Advanced Structural Analysis

Advanced Structural Analysis

Devdas Menon

Alpha Science International Ltd.
Oxford, U.K.

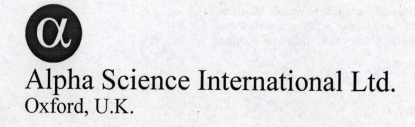

Advanced Structural Analysis

700 pgs. | 5 tbls. | 290 figs.

Devdas Menon
Professor
Department of Civil Engineering
Indian Institute of Technology, Madras
Chennai, India

ALPHA SCIENCE INTERNATIONAL LTD.

7200 The Quorum, Oxford Business Park North
Garsington Road, Oxford OX4 2JZ, U.K.

www.alphasci.com

ISBN 978-1-84265-497-0

Printed in India

To
my alma mater,
IIT Madras

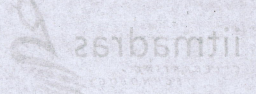

Preface

There is an increasing tendency among modern structural engineers to lean heavily on software packages for everything. This induces a false sense of knowledge, security and power. The computer is indeed a powerful tool and an asset for any structural engineer, for carrying out repetitive work and for generating quick solutions to complex problems. It is dangerous, however, to make the tool one's master, and to make it a convenient substitute for human knowledge, experience and creative thinking.

Young engineers today are rapidly losing the ability to do structural analysis without the assistance of the digital computer, and this is happening worldwide. This problem is not being addressed adequately in education, so that graduating students develop self-confidence in their analytical abilities. As a result, experienced structural engineers are alarmed at the lack of academic preparedness in the graduating students.

The problem is not so much with limitations in the intellectual abilities of the students, as it has to do with the very process of learning. The student must not only be empowered to carry out structural analysis of small structures by simple manual methods, but must also understand the basics of how a software package works. There is a need for a systematic approach to learning both manual analysis (using classical methods) and computer-aided methods (based on finite element analysis), with clarity of understanding and joy of learning.

One of the areas of structural analysis that is not adequately covered in present-day education is the topic of *matrix methods*, which provides a bridge between traditional methods and modern computer-aided methods. This book, which is a natural extension of my previous book on Structural Analysis, attempts to fill this gap. It aims to equip students and practising engineers with a thorough understanding of the laws underlying the mechanics of structures in the mathematical framework of matrices.

The initial three chapters review the basic concepts in structural analysis and matrix algebra. The next three chapters discuss in detail and demonstrate through many examples how matrix methods can be applied to linear static analysis of skeletal structures (plane and space trusses; beams and grids; plane and space frames) by the stiffness method. Also, it is shown how simple structures can be conveniently solved using a reduced stiffness formulation, involving far less computational effort. The flexibility method, which is not adopted in software packages, is also discussed in some detail in this book, because it offers good insights for a complete understanding of the subject. Finally, in the seventh chapter, analysis of elastic instability and second-order response is covered.

The main objective is to enable the reader (student / teacher / practising engineer) to have a good grasp of all the fundamental issues in these advanced topics in Structural Analysis, besides enjoying the learning process, and developing analytical and intuitive skills. With these strong fundamentals, the reader will be well prepared to understand what happens behind the 'black box' of the software package commonly used for structural analysis. Numerous examples are demonstrated in this book to show, step-by-step, how the solution is generated, with the help of simple computational tools like MATLAB for handling matrices. No computer program or CD-ROM is supplied to the reader along with this book, because its very objective is to engage the reader in a process of self-learning, rather than simply learning to feed input data mechanically into a program developed by somebody else. For a complete understanding, the reader is encouraged to answer the review questions and problems posed at the end of various chapters.

With this understanding, the reader will be well prepared to dwell into further topics like Finite Element Analysis (including nonlinear analysis and dynamics), as applied to structural engineering, and also to knowledgeably use software packages that are widely available.

Devdas Menon

Professor, Department of Civil Engineering
Indian Institute of Technology, Madras, Chennai 600036, India
(email: dmenon@iitm.ac.in)

Acknowledgement

I wish to gratefully acknowledge the support received from Indian Institute of Technology, Madras, in making this book a reality, as part of the Institute's Golden Jubilee celebrations. The sabbatical sanctioned to me by the Institute and the support and encouragement given by the Centre for Continuing Education and the Department of Civil Engineering at IIT Madras have enabled this book to come out in a relatively short time.

I am grateful to all my teachers, faculty colleagues and students for their support. In particular, I wish to thank Professors Meher Prasad and V Kalyanaraman for their assistance and encouragement. I am deeply indebted to A S Balu, PhD scholar and Dr S Latheswary for meticulously scrutinising the manuscript and for their suggestions. I also wish to thank my students, K Girija, S Arun, V Raju, Vipin Unnithan, Robin Davis, Maganti Janardhana and Ambili Thampi for their help and feedback. I thank Smitha, Sankari, Sivagami and Prabheesh for their secretarial assistance.

I thank Mr N K Mehra and Ms Anupama Jauhry of Narosa Publishing House for bringing out this publication in a very professional manner.

Finally, I wish to thank my dear wife, Roshni, for her patience, forbearance and tremendous support, without which this book would not have been possible. I am ever grateful to my parents for their blessings and constant encouragement.

Devdas Menon

Notation

The following symbols have been used commonly in various chapters in this book. They have also been defined in the text whenever they first appear. Other symbols, specific to individual chapters, are not included in the list below.

$\mathbf{D}$	Displacement vector (global coordinates); D_j denotes the jth element of this vector
$\mathbf{D_A}$	Displacement vector, corresponding to *active* global coordinates
$\mathbf{D_R}$	Displacement vector, corresponding to *restrained* global coordinates
$\mathbf{D_*^i}$	Displacement vector (local coordinates) of ith element (comprising $D_{1*}^i, D_{2*}^i, \ldots$)
$\mathbf{D^i}$	Displacement vector (in global axes system) of ith element (comprising $D_1^i, D_2^i, \ldots$)
$\mathbf{f}$	Structure flexibility matrix (global coordinates), relating $\mathbf{D}$ to $\mathbf{F}$; this square, symmetric matrix can have partitions, corresponding to *active* and *redundant* global coordinates, denoted by the subscripts, A and X, respectively
$\mathbf{f_*^i}$	Element flexibility matrix (local coordinates) of the ith element, relating $\mathbf{D_*^i}$ to $\mathbf{F_*^i}$
$\mathbf{F}$	Force vector (global coordinates); F_j denotes the jth element of this vector
$\mathbf{F_e}$	Equivalent joint load vector (global coordinates)
$\mathbf{F_f}$	Fixed end force vector; this matrix can have partitions, corresponding to *active* and *restrained* global coordinates, denoted by the subscripts, A and R, respectively
$\mathbf{F_A}$	Force vector (load vector), corresponding to *active* global coordinates
$\mathbf{F_R}$	Force vector (support reactions), corresponding to *restrained* global coordinates
$\mathbf{F_X}$	Redundant force vector, corresponding to *redundant* locations (flexibility method)
$\mathbf{F_*^i}$	Force vector (local coordinates) of ith element (comprising $F_{1*}^i, F_{2*}^i, \ldots$)
$\mathbf{F_{*f}^i}$	Fixed end force vector (local coordinates) of ith element (comprising $F_{1*f}^i, F_{2*f}^i, \ldots$)
$\mathbf{F^i}$	Force vector (in global axes system) of ith element (comprising $F_1^i, F_2^i, \ldots$)
$\mathbf{k}$	Structure stiffness matrix (global coordinates), relating $\mathbf{F}$ to $\mathbf{D}$; this square, symmetric matrix can have partitions, corresponding to *active* and *restrained* global coordinates, denoted by the subscripts, A and R, respectively
$\tilde{\mathbf{k}}$	Reduced structure stiffness matrix (global coordinates)
$\mathbf{k_*^i}$	Element stiffness matrix (local coordinates) of the ith element, relating $\mathbf{F_*^i}$ to $\mathbf{D_*^i}$
$\tilde{\mathbf{k}}_*^i$	Reduced element stiffness matrix (local coordinates) of the ith element
$\mathbf{k^i}$	Element stiffness matrix (with element coordinates expressed in global axes system) of the ith element, relating $\mathbf{F^i}$ to $\mathbf{D^i}$
$\mathbf{T^i}$	Transformation matrix (square matrix) of the ith element, relating the coordinates defined in the local axes system $(1*, 2*, \ldots)$ to the global axes system $(1, 2, \ldots)$

$\mathbf{T}_\mathrm{D}^\mathrm{i}$	Displacement transformation matrix, relating the displacements at the local coordinates of the ith element $(1^*, 2^*, ...)$ to the global coordinates $(1, 2, 3, ...)$ of the structure; this matrix can have partitions, corresponding to *active* and *restrained* global coordinates, denoted by the subscripts, A and R, respectively
$\mathbf{T}_\mathrm{F}^\mathrm{i}$	Force transformation matrix (used in flexibility method), relating forces at the local coordinates of the ith element $(1^*, 2^*, ...)$ to the global coordinates $(1, 2, 3, ...)$ of the structure; this matrix can have partitions, corresponding to *active* and *redundant* global coordinates, denoted by the subscripts, A and X, respectively
A	Area of a cross-section
E	Modulus of elasticity
EI	Flexural rigidity of a prismatic element
EA	Axial rigidity of a prismatic element
GJ	Torsional rigidity of a prismatic element
G	Modulus of rigidity
I	Second moment of area (moment of inertia) of a section
J	Torsion constant (or polar moment of inertia in circular/tubular sections)
L, l	Length
M	Bending moment at a section or moment applied at a node
P, W	Concentrated force (load)
m, n, r	integers denoting number of members, number of active degrees of freedom and number of restrained coordinates respectively
i, j, k	indices (integers), used as subscripts, denoting coordinate locations or elements
N	Axial force at a section in an element
S	Shear force at a section in an element
q	load intensity per unit length
x, y, z	Global axes system (Cartesian coordinates) for the structure
x^*, y^*, z^*	Local axes system (Cartesian coordinates) for an element
Δ	Deflection at a node
ε	Strain
φ	Curvature at a node
ν	Poisson's ratio
θ	Rotation at a node
θ_x^i	angle between local x^* (longitudinal) axis of the ith element and global x axis
σ	Stress
{ }	Curly braces indicate a vector (matrix with one column)
[ ]	Box brackets indicate a matrix (square or rectangular)

Contents

1

Review of Basic Concepts

1.1 INTRODUCTION

An advanced course in any stream of knowledge is like a structure. It needs a solid foundation, for it to meaningfully serve its purpose. *Advanced Structural Analysis* can be learnt easily, if the student has a good understanding of the fundamentals of *Structural Analysis*. However, such understanding usually takes time and repeated exposure. There is a simple but certain formula to accelerating this process of learning. Whenever a lack of clarity due to poor background knowledge is detected, it is advisable to go back to the basics and set things right at the fundamental level. If this is not done, and an attempt is made to somehow erect the superstructure on a shaky foundation, the learning process will not be fulfilling and enjoyable. In today's world, where structural analysis is routinely done using computer software, it is all the more important for structural engineers to have the necessary basic skills to check and interpret meaningfully the solutions generated by the computer.

In this Chapter, an overview of some of the fundamental concepts of structural analysis is given, based on the material given in *Structural Analysis*, a textbook written by the same author. We begin with a simple description of 'structural analysis'.

> **Structural analysis** is the application of solid mechanics to predict the **response** (in terms of forces and displacements) of a given **structure** (existing or proposed) subject to specified **loads**.

The structure may be viewed as a 'system', on which we apply loads as 'input', with the objective of finding the response, which is the desired 'output'.

1.2 STRUCTURE

A **structure** may be defined as an assemblage of load-bearing elements in a construction. The non-load-bearing elements (such as the cladding, partitions and finishes) are considered to be 'non-structural'. The structural system defines the manner in which various forces acting on the structure get transmitted. By making appropriate idealisations, we reduce the structural system to separate sub-systems, each of which is made of structural elements. While separating out the sub-system from the main system, and possibly, individual **elements** from the sub-system, we must be careful to model the inter-connections (**joints** and **supports**) appropriately.

1.2.1 Structural Elements

When two of the dimensions (cross-sectional dimensions) are very small in comparison with the third (the 'length'), the element is modelled as a one-dimensional **line element** (refer Fig. 1.1). On the other hand, when only one of the dimensions (viz., the 'thickness') is very small compared to the other two, the element is modelled as a two-dimensional **surface element**†. The **space frame element** [Fig. 1.1f] (commonly encountered in building construction) is perhaps the most generalized element, of which the other elements (shown in Fig. 1.1) happen to be special cases. As indicated, six independent internal force resultants can act at any section of this element: bending moments (M_z and M_y), twisting moment (M_x), flexural shear forces (S_z and S_y), and axial force (N_x). When this element lies in a Cartesian reference plane (such as xy) and is not subject to torsion, it reduces to a **plane frame element** [Fig. 1.1d], in which the number of independent internal force resultants reduce to three, viz., M_z, S_y and N_x (or simply M, S and N). The plane frame element reduces to the beam element [Fig. 1.1c] in the absence of axial force (only M and S operate), and to a shaft element [Fig. 1.1b] when only twisting moment (T) operates, and to a truss element [Fig. 1.1a] when only the axial force (N) is to be considered. The space frame element reduces to a **grid element** [Fig. 1.1e] when it is located in a Cartesian plane (usually the horizontal xz plane), and can be subject to twisting moment, in addition to bending moment and shear force.

Assemblages of line elements are called 'skeletal' structures or 'framed' structures. When all the line elements of a skeletal structure lie in a single plane, and the loading is also in this plane (i.e., no out-of-plane loading), the structure is called a **planar** structure; otherwise, it is called a **space** structure. Typical examples of planar skeletal structures are

† Plates and shells are examples of surface elements. The main scope of this book is limited to skeletal structures, and does not include surface elements.

beams, plane trusses, plane frames, cables and arches. Typical examples of space skeletal structures are grids, space trusses and space frames.

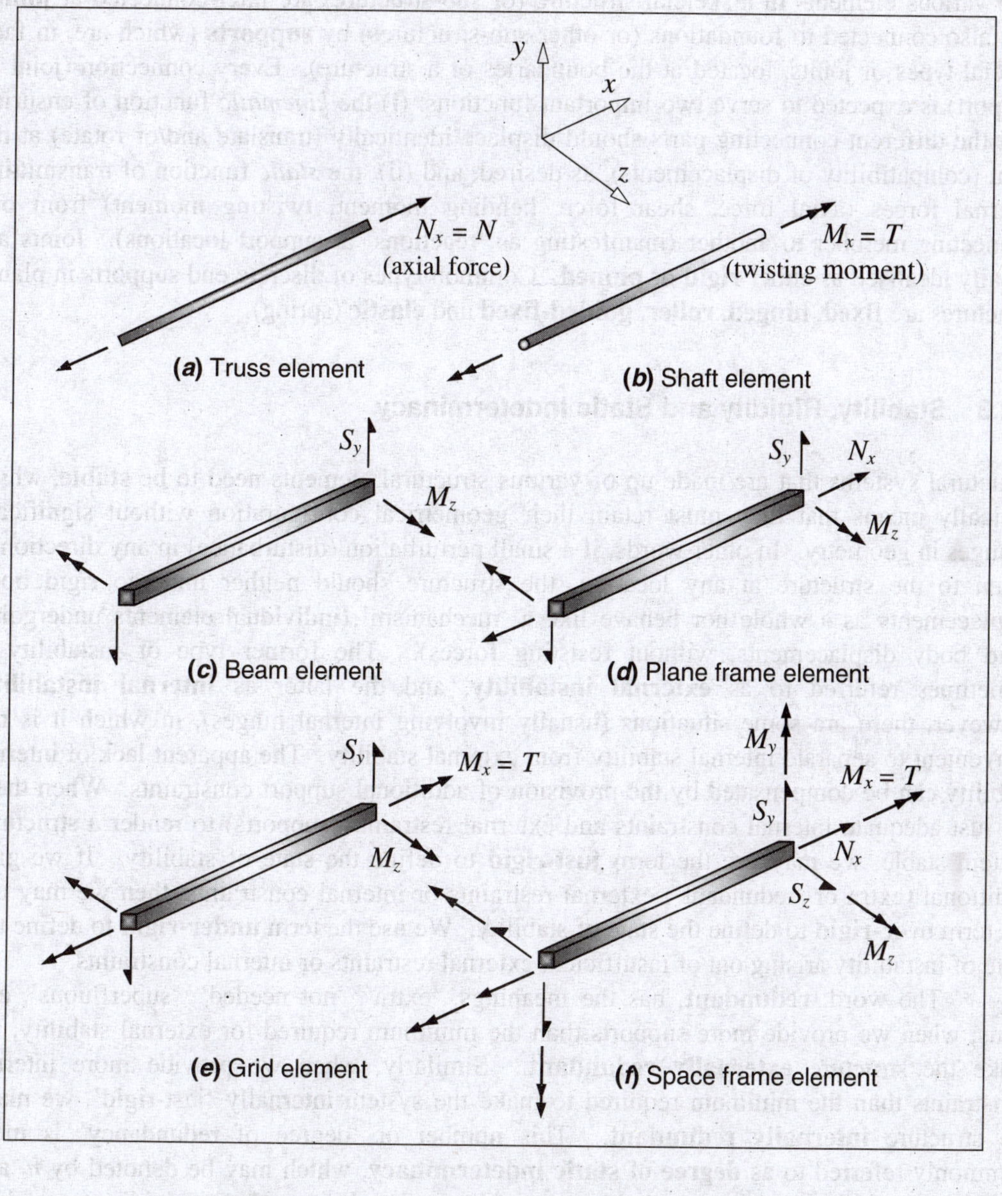

Figure 1.1 One-dimensional 'line' elements

1.2.2 Joints and Supports

The various elements in a skeletal structure (or sub-structure) are inter-connected at **joints**, and also connected to foundations (or other sub-structures) by **supports** (which are, in fact, special types of joints, located at the boundaries of a structure). Every connection (joint or support) is expected to serve two important functions: (i) the *kinematic* function of ensuring that the different connecting parts should displace identically (translate and/or rotate) at the joint (compatibility of displacements), as desired, and (ii) the *static* function of transmitting internal forces (axial force, shear force, bending moment, twisting moment) from one connecting member to another (manifesting as 'reactions' at support locations). Joints are usually idealised as either **rigid** or **pinned**. Common types of discrete end supports in planar structures are **fixed**, **hinged**, **roller**, **guided-fixed** and **elastic** (spring).

1.2.3 Stability, Rigidity and Static Indeterminacy

Structural systems that are made up of various structural elements need to be **stable**, which basically means that they must retain their geometrical configuration without significant changes in geometry. In other words, if a small perturbation (disturbance) in any direction is given to the structure at any location, the structure should neither undergo rigid body displacements as a whole nor behave like a 'mechanism' (individual elements undergoing rigid body displacements, without resisting forces). The former type of instability is sometimes referred to as **external instability**, and the latter as **internal instability**. However, there are some situations (usually involving internal hinges), in which it is not convenient to separate internal stability from external stability. The apparent lack of internal stability can be compensated by the provision of additional support constraints. When there are just adequate internal constraints and external restraints (supports) to render a structural system stable, we may use the term **just-rigid** to define the state of stability. If we give additional (extra or 'redundant') external restraints or internal constraints, then we may use the term **over-rigid** to define the state of stability. We use the term **under-rigid** to define the state of instability arising out of insufficient external restraints or internal constraints.

The word, **redundant**, has the meanings, 'extra', 'not needed', 'superfluous', etc. Thus, when we provide more supports than the minimum required for external stability, we make the structure **externally redundant**. Similarly, when we provide more internal constraints than the minimum required to make the system internally 'just-rigid', we make the structure **internally redundant**. This number or 'degree of redundancy' is more commonly referred to as **degree of static indeterminacy**, which may be denoted by n_s and can be obtained by summing up the degrees of internal and external over-rigidity (n_{si} and n_{se} respectively). Over-rigid structures are said to be **statically indeterminate**, because the use of statics (equations of static equilibrium) is not sufficient to find the force response quantities. The degree of static indeterminacy is given by the number of unknown forces (internal forces or support reactions) that exceed the number of available independent equations of equilibrium. Typical examples are illustrated in Fig. 1.2.

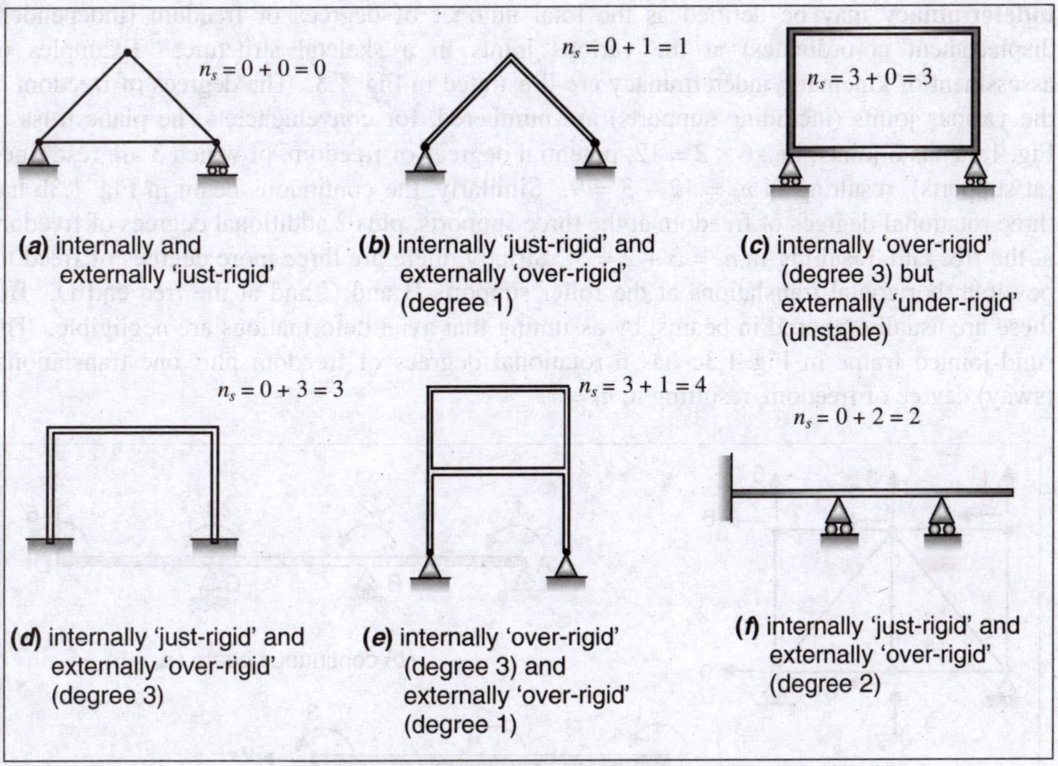

Figure 1.2 Assessment of degree of static indeterminacy, n_s

1.2.4 Kinematic Indeterminacy

There is concept allied to 'static indeterminacy' in structural analysis called **kinematic indeterminacy**. Whereas the term 'statics' is associated with forces, the term 'kinematics' is associated with displacements. The concept of kinematic indeterminacy is usually associated with 'degrees of freedom', or independent co-ordinates required to locate the structure in the displaced configuration, relative to the original configuration. It would appear that a structure has infinite degrees of freedom, because it is a continuum. But *under static loading*, it can be shown that the actual degrees of freedom in a skeletal structure are limited to the ones at the joints where different structural elements converge.

If the displacements (translations and rotations) at the two ends of a particular skeletal element are known, then the displacement at any intermediate point of the element can be derived†, by applying appropriate principles of mechanics (involving the distribution strains along the length of member under a given loading). The **degree of kinematic indeterminacy** may be defined as the total number of degrees of freedom (independent displacement co-ordinates) at the various joints in a skeletal structure. Examples of assessment of kinematic indeterminacy are illustrated in Fig. 1.3. The degrees of freedom at the various joints (including supports) are numbered, for convenience. The plane truss in Fig. 1.3a has 6 joints, i.e., $6 \times 2 = 12$, potential degrees of freedom, of which 3 are restrained (at supports), resulting in $n_k = 12 - 3 = 9$. Similarly, the continuous beam in Fig. 1.3b has three rotational degrees of freedom at the three supports, plus 2 additional degrees of freedom at the free end, resulting in $n_k = 3 + 2 = 5$. Strictly, there are three more degrees of freedom possible: horizontal translations at the roller supports B and C and at the free end D. But these are usually ignored in beams, by assuming that axial deformations are negligible. The rigid-jointed frame in Fig. 1.3c has 6 rotational degrees of freedom plus one translational (sway) degree of freedom, resulting in $n_k = 7$.

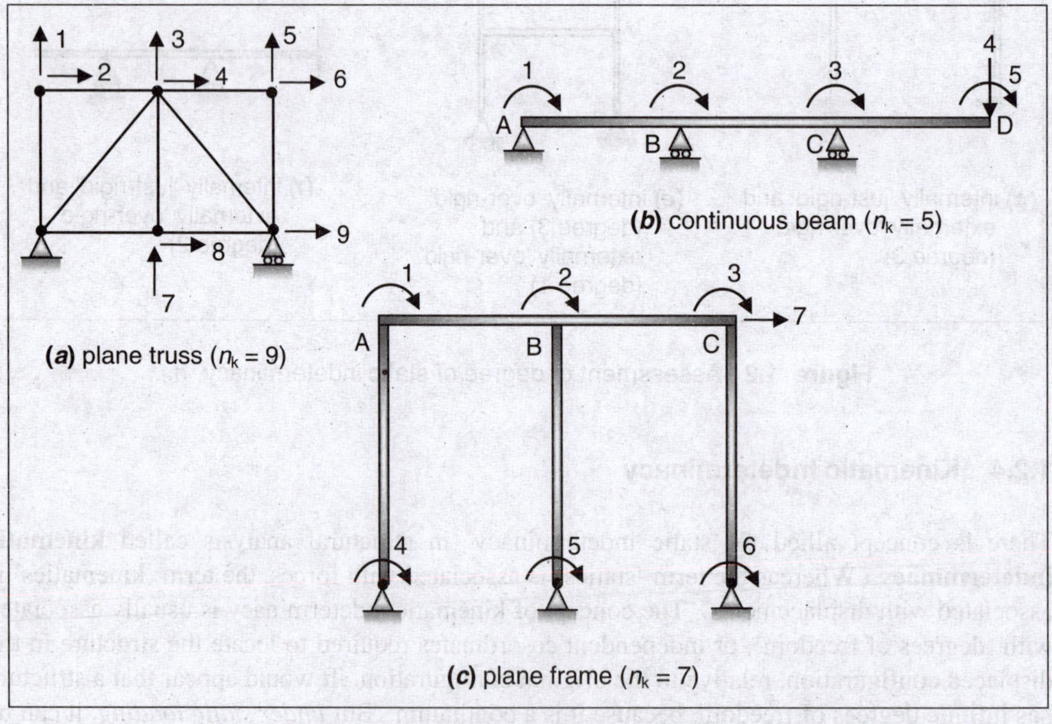

Figure 1.3 Assessment of degree of kinematic indeterminacy, n_k

† This is easy to understand with regard to a truss element. The deflections at an intermediate location in the element can be obtained by a linear interpolation of the corresponding deflection components at the two ends (assuming a constant axial strain at all points of the element).

1.3 LOADS

Loads act on a structure, causing it to undergo internal stresses and displacements, which the structure should be able to withstand satisfactorily, meeting the requirements of stability, strength and serviceability. Loads take the form of either **direct actions** or **indirect loading**.

1.3.1 Direct Actions

The most common source of loading is by 'direct actions', meaning forces and/or moments applied directly on a structure (or a part of it). These loads may be caused by the action of gravity, the natural environment or other sources. They include **dead loads** (due to self-weight), **live loads** (due to human occupancy, storage, vehicular traffic, etc.), **wind loads, snow loads, earthquake loads, soil pressures, water pressures, wave loads, impact loads** and **blast loads**. Most of these loads (except dead loads) are difficult to quantify, because they are random in nature, both temporally (i.e., in time) and spatially (i.e., in space). Building codes often prescribe the magnitudes of loads to be assumed, on the basis of statistical data available and acceptable risk of exceedance. The distributions of loads are also idealised, for convenience, in the form of 'equivalent' loads (say, uniformly distributed or linearly varying), such that they generally result in 'safe' designs. Loads, which are 'dynamic' in nature, are often replaced by 'equivalent static' loads, for convenience in structural analysis. The static loads have to be multiplied by an 'impact factor' (or 'dynamic magnification factor'), to account for the maximum amplified response due to the dynamic effect.

Direct actions (directly applied external loads) usually take the form of either concentrated or distributed forces or moments.

1.3.2 Indirect Loading

It is possible that a response (*force response* as well as *displacement response*) may be induced in a structure without being caused by externally applied loads ('direct actions'). We may use the term 'indirect loading' to refer to the cause of such load effects. Usually, this type of loading occurs in addition to direct actions, and if not anticipated in design, may result in serious consequences.

The sources of such indirect static loading may be **support displacements** (settlements or rotational slips), **constructional errors** (accidental eccentricities and 'lack of fit') and **environmental changes** (temperature and shrinkage changes, creep effects).

1.4 RESPONSE

When a structure is loaded, it *responds* by (i) developing internal force resultants and support reactions (**force response**) and (ii) undergoing displacements in the form of deflections, rotations, curvatures, etc. (**displacement response**). For a stable, linearly elastic structure, it can be proved that there exists a unique response that satisfies the following three requirements:

1. Equilibrium of forces;
2. Compatibility of displacements; and
3. Force-displacement relations.

1.4.1 Equilibrium of Forces

From Newton's first law of motion, it follows that a body will remain in a state of rest or of uniform motion, unless there is a non-zero resultant force (or moment) acting on it. We use the term static equilibrium to define this 'state of rest or of uniform motion'. In Cartesian coordinates, the equations of equilibrium may be expressed as follows:

$$\sum F_x = 0; \ \sum F_y = 0; \ \sum F_z = 0 \tag{1.1}$$

$$\sum M_x = 0; \ \sum M_y = 0; \ \sum M_z = 0 \tag{1.2}$$

For a planar structure, assuming all the forces and moment vectors to lie in the *x-y* plane, the above equations reduce to the following three.

$$\sum F_x = 0; \ \sum F_y = 0; \ \sum M_z = 0 \tag{1.3}$$

Alternative approaches to expressing static equilibrium using work and energy concepts are possible. This is discussed in Sections 1.6 and 1.7.

If a structure is in a state of static equilibrium, every part of it (i.e., every 'free-body') must also satisfy the requirements of static equilibrium. In a planar structure, there are three independent equations of equilibrium for the overall free-body (in which the support reactions are exposed), plus three additional independent equations for every cut free-body of the structure (in which the internal force resultants at the cut section are exposed). For a space structure, the number will be six, plus an additional six for every cut free-body.

The loading and response on a structure can be collectively described in terms of a 'force field' and 'displacement field'. Free-bodies enable us to access the force field. The mere fact that we have a complete force field (i.e., loads, support reactions, internal force resultants at every section), which fully satisfies the conditions of static equilibrium, does not guarantee that we are in possession of the true force field. It is possible, in the case of statically indeterminate ('over-rigid') structures, to concoct a large number of complete force fields, which all satisfy the conditions of static equilibrium. Such force fields are said to be

statically admissible. Any choice of redundants from the force field of a statically indeterminate structure is admissible, provided the redundants themselves are statically indeterminate. Thus, while internal force resultants can be chosen as redundants in externally 'over-rigid' structures, support reactions cannot be chosen as redundants in structures that are internally 'over-rigid', but externally 'just-rigid'. The correct solutions to the redundants can be obtained by invoking appropriate conditions of compatibility and force-displacement relations.

1.4.2 Compatibility of Displacements

Associated with every 'force field', there is a corresponding 'displacement field'. A **kinematically admissible** displacement field is one that preserves continuity within the structure and satisfies the specified kinematic (or geometric) boundary conditions. These requirements of the displacement field are referred to as *compatibility* or *geometric consistency* requirements. Just as it is necessary for a statically admissible force field to satisfy conditions of equilibrium, so it is necessary for a kinematically admissible displacement field to satisfy conditions of compatibility. And just as displacements are irrelevant in the context of static admissibility (in 'first-order' structural analysis), forces are irrelevant in the context of kinematic admissibility.

Compatibility conditions are typically invoked at joint and support locations. For example, compatibility requires that all the segments inter-connecting at a rigid joint undergo identical translations and rotations at the joint location.

1.4.3 Force-Displacement Relationships

The force field and displacement field in any structure are inter-related by means of force-displacement relationships. These relationships can be expressed either in a **stiffness** format or a **flexibility** format. If **linear elastic** behaviour of the structure is assumed, the force quantities are directly proportional to the displacement quantities. It is easy to establish these relationships at the element level. For example, the **axial stiffness** of a truss element, defined as the axial force (tension) required to cause unit axial deformation (elongation) is given by the ratio of the axial rigidity (product of elastic modulus E and cross-sectional area A) to the length L of the element. The reciprocal of the axial stiffness is the **axial flexibility**, L/EA, defined as the axial deformation caused by unit axial force. Similarly, measures of flexural stiffness, shear stiffness and torsional stiffness can be defined. In each of these measures, a rigidity term (flexural rigidity or torsional rigidity or shear rigidity) appears in the numerator and the length term appears in the denominator. The rigidity term is a product of an elastic property (such as elastic modulus or shear modulus) and a cross-sectional property (such as second moment of area). The higher the rigidity, the greater is the stiffness, and conversely, the smaller the flexibility of the element under consideration. Similarly, longer elements have lesser stiffness.

A structure is typically made up of several elements. We can define measures of stiffness and flexibility, considering the structure as a whole. When a unit load is applied at a point on the structure in a certain direction (at coordinate *j*), then the displacement measured at any other point (at coordinate *i*) due to this action, with no other loads acting, is referred to as the **flexibility coefficient** f_{ij}. Conversely, when all the degrees of freedom in a structure are arrested and only one displacement with unit value (at coordinate *j*) is allowed to occur, then the force generated at some location (coordinate *i*) is referred to as the **stiffness coefficient** k_{ij}. High stiffness values of individual elements in a structure contribute to higher values of stiffness coefficients. When all the stiffness coefficients, corresponding to various degrees of freedom, are assembled in a matrix form the resulting matrix is called the **stiffness matrix** of the structure. Similarly, the **flexibility matrix** is an assemblage of flexibility coefficients. In linear elastic stable structures, the stiffness matrix and flexibility matrix are square symmetric matrices, one being the inverse of the other. In this book, we shall discuss in detail as to how these matrices can be generated and used in analysing structures. For the present, it suffices to observe that in these matrices are embedded the fundamental laws that govern the behaviour of the structure. These matrices provide the basic relationships between the force field and the associated displacement field.

1.4.4 Levels of Structural Analysis

There are various levels of rigour to which structural analysis can be taken. The higher the level, the closer the mathematical model will be to physical reality. The level of structural analysis is determined by simplifications related to the following aspects:

- degree of indeterminacy
- static or dynamic
- linear or nonlinear
- deterministic or stochastic

In reality, all problems in mechanics are highly indeterminate, if we are interested in finding the stresses and strains at all points in the body under consideration. By limiting our scope to skeletal structures and our interest to internal force resultants (rather than stresses) and deflections and rotations (rather than strains), we reduce the indeterminacy significantly in structural analysis. We also reduce indeterminacy in framed structures by often assuming that axial and shear deformations are negligible, in comparison with flexural deformations. The less the indeterminacy, the less is the number of unknowns, and hence the easier the analysis. Also, if the static indeterminacy is less than the kinematic indeterminacy, it makes sense to prefer the **force method** of analysis (in which the basic unknowns are forces); otherwise, it is appropriate to adopt the **displacement method** of analysis (in which the basic unknowns are displacements). Through proper modelling and structural idealisation, including taking advantage of simplifying assumptions and symmetry or anti-symmetry, if any, in the response, it is possible to reduce the indeterminacy and thereby reduce the computational effort in analysis.

Static versus Dynamic Analysis

When we say that a structure is in a 'static' condition, we mean that it is in a state of rest (or uniform motion); i.e., there are no accelerations in any part of the structure. We expect such a static structural response, when the applied loading is static and the system characteristics (such as stiffness) remain invariant with time. Loads (typically, 'dead loads') that are fixed in position and do not vary with respect to time, in magnitude or direction, are called 'static loads' and result in a static response. We use simple static analysis to determine this static response. However, in practice, we encounter many loads (such as live loads, wind or earthquake loads, impact loads, wave loads) that vary in magnitude and/or position with time, and these usually result in a time-varying response. We need to perform **dynamic analysis** to predict such a response. As the structure is not in a state of 'static equilibrium', we need to account for the additional forces associated with the accelerations. D'Alembert proposed the concept of 'dynamic equilibrium', wherein the equations of static equilibrium can still be used, but including the so-called 'inertial forces' (products of masses and corresponding accelerations). If the inertial forces are relatively small in magnitude, the dynamic problem can be approximated as a static problem, and it suffices to do a static analysis, which is much simpler than dynamic analysis. Because of the difficulties associated with dynamic analysis, codes of practice permit structural designers to resort to 'quasi-static analysis'. The objective is to arrive at the maximum response of the structure under the given dynamic loads. It therefore suffices to find some equivalent static loading that results (approximately) in this maximum response. This is usually achieved by applying some 'dynamic amplification factor' (also called 'impact factor' in the case of bridges, or 'gust factor' in wind-resistant design). The static condition is, of course, a special case of the more complex dynamic condition. In the general dynamic condition, there is a force equilibrium maintained at any instant of time between the external applied load, if any, and the internal forces comprising elastic forces, damping forces and inertial forces. Dynamic analysis is unavoidable in some cases, such as asymmetric or irregular structures subject to earthquake loading.

Linear versus Nonlinear Analysis

In structural analysis, we commonly make the assumption of linear elastic behaviour. This enables us to invoke the *Principle of Superposition*, which states that the responses due to several loads on a structure can be obtained by superposing, algebraically, the individual responses due to the various loads acting one at a time. This principle could be used in reference to forces, displacements or both. However, this assumption of linearity can result in significant error (usually under-estimation of the response), when (i) there are relatively large deformations involved or there are axial forces inducing so-called *P-delta* effects, requiring considerations of equilibrium in the deformed configuration, and (ii) the stress-strain behaviour of the structural material is non-linear or exceeds the elastic limit. The former induces **geometric nonlinearity**, while the latter induces **material nonlinearity**, calling for a nonlinear analysis of the structure, which requires a step-by-step iterative solution procedure, which is difficult and sometimes has problems related to a lack of convergence and non-uniqueness in the response. If the nonlinearity is not significant, we can avoid a **nonlinear analysis** and manage with a linear analysis, which is a special case of the nonlinear analysis. However, when deformations are large and the material yields, nonlinear analysis must be carried out for an accurate solution.

Deterministic versus Probabilistic Analysis

Every structure has an inherent probability of failure (however, small that it may be), i.e., there is a chance that a structure may not carry out its intended function during its lifetime. Thus, no structure is fully safe. This is due to uncertainty in all physical phenomena. This uncertainty (randomness) could be a characteristic in the input (loads) and also in the system characteristics (stiffness, damping, etc.). Modelling of these uncertainties can be carried out in a 'probabilistic' framework in structural analysis and structural design, provided we have adequate statistical data to model the various uncertainties. The uncertainties are implicitly accounted for in a deterministic framework by adopting nominal values for the random parameters and applying factors of safety. In such an approach, the resistance parameters are under-estimated (by using partial safety factors) and the loads over-estimated (by using partial load and load combination factors). However, for a more accurate structural analysis, we need to carry out a **probabilistic (stochastic) analysis**, which is generally very difficult.

Thus, various levels of structural analysis are possible, ranging from simplified linear, static and deterministic analysis to nonlinear, dynamic and probabilistic analysis. In this book, the emphasis is primarily on linear static deterministic analysis of skeletal structures. However, some aspects related to geometric nonlinearity are covered in this book, within the framework of linear analysis (refer Chapter 7).

1.5 FORCE RESPONSE IN STATICALLY DETERMINATE STRUCTURES

In basic structural analysis, the first topics to be covered relate to finding the force response, and subsequently, the displacement response of simple, just-rigid, statically determinate structures, such as beams, trusses, cables, arches and frames.

1.5.1 Support Reactions

Finding the support reactions is usually a necessary first step in structural analysis, one that helps us determine the internal force resultants in statically determinate structures. In practice, we also need to determine support reactions in order to account for the loads transmitted to the supporting sub-structures, so that we can design these supports (such as foundations) appropriately. The support reactions are in the form of forces and moments in various directions. The simplest way is usually by direct application of equilibrium equations on free-bodies of the structure. In some situations involving internal hinges, it may also be convenient to apply the Principle of Virtual Displacements.

For convenience in calculations, we can replace distributed loads by their statically equivalent resultant force acting at the appropriate line of action (having the same moment effect), for the limited purpose of finding support reactions. In the calculation of support reactions, it is the geometry of critical points in the overall structure (including locations of internal hinges, if any) and the magnitudes of the statically equivalent loads and their respective lines of actions that influence the calculations. Hence, it is possible for structures that apparently look different, but share the above characteristics, to have identical values of support reactions. This is demonstrated in Fig. 1.4, which shows four different structures (two three-hinged frames, a three-hinged arch and a cable).

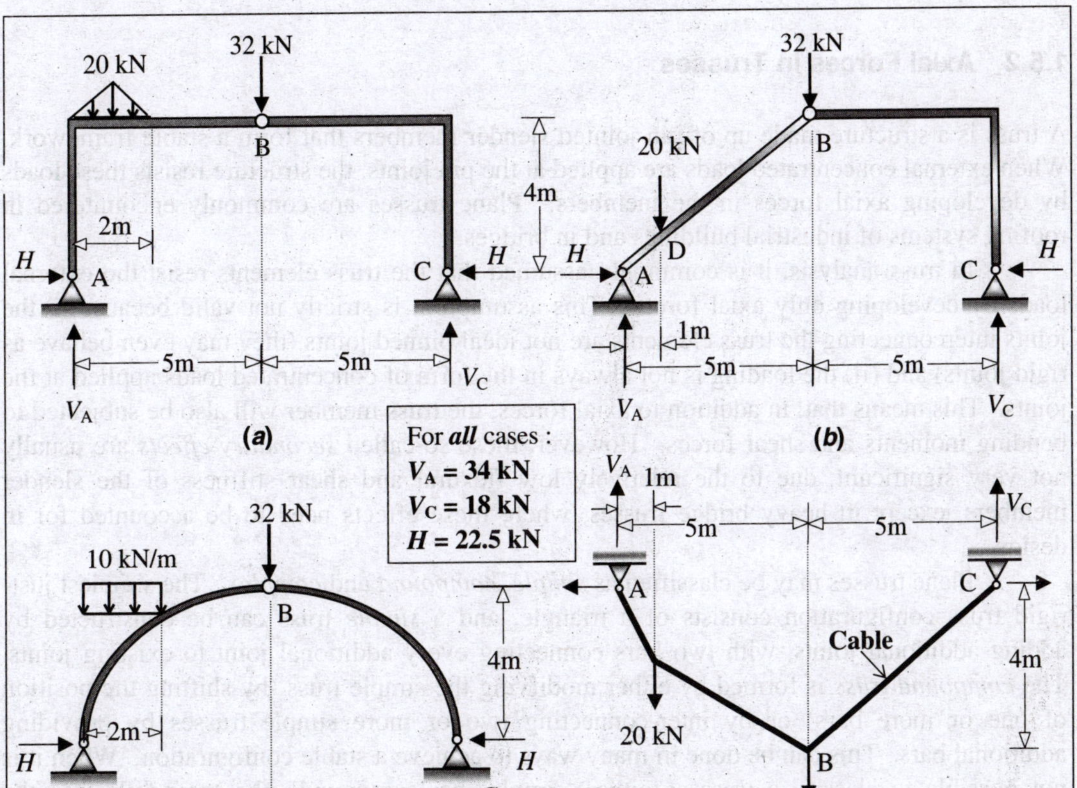

Figure 1.4 Structures that are different, but have identical support reactions

All the structures in Fig. 1.4 have the same span of 10m and resultant loads of 20kN and 32kN at 1m and 5m respectively from the left support A. As the two supports at A and C are at the same level, the vertical reaction components are easily obtainable from moment equilibrium considerations ($\sum M_A = 0$ and $\sum M_C = 0$), whereby the 20kN load is transmitted to V_A and V_C in the ratio 1/10 and 9/10 respectively, while the 32kN load acting at mid-span is transmitted equally to the two supports. After determining $V_A = 34$kN and $V_C = 18$kN, the horizontal reaction component H can be calculated by considering the free-body AB or BC and applying the equilibrium condition that the bending moment in the frame at the internal hinge B† is zero ($M_B = 0$ kNm), whereby $H = 22.5$ kN. Alternatively, we can invoke the Principle of Virtual Displacements to find the support reactions, as demonstrated in Section 1.6.1.

1.5.2 Axial Forces in Trusses

A truss is a structure made up of pin-jointed slender members that form a stable framework. When external concentrated loads are applied at the pin joints, the structure resists these loads by developing axial forces in the members. Plane trusses are commonly encountered in roofing systems of industrial buildings and in bridges.

In truss analysis, it is commonly assumed that the truss elements resist the external loads by developing only axial forces. This assumption is strictly not valid because (i) the joints interconnecting the truss elements are not ideal pinned joints (they may even behave as rigid joints) and (ii) the loading is not always in the form of concentrated loads applied at the joints. This means that, in addition to axial forces, the truss member will also be subjected to bending moments and shear forces. However, these so-called *secondary effects* are usually not very significant, due to the relatively low flexural and shear stiffness of the slender members, except in heavy bridge trusses, where these effects need to be accounted for in design.

Plane trusses may be classified as *simple*, *compound* and *complex*. The simplest just-rigid truss configuration consists of a triangle, and a *simple* truss can be constructed by adding additional joints, with two bars connecting every additional joint to existing joints. The *compound truss* is formed by either modifying the simple truss (by shifting the position of one or more bars) or by inter-connecting two or more simple trusses by providing additional bars. This can be done in many ways to achieve a stable configuration. When it is not possible to classify a truss as either 'simple' or 'compound', the truss falls into the category of *complex truss*. As the name suggests, the analysis of the force response of such trusses is complex, as the equations of equilibrium related to the unknown forces are generally heavily coupled.

† It is important to note that the condition $M_B = 0$ kNm is the additional equation of equilibrium to be considered here. This is different from the moment equilibrium equation, $\sum M_B = 0$, which does not result in any additional independent equation.

The total number of unknown forces is given by the number of member forces (equal to *m*) plus the number of reaction components (equal to *r*). The number of available independent equations of equilibrium for a plane truss is equal to two per joint ($\sum F_x = 0$ and $\sum F_y = 0$), and hence the total number of available equations of equilibrium is equal to $2j$. When the condition $m + r = 2j$ is satisfied, the truss is likely to be stable, just-rigid and statically determinate. When $m + r$ takes on a value less than $2j$, implying that there are not enough members for internal stability or not enough support restraints for external stability, we clearly have a case of an under-rigid and unstable truss. On the other hand, when $m + r$ exceeds $2j$, we have more unknown forces that the available equations of equilibrium, indicating that the truss is likely to be over-rigid and statically indeterminate. In the case of a space truss, the simplest stable configuration is that of a tetrahedron, and the requirement is that $m + r$ should not be less than $3j$ for the space truss to be stable.

However, the number count, $m + r \geq 2j$ in the case of a plane truss, and $m + r \geq 3j$ in the case of a space truss, is but a necessary condition and not a sufficient condition to guarantee stability. Internal instability may arise due to improper positioning of members and external instability may arise due to reaction components being either concurrent or not adequately oriented. Also, there may be *critical forms* of a truss, which may be unstable. Such instability can be detected by carrying out the "zero load test", whereby it is possible to conceive of a set of statically admissible internal bar forces even in the absence of external loading in a truss that satisfies the condition of just-rigidity by count.

The bar forces in simple statically determinate plane trusses can be analysed by the **method of joints**. In this method, the forces in the members are determined by the application of the equations $\sum F_x = 0$ and $\sum F_y = 0$ at the joints. We can conveniently carry out the analysis at joints in a plane truss where there are not more than two unknown bar forces (in the case of a plane truss), which can be easily solved by applying $\sum F_x = 0$ and $\sum F_y = 0$ (and also $\sum F_z = 0$ in the case of space trusses) at that joint. Thus, we proceed from one joint to another. A good analyst will try to first spot the bar forces whose magnitude can be easily determined by considering simple force equilibrium at joints, which have some special features. Typically, if there are two pairs of collinear forces, meeting at a joint and no other force, the magnitude of the forces in each pair must be equal and opposite, as indicated in Fig. 1.5a. Such a situation may be encountered when four bars meet at a joint, and no external load acts at that joint (Fig. 1.5b), or when three bars meet at a joint and there is an external force acting at the joint as indicated in Fig. 1.5c. If the value of this external force *F* is equal to zero, it follows that the bar force N_1 in Fig. 1.5c and Fig. 1.5d is also zero.

The method of joints turns out to be cumbersome when we encounter just-rigid compound and complex trusses in which there are more than two unknown bar forces at every joint in a plane truss. Although the number of independent equations of equilibrium are adequate to solve for the unknown forces, there is significant coupling among the equations. In such cases, it is more convenient to apply the **tension coefficient method**, which is a variation of the method of joints. It provides us an elegant and systematic procedure of formulating the various simultaneous equations, and is particularly suitable for solution using the computer. This method is especially suitable for solving space trusses.

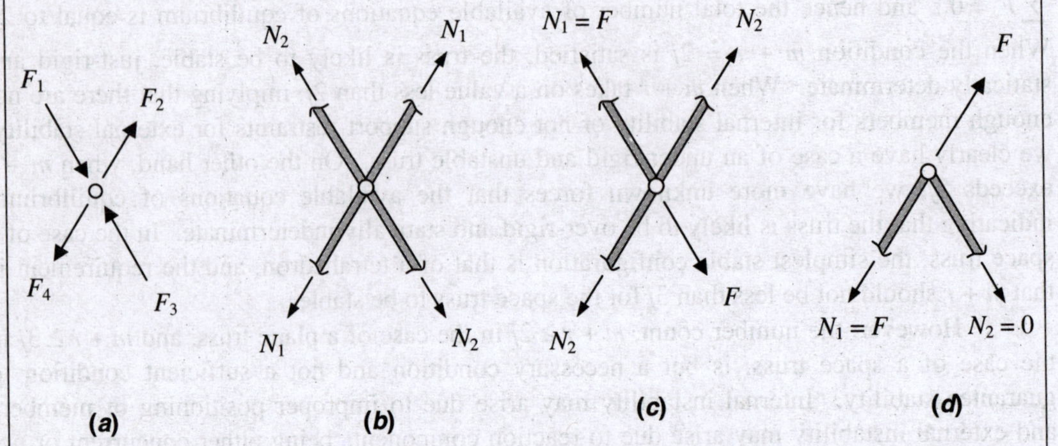

Figure 1.5 Two pairs of collinear forces

Consider a typical joint A, to which several members are connected, as shown in Fig. 1.6. The system of forces acting at the joint consists of the various bar forces (N_1, N_2, …) and the components, F_{Ax}, F_{Ay} *and* F_{Az}, of the external force (or reaction) along the three co-ordinate axes acting at the joint. We can identify the components of the bar forces along the specified Cartesian co-ordinates by defining the *direction cosines* of the various bars. For example, if the ith bar connects joint A (having co-ordinates x_A, y_A, and z_A with respect to some origin) to the joint i (having the co-ordinates x_i, y_i, and z_i), the length is given by $L_i = \sqrt{(x_i - x_A)^2 + (y_i - y_A)^2 + (z_i - z_A)^2}$. Considering the bar force N_i to point outwards from A, the direction cosines (l_i, m_i, n_i) of N_i are given by $l_i = \dfrac{(x_i - x_A)}{L_i}$, $m_i = \dfrac{(y_i - y_A)}{L_i}$ and $n_i = \dfrac{(z_i - z_A)}{L_i}$. Defining the 'tension co-efficient' T_i of the ith bar as N_i/L_i (tensile force per unit length), the equations of equilibrium at joint A can be expressed as:

$$\sum F_x = 0 \;\Rightarrow\; \sum_i T_i(x_i - x_A) + F_{Ax} = 0 \tag{1.4}$$

$$\sum F_y = 0 \;\Rightarrow\; \sum_i T_i(y_i - y_A) + F_{Ay} = 0 \tag{1.5}$$

$$\sum F_z = 0 \;\Rightarrow\; \sum_i T_i(z_i - z_A) + F_{Az} = 0 \tag{1.6}$$

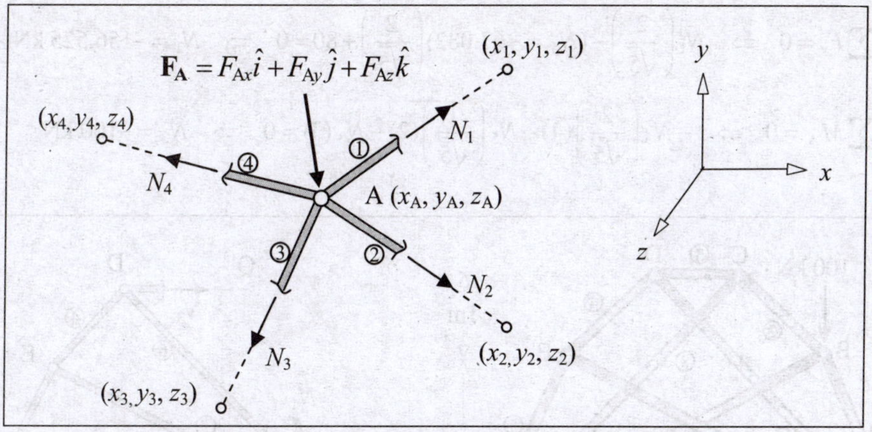

Figure 1.6 System of forces acting at a joint

When our interest is limited to finding forces in selected bars (such as the bars with the maximum compression/tension in the top/bottom chords), then the 'method of joints' and the 'tension coefficient method' may require us to solve equilibrium equations at many joints before we arrive at the desired solution. In such cases, the **method of sections** provides a powerful alternative. In this method (as applied to plane trusses), the truss is ideally cut through a section which exposes not more than three unknown bar forces, which can be solved for by applying three equilibrium equations on the cut free-body. For example, by applying $\sum M = 0$ about any joint (which may even lie outside the cut free-body) where two of the cut bars converge, we can easily find the bar force in the third bar. This method is also sometimes convenient to apply in compound trusses, as demonstrated in Fig. 1.7.

The method of joints is not suitable here to analyse the truss in Fig. 1.7a because there are three members (and hence three unknown bar forces) at any joint. We can however solve the problem by making intelligent use of the method of sections. The trick lies in isolating a suitable free-body of the truss, in which only three bar forces are exposed. We note that this is impossible to achieve if we cut a single continuous section. We can, however, achieve this by cutting more than one section and separating out a free-body in which only three bars are cut, as indicated in Fig. 1.7b, thereby exposing only N_1, N_3 and N_5, which can now be easily solved.

$$\sum M_Q = 0 \;\Rightarrow\; N_5\left(\frac{2}{\sqrt{5}}\right)(2.5) - N_5\left(\frac{1}{\sqrt{5}}\right)(1.5) + 80\,(1.5) = 0 \;\Rightarrow\; N_5 = -67.082 \text{ kN}$$

$$\sum F_y = 0 \;\Rightarrow\; N_1\left(\frac{2}{\sqrt{5}}\right) - (N_5 = -67.082)\left(\frac{2}{\sqrt{5}}\right) + 80 = 0 \;\Rightarrow\; N_1 = -156.525 \text{ kN}$$

$$\sum M_A = 0 \;\Rightarrow\; -N_5\left(\frac{2}{\sqrt{5}}\right)(4) - N_5\left(\frac{1}{\sqrt{5}}\right)(2) - N_3\,(3) = 0 \;\Rightarrow\; N_3 = -100 \text{ kN}$$

$$\sum F_y = 0 \;\Rightarrow\; N_1\left(\frac{2}{\sqrt{5}}\right) - (N_5 = -67.082)\left(\frac{2}{\sqrt{5}}\right) + 80 = 0 \;\Rightarrow\; N_1 = -156.525 \text{ kN}$$

$$\sum M_A = 0 \;\Rightarrow\; -N_5\left(\frac{2}{\sqrt{5}}\right)(4) - N_5\left(\frac{1}{\sqrt{5}}\right)(2) - N_3(3) = 0 \;\Rightarrow\; N_3 = -100 \text{ kN}$$

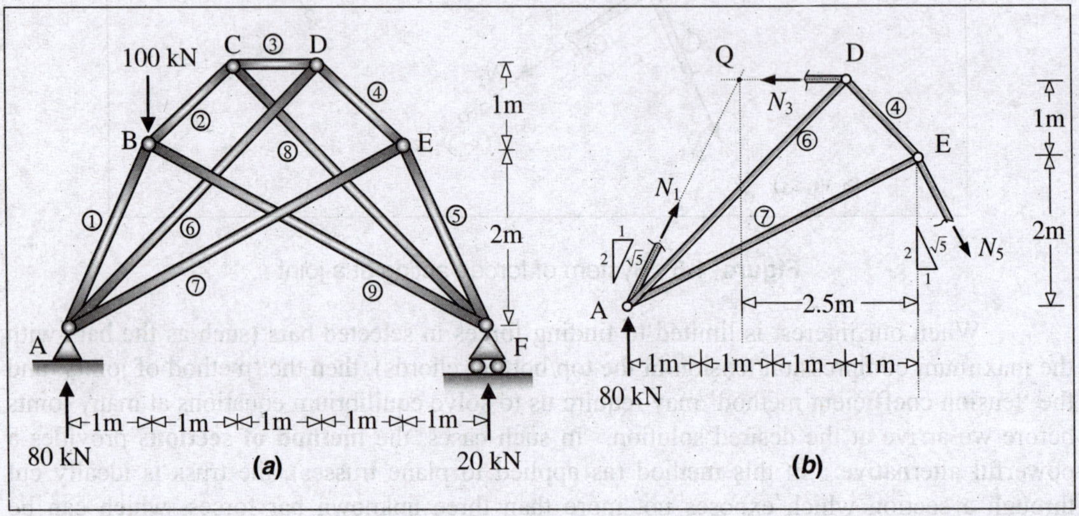

Figure 1.7 Clever use of method of sections to analyse a compound truss

There are many other methods of solution possible, including graphical methods and application of the Principle of Virtual Displacements.

1.5.3 Internal Forces and Deformations in Beams

The *beam* may be defined as a skeletal structural member that resists externally applied loads essentially by bending. Usually, the external forces are applied in a direction that is perpendicular to the longitudinal axis of the beam. The overall force equilibrium in the beam (as a free-body) is maintained by appropriate support reactions equilibrating with the externally applied loads. The main internal force resultant in a beam is the so-called **bending moment**, $M(x)$, which may be viewed as a couple comprising two normal forces (one compression and the other tension), separated by a lever arm. Its magnitude is obtained by considering moment equilibrium, $\sum M = 0$, on any free-body (cut-section) exposing the section at location x along the longitudinal axis of the beam. This bending moment may vary along the length of the beam, and when this happens, equilibrium demands the presence of another force resultant, namely, **shear force**, $S(x)$, which is also given by considering force equilibrium, $\sum F_y = 0$, on the free-body.

Bending moment causes *bending* in the form of a change in *curvature*, while shear force induces *shear strains*. Fig. 1.8 shows the sign conventions adopted in this book for positive bending moment M, curvature φ, shear force S and shear strain γ. A positive sign for S is assigned when the moment produced by the shear forces on an element is in the clockwise sense (Fig. 1.8a). The associated shear strain γ (change in angle per unit length) is also indicated in Figs 1.8a and b. The bending moment M, as well as the corresponding curvature φ, is assigned a positive sign when the beam undergoes *sagging* (Fig. 1.8c) and a negative sign when the beam undergoes *hogging* (Fig. 1.8d). Under linear elastic behaviour, a constant shear force in a prismatic beam segment will induce a constant shear strain, which is given by the shear force divided by the **shear rigidity**, GA', where G is the modulus of rigidity and A' is the modified area of cross-section of the beam (modified to account for the non-uniform variation of shear stress under flexural shear). Similarly, a constant bending moment will make the beam segment bend into a circular segment of radius $R = 1/\varphi$, with the curvature φ given by the bending moment divided by the **flexural rigidity**, EI, where E is the modulus of elasticity and I is the second moment of area of the cross-section.

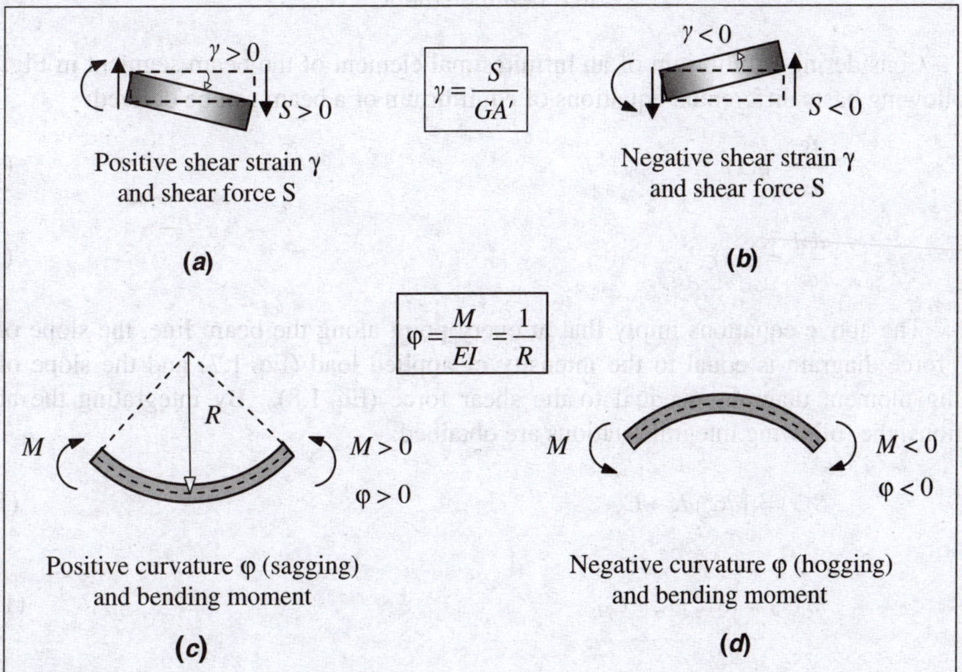

Figure 1.8 Sign conventions for beam analysis

Consider a free-body of a beam element (along the horizontal x-axis), as shown in Fig. 1.9. The load intensity, $q(x)$, is considered positive while acting in a vertically upward y-direction. The shear force $S(x)$ produces a clockwise moment with respect to the origin (located to the left of the section at x) and thus is assigned a positive sign. The bending moment $M(x)$, as indicated, causes sagging of the beam element and thus is assigned a positive sign. The directions of q, S and M indicated at the origin O are also positive.

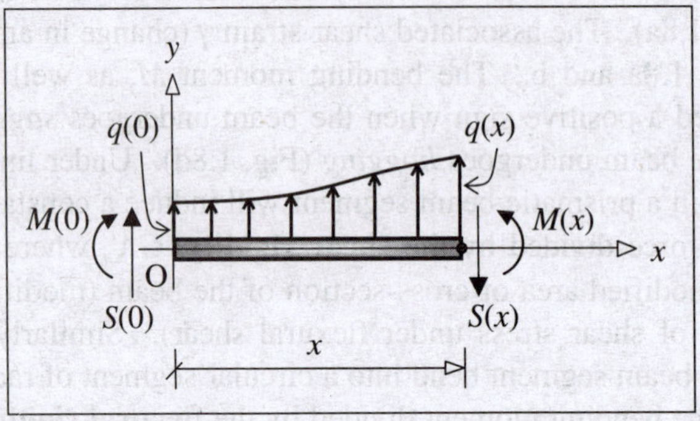

Figure 1.9 Assumed positive directions of loading, shear force and bending moment in a beam element

Considering equilibrium of an infinitesimal element of the beam segment in Fig. 1.9, the following basic differential equations of equilibrium of a beam can be derived:

$$\frac{dS}{dx} = q(x) \tag{1.7}$$

$$\frac{dM}{dx} = S(x) \tag{1.8}$$

The above equations imply that at every point along the beam line, the slope of the shear force diagram is equal to the intensity of applied load (Eq. 1.7) and the slope of the bending moment diagram is equal to the shear force (Eq. 1.8). By integrating the above equations, the following integral relations are obtained:

$$S(x) = \int_a^x q(\xi)d\xi + C_S \tag{1.9}$$

$$M(x) = \int_a^x S(\xi)d\xi + C_M \tag{1.10}$$

where C_S and C_M are constants (having the dimensions of shear force and bending moment respectively), whose values are determinable by satisfying appropriate *static* boundary conditions.

Eq. 1.9 implies that the difference between the values of shear force at two points on the beam line is equal to the area of the loading diagram between the two points. Similarly, Eq. 1.10 implies that the difference between the values of bending moment at two points on the beam line is equal to the area of the shear force diagram between the two points.

The above relationships between the three *static* variables, viz., load intensity $q(x)$, shear force $S(x)$ and bending moment $M(x)$ resemble the relationships between the three *kinematic* variables, viz., curvature $\varphi(x)$, slope $\theta(x)$ and deflection $\Delta(x)$, depicted in Fig. 1.10. Referring to the figure, we note that the slope $\theta(x)$ is the gradient of the deflection $\Delta(x)$ at any location x:

$$\theta(x) = \frac{d\Delta}{dx} \tag{1.11}$$

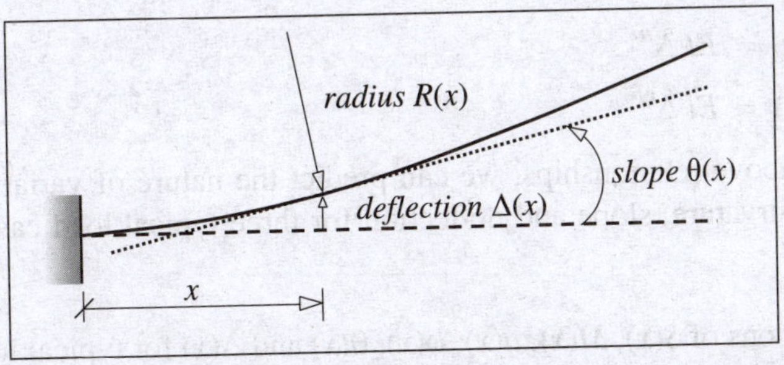

Figure 1.10 Displacement response quantities in a beam element

The curvature, $\varphi(x)$, defined as the reciprocal of the radius of curvature R at any location x, is obtained from differential calculus as

$$\varphi(x) = \frac{\dfrac{d^2\Delta}{dx^2}}{\left[1+\left(\dfrac{d\Delta}{dx}\right)^2\right]^{3/2}} \tag{1.12}$$

This complicated expression can be simplified by noting that the deflections and slopes in most real structures are very small. Then, the term $\left(\frac{d\Delta}{dx}\right)^2$ will turn out to be negligible, whereby Eq. 1.12 reduces to a simple expression for curvature as the second derivative of the deflection. Invoking Eq. 1.11, curvature $\varphi(x)$ can be approximated as the gradient of the slope $\theta(x)$:

$$\varphi(x) \approx \frac{d^2\Delta}{dx^2} \approx \frac{d\theta}{dx} \tag{1.13}$$

The relevant equations to estimate the slope $\theta(x)$ and deflection $\Delta(x)$ from curvature $\varphi(x)$ at any location x can now be expressed in an integral form, similar to Eqns 1.9 and 1.10:

$$\theta(x) = \int \varphi(x) + C_\theta \tag{1.14}$$

$$\Delta(x) = \int \theta(x) + C_\Delta \tag{1.15}$$

where C_θ and C_Δ are constants (having the dimensions of slope and deflection respectively), whose values are determinable by satisfying appropriate *kinematic* boundary conditions.

Invoking the linear relationship between bending moment and curvature, in terms of flexural rigidity *EI*, the following relationships between the force response variables and deflection $\Delta(x)$ can be generated, assuming linear elastic behaviour:

$$M(x) = EI\,\Delta'' \tag{1.16}$$

$$S(x) = EI\,\Delta''' \tag{1.17}$$

$$q(x) = EI\,\Delta^{\text{IV}} \tag{1.18}$$

Using the above relationships, we can predict the nature of variation of shear force, bending moment, curvature, slope and deflection for three typical load cases, as indicated in Table 1.1.

Table 1.1 Variations of $S(x)$, $M(x)$, $q(x)$, $\varphi(x)$, $\theta(x)$ and $\Delta(x)$ for typical loading conditions

Load case / Response variable	(1) Linearly varying $q(x)$ 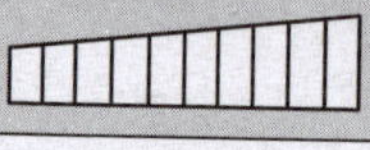	(2) Uniformly distributed q 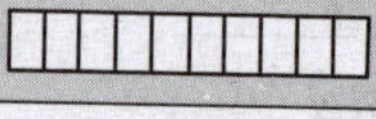	(3) Concentrated Q 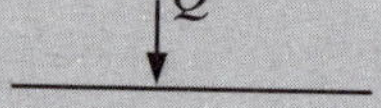
Shear force $S(x)$	Quadratic (2nd order)	Linear (1st order)	Uniform
Bending moment $M(x)$	Cubic (3rd order)	Quadratic (2nd order)	Linear (1st order)
Curvature $\varphi(x)$	Cubic (3rd order)	Quadratic (2nd order)	Linear (1st order)
Slope $\theta(x)$	Quartic (4th order)	Cubic (3rd order)	Quadratic (2nd order)
Deflection $\Delta(x)$	Quintic (5th order)	Quartic (4th order)	Cubic (3rd order)

The slope $\theta(x)$ and deflection $\Delta(x)$ are related to the curvature $\varphi(x)$ in exactly the same way as the shear force $S(x)$ and the bending moment $M(x)$ are related to the load intensity $q(x)$. This close correspondence between the kinematic and static variables has given birth to an ingenious method called the **conjugate beam method**, which is similar to the **moment area method** and can be conveniently used to find slopes and deflections in beams with known curvature distributions and boundary conditions. If we visualise an equivalent beam, called *conjugate beam*, with appropriate boundary conditions, with a distributed loading having intensity $\tilde{q}(x)$ equal to the curvature $\varphi(x)$ acting on the conjugate beam, then at any location x, the shear force $\tilde{S}(x)$ will be equal to the slope $\theta(x)$ in the real beam and the bending moment $\tilde{M}(x)$ will be equal to the deflection $\Delta(x)$. The fixed end becomes free and the free end becomes fixed in the conjugate beam; The hinged / roller support remains unchanged in the conjugate beam; a continuous beam support is replaced by an internal hinge in the conjugate beam, and an internal hinge is replaced by a continuous (intermediate) beam support in the conjugate beam.

Slopes and deflections in beams can alternatively be calculated using the method of direct integration. The Principle of Virtual Forces (or Unit Load Method) also provides a very powerful method for finding displacements in structures, as described in Section 1.6.2.

1.5.4 Axial Forces in Cables and Funicular Arches

Cables are structural elements that resist loads by developing axial tension. The main loads constitute gravity loads (concentrated or distributed) in cables that support suspension bridges, cable stayed bridges, suspended roof systems, cable car systems, etc. In such cases, on the cables are usually gravity loads (either concentrated or distributed) transmitted from the suspended system (deck slab, roof or trolley). Compared to these heavy loads, the self weight of the cable is usually negligible, and hence is generally ignored in the force analysis. In the force analysis of the cable, we commonly make the assumptions that the cable is perfectly flexible and inextensible.

The cable system has a unique feature that makes it different from other structural components, such as beams, trusses and frames. By virtue of its flexibility, the geometric configuration of the cable (cable profile) changes with the applied load. This happens simply because the equilibrium of forces has to be satisfied at all locations in the cable, with the resultant internal force at any point in the cable being an axial tension (zero shear force and bending moment). It thus becomes obvious that the cable profile is made of straight line segments when it is subject to concentrated loads, and will have a smooth curve when it is subject to distributed loads using its length.

Arches, unlike cables, resist external loads predominantly through axial compression, instead of axial tension. Indeed, if the cable profiles, with the same system of gravity loads are inverted, we have ideal arches (also called *funicular arches*) in which only axial compressive forces are present. However, unlike cables, arches are relatively rigid structures, which do not change shape when the applied loads are changed. In other words, a given arch configuration will be ideal (funicular) only for a particular type of loading, and when other loads act on the arch, the axial compressive forces will be accompanied by bending moments and shear forces.

In most practical problems involving cables and arches, the loads applied are gravity loads. For simplicity, let us consider the case where the two hinged supports are at the same elevation, and only two concentrated loads are involved as shown in Fig. 1.11. The system in Fig. 1.11a refers to a cable, whereas the one in Fig. 1.11b refers to a funicular arch, one being the inverted form of the other. We would expect the internal forces in the segments AC, CD and DB to identical in the two cases, the only difference being that these forces are tensile in the cable, whereas they are compressive in the arch. The vertical reaction components V_A and V_B will also be identical in the two cases (considering moment equilibrium with respect to the support A or B). These vertical reaction components will be identical to those in a simply supported beam (or rigid body), in which the same system of vertical load is applied (Fig. 1.11e). The horizontal reaction component H in the funicular systems (pointing outward in the case of the cable and inward in the case of the funicular arch) will have a magnitude such that the resultant reactions at supports A and B will have angles of inclination that correspond to those of the funicular segments AC and DB.

If we cut any section through the cable or funicular arch, as indicated in Figs 1.11c and 1.11d, we note that the horizontal component of the axial force in the cable will be equal to H, in order to satisfy horizontal force equilibrium ($\Sigma F_x = 0$). This is true for all segments, provided the external loads have no horizontal components. For this reason, the horizontal force H, which is a constant, is sometimes referred to as **horizontal tension** (short for horizontal component of axial tension) in the case of the cable, and **horizontal thrust** (short for horizontal component of axial thrust) in the case of the arch.

The resultant axial force $N(x)$ and its vertical and horizontal components $V(x)$ and H respectively at any location x along the horizontal span form a triangle of forces such that:

$$N(x) = \sqrt{V(x)^2 + H^2} \tag{1.19}$$

$$V(x) = H \tan\theta(x) \tag{1.20}$$

$$H = N(x)\cos\theta(x) \tag{1.21}$$

where $\theta(x)$ is the angle of inclination of the tangent to the cable at x with respect to the horizontal. Force equilibrium and comparison with the free-body of equivalent simply supported beam in Fig. 1.11f also reveal that $V(x)$ is numerically equal to the shear force $S(x)$ in the beam. Also, if we consider any free-body of the cable or funicular arch (Figs 1.11c and 1.11d), we note that the pair of equal and opposite horizontal force component H form a couple, generating a hogging moment that neutralizes the sagging moment $M^{o}(x)$ generated by the system of vertical forces. If $y(x)$ is the difference in elevation between the left end support and the cable or arch section under consideration at a distance x, then the condition of moment equilibrium requires:

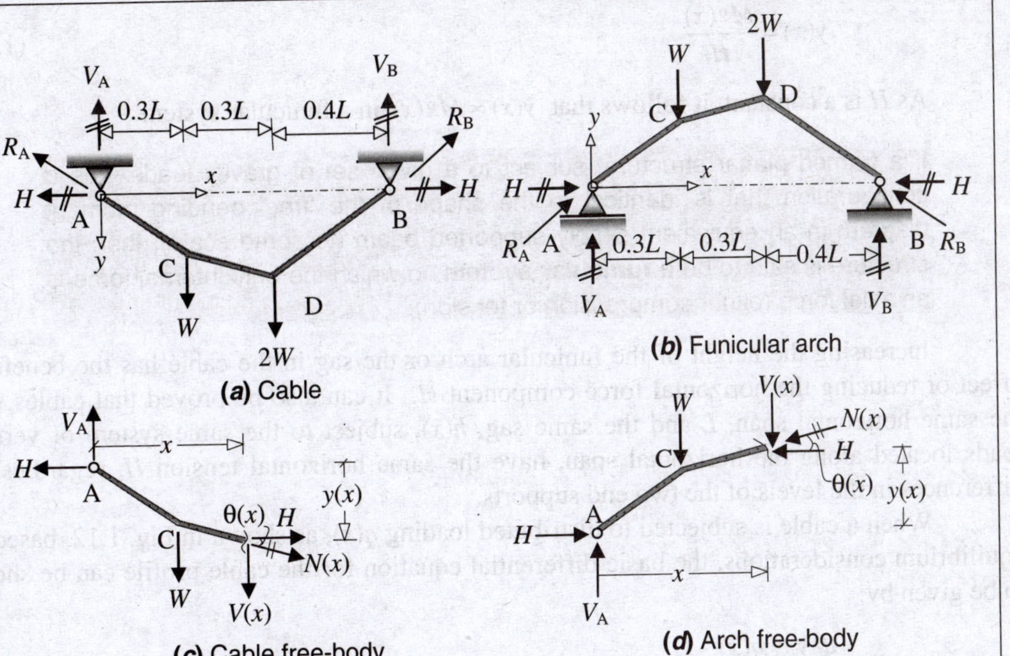

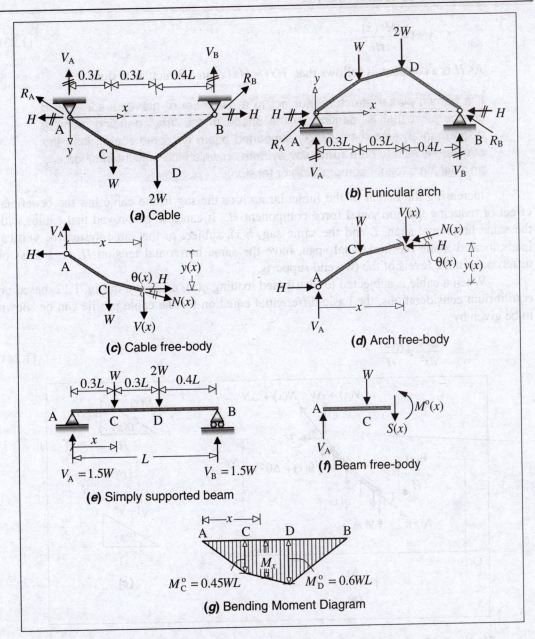

Figure 1.11 Funicular cable and arch profiles and free bending moment diagram

$$M(x) = M^{o}(x) - Hy(x) \tag{1.22}$$

where $M(x)$ is the net moment in the arch or cable section at x. In the case of funicular systems, $M(x) = 0$ at all locations, whereby

$$y(x) = \frac{M^{\circ}(x)}{H} \tag{1.23}$$

As H is a constant, it follows that $y(x) \propto M^{\circ}(x)$ in a funicular system.

If a framed planar structure, subject to a given set of gravity loads, has a configuration that is identical to the shape of the 'free' bending moment diagram in an equivalent simply supported beam (to some scale), then the structure is said to be a **funicular system**, in which the only internal force is an axial force (either compression or tension).

Increasing the height of the funicular arch or the sag in the cable has the beneficial effect of reducing the horizontal force component H. It can also be proved that cables with the same horizontal span, L and the same sag, $h(x)$, subject to the same system of vertical loads located along the horizontal span, have the same horizontal tension H, regardless of differences in the levels of the two end supports.

When a cable is subjected to distributed loading $q(x)$, as shown in Fig. 1.12, based on equilibrium considerations, the basic differential equation for the cable profile can be shown to be given by:

$$\frac{d^2y}{dx^2} = \frac{q(x)}{H} \tag{1.24}$$

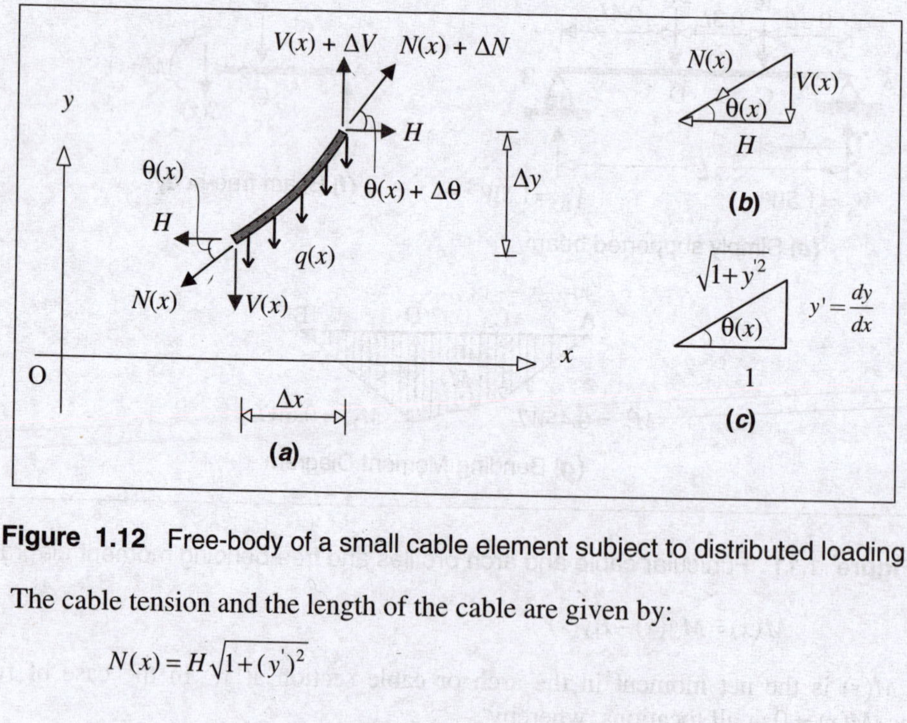

Figure 1.12 Free-body of a small cable element subject to distributed loading

The cable tension and the length of the cable are given by:

$$N(x) = H\sqrt{1+(y')^2} \tag{1.25}$$

$$S \approx \int_0^L \left[1 + \frac{y'(x)^2}{2} \right] dx \tag{1.26}$$

Using the above equations, for the simplest case of a cable supporting a uniformly distributed load, q_o, which is constant along the horizontal span, it can be shown that the cable profile is parabolic. For a symmetrical cable, with span L and sag at mid-span h, the horizontal tension is easily given by:

$$H = \frac{q_o L^2}{8h} \tag{1.27}$$

When the cable is subject to uniformly distributed loading along the curved length (as in the case of self-weight), the profile is that of a *catenary*. It follows, therefore, that when a catenary arch is subject to self-weight, there will be only axial compressive forces in the arch. Similarly, the parabolic profile is the funicular arch profile corresponding to gravity loading that is uniformly distributed along the horizontal span, corresponding to which the horizontal thrust is given by $H = \frac{q_o L^2}{8h}$ for a symmetrical arch (refer Eq. 1.27). If the loading consists of a concentrated load, the funicular profile, which corresponds to the 'free' bending moment diagram, is made up of two straight lines ('linear arch').

Sometimes, it may be possible to use the results of funicular arches, even to solve non-funicular problems. Consider, for example, a two-hinged parabolic arch subject to uniformly distributed loading on one-half of the span, as shown in Fig. 1.13a. Such an arch would normally be treated as a statically indeterminate arch with the horizontal thrust H as the redundant. However, we note that the arch in Fig. 1.13b, with the loading applied on the other half-span, is a mirror image of the arch in Fig. 1.13a.

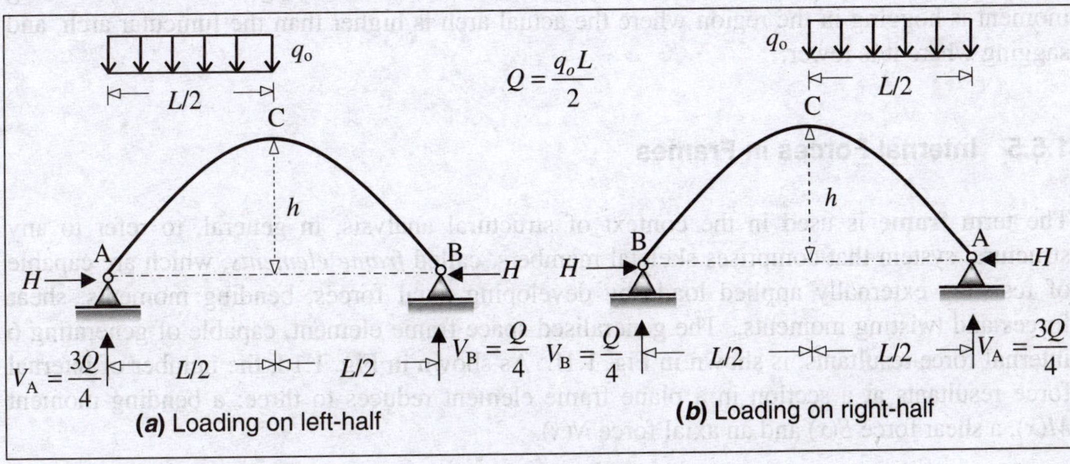

Figure 1.13 Demonstration of Principle of Parity

Figs 1.13a and b actually represent the same arch viewed from two opposite sides, whereby the horizontal reaction H must be the same. This is sometimes referred to as the **Principle of Parity**†. By applying the Principle of Superposition to the two loading cases in Fig. 1.13 and observing that for the resulting loading case the parabolic arch profile is funicular with a horizontal reaction equal to $\frac{q_o L^2}{8h}$, it follows that $H = \frac{q_o L^2}{16h}$ for the arches in Figs 1.13a and b. These arches are non-funicular and the shear force and bending moment can be derived at any section, in addition to the axial force.

For a non-funicular arch, with known values of the vertical force distribution $V(x)$ and horizontal thrust H, the normal thrust $N(x)$ and shear force $S(x)$ can be obtained by means of the following coordinate transformation in terms of the gradient $y' = \frac{dy}{dx}$ at x:

$$\begin{Bmatrix} N(x) \\ S(x) \end{Bmatrix} = \frac{1}{\sqrt{1+(y')^2}} \begin{bmatrix} 1 & y' \\ -y' & 1 \end{bmatrix} \begin{Bmatrix} H \\ V(x) \end{Bmatrix} \tag{1.28}$$

Furthermore, it can be proved that the bending moment at any section of an arch is proportional to the vertical intercept (the difference in elevation) between the point of the centre-line of the given arch and the corresponding point (vertically above or below) on the corresponding funicular arch having the same span and horizontal thrust for a given system of loads. This is sometimes referred to as **Eddy's Theorem**. Using this concept, the bending moment distribution in a non-funicular statically determinate arch can be easily generated, as demonstrated in Fig. 1.14 for a three-hinged arch subject to a concentrated load.

The funicular arch shown in Fig. 1.14b has the same horizontal thrust $H = M^o{}_B / y_B$ as the three-hinged arch in Fig. 1.14b. The bending moment diagram is generated by superimposing the centre-lines of these two arches, as shown in Fig. 1.14c. The bending moment is hogging in the region where the actual arch is higher than the funicular arch, and sagging where it is lower.

1.5.5 Internal Forces in Frames

The term **frame** is used in the context of structural analysis, in general, to refer to any structural system that comprises skeletal members, called *frame elements*, which are capable of resisting externally applied loads by developing axial forces, bending moments, shear forces and twisting moments. The generalised space frame element, capable of generating 6 internal force resultants, is shown in Fig. 1.1f. As shown in Fig. 1.1d, the number of internal force resultants at a section in a plane frame element reduces to three: a bending moment $M(x)$, a shear force $S(x)$ and an axial force $N(x)$.

† Applying this principle, we also note that the values of the response quantities at the crown C (bending moment M_C, shear force S_C and deflection Δ_C) will be identical in the two cases.

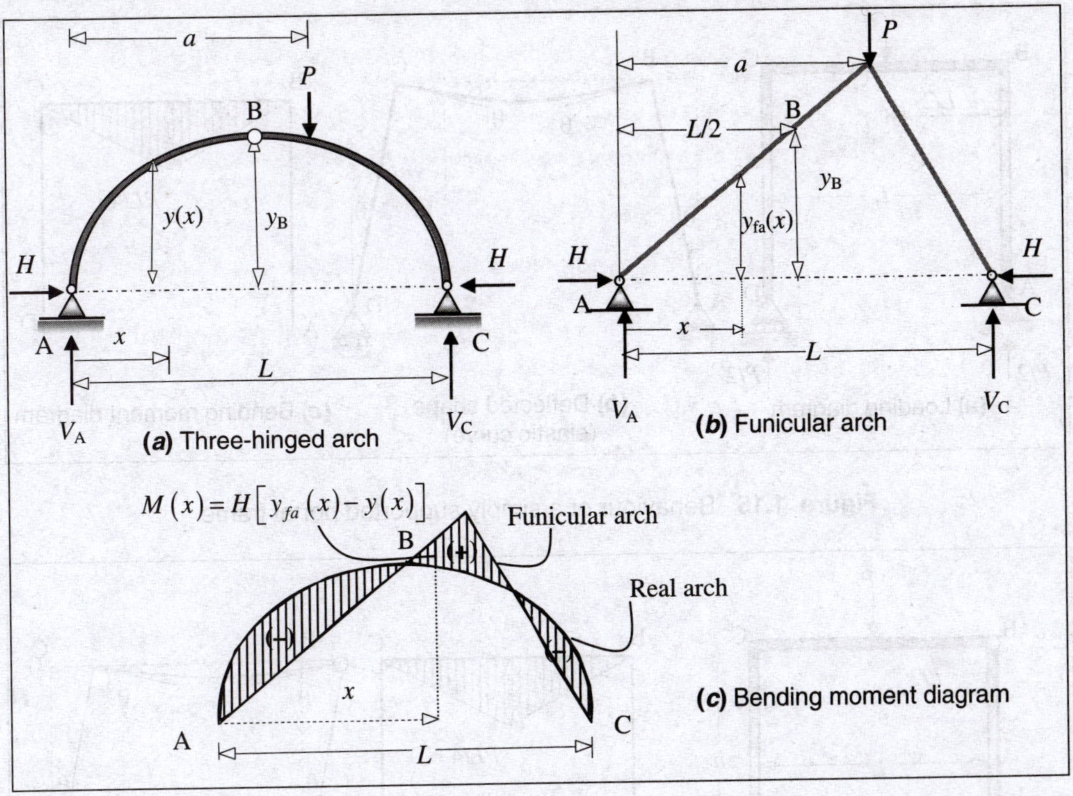

Figure 1.14 Eddy's Theorem to find the bending moment diagram in a three-hinged arch

If there are no intermediate external loads acting on a frame element, the shear force and axial force will be constant along the length of the member and the bending moment will vary linearly (or will be equal to zero). Flexural (bending) action is usually predominant in frame elements, while axial and shear deformations are generally considered to be negligibly small in most cases. It is instructive to draw the deflected shape of the frame under any given loading, and to relate the curvatures to the bending moment diagram, as in the case of the beam. This is demonstrated in Figs 1.15 and 1.16 for typical portal frames. The axial force and shear force distributions can also be drawn.

1.5.6 Influence Lines

An *influence line diagram* (or simply, *influence line*) is a graph which depicts the variation of a particular force response function (such as a support reaction or an internal force at a given section) for different positions of a unit concentrated load on the span of a structure. The ordinate of the influence line at any point on the span gives the value of the response function due to a *unit load* at that point.

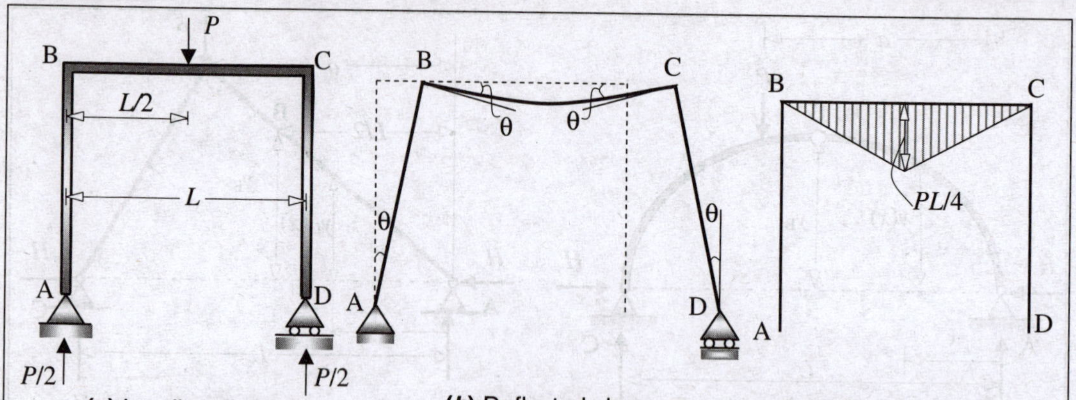

(*a*) Loading diagram (*b*) Deflected shape (elastic curve) (*c*) Bending moment diagram

Figure 1.15 Behaviour of a simply supported portal frame

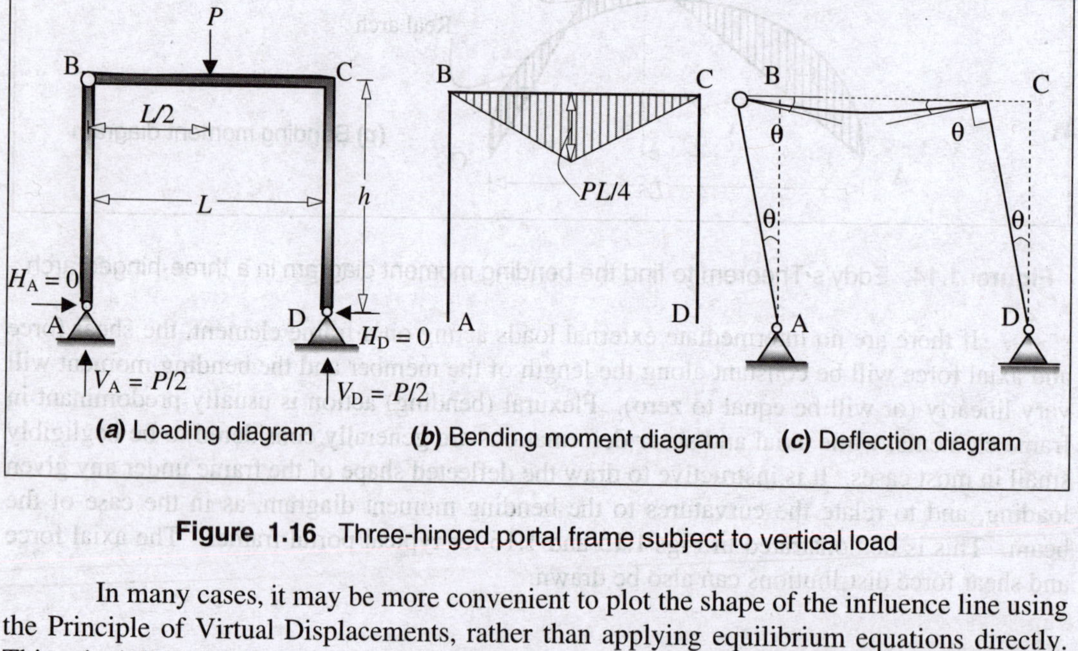

(*a*) Loading diagram (*b*) Bending moment diagram (*c*) Deflection diagram

Figure 1.16 Three-hinged portal frame subject to vertical load

In many cases, it may be more convenient to plot the shape of the influence line using the Principle of Virtual Displacements, rather than applying equilibrium equations directly. This principle, when applied to influence lines, is also known as **Müller-Breslau's Principle**, and states that the influence line for any force response function in a structure is given by the deflected shape of the structure resulting from a unit displacement corresponding to the force under consideration. Figs 1.17 and 1.18 illustrate typical influence lines for various force response quantities in a cantilever beam and a simply supported beam with overhangs respectively.

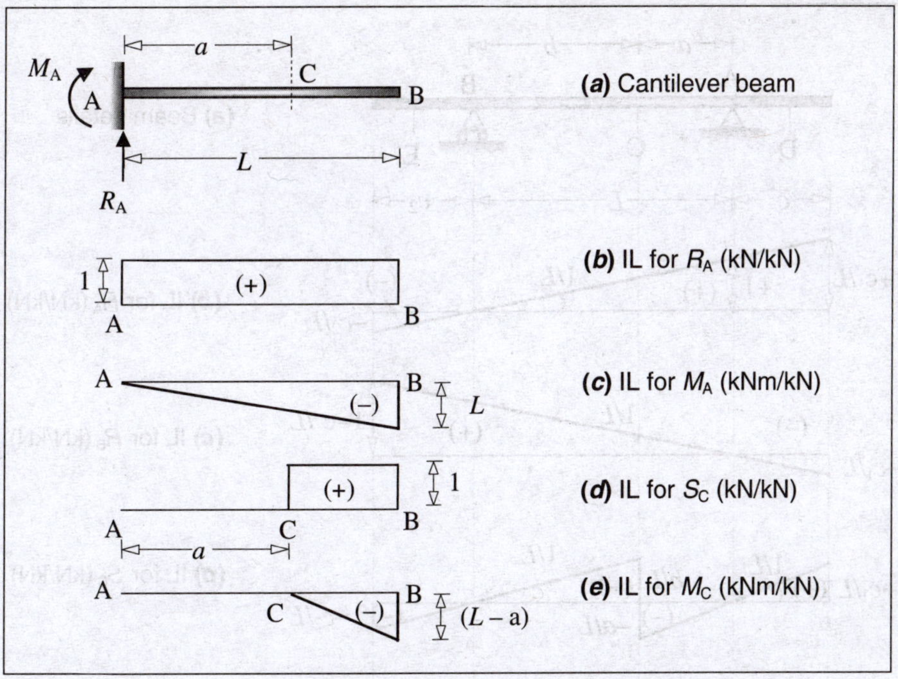

Figure 1.17 Typical influence lines for a cantilever beam

The influence line for a vertical support reaction in a beam is given by the deflected shape of the beam with a unit upward deflection of the support. At a fixed end support, the influence line for the fixed end moment is obtained by giving the beam a unit rotation (clockwise or anti-clockwise) at the end under consideration.

If the influence line for bending moment at a section is required, we imagine an internal hinge at this location and give a deflection to the beam at this point such that the change in angle between the left segment and right segment is equal to unity. Similarly, if the influence line for shear force at a section is required, we visualise a cut in the beam at this location and give a relative vertical separation of unit magnitude, with the same inclination to the left segment and right segments (so that there is no relative rotation). For statically determinate structures, all these deflected shapes, and hence, the influence lines, comprise straight lines (as the structure is now rendered under-rigid), and are relatively easy to sketch.

Influence lines can be conveniently used to find the maximum values of any force response function, provided we locate the loading (which may comprise a train of concentrated loads or distributed loads) appropriately.

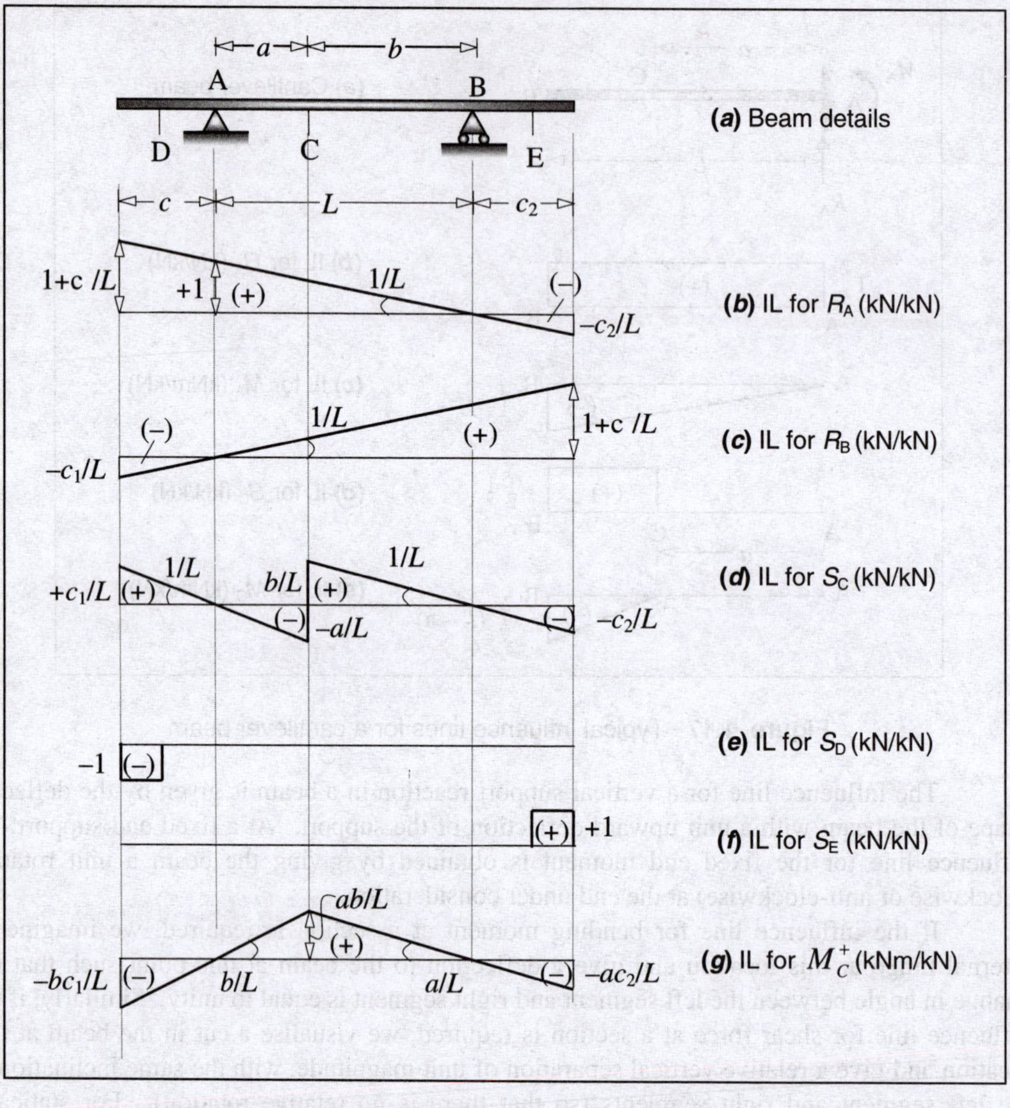

Figure 1.18 Typical influence lines for a simply supported beam with overhangs

It can be proved that the maximum bending moment at a particular section in a simply supported beam under a train of concentrated loads occurs when a particular load is located at the section, such that when this load crosses over the section, the average load on the segment to the left of the section minus the average load on the right segment changes sign. The maximum bending moment under any given concentrated load, which forms part of a train of loads moving across a simply supported beam, occurs when the mid-span point of the beam bisects the distance between the centre of gravity of the load system and the given load.

Influence line diagrams can conveniently be plotted for all types of statically determinate structures, notably beams and trusses used for bridge decks. Fig. 1.19 depicts influence lines for axial forces in typical members of a Warren truss in a 'through type' bridge deck.

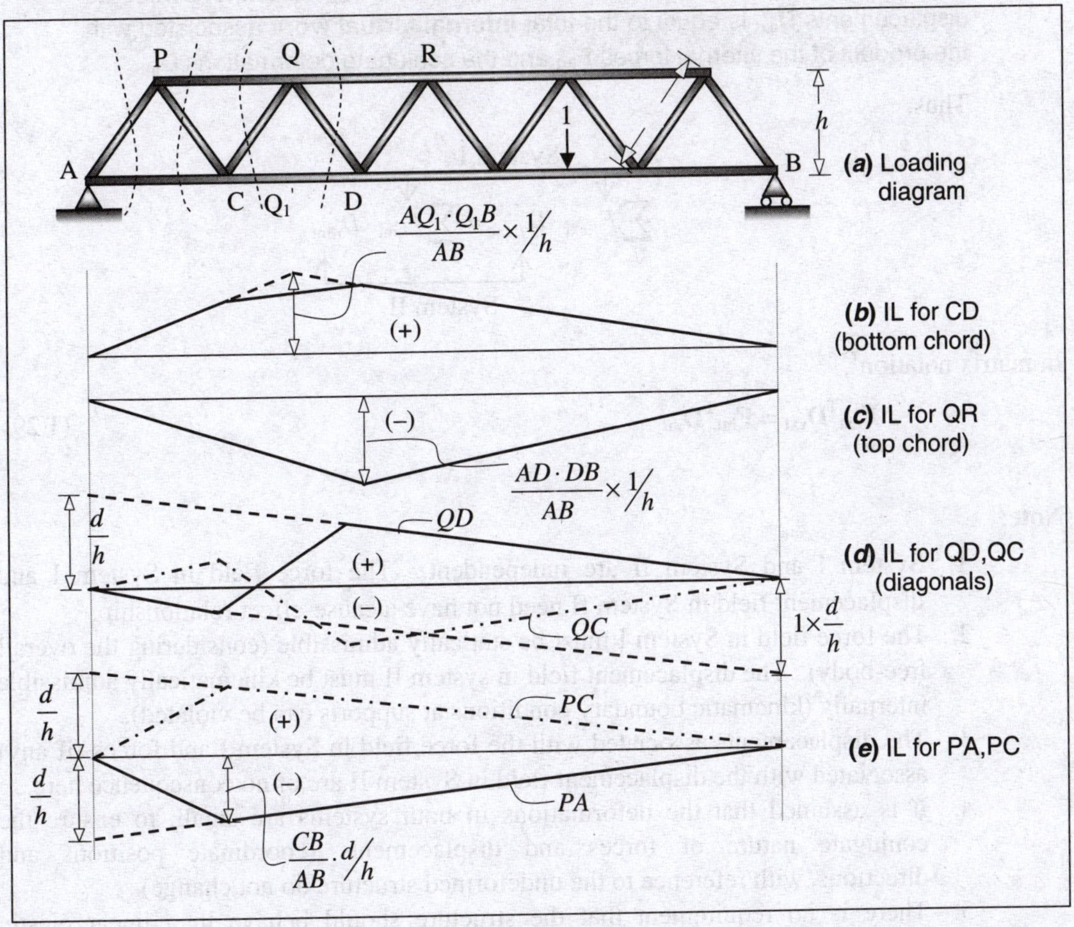

Figure 1.19 Influence lines for Warren truss (through type)

1.6 PRINCIPLE OF VIRTUAL WORK

Traditionally, the Principle of Virtual Work has been used to find unknown forces or displacements in loaded elastic bodies, by imparting appropriate perturbations (of small magnitude) to the force field or displacement field (which are inter-related by a cause-effect relationship). However, in its more generalised form, such cause-effect relationships are not required to be considered, and in this versatile form, the Principle of Virtual Work provides a very powerful tool to determine unknown quantities in the force and displacement fields in a structure.

> According to the Principle of Virtual Work, the total **external virtual work** associated with the products of external forces $\mathbf{F_{ext}}$ and the conjugate displacements $\mathbf{D_{ext}}$ is equal to the total **internal virtual work** associated with the product of the internal forces $\mathbf{F_{int}}$ and the conjugate deformations $\mathbf{D_{int}}$.

Thus,

System I

$$\sum_j F_{j,\text{ext}} \cdot D_{j,\text{ext}} = \sum_i F_{i,\text{int}} \cdot D_{i,\text{int}}$$

System II

In matrix notation[†],

$$\mathbf{F_{ext}}^{\mathbf{T}}\mathbf{D_{ext}} = \mathbf{F_{int}}^{\mathbf{T}}\mathbf{D_{int}} \tag{1.29}$$

Note:

1. System I and System II are independent. The force field in System I and displacement field in System II need not have a cause-effect relationship.
2. The force field in System I must be statically admissible (considering the overall free-body). The displacement field in system II must be kinematically admissible internally (kinematic boundary conditions at supports can be violated).
3. The displacements associated with the force field in System I and forces (if any) associated with the displacement field in System II are of no consequence here.
4. It is assumed that the deformations in both systems are small, to ensure the conjugate nature of forces and displacements (coordinate positions and directions, with reference to the undeformed structure do not change).
5. There is no requirement that the structure should behave in a linear elastic manner; even elastic behaviour is not assumed here.
6. It does not matter whether the two independent systems are real or virtual. If system I is real and system II virtual, the principle is called the *Principle of Virtual Displacements*. If the converse is true, it is referred to as the *Principle of Virtual Forces*.

[†] The force and displacement vectors, **F** and **D**, are matrices, each comprising a single column of size n. They may be alternatively represented as $\{F\}$ and $\{D\}$ respectively. Work W is a scalar, obtainable by multiplying these two vectors suitably. For this purpose, we perform a simple transposition operation on **F**, converting the vector from a column matrix of size $n\times1$ to a row matrix $\mathbf{F}^{\mathbf{T}}$ of size $1\times n$.

1.6.1 Principle of Virtual Displacements

We can invoke the Principle of Virtual Displacements to determine unknown quantities in the force field of a structure. This procedure provides an alternative to applying the conventional equations of static equilibrium, and is useful in finding support reactions and internal forces in just-rigid (statically determinate) structures.

We consider the overall free-body of the structure, and treat the various forces acting on it (including the unknown force quantity, say F_A) as system I (real force field in equilibrium). Now, we cleverly choose a separate displacement field (system II) such that, when we apply the principle, we obtain the work equation (equilibrium equation) in which only one unknown, viz., F_A, is involved. For this, there should be a finite displacement, D_A, conjugate with F_A, in the virtual displaced configuration in system II, with zero displacements at locations corresponding to other unknown forces in system I. Also, we should be able to express the displacements conjugate with the various known forces (such as given loads) in terms of D_A, and thereby obtain the external work components associated with these forces. This becomes easy when we deal with statically determinate systems, because when we release the restraint at A to impart the displacement D_A, the just-rigid structure is now rendered unstable, and the displacement imparted will be a **rigid-body displacement**. The instability of the imaginary system II is, of course, of no consequence here. The rigid body movement serves to not only make the displacements $\mathbf{D_{ext}}$ well-defined (in terms of D_A), but also implies that there are no deformations ($\mathbf{D_{int}}$) in system II. Thus, although there are internal forces ($\mathbf{F_{int}}$) in system I, the internal virtual work is invariably zero in statically determinate structures (which may be treated as rigid bodies for the purpose of finding reactions and internal forces), whereby the work Eq. 1.29 reduces to:

$$\sum_j F_{j,\text{ext}} \cdot D_{j,\text{ext}} = 0 \tag{1.30}$$

We note that the dummy displacement term D_A enters all the virtual work components in Eq. 1.30, and hence gets eliminated, resulting in an equation that involves only the unknown F_A. For this reason, this method of applying the Principle of Virtual Displacements is sometimes referred to as the **dummy displacement method**. Alternatively, we can assign a unit value to the dummy displacement D_A, and when this technique is employed, the method is sometimes referred to as the **unit displacement method**. This procedure can be used effectively to determine, one at a time, support reactions as well as internal force resultants (such as axial force in a truss member or cable, bending moment or shear force at a beam section) in statically determinate structures.

For example, to find the horizontal reaction H in the frame shown in Fig. 1.4a, we can visualise a virtual displacement field in which each of the two supports at A and B move inward horizontally by δ, but with no vertical translation, as indicated in Fig 1.20. This causes the L-shaped segments AB and BC to undergo rigid body rotations of δ/4, as shown. The corresponding deflections at the resultant load points are δ/4 and 5δ/4 (both upward) at the 20kN and 32kN load locations in Fig. 1.4a. Applying the Principle of Virtual Work (Eq. 1.30) to the real force field in Fig. 1.4a and the virtual displacement field in Fig. 1.20,

$$2(H)(\delta) - (20)(\delta/4) - (32)(5\delta/4) = 0 \Rightarrow H = 22.5 \text{ kN}$$

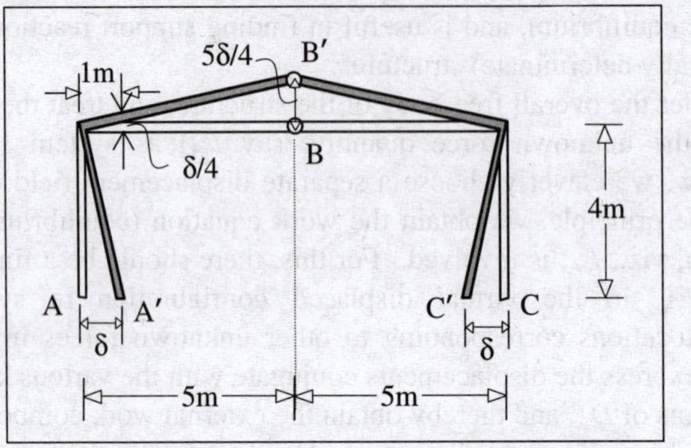

Figure 1.20 Virtual displacement field to find the horizontal support reaction in the frame in Fig. 1.4a

1.6.2 Principle of Virtual Forces

We can invoke the Principle of Virtual Forces to determine unknown quantities in the displacement field of a structure. This procedure invokes the assistance of statics to solve problems of kinematics (small changes in geometry in a structure). This is a very useful way of establishing compatibility relationships in the displacement field, and is especially useful in finding unknown displacements (deflections and rotations) in all kinds of structures, statically determinate or indeterminate. The displacement field, satisfying internal compatibility, need not be caused by external loads, and may be caused by environmental effects or constructional errors. We can invoke the Principle of Virtual Forces to find one unknown displacement at a time, by visualising an imaginary force field in which we apply a dummy load that is conjugate with the desired displacement (at the same location and with the same direction). The dummy load gets eliminated while applying Eq. 1.34, and in this form the method is called **dummy load method**. It is more convenient to consider a unit value of the dummy load, and in this form, the method is called **unit load method**.

Let us consider, for example, a displacement field in a truss whose bar elongations $\{e\}$ are known, and we need to find the horizontal deflection D_j at a joint coordinate j that is compatible with the bar deformations. These changes in bar lengths may be arbitrary, if caused by temperature effects or 'lack of fit' in statically determinate trusses (in which case, the real displacement field is not accompanied by any real force field). We now visualise a virtual force field in which the truss is subject to a unit horizontal force, $F_j = 1$, and determine a set of bar forces that is statically admissible. If the truss happens to be statically indeterminate, we can conveniently assign zero values to the redundants, and thereby arrive at any set of bar forces that satisfies force equilibrium (no need to satisfy compatibility of displacements). Denoting the axial force in the ith bar as n_{ij}, and applying the Principle of Virtual Work (Eq. 1.29),

$$D_j = \sum_{\text{all bars}} n_{ij} e_i \tag{1.31}$$

If the displacement field is caused by external loads (or by temperature effects or lack of fit in statically indeterminate trusses), then we need to first analyse the bar forces $\{N\}$ and thereby compute the bar elongations $\{e\}$. If the behaviour is linear elastic, the bar elongation e_i is easily given by the product of the bar force N_i and the axial flexibility $f_i = \frac{L_i}{(EA)_i}$, where $(EA)_i$ and L_i are respectively the axial rigidity and length of the ith bar.

$$D_j = \sum_{\text{all bars}} n_{ij} \frac{N_i L_i}{(EA)_i} \tag{1.32}$$

Similarly, if we know the variation of curvature $\varphi(x)$ along the length of a beam or frame and it is desired to find a deflection or a rotation D_j at the jth coordinate, then we analyse the same structure under a virtual load, $F_j = 1$, and determine a distribution of bending moments $m_j(x)$ that is statically admissible. If the structure happens to be statically indeterminate, we can conveniently assign zero values to the redundants, and thereby arrive at any bending moment diagram that satisfies equilibrium (no need to satisfy compatibility of displacements). Applying the Principle of Virtual Work (Eq. 1.29), and ignoring the work associated with shear and axial forces (which are generally negligible),

$$D_j = \int m_j(x)\, \varphi(x)\, dx \tag{1.33}$$

If the displacement field is caused by external loads (or by temperature effects or support movements in statically indeterminate beams and frames), then we need to first analyse the bending moment $M(x)$ due to this loading and thereby compute the curvature $\varphi(x)$. If the behaviour is linear elastic, the curvature $\varphi(x)$ is easily given by the ratio of the bending moment $M(x)$ and flexural rigidity EI, whereby

$$D_j = \int m_j(x) \frac{M(x)}{EI} dx \tag{1.34}$$

Eqns 1.33 and 1.34 can be more easily solved by the 'area multiplication' technique (instead of explicit integration), by recognising that the integrand involves the product of two continuous variables, $m_j(x)$, which is invariably linearly distributed, and $\varphi(x)$ or $M(x)/EI$. It can be shown that the 'volume integral' is easily obtained is given by the product of the area of the $\varphi(x)$ diagram and the value of $m(x)$ at the centroidal location $\bar{x}$ of the area of the $\varphi(x)$ diagram, i.e.,

$$\int \varphi(x) m(x)\, dx = [\text{Area of } \varphi(x) \text{ diagram}] \times m(\bar{x}) \tag{1.35}$$

Note that if $\varphi(x)$ is also linearly distributed, the same volume integral can be obtained as the product of the area of the $m(x)$ diagram and the value of $\varphi(x)$ at the centroid of the area of the $m(x)$ diagram. Formulas for area and centroidal distances for triangular, trapezoidal and parabolic shapes of the curvature diagram are shown in Table 1.2.

Table 1.2 Areas and centroidal distances

Shape	Area	Centroidal distance
(a) 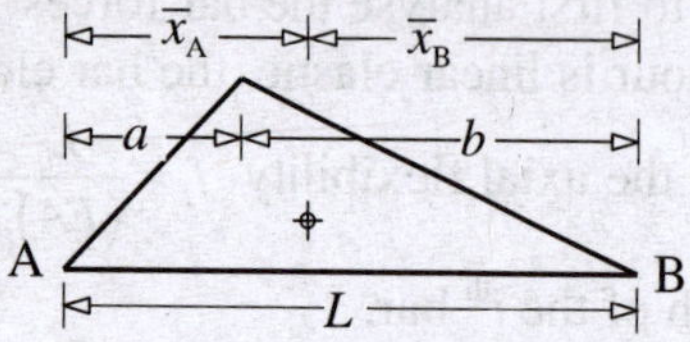	$\dfrac{\varphi L}{2}$	$\bar{x}_A = \dfrac{(a+L)}{3}$ $\bar{x}_B = \dfrac{(b+L)}{3}$
(b) 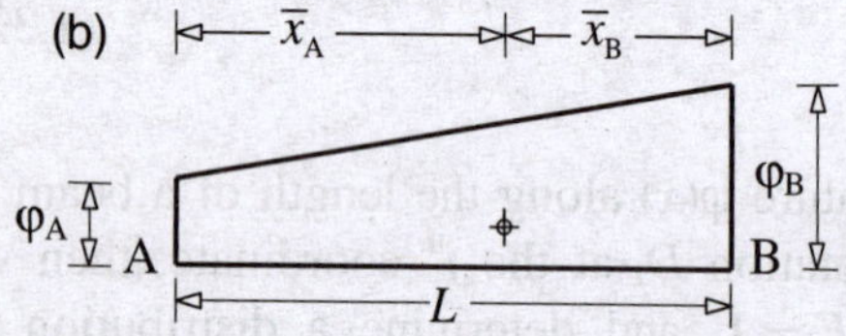	$\dfrac{\varphi_A + \varphi_B}{2}(L)$	$\bar{x}_A = \dfrac{L}{3}\left[1+\dfrac{\varphi_B}{\varphi_A+\varphi_B}\right]$ $\bar{x}_B = \dfrac{L}{3}\left[2-\dfrac{\varphi_B}{\varphi_A+\varphi_B}\right]$
(c) 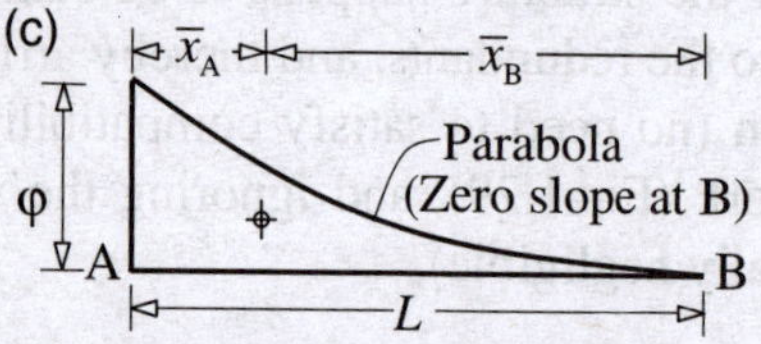	$\dfrac{1}{3}\varphi L$	$\bar{x}_A = \dfrac{L}{4}$ $\bar{x}_B = \dfrac{3L}{4}$
(d) 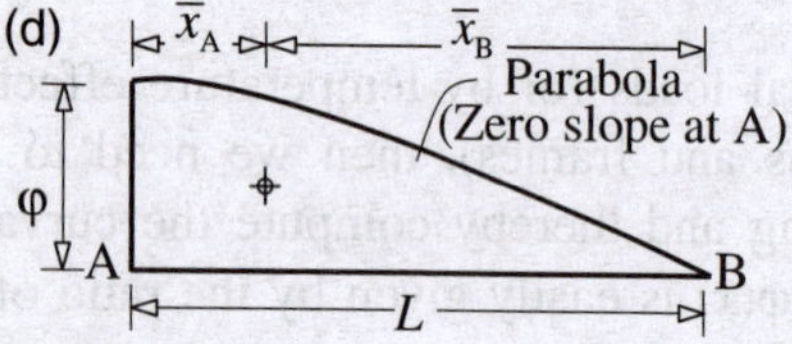	$\dfrac{2}{3}\varphi L$	$\bar{x}_A = \dfrac{3}{8}L$ $\bar{x}_B = \dfrac{5}{8}L$

If both areas happen to be trapeziums, as shown in Fig. 1.21, the volume integral is given by

$$\int \varphi(x)\, m(x)\,dx = \frac{L}{6}\left[\varphi_1(2m_1+m_2)+\varphi_2(2m_2+m_1)\right] \tag{1.36a}$$

or,

$$\int \varphi(x)\, m(x)\,dx = \frac{L}{6}\left[m_1(2\varphi_1+\varphi_2)+m_2(2\varphi_2+\varphi_1)\right] \tag{1.36b}$$

(a) Linearly distributed $\varphi(x)$ and $m(x)$

$$A_1 = \int \varphi(x)\,dx$$

$$A_2 = \int m(x)\,dx$$

$$\boxed{V = A_1 m(\bar{x}_1) = A_2 \varphi(\bar{x}_2))}$$

(b) Solids generated by $\int \varphi(x)m(x)\,dx$

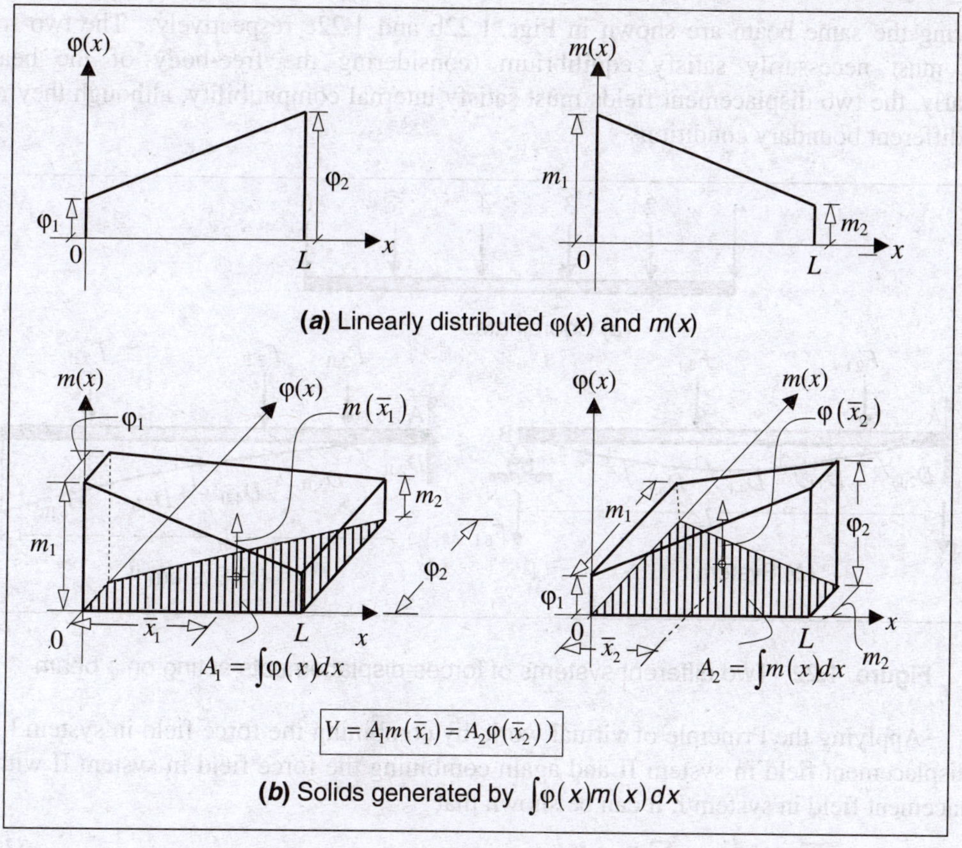

Figure 1.21   Volume integrals for linearly distributed $\varphi(x)$

1.6.3   Reciprocal Theorems

The Principle of Virtual Work provides the basis for many other principles and theorems. Energy-related principles are discussed in Section 1.7. In this Section, we review the so-called *Reciprocal Theorems* and the Müller-Breslau's Principle, which are applicable for linear elastic structures. The Reciprocal Theorems are of particular historical importance, because they provided the first break through in the solution of statically indeterminate problems.

Consider a linear elastic structure, subject to two different force-displacement conditions, which we may refer to as systems I and II. Let $\{F_\mathrm{I}\}$ and $\{F_\mathrm{II}\}$ denote the sets of external forces acting on these two systems, and let $\{D_\mathrm{I}\}$ and $\{D_\mathrm{II}\}$ denote their conjugate displacements. This is illustrated in Fig. 1.22 for the case of a simple flexural member. Six coordinate points are identified, for convenience, as shown in Fig. 1.22a. The external forces (applied loads and support reactions), as well as deflections, in two different systems involving the same beam are shown in Figs. 1.22b and 1.22c respectively. The two force fields must necessarily satisfy equilibrium (considering the free-body of the beam). Similarly, the two displacement fields must satisfy internal compatibility, although they may have different boundary conditions.

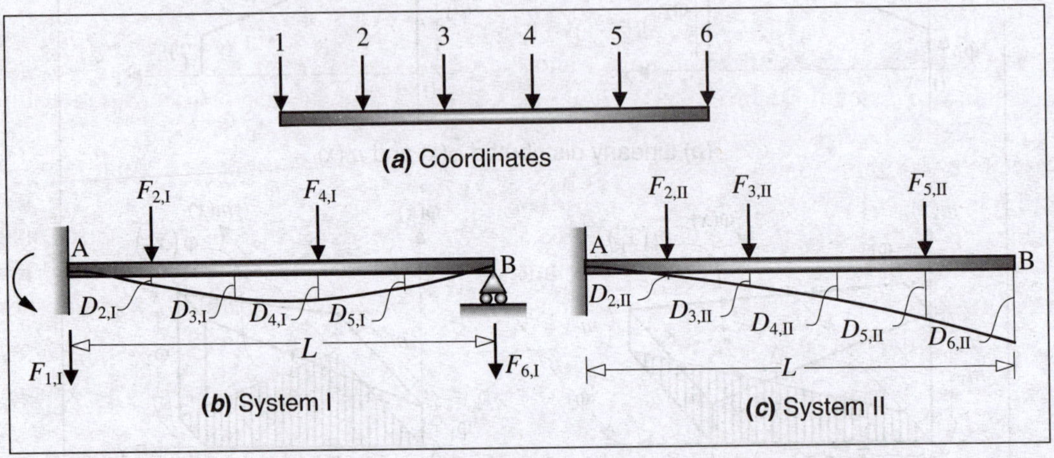

Figure 1.22 Two different systems of forces-displacements acting on a beam

Applying the Principle of virtual work, by combining the force field in system I with the displacement field in system II and again combining the force field in system II with the displacement field in system I, it can be shown that

$$\sum_j F_{j,\mathrm{I}} \cdot D_{j,\mathrm{II}} = \sum_j F_{j,\mathrm{II}} \cdot D_{j,\mathrm{I}} \tag{1.37}$$

This reciprocal relationship between the force and displacement fields in the two systems is called **Betti's Theorem**.

Betti's theorem states that the total external virtual work associated with forces $\{F_{\mathrm{I}}\}$ in system I and the conjugate displacements $\{D_{\mathrm{II}}\}$ in system II is equal to that associated with forces $\{F_{\mathrm{II}}\}$ in system II and conjugate displacements $\{D_{\mathrm{I}}\}$ in system I in a linear elastic structure.

When considerations are limited to a single load P acting in each of the two systems, Eq. 1.37 establishes that the displacement must also be identical in the two systems. In this form, the reciprocal theorem is referred to as **Maxwell's Theorem**, which in fact historically preceded Betti's Theorem.

Maxwell's reciprocal theorem states that in a linear elastic structure, if a load *F* acting at some coordinate location '1' causes a displacement *D* at some other coordinate location '2', then the same displacement *D* will occur at '1', if the load *F* acts at '2' in a separate loading condition.

It may be noted that the load *F* may be either a concentrated force (with N units) or a concentrated moment (with Nmm units), and the displacement could be a translation (with mm units) or a rotation (in radians). It is also possible to have a reciprocal relationship between a force / displacement and moment / rotation. This can be easily understood by the concept of flexibility coefficient, which involves the application of a 'unit load' (unit force or unit moment). Referring to Fig. 1.23, Maxwell's reciprocal theorem establishes that the clockwise rotation at B, f_{ji}, due to a unit force acting downward at A ($F_i = 1$) is exactly equal to downward deflection at A, f_{ij}, due to a unit moment applied at B ($F_j = 1$), provided the structure behaves in a linear elastic manner. This reciprocal relationship establishes symmetry in the **flexibility matrix**:

$$f_{ij} = f_{ji} \tag{1.38}$$

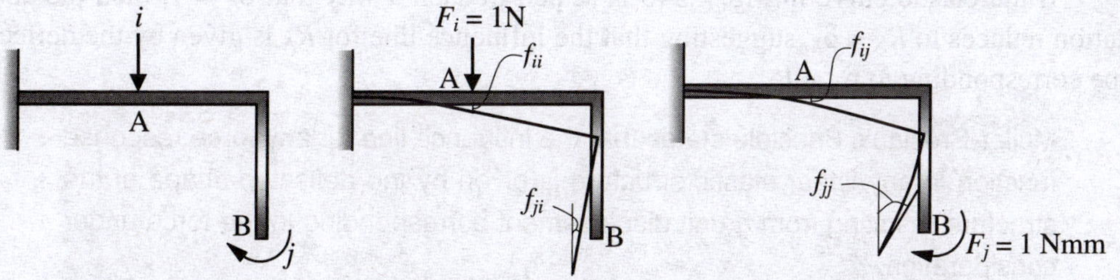

(***a***) Coordinates (***b***) Unit load (force) at *i* (***c***) Unit load (moment) at *j*

Figure 1.23 Flexibility coefficients in a linear elastic structure

Finally, we note that *Müller-Breslau's Principle* can be proved for linear elastic structures (statically determinate or indeterminate) using Betti's Theorem. Consider, for example, a continuous beam subject to a unit load that can be placed anywhere on the span, as shown in Fig. 1.24a. We are interested in generating the influence lines of the force response functions, say for the support reaction R_A, as shown in Fig. 1.24a (system I). Consider, separately, a system II involving the same beam, but with the constraint at A removed, and a force F_A (corresponding to R_A) applied so as to induce a deflection δ_A (Fig. 1.24b) at A. Applying Betti's theorem (Eq. 1.37) to the systems in Figs 1.24a and b, $(R_A)\,(\delta_A) - (1)\,(\delta_X) = (F_A)\,(0)$.

$$\Rightarrow R_A = \frac{\delta_X}{\delta_A} \tag{1.39}$$

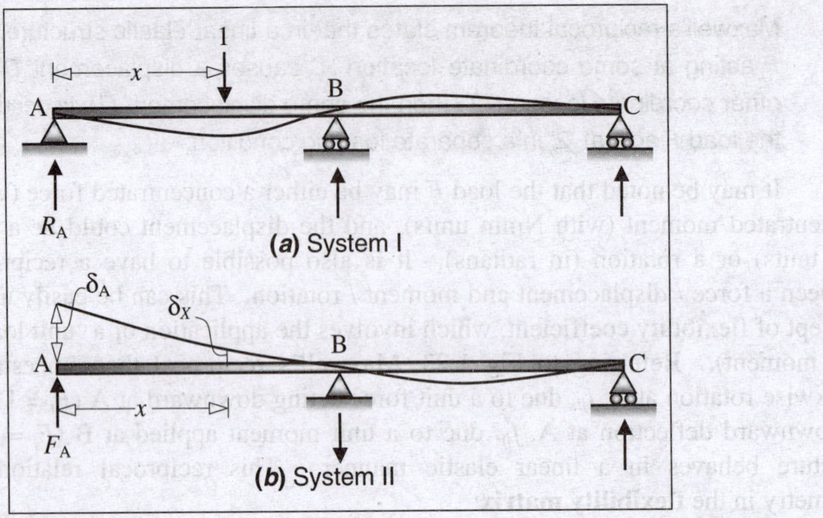

Figure 1.24 Derivation of Müller-Breslau's Principle for a continuous beam

If the elastic curve in Fig. 1.24b is scaled in such a way that $\delta_A = 1$, then the above equation reduces to $R_A = \delta_X$, suggesting that the influence line for R_A is given by the deflected shape corresponding to $\delta_A = 1$.

Müller-Breslau's Principle states that the influence line for any force response function in any linear elastic structure is given by the deflected shape of the structure resulting from a unit displacement corresponding to the force under consideration.

1.7 ENERGY PRINCIPLES

Energy principles provide powerful methods that have many uses in structural analysis. They are particularly advantageous in analysing indeterminate structures, formulating flexibility and stiffness coefficients, and in structural dynamics. Like the two forms of the Principle of Virtual Work, viz., the Principle of Virtual Displacements (PVD) and the Principle of Virtual Forces (PVF), there are two different categories of energy methods, viz. displacement-based methods and force-based methods. These methods are applicable to **elastic** structures, and are expressed in terms of energy concepts, such as *strain energy* and *complementary strain energy*, *load potential energy*, *total potential energy* and *complementary total potential energy*, which become easy to understand with increasing familiarity.

1.7.1 Potential and Kinetic Energy

Energy may be described as the ability to do work and thereby bring about change. Quantitatively, it is treated as equivalent to work, and has the same units as work (Nm or Joule in SI system). Energy exists in many forms, such as mechanical, thermal (heat), light, sound, chemical and electrical. Here, we limit our interest to mechanical energy, as observed in structural systems.

Energy may also be visualised as being of two different kinds: *potential* and *kinetic*. **Potential energy**, as the name suggests, is energy that has not yet been used and thus has the potential for being used in the future. **Kinetic energy**, on the other hand, is energy that is in use (or in motion). In structural mechanics, we associate the concept of potential energy with all forces, because of their potential to do work. Some of these forces are *conservative*, which means that their potential to do work can be retrieved (i.e., the energy is conserved, without getting lost), when their displacements are reversed. The *force of gravity*, which is inherent in the 'gravitational field' of the earth and acts on any mass located in this field, and the *elastic force*, which is inherent in an elastic structure (which behaves like a spring) are examples of conservative forces, and the potential energy associated with them are called *gravitational potential energy* and *elastic potential energy* (or *strain energy*) respectively. In *conservative* systems, when there is a movement from one position to another, the resulting change in potential energy depends only on the initial and final positions, and not on the path taken.

In general, if we consider a single rigid mass m, initially at rest, acted upon by a resultant force F, causing it to translate by a distance x in the direction of the force, then, we observe that the kinetic energy, denoted conventionally by the symbol T, imparted to the mass, is equal to Fx. From Newton's second law of motion, we also observe that the applied force F (in N) is equal to the product of the mass m (in kg) and the acceleration $\ddot{x}$ (in m/s^2); i.e., $F = m\ddot{x}$. We note that the distance x (in m) travelled linearly by the mass (assumed to have a constant acceleration) is given by $x = \frac{1}{2}\ddot{x}t^2$, where t (in s) is the time interval. Substituting these expressions in $T = Fx$, and noting that the product $\ddot{x}t$ is equal to velocity $\dot{x}$ (also denoted as v), we arrive at the well-known expression for kinetic energy, T of any mass m moving with a velocity v at any instant of time:

$$T = \tfrac{1}{2} mv^2 \tag{1.40}$$

Thus, considering the case of a mass m, initially at rest, falling freely through a height h, and equating the work done by the gravitational force, mgh (equal to the loss in gravitational potential energy) to the gain in kinetic energy, $T = \tfrac{1}{2} mv^2$, we observe that the velocity gained by the mass is equal to $\sqrt{2gh}$, using the *Principle of Conservation of Energy*, which is applicable for a conservative system.

Let us now consider the mass m to be located at the free end of a cantilever beam, having a span L and uniform flexural rigidity EI, as shown in Fig. 1.25a. Let us assume the beam to be otherwise weightless (negligible self-weight), with its original configuration being perfectly horizontal. Let the vertical (downward) deflection at the free end be denoted by Δ. For convenience, we may model the beam as a spring, as shown in Fig. 1.25b, with Δ being the displacement from the undeformed configuration of the spring. The stiffness, *k*, of the spring, defined as the force required to cause unit displacement (*P*/Δ), can be shown to be equal to $3EI/L^3$. We know, from experience, that the beam (or spring) will vibrate and eventually settle to a steady (static) configuration, with a maximum deflection, Δ_{static}, as marked in Fig. 1.25.

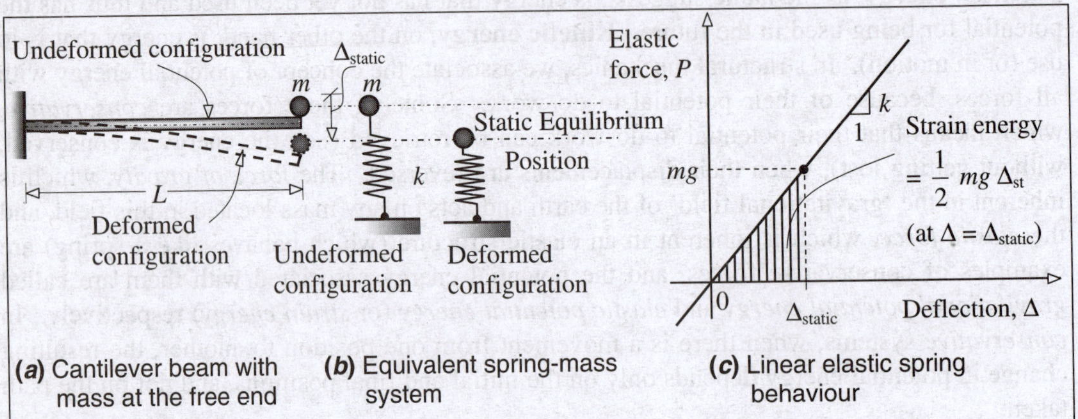

Figure 1.25 Behaviour of a spring-mass system

Now, let us look at the energy balance in this system, in terms of mechanical energy, and compare the total energy in the original configuration with that in the final deflected configuration. In both these configurations, the mass is at rest; hence, there is no kinetic energy. The mass has moved downwards by Δ_{static}. Hence, the loss in its gravitational potential energy in this process is equal to $mg\Delta_{static}$. Corresponding to an internal force, equal to *mg*, the deflection in the spring is given by $\Delta_{static} = mg/k$, and the corresponding internal (real) work done (strain energy) will be equal to only ½ $mg\Delta_{static}$, as depicted by the shaded area in the force-deformation plot of the spring in Fig. 1.25c.

Had the system been truly conservative, the mass in the ideal spring-mass system is expected to oscillate up and down, executing 'free vibrations' (similar to a pendulum) endlessly, always maintaining a balance of mechanical energy. Of course, we know that in actual practice, this unending motion does not take place, and that the vibrations get 'damped' out fairly quickly, with the mass attaining a state of rest at the static equilibrium position, corresponding to which $\Delta = \Delta_{static}$. We need to acknowledge that real-life systems are *non-conservative*. Energy gets *dissipated* in such systems through *damping*. A simple way of including this in our spring-mass model is by including a viscous damper, by means of a 'dashpot'. However, the dynamics of motion need not concern us when we are dealing with problems that are essentially 'static'. There are intelligent ways of dealing with this problem of energy dissipation within an energy framework, by using such concepts as *complementary strain energy* and *total potential energy*. Under static loading conditions, it is convenient to assume that the loads are applied, not suddenly, but very slowly and gradually, such that the structure also deflects very gradually to the configuration corresponding to static equilibrium, without any acceleration or vibration.

1.7.2 Strain Energy and Complementary Strain Energy

When a body (or structure) is deformed, by the action of some loading, it undergoes strains and changes its configuration. Internal work is performed in this process by the stresses corresponding to the strains. We can visualise this internal energy as the work we need to perform in order to lead the system, gradually, from the undeformed configuration to the deformed configuration. If this work is path-independent, and depends only on the initial and final configurations, then the body is said to be *elastic* and the internal energy is called **strain energy**. Conversely, we can visualise strain energy as the energy that gets released when the displacements are gradually and completely reversed and the body returns to its original undeformed configuration. This energy released will be exactly equal to the work done in bringing the system to the deformed configuration in the first place, if the body is truly elastic.

The conservative nature of the system is preserved, as we have seen, in dynamics under ideal conditions of undamped vibrations (generating kinetic energy). In statics also, the conservative nature of the system can also be considered to be preserved when we assume that the loading is applied isothermally. However, in this case, we need to invoke the assistance of some imaginary agency (a friendly genie, perhaps), which does the additional work in ensuring that the loading and hence the response indeed take place gradually. The work done by this agency is complementary to the strain energy stored inside the elastic structure; hence, it is given the name, **complementary strain energy**.

The concept of strain energy in structural elements owes its origin to the more fundamental concept of **strain energy density**, applicable for an elastic material with a well-defined constitutive (stress-strain) law. Consider a small piece of the elastic material, having unit volume, subject to uniaxial strain, ε, increasing gradually from zero to a maximum value ε_{max}. The uniaxial stress, σ, will also build up correspondingly, following a linear path or the more general nonlinear path shown in Fig. 1.26. The strain energy density, μ, is defined as *the internal work done per unit volume in an elastic material that is subject to gradually increasing strain under adiabatic conditions* (no heat flow). It is a 'point function', whose value depends on ε_{max}. It is also a 'positive definite' function, which is always non-negative and takes on a value of zero when $\varepsilon = 0$. Thus, corresponding to any defined stress-strain law for an elastic material, we can define a strain energy density function, μ, as follows:

$$\mu = \int_0^{\varepsilon_{max}} \sigma(\varepsilon) d\varepsilon \tag{1.41}$$

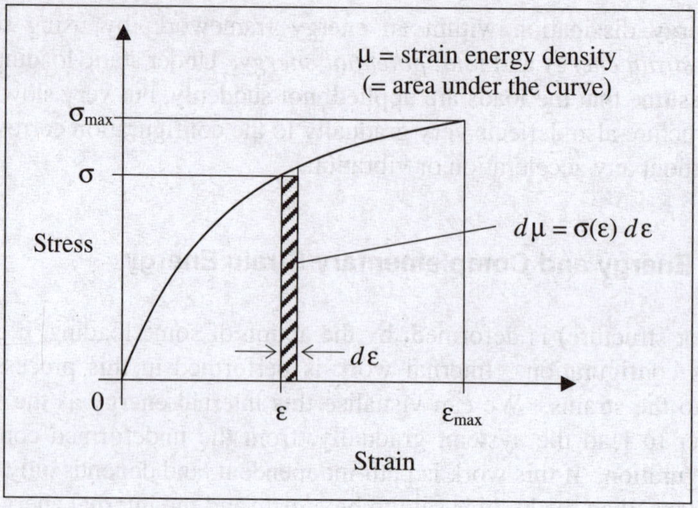

Figure 1.26 Strain energy density

Under linear elastic behaviour, such as uniaxial stress and strain in a material governed by Hooke's law, $\sigma = E\varepsilon$, whereby $\mu_{axial} = \frac{1}{2}E\varepsilon^2 = \frac{1}{2}\frac{\sigma^2}{E}$.

The strain energy for any structural member *m* can be computed by integrating the strain energy density over the volume *V* of the element:

$$U_m = \iiint_V \mu dxdydz \tag{1.42}$$

We can use Eq. 1.42 to derive expressions for strain energy in skeletal structural members, subject to different types of structural actions in linear elastic structural elements. For example, considering an axial element, in which the axial strains are assumed to be uniform across the entire cross-sectional area *A*, the volume integral in Eq. 1.42 reduces to the following line integral, providing a measure of *axial strain energy*:

$$U_{axial} = \int \frac{1}{2}\mu_{axial} A dx = \frac{1}{2}\int N(x)\varepsilon(x)dx \tag{1.43}$$

where $N(x) = \sigma(x)A$ is the axial (tensile) force acting at any location *x* in the element, and $\varepsilon(x)$ is the conjugate axial strain, such that $\varepsilon(x)dx$ is equal to the elemental elongation, *du*, in an infinitesimal element of length *dx*, located at *x*.

Similarly, considering the linearly varying normal strains generated across a beam cross-section at location *x*, which is subject to a bending moment $M(x)$ and conjugate curvature $\varphi(x)$, noting that the normal strain $\varepsilon(x,y)$ and conjugate stress $\sigma(x,y)$ at any depth *y* below the centroidal (neutral) axis are given by $\varphi(x)y$ and $M(x)y/I$ respectively, with $I = \iint_A y^2 dydz$, the expression for *flexural strain energy* can be generated from Eq. 1.42:

$$U_{bending} = \frac{1}{2}\int M(x)\varphi(x)dx \tag{1.44}$$

Analogous to the concept of strain energy density, as defined by Eq. 1.41 (refer Fig. 1.26), we can define **complementary strain energy density,** μ^*, as the complementary strain energy per unit volume in an elastic body, as follows:

$$\mu^* = \int_0^{\sigma_{max}} \varepsilon(\sigma)d\sigma \tag{1.45}$$

$$U_m^* = \iiint_V \mu dxdydz \tag{1.46}$$

The sum of the strain energy, U, and complementary strain energy, U_m^*, will be equal to the internal virtual work product. It may be noted that unlike U_m and U_m^*, the internal virtual work product is independent of the load-deformation path in an elastic member. Thus,

$$U_m + U_m^* = F_m D_m \tag{1.47}$$

This complementary nature is easily seen in Figs 1.27a and b, in which the two types of internal energy, U_m and U_m^*, are superimposed, for the nonlinear elastic and linear elastic situations respectively. In the case of linear elastic behaviour, $U_m = U_m^* = \frac{1}{2}F_m D_m$, making it convenient to inter-change complementary strain energy with strain energy.

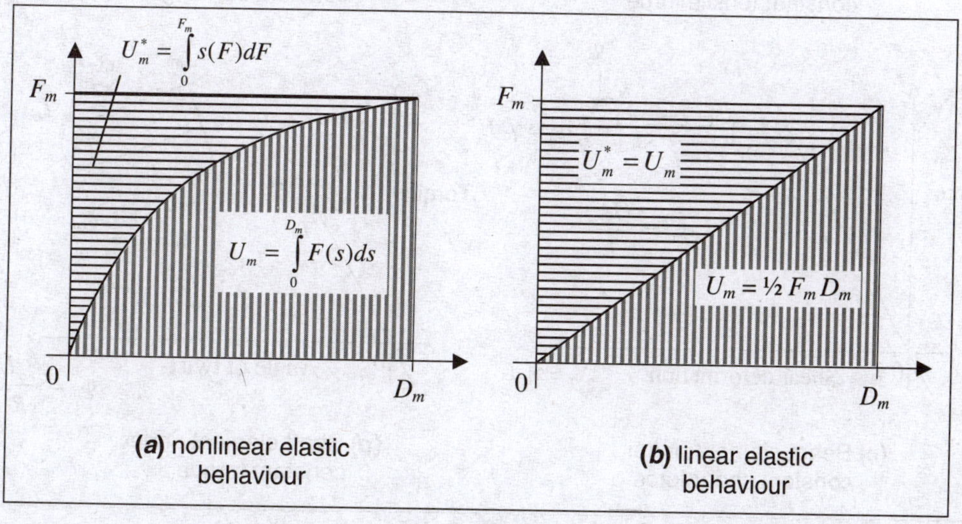

Figure 1.27 Strain energy and complementary strain energy in an elastic member

Axial strain energy in an element (length L, axial rigidity EA) subject to tensile force $N(x)$ and conjugate axial strain $\varepsilon(x) = N(x)/EA$ is given by:

$$U_{\text{axial}} = \frac{1}{2}\int_0^L EA\varepsilon(x)^2 dx = U^*_{\text{axial}} = \frac{1}{2}\int_0^L \frac{N(x)^2}{EA} dx \tag{1.48}$$

If the axial force is constant, and the element (for example the ith bar in a truss) has an **axial stiffness** $k_i = (EA/L)_i$, which also implies an axial flexibility $f_i = (L/EA)_i$, then the bar will have elongation $e_i = f_i N_i$ corresponding to a tensile force N_i (refer Fig. 1.28a). The total axial strain energy of the truss is given by:

$$U_{\text{truss}} = \frac{1}{2}\sum_i \left(\frac{EA}{L}\right)_i e_i^2 = U^*_{\text{truss}} = \frac{1}{2}\sum_i \left(\frac{L}{EA}\right)_i N_i^2 \tag{1.49}$$

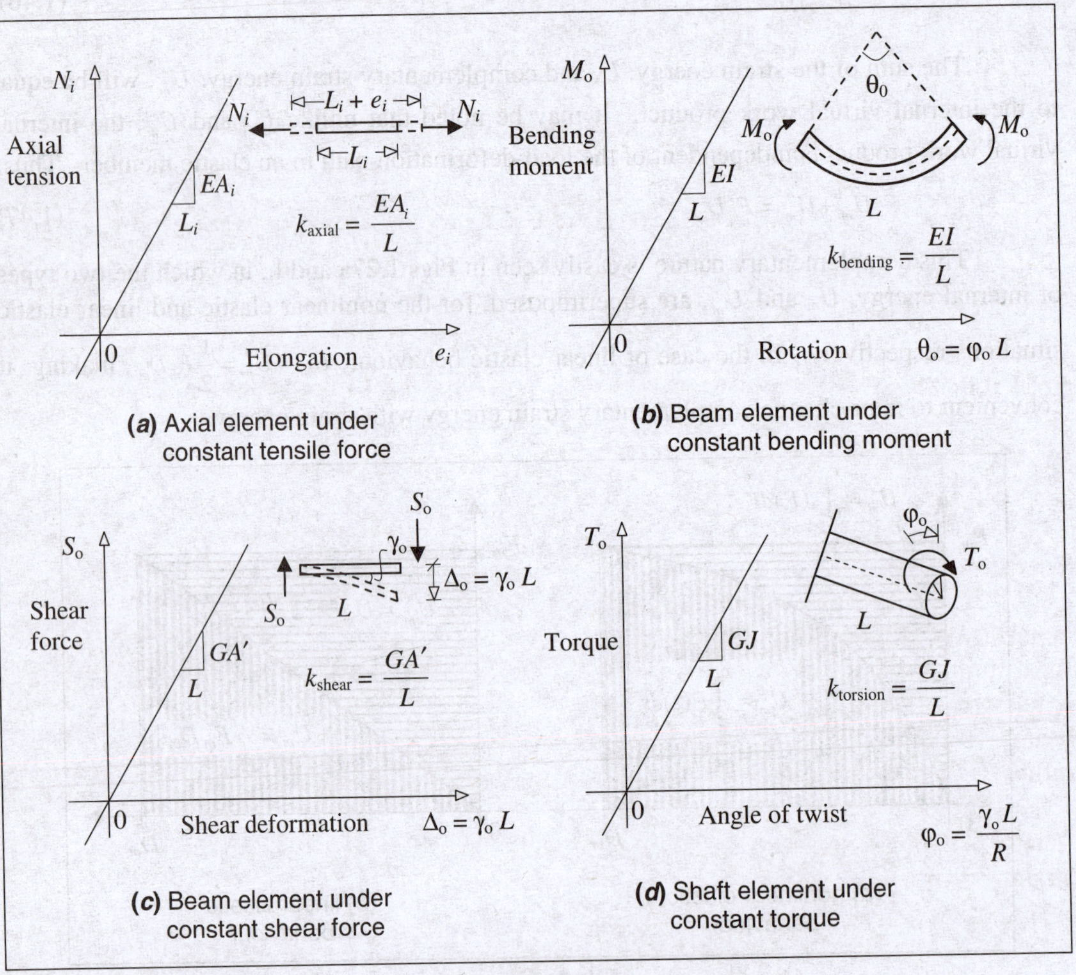

Figure 1.28 Linear elastic behaviour of typical line elements

Flexural (bending) strain energy in a beam element (length *L*, flexural rigidity *EI*) subject to bending moment $M(x)$ and conjugate curvature $\varphi(x) = M(x)/EI$ is given by:

$$U_{\text{bending}} = \frac{1}{2}\int_0^L EI\varphi(x)^2 dx = U^*_{\text{bending}} = \frac{1}{2}\int_0^L \frac{M(x)^2}{EI} dx \qquad (1.50)$$

If the bending moment is constant (M_o), the beam will undergo a uniform curvature, $\varphi_o = M_o/EI$, with its ends undergoing a relative rotation $\theta_o = \varphi_o L = M_o L/EI$, the slope of the M_o-θ_o line will give a measure of the **flexural (bending) stiffness**, $k_{bending} = EI/L$, as depicted in Fig. 1.28b. The flexural strain energy of the beam element is given by:

$$U_{\text{bending}} = \frac{1}{2} EIL\varphi_o^2 = U^*_{\text{bending}} = \frac{1}{2}\frac{M_o^2 L}{EI} \qquad (1.51)$$

Shear strain energy in an element (length *L*, effective shear rigidity *GA′*) subject to shear force $S(x)$ and conjugate shear strain $\gamma(x) = S(x)/GA'$ is given by:

$$U_{\text{shear}} = \frac{1}{2}\int_0^L GA'\gamma(x)^2 dx = U^*_{\text{shear}} = \frac{1}{2}\int_0^L \frac{S(x)^2}{GA'} dx \qquad (1.52)$$

If the shear force in a beam element is constant (S_o), the beam segment will undergo a uniform shear strain, $\gamma_o = S_o/GA'$, with its ends undergoing a relative translation (with a linear profile on account of shear deformations) $\Delta_o = \gamma_o L = S_o L/GA'$, the slope of the S_o-Δ_o line will give a measure of the **shear stiffness**, $k_{shear} = GA'/L$, as depicted in Fig. 1.28c. The shear strain energy of the beam element is given by:

$$U_{\text{shear}} = \frac{1}{2} GA'L\gamma_o^2 = U^*_{\text{shear}} = \frac{1}{2}\frac{S_o^2 L}{GA'} \qquad (1.53)$$

It is important to note that flexural shear is invariably accompanied by bending moments, and so the total strain energy in a beam is made up of bending strain energy plus shear strain energy. For normal, well-proportioned beams, the shear strain energy component is relatively small, and is generally neglected. But in the case of relatively short and deep beams, shear strain energy is comparable to flexural strain energy and hence strongly influences the displacement response.

Torsional strain energy in a shaft element, assumed to be a solid or hollow cylinder (length *L*, torsional rigidity *GJ*) subject to twisting moment $T(x)$ and conjugate torsional shear strain $\gamma(x) = T(x)/GJ$ is given by:

$$U_{\text{torsion}} = \frac{1}{2}\int_0^L GJ\gamma(x)^2 dx = U^*_{\text{torsion}} = \frac{1}{2}\int_0^L \frac{T(x)^2}{GJ} dx \qquad (1.54)$$

If the twisting moment in the shaft element (assumed to be cylindrical with radius *R*) is constant (T_o), it will undergo a uniform conjugate torsional shear strain, $\gamma_o = T_o R/GJ$, with its ends undergoing a relative rotation (angle of twist), $\phi_o = \gamma_o L/R = T_o L/GJ$, the slope of the

T_o-ϕ_o line will give a measure of the **torsional stiffness**, $k_{torsion} = GJ/L$, as depicted in Fig. 1.28d. The torsional strain energy of the shaft element is given by:

$$U_{\text{torsion}} = \frac{1}{2} GJL\gamma_o^{\ 2} = U^*_{\text{torsion}} = \frac{1}{2}\frac{T_o^2 L}{GJ} \tag{1.55}$$

An important point to bear in mind is that we cannot apply the Principle of Superposition to strain energies, even if the structure happens to be linearly elastic! In other words, if a force F_1 generates a strain energy U_1, and another force F_2 generates a strain energy U_2 in the same structure, but in a separate loading condition, then when both F_1 and F_2 act together on the structure, the total strain energy, U, will not be equal to $U_1 + U_2$ (it will exceed this by an amount equal to $2\sqrt{U_1 U_2}$)! This is because U is nonlinearly related to F.

1.7.3 Strain Energy = External Work

Applying the Principle of Virtual Work to the real force field and real displacement field associated with the same loaded structure; equating the total external virtual work to the total internal virtual work: $\sum_j F_{j,\text{ext}} D_{j,\text{ext}} = \sum_i F_{i,\text{int}} D_{i,\text{int}}$

The work is considered 'virtual' because it merely involves a product of a force and a conjugate displacement, without accounting for any relationship between the two. Real work, on the other hand, would account for this possible dependence. Indeed, in elastic structures, there is a very strong dependence, and if the loading process is done gradually, and if the structural behaviour is linearly elastic, then the total internal virtual work, when multiplied by ½, will qualify to be called *strain energy*, U, provided it is initially in an undeformed condition. If we multiply both sides of the virtual work equation by ½, we arrive at the following interesting result:

$$U = \frac{1}{2}\sum_j F_{j,\text{ext}} D_{j,\text{ext}} \tag{1.56}$$

The term on the right-hand side of the above equation, under the condition of gradual loading, qualifies to be called external real work, or simply external work. Eq. 1.56 provides the basis for an 'energy method' to calculate a displacement D in a statically determinate structure, where only one load F acts on the structure and D is conjugate with F.

It is important to note that if, prior to the application of the forces, the structure had some initial strain energy U_{initial} on account of some other loading (caused by external forces), then as explained earlier, Eq. 1.56 needs to be modified as follows:

$$U - U_{\text{initial}} = \frac{1}{2}\sum_j F_{j,\text{ext}} D_{j,\text{ext}} + \sum_j F_{j,\text{initial}} D_{j,\text{ext}} \tag{1.57}$$

Using this energy method, it is possible to derive the Reciprocal Theorems, derived earlier using the Principle of Virtual Work.

1.7.4 Total Potential Energy and Total Complementary Potential Energy

We have observed in Section 1.7.1 that the external work done by the force of gravity uses up a potential energy, called *gravitational potential energy*, which is path-independent. Extending this concept to all kinds of external forces acting on a structure (not necessarily due to gravity), we can talk of a **load potential energy**. Work is done when these forces undergo displacements, in which process their 'potential' gets used up. For this reason, a negative sign is applied on the external work product, while defining load potential energy. Thus, using the notations, $F_{j,ext}$ and $D_{j,ext}$ to denote the external force and conjugate displacement, the expression for load potential energy, denoted by the symbol V, is given by:

$$V = -\sum_j F_{j,\text{ext}} D_{j,\text{ext}} \tag{1.58}$$

The **total potential energy** (TPE) of an elastic structure is defined as the sum of the strain energy U and the load potential energy V. It is denoted by the symbol, Π.

$$\Pi = U + V \tag{1.59}$$

Similarly, the **total complementary potential energy** (TCPE) of an elastic structure is defined as the sum of the complementary strain energy U^* and the load potential energy V. It is denoted by the symbol, Π^*

$$\Pi^* = U^* + V \tag{1.60}$$

The absolute values of TPE and TCPE are not very meaningful. What is of interest to us is how these values change when there is a small 'perturbation' given to the force field or the displacement field. By applying the Principle of Virtual Work, we can derive a set of useful principles, associated with TPE and TCPE, and use these principles to find the force and displacement responses of elastic structures.

1.7.5 Displacement-based Energy Principles

Principle of Stationary Total Potential Energy

Let us consider giving an imaginary and small *perturbation* to the displacement field, following the displacement-based approach. This means introducing slight and arbitrary virtual changes in the displacements in the elastic structure (for example, a truss), without violating internal compatibility (maintaining the kinematic relationships between bar elongations and joint displacements in the truss example) and without disturbing the force field, which is in a state of static equilibrium.

Let us denote by the symbol, δD_j, the virtual displacement imparted at the jth coordinate and let δe_i denote the virtual elongation in the ith truss member. The perturbation in the displacement field will result in a 'variation' in the strain energy, given by $\delta U = \Sigma N_i \, \delta e_i$, and a variation in the load potential energy, given by $\delta V = -\Sigma F_j \, \delta D_j$. Only the 'first variation' is considered here. Invoking the Principle of Virtual Displacements, and transposing terms, $\delta U + \delta V = 0$.

$$\delta(U + V) = \delta\Pi = 0 \tag{1.61}$$

The **Principle of Stationary Total Potential Energy** (PSTPE) states that, when the displacement field in a loaded elastic structure is given a small and arbitrary perturbation, maintaining compatibility and without disturbing the associated force field, then the first variation of the total potential energy is equal to zero, if the forces are in a state of static equilibrium.

This principle can be used to find unknown reactions and internal forces (axial forces, bending moments and shear forces, for example) in statically determinate structures in more-or-less the same way as the Principle of Virtual Displacements.

We can consider the total potential energy, Π, to be a function of the displacement parameters identified in the displacement field. The number of such displacement parameters is equal to the degrees of freedom (n) identified, which is generally equal to the degree of kinematic indeterminacy n_k. Thus, we note that Eq. 1.61 can be more precisely stated as $\delta\Pi\,(\delta D_1, \delta D_2, \ldots, \delta D_n) = 0$. Expanding the expression for $\delta\Pi$, in terms of the partial derivatives of Π, by invoking the 'chain rule' in calculus, and noting that the virtual displacements, $\delta D_1, \delta D_2, \ldots, \delta D_n$ are, by definition, arbitrary and independent of one another, Eq. 1.61 can be satisfied if and only if

$$\frac{\partial \Pi}{\partial D_j} = 0 \qquad \text{for } j = 1, 2, \ldots, n \tag{1.62}$$

Whereas the original form of PSTPE (Eq. 1.61) merely ensures that equilibrium in the force field will be satisfied if $\delta\Pi = 0$, Eq. 1.62 makes a stronger statement of equilibrium, by requiring all the independent displacement parameters to render the total potential energy stationary.

The alternative form of the **Principle of Stationary Total Potential Energy** states that the total potential energy, Π, in a loaded elastic structure, expressed as a function of n independent displacements, $D_1, D_2, \ldots, D_n$, in a compatible displacement field must be rendered stationary, with the partial derivative of Π with respect to every D_j being equal to zero, if the associated force field is to be in a state of static equilibrium.

It can be proved that for stable structures that exhibit linear elastic behaviour, this corresponds to a condition of *minimum* total potential energy, and for this reason, for such structures, PSTPE is also referred to as the **Principle of Minimum Total Potential Energy (PMTPE).** Its formal statement is identical to the one given for PSTPE, except that the term "elastic" is replaced with "linear elastic", and the term "stationary" with "minimum".

Castigliano's Theorem – Part I

The alternative form of PSTPE, represented by the set of differential equations in Eq. 1.62, can be further simplified, by expanding the total potential energy in terms of strain energy and load potential energy, as $\Pi = U - \sum_{j=1}^{n} F_j D_j$. Substituting this expression for Π in Eq. 1.62, and transposing terms, we obtain a beautiful and simple relationship between F_j and D_j:

$$\frac{\partial U}{\partial D_j} = F_j \qquad \text{for } j = 1, 2, \ldots, n \tag{1.63}$$

This equation, attributed to an Italian railway engineer called Castigliano in 1879, is commonly referred to as *Castigliano's Theorem (Part I).*

Castigliano's Theorem (Part I) states that, if the strain energy, U, in an elastic structure, subjected to a system of external forces in static equilibrium, can be expressed as a function of n independent displacements, D_1, D_2,..., D_n, satisfying compatibility, then the partial derivative of U with respect to every D_j will be equal to the value of the conjugate force, F_j.

Although Castigliano, in his time, had applied this theorem to linear static behaviour, it is clear from our derivation based on PSTPE, that there is actually no requirement that the elastic behaviour should necessarily be linear. It is equally applicable to structures that exhibit nonlinear elastic behaviour.

Deriving Stiffness Coefficients

We define the **stiffness coefficient,** k_{ij}, as the force at the ith coordinate due to a unit displacement applied at the jth coordinate, and with all other identified degrees of freedom arrested (using imaginary restraints) in the structure. Using this definition, and applying the Principle of Superposition, which is valid for linear elastic structures, we can inter-relate all the forces and displacements in the structure, corresponding to the identified n degrees of freedom, as follows:

$$F_i = k_{i1}D_1 + k_{i2}D_2 + \ldots + k_{in}D_n = \sum_{j=1}^{n} k_{ij}D_j \tag{1.64}$$

Differentiating Eq. 1.64 with respect to any particular displacement, D_j, and noting that all the displacement parameters are, by definition, independent of one another, and all the stiffness coefficients are constant in a linearly elastic structure, we arrive at an interesting result: $\frac{\partial F_i}{\partial D_j} = k_{ij}$. Invoking Castigliano's Theorem I ($F_i = \partial U/\partial D_i$), we obtain:

$$k_{ij} = \frac{\partial^2 U}{\partial D_i \partial D_j} \tag{1.65}$$

The **stiffness coefficient, k_{ij}**, in a linearly elastic structure can be obtained by expressing the strain energy U as a function of n independent displacements (D_1, D_2,..., D_n) associated with a kinematically admissible displacement field, and finding the mixed partial derivative of U with respect to displacements D_i and D_j.

The use of this energy principle in finding stiffness coefficients is demonstrated later. It is noteworthy, from the nature of Eq. 1.65, that the stiffness coefficient, k_{ji} will be identical to k_{ij}. This proves that the stiffness matrix **k** is symmetric, just like the flexibility matrix **f**.

1.7.6 Force-based Energy Principles

Principle of Stationary Total Complementary Potential Energy

Let us now consider giving an imaginary and small *perturbation* to the force field, following the force-based approach. This means giving a slight virtual change to the force field, maintaining the state of static equilibrium and without disturbing the displacement field, which is in a state of internal compatibility. This also means that we will be dealing with *complementary strain energy* (instead of strain energy), and thereby *complementary total potential energy* (instead of total potential energy).

Let us denote by the symbol, δF_j, the virtual force perturbation imparted at the jth coordinate and let δN_i denote the corresponding virtual internal force in the ith truss member (maintaining equilibrium). The perturbation in the force field will result in a 'variation' in the complementary strain energy, given by $\delta U^* = \Sigma e_i \, \delta N_i$, and a variation in the load potential energy, given by $\delta V = -\Sigma D_j \, \delta F_j$. Only the 'first variation' is considered here. Invoking the Principle of Virtual Forces, and transposing terms, $\delta U^* + \delta V = 0$.

$$\delta(U^* + V) = \Pi^* = 0 \tag{1.66}$$

The **Principle of Stationary Total Complementary Potential Energy** states that, when the force field in a loaded elastic structure is given a small and arbitrary perturbation, maintaining equilibrium and without disturbing the associated displacement field, then the first variation of the total complementary potential energy is equal to zero, if the displacements satisfy compatibility.

This principle can be used to find unknown displacements in more-or-less the same way demonstrated by the application of the Principle of Virtual Forces for statically determinate structures.

As with PSTPE, in order to understand why the term *stationary* is used in PSTCPE, we need to move from the domain of variational calculus to that of differential calculus. We can consider the total complementary potential energy, Π^*, to be a function of the unknown forces identified in the force field: $\delta\Pi^*(\delta F_1, \delta F_2, \ldots, \delta F_n) = 0$. Expanding the expression for $\delta\Pi^*$, in terms of the partial derivatives of Π^*, by invoking the 'chain rule' in calculus, and noting that because the virtual displacements, $\delta F_1, \delta F_2, \ldots, \delta F_n$ are, by definition, arbitrary and independent of one another, Eq. 1.66 can be satisfied if and only if

$$\frac{\partial \Pi^*}{\partial F_j} = 0 \qquad \text{for } j = 1, 2, \ldots, n \tag{1.67}$$

Whereas the original form of PSTPE (Eq. 1.66) merely ensures that compatibility in the displacement field will be satisfied if $\delta\Pi^* = 0$, Eq. 1.67 makes a stronger statement of compatibility, by requiring all the independent force parameters in the force field to render the total complementary potential energy stationary.

The alternative form of the **Principle of Stationary Total Complementary Potential Energy** states that the total complementary potential energy, Π^*, in a loaded elastic structure, expressed as a function of n independent forces, $F_1, F_2, \ldots, F_n$, in a statically admissible force field must be rendered stationary, with the partial derivative of Π^* with respect to every F_j being equal to zero, if the associated displacement field is to satisfy compatibility.

It can be proved that for stable structures that exhibit linear elastic behaviour, this corresponds to a condition of *minimum* total complementary potential energy, which is equal to the total potential energy (complementary strain energy and strain energy are the same for linear elastic structures). In such situations, PSTCPE is also referred to as the **Principle of Minimum Total Complementary Potential Energy** (PMTCPE). Its formal statement is identical to the one given for PSTCPE, except that the term "elastic" is replaced with "linear elastic", and the term "stationary" with "minimum".

Castigliano's Theorem – Part II

The alternative form of PSTCPE, represented by the set of differential equations in Eq. 1.67, can be further simplified, by expanding the total complementary potential energy in terms of strain energy and load potential energy, as $\Pi^* = U^* - \sum_{j=1}^{n} F_j D_j$. Substituting this expression for Π^* in Eq. 1.67, and transposing terms, we obtain a beautiful and simple relationship between D_j and F_j, which is analogous to Eq. 1.63:

$$\frac{\partial U^*}{\partial F_j} = D_j \qquad \text{for } j = 1, 2, \ldots, n \tag{1.68}$$

The derivation of this equation is generally attributed to Engesser, although it was Castigliano who in 1879 first published this relationship (as Part II of his paper), limiting considerations to linear elastic behaviour (with strain energy, U, substituted in place of complementary strain energy, U^*):

$$\frac{\partial U}{\partial F_j} = D_j \qquad \text{for } j = 1, 2, \ldots, n \tag{1.69}$$

Castigliano's Theorem (Part II) states that, if the complementary strain energy, U^*, in an elastic structure, with a given kinematically admissible displacement field, is expressed as a function of n independent external forces, $F_1, F_2, \ldots, F_n$, satisfying equilibrium, then the partial derivative of U^* with respect to every F_j will be equal to the value of the conjugate displacement, D_j. If the behaviour is linear elastic, U^* can be replaced by the strain energy function U.

Castigliano's Theorem II is, by far, the most useful energy method available, for finding displacements in statically determinate structures. In such structures, the force response is statically determinable, whereby the complementary strain energy is determinable. In order to simplify calculations, however, the explicit formulation of U^* can be completely avoided, using the following simplified versions of Castigliano's Theorem II for flexural members and trusses respectively:

$$D_j = \int \frac{1}{EI}\left(M(x)\big|_{F_1,F_2,\ldots} \right)\left(\frac{\partial M}{\partial F_j} \right) dx \qquad \text{for } j = 1, 2, \ldots, n \tag{1.70}$$

$$D_j = \sum_i \left(\frac{L}{EA} \right)_i \left(N_i\big|_{F_1,F_2,\ldots} \right)\left(\frac{\partial N_i}{\partial F_j} \right) \qquad \text{for } j = 1, 2, \ldots, n \tag{1.71}$$

Deriving Flexibility Coefficients

We define the **flexibility coefficient,** f_{ij}, as the displacement at the ith coordinate due to a unit force applied at the jth coordinate, and with no other forces on the structure. Using this definition, and applying the Principle of Superposition, which is valid for linear elastic structures, we can inter-relate all the forces and displacements in the structure, corresponding to the identified n degrees of freedom, as follows:

$$D_i = f_{i1}F_1 + f_{i2}F_2 + \ldots + f_{in}F_n = \sum_{j=1}^{n} f_{ij}F_j \tag{1.72}$$

Differentiating Eq. 1.72 with respect to any particular force, F_j, and noting that all the forces are, by definition, independent of one another, and all the flexibility coefficients are constant in a linearly elastic structure, we arrive at an interesting result: $\frac{\partial D_i}{\partial F_j} = f_{ij}$. Invoking Castigliano's Theorem II ($D_i = \partial U^* / \partial F_i$) , we obtain:

$$f_{ij} = \frac{\partial^2 U^*}{\partial F_i \partial F_j} \tag{1.73}$$

The **flexibility coefficient, f_{ij}**, in a linearly elastic structure can be obtained by expressing the complementary strain energy U^* as a function of n independent forces (F_1, F_2,..., F_n) associated with a statically admissible force field, and finding the mixed partial derivative of U^* with respect to displacements F_i and F_j.

The use of this energy principle in finding flexibility coefficients is demonstrated later. It is noteworthy, from the nature of Eq. 1.73, that the stiffness coefficient, f_{ji} will be identical to k_{ij}. This proves that the flexibility matrix **f** is symmetric, just like the flexibility matrix **k**.

Theorem of Least Work

Consider a statically indeterminate structure and let the unknown redundant forces be denoted as X_1, X_2,.., X_n, where $n = n_s$, the degree of static indeterminacy in the structure. There are many possible choices of the redundants, but their total number is fixed and equal to n_s. According to the PMTCPE, the choices of the redundants must be such that they render the total complementary energy, Π^*, in the structure *minimum*, if the structure is linear elastic and stable. Applying the compatibility conditions given by Eq. 1.69 (due to Engesser / Castigliano), to the statically indeterminate problem involving redundant forces, and noting that, the displacements conjugate with the redundants are all equal to zero (except possibly under conditions of indirect loading),

$$\frac{\partial U^*}{\partial X_j} = D_j = 0 \qquad \text{for } j = 1, 2, \ldots, n_s \tag{1.74}$$

The **Theorem of Least Work** states that, of all the possible values that the redundants in a statically indeterminate and linearly elastic structure can assume, the true solutions, ensuring compatibility in the displacement field, correspond to the condition of minimum complementary strain energy U^* (or minimum strain energy U).

1.8 FORCE METHODS — STATICALLY INDETERMINATE STRUCTURES

Historically, it appears that the problem of analysing statically indeterminate structures remained unsolved for many years, until the second half of the 19th century, and the earliest breakthroughs came with the emergence of the **Reciprocal Theorems** and **Clapeyron's Theorem of Three Moments**, as applied to simple statically indeterminate beams. Until then, statically indeterminate problems posed a formidable challenge. Builders were wary of constructing over-rigid structures, and so most structures were deliberately made statically determinate, such as post-and-lintel construction, simply supported bridges, three-hinged arches, etc. It was also recognised that although over-rigid structures possessed higher capacity, by virtue of the 'redundancy' in the load-transmission, and also higher stiffness, there were problems associated with 'self-straining' effects, which were absent in just-rigid structures.

In practice today, many structures are *over-rigid*, and hence **statically indeterminate**. This means that the number of unknown forces (support reactions plus internal forces) in the structure exceed the number of available independent equations of static equilibrium. The structure has more restraints and/or internal constraints than the minimum required to make it just-rigid (stable), as discussed earlier. The number by which the unknown forces exceed the available equilibrium equations is called the **degree of static indeterminacy**, denoted by n_s. When the structure is statically indeterminate, we are compelled to also account for the deformations in the structure and thereby account for **compatibility** conditions in the displacement field. These compatibility equations provide the necessary extra equations needed to solve the problem. The number of such compatibility equations to be invoked is therefore exactly equal to the degree of static indeterminacy, n_s. There are different ways in which these compatibility conditions can be invoked, and herein lies the main difference among the various **force methods** available for finding the force response in statically determinate structures. In any one force method, there is considerable choice involved with regard to the choice of redundants.

Once the redundants (X_1, X_2,.., X_n, where $n = n_s$) are chosen, the compatibility requirements, corresponding to the displacements conjugate with these forces, are fixed, depending on the nature of the structure. In general, these displacements are zero, and we can invoke the **Theorem of Least Work**, for example, to solve for the redundants. Alternatively, we can work with the **primary structure**, which is an imaginary just-rigid structure that is left behind when we imagine all the redundants (forces and associated displacement constraints) to vanish. We then find the conjugate displacements at the redundant locations, first due to the applied loading, and then (separately) due to the redundants. Applying the Principle of Superposition, we can invoke the appropriate compatibility conditions, consistent with the problem. This method is known as the **Method of Consistent Deformations**. We can formulate this method in a convenient matrix format, making use of the flexibility matrix using an element approach, and in this form, it is referred to as the **Flexbility Method** (to be explored at length in this book). Other, rather ingenious methods include the **Column Analogy Method**, which is especially useful for solving non-prismatic beams and frames.

It is instructive to note the following key features, commonly shared by all the major force methods, in some form or other, to determine the force response in statically indeterminate structures:

1. Find the **degree of static indeterminacy**, n_s.
2. Select the **redundants**, X_1, X_2,.., X_n.
3. Identify the required **conditions of compatibility** of the displacements conjugate with the redundants.
4. Express the required displacements in terms of the redundants using a **flexibility format**.
5. Analyse the force response of the primary structure, using **conditions of static equilibrium**, to provide the input needed to express the compatibility equations.
6. Solve the compatibility equations and find the unknown redundants.
7. Treating these known redundants as additional loads on the structure (now statically determinate), find the force response.

Essentially, we are reducing the indeterminate structure to a determinate one (primary structure), and then applying the Principle of Superposition (to both the force field and the displacement field) to account for the contributions of the applied loading and the redundants, satisfying compatibility. This technique, therefore, can be applied only to linear elastic problems.

1.8.1 Method of Consistent Deformations

Consider, for example, the three-span continuous beam ABCD, with a fixed end at A and a simply supported end at D, as shown in Fig. 1.29a. This structure is statically indeterminate to the third degree. Let the structure be subject to arbitrary external loads, and let there also be some displacement loading in the form of support displacements (such as rotational slip θ_A and support settlement δ_C in Fig. 1.29a). We can conveniently consider the cantilever beam in Fig. 1.29b as the primary structure, by choosing the support reactions at B, C and D as the redundants, X_1, X_2 and X_3 respectively. This implies releasing the restraints against vertical translation at these support locations. For convenience, it is advantageous to consider each loading condition separately, and then apply the Principle of Superposition to arrive at the desired expressions for displacements, $D_1, D_2, \ldots, D_n$, on account of their combined effect. In general, these displacements are zero. However, there may be exceptional situations, involving support displacements (or environmental changes or constructional errors), where some of these values are non-zero, but of known magnitude. If there are support displacements specified at locations other than those corresponding to the selected redundants (such as θ_A in Fig. 1.29a), then the effect of such displacement loading input must be accounted for in the expressions for the displacements, $D_1, D_2, \ldots, D_n$, as shown in Fig. 1.29c. The action of unit loads, corresponding to $X_1 = 1$, $X_2 = 1$ and $X_3 = 1$, are shown in Figs 1.29 d, e and f respectively.

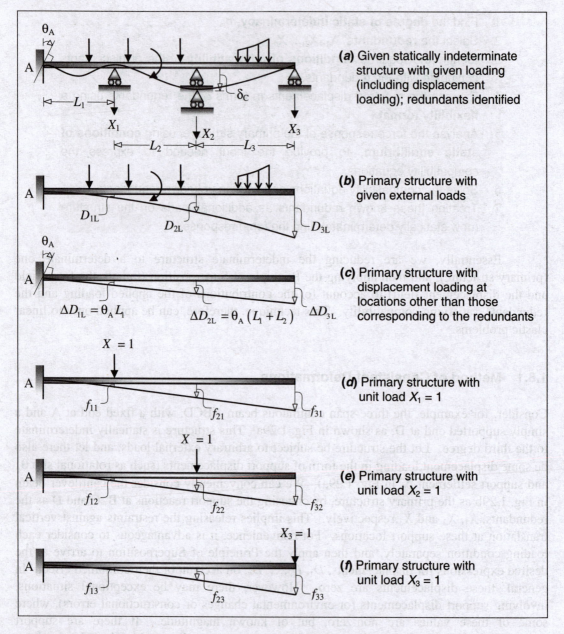

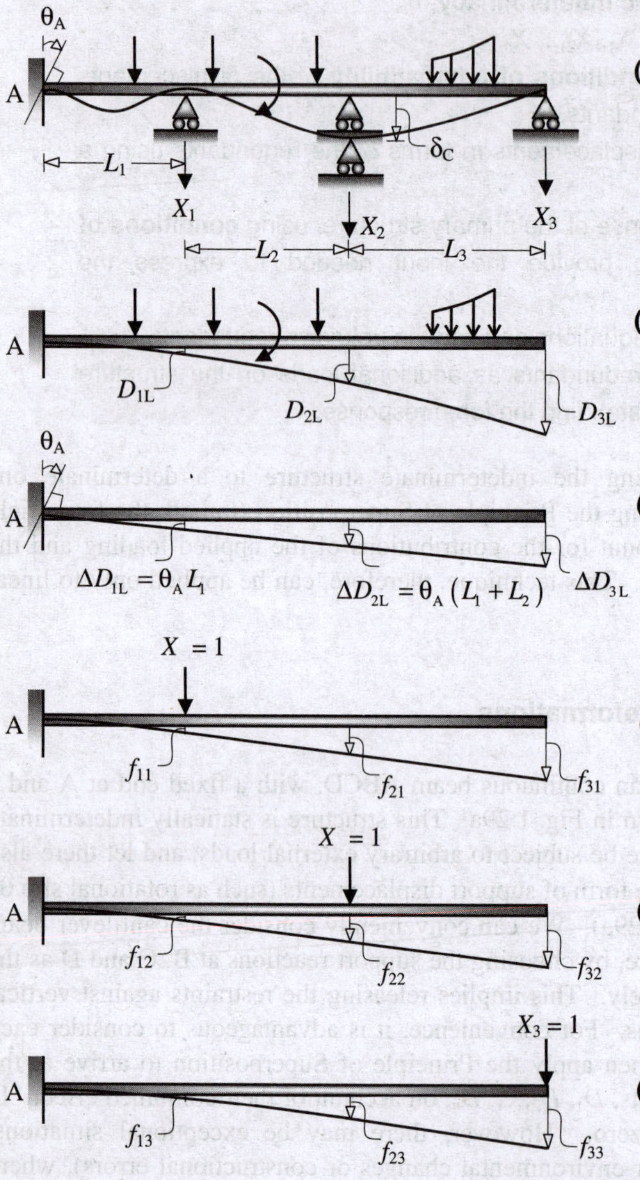

(*a*) Given statically indeterminate structure with given loading (including displacement loading); redundants identified

(*b*) Primary structure with given external loads

(*c*) Primary structure with displacement loading at locations other than those corresponding to the redundants

(*d*) Primary structure with unit load $X_1 = 1$

(*e*) Primary structure with unit load $X_2 = 1$

(*f*) Primary structure with unit load $X_3 = 1$

Figure 1.29 Continuous beam (with $n_s = 3$) subject to arbitrary loading and primary structure subject to applied loading and unit loads ($X_j = 1$)

Considering the primary structure, the sum of the displacements at the redundant locations due to (i) the external loads (D_{jL}), (ii) additional displacements (ΔD_{jL}) caused by displacement loading (due to input at non-redundant locations), (iii) the redundants (D_{jX}) must be equal to the net (final) values (D_X), consistent with the given problem.

Applying these compatibility conditions to the continuous beam example shown in Fig. 1.29,

$$\begin{Bmatrix} D_{1L} + \Delta D_{1L} \\ D_{2L} + \Delta D_{2L} \\ D_{3L} + \Delta D_{3L} \end{Bmatrix} + \begin{bmatrix} f_{11} & f_{12} & f_{13} \\ f_{21} & f_{22} & f_{23} \\ f_{31} & f_{32} & f_{33} \end{bmatrix} \begin{Bmatrix} X_1 \\ X_2 \\ X_3 \end{Bmatrix} = \begin{Bmatrix} D_1 = 0 \\ D_2 = \delta_C \\ D_3 = 0 \end{Bmatrix}$$

where, applying the Principle of Virtual Forces (Unit Load Method), and denoting $M_L(x)$ and $m_j(x)$ as the bending moments at x due to the applied loading and $X_j = 1$ respectively,

$$1 \times D_{jL} = \int \frac{M_L(x)}{EI} m_j(x) dx \tag{1.75}$$

$$f_{jk} = f_{kj} = \int \frac{1}{EI} m_j(x) m_k(x) dx \tag{1.76}$$

The general solution for the redundants $\{X\}$ is given by solving the set of simultaneous equations as follows:

$$\{X\} = [f]^{-1}\{\{D\} - \{D_L + \Delta D_L\}\} \tag{1.77}$$

where $[f]$ denotes the flexibility matrix, which is a square, symmetric matrix. Eq. 1.77 provides a compact form of the solution for the redundants. Of course, it is not necessary to invert the flexibility matrix in all cases; alternative solution techniques, such as Gauss elimination method, work just as well for solving the simultaneous equations.

After the values of the redundants $\{X\}$ are obtained, the complete force response of the structure can be generated. This can be done by superimposing the force responses of the primary structure subject to the different load cases, as obtained earlier, and factoring in the values of the redundants appropriately. For example, the expression for bending moment $M(x)$ at any section in a statically indeterminate beam or framed structure, is given by:

$$M(x) = M_L(x) + X_1 m_1(x) + X_2 m_2(x) + \ldots + X_n m_n(x) \tag{1.78}$$

1.8.2 Theorem of Least Work

As mentioned earlier, the Theorem of Least Work is applicable in situations where there is no displacement loading (such as temperature effects and support settlements). After identifying the redundants, at the very outset, the required compatibility equations can be expressed in terms of minimising the complementary strain energy, U^*, in the structure. The overall expression for complementary strain energy U^* in a framed structure is given by:

$$U^{*} = \frac{1}{2}\int \frac{M(x)^2}{EI}dx + \frac{1}{2}\int \frac{N(x)^2}{EA}dx + \frac{1}{2}\int \frac{S(x)^2}{GA'}dx \tag{1.79}$$

where $M(x)$, $N(x)$ and $S(x)$ denote the bending moment, axial force and shear force respectively at any section x, and EI, EA and GA' denote the flexural, axial and shear rigidities respectively.

Accordingly, the Theorem of Least Work (Eq. 1.74) takes the following form in applications to beams and frames:

$$\frac{\partial U^{*}}{\partial X_j} = \int \frac{M(x)}{EI}\frac{\partial M}{\partial X_j}dx + \int \frac{N(x)}{EA}\frac{\partial N}{\partial X_j}dx + \int \frac{S(x)}{GA'}\frac{\partial S}{\partial X_j}dx = 0 \tag{1.80}$$

In most cases of well-proportioned frame and beam members, axial and shear energy terms can be ignored in the above equation. In applications to truss members, with axial flexibility $f_i = L_i/(EA_i)$, the Theorem of Least Work takes the following form:

$$\frac{\partial U^{*}}{\partial X_j} = \sum_i f_i N_i \frac{dN_i}{dX_j} = 0 \tag{1.81}$$

In two-hinged symmetrical arches subject to arbitrary loading, the application of the Theorem of Least Work results in the following expression for the horizontal thrust H:

$$H = \frac{\int_0^S \frac{M^{o}(s)y(s)}{EI(s)}ds}{\int_0^S \frac{y^2(s)}{EI(s)}ds} \tag{1.82}$$

where $M^{o}(x)$ is the bending moment in the simply supported arch (primary structure). In symmetrically loaded parabolic arches (with $y(x) = \frac{4h}{L^2}x(L-x)$), assuming $I(x) = I_o\sqrt{1+y'^2}$, Eq. 1.82 simplifies to:

$$H = \frac{\int_0^{L/2} M^{o}(x)y(x)dx}{\int_0^{L/2} y^2(x)dx} \tag{1.83}$$

If the loading is unsymmetric (such as by the action of an eccentric concentrated load W), the load may be apply symmetrically on the arch (on both sides), and by invoking the Principle of Parity and the Principle of Superposition, the value of the horizontal thrust in the given arch taken as half the value obtained by applying Eq. 1.83 to the symmetrically loaded arch. Similarly, for a symmetrically loaded fixed parabolic arch, the application of the Theorem of Least Work results in the following expressions for the horizontal thrust H and the fixed end moment M^{F} (refer Fig. 1.30).

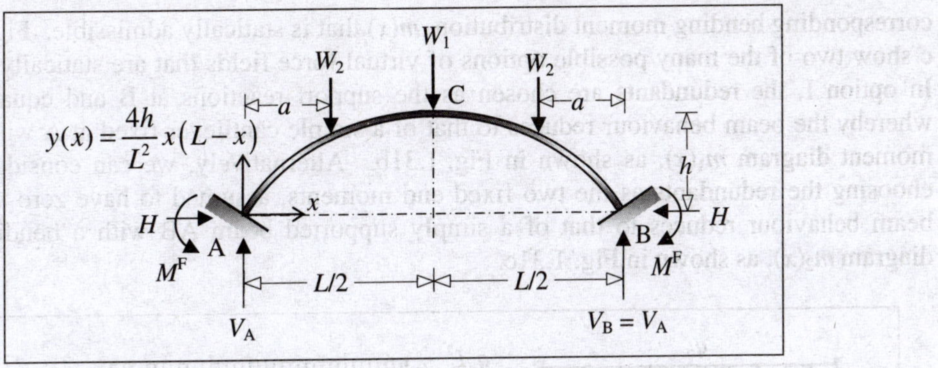

Figure 1.30 Symmetrically loaded fixed parabolic arch

$$\begin{bmatrix} \int_0^{L/2} y^2 dx & \int_0^{L/2} y dx \\ \int_0^{L/2} y dx & \int_0^{L/2} dx \end{bmatrix} \begin{Bmatrix} H \\ M^F \end{Bmatrix} = \begin{Bmatrix} \int_0^{L/2} M^o y dx \\ \int_0^{L/2} M^o dx \end{Bmatrix} \tag{1.84}$$

1.8.3 Calculating Displacements in Statically Indeterminate Structures

It is also noteworthy that displacements in statically indeterminate structures can be conveniently calculated by invoking the power of the Principle of Virtual Forces (Unit Load Method). As the assumed virtual force field is required to only satisfy equilibrium, any statically admissible solution is acceptable; hence all the redundants chosen can be conveniently equated to zero. For example, consider the problem of finding the maximum deflection in a prismatic fixed beam (span L, flexural rigidity EI) subject to a uniformly distributed gravity load q_o per unit length (Fig. 1.31a). As the beam is symmetrically loaded, the bending moment diagram will be symmetric, and can be visualised as the superposition of a parabolic sagging moment diagram with a maximum value of $q_oL^2/8$ at mid-span (due to simply supported action) and a hogging moment diagram of constant magnitude M^F. By applying the Conjugate Beam Method, it can be easily shown that, for equilibrium of the conjugate beam (with free-free boundary conditions), the area of the sagging curvature diagram must be equal to that of the hogging curvature diagram, whereby, $\frac{M^F L}{EI} = \frac{q_o L^2}{8EI} \times L \times \frac{2}{3} \Rightarrow M^F = \frac{q_o L^2}{12}$.

We now have a (real) displacement field, in which we know the distribution of curvatures, given by the superposition of two bending moment diagrams (divided by EI), as shown in Fig. 1.31a. To find the maximum deflection (at mid-span) that is compatible with the curvature (M/EI) diagram and the fixed end boundary conditions of the beam, we visualise a unit virtual load acting at the mid-span location of the beam and obtain a corresponding bending moment distribution, $m(x)$ that is statically admissible. Figs 1.31b and c show two of the many possible options of virtual force fields that are statically admissible. In option I, the redundants are chosen as the support reactions at B and equated to zero, whereby the beam behaviour reduces to that of a simple cantilever fixed at A with a bending moment diagram $m_1(x)$, as shown in Fig. 1.31b. Alternatively, we can consider option II, choosing the redundants as the two fixed end moments, assumed to have zero values. The beam behaviour reduces to that of a simply supported beam AB with a bending moment diagram $m_2(x)$, as shown in Fig. 1.31c.

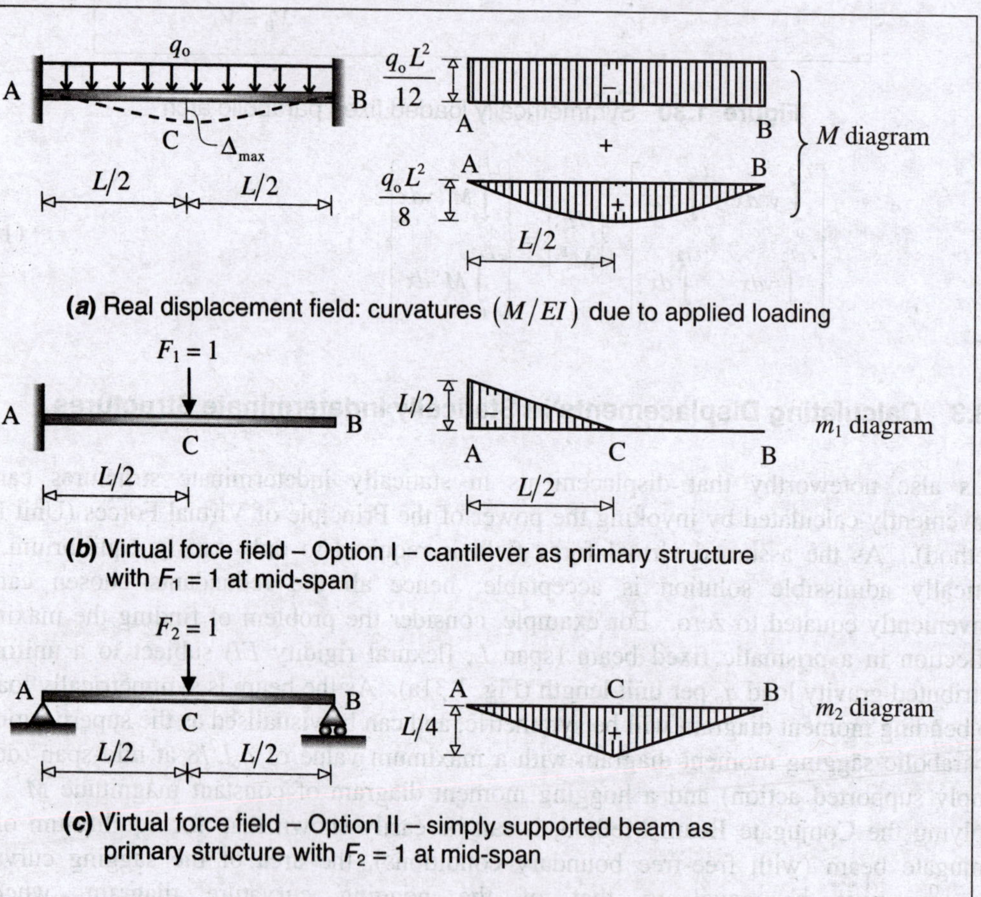
q_o
A
B
C
Δ_{max}
L/2
L/2
$\frac{q_o L^2}{12}$
$\frac{q_o L^2}{8}$
A
B
+
M diagram
L/2
(a) Real displacement field: curvatures (M/EI) due to applied loading
$F_1 = 1$
A
B
C
L/2
L/2
A
C
B
m_1 diagram
(b) Virtual force field – Option I – cantilever as primary structure with $F_1 = 1$ at mid-span
$F_2 = 1$
A
B
C
L/2
L/2
L/4
m_2 diagram
(c) Virtual force field – Option II – simply supported beam as primary structure with $F_2 = 1$ at mid-span

Figure 1.31 Deflection in a fixed beam by Principle of Virtual Forces

Applying the Principle of Virtual Forces (Eq. 1.34), and using the area-multiplication technique (Eq. 1.35), we get the desired expression for the mid-span deflection. Both options yield the same solution.

$$\text{Option I:} \quad \Delta_{max} = \int_0^{L/2} m_1 \frac{M}{EI} dx = \frac{L}{2}\left[\left\{\frac{1}{2}\frac{L}{2} \times \frac{q_o L^2}{12EI}\right\} - \left\{\frac{2}{3}\frac{q_o L^2}{8EI} \times \frac{3}{8}\frac{L}{2}\right\}\right] = \frac{1}{384}\frac{q_o L^4}{EI}$$

$$\text{Option II:} \quad \Delta_{max} = 2\int_0^{L/2} m_2 \frac{M}{EI} dx = 2\left(\frac{L}{2}\right)\left[-\left\{\frac{1}{2}\frac{L}{4} \times \frac{q_o L^2}{12EI}\right\} + \left\{\frac{2}{3}\frac{q_o L^2}{8EI} \times \frac{5}{8}\frac{L}{4}\right\}\right] = \frac{1}{384}\frac{q_o L^4}{EI}$$

1.9 DISPLACEMENT METHODS — KINEMATICALLY INDETERMINATE STRUCTURES

As mentioned earlier, there are basically two different approaches in solving indeterminate problems in structural analysis: the *force* (or *flexibility*) approach and the *displacement* (or *stiffness*) approach. These two approaches are completely different, and yet give identical results, in terms of the force and displacement response of any linear elastic structure subject to prescribed loading.

Let us briefly overview here the main differences between displacement and force methods of analysis. The basic unknowns are *displacements* in displacement methods, as the name suggests, whereas these are *forces* in the force methods. For this reason, the type of indeterminacy involved is *kinematic* in displacement methods, whereas it is *static* indeterminacy in the force methods. The unknown displacements in displacement methods are decided by the *degrees of freedom* identified in the structure (although we can in some instances reduce this number), whereas the unknown redundant forces are not pre-defined in force methods, and we have much choice in selecting the *redundants*. *Compatibility* conditions are not adequate to solve for the unknown displacements in displacement methods, and *equilibrium* conditions are not adequate to solve for the unknown forces in force methods. We solve for the unknown displacements in displacement methods by invoking appropriate *equilibrium equations*, whereas we use *compatibility equations* to solve for the unknown forces in force methods. In order to do this, we express the force-displacement relations in displacement methods in a *stiffness format*, whereas we do this in a *flexibility format* in force methods. After solving and finding the unknown displacements, we find the desired force response in the displacement methods, whereas we get the force response more directly in the force methods after finding the unknown redundants. However, in some displacement methods (**Moment Distribution Method** and **Kani's Method**) applicable to beams and frames, it is possible to by-pass the step of calculating explicitly the unknown displacements, and we can directly get the end moments in the various beam and frame members. In both force and displacement methods, we make use of the principle of superposition, with respect to forces and displacements.

The basic displacement method originated with the **Slope-deflection Method**, proposed in 1915 by George Maney. This method, applicable to beams and frames, was later generalised in a matrix format suitable for computer use, in the form of the **Stiffness Method**, which is explored in detail in this book, and which is applicable to all kinds of kinematically indeterminate structures.

1.9.1 Basic Approach of Displacement Methods

Consider, for example, the three-span continuous beam ABCD shown in Fig. 1.32. There are only two unknown joint displacements, namely $D_1 = \theta_B$ and $D_2 = \theta_C$ (assumed clockwise positive). The *primary structure* (kinematically determinate) is obtained by arresting the displacements D_1 and D_2. This implies rotational fixity at B and C, whereby the beam segments AB, BC and CD can be separated out as three fixed beams, as shown in Fig. 1.32b.

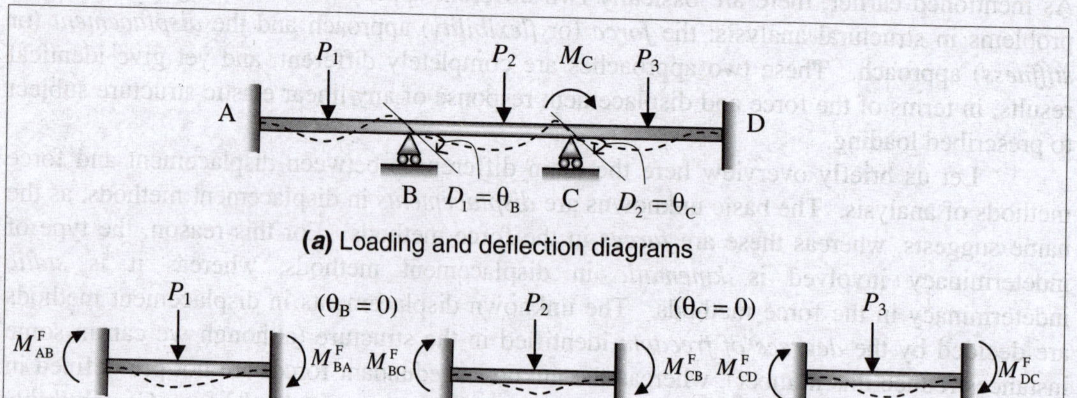

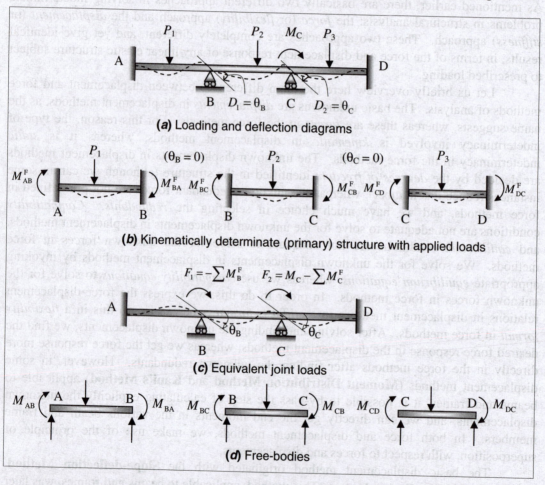

Figure 1.32 Application of displacement method to a three-span continuous beam

The fixed end moments (M_{AB}^{F}, M_{BA}^{F}, etc., all assumed clockwise positive), for these beams, subject to any arbitrary loading must be known (using appropriate tables or methods like the *conjugate beam method* or *column analogy method*). Table 1.3 summarises the values of fixed end moments for some typical load cases in prismatic fixed beams.

By superposing the end moments in the three beams at the common supports B and C, we note that there are resultant forces (moments), $\sum M_{B}^{F} = M_{BA}^{F} + M_{BC}^{F}$ and $\sum M_{C}^{F} = M_{CB}^{F} + M_{CD}^{F}$ accumulating at B and C. In order to satisfy equilibrium conditions, these forces should be eliminated and other nodal forces acting at the joints should be accounted for. This is achieved by applying these nodal forces (moments) in an opposite direction, as shown in Fig. 1.32c, and by including any nodal force (moment) acting in the original loading diagram (such as the clockwise moment M_C in Fig. 1.32a). The final nodal loads to be considered in the analysis ($F_1 = -\sum M_{B}^{F}$ and $F_2 = M_C - \sum M_{C}^{F}$ Fig. 1.32c) are called *equivalent joint loads*. The results of analysis of Fig. 1.32c can be superposed on those obtained in Fig. 1.32b. We note that, as the structure is assumed to be linear elastic, applying the principle of superposition (adding algebraically the responses in Figures 1.32b and 1.32c), we get not only the force response, but also the displacement response of Fig. 1.32a. The rotations at B and C in Fig. 1.32c will be exactly equal to the ones in Fig. 1.32a, because $\theta_B = \theta_C = 0$ in Fig. 1.32b.

This is very interesting. The beam system in Fig. 1.32c with the equivalent joint loads (moments), F_1 and F_2, will have exactly the same displacements (rotations), D_1 and D_2, at the degrees of freedom identified in the kinematically indeterminate system with arbitrary loading shown in Fig. 1.32a. In displacement methods of analysis, we need to go through this exercise (find equivalent joint loads) if the applied loads on the beam (or frame) elements act at intermediate locations (not at the joints), in order to find the unknown displacements.

As mentioned earlier, there exist definite stiffness relations inter-relating the forces F_1 and F_2, to the conjugate displacements, D_1 and D_2, in any linear elastic structure: $\{F\} = [k]\{D\}$. With reference to the continuous beam system in Fig. 1.32, for example, the stiffness coefficients k_{11} and k_{12} can be obtained as the moments generated at the imaginary rotational restraints at coordinates 1 and 2 (at B and C) respectively on account of a rotational slip, $D_1 = 1$, with $D_2 = 0$. The above set of *displacement-method based equilibrium* equations can be solved simultaneously, for any given load vector $\{F\}$ by any suitable method, such as matrix inversion or Gauss elimination, to determine the unknown displacements $\{D\}$. After finding these displacements, the individual beam elements can be analysed separately and the final end moments in each beam element can be obtained by applying the Principle of Superposition. Equations relating the final end moments to the fixed-end moments and the beam end displacements are called *slope-deflection equations*.

It is instructive to note the following key features, commonly shared by all the major displacement methods, in some form or other, to determine the force response in kinematically indeterminate structures.

Table 1.3 Fixed end moments in a prismatic fixed beam of span L (subject to total load W)

<table>
<tr><td>1</td><td>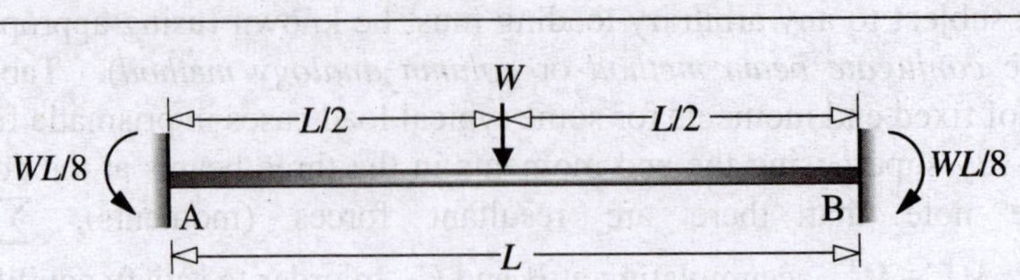
</td></tr>
<tr><td>2</td><td>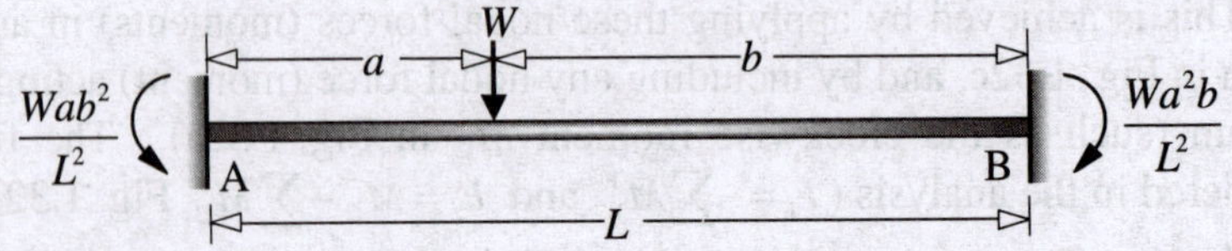
</td></tr>
<tr><td>3</td><td>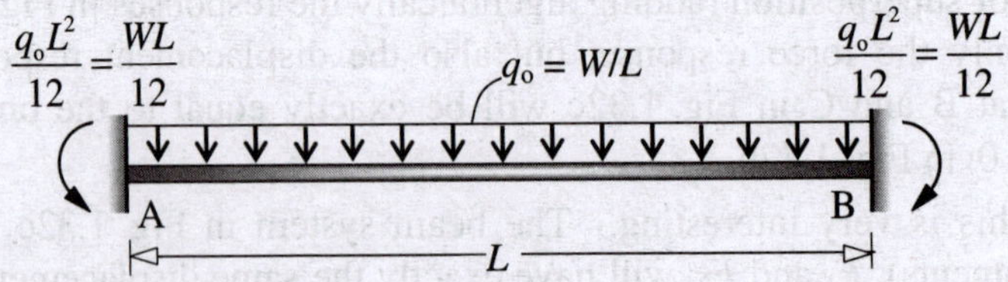
</td></tr>
<tr><td>4</td><td>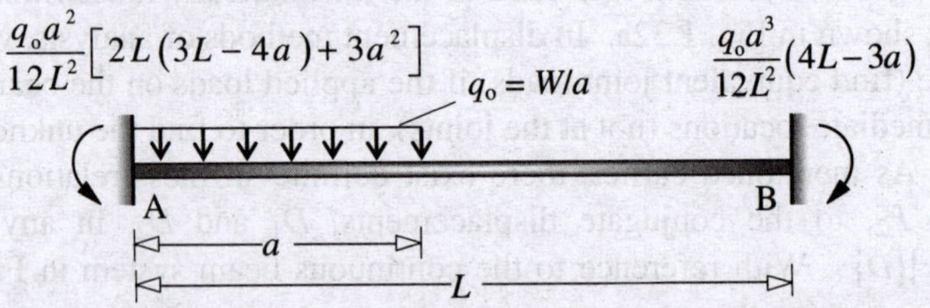
</td></tr>
<tr><td>5</td><td>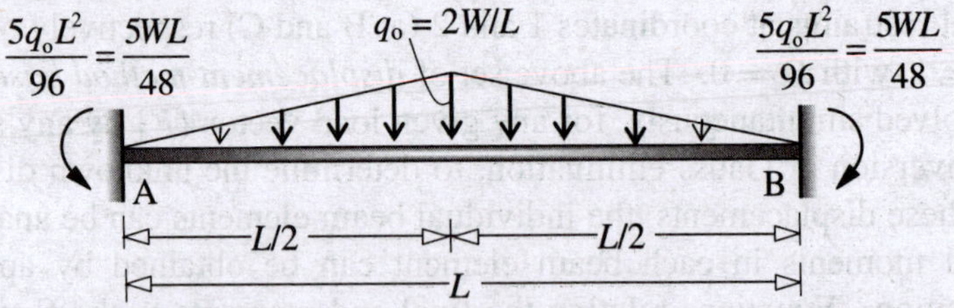
</td></tr>
</table>

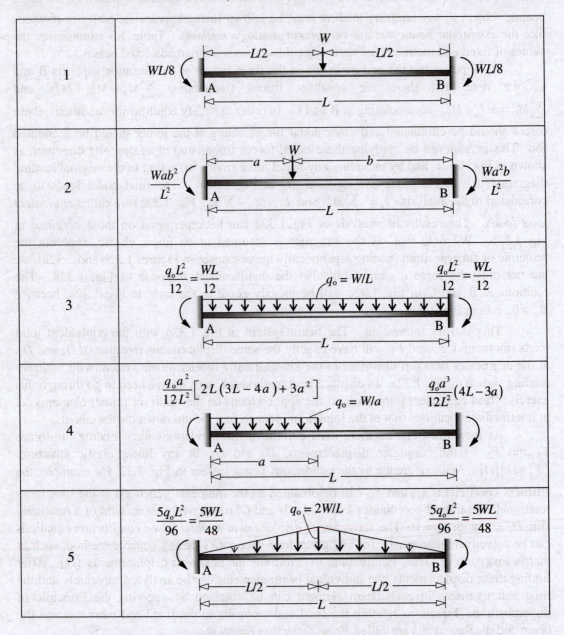

6	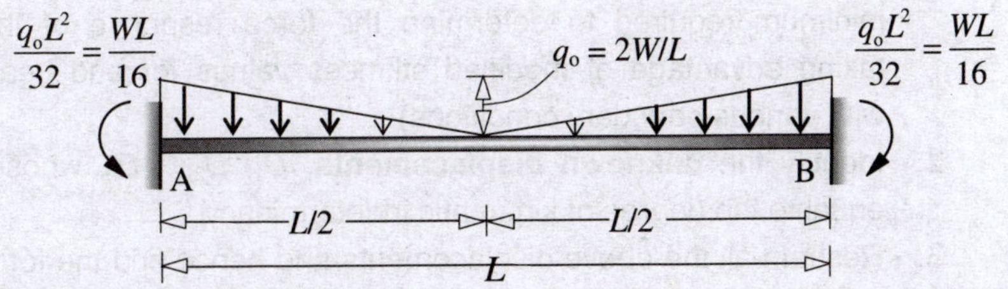$\frac{q_o L^2}{32} = \frac{WL}{16}$ $q_o = 2W/L$ $\frac{q_o L^2}{32} = \frac{WL}{16}$ A B L/2 L/2 L
7	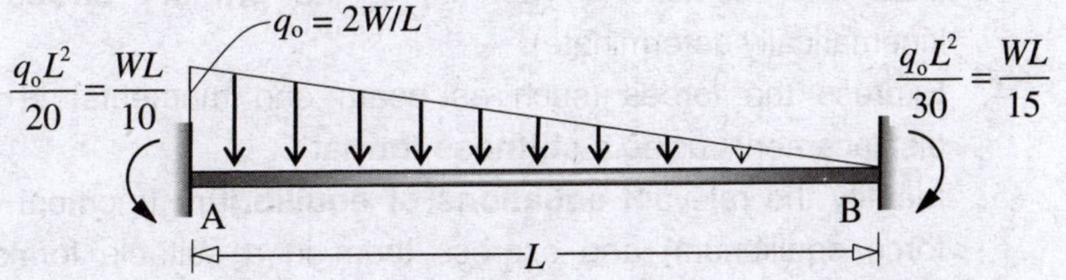$q_o = 2W/L$ $\frac{q_o L^2}{20} = \frac{WL}{10}$ $\frac{q_o L^2}{30} = \frac{WL}{15}$ A B L
8	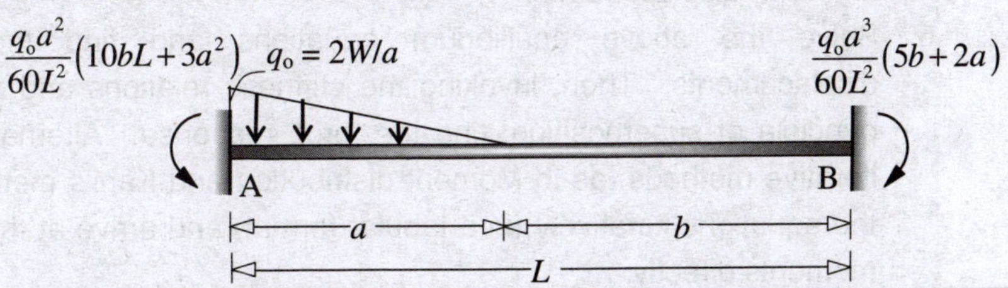$\frac{q_o a^2}{60L^2}(10bL + 3a^2)$ $q_o = 2W/a$ $\frac{q_o a^3}{60L^2}(5b + 2a)$ A B a b L
9	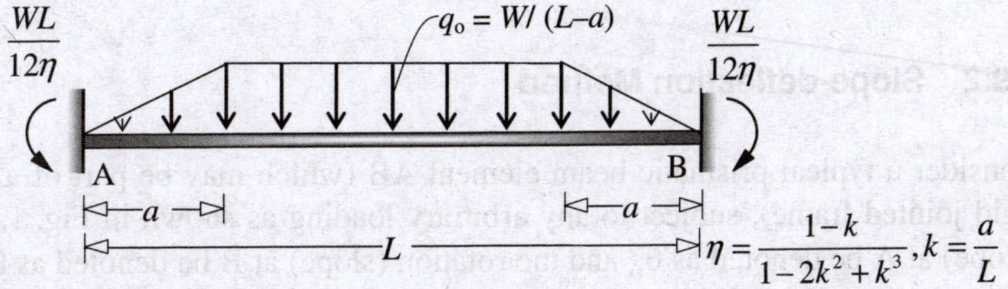$\frac{WL}{12\eta}$ $q_o = W/(L-a)$ $\frac{WL}{12\eta}$ A B a a L $\eta = \frac{1-k}{1-2k^2+k^3}, k = \frac{a}{L}$
10	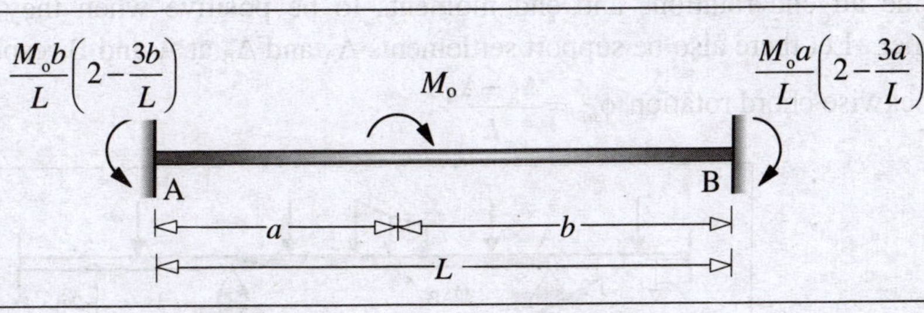$\frac{M_o b}{L}\left(2 - \frac{3b}{L}\right)$ M_o $\frac{M_o a}{L}\left(2 - \frac{3a}{L}\right)$ A B a b L

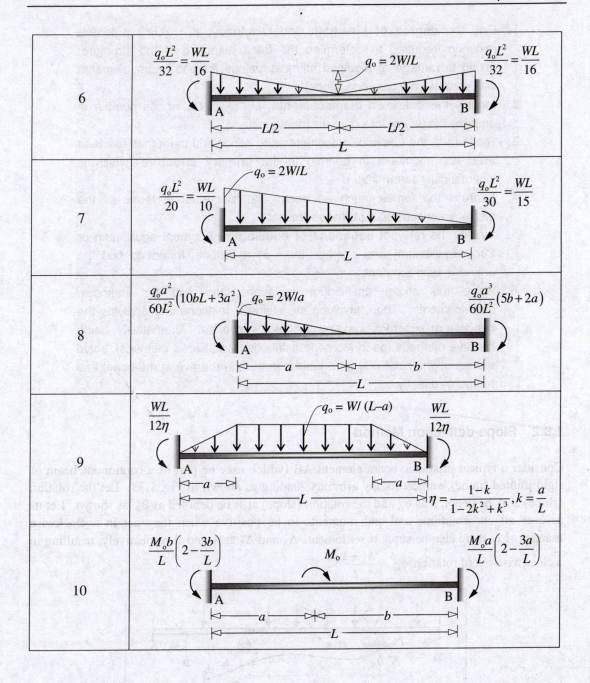

1. Find the **degree of kinematic indeterminacy**, n_k. (Limit it to the minimum required to determine the force response of the structure, taking advantage of modified stiffness values for end beam elements with various boundary conditions).
2. Identify the **unknown displacements**, D_1, D_2,.., D_n, whose number is equal to the degree of kinematic indeterminacy.
3. Restrain all the above displacements and hence find the forces (such as fixed end moments in beams) in the **primary structure** (which is kinematically determinate).
4. Express the forces (such as beam end moments) in terms of the displacements using a **stiffness format**.
5. Identify the relevant **equations of equilibrium** (moment equilibrium or force equilibrium) and express them in a suitable format to find the unknown displacements.
6. Solve the above equilibrium equations and find the unknown displacements. Then, invoking the stiffness relations and applying the principle of superposition, find the force response. Alternatively, using iterative methods (as in Moment distribution and Kani's methods), solve the equations iteratively (in a tabular format) and arrive at the beam end moments directly.

1.9.2 Slope-deflection Method

Consider a typical prismatic beam element AB (which may be part of a continuous beam or rigid jointed frame), subject to any arbitrary loading as shown in Fig. 1.33. Let the rotation (slope) at A be denoted as θ_A and the rotation (slope) at B be denoted as θ_B, as shown. Let us assume all end rotations and end moments to be positive when they act in a clockwise manner. Let there also be support settlements Δ_A and Δ_B at A and B respectively, resulting in a clockwise chord rotation $\phi_{AB} = \dfrac{\Delta_B - \Delta_A}{L}$.

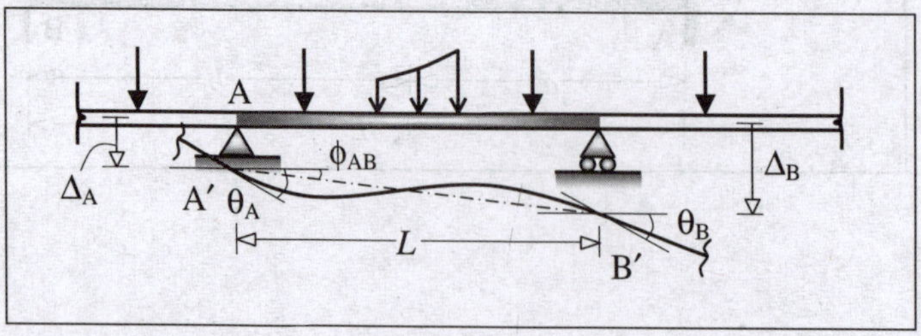

Figure 1.33 Typical prismatic beam element

The following slope-deflection equations can be easily derived for this beam element:

$$\begin{Bmatrix} M_{AB} \\ M_{BA} \end{Bmatrix} = \begin{Bmatrix} M^F_{AB} \\ M^F_{BA} \end{Bmatrix} + \left(\frac{EI}{L}\right)\begin{bmatrix} 4 & 2 \\ 2 & 4 \end{bmatrix}\begin{Bmatrix} \theta_A - \phi_{AB} \\ \theta_B - \phi_{AB} \end{Bmatrix} \tag{1.85}$$

The above equations can be modified suitably for beam elements with one of the beam ends having a hinged/roller support or guided fixed support condition. Such modifications are helpful in reducing the degree of kinematic indeterminacy in the structure for the purpose of simplified analysis. For example, if the end B is a hinged/roller support end, then we can avoid considering the end rotation θ_B as an unknown displacement. We must avoid arresting this degree of freedom in the primary structure, and the fixed-end moment to be considered is that of a propped cantilever (with end B propped), $M^{F_o}_{AB}$ at end A. This can be easily derived from the conventional fixed end moments $M^F{}_{AB}$ and $M^F{}_{BA}$ (for the identically loaded fixed beam) as follows, assuming all moments to be clockwise-positive:

$$M^{F_o}_{AB} = M^F_{AB} - M^F_{BA}/2 \tag{1.86}$$

For this case (end B hinged), the modified slope-deflection equation is given by:

$$M_{AB} = M^{F_o}_{AB} + \frac{3EI}{L}(\theta_A - \phi_{AB}) \tag{1.87}$$

After identifying the minimum degree of kinematic indeterminacy corresponding to the unknown displacements in the structure, and writing down the slope-deflection equations for all the beam elements, appropriate equations of equilibrium have to be formulated. Thus, corresponding to an unknown joint rotation θ_j, the equilibrium equation takes the form, $\sum M_{\text{beam end},j} = M_j$, where M_j is the concentrated moment (acting clockwise), if any, acting at the joint under consideration. In 'sway-type' problems, it is desirable to express the various chord rotations correctly in terms of the identified unknown translations. The corresponding force equilibrium equations to be applied will be in the direction corresponding to the translations. The equilibrium equations, expressed in terms of the unknown rotations and translations, are then solved simultaneously. Substituting the values of the displacements in the various slope-deflection equations, the beam-end moments are obtained.

1.9.3 Moment Distribution Method

The *Moment distribution method* is an ingenious method, originally proposed by Hardy Cross in 1930, as an alternative to the Slope-deflection method, with the advantage of directly yielding the beam end moments without explicitly finding the unknown displacements. The method was originally conceived as a relaxation technique of solving the equilibrium equations. Later, Hardy Cross discovered that there is a physical significance to every step in the mathematical procedure of analysis, which involved fixing the unknown displacements and releasing them one at a time. The process of fixity results in moments accumulating at various joints, which need to be released (or "balanced"). With every release, it is possible to use stiffness relations to assess how the released moment at a joint gets "distributed" to the various connecting elements and how some of the end moments get "carried over" to the remote ends of the various elements. Some of these carried over moments create imbalance and these now have to be released. This alternating procedure of fixing and releasing (balancing) is iteratively carried out until the residual unbalanced moments at all joints become negligible.

In simple problems, involving only one rotational degree of freedom, the exact solution is obtained in a single step. In other problems, involving two or more rotational degrees of freedom, several iterative steps are required until the final solution is obtained. In practice, we terminate the iterative procedure after a few cycles, and accept the solution which is considered approximate. For improved accuracy, we can continue with further cycles until the desired level of convergence is attained. The iteration can be conveniently done in a tabular format, referred to as the *moment distribution table*.

When a concentrated moment is applied at a rigid joint that is connected to several beam elements, the applied moment gets apportioned to the individual beam ends in direct proportion to their relative flexural stiffnesses. The joint rotational stiffness is equal to the sum of the individual flexural stiffnesses of the various connecting elements. The relative stiffness ($k_{io}/\Sigma k_{io}$) for the ith element at a joint O is called *distribution factor*. The stiffness measures for beam elements under various boundary conditions are summarised in Fig. 1.34.

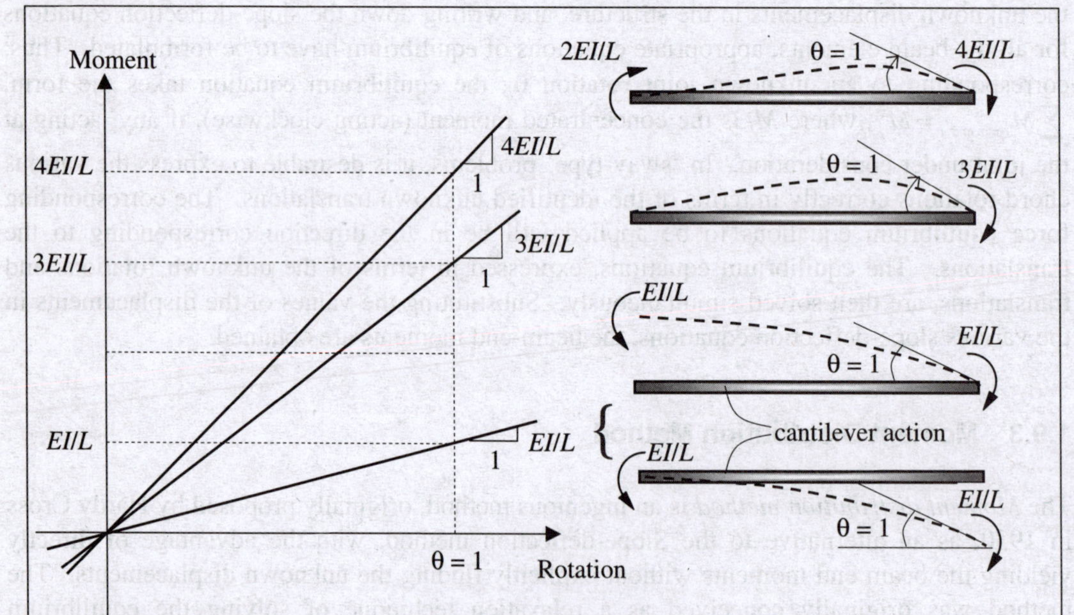

Figure 1.34 Moment-rotation plots showing flexural stiffnesses of beam elements with different boundary conditions

Some portion of the moment distributed at one end of a beam gets transmitted to the other end, and this fraction is referred to as *carry-over factor*. As can be seen from Fig. 1.34, the carry-over factor is equal to +½ when the far end is a fully fixed end, and equal to –1 when it is a guided-fixed support (cantilever action)

The Moment distribution method ceases to be an attractive option in problems involving unknown "sway" degrees of freedom. It becomes necessary to explicitly solve equilibrium equations and to resort to multiple moment distribution tables. In this regard, Kani's method is better suited for handling structures, such as unsymmetric multi-storeyed frames, in which rotations and sway degrees of freedom have to be considered.

1.9.4 Kani's Method

Kani's method, proposed by Gasper Kani of East Germany in 1957, is another ingenious method (suitable for manual analysis) that specifically addresses the shortcoming of the Moment distribution method, when applied to sway type problems, and manages the solution within a single iteration table. Although originally proposed for the analysis of multi-storeyed frames, this method can be used for all types of beams and plane frames. It combines the use of slope-deflection equations (in a different format) with the iterative procedure underlying the Moment distribution method (although in a different form). The unknown quantities in Kani's method are "rotation moments" and "sway moments", and these are solved for using appropriate moment equilibrium and force equilibrium equations, similar to the Slope-deflection method, using a numerical technique that resembles the Gauss-Seidel method of solving linear simultaneous equations. After we obtain the rotation and sway moments, we can easily substitute these values in the slope-deflection equations to obtain the final end moments, as in the Slope-deflection method. In this respect, Kani's method involves an additional step in "non-sway" problems (involving only rotational degrees of freedom), compared to the Moment distribution method, as the final end moments are not obtained directly at the end of the iterative cycle. However, in "sway" type problems, Kani's method turns out to be much more efficient compared to Moment distribution method, as the iterative procedure, although more involved than for non-sway problems, involves a single algorithm that can be accommodated in a single tabular format.

Kani's method uses the concept of *rotation factor* in the place of the *distribution factor* (in Moment distribution method). As in the case of the Moment distribution method, Kani's method is simple and accurate to apply to problems that involve only one rotational degree of freedom, although the solution (which is exact) requires an additional step, compared to the Moment distribution method. In problems involving unknown "sway" degrees of freedom, the unknown sway moments are accounted by using *displacement factors* in addition to rotation factors. The iterative solution is obtained in a single table, although convergence is slow in the presence of sway. However, it does not serve much purpose today to explore Kani's method for large structures, for which the Stiffness Method, is better suited.

REVIEW QUESTIONS

1.1 Distinguish between *statics* and *kinematics* in structural analysis.

1.2 Distinguish between a *planar structure* and a *space structure*.

1.3 How do you justify modelling a three-dimensional beam as a *line element* in structural analysis?

1.4 Distinguish between *pinned joint* and *rigid joint*, in terms of their kinematic and static functions.

1.5 Distinguish between *internal* and *external stability*. Under each of the two categories, give examples of *under-rigid*, *just-rigid* and *over-rigid* systems.

1.6 How is the concept of static determinacy and indeterminacy related to the kinematic concept of rigidity (*just-rigid* and *over-rigid* conditions)?

1.7 Distinguish between *static* and *kinematic* indeterminacy.

1.8 Distinguish between *direct actions* and *indirect loading*.

1.9 Why does 'lack of fit' in trusses induce stresses in *internally over-rigid* systems, but not in *just-rigid* systems?

1.10 Cite the three main requirements to be satisfied in determining the structural response of any stable, linearly elastic structure.

1.11 Discuss the importance of the *free-body*. How many independent equations of equilibrium are obtainable from every cut free-body of a *spatial* structure?

1.12 Distinguish between *statically admissible* and *kinematically admissible* structural systems. Give examples.

1.13 What is meant by *linear elastic* behaviour? Discuss its implications in structural analysis.

1.14 In statically determinate problems, it is sufficient to ensure that *equilibrium* conditions are satisfied to arrive at the correct force response. There is no need to explicitly satisfy *compatibility* requirements. Does this mean that compatibility is automatically satisfied? Discuss!

1.15 Distinguish between (a) *static* and *dynamic* analysis (b) *linear* and *nonlinear* analysis (c) *deterministic* and *stochastic* analysis. Show that one is a special case of the other.

1.16 'A system of loads acting on a structure may be replaced by a statically equivalent resultant force'. Is this justified for the purpose of calculating (a) support reactions (b) internal force resultants? Explain.

1.17 Consider a prismatic cantilever beam AB, fixed at A and free at B, with span L and flexural rigidity EI. Sketch the variation of curvature along the length of the beam given that the beam is subject to (i) a concentrated moment M applied clockwise at B and (ii) a concentrated load P applied at the mid-span location. For the case (ii), sketch the variation of the deflected shape of the beam due to (a) flexural deformations alone and (b) shear deformations alone.

1.18 Derive the basic differential equations of equilibrium, involving shear force $S(x)$ and bending moment $M(x)$, in a beam subject to a distributed force (acting upwards) $q(x)$ per unit length and a distributed moment (acting clockwise) $m(x)$ per unit length. Hence, find $S(x)$ and $M(x)$, given $q(x)$ = 0 kN/m and $m(x)$ = 10kN/m in a simply supported beam of 5m span. Draw the deflection, shear force and bending moment diagrams.

1.19 What are the assumptions commonly made in the analysis of forces in cable systems?

1.20 Show how the axial forces in cables and funicular arches are related to the distribution of bending moments in equivalent simply supported beams having the same span and vertical load distribution.

1.21 If the arch configuration is funicular under a given loading, does it matter whether the arch is three-hinged, two-hinged or fixed? Explain.

1.22 State and prove Eddy's Theorem.

1.23 Sketch the shapes of the bending moment and deflection diagrams for a typical rectangular portal frame ABCD (members AB, BC and CD having equal length), with a hinged support at A and a roller support at D, subject to a uniformly distributed gravity load acting on beam BC. (No calculations required; only sketches).

1.24 Consider a symmetric parabolic arch ABC, with a hinged support at A, roller support at B and crown at C, having a span L and rise h, subject to a horizontal force H, acting at B, directed towards A. Derive an expression for H and draw the bending moment diagram.

1.25 Distinguish between "bending moment at a section" and "influence line for bending moment at a section".

1.26 State and prove Müller-Breslau's Principle, as applied to statically determinate beams.

1.27 Why are influence lines for statically determinate beams invariably made of straight lines?

1.28 Consider a gantry girder with a simply supported span of 6m, subject to a crane loading comprising two wheel loads of 25 kN each with a wheel spacing of 0.9m. Find the maximum bending moment in the gantry girder on account of the crane loading.

1.29 Consider a simply supported beam of span 10m, with an overhang of 2.5m on one side, subject to a dead load of 10 kN/m and live load (of any length) of 30 kN/m.

Using influence lines, find the design values of shear force and bending moment at the mid-span location of the simply supported span.

1.30 For what kind of problems is it easier to determine support reactions by applying the *Principle of Virtual Displacements*?

1.31 Why is the *internal virtual work* equal to zero in the application of the *Principle of Virtual Displacements* to just-rigid structures?

1.32 Consider an axial element of length L, subject to an arbitrary but kinematically admissible displacement field $u(x)$, with translations $u(0) = D_1$ and $u(L) = D_2$ at the two ends. Consider the same axial element, subject separately to an arbitrary but statically admissible force field involving axial tension $N(x)$ and forces F_1 and F_2 conjugate with D_1 and D_2 respectively. Prove the validity of the Principle of Virtual Forces for this situation.

1.33 Consider a pin-jointed frame, ABC, which is an equilateral triangle of side 1m. The bar BC alone is heated and undergoes an increase in length of 3mm. Use the Principle of Virtual Forces to find the movement of the joint A (towards BC).

1.34 Using the Unit Load Method, derive expressions to find the slope and deflection at the free end of a prismatic cantilever beam of span L and flexural rigidity EI, subject to a uniformly distributed load of intensity q_o per unit length along the entire span.

1.35 State and prove Maxwell's reciprocal theorem.

1.36 State and prove Müller-Breslau's Principle, as applied to statically indeterminate beams.

1.37 What is meant by flexibility coefficient f_{ij}? Derive suitable expressions to find the flexibility coefficient in (a) a truss and (b) a rigid-jointed frame (in which axial and shear deformations are considered negligible).

1.38 A beam is loaded at three coordinate locations such that $F_1 = +60$ kN, $F_2 = -100$ kN, $F_3 = +40$ kNm, and the corresponding displacements at these coordinates are $D_1 = 2.5$ mm, $D_2 = -5$ mm and $D_3 = -0.0015$ radian. If, in a separate loading condition in which F_3 alone acts, the observed deflections are $D_1 = -1$ mm, $D_2 = +3$ mm and $D_3 = +0.0025$ radian, find out the value of F_3.

1.39 Distinguish the terms, *strain energy* and *complementary strain energy*. Under what condition can these quantities be inter-changed, without error in calculations?

1.40 A load F_1 acting on a linear elastic structure induces a strain energy of 80 Nm. When another load F_2 is placed on the structure, the strain energy increases to 180 kNm. If the load F_1 is now removed from the structure, what would be the residual strain energy?

1.41 Prove Maxwell's Reciprocal Theorem, using an energy method.

1.42 Match the following:

(A) Principle of Stationary Complementary Potential Energy	(1) Dummy Displacement Method
(B) Principle of Stationary Total Potential Energy	(2) Principle of Virtual Displacements
(C) Castigliano's Theorem (Part I)	(3) Dummy (Unit) Load Method
(D) Castigliano's Theorem (Part II)	(4) Principle of Virtual Forces

1.43 What are the main differences between the displacement-based and force-based energy methods?

1.44 How do the stationary energy principles get the name *stationary*?

1.45 State the two Castigliano Theorems. What are their uses?

1.46 How can energy methods be used to find stiffness and flexibility coefficients?

1.47 Enumerate the basic features of the force methods of analysis of statically indeterminate structures.

1.48 Show how Betti's Theorem can be used to find directly the moment at the fixed end of a propped cantilever beam, subject to arbitrary gravity loading.

1.49 Explain how the conjugate beam corresponding to a statically indeterminate beam is an unstable structure. How is it possible to apply the Conjugate Beam Method in spite of this instability?

1.50 Why is it that computer-aided analysis procedures for indeterminate structures are based on the stiffness approach rather than the flexibility approach?

1.51 Consider a rectangular portal frame ABCD, with hinged supports at A and D, subject to some gravity loading acting on the horizontal element BC. Can one of the two vertical reactions be treated as a redundant? Explain.

1.52 Consider a typical two-span continuous beam ABC, with the end A fixed and end C simply supported. Draw the probable shapes of the deflection and bending moment diagrams, corresponding to (a) a clockwise rotational slip θ_A at end A, and (b) a settlement δ_C at the support C.

1.53 Consider a rectangular portal frame ABCD, with hinged supports at A and D. How would you analyse this problem considering a uniform rise in temperature in the frame? Sketch probable shapes of the deflection and bending moment diagrams.

1.54 Can the Theorem of Least Work be applied to solve statically indeterminate problems with support settlements? Discuss.

1.55 Consider a symmetric two-hinged parabolic arch ABC, with hinged supports at A and B and crown at C, having a span L and rise h, subject to a distributed gravity load of total magnitude W, which varies linearly along the span, with zero intensity at A and maximum value at B. Derive an expression for the horizontal thrust H and sketch the approximate shape of the bending moment diagram.

1.56 Briefly explain how the Theorem of Least Work can be applied to find the redundant reactions in a fixed arch subject to symmetric gravity loading.

1.57 Demonstrate, with the help of an example, that *any* statically admissible bending moment diagram can be used to find the mid-span deflection in a prismatic fixed beam (span L, flexural rigidity EI), subject to a concentrated gravity load W at mid-span, while applying the Principle of Virtual Work.

1.58 Consider a typical beam of span 6m in a multi-storeyed frame, subject to a uniformly distributed load of 30 kN/m. If, based on a gravity load analysis of the entire frame, it is found that the hogging moments at the two ends of the beam are equal to 110.5 kNm and 125 kNm, and the flexural rigidity of the beam is 75 000 kNm^2, find the deflection at the mid-span of the beam.

1.59 What are the main differences between displacement and force methods of analysis?

1.60 Explain the basis of the displacement method of analysis with reference to a continuous beam.

1.61 Given the fixed end moments of a prismatic beam element subject to arbitrary gravity loading, how will you find the fixed end moment of the same beam when one of the fixed ends is made a hinged end.

1.62 Derive the basic slope-deflection equations for a typical prismatic beam element AB, having span L and flexural rigidity EI, subject to arbitrary loading, in terms of the fixed end moments, end rotations and chord rotation (all assumed clockwise).

1.63 How can the Slope-deflection method of analysis be used to solve problems involving (a) rotations of fixed end supports and (b) differential settlements in beams?

1.64 How can the Moment distribution method be categorised as a *displacement method*, when there are no unknown displacements to determine? The unknown quantities seem to be bending moments, and for this reason, should not this method be categorised as a *force method* of analysis?

1.65 Are the Moment distribution and Kani's methods of analysis "exact" methods (like the Slope-deflection method) or approximate methods? Discuss.

1.66 Discuss the merits and demerits of the Moment distribution method vis-à-vis Kani's method.

2

Matrix Concepts & Algebra

2.1 INTRODUCTION

As discussed in Chapter 1, the main objective of Structural Analysis is to determine the force response and displacement response of a given structure that is subject to specified loading. We deal with a mathematical model of the structure, with given system characteristics, and by applying the laws pertaining to the structural behaviour, we predict the response (output) corresponding to a given load input. The load may constitute direct actions (forces) or indirect loading (including displacements).

We can visualise all the forces in the structure (loads, reactions, internal force resultants) to constitute a 'force field'. Similarly, we can visualise all the displacements (including member deformations) to constitute a 'displacement field'. Some of these forces and displacements are known at the beginning of the analysis (load input), and our task is to find the remaining unknown forces and displacements (response output) that are of interest to us. For a correct solution, we know that three basic requirements need to be satisfied, viz., (i) equilibrium in the force field, (ii) compatibility in the displacement field and (iii) force-displacement relationships that account for the system characteristics (including constitutive relationships for the materials). If linear elastic behaviour is assumed and geometric nonlinearity is ignored, as is commonly done, the system characteristics, such as stiffness coefficients, are treated as constants, resulting in a definite set of linear relationships linking the various variables in the force and displacement fields.

Although many variables in the force and displacement fields (such as bending moments and deflections in beam elements) are *continuous variables*, it is possible to limit considerations of these variables to specific *coordinate* locations (such as joint locations), where they appear as *discrete variables*. These variables take on specific values (real numbers) in practical structural analysis, and these numbers can be arranged conveniently in a *matrix* format. Such a format is particularly desirable and suitable for computer application, whereby the computer does the required "number crunching" to yield the numerical output of the response variables. The basic variables, such as *force* and *displacement vectors*, and the system characteristics such as the *stiffness* or *flexibility matrices*, are inter-related by the basic laws of structural analysis. We can invoke the mathematical concepts of linear algebra to formulate these relationships and to solve for the desired unknowns.

Matrix methods are simple, yet powerful, mathematical concepts that have a wide range of applications in many diverse fields. The power and beauty of matrix algebra lie in its ability to convey, in a very compact form, mathematical expressions that would appear laborious and somewhat messy, in the conventional scalar algebra. In this Chapter we give an overview of the basic matrix concepts and matrix algebra that we adopt in the application of matrix methods in structural analysis, as described in this book.

Following this Introduction, in Sections 2.2 and 2.3, the basic definitions of a **matrix** and a **vector** are explained. In the remaining Sections of this Chapter, we show the essential mathematical operations and algebra involving matrices and vectors. In Section 2.4, we define the **elementary matrix operations**. This is followed in Section 2.5 by a description of the important operation of **matrix multiplication**. In Sections 2.6 and 2.7, we discuss respectively the concepts of **transpose** and **rank** of a matrix. In Section 2.8, we discuss the possibilities of finding **solutions to linear simultaneous equations** using matrix methods. In Section 2.9, we discuss the **inverse** of a square matrix and in Section 2.10, we cover the important topic of **eigenvalues and eigenvectors** of a square matrix.

2.2 MATRIX

By definition, a **matrix** is a rectangular array of elements arranged in horizontal rows and vertical columns. The entries of a matrix, called **elements**, are *scalar* quantities, which are variables described by single numbers. The scalar entries are commonly numbers, but they may also be functions, operators or even matrices (called sub-matrices) themselves.

The size of the rectangular array defines the **order** of the matrix. If the array comprises m horizontal rows and n vertical columns, then the matrix is said to have an order, $m \times n$. If m is equal to n, the rectangle becomes a square, and we have **square matrix** (of order n).

In general, a matrix is designated by an uppercase bold-face letter, such as **A**, or with the letter encased in square parentheses, such as $[A]$. The element located in the ith row and j^{th} column, is designated with the corresponding lowercase letter and with subscripts, ij. The row index i is always placed before the column index j. Thus, the *scalar* quantity, a_{ij} denotes a typical element of the matrix **A**. Sometimes, the matrix itself is designated as $[a_{ij}]$. The order of the matrix, $m \times n$, can also be indicated as a subscript after the right parenthesis. In summary, the matrix **A** can be shown in any of the following equivalent ways:

$$\mathbf{A} = [A] = [A]_{m\times n} = \left[a_{ij}\right] = \left[a_{ij}\right]_{m\times n} = \begin{bmatrix} a_{11} & a_{12} & a_{13} & \cdots & a_{1n} \\ a_{21} & a_{22} & a_{23} & \cdots & a_{2n} \\ a_{31} & a_{32} & a_{33} & \cdots & a_{3n} \\ \vdots & \vdots & \vdots & & \vdots \\ a_{m1} & a_{m2} & a_{m3} & \cdots & a_{mn} \end{bmatrix} \tag{2.1}$$

Elements located on the principal diagonal of the matrix, where the row index matches the column index (i.e., $i = j$), are called **diagonal elements**. A **diagonal matrix** is a square matrix, whose elements are all zero, except possibly the diagonal elements; i.e., the *off-diagonal* elements in a diagonal matrix are zero. The **identity matrix** (or **unit matrix**), **I**, is a diagonal matrix, all of whose diagonal elements are equal to unity; i.e., $i_{11} = i_{22} = ... = i_{nn} = 1$ and $i_{j\neq k} = 0$.

A **zero row** in a matrix is one whose elements are all zeros. When all the rows in a matrix are zero rows, the matrix is called a **null matrix** (or **zero matrix**), denoted by the symbol, **O**. A **sparse matrix** is one in which most of the elements are zero. The *stiffness matrix* of a large structure is typically a *sparse* matrix, which is also *symmetric*, with the non-zero elements lying within a band parallel to the principal diagonal. Such a matrix, in which the elements outside the diagonal band, are all zero, is called a **banded matrix**. While carrying out computations on a digital computer, it is convenient to limit the storage and operations to the non-zero elements in the banded region.

If all the elements above the main diagonal of a matrix **A** are zero, i.e., $a_{ij} = 0$ for $j > i$, the matrix is said to be **lower triangular**. Conversely, if all the elements below the main diagonal of a matrix **A** are zero, i.e., $a_{ij} = 0$ for $i > j$, the matrix is said to be **upper triangular**.

A **sub-matrix** of a given matrix **A** is any matrix that can be obtained from **A** by the removal of any number of rows and columns. This is akin to the definition of a *sub-set*. When the sub-matrices are formed by dividing the matrix using horizontal and vertical lines between rows and columns (a procedure called *partitioning*), the resulting sub-matrices are said to be **partitioned**. By judicious partitioning, we can simplify matrix operations, dealing with limited matrix sizes (containing sub-matrices of interest to us), rather than with large entire matrices. Thus, for example, the matrix **A**, as described in Eq. 2.1 can be partitioned into four sub-matrices as follows:

$$\mathbf{A} = \left[a_{ij}\right]_{m\times n} = \left[\begin{array}{c|c} \left[b_{ij}\right]_{l\times p} & \left[c_{ij}\right]_{l\times(n-p)} \\ \hline \left[d_{ij}\right]_{(m-l)\times p} & \left[e_{ij}\right]_{(m-l)\times(n-p)} \end{array}\right]_{m\times n} = \left[\begin{array}{c|c} \mathbf{B} & \mathbf{C} \\ \hline \mathbf{D} & \mathbf{E} \end{array}\right] \tag{2.2}$$

Any *banded* matrix **A** can be conveniently partitioned into the following **block diagonal** form, in which there are k sub-matrices $\mathbf{A}_i$ (containing mostly non-zero elements) along the diagonal band:

$$\mathbf{A} = \begin{bmatrix} \mathbf{A}_1 & & & & \\ & \mathbf{A}_2 & & \mathbf{O} & \\ & & \mathbf{A}_3 & & \\ & \mathbf{O} & & \ddots & \\ & & & \cdots & \mathbf{A}_k \end{bmatrix} \tag{2.3}$$

2.3 VECTOR

A **vector** is a simple array of scalar quantities, typically arranged in a vertical column. Hence, the vector can be visualised as a matrix of order $m \times 1$, where the number m is called the **dimension** of the vector. The scalar entries of a vector are called **components** of the vector.

In traditional mathematics, the vector is designated by a lowercase bold-face letter, such as **v**. However, in structural analysis, we make a minor departure from this convention, and treat the vector as another matrix (column matrix), using the uppercase letter (such as **V**) to designate both matrices and vectors. The *force vector* **F** and the *displacement vector* **D** are typical examples. However, in order to indicate that these symbols specifically denote vectors (column arrays) and not matrices (rectangular arrays), we shall use the notations $\{F\}$ and $\{D\}$ in lieu of **F** and **D** respectively.

A typical vector **V** of dimension m can be represented in any of the following equivalent ways:

$$\mathbf{V} = \{V\}_m = \{v_i\} = [V]_{m\times 1} = \left[v_{ij}\right]_{m\times 1} = \begin{Bmatrix} v_1 \\ v_2 \\ v_3 \\ \vdots \\ v_m \end{Bmatrix} \tag{2.4}$$

Thus, a rectangular matrix of order $m \times n$ can be visualised as being made up of a row of n vectors, each of dimension m. The matrix reduces to a vector when there is only one column, i.e., $n = 1$. Furthermore, this vector reduces to a scalar when it has only one component (i.e., $m = n = 1$).

It is also possible to depict the vector in the form of a horizontal row, rather than a vertical column. However, in this book, we reserve the term 'vector' to refer to the 'column matrix' and use the term 'row matrix' or 'row vector', while dealing with the single horizontal row representation.

We can visualise a multi-dimensional **linear vector space,** $\Re^m$, whose **dimension** m is given by the minimum number of **linearly independent vectors** (with *real* components) required to **span**[†] the space. (The formal definition of linear independence is given later in Section 2.6). Any set of vectors that are linearly independent and also span the vector space is called a **basis** of the vector space. It is easy for us to picture and draw vectors in spaces with dimensions up to three; for higher dimensions, we need to stretch our imagination and visualise an extended 'hyper-space'. Any vector in this space can then be visualised as a line (with an arrow-head) directed from the origin to a point in the vector space, such that the line has the various components of the vector when projected on to the respective orthogonal axes. Thus, the vector is conceived of having a **direction** in the linear vector space. The vector also has a **magnitude** or *length*, which is a non-negative scalar quantity, equal to the square root of the sum of the squares of all the components. Thus, the magnitude of the vector **V**, which may be denoted by $\|\mathbf{V}\|$, is given by $\|\mathbf{V}\| = \sqrt{\sum_{i=1}^{m} v_i^2}$. For example, the vector $\mathbf{V} = \begin{bmatrix} 2 \\ -1 \\ 3 \end{bmatrix}$ has a magnitude of $\sqrt{14}$ and can be visualised in $\Re^3$ vector space as $\mathbf{V} = 2\hat{\mathbf{i}} - \hat{\mathbf{j}} + 3\hat{\mathbf{k}}$, where $\hat{\mathbf{i}}, \hat{\mathbf{j}}, \hat{\mathbf{k}}$ denote **unit vectors** (mutually orthogonal) along the x-, y-, z- Cartesian coordinate axes, represented by $\begin{bmatrix} 1 \\ 0 \\ 0 \end{bmatrix}$, $\begin{bmatrix} 0 \\ 1 \\ 0 \end{bmatrix}$ and $\begin{bmatrix} 0 \\ 0 \\ 1 \end{bmatrix}$ respectively, and forming an **orthogonal basis** for $\Re^3$.

When the components of any vector are divided by the magnitude of the vector, the resulting vector has a magnitude equal to one. Such a vector is sometimes referred to as a **normalised vector**. A set of **orthogonal** (mutually perpendicular) non-zero vectors are invariably linearly independent. When they are also normalised, these vectors are called **orthonormal vectors**. Any set of m orthonormal vectors form an **orthonormal basis** for the $\Re^m$ vector space. The simplest set, of course, corresponds to the one in which each unit vector points in the direction of a coordinate axis. By rotating the axes, we can generate a new orthonormal basis. We shall refer to such **coordinate transformations** in Chapter 3.

The identity matrix **I** of order $m \times m$ may be visualised as being composed of a set of m independent and mutually orthogonal unit vectors, each of dimension m in the $\Re^m$ vector space. It provides a *basis* of the vector space, whereby any vector can be defined and operated upon, applying the laws of vector algebra, with the resulting vector remaining in the same $\Re^m$ vector space.

[†] Vectors are said to *span* a vector space, if the space consists of *all* possible linear combinations of those vectors.

2.4 ELEMENTARY MATRIX OPERATIONS

The algebra for symbolic operations on matrices (and vectors) is different from the algebra for operations on scalars or single numbers. It is possible to express the exact equivalent of matrix algebra equations in terms of scalar algebra expressions, and from the resulting expressions we can appreciate the power and compactness of matrix notations and algebra. The rules that apply to matrices also apply to vectors (which are column matrices).

It is interesting to note that, except for some simple operations, such as addition and scalar multiplication, the simple and intuitively obvious rules that apply to scalars (or numbers) are not generally applicable in an analogous manner to the elements of matrices.

2.4.1 Equality

Two matrices, **A** and **B**, are said to be **equal** if they have the same order and if their corresponding elements are equal (i.e., $a_{ij} = b_{ij}$). This is intuitively obvious.

2.4.2 Scalar Multiplication

When a matrix **A** is multiplied by a scalar, λ, the operation results a matrix of the same order, denoted as $\lambda\mathbf{A}$, whose elements have values of λ times a_{ij}. In other words, the multiplication is to be simply carried out element-wise, as is intuitively obvious.

$$\lambda\mathbf{A} = \lambda\left[a_{ij}\right]_{m\times n} = \left[\lambda a_{ij}\right]_{m\times n} = \mathbf{A}\lambda \tag{2.5}$$

If **V** is an m-dimensional vector, then the scalar product $\lambda\mathbf{V}$ is another vector that has the same direction as **V** and a magnitude equal to $\lambda\|\mathbf{V}\|$. The line containing all possibilities of the vector, $\lambda\mathbf{V}$, extending from $-\infty$ to $+\infty$ and passing through the origin (the **zero vector** being a special case of $\lambda\mathbf{V}$, corresponding to $\lambda = 0$), is a **subspace** of the vector space $\Re^m$ that contains **V**.

2.4.3 Matrix Addition

Addition of matrices is possible if they have the same order. The addition results in a matrix whose elements are given by the algebraic sum of the corresponding elements in the matrices being added. Thus,

$$\mathbf{A}+\mathbf{B} = \left[a_{ij}\right]_{m\times n} + \left[b_{ij}\right]_{m\times n} = \left[a_{ij} + b_{ij}\right]_{m\times n} \tag{2.6}$$

The addition operation is obviously *commutative*, i.e., if matrices **A** and **B** have the same order,

$$\mathbf{A}+\mathbf{B}=\mathbf{B}+\mathbf{A} \tag{2.7}$$

If the matrix **B** in Eq. 2.7 is a null matrix (i.e., **B** = **O**), it follows that $\mathbf{A}+\mathbf{O}=\mathbf{A}$.

The addition operation is obviously also *associative*, i.e., if matrices **A**, **B** and **C** have the same order,

$$\mathbf{A}+(\mathbf{B}+\mathbf{C})=(\mathbf{A}+\mathbf{B})+\mathbf{C} \tag{2.8}$$

The operation of **subtraction** is analogous to the operation of addition, with the plus sign in Eq. 2.6 replaced by a negative sign. This can also be proved by invoking the property of scalar multiplication: $\mathbf{A}-\mathbf{B}=\left[a_{ij}\right]_{m\times n}+(-1)\left[b_{ij}\right]_{m\times n}=\left[a_{ij}-b_{ij}\right]_{m\times n}$.

A linear combination of two *m*-dimensional vectors, $\mathbf{V}_1$ and $\mathbf{V}_2$ in $\Re^m$ vector space results in another *m*-dimensional vector, $\lambda_1\mathbf{V}_1+\lambda_2\mathbf{V}_2$, in the same space, which can be visualised in a manner akin to the *resultant force* in the *triangle of forces* that we are familiar with in structural analysis. This visualisation also conveys that the plane (or hyper-plane) containing the vectors $\mathbf{V}_1$ and $\mathbf{V}_2$ in $\Re^m$ is the *subspace* that contains the vectors generated by all possible linear combinations of $\mathbf{V}_1$ and $\mathbf{V}_2$. This *plane* reduces to a *line* if $\mathbf{V}_1$ and $\mathbf{V}_2$ have the same direction, and reduces to a *point* at the origin (**zero vector**) if the scalar multipliers, λ_1 and λ_2, are both zero or if $\mathbf{V}_1=\mathbf{V}_2=\mathbf{O}$. A set of vectors, $\{\mathbf{V}_1, \mathbf{V}_2, \ldots, \mathbf{V}_n\}$, having the same dimension *n*, is said to be **linearly independent** if no linear combination of them (other than the zero combination) results in a zero vector; i.e., the only set of scalars, $c_1, c_2, \ldots, c_n$, that satisfy the equation, $\sum_{i=1}^{n} c_i\mathbf{V}_i=\mathbf{O}$, is the set, $c_1 = c_2 = \ldots = c_n = 0$.

Non-zero vectors that are mutually **orthogonal** (perpendicular) are invariably linearly independent.

2.5 MATRIX MULTIPLICATION

The intuitive concept that operations on matrices can be carried out element-wise, which served us well for matrix addition and scalar multiplication, unfortunately, fails when we try to apply it to matrix multiplication. Furthermore, the elements of the product of two matrices, **A** and **B**, which is denoted as **AB** or [**A**][**B**], is not possible in general if both have the same order (except if they are square matrices). Also, the commutative property, **AB** = **BA**, is not valid in general, and it may not be possible to define **BA**, even if the matrix **AB** can be defined! Another intuition-defying fact is that if **AB** = **AC**, it does not generally follow that **B** is equal to **C**! Furthermore, if the product **AB** = **O** (null matrix), it does not generally follow that either **A** = **O** or **B** = **O** or both **A** and **B** are null matrices! Hence, we need to proceed with caution, while dealing with the algebra related to matrix multiplication.

The product **AB** of two matrices, **A** and **B**, can be defined if the number of columns of **A** is equal to the number of rows of **B**. If **A** has an order $m \times n$ and **B** an order $n \times p$, then the product **AB** will have an order $m \times p$. (Clearly, by this rule, the product **BA** can exist as a matrix, only if $m = p$). When the product is written as **AB**, **A** is said to *pre-multiply* **B**. Alternatively, it may be said that **B** *post-multiplies* **A**.

Let the product **AB** = **C**; i.e., $\left[a_{ij}\right]_{m\times n}\left[b_{ij}\right]_{n\times p} = \left[c_{ij}\right]_{m\times p}$. The element c_{ij} is obtained by multiplying each of the n elements in the ith row of **A** with the corresponding element in the jth column of **B** and summing up the results. Thus,

$$c_{ij} = \sum_{k=1}^{n} a_{ik}b_{kj} = a_{i1}b_{1j} + a_{i2}b_{2j} + \cdots + a_{in}b_{nj} \qquad (i = 1, 2, \ldots, m;\ j = 1, 2, \ldots, p) \qquad (2.9)$$

While carrying out the matrix multiplication operation to determine any element c_{ij} of the product **C**, it is necessary and sufficient to focus attention only the ith row of **A** and j^{th} column of **B**, (ignoring other rows in **A** and other columns in **B**) as indicated schematically in Fig. 2.1, and then carrying out the multiplications and summation described by Eq. 2.9. Eq. 2.9 can therefore be alternatively expressed as a scalar product of a row vector (i^{th} row of **A**) and a column vector (j^{th} column of **B**): $\begin{bmatrix} a_{i1} & a_{i2} & a_{i3} & \cdots & a_{in} \end{bmatrix}_{1\times n}\begin{bmatrix} b_{1j} \\ b_{2j} \\ \vdots \\ b_{nj} \end{bmatrix}_{n\times 1} = c_{ij}$

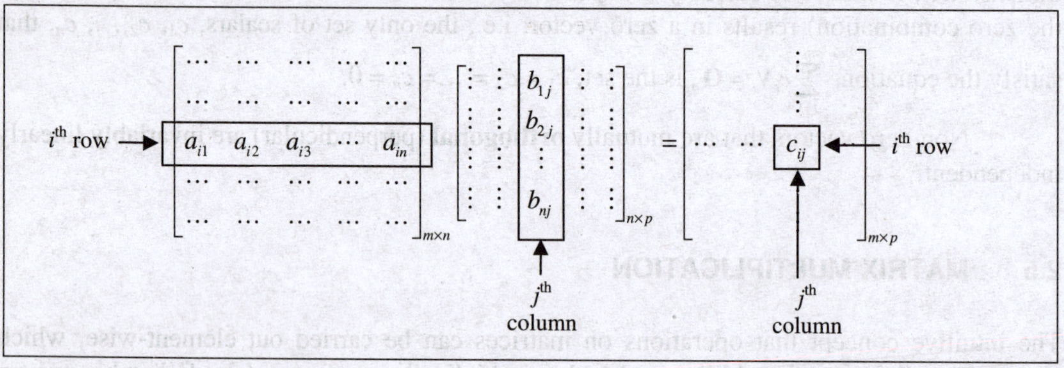

Figure 2.1 Matrix multiplication: focus on i^{th} row of **A** and j^{th} column of **B** to find c_{ij}

Thus, the elements that make up the jth column of **C**, i.e., $c_{1j}, c_{2j}, \ldots, c_{mj}$ are generated by taking, one at a time, the scalar products of the row vectors at rows 1, 2,…, m of **A** respectively with the jth column vector of **B**. It turns out that the jth column vector of **C** is a product of the entire matrix **A** and the jth column vector of **B**, and can be generated by b_{1j} times the first column vector of **A** plus b_{2j} times the second column vector of **A** and so on until we reach b_{nj} times the last (n^{th}) column vector of **A**. In other words, the jth column vector of **C** is a linear combination of the column vectors of **A**. Similarly, it can be shown that the ith row vector of **C** is a linear combination of the row vectors of **B**. By partitioning **A** into columns and **B** into rows, we can generate the matrix **C** as the addition of the products of the column vectors of **A** and the respective row vectors of **B**.

> The product $[C]_{m\times p}$ of two matrices $[A]_{m\times n}$ and $[B]_{n\times p}$ is such that every column vector of **C** is a linear combination of the column vectors of the pre-multiplying matrix **A**, and every row vector of **C** is a linear combination of the row vectors of the post-multiplying matrix **B**. Furthermore, the matrix **C** can be generated as the sum of the products of the column vectors of **A** and the respective row vectors of **B**.

These concepts (not commonly understood) underlying matrix multiplication can be observed in the following example:

$$\begin{bmatrix} 2 & -1 \\ 3 & 4 \\ 1 & 2 \end{bmatrix}_{3\times2} \begin{bmatrix} 1 & 4 \\ 2 & 0 \end{bmatrix}_{2\times2} = \begin{bmatrix} 2 \\ 3 \\ 1 \end{bmatrix} [1 \;\; 4] + \begin{bmatrix} -1 \\ 4 \\ 2 \end{bmatrix} [2 \;\; 0] = \begin{bmatrix} 2 & 8 \\ 3 & 12 \\ 1 & 4 \end{bmatrix}_{3\times2} + \begin{bmatrix} -2 & 0 \\ 8 & 0 \\ 4 & 0 \end{bmatrix}_{3\times2} = \begin{bmatrix} 0 & 8 \\ 11 & 12 \\ 5 & 4 \end{bmatrix}_{3\times2} \quad \text{(a)}$$

In general, it can easily be shown that the matrix multiplication operation does not possess the property of *commutativity*; i.e., in general, **AB** is not equal to **BA**, even if the matrices **A** and **B** have the correct orders ($m \times n$ and $n \times m$) to make such products possible. This lack of the commutative property of matrix multiplication should caution us against making any inferences with regard to the product of matrix multiplication, as pointed out earlier. For example, if $\mathbf{A} = \begin{bmatrix} 4 & 2 \\ 2 & 1 \end{bmatrix}$, $\mathbf{B} = \begin{bmatrix} 2 & 2 \\ 4 & 2 \end{bmatrix}$ and $\mathbf{C} = \begin{bmatrix} 4 & 4 \\ 0 & -2 \end{bmatrix}$, it follows that the product $\mathbf{AB} = \begin{bmatrix} 16 & 12 \\ 8 & 6 \end{bmatrix}$ and also $\mathbf{AC} = \begin{bmatrix} 16 & 12 \\ 8 & 6 \end{bmatrix}$. As **AB** = **AC**, we may be tempted to "cancel" out the common matrix **A** on both sides of the equation, and thereby we may wrongly conclude that **B** is equal to **C**, which is clearly not the case.

However, matrix multiplication does have the *associative* property:

$$\mathbf{A(BC)} = \mathbf{(AB)C} \tag{2.10}$$

It also possesses the *left distributive* and *right distributive* properties:

$$\mathbf{A(B+C)} = \mathbf{AB} + \mathbf{AC} \tag{2.11}$$

$$\mathbf{(B+C)A} = \mathbf{BA} + \mathbf{CA} \tag{2.12}$$

It is assumed in the above equations that the matrices **A**, **B** and **C**, have the appropriate orders to make possible the multiplications and additions. We may use the term *conformable* to indicate the validity of such operations.

The product of any matrix **A** with a conformable null matrix **O** results in a null matrix:

$$\mathbf{AO} = \mathbf{O} \quad \text{and} \quad \mathbf{OA} = \mathbf{O} \tag{2.13}$$

The *identity matrix* **I** (a *diagonal matrix* with diagonal elements equal to unity) has a useful and unique property: its product with any conformable matrix **A** will be equal to **A**:

$$\mathbf{AI} = \mathbf{A} \quad \text{and} \quad \mathbf{IA} = \mathbf{A} \tag{2.14}$$

2.6 TRANSPOSE OF A MATRIX

Transposition is an operation in which the rectangular array of the matrix is rearranged ("transposed") such that the order of the matrix changes from $m \times n$ to $n \times m$, with the rows changed into columns, preserving the order. If the original matrix is $\mathbf{A} = \left[a_{ij}\right]_{m\times n}$, then the **transpose of A**, which is denoted as $\mathbf{A}^{\mathrm{T}}$ (read as "A-transpose"), is given by $\mathbf{A}^{\mathrm{T}} = \left[a_{ij}\right]^{\mathrm{T}} = \left[a_{ji}\right]_{n\times m}$.

Some important properties of the transpose, involving transposition of (i) a transpose, (ii) a scalar multiple of a matrix, (iii) a sum of two matrices and (iv) a product of two matrices, are given below:

$$\left(\mathbf{A}^{\mathrm{T}}\right)^{\mathrm{T}} = \mathbf{A} \tag{2.15}$$

$$(\lambda\mathbf{A})^{\mathrm{T}} = \lambda\mathbf{A}^{\mathrm{T}} \tag{2.16}$$

$$(\mathbf{A}+\mathbf{B})^{\mathrm{T}} = \mathbf{A}^{\mathrm{T}} + \mathbf{B}^{\mathrm{T}} \tag{2.17}$$

$$(\mathbf{AB})^{\mathrm{T}} = \mathbf{B}^{\mathrm{T}}\mathbf{A}^{\mathrm{T}} \tag{2.18}$$

Eq. 2.18 is especially noteworthy: the transpose of a product is not the product of the transposes (as one might be inclined to believe), but the *commuted* product of the transposes. Let us consider the product, $\mathbf{S} = \mathbf{A}^{\mathrm{T}}\mathbf{A}$, involving any rectangular matrix $\mathbf{A} = \left[a_{ij}\right]_{m\times n}$ and its transpose $\mathbf{A}^{\mathrm{T}} = \left[a_{ji}\right]_{n\times m}$. This will obviously be a *square matrix* of order n. If we now consider the transpose of **S**, and invoke Eqns 2.18 and 2.15, we observe that $\mathbf{S}^{\mathrm{T}} = \mathbf{A}^{\mathrm{T}}\mathbf{A} = \mathbf{S}$. Such a square matrix, whose transpose is equal to itself ($\mathbf{S}^{\mathrm{T}} = \mathbf{S}$), will reflect symmetry in the elements with respect to the principal diagonal ($a_{ji} = a_{ij}$), and hence is called a **symmetric matrix**. In structural analysis, *flexibility* and *stiffness* matrices belong to this category. There are many advantages in dealing with such *symmetric* matrices, while solving linear simultaneous equations to find the structural response. When the transpose is equal to the negative of the matrix, i.e., $\mathbf{A}^{\mathrm{T}} = -\mathbf{A}$ (i.e., $a_{ji} = -a_{ij}$), the matrix is said to be **skew-symmetric**.

Using transposition, a vector (column matrix) can be conveniently converted into a row matrix. If we have two vectors, say $\{F\}$ and $\{D\}$, both having real (not complex) components with the same dimension m, and if we are interested in their product, then it is necessary to consider the transpose of one of them. The product $\{F\}^T\{D\}$ results in a matrix of order 1×1, which is nothing but a scalar. Such a product, which is sometimes denoted as $\langle \mathbf{F}, \mathbf{D} \rangle$, is called **inner product**, which is an extension of the *dot product* in vector algebra (involving two- and three-dimensional vectors) to real vectors of all dimensions. This product essentially reflects the projected length of one vector along the direction of the other vector. The *magnitude* of any vector **V**, defined earlier as $\|\mathbf{V}\| = \sqrt{\sum_{i=1}^{m} v_i^2}$, may be viewed as the inner product of **V** with itself, i.e., $\langle \mathbf{V}, \mathbf{V} \rangle = \mathbf{V}^T\mathbf{V}$. Also, when the inner product of two vectors, **V** and **W**, turns out to be zero, i.e., $\langle \mathbf{V}, \mathbf{W} \rangle = \mathbf{V}^T\mathbf{W} = 0$, it follows that the two vectors are aligned in directions that are mutually orthogonal (perpendicular), whereby they are called **orthogonal vectors**. If all the vectors in a set of orthogonal vectors are *normalised* (by dividing the components of each vector by its magnitude) such that they become *unit vectors*, then they are referred to as **orthonormal vectors**. If $\hat{\mathbf{X}}_i$ and $\hat{\mathbf{X}}_j$, denote two such vectors from an orthonormal set, then their inner product is given by the *Kronecker delta* δ_{ij}:

$$\hat{\mathbf{X}}_i{}^T\hat{\mathbf{X}}_j = \delta_{ij} = \begin{cases} 0 & \text{if } i = j \\ 1 & \text{if } i \neq j \end{cases} \tag{2.19}$$

If the vectors, $\{F\}$ and $\{D\}$ denote force and displacement vectors in a structure, then the scalar quantity represented by the inner product, $\{F\}^T\{D\} = \lambda$, is none other than *virtual work*. If, on the other hand, we consider the product, $\{F\}\{D\}^T$, called the **outer product**, the resulting matrix is a *square matrix* of order n. Such a product would not be meaningful in the context of virtual work. Hence, if we know the outcome to be a scalar, we must always choose the inner product. In the case of virtual work, an alternative choice of the inner product is given by $\{D\}^T\{F\} = \lambda$. This reflects the general *commutative* property of the inner product, whereby $\mathbf{F}^T\mathbf{D} = \mathbf{D}^T\mathbf{F}$. We also observe this from the expansion of the inner product of any two vectors non-zero, **V** and **W**, in terms of the included angle θ, as $\mathbf{V}^T\mathbf{W} = \|\mathbf{V}\|\|\mathbf{W}\|\cos\theta = \mathbf{W}^T\mathbf{V}$. The cosine of the angle between the two vectors is given by $\cos\theta = \dfrac{\mathbf{V}^T\mathbf{W}}{\|\mathbf{V}\|\|\mathbf{W}\|} = \dfrac{\mathbf{W}^T\mathbf{V}}{\|\mathbf{V}\|\|\mathbf{W}\|}$.

Interestingly, if we have three m-dimensional vectors, **V**, **W** and **X**, which satisfy the equality, $\langle \mathbf{V}, \mathbf{W} \rangle = \langle \mathbf{V}, \mathbf{X} \rangle = \lambda$, it follows that the lengths of the projections of the vectors **W** and **X** along the direction of **V** are both equal (equal to λ), suggesting that there are infinite possibilities of **W** and **X**, which need not be equal. However, if this equality is valid for any non-zero and arbitrary choice of the vector **A**, then the condition of equality can be satisfied if and only if **W** and **X** are identically equal, i.e., $\mathbf{W} = \mathbf{X}$. We will find this property, which is an offshoot of the commutative property of the inner product of vectors, of immense use in establishing the *Principle of Contra-gradient* in structural analysis (refer Chapter 3).

Furthermore, if two *m*-dimensional vectors **V** and **W** are inter-related by means of a square symmetric matrix **A** (of order *n*), such that **W** = **AV**, then the inner product of **V** and **W** = **AV** can be expressed as $\langle \mathbf{V}, \mathbf{W} \rangle = \{V\}^{\mathrm{T}}[A]\{V\} = \lambda$. Clearly, the transpose of $\{V\}^{\mathrm{T}}[A]\{V\}$ must also be equal to the scalar λ. Invoking Eq. 2.18 (extending to three matrices), this implies:

$$\left(\{V\}^{\mathrm{T}}[A]\{V\}\right)^{\mathrm{T}} = \{V\}^{\mathrm{T}}[A]^{\mathrm{T}}\{V\} = \lambda = \{V\}^{\mathrm{T}}[A]\{V\} \qquad \text{(b)}$$

whereby, it follows that $\mathbf{A}^{\mathrm{T}} = \mathbf{A}$; i.e., the matrix **A** is a **symmetric matrix**. This property is useful in establishing the symmetric nature of the structure flexibility and stiffness matrices (refer Chapter 3).

2.7 RANK OF A MATRIX

Perhaps the most important use of matrices is in solving equations. Typically, we need to solve a set of *linear simultaneous equations*, which may be expressed in scalar form as follows:

$$\sum_{j=1}^{n} a_{ij} X_j = c_i \qquad (i = 1, 2, \ldots, m) \qquad (2.20)$$

or, in matrix format, as follows:

$$[A]\{X\} = \{C\} \qquad (2.20a)$$

where **A** is a matrix, called **coefficient matrix** of order $m \times n$; **X** is an *n*-dimensional vector (comprising unknown variables, $X_1, X_2, \ldots, X_n$) and **C** is an *m*-dimensional vector (comprising known constants, $c_1, c_2, \ldots, c_m$). There is no guarantee that the above set of simultaneous equations has a solution for **X**, and in case there is a solution, it is possible that the solution may be either unique or non-unique (multiple solutions). All these aspects related to the *solvability* of the equations can be discovered by finding a property called the **rank** of the coefficient matrix **A**.

The **rank** of a matrix is equal to the number of **linearly independent column vectors** of the matrix, and this number is identical to the number of **linearly independent row vectors**. The maximum value of the rank *r* of any matrix of order $m \times n$ is given by *m* or *n* (whichever is lower), and the minimum value is 1. If $r = m$, the matrix is said to have **full row rank**, and if $r = n$, the matrix is said to have **full column rank**.

When the constant vector **C** on the right-hand side of Eq. 2.20a is a *zero vector* (i.e., **C** = **O**), the equations are referred to as **homogeneous equations** and the vector space containing all possible solutions of **X** to the equation **AX** = **O** is a property of the matrix **A**, called the **nullspace** of A.

As **C** is effectively a product of two matrices, **A** and **X**, it follows that it is necessarily a linear combination of the column vectors of **A**. It is possible that some of the *n* column vectors in **A** may lack linear independence; i.e., they be multiples or linear combinations of the other column vectors in **A**. The subspace in the vector space $\Re^m$ containing all linear combinations of the independent column vectors (with dimension *m*) is called the **column space** of **A**. Similarly, the row vectors in **A** (which happen to be the column vectors of the transpose of **A**, with dimension *n*) may lack linear independence. The subspace in the vector space $\Re^n$ containing all linear combinations of the independent row vectors is called the **row space** of **A** (and are orthogonal to the nullspace of **A**). Both the column space and row space are sub-spaces having the same dimension, equal to the rank *r* of the matrix **A**.

For example, consider the matrix $\mathbf{A} = \begin{bmatrix} 1 & 2 & 3 \\ 2 & 1 & 3 \\ 3 & 1 & 4 \\ 4 & 2 & 6 \end{bmatrix}$. The third column vector is simply the sum of the first two columns, and on account of this linear dependence, we can intuit that the rank of the matrix is 2 (and that the column space is a linear combination of the first two column vectors). This also means that there can be only 2 independent rows out of the 4 rows in **A**, although this may not be immediately apparent.

2.7.1 Row Reduced Echelon Form

A relatively easy and certain way of determining the rank of a matrix is by reducing the matrix to a *row reduced echelon form* **R** through a process of *elimination*. This involves carrying out a series of operations (scalar multiplication and addition) on the various rows of the matrix, identifying suitable (non-zero) *pivot* elements, with the objective of transforming the matrix **A** as closely as possible to an identity matrix **I** in the upper left corner. In the general case of a rectangular matrix of order $m \times n$, the row reduced echelon form takes the following generalised form (shown partitioned):

$$[R]_{m\times n} = \left[\begin{array}{c|c} [I]_{r\times r} & [F]_{r\times(n-r)} \\ \hline [O]_{(m-r)\times r} & [O]_{(m-r)\times(n-r)} \end{array}\right] \qquad (2.21)$$

The maximum size of the identity matrix possible, which is equal to the number of *pivots*, provides the measure of the rank *r*. If the row vectors are not linearly independent, i.e., if the matrix does not have a full row rank ($r < m$), then the elimination procedure will reveal that the vectors in $m - r$ rows at the bottom will reduce to *zero row vectors*, as indicated in Eq. 2.21. If the column vectors are not linearly independent, i.e., if the matrix does not have a full column rank ($r < n$), then the elimination procedure will reveal this through the absence of pivots in these $n-r$ columns located on the right hand side and denoted by the sub-matrix $\left[\begin{array}{c}\mathbf{F}\\ \hline \mathbf{O}\end{array}\right]$, as indicated in Eq. 2.21. The sub-matrix **F** may be called the *free variable coefficient matrix.*

Let us demonstrate this for the 4×3 matrix discussed earlier. The application of the elimination procedure to achieve the row reduced echelon form is shown below:

$$\mathbf{A} = \begin{bmatrix} \boxed{1} & 2 & 3 \\ 2 & 1 & 3 \\ 3 & 1 & 4 \\ 4 & 2 & 6 \end{bmatrix} \rightarrow \begin{bmatrix} 1 & 2 & 3 \\ 0 & \boxed{3} & 3 \\ 0 & 5 & 5 \\ 0 & 6 & 6 \end{bmatrix} \rightarrow \begin{bmatrix} 1 & 2 & 3 \\ 0 & 1 & 1 \\ 0 & 0 & 0 \\ 0 & 0 & 0 \end{bmatrix} \rightarrow \left[\begin{array}{cc|c} 1 & 0 & 1 \\ 0 & 1 & 1 \\ \hline 0 & 0 & 0 \\ 0 & 0 & 0 \end{array}\right] = \mathbf{R} = \left[\begin{array}{c|c} \mathbf{I} & \mathbf{F} \\ \hline \mathbf{O} & \mathbf{O} \end{array}\right] \quad \text{(c)}$$

In the first operation, using the first element $a_{11} = 1$ as the first pivot (marked with a box), we reduce all the elements (below the first row) in the first column to zero (multiply row 1 by 2 and subtract row 2; multiply row 1 by 3 and subtract row 3; multiply row 1 by 4 and subtract row 4). In the resulting reduced matrix, we identify the first non-zero element in the second column below the first row as the second pivot (this occurs in the second row, and has a value equal to 3, marked with a box)†. In the next operation, we divide the second row by the pivot value 3 to reduce this pivot to unity. Also, we reduce the elements in the lower two rows in the second column to zero (multiply row 2 by 5 and subtract row 3; multiply row 2 by 6 and subtract row 4). We observe that this results in zero row vectors in rows 3 and 4. Finally, we reduce the second column element in the first row to zero (multiply row 2 by 2 and subtract from row 1), in order to generate the identity matrix (of order 2), as indicated. Thus the row reduced echelon matrix **R** is generated. The rank of this matrix is clearly equal to 2.

2.8 LINEAR SIMULTANEOUS EQUATIONS

The row reduced echelon matrix helps us to understand thoroughly the types of solution possible for any set of equations, $\mathbf{AX} = \mathbf{C}$ or the homogeneous set, $\mathbf{AX} = \mathbf{O}$. The solution to the latter, referred to as the *nullspace solution*, always contains the zero vector (null vector) solution $\mathbf{X} = \mathbf{O}$, but may also contain additional vectors. The solution to the former depends on (i) whether the rank r of **A** is less than or equal to the number of unknowns (n components of **X**) and (ii) whether it is less than or equal to the number of constants (given by m) and (iii) whether the equations, when provided in excess of the unknowns, are *consistent.* If the equations are consistent, we can find a *particular solution* $\mathbf{X_p}$ that satisfies $\mathbf{AX} = \mathbf{C}$.

† If the pivot cannot be found in the second row, then we search in the third row, and so on. We then exchange the rows (permutation operation), ensuring that the rows on top (and columns on the left) are always ones with pivots.

Superimposing this solution on the *nullspace solution* $\mathbf{X_n}$, the *complete solution* to the set of linear equations is obtainable:

$$\mathbf{X} = \mathbf{X_p} + \mathbf{X_n} \tag{2.22}$$

We need to carry out the same elimination operations on the constant vector **C**, so that the equations **AX** = **C** in the reduced form† become **RX** = **D**, which may be expanded as:

$$\begin{bmatrix} [I]_{r\times r} & [F]_{r\times(n-r)} \\ [O]_{(m-r)\times r} & [O]_{(m-r)\times(n-r)} \end{bmatrix}_{m\times n} \begin{bmatrix} [X_{pivot}]_{r\times 1} \\ [X_{free}]_{(n-r)\times 1} \end{bmatrix}_{n\times 1} = \begin{bmatrix} [D_{pivot}]_{r\times 1} \\ [D_{zero}]_{(m-r)\times 1} \end{bmatrix}_{m\times 1} \tag{2.23}$$

where the vector **X** may be partitioned into (i) a set of **pivot variables**, denoted as $\{X_{pivot}\}$ having a dimension equal to the rank of **A**, and (ii) a set of **free variables**, denoted as $\{X_{free}\}$ with a dimension equal to $n - r$. Similarly, the reduced constant vector, **D**, gets partitioned into (i) a set of *pivot row constants*, denoted as $\{D_{pivot}\}$ having a dimension equal to the rank of **A**, and (ii) a set of *zero row constants*, denoted as $\{D_{zero}\}$ with a dimension equal to $m - r$.

Carrying out the matrix multiplication, Eq. 2.23 simplifies to the following equations:

$$\{X_{pivot}\} = \{D_{pivot}\} - [F]\{X_{free}\} \tag{2.24}$$

$$\{D_{zero}\} = \{O\} \tag{2.25}$$

It may be noted that when **A** has a full row rank ($m = r$), the constant vector **D** comprises only pivot row constants (no $\mathbf{D_{zero}}$), whereby Eq. 2.25 has no relevance. Similarly, when **A** has a full column rank ($n = r$), the vector of unknowns **X** comprises only pivot variables (no $\mathbf{X_{free}}$ and no **F**), whereby Eq. 2.24 simplifies to $\mathbf{X_p} = \mathbf{D}$. A particular solution $\mathbf{X_p}$ can be conveniently obtained by setting all the free variables to zero ($\mathbf{X_{free}} = \mathbf{O}$) and then solving for the pivot variables, $\mathbf{X_{pivot}}$.

Case 1*: *r* < *m* and *r* < *n

In this case, there are linearly dependent rows ($r < m$) as well as linearly dependent columns ($r < n$) in the coefficient matrix **A**. A particular solution $\mathbf{X_p}$ is possible only if the constant vector **C** is such that, after transformation to **D** (through elimination), Eq. 2.25 satisfied. In other words, a solution for any arbitrary constant vector **C** is not possible. Considering, for example, the same 4×3 matrix (with $r = 2$) discussed earlier, and performing the row reduction operations on the augmented matrix $[\mathbf{A} \mid \mathbf{C}]$, we obtain the following reduced $[\mathbf{R} \mid \mathbf{D}]$ form:

† This can be conveniently done by performing the row reduction operations on the **augmented matrix** $[\mathbf{A} \mid \mathbf{C}]$ which gets transformed into $[\mathbf{R} \mid \mathbf{D}]$. This is demonstrated later with an example.

$$[\mathbf{A} \mid \mathbf{C}] = \left[\begin{array}{ccc|c} \boxed{1} & 2 & 3 & c_1 \\ 2 & 1 & 3 & c_2 \\ 3 & 1 & 4 & c_3 \\ 4 & 2 & 6 & c_4 \end{array}\right] \rightarrow \left[\begin{array}{ccc|c} \boxed{1} & 2 & 3 & c_1 \\ 0 & \boxed{3} & 3 & 2c_1 - c_2 \\ 0 & 5 & 5 & 3c_1 - c_3 \\ 0 & 6 & 6 & 4c_1 - c_4 \end{array}\right] \rightarrow \left[\begin{array}{ccc|c} 1 & 0 & 1 & (-c_1 + 2c_2)/3 \\ 0 & 1 & 1 & (2c_1 - c_2)/3 \\ 0 & 0 & 0 & (c_1 - 5c_2)/3 + c_3 \\ 0 & 0 & 0 & (-2c_2 + c_4) \end{array}\right] = [\mathbf{R} \mid \mathbf{D}] \quad \text{(d)}$$

whereby the transformed set of equations $\mathbf{RX} = \mathbf{D}$ takes the following form:

$$\mathbf{RX} = \left[\begin{array}{c|c} \mathbf{I} & \mathbf{F} \\ \hline \mathbf{O} & \mathbf{O} \end{array}\right] \left[\begin{array}{c} \mathbf{X}_{\mathbf{pivot}} \\ \hline \mathbf{X}_{\mathbf{free}} \end{array}\right] = \left[\begin{array}{cc|c} 1 & 0 & 1 \\ 0 & 1 & 1 \\ \hline 0 & 0 & 0 \\ 0 & 0 & 0 \end{array}\right] \left[\begin{array}{c} X_1 \\ X_2 \\ \hline X_3 \end{array}\right] = \left[\begin{array}{c} \mathbf{D}_{\mathbf{pivot}} \\ \hline \mathbf{D}_{\mathbf{zero}} \end{array}\right] = \left[\begin{array}{c} (-c_1 + 2c_2)/3 \\ (2c_1 - c_2)/3 \\ \hline c_3 + (c_1 - 5c_2)/3 \\ (-2c_2 + c_4) \end{array}\right] \quad \text{(e)}$$

Clearly, a solution is possible only if $\mathbf{D}_{\mathbf{zero}} = \mathbf{O}$ (Eq. 2.25), which requires the components of the constant vector **C** to satisfy the following equations to ensure consistency in the set of equations:

$$c_3 + (c_1 - 5c_2)/3 = 0 \text{ and } (-2c_2 + c_4) = 0 \quad \Rightarrow \quad c_3 = (5c_2 - c_1)/3 \text{ and } c_4 = 2c_2$$

Hence, in this example, a solution exists only if the constant vector has the form, $\mathbf{C} = \begin{bmatrix} c_1 \\ c_2 \\ (5c_2 - c_1)/3 \\ 2c_2 \end{bmatrix}$. For example, $\mathbf{C} = \begin{bmatrix} 0 \\ 3 \\ 5 \\ 6 \end{bmatrix}$, which is obtainable by subtracting the second column in **A** from twice the first column (or subtracting the third column from thrice the first column, or other combinations), is acceptable. We can deduce a particular solution such as $\mathbf{X}_{\mathbf{p}} = \begin{bmatrix} 2 \\ -1 \\ 0 \end{bmatrix}$, which can also be generated by solving Eq. 2.24, after assuming a choice of the free variable, $X_3 = 0$: $\{X_{pivot}\} = \{D_{pivot}\} - [F]\{X_{free}\} = \{D_{pivot}\}$

$$\Rightarrow X_1 = (-c_1 + 2c_2)/3 = 2 \quad \text{and} \quad X_2 = (2c_1 - c_2)/3 = -1$$

Now, to get the complete solution, we need to find the nullspace solution $\mathbf{X}_{\mathbf{n}}$, corresponding to $\mathbf{C} = \mathbf{O}$. What would be a general solution to the homogeneous set of equations? As we know that the third column is given by the sum of the first two columns in the coefficient matrix **A**, and we require the linear combination of all three column vectors to result in a null vector, we can deduce that $\mathbf{X}_{\mathbf{n}} = \begin{bmatrix} -\lambda \\ -\lambda \\ \lambda \end{bmatrix}$, where λ is a scalar that can have any value, including zero (i.e., infinite solutions are possible). This can also be generated by solving Eq. 2.24, after assuming a choice of the free variable, $X_3 = \lambda$ and noting that $\mathbf{D} = \mathbf{O}$:

$$\{X_{pivot}\}=\{D_{pivot}\}-[F]\{X_{free}\}=-[F]\{X_{free}\} \Rightarrow \begin{bmatrix} X_1 \\ X_2 \end{bmatrix} = -\begin{bmatrix} 1 \\ 1 \end{bmatrix}\lambda = \begin{bmatrix} -\lambda \\ -\lambda \end{bmatrix}$$

The complete solution (Eq. 2.22) to the set of equations in this example is as follows:

$$\mathbf{X}=\mathbf{X_p}+\mathbf{X_n}=\begin{bmatrix} 2 \\ -1 \\ 0 \end{bmatrix}+\begin{bmatrix} -\lambda \\ -\lambda \\ \lambda \end{bmatrix}=\begin{bmatrix} 2-\lambda \\ -1-\lambda \\ \lambda \end{bmatrix} \text{ is a complete solution to } \begin{bmatrix} 1 & 2 & 3 \\ 2 & 1 & 3 \\ 3 & 1 & 4 \\ 4 & 2 & 6 \end{bmatrix}\begin{bmatrix} X_1 \\ X_2 \\ X_3 \end{bmatrix}=\begin{bmatrix} 0 \\ 3 \\ 5 \\ 6 \end{bmatrix} \quad \text{(f)}$$

To summarise, when $r < m$ and $r < n$, the matrix **A** has the reduced row echelon form, $\mathbf{R}=\left[\begin{array}{c|c} \mathbf{I} & \mathbf{F} \\ \hline \mathbf{O} & \mathbf{O} \end{array}\right]$, whereby either no solution or infinite solutions are possible.

***Case 2:* $r = n < m$**

In this case, the coefficient matrix **A** has a full column rank ($r = n$), but there are linearly dependent rows ($r < m$), which also implies that the number of rows exceeds the number of columns ($m > n$). We have more equations than unknowns, and we need to ensure that the equations are consistent (linearly dependent) for a solution to be possible. The reduced row echelon form of the matrix **A** has the generic form, $\mathbf{R}=\left[\begin{array}{c} \mathbf{I} \\ \hline \mathbf{O} \end{array}\right]$, whereby it follows that for a solution to exist, the condition $\mathbf{D_{zero}} = \mathbf{O}$ (Eq. 2.25) has to be satisfied (as in Case 1). But, as there are no zero variables (i.e., only pivot variables), **F** (refer Eq. 2.21) does not exist, the nullspace solution is a null vector ($\mathbf{X_n} = \mathbf{O}$), and the complete solution is directly given by $\mathbf{X} = \mathbf{D_{pivot}}$.

Let us illustrate this by considering the 4×3 matrix **A** of the previous example, with the third column slightly modified to eliminate linear dependence (the last entry alone is changed from '6' to '8'), thereby ensuring full column rank. The elimination operation on the augmented matrix [Eq. (d)] gets modified as follows:

$$[\mathbf{A} \mid \mathbf{C}]=\left[\begin{array}{ccc|c} \boxed{1} & 2 & 3 & c_1 \\ 2 & 1 & 3 & c_2 \\ 3 & 1 & 4 & c_3 \\ 4 & 2 & 8 & c_4 \end{array}\right] \rightarrow \left[\begin{array}{ccc|c} \boxed{1} & 2 & 3 & c_1 \\ 0 & \boxed{3} & 3 & 2c_1-c_2 \\ 0 & 5 & 5 & 3c_1-c_3 \\ 0 & 6 & 4 & 4c_1-c_4 \end{array}\right] \rightarrow \left[\begin{array}{ccc|c} 1 & 0 & 1 & (-c_1+2c_2)/3 \\ 0 & 1 & 1 & (2c_1-c_2)/3 \\ 0 & 0 & \boxed{2} & (-2c_2+c_4) \\ 0 & 0 & 0 & c_3+(c_1-5c_2)/3 \end{array}\right]$$

$$\rightarrow \left[\begin{array}{ccc|c} 1 & 0 & 0 & (-c_1+5c_2)/3-c_4/2 \\ 0 & 1 & 0 & (2c_1+2c_2)/3-c_4/2 \\ 0 & 0 & 1 & c_4/2-c_2 \\ \hline 0 & 0 & 0 & c_3+(c_1-5c_2)/3 \end{array}\right]=[\mathbf{R} \mid \mathbf{D}] \quad \text{(g)}$$

where the third and fourth rows have been inter-changed in order to locate the zero row vector at the bottom. Clearly, there are no free variables here ($r = n = 3$), and **F** (refer

Eq. 2.21) does not exist. A solution is possible only if $\mathbf{D}_{\text{zero}} = \mathbf{O}$ (Eq. 2.25), whereby $c_3 = -(c_1 - 5c_2)/3$ (as in the earlier example). Considering the same **C** as in the previous example, the complete solution (with $\mathbf{X_n} = 0$) is given by:

$$\mathbf{AX} = \mathbf{C} \rightarrow \begin{bmatrix} 1 & 2 & 3 \\ 2 & 1 & 3 \\ 3 & 1 & 4 \\ 4 & 2 & 8 \end{bmatrix} \begin{bmatrix} X_1 \\ X_2 \\ X_3 \end{bmatrix} = \begin{bmatrix} 0 \\ 3 \\ 5 \\ 6 \end{bmatrix} \rightarrow \begin{bmatrix} 1 & 0 & 0 \\ 0 & 1 & 0 \\ 0 & 0 & 1 \\ 0 & 0 & 0 \end{bmatrix} \begin{bmatrix} X_1 \\ X_2 \\ X_3 \end{bmatrix} = \begin{bmatrix} 2 \\ -1 \\ 0 \\ 0 \end{bmatrix} \rightarrow \mathbf{RX} = \mathbf{D} \Rightarrow \mathbf{X} = \begin{bmatrix} 2 \\ -1 \\ 0 \end{bmatrix} \quad \text{(h)}$$

Case 3: r = m < n

In this case, the coefficient matrix **A** has a full row rank ($r = m$), but there are linearly dependent columns ($r < n$), which also implies that the number of columns exceeds the number of rows ($n > m$). We have more unknowns than equations, which is a situation we encounter in structural analysis, while dealing with equations of equilibrium in statically indeterminate structures. The reduced row echelon form of the matrix **A** has the generic form, $\mathbf{R} = [\mathbf{I} \mid \mathbf{F}]$. Owing to the absence of zero row vectors, there is no constraint on the constant vector **C**, and a solution is certainly possible. However, as there are free variables present, the nullspace solution $\mathbf{X_n}$ has infinite possibilities (as in Case 1) in the complete solution.

Let us illustrate this by considering the transpose of the 4×3 matrix **A** of the previous example (Case 2), thereby ensuring full row rank. The elimination operation on the augmented matrix is as follows:

$$[\mathbf{A} \mid \mathbf{C}] = \left[\begin{array}{cccc|c} \boxed{1} & 2 & 3 & 4 & c_1 \\ 2 & 1 & 1 & 2 & c_2 \\ 3 & 3 & 4 & 8 & c_3 \end{array}\right] \rightarrow \left[\begin{array}{cccc|c} 1 & 2 & 3 & 4 & c_1 \\ 0 & \boxed{3} & 5 & 6 & 2c_1 - c_2 \\ 0 & 3 & 5 & 4 & 3c_1 - c_3 \end{array}\right] \rightarrow \left[\begin{array}{cccc|c} 1 & 0 & -1/3 & 0 & (-c_1 + 2c_2)/3 \\ 0 & 1 & 5/3 & 2 & (2c_1 - c_2)/3 \\ 0 & 0 & 0 & \boxed{2} & (-c_1 - c_2 + c_3) \end{array}\right]$$

$$\Rightarrow \left[\begin{array}{ccc|c} 1 & 0 & 0 & -1/3 \\ 0 & 1 & 0 & 5/3 \\ 0 & 0 & 1 & 0 \end{array}\right] \begin{bmatrix} X_1 \\ X_2 \\ X_4 \\ \hline X_3 \end{bmatrix} = \begin{bmatrix} (-c_1 + 2c_2)/3 \\ (5c_1 + 2c_2)/3 - c_3 \\ (-c_1 - c_2 + c_3)/2 \end{bmatrix} = \begin{bmatrix} d_1 \\ d_2 \\ d_3 \end{bmatrix} \rightarrow [\mathbf{I} \mid \mathbf{F}]\tilde{\mathbf{X}} = \mathbf{D} \quad \text{(i)}$$

where the third and fourth columns have been inter-changed in order to preserve the identity matrix in left-hand side of the **R** matrix, whereby the sequence of component placement in **X** gets modified (shown as $\tilde{\mathbf{X}}$). As there are no zero row vectors ($r = m = 3$), there is no restriction on the constant vector **C** for solvability of the equations. However, we have more unknowns ($n = 4$) than available equations ($r = 3$), whereby multiple solutions are possible. We can easily find a particular solution by assigning a zero value to the free variable ($X_3 = 0$), whereby the pivot variables are given directly by the components of **D**. For example, considering $\mathbf{C}=\begin{bmatrix}3\\6\\2\end{bmatrix} \Rightarrow \mathbf{D}=\begin{bmatrix}3\\7\\-3.5\end{bmatrix} \Rightarrow \mathbf{X_p}=\begin{bmatrix}3\\7\\0\\-3.5\end{bmatrix}$

Now, to get the complete solution, we need to find the nullspace solution $\mathbf{X_n}$, corresponding to $\mathbf{C}=\mathbf{O}$. This can be generated by solving Eq. 2.24, after assuming a choice of the free variable, $X_3 = \lambda$ and noting that $\mathbf{D}=\mathbf{O}$:

$$\{X_{pivot}\}=\{D_{pivot}\}-[F]\{X_{free}\}=-[F]\{X_{free}\} \Rightarrow \begin{bmatrix}X_1\\X_2\\X_4\end{bmatrix}=-\begin{bmatrix}-1/3\\5/3\\0\end{bmatrix}\lambda \Rightarrow \mathbf{X_n}=\begin{bmatrix}\lambda/3\\-5\lambda/3\\\lambda\\0\end{bmatrix}$$

where λ is a scalar that can take any value. The nullspace includes the null vector.

The complete solution (Eq. 2.22) to the set of equations in this example is as follows:

$$\Rightarrow \mathbf{X}=\mathbf{X_p}+\mathbf{X_n}=\begin{bmatrix}3\\7\\0\\-3.5\end{bmatrix}+\begin{bmatrix}\lambda/3\\-5\lambda/3\\\lambda\\0\end{bmatrix}=\begin{bmatrix}3+\lambda/3\\7-5\lambda/3\\\lambda\\-3.5\end{bmatrix} \text{ satisfies } \begin{bmatrix}1&2&3&4\\2&1&1&2\\3&3&4&8\end{bmatrix}\begin{bmatrix}X_1\\X_2\\X_3\\X_4\end{bmatrix}=\begin{bmatrix}3\\6\\2\end{bmatrix} \quad \text{(j)}$$

It is evident that infinite solutions are possible.

***Case 4:* $r = m = n$**

In this case, the coefficient matrix **A** has a full column rank ($r = n$) as well as a full row rank ($r = m$), which also implies that the matrix is a square matrix ($m = n$). Such a matrix is said to be **invertible** or **non-singular**. The reduced row echelon form of the matrix **A** takes the generic form, $\mathbf{R}=\mathbf{I}$, indicating that the set of equations, $\mathbf{AX}=\mathbf{C}$, can (by elimination) be reduced to $\mathbf{IX}=\mathbf{D}$, whereby it follows that there is a *unique solution*, $\mathbf{X}=\mathbf{D}$ for any given constant vector **C**. The nullspace (solution corresponding to $\mathbf{C}=\mathbf{O}$) is a null vector ($\mathbf{X_n}=\mathbf{O}$), which does not add anything to the unique particular solution.

Given a set of linear simultaneous equations, **AX** = **C**, a unique solution exists only if the coefficient matrix **A** is a square matrix whose row reduced echelon form **R** is an identity matrix (**R** = **I**). Such a square matrix (of order $n \times n$) has a full rank ($r = n$), and is said to be **invertible** or **non-singular**. If, however, there is linear dependence in the column vectors or the row vectors (or both) in the matrix **A**, whether square or rectangular ($m \times n$), then this will be reflected in the rank of the matrix which will be less than the number of rows ($r < m$) or number of columns ($r < n$) or both. If the number of rows exceeds the number of linearly independent equations ($r < m$), then elimination will reveal $(m - r)$ zero row vectors. A solution is possible only if the constants in **C** reflect the same linear dependence as in the rows of **A** (i.e., the equations are **consistent**)†. This solution will be unique if **A** has a full column rank ($r = n < m$). If, however, there is linear dependence in the column vectors of **A**, whereby the number of unknown components in **X** exceeds the number of linearly independent equations ($r < n$), infinite solutions are possible, provided the equations are consistent.

2.9 MATRIX INVERSION

Unlike multiplication, division of matrices is not possible. In matrix algebra, the nearest operation to division (as applied in scalar algebra) is the operation of *inversion*. For example, let us consider the following set of linear equations in scalar algebra:

$$a_i X_i = c_i \qquad (i = 1, 2, \ldots, m) \tag{2.26}$$

where a_i and c_i are constants. These m equations are "uncoupled" and hence we can solve for the unknowns X_i by applying the simple operation of division:

$$X_i = \frac{c_i}{a_i} = (a_i)^{-1}(c_i) \qquad (i = 1, 2, \ldots, m) \tag{2.27}$$

In a more general situation, the set of linear equations may be coupled, as shown earlier in Eq. 2.20. These equations, which, in matrix form, are represented by **AX** = **C**, need to be solved simultaneously. As discussed earlier, these equations can be solved by the matrix operation of *elimination*. In the traditional Gauss elimination procedure, it is sufficient to reduce the coefficient matrix **A** to an upper triangular form **U** for this purpose, while carrying out the elementary matrix operations on the augmented matrix $[\mathbf{A} \mid \mathbf{C}]$. If **A** is a square matrix and the equations are consistent, the unique solution can be obtained by *back substitution* (i.e., solving the last row for X_m, and using this value, for X_{m-1}, and so on, till X_1). However, by going a few steps further, as shown in Section 2.7, the **A** matrix can be reduced to the row reduced echelon form **R**, and the complete solution, if any, can be directly obtained.

When the matrix **A** is square and of full rank, an alternative approach to solving the equations is by the operation of *inversion*. In a manner analogous to Eq. 2.27, the unknown vector **X** is obtained by "pre-multiplying" the constant vector **C** with the **inverse** of the matrix **A**:

$$\{X\} = [A]^{-1}\{C\} \tag{2.28}$$

† Otherwise, pre-multiplying the equation **AX** = **C** on both sides by $\mathbf{A}^T$, we can find a solution to the set of equations, $(\mathbf{A}^T\mathbf{A})\mathbf{X} = \mathbf{A}^T\mathbf{C}$, where $(\mathbf{A}^T\mathbf{A})$ is a square matrix of order $n \times n$.

While **A** and **C** are known (constants), **X** is unknown. If the equations are uncoupled, as indicated in Eq. 2.26, the matrix **A** becomes a *diagonal matrix* whose diagonal elements are given by $a_{ii} = a_i$. In order to find a solution for **X**, we need to understand the meaning of the **inverse** of the matrix **A**.

It is important to note that matrix inversion is possible only for *square* matrices, but not all square matrices are invertible! As mentioned earlier, the matrix of order $n \times n$ is said to be **invertible** or **non-singular** if all the column vectors are linearly independent (or alternatively, all the row vectors are linearly independent), whereby the rank r of the matrix is equal to n. The inverse of the non-singular square matrix **A**, which is denoted as $\mathbf{A}^{-1}$ (and read as "A-inverse"), and which has the same order $n \times n$ as **A**, is defined such that the product of **A** and $\mathbf{A}^{-1}$ is an *identity matrix* (a diagonal matrix having unit values along the main diagonal).

$$\mathbf{AA}^{-1} = \mathbf{A}^{-1}\mathbf{A} = \mathbf{I} \tag{2.29}$$

Thus, pre-multiplying both sides of Eq. 2.20a by the inverse matrix, $\mathbf{A}^{-1}$, $\mathbf{A}^{-1}\mathbf{AX} = \mathbf{A}^{-1}\mathbf{C}$, and, invoking the property of the inverse (Eq. 2.29), Eq. 2.28 is derived.

When the order of **A** is one, the matrix reduces to a scalar a, whereby the inverse $\mathbf{A}^{-1}$ is also a scalar, whose value is the reciprocal of a (i.e., $1/a$). Evidently, the inverse can be defined only if a is not zero. If the square matrix **A** is a *diagonal* matrix of order n, then the inverse will also be a diagonal matrix, whose diagonal elements are the reciprocals of the corresponding elements of **A**; i.e., $\left(a^{-1}\right)_{ii} = 1/a_{ii}$ for $i = 1, 2,\ldots, n$. Clearly, if even one of the diagonal elements of **A** is zero, the matrix becomes *singular*, and its inverse cannot be defined.

2.9.1 Basic Properties of Inverse of a Matrix

Some important properties of the inverse, involving inversion of (i) an inverse, (ii) a transpose, (iii) a scalar multiple of a matrix, and (iv) a product of two matrices, are given below:

$$\left(\mathbf{A}^{-1}\right)^{-1} = \mathbf{A} \tag{2.30}$$

$$\left(\mathbf{A}^{\mathrm{T}}\right)^{-1} = \left(\mathbf{A}^{-1}\right)^{\mathrm{T}} \tag{2.31}$$

$$(\lambda\mathbf{A})^{-1} = \frac{1}{\lambda}\mathbf{A}^{-1} \tag{2.32}$$

$$(\mathbf{AB})^{-1} = \mathbf{B}^{-1}\mathbf{A}^{-1} \tag{2.33}$$

Eq. 2.33 is especially noteworthy: the inverse of a product is not the product of the inversed matrices (as one might be inclined to believe), but the *commuted* product of the inverses†.

2.9.2 Determinants and Cofactors

In general, it can be proved that the inverse exists, if a scalar property called the **determinant** of the square matrix **A**, denoted as $|\mathbf{A}|$ or det **A** or $\Delta(\mathbf{A})$, is not equal to zero. For a diagonal matrix, the determinant is given by the product of all the diagonal elements. For a matrix **A** of order 2 × 2, the determinant is given by:

$$\det \mathbf{A} = \begin{vmatrix} a_{11} & a_{12} \\ a_{21} & a_{22} \end{vmatrix} = (a_{11}a_{22} - a_{21}a_{12}) \tag{2.34}$$

It is possible to write out a general formula for finding the determinant of a matrix square matrix **A** of any order $n \times n$; however, this would be rather complicated. Perhaps the simplest way of finding the determinant of any square matrix is by applying the process of elimination, reducing the matrix to an upper diagonal form **U** (from which it is further possible to reduce to a diagonal form). The determinant is directly obtainable as the product of the *pivot* elements in the diagonal, if there are no row exchanges needed or if an even number of row exchanges are involved. If, however, an odd number of row exchanges are needed for this elimination procedure, then a negative sign has to be attached to the product of the pivots. Clearly, a non-zero determinant is possible for a given square matrix **A** only if there are pivot elements in *all* the rows of the matrix; i.e., the matrix has to have a *full rank* for it to be *non-singular* and *invertible*.

For example, we can reduce the 2 × 2 matrix in Eq. 2.34 easily, in one step, without any row exchange, to reduce it to an upper diagonal form and thereby find the determinant as the product of the two pivots. For the matrix to be non-singular, both the pivots should be non-zero.

$$\det \mathbf{A} = \begin{vmatrix} \boxed{a_{11}} & a_{12} \\ a_{21} & a_{22} \end{vmatrix} = \begin{vmatrix} \boxed{a_{11}} & a_{12} \\ 0 & \boxed{a_{22} - \dfrac{a_{21}}{a_{11}} a_{12}} \end{vmatrix} = (a_{11})\left(a_{22} - \frac{a_{21}}{a_{11}} a_{12}\right) = (a_{11}a_{22} - a_{21}a_{12}) \tag{2.34a}$$

It can be shown that the determinant of a square matrix **A** of order $n \times n$ is the algebraic sum of $n!$ terms formed from the elements a_{ij} such that one and only one element from each row i and each column j appears in each term and each term contains n elements:

† In connection with the relation $(\mathbf{AB})^{-1} = \mathbf{B}^{-1}\mathbf{A}^{-1}$, a striking and witty observation made is this: *The process of wearing shoes involves putting on socks prior to putting on shoes. However, in the reverse process of taking off the shoes, the shoes have to be removed first, and then the socks. The sequence gets inverted!*

$$\det \mathbf{A} = \sum_{j=1}^{n} a_{ij}\alpha_{ij} \qquad \text{for } i = 1, 2, 3, \ldots, n \tag{2.35}$$

where $\alpha_{ij} = -(1)^{i+j}\beta_{ij}$ is the **cofactor** of the element a_{ij}, whose value is given by the determinant β_{ij} of the $(n-1)\times(n-1)$ matrix obtained by deleting the ith row and j^{th} column of the matrix **A**, with the appropriate sign (positive or negative).

An expansion, based on the elements of the first row, is given by

$$\det \mathbf{A} = a_{11}\alpha_{11} + a_{12}\alpha_{12} + \ldots + a_{1n}\alpha_{1n} \tag{2.35a}$$

Thus, for example, the determinant of a 3 × 3 matrix **A** can be generated using cofactors as follows:

$$\det \mathbf{A} = a_{11}\begin{vmatrix} a_{22} & a_{23} \\ a_{32} & a_{33} \end{vmatrix} - a_{12}\begin{vmatrix} a_{21} & a_{23} \\ a_{31} & a_{33} \end{vmatrix} + a_{13}\begin{vmatrix} a_{21} & a_{22} \\ a_{31} & a_{32} \end{vmatrix} \tag{2.36}$$

Determinants and cofactors have many interesting uses. For example, the absolute value of the determinant of the 2 × 2 matrix in Eq. 2.34 yields a very useful formula to find the area of a parallelogram in terms of the coordinates of the 4 corner points, viz., (0,0), (a_{11},a_{12}), (a_{21},a_{22}) and $(a_{11}+a_{21},a_{12}+a_{22})$. Similarly, the volume of a parallelepiped is given by $|\det \mathbf{A}|$ for a 3 × 3 matrix (Eq. 2.36). Cofactors are useful in finding matrix inverses and solving linear simultaneous equations, as discussed later.

There are some properties of determinants that are especially noteworthy. The determinant of the product of two square matrices is equal to the product of the two determinants:

$$\det \mathbf{AB} = (\det \mathbf{A})(\det \mathbf{B}) \tag{2.37}$$

whereby, it can be shown that:

$$\det \mathbf{A}^m = (\det \mathbf{A})^m \tag{2.37a}$$

$$\det \lambda\mathbf{A} = \lambda^n (\det \mathbf{A}) \qquad \text{for } [A]_{n\times n} \tag{2.37b}$$

The determinants of a square matrix and its transpose are identical:

$$\det \mathbf{A}^{\mathrm{T}} = \det \mathbf{A} \tag{2.38}$$

2.9.3 Adjoint Method of Finding Inverse

In general, for any square matrix of order $n \times n$, provided, $\det \mathbf{A} \neq 0$, it can be shown that

$$\mathbf{A}^{-1} = \frac{1}{\det \mathbf{A}}\tilde{\mathbf{A}} \tag{2.39}$$

where $\tilde{\mathbf{A}}$ is called the **adjoint matrix** of **A**, which is the transpose of a matrix whose elements comprise the cofactors α_{ij} of **A**. This technique, sometimes called the **adjoint method** of finding the inverse, however, becomes cumbersome when the order of the matrix exceeds three or four.

Using the adjoint method, it can be shown that inverse of a non-singular square matrix of order 2×2 is given by

$$\begin{bmatrix} a & b \\ c & d \end{bmatrix}^{-1} = \frac{1}{ad - bc}\begin{bmatrix} d & -b \\ -c & a \end{bmatrix} \tag{2.39a}$$

If the matrix is *symmetric* (as in the case of stiffness and flexibility matrices), the inverse is also symmetric. For a symmetric non-singular 3×3 matrix, the following formula (with elements shown only in the lower triangular segment) can be derived using Eq. 2.39:

$$\begin{bmatrix} a & & \text{sym} \\ b & c & \\ d & e & f \end{bmatrix}^{-1} = \frac{1}{a(cf - e^2) - b(bf - de) + d(be - cd)}\begin{bmatrix} (cf - e^2) & & \text{sym} \\ (de - bf) & (fa - d^2) & \\ (be - cd) & (bd - ae) & (ac - b^2) \end{bmatrix} \tag{2.39b}$$

The application of the above formula is illustrated below for a typical 3×3 symmetric matrix:

$$\begin{bmatrix} 7/3 & 2/3 & -3/8 \\ 2/3 & 7/3 & -3/8 \\ -3/8 & -3/8 & 3/8 \end{bmatrix}^{-1} = \frac{1}{1.40625}\begin{bmatrix} 0.734375 & -0.109375 & 0.625 \\ -0.109375 & 0.734375 & 0.625 \\ 0.625 & 0.625 & 5.0 \end{bmatrix} \tag{k}$$

2.9.4 Cramer's Rule

The concepts of cofactors and determinants can be extended to solving linear simultaneous equations (Eq. 2.20) directly, without the need to find the inverse of the coefficient matrix **A**. Combining Eqns 2.28 and 2.39, the solution to a set of consistent equations, $[A]\{X\} = \{C\}$, can be shown to be given by:

$$X_i = \frac{\begin{vmatrix} a_{11} & \cdots & a_{1,j-1} & c_1 & \cdots & a_{1n} \\ \vdots & & \vdots & \vdots & & \vdots \\ a_{n1} & & a_{n,j-1} & c_n & & a_{nn} \end{vmatrix}}{\begin{vmatrix} a_{11} & \cdots & a_{1,j-1} & a_{1j} & \cdots & a_{1n} \\ \vdots & & \vdots & \vdots & & \vdots \\ a_{n1} & & a_{n,j-1} & a_{nj} & & a_{nn} \end{vmatrix}} \qquad (i = 1, 2, \ldots, n) \tag{2.40}$$

The above simplified expression for the unknown X_i as the ratio of two determinants, with $\Delta(\mathbf{A}) \neq 0$ in the denominator, and in the numerator, the determinant of a matrix having the same columns as **A** but with the jth column is replaced by the constant vector $\{C\}$, is known as **Cramer's rule**, which is suitable for solving a small number of simultaneous equations. It requires generation of $n + 1$ determinants, which is cumbersome by algebraic formulation. Elimination-based algorithmic methods are much better suited for computer application.

2.9.5 Condition of a Matrix

There are many other techniques of finding the inverse of a matrix, most of which are iterative in nature. Computationally, the effort required increases significantly as the order of the matrix increases. The stability of iterative solution procedures and the accuracy of the end result, however, are dependent on the **condition** of the matrix, which is a measure of its non-singularity. A matrix is said to be **ill-conditioned** (or *near-singular*) if its determinant is very small in comparison with the value of its average element, and such a matrix is vulnerable to erroneous estimation of its inverse by iterative solvers. In the force method of structural analysis, it is seen that certain choices of redundants in a statically indeterminate structure may result in ill-conditioned flexibility matrices. On the other hand, stiffness matrices are relatively **well-conditioned** and have the property of **positive definiteness**†, whereby their inverses can be stably and accurately generated. In particular, when there is **diagonal dominance** in the matrix (the diagonal element in any row or column has a magnitude that exceeds the sum of the absolute values of all other elements in that row or column), as in *non-sway* type kinematically indeterminate frames (where only rotational degrees of freedom exist), the iterative procedures to calculate the inverse of the stiffness matrix are robust.

2.9.6 Gauss-Jordan Elimination Method of Finding Inverse

Here, we present a generalised *elimination* procedure, called the **Gauss-Jordan method**, to find the inverse of any non-singular square matrix **A**. This method is based on the property of any non-singular matrix for reduction, by means of elementary row operations, to an identity matrix. In the Gauss-Jordan procedure, an identity matrix **I** of the same order n as **A** is placed next to the matrix **A** in an augmented form, [**A** | **I**]. Using elementary row operations the matrix **A** is transformed into **I**, carrying out the same operation to all the rows of the augmented matrix. This is achieved through a sequence of steps that are aimed at arriving at (i) a value of unity in the diagonal element in any row (by dividing the row with the current value at the diagonal location) and (ii) a zero value in all other non-diagonal elements in any

---

† The square matrix **A** of order n is said to be *positive definite*, if for any arbitrary choice of an n-dimensional vector **X**, the product $\mathbf{X}^T\mathbf{A}\mathbf{X}$ yields a scalar quantity that is invariably positive. All the *eigenvalues* of such a matrix are real, distinct and spaced well apart.

row (through multiplication of the appropriate row with a unit diagonal value and thereafter addition or subtraction of the row under consideration). This sequence of elimination operations on the augmented matrix that could be conveniently summarised in matrix form as follows:

$$\mathbf{E}_q\left(\mathbf{E}_{q-1}\left(\ldots\left(\mathbf{E}_2\left(\mathbf{E}_1[\mathbf{A} \mid \mathbf{I}]\right)\right)\right)\right)=\mathbf{E}[\mathbf{A} \mid \mathbf{I}]=[\mathbf{I} \mid ?]=\left[\mathbf{I} \mid \mathbf{A}^{-1}\right] \tag{l}$$

If $\mathbf{E}=\mathbf{E}_q\mathbf{E}_{q-1}\ldots\mathbf{E}_2\mathbf{E}_1$ denotes the *resultant elimination matrix*, such that $\mathbf{EA}=\mathbf{I}$, as indicated in the above equation, then it follows that $\mathbf{E}=\mathbf{A}^{-1}$, whereby the sub-matrix on the right-hand side of the final augmented matrix is $\mathbf{EI}=\mathbf{E}=\mathbf{A}^{-1}$. Thus, the final augmented matrix takes the form, $[\mathbf{I} \mid \mathbf{A}^{-1}]$, from which the desired matrix inverse $\mathbf{A}^{-1}$ can be extracted.

The application of this technique is illustrated with respect to the same 3×3 matrix whose inverse we had found using the adjoint method [Eq. (k)]. We begin first by transforming the first element in the first row of **A** to unity (by dividing the entire first row of the augmented matrix, $[\mathbf{A} \mid \mathbf{I}]$ by $7/3$. Then, multiplying this transformed first row (i) by $-2/3$ and adding the results element-wise to the second row, and (ii) by $3/8$ and adding the results element-wise to the third row, we achieve the following transformation:

$$\left[\begin{array}{ccc|ccc} 7/3 & 2/3 & -3/8 & 1 & 0 & 0 \\ 2/3 & 7/3 & -3/8 & 0 & 1 & 0 \\ -3/8 & -3/8 & 3/8 & 0 & 0 & 1 \end{array}\right] \longrightarrow \left[\begin{array}{ccc|ccc} 1 & 2/7 & -9/56 & 3/7 & 0 & 0 \\ 0 & 15/7 & -15/56 & -2/7 & 1 & 0 \\ 0 & -15/56 & 141/448 & 9/56 & 0 & 1 \end{array}\right] \tag{m}$$

Next, we work with the second column of the modified **A** matrix and reduce it to a vector with a unit value in the second row (diagonal element), by dividing the entire second row of the augmented matrix by $15/7$, and with zero values in the other two rows (through multiplication and addition). Finally, we repeat this procedure with the third column of the modified **A** matrix, thereby transforming this matrix to an identity matrix. The resulting 3×3 matrix on the right side of the augmented matrix is then the desired inverse of **A**, as indicated below.

$$\longrightarrow \left[\begin{array}{ccc|ccc} 1 & 0 & -1/8 & 7/15 & -2/15 & 0 \\ 0 & 1 & -1/8 & -2/15 & 7/15 & 0 \\ 0 & 0 & 9/32 & 1/8 & 1/8 & 1 \end{array}\right] \longrightarrow \left[\begin{array}{ccc|ccc} 1 & 0 & 0 & 47/90 & -7/90 & 4/9 \\ 0 & 1 & 0 & -7/90 & 47/90 & 4/9 \\ 0 & 0 & 1 & 4/9 & 4/9 & 32/9 \end{array}\right] \tag{n}$$

$$\Rightarrow \begin{bmatrix} 7/3 & 2/3 & -3/8 \\ 2/3 & 7/3 & -3/8 \\ -3/8 & -3/8 & 3/8 \end{bmatrix}^{-1} = \begin{bmatrix} 47/90 & -7/90 & 4/9 \\ -7/90 & 47/90 & 4/9 \\ 4/9 & 4/9 & 32/9 \end{bmatrix} \tag{o}$$

The value of the inverse obtained in Eq. (o) by the Gauss-Jordan elimination method is identical to the one obtained earlier in Eq. (k) by the adjoint method. It may be noted that the inverse of the symmetric matrix **A** (as demonstrated in the above example) is also an inverse, but the generalised Gauss-Jordan elimination method is not able to take advantage of this, by reducing the calculations involved. However, there are other methods, notably the **LDLT decomposition method**†, which is specially suited for finding the inverse of a symmetric matrix. In this method, the square non-singular symmetric matrix **A** is *decomposed* (or *factorised*) into a form $\mathbf{LDL^T}$, where **L** is a product of a lower triangular matrix and a 'permutation matrix', with a transpose $\mathbf{L^T}$, and **I** is a diagonal (or block diagonal) matrix.

2.9.7 Cholesky Decomposition Method of Finding Inverse

A simpler and faster version of the LDLT decomposition method is the **Chlolesky decomposition method** which is applicable when the symmetric matrix **A** is **positive definite** (as is commonly the case with stiffness matrices in structural analysis). In this method, **A** is *decomposed* (or *factorised*) into a product of a lower triangular matrix **L** and its transpose $\mathbf{L^T}$, which is an upper triangular matrix. This factorisation $\mathbf{A = LL^T}$ is unique, provided **A** is positive definite. (Conversely, it can be proved that, if Cholesky decomposition is possible, **A** must be positive definite). There are several algorithms available for Cholesky decomposition. The Cholesky-Crout algorithm is briefly described here. The Cholesky factorisation for a typical 3 × 3 matrix takes the following form:

$$\mathbf{A} = \mathbf{LL}^{\mathrm{T}} = \begin{bmatrix} l_{11} & 0 & 0 \\ l_{21} & l_{22} & 0 \\ l_{31} & l_{32} & l_{33} \end{bmatrix} \begin{bmatrix} l_{11} & l_{21} & l_{31} \\ 0 & l_{22} & l_{32} \\ 0 & 0 & l_{33} \end{bmatrix} = \begin{bmatrix} l_{11}^2 & l_{11}l_{21} & l_{11}l_{31} \\ l_{11}l_{21} & l_{21}^2 + l_{22}^2 & l_{21}l_{31} + l_{22}l_{32} \\ l_{11}l_{31} & l_{21}l_{31} + l_{22}l_{32} & l_{31}^2 + l_{32}^2 + l_{33}^2 \end{bmatrix} = \begin{bmatrix} a_{11} & & \text{sym} \\ a_{21} & a_{22} & \\ a_{31} & a_{32} & a_{33} \end{bmatrix} \quad \text{(p)}$$

Thus, the first entry in the **L** matrix is easily obtained as: $l_{11} = \sqrt{a_{11}}$. The next entries in the first column can now be generated as $l_{i1} = a_{i1} / l_{i1}$ (for $i = 1, 2, \ldots, n$). Thus, we can proceed to the next column, starting with diagonal element, $l_{22} = \sqrt{a_{22} - l_{21}^2}$, and then, $l_{32} = (a_{32} - l_{21}l_{31})/l_{22}$, and so on. The general algorithm to derive the elements of the **L** matrix can be written as follows:

$$l_{ii} = \sqrt{a_{ii} - \sum_{j=1}^{i-1} l_{ij}^2} \qquad \text{for } i = 1, 2, \ldots, n \tag{2.41}$$

$$l_{ij} = \frac{1}{l_{jj}}\left(a_{ij} - \sum_{k=1}^{j-1} l_{ik} l_{jk} \right) \qquad \text{for } i = 1, 2, \ldots, n \text{ and } i > j \tag{2.41a}$$

It may be noted that the diagonal elements l_{ii} are invariably positive, which is possible because the quantity under the square root in Eq. 2.41 is positive for positive definite

† The LDLT decomposition method is faster than the more generalised LU decomposition method (which is applicable to non-symmetric matrices that are decomposed into a lower triangular matrix **L** with unit diagonal elements and upper triangular matrix **U** with zero diagonal elements).

matrices. Applying the above procedure, it can be shown that the 3×3 symmetric matrix considered in the earlier example can be factorised as follows:

$$\begin{bmatrix} 7/3 & 2/3 & -3/8 \\ 2/3 & 7/3 & -3/8 \\ -3/8 & -3/8 & 3/8 \end{bmatrix} = \begin{bmatrix} 1.527525 & 0 & 0 \\ 0.436436 & 1.463850 & 0 \\ -0.245495 & -0.182981 & 0.530330 \end{bmatrix} \begin{bmatrix} 1.527525 & 0.436436 & -0.245495 \\ 0 & 1.463850 & -0.182981 \\ 0 & -0.182981 & 0.530330 \end{bmatrix} \quad \text{(q)}$$

After the Cholesky decomposition, $\mathbf{A} = \mathbf{LL}^\mathrm{T}$, is achieved, the inverse matrix can be generated by invoking the following relationship (obtained by inverting both sides of $\mathbf{A} = \mathbf{LL}^\mathrm{T}$ and post-multiplying by **L**):

$$\mathbf{A} = \mathbf{LL}^\mathrm{T} \quad \Rightarrow \quad \mathbf{A}^{-1}\mathbf{L} = \left(\mathbf{L}^\mathrm{T}\right)^{-1} \tag{2.42}$$

Invoking the matrix property that the elements of the main diagonal of $\left(\mathbf{L}^\mathrm{T}\right)^{-1}$ are given by the corresponding reciprocals of the main diagonal of $\mathbf{L}^\mathrm{T}$ (same as **L**), and denoting $\mathbf{A}^{-1} = \mathbf{B}$, Eq. 2.42 may be expanded as follows, for the case of a typical 3×3 matrix:

$$\begin{bmatrix} b_{11} & b_{21} & b_{31} \\ b_{21} & b_{22} & b_{32} \\ b_{31} & b_{32} & b_{33} \end{bmatrix} \begin{bmatrix} l_{11} & 0 & 0 \\ l_{21} & l_{22} & 0 \\ l_{31} & l_{32} & l_{33} \end{bmatrix} = \begin{bmatrix} 1/l_{11} & ? & ? \\ 0 & 1/l_{22} & ? \\ 0 & 0 & 1/l_{33} \end{bmatrix} \quad \text{(r)}$$

where the six unknowns in the symmetric matrix $\mathbf{A}^{-1} = \mathbf{B}$ can be determined by solving the six available simultaneous equations pertaining to the lower triangle, using the method of 'backward substitution', starting with the last row and with the diagonal element: $b_{33} = 1/l_{33}^2$, $b_{32} = -b_{33}l_{32}/l_{22}$ and $b_{31} = -(b_{32}l_{21} + b_{33}l_{31})/l_{11}$. Next, moving upwards, $b_{22} = 1/l_{22}^2 - b_{32}l_{32}/l_{22}$, $b_{21} = -(b_{22}l_{21} + b_{32}l_{31})/l_{11}$ and $b_{11} = 1/l_{11}^2 - (b_{21}l_{21} + b_{31}l_{31})/l_{11}$.

In actual practice, however, the main application of Cholesky decomposition is not in explicitly finding the inverse of the square symmetric matrix **A**, but in directly solving a set of simultaneous equations, $[A]\{X\} = \{C\}$, as indicated in Eq. 2.20, to determine the unknown quantities in the vector **X**, given the *coefficient matrix* **A** and the *constant vector* **C** (all of order *n*), **A** being a positive definite symmetric matrix. To achieve this objective, let us define a vector **Y** as follows:

$$\mathbf{Y} = \mathbf{L}^\mathrm{T}\mathbf{X} \tag{2.43}$$

Observing that $\mathbf{A} = \mathbf{LL}^\mathrm{T}$ and substituting in $\mathbf{AX} = \mathbf{C}$,

$$\mathbf{LY} = \mathbf{C} \tag{2.44}$$

which may be expanded as follows, for the case of a typical 3×3 matrix **A**:

$$\begin{bmatrix} l_{11} & 0 & 0 \\ l_{21} & l_{22} & 0 \\ l_{31} & l_{32} & l_{33} \end{bmatrix} \begin{bmatrix} Y_1 \\ Y_2 \\ Y_3 \end{bmatrix} = \begin{bmatrix} c_1 \\ c_2 \\ c_3 \end{bmatrix} \tag{s}$$

The above set of equations can be solved easily for **Y** using the method of 'forward substitution', starting with the first row: $Y_1 = c_1 / l_{11}$, $Y_2 = (c_2 - l_{21}Y_1)/l_{22}$ and $Y_3 = (c_3 - l_{31}Y_1 - l_{32}Y_2)/l_{33}$. In general, for a vector of dimension n,

$$Y_j = \left(c_j - \sum_{i=1}^{j} l_{ji} Y_i \right) \Big/ l_{jj} \qquad j = 1, 2, ..., n \tag{2.45}$$

We can now solve for **X** invoking the linear transformation relating **X** with **Y**. Expanding Eq. 2.42 for the case of a typical 3 × 3 matrix **A**:

$$\begin{bmatrix} l_{11} & l_{21} & l_{31} \\ 0 & l_{22} & l_{32} \\ 0 & 0 & l_{33} \end{bmatrix} \begin{bmatrix} X_1 \\ X_2 \\ X_3 \end{bmatrix} = \begin{bmatrix} Y_1 \\ Y_2 \\ Y_3 \end{bmatrix} \tag{t}$$

Using the method of 'backward substitution', starting with the last row: $X_3 = Y_3 / l_{33}$, $X_2 = (Y_2 - l_{32}X_3)/l_{22}$ and $X_1 = (Y_1 - l_{21}X_2 - l_{31}X_3)/l_{11}$. In general, for a vector of dimension n,

$$X_j = \left(Y_j - \sum_{i=j+1}^{n} l_{ij} X_i \right) \Big/ l_{jj} \qquad j = n, n-1, ..., 2, 1 \tag{2.46}$$

Let us demonstrate this using the same 3 × 3 matrix **A** we had factorised in Eq. (q) by solving the following set of equations, which is in the form, $[A]\{X\} = \{C\}$:

$$\begin{bmatrix} 7/3 & 2/3 & -3/8 \\ 2/3 & 7/3 & -3/8 \\ -3/8 & -3/8 & 3/8 \end{bmatrix} \begin{bmatrix} X_1 \\ X_2 \\ X_3 \end{bmatrix} = \begin{bmatrix} 30 \\ 90 \\ -45 \end{bmatrix} \tag{u}$$

Applying Eq. 2.45,

$$Y_1 = c_1 / l_{11} = 30/1.527525 = 19.63961$$

$$Y_2 = (c_2 - l_{21}Y_1)/l_{22} = \frac{90 - (0.436436)(19.63961)}{1.463850} = 55.62631$$

$$Y_3 = (c_3 - l_{31}Y_1 - l_{32}Y_2)/l_{33} = \frac{-45 - (-0.245495)(19.63961) - (-0.182981)(55.62631)}{0.530330} = -56.56858 \tag{v}$$

Applying Eq. 2.46,

$$X_3 = Y_3 / l_{33} = -56.56858 / 0.530330 = -106.667$$

$$X_2 = (Y_2 - l_{32} X_3)/l_{22} = \frac{55.62631 - (-0.182981)(-106.667)}{1.463850} = 24.667 \tag{w}$$

$$X_1 = (Y_1 - l_{21} X_2 - l_{31} X_3)/l_{11} = \frac{19.63961 - (0.436436)(24.667) - (-0.245495)(-106.667)}{1.527525} = -11.333$$

We can verify this solution by substituting in Eq. (u). Alternatively, we can find the solution, taking advantage of the inverse of **A** that we have already generated in Eq. (o):

$$\mathbf{X} = \mathbf{A}^{-1}\mathbf{C} = \begin{bmatrix} 47/90 & -7/90 & 4/9 \\ -7/90 & 47/90 & 4/9 \\ 4/9 & 4/9 & 32/9 \end{bmatrix} \begin{bmatrix} 30 \\ 90 \\ -45 \end{bmatrix} = \begin{bmatrix} -11.333 \\ 24.667 \\ -106.667 \end{bmatrix} \tag{x}$$

2.10 EIGENVALUES & EIGENVECTORS

As seen earlier, any given matrix **A** can be visualised as a device that is capable of transforming a vector **X** (whose components are unknown) to another known vector **C** through a *linear* transformation, $\mathbf{AX} = \mathbf{C}$. In this Section, we look at certain special types of vectors **X**, called **eigenvectors**, associated with any *square* matrix **A**, such that the transformation results in vectors that are *parallel* to **X**. This means that the product $\mathbf{AX}$ is a scalar multiple of **X**, whereby the transformation takes the following form:

$$\mathbf{AX} = \lambda \mathbf{X} \tag{2.47}$$

where λ, the *scalar multiplier*, is a unique property of the matrix **A**, called **eigenvalue**. This linear transformation results in the direction of the eigenvector remaining unchanged (for positive eigenvalues) or getting reversed (for negative eigenvalues). Indeed, most vectors will not satisfy this equation; they will change direction when transformed.

Eigenvalues are important features of any dynamical system; they correspond to the *natural frequencies* of the system. It is necessary to know their values, as they govern the dynamic response of the system. In most engineering situations, we try to keep the eigen-frequencies (of systems such as bridge structures and rockets) away from the "exciting" frequencies of the applied loading (wind or sloshing of rocket fuel), so that the response does not get amplified (*resonance*). On the other hand, the stock-broker aims to get in line with the natural frequencies of the stock market in order to maximise his profits!

Clearly, $\mathbf{X} = \mathbf{O}$ is a possible solution to Eq. 2.47. But this is practically useless, and is referred to as a *trivial* solution. We are looking here for a non-trivial solution, whereby the vector **X** must be a non-zero vector in order to qualify to be an *eigenvector*. We also observe that $\lambda = 0$ is a possible eigenvalue, satisfying $\mathbf{AX} = \mathbf{O}$, but only if **A** is *singular*. We are interested in the general case of a square non-singular matrix **A** of order $n \times n$, for which there will be as many as n eigenvalues. Some of these n eigenvalues may be *repeated* values

($\lambda_1 = \lambda_2$); some may involve *complex* values ($\lambda = c \pm id$), corresponding to which, it is possible that we may not be able to define uniquely independent eigenvectors. Fortunately, for the practical problems we encounter in structural analysis, in stability and dynamics, the square matrix **A** is usually *symmetric*, and this implies that all the eigenvalues will be *real* (not complex), and the corresponding n eigenvectors, associated with distinct eigenvalues, will be not only *linearly independent* but also mutually orthogonal.

The identity matrix **I** (of order $n \times n$) is an interesting and exceptional example of a symmetric matrix with n repeated eigenvalues (all equal to 1!). Every n-dimensional vector **X** turns out to be an eigenvector of the identity matrix **I**! This is because **IX** = **X**. These eigenvectors need not be mutually independent.

> Every symmetric (or Hermitian†) matrix has real eigenvalues and corresponding real eigenvectors. If the eigenvectors come from different eigenvalues (not repeated), then they are mutually orthogonal and can be chosen to be orthonormal.

Some other useful properties of eigenvalues are highlighted below.

> The sum of the n eigenvalues of a square non-singular matrix **A** of order $n \times n$ is equal to the 'trace' (sum of the diagonal elements) of **A** and the product of the n eigenvalues is equal to the determinant of **A**.

$$\lambda_1 + \lambda_2 + \ldots + \lambda_n = a_{11} + a_{22} + \ldots + a_{nn} \tag{2.48}$$

$$\lambda_1 \lambda_2 \ldots \lambda_n = \det \mathbf{A} \tag{2.49}$$

Eqns 2.48 and 2.49 provide a useful check on the eigenvalues generated for a given matrix. Indeed, they are sufficient to find the eigenvalues of a 2×2 matrix. For example, let us consider the square matrix, $\mathbf{A} = \begin{bmatrix} 3 & 1 \\ 1 & 3 \end{bmatrix}$, for which $\lambda_1 + \lambda_2 = 3 + 3 = 6$ and $\lambda_1 \lambda_2 = \det \mathbf{A} = 8$. The resulting quadratic equation may be solved easily to yield the two eigenvalues:

$$\lambda^2 - 6\lambda + 8 = (\lambda - 4)(\lambda - 2) = 0 \;\Rightarrow\; \lambda_1 = 4,\ \lambda_2 = 2 \;\text{ or }\; \lambda_1 = 2,\ \lambda_2 = 4 \tag{y}$$

† A *Hermitian* matrix is one whose diagonal entries are real and each off-diagonal element, $a + ib$, is matched by its complex conjugate, $a - ib$, as its mirror image across the main diagonal. A real symmetric matrix is a special case of a Hermitian matrix.

2.10.1 Characteristic Equation

In general, the *eigenvalue problem* involves finding a set of eigenvalues ($\lambda_1, \lambda_2, ..., \lambda_n$) and corresponding eigenvectors ($\mathbf{X_1}, \mathbf{X_2}, ..., \mathbf{X_n}$) that satisfy Eq. 2.47, for a given square non-singular matrix **A** of order $n \times n$. In order to solve the eigenvalue problem, which involves two sets of unknowns, given one set of equations, we can re-write Eq. 2.47 as follows:

$$(\mathbf{A} - \lambda\mathbf{I})\mathbf{X} = \mathbf{O} \tag{2.50}$$

We observe that the matrix **A** is now modified as $\mathbf{A} - \lambda\mathbf{I}$, suggesting that the diagonal elements of **A** are "shifted" by λ, with off-diagonal terms remaining unchanged:

$$\begin{bmatrix} (a_{11}-\lambda) & a_{12} & \cdots & a_{1n} \\ a_{21} & (a_{22}-\lambda) & \cdots & a_{2n} \\ \vdots & \vdots & \ddots & \vdots \\ a_{n1} & a_{n2} & \cdots & (a_{nn}-\lambda) \end{bmatrix} \begin{bmatrix} x_1 \\ x_2 \\ \vdots \\ x_n \end{bmatrix} = \begin{bmatrix} 0 \\ 0 \\ \vdots \\ 0 \end{bmatrix} \tag{2.50a}$$

Clearly, for a *non-trivial* solution ($\mathbf{X} \neq \mathbf{O}$) to exist, the matrix $\mathbf{A} - \lambda\mathbf{I}$ has to be *singular*, whereby its determinant must be equal to zero:

$$|\mathbf{A} - \lambda\mathbf{I}| = 0 \tag{2.51}$$

The above equation, which is a polynomial equation of order *n*, in terms of λ, is known as the **characteristic equation**. The roots of this equation yield the desired *n* eigenvalues. After finding these eigenvalues, using a suitable numerical scheme, and substituting them one at a time in Eq. 2.50, we can find the eigenvector $\mathbf{X_j}$ corresponding to λ_j. This eigenvector corresponds to the *nullspace* of the matrix $\mathbf{A} - \lambda_j\mathbf{I}$, as discussed earlier (refer Section 2.7). As the $\mathbf{A} - \lambda_j\mathbf{I}$ is a *singular* matrix, its rank is such that $r < n$. The column vectors of this matrix are linearly dependent in some manner. The components of the eigenvector $\mathbf{X_j}$ should be selected in such a way that a linear combination of the column vectors of $\mathbf{A} - \lambda_j\mathbf{I}$ results in a null vector. In general, the solution can be conveniently determined by reducing the matrix $\mathbf{A} - \lambda_j\mathbf{I}$ to its row reduced form, $\mathbf{R} = \left[\begin{array}{c|c} \mathbf{I} & \mathbf{F} \\ \hline \mathbf{O} & \mathbf{O} \end{array}\right]$ through a process of elimination, whereby $(\mathbf{A} - \lambda_j\mathbf{I})\mathbf{X_j} = \mathbf{O}$ becomes $\mathbf{RX_j} = \left[\begin{array}{c|c} \mathbf{I} & \mathbf{F} \\ \hline \mathbf{O} & \mathbf{O} \end{array}\right] \left[\begin{array}{c} \mathbf{X}_{\mathbf{j},pivot} \\ \hline \mathbf{X}_{\mathbf{j},free} \end{array}\right] = \left[\begin{array}{c} \mathbf{O} \\ \hline \mathbf{O} \end{array}\right]$. By assigning suitable (arbitrary) values to the *free variables* $\mathbf{X}_{\mathbf{j},free}$, the remaining components in the eigenvector can be directly obtained as $\mathbf{X}_{\mathbf{j},pivot} = -\mathbf{F}\mathbf{X}_{\mathbf{j},free}$. Usually, it is convenient to assign unit values to the free variables, and thus find the pivot variables. Clearly, multiple solutions are possible, all these vectors are scalar multiples of one another, and it is sufficient to choose any one vector (with one of the components having a unit value). The eigenvector can be suitably normalised in various ways. The standard way of normalising is by dividing the components of the eigenvector by the length of the vector, thereby reducing it to a unit vector.

For example, considering the symmetric 2×2 matrix discussed earlier, the characteristic equation is given by $\begin{vmatrix} (3-\lambda) & 1 \\ 1 & (3-\lambda) \end{vmatrix} = (3-\lambda)^2 - 1 = 0$, resulting in the same quadratic equation and solutions for λ, as in Eq. (y). Now, we need to find the two eigenvectors. Corresponding to $\lambda_1 = 4$, the matrix $\mathbf{A} - \lambda_1\mathbf{I} = \begin{bmatrix} -1 & 1 \\ 1 & -1 \end{bmatrix}$, which is obviously singular, with the first column vector being the negative of the second column vector. The linear combination of the two column vectors that results in a null vector is obviously one in which the multipliers are the same for both vectors (any scalar c). Thus, the eigenvector, corresponding to $\lambda_1 = 4$, is given by $\mathbf{X}_1 = \begin{bmatrix} c \\ c \end{bmatrix}$, or simply, $\mathbf{X}_1 = \begin{bmatrix} 1 \\ 1 \end{bmatrix}$. Alternatively, through elimination, we observe that $\mathbf{A} - \lambda_1\mathbf{I}$ has the row reduced echelon form, $\mathbf{R} = \left[\begin{array}{c|c} 1 & -1 \\ \hline 0 & 0 \end{array}\right]$. Assigning a unit value to the free variable (X_2), we observe that the pivot variable (X_1) is also equal to 1.

Similarly, corresponding to $\lambda_2 = 2$, the matrix $\mathbf{A} - \lambda_2\mathbf{I} = \begin{bmatrix} 1 & 1 \\ 1 & 1 \end{bmatrix}$, in which the first column vector is identical to the second column vector. Clearly, the eigenvector is given by $\mathbf{X}_2 = \begin{bmatrix} -c \\ c \end{bmatrix}$, or simply, $\mathbf{X}_2 = \begin{bmatrix} -1 \\ 1 \end{bmatrix}$. Alternatively, through elimination, we observe that $\mathbf{A} - \lambda_2\mathbf{I}$ has the row reduced echelon form, $\mathbf{R} = \left[\begin{array}{c|c} 1 & 0 \\ \hline 0 & 0 \end{array}\right]$. Assigning a unit value to the free variable (X_2), we observe that the pivot variable (X_1) is equal to -1.

Let us now demonstrate the solution to the eigenvalue problem with reference to a 3×3 symmetric matrix, $\mathbf{A} = \begin{bmatrix} 7 & -2 & 0 \\ -2 & 6 & -2 \\ 0 & -2 & 5 \end{bmatrix}$, which is typical of the kind we encounter in structural analysis problems (dynamics and stability) involving three degrees of freedom. The *characteristic equation*, which is a cubic equation, is given by:

$$|\mathbf{A} - \lambda\mathbf{I}| = \begin{vmatrix} (7-\lambda) & -2 & 0 \\ -2 & (6-\lambda) & -2 \\ 0 & -2 & (5-\lambda) \end{vmatrix} = 0 \tag{z}$$

$$\Rightarrow \quad -\lambda^3 + 18\lambda^2 - 99\lambda + 162 = 0$$

which may be conveniently factorised (in this particular case), or solved by a suitable numerical scheme, to yield the following roots: $\lambda_1 = 3,\ \lambda_2 = 6,\ \lambda_3 = 9$. We may verify that these solutions satisfy Eqns 2.48 and 2.49.

Corresponding to $\lambda_1 = 3$, the matrix $\mathbf{A} - \lambda_1\mathbf{I} = \begin{bmatrix} 4 & -2 & 0 \\ -2 & 3 & -2 \\ 0 & -2 & 2 \end{bmatrix}$, whose row reduced echelon form can be shown to be $\mathbf{R} = \left[\begin{array}{cc|c} 1 & 0 & -1/2 \\ 0 & 1 & -1 \\ \hline 0 & 0 & 0 \end{array}\right]$. Assigning a unit value to the free variable (X_3), we observe that the pivot variables are given by:

$$\mathbf{X}_{1,pivot} = -\mathbf{F}\mathbf{X}_{1,free} = -\begin{bmatrix} (-1/2)(1) \\ (-1)(1) \end{bmatrix} = \begin{bmatrix} 1/2 \\ 1 \end{bmatrix} \Rightarrow \mathbf{X}_1 = \begin{bmatrix} 1/2 \\ 1 \\ 1 \end{bmatrix} \text{ or } \begin{bmatrix} c/2 \\ c \\ c \end{bmatrix}.$$

Similarly, we can show that, corresponding to $\lambda_2 = 6$, $\mathbf{X}_2 = \begin{bmatrix} -1 \\ -1/2 \\ 1 \end{bmatrix}$ or $\begin{bmatrix} c \\ c/2 \\ -c \end{bmatrix}$, and corresponding to $\lambda_3 = 9$, $\mathbf{X}_3 = \begin{bmatrix} 1 \\ -1 \\ 1/2 \end{bmatrix}$ or $\begin{bmatrix} c \\ -c \\ c/2 \end{bmatrix}$. We may normalise all the eigenvectors, as described earlier, reducing them to unit vectors (with the first component always having a positive value) as follows: $\hat{\mathbf{X}}_1 = \begin{bmatrix} 1/3 \\ 2/3 \\ 2/3 \end{bmatrix}$, $\hat{\mathbf{X}}_2 = \begin{bmatrix} 2/3 \\ 1/3 \\ -2/3 \end{bmatrix}$ and $\hat{\mathbf{X}}_3 = \begin{bmatrix} 2/3 \\ -2/3 \\ 1/3 \end{bmatrix}$. Such real normalised eigenvectors of a symmetric matrix are invariably mutually *orthogonal*, and are referred to as **orthonormal eigenvectors**.

2.10.2 Properties of Orthogonal Eigenvector Matrix

All the n distinct orthonormal eigenvectors, $\hat{\mathbf{X}}_1, \hat{\mathbf{X}}_2, \ldots, \hat{\mathbf{X}}_n$, of a real symmetric matrix $\mathbf{A}$ of order $n \times n$ can be assembled sequentially as column vectors in a matrix, called **orthonormal eigenvector matrix**, which may be denoted as $\mathbf{Q}$. This is an **orthogonal matrix**, which has some special and useful features.

Firstly, owing to the orthogonal nature of the orthonormal eigenvectors of **A**, the transpose $\mathbf{Q}^{\mathbf{T}}$ and its inverse $\mathbf{Q}^{-1}$ turn out to be equal. This is the fundamental feature of an orthogonal matrix†, whose determinant is always equal to ± 1.

$$\mathbf{Q}^{\mathbf{T}} = \mathbf{Q}^{-1} \tag{2.52}$$

whereby:

$$\mathbf{Q}^{\mathbf{T}}\mathbf{Q}^{-1} = \mathbf{I} \tag{2.52a}$$

We can verify this with reference to the 3×3 symmetric matrix example, whose orthonormal eigenvector matrix is $\mathbf{Q} = \begin{bmatrix} 1/3 & 2/3 & 2/3 \\ 2/3 & 1/3 & -2/3 \\ 2/3 & -2/3 & 1/3 \end{bmatrix}$, which, incidentally, also happens to be symmetric in this particular example, owing to the manner in which the components of the eigenvectors have been chosen.

Secondly, there are interesting relationships that relate the real symmetric matrix **A** with its *orthonormal eigenvector matrix* $\mathbf{Q} = \left[\{\hat{X}_1\} \ \{\hat{X}_2\} \ \cdots \ \{\hat{X}_n\}\right]$ and the respective eigenvalues $\lambda_1, \lambda_2, ..., \lambda_n$. These eigenvalues can be arranged sequentially as diagonal elements of a diagonal matrix called **eigenvalue matrix**, denoted as $\mathbf{\Lambda}$ (capital lambda). If $\mathbf{\Lambda}$ is pre-multiplied by **Q** and post-multiplied by its transpose $\mathbf{Q}^{\mathbf{T}}$ (or its inverse, $\mathbf{Q}^{-1}$), then the resulting matrix is none other than the original matrix **A**! In other words, the matrix **A** can be factorised or decomposed as $\mathbf{Q\Lambda Q^T}$. This "eigen-decomposition" is sometimes referred to as the *spectral theorem* in linear algebra.

$$\mathbf{A} = \mathbf{Q\Lambda Q^T} = \lambda_1 \mathbf{X_1 X_1^T} + \lambda_2 \mathbf{X_2 X_2^T} + ... + \lambda_n \mathbf{X_n X_n^T} \tag{2.53}$$

We can verify this in our example problem:

$$\mathbf{Q\Lambda Q^T} = \begin{bmatrix} 1/3 & 2/3 & 2/3 \\ 2/3 & 1/3 & -2/3 \\ 2/3 & -2/3 & 1/3 \end{bmatrix} \begin{bmatrix} 3 & 0 & 0 \\ 0 & 6 & 0 \\ 0 & 0 & 9 \end{bmatrix} \begin{bmatrix} 1/3 & 2/3 & 2/3 \\ 2/3 & 1/3 & -2/3 \\ 2/3 & -2/3 & 1/3 \end{bmatrix} = \begin{bmatrix} 7 & -2 & 0 \\ -2 & 6 & -2 \\ 0 & -2 & 5 \end{bmatrix} = \mathbf{A}$$

If, instead of the *orthonormal eigenvector matrix* **Q**, we use any other eigenvector matrix $\mathbf{S} = \left[\{X_1\} \ \{X_2\} \ \cdots \ \{X_n\}\right]$ (containing *linearly independent eigenvectors* that need not be normalised to make them mutually orthogonal), then also factorisation of the matrix **A** is possible, as follows:

$$\mathbf{A} = \mathbf{S\Lambda S^{-1}} \tag{2.54}$$

† A real matrix is orthogonal if and only if its columns (and rows), considered as vectors, form an orthonormal set.

A real symmetric matrix $\mathbf{A}$ order $n \times n$ can be factored as $\mathbf{A} = \mathbf{S\Lambda S}^{-1}$, where $\mathbf{S}$ is the matrix containing n linearly independent eigenvectors of $\mathbf{A}$, and $\mathbf{\Lambda}$ is a diagonal matrix containing the corresponding eigenvalues of $\mathbf{A}$. If the orthonormal eigenvector matrix $\mathbf{Q}$ is used in the place of $\mathbf{S}$, then the factorisation can be carried out more conveniently as $\mathbf{A} = \mathbf{Q\Lambda Q}^{T}$.

Alternatively, if the matrix $\mathbf{A}$ is post-multiplied by $\mathbf{Q}$ and pre-multiplied by its transpose $\mathbf{Q}^{T}$ (or its inverse, $\mathbf{Q}^{-1}$), then the resulting matrix is none other than the diagonal eigenvalue matrix $\mathbf{\Lambda}$! In other words, the matrix $\mathbf{A}$ can be *diagonalised* with the help of the eigenvector matrix:

$$\mathbf{Q}^{T}\mathbf{AQ} = \mathbf{\Lambda} \tag{2.55}$$

Diagonalisation of $\mathbf{A}$ is also possible by means of the eigenvector matrix $\mathbf{S} = \left[\{X_1\} \ \{X_2\} \ \cdots \ \{X_n\}\right]$ (containing linearly independent eigenvectors), instead of the orthonormal eigenvector matrix, $\mathbf{Q}$, as follows:

$$\mathbf{S}^{-1}\mathbf{AS} = \mathbf{\Lambda} \tag{2.56}$$

If a real symmetric matrix $\mathbf{A}$ of order $n \times n$ has n linearly independent eigenvectors which constitute the eigenvector matrix $\mathbf{S}$ of order $n \times n$, then by pre-multiplying $\mathbf{A}$ with the inverse of $\mathbf{S}$ and post-multiplying with $\mathbf{S}$, the resulting matrix† is a diagonal matrix $\mathbf{\Lambda}$ containing the eigenvalues of $\mathbf{A}$: $\mathbf{SAS}^{-1} = \mathbf{\Lambda}$. If the orthonormal eigenvector matrix $\mathbf{Q}$ is used in the place of $\mathbf{S}$, then the diagonalisation can be carried out more conveniently as $\mathbf{Q}^{T}\mathbf{AQ} = \mathbf{\Lambda}$.

We can easily verify these properties in our example problem. Clearly, making the eigenvectors orthonormal simplifies the calculations. The orthonormal eigenvectors define an **orthogonal basis** in the "eigenspace", which facilitates transformations of vectors conveniently, as demonstrated in various applications in structural analysis (such as problems related to dynamics and stability). It may be noted that any vector in the eigenspace can be expressed as a linear combination of the eigenvectors.

2.10.3 Practical Applications in Structural Analysis

In practical applications in structural analysis, the eigenvalues may correspond to the *critical buckling loads* (in stability analysis) or the *natural frequencies* (in dynamic analysis) of a

† This is a special case of *similarity transformation* of the type, $\mathbf{M}^{-1}\mathbf{AM}$, in which *any* invertible matrix $\mathbf{M}$ can be used in the place of the eigenvector matrix $\mathbf{S}$. The family of matrices thus generated are "similar", which means that they all share the same eigenvalues.

system with multiple degrees of freedom. The eigenvectors define the shape of the *buckling mode* of the structure (in stability analysis) or the *natural mode of vibration* (in dynamic analysis) of the structure. This shape is commonly referred to as *mode shape*. Associated with every mode of buckling or free vibration, there is a corresponding eigenvalue. In problems involving continuous systems, with infinite degrees of freedom, and in such cases, the eigenvector becomes an *eigenfunction*, as we shall see later.

These eigenvectors are mutually orthogonal, whereby they form an *orthogonal basis* for defining a vector space in which any displaced configuration of the structure can be defined. As we shall see later, this basis is very useful in solving problems involving multiple degrees of freedom using the so-called *mode superposition method.* In solving any linear system of coupled differential equations (such as equations of motion in a structure with multiple degrees of freedom), it is possible to achieve a transformation of coordinates, such that the equations become uncoupled in the new coordinate system, and so can be easily solved. This implies that the coefficient matrix **A**, which is generally a square symmetric matrix, becomes diagonalised. The *orthonormal eigenvector matrix* **Q** of **A** provides the necessary basis for such a transformation. These concepts are also adopted in finding principal stresses at a point in a body, whereby the general stress tensor (with nine components) gets diagonalised, with the diagonal terms representing the three principal stresses (off-diagonal terms representing shear stresses are zero). Similarly, by finding the principal axes of areas and masses, we are able to find the principal moments of inertia.

In Chapter 7, we come across the eigenvalue problem in finding the critical load that causes elastic instability in a structure ('bifurcation buckling'). The eigenvalue problem formulation takes on a different (more generalised) form compared to the elementary form shown in Eq. 2.47. The generalised form is of the following type:

$$\mathbf{AX} = \lambda \mathbf{BX} \tag{2.57}$$

where **X** corresponds to the displacement vector, **A** corresponds to the primary ('first-order') stiffness matrix of the structure and $\lambda\mathbf{B}$ corresponds to the 'geometric' stiffness matrix of the structure. **A** and **B** are symmetric matrices of order $n \times n$, where n is given by the degrees of freedom in the structure. Eq. 2.57 can be expressed in a form similar to Eq. 2.50 as follows:

$$(\mathbf{A} - \lambda\mathbf{B})\mathbf{X} = \mathbf{O} \tag{2.58}$$

A non-trivial solution is possible only when the determinant of $|\mathbf{A} - \lambda\mathbf{B}|$ vanishes, and this physically manifests in the structure becoming unstable (overall stiffness matrix becomes singular). Of the n eigenvalues (values of λ), possible, the lowest one, corresponding to the critical buckling load, is alone of practical significance. The corresponding eigenvector reflects the 'fundamental' *mode shape* in which the buckling of the structure occurs.

It is possible to arrive at the solution to the above eigenvalue problem using simple numerical techniques, as demonstrated in Chapter 7. However, for a more generalised algorithmic-based solution, it is desirable to cast Eq. 2.57 in the same form as Eq. 2.47 by means of the following transformation:

$$(\mathbf{B}^{-1}\mathbf{A})\mathbf{X} = \lambda\mathbf{X} \tag{2.59}$$

There are several iterative procedures available to solve the eigenvalue problem.

2.10.4 Iterative Methods

Finding the eigenvalues as roots of the characteristic equation, and thereafter finding the eigenvectors, is not a procedure that is suitable while dealing with square matrices of order greater than 3×3. In such cases, we need an iterative (algorithmic) procedure, rather than an algebraic one, to render it suitable for computer application. Many iterative schemes are available for this purpose. In general, the objective is to converge upon an eigenvalue and corresponding eigenvector, by solving the basic eigenvalue equation, $\mathbf{AX} = \lambda\mathbf{X}$, each iteration yielding a new and improved approximation.

The **power method** is an algorithm especially suitable for finding the dominant eigenvalue (which is larger than all other eigenvalues: $|\lambda_1| \leq |\lambda_2| ... \leq |\lambda_{n-1}| < |\lambda_n|$) and corresponding eigenvector of any real matrix **A** of order $n \times n$ that has n linearly independent eigenvectors. The ordinary power method starts with an initial guess of an eigenvector $\mathbf{X}_0$ (usually with all components equal to 1) and then proceeds as $\mathbf{X}_1 = \mathbf{AX}_0$, $\mathbf{X}_2 = \mathbf{AX}_1$, until the kth step, whereby, effectively, $\mathbf{X}_k = \mathbf{A}^k\mathbf{X}_0$. As k gets larger, $\mathbf{X}_k$ approaches the eigenvector $\mathbf{X}_\mathbf{n}$ corresponding to the dominant eigenvalue λ_n. If $\mathbf{X}_k$ is scaled such that its largest component is unity, then the component of $\mathbf{X}_{k+1} = \mathbf{AX}_k$ having the largest absolute value must be λ_n. Convergence is slow if the dominance of the largest eigenvalue is not significant, i.e., if $|\lambda_{n-1}/\lambda_n|$ is close to 1. The convergence can be significantly expedited by various improvements on the power method. If we are interested in the smallest eigenvalue λ_1, as is often the case in many structural analysis problems, such as finding the *fundamental natural frequency* or the *fundamental buckling mode*, we can conveniently re-formulate the problem, operating on $\mathbf{A}^{-1}$ (instead of **A**). This is known as the **inverse power method**. The power methods, including its many variants, are suitable when the matrix **A** is large and sparse.

There are other iterative methods, such as **Jacobi's method**, that are capable of generating all the eigenvalues and eigenvectors. In practical problems, however, our interest is limited to finding only a few eigenvalues and eigenvectors, corresponding to the first few modes, as the participation of higher modes may not be significant. The **Stodola-Vianello method** is suitable in many boundary value problems of interest in mathematical physics, such as the free vibration problem in structural dynamics.

There are algorithms available in packages like MATLAB that can be conveniently used to solve the eigenvalue problem. However, one must exercise caution, as the results are strongly dependent on the conditioning of the matrices, and significant numerical errors are possible. It is, therefore, necessary to use alternative methods of verifying the results (at least, in terms of order of magnitude), from a practical perspective.

REVIEW QUESTIONS

2.1 What is a *matrix*? Explain the relevance of *matrix methods* in structural analysis?

2.2 What is a vector? What is meant by *linear vector space*?

2.3 Show that the product of two matrices, if possible, results in a linear combination of column vectors or row vectors of the multiplying matrices.

2.4 What are the characteristics of a *symmetric* matrix and a *skew-symmetric* matrix?

2.5 What is meant by the *rank* of a matrix? Explain how the solvability of a system of linear simultaneous equations can be assessed in terms of the rank of the coefficient matrix.

2.6 What is meant by the *row reduced echelon form* of a matrix? Demonstrate, by means of an example, how this can be generated by a process of *elimination.*

2.7 What conditions need to be satisfied for a matrix **A** to have an *inverse*?

2.8 What is meant by the *determinant* of a matrix? What do the terms, *cofactor* and *adjoint*, mean?

2.9 Derive the basis for *Cramer's Rule.*

2.10 What is meant by the *condition* of a matrix?

2.11 Under which circumstances is the Cholesky decomposition method of finding the inverse of a matrix advantageous? Describe briefly the basis of this method.

2.12 What is meant by the terms, *eigenvalue* and *eigenvector*? Cite some applications in structural analysis.

2.13 What is meant by *characteristic equation* in the context of the eigenvalue problem?

2.14 What are the properties of the *orthogonal eigenvector matrix*?

2.15 Briefly describe some common iterative methods of finding eigenvalues and eigenvectors.

3

Matrix Analysis of Structures — Basic Concepts

3.1 INTRODUCTION

In this Chapter, we take an overview of the basic matrix formulations involved in structural analysis. We begin by recalling that the objective of structural analysis is to find the force and displacement responses of a given structure subject to prescribed loading (comprising direct actions or indirect loads such as due to support movements and temperature effects). Matrix methods can be employed for this purpose, if we can suitably express all the relevant force and displacement quantities in matrix (vector) form and if we can establish their linear inter-relationships through appropriate matrices (stiffness /flexibility matrix, force /displacement transformation matrix). Indeed, this kind of matrix approach to structural analysis has become a necessity with the advent of the digital computer, facilitating the use of software to carry out structural analysis.

The first step in the analysis is to define the idealised model of the structure. In the case of skeletal structures, this implies defining the geometry of the structure, identifying and labelling the various line elements and their degrees of freedom, and identifying the various joints and supports. The geometric and material properties of the various structural elements also form part of the input data in the analysis. Next, the load data, in an appropriate format, has to be input. In order to capture all the relevant data, both input (model, loading) and output (response), we need a suitable framework. This is provided by defining *global* and *local* coordinates, which provide a basis for identifying the various force and displacement variables. This basic framework for modelling the structure, the loads and the response is discussed in Section 3.2.

In Section 3.2, we define **global and local coordinates** that are needed to represent forces and displacements at the structure and element levels. This is followed by a discussion in Section 3.3 on how it is possible to inter-relate global and local coordinates by means of **transformation matrices**. In Sections 3.4 and 3.5, concepts related to **flexibility matrix** and **stiffness matrix**, inter-relating force and displacement vectors at the structure (global) and element (local) levels are covered. In Section 3.6, we discuss how intermediate loads in frame elements can be converted to **equivalent joint loads**, to facilitate application of matrix methods. Finally, in Section 3.7, we take an overview of the steps involved in the **stiffness and flexibility methods of analysis**, as well as comparison of the two methods.

3.2 COORDINATE SYSTEMS

3.2.1 Global Coordinates

Nodes

We use the term *node* to define any point on the structure, where we are interested in prescribing or finding forces (loads or reactions) or displacements (translations and rotations). *Joints*, where two or more prismatic elements get inter-connected, and *supports*, where displacements are restrained, are examples of nodes. However, a node can also be located at any intermediate location in an element of a structure where a concentrated load is applied or where we are interested in finding a displacement. In short, a node is a point location of interest to us in the model of the skeletal structure, in terms of one or more concentrated forces (including moments) or displacements at that point.

A question that may arise at this stage is in relation to distributed loads. *How can we accommodate forces that are continuously distributed and would require infinite nodes*? The answer is obvious: *we cannot*! By definition, a matrix is a means of storing data that is discrete and not a continuum. Hence, we cannot directly deal with distributed loads in matrix operations in structural analysis. We need to somehow convert these distributed loads into equivalent concentrated loads for this purpose. This is discussed in Section 3.7. If we can handle distributed loads that act at intermediate locations in a structure (in between joints) through the device of *equivalent loads* (acting at joints), then we can do the same with even concentrated loads that act at intermediate locations. This is usually done in matrix methods, and for this reason, the nodes are invariably located only at joints and supports; however, additional nodes also need to be considered at locations where displacements are desired.

In addition to location, we also need to define the directions of the force or displacement vector. The force and displacement vectors at a particular node location form a *conjugate* pair. When we define the force vector at a node in terms of its scalar components (in a Cartesian coordinate framework, for example), and assign directions by means of arrows (touching the structure at a particular node), we are also simultaneously defining the components of the conjugate displacement vector, and the same arrows are used for physical representation of displacement component at the nodes. Thus, F_j and D_j form a conjugate pair, denoting the force and displacement components respectively of the force and displacement vectors in the j direction. If we are following the flexibility (force) method of analysis, we give primary importance to forces (such as F_j). On the other hand, if we are following the stiffness (displacement) method of analysis, we give primary importance to displacements (such as D_j).

Let us begin by considering the basic framework required in the stiffness (displacement) method, which is better suited than the flexibility (force) method of analysis for application of matrix methods. Associated with every node, we can define Cartesian coordinate components, corresponding to independent degrees of freedom (displacements) at any node. We use the term *global coordinates* (defined later) to identify these degrees of freedom. The coordinate axes used for this purpose are called *global axes*. This provides the necessary framework to define the displacement field (and the conjugate force field) in a structure in a *matrix* representation.

Global Axes

Consider a typical rigid-jointed plane frame structure, as shown in Fig. 3.1a. We can define the geometry of the structure by identifying the various joint locations (A, B, C and D), in terms of their Cartesian coordinate locations (x_A,y_A; x_B,y_B; x_C,y_C; x_D,y_D) with reference to some origin O. Next, we need to identify the members (or elements) that constitute the structure (plane frame elements, '1', '2' and '3') and relate them to the joints previously identified. For example, member '2' is a prismatic frame element inter-connecting joints B and C in Fig. 3.1a.

This reference framework, applicable for the entire structure, is called a *global coordinate system*, and in a general case involving a three-dimensional framed structure, must have a single well-defined origin O, with the three orthogonal axes (x-, y- and z-) with their directions clearly identified. The axes are called *global axes*. The directions of these axes should be chosen such that they satisfy the vector cross-product (of the respective unit vectors): $\hat{i} \times \hat{j} = \hat{k}$. Thus, if the positive x- and y-axes directions are assumed to point to the right and upwards respectively, it would be appropriate to consider the positive z-axis direction to point outwards from the x-y plane as shown in Fig. 3.1a (with only the tip of the arrow-head visible at O). This implies that rotations and moments in the x-y plane (i.e., with respect to the z-axis) are treated as positive when they act in an *anti-clockwise* direction. In practice, it is convenient to locate the origin O at a lower left-extreme location (including the joint A), so that the various joint coordinates have positive values.

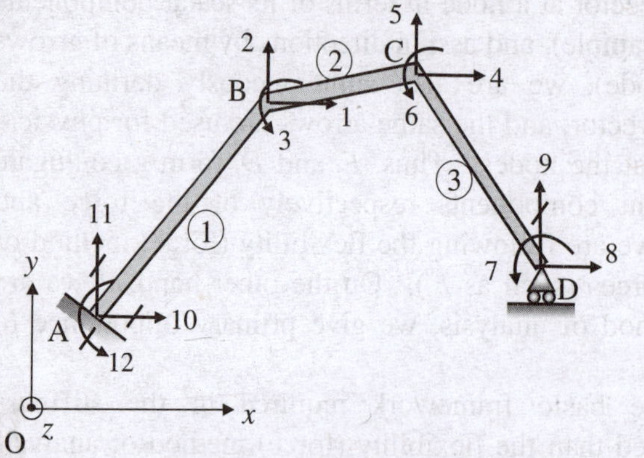

Displacement vector:

$$\{D\} = \begin{Bmatrix} D_A \\ \cdots \\ D_R \end{Bmatrix} = \begin{Bmatrix} D_1 = \Delta_{Bx} \\ D_2 = \Delta_{By} \\ D_3 = \theta_{Bz} \\ D_4 = \Delta_{Cx} \\ D_5 = \Delta_{Cy} \\ D_6 = \theta_{Cz} \\ D_7 = \theta_{Dz} \\ D_8 = \Delta_{Dx} \\ \cdots\cdots \\ D_9 = 0 \\ D_{10} = 0 \\ D_{11} = 0 \\ D_{12} = 0 \end{Bmatrix}$$

(*a*) Global coordinates (8 active; 4 restrained)

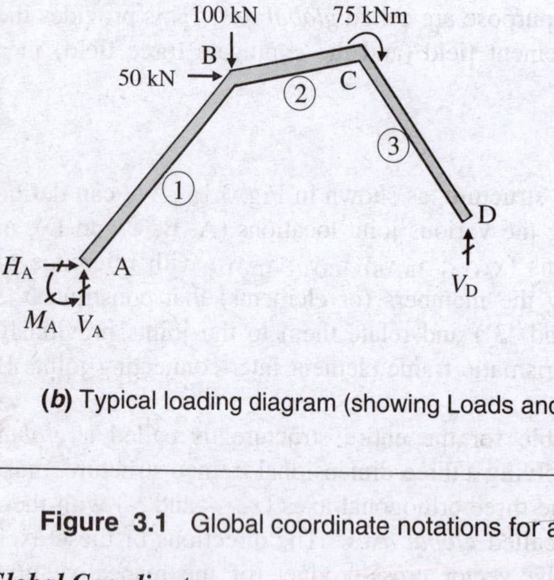

Force vector:

$$\{F\} = \begin{Bmatrix} F_A \\ \cdots \\ F_R \end{Bmatrix} = \begin{Bmatrix} F_1 = +50\text{kN} \\ F_2 = -100\text{kN} \\ F_3 = 0\text{ kNm} \\ F_4 = 0\text{ kN} \\ F_5 = 0\text{ kN} \\ F_6 = -75\text{kNm} \\ F_7 = 0\text{ kNm} \\ F_8 = 0\text{ kN} \\ \cdots\cdots \\ F_9 = V_D \\ F_{10} = H_A \\ F_{11} = V_A \\ F_{12} = M_A \end{Bmatrix}$$

(*b*) Typical loading diagram (showing Loads and Reactions)

Figure 3.1 Global coordinate notations for a typical plane frame (stiffness method)

Global Coordinates

In order to identify the displacements and conjugate forces at any node in the structure, we need to invoke the concept of *global coordinates*. In general, any node in space has six independent degrees of freedom, and the corresponding displacements in Cartesian coordinates can be expressed in terms of three translations (deflections), Δ_x, Δ_y and Δ_z, and three rotations, θ_x, θ_y and θ_z. We can conveniently number these degrees of freedom in some sequence, starting with any one node, using integers, ('1', '2', etc.), and thereby sequentially cover all the degrees of freedom at all the relevant nodes.

As mentioned earlier, every indexing number j refers to a global coordinate which identifies the location and direction of the scalar component D_j of the displacement vector, and also that of the conjugate force (including moment) vector component, F_j. This is denoted by means of an arrow, which may be straight (if representing a linear translation or force) or curved (if representing a rotation or moment), as illustrated in the plane frame example in Fig. 3.1a. Thus, for example, in Fig. 3.1a, with reference to the straight arrow marked '2' (acting at B and pointing upwards), the term D_2 refers to a upward deflection at B, while the term F_2 refers to an upward force (concentrated load), if any, acting at B. Similarly, with reference to the curved arrow marked '3' (acting at B in an anti-clockwise direction), the term D_3 refers to an anti-clockwise rotation of the rigid joint B, while the term F_3 refers to a concentrated moment (load), if any, acting anti-clockwise at the rigid joint B. The various displacements at all the nodes of the structure constitute the **displacement vector**, denoted as $\{D\}$, while all the conjugate forces constitute the **force vector**, denoted as $\{F\}$.

It is tacitly assumed that the origin of the global axes gets shifted to the node under consideration, for the purpose of defining the displacement and force vectors applicable at that node†.

Displacements: Active and Restrained Coordinates

It is necessary to make a distinction between global coordinates located at joints (where displacements can occur freely) and ones located at supports (where displacements are arrested). For convenience, while assigning numbers (integers) to the global coordinates, we can list the global coordinates corresponding to the *active* degrees of freedom at the beginning and the *restrained* coordinates at the end of the list. Thus, in the stiffness (displacement) approach, we can conveniently partition the displacement vector **D** as follows:

$$\mathbf{D} = \left[\frac{\mathbf{D_A}}{\mathbf{D_R}}\right] \tag{3.1}$$

where the subscripts, *A* and *R*, denote *active* and *restrained* conditions of displacements respectively. $\mathbf{D_A}$ and $\mathbf{D_R}$ may be called **active displacement vector** and **restrained displacement vector** respectively. In general, therefore, $\{D_R\} = \{0\}$, except in situations where there are support movements‡ whose magnitudes are known *a priori*. At any rate, $\mathbf{D_R}$ is known, while $\mathbf{D_A}$ is unknown.

This is illustrated in the plane frame example in Fig. 3.1a, where there are four nodes (A, B, C and D) with three possible degrees of freedom at each node, thus generating 12 global coordinates. The numbering of these coordinates is sequenced in such a manner that the active degrees of freedom at the rigid joints B and C are listed first. At the roller support

† The magnitude of the (displacement or force) vector is given by the length of the vector with reference to the node (to which it pertains) as origin; this is different from the common origin O (shown in Fig. 3.1a) that is used for defining the geometrical coordinates of the nodes in the global coordinate system.

‡ *Elastic supports* cannot be treated under this category, as the support movements are unknown and load-dependent. They need to be modelled using springs (which are structural elements). The ends of these springs can be treated as restrained supports.

location at D, there are two active degrees of freedom, while the third degree of freedom (vertical translation) is restrained. At the fixed support at A, all three degrees of freedom are restrained. Thus, of the 12 global coordinates, 8 correspond to active coordinates while the remaining 4 correspond to restrained coordinates, as indicated. The arrows representing the restrained global coordinates ('9', '10', '11' and '12' in Fig. 3.1a) are marked with a slash, to facilitate easy recognition. The degree of kinematic indeterminacy, n_k, is given by the number of active coordinates, and in this instance, $n_k = 8$. In the stiffness (displacement) method of analysis, the primary unknowns are the active displacements, the components of $\mathbf{D_A}$. It is after finding these unknown displacements, that we are able to find the other unknowns in the response: support reactions (global coordinates) and internal force resultants (local coordinates).

It may be noted that the identification of restrained degrees of freedom is particularly useful in the stiffness (displacement) method of analysis, as it enables the identification and determination of support reactions that are conjugate with these coordinate locations. Furthermore, they facilitate handling of problems involving support displacements, whereby $\mathbf{D_R}$ is a non-zero vector. This, however, (as explained later) is not facilitated in the flexibility (force) method of analysis. We need to invoke an alternative approach to find support reactions and handle support displacements.

Forces: Actions and Reactions

The global coordinates corresponding to the active degrees of freedom also serve as potential *action* (concentrated load) locations, while those coordinates corresponding to the restrained coordinate locations serve as support *reaction* locations. Thus, the subscripts, *A* and *R*, can be conveniently interpreted, in the context of forces, as *action* and *reaction* respectively.

Conjugate with the *active displacement vector*, $\{D_A\}$, we can define an **action vector** or **load vector**, which may be denoted as $\{F_A\}$. Similarly, conjugate with the *restrained displacement vector*, $\{D_R\}$, we can define a **reaction vector**, which may be denoted as $\{F_R\}$. In general, $\mathbf{F_A}$ is known, while $\mathbf{F_R}$ is unknown.

Thus, we can partition the force vector $\mathbf{F}$ in the manner analogous to the partitioning of the displacement vector $\mathbf{D}$ (Eq. 3.1) as follows:

$$\mathbf{F} = \left[\begin{array}{c}\mathbf{F_A} \\ \hline \mathbf{F_R}\end{array}\right] \tag{3.2}$$

We will see later (Section 3.5) that, by the stiffness method, the force vector $\mathbf{F}$ is linearly related to the displacement vector $\mathbf{D}$ by means of a *structure stiffness matrix* $\mathbf{k}$, which can also be partitioned (separating active and reaction coordinates). We can then find the unknown displacement components of $\mathbf{D_A}$ by solving the set of stiffness equations (linear simultaneous equations), $\mathbf{F_A} = \mathbf{k_{AA}D_A} + \mathbf{k_{AR}D_R}$, involving the known load vector $\mathbf{F_A}$ and known restrained displacement vector $\mathbf{D_R}$. This essentially invokes the *equilibrium conditions* required to find the unknown displacements. After finding $\mathbf{D_A}$, we can easily find the unknown reaction vector $\mathbf{F_R}$ by directly applying another set of stiffness relations, $\mathbf{F_R} = \mathbf{k}_{RA}\mathbf{D}_A + \mathbf{k}_{RR}\mathbf{D}_R$.

Consider the plane frame example in Fig. 3.1b, where a typical set of loads is shown. Corresponding to this loading, out of the 8 potential load locations, only 3 have non-zero actions, as shown. There are 4 possible reaction components, corresponding to the coordinates, '9', '10', '11' and '12' (in Fig. 3.1a), representing V_D, H_A, V_A and M_A respectively, as shown with slashed arrows in Fig. 3.1b. Thus, in this example, $\mathbf{F_A}$ (like $\mathbf{D_A}$) is a vector with a dimension equal to 8, while $\mathbf{F_R}$ (like $\mathbf{D_R}$) is a vector with a dimension equal to 4.

Flexibility Method: Active and Redundant Global Coordinates

From the perspective of the flexibility (force method), the plane frame structure in Fig. 3.1 is 'externally over-rigid' to the first degree, and hence, is *statically indeterminate* ($n_s = 1$). According to this method of analysis, we need to identify a 'redundant' and by releasing the constraint associated with it, we obtain a *statically determinate 'primary' structure*, which becomes our basic model for carrying out structural analysis.

Thus, in order to designate the redundant forces and apply appropriate compatibility equations (to find the redundant forces), we need to identify global coordinates associated with the redundant locations. We shall refer to such coordinates as **redundant coordinates**. The number of redundant global coordinates will be equal to the degree of static indeterminacy of the structure, n_s. It is important to note that many alternative choices of redundant coordinates are possible for any given statically indeterminate structure.

We may recall that, unlike the stiffness method, the basic unknowns in the flexibility method are *forces* (and not *displacements*). The unknown forces, acting on a statically determinate structure, include the redundant forces $\mathbf{F_X}$ acting at n_s global coordinates and the internal force resultants that are described in terms of local coordinates (refer Section 3.2.2). In global coordinates, the force vector $\mathbf{F}$ is of primary importance here, and it should be possible for the components of $\mathbf{F}$ to assume any arbitrary set of values, without violating equilibrium. For this reason, we cannot use the same $\mathbf{F}$ (defined by Eq. 3.2) we had adopted as a derived quantity in the stiffness formulation, because it contains the reaction vector $\mathbf{F_R}$ which is linearly dependent on the load vector $\mathbf{F_A}$. It is appropriate to replace $\mathbf{F_R}$ with the **redundant vector** $\mathbf{F_X}$ in the flexibility formulation, because, as we have seen earlier (refer Section 1.8), any choice of redundants is statically admissible. Accordingly, the force vector in the flexibility method may be defined for a statically indeterminate structure, in the global coordinate system, as comprising the load vector $\mathbf{F_A}$ and the redundant vector $\mathbf{F_X}$:

$$\mathbf{F} = \begin{bmatrix} \mathbf{F_A} \\ \hline \mathbf{F_X} \end{bmatrix} \tag{3.3}$$

Similarly, the displacement vector, $\mathbf{D}$, which is conjugate with $\mathbf{F}$, is defined in the flexibility method (in lieu of Eq. 3.1), as comprising the *active displacement vector* $\mathbf{D_A}$ and the *redundant displacement vector* $\mathbf{D_X}$:

$$D = \begin{bmatrix} D_A \\ \hline D_X \end{bmatrix} \quad (3.4)$$

The dimension of the force and displacement vectors above is given by the sum of the active degrees of freedom in the statically determinate structure and the degree of static indeterminacy, $n + n_s$. If the structure happens to be statically determinate, obviously, we should not include F_X in the force vector and D_X in the displacement vector. For the plane frame example of Fig. 3.1, with $n_s = 1$, the vertical support reaction at D, corresponding to the global coordinate '9', may be selected as the redundant force X_1, as shown in Fig. 3.2. The primary structure here is a cantilever frame, and the appropriate compatibility conditione to be applied to solve for X_1 is given by $\{D_X\} = D_9 = 0$.

Force vector: $\{F\} = \left\{ \begin{array}{c} F_A \\ \cdots \\ F_X \end{array} \right\} = \left\{ \begin{array}{l} F_1 = +50\text{kN} \\ F_2 = -100\text{kN} \\ F_3 = 0 \text{ kNm} \\ F_4 = 0 \text{ kN} \\ F_5 = 0 \text{ kN} \\ F_6 = -75\text{kNm} \\ F_7 = 0 \text{ kNm} \\ F_8 = 0 \text{ kNm} \\ \cdots\cdots \\ F_9 = X_1 \text{(kN)} \end{array} \right.$

Compatibility condition $\{D_X\} = D_9 = 0$

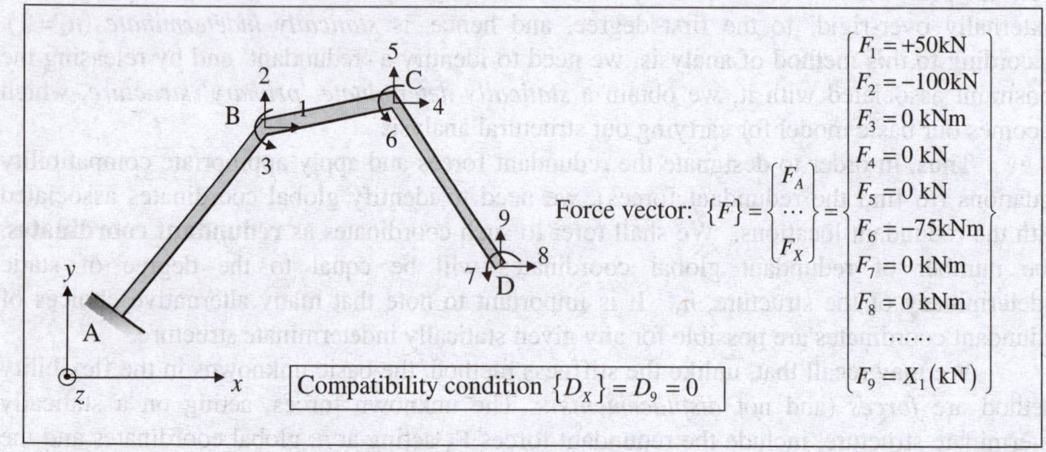

Figure 3.2 Global coordinate notation for the plane frame in Fig. 3.1 (flexibility method)

We will see later (Section 3.6) that, by the flexibility method, the displacement vector **D** is linearly related to the force vector **F** by means of a *structure flexibility matrix* **f**, which can also be partitioned (separating active and redundant coordinates). We can then find the unknown redundant force components of F_X by solving the set of flexibility equations (linear simultaneous equations), $D_X = f_{XA} F_A + f_{XX} F_X$, involving the known load vector F_A and known redundant displacement vector D_X. This essentially invokes the *compatibility conditions* required to find the redundants. After finding F_X, if required, we can easily find the unknown displacement vector D_A by directly applying another set of flexibility relations, $D_A = f_{AA} F_A + f_{AX} F_X$.

The questions that remain unaddressed, with regard to the flexibility method are twofold. Firstly, *having eliminated F_R from the force vector, how do we find support reactions?* Secondly, *having eliminated D_R from the displacement vector, how do we account for indirect loading caused by support movements?*

A convenient way of addressing both of the above issues is by inserting appropriate rigid links (translational or rotational springs with infinite stiffness) at the support locations. The support reaction manifests as the internal force in the rigid link. In this way, we shift support reactions from the 'external' force vector to the 'internal' force vector; i.e., from global coordinates to local coordinates (described in Section 3.2.2). Using the same device, we can impart support movements (translations and rotations), if any, in the form of equivalent initial deformations (similar to temperature changes) in the rigid links. Thus, we shift the restrained global coordinates to appropriate local coordinates in the flexibility (force) method of analysis. This is demonstrated later through examples.

In the flexibility method of matrix analysis, we can replace supports by appropriate rigid links, and thereby find support reactions as the internal forces in these links. Similarly, we can assign support displacements, if required, by way of initial deformations in the rigid links.

3.2.2 Local Coordinates

Whereas *global* axes and coordinates help us identify to forces and displacements at the level of the entire structure, *local* axes and coordinates help us in identifying forces and displacements at the element level. Corresponding to every node in the overall structure having global coordinates, we also need to define local coordinates. Corresponding to every node, there will be as many sets of local coordinates as there are elements converging at the node under consideration. If, for some reason, a node is chosen at an intermediate location of a particular element in the overall structure, then it is necessary to divide the element under consideration into two elements at the node, resulting in two sets of local coordinates corresponding to the global coordinates at this node.

The number of local coordinates to be considered, as in the case of global coordinates, depends on whether we are adopting the stiffness method or the flexibility method. In the stiffness method, we give primacy to displacements, whereas in the flexibility method, we give primacy to forces.

Stiffness Method

With reference to the rigid-jointed plane frame example in Fig. 3.1, there are three frame elements, each of which can have independent frames of reference in a *local coordinate* system, as shown in Fig. 3.3. Typically, the local axes for each element are aligned in the direction of the element, with the left end of the element as the origin. In order to distinguish the local Cartesian coordinates from the global Cartesian coordinates, an asterisk is marked while denoting the *local axes* (i.e., as x^*, y^* and z^*), with the x^* axis always pointing along the longitudinal direction of the element, from the "start node" (usually at the lower left end) to the "end node" (usually at the upper right end), making an angle of θ_x with reference to the global x-axis, as shown in Fig. 3.3.

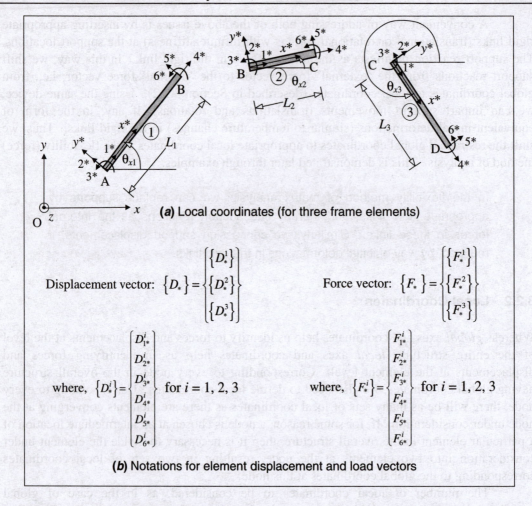

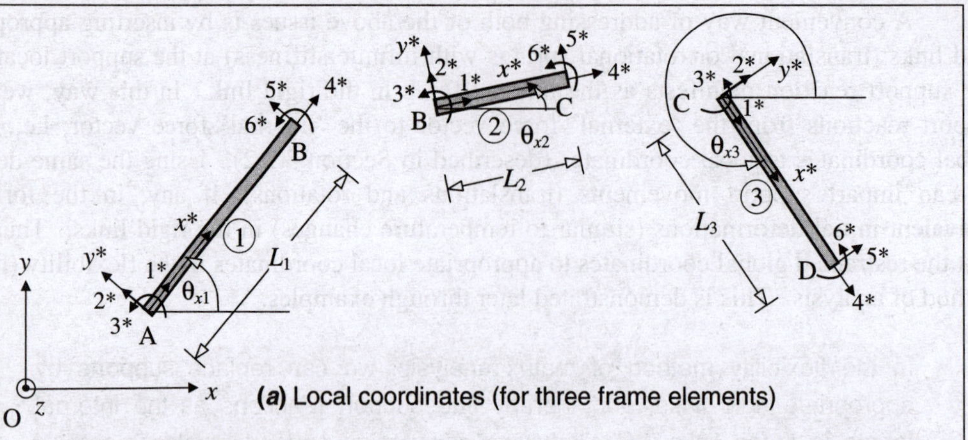

(*a*) Local coordinates (for three frame elements)

Displacement vector: $\{D_*\} = \begin{Bmatrix} \{D_*^1\} \\ \{D_*^2\} \\ \{D_*^3\} \end{Bmatrix}$ Force vector: $\{F_*\} = \begin{Bmatrix} \{F_*^1\} \\ \{F_*^2\} \\ \{F_*^3\} \end{Bmatrix}$

where, $\{D_*^i\} = \begin{Bmatrix} D_{1*}^i \\ D_{2*}^i \\ D_{3*}^i \\ D_{4*}^i \\ D_{5*}^i \\ D_{6*}^i \end{Bmatrix}$ for $i = 1, 2, 3$ where, $\{F_*^i\} = \begin{Bmatrix} F_{1*}^i \\ F_{2*}^i \\ F_{3*}^i \\ F_{4*}^i \\ F_{5*}^i \\ F_{6*}^i \end{Bmatrix}$ for $i = 1, 2, 3$

(*b*) Notations for element displacement and load vectors

Figure 3.3 Local coordinate notations for the plane frame in Fig. 3.1 (stiffness method)

Every element will have, in general, at either end of it, an independent set of local coordinates that correspond to the degrees of freedom of the element. We can conveniently number these degrees of freedom in some sequence, starting with the left end and using integers with an asterisk attached ('1*', '2*', etc.), and thereby sequentially cover all the degrees of freedom at all the relevant nodes. For each of the plane frame elements in Fig. 3.3, there are six degrees of freedom (three at either end), as indicated. It is good practice to follow the same sequence of numbering (and same direction of axes) at the element level as the one adopted at the structure level. For a typical plane frame element (i^{th} element), according to the stiffness (displacement) method, the **element displacement vector**, $\{D_*^i\}$, and **element force vector**, $\{F_*^i\}$, take the following form:

$$\mathbf{D}_*^i = \begin{bmatrix} D_{1*}^i \\ D_{2*}^i \\ D_{3*}^i \\ D_{4*}^i \\ D_{5*}^i \\ D_{6*}^i \end{bmatrix} \quad \text{and} \quad \mathbf{F}_*^i = \begin{bmatrix} F_{1*}^i \\ F_{2*}^i \\ F_{3*}^i \\ F_{4*}^i \\ F_{5*}^i \\ F_{6*}^i \end{bmatrix} \tag{3.5}$$

In the above example, D_{1*}^i and D_{4*}^i refer to **axial translations** (in the longitudinal direction) at the left end and right end respectively, with $e_i = D_{4*}^i - D_{1*}^i$ denoting the **axial elongation** in the ith element. D_{2*}^i and D_{5*}^i refer to **deflections** (in the transverse direction, as shown), while D_{3*}^i and D_{6*}^i refer to **rotations** (anti-clockwise) at the left end and right end respectively. Similarly, F_{1*}^i and F_{4*}^i refer to **axial forces**; F_{2*}^i and F_{5*}^i refer to **shear forces**; F_{3*}^i and F_{6*}^i refer to **bending moments** at the left end and right end respectively of the ith element, with directions as indicated.

It is tacitly assumed that the origin of the local axes gets shifted from the "start node" to the "end node", for the purpose of defining the displacement and force vectors applicable at the end node (local coordinates labelled as 4*, 5* and 6* in Fig. 3.3a).

It is also possible to express the element displacement and force vectors at the start and end nodes, with reference to the global axes (*x*-, *y*-, *z*- axes, with directions common to all elements), instead of the various local axes (*x**-, *y**-, *z**- axes, with directions that depend on the orientation of the element), by means of a simple *coordinate transformation*. This is discussed in Section 3.4.

Flexibility Method and Reduced Stiffness Method

It is interesting to note that although the components of the displacement vector, $\mathbf{D}_*^i$, are, in general[†], independent of one another (as, by definition, they correspond to independent degrees of freedom), the elements of the force vector, $\mathbf{F}_*^i$, are not all independent. There are three independent equations of equilibrium that need to be satisfied by the free-body of the plane frame element. Hence, in the flexibility approach, we need to restrict the number of local coordinates.

Referring to the various skeletal elements shown in Fig. 1.1, we observe that the most general element (of which all others are special cases) is the *space frame element* (Fig. 1.1f), having 6 degrees of freedom (3 translational plus 3 rotational) at either end, whereby, according to the stiffness method, $\mathbf{D}_*^i$ (and hence, $\mathbf{F}_*^i$ also) has a dimension $q = 12$. As there are 6 independent equations of equilibrium, inter-relating the various member end forces, it follows that, according to the flexibility method, the dimension of $\mathbf{F}_*^i$ (and hence, $\mathbf{D}_*^i$) is

---

[†] Some degrees of freedom, corresponding to support locations, may be restrained. The displacements at these coordinates will be non-zero (have arbitrary values) if there are support movements.

given by $\tilde{q} = 6$. We can also detect this by inspecting the nature of the **element stiffness matrix**, $\mathbf{k}_*^i$, and its inverse, the **element flexibility matrix**, $\mathbf{f}_*^i$, inter-relating $\mathbf{F}_*^i$ and $\mathbf{D}_*^i$ (described later in Sections 3.5 and 3.6). We will see that, although the stiffness matrix for the space frame element has an order 12×12, there is significant linear dependence among the column vectors (and row vectors) of the matrix $\mathbf{k}_*^i$, revealing that this matrix is not invertible; i.e., a flexibility matrix $\left[f_*^i\right]_{12\times12} = \left[k_*^i\right]_{12\times12}^{-1}$ cannot exist. The size of the flexibility matrix is given by the *rank* of the 12×12 stiffness matrix, which turns out to be equal to $\tilde{q} = 6$. In other words, the appropriate order of the flexibility matrix $\mathbf{f}_*^i$ for a space frame element is 6×6. The number of linearly independent vectors in the flexibility matrix also turns out to be a measure of the number of independent equations of equilibrium of forces. Also, interestingly, as the flexibility matrix $\left[f_*^i\right]_{6\times6}$ has a full rank, equal to 6, it is invertible, and the inverse of this matrix, $\left[\tilde{k}_*^i\right]_{6\times6} = \left[f_*^i\right]_{6\times6}^{-1}$ does exist, and may be referred to as the **reduced element stiffness matrix**, denoted as $\tilde{\mathbf{k}}_*^i$, having reduced degrees of freedom, $\tilde{q} = q/2 = 6$. Incidentally, this reduced stiffness matrix $\tilde{\mathbf{k}}_*^i$ can alternatively be obtained from the full stiffness matrix $\mathbf{k}_*^i$ through row reduction elimination techniques. It is possible to analyse space frames by the stiffness method using the reduced stiffness matrix, although the versatile and generalised nature of the stiffness formulation is diminished.

By eliminating irrelevant degrees of freedom from the general space frame element, we can arrive at the dimension q of other element types (shown in Fig. 1.1), applicable in the conventional stiffness method. Thus, for example, by eliminating the displacements that lie outside the local x^*-y^* plane (translations in the local z^*-direction and rotations with respect to the local x^*- and y^*-axes directions), we have 3 degrees of freedom at each end of a *plane frame element* (Fig. 1.1d). Thus, according to the regular stiffness method, $\mathbf{D}_*^i$ (and hence, $\mathbf{F}_*^i$ also) has a dimension $q = 6$. However, there are only 3 independent equations of equilibrium available for the element, whereby according to the flexibility method, the dimension of $\mathbf{F}_*^i$ (and hence, $\mathbf{D}_*^i$) for a plane frame element is given by $\tilde{q} = q/2 = 3$. This also happens to be the rank of the 6×6 stiffness matrix. By inverting the 3×3 flexibility matrix, or alternatively, by applying row reduction elimination on the 6×6 stiffness matrix, we can generate the reduced 3×3 stiffness matrix $\tilde{\mathbf{k}}_*^i$ for the plane frame element. In other words, by the flexibility method, we have to necessarily assign a dimension $\tilde{q} = 3$ to the element force and displacement vectors of a plane frame element; however, for the same element, we need to assign $q = 6$ in the regular stiffness formulation, or we may choose to assign $\tilde{q} = 3$ in the reduced stiffness formulation.

Further simplifications in both stiffness and flexibility formulations are possible, if we ignore the effect of axial deformations, as is often done in structural analysis of frames; but this simplification is an approximation. Such simplifications are not adopted for the purpose of developing general-purpose software for structural analysis, which are invariably based on the regular stiffness method (and not the reduced stiffness method or the flexibility method). We shall demonstrate later how these simplifications are useful for the limited purpose of handling problems with limited indeterminacy by using manual methods.

Combined Element Displacement and Force Vectors

It is sometimes convenient to assemble the element-level forces and displacements of all the elements of the structure in a consolidated manner. Thus, if *m* elements are involved, we can define the **combined element displacement vector**, $\{D_*\}$, and **combined element force vector**, $\{F_*\}$, as follows:

$$\mathbf{D}_* = \begin{bmatrix} \mathbf{D}_*^1 \\ \mathbf{D}_*^2 \\ \vdots \\ \vdots \\ \mathbf{D}_*^m \end{bmatrix} \quad \text{and} \quad \mathbf{F}_* = \begin{bmatrix} \mathbf{F}_*^1 \\ \mathbf{F}_*^2 \\ \vdots \\ \vdots \\ \mathbf{F}_*^m \end{bmatrix} \tag{3.6}$$

These vectors should include components pertaining to all possible elements in the structure, including spring supports, if any. In the flexibility method, we can also include the rigid links (to simulate the support reactions) as members. Every support condition, whether rigid or elastic, can be conveniently modelled using a spring (translational or rotational) with known spring stiffness or flexibility (infinite stiffness in the case of a rigid link).

3.3 TRANSFORMATION MATRICES

Having defined a single set of *global coordinates* at the level of the entire structure, and multiple sets of *local coordinates* at the various element levels, it is now necessary to describe the inter-relationships between these two coordinate systems. Various types of *coordinate transformations* are possible to facilitate this.

3.3.1 Element Transformation Matrix (Stiffness Method)

The simplest transformation is one in which the two Cartesian coordinate systems (local axes and global axes) are inter-related through an **element transformation matrix**. This enables us to transform the element displacement and force vectors from the local coordinate system (x^*-, y^*-, z^*- axes, different for every element) to the common global coordinate system (x-, y-, z- axes) and vice versa. Let us use the symbols, $\{D^i\}$ and $\{F^i\}$, to denote respectively, the displacement and force vectors of the element *i* in the global axes system. Let the *element transformation matrix* $\mathbf{T}^\mathbf{i}$ denote the matrix that transforms $\{D^i\}$ and $\{F^i\}$ to $\{D_*^i\}$ and $\{F_*^i\}$ respectively, i.e., from the global axes system to the local axes system, for the element *i*:

$$\left\{D_*^i\right\}_{q\times1} = \left[T^i\right]_{q\times q}\left\{D^i\right\}_{q\times1} \tag{3.7a}$$

$$\left\{F_*^i\right\}_{q\times1} = \left[T^i\right]_{q\times q}\left\{F^i\right\}_{q\times1} \tag{3.7b}$$

The transformation matrix $\mathbf{T}^i$, as defined above, is a square matrix of order $q \times q$, where q is the number of degrees of freedom of the element (dimension of $\mathbf{D}_*^i$ and $\mathbf{F}_*^i$, applicable in the stiffness method). For example, let us consider a typical element '2' in the plane frame example of Fig. 3.1, as depicted in Fig. 3.4. The member-end displacements and forces for this element, expressed in the local axes system, are shown in Fig. 3.4a (as shown earlier in Fig. 3.3); these displacements and forces, expressed in the global axes system, are shown in Fig. 3.4b.

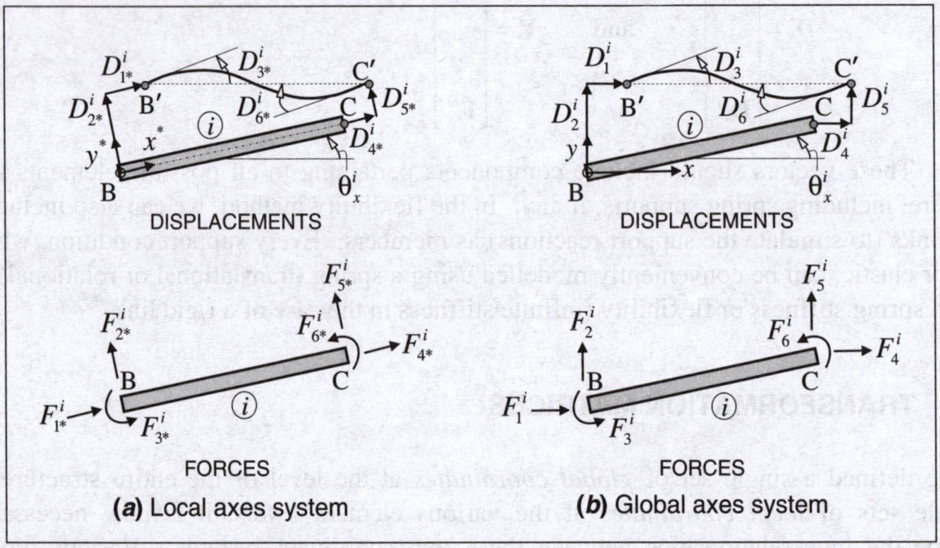

Figure 3.4 Member-end displacements and forces in a typical frame element

Let us first derive the components of the transformation matrix $\mathbf{T}^2$ pertaining to the coordinates at the "start node". This transformation involves a simple (anti-clockwise) rotation of the coordinate axes in the *x-y* plane by an angle θ_x^2, without shifting the origin. Such a transformation clearly does not affect the member-end rotation ($D_{3*}^2 = D_3^2$) and moment ($F_{3*}^2 = F_3^2$). The translation (or force) along the local x^* axis is obtained by considering the projections of the corresponding vectors (along global *x*- and *y*- axes) on the x^* axis. Thus, with regard to the components, D_{1*}^2 and F_{1*}^2, we can easily† show that:

† Our training in *statics* is generally strong, compared to our feel for *kinematics*. For this reason, it is perhaps more easy to visualise the components of the force vector, rather than the displacement vector.

$$D_{1*}^2 = D_1^2 \cos\theta_x^2 + D_2^2 \sin\theta_x^2 \quad \text{and} \quad F_{1*}^2 = F_1^2 \cos\theta_x^2 + F_2^2 \sin\theta_x^2$$

Similarly, the translation (or force) along the local y^* axis is obtained by considering the projections of the corresponding vectors (along global x- and y- axes) on the y^* axis. Thus, with regard to the components, D_{2*}^2 and F_{2*}^2, we can show that:

$$D_{2*}^2 = -D_1^2 \sin\theta_x^2 + D_2^2 \cos\theta_x^2 \quad \text{and} \quad F_{2*}^2 = -F_1^2 \sin\theta_x^2 + F_2^2 \cos\theta_x^2$$

To summarise,

$$\begin{bmatrix} D_{1*}^2 \\ D_{2*}^2 \\ D_{3*}^2 \end{bmatrix} = \begin{bmatrix} \cos\theta_x^2 & \sin\theta_x^2 & 0 \\ -\sin\theta_x^2 & \cos\theta_x^2 & 0 \\ 0 & 0 & 1 \end{bmatrix} \begin{bmatrix} D_1^2 \\ D_2^2 \\ D_3^2 \end{bmatrix} \quad \text{and} \quad \begin{bmatrix} F_{1*}^2 \\ F_{2*}^2 \\ F_{3*}^2 \end{bmatrix} = \begin{bmatrix} \cos\theta_x^2 & \sin\theta_x^2 & 0 \\ -\sin\theta_x^2 & \cos\theta_x^2 & 0 \\ 0 & 0 & 1 \end{bmatrix} \begin{bmatrix} F_1^2 \\ F_2^2 \\ F_3^2 \end{bmatrix} \quad \text{(a)}$$

Similarly, shifting the origin to the "end node" of element '2', we can deduce:

$$\begin{bmatrix} D_{4*}^2 \\ D_{5*}^2 \\ D_{6*}^2 \end{bmatrix} = \begin{bmatrix} \cos\theta_x^2 & \sin\theta_x^2 & 0 \\ -\sin\theta_x^2 & \cos\theta_x^2 & 0 \\ 0 & 0 & 1 \end{bmatrix} \begin{bmatrix} D_4^2 \\ D_5^2 \\ D_6^2 \end{bmatrix} \quad \text{and} \quad \begin{bmatrix} F_{4*}^2 \\ F_{5*}^2 \\ F_{6*}^2 \end{bmatrix} = \begin{bmatrix} \cos\theta_x^2 & \sin\theta_x^2 & 0 \\ -\sin\theta_x^2 & \cos\theta_x^2 & 0 \\ 0 & 0 & 1 \end{bmatrix} \begin{bmatrix} F_4^2 \\ F_5^2 \\ F_6^2 \end{bmatrix} \quad \text{(b)}$$

In this manner, we can generate the 6 × 6 transformation matrix $\mathbf{T}^i$ for a plane frame element, combining the contributions from the "start node" and "end node", as a block diagonal matrix. For convenience in understanding, at the top of each column vector, the 'cause' of the vector is labelled. Unit displacements, expressed in the global axes system, are indicated in this row above the matrix. We observe that the column vectors are not only linearly independent, but are also normalised (magnitude equal to 1). This transformation matrix is an orthogonal matrix, consisting of a set of orthonormal eigenvectors associated with the coordinate transformation:

$$\begin{matrix} & (D_1^i=1) & (D_2^i=1) & (D_3^i=1) & (D_4^i=1) & (D_5^i=1) & (D_6^i=1) \end{matrix}$$

$$\mathbf{T}^i = \begin{bmatrix} \cos\theta_x^i & \sin\theta_x^i & 0 & 0 & 0 & 0 \\ -\sin\theta_x^i & \cos\theta_x^i & 0 & 0 & 0 & 0 \\ 0 & 0 & 1 & 0 & 0 & 0 \\ 0 & 0 & 0 & \cos\theta_x^i & \sin\theta_x^i & 0 \\ 0 & 0 & 0 & -\sin\theta_x^i & \cos\theta_x^i & 0 \\ 0 & 0 & 0 & 0 & 0 & 1 \end{bmatrix} \quad (3.8)$$

As discussed earlier, the element transformation matrix, which is a square, symmetric matrix, can be used to transform both displacements and forces at the ends of a plane frame element from the global axes system to the local axes system (refer Eqns 3.7a and b). It is also possible to do the inverse transformation, i.e., from the local axes system to the global axes system, using the inverse of the transformation matrix. As $\mathbf{T}^i$ is an orthogonal matrix, its inverse $(\mathbf{T}^i)^{-1}$ is equal to its transpose $(\mathbf{T}^i)^{\mathrm{T}}$, which is easy to deduce from $\mathbf{T}^i$. Thus, we have:

$$\{D^i\}_{q\times 1} = [T^i]^T_{q\times q}\{D^i_*\}_{q\times 1} \tag{3.9a}$$

$$\{F^i\}_{q\times 1} = [T^i]^T_{q\times q}\{F^i_*\}_{q\times 1} \tag{3.9b}$$

The transformations described by Eqns 3.8 and 3.9, however, do not relate the *local coordinates* to the *global coordinates*. Eqns 3.9a and 3.9b enable us to shift from local coordinates (expressed in the local axes system) to an equivalent description in the global axes system. It is now possible to work with all the element-level displacements (and forces) as well as the structure-level displacements (and forces) in a common global-axes framework. However, in order to proceed further, we need to link the element-level displacements $\mathbf{D^i}$ to the structure-level displacements $\mathbf{D}$, by invoking appropriate *compatibility conditions*. This linkage is simple, as it only requires matching the number codes of the member-end displacements with the corresponding global coordinates (described later). Global coordinates at nodes other than the ones connected to the two ends of a member are not involved in this linkage, as far as displacements are concerned. However, loads applied on the structure at any global coordinate location can indeed affect the member-end forces in any element. Thus, unlike displacements, there is no simple relationship between element-level forces $\mathbf{F^i}$ and the structure-level forces $\mathbf{F}$. We need to apply appropriate *equilibrium equations* at the various nodes, for this purpose. In the stiffness method of analysis, these are the governing equations, which are expressed in terms of the structure stiffness equation $\mathbf{F = kD}$, involving the global coordinates. After finding the structure displacements $\mathbf{D}$, we can relate these to the element-level displacements $\mathbf{D^i}$, and applying Eq. 3.7a, we can find all the member-end displacements $\mathbf{D^i_*}$ expressed in the respective local coordinates. Using the *member stiffness* relations, $\mathbf{F^i_* = k^i_* D^i_*}$, we can then find the desired member-end forces. As we shall see later, we can derive the structure stiffness matrix $\mathbf{k}$ from the various element stiffness matrices and the corresponding transformation matrices.

The **element transformation matrix** $\mathbf{T^i}$ is a square, orthogonal matrix, which enables member-end displacements and forces, expressed in a global axes framework, to be transformed to the local axes system ($\mathbf{D^i_* = T^i D^i}$ and $\mathbf{F^i_* = T^i F^i}$). Transformation from the local axes system to the global axes system is enabled by the transpose of this matrix ($\mathbf{D^i = T^{i^T} D^i_*}$ and $\mathbf{F^i = T^{i^T} F^i_*}$).

3.3.2 Transformations from Global to Local Coordinates

It is possible to directly relate the member-end displacements $\mathbf{D^i_*}$ to the global displacements $\mathbf{D}$ (including displacements at all structure nodes) by means of an **element displacement**

transformation matrix, $\mathbf{T}_D^i$, by applying appropriate *compatibility conditions*. We can also store the displacement transformation matrices for all the elements in a single vector, $\mathbf{T}_D$. This makes the formulation simpler and easier to understand, although it may not be suitable for digital computation, as it requires more data storage (storing a lot of zero elements in a large transformation matrix). This is discussed in Section 3.3.3.

In the flexibility (force) method of analysis, we give primacy to forces, and not displacements. Here, instead of the *element displacement transformation matrix*, we use the **element force transformation matrix**, $\mathbf{T}_F^i$, by applying appropriate *equilibrium conditions*. We can store the force transformation matrices for all the elements in a single vector, $\mathbf{T}_F$. This is discussed in Section 3.3.3.

These transformations enable global coordinates to be converted to local coordinates, as shown schematically in Fig. 3.5. The transformation matrices $\mathbf{T}_D^i$ and $\mathbf{T}_F^i$ are, in general, rectangular matrices, unlike the $\mathbf{T}^i$ matrix described earlier, because they involve all relevant global coordinates of the structure (not just the global coordinates located at the two ends of the element *i*).

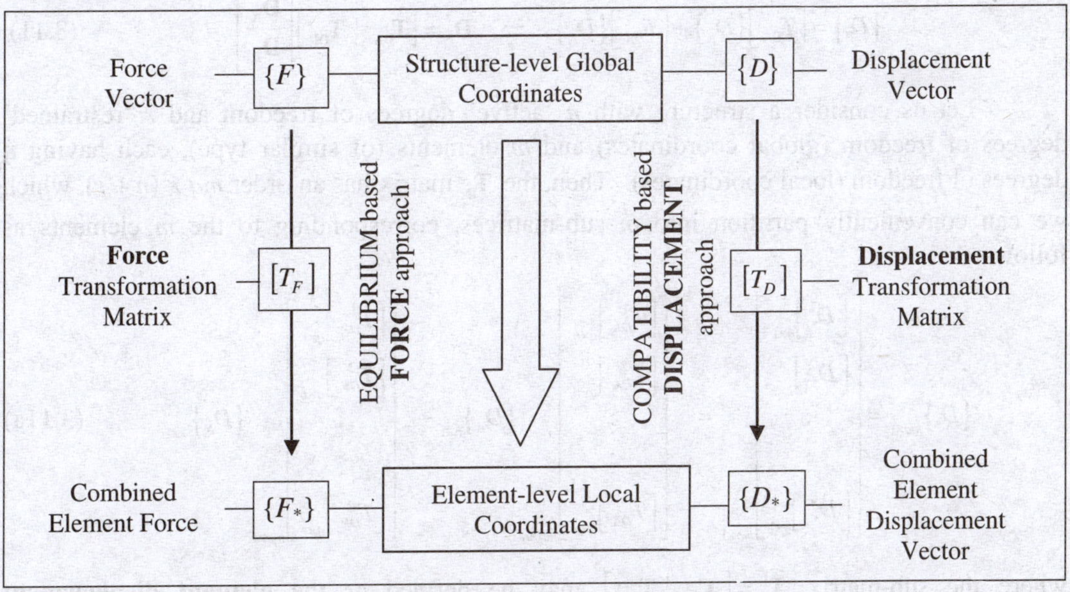

Figure 3.5 Transformations from global to local coordinates

These transformations are essentially *linear transformations*, which implicitly invoke the Principle of Superposition. Using this concept, we can understand the physical meaning underlying these transformations, in terms of either the response of a unit displacement in a kinematically determinate structure (stiffness method) or that of a unit force in a statically determinate structure (flexibility method).

3.3.3 Displacement Transformation Matrix (Stiffness Method)

In the stiffness method, the *displacement transformation matrix* enables a transformation of the structure-level displacements (nodal displacements at the *global* coordinates) to element-level displacements (deformations at the *local* coordinates) using compatibility conditions. Let us denote this matrix as $\mathbf{T_D}$. This implies that the *combined element displacement vector*, $\{D_*\}$, is obtainable by pre-multiplying the *displacement vector*, $\{D\}$, with the **displacement transformation matrix** $[T_D]$, as indicated in Fig. 3.5.

$$\mathbf{D_* = T_D D} \tag{3.10}$$

$\mathbf{T_D}$ may be conveniently partitioned into two sub-matrices, comprising the **active displacement transformation matrix** $[T_{DA}]$ (corresponding to the *active* degrees of freedom) and the **restrained displacement transformation matrix** $[T_{DR}]$ (corresponding to the restrained degrees of freedom):

$$\{D_*\} = [T_{DA}]\{D_A\} + [T_{DR}]\{D_R\} \quad \Rightarrow \quad \mathbf{D_*} = [\mathbf{T_{DA}} \mid \mathbf{T_{DR}}]\left[\frac{\mathbf{D_A}}{\mathbf{D_R}}\right] \tag{3.11}$$

Let us consider a structure with n 'active' degrees of freedom and r 'restrained' degrees of freedom (global coordinates) and m elements (of similar type), each having q degrees of freedom (local coordinates). Then, the $\mathbf{T_D}$ matrix has an order $mq \times (n + r)$, which we can conveniently partition into m sub-matrices, corresponding to the m elements as follows:

$$\{D_*\}_{mq\times1} = \begin{bmatrix} [D_*^1]_{q\times1} \\ [D_*^2]_{q\times1} \\ \vdots \\ \vdots \\ [D_*^m]_{q\times1} \end{bmatrix}_{mq\times1} = \begin{bmatrix} [T_{DA}^1]_{q\times n} \\ [T_{DA}^2]_{q\times n} \\ \vdots \\ \vdots \\ [T_{DA}^m]_{q\times n} \end{bmatrix}_{mq\times n} \{D_A\}_{n\times1} + \begin{bmatrix} [T_{DR}^1]_{q\times r} \\ [T_{DR}^2]_{q\times r} \\ \vdots \\ \vdots \\ [T_{DR}^m]_{q\times r} \end{bmatrix}_{mq\times r} \{D_R\}_{r\times1} \tag{3.11a}$$

where the sub-matrix $\mathbf{T_D^i} = [\mathbf{T_{DA}^i} \mid \mathbf{T_{DR}^i}]$ may be defined as the **element displacement transformation matrix** (for the ith *element*), which enables a transformation of the *displacement vector*, $\{D\}$, to the ith *element displacement vector*, $\{D_*^i\}$, as follows:

$$\mathbf{D_*^i} = \mathbf{T_D^i D} = [\mathbf{T_{DA}^i} \mid \mathbf{T_{DR}^i}]\left[\frac{\mathbf{D_A}}{\mathbf{D_R}}\right] \tag{3.12}$$

Defining the **displacement transfer coefficient**, $T^i_{D-l^*,j}$ as the displacement at the local coordinate l^* in the element i due to a unit displacement, $D_j = 1$, with all other global coordinates restrained ($D_{k\neq j} = 0$), Eq. 3.10 may be expanded in terms of the coefficients, $T^i_{D-l^*,j}$, with an additional subscript, A or R, depending on whether D_j relates to an *active* or *restrained* global coordinate, as follows:

$$\begin{bmatrix} D^i_{1*} \\ D^i_{2*} \\ \cdot \\ \cdot \\ D^i_{q*} \end{bmatrix} = \begin{bmatrix} T^i_{DA-1*1} & T^i_{DA-1*2} & \cdot\ \cdot & T^i_{DA-1*n} \\ T^i_{DA-2*1} & T^i_{DA-2*2} & \cdot\ \cdot & T^i_{DA-2*n} \\ \cdot & \cdot & \cdot\ \cdot & \cdot \\ \cdot & \cdot & \cdot\ \cdot & \cdot \\ T^i_{DA-q*1} & T^i_{DA-q*2} & \cdot\ \cdot & T^i_{DA-q*n} \end{bmatrix} \begin{bmatrix} D_1 \\ D_2 \\ \cdot \\ \cdot \\ D_n \end{bmatrix} + \begin{bmatrix} T^i_{DR-1*n+1} & \cdots & T^i_{DR-1*n+r} \\ T^i_{DR-2*n+1} & \cdots & T^i_{DR-2*n+r} \\ \cdot & & \cdot \\ \cdot & & \cdot \\ T^i_{DR-m*n+1} & \cdots & T^i_{DR-m*n+r} \end{bmatrix} \begin{bmatrix} D_{n+1} \\ \vdots \\ D_{n+r} \end{bmatrix} \quad (3.13)$$

$$\Rightarrow \quad D^i_{l*} = \sum_{j=1}^{n+r} T^i_{D-l*j} D_j = \sum_{j=1}^{n} T^i_{DA-l*j} D_j + \sum_{j=n+1}^{n+r} T^i_{DR-l*j} D_j \qquad \text{for } l = 1, 2, \ldots, q \quad (3.13a)$$

We can generate the elements in a typical j^{th} column of the $\mathbf{T}^i_D$ matrix by visualising the displacement field of the structure, defined by the conditions, $D_j = 1$ and $D_{k\neq j} = 0$. This corresponds to the classical *stiffness* (or *displacement*) approach in structural analysis, in which the primary structure is a fully restrained (kinematically determinate) structure (as shown in Fig. 3.6a), and we impart only one displacement (of unit value) at a time. It is as though the restraint (artificial, in the case of an active degree of freedom) corresponding to the jth global coordinate undergoes a translational or rotational slip, $D_j = 1$. By visualising the deflected shape, it is easy to find the corresponding element-level displacements. This effect is a very local effect, inducing a non-zero displacement response only at a few local coordinates in elements connected to the node under consideration. Transformations involving *translational* displacements will obviously be in terms of the *direction cosines* of the elements, based on compatibility considerations.

> The **displacement transformation matrix** $\mathbf{T}_D$ is a unique matrix, satisfying compatibility requirements, for any given structure. A typical element, $T^i_{D-l^*,j}$, of this matrix may be visualised as the displacement being "transferred" on account of a unit displacement at the jth global coordinate to the local l^* coordinate of the ith element in the structure, with all other degrees of freedom restrained.

Here, we show, with reference to the plane frame example in Fig. 3.1, how the active displacement transformation matrix for the element '2', $\mathbf{T}^2_{DA}$, can be generated, by applying unit displacements $D_j = 1$ for j = 1, 2,..., 8 to the primary structure (Fig. 3.6a) in separate cases. Fig. 3.6b shows the deflected shape, corresponding to a unit horizontal translation, $D_1 = 1$, with all other degrees of freedom restrained. It is clear that non-zero displacements in the local coordinate system will be realised only with respect to D_{4*}^1, D_{5*}^1, D_{1*}^2 and D_{2*}^2 (refer Fig. 3.3a).

(a) Primary (kinematically determinate) structure
(All joint displacements are restrained)

(b) $D_1 = 1$ and $D_{k \neq 1} = 0$
($\theta_A = \theta_B = \theta_C = \theta_D = 0$)

Detail 'Y'
(showing end displacements in element '2' only)

Detail 'X'
(showing end displacements in element '2' only)

(c) $D_2 = 1$ and $D_{k \neq 2} = 0$
($\theta_A = \theta_B = \theta_C = \theta_D = 0$)

(d) $D_3 = 1$ and $D_{k \neq 3} = 0$
($\theta_A = \theta_C = \theta_D = 0$)

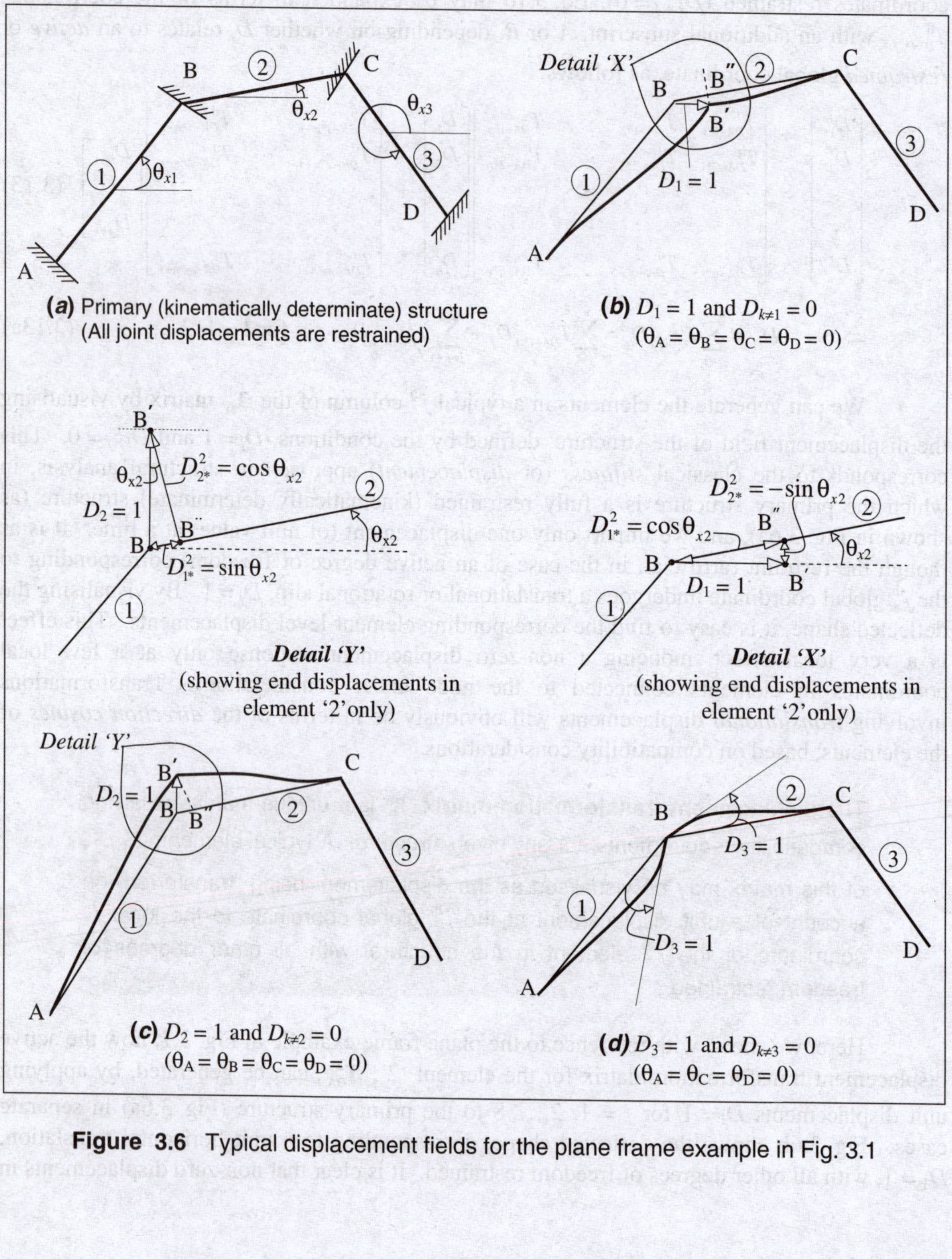

Figure 3.6 Typical displacement fields for the plane frame example in Fig. 3.1

Based on compatibility considerations, as shown in Fig. 3.6b (refer detail 'X', where the axial and transverse displacements at end B in element '2' are given by BB″ and B″B′ in the right-angled triangle[†] BB′B″), it is seen that $D_1 = 1$ implies $D^2_{1*} = \cos\theta_{x2}$ and $D^2_{2*} = -\sin\theta_{x2}$. This means that in the first column of the $\mathbf{T}^2_{\mathbf{DA}}$ matrix, comprising six values of D^2_{l*1}, the first two values will be equal to $\cos\theta_{x2}$ and $-\sin\theta_{x2}$ respectively, while the remaining four values will be equal to zero (for l = 3, 4, 5 and 6).

Similarly, referring to Fig. 3.6c, it is seen that $D_2 = 1$ implies $D^2_{1*} = \sin\theta_{x2}$ and $D^2_{2*} = \cos\theta_{x2}$ for the element '2' (refer detail 'Y'). This means that in the second column of the $\mathbf{T}^2_{\mathbf{DA}}$ matrix, comprising six values of D^2_{l*2}, the first two values will be equal to $\sin\theta_{x2}$ and $\cos\theta_{x2}$ respectively, while the remaining four values will be equal to zero (for l = 3, 4, 5 and 6). Referring to Fig. 3.6d, it is seen that a unit anti-clockwise rotation, $D_3 = 1$, implies $D^1_{6*} = 1$ and $D^2_{3*} = 1$, with all other values in the third column of the $\mathbf{T}^2_{\mathbf{DA}}$ matrix equal to zero. With the above results, we can easily fill up the first three columns of the $\mathbf{T}^2_{\mathbf{DA}}$ matrix. We observe that the next three columns (corresponding to $D_4 = 1$, $D_5 = 1$ and $D_6 = 1$ respectively) can be filled in a similar way. The last two columns, corresponding to $D_7 = 1$ and $D_8 = 1$ will comprise null vectors, because these displacements that occur at the joint D in the frame will not affect the element '2' in the restrained primary structure.

Thus, it can be shown that for the element '2' in this example, the active displacement transformation matrix $\mathbf{T}^2_{\mathbf{DA}}$, of order 6 × 8, is given by:

$$\begin{array}{cccccccc} (D_1=1) & (D_2=1) & (D_3=1) & (D_4=1) & (D_5=1) & (D_6=1) & (D_7=1) & (D_8=1) \end{array}$$
$$\mathbf{T}^2_{\mathbf{DA}} = \begin{bmatrix} \cos\theta_{x2} & \sin\theta_{x2} & 0 & 0 & 0 & 0 & 0 & 0 \\ -\sin\theta_{x2} & \cos\theta_{x2} & 0 & 0 & 0 & 0 & 0 & 0 \\ 0 & 0 & 1 & 0 & 0 & 0 & 0 & 0 \\ 0 & 0 & 0 & \cos\theta_{x2} & \sin\theta_{x2} & 0 & 0 & 0 \\ 0 & 0 & 0 & -\sin\theta_{x2} & \cos\theta_{x2} & 0 & 0 & 0 \\ 0 & 0 & 0 & 0 & 0 & 1 & 0 & 0 \end{bmatrix} \quad \text{(c)}$$

Similarly, it is possible to derive the restrained displacement transformation sub-matrix $\mathbf{T}^2_{\mathbf{DR}}$, of order 6 × 4 for the element '2'. It is convenient to write a simple algorithm to automate the generation of the displacement transformation matrices for various elements in a structure, once the geometrical data input of the structure, identification of elements and global coordinates and selection of element types and local coordinates are carried out.

If, for some reason, our interest is limited to considerations of only a few active or restrained displacement coordinates (i.e., not all degrees of freedom in the structure), then we

[†] In detail 'X', the deflected position B′ of the joint B is located horizontally to the right of B by a unit distance. B″ denotes the projection of B′ on the line BC (of element '2'). In detail 'Y', B′ is located vertically above B.

can accordingly limit the dimension of the $\{D\}$ vector, and hence the order of the $\mathbf{T}_D^i$ matrix (as given by Eq. 3.10) for this limited purpose. For example, if our interest is limited to finding the effect of only D_1, D_2 and D_3 in the above example, then the $\mathbf{T}_{DA}^2$ matrix in Eq. (c) will have a order of only 6×3, given by the first three columns of the full 6×8 matrix shown above.

We observe that the element transformation matrix $\mathbf{T}^2$ (square matrix) defined earlier (refer Eqns (a), (b) and 3.8) is a part of the rectangular $\mathbf{T}_D^2 = \left[\mathbf{T}_{DA}^2 \;\vdots\; \mathbf{T}_{DR}^2\right]$ matrix, which in turn is part of the *displacement transformation matrix* of the plane frame structure, $\mathbf{T}_D = \begin{bmatrix} \mathbf{T}_D^1 \\ \mathbf{T}_D^2 \\ \mathbf{T}_D^3 \end{bmatrix} = \begin{bmatrix} \mathbf{T}_{DA}^1 & \mathbf{T}_{DR}^1 \\ \mathbf{T}_{DA}^2 & \mathbf{T}_{DR}^2 \\ \mathbf{T}_{DA}^3 & \mathbf{T}_{DR}^3 \end{bmatrix}_{18\times12}$. Many elements of this large displacement matrix are zero. Therefore, from a computational point of view, it is advantageous in the stiffness method to deal with the three separate transformation matrices, $\mathbf{T}^1$, $\mathbf{T}^2$ and $\mathbf{T}^3$, each of order 6×6, than the large $\mathbf{T}_D$ matrix of order 18×12. However, in order to take advantage of this, we need to also have a "code" identification that links the local coordinates for each element with the global coordinates. This is described in detail later.

3.3.4 Force Transformation Matrix (Flexibility Method)

In the flexibility method, the *force transformation matrix* enables a transformation of the structure-level loads (applied actions at the global coordinates) to element-level forces (member-end forces at the *local* coordinates, using equilibrium conditions, in a statically determinate structure. Let us denote this matrix as $\mathbf{T}_F$, as indicated in Fig. 3.5. This implies that the *combined element force vector*, $\{F_*\}$, is obtainable by pre-multiplying the *force vector*, $\{F\}$, with the **force transformation matrix** $[T_F]$, in a manner that is somewhat analogous to the displacement transformation described earlier:

$$\mathbf{F}_* = \mathbf{T}_F \mathbf{F} \tag{3.14}$$

Referring to Eq. 3.3, for a typical statically indeterminate structure, $\mathbf{T}_F$ may be conveniently partitioned into two sub-matrices, comprising the **load transformation matrix** $[T_{FA}]$ (corresponding to the *active* global coordinates) and the **redundant force transformation matrix** $[T_{FX}]$ (corresponding to the *redundant* global coordinates):

$$\{F_*\} = [T_{FA}]\{F_A\} + [T_{FX}]\{F_X\} \;\Rightarrow\; \mathbf{F}_* = [\mathbf{T}_{FA} \;\vdots\; \mathbf{T}_{FR}]\begin{bmatrix} \mathbf{F}_A \\ \mathbf{F}_X \end{bmatrix} \tag{3.15}$$

Let us consider a typical statically indeterminate structure with n 'active' global coordinates (potential load locations) and n_s redundant global coordinates and m elements (of similar type), each having $\tilde{q}$ independent equations of equilibrium (local coordinates). Then, the $\mathbf{T_F}$ matrix, for the selected 'primary' structure, has an order $m\tilde{q}\times(n+n_s)$, which we can conveniently partition into m sub-matrices, corresponding to the m elements as follows:

$$[F_*]_{m\tilde{q}\times 1} = \begin{bmatrix} [F_*^1]_{\tilde{q}\times 1} \\ [F_*^2]_{\tilde{q}\times 1} \\ \vdots \\ \vdots \\ [F_*^m]_{\tilde{q}\times 1} \end{bmatrix}_{m\tilde{q}\times 1} = \begin{bmatrix} [T_{FA}^1]_{\tilde{q}\times n} \\ [T_{FA}^2]_{\tilde{q}\times n} \\ \vdots \\ \vdots \\ [T_{FA}^m]_{\tilde{q}\times n} \end{bmatrix}_{m\tilde{q}\times n} [F_A]_{n\times 1} + \begin{bmatrix} [T_{FX}^1]_{\tilde{q}\times n_s} \\ [T_{FX}^2]_{\tilde{q}\times n_s} \\ \vdots \\ \vdots \\ [T_{FX}^m]_{\tilde{q}\times n_s} \end{bmatrix}_{m\tilde{q}\times n_s} [F_X]_{n_s\times 1} \tag{3.15a}$$

where the sub-matrix $\mathbf{T_F^i} = \left[\mathbf{T_{FA}^i} \mid \mathbf{T_{FX}^i}\right]$ may be defined as the **force transformation matrix for the ith *element*,** which enables a transformation of the *force vector*, $\{F\}$, to the ith *element force vector*, $\{F_*^i\}$, as follows:

$$\mathbf{F_*^i} = \mathbf{T_F^i F} = \left[\mathbf{T_{FA}^i} \mid \mathbf{T_{FX}^i}\right] \begin{bmatrix} \mathbf{F_A} \\ \hline \mathbf{F_X} \end{bmatrix} \tag{3.16}$$

If the structure is statically determinate ($n_s = 0$), there is no need to partition the $\mathbf{T_F}$ matrix ($\mathbf{T_F} = \mathbf{T_{FA}}$). As mentioned earlier, in the force approach, unlike the displacement approach, it is not meaningful to include the reaction coordinates in the force transformation. However, it is possible to include the support reactions through additional rigid link[†] members in the $\{F_*\}$ vector.

We will also observe, through various applications in subsequent Chapters, that the full force transformation matrix $\mathbf{T_F}$ for a 'just-rigid' statically determinate structure is invariably a *square* matrix, as the dimension of the 'external' force vector turns out to be identical to that of the 'internal' force vector, $\mathbf{F_*}$. For example, in a 'just-rigid' statically determinate plane truss, we recall that the condition, $m + r = 2j$ ("total number of unknown forces must be equal to total number of independent equations"), has to be satisfied. If we include r rigid links (to simulate r support reactions), then we must also add r additional

[†] For example, a vertical support reaction F_j acting upward (say, in the global y-axis direction), can be simulated as the internal force N_i in a rigid link (a small, infinitely stiff axial element) inserted at the support location between the restrained end and the connecting structural element (refer Fig. 3.7). Depending on whether the axial force is tensile or compressive, and depending on whether the restrained end is below or above, the appropriate sign (positive or negative) has to be assigned $\left(F_j = \pm N_i\right)$.

active global coordinates (arising from inserting the rigid link members) to the count of $n = 2j - r$ active degrees of freedom. Thus, the quantity, $m + r$, represents the total number of (independent) unknown element forces in the structure, while $2j = n + r$ represents the total number of active global coordinates. If we do not wish to include support reactions, as is often the case in manual analysis (where we can infer the values from free-bodies), then, by deducting r from both sides of the equation, $m + r = n + r$, we observe that, in a statically determinate plane truss (with statically determinate support reactions), the condition, $m = n$, has to be invariably satisfied; i.e., the number of truss members must be equal to the number of potential load locations (active degrees of freedom). In the truss example, we have one unknown force per member, i.e., $\tilde{q} = 1$. In a general 'just-rigid' structure, involving m elements (of similar type), each having $\tilde{q} = 1, 2, 3$ or 6 independent equations of equilibrium, the condition, $m\tilde{q} = n$, has to be invariably satisfied, provided the structure is externally statically determinate.

In practice, we may also choose to restrict the dimension of the $\mathbf{F_A}$ vector to the number of global coordinates of interest to us (where non-zero loads act or where we need to find displacements), and in such cases, the $\mathbf{T_F}$ matrix becomes rectangular.

Defining the **force transfer coefficient**, $T^i_{Fl^*,j}$ as the force at the local coordinate l^* in the element i due to a unit load, $F_j = 1$, with no other loads acting ($F_{k \neq j} = 0$), the element force $F^i_{l^*}$ may be expressed in terms of the coefficients, $T^i_{F-l^*j}$ and the components F_j of the force vector, as follows:

$$\Rightarrow \quad F^i_{l^*} = \sum_{j=1}^{n} T^i_{F-l^*j} F_j \quad \text{for } l = 1, 2, \ldots, \tilde{q} \tag{3.17}$$

Considering a structure made up with m elements (such as *beam elements*) with $\tilde{q} = 2$, Eq. 3.17 may be expanded as follows:

$$\begin{bmatrix} \begin{Bmatrix} F^1_{1^*} \\ F^1_{2^*} \end{Bmatrix} \\ \begin{Bmatrix} F^2_{1^*} \\ F^2_{2^*} \end{Bmatrix} \\ \vdots \\ \begin{Bmatrix} F^m_{1^*} \\ F^m_{2^*} \end{Bmatrix} \end{bmatrix} = \Big[\{T_{F-*1}\} \quad \{T_{F-*2}\} \quad \cdots \quad \cdots \quad \{T_{F-*n}\} \Big]_{2m \times n} \begin{bmatrix} F_1 \\ F_2 \\ \cdot \\ F_n \end{bmatrix} \quad \text{for } \tilde{q} = 2 \tag{3.18}$$

We can generate the elements in a typical j^{th} column of the $\mathbf{T_F}$ matrix, $\{T_{F-*j}\}$, by visualising the force field of the structure, defined by the conditions, $F_j = 1$ and $F_{k \neq j} = 0$. This corresponds to the classical *flexibility* (or *force*) approach in structural analysis, in which the primary structure is a just-rigid (statically determinate) structure, and we impart only a *unit* *load* at a time. By applying equilibrium equations, it is easy to find the various element-level (member end) forces corresponding to the action of $F_j = 1$ on this primary structure. We can visualise $T^i_{F-l^*,j}$ as the force being transferred from $F_j = 1$ to $F^i_{l^*}$; hence, the term *force transfer coefficient* is meaningful.

The force transfer coefficients are constants that define certain inherent properties of the structure, enabling us to determine the unknown internal force resultants due to any external loading (in the form of direct actions).

As an illustrative example, let us consider the simple plane truss in Fig. 3.7a(i), which is made up of equilateral triangles of side $2a$.

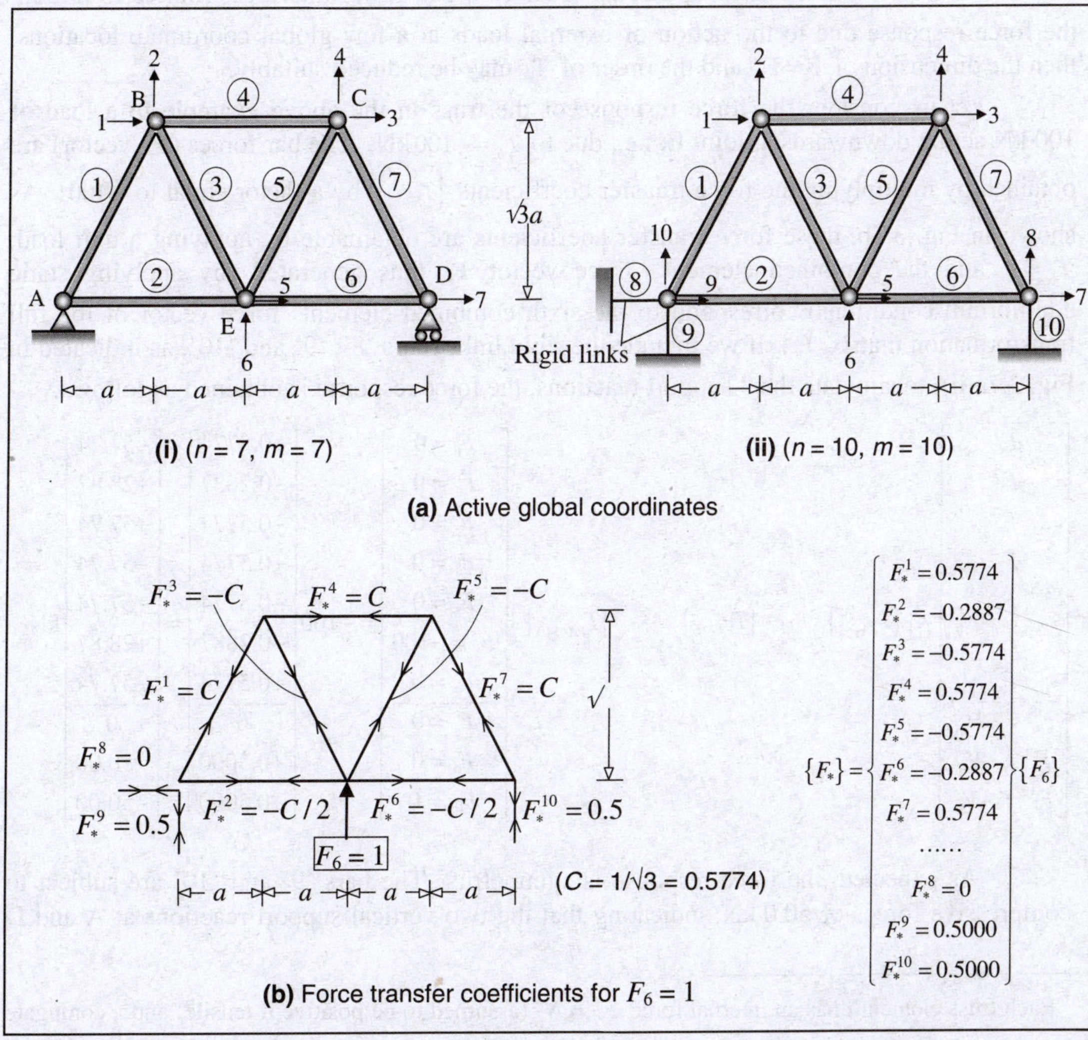

Figure 3.7 Example of force transformation in a simple plane truss (flexibility method)

The truss, which has 7 bars ($m = 7$) and 5 joints ($j = 5$), is simply supported. Of the $2j = 10$ potential (translational) degrees of freedom, 3 degrees of freedom are restrained. In the flexibility method, we may choose to ignore the support reactions ($r = 3$), and work with 7 active global coordinates (load locations) and 7 bar forces (corresponding to local coordinates, ; $\tilde{q} = 1$[†]), as indicated in Fig. 3.7a, whereby $\mathbf{F_A}$ and $\mathbf{F_*}$ both have dimensions of $n = m = 7$ and the full force transformation matrix $\mathbf{T_F} = \mathbf{T_{FA}}$ has an order 7 × 7.

We may also choose to include the support reactions by providing 3 rigid links, as also shown in Fig. 3.7a(ii). This increases the number of members as well as the number of active global coordinates (load locations) by $r = 3$; thus, in this case, $n = m = 10$, whereby the full force transformation matrix $\mathbf{T_F}$ has an order 10 × 10. If our interest is limited to finding the force response due to the action of external loads at a few global coordinate locations, then the dimension of $\mathbf{F} = \mathbf{F_A}$ and the order of $\mathbf{T_F}$ may be reduced suitably.

Let us consider the force response of the truss in the above example to a load of 100 kN acting downwards at joint E; i.e., due to $F_6 = -100\,\text{kN}$. The bar forces ($\mathbf{F_*}$ vector) are obtained by multiplying the force transfer coefficients $\{T_{F-*6}\}$ by a factor equal to -100. As shown in Fig. 3.7b, these force transfer coefficients are obtainable by applying a unit load, $F_6 = 1$, and the combined element force vector $\mathbf{F_*}$ thus generated (by applying static equilibrium conditions) correspond to the sixth combined element force vector of the full transformation matrix $\mathbf{T_F}$. If we include the rigid links (bars '8', '9' and '10', as indicated in Fig. 3.7a(ii)) to simulate the 3 support reactions, the force response is obtained as follows:

$$\begin{bmatrix} F_*^1 \\ F_*^2 \\ F_*^3 \\ F_*^4 \\ F_*^5 \\ F_*^6 \\ F_*^7 \\ \hdashline F_*^8 = -H_A \\ F_*^9 = -V_A \\ F_*^{10} = -V_D \end{bmatrix} = \left[\{T_{FA-*1}\} \quad \cdots \quad \{T_{FA-*6}\} \quad \cdots \quad \{T_{FA-*10}\} \right] \begin{bmatrix} F_1 = 0 \\ F_2 = 0 \\ F_3 = 0 \\ F_4 = 0 \\ F_5 = 0 \\ F_6 = -100 \\ F_7 = 0 \\ \hdashline F_8 = 0 \\ F_9 = 0 \\ F_{10} = 0 \end{bmatrix} = -100 \begin{bmatrix} +0.5774 \\ -0.2887 \\ -0.5774 \\ +0.5774 \\ -0.5774 \\ -0.2887 \\ +0.5774 \\ \hdashline 0 \\ +0.5000 \\ +0.5000 \end{bmatrix} = \begin{bmatrix} -57.74 \\ +28.87 \\ +57.74 \\ -57.74 \\ +57.74 \\ +28.87 \\ -57.74 \\ \hdashline 0 \\ -50.00 \\ -50.00 \end{bmatrix} \text{kN}$$

As expected, the force response is symmetric. The bars '9' and '10' are subject to compressive forces of 50.0 kN, indicating that the two vertical support reactions at A and D

[†] Each truss element i has an internal force $F_*^i = N_i$ (assumed to be positive if tensile) and a conjugate deformation $D_*^i = e_i$ (assumed to be positive if it is an elongation). For convenience, we use the symbol F_*^i instead of F_{1*}^i (refer Chapter 4 for further details on truss elements).

are equal to 50.0 kN (acting upwards); the horizontal reaction at A (equal to $-F_*^8$) is equal to zero.

The **force transformation matrix** is a unique matrix, satisfying static equilibrium, enabling structure-level forces (loads) to be "transferred" to element-level forces, in a given statically determinate structure. In the case of a statically indeterminate structure, it is not possible to define such a unique matrix, because multiple statically admissible solutions are possible. However, for a chosen primary structure (statically determinate), the force transformation matrix is unique. Thus, the $\mathbf{T_F}$ matrix in a statically indeterminate structure depends on the choice of redundants.

3.3.5 Principle of Contra-gradient

In Section 3.3.3, we observed how, in any structure, a transformation is possible from global (structure-level) coordinates to local (element-level) coordinates in the stiffness (displacement) method, by means of the *displacement transformation matrix* $\mathbf{T_D}$, which invokes *compatibility* conditions. Similarly, in Section 3.3.4, we observed how, in a statically determinate structure, a transformation is possible from global (structure-level) coordinates to local (element-level) coordinates in the flexibility (force) method, by means of the *force transformation matrix* $\mathbf{T_F}$, which invokes *equilibrium* conditions. Both these transformations involve a shift from global to local coordinates. We will now explore the possibility of a shift in the *reverse direction* ("contra-gradient"), i.e., from local coordinates to global coordinates. We know, from Section 3.3.1, that such a transformation is possible with respect to the element transformation matrix $\mathbf{T}^i$, which is a square orthogonal matrix, with $(\mathbf{T}^i)^{-1} = (\mathbf{T}^i)^T$. However, as mentioned earlier, this matrix does not deal with all the global coordinates in the structure.

Contra- displacement Transformation Matrix

The displacement transformation (stiffness method) is of the form, $\mathbf{D}_* = \mathbf{T_D D}$ (Eq. 3.10), enabling a shift from global displacement coordinates (in the structure) to local displacement coordinates (in the various elements). Let us now see whether a "contra-gradient" displacement transformation in the form, $\mathbf{D} = \mathbf{T_{D\text{-}contra}} \mathbf{D}_*$, is possible and meaningful. Clearly, such a "contra-transformation" is possible only if we can generate a **contra-displacement transformation matrix**, which we may tentatively denote as $\mathbf{T_{D\text{-}contra}}$, such that unique displacement vectors $\mathbf{D}$ (global coordinates) can be generated, corresponding to a unit displacement at a member-end (local coordinate) with all other local degrees of freedom restrained. In other words, if there is only a single member deformation at the element level, is it possible to predict the entire displacement field (all nodal displacements), based solely on compatibility considerations?

We may recall, from our knowledge of force methods, that such a transformation is possible only in the case of internally 'just-rigid' (statically determinate) structures. For example, it is possible to define all the joint displacements in a truss (using the Principle of Virtual Forces), due to a unit deformation in one bar alone (due to 'lack of fit' or temperature change). If the truss were to be internally 'over-rigid' (statically indeterminate), however, it is not possible in general for one bar alone to have an arbitrary deformation; there will be deformations in other connecting bars, as well as internal forces, and it will not be possible to find the joint displacements using compatibility conditions alone[†]. In an internally 'just-rigid' (statically determinate) structure, the number of active degrees of freedom (*n* global coordinates) match the number of independent equations of static equilibrium available for all the elements. As mentioned earlier, we can also include rigid links as additional members in the combined element displacement vector, and correspondingly include restrained degrees of freedom in the structure-level displacement vector. For such just-rigid structures, the dimension of the combined element displacement vector $\mathbf{D}_*$ (involving $m\tilde{q}+r$ local coordinates) is identical to that of the displacement vector $\mathbf{D}=\mathbf{D}_\mathrm{A}$ (involving $n+r$ global coordinates), whereby the *contra-displacement transformation matrix*, $\mathbf{T}_{\text{D-contra}}$ turns out to be a square matrix of order $(n+r)\times(n+r)$. This transformation takes the following form:

$$\{D_A\}_{(n+r)\times 1} = [T_{D-contra}]_{(n+r)\times(n+r)} \{D_*\}_{(n+r)\times 1} \tag{3.19}$$

The obvious requirement for such a transformation is that the structure should be 'just-rigid' in an overall sense (i.e., 'external' plus 'internal'). In such structures, the above transformation serves the useful purpose of providing compatibility relationships that enable us to find joint displacements in just-rigid structures on account of initial member deformations (due to temperature effects or constructional errors) or support movements (indirect loading). We observe that the full $\mathbf{T}_{\text{D-contra}}$ matrix enables us to generate *all* the nodal displacements in a structure due to any arbitrary set of element deformations in a just-rigid structure. If, however, our interest is limited to finding only a few nodal displacements, we can limit the size of the transformation matrix accordingly. We recall that, by invoking the Principle of Virtual Forces (refer Section 1.6.2), we can find one displacement at a time, by applying an appropriate *unit load*. We can handle multiple displacements, using a matrix approach. Clearly, the *contra-displacement transformation matrix*, $\mathbf{T}_{\text{D-contra}}$, is based on a flexibility (force) approach, in which the basic transformation matrix is the force transformation matrix $\mathbf{T}_\mathrm{F}$. It now remains to be seen as to how $\mathbf{T}_{\text{D-contra}}$ is related to $\mathbf{T}_\mathrm{F}$.

Invoking the Principle of Virtual Work, considering the force field and displacement field of the same statically determinate structure (including rigid links to simulate support

[†] The joint displacements (as well as the final deformations and internal forces induced in the various bars) will depend on the axial stiffnesses of the bars.

reactions), we observe that the total external virtual work is equal to the total internal virtual work. Each of these two quantities are scalars, obtained as the *inner product* of a row vector of order $1 \times (n + r)^{\dagger}$ by a conjugate column vector of order $(n + r) \times 1$ (refer Section 2.5). We can convert either the force or displacement column vector at the structure and element levels as row vectors by taking the transpose of the vector. Thus, for example, considering the external virtual work to be given by $\{F\}^{\mathrm{T}}\{D\} = \{F_A\}^{\mathrm{T}}\{D_A\}$ and the internal virtual work to be given by $\{F_*\}^{\mathrm{T}}\{D_*\}$, and applying the Principle of Virtual Work,

$$\{F_A\}^{\mathrm{T}}\{D_A\} = \{F_*\}^{\mathrm{T}}\{D_*\} \tag{d}$$

Substituting $\mathbf{F_* = T_F F_A}$ (Eq. 3.14, with $\mathbf{T_F = T_{FA}}$ for a statically determinate structure), and performing the transpose operation:

$$\{F_*\}^{\mathrm{T}} = \left([T_F]\{F_A\}\right)^{\mathrm{T}} = \{F_A\}^{\mathrm{T}}[T_F]^{\mathrm{T}}$$

Substituting the above expression in Eq. (d),

$$\{F_A\}^{\mathrm{T}}\{D_A\} = \{F_A\}^{\mathrm{T}}[T_F]^{\mathrm{T}}\{D_*\}$$

Observing that the quantity on either side of the equation is a scalar, we can conclude by comparing the terms on both sides of the equation (eliminating $\{F_A\}^{\mathrm{T}}$),

$$\{D_A\}_{(n+r)\times 1} = [T_F]^{\mathrm{T}}_{(n+r)\times(n+r)}\{D_*\}_{(n+r)\times 1} \tag{3.20}$$

The above equation clearly establishes a compatibility relationship between the global coordinates and the local coordinates, in terms of displacements. Two points are noteworthy. Firstly, the transformation matrix involved here is none other than the force transformation matrix (in transpose form), which is based on static equilibrium conditions. Secondly, the transformation here is in the reverse direction ("contra-gradient"), from the local coordinates to the global coordinates. Comparing Eq. 3.20 with Eq. 3.19, we observe that the $\mathbf{T_{D\text{-}contra}}$ matrix we are seeking is given by the $\mathbf{T_F}^{\mathrm{T}}$ matrix.

---

† We can account for the external virtual work associated with possible support movements by including the r restrained degrees of freedom in the global coordinates, i.e., $\mathbf{D} = \begin{bmatrix} \mathbf{D_A} \\ \hline \mathbf{D_R} \end{bmatrix}$ and $\mathbf{F} = \begin{bmatrix} \mathbf{F_A} \\ \hline \mathbf{F_R} \end{bmatrix}$. At the same time, in the force approach, our practice is to include r rigid links, whereby the dimension of $\mathbf{D_*}$ and $\mathbf{F_*}$ vectors is equal to $m\tilde{q} + r$, which is equal to the dimension, $n + r$, of $\mathbf{D} = \mathbf{D_A}$ and $\mathbf{F} = \mathbf{F_A}$, for a just-rigid (statically determinate) structure (refer Fig. 3.7a(ii)).

The **Principle of Contra-gradient**, as applied in the flexibility method of analysis, states that if in a statically determinate structure, a force transformation matrix $\mathbf{T_F}$ can be established, based on equilibrium conditions, to transform the loads on the structure to member forces ($\mathbf{F_* = T_F F_A}$), then the transpose of $\mathbf{T_F}$ serves to transform the member deformations to the active structure displacements ($\mathbf{D_A = T_F{}^T D_*}$). This implies that the same equilibrium conditions serve as compatibility conditions in a just-rigid structure, with the difference that $\mathbf{T_F}$ pertains to a transformation from the global coordinates to the local coordinates, while $\mathbf{T_F{}^T}$ pertains to a transformation in the "contra" direction, i.e., from local coordinates to global coordinates, relating displacements. Hence, we may refer to $\mathbf{T_F{}^T}$ as the **contra-displacement transformation matrix.**

This Principle is depicted schematically in Fig. 3.8.

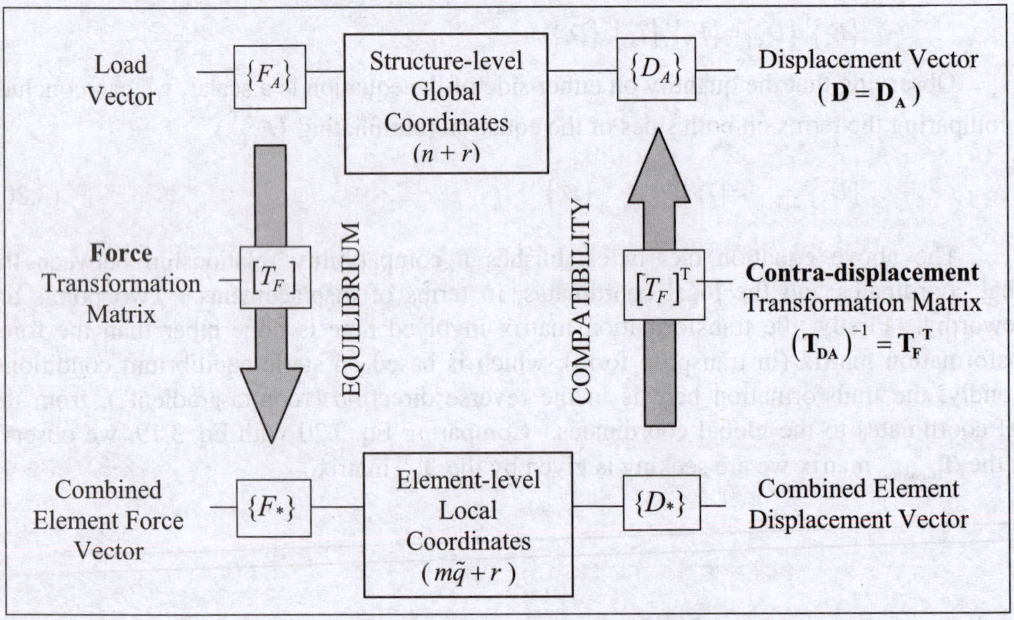

Figure 3.8 Principle of Contra-gradient applied to a statically determinate (just-rigid) structure including rigid links to simulate reactions (flexibility method)

Clearly, the contra-displacement matrix is meaningful in the flexibility method, while dealing with statically determinate (just-rigid) structures. We can also try to inter-relate this matrix with the *displacement transformation matrix*, provided we have a common framework of coordinates. This implies that we deal with just-rigid structure, with 'rigid links' to simulate the support reactions (as shown in Fig. 3.7a(ii), whereby $\mathbf{D} = \mathbf{D_A}$ (with a dimension of $n + r$), and the combined element displacement vector $\mathbf{D_*}$ has a dimension $m\tilde{q} + r = n + r$. This corresponds to the *reduced stiffness* formulation discussed earlier. We may also choose to ignore the support reactions, and thereby limit the dimension of $\mathbf{D_*}$ to $m\tilde{q} = n$, whereby the $\mathbf{T_F}$ matrix (as well as the $\mathbf{T_{DA}}$ matrix) has an order of $n \times n$.

According to the stiffness approach, compatibility for this just-rigid structure is defined by $\mathbf{D_*} = \mathbf{T_{DA}}\mathbf{D_A}$ (refer Eq. 3.11), whereas according to the flexibility approach, the same is described (in the "contra" direction) by $\mathbf{D_A} = \mathbf{T_F}^{\mathrm{T}}\mathbf{D_*}$. Both $\mathbf{T_{DA}}$ and $\mathbf{T_F}^{\mathrm{T}}$ are square matrices of order $n \times n$. These two square matrices will generally be non-singular and invertible, whereby it follows that one is necessarily the inverse of the other:

$$\left(\mathbf{T_F}^{\mathrm{T}}\right)^{-1} = \mathbf{T_{DA}} \tag{3.21a}$$

$$\left(\mathbf{T_{DA}}\right)^{-1} = \mathbf{T_F}^{\mathrm{T}} \tag{3.21b}$$

This is also indicated in Fig. 3.8. We shall explore examples to illustrate this in the Chapters to follow.

The **contra-displacement matrix** $\mathbf{T_F}^{\mathrm{T}}$ (which is based on a flexibility formulation) for a statically determinate (just-rigid) structure and the **active displacement matrix** $\mathbf{T_{DA}}$ (which is based on a reduced stiffness formulation) for the same structure, are both square matrices describing compatibility relationships between the local and global coordinates, one matrix being the inverse of the other.

Applying the Principle of Contra-gradient, we may re-interpret the **force transfer coefficient**, $T^i_{F-l^*,j}$ as the displacement D_j transferred to the (active) global coordinate j on account of a unit displacement (deformation), $D^i_{l^*} = 1$ at the local coordinate l^* in the element i of a just-rigid structure. Eq. 3.20 may be accordingly expanded as follows:

$$D_j = \sum_{i=1}^{m}\sum_{l=1}^{\tilde{q}} T^i_{F-l^*,j} D^i_{l^*} \qquad \text{for } j = 1, 2, \ldots, n \tag{3.22}$$

For example, referring to the plane truss shown in Fig. 3.7, we can apply the above equation to find the various joint displacements due to a 'lack of fit' or changes in lengths in various members due to temperature effects or even support movements. In this truss example, $\tilde{q} = 1$, and $m = 10$ (including $r = 3$). Using the force transfer coefficients derived, earlier, and considering a situation involving an increase in length in the top chord members ('1', '4' and '7') by 2mm each on account of a temperature rise, we can find the resulting deflection at joint E, applying Eq. 3.20:

$$D_6 = \left[T_{F-*,6}\right]^{\mathrm{T}}_{1\times10}\left[D^i_*\right]_{10\times1} = \left(T^1_{F-*,6}\right)\left(D^1_*\right)+\left(T^4_{F-*,6}\right)\left(D^4_*\right)+\left(T^7_{F-*,6}\right)\left(D^7_*\right)$$

$$= 3\times(+0.5774)(+2.0) = +3.46 \text{ mm } (\uparrow)$$

Similarly, considering the effect of a support settlement of 5mm at D, by means of a reduction in length ($D^i_* = 5$ mm) in the rigid link at D (member '10'), the resulting deflection at joint E is given by:

$$D_6 = \left[T_{F-*,6}\right]^{\mathrm{T}}_{1\times10}\left[D^i_*\right]_{10\times1} = \left(T^{10}_{F-*,6}\right)\left(D^{10}_*\right) = (+0.5000)(-5.0) = -2.50 \text{ mm } (\downarrow)$$

The **force transfer coefficient**, $T^i_{F-l*,j}$, which is an element of the force transformation matrix $\mathbf{T_F}$ of a statically determinate structure, may be visualised as the force being "transferred" to the local *l** coordinate of the *i* th element on account of a unit load acting at the *j* th global coordinate ($F_j = 1$), satisfying conditions of static equilibrium. The same transfer coefficient, which is also an element of the contra-displacement transformation matrix $\mathbf{T_F}^{\mathrm{T}}$ of the 'just-rigid' structure can be visualised as the displacement being "transferred" (in the "contra" direction) to the *j* th global coordinate on account of a unit displacement (deformation) at the local *l** coordinate of the *i* th element ($D^i_{l*} = 1$), satisfying conditions of compatibility.

Contra- force Transformation Matrix

The force transformation (flexibility method) is of the form, $\mathbf{F_* = T_F F}$ (Eq. 3.14), enabling a shift from global force coordinates (in the structure) to local force coordinates (in the various elements). Let us now see whether a "contra-gradient" force transformation in the form, $\mathbf{F = T_{F\text{-contra}} F_*}$, is possible and meaningful. Clearly, such a "contra-transformation" is possible only if we can generate a **contra-force transformation matrix**, which we may tentatively denote as $\mathbf{T_{F\text{-contra}}}$, such that unique force vectors $\mathbf{F}$ (global coordinates) can be generated, corresponding to a unit force at a member-end (local coordinate) with no other element-level force acting. In the *reduced stiffness formulation*, the action of $F^i_{l*} = 1$ will invariably generate an internal force that causes stresses in the element *i*, and which calls for appropriate equilibrating forces $\mathbf{F}$ at the global coordinates connected to the two ends of the member. In the conventional stiffness formulation, however, as all the *q* components of the element force vector cannot assume arbitrary zero values (to satisfy equilibrium), the elements remain unstressed, and the action of $F^i_{l*} = 1$ must remain 'external'. In this situation, the $\mathbf{T_{F\text{-contra}}}$ matrix only serves to *resolve* the element-end forces (expressed in local coordinates) to the structure global coordinates (similar to the $\mathbf{T}^{i^{\mathrm{T}}}$ matrix discussed in Section 3.3.1, except that

$\mathbf{T_{F\text{-}contra}}$ contains contributions from all elements and the linkage is with the *structure* global coordinates).

In the reduced stiffness formulation, this transformation takes the following form for any structure (kinematically indeterminate), assumed to be made of *m* elements of the same type:

$$\{F\}_{(n+r)\times 1} = [T_{F-contra}]_{(n+r)\times m\tilde{q}}\{F_*\}_{m\tilde{q}\times 1} \tag{3.23}$$

In the conventional stiffness formulation, the dimension $\tilde{q}$ is to be replaced by q.

We observe that the full $\mathbf{T_{F\text{-}contra}}$ matrix, which is in general a rectangular matrix, enables us to generate a unique and statically admissible set of nodal forces (acting at the active and restrained global coordinates) in a structure due to any arbitrary set of element forces in the structure. The *contra-force transformation matrix*, $\mathbf{T_{F\text{-}contra}}$, is based on a modified stiffness (force) approach, in which the basic transformation matrix is the *displacement transformation matrix* $\mathbf{T_D}$. It now remains to be seen as to how $\mathbf{T_{F\text{-}contra}}$ is related to $\mathbf{T_D}$.

Invoking the Principle of Virtual Work, considering the force field and displacement field of the same structure, we observe that the total external virtual work is equal to the total internal virtual work. Considering the external virtual work to be given by $\{D\}^{\mathrm{T}}\{F\}$ and the internal virtual work to be given by $\{D_*\}^{\mathrm{T}}\{F_*\}$, and applying the Principle of Virtual Work,

$$\{D\}^{\mathrm{T}}\{F\} = \{D_*\}^{\mathrm{T}}\{F_*\} \tag{e}$$

Substituting $\mathbf{D_* = T_D D}$ (Eq. 3.10), and performing the transpose operation:

$$\{D_*\}^{\mathrm{T}} = ([T_D]\{D\})^{\mathrm{T}} = \{D\}^{\mathrm{T}}[T_D]^{\mathrm{T}}$$

Substituting the above expression in Eq. (e),

$$\mathbf{D^T F = D^T T_D{}^T F_*} \tag{f}$$

Observing that the quantity on either side of the equation is a scalar, we can conclude by comparing the terms on both sides of the equation (eliminating $\mathbf{D^T}$),

$$\mathbf{F = T_D{}^T F_*} \tag{3.24}$$

The above equation reflects a statically admissible relationship between the global coordinates and the local coordinates, in terms of forces. Two points are noteworthy. Firstly, the transformation matrix involved here is none other than the displacement transformation matrix (in transpose form), which is based on compatibility conditions. Secondly, the transformation here is in the reverse direction ("contra-gradient"), from the local coordinates to the global coordinates. Comparing Eq. 3.24 with Eq. 3.23, we observe that the $\mathbf{T_{F\text{-}contra}}$ matrix we are seeking is given by the $\mathbf{T_D{}^T}$ matrix. This is depicted schematically in Fig. 3.9.

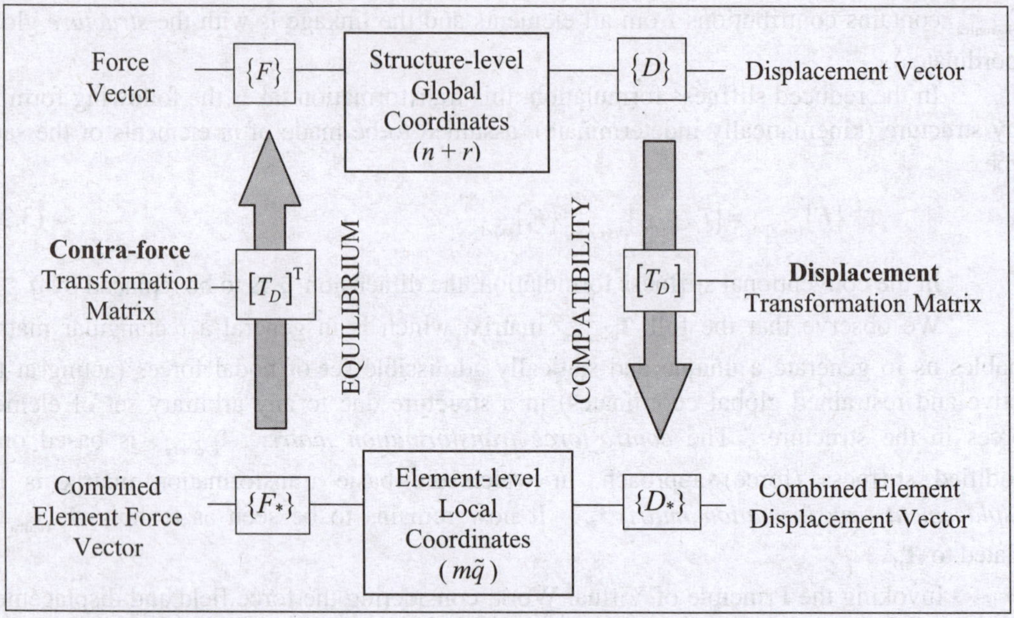

Figure 3.9 Principle of Contra-gradient, as applied in the reduced stiffness formulation

The **Principle of Contra-gradient**, as applied in the stiffness method of analysis, states that if in any structure, a displacement transformation matrix $\mathbf{T_D}$ can be established, based on compatibility conditions, to transform the nodal displacements in the structure to member displacements ($\mathbf{D_*} = \mathbf{T_D D}$), then the transpose of $\mathbf{T_D}$ serves to transform the member forces to the nodal forces acting in the form of loads and reactions on the structure ($\mathbf{F} = \mathbf{T_D}^{\mathrm{T}} \mathbf{F_*}$). Whereas $\mathbf{T_D}$ pertains to a transformation from the global coordinates to the local coordinates, $\mathbf{T_D}^{\mathrm{T}}$ pertains to a transformation in the "contra" direction, i.e., from local coordinates to global coordinates, relating forces. Hence, we may refer to $\mathbf{T_D}^{\mathrm{T}}$ as the **contra-force transformation matrix**. In the regular stiffness method, this transformation resolves the member-end forces to equivalent nodal forces in the structure. However, in the reduced stiffness method (where the $\mathbf{F_*}$ vector describes the internal force distributions in the elements), this transformation satisfies conditions of force equilibrium between the structure-level nodal forces $\mathbf{F}$ and the element-level forces $\mathbf{F_*}$ in the primary (kinematically determinate) structure.

As the contra-force transformation matrix $\mathbf{T_D}^{\mathrm{T}}$ is, in general, a rectangular matrix (unlike its counterpart, the contra-displacement transformation matrix $\mathbf{T_F}^{\mathrm{T}}$ which is a square matrix applicable to a statically determinate structure), we cannot, in general, take the inverse of $\mathbf{T_D}^\mathrm{T}$ and thereby relate it to $\mathbf{T_F}$.

Applying the Principle of Contra-gradient, we may re-interpret the **displacement transfer coefficient**, $T^i_{D-l^*,j}$ as the force F_j transferred to the (active) global coordinate j on account of a unit force, $F^i_{l^*} = 1$ at the local coordinate l^* in the element i of any structure. Eq. 3.24 may be accordingly expanded as follows:

$$F_j = \sum_{i=1}^{m}\sum_{l=1}^{\tilde{q}} T^i_{D-l^*,j} F^i_{l^*} \qquad \text{for } j = 1, 2, \ldots, n \qquad (3.25)$$

Thus, in the modified stiffness method, we can apply this Principle as an alternative means (using *statics*, instead of *kinematics*) to set up the $\mathbf{T_D}$ matrix. For example, let us consider the plane truss shown in Fig. 3.7a(i), corresponding to which, $\tilde{q} = 1$, $m = 7$, $r = 3$. Fig. 3.10a shows the same truss, with the restrained global coordinates ('8', '9' and '10') also included. The matrix $\mathbf{T_D}^\mathrm{T}$ for this truss has an order 10 × 7.

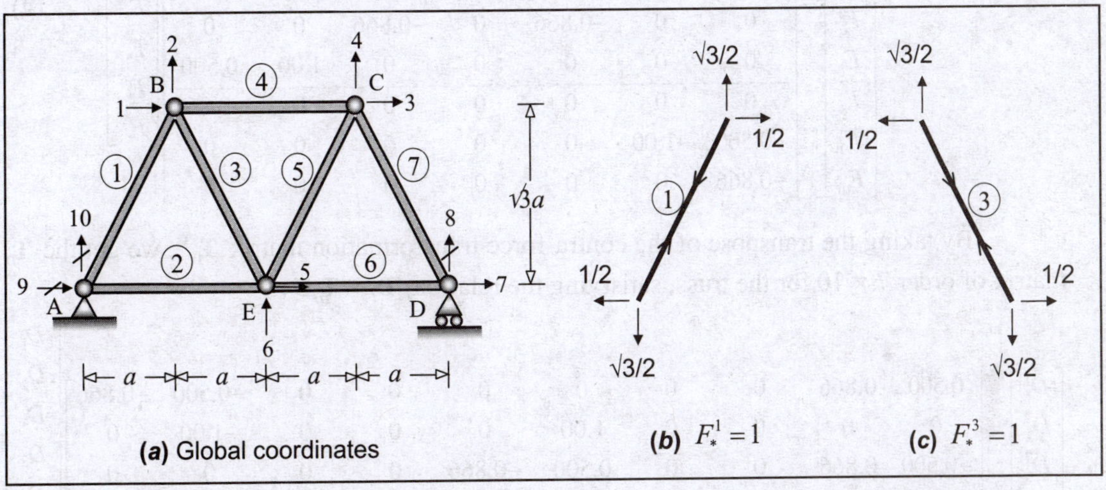

Figure 3.10 Example of force transformation in a simple plane truss (flexibility method)

Let us generate the $\mathbf{T_D}^\mathrm{T}$ matrix for this truss. For example, the first column vector corresponds to a condition, $F^1_* = 1$ (only element '1' has a unit tensile force; all other bars have zero force in the primary structure)†. As indicated in Fig. 3.10b, this condition requires a pair of equal and opposite forces acting at joints A and B, in order to satisfy equilibrium.

† As there is only one degree of freedom ($\tilde{q} = 1$) considered for the truss element in the reduced stiffness formulation, as in the flexibility method (refer Section 3.4.5 for further details), for convenience, we may use the symbol F^i_* (instead of F^i_{1*}) to designate i^{th} the bar force.

Corresponding to the global coordinate directions shown in Fig. 3.10a, this implies joint forces $F_1 = +0.500,\ F_2 = +0.866,\ F_9 = -0.500,\ F_{10} = -0.866$, with all other joint forces being equal to zero. Similarly, to generate the third column vector of $\mathbf{T_D}^\mathbf{T}$, we need to find the force vector that will equilibrate with $F_*^3 = 1$ (Fig. 3.10c). This implies joint forces $F_1 = -0.500,\ F_2 = +0.866,\ F_5 = +0.500,\ F_6 = -0.866$, with all other joint forces being equal to zero. Similarly, we can generate all the columns of the $\mathbf{T_D}^\mathbf{T}$ matrix. The sum of all the elements in any column should add up to zero, in order to satisfy force equilibrium.

Accordingly, for this example truss, with $n = 7,\ r = 3,\ m = 7$, the relation $\mathbf{F} = \mathbf{T_D}^\mathbf{T}\mathbf{F_*}$ takes the following form:

$$\begin{bmatrix} F_1 \\ F_2 \\ F_3 \\ F_4 \\ F_5 \\ F_6 \\ F_7 \\ \hline F_8 \\ F_9 \\ F_{10} \end{bmatrix} = \begin{bmatrix} 0.500 & 0 & -0.500 & -1.00 & 0 & 0 & 0 \\ 0.866 & 0 & 0.866 & 0 & 0 & 0 & 0 \\ 0 & 0 & 0 & 1.00 & 0.500 & 0 & -0.500 \\ 0 & 0 & 0 & 0 & 0.866 & 0 & 0.866 \\ 0 & 1.00 & 0.500 & 0 & -0.500 & -1.00 & 0 \\ 0 & 0 & -0.866 & 0 & -0.866 & 0 & 0 \\ 0 & 0 & 0 & 0 & 0 & 1.00 & 0.500 \\ \hline 0 & 0 & 0 & 0 & 0 & 0 & -0.866 \\ -0.500 & -1.00 & 0 & 0 & 0 & 0 & 0 \\ -0.866 & 0 & 0 & 0 & 0 & 0 & 0 \end{bmatrix} \begin{bmatrix} F_*^1 \\ F_*^2 \\ F_*^3 \\ F_*^4 \\ F_*^5 \\ F_*^6 \\ F_*^7 \end{bmatrix} \qquad \text{(g)}$$

By taking the transpose of the contra-force transformation matrix $\mathbf{T_D}^\mathbf{T}$, we get the $\mathbf{T_D}$ matrix of order 7 × 10 for the truss, satisfying the relation, $\mathbf{D_*} = \mathbf{T_D}\mathbf{D}$:

$$\begin{bmatrix} D_*^1 \\ D_*^2 \\ D_*^3 \\ D_*^4 \\ D_*^5 \\ D_*^6 \\ D_*^7 \end{bmatrix} = \left[\begin{array}{ccccccc|ccc} 0.500 & 0.866 & 0 & 0 & 0 & 0 & 0 & 0 & -0.500 & -0.866 \\ 0 & 0 & 0 & 0 & 1.00 & 0 & 0 & 0 & -1.00 & 0 \\ -0.500 & 0.866 & 0 & 0 & 0.500 & -0.866 & 0 & 0 & 0 & 0 \\ -1.00 & 0 & 1.00 & 0 & 0 & 0 & 0 & 0 & 0 & 0 \\ 0 & 0 & 0.500 & 0.866 & -0.500 & -0.866 & 0 & 0 & 0 & 0 \\ 0 & 0 & 0 & 0 & -1.00 & 0 & 1.00 & 0 & 0 & 0 \\ 0 & 0 & -0.500 & 0.866 & 0 & 0 & 0.500 & -0.866 & 0 & 0 \end{array}\right] \begin{bmatrix} D_1 \\ D_2 \\ D_3 \\ D_4 \\ D_5 \\ D_6 \\ D_7 \\ \hline D_8 \\ D_9 \\ D_{10} \end{bmatrix}$$

The first column vector in the above $\mathbf{T_D}$ matrix corresponds to the condition $D_1 = 1$ in the primary (kinematically restrained) structure. We can easily visualise the corresponding compatible displacement field, wherein element '1' elongates by 0.500, element '3' contracts by 0.500, and element '4' contracts by 1.00, as indicated.

The $\mathbf{T_D}^T$ matrix has other uses in the stiffness method. For example, it enables us to deal with problems involving indirect loading cases, such as due to temperature effects and 'lack of fit' in trusses. In such situations, we visualise the response of the structure, subject to this loading, in the restrained (kinematically determinate) condition. It is then possible to find the "initial" member displacements, and using appropriate stiffness relations, the corresponding "initial" combined element force vector $\mathbf{F}_{*\text{initial}}$, and thereby the nodal forces to be considered for further analysis, $\mathbf{F}_{\text{initial}} = \mathbf{T_D}^T\mathbf{F}_{*\text{initial}}$. This is discussed in the Chapters to follow.

The **displacement transfer coefficient**, $T^i_{D-l*,j}$, which is an element of the displacement transformation matrix $\mathbf{T_D}$, may be visualised not as the displacement being "transferred" to the local *l** coordinate of the *i* th element on account of a unit displacement at the *j* th global coordinate ($D_j = 1$), with all other degrees of freedom restrained, satisfying conditions of compatibility. Applying the Principle of Contra-gradient in the reduced stiffness formulation, the same transfer coefficient, which is also an element of the contra-force transformation matrix $\mathbf{T_D}^T$ can be visualised as the force being "transferred" (in the "contra" direction) to the *j* th global coordinate on account of a unit (internal) force at the local *l** coordinate of the *i* th element ($F^i_{l*} = 1$), satisfying conditions of static equilibrium.

3.4 STIFFNESS MATRIX

The **stiffness matrix** is another type of transformation matrix, which transforms the displacement vector to the force vector. It establishes the force-displacement relationships underlying the linear behaviour of the structure. The components of this matrix are called **stiffness coefficients**.

The **stiffness coefficient** k_{jk} may be defined as the force F_j generated at the coordinate *j* on account of a unit displacement at the coordinate *k* $(D_k = 1)$, with all other degrees of freedom restrained $(D_{l \neq k} = 0)$.

We begin by first defining, in Section 3.4.1, the *element stiffness matrix* $\mathbf{k}^i_*$, involving a linear relationship between $\mathbf{F}^i_*$ and $\mathbf{D}^i_*$ for the element *i*, in the local coordinate system. Next, in Section 3.4.2, we shall see how this matrix can be transformed to an equivalent matrix $\mathbf{k}^i$ in the global axes system, making use of the *element transformation matrix* $\mathbf{T}^i$ (defined by Eq. 3.7). In Section 3.4.3, we shall see how the *structure stiffness matrix* $\mathbf{k}$, relating $\mathbf{F}$ and $\mathbf{D}$, can be generated from $\mathbf{k}^i$ derived for the various elements that make up the structure. In Section 3.4.4, we shall see alternative ways of generating the structure stiffness matrix $\mathbf{k}$ from (i) the *unassembled stiffness matrix* $\mathbf{k}_*$, making use of the *displacement transformation matrix* $\mathbf{T}_D$, and (ii) the *element stiffness matrix* $\mathbf{k}_*^i$, making use of the *element displacement transformation matrix* $\mathbf{T}_D^i$ by the *direct stiffness method*. Finally, in Section 3.4.5, we shall see how the *reduced element stiffness matrix* $\tilde{\mathbf{k}}_*^i$ (or the *unassembled reduced stiffness matrix* $\tilde{\mathbf{k}}_*$) and the *element displacement transformation matrix* $\mathbf{T}_D^i$ (or the *displacement transformation matrix* $\mathbf{T}_D$) can be used to generate the *structure stiffness matrix* $\mathbf{k}$, in the reduced stiffness formulation, involving lesser degrees of freedom at the element level.

3.4.1 Element Stiffness Matrix (Local Coordinates)

The stiffness relation for a typical element *i*, expressed in the local coordinate system, relating the element force vector $\mathbf{F}_*^i$ to the element displacement vector $\mathbf{D}_*^i$ is as follows:

$$\{F_*^i\}=[k_*^i]\{D_*^i\} \tag{3.26}$$

where $\mathbf{k}_*^i$ is the **element stiffness matrix** of the element *i*, of order $q \times q$. The dimension q of the displacement and force vectors, $\mathbf{D}_*^i$ and $\mathbf{F}_*^i$, is given by the degrees of freedom at the two ends of the element. As mentioned earlier, $q=12$ in a space frame element. This dimension reduces to $q=6$ in a *plane frame element* (refer Eq. 3.5) as shown in Fig. 3.11a, and further to $q=4$ in a *plane truss element* (Fig. 3.11b). In a continuous beam system, we can conveniently align the local axes system and global axes system in the same direction, with $\theta_x^i=0$. Such a beam element may also be treated as a special case of the plane frame element, in which axial degrees of freedom are eliminated), whereby $q=4$ (Fig. 3.11c). If the beam is inclined to the horizontal axis ($\theta_x^i \neq 0$), it would be appropriate to treat it as a plane frame element (Fig. 3.11a). If axial deformations are to be considered negligible, we simply assign a very high value to the axial rigidity of the plane frame element.

For a typical plane frame element (Fig. 3.11a), Eq. 3.26 may be expanded as follows:

$$\begin{matrix} & & (D_{1*}^i=1) & (D_{2*}^i=1) & (D_{3*}^i=1) & (D_{4*}^i=1) & (D_{5*}^i=1) & (D_{6*}^i=1) & \end{matrix}$$

$$\begin{bmatrix} F_{1*}^i \\ F_{2*}^i \\ F_{3*}^i \\ \hline F_{4*}^i \\ F_{5*}^i \\ F_{6*}^i \end{bmatrix} = \left[\begin{array}{ccc|ccc} k_{1*1*}^i & k_{1*2*}^i & k_{1*3*}^i & k_{1*4*}^i & k_{1*5*}^i & k_{1*6*}^i \\ k_{2*1*}^i & k_{2*2*}^i & k_{2*3*}^i & k_{2*4*}^i & k_{2*5*}^i & k_{2*6*}^i \\ k_{3*1*}^i & k_{3*2*}^i & k_{3*3*}^i & k_{3*4*}^i & k_{3*5*}^i & k_{3*6*}^i \\ \hline k_{4*1*}^i & k_{4*2*}^i & k_{4*3*}^i & k_{4*4*}^i & k_{4*5*}^i & k_{4*6*}^i \\ k_{5*1*}^i & k_{5*2*}^i & k_{5*3*}^i & k_{5*4*}^i & k_{5*5*}^i & k_{5*6*}^i \\ k_{6*1*}^i & k_{6*2*}^i & k_{6*3*}^i & k_{6*4*}^i & k_{6*5*}^i & k_{6*6*}^i \end{array}\right] \begin{bmatrix} D_{1*}^i \\ D_{2*}^i \\ D_{3*}^i \\ \hline D_{4*}^i \\ D_{5*}^i \\ D_{6*}^i \end{bmatrix} \tag{3.26a}$$

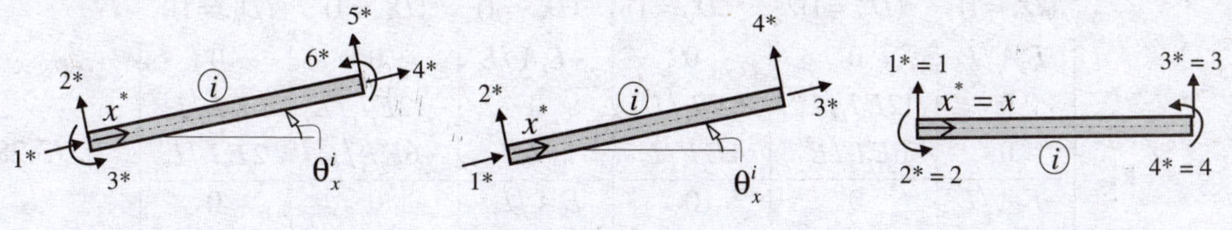

(a) Plane frame element ($q = 6$)

(b) Plane truss element ($q = 4$)

(c) Beam element ($q = 4$)

Figure 3.11 Local coordinates in typical planar elements (stiffness method)

In Eq. 3.26a, vertical partitions separating the 6 column vectors are shown to indicate that any column vector (say, j^{th} column) of the stiffness matrix can be generated by considering the force response caused by a unit displacement, $D^i_{j*} = 1$, with all other degrees of freedom restrained ($D^i_{k*\neq j*} = 0$), as indicated in parenthesis in the top row. Also, for convenience in understanding, the vectors, $\mathbf{D^i_*}$ and $\mathbf{F^i_*}$ may be partitioned, to distinguish between the local coordinates (1*, 2*, 3*) at the *start node* and those at the *end node* (4*, 5*, 6*) of the element.

The stiffness coefficients can be generated from first principles, as shown in the Chapters to follow. In the classical stiffness approach, we assume a suitable displacement function, in terms of the member-end displacements, as the basis for derivation. For example, using an energy approach (refer Eqns 1.64 and 1.65), and expressing the strain energy U_i in the element as a function of the components of $\mathbf{D^i_*}$,

$$k^i_{j*k*} = \frac{\partial^2 U_i}{\partial D^i_{j*} \partial D^i_{k*}} = k^i_{k*j*} \tag{3.27}$$

The above equation also reveals that the stiffness matrix is a *symmetric* matrix $\left(k^i_{j*k*} = k^i_{k*j*}\right)$.

Prismatic Plane Frame Element

For a *prismatic* plane frame element, with length L_i, flexural rigidity E_iI_i and axial rigidity E_iA_i (ignoring shear deformations), we can easily prove (as shown in Chapter 6) that the element stiffness matrix (in Eq. 3.26), with the local coordinates as shown in Fig. 3.11a, takes the following form:

$$
\mathbf{k}_*^{\mathbf{i}} = \begin{array}{c} \\ \left[\begin{array}{ccc|ccc} (D_{1*}^i=1) & (D_{2*}^i=1) & (D_{3*}^i=1) & (D_{4*}^i=1) & (D_{5*}^i=1) & (D_{6*}^i=1) \\ E_iA_i/L_i & 0 & 0 & -E_iA_i/L_i & 0 & 0 \\ 0 & 12E_iI_i/L_i^3 & 6E_iI_i/L_i^2 & 0 & -12E_iI_i/L_i^3 & 6E_iI_i/L_i^2 \\ 0 & 6E_iI_i/L_i^2 & 4E_iI_i/L_i & 0 & -6E_iI_i/L_i^2 & 2E_iI_i/L_i \\ \hline -E_iA_i/L_i & 0 & 0 & E_iA_i/L_i & 0 & 0 \\ 0 & -12E_iI_i/L_i^3 & -6E_iI_i/L_i^2 & 0 & 12E_iI_i/L_i^3 & -6E_iI_i/L_i^2 \\ 0 & 6E_iI_i/L_i^2 & 2E_iI_i/L_i & 0 & -6E_iI_i/L_i^2 & 4E_iI_i/L_i \end{array}\right] \end{array} \qquad (3.28)
$$

Prismatic Plane Truss Element

The truss element resists external loads by developing axial forces alone. Hence, only axial stiffness components of the stiffness matrix can have non-zero values. As rotational degrees of freedom are absent, we can eliminate the third and sixth rows and columns in the matrix in Eq. 3.28. Member-end translations in the local y^*- axis direction are possible, but these are rigid body translations (assuming small deformations). As the element cannot develop shear forces, the stiffness coefficients corresponding to local coordinates 2* and 4* in Fig. 3.11c (second and fourth rows and columns in the 4×4 stiffness matrix) must be all zero. Accordingly, the stiffness matrix for the plane truss element, with local coordinates as shown in Fig. 3.11b, takes the following form:

$$
\mathbf{k}_*^{\mathbf{i}} = \frac{E_iA_i}{L_i}\left[\begin{array}{cc|cc} (D_{1*}^i=1) & (D_{2*}^i=1) & (D_{3*}^i=1) & (D_{4*}^i=1) \\ 1 & 0 & -1 & 0 \\ 0 & 0 & 0 & 0 \\ \hline -1 & 0 & 1 & 0 \\ 0 & 0 & 0 & 0 \end{array}\right] \qquad (3.29)
$$

Prismatic Beam Element

By eliminating the rows and columns (first and fourth) corresponding to the axial degrees of freedom from the matrix in Eq. 3.28, we can generate the element stiffness matrix for the prismatic beam element, with local coordinates as shown in Fig. 3.11c:

$$
\mathbf{k}_*^{\mathbf{i}} = \frac{E_iI_i}{L_i}\left[\begin{array}{cc|cc} (D_{1*}^i=1) & (D_{2*}^i=1) & (D_{3*}^i=1) & (D_{4*}^i=1) \\ 12/L_i^2 & 6/L_i & -12/L_i^2 & 6/L_i \\ 6/L_i & 4 & -6/L_i & 2 \\ \hline -12/L_i^2 & -6/L_i & 12/L_i^2 & -6/L_i \\ 6/L_i & 2 & -6/L_i & 4 \end{array}\right] \qquad (3.30)
$$

3.4.2 Element Stiffness Matrix in Global Axes System

In Section 3.3.1, we found a way of transforming the element displacement and force vectors, $\mathbf{D}_*^i$ and $\mathbf{F}_*^i$ from local coordinates to the global axes system, where they are re-designated as $\mathbf{D}^i$ and $\mathbf{F}^i$ respectively. We first defined the element transformation matrix $\mathbf{T}^i$ as the means to transform $\mathbf{D}^i$ and $\mathbf{F}^i$ (in global axes system) to $\mathbf{D}_*^i$ and $\mathbf{F}_*^i$ (in local axes system): ($\mathbf{D}_*^i = \mathbf{T}^i\mathbf{D}^i$ and $\mathbf{F}_*^i = \mathbf{T}^i\mathbf{F}^i$). Then we showed that $\mathbf{T}^i$ being a square, orthogonal matrix can be inverted, the inverse being the same as the transpose. Thus, we may transform from the local axes system to the global axes system, using the relations, $\mathbf{D}^i = \mathbf{T}^{i^T}\mathbf{D}_*^i$ and $\mathbf{F}^i = \mathbf{T}^{i^T}\mathbf{F}_*^i$.

We can now use these concepts to transform the element stiffness matrix $\mathbf{k}_*^i$ (defined by Eq. 3.26) from the local axes system to the global axes system. Substituting Eq. 3.9b and Eq. 3.7a in Eq. 3.26,

$$\mathbf{F}^i = \mathbf{T}^{i^T}\mathbf{F}_*^i = \mathbf{T}^{i^T}\left(\mathbf{k}_*^i\mathbf{D}_*^i\right) = \mathbf{T}^{i^T}\mathbf{k}_*^i\left(\mathbf{T}^i\mathbf{D}^i\right) = \left(\mathbf{T}^{i^T}\mathbf{k}_*^i\mathbf{T}^i\right)\mathbf{D}^i \tag{h}$$

Defining $\mathbf{k}^i$ as the **element stiffness matrix in global axes system** of the element *i*, of order $q \times q$, we have:

$$\left\{F^i\right\} = \left[k^i\right]\left\{D^i\right\} \tag{3.31}$$

Comparing this equation with Eq. (h), we arrive at the desired means of transforming the element stiffness matrix $\mathbf{k}_*^i$ (local axes system) to $\mathbf{k}^i$ (global axes system):

$$\left[k^i\right] = \left[T^i\right]^T\left[k_*^i\right]\left[T^i\right] \tag{3.32}$$

We can also understand the basis for this transformation by looking at the schematic diagram shown in Fig. 3.12, where various transformations are illustrated. The basic flow is depicted in a clockwise direction, beginning with a transformation of the element displacement vector to local coordinates, $\mathbf{D}_*^i = \mathbf{T}^i\mathbf{D}^i$, followed by an element stiffness transformation in local coordinates, $\mathbf{F}_*^i = \mathbf{k}_*^i\mathbf{D}_*^i$, which in turn is followed by a transformation of the element force vector from local coordinates to the global axes system, $\mathbf{F}^i = \mathbf{T}^{i^T}\mathbf{F}_*^i$. Using this diagram, we can easily work out other relationships, including the stiffness relation, $\mathbf{F}_*^i = \mathbf{k}_*^i\mathbf{D}_*^i = \mathbf{k}_*^i\mathbf{T}^i\mathbf{D}^i$, which is a kind of resultant relation (directed in the anti-clockwise direction in Fig. 3.12). Another useful relation is to be found in the diagonal direction, whereby a direct transformation of the element displacement vector in global axes system $\mathbf{D}^i$ to element force vector $\mathbf{F}_*^i$ in local coordinates is possible:

$$\left\{F_*^i\right\} = \left[k_*^i\right]\left[T^i\right]\left\{D^i\right\} \tag{3.33}$$

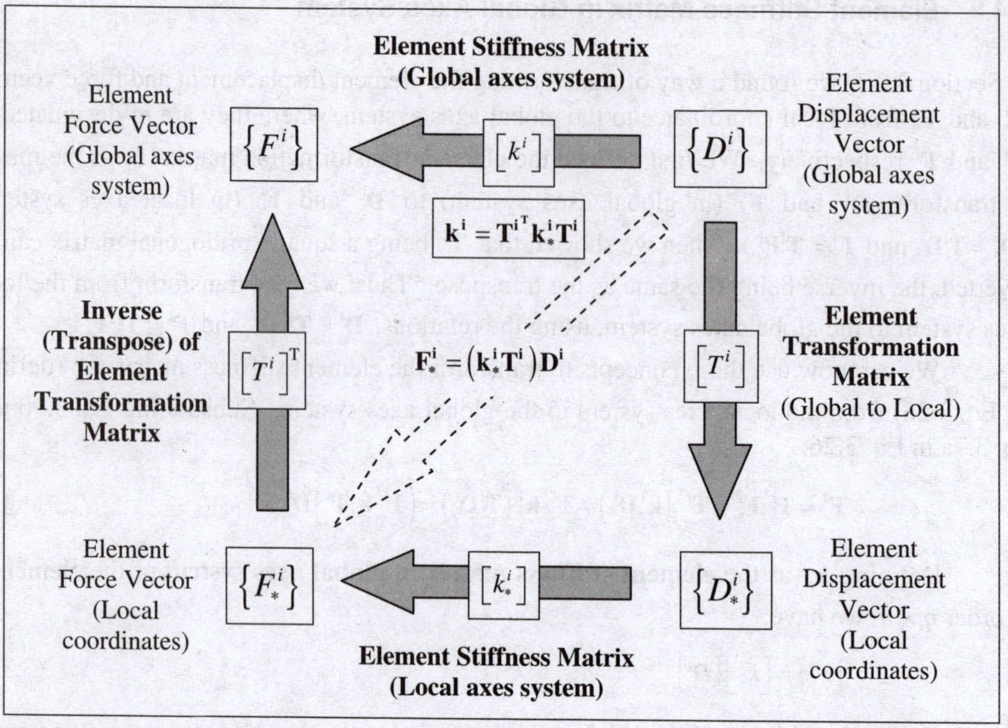

Figure 3.12 Transformations inter-relating element force and displacement vectors in local and global axes systems (stiffness method)

Eq. 3.32 reveals that, as $\mathbf{k}^i_*$ is a square symmetric matrix, the transformed matrix $\mathbf{k}^i$ must also be a square symmetric matrix of the same order $q \times q$. For the three typical types of planar elements discussed earlier (Fig. 13.11), we can generate the $\mathbf{k}^i$ respective matrices, applying Eq. 3.32. The element coordinates for the plane frame, plane truss and beam elements, expressed in the global axes system, are shown in Fig. 3.13. The element transformation matrix $\mathbf{T}^i$ for the plane frame element (Fig. 3.13a) had been derived earlier (refer Eq. 3.8). From this general transformation matrix, by eliminating the third and sixth rows and columns (corresponding to the rotational degrees of freedom), we can generate the element transformation matrix for the plane truss element (Fig. 3.13b). The element transformation matrix for a beam element will be the identity matrix $\mathbf{T}^i = \mathbf{I}$, as the global axes directions coincide with the local axes directions, as discussed earlier (refer Fig. 3.11c); in other words, $\mathbf{k}^i = \mathbf{k}^i_*$ (Eq. 3.30) for such a matrix. All the element transformation matrices are *orthogonal* matrices.

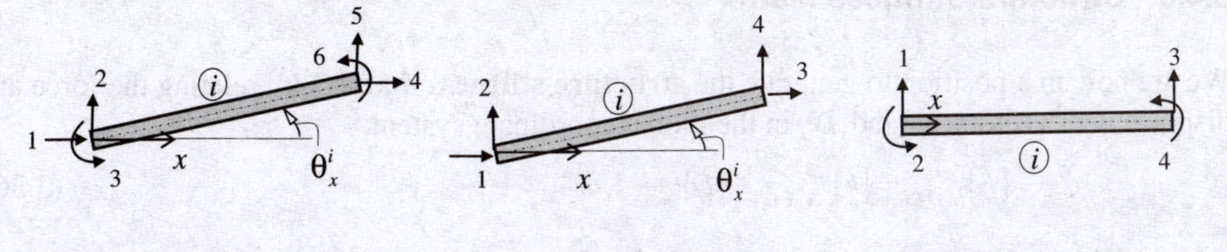

(a) Plane frame element ($q = 6$) **(b)** Plane truss element ($q = 4$) **(c)** Beam element ($q = 4$)

Figure 3.13 Degrees of freedom in typical planar elements, expressed in global axes system (stiffness method)

Thus, applying Eq. 3.32, we can show that the $\mathbf{k^i}$ matrix for a **plane frame element** is given by:

$$\mathbf{k^i} = \begin{array}{cccccc} (D_1^i=1) & (D_2^i=1) & (D_3^i=1) & (D_4^i=1) & (D_5^i=1) & (D_6^i=1) \end{array}$$

$$\mathbf{k^i} = \left[\begin{array}{ccc|ccc} (\alpha c^2+\beta s^2) & (\alpha-\beta)(cs) & -\chi s & -(\alpha c^2+\beta s^2) & -(\alpha-\beta)(cs) & -\chi s \\ (\alpha-\beta)(cs) & (\alpha s^2+\beta c^2) & \chi c & -(\alpha-\beta)(cs) & -(\alpha s^2+\beta c^2) & \chi c \\ -\chi s & \chi c & 4\delta & \chi s & -\chi c & 2\delta \\ \hline -(\alpha c^2+\beta s^2) & -(\alpha-\beta)(cs) & \chi s & (\alpha c^2+\beta s^2) & (\alpha-\beta)(cs) & \chi s \\ -(\alpha-\beta)(cs) & -(\alpha s^2+\beta c^2) & -\chi c & (\alpha-\beta)(cs) & (\alpha s^2+\beta c^2) & -\chi c \\ -\chi s & \chi c & 2\delta & \chi s & -\chi c & 4\delta \end{array}\right]$$

$$\alpha = \frac{E_i A_i}{L_i} \;;\; \beta = \frac{12 E_i I_i}{L_i^3} \;;\; \chi = \frac{6 E_i I_i}{L_i^2} \;;\; \delta = \frac{E_i I_i}{L_i} \;;\; c = \cos\theta_x^i \;;\; s = \sin\theta_x^i \qquad (3.34)$$

Similarly, we can derive the $\mathbf{k^i}$ matrix for a **plane truss element**:

$$\begin{array}{cccc} (D_1^i=1) & (D_2^i=1) & (D_3^i=1) & (D_4^i=1) \end{array}$$

$$\mathbf{k^i} = \frac{E_i A_i}{L_i}\left[\begin{array}{cc|cc} c^2 & cs & -c^2 & -cs \\ cs & s^2 & -cs & -s^2 \\ \hline -c^2 & -cs & c^2 & cs \\ -cs & -s^2 & cs & s^2 \end{array}\right] \qquad (3.35)$$

$$c = \cos\theta_x^i \;;\; s = \sin\theta_x^i$$

It is instructive to understand the physical significance of the stiffness coefficients in the $\mathbf{k^i}$ matrix. This is discussed in the Chapters to follow.

3.4.3 Structure Stiffness Matrix

We are now in a position to generate the **structure stiffness matrix, k**, relating the force and displacement vectors, **F** and **D**, in the global coordinate system.

$$\{F\}_{(n+r)\times 1} = [k]_{(n+r)\times(n+r)}\{D\}_{(n+r)\times 1} \tag{3.36}$$

We recall that, any component of **k**, which is the stiffness coefficient k_{jk}, is defined as the force F_j generated at the global coordinate *j* due to a unit displacement, $(D_k = 1)$, at the global coordinate *k*, all other degrees of freedom being restrained $(D_{l\neq k} = 0)$. We may visualise this displacement field as the outcome of a translational or rotational slip at the joint under consideration in a fully restrained (kinematically determinate) structure. Also, as discussed earlier, the unit displacement $(D_k = 1)$ is imparted to all those elements that are connected to the global coordinate *k*. Indeed only these elements will be subject to internal forces, while elements at other remote locations will remain unstressed. This is illustrated in the plane frame example shown earlier in Fig. 3.6. If any of the stressed elements happen to be connected to the global coordinate *j*, then they will contribute to F_j, and thereby, to k_{jk}. Indeed, in order to satisfy force equilibrium, we find that k_{jk} is equal to the algebraic sum of the corresponding element stiffness coefficients $k^i_{\tilde{j}\tilde{k}}$, expressed in the global axes system, $\tilde{j}$ and $\tilde{k}$ denoting the element coordinate numbers identified for the element *i* that match the global coordinates *j* and *k* ($D^i_{\tilde{j}} = D_j$ and $D^i_{\tilde{k}} = D_k$). Thus, each entry in the element stiffness matrix $\mathbf{k^i}$ is also an entry somewhere in the structure stiffness matrix **k**. The main task involved in assembling the matrix **k** is in identifying this location precisely, using a rule.

$$k_{jk} = \sum_i k^i_{\tilde{j}\tilde{k}} \quad \text{satisfying } D^i_{\tilde{j}} = D_j \text{ and } D^i_{\tilde{k}} = D_k \tag{3.37}$$

The displacement compatibility requirements (matching of structure and element coordinates) is illustrated for the case of the plane frame example in Fig. 3.14. Thus, in this example, after generating the three element stiffness matrices in the global axes system, $\mathbf{k^1}$, $\mathbf{k^2}$ and $\mathbf{k^3}$ (using Eq. 3.34), we can assign the various stiffness components to the relevant locations in the structure stiffness matrix, by matching the coordinate indices. For convenience, let us mark the matching structure global coordinates in parenthesis above the first row and to the right of the last column of each of the 3 element stiffness matrices, referring to the displacement compatibility relations spelt out in Fig. 3.14. Next to each element stiffness matrix $\mathbf{k^i}$, we indicate the “block” locations of the structure stiffness matrix **k**, to which the four quadrants of each $\mathbf{k^i}$ matrix contributes as entries.

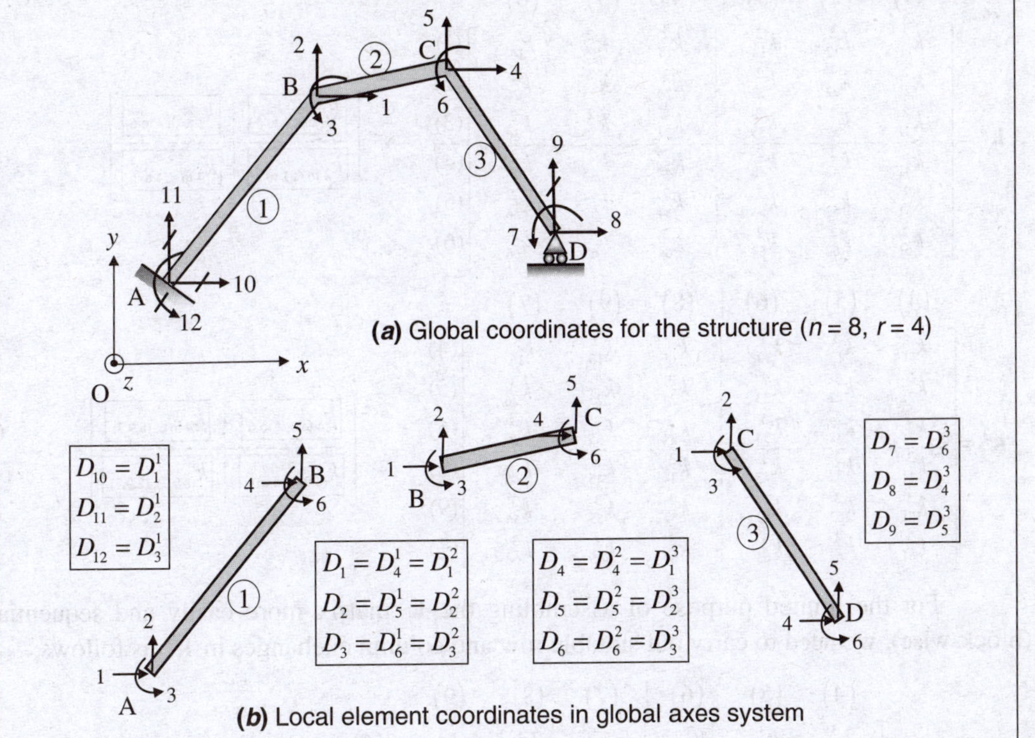

(a) Global coordinates for the structure ($n = 8$, $r = 4$)

(b) Local element coordinates in global axes system

Figure 3.14 Displacement compatibility between structure and element coordinates in the global axes system (plane frame example in Fig. 3.1)

Assigning "blocks" of $\mathbf{k}^i$ to $\mathbf{k}$ is easy and convenient, provided the numbering in each block follows an incremental sequence. In this example, we find that, except for the global coordinates in the block, "7 to 9", the numbering sequence is incremental. In the block "7 to 9", we must take care to shuffle the entries of $\mathbf{k}^3$ such that the element coordinates, (6, 4, 5) correspond to the global coordinates, (7, 8, 9), as indicated in Fig. 3.14 (elements '1' and '2' make zero contributions to the column vectors, corresponding to $D_7 = 1$, $D_8 = 1$, $D_9 = 1$).

$$\mathbf{k}^1 = \begin{array}{c} \begin{array}{cccccc} (10) & (11) & (12) & (1) & (2) & (3) \end{array} \\ \left[\begin{array}{ccc|ccc} k_{11}^1 & k_{12}^1 & k_{13}^1 & k_{14}^1 & k_{15}^1 & k_{16}^1 \\ k_{21}^1 & k_{22}^1 & k_{23}^1 & k_{24}^1 & k_{25}^1 & k_{26}^1 \\ k_{31}^1 & k_{32}^1 & k_{33}^1 & k_{34}^1 & k_{35}^1 & k_{36}^1 \\ \hline k_{41}^1 & k_{42}^1 & k_{43}^1 & k_{44}^1 & k_{45}^1 & k_{46}^1 \\ k_{51}^1 & k_{52}^1 & k_{53}^1 & k_{54}^1 & k_{55}^1 & k_{56}^1 \\ k_{61}^1 & k_{62}^1 & k_{63}^1 & k_{64}^1 & k_{65}^1 & k_{66}^1 \end{array}\right] \end{array} \begin{array}{c} \\ (10) \\ (11) \\ (12) \\ (1) \\ (2) \\ (3) \end{array} \rightarrow \left[\begin{array}{c|c} \boxed{k_{10\text{ to }12,\,10\text{ to }12}} & \boxed{k_{10\text{ to }12,\,1\text{ to }3}} \\ \hline \boxed{k_{1\text{ to }3,\,10\text{ to }12}} & \boxed{k_{1\text{ to }3,\,1\text{ to }3}} \end{array}\right] \quad \text{(i)}$$

$$
\mathbf{k}^2 = \begin{array}{c} \begin{array}{cccccc} (1) & (2) & (3) & (4) & (5) & (6) \end{array} \\ \left[\begin{array}{ccc|ccc} k_{11}^2 & k_{12}^2 & k_{13}^2 & k_{14}^2 & k_{15}^2 & k_{16}^2 \\ k_{21}^2 & k_{22}^2 & k_{23}^2 & k_{24}^2 & k_{25}^2 & k_{26}^2 \\ k_{31}^2 & k_{32}^2 & k_{33}^2 & k_{34}^2 & k_{35}^2 & k_{36}^2 \\ \hline k_{41}^2 & k_{42}^2 & k_{43}^2 & k_{44}^2 & k_{45}^2 & k_{46}^2 \\ k_{51}^2 & k_{52}^2 & k_{53}^2 & k_{54}^2 & k_{55}^2 & k_{56}^2 \\ k_{61}^2 & k_{62}^2 & k_{63}^2 & k_{64}^2 & k_{65}^2 & k_{66}^2 \end{array}\right] \end{array} \begin{array}{c} \\ (1) \\ (2) \\ (3) \\ (4) \\ (5) \\ (6) \end{array} \rightarrow \left[\begin{array}{c|c} \boxed{k_{1\text{ to }3,\,1\text{ to }3}} & \boxed{k_{1\text{ to }3,\,4\text{ to }6}} \\ \hline \boxed{k_{4\text{ to }6,\,1\text{ to }3}} & \boxed{k_{4\text{ to }6,\,4\text{ to }6}} \end{array}\right] \qquad \text{(j)}
$$

$$
\mathbf{k}^3 = \begin{array}{c} \begin{array}{cccccc} (4) & (5) & (6) & (8) & (9) & (7) \end{array} \\ \left[\begin{array}{ccc|ccc} k_{11}^3 & k_{12}^3 & k_{13}^3 & k_{14}^3 & k_{15}^3 & k_{16}^3 \\ k_{21}^3 & k_{22}^3 & k_{23}^3 & k_{24}^3 & k_{25}^3 & k_{26}^3 \\ k_{31}^3 & k_{32}^3 & k_{33}^3 & k_{34}^3 & k_{35}^3 & k_{36}^3 \\ \hline k_{41}^3 & k_{42}^3 & k_{43}^3 & k_{44}^3 & k_{45}^3 & k_{46}^3 \\ k_{51}^3 & k_{52}^3 & k_{53}^3 & k_{54}^3 & k_{55}^3 & k_{56}^3 \\ k_{61}^3 & k_{62}^3 & k_{63}^3 & k_{64}^3 & k_{65}^3 & k_{66}^3 \end{array}\right] \end{array} \begin{array}{c} \\ (4) \\ (5) \\ (6) \\ (8) \\ (9) \\ (7) \end{array} \rightarrow \left[\begin{array}{c|c} \boxed{k_{4\text{ to }6,\,4\text{ to }6}} & \boxed{k_{4\text{ to }6,\,(8,9,7)}} \\ \hline \boxed{k_{(8,9,7),\,4\text{ to }6}} & \boxed{k_{(8,9,7),\,(8,9,7)}} \end{array}\right] \qquad \text{(k)}
$$

For the limited purpose of assembling the $\mathbf{k}$ matrix more easily and sequentially (block-wise), we need to carry out suitable row and column exchanges in $\mathbf{k}^3$ as follows:

$$
\left(\mathbf{k}^3\right)_{\text{modified}} = \begin{array}{c} \begin{array}{cccccc} (4) & (5) & (6) & (7) & (8) & (9) \end{array} \\ \left[\begin{array}{ccc|ccc} k_{11}^3 & k_{12}^3 & k_{13}^3 & k_{16}^3 & k_{14}^3 & k_{15}^3 \\ k_{21}^3 & k_{22}^3 & k_{23}^3 & k_{26}^3 & k_{24}^3 & k_{25}^3 \\ k_{31}^3 & k_{32}^3 & k_{33}^3 & k_{36}^3 & k_{34}^3 & k_{35}^3 \\ \hline k_{61}^3 & k_{62}^3 & k_{63}^3 & k_{66}^3 & k_{64}^3 & k_{65}^3 \\ k_{41}^3 & k_{42}^3 & k_{43}^3 & k_{46}^3 & k_{44}^3 & k_{45}^3 \\ k_{51}^3 & k_{52}^3 & k_{53}^3 & k_{56}^3 & k_{54}^3 & k_{55}^3 \end{array}\right] \end{array} \begin{array}{c} \\ (4) \\ (5) \\ (6) \\ (7) \\ (8) \\ (9) \end{array} \rightarrow \left[\begin{array}{c|c} \boxed{k_{4\text{ to }6,\,4\text{ to }6}} & \boxed{k_{4\text{ to }6,\,7\text{ to }9}} \\ \hline \boxed{k_{7\text{ to }9,\,4\text{ to }6}} & \boxed{k_{7\text{ to }9,\,7\text{ to }9}} \end{array}\right] \qquad \text{(l)}
$$

We observe that by combining Eqns (i), (j) and (l), the $\mathbf{k}$ matrix can be easily assembled block-wise. There are 4 nodes (joints) in the structure, with 3 degrees of freedom associated with each .node. Thus, we may visualise the $\mathbf{k}$ matrix of order 12×12 to comprise $4 \times 4 = 16$ blocks, each of order 3×3. The resulting matrix is a square, symmetric matrix, with some of the off-diagonal blocks comprising zero elements:

$$
\mathbf{k} = \left[\begin{array}{ccc|c} \boxed{k_{1\text{ to }3,\,1\text{ to }3}} & \boxed{k_{1\text{ to }3,\,4\text{ to }6}} & \boxed{0} & \boxed{k_{1\text{ to }3,\,10\text{ to }12}} \\ \boxed{k_{4\text{ to }6,\,1\text{ to }3}} & \boxed{k_{4\text{ to }6,\,4\text{ to }6}} & \boxed{k_{4\text{ to }6,\,7\text{ to }9}} & \boxed{0} \\ \boxed{0} & \boxed{k_{7\text{ to }9,\,4\text{ to }6}} & \boxed{k_{7\text{ to }9,\,7\text{ to }9}} & \boxed{0} \\ \hline \boxed{k_{10\text{ to }12,\,1\text{ to }3}} & \boxed{0} & \boxed{0} & \boxed{k_{10\text{ to }12,\,10\text{ to }12}} \end{array}\right] \qquad \text{(m)}
$$

where, taking advantage of symmetry, and summing up the contributions from different elements (Eq. 3.37),

$$\boxed{k_{1\text{ to }3,\,1\text{ to }3}} = \begin{bmatrix} k_{44}^{1}+k_{11}^{2} & & -\text{sym}- \\ k_{54}^{1}+k_{21}^{2} & k_{55}^{1}+k_{22}^{2} & \\ k_{64}^{1}+k_{31}^{2} & k_{65}^{1}+k_{32}^{2} & k_{66}^{1}+k_{33}^{2} \end{bmatrix}; \quad \boxed{k_{4\text{ to }6,\,1\text{ to }3}} = \boxed{k_{1\text{ to }3,\,4\text{ to }6}}^{T} = \begin{bmatrix} k_{41}^{2} & & -\text{sym}- \\ k_{51}^{2} & k_{52}^{2} & \\ k_{61}^{2} & k_{62}^{2} & k_{63}^{2} \end{bmatrix}$$

$$\boxed{k_{4\text{ to }6,\,4\text{ to }6}} = \begin{bmatrix} k_{44}^{2}+k_{11}^{3} & & -\text{sym}- \\ k_{54}^{2}+k_{21}^{3} & k_{55}^{2}+k_{22}^{3} & \\ k_{64}^{2}+k_{31}^{3} & k_{65}^{2}+k_{32}^{3} & k_{66}^{2}+k_{33}^{3} \end{bmatrix}; \quad \boxed{k_{7\text{ to }9,\,4\text{ to }6}} = \boxed{k_{4\text{ to }6,\,7\text{ to }9}}^{T} = \begin{bmatrix} k_{61}^{3} & & -\text{sym}- \\ k_{41}^{3} & k_{42}^{3} & \\ k_{51}^{3} & k_{52}^{3} & k_{53}^{3} \end{bmatrix}$$

(n)

$$\boxed{k_{7\text{ to }9,\,7\text{ to }9}} = \begin{bmatrix} k_{66}^{3} & -\text{sym}- & \\ k_{46}^{3} & k_{44}^{3} & \\ k_{56}^{3} & k_{54}^{3} & k_{55}^{3} \end{bmatrix}; \quad \boxed{k_{10\text{ to }12,\,10\text{ to }12}} = \begin{bmatrix} k_{11}^{1} & -\text{sym}- & \\ k_{21}^{1} & k_{22}^{1} & \\ k_{31}^{1} & k_{32}^{1} & k_{33}^{1} \end{bmatrix}$$

$$\boxed{k_{10\text{ to }12,\,1\text{ to }3}} = \boxed{k_{1\text{ to }3,\,10\text{ to }12}}^{T} = \begin{bmatrix} k_{41}^{1} & & -\text{sym}- \\ k_{51}^{1} & k_{52}^{1} & \\ k_{61}^{1} & k_{62}^{1} & k_{63}^{1} \end{bmatrix}$$

For convenience, we may separate out the active and restrained degrees of freedom (refer Eqns 3.1 and 3.2), by suitably partitioning the structure stiffness matrix $\mathbf{k}$, as indicated in Eq. (m). Accordingly, we may expand the basic stiffness relation (Eq. 3.36) as follows:

$$\begin{bmatrix} \mathbf{F_A} \\ \hline \mathbf{F_R} \end{bmatrix} = \left[\begin{array}{c|c} \mathbf{k_{AA}} & \mathbf{k_{AR}} \\ \hline \mathbf{k_{RA}} & \mathbf{k_{RR}} \end{array}\right] \begin{bmatrix} \mathbf{D_A} \\ \hline \mathbf{D_R} \end{bmatrix} \tag{3.38}$$

$$\Rightarrow \quad \mathbf{F_A} = \mathbf{k_{AA}}\mathbf{D_A} + \mathbf{k_{AR}}\mathbf{D_R} \tag{3.38a}$$

$$\mathbf{F_R} = \mathbf{k_{RA}}\mathbf{D_A} + \mathbf{k_{RR}}\mathbf{D_R} \tag{3.38b}$$

The matrix $\mathbf{k_{AA}}$, pertaining to the active global coordinates, is not only square and symmetric, but is generally *banded* as well, taking a *block diagonal* form. A solution to the unknown displacement vector $\mathbf{D_A}$ is possible only if this matrix is *non-singular*. A zero determinant indicates that the structure that has been modelled is unstable (under-rigid). We will see later that the inverse of the $\mathbf{k_{AA}}$ matrix is the *structure flexibility matrix*. The matrix $\mathbf{k_{AA}}$ has an order $n \times n$, where n is the number of active degrees of freedom in the structure, which is also equal to the degree of kinematic indeterminacy, n_k. The other matrices, $\mathbf{k_{AR}}$, $\mathbf{k_{RA}}$ and $\mathbf{k_{RR}}$, which involve r restrained degrees of freedom, are of interest to us only if there are support movements or if we are interested in finding the support reactions.

In the stiffness method, we need to correctly determine the load vector $\mathbf{F_A}$, which should include equivalent joint loads (due to intermediate loading and indirect loads), as discussed in Section 3.7. Corresponding to any load vector $\mathbf{F_A}$, we can solve Eq. 3.38a, if a solution exists, and thus find the unknown displacement vector $\mathbf{D_A}$ (*displacement response*) to the prescribed loading, which may include support movements, as defined by $\mathbf{D_R}$. Subsequently we can find the *force response*, in terms of support reactions (applying Eq. 3.38b) and member-end forces in local coordinates (applying Eq. 3.33), taking care to include any initial fixed-end forces.

3.4.4 Structure Stiffness Matrix Using Displacement Transformation Matrix

An alternative way of generating the structure stiffness matrix $\mathbf{k}$ from the element stiffness matrices $\mathbf{k_*^i}$ is by making use of the element displacement transformation matrices $\mathbf{T_D^i}$. This method has the advantage that we need to do anything special to keep track of the linkages between the member-end coordinates and the global coordinates. We also do not need to convert the element stiffness matrix $\mathbf{k_*^i}$ to the global axes system. All this is conveniently accounted for by the displacement transformation matrix $\mathbf{T_D^i}$ and its transpose. We may also make the formulation more compact by referring to the *combined* element displacement and force vectors, $\mathbf{D_*}$ and $\mathbf{F_*}$ (which include all the individual element vectors, $\mathbf{D_*^i}$ and $\mathbf{F_*^i}$, respectively) in combination with the displacement transformation matrix $\mathbf{T_D}$ (which includes all the individual element $\mathbf{T_D^i}$ matrices).

Let us now express the stiffness relation between the combined element force and displacement vectors, $\mathbf{F_*}$ and $\mathbf{D_*}$, as follows:

$$\{F_*\}=[k_*]\{D_*\} \tag{3.39}$$

which may be expanded as follows:

$$\begin{bmatrix} \{F_*^1\} \\ \{F_*^2\} \\ \vdots \\ \{F_*^m\} \end{bmatrix}_{mq\times 1} = \begin{bmatrix} [k_*^1] & & & \\ & [k_*^2] & & \\ & & \ddots & \\ & & & [k_*^m] \end{bmatrix}_{mq\times mq} \begin{bmatrix} \{D_*^1\} \\ \{D_*^2\} \\ \vdots \\ \{D_*^m\} \end{bmatrix}_{mq\times 1} \tag{3.39a}$$

where $\mathbf{k_*}$ is a square block diagonal matrix, whose elements are the various element stiffness matrices. If the structure comprises m elements, each with q degrees of freedom, then the $\mathbf{k_*}$ matrix, which is called the **unassembled**† **stiffness matrix**, has an order $mq \times mq$.

Fig. 3.15 shows, by means of a schematic diagram (similar to Fig. 3.12), how the $\mathbf{k}_*$ matrix can be transformed into the structure stiffness matrix $\mathbf{k}$. The basic flow is depicted in a clockwise direction, beginning with a transformation of the (structure) displacement vector to the combined displacement vector, $\mathbf{D}_* = \mathbf{T}_\mathbf{D}\mathbf{D}$, followed by an element stiffness transformation in local coordinates, $\mathbf{F}_* = \mathbf{k}_*\mathbf{D}_*$, which in turn is followed by a transformation of the combined element force vector to the (structure) force vector, $\mathbf{F} = \mathbf{T}_\mathbf{D}{}^\mathrm{T}\mathbf{F}_*$.

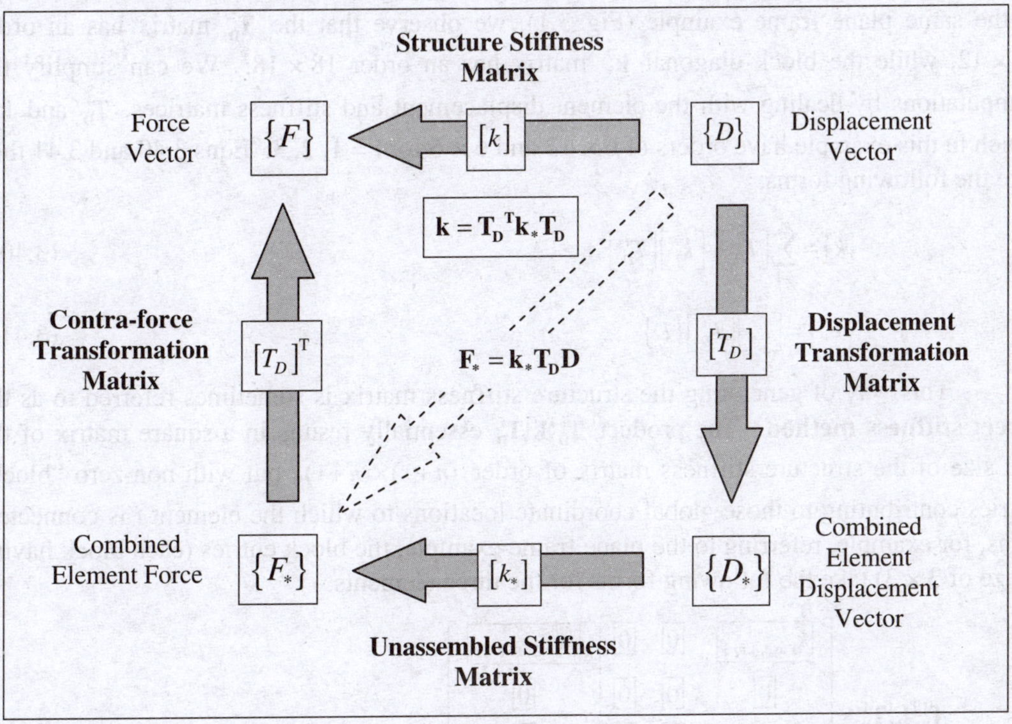

Figure 3.15 Transforming unassembled stiffness matrix to structure stiffness matrix

Thus, we can show:

$$\mathbf{F} = \mathbf{T}_\mathbf{D}{}^\mathrm{T}\mathbf{F}_* = \mathbf{T}_\mathbf{D}{}^\mathrm{T}\left(\mathbf{k}_*\mathbf{D}_*\right) = \mathbf{T}_\mathbf{D}{}^\mathrm{T}\mathbf{k}_*\left(\mathbf{T}_\mathbf{D}\mathbf{D}\right) = \left(\mathbf{T}_\mathbf{D}{}^\mathrm{T}\mathbf{k}_*\mathbf{T}_\mathbf{D}\right)\mathbf{D} \tag{o}$$

$$[k] = [T_D]^\mathrm{T}\left[k_*\right][T_D] \tag{3.40}$$

† The term "unassembled" is used here in the sense that the element stiffnesses of the various elements in a structure are simply put together in a matrix in a mutually disconnected manner without accounting for the connections between them. In the *structure* stiffness matrix, on the other hand, the assemblage of element stiffnesses is complete.

Another useful relation is to be found in the diagonal direction, whereby a direct transformation of the displacement vector **D** to the combined element force vector $\mathbf{F}_*$ in local coordinates is possible:

$$\{F_*\} = [k_*][T_D]\{D\} \tag{3.41}$$

Although Eqns 3.40 and 3.41 present a compact formulation, in actual implementation, this procedure requires dealing with large matrices. For example, referring to the same plane frame example (Fig. 3.1), we observe that the $\mathbf{T_D}$ matrix has an order 18 × 12, while the block diagonal $\mathbf{k}_*$ matrix has an order 18 × 18: We can simplify the computations by dealing with the element displacement and stiffness matrices, $\mathbf{T_D^i}$ and $\mathbf{k_*^i}$, which in this example have orders of 6 × 12 and 6 × 6 for $i = 1, 2, 3$. Eqns 3.40 and 3.41 then take the following forms:

$$[k] = \sum_{i=1}^{m} \left[T_D^i\right]^{\mathrm{T}} \left[k_*^i\right]\left[T_D^i\right] \tag{3.40a}$$

$$\{F_*^i\} = \left[k_*^i\right]\left[T_D^i\right]\{D\} \tag{3.41a}$$

This way of generating the structure stiffness matrix is sometimes referred to as the **direct stiffness method**. The product $\mathbf{T_D^{i\,T} k_*^i T_D^i}$ essentially results in a square matrix of the full size of the structure stiffness matrix of order $(n+r)\times(n+r)$, but with non-zero "block" entries contributing to those global coordinate locations to which the element i is connected. Thus, for example, referring to the plane frame example, the block entries (each block having a size of 3 × 3) take the following forms for the three elements:

$$\mathbf{T_D^{1T} k_*^1 T_D^1} = \left[\begin{array}{ccc|c} k_{1\text{ to }3,\,1\text{ to }3} & 0 & 0 & k_{1\text{ to }3,\,10\text{ to }12} \\ 0 & 0 & 0 & 0 \\ 0 & 0 & 0 & 0 \\ \hline k_{10\text{ to }12,\,1\text{ to }3} & 0 & 0 & k_{10\text{ to }12,\,10\text{ to }12} \end{array}\right]$$

$$\mathbf{T_D^{2T} k_*^2 T_D^2} = \left[\begin{array}{ccc|c} k_{1\text{ to }3,\,1\text{ to }3} & k_{1\text{ to }3,\,4\text{ to }6} & 0 & 0 \\ k_{4\text{ to }6,\,1\text{ to }3} & k_{4\text{ to }6,\,4\text{ to }6} & 0 & 0 \\ 0 & 0 & 0 & 0 \\ \hline 0 & 0 & 0 & 0 \end{array}\right]$$

$$\mathbf{T}_D^{3T}\mathbf{k}_*^3\mathbf{T}_D^3 = \left[\begin{array}{ccc:c} \boxed{0} & \boxed{0} & \boxed{0} & \boxed{0} \\ \boxed{0} & \boxed{k_{4\text{ to }6,\,4\text{ to }6}} & \boxed{k_{4\text{ to }6,\,7\text{ to }9}} & \boxed{0} \\ \boxed{0} & \boxed{k_{7\text{ to }9,\,4\text{ to }6}} & \boxed{k_{7\text{ to }9,\,7\text{ to }9}} & \boxed{0} \\ \hdashline \boxed{0} & \boxed{0} & \boxed{0} & \boxed{0} \end{array}\right] \qquad \text{(p)}$$

Summing up the contributions from the three elements (applying Eq. 3.40a), we get exactly the same result for the structure stiffness matrix $\mathbf{k}$, as indicated by Eq. (m). The matrix is shown partitioned into sectors, $\mathbf{k}_{AA}$, $\mathbf{k}_{AR}$, $\mathbf{k}_{RA}$ and $\mathbf{k}_{RR}$, as depicted in Eq. 3.38.

3.4.5 Reduced Element Stiffness Matrix

As mentioned earlier (in Section 3.2.2), the dimension, $\tilde{q}$, of the element force vector, $\mathbf{F}_*^i$, which is given primary importance in the flexibility method, cannot be the same (q) as that possible in the stiffness method, where the element displacement vector, $\mathbf{D}_*^i$, has primary importance. Whereas the dimension q in the conventional stiffness method is governed by the degrees of freedom (independent member-end displacements), the dimension $\tilde{q}$ in the flexibility method is governed by the number of independent member-end forces, which is equal to the number of independent equations of equilibrium and can be further simplified by making appropriate assumptions, such as ignoring axial deformations in frames. Thus, for example, for a plane frame element, $q = 6$ reducible to $\tilde{q} = 2$ (same as for beam element), while for a plane truss element, $q = 4$ reducible to $\tilde{q} = 1$. The element stiffness matrix $\mathbf{k}_*^i$ of order $q \times q$ is singular and therefore cannot be inverted, while the element flexibility matrix $\mathbf{f}_*^i$ of order $\tilde{q}\times\tilde{q}$ is non-singular and therefore invertible. The inverse of this matrix may be called **the reduced element stiffness matrix**, $\tilde{\mathbf{k}}_*^i$, of order $\tilde{q}\times\tilde{q}$. This matrix can alternatively be generated from the full stiffness matrix $\mathbf{k}_*^i$ through row reduction elimination techniques. By assembling the various $\tilde{\mathbf{k}}_*^i$ matrices in a *block diagonal* form, we obtain the **reduced unassembled stiffness matrix,** $\tilde{\mathbf{k}}_*$.

> The **reduced element stiffness formulation** has the advantage of enabling us to deal with transformation and stiffness matrices of small sizes. It is especially suitable when we wish to make simplifying assumptions in structural analysis, such as ignoring axial deformations in frame elements. However, it should be used with caution, while dealing with "sway" type problems and with sloping members in frames.

The schematic diagram shown in Fig. 3.15 is equally applicable in the reduced element stiffness formulation. All we need to do is to substitute $\tilde{\mathbf{k}}_*$ in the place of $\mathbf{k}_*$, $\tilde{\mathbf{k}}_*^i$ in the place of $\mathbf{k}_*^i$, and $\tilde{q}$ in the place of q, while applying Eqns 3.39 to 3.41.

Let us generate the element stiffness matrices for typical prismatic truss, beam and plane frame elements. The local coordinates for these three element types are shown in Fig. 3.16. The member stiffness relations are also indicated in the figure. A detailed derivation of the stiffness coefficients for these three elements, along with various applications, is given in the Chapters to follow.

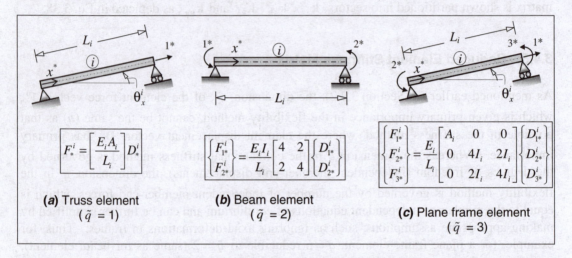

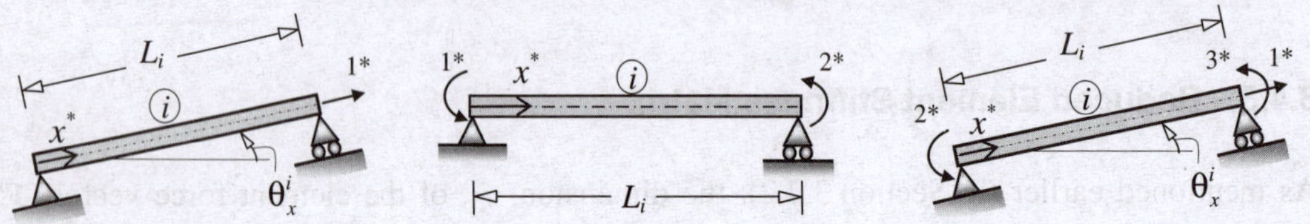

$$F_*^i = \left[\frac{E_i A_i}{L_i}\right] D_*^i$$

(*a*) Truss element ($\tilde{q} = 1$)

$$\begin{Bmatrix} F_{1*}^i \\ F_{2*}^i \end{Bmatrix} = \frac{E_i I_i}{L_i} \begin{bmatrix} 4 & 2 \\ 2 & 4 \end{bmatrix} \begin{Bmatrix} D_{1*}^i \\ D_{2*}^i \end{Bmatrix}$$

(*b*) Beam element ($\tilde{q} = 2$)

$$\begin{Bmatrix} F_{1*}^i \\ F_{2*}^i \\ F_{3*}^i \end{Bmatrix} = \frac{E_i}{L_i} \begin{bmatrix} A_i & 0 & 0 \\ 0 & 4I_i & 2I_i \\ 0 & 2I_i & 4I_i \end{bmatrix} \begin{Bmatrix} D_{1*}^i \\ D_{2*}^i \\ D_{3*}^i \end{Bmatrix}$$

(*c*) Plane frame element ($\tilde{q} = 3$)

Figure 3.16 Local coordinates and member stiffness relations in prismatic truss, beam and plane frame elements (reduced stiffness method)

Prismatic Truss Element

In the truss element (Fig. 3.16a), there is only one effective local coordinate, corresponding to the axial degree of freedom (local coordinate 1*). We may define it as the relative translation in the axial direction (local x^* axis) between the two ends of the element (by providing two equal and opposite arrows). Alternatively, and perhaps more conveniently, we can arrest one end, whereby the same axial deformation in the element is given by the translation at the other end. It is assumed to be positive when it is an *elongation*. As there is only one degree of freedom involved, for convenience, we may designate the displacement as D_*^i (rather than D_{1*}^i) or simply e_i for the element i. Likewise, the conjugate internal force is the axial force in the element (assumed to be tensile if positive) may be designated as F_*^i, or simply N_i (as in Chapter 1). For a *prismatic* truss element, with length L_i and axial rigidity $E_i A_i$, the reduced element stiffness matrix has an order 1 × 1, which takes the following form:

$$\tilde{\mathbf{k}}_*^{\mathbf{i}} = \left[\frac{E_i A_i}{L_i} \right] \tag{3.42}$$

Incidentally, this constitutes the first element (k_{11}^i) of the $\mathbf{k}_*^{\mathbf{i}}$ matrix of order 4×4 shown in Eq. 3.29 as well as the $\mathbf{k}_*^{\mathbf{i}}$ matrix of order 6×6 shown in Eq. 3.28. In the reduced stiffness formulation, the same truss element can be used for one-dimensional (axial) systems, plane trusses and space trusses.

Prismatic Beam Element

In the beam element (Fig. 3.16b), we may identify any two of the four degrees of freedom adopted in the regular stiffness formulation (Fig. 3.11c) as our reduced degrees of freedom. It is convenient to work with member-end rotations and moments (as in the *slope-deflection method*). Thus, we may choose the two member-end rotations, D_{1*}^i and D_{2*}^i, as the basic unknowns in the reduced stiffness formulation for beam elements. Their positive directions are assumed in the anti-clockwise sense (in keeping with the earlier convention shown in Fig. 3.11c, although this is in contrast to the convention traditionally adopted in the slope-deflection method). We will see later (in Chapters 5 and 6) that we can even deal with relative translations at the member ends ("sway" degree of freedom) by means of appropriate *chord rotations*, expressed in terms of D_{1*}^i and D_{2*}^i. The conjugate forces, F_{1*}^i and F_{2*}^i, corresponding to D_{1*}^i and D_{2*}^i, are the member-end moments (assumed anti-clockwise positive).

For a *prismatic* beam element (Fig. 3.16b), with length L_i and flexural rigidity $E_i I_i$, the reduced element stiffness matrix has an order 2×2, and takes the following form (derived in Chapter 5):

$$\tilde{\mathbf{k}}_*^{\mathbf{i}} = \frac{E_i I_i}{L_i} \begin{bmatrix} 4 & 2 \\ 2 & 4 \end{bmatrix} \tag{3.43}$$

Incidentally, the four elements of the above reduced stiffness matrix are obtainable from the 4×4 $\mathbf{k}_*^{\mathbf{i}}$ matrix shown in Eq. 3.30, deleting the rows and columns (first and third) corresponding to the two translational degrees of freedom.

Prismatic Plane Frame Element

The plane frame element may be visualised as a combination of the truss element and the beam element, involving three degrees of freedom. Thus, we may choose the relative translation at the two ends of the element in the axial direction as D_{1*}^i and the two member-end rotations, D_{2*}^i and D_{3*}^i, as the basic unknowns in the reduced stiffness formulation as shown in Fig. 3.16c. The conjugate forces, F_{1*}^i, F_{2*}^i and F_{3*}^i refer respectively to the axial force (tension positive), and the two member-end moments (assumed anti-clockwise positive).

For a *prismatic* plane frame element (Fig. 3.16c), with length L_i, axial rigidity E_iA_i and flexural rigidity E_iI_i, the reduced element stiffness matrix has an order 3×3, and takes the following form, which is obtainable by combining Eqns 3.42 and 3.43, and assuming that there is no coupling[†] between axial and flexural stiffnesses:

$$\tilde{\mathbf{k}}_*^{\mathbf{i}} = \frac{E_i}{L_i}\begin{bmatrix} A_i & 0 & 0 \\ 0 & 4I_i & 2I_i \\ 0 & 2I_i & 4I_i \end{bmatrix} \tag{3.44}$$

Incidentally, the four elements of the above reduced stiffness matrix are obtainable from the $\mathbf{k}_*^{\mathbf{i}}$ matrix of order 6×6 shown in Eq. 3.28.

Assembly of Structure Stiffness Matrix

As outlined in Section 3.4.4, the structure stiffness matrix $\mathbf{k}$ can be generated from the element stiffness matrices $\tilde{\mathbf{k}}_*^{\mathbf{i}}$ by making use of the element displacement transformation matrices $\mathbf{T}_{\mathbf{D}}^{\mathbf{i}}$ (*direct stiffness method*). Alternatively, we can work with the unassembled stiffness matrix $\tilde{\mathbf{k}}_*$ and the displacement transformation matrix $\mathbf{T}_{\mathbf{D}}$, which include the contributions of all the elements in the structure.

Let us demonstrate the application of the reduced stiffness method with reference to the 7-bar plane truss example shown earlier (Fig. 3.10). Let us assume that all the bars have not only the same length (2*a*) but also the same axial rigidity, *EA*. Hence,

$$\tilde{\mathbf{k}}_*^{\mathbf{i}} = \frac{EA}{2a} \qquad i = 1, 2, \ldots, 7 \tag{q}$$

The reduced unassembled stiffness matrix is accordingly given by the diagonal matrix:

$$\tilde{\mathbf{k}}_* = \frac{EA}{2a}\begin{bmatrix} 1 & & & & & & \\ & 1 & & & \mathbf{O} & & \\ & & 1 & & & & \\ & & & 1 & & & \\ & \mathbf{O} & & & 1 & & \\ & & & & & 1 & \\ & & & & & & 1 \end{bmatrix} \tag{r}$$

We had generated the $\mathbf{T}_{\mathbf{D}}$ matrix for this truss in Eq. (g); the various displacement transformation matrices $\mathbf{T}_{\mathbf{D}}^{\mathbf{i}}$ for the 7 elements are easily obtainable as follows:

[†] This assumption is implicit in *linear* first-order structural analysis.

$$
\begin{aligned}
\mathbf{T}_\mathbf{D}^1 &= [0.500\ \ 0.866\ \ 0\ \ 0\ \ 0\ \ 0\ \ 0 \;\vdots\; 0\ \ -0.500\ \ -0.866] \\
\mathbf{T}_\mathbf{D}^2 &= [0\ \ 0\ \ 0\ \ 0\ \ 1.00\ \ 0\ \ 0 \;\vdots\; 0\ \ -1.00\ \ 0] \\
\mathbf{T}_\mathbf{D}^3 &= [-0.500\ \ 0.866\ \ 0\ \ 0\ \ 0.500\ \ -0.866\ \ 0 \;\vdots\; 0\ \ 0\ \ 0] \\
\mathbf{T}_\mathbf{D}^4 &= [-1.00\ \ 0\ \ 1.000\ \ 0\ \ 0\ \ 0\ \ 0 \;\vdots\; 0\ \ 0\ \ 0] \\
\mathbf{T}_\mathbf{D}^5 &= [0\ \ 0\ \ 0.500\ \ 0.866\ \ -0.500\ \ -0.866\ \ 0 \;\vdots\; 0\ \ 0\ \ 0] \\
\mathbf{T}_\mathbf{D}^6 &= [0\ \ 0\ \ 0\ \ 0\ \ -1.00\ \ 0\ \ 1.00 \;\vdots\; 0\ \ 0\ \ 0] \\
\mathbf{T}_\mathbf{D}^7 &= [0\ \ 0\ \ -0.500\ \ 0.866\ \ 0\ \ 0\ \ 0.500 \;\vdots\; -0.866\ \ 0\ \ 0]
\end{aligned}
\qquad \text{(s)}
$$

The structure stiffness matrix may be generated by summing up the individual contributions of the various elements, $[k] = \sum_{i=1}^{7} \left[T_D^i\right]^\mathrm{T} \left[\tilde{k}_*^i\right] \left[T_D^i\right] = \frac{EA}{2a} \sum_{i=1}^{7} \left[T_D^i\right]^\mathrm{T} \left[T_D^i\right]$ (refer Eq. 3.40a), or by the operation, $\mathbf{k} = \mathbf{T_D}^\mathrm{T} \tilde{\mathbf{k}}_* \mathbf{T_D} = \frac{EA}{2a} \mathbf{T_D}^\mathrm{T} \mathbf{T_D}$. The matrix operations can be conveniently done using softwares like MATLAB or SKYLAB.

$$
\mathbf{k} = \frac{EA}{2a}
\left[\begin{array}{ccccccc|ccc}
1.50 & 0 & -1.00 & 0 & -0.25 & 0.433 & 0 & 0 & -0.25 & -0.433 \\
0 & 1.50 & 0 & 0 & 0.433 & -0.75 & 0 & 0 & -0.433 & -0.75 \\
-1.00 & 0 & 1.50 & 0 & -0.25 & -0.433 & -0.25 & 0.433 & 0 & 0 \\
0 & 0 & 0 & 1.50 & -0.433 & -0.75 & 0.433 & -0.75 & 0 & 0 \\
-0.25 & 0.433 & -0.25 & -0.433 & 2.50 & 0 & -1.00 & 0 & -1.00 & 0 \\
0.433 & -0.75 & -0.433 & -0.75 & 0 & 1.50 & 0 & 0 & 0 & 0 \\
0 & 0 & -0.25 & 0.433 & -1.00 & 0 & 1.25 & -0.433 & 0 & 0 \\
\hline
0 & 0 & 0.433 & -0.75 & 0 & 0 & -0.433 & 0.75 & 0 & 0 \\
-0.25 & -0.433 & 0 & 0 & -1.00 & 0 & 0 & 0 & 1.25 & 0.433 \\
-0.433 & -0.75 & 0 & 0 & 0 & 0 & 0 & 0 & 0.433 & 0.75
\end{array}\right]
\qquad \text{(t)}
$$

It may be noted that (i) the matrix generated is symmetric, with all the diagonal elements positive and (ii) the sum of all the elements in any column (and hence, any row) must add up to zero, in order to satisfy equilibrium. This matrix has the form $\mathbf{k} = \left[\begin{array}{c|c} \mathbf{k_{AA}} & \mathbf{k_{AR}} \\ \hline \mathbf{k_{RA}} & \mathbf{k_{RR}} \end{array}\right]$, as mentioned earlier.

If our objective is limited to finding the bar forces alone due to a given set of loads, it is sufficient to generate the 7×7 square and symmetric sub-matrix $\mathbf{k_{AA}}$, involving only the 7 active global coordinates. However, we also need the 7×3 sub-matrix $\mathbf{k_{AR}}$, in case there are support movements to be accounted for (refer Eq. 3.38a). After solving for the unknown displacements $\mathbf{D_A}$, we obtain the bar forces through the transformation, $\mathbf{F}_*^\mathbf{i} = \left(\tilde{\mathbf{k}}_*^\mathbf{i} \mathbf{T}_*^\mathbf{i}\right) \mathbf{D_A}$. To find the support reactions by the stiffness method, we also need the 3×7 sub-matrix $\mathbf{k_{RA}}$, and in case there are support movements, also the 3 × 3 sub-matrix $\mathbf{k}_{\mathbf{RR}}$. If there are indirect load effects to be considered (due to 'lack of fit' or temperature changes), then these effects have to be superimposed, as discussed in Section 3.7 and Chapter 4.

It may be noted that we can arrive at exactly the same stiffness matrix **k** by considering 4 degrees of freedom for each truss element, instead of $\tilde{q} = 1$. This is the way it is done in standard softwares for structural analysis.

3.5 FLEXIBILITY MATRIX

The **flexibility matrix** is a type of transformation matrix, which transforms the force vector to the displacement vector. It establishes the force-displacement relationships underlying the linear behaviour of the structure in the force method of analysis. The components of this matrix are called **flexibility coefficients.**

> The **flexibility coefficient** f_{jk} may be defined as the displacement D_j generated at the coordinate *j* on account of a unit load at the coordinate *k* $(F_k = 1)$, with no other load acting on the structure $(F_{j \neq k} = 0)$.

We begin by first defining, in Section 3.5.1, the *element flexibility matrix* $\mathbf{f}_*^{\mathbf{i}}$, involving a linear relationship between $\mathbf{D}_*^{\mathbf{i}}$ and $\mathbf{F}_*^{\mathbf{i}}$ for the element *i*, in the local coordinate system. Next, in Section 3.5.2, we shall see how the *structure flexibility matrix* **f** for any statically determinate structure, relating the structure displacements $\mathbf{D} = \mathbf{D}_{\mathbf{A}}$ and forces $\mathbf{F} = \mathbf{F}_{\mathbf{A}}$, can be generated from (i) the *unassembled flexibility matrix* $\mathbf{f}_*$, making use of the *force transformation matrix* $\mathbf{T}_{\mathbf{F}}$, and (ii) the *element flexibility matrix* $\mathbf{f}_*^{\mathbf{i}}$, making use of the *element force transformation matrix* $\mathbf{T}_{\mathbf{F}}^{\mathbf{i}}$ derived for the various elements that make up the structure. Finally, in Section 3.5.3, how these concepts can be extended to statically indeterminate structures, for which it is necessary to make a selection of the redundants, based on which the structure flexibility matrix $\mathbf{f} = \left[\begin{array}{c|c} \mathbf{f}_{\mathbf{AA}} & \mathbf{f}_{\mathbf{AX}} \\ \hline \mathbf{f}_{\mathbf{XA}} & \mathbf{f}_{\mathbf{XX}} \end{array}\right]$, relating $\mathbf{D} = \left[\begin{array}{c} \mathbf{D}_{\mathbf{A}} \\ \hline \mathbf{D}_{\mathbf{X}} \end{array}\right]$ and $\mathbf{F} = \left[\begin{array}{c} \mathbf{F}_{\mathbf{A}} \\ \hline \mathbf{F}_{\mathbf{X}} \end{array}\right]$, can be generated.

3.5.1 Element Flexibility Matrix (Local Coordinates)

The flexibility relation for a typical element *i*, expressed in the local coordinate system, relating the element force vector $\mathbf{F}_*^{\mathbf{i}}$ to the element displacement vector $\mathbf{D}_*^{\mathbf{i}}$ is as follows:

$$\{D_*^i\} = [f_*^i]\{F_*^i\} \tag{3.45}$$

where $\mathbf{f}_*^i$ is the **element flexibility matrix** of the element i, of order $\tilde{q}\times\tilde{q}$, and is the inverse of the *reduced element stiffness matrix* $\tilde{\mathbf{k}}_*^i$, defined in Section 3.4.5. The dimension $\tilde{q}$ of the force and displacement, $\mathbf{F}_*^i$ and $\mathbf{D}_*^i$, is given by the number of independent equations available for the element. As mentioned earlier, $\tilde{q}=6$ in a space frame element. This dimension reduces to $\tilde{q}=3$ in a *plane frame element*, and further to $\tilde{q}=2$ in a *beam element* and to $\tilde{q}=1$ in a *truss element*.

For a typical plane frame element (Fig. 3.16c), Eq. 3.45 may be expanded as follows:

$$\begin{array}{c} \\ \begin{bmatrix} D_{1*}^i \\ D_{2*}^i \\ D_{3*}^i \end{bmatrix} \end{array} = \begin{array}{c} (F_{1*}^i=1) \quad (F_{2*}^i=1) \quad (F_{3*}^i=1) \\ \left[\begin{array}{c|c|c} f_{1*1*}^i & f_{1*2*}^i & f_{1*3*}^i \\ f_{2*1*}^i & f_{2*2*}^i & f_{2*3*}^i \\ f_{3*1*}^i & f_{3*2*}^i & f_{3*3*}^i \end{array}\right] \end{array} \begin{array}{c} \\ \begin{bmatrix} F_{1*}^i \\ F_{2*}^i \\ F_{3*}^i \end{bmatrix} \end{array} \tag{3.45a}$$

In Eq. 3.45a, vertical partitions separating the 3 column vectors are shown to indicate that any column vector (say, j^{th} column) of the flexibility matrix can be generated by considering the displacement response caused by a unit load, $F_{j*}^i=1$, with no other loads acting ($F_{k*\neq j*}^i=0$), as indicated in parenthesis in the top row. In the flexibility method, unlike the stiffness method, the basic element must be stable and statically determinate, for us to be able to generate the element flexibility matrix. The element flexibility matrix will depend on the support conditions assumed and the choice of the local coordinates. For example, two alternative choices for a beam element are shown in Fig. 3.17, corresponding to which the element flexibility matrices are indicated in the figure (for a prismatic element).

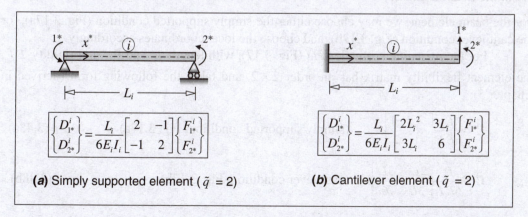

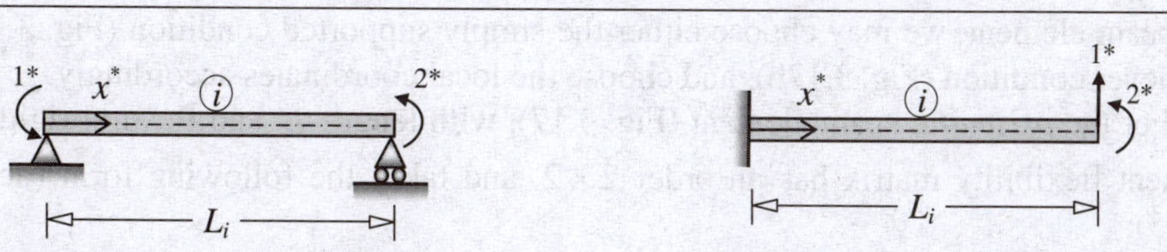

$$\begin{Bmatrix} D_{1*}^i \\ D_{2*}^i \end{Bmatrix} = \frac{L_i}{6E_iI_i}\begin{bmatrix} 2 & -1 \\ -1 & 2 \end{bmatrix}\begin{Bmatrix} F_{1*}^i \\ F_{2*}^i \end{Bmatrix}$$

(a) Simply supported element ($\tilde{q}$ = 2)

$$\begin{Bmatrix} D_{1*}^i \\ D_{2*}^i \end{Bmatrix} = \frac{L_i}{6E_iI_i}\begin{bmatrix} 2L_i^2 & 3L_i \\ 3L_i & 6 \end{bmatrix}\begin{Bmatrix} F_{1*}^i \\ F_{2*}^i \end{Bmatrix}$$

(b) Cantilever element ($\tilde{q}$ = 2)

Figure 3.17 Alternative choices of local coordinates for a prismatic beam element (flexibility method)

The flexibility coefficients can be generated from first principles, as shown in the Chapters to follow. For example, using an energy approach, and expressing the complementary strain energy U_*^i in the element as a function of the components of $\mathbf{F}_*^i$,

$$f_{j*k*}^i = \frac{\partial^2 U_i^*}{\partial F_{j*}^i \partial F_{k*}^i} = f_{k*j*}^i \tag{3.46}$$

The above equation also reveals that the flexibility matrix is a *symmetric* matrix $\left(f_{j*k*}^i = f_{k*j*}^i \right)$.

Let us generate the element stiffness matrices for typical prismatic truss, beam and plane frame elements.

Prismatic Truss Element

In the truss element (Fig. 3.16a), there is only one effective local coordinate, corresponding to the axial degree of freedom (local coordinate 1*). As explained earlier, the axial flexibility of this element is the reciprocal of the axial stiffness. For a *prismatic* truss element, with length L_i and axial rigidity $E_i A_i$, the element flexibility matrix $\mathbf{f}_*^i$ has an order 1×1, which takes the following form:

$$\mathbf{f}_*^i = \left[\frac{L_i}{E_i A_i} \right] \tag{3.47}$$

This matrix is the inverse of the $\tilde{\mathbf{k}}_*^i$ matrix defined in Eq. 3.42. In the flexibility formulation, the same truss element can be used for one-dimensional (axial) systems, plane trusses and space trusses.

Prismatic Beam Element

For the beam element, we may choose either the simply supported condition (Fig. 3.17a), or the cantilever condition (Fig. 3.17b), and choose the local coordinates accordingly.

For the *prismatic* beam element (Fig. 3.17), with length L_i and flexural rigidity $E_i I_i$, the element flexibility matrix has an order 2×2, and takes the following form (derived in Chapter 5):

$$\mathbf{f}_*^i = \frac{L_i}{6E_i I_i} \begin{bmatrix} 2 & -1 \\ -1 & 2 \end{bmatrix} \quad \text{(simply supported condition, Fig. 3.17a)} \tag{3.48a}$$

$$\mathbf{f}_*^i = \frac{L_i}{6E_i I_i} \begin{bmatrix} 2L_i^2 & 3L_i \\ 3L_i & 6 \end{bmatrix} \quad \text{(cantilever condition, Fig. 3.17b)} \tag{3.48b}$$

Incidentally, the matrix shown in Eq. 3.48a is the inverse of the $\tilde{\mathbf{k}}_*^i$ matrix defined in Eq. 3.43, as both matrices refer to the same set of local coordinates.

Prismatic Plane Frame Element

The plane frame element may be visualised as a combination of the truss element and the beam element, involving three local coordinates. We may choose either a simply supported or a cantilever condition, as shown in Fig. 3.18, as in the case of the beam element.

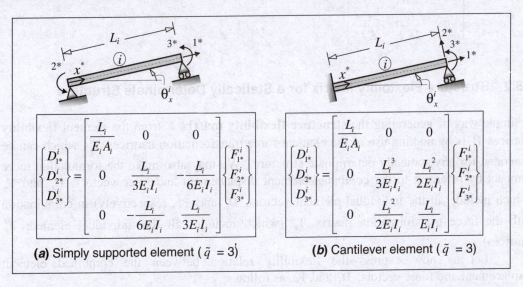

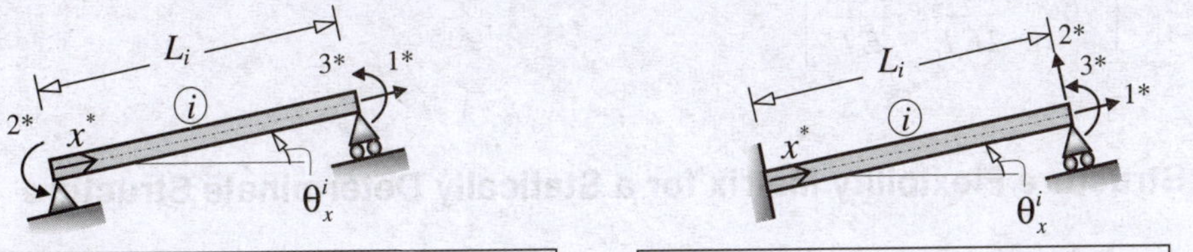

$$\begin{Bmatrix} D_{1*}^i \\ D_{2*}^i \\ D_{3*}^i \end{Bmatrix} = \begin{bmatrix} \dfrac{L_i}{E_iA_i} & 0 & 0 \\ 0 & \dfrac{L_i}{3E_iI_i} & -\dfrac{L_i}{6E_iI_i} \\ 0 & -\dfrac{L_i}{6E_iI_i} & \dfrac{L_i}{3E_iI_i} \end{bmatrix} \begin{Bmatrix} F_{1*}^i \\ F_{2*}^i \\ F_{3*}^i \end{Bmatrix}$$

(a) Simply supported element ($\tilde{q}$ = 3)

$$\begin{Bmatrix} D_{1*}^i \\ D_{2*}^i \\ D_{3*}^i \end{Bmatrix} = \begin{bmatrix} \dfrac{L_i}{E_iA_i} & 0 & 0 \\ 0 & \dfrac{L_i^3}{3E_iI_i} & \dfrac{L_i^2}{2E_iI_i} \\ 0 & \dfrac{L_i^2}{2E_iI_i} & \dfrac{L_i}{E_iI_i} \end{bmatrix} \begin{Bmatrix} F_{1*}^i \\ F_{2*}^i \\ F_{3*}^i \end{Bmatrix}$$

(b) Cantilever element ($\tilde{q}$ = 3)

Figure 3.18 Alternative choices of local coordinates for a prismatic plane frame element (flexibility method)

For a *prismatic* plane frame element (Fig. 3.18), with length L_i, axial rigidity E_iA_i and flexural rigidity E_iI_i, the element flexibility matrix has an order 3×3, and takes the following form, assuming that there is no coupling between axial and flexural flexibilities:

$$\mathbf{f}_*^i = \begin{bmatrix} \dfrac{L_i}{E_iA_i} & 0 & 0 \\ 0 & \dfrac{L_i}{3E_iI_i} & -\dfrac{L_i}{6E_iI_i} \\ 0 & -\dfrac{L_i}{6E_iI_i} & \dfrac{L_i}{3E_iI_i} \end{bmatrix} \quad \text{(simply supported condition, Fig. 3.18a)} \qquad (3.49a)$$

Incidentally, the matrix shown in Eq. 3.49a is the inverse of the $\tilde{\mathbf{k}}_*^i$ matrix defined in Eq. 3.44, as both matrices refer to the same set of local coordinates.

$$\mathbf{f}_*^{\mathbf{i}} = \begin{bmatrix} \frac{L_i}{E_i A_i} & 0 & 0 \\ 0 & \frac{L_i^3}{3E_i I_i} & \frac{L_i^2}{2E_i I_i} \\ 0 & \frac{L_i^2}{2E_i I_i} & \frac{L_i}{E_i I_i} \end{bmatrix} \quad \text{(cantilever condition, Fig. 3.18b)} \qquad (3.49b)$$

3.5.2 Structure Flexibility Matrix for a Statically Determinate Structure

A simple way of generating the **structure flexibility matrix** $\mathbf{f}$ from the element flexibility matrices $\mathbf{f}_*^{\mathbf{i}}$ is by making use of the element force transformation matrices $\mathbf{T}_{\mathbf{F}}^{\mathbf{i}}$, which can be generated for any statically determinate structure. We may also make the formulation more compact by referring to the *combined* element displacement and force vectors, $\mathbf{D}_*$ and $\mathbf{F}_*$ (which include all the individual element vectors, $\mathbf{D}_*^{\mathbf{i}}$ and $\mathbf{F}_*^{\mathbf{i}}$, respectively) in combination with the force transformation matrix $\mathbf{T}_{\mathbf{F}}$ (which includes all the individual element $\mathbf{T}_{\mathbf{F}}^{\mathbf{i}}$ matrices).

Let us now express the flexibility relation between the combined element displacement and force vectors, $\mathbf{D}_*$ and $\mathbf{F}_*$, as follows:

$$\{D_*\} = [f_*]\{F_*\} \qquad (3.50)$$

which may be expanded as follows:

$$\begin{bmatrix} \{D_*^1\} \\ \{D_*^2\} \\ \vdots \\ \{D_*^m\} \end{bmatrix}_{m\tilde{q}\times 1} = \begin{bmatrix} [f_*^1] & & & \\ & [f_*^2] & & \\ & & \ddots & \\ & & & [f_*^m] \end{bmatrix}_{m\tilde{q}\times m\tilde{q}} \begin{bmatrix} \{F_*^1\} \\ \{F_*^2\} \\ \vdots \\ \{F_*^m\} \end{bmatrix}_{m\tilde{q}\times 1} \qquad (3.50a)$$

where $\mathbf{f}_*$ is a square block diagonal matrix, whose elements are the various element flexibility matrices. If the structure comprises m elements, each with $\tilde{q}$ degrees of freedom, then the $\mathbf{f}_*$ matrix, which is called the **unassembled flexibility matrix**, has an order $m\tilde{q}\times m\tilde{q}$.

Fig. 3.19 shows, by means of a schematic diagram, how the $\mathbf{f}_*$ matrix can be transformed into the structure flexibility matrix $\mathbf{f}$. The basic flow is depicted in an anti-clockwise direction, beginning with a transformation of the (structure) force vector to the combined force vector, $\mathbf{F}_* = \mathbf{T}_{\mathbf{F}}\mathbf{F}$, followed by an element flexibility transformation in local coordinates, $\mathbf{D}_* = \mathbf{f}_*\mathbf{F}_*$, which in turn is followed by a transformation of the combined element displacement vector to the (structure) displacement vector, $\mathbf{D} = \mathbf{T_F}^\mathbf{T}\mathbf{D_*}$. Thus, we can show:

$$\mathbf{D} = \mathbf{T_F}^\mathbf{T}\mathbf{D_*} = \mathbf{T_F}^\mathbf{T}\left(\mathbf{f_*}\mathbf{F_*}\right) = \mathbf{T_F}^\mathbf{T}\mathbf{f_*}\left(\mathbf{T_F}\mathbf{F}\right) = \left(\mathbf{T_F}^\mathbf{T}\mathbf{f_*}\mathbf{T_F}\right)\mathbf{F} \tag{u}$$

$$\Rightarrow \quad \{D\} = [f]\{F\} \tag{3.51}$$

$$[f] = [T_F]^\mathrm{T}[f_*][T_F] \tag{3.52}$$

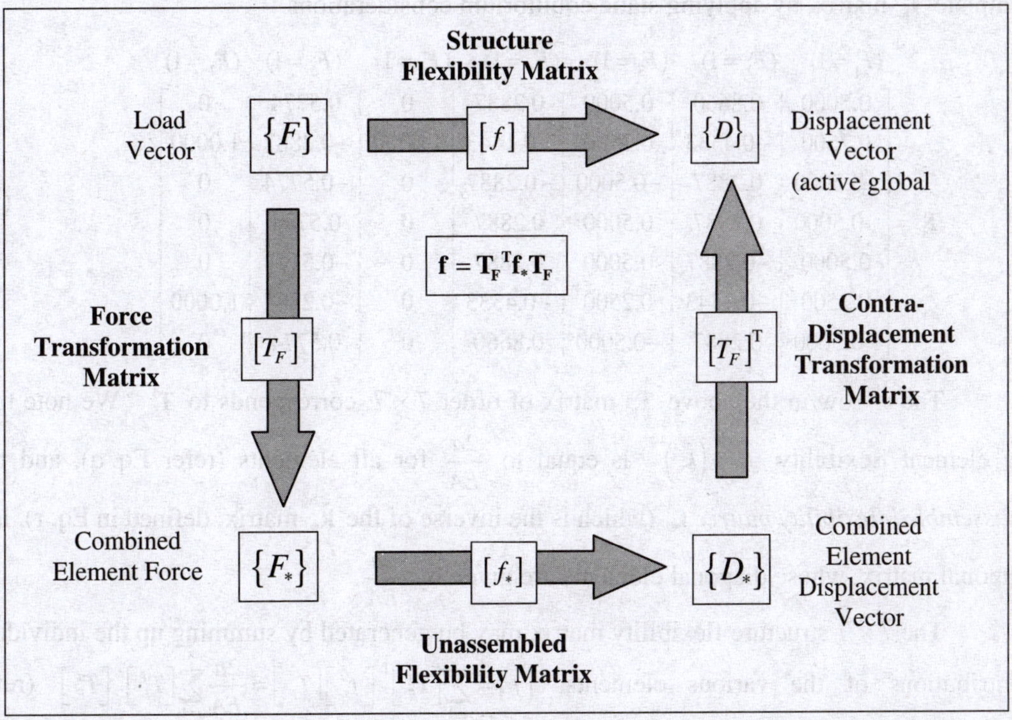

Figure 3.19 Transforming unassembled flexibility matrix to structure flexibility matrix

Although Eq. 3.51 presents a compact formulation, in actual implementation, this procedure may require dealing with large matrices. We can simplify the computations by dealing with the element force and flexibility matrices, $\mathbf{T_F^i}$ and $\mathbf{f_*^i}$. Eq. 3.52 then takes the following form:

$$[f] = \sum_{i=1}^{m}\left[T_F^i\right]^\mathrm{T}\left[f_*^i\right]\left[T_F^i\right] \tag{3.52a}$$

The product $\mathbf{T_F^i}^\mathbf{T}\mathbf{f_*^i}\mathbf{T_F^i}$ essentially results in a square matrix of the full size of the structure flexibility matrix, which is of order $(n+r)\times(n+r)$ if r "rigid link" elements are introduced in order to simulate the support reaction forces; otherwise, the **f** matrix will be of order $n \times n$, where n is equal to the number of active global coordinates. In the latter case, the **f** matrix will be the reciprocal of the $\mathbf{k}_{AA}$ matrix (structure stiffness matrix, involving active coordinates alone) for the same statically determinate (just-rigid) structure.

For example, let us consider the same plane truss example, for which we had derived the reduced element structure stiffness matrix in Section 3.4.5, and for which we had derived the sixth column vector of the $\mathbf{T_F}$ matrix (refer Fig. 3.7). Now, ignoring the support reactions, and taking advantage of the symmetry in the structure, we can generate the complete $\mathbf{T_F}$ matrix, by applying static equilibrium considerations[†]:

$$\mathbf{T_F} = \begin{array}{c} \begin{matrix} (F_1=1) & (F_2=1) & (F_3=1) & (F_4=1) & (F_5=1) & (F_6=1) & (F_7=1) \end{matrix} \\ \left[\begin{array}{c|c|c|c|c|c|c} 0.5000 & 0.8660 & 0.5000 & 0.2887 & 0 & 0.5774 & 0 \\ 0.7500 & -0.4333 & 0.7500 & -0.1443 & 1.0000 & -0.2887 & 1.0000 \\ -0.5000 & 0.2887 & -0.5000 & -0.2887 & 0 & -0.5774 & 0 \\ -0.5000 & 0.2887 & 0.5000 & 0.2887 & 0 & 0.5774 & 0 \\ 0.5000 & -0.2887 & 0.5000 & 0.2887 & 0 & -0.5774 & 0 \\ 0.2500 & -0.1443 & 0.2500 & -0.4333 & 0 & -0.2887 & 1.0000 \\ -0.5000 & 0.2887 & -0.5000 & 0.8660 & 0 & 0.5774 & 0 \end{array}\right] \end{array} \quad \text{(v)}$$

The ith row in the above $\mathbf{T_F}$ matrix of order 7×7, corresponds to $\mathbf{T_F^i}$. We note that the element flexibility $f_*^i = \left(\tilde{k}_*^i\right)^{-1}$ is equal to $\frac{2a}{EA}$ for all elements (refer Eq. q), and the *unassembled flexibility matrix* $\mathbf{f}_*$ (which is the inverse of the $\tilde{\mathbf{k}}_*$ matrix, defined in Eq. r), is a diagonal matrix, whose diagonal elements are given by $\frac{2a}{EA}$.

The 7×7 structure flexibility matrix may be generated by summing up the individual contributions of the various elements, $[f] = \sum_{i=1}^{7}\left[T_F^i\right]^{\mathrm{T}}\left[f_*^i\right]\left[T_F^i\right] = \frac{2a}{EA}\sum_{i=1}^{7}\left[T_F^i\right]^{\mathrm{T}}\left[T_F^i\right]$ (refer Eq. 3.52a), or by applying Eq. 3.52, whereby $\mathbf{f} = \mathbf{T_F}^{\mathrm{T}}\mathbf{f}_*\mathbf{T_F} = \frac{2a}{EA}\mathbf{T_F}^{\mathrm{T}}\mathbf{T_F}$:

[†] The reader is invited to carry out this exercise and verify the results.

$$\mathbf{f} = \frac{2a}{EA} \begin{array}{c} \begin{array}{ccccccc} (F_1=1) & (F_2=1) & (F_3=1) & (F_4=1) & (F_5=1) & (F_6=1) & (F_7=1) \end{array} \\ \left[\begin{array}{c:c:c:c:c:c:c} 1.8750 & -0.5055 & 1.3750 & -0.3609 & 0.7500 & -0.5774 & 1.0000 \\ -0.5055 & 1.2919 & -0.2168 & 0.5417 & -0.4333 & 1.0002 & -0.5776 \\ 1.3750 & -0.2168 & 1.8750 & -0.0721 & 0.7500 & 0.0000 & 1.0000 \\ -0.3609 & 0.5417 & -0.0721 & 1.2919 & -0.1443 & 1.0002 & -0.5776 \\ 0.7500 & -0.4333 & 0.7500 & -0.1443 & 1.0000 & -0.2887 & 1.0000 \\ -0.5774 & 1.0002 & 0.0000 & 1.0002 & -0.2887 & 1.8336 & -0.5774 \\ 1.0000 & -0.5776 & 1.0000 & -0.5776 & 1.0000 & -0.5774 & 2.0000 \end{array} \right] \end{array} \qquad \text{(w)}$$

The jth column in the above 7×7 structure flexibility matrix describes the various bar elongations (D_*^i) due to the action of a unit load applied at the coordinate j $\left(F_j = 1\right)$. The matrix is symmetric and invertible, with all the diagonal elements positive.

> In statically determinate structures, the generation of the **structure flexibility matrix** $\mathbf{f}$ is meaningful and worthwhile only if we are interested in finding the displacement vector, $\mathbf{D} = \mathbf{D}_\mathrm{A} = \mathbf{f}\mathbf{F}_\mathrm{A}$. There are no simultaneous equations to solve here for this purpose. The element forces are directly obtainable from the load vector $\mathbf{F}_\mathrm{A}$ by the force transformation $\mathbf{F}_*^\mathrm{i} = \mathbf{T}_\mathrm{F}^\mathrm{i}\mathbf{F}_\mathrm{A}$. The same structure, however, is kinematically indeterminate, and by the stiffness method, there is a need to generate the structure stiffness matrix $\mathbf{k}_\mathrm{AA}$ (at the active global coordinates) and to solve linear simultaneous equations, $\mathbf{F}_\mathrm{A} = \mathbf{k}_\mathrm{AA}\mathbf{D}_\mathrm{A}$, in order to find the unknown displacements $\mathbf{D}_\mathrm{A}$, and thereby the element forces through the transformation, $\mathbf{F}_*^\mathrm{i} = \mathbf{k}_*^\mathrm{i}\mathbf{T}_\mathrm{D}^\mathrm{i}\mathbf{D}_\mathrm{A}$. Furthermore, if there are indirect loads (such as support movements or temperature effects), these need to be also accounted for in finding the response by the stiffness method by additional effort. In contrast, by the flexibility method, the element forces remain unaffected by indirect loads. Hence, the flexibility method of analysis is the most efficient solution for statically determinate structures, although software packages are designed to apply the stiffness method to even such structures (such as just-rigid trusses), which may have a high degree of kinematic indeterminacy and zero degree of static indeterminacy.

The inverse of the structure flexibility matrix given in Eq. (w), which can be easily generated using software like MATLAB, yields a 7×7 structure stiffness matrix, which is none other than $\mathbf{k}_\mathrm{AA}$, involving the active degrees of freedom.

$$\mathbf{f}^{-1} = \mathbf{k}_{AA} = \frac{EA}{2a}\begin{bmatrix} 1.5000 & 0.0001 & -1.0000 & 0.0000 & -0.2500 & 0.4329 & 0.0000 \\ 0.0001 & 1.5000 & 0.0000 & 0.0000 & 0.4335 & -0.7499 & -0.0001 \\ -1.0000 & 0.0000 & 1.5000 & -0.0001 & -0.2500 & -0.4329 & -0.2500 \\ 0.0000 & 0.0000 & -0.0001 & 1.5000 & -0.4335 & -0.7499 & 0.4335 \\ -0.2500 & 0.4335 & -0.2500 & -0.4335 & 2.5005 & 0.0000 & -1.0003 \\ 0.4329 & -0.7499 & -0.4329 & -0.7499 & 0.0000 & 1.4997 & -0.0002 \\ 0.0000 & -0.0001 & -0.2500 & 0.4335 & -1.0003 & -0.0002 & 1.2503 \end{bmatrix} \quad \text{(x)}$$

Comparing the results generated above with the $\mathbf{k}_{AA}$ matrix generated earlier by the stiffness method (refer Eq. t) shows that the values match very closely, except for rounding off errors (in the fourth significant place). We could have, alternatively, generated the structure flexibility matrix as the inverse of the $\mathbf{k}_{AA}$ matrix:

$$\mathbf{k}_{AA} = \mathbf{f}^{-1} \quad \Rightarrow \quad \mathbf{f} = \mathbf{k}_{AA}^{-1} \quad (3.53)$$

3.5.3 Structure Flexibility Matrix for a Statically Indeterminate Structure

In the flexibility (force) method of analysis, statically indeterminate problems are handled by identifying appropriate *redundant forces* (equal in number to the degree of static indeterminacy, n_s), which are treated as unknown forces acting on a statically determinate *primary* structure, along with the external loads.

As explained in Section 3.3.4, the force vector, for such structures is given by $\mathbf{F} = \begin{bmatrix} \mathbf{F}_A \\ \hline \mathbf{F}_X \end{bmatrix}$, and the conjugate displacement vector is $\mathbf{D} = \begin{bmatrix} \mathbf{D}_A \\ \hline \mathbf{D}_X \end{bmatrix}$. The dimension of the *active* force and displacement vectors, $\mathbf{F}_A$ and $\mathbf{D}_A$, is equal to the number of active degrees of freedom, n, in the given statically indeterminate structure, if 'rigid links' are not introduced to simulate the support reactions; otherwise, it gets enhanced to $n + r$. The dimension of the force and displacement vectors, $\mathbf{F}_X$ and $\mathbf{D}_X$, corresponding to the *redundant* coordinate locations in the structure, is equal to the degree of static indeterminacy, n_s. Typically, $\mathbf{F}_A$ and $\mathbf{D}_X$ are unknown, and our task is to find $\mathbf{F}_X$ (and thereby the element-level forces), and if required, $\mathbf{D}_A$ also. It is convenient to number the redundant coordinates sequentially after the numbering of the active global coordinates.

If there are indirect loads involved (other than support displacements at the redundant coordinates, $\mathbf{D}_X$), these can be expressed as "initial displacements" in terms of changes in element lengths in the 'free' condition, such as due to temperature changes or 'lack of fit' or due to support movements at non-redundant support locations (indicated as changes in lengths in the rigid links). The **initial element displacement vector**, $\mathbf{D}^i_{*initial}$, can be transformed to the global coordinates, to generate the initial displacement vector $\mathbf{D}_{\text{initial}} = \left[\begin{array}{c}\mathbf{D}_{\text{A,initial}} \\ \hdashline \mathbf{D}_{\text{X,initial}}\end{array}\right]$ as follows:

$$\mathbf{D}_{\text{initial}} = \left[\begin{array}{c}\mathbf{D}_{\text{A,initial}} \\ \hdashline \mathbf{D}_{\text{X,initial}}\end{array}\right] = \left[\begin{array}{c}\mathbf{T}_{\text{FA}}^{\text{T}} \\ \hdashline \mathbf{T}_{\text{FX}}^{\text{T}}\end{array}\right] \mathbf{D}_{*\text{initial}} \tag{3.54}$$

The basic flexibility relationship at the structure level (Eq. 3.51), accordingly, takes the following form, applying the Principle of Superposition, in terms of the **primary structure flexibility matrix, f** :

$$\left[\begin{array}{c}\mathbf{D}_{\text{A}} \\ \hdashline \mathbf{D}_{\text{X}}\end{array}\right] = \left[\begin{array}{c}\mathbf{D}_{\text{A,initial}} \\ \hdashline \mathbf{D}_{\text{X,initial}}\end{array}\right] + \left[\begin{array}{c:c}\mathbf{f}_{\text{AA}} & \mathbf{f}_{\text{AX}} \\ \hdashline \mathbf{f}_{\text{XA}} & \mathbf{f}_{\text{XX}}\end{array}\right] \left[\begin{array}{c}\mathbf{F}_{\text{A}} \\ \hdashline \mathbf{F}_{\text{X}}\end{array}\right] \tag{3.55}$$

$$\Rightarrow \qquad \{D_A\} = \{D_{A,\text{initial}}\} + [f_{AA}]\{F_A\} + [f_{AX}]\{F_X\} \tag{3.55a}$$

$$\{D_X\} = \{D_{X,\text{initial}}\} + [f_{XA}]\{F_A\} + [f_{XX}]\{F_X\} \tag{3.55b}$$

$$\Rightarrow \qquad \{F_X\} = [f_{XX}]^{-1}\left(\{D_X\} - \{D_{X,\text{initial}}\} - [f_{XA}]\{F_A\}\right) \tag{3.56}$$

Under direct loading conditions, in general, the compatibility conditions required to solve Eq. 3.55, are given by $\mathbf{D}_{\mathbf{X}} = \mathbf{D}_{\mathbf{X,initial}} = \mathbf{O}$.

$$\mathbf{D}_{\mathbf{X}} = \mathbf{O} \qquad \Rightarrow \qquad \{F_X\} = -[f_{XX}]^{-1}[f_{XA}]\{F_A\} \tag{3.56a}$$

Substituting Eq. 3.56a in Eq. 3.55a, and simplifying (*condensing out* the redundant coordinates in a process called *static condensation*), we arrive at a structure flexibility relation for the statically indeterminate structure. For convenience, we may replace the notations $\mathbf{F}_{\mathbf{A}}$ and $\mathbf{D}_{\mathbf{A}}$ (applicable for the primary structure) with **F** and **D** for the given statically indeterminate structure. Accordingly, Eq. 3.51 simplifies to:

$$\{D\} = \left[\bar{f}\right]\{F\} \tag{3.57}$$

where $\bar{\mathbf{f}}$ is the **structure flexibility matrix** of order $n \times n$ for the statically indeterminate structure, given by:

$$\left[\bar{f}\right] = [f_{AA}] - [f_{AX}][f_{XX}]^{-1}[f_{XA}] \tag{3.58}$$

The **primary structure flexibility matrix** $\mathbf{f} = \left[\begin{array}{c:c}\mathbf{f}_{\text{AA}} & \mathbf{f}_{\text{AX}} \\ \hdashline \mathbf{f}_{\text{XA}} & \mathbf{f}_{\text{XX}}\end{array}\right]$, on the other hand, contains components pertaining to both the active global coordinates and the redundant coordinates. It is a square symmetric matrix of order $(n+n_s)\times(n+n_s)$ if no 'rigid links' are included in the model to simulate the support reactions. Otherwise, if the 'rigid links' are included, then the order of the matrix gets expanded to $(n+r+n_s)\times(n+r+n_s)$.

We can also apply static condensation to eliminate the redundant coordinates from the force transformation matrix relation, $\mathbf{F_* = T_{FA}F_A + T_{FX}F_X}$ (substituting Eq. 3.56), as follows:

$$\{F_*\} = \left[\bar{T}_F\right]\{F\} \tag{3.59}$$

$$\{F_*^i\} = \left[\bar{T}_F^i\right]\{F\} \tag{3.59a}$$

where $\mathbf{\bar{T}_F^i}$ is the **element force transformation matrix** for the ith element, while $\mathbf{\bar{T}_F}$ is the overall **force transformation matrix** for the statically indeterminate structure, which includes considerations of displacement compatibility and flexibility relations, unlike the force transformation matrix defined for a statically determinate structure (where only force equilibrium is considered).

$$\left[\bar{T}_F\right] = [T_{FA}] - [T_{FX}][f_{XX}]^{-1}[f_{XA}] \tag{3.60}$$

$$\left[\bar{T}_F^i\right] = \left[T_{FA}^i\right] - \left[T_{FX}^i\right][f_{XX}]^{-1}[f_{XA}] \tag{3.60a}$$

In order to generate the matrix relations to find the force response in statically indeterminate structures, we need to first select the redundants and thereby identify the primary structure. For this statically determinate primary structure, we can set up the element force transformation matrices $\mathbf{T_F^i} = \left[\mathbf{T_{FA}^i} \mid \mathbf{T_{FX}^i}\right]$ and generate the primary structure flexibility matrix $\mathbf{f} = \begin{bmatrix} \mathbf{f_{AA}} & \mathbf{f_{AX}} \\ \mathbf{f_{XA}} & \mathbf{f_{XX}} \end{bmatrix}$. Using these matrices (which depend on the choice of the redundants), the element force transformation matrix, $\mathbf{\bar{T}_F^i = T_{FA}^i - T_{FX}^i f_{XX}^{-1} f_{XA}}$, and the structure flexibility matrix, $\mathbf{\bar{f} = f_{AA} - f_{AX} f_{XX}^{-1} f_{XA}}$, which are unique for the given statically indeterminate structure, can be generated. Thereby, for any given load vector, $\mathbf{F}$, the element forces, $\mathbf{F_*^i = \bar{T}_F^i F}$, and the nodal displacements, $\mathbf{D = \bar{f}\,F}$, can be generated.

Let us illustrate this procedure with reference to the plane truss in Fig. 3.20a, which is the same example truss we had seen earlier (Fig. 3.7), except that additional support reactions (vertical reaction at E and horizontal reaction at D) have been introduced, making the structure statically indeterminate (to the second degree). Choosing these two reactions as the redundants, $\mathbf{F_X} = \begin{Bmatrix} X_1 = F_6 \\ X_2 = F_7 \end{Bmatrix}$, the primary (statically determinate) structure, shown in

Fig. 3.20b is identical to the truss we had analysed earlier, with the same global coordinates. The compatibility conditions are given by $\mathbf{D_X} = \begin{Bmatrix} D_6 = 0 \\ D_7 = 0 \end{Bmatrix}$.

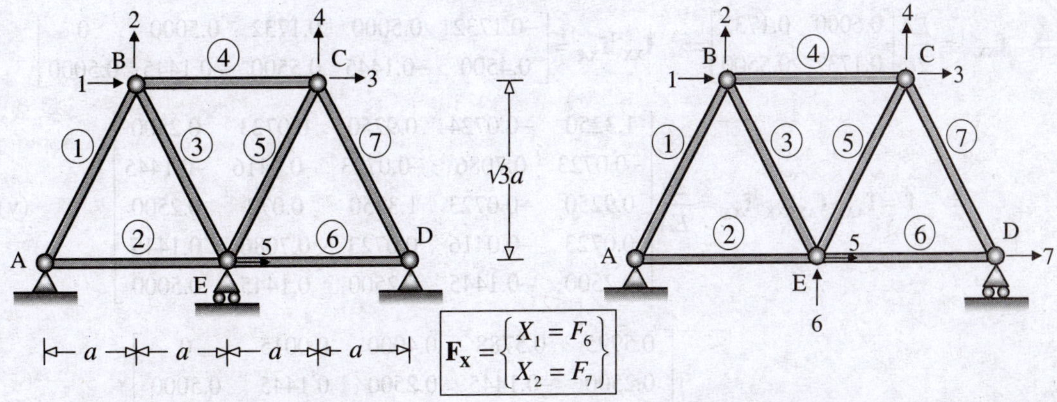

(a) Statically indeterminate truss ($n = 5$, $n_s = 2$, $m = 7$)

(b) Statically determinate primary structure ($n = 5 + 2$, $m = 7$)

Figure 3.20 Choice of redundants in a statically indeterminate truss (flexibility method)

The force transformation matrix, $\mathbf{T_F}$, is the same as the one derived in Eq. (v), and the structure flexibility matrix is also the same as derived in Eq. (w). By partitioning of these matrices as $\mathbf{T_F^i} = \left[\mathbf{T_{FA}^i} \mid \mathbf{T_{FX}^i} \right]$ and $\mathbf{f} = \begin{bmatrix} \mathbf{f_{AA}} & \mathbf{f_{AX}} \\ \mathbf{f_{XA}} & \mathbf{f_{XX}} \end{bmatrix}$, we obtain:

$$\mathbf{T_{FA}} = \begin{bmatrix} 0.5000 & 0.8660 & 0.5000 & 0.2887 & 0 \\ 0.7500 & -0.4333 & 0.7500 & -0.1443 & 1.0000 \\ -0.5000 & 0.2887 & -0.5000 & -0.2887 & 0 \\ -0.5000 & 0.2887 & 0.5000 & 0.2887 & 0 \\ 0.5000 & -0.2887 & 0.5000 & 0.2887 & 0 \\ 0.2500 & -0.1443 & 0.2500 & -0.4333 & 0 \\ -0.5000 & 0.2887 & -0.5000 & 0.8660 & 0 \end{bmatrix}; \quad \mathbf{T_{FX}} = \begin{bmatrix} 0.5774 & 0 \\ -0.2887 & 1.0000 \\ -0.5774 & 0 \\ 0.5774 & 0 \\ -0.5774 & 0 \\ -0.2887 & 1.0000 \\ 0.5774 & 0 \end{bmatrix}$$

$$\mathbf{f_{AA}} = \frac{2a}{EA} \begin{bmatrix} 1.8750 & -0.5055 & 1.3750 & -0.3609 & 0.7500 \\ -0.5055 & 1.2919 & -0.2168 & 0.5417 & -0.4333 \\ 1.3750 & -0.2168 & 1.8750 & -0.0721 & 0.7500 \\ -0.3609 & 0.5417 & -0.0721 & 1.2919 & -0.1443 \\ 0.7500 & -0.4333 & 0.7500 & -0.1443 & 1.0000 \end{bmatrix}; \quad \mathbf{f_{AX}} = \frac{2a}{EA} \begin{bmatrix} -0.5774 & 1.0000 \\ 1.0000 & -0.5776 \\ 0 & 1.0000 \\ 1.0000 & -0.5776 \\ -0.2887 & 1.0000 \end{bmatrix}$$

$$\mathbf{f}_{XA} = \frac{2a}{EA}\begin{bmatrix} -0.5774 & 1 & 0 & 1 & -0.2887 \\ 1 & -0.5774 & 1 & -0.5774 & 1 \end{bmatrix} \; ; \; \mathbf{f}_{XX} = \frac{2a}{EA}\begin{bmatrix} 1.8336 & -0.5774 \\ -0.5774 & 2.0000 \end{bmatrix}$$

We may now carry out the matrix operations (preferably, using alogorithms such as those available in MATLAB), to generate $\bar{\mathbf{f}}$ and $\bar{\mathbf{T}}_F$:

$$\Rightarrow \mathbf{f}_{XX}^{-1} = \frac{EA}{2a}\begin{bmatrix} 0.6000 & 0.1732 \\ 0.1732 & 0.5500 \end{bmatrix} \Rightarrow \mathbf{f}_{XX}^{-1}\mathbf{f}_{XA} = \begin{bmatrix} -0.1732 & 0.5000 & 0.1732 & 0.5000 & 0 \\ 0.4500 & -0.1445 & 0.5500 & -0.1445 & 0.5000 \end{bmatrix}$$

$$\Rightarrow \bar{\mathbf{f}} = \mathbf{f}_{AA} - \mathbf{f}_{AX}\mathbf{f}_{XX}^{-1}\mathbf{f}_{XA} = \frac{2a}{EA}\begin{bmatrix} 1.3250 & -0.0724 & 0.9250 & 0.0723 & 0.2500 \\ -0.0723 & 0.7086 & -0.0723 & -0.0416 & -0.1445 \\ 0.9250 & -0.0723 & 1.3250 & 0.072 & 0.2500 \\ 0.0723 & -0.0416 & 0.0723 & 0.7086 & 0.1445 \\ 0.2500 & -0.1445 & 0.2500 & 0.1445 & 0.5000 \end{bmatrix} \quad \text{(y)}$$

$$\Rightarrow \bar{\mathbf{T}}_F = \mathbf{T}_{FA} - \mathbf{T}_{FX}\mathbf{f}_{XX}^{-1}\mathbf{f}_{XA} = \begin{bmatrix} 0.5995 & 0.5788 & 0.4000 & 0.0015 & 0 \\ 0.2500 & -0.1445 & 0.2500 & 0.1445 & 0.5000 \\ -0.6000 & 0.5774 & -0.4000 & 0 & 0 \\ -0.4000 & 0 & 0.4000 & 0 & 0 \\ 0.4000 & 0 & 0.6000 & 0.5774 & 0 \\ -0.2500 & 0.1445 & -0.2500 & -0.1445 & -0.5000 \\ -0.4000 & 0 & -0.6000 & 0.5774 & 0 \end{bmatrix} \quad \text{(z)}$$

Let us consider a loading case, $F_1 = F_2 = F_3 = F_4 = F_5 = -P$ in the statically indeterminate structure (Fig. 3.21a). The bar forces $\mathbf{F}_*$ and joint deflections $\mathbf{D}$ can now be easily generated.

$$\mathbf{F}_* = \bar{\mathbf{T}}_F\mathbf{F} = \begin{bmatrix} 0.5995 & 0.5788 & 0.4000 & 0.0015 & 0 \\ 0.2500 & -0.1445 & 0.2500 & 0.1445 & 0.5000 \\ -0.6000 & 0.5774 & -0.4000 & 0 & 0 \\ -0.4000 & 0 & 0.4000 & 0 & 0 \\ 0.4000 & 0 & 0.6000 & 0.5774 & 0 \\ -0.2500 & 0.1445 & -0.2500 & -0.1445 & -0.5000 \\ -0.4000 & 0 & -0.6000 & 0.5774 & 0 \end{bmatrix}\begin{bmatrix} -P \\ -P \\ -P \\ -P \\ -P \end{bmatrix} = \begin{bmatrix} -1.5803P \\ -P \\ 0.4226P \\ 0 \\ -1.5774P \\ P \\ 0.4226P \end{bmatrix}$$

$$\mathbf{D} = \bar{\mathbf{f}}\,\mathbf{F} = \frac{2a}{EA}\begin{bmatrix} 1.3250 & -0.0724 & 0.9250 & 0.0723 & 0.2500 \\ -0.0723 & 0.7086 & -0.0723 & -0.0416 & -0.1445 \\ 0.9250 & -0.0723 & 1.3250 & 0.072 & 0.2500 \\ 0.0723 & -0.0416 & 0.0723 & 0.7086 & 0.1445 \\ 0.2500 & -0.1445 & 0.2500 & 0.1445 & 0.5000 \end{bmatrix}\begin{bmatrix} -P \\ -P \\ -P \\ -P \\ -P \end{bmatrix} = \frac{2Pa}{EA}\begin{bmatrix} -2.500 \\ -0.378 \\ -2.500 \\ -0.956 \\ -1 \end{bmatrix}$$

The force response and displacement response are shown in Figs 3.21a and b respectively, based on the above output.

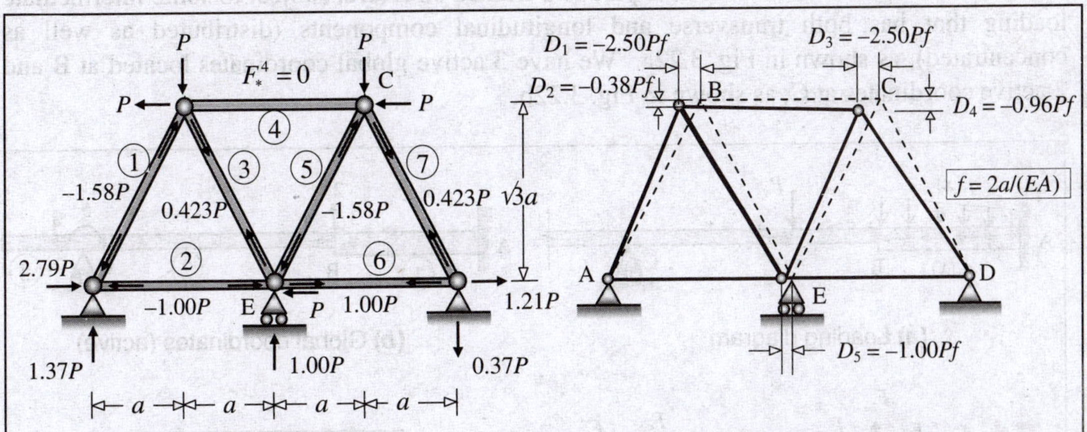

(*a*) Loading diagram (force response also shown)

(*b*) Displacement response

Figure 3.21 Response of a statically indeterminate truss to direct actions (flexibility method)

3.6 EQUIVALENT JOINT LOADS & FIXED END FORCES

3.6.1 Need for Equivalent Joint Loads

As mentioned earlier, the applicability of matrix methods in structural analysis is rooted in the assumption that we can express all the variables in the force and displacement fields in the structure (including all the elements) in terms of their coordinate locations. However, in the case of loading, this implies that the loads have to be treated as discrete concentrated loads acting at the active global coordinate locations (usually joint locations), so that they can be conveniently listed in the load vector $\mathbf{F_A}$. In practice, loads may act not only at the joint locations, but at intermediate locations (in between joints) and may be distributed rather than concentrated.

In order to handle such *intermediate loads* by matrix methods, we need to convert the applied loading to appropriate **equivalent joint loads**. The concept underlying equivalent joint loads is best understood by understanding the basis of the displacement method as applied to beams and frames, in which intermediate loads are commonly encountered. This has been discussed in Section 1.9.1 (refer Fig. 1.32). The same concept can be extended to intermediate axial loads in frame elements (or axial elements).

3.6.2 Equivalent Joint Loads from Fixed End Forces

Consider the segment ABC, which is part of a framed structure, subject to some intermediate loading that has both transverse and longitudinal components (distributed as well as concentrated), as shown in Fig. 3.22a. We have 3 active global coordinates located at B and 2 active coordinates at C, as shown in Fig. 3.22b.

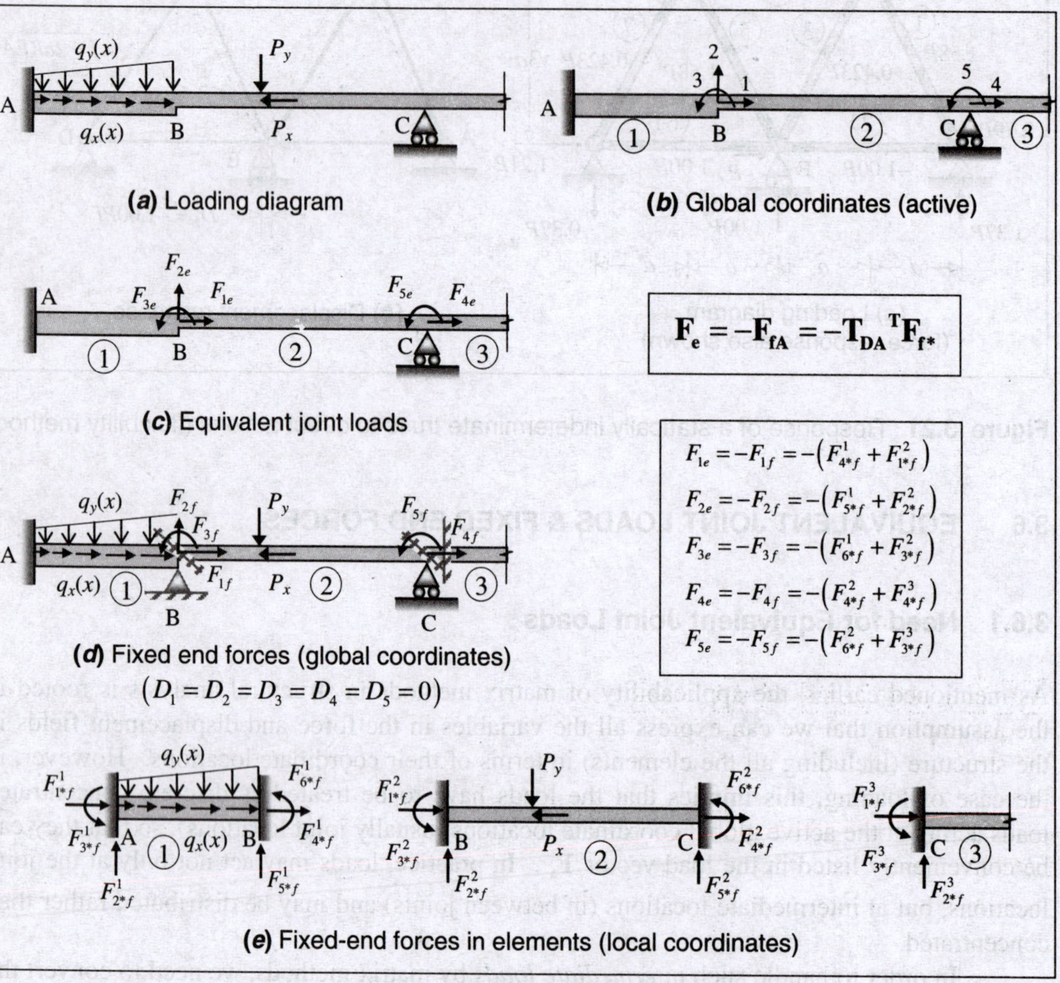

$$\mathbf{F_e} = -\mathbf{F_{fA}} = -\mathbf{T_{DA}}^{\mathbf{T}}\mathbf{F_{f*}}$$

$$F_{1e} = -F_{1f} = -\left(F^1_{4*f} + F^2_{1*f}\right)$$
$$F_{2e} = -F_{2f} = -\left(F^1_{5*f} + F^2_{2*f}\right)$$
$$F_{3e} = -F_{3f} = -\left(F^1_{6*f} + F^2_{3*f}\right)$$
$$F_{4e} = -F_{4f} = -\left(F^2_{4*f} + F^3_{4*f}\right)$$
$$F_{5e} = -F_{5f} = -\left(F^2_{6*f} + F^3_{3*f}\right)$$

$$\left(D_1 = D_2 = D_3 = D_4 = D_5 = 0\right)$$

Figure 3.22 Finding equivalent joint loads in a structure

Our objective is to replace the intermediate loads with a set of equivalent concentrated forces $(F_{1e}, F_{2e}, F_{3e}, F_{4e}, F_{5e})$ acting at the joints B and C, as shown in Fig. 3.22c. The basis for establishing equivalence is that the conjugate displacements $(D_1, D_2, D_3, D_4, D_5)$ at the active degrees of freedom due to the equivalent joint loads must be identical to those due to the given intermediate loads at these locations. By arresting these degrees of freedom $(D_1 = D_2 = D_3 = D_4 = D_5 = 0)$, we generate forces $(F_{1f}, F_{2f}, F_{3f}, F_{4f}, F_{5f})$ at the artificial restraints provided, on account of the applied loading acting on the kinematically determinate structure, as shown in Fig. 3.22d. We can conveniently separate out the various elements, whose ends are fully restrained, and find out the various **element fixed end forces** $\mathbf{F}^{i}_{*f}$ in the individual elements (in local coordinates), as indicated in Fig. 3.22e. For this, we need to be able to calculate the fixed end forces in prismatic members subject to intermediate loading. Table 1.3 can be used conveniently to find *fixed end moments* in prismatic beam elements subject to some common types of intermediate loads.

After finding the element fixed end forces in the local coordinate system (Fig. 3.22e), we can transform these to global coordinates, and sum up the components contributed by the connecting elements. This may be done by using appropriate transformation matrices, as discussed in Section 3.3. Thus, we arrive at the **fixed end force vector**, $\mathbf{F}_f = \left[\begin{array}{c}\mathbf{F}_{fA} \\ \hline \mathbf{F}_{fR}\end{array}\right]$, for the structure at the active global coordinates ($\mathbf{F}_{fA}$) as well as the restrained global coordinates ($\mathbf{F}_{fR}$). The fixed end forces indicated in Fig. 3.22d correspond to the active global coordinates ($\mathbf{F}_{fA}$).

$$\mathbf{F}^{i}_{f} = \mathbf{T}^{iT}\mathbf{F}^{i}_{*f} \tag{3.61}$$

$$\mathbf{F}_f = \left[\begin{array}{c}\mathbf{F}_{fA} \\ \hline \mathbf{F}_{fR}\end{array}\right] = \left[\begin{array}{c}[\mathbf{T}_{DA}]^T \\ \hline [\mathbf{T}_{DR}]^T\end{array}\right]\mathbf{F}_{f*} \tag{3.62}$$

Combining Figs 3.22c and d, and applying the Principle of Superposition (assuming linear behaviour), we observe that this superposition must yield the force and displacement fields in Fig. 3.22a, in terms of forces as well as displacements. First, looking at the displacements at the active global coordinates, and noting that all these displacements are zero in the restrained structure in Fig. 3.22d, it follows that the displacements $(D_1, D_2, D_3, D_4, D_5)$ in Fig. 3.22a must necessarily be equal to the ones in Fig. 3.22c (due to the equivalent joint loads) at the respective locations. Next, to satisfy the superposition of forces, we note that the sum of the **equivalent joint load vector** $\mathbf{F}_e$ (in Fig. 3.22c) and the fixed end force vector at the active global coordinates, $\mathbf{F}_{fA} = \mathbf{T}_{DA}{}^{T}\mathbf{F}_{f*}$ (in Fig. 3.22d) must be equal to zero, so that we are left with only the intermediate loads shown in Fig. 3.22a. Hence, the equivalent joint loads are equal and opposite to the fixed end forces generated at the active global coordinates.

$$\mathbf{F}_e = -\mathbf{F}_{fA} = -\mathbf{T}_{DA}{}^{T}\mathbf{F}_{f*} \tag{3.63}$$

It is possible to formulate the equivalent joint loads from the element fixed end forces directly, without invoking the transformation shown in Eq. 3.63. Indeed, this becomes necessary in some cases in the reduced stiffness method and the flexibility method, where we work with limited degrees of freedom per element ($\tilde{q} < q$), for which the transformation indicated by Eq. 3.63 may not be applicable. This is discussed in the Chapters to follow.

When intermediate loads act in between the joints of a structure, they can be converted to **equivalent joint loads**, to facilitate formulating the load vector in matrix analysis. The basis for equivalence is that the equivalent joint loads should generate the same conjugate displacements at the joints on which they act as the original intermediate loads at these locations[†]. If the displacements at these joints are arrested in the original structure, then the resultant **fixed end forces** generated at the artifical restraints, when applied in an opposite direction, correspond to the equivalent joint loads.

Let us illustrate how the equivalent joint loads can be directly evaluated by a numerical example. Consider the plane frame in Fig. 3.23a, comprising 4 prismatic elements ($m = 4$) and 5 nodes with 3 potential degrees of freedom per node, of which 9 degrees are active ($n = 9$), while 6 are restrained ($r = 6$), as indicated. The loading diagram is shown in Fig. 3.23b. By arresting the active degrees of freedom, the various elements can be separated out and the fixed end forces on account of the intermediate loads can be calculated, as shown in Fig. 3.23c. By summing up the respective forces at the active coordinate locations, the fixed end forces in the structure, $\mathbf{F}_{fA}$, can be easily computed and the equivalent joint loads, $\mathbf{F}_e = -\mathbf{F}_{fA}$, generated, as indicated in Fig. 3.23d.

The equivalent joint loads need to be added to any nodal loads $\mathbf{F}_A$ that may be prescribed at the appropriate active global coordinates (refer Figs 3.23b and d).

3.6.3 Equivalent Joint Loads due to Indirect Loading (Stiffness Method)

In the stiffness method, we need to invoke the concept of equivalent joint loads not only when we encounter intermediate loads, but also when we deal with situations of indirect loading associated with initial prescribed changes in lengths in members, $\mathbf{D}^i_{*initial}$, due to temperature changes or 'lack of fit'. These introduce additional fixed end forces, $\Delta\mathbf{F}^i_{*f} = \mathbf{k}^i_*\mathbf{D}^i_{*initial}$, in the various elements as well as in the primary structure ($\Delta\mathbf{F}_f$). Eqns 3.61 and 3.62 may be modified accordingly as follows:

$$\mathbf{F}^i_f = \mathbf{T}^{iT}\left(\mathbf{F}^i_{*f} + \Delta\mathbf{F}^i_{*f}\right) \tag{3.64}$$

$$\mathbf{F}_f = \begin{bmatrix} \mathbf{F}_{fA} \\ \hline \mathbf{F}_{fR} \end{bmatrix} = \begin{bmatrix} [\mathbf{T}_{DA}]^T \\ \hline [\mathbf{T}_{DR}]^T \end{bmatrix}\left(\mathbf{F}_{f*} + \Delta\mathbf{F}^i_{*f}\right) \tag{3.65}$$

[†] At other locations, the displacements are likely to be different.

$$\Delta \mathbf{F}_{*f}^i = \mathbf{k}_{*}^i \mathbf{D}_{*initial}^i \tag{3.65a}$$

(a) Global coordinates
(n = 9, r = 6)

(b) Loading diagram

[Nodal loads:
$F_1 = 30$ kN, $F_9 = -30$ kNm]

(c) Fixed-end forces due to intermediate loads in various elements

$F_{1e} = -F_{1f} + F = -(-27) + 30 = +57$ kN

$F_{2e} = -F_{2f} = -(60 + 36) = -96$ kN

$F_{3e} = -F_{3f} = -(-36 + 18) = +18$ kNm

$F_{4e} = -F_{4f} = -(0) = 0$ kN

$F_{5e} = -F_{5f} = -(36) = -36$ kN

$F_{6e} = -F_{6f} = -(-18) = +18$ kNm

$F_{7e} = -F_{7f} = -(0) = 0$ kNm

$F_{8e} = -F_{8f} = -(30) = -30$ kNm

$F_{9e} = -F_{9f} + F = -(-30) - 50 = -20$ kNm

(d) Equivalent joint loads + Nodal loads
(at active global coordinates)

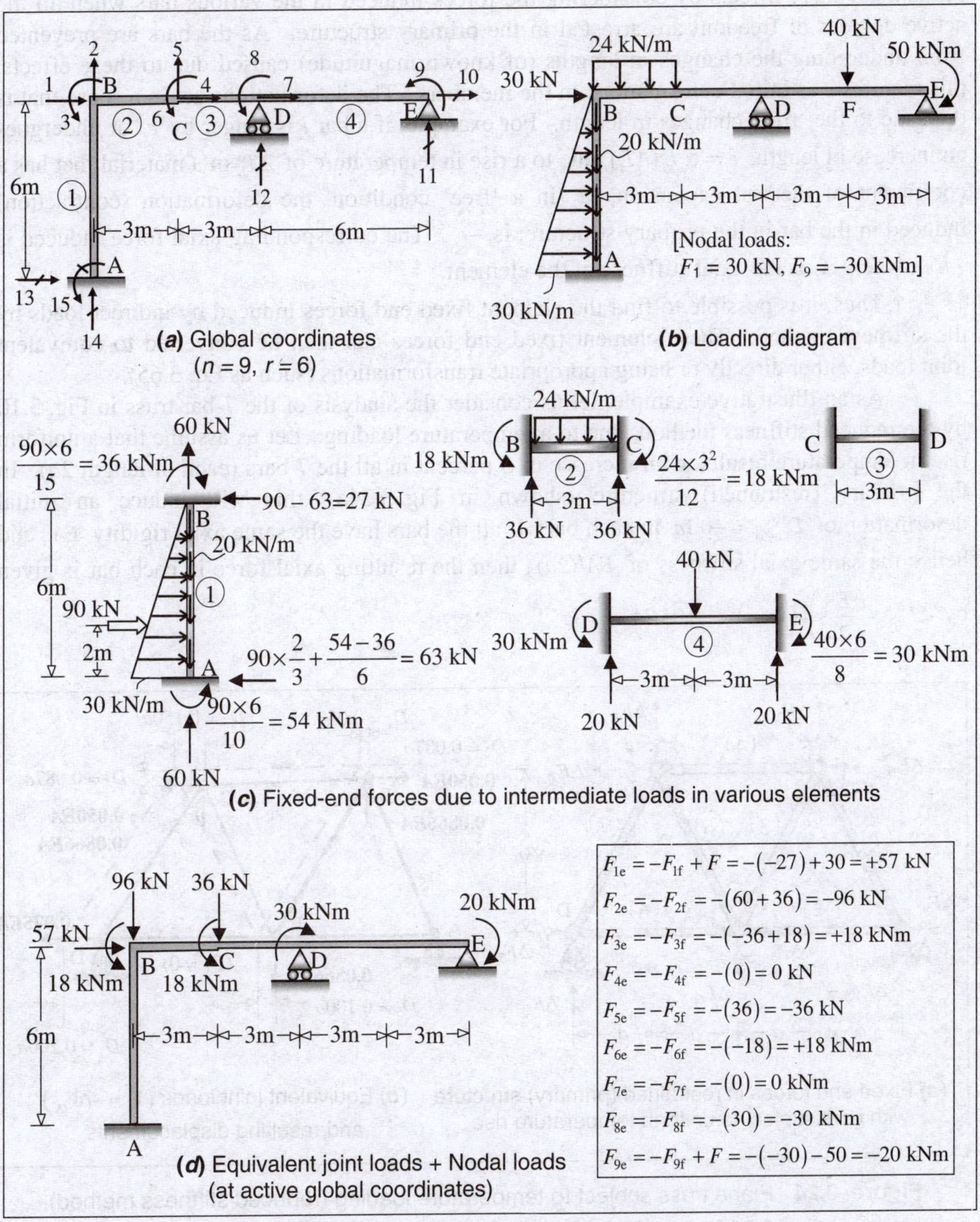

Figure 3.23 Assessment of equivalent joint loads in a plane frame

For example, when there is 'lack of fit' or temperature changes in truss members, we account for these effects by considering the forces induced in the various bars when all the active degrees of freedom are arrested in the primary structure. As the bars are prevented from undergoing the changes in lengths (of known magnitude) caused due to these effects, this manifests as initial deformations in the members. The deformations act in a sense that is opposite to the 'free' changes in length. For example, if a bar i is longer by e_i or undergoes an increase in length, $e_i = \alpha_i L_i (\Delta T_i)$ due to a rise in temperature of ΔT_i in a material that has a coefficient of thermal expansion, α_i in a 'free' condition, the deformation (contraction) induced in the bar in the primary structure, is $-e_i$. The corresponding axial force induced is $-k_i e_i$, where k_i is the axial stiffness of the element.

Thus, it is possible to find the element fixed end forces induced by indirect loads by the stiffness method. These element fixed end forces can then be converted to equivalent joint loads, either directly or using appropriate transformations (such as Eq. 3.65).

As an illustrative example, let us consider the analysis of the 7-bar truss in Fig. 3.10 by the reduced stiffness method, due to a temperature loading. Let us assume that a uniform rise in temperature results in an increase of 5 percent in all the 7 bars (each of length $2a$). In the primary (restrained) structure (shown in Fig. 3.24a), this will induce an initial deformation of $D^i_{*\text{initial}} = -0.1a$ in each bar. If all the bars have the same axial rigidity EA, and hence the same axial stiffness of $EA/(2a)$, then the resulting axial force in each bar is given by $\Delta F^i_{*f} = \left(\dfrac{EA}{2a}\right)(-0.1a) = -0.05EA$.

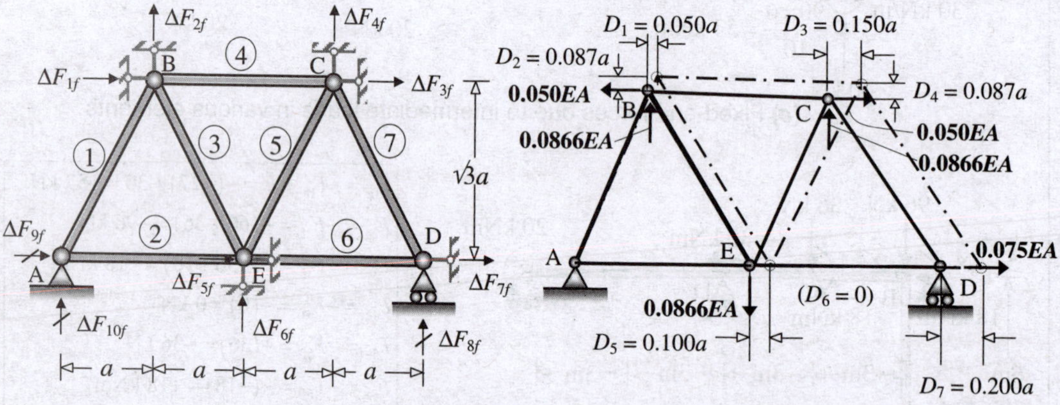

(a) Fixed end forces in restrained (primary) structure with bars subject to uniform temperature rise

(b) Equivalent joint loads ($\mathbf{F_e} = -\Delta \mathbf{F_{fA}}$) and resulting displacements

Figure 3.24 Plane truss subject to temperature loading (reduced stiffness method)

Using the $\mathbf{T_D}^{\mathrm{T}}$ matrix derived in Eq. (g), the fixed end forces, $\mathbf{F_f} = \mathbf{T_D}^{\mathrm{T}}(\Delta\mathbf{F_{f*}})$, in global coordinates can be determined. The resulting equivalent joint loads (at the active coordinates), are shown in Fig. 3.24b.

$$\begin{bmatrix} \Delta F_{1f} \\ \Delta F_{2f} \\ \Delta F_{3f} \\ \Delta F_{4f} \\ \Delta F_{5f} \\ \Delta F_{6f} \\ \Delta F_{7f} \\ \hline \Delta F_{8f} \\ \Delta F_{9f} \\ \Delta F_{10f} \end{bmatrix} = \begin{bmatrix} 0.500 & 0 & -0.500 & -1.00 & 0 & 0 & 0 \\ 0.866 & 0 & 0.866 & 0 & 0 & 0 & 0 \\ 0 & 0 & 0 & 1.00 & 0.500 & 0 & -0.500 \\ 0 & 0 & 0 & 0 & 0.866 & 0 & 0.866 \\ 0 & 1.00 & 0.500 & 0 & -0.500 & -1.00 & 0 \\ 0 & 0 & -0.866 & 0 & -0.866 & 0 & 0 \\ 0 & 0 & 0 & 0 & 0 & 1.00 & 0.500 \\ \hline 0 & 0 & 0 & 0 & 0 & 0 & -0.866 \\ -0.500 & -1.00 & 0 & 0 & 0 & 0 & 0 \\ -0.866 & 0 & 0 & 0 & 0 & 0 & 0 \end{bmatrix} \begin{bmatrix} -0.05EA \\ -0.05EA \\ -0.05EA \\ -0.05EA \\ -0.05EA \\ -0.05EA \\ -0.05EA \end{bmatrix} = \begin{bmatrix} 0.0500EA \\ -0.0866EA \\ -0.0500EA \\ -0.0866EA \\ 0 \\ 0.0866EA \\ -0.0750EA \\ \hline 0.0433EA \\ 0.0750EA \\ 0.0433EA \end{bmatrix} \quad (1)$$

3.6.4 Modified Stiffness Relations Including Fixed End Forces

The structure stiffness relation, given by Eqns 3.36 and 3.38 needs to be modified to include the contribution of the fixed end forces $\mathbf{F_f}$ and $\Delta\mathbf{F_f}$, as follows:

$$\mathbf{F} - (\mathbf{F_f} + \Delta\mathbf{F_f}) = \mathbf{kD} \tag{3.66}$$

$$\Rightarrow \begin{bmatrix} \mathbf{F_A} \\ \hline \mathbf{F_R} \end{bmatrix} - \begin{bmatrix} \mathbf{F_{fA}} + \Delta\mathbf{F_{fA}} \\ \hline \mathbf{F_{fR}} + \Delta\mathbf{F_{fR}} \end{bmatrix} = \left[\begin{array}{c|c} \mathbf{k_{AA}} & \mathbf{k_{AR}} \\ \hline \mathbf{k_{RA}} & \mathbf{k_{RR}} \end{array}\right] \begin{bmatrix} \mathbf{D_A} \\ \hline \mathbf{D_R} \end{bmatrix} \tag{3.67}$$

$$\Rightarrow \boxed{\mathbf{D_A} = [\mathbf{k_{AA}}]^{-1}[\mathbf{F_A} - (\mathbf{F_{fA}} + \Delta\mathbf{F_{fA}} + \mathbf{k_{AR}}\mathbf{D_R})]} \tag{3.68}$$

$$\boxed{\mathbf{F_R} = \mathbf{F_{fR}} + \Delta\mathbf{F_{fR}} + \mathbf{k_{RA}}\mathbf{D_A} + \mathbf{k_{RR}}\mathbf{D_R}} \tag{3.69}$$

Eq. 3.68 turns out to be a comprehensive governing equation that enables us to find the unknown displacements in any structure by the stiffness method. After finding these displacements, we can find the support reactions using Eq. 3.69. We can also find the various element level forces, including the effect of the fixed end forces, by modifying Eq. 3.41 as follows:

$$\boxed{\mathbf{F_*} = \mathbf{F_{*f}} + \Delta\mathbf{F_{*f}} + \mathbf{k_*}\mathbf{T_D}\mathbf{D}} \tag{3.70}$$

The above equation has the form of the slope-deflection equations we had used in the slope-deflection method of displacement analysis of beams and frames.

If $\mathbf{D_R} = \mathbf{O}$, then Eq. 3.70 reduces to:

$$\mathbf{F}_* = \mathbf{F}_{*f} + \Delta\mathbf{F}_{*f} + \mathbf{k}_*\mathbf{T}_{DA}\mathbf{D}_A \tag{3.70a}$$

If the reduced stiffness method is used instead of the regular stiffness method, then the symbol $\mathbf{k}_*$ should be substituted with $\tilde{\mathbf{k}}_*$ in Eqns 3.66 and 3.67.

For example, referring to the truss in Fig. 3.24, and observing that the loading is entirely due to temperature effects alone $(\mathbf{F}_A = \mathbf{F}_f = \mathbf{O})$, Eq. 3.68 simplifies to $\mathbf{D}_A = [\mathbf{k}_{AA}]^{-1}(-\Delta\mathbf{F}_{fA})$, where $[\mathbf{k}_{AA}]^{-1} = \mathbf{f}$ is the structure flexibility matrix we had derived earlier (Eq. w) and the components of $\Delta\mathbf{F}_{fA}$ are given by the first 7 components in the $\Delta\mathbf{F}_f$ vector we had derived earlier. Accordingly, $\mathbf{D}_A = -\mathbf{f}(\Delta\mathbf{F}_{fA}) \Rightarrow$

$$\begin{bmatrix} D_1 \\ D_2 \\ D_3 \\ D_4 \\ D_5 \\ D_6 \\ D_7 \end{bmatrix} = \frac{2a}{EA}\begin{bmatrix} 1.8750 & -0.5055 & 1.3750 & -0.3609 & 0.7500 & -0.5774 & 1.0000 \\ -0.5055 & 1.2919 & -0.2168 & 0.5417 & -0.4333 & 1.0000 & -0.5774 \\ 1.3750 & -0.2168 & 1.8750 & -0.0721 & 0.7500 & 0 & 1.0000 \\ -0.3609 & 0.5417 & -0.0721 & 1.2919 & -0.1443 & 1.0000 & -0.5774 \\ 0.7500 & -0.4333 & 0.7500 & -0.1443 & 1.0000 & -0.2887 & 1.0000 \\ -0.5774 & 1.0000 & 0 & 1.0000 & -0.2887 & 1.8336 & -0.5774 \\ 1.0000 & -0.5774 & 1.0000 & -0.5774 & 1.0000 & -0.5774 & 2.0000 \end{bmatrix}\begin{bmatrix} -0.0500EA \\ 0.0866EA \\ 0.0500EA \\ 0.0866EA \\ 0 \\ -0.0866EA \\ 0.0750EA \end{bmatrix}$$

$$\Rightarrow \begin{bmatrix} D_1 \\ D_2 \\ D_3 \\ D_4 \\ D_5 \\ D_6 \\ D_7 \end{bmatrix} = \begin{bmatrix} 0.0500a \\ 0.0866a \\ 0.1500a \\ 0.0866a \\ 0.1000a \\ 0 \\ 0.2000a \end{bmatrix} \tag{2}$$

These joint displacements (which are in fact predictable) are indicated in Fig. 3.24b. If we now calculate the resultant bar forces using Eq. 3.70a, and observing that for this truss example, $\tilde{\mathbf{k}}_* = \frac{EA}{2a}\mathbf{I}$, we get the following interesting result:

$$\begin{bmatrix} F_*^1 \\ F_*^2 \\ F_*^3 \\ F_*^4 \\ F_*^5 \\ F_*^6 \\ F_*^7 \end{bmatrix} = \begin{bmatrix} -0.05EA \\ -0.05EA \\ -0.05EA \\ -0.05EA \\ -0.05EA \\ -0.05EA \\ -0.05EA \end{bmatrix} + \frac{EA}{2a}\begin{bmatrix} 0.5 & 0.866 & 0 & 0 & 0 & 0 & 0 \\ 0 & 0 & 0 & 0 & 1 & 0 & 0 \\ -0.5 & 0.866 & 0 & 0 & 0.5 & -0.866 & 0 \\ -1 & 0 & 1 & 0 & 0 & 0 & 0 \\ 0 & 0 & 0.5 & 0.866 & -0.5 & -0.866 & 0 \\ 0 & 0 & 0 & 0 & -1 & 0 & 1 \\ 0 & 0 & -0.5 & 0.866 & 0 & 0 & 0.5 \end{bmatrix}\begin{bmatrix} 0.0500a \\ 0.0866a \\ 0.1500a \\ 0.0866a \\ 0.1000a \\ 0 \\ 0.2000a \end{bmatrix} = \begin{bmatrix} 0 \\ 0 \\ 0 \\ 0 \\ 0 \\ 0 \\ 0 \end{bmatrix} \tag{3}$$

All the bar forces turn out to be zero! This is only to be expected, as we know that the force response of a just-rigid structure, subject to indirect loading (such as temperature changes) will be zero. We can also verify, using Eq. 3.69, which simplifies to $\mathbf{F_R} = \Delta\mathbf{F_{fR}} + \mathbf{k_{RA}}\mathbf{D_A}$ that the support reactions will also be zero.

3.6.5 Flexibility Relations Including Fixed End Forces

The concept of equivalent joint loads, with its basis emerging from the fixed end forces in a kinematically restrained structure, is essentially a concept from the stiffness method. We need to invoke the same concept in the flexibility method, if we wish to apply a matrix approach to dealing with intermediate loads. In this process, however, the classical flexibility method (which we had discussed in Chapter 1) loses its elegance and becomes somewhat cumbersome. It is also somewhat ironical that we need to first know the statically indeterminate solution pertaining to fixed end forces due to the action of intermediate loads, in order to carry out the analysis by the flexibility method for even statically determinate structures.

Nevertheless, if we choose to use the concept of equivalent joint loads in the flexibility method, then we need to transform the fixed end forces at the element level to the global coordinates by generating the fixed end force vector $\mathbf{F_f}$; this is best done manually, as we do not have a $\mathbf{T_D}^{\mathrm{T}}$ matrix in the flexibility method.

We also need to modify the primary structure flexibility relation, given by Eq. 3.55, as follows:

$$\begin{bmatrix} \mathbf{D_A} \\ \hline \mathbf{D_X} \end{bmatrix} = \begin{bmatrix} \mathbf{D_{A,initial}} \\ \hline \mathbf{D_{X,initial}} \end{bmatrix} + \left[\begin{array}{c|c} \mathbf{f_{AA}} & \mathbf{f_{AX}} \\ \hline \mathbf{f_{XA}} & \mathbf{f_{XX}} \end{array}\right] \begin{bmatrix} \mathbf{F_A} - \mathbf{F_{fA}} \\ \hline \mathbf{F_X} - \mathbf{F_{fX}} \end{bmatrix} \tag{3.71}$$

$$\Rightarrow \quad \boxed{\mathbf{F_X} = \mathbf{F_{fX}} + [\mathbf{f_{XX}}]^{-1}\left[\mathbf{D_X} - \mathbf{D_{X,initial}} - \mathbf{f_{XA}}(\mathbf{F_A} - \mathbf{F_{fA}})\right]} \tag{3.72}$$

$$\boxed{\mathbf{D_A} = \mathbf{D_{A,initial}} + \mathbf{f_{AA}}(\mathbf{F_A} - \mathbf{F_{fA}}) + \mathbf{f_{AX}}(\mathbf{F_X} - \mathbf{F_{fX}})} \tag{3.73}$$

The element forces are obtainable as follows:

$$\boxed{\mathbf{F_*^i} = \mathbf{F_{*f}^i} + \mathbf{T_{FA}^i}\mathbf{F_A} + \mathbf{T_{FX}^i}\mathbf{F_X}} \tag{3.74}$$

Alternatively, in the absence of indirect loading, we can deal with the kinematically indeterminate structure directly, by considering the load vector, $\mathbf{F} = \mathbf{F_A}$, and computing the fixed end forces at the global coordinates ($\mathbf{F_f}$), and avoid dealing with the redundants altogether, as indicated by Eqns 3.57 and 3.59, which need to be modified as follows:

$$\boxed{\mathbf{D} = \overline{\mathbf{f}}(\mathbf{F} - \mathbf{F_f})} \tag{3.75}$$

where $\mathbf{D} = \mathbf{D}_\mathbf{A}$ and $\overline{\mathbf{f}}$ is the **structure flexibility matrix** of order $n \times n$ (expandable to $(n+r)\times(n+r)$ if rigid links are included to simulate the support reactions) for the statically indeterminate structure, given by Eq. 3.58.

$$\boxed{\mathbf{F}_* = \mathbf{F}_{*\mathbf{f}} + \overline{\mathbf{T}}_\mathbf{F}\mathbf{F}} \tag{3.76}$$

where $\overline{\mathbf{T}}_\mathbf{F}$ is the **force transformation matrix** of order $n \times n$ (expandable to $(n+r)\times(n+r)$ if rigid links are included to simulate the support reactions) for the statically indeterminate structure, given by Eq. 3.60.

We shall demonstrate the application of this to a few problems in the Chapters to follow. In general, however, we note that the matrix approach to the flexibility method is not advocated when intermediate loads are involved; the classical approach (discussed in Section 1.8.1) is more convenient till the stage of formulating the compatibility equations.

3.7 STIFFNESS & FLEXIBILITY METHODS

In this Section, we shall summarise the basic steps involved in the stiffness and flexibility methods of matrix analysis of skeletal structures. The procedures are essentially algorithmic in nature, and suited for computer programming. The stiffness method is discussed in Section 3.7.1, followed by the flexibility method in Section 3.7.2. In the concluding Section 3.7.3, we shall compare the two approaches and see why the stiffness method is the one that is favoured for general purpose software development.

In general, in both methods, there are input modules (giving input data regarading the structure and the loading) as well as analysis modules in which the input data is processed using matrix methods to generate the desired force and displacement responses.

3.7.1 Stiffness Method

The basic steps to be followed in the stiffness method comprise the following:

1. Give the **structure input data** in terms of:
   (a) joint coordinates (such as x_j, y_j, z_j) in the global axes system at suitably labelled joints (A, B,… or '1', '2',…);
   (b) global coordinates, corresponding to the independent degrees of freedom at the joints and supports, numbered using integers, proceeding sequentially from one joint to another, with the *restrained* degrees of freedom numbered at the end of the list (as shown in Fig. 3.1a);
   (c) elements (numerically labelled) inter-connecting various joints, indicating member incidence ('start node' and 'end node') and element type (such as

‘plane frame element’, etc.), which automatically defines the local coordinates for each element;

(d) element material and cross-sectional properties.

2. Generate the **element stiffness matrix** $\mathbf{k}_*^i$ and the element transformation matrix $\mathbf{T}^i$ and compute $\mathbf{k}_*^i\mathbf{T}^i$ and $\mathbf{k}^i = \mathbf{T}^{i^T}\mathbf{k}_*^i\mathbf{T}^i$, including the linkages to the global coordinate numbers, for all the elements. Alternatively, find the element displacement transformation matrix $\mathbf{T}_D^i$ and compute $\mathbf{k}_*^i\mathbf{T}_D^i$ and $\mathbf{T}_D^{i\,T}\mathbf{k}_*^i\mathbf{T}_D^i$ for all the elements. (In the case of the reduced stiffness method, use $\tilde{\mathbf{k}}_*^i$ in place of $\mathbf{k}_*^i$).

3. Generate the **structure stiffness matrix** $\mathbf{k} = \left[\begin{array}{c|c} \mathbf{k}_{AA} & \mathbf{k}_{AR} \\ \hline \mathbf{k}_{RA} & \mathbf{k}_{RR} \end{array}\right]$ by summing up the contributions from all elements appropriately ($\mathbf{T}^{i^T}\mathbf{k}_*^i\mathbf{T}^i$ or $\mathbf{T}_D^{i\,T}\mathbf{k}_*^i\mathbf{T}_D^i$ or $\mathbf{T}_D^{i\,T}\tilde{\mathbf{k}}_*^i\mathbf{T}_D^i$). Also, generate $[\mathbf{k}_{AA}]^{-1}$.

4. Give the **load input data** in terms of:

   (a) nodal load data, corresponding to the active coordinates ($\mathbf{F}_A$)

   (b) intermediate load data, if any, in the form of element fixed end forces ($\mathbf{F}_{*f}^i$) for all the elements

   (c) indirect load data, if any, in terms of support displacements, corresponding to the restrained coordinates ($\mathbf{D}_R$), and changes in element lengths in the ‘free’ condition, such as due to temperature changes or ‘lack of fit’ ($\mathbf{D}_{*\text{initial}}^i$), which can be transformed into additional element fixed end forces as $\Delta\mathbf{F}_{*f}^i = \mathbf{k}_*^i\mathbf{D}_{*\text{initial}}^i$.

5. Generate the **fixed end force vector,** $\mathbf{F}_f + \Delta\mathbf{F}_f = \left[\begin{array}{c} \mathbf{F}_{fA} + \Delta\mathbf{F}_{fA} \\ \hline \mathbf{F}_{fR} + \Delta\mathbf{F}_{fR} \end{array}\right]$ by summing up the fixed end force contributions from all elements appropriately, considering $\mathbf{T}^{i^T}\left(\mathbf{F}_{*f}^i + \Delta\mathbf{F}_{*f}^i\right)$ or $\mathbf{T}_D^{i\,T}\left(\mathbf{F}_{*f}^i + \Delta\mathbf{F}_{*f}^i\right)$.

6. Compute the unknown displacements, $\mathbf{D}_A = [\mathbf{k}_{AA}]^{-1}\left[\mathbf{F}_A - (\mathbf{F}_{fA} + \Delta\mathbf{F}_{fA} + \mathbf{k}_{AR}\mathbf{D}_R)\right]$.

7. Compute the element forces, $\mathbf{F}_*^i = \mathbf{F}_{*f}^i + \Delta\mathbf{F}_{*f}^i + \mathbf{k}^i\mathbf{T}^i\mathbf{D}$ (or, in lieu of $\mathbf{k}^i\mathbf{T}^i\mathbf{D}$, use $\mathbf{k}_*^i\mathbf{T}_D^i\mathbf{D}$ or $\tilde{\mathbf{k}}_*^i\mathbf{T}_D^i\mathbf{D}$).

8. Compute the support reactions, $\mathbf{F}_R = \mathbf{F}_{fR} + \Delta\mathbf{F}_{fR} + \mathbf{k}_{RA}\mathbf{D}_A + \mathbf{k}_{RR}\mathbf{D}_R$.

9. If the structure has to be analysed for other loading conditions, repeat steps 4 to 8.

3.7.2 Flexibility Method

The basic steps to be followed in the flexibility method comprise the following:

1. Give the **structure input data** in terms of:
   (a) joint coordinates (such as x_j, y_j, z_j) in the global axes system at suitably labelled joints (A, B,... or '1', '2',...), including additional joints at supports, if rigid links are included to model support reactions;
   (b) global coordinates, corresponding to the independent *active* degrees of freedom at the joints, numbered using integers, proceeding sequentially from one joint to another; with the *redundant* coordinates (for suitably chosen redundants in the case of a statically indeterminate structure), numbered at the end of the list (as shown in Fig. 3.2);
   (c) elements (numerically labelled) inter-connecting various joints, indicating member incidence ('start node' and 'end node') and element type (such as 'plane frame element', etc.), which automatically defines the local coordinates for each element (rigid links to be included);
   (d) element material and cross-sectional properties.
2. Generate the **element flexibility matrix** $\mathbf{f}_*^i$ and the element force transformation matrix $\mathbf{T}_F^i$ and compute $\mathbf{f}_*^i\mathbf{T}_F^i$ and $\mathbf{T}_F^{iT}\mathbf{f}_*^i\mathbf{T}_F^i$ for all the elements in the primary structure.
3. If the structure is statically indeterminate generate the **primary structure flexibility matrix** $\mathbf{f} = \begin{bmatrix} \mathbf{f}_{AA} & \mathbf{f}_{AX} \\ \mathbf{f}_{XA} & \mathbf{f}_{XX} \end{bmatrix}$ by summing up the contributions from all elements appropriately ($\mathbf{T}_F^{iT}\mathbf{f}_*^i\mathbf{T}_F^i$). Also, generate $[\mathbf{f}_{XX}]^{-1}$. [If there are no indirect loads, to be considered, compute the **structure flexibility matrix** $\overline{\mathbf{f}} = \mathbf{f}_{AA} - \mathbf{f}_{AX}\mathbf{f}_{XX}^{-1}\mathbf{f}_{XA}$ and element transformation matrix $\overline{\mathbf{T}}_F^i = \mathbf{T}_{FA}^i - \mathbf{T}_{FX}^i\mathbf{f}_{XX}^{-1}\mathbf{f}_{XA}$ for the statically indeterminate structure]. If the structure is statically determinate, simply compute $\mathbf{f} = \mathbf{f}_{AA}$.
4. Give the **load input data** in terms of:
   (a) nodal load data, corresponding to the active coordinates ($\mathbf{F}_A$)
   (b) intermediate load data, if any, in the form of element fixed end forces ($\mathbf{F}_{*f}^i$) for all the elements and the the **fixed end force vector,** $\mathbf{F}_f = \begin{bmatrix} \mathbf{F}_{fA} \\ \mathbf{F}_{fX} \end{bmatrix}$;
   (c) indirect load data, if any, in terms of support displacements, corresponding to the redundant coordinates ($\mathbf{D}_X$), and changes in element lengths in the 'free' condition, such as due to temperature changes or 'lack of fit' or support movements in non-redundant support locations ($\mathbf{D}_{*initial}^i$), which are transformed to the global coordinates as $\begin{bmatrix} \mathbf{D}_{A,initial} \\ \mathbf{D}_{X,initial} \end{bmatrix} = \begin{bmatrix} \mathbf{T}_{FA}^T \\ \mathbf{T}_{FX}^T \end{bmatrix} \mathbf{D}_{*initial}^i$.

5. In the case of statically indeterminate structures, compute the unknown redundants, $\mathbf{F_X} = \mathbf{F_{fX}} + [\mathbf{f_{XX}}]^{-1}\left[\mathbf{D_X} - \mathbf{D_{X,initial}} - \mathbf{f_{XA}}(\mathbf{F_A} - \mathbf{F_{fA}})\right]$; otherwise, proceed to step 6 in the case of statically determinate structures (with $\mathbf{F_X} = \mathbf{O}$).
6. Compute the element forces, $\mathbf{F_*^i} = \mathbf{F_{*f}^i} + \mathbf{T_{FA}^i}\mathbf{F_A} + \mathbf{T_{FX}^i}\mathbf{F_X}$, and extract the support reactions from the forces in the rigid links. [Alternatively, if there are no indirect loads, to be considered, compute $\mathbf{F_*^i} = \mathbf{F_{*f}^i} + \bar{\mathbf{T}}_{\mathbf{F}}^{\mathbf{i}}\mathbf{F}$].
7. Compute the joint displacements, $\mathbf{D_A} = \mathbf{D_{A,initial}} + \mathbf{f_{AA}}(\mathbf{F_A} - \mathbf{F_{fA}}) + \mathbf{f_{AX}}(\mathbf{F_X} - \mathbf{F_{fX}})$. [Alternatively, if there are no indirect loads, to be considered, compute $\mathbf{D} = \bar{\mathbf{f}}\ (\mathbf{F} - \mathbf{F_f})$].
8. If the structure has to be analysed for other loading conditions, repeat steps 4 to 7.

3.7.3 Comparison between Stiffness and Flexibility Methods

The main distinctions between the stiffness and flexibility methods may be summarised as follows.

	Aspect under consideration	Stiffness Method	Flexibility Method
1	Basic Unknowns?	Displacements	Forces
2	Type of Indeterminacy?	Kinematic	Static
3	Zero indeterminacy possible?	Not possible	Possible (different procedure for statically determinate structures)
4	Choice of primary structure?	No choice	Multiple choices possible in statically indeterminate structures
5	Format of force-dispacement relation	Stiffness format	Flexibility format
6	Condition of structure stiffness matrix	Well-conditioned	May be ill-conditioned
7	Basis of handling intermediate loads	Fixed end forces	Fixed end forces†
8	Basis of governing equations to solve for unknowns	Equilibrium	Compatibility

† This concept is borrowed from the stiffness method, and assumes that we know the force response of the statically indeterminate element with fixed ends, subject to arbitrary intermediate loads.

The main reasons as to why the stiffness method is preferred to the flexibility method for computer-based structural analysis (general purpose software) are the following:

1. The solution procedure is unique, and does not depend on the user for the choice of the primary structure. Hence, programming is simplified.
2. The structure stiffness matrix for stable linear structures is invariably a well-conditioned matrix, thereby facilitating accurate solutions† of the response.

From a human effort perspective (manual solution), in the absence of any software, the method to be preferred is governed by the size of the matrix (stiffness or flexibility) that needs to be inverted, in order to generate the response. Clearly, if the static indeterminacy is less than the kinematic indeterminacy, we should prefer the flexibility method to the stiffness method, and it is possible to multiply and invert matrices using software like MATLAB. We encounter this situation, for example, in roof and bridge trusses and in transmission towers, which are practically statically determinate and whose kinematic indeterminacy is very high. It would be absurd to prefer the stiffness method to the flexibility method in such cases; nevertheless, in practice, this is exactly what many designers end up doing, by using readily available software packages! Designers should at least have the means of doing quick checks on the solutions generated using simple manual methods.

The reduced stiffness method offers substantial savings in computational effort. However, they need to be used with caution, and they lack the versatility and generalised applications that are possible in the conventional stiffness method adopted in software packages. Nevertheless, the reduced stiffness method and the flexibility method are exciting methods for manual analysis, and for these reasons, they are also covered in the Chapters to follow.

Developing efficient algorithms to reduce computational effort and storage size of matrices, and to achieve accuracy in solving equations, has been an area of much research. This is particularly important in developing software to analyse large structures by the stiffness method. The stiffness matrix of such large structures is usually *sparse* (i.e., containing a large number of zero elements) and *banded* (i.e., with the non-zero elements located within a symmetric band centred along the principal diagonal). It has been observed that the number of arithmetic operations involved in solving the set of equilibrium equations to find the unknown displacements, $\mathbf{D_A}$, increases linearly with the number of active degrees of freedom (n) and with the square of the *half-band width* (n_b). Hence, there is much to be gained in reducing the *bandwidth* of the stiffness matrix, not only in terms of reduced data storage, but, more important, in terms of much less computational effort, while dealing with large structures. Ideally, the numbering of the various coordinates in the structure should be such that the band along the principal diagonal contains practically no zero elements. In

† The structure flexibility matrix, on the other hand, may be ill-conditioned, especially in structures with a high degree of static indeterminacy. Consequently, inversion of the $\mathbf{f_{XX}}$ matrix can result in significant errors and thereby give inaccurate results.

general, this requires the sequencing of the numbering scheme to be done along the shorter direction of the structure. Structural analysis software programs usually have an in-built feature that does an automatic re-numbering of the coordinates specified by the user, in order to achieve this.

Developing efficient computer programs to perform matrix analysis of structures, however, is a topic that lies outside the scope of this book, whose main purpose is to enable an understanding of the basic concepts of matrix analysis of structures. These concepts are demonstrated by means of several illustrative examples, dealing with relatively small structures, and requiring no more that simple MATLAB matrix operations.

REVIEW QUESTIONS

3.1 Distinguish between *global* and *local* coordinates. Demonstrate with an example plane truss.

3.2 Can the same local coordinates be adopted in both stiffness and flexibility methods of analysis? Discuss.

3.3 How do we find support reactions in matrix methods based on (a) stiffness method and (b) flexibility method?

3.4 What are the different types of *transformation matrices* encountered in matrix methods of structural analysis? Discuss.

3.5 Define *displacement transfer coefficient* and *force transfer coefficient*.

3.6 Can a unique *force transformation matrix* be defined for a statically indeterminate structure? Discuss.

3.7 What is the *Principle of Contra-gradient*? Discuss.

3.8 Prove that the force response obtained by applying static equilibrium concepts in a statically determinate structure also satisfies compatibility of displacements.

3.9 Define *stiffness coefficient* and *stiffness matrix*. How is the *structure stiffness matrix* related to the *element stiffness matrix*?

3.10 Define *flexibility coefficient* and *flexibility matrix*. How is the *structure flexibility matrix* related to the *element flexibility matrix*?

3.11 Prove that the stiffness matrix for a linear elastic structure is symmetric. What are the other typical characteristics of the structure stiffness matrix?

3.12 Prove that the flexibility matrix for a linear elastic structure is symmetric. Comment on the *condition* of the structure flexibility matrix.

3.13 Are the element stiffness matrix and element flexibility matrix always non-singular?

3.14 What is the advantage in the reduced element stiffness formulation? Will it result in the same structure stiffness matrix as the conventional element stiffness formulation?

3.15 Is it possible to derive the element stiffness and flexibility matrices for a non-prismatic element? How?

3.16 How do we account for indirect loads due to temperature effects in (a) stiffness method and (b) flexibility method?

3.17 How do we account for indirect loads due to support movements in (a) stiffness method and (b) flexibility method?

3.18 What is meant by equivalent joint loads? What purpose do they serve? Is it necessary to generate equivalent joint loads in the flexibility method (matrix approach)?

3.19 Enumerate the steps involved in finding the force and displacement responses by the stiffness method of analysis, under the action of both direct and indirect loads.

3.20 Enumerate the steps involved in finding the force and displacement responses by the flexibility method of analysis, under the action of both direct and indirect loads.

3.21 Compare and contrast the stiffness and flexibility methods of matrix analysis of structures. Hence, explain why the stiffness method is preferred by softwares for structural analysis?

4

Matrix Analysis of Structures with Axial Elements

4.1 INTRODUCTION

In this Chapter, we will see how matrix methods of analysis can be applied to one-dimensional axial structures, plane trusses and space trusses. In subsequent Chapters, we will show applications to beams, grids and rigid-jointed frames, which involve dealing with larger stiffness and flexibility matrices at the element level.

The pin-jointed truss element is an axial element. Typically, the internal force in such an element is an axial force (either tensile or compressive) and the corresponding deformation is an axial elongation or contraction. In the simplest formulation possible (reduced element stiffness method or flexibility method), we need to consider only a single degree of freedom for each prismatic element, corresponding to the axial deformation. Thus, for the ith element, we can define the axial stiffness, k_i, and its reciprocal, the axial flexibility, f_i as simple scalar quantities. However, for a more generic and versatile formulation, as is possible in the conventional stiffness method, it is desirable to identify degrees of freedom at the two ends of the element.

The simplest case is that of straight axial elements aligned in one direction. In such *one-dimensional structures*, we have two axial degrees of freedom (one at each end) per element in the regular stiffness method, and the stiffness matrix of the element (in local coordinates) is of order 2×2. In two-dimensional structures (*plane trusses*), in general, we can conceive of four translational degrees of freedom (two at each end) per element and in three-dimensional structures (space trusses), we can conceive of six degrees of freedom (three at each end). However, only the axial degrees of freedom at the two ends (along the local x^*-axis of the element) are significant, in terms of causing elastic deformations in the truss element. The axial deformation, e_i, of the element is given by the relative displacement in the axial direction between the two end axial displacements, D_{1*}^i and D_{2*}^i, as explained in Section 3.4.1. The axial force in the element, N_i, is given directly by the product of the axial deformation, e_i, and the axial stiffness, k_i, of the element, and hence the member-end forces, F_{1*}^i and F_{2*}^i, which are equal to $\pm N_i$, are dependent only on the axial displacements, D_{1*}^i and D_{2*}^i, and not on displacements in the transverse directions. In other words, the truss member has zero stiffness, corresponding to degrees of freedom along the local y^*- and z^*-axes of the element. In a space truss, it is possible for the element to have displacements at the two ends in two orthogonal transverse directions, but this will not be accompanied by conjugate forces in these directions. Such rigid-body displacements (of small magnitude) are not of significance at the element level, and for this reason, the same simple truss element model with two axial degrees of freedom can be adopted for one-dimensional systems, plane trusses and space trusses. It is also possible to work with a reduced stiffness formulation, involving only one axial degree of freedom. The reciprocal of this axial stiffness is the axial flexibility to be considered in the flexibility method of analysis of trusses. The stiffness and flexibility matrices for a typical prismatic truss element are discussed in Section 4.2.

The loading on the pin-jointed frame may be in the form of direct actions or in the form of indirect loading (self-straining due to temperature effects or 'lack of fit'). Direct actions are usually in the form of concentrated loads applied at the pin joint locations, but may occasionally act as distributed loads (in one-dimensional structures). In the latter case, it is necessary to invoke the concept of *equivalent joint loads*.

The application of matrix methods to one-dimensional axially loaded systems is presented in Section 4.3. Sections 4.4 and 4.5 deal with the matrix analysis of plane trusses and space trusses respectively. Through different examples, it is shown how matrix methods can be adopted to find the complete force and displacement responses of trusses.

It may be noted that cables, which are also axial elements, can be treated as axial (truss) elements, provided they are not laterally loaded, remain straight and in a state of axial tension (act as ties). Otherwise, they need a different and special treatment, which accounts for the non-linear behaviour associated with large displacements (changes in cable profile) and possible slackening (loss of axial stiffness).

4.2 STIFFNESS AND FLEXIBILITY MATRICES FOR A TRUSS ELEMENT

A *truss (axial) element* is a skeletal element that resists external loads by developing internal axial forces that may be tensile or compressive, undergoing axial deformations.

4.2.1 Truss Element with Single Degree of Freedom

Consider a prismatic truss (axial) element i, having a length L_i and a cross-sectional area A_i. Let the axial force (assumed positive, if tensile) be N_i and the corresponding total elastic deformation (elongation) be e_i, as shown in Fig. 4.1a. Assuming linear elastic behaviour, we can invoke Hooke's law to find a linear relationship between N_i and e_i in terms of the elastic modulus E_i of the material. If $u(x)$ denotes the axial displacement of any point located at x from the left end (origin), then the axial strain at this point is given by $\varepsilon_{axial} = \frac{du}{dx^*}$. As the axial force N_i is assumed to be constant throughout the length and the cross-sectional area A_i is also constant, the normal (axial) stress $\sigma_{axial} = \frac{N_i}{A_i} = E_i \varepsilon_{axial}$ will also be constant throughout the length. This implies that the axial strain is also constant throughout the length of the element, $\varepsilon_{axial} = \frac{du}{dx^*} = \frac{e_i}{L_i}$, whereby $\frac{N_i}{A_i} = E_i \frac{e_i}{L_i} \Rightarrow N_i = \frac{E_i A_i}{L_i} e_i$.

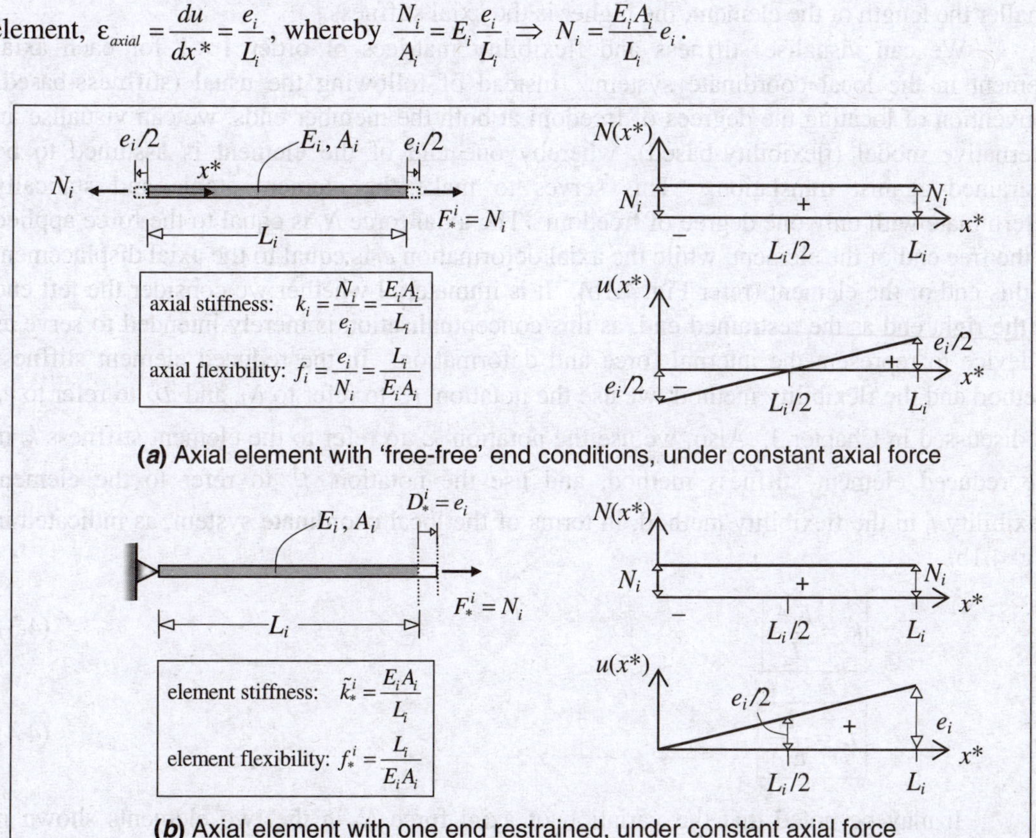

(a) Axial element with 'free-free' end conditions, under constant axial force

(b) Axial element with one end restrained, under constant axial force

Figure 4.1 Prismatic truss (axial) element with one degree of freedom

The **axial stiffness** k_i of such an element, defined as the axial force N_i associated with unit axial deformation (elongation $e_i = 1$), assuming constant axial strain, is therefore, given by:

$$k_i = \frac{N_i}{e_i} = \frac{E_i A_i}{L_i} \tag{4.1}$$

Conversely, we can define the **axial flexibility** f_i of the element as the axial deformation caused by a unit axial force:

$$f_i = \frac{e_i}{N_i} = \frac{L_i}{E_i A_i} \tag{4.2}$$

The axial flexibility is the reciprocal of the axial stiffness. The quantity, E_iA_i, appearing in these two terms, is called the **axial rigidity**. The higher the axial rigidity and the smaller the length of the element, the higher is the axial stiffness.

We can visualise stiffness and flexibility matrices of order 1×1 for each axial element in the local coordinate system. Instead of following the usual (stiffness-based) convention of locating the degrees of freedom at both the member ends, we can visualise an alternative model (flexibility-based), whereby one end of the element is assumed to be restrained against translation. This serves to make the element stable and statically determinate with only one degree of freedom. The axial force N_i is equal to the force applied at the free end of the element, while the axial deformation e_i is equal to the axial displacement at this end of the element (refer Fig. 4.1b). It is immaterial whether we consider the left end or the right end as the restrained end, as this conceptualisation is merely intended to serve as a device to represent the internal force and deformation. In the reduced element stiffness method and the flexibility method, we use the notation F_*^i to refer to N_i, and D_*^i to refer to e_i, as discussed in Chapter 3. Also, we use the notation $\tilde{k}_*^i$ to refer to the element stiffness k_i in the reduced element stiffness method, and use the notation f_*^i to refer to the element flexibility f_i in the flexibility method, in terms of the local coordinate system, as indicated in Fig. 4.1b.

$$\boxed{k_*^i = \frac{E_i A_i}{L_i}} \tag{4.3}$$

$$\boxed{f_*^i = \frac{L_i}{E_i A_i}} \tag{4.4}$$

It may be noted that the variation of axial force N_i in the two elements shown in Figs 4.1a and b are identical (constant throughout the length), whereby the axial strain, $\frac{du}{dx^*} = \frac{e_i}{L_i}$, is also constant, and hence the axial displacement, $u(x^*)$ varies linearly across the length of the element with the same slope, as shown in Fig. 4.1; the difference in the two displacement distributions lies in the different boundary conditions.

$$\frac{du}{dx^*} = \frac{e_i}{L_i} \qquad \Rightarrow \qquad u(x^*) = \int \frac{e_i}{L_i} dx^* = \frac{e_i}{L_i} x^* + C \tag{a}$$

In Fig. 4.1a, as the applied loading is symmetric, it is obvious that the displacement at the mid-point must be zero; i.e., $u(L_i/2) = 0 \Rightarrow C = -e_i/2$. On the other hand, in Fig. 4.1b, the displacement is zero at the left end (kinematic boundary condition), i.e., $u(0) = 0 \Rightarrow C = 0$. Accordingly,

$$u(x^*) = \begin{cases} e_i\left(\dfrac{x^*}{L_i} - \dfrac{1}{2}\right) & \text{for element in Fig. 4.1a} \\ e_i\left(\dfrac{x^*}{L_i}\right) & \text{for element in Fig. 4.1b} \end{cases} \tag{b}$$

The distributions of $u(x^*)$ are plotted in Fig. 4.1 for the two elements.

The truss (axial) element with a single degree of freedom can be used in the reduced element stiffness method and the flexibility method, as applied to one-dimensional axial (line) structures, plane trusses and space trusses, in which the elements undergo only axial deformations and axial forces.

The displacement transformation matrix $\mathbf{T_D}$ (in the reduced element stiffness method) and the force transformation matrix $\mathbf{T_F}$ (in the flexibility method), need to be appropriately generated, using compatibility and equilibrium conditions, as discussed later in this Chapter. Using the $\mathbf{T_D}$ matrix, along with the reduced unassembled stiffness matrix $\tilde{\mathbf{k}}_*$, the structure stiffness matrix $\mathbf{k}$ can be generated for application in the reduced element stiffness method. Similarly, using the $\mathbf{T_F}$ matrix, along with the unassembled flexibility matrix $\mathbf{f}_*$, the structure flexibility matrix $\mathbf{f}$ can be generated for application in the flexibility method.

4.2.2 Axial Element with Two Degrees of Freedom (Stiffness Method)

The axial element with a single degree of freedom is not suitable for application in the conventional stiffness method, even when applied in a one-dimensional (axial) structure comprising many such elements. This is because there are actually two independent degrees of freedom (at the two ends) of each element We need to define the axial stiffness with reference to local coordinates defined at the two ends of the axial element (refer Fig. 4.2a). We can do this easily extending the stiffness concept defined in the previous Section to two degrees of freedom. Each end of the element has an axial degree of freedom, which is shown directed along the local x^*-axis. Following the convention described earlier (Section 3.4.1), the coordinates are denoted as 1* and 2* at the "start node" (left end) and "end node" (right end) respectively. Based on simple compatibility requirements, the expression for member elongation e_i in terms of the member end translations, D_{1*}^i and D_{2*}^i, is easily obtained as:

$$e_i = D_{2*}^i - D_{1*}^i \tag{4.5}$$

In the stiffness approach, the axial tension N_i in the member can be easily obtained as a product of the elongation e_i (given by Eq. 4.5) and the axial stiffness k_i (given by Eq. 4.1):

$$N_i = k_i e_i = \frac{E_i A_i}{L_i} e_i \tag{4.6}$$

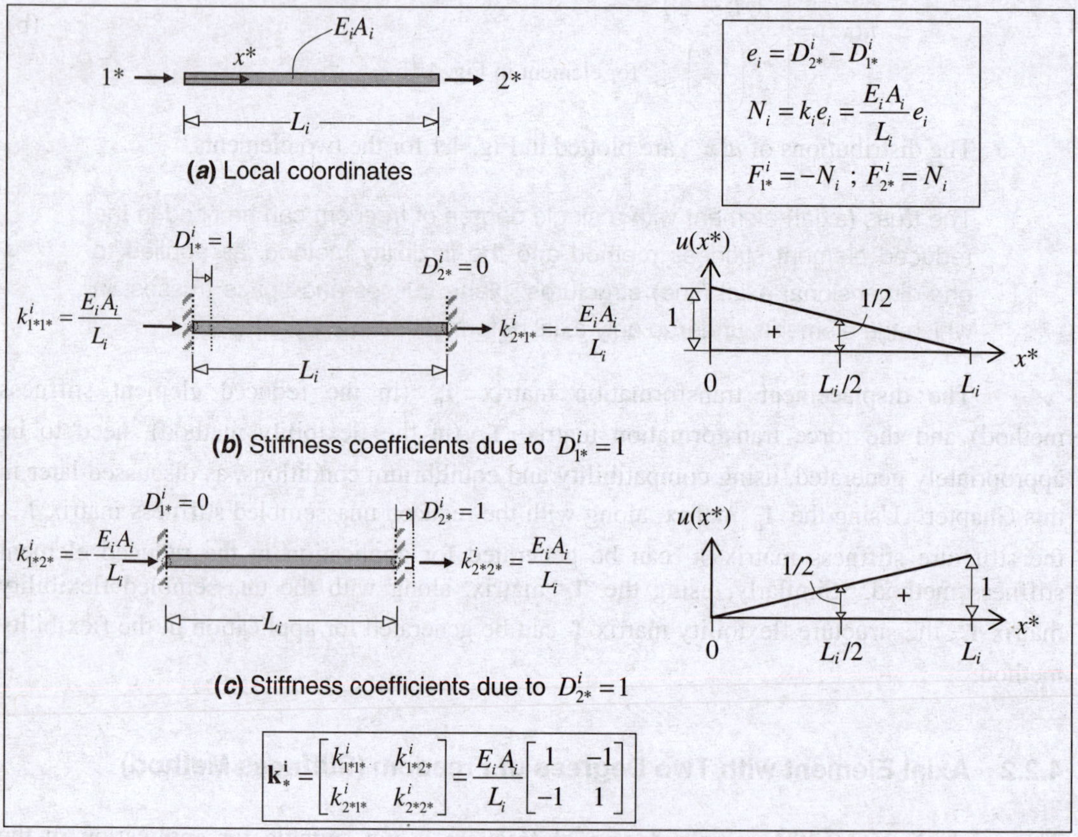

Figure 4.2 Generation of 2 × 2 stiffness matrix for an axial element

Considering any section through the element (Fig. 4.2a) and applying the equilibrium equation, $\sum F_{x^*} = 0$, to the two resulting free-bodies, the expressions for the member end forces F_{1*}^i and F_{2*}^i can be easily obtained in terms of the axial tension N_i:

$$F_{1*}^{i} = -N_i \quad ; \quad F_{2*}^{i} = +N_i \tag{4.7}$$

The stiffness matrix $\mathbf{k}_*^\mathbf{i}$ for this element is a 2×2 matrix. The primary (kinematically determinate) system is one in which both degrees of freedom are arrested. The forces (reactions) induced at the two coordinates on account of the displacement fields, $\{D_{1*}^{i} = 1;\ D_{2*}^{i} = 0\}$ and $\{D_{1*}^{i} = 0;\ D_{2*}^{i} = 1\}$ are depicted in Figs 4.2b and c respectively. We visualise the restrained system to have somehow undergone the prescribed displacement field. For example, $\{D_{1*}^{i} = 1; D_{2*}^{i} = 0\}$ corresponds to a translational slip of unit magnitude at the left end, as shown in Fig. 4.2b. This induces a unit axial deformation, $e_i = -1$ (Eq. 4.5), causing an axial force $N_i = k_i e_i = -\frac{E_i A_i}{L_i}$ (Eq. 4.6), which in turn equilibrates with the reactions at the two ends, whereby, $F_{1*}^{i} = k_{1*1*}^{i} = \frac{E_i A_i}{L_i}$ and $F_{2*}^{i} = k_{2*1*}^{i} = -\frac{E_i A_i}{L_i}$ (Eq. 4.7). Similarly, for the condition shown in Fig. 4.2c, we can derive $k_{1*2*}^{i} = -\frac{E_i A_i}{L_i}$ and $k_{2*2*}^{i} = +\frac{E_i A_i}{L_i}$.

The element stiffness matrix $\mathbf{k}_*^\mathbf{i}$ is accordingly generated as follows:

$$\mathbf{k}_*^\mathbf{i} = \begin{bmatrix} k_{1*1*}^{i} & k_{1*2*}^{i} \\ k_{2*1*}^{i} & k_{2*2*}^{i} \end{bmatrix} = \frac{E_i A_i}{L_i} \begin{bmatrix} 1 & -1 \\ -1 & 1 \end{bmatrix} \tag{4.8}$$

$$\Rightarrow \quad \boxed{\begin{bmatrix} F_{1*}^{i} \\ F_{2*}^{i} \end{bmatrix} = \frac{E_i A_i}{L_i} \begin{bmatrix} 1 & -1 \\ -1 & 1 \end{bmatrix} \begin{bmatrix} D_{1*}^{i} \\ D_{2*}^{i} \end{bmatrix}} \tag{4.8a}$$

The variation of axial translation along the length of the element can be easily derived, as shown earlier (refer Eqns a and b):

$$u(x^*) = \begin{cases} \left(1 - \frac{x^*}{L_i}\right) & \text{for element in Fig. 4.2b} \\ \left(\frac{x^*}{L_i}\right) & \text{for element in Fig. 4.2c} \end{cases} \tag{c}$$

The distributions of $u(x^*)$ are plotted in Figs 4.2b and c for the two elements.

Energy Formulation

We can, alternatively, derive the stiffness coefficients from first principles using the energy formulation described in Section 1.7.5. For this purpose, we need to formulate an expression for the strain energy U_i in the element, in terms of the end displacements, D_{1*}^{i} and D_{2*}^{i} (Fig. 4.2a).

The axial strain, assumed to be constant throughout the length of the element, can be expressed as the gradient of the axial displacement $u(x^*)$ at any intermediate location (at a distance x^* from the left end) as $\varepsilon_{axial} = \frac{du}{dx^*} = \frac{e_i}{L_i} = \frac{D^i_{2*} - D^i_{1*}}{L_i}$. The axial stress is given by $\sigma_{axial} = \frac{N_i}{A_i} = E_i \varepsilon_{axial}$, whereby $N_i = \frac{E_i A_i}{L_i}\left(D^i_{2*} - D^i_{1*}\right)$. The expression for strain energy, accordingly, takes the following form:

$$U_i = \frac{1}{2}\int_0^{L_i} N(x^*)\frac{du}{dx^*}dx = \frac{E_i A_i}{2L_i^2}\int_0^{L_i}\left(D^i_{2*} - D^i_{1*}\right)^2 dx = \frac{E_i A_i}{2L_i}\left[\left(D^i_{1*}\right)^2 + \left(D^i_{2*}\right)^2 - 2\left(D^i_{1*}\right)\left(D^i_{2*}\right)\right] \quad \text{(d)}$$

whereby, applying Eq. 1.65,

$$\begin{aligned}
\frac{\partial U_i}{\partial D^i_{1*}} &= \frac{E_i A_i}{2L_i}\left[2\left(D^i_{1*}\right) - 2\left(D^i_{2*}\right)\right] \Rightarrow k_{1*1*} = \frac{\partial^2 U_i}{\partial {D^i_{1*}}^2} = +\frac{E_i A_i}{L_i} \text{ and } k_{1*2*} = \frac{\partial^2 U_i}{\partial D^i_{1*}\partial D^i_{2*}} = -\frac{E_i A_i}{L_i} \\
\frac{\partial U_i}{\partial D^i_{2*}} &= \frac{E_i A_i}{2L_i}\left[2\left(D^i_{2*}\right) - 2\left(D^i_{1*}\right)\right] \Rightarrow k_{2*2*} = \frac{\partial^2 U_i}{\partial {D^i_{2*}}^2} = +\frac{E_i A_i}{L_i} \text{ and } k_{2*1*} = \frac{\partial^2 U_i}{\partial D^i_{2*}\partial D^i_{1*}} = -\frac{E_i A_i}{L_i}
\end{aligned} \quad \text{(e)}$$

The resulting element stiffness matrix is identical to the one obtained earlier (Eq. 4.8).

2 × 2 Flexibility Matrix not Possible!

Although we could generate both the stiffness matrix and flexibility matrix, considering a single degree of freedom for the axial element (with $\mathbf{f}^i_* = \left[\tilde{\mathbf{k}}^i_*\right]^{-1} = \frac{L_i}{E_i A_i}$), we will find that the flexibility matrix cannot be formulated for the same element with two degrees of freedom.

It is interesting to note that the determinant of the 2×2 stiffness matrix $\mathbf{k}^i_*$ in Eq. 4.8 is zero (rank = 1), whereby the matrix is *not invertible*. This means that a 2×2 flexibility matrix (inverse of the stiffness matrix) for the element shown in Fig. 4.2a cannot be defined. This is so, because this element is *unstable* and cannot resist the action of any arbitrary load (forces F^i_{1*} and F^i_{2*}). In order to have a valid flexibility matrix, it is necessary to arrest one degree of freedom. The resulting flexibility matrix is a 1×1 matrix, as defined earlier (Eq. 4.4).

Element Stiffness Matrix in Global Axes System

In a one-dimensional structure, the local coordinate x^*-axis of an axial element (with 2 degrees of freedom) can be conveniently chosen to be aligned in the same direction as the global x-axis for the structure. Thus, the 2×2 element transformation matrix $\mathbf{T}^i$, relating the displacement and force vector components, (D^i_1, D^i_2) and (F^{i}_1, F^i_2), in the global axes system to the local coordinate system, as defined in Section 3.3.1, will be an *identity* matrix ($\mathbf{T}^i = \mathbf{I} = \mathbf{T}^{i^T}$):

$$\begin{bmatrix} D_{1*}^i \\ D_{2*}^i \end{bmatrix} = \begin{bmatrix} 1 & 0 \\ 0 & 1 \end{bmatrix} \begin{bmatrix} D_1^i \\ D_2^i \end{bmatrix} = \begin{bmatrix} D_1^i \\ D_2^i \end{bmatrix} \quad \text{and} \quad \begin{bmatrix} F_{1*}^i \\ F_{2*}^i \end{bmatrix} = \begin{bmatrix} F_1^i \\ F_2^i \end{bmatrix} \tag{4.9}$$

The element stiffness matrix in the global axes system is given by the transformation $\mathbf{k}^i = \mathbf{T}^{i^T}\mathbf{k}_*^i\mathbf{T}^i$ (refer Eq. 3.32), which in this case, reduces to $\mathbf{k}^i = \mathbf{k}_*^i$, whereby:

$$\begin{bmatrix} F_1^i \\ F_2^i \end{bmatrix} = \frac{E_i A_i}{L_i} \begin{bmatrix} 1 & -1 \\ -1 & 1 \end{bmatrix} \begin{bmatrix} D_1^i \\ D_2^i \end{bmatrix} \tag{4.10}$$

However, in two- or three-dimensional trusses, the truss element with 2 degrees of freedom will not be suitable for application in the conventional stiffness method, because there will be displacements in the transverse directions that need to be accounted for. This is discussed in the Sections 4.2.3 and 4.2.4.

4.2.3 Plane Truss Element with Four Degrees of Freedom (Stiffness Method)

In a plane truss, the truss elements can have any orientation. Unlike the one-dimensional structure, the local coordinate axes (x^*-axis and y^*-axis) of a plane truss element will generally not be aligned in the same direction as the global axes (x-axis and y-axis) for the structure, although they are in the same (usually vertical) xy plane. In order to define the displaced configuration of the plane truss element, with reference to its original configuration, it is necessary to define two degrees of freedom at each end of the element. Thus, there are four degrees of freedom to be considered, as shown in Fig. 4.3.

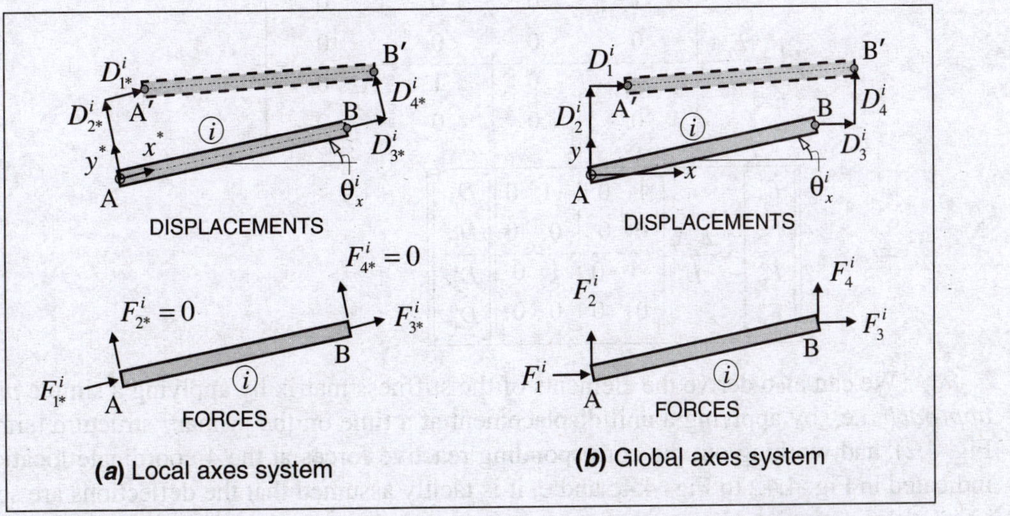

Figure 4.3 Member-end displacements and forces in a plane truss element

Following the convention described in Chapter 3, the local coordinates are denoted as 1* and 2* in the x^*- and y^*- directions respectively at the "start node" (left end) and as 3* and 4* in the x^*- and y^*- directions respectively at the "end node" (right end). Elastic deformations are possible only along the longitudinal axis direction (x^*-axis) in a truss element, whereby based on simple compatibility requirements, the expression for member elongation e_i in terms of the member end axial translations, D^i_{1*} and D^i_{3*}, is easily obtained as:

$$e_i = D^i_{3*} - D^i_{1*} \tag{4.11}$$

The axial tension N_i in the element is given by $k_i e_i = \frac{E_i A_i}{L_i} e_i$ (Eq. 4.1), and applying equilibrium equations, as shown in Section 4.2.2,

$$F^i_{1*} = -N_i = -\frac{E_i A_i}{L_i} e_i \quad ; \quad F^i_{3*} = +N_i = \frac{E_i A_i}{L_i} e_i \tag{4.12}$$

The translations, D^i_{2*} and D^i_{4*}, at the two ends of the element in transverse direction (y^*-axis) are to be viewed as rigid body displacements, and as the element has zero stiffness in this direction, the conjugate forces[†], F^i_{2*} and F^i_{4*}, will be zero:

$$F^i_{2*} = F^i_{4*} = 0 \tag{4.12a}$$

Thus, combining Eqns 4.11, 4.12 and 4.12a, we can generate the 4×4 element stiffness matrix $\mathbf{k}^i_*$ for the plane truss element (derived earlier in Chapter 3, refer Eq. 3.29):

$$\begin{array}{cccc} (D^i_{1*}=1) & (D^i_{2*}=1) & (D^i_{3*}=1) & (D^i_{4*}=1) \end{array}$$

$$\mathbf{k}^i_* = \frac{E_i A_i}{L_i} \left[\begin{array}{cc|cc} 1 & 0 & -1 & 0 \\ 0 & 0 & 0 & 0 \\ \hline -1 & 0 & 1 & 0 \\ 0 & 0 & 0 & 0 \end{array} \right] \tag{4.13}$$

$$\Rightarrow \boxed{\begin{bmatrix} F^i_{1*} \\ F^i_{2*} \\ \hline F^i_{3*} \\ F^i_{4*} \end{bmatrix} = \frac{E_i A_i}{L_i} \left[\begin{array}{cc|cc} 1 & 0 & -1 & 0 \\ 0 & 0 & 0 & 0 \\ \hline -1 & 0 & 1 & 0 \\ 0 & 0 & 0 & 0 \end{array} \right] \begin{bmatrix} D^i_{1*} \\ D^i_{2*} \\ \hline D^i_{3*} \\ D^i_{4*} \end{bmatrix}} \tag{4.13a}$$

We can also derive the elements of the stiffness matrix by applying a simple *physical approach*, i.e., by applying a unit displacement at a time on the primary structure (similar to Fig. 4.2), and working out the corresponding reactive forces at the 4 coordinate locations, as indicated in Fig. 4.4. In Figs 4.4c and e, it is tacitly assumed that the deflections are so small

[†] These forces correspond to shear forces, which are assumed to be zero in a truss element.

that the element length in the inclined position is practically the same as that in the original (horizontal) position, without inducing any axial deformation. These displacements correspond to 'rigid' body displacements, whereby the stiffness coefficients (corresponding to $D^i_{2*} = 1$ and $D^i_{4*} = 1$) are all equal to zero.

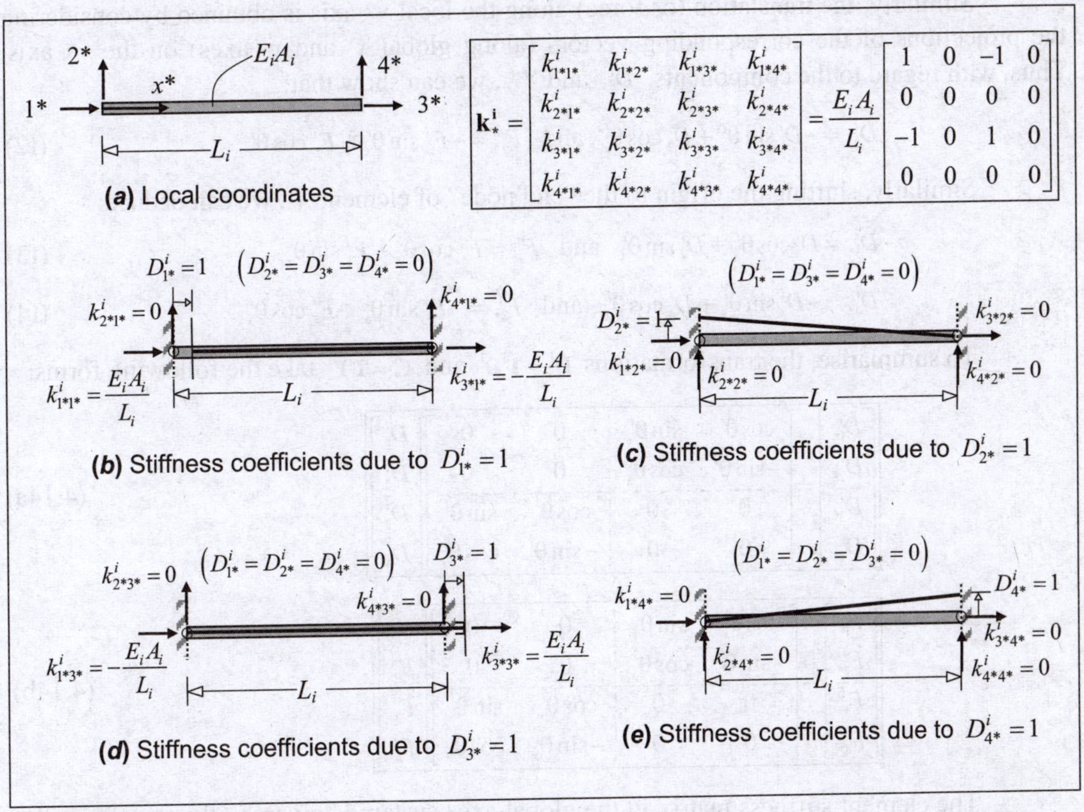

Figure 4.4 Generation of 4 × 4 stiffness matrix for a plane truss element (local coordinate system) by physical approach

Element Stiffness Matrix in Global Axes System

It is interesting to note that although the stiffnesses and forces in the transverse direction are zero in the local coordinate system (Fig. 4.3a), there will be non-zero components of these stiffnesses and forces in the global axes system (Fig. 4.3b).

Let us first derive the components of the transformation matrix $\mathbf{T}^i$ pertaining to the coordinates at the "start node". This transformation involves a simple (anti-clockwise) rotation of the coordinate axes in the *x-y* plane by an angle θ^i_x, without shifting the origin (refer Figs 4.3a and b). The translation (or force) along the local *x** axis is obtained by considering the projections of the corresponding vectors (along global *x*- and *y*- axes) on the *x** axis. Thus, with regard to the components, D_{1*}^i and F_{1*}^i, we can easily show that:

$$D_{1*}^i = D_1^i \cos\theta_x^i + D_2^i \sin\theta_x^i \quad \text{and} \quad F_{1*}^i = F_1^i \cos\theta_x^i + F_2^i \sin\theta_x^i \tag{f1}$$

Similarly, the translation (or force) along the local *y** axis is obtained by considering the projections of the corresponding vectors (along global *x*- and *y*- axes) on the *y** axis. Thus, with regard to the components, D_{2*}^i and F_{2*}^i, we can show that:

$$D_{2*}^i = -D_1^i \sin\theta_x^i + D_2^i \cos\theta_x^i \quad \text{and} \quad F_{2*}^i = -F_1^i \sin\theta_x^i + F_2^i \cos\theta_x^i \tag{f2}$$

Similarly, shifting the origin to the "end node" of element '*i*', we can deduce:

$$D_{3*}^i = D_3^i \cos\theta_x^i + D_4^i \sin\theta_x^i \quad \text{and} \quad F_{3*}^i = F_3^i \cos\theta_x^i + F_4^i \sin\theta_x^i \tag{f3}$$

$$D_{4*}^i = -D_3^i \sin\theta_x^i + D_4^i \cos\theta_x^i \quad \text{and} \quad F_{4*}^i = -F_3^i \sin\theta_x^i + F_4^i \cos\theta_x^i \tag{f4}$$

To summarise, the transformations $\mathbf{D_*^i} = \mathbf{T^i D^i}$ and $\mathbf{F_*^i} = \mathbf{T^i F^i}$ take the following forms:

$$\begin{bmatrix} D_{1*}^i \\ D_{2*}^i \\ \hline D_{3*}^i \\ D_{4*}^i \end{bmatrix} = \left[\begin{array}{cc|cc} \cos\theta_x^i & \sin\theta_x^i & 0 & 0 \\ -\sin\theta_x^i & \cos\theta_x^i & 0 & 0 \\ \hline 0 & 0 & \cos\theta_x^i & \sin\theta_x^i \\ 0 & 0 & -\sin\theta_x^i & \cos\theta_x^i \end{array}\right] \begin{bmatrix} D_1^i \\ D_2^i \\ \hline D_3^i \\ D_4^i \end{bmatrix} \tag{4.14a}$$

$$\begin{bmatrix} F_{1*}^i \\ F_{2*}^i \\ \hline F_{3*}^i \\ F_{4*}^i \end{bmatrix} = \left[\begin{array}{cc|cc} \cos\theta_x^i & \sin\theta_x^i & 0 & 0 \\ -\sin\theta_x^i & \cos\theta_x^i & 0 & 0 \\ \hline 0 & 0 & \cos\theta_x^i & \sin\theta_x^i \\ 0 & 0 & -\sin\theta_x^i & \cos\theta_x^i \end{array}\right] \begin{bmatrix} F_1^i \\ F_2^i \\ \hline F_3^i \\ F_4^i \end{bmatrix} \tag{4.14b}$$

The element stiffness matrix in the global axes system is given by the transformation $\mathbf{k^i} = \mathbf{T^{i^T} k_*^i T^i}$ (refer Eq. 3.32), whereby:

$$\begin{array}{c} \quad (D_1^i = 1) \quad (D_2^i = 1) \;|\; (D_3^i = 1) \quad (D_4^i = 1) \\ \mathbf{k^i} = \dfrac{E_i A_i}{L_i} \left[\begin{array}{cc|cc} c_i^2 & c_i s_i & -c_i^2 & -c_i s_i \\ c_i s_i & s_i^2 & -c_i s_i & -s_i^2 \\ \hline -c_i^2 & -c_i s_i & c_i^2 & c_i s_i \\ -c_i s_i & -s_i^2 & c_i s_i & s_i^2 \end{array}\right] \Rightarrow \begin{bmatrix} F_1^i \\ F_2^i \\ \hline F_3^i \\ F_4^i \end{bmatrix} = \dfrac{E_i A_i}{L_i} \left[\begin{array}{cc|cc} c_i^2 & c_i s_i & -c_i^2 & -c_i s_i \\ c_i s_i & s_i^2 & -c_i s_i & -s_i^2 \\ \hline -c_i^2 & -c_i s_i & c_i^2 & c_i s_i \\ -c_i s_i & -s_i^2 & c_i s_i & s_i^2 \end{array}\right] \begin{bmatrix} D_1^i \\ D_2^i \\ \hline D_3^i \\ D_4^i \end{bmatrix} \\ c_i = \cos\theta_x^i \;;\; s_i = \sin\theta_x^i \end{array} \tag{4.15}$$

We can also derive the elements of the stiffness matrix $\mathbf{k^i}$ by applying a simple *physical approach*, i.e., by applying a unit displacement at a time on the primary structure (similar to Fig. 4.4), and working out the corresponding reactive forces at the 4 coordinate locations, as indicated in Fig. 4.5. In order to do this, we first find the component of the unit displacement in the longitudinal direction; this gives the axial deformation, e_i, which when multiplied by the axial stiffness, $k_i = E_i A_i / L_i$, gives the axial force, N_i, in the element. By resolving the equilibrating forces at the two ends, we can generate the stiffness components, which are identical to the ones derived in Eq. 4.15.

$$\mathbf{k^i} = \begin{bmatrix} k_{11}^i & k_{12}^i & k_{13}^i & k_{14}^i \\ k_{21}^i & k_{22}^i & k_{23}^i & k_{24}^i \\ k_{31}^i & k_{32}^i & k_{33}^i & k_{34}^i \\ k_{41}^i & k_{42}^i & k_{43}^i & k_{44}^i \end{bmatrix} = k_i \begin{bmatrix} c_i^2 & c_i s_i & -c_i^2 & -c_i s_i \\ c_i s_i & s_i^2 & -c_i s_i & -s_i^2 \\ -c_i^2 & -c_i s_i & c_i^2 & c_i s_i \\ -c_i s_i & -s_i^2 & c_i s_i & s_i^2 \end{bmatrix}$$

$$c_i = \cos\theta_x^i$$
$$s_i = \sin\theta_x^i$$
$$k_i = \frac{E_i A_i}{L_i}$$

(a) Local coordinates in global axes system

(b) Stiffness coefficients due to $D_1^i = 1$

(c) Stiffness coefficients due to $D_2^i = 1$

(d) Stiffness coefficients due to $D_3^i = 1$

(e) Stiffness coefficients due to $D_4^i = 1$

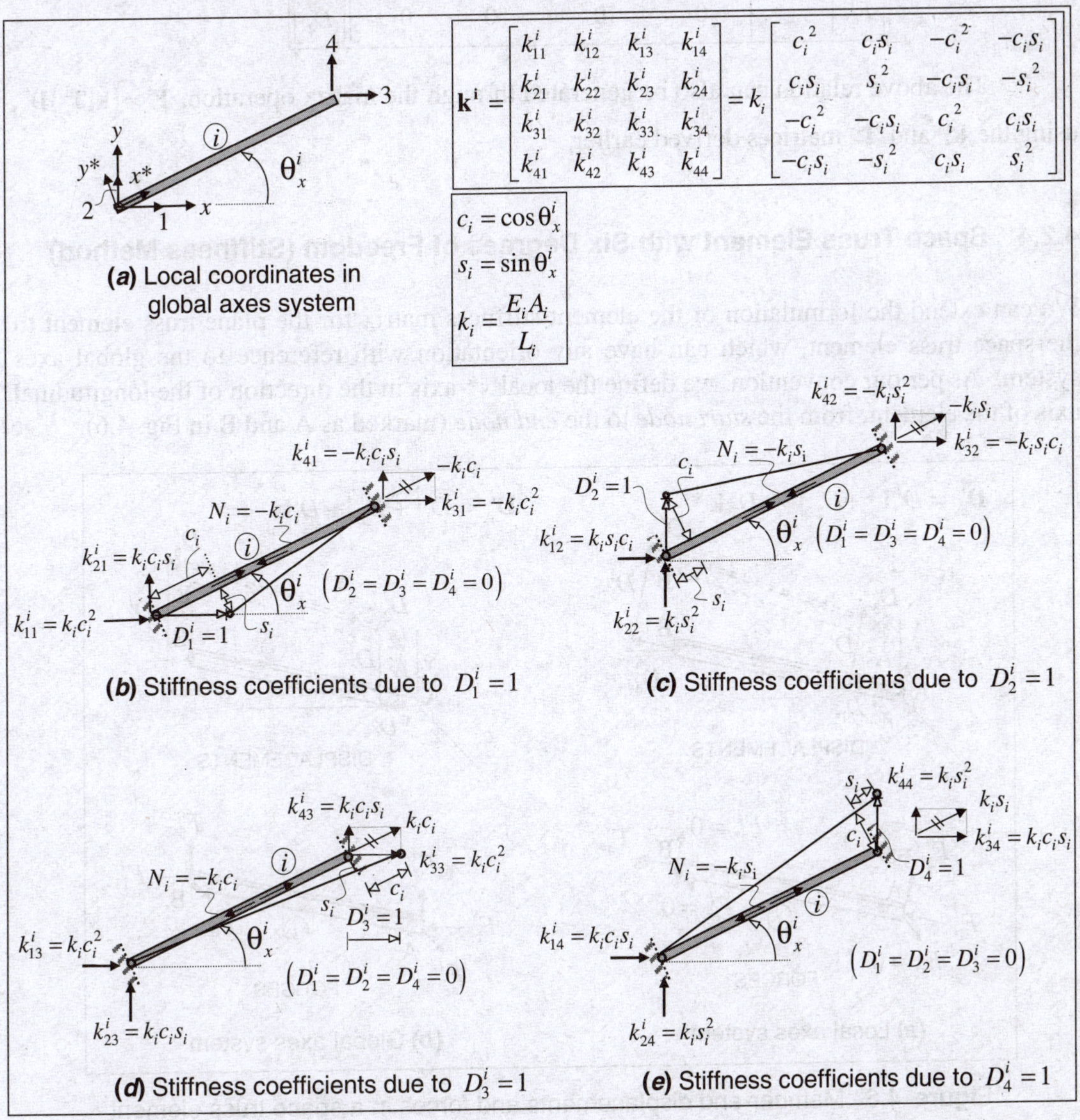

Figure 4.5 Generation of 4×4 stiffness matrix for a plane truss element (global axes system) by physical approach

Incidentally, we also observe from Fig. 4.5 that the element end forces in the local coordinate system are obtainable as:

$$\Rightarrow \begin{bmatrix} F_{1*}^i \\ F_{2*}^i \\ F_{3*}^i \\ F_{4*}^i \end{bmatrix} = \frac{E_i A_i}{L_i} \left[\begin{array}{cc|cc} \cos\theta_x^i & \sin\theta_x^i & -\cos\theta_x^i & -\sin\theta_x^i \\ 0 & 0 & 0 & 0 \\ \hline -\cos\theta_x^i & -\sin\theta_x^i & \cos\theta_x^i & \sin\theta_x^i \\ 0 & 0 & 0 & 0 \end{array} \right] \begin{bmatrix} D_1^i \\ D_2^i \\ D_3^i \\ D_4^i \end{bmatrix} \tag{4.16}$$

The above relation can also be generated through the matrix operation, $\mathbf{F_*^i} = \left(\mathbf{k_*^i T^i}\right)\mathbf{D^i}$, using the $\mathbf{k_*^i}$ and $\mathbf{T^i}$ matrices derived earlier.

4.2.4 Space Truss Element with Six Degrees of Freedom (Stiffness Method)

We can extend the formulation of the element stiffness matrix for the plane truss element to the space truss element, which can have any orientation with reference to the global axes system. As per our convention, we define the local x^*-axis in the direction of the longitudinal axis of the element, from the *start node* to the *end node* (marked as A and B in Fig. 4.6).

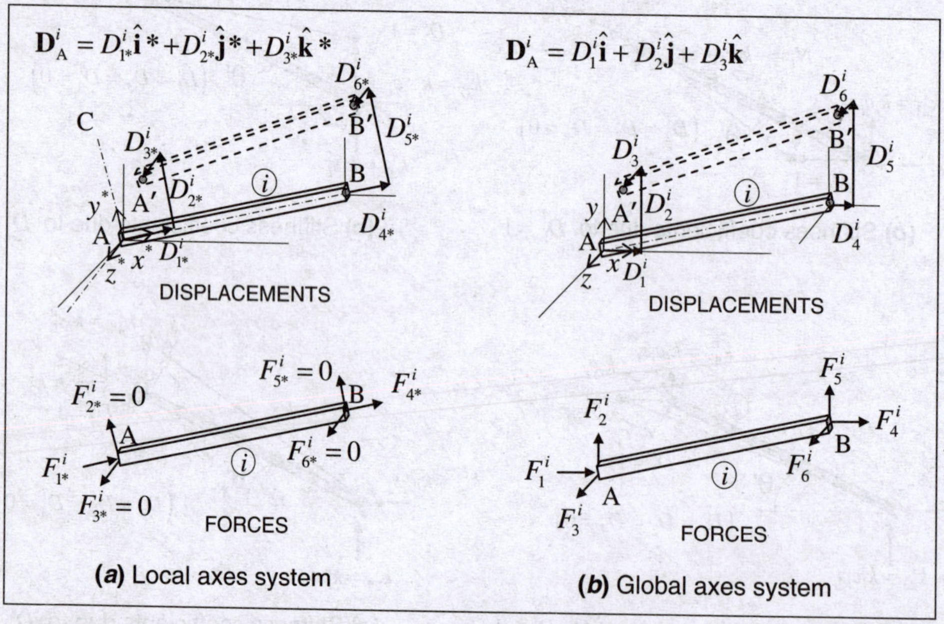

Figure 4.6 Member-end displacements and forces in a space truss element

We are free to choose the other two local coordinate axes (y^*- and z^*-axes) as any two mutually orthogonal directions that are both perpendicular to the x^*-axis. These may be conveniently oriented along the principal axes directions of the cross-section of the element, and for this purpose, it is necessary to define the coordinates of a reference point (such as C in Fig. 4.6a, such that AC points in the positive y*-direction. Then, the cross-product, $\hat{\mathbf{i}}^* \times \hat{\mathbf{j}}^* = \hat{\mathbf{k}}^*$ defines the direction of the base vector in the positive z*-direction.†

In order to define the displaced configuration of the space truss element, with reference to its original configuration, it is necessary to define three degrees of freedom at each end of the element. In addition to translations along the axial (longitudinal) x*-direction and transverse y*- direction shown in Fig. 4.3a, we need to consider the transverse z*- direction in the local axes system. Clearly, the forces in the two transverse directions (y*- and z*-), which correspond to shear forces, are always zero, as the truss element, by definition is capable of resisting forces only in the axial direction. Accordingly, the translations in these two transverse directions, $D^i_{2*}, D^i_{3*}, D^i_{5*}$ and D^i_{6*}, are to be treated as rigid body displacements. Extending the derivation of the element stiffness matrix for the plane truss element to the space truss element (Fig. 4.6a), we can easily establish:

$$\mathbf{k}^{\mathbf{i}}_{*} = \frac{E_i A_i}{L_i}\left[\begin{array}{ccc|ccc} 1 & 0 & 0 & -1 & 0 & 0 \\ 0 & 0 & 0 & 0 & 0 & 0 \\ 0 & 0 & 0 & 0 & 0 & 0 \\ \hline -1 & 0 & 0 & 1 & 0 & 0 \\ 0 & 0 & 0 & 0 & 0 & 0 \\ 0 & 0 & 0 & 0 & 0 & 0 \end{array}\right] \qquad (4.17)$$

$$\Rightarrow \begin{bmatrix} F^i_{1*} \\ F^i_{2*} \\ F^i_{3*} \\ \hline F^i_{4*} \\ F^i_{5*} \\ F^i_{6*} \end{bmatrix} = \frac{E_i A_i}{L_i}\left[\begin{array}{ccc|ccc} 1 & 0 & 0 & -1 & 0 & 0 \\ 0 & 0 & 0 & 0 & 0 & 0 \\ 0 & 0 & 0 & 0 & 0 & 0 \\ \hline -1 & 0 & 0 & 1 & 0 & 0 \\ 0 & 0 & 0 & 0 & 0 & 0 \\ 0 & 0 & 0 & 0 & 0 & 0 \end{array}\right] \begin{bmatrix} D^i_{1*} \\ D^i_{2*} \\ D^i_{3*} \\ \hline D^i_{4*} \\ D^i_{5*} \\ D^i_{6*} \end{bmatrix} \qquad (4.17a)$$

Element Stiffness Matrix in Global Axes System

In order to derive the element stiffness matrix $\mathbf{k}^{\mathbf{i}}$ in the global axes system, we need to first generate the transformation matrix $\mathbf{T}^{\mathbf{i}}$, which is an *orthogonal* transformation matrix, that enables a displacement or force vector, defined in the global axes system (Fig. 4.6b) to be transformed to the local coordinate system (Fig. 4.6a), without change in the length of the vector. For example, the displacement vector, $\mathbf{D}^{\mathbf{i}}_{*}$, can be expressed as follows, for the element i whose *start node* is A and *end node* is B:

† Alternatively, it is convenient to choose the z*-axis on the horizontal xz plane. Then the y*-axis is located in a vertical plane passing through the x*- and y-axes.

$$\mathbf{D}_*^{\mathbf{i}} = \begin{Bmatrix} \mathbf{D}_{\mathbf{A}*}^{\mathbf{i}} \\ \hline \mathbf{D}_{\mathbf{B}*}^{\mathbf{i}} \end{Bmatrix} = \left[\begin{array}{c|c} \mathbf{R}^{\mathbf{i}} & \mathbf{O} \\ \hline \mathbf{O} & \mathbf{R}^{\mathbf{i}} \end{array}\right] \begin{Bmatrix} \mathbf{D}_{\mathrm{A}} \\ \mathbf{D}_{\mathrm{B}} \end{Bmatrix} \tag{g}$$

where $\mathbf{R}^{\mathbf{i}}$ is a 3×3 orthogonal matrix, called *rotation matrix*, comprising 9 *direction cosines*, linking each of the unit vectors, $\hat{\mathbf{i}}^*, \hat{\mathbf{j}}^*, \hat{\mathbf{k}}^*$ in the local axes system to the unit vectors $\hat{\mathbf{i}}, \hat{\mathbf{j}}, \hat{\mathbf{k}}$ in the global axes system:

$$\begin{bmatrix} \hat{\mathbf{i}}^* \\ \hat{\mathbf{j}}^* \\ \hat{\mathbf{k}}^* \end{bmatrix} = \begin{bmatrix} \cos\theta^i_{x^*x} & \cos\theta^i_{x^*y} & \cos\theta^i_{x^*z} \\ \cos\theta^i_{y^*x} & \cos\theta^i_{y^*y} & \cos\theta^i_{y^*z} \\ \cos\theta^i_{z^*x} & \cos\theta^i_{z^*y} & \cos\theta^i_{z^*z} \end{bmatrix} \begin{bmatrix} \hat{\mathbf{i}} \\ \hat{\mathbf{j}} \\ \hat{\mathbf{k}} \end{bmatrix} = \begin{bmatrix} c_{x^*x} \\ c_{y^*x} \\ c_{z^*x} \end{bmatrix} \hat{\mathbf{i}} + \begin{bmatrix} c_{x^*y} \\ c_{y^*y} \\ c_{z^*y} \end{bmatrix} \hat{\mathbf{j}} + \begin{bmatrix} c_{x^*z} \\ c_{y^*z} \\ c_{z^*z} \end{bmatrix} \hat{\mathbf{k}} \tag{h}$$

where, for example, the direction cosine, $\cos\theta^i_{x^*y} = \cos\theta^i_{yx^*} = \hat{\mathbf{i}}^* . \hat{\mathbf{j}}$, is expressed in terms of the angle between the x^*-axis (local) and the y-axis (global). The direction cosines in the first row[†] of the $\mathbf{R}^{\mathbf{i}}$ matrix can be conveniently expressed in terms of the global axes coordinates at the start node A ($x_{\mathrm{A}}, y_{\mathrm{A}}, z_{\mathrm{A}}$) and end node B ($x_{\mathrm{B}}, y_{\mathrm{B}}, z_{\mathrm{B}}$):

$$\begin{aligned} c_{x^*x} &= \cos\theta^i_{x^*x} = \frac{x_{\mathrm{B}} - x_{\mathrm{A}}}{L_i} = c_x \\ c_{x^*y} &= \cos\theta^i_{x^*y} = \frac{y_{\mathrm{B}} - y_{\mathrm{A}}}{L_i} = c_y \\ c_{x^*z} &= \cos\theta^i_{x^*z} = \frac{z_{\mathrm{B}} - z_{\mathrm{A}}}{L_i} = c_z \end{aligned} \tag{4.18}$$

where L_i is the length of the element, given by:

$$L_i = \sqrt{(x_{\mathrm{B}} - x_{\mathrm{A}})^2 + (y_{\mathrm{B}} - y_{\mathrm{A}})^2 + (z_{\mathrm{B}} - z_{\mathrm{A}})^2} \tag{4.18a}$$

Thus, the transformation matrix $\mathbf{T}^{\mathbf{i}}$ to be used in the transformations, $\mathbf{D}_*^{\mathbf{i}} = \mathbf{T}^{\mathbf{i}}\mathbf{D}^{\mathbf{i}}$ and $\mathbf{F}_*^{\mathbf{i}} = \mathbf{T}^{\mathbf{i}}\mathbf{F}^{\mathbf{i}}$, takes the following form:

$$\mathbf{T}^{\mathbf{i}} = \left[\begin{array}{ccc|ccc} c_x & c_y & c_z & 0 & 0 & 0 \\ c_{y^*x} & c_{y^*y} & c_{y^*z} & 0 & 0 & 0 \\ c_{z^*x} & c_{z^*y} & c_{z^*z} & 0 & 0 & 0 \\ \hline 0 & 0 & 0 & c_x & c_y & c_z \\ 0 & 0 & 0 & c_{y^*x} & c_{y^*y} & c_{y^*z} \\ 0 & 0 & 0 & c_{z^*x} & c_{z^*y} & c_{z^*z} \end{array}\right] \tag{4.19}$$

[†] The other 6 direction cosines, which depend on the choice of the orientations of the y^*- and z^*-axes, are not actually required to be calculated for the purpose of generating the structure stiffness matrix and for finding the member-end forces (refer Eqns 4.20 and 4.21).

After finding the member-end displacements in the global axes system, we can find the member-end forces, using the transformation, $\mathbf{F}_*^i = \left(\mathbf{k}_*^i\mathbf{T}^i\right)\mathbf{D}^i$, which takes the following form (combining Eqns 4.17 and 4.19), and also generate the structure stiffness matrix, $\mathbf{k}^i = \mathbf{T}^{i^T}\mathbf{k}_*^i\mathbf{T}^i$. As expected, the member-end forces in the transverse directions are always zero.

$$\begin{bmatrix} F_{1*}^i \\ F_{2*}^i \\ F_{3*}^i \\ \hline F_{4*}^i \\ F_{5*}^i \\ F_{6*}^i \end{bmatrix} = \frac{E_iA_i}{L_i}\left[\begin{array}{ccc|ccc} c_x & c_y & c_z & -c_x & -c_y & -c_z \\ 0 & 0 & 0 & 0 & 0 & 0 \\ 0 & 0 & 0 & 0 & 0 & 0 \\ \hline -c_x & -c_y & -c_z & c_x & c_y & c_z \\ 0 & 0 & 0 & 0 & 0 & 0 \\ 0 & 0 & 0 & 0 & 0 & 0 \end{array}\right]\begin{bmatrix} D_1^i \\ D_2^i \\ D_3^i \\ \hline D_4^i \\ D_5^i \\ D_6^i \end{bmatrix} \tag{4.20}$$

$$\mathbf{k}^i = \frac{E_iA_i}{L_i}\left[\begin{array}{ccc|ccc} c_x^2 & c_xc_y & c_xc_z & -c_x^2 & -c_xc_y & -c_xc_z \\ c_xc_y & c_y^2 & c_yc_z & -c_xc_y & -c_y^2 & -c_yc_z \\ c_xc_z & c_yc_z & c_z^2 & -c_xc_z & -c_yc_z & -c_z^2 \\ \hline -c_x^2 & -c_xc_y & -c_xc_z & c_x^2 & c_xc_y & c_xc_z \\ -c_xc_y & -c_y^2 & -c_yc_z & c_xc_y & c_y^2 & c_yc_z \\ -c_xc_z & -c_yc_z & -c_z^2 & c_xc_z & c_yc_z & c_z^2 \end{array}\right] \tag{4.21}$$

$$\Rightarrow \begin{bmatrix} F_1^i \\ F_2^i \\ F_3^i \\ \hline F_4^i \\ F_5^i \\ F_6^i \end{bmatrix} = \frac{E_iA_i}{L_i}\left[\begin{array}{ccc|ccc} c_x^2 & c_xc_y & c_xc_z & -c_x^2 & -c_xc_y & -c_xc_z \\ c_xc_y & c_y^2 & c_yc_z & -c_xc_y & -c_y^2 & -c_yc_z \\ c_xc_z & c_yc_z & c_z^2 & -c_xc_z & -c_yc_z & -c_z^2 \\ \hline -c_x^2 & -c_xc_y & -c_xc_z & c_x^2 & c_xc_y & c_xc_z \\ -c_xc_y & -c_y^2 & -c_yc_z & c_xc_y & c_y^2 & c_yc_z \\ -c_xc_z & -c_yc_z & -c_z^2 & c_xc_z & c_yc_z & c_z^2 \end{array}\right]\begin{bmatrix} D_1^i \\ D_2^i \\ D_3^i \\ \hline D_4^i \\ D_5^i \\ D_6^i \end{bmatrix} \tag{4.21a}$$

4.3 ONE-DIMENSIONAL AXIAL STRUCTURES

In this Section, we shall learn to analyse simple one-dimensional structures, made up of m straight prismatic axial elements, for which we have derived element stiffness and flexibility matrix relations in Section 4.2. The formulation presented here is also applicable for a set of m linear springs (with spring constants, $k_1, k_2, ..., k_m$), connected in series, with restraints against translation at both ends (or one end). We shall first review the two stiffness method approaches (using the conventional stiffness method and the reduced element stiffness method) and illustrate the application of these methods through various examples, involving both direct loading and indirect loading. Next, we shall see how these same examples can be solved using the flexibility method. The basic steps to be followed in these matrix methods of analysis have already been outlined in Section 3.7.

4.3.1 Analysis by Conventional Stiffness Method

Consider the one-dimensional structure in Fig. 4.7a, made up of m straight prismatic axial elements, integrally connected in series to one another, and axially restrained at the two extreme ends. We may number the *active* global coordinates sequentially, from '1' to '$m-1$', and the two restrained coordinates (at the right support B and left support A respectively) as 'm' and '$m+1$', as indicated in Fig. 4.7b. In case there is no restraint against translation (no support) at the end B, the degree of freedom marked as 'm' will serve as another active global coordinate, while the '$m+1$' coordinate (at A) will remain restrained.

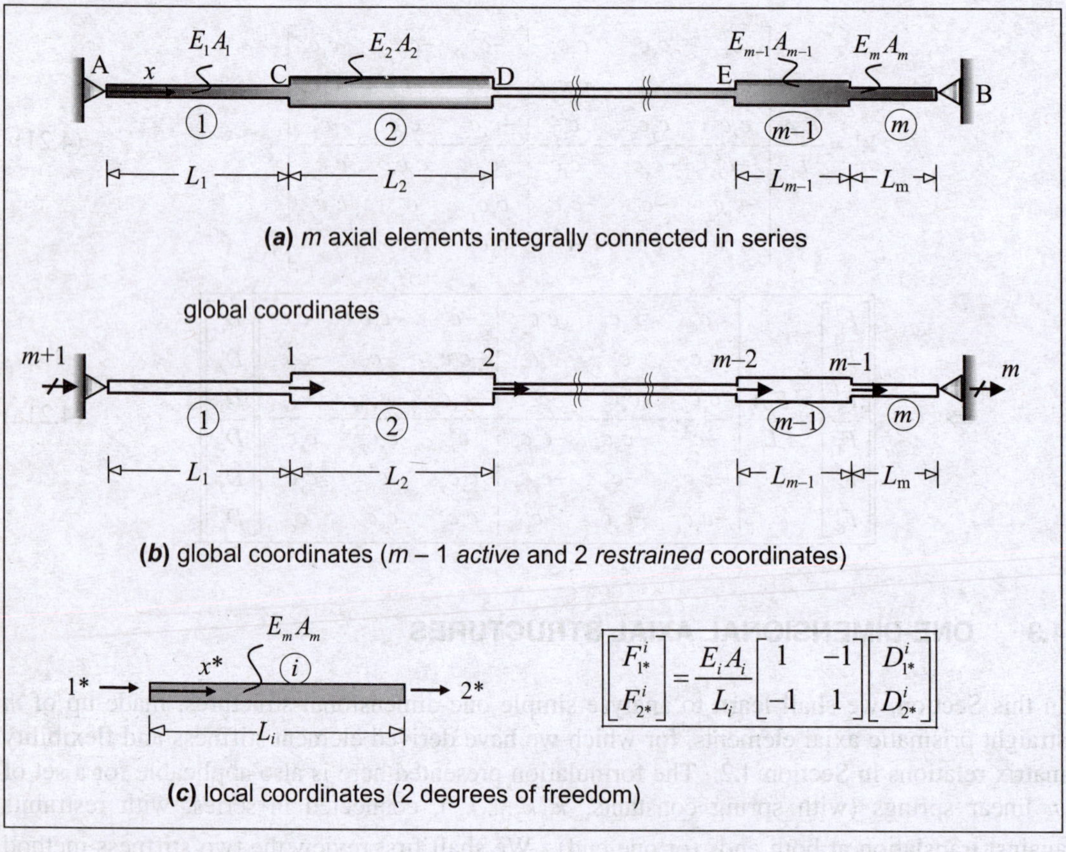

Figure 4.7 One-dimensional structure comprising *m* prismatic axial elements

In the conventional stiffness method, we consider each element to have two degrees of freedom (with local coordinates as shown in Fig. 4.7c), with the element stiffness as defined by Eq. 4.8.

Unassembled Element Stiffness Matrix

For the one-dimensional axial structure shown in Fig. 4.7a, the $2m$-dimensional *combined element displacement vector*, $\{D_*\}$, and *combined element force vector*, $\{F_*\}$, in local coordinates, are inter-related by the ***unassembled stiffness matrix***, $\mathbf{k}_*$, as explained in Section 3.4.4. This matrix is a block diagonal matrix of order $2m \times 2m$:

$$\mathbf{F}_* = \mathbf{k}_* \mathbf{D}_* \;\Rightarrow\; \begin{bmatrix} \begin{Bmatrix} F_{1*}^1 \\ F_{2*}^1 \end{Bmatrix} \\ \begin{Bmatrix} F_{1*}^2 \\ F_{2*}^2 \end{Bmatrix} \\ \vdots \\ \begin{Bmatrix} F_{1*}^m \\ F_{2*}^m \end{Bmatrix} \end{bmatrix} = \begin{bmatrix} \begin{bmatrix} k_1 & -k_1 \\ -k_1 & k_1 \end{bmatrix} & \begin{bmatrix} 0 & 0 \\ 0 & 0 \end{bmatrix} & \cdots & \begin{bmatrix} 0 & 0 \\ 0 & 0 \end{bmatrix} \\ \begin{bmatrix} 0 & 0 \\ 0 & 0 \end{bmatrix} & \begin{bmatrix} k_2 & -k_2 \\ -k_2 & k_2 \end{bmatrix} & \cdots & \begin{bmatrix} 0 & 0 \\ 0 & 0 \end{bmatrix} \\ \vdots & \vdots & & \vdots \\ \begin{bmatrix} 0 & 0 \\ 0 & 0 \end{bmatrix} & \begin{bmatrix} 0 & 0 \\ 0 & 0 \end{bmatrix} & \cdots & \begin{bmatrix} k_m & -k_m \\ -k_m & k_m \end{bmatrix} \end{bmatrix} \begin{bmatrix} \begin{Bmatrix} D_{1*}^1 \\ D_{2*}^1 \end{Bmatrix} \\ \begin{Bmatrix} D_{1*}^2 \\ D_{2*}^2 \end{Bmatrix} \\ \vdots \\ \begin{Bmatrix} D_{1*}^m \\ D_{2*}^m \end{Bmatrix} \end{bmatrix} ;\; k_i = \frac{E_i A_i}{L_i} \quad (4.22)$$

Assembling Structure Stiffness Matrix

As the local x^*-axis of the element can be conveniently aligned in the same direction as the global x-axis, coordinate transformations become simplified. The structure stiffness matrix $\mathbf{k}$ can be assembled in several ways, such as by:

(i) using the element transformation matrix, $\mathbf{T}^i = \mathbf{I}$, relating the global axis to the local axis, and summing up the contributions, $\mathbf{k}^i = \mathbf{T}^{i\mathrm{T}} \mathbf{k}_*^i \mathbf{T}^i = \mathbf{k}_*^i$ to $\mathbf{k}$, keeping track of the linkages with the global coordinate numbers;

(ii) using the displacement transformation matrix $\mathbf{T_D}$, relating the displacement vector $\mathbf{D}$ to the combined element displacement vector, $\mathbf{D}_*$, and applying $\mathbf{k} = \mathbf{T_D}^{\mathrm{T}} \mathbf{k}_* \mathbf{T_D}$;

(iii) using a 'physical approach', applying a unit displacement, $D_j = 1$, at a time on the restrained primary structure.

Let us consider here the method involving $\mathbf{T_D}$, which has an order of $2m \times (m+1)$, and is given by the following relationship, $[D_*]_{2m\times 1} = [T_D]_{2m\times(m+1)} [D]_{(m+1)\times 1}$:

$$\mathbf{D}_* = \left[\mathbf{T}_{\mathbf{DA}} \mid \mathbf{T}_{\mathbf{DR}}\right]\left[\frac{\mathbf{D}_\mathbf{A}}{\mathbf{D}_\mathbf{R}}\right] \Rightarrow \begin{bmatrix} \begin{Bmatrix} D_{1*}^1 \\ D_{2*}^1 \end{Bmatrix} \\ \begin{Bmatrix} D_{1*}^2 \\ D_{2*}^2 \end{Bmatrix} \\ \vdots \\ \begin{Bmatrix} D_{1*}^m \\ D_{2*}^m \end{Bmatrix} \end{bmatrix} = \left[\begin{array}{cccc|cc} 0 & 0 & \cdots & 0 & 0 & 1 \\ 1 & 0 & & 0 & 0 & 0 \\ \hline 1 & 0 & \cdots & 0 & 0 & 0 \\ 0 & 1 & & 0 & 0 & 0 \\ \hline \vdots & \vdots & & \vdots & \vdots & \vdots \\ \hline 0 & 0 & \cdots & 1 & 0 & 0 \\ 0 & 0 & & 0 & 1 & 0 \end{array}\right] \begin{bmatrix} D_1 \\ D_2 \\ \vdots \\ D_{m-1} \\ \hline D_m \\ D_{m+1} \end{bmatrix} \tag{4.23}$$

For example, to generate the first column in the $\mathbf{T}_\mathbf{D}$ matrix, we apply $D_1 = 1$ on the restrained (primary) structure, with $D_{j\neq 1} = 0$, and observe that this causes the right end of element '1' and the left end of element '2' to also undergo unit displacement, whereby $D_{2*}^1 = D_{1*}^2 = 1$, with all other member-end displacements equal to zero (compatibility requirement). Thus, we observe that the $\mathbf{T}_\mathbf{D}$ matrix contain mostly zeroes, except for two elements (equal to 1) in each column in the *active* coordinate zone ($\mathbf{T}_{\mathbf{DA}}$), and just one element equal to 1 in the last two columns, corresponding to the *restrained* coordinate zone ($\mathbf{T}_{\mathbf{DA}}$), as shown.

We next compute the product $\mathbf{k}_*\mathbf{T}_\mathbf{D}$, which is needed later to compute the element-end forces ($\mathbf{F}_* = (\mathbf{k}_*\mathbf{T}_\mathbf{D})\mathbf{D}$), as indicated below:

$$\mathbf{F}_* = \left[\mathbf{k}_*\mathbf{T}_{\mathbf{DA}} \mid \mathbf{k}_*\mathbf{T}_{\mathbf{DR}}\right]\left[\frac{\mathbf{D}_\mathbf{A}}{\mathbf{D}_\mathbf{R}}\right] \Rightarrow \begin{bmatrix} \begin{Bmatrix} F_{1*}^1 \\ F_{2*}^1 \end{Bmatrix} \\ \begin{Bmatrix} F_{1*}^2 \\ F_{2*}^2 \end{Bmatrix} \\ \vdots \\ \begin{Bmatrix} F_{1*}^m \\ F_{2*}^m \end{Bmatrix} \end{bmatrix} = \left[\begin{array}{cccc|cc} -k_1 & 0 & \cdots & 0 & 0 & k_1 \\ k_1 & 0 & & 0 & 0 & -k_1 \\ \hline k_2 & -k_2 & \cdots & 0 & 0 & 0 \\ -k_2 & k_2 & & 0 & 0 & 0 \\ \hline \vdots & \vdots & \ddots & \vdots & \vdots & \vdots \\ \hline 0 & 0 & \cdots & k_m & -k_m & 0 \\ 0 & 0 & & -k_m & k_m & 0 \end{array}\right] \begin{bmatrix} D_1 \\ D_2 \\ \vdots \\ D_{m-1} \\ \hline D_m \\ D_{m+1} \end{bmatrix} \tag{4.24}$$

Finally, we generate the *structure stiffness matrix*, $\mathbf{k} = \mathbf{T}_\mathbf{D}^\mathrm{T}\mathbf{k}_*\mathbf{T}_\mathbf{D}$, of order $(m+1)\times(m+1)$, making use of the $\mathbf{k}_*\mathbf{T}_\mathbf{D}$ matrix in Eq. 4.24:

$$\mathbf{k} = \mathbf{T}_\mathbf{D}^\mathrm{T}\left(\mathbf{k}_*\mathbf{T}_\mathbf{D}\right) = \left[\begin{array}{cc|cc|c|cc} 0 & 1 & 1 & 0 & \cdots & 0 & 0 \\ 0 & 0 & 0 & 1 & \cdots & 0 & 0 \\ \vdots & & \vdots & & \ddots & & \vdots \\ 0 & 0 & 0 & 0 & \cdots & 1 & 0 \\ \hline 0 & 0 & 0 & 0 & \cdots & 0 & 1 \\ 1 & 0 & 0 & 0 & \cdots & 0 & 0 \end{array}\right] \left[\begin{array}{cccc|cc} -k_1 & 0 & \cdots & 0 & 0 & k_1 \\ k_1 & 0 & & 0 & 0 & -k_1 \\ \hline k_2 & -k_2 & \cdots & 0 & 0 & 0 \\ -k_2 & k_2 & & 0 & 0 & 0 \\ \hline \vdots & \vdots & \ddots & \vdots & \vdots & \vdots \\ \hline 0 & 0 & \cdots & k_m & -k_m & 0 \\ 0 & 0 & & -k_m & k_m & 0 \end{array}\right] \Rightarrow$$

$$\left[\frac{\mathbf{F_A}}{\mathbf{F_R}}\right]=\left[\begin{array}{c|c}\mathbf{k_{AA}} & \mathbf{k_{AR}}\\ \hline \mathbf{k_{RA}} & \mathbf{k_{RR}}\end{array}\right]\left[\frac{\mathbf{D_A}}{\mathbf{D_R}}\right] \Rightarrow \begin{bmatrix} F_1 \\ F_2 \\ \vdots \\ F_{m-1} \\ \hline F_m \\ F_{m+1}\end{bmatrix} = \left[\begin{array}{cccc|cc} k_1+k_2 & -k_2 & \cdots & 0 & 0 & -k_1 \\ -k_2 & k_2+k_3 & \cdots & 0 & 0 & 0 \\ \vdots & \vdots & \ddots & \vdots & \vdots & \vdots \\ 0 & 0 & \cdots & k_{m-1}+k_m & -k_m & 0 \\ \hline 0 & 0 & \cdots & -k_m & k_m & 0 \\ -k_1 & 0 & \cdots & 0 & 0 & k_1 \end{array}\right] \begin{bmatrix} D_1 \\ D_2 \\ \vdots \\ D_{m-1} \\ \hline D_m \\ D_{m+1}\end{bmatrix} \quad (4.25)$$

We can verify the results (Eqns 4.24 and 4.25), by employing the 'physical approach' and working from first principles. For example, let us consider the effect of $D_1 = 1$ on the restrained (primary) structure, with $D_{j\neq 1} = 0$ (refer Fig. 4.7b). This will induce axial forces of $N_1 = +k_1$ and $N_2 = -k_2$ respectively in elements '1' and '2', causing $F_{1*}^1 = -k_1, F_{2*}^1 = +k_1$ and $F_{1*}^2 = k_2, F_{2*}^2 = -k_2$, and no other member-end forces, as indicated in the first column of the $\mathbf{k_* T_D}$ matrix in Eq. 4.24. Furthermore, to satisfy force equilibrium, the resultant force at the node '1' must be equal to $k_{11} = k_1 + k_2$, with equal and opposite reactive forces $k_{m+1,1} = -k_1$ at the node 'm + 1' and $k_{21} = -k_2$ at the node '2'. These values are reflected in the first column of the structure stiffness matrix in Eq. 4.25. All the elements in any column of the $\mathbf{k_* T_D}$ matrix (in Eq. 4.24) and $\mathbf{k}$ matrix (in Eq. 4.25) must add up to zero in order to satisfy force equilibrium ($\sum F_x = 0$). Furthermore, all the diagonal elements in the $\mathbf{k}$ matrix (in Eq. 4.25) will have positive entries. The stiffness k_{jj} at node 'j' will be the sum of the axial stiffness of the elements connected directly to j, whereby, typically, for an intermediate node, $k_{jj} = k_j + k_{j+1}$.

In case there is no restraint at B (in Fig. 4.1a), then the matrices in Eqns 4.24 and 4.25 remain unchanged, except for the partition line locations, which shift from in between the global coordinates $m-1$ and m to in between m and $m+1$.

Fixed End Forces and Axial Displacements

In case there are intermediate loads acting in any element, or if the element is subject to indirect loading (due to temperature effects or 'lack of fit'), we need to find the initial force and displacement responses in the 'fixed end' element (which is part of the primary structure) and thus arrive at the *equivalent joint loads*. We shall consider here three such cases in a prismatic axial element.

<u>**Case 1**: Intermediate axial point load</u>

Consider the action of an intermediate axial point load P acting at B, located at a distance a from the left end A of a fixed prismatic axial element ABC of length L and axial rigidity EA, as shown in Fig. 4.8a. The segment AB has an axial stiffness $k_1 = \frac{EA}{a}$, while the segment BC has an axial stiffness $k_2 = \frac{EA}{b}$. If we treat these two integrally connected axial segments as two springs in series, it follows that the axial stiffness of the element as a whole, defined as $k = \dfrac{P}{\Delta_B}$, is given by $k = k_1 + k_2 = \dfrac{EAL}{ab}$, whereby $\Rightarrow \Delta_B = \dfrac{P}{k_1 + k_2} = \dfrac{Pab}{EAL}$ (refer Fig. 4.8b).

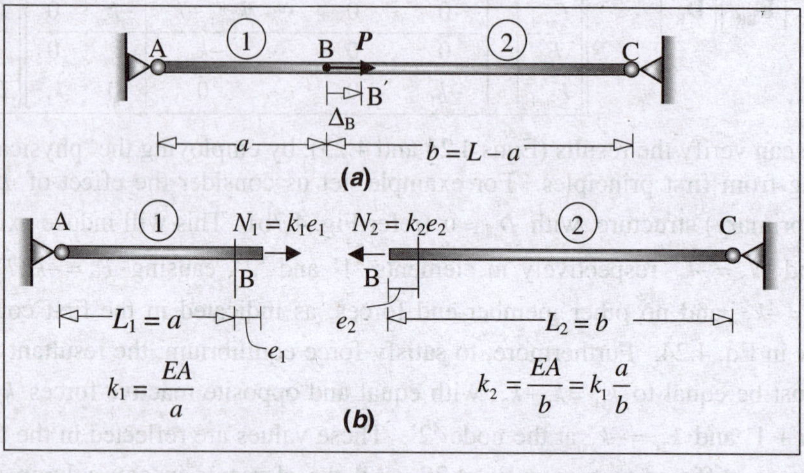

Figure 4.8 Prismatic axial element subject to an intermediate concentrated load

The compatibility requirement is that the elongation in AB and the contraction in BC must be equal to the translation of the point B, Δ_B, whereby the axial forces in the two segments are obtainable as:

$$N_1 = k_1 e_1 = k_1 \Delta_B = \frac{EA}{a} \cdot \frac{Pab}{EAL} = P\frac{b}{L} \text{ (tension)}$$

$$N_2 = k_2 e_2 = k_2 (-\Delta_B) = \frac{EA}{b} \cdot \left(-\frac{Pab}{EAL} \right) = -P\frac{a}{L} \text{ (compression)}$$

The corresponding axial force distribution is shown in Fig. 4.9b. The fixed end forces are marked in Fig. 4.9a, whereby, following our standard notations,

$$F^i_{1*f} = -P\frac{b}{L_i};\ F^i_{2*f} = -P\frac{a}{L_i} \quad \text{and} \quad u_{if,\max} = u_{if}(x^* = a) = \frac{Pab}{E_i A_i L_i} \qquad (4.26)$$

As the axial force $N(x^*)$ is constant in each of the two segments, the axial displacement $u(x^*)$ will vary linearly from zero at the two fixed ends to a peak value $\Delta_B = \dfrac{Pab}{EAL}$ at B, as shown in Fig. 4.9c. Thus, in Fig. 4.9, we have the complete force and displacement response of the axial element to the given loading.

In the special case, when the load is applied at the mid-point of the element ($a = b = L/2$),

$$F^i_{1*f} = F^i_{2*f} = -P/2 \quad \text{and} \quad u_{if,\max} = u_{if}\left(x^* = \frac{L_i}{2}\right) = \frac{PL_i}{4E_i A_i}. \qquad (4.26a)$$

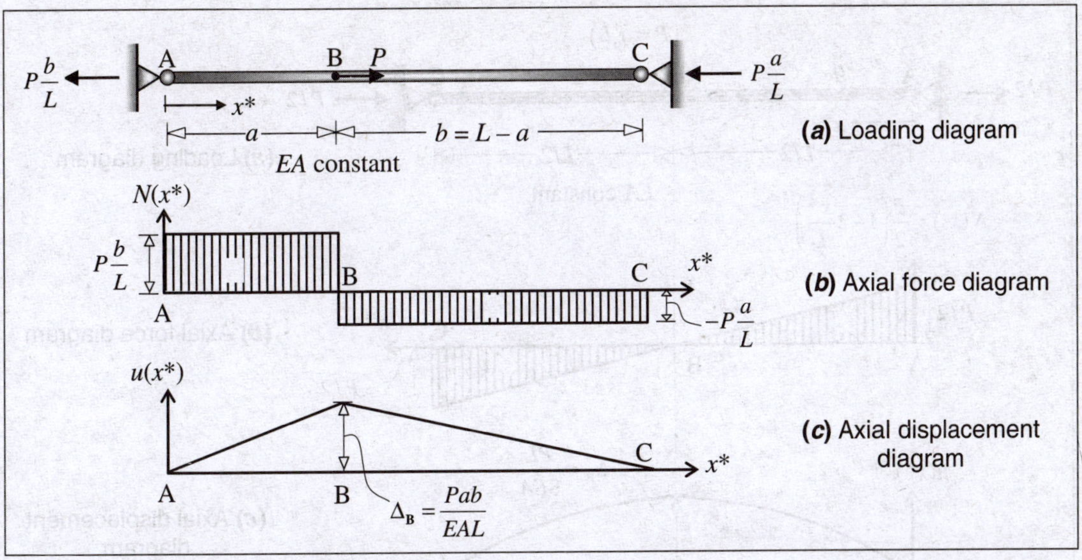

Figure 4.9 Force and displacement responses in a prismatic axial element subject to an intermediate axial point load

Case 2: Uniformly distributed axial load

Consider the action of a uniformly distributed axial load (q_o per unit length) of total magnitude $P = q_o L$ acting on a fixed prismatic axial element ABC of overall length L and axial rigidity EA, as shown in Fig. 4.10a. On account of the symmetry in the structure, applying the Principles of Parity and Superposition, it follows that the two support reactions will be equal to $-P/2$. The resulting axial force distribution will be linearly varying, with a zero value at the mid-point B, and with tension in the segment AB and compression in the segment BC, as shown in Fig. 4.10b.

Accordingly, the axial stress, and hence the axial strain, will also be varying linearly, as follows:

$$\frac{du}{dx^*} = \frac{N(x^*)/A}{E} = \frac{P}{2EA}\left(1 - 2\frac{x^*}{L}\right)$$

Hence, the axial displacement will have a parabolic variation, as shown in Fig. 4.10c, with a peak value, Δ_B, at the mid-point location, whose magnitude can be derived as follows:

$$\Rightarrow \qquad \Delta_B = u(x^* = L/2) = \frac{P}{2EA}\int_0^{L/2}\left(1 - 2\frac{x^*}{L}\right)dx^* = \frac{PL}{8EA}$$

$$F^i_{1^*f} = F^i_{2^*f} = -P/2 \qquad \text{and} \qquad u_{if,\max} = u_{if}(x^* = \frac{L_i}{2}) = \frac{PL_i}{8E_iA_i} \qquad (4.27)$$

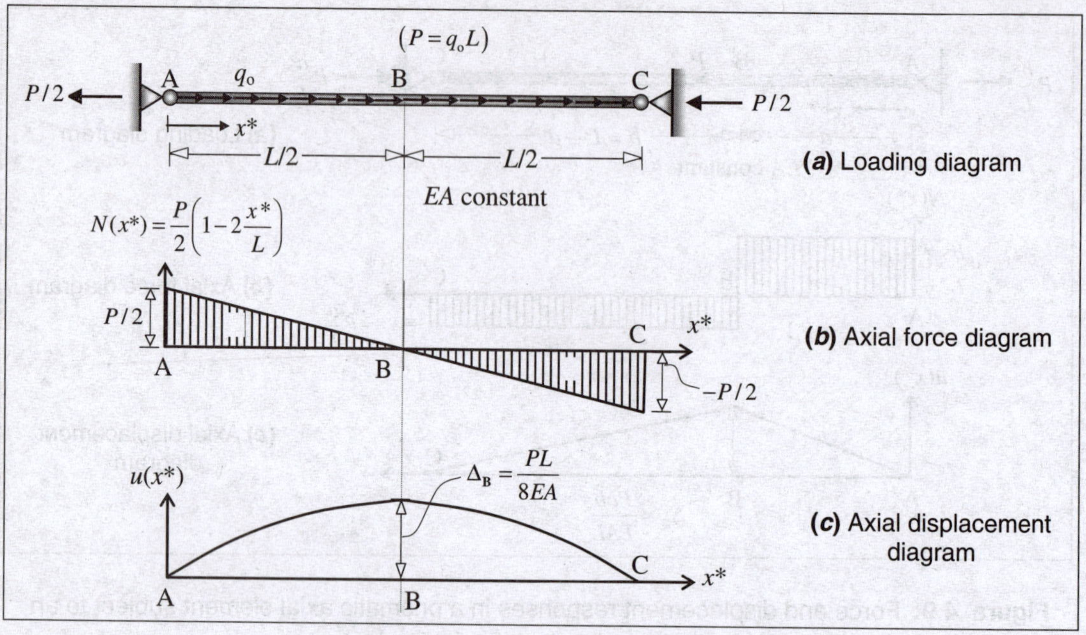

Figure 4.10 Force and displacement responses in a prismatic axial element subject to a uniformly distributed axial load

Case 3: Uniform temperature rise (or 'lack of fit')

Consider the action of a uniform increase in temperature of ΔT_i along the length of the element i, which has a length L_i and is made of a material with a coefficient of thermal expansion, α_i. Had the element been free to change its length, then this 'free length' would be given by $e_{io} = \alpha_i L_i (\Delta T_i)$. A similar situation arises in the 'lack of fit' condition, where e_{io} denotes the increase in length of the element owing to constructional error. Under the 'fixed end' condition, the element will be subject to a uniform (compressive) axial force, $\Delta N_{if} = -\frac{E_i A_i}{L_i} e_{io}$, resulting in fixed end forces:

$$\Delta F^i_{1^*f} = \frac{E_i A_i}{L_i} e_{io} \quad \text{and} \quad \Delta F^i_{2^*f} = -\frac{E_i A_i}{L_i} e_{io}. \tag{4.28}$$

The axial displacement will be zero at all points in the fixed end condition.

> The axial displacements in a prismatic axial element will be piece-wise linear under the action of intermediate point loads, parabolic under uniformly distributed loading and zero under temperature loading (or 'lack of fit') in the primary (restrained) structure. Under the action of equivalent joint loads and nodal loads, these displacements will have a linear variation between joints.

EXAMPLE 4.1

Analyse the non-prismatic axially loaded structural system shown in Fig. 4.11 by the Stiffness Method. Find the complete force and displacement responses. Assume the segment BCD to have an axial rigidity, $EA = 5000\,\text{kN}$ and the segment AB to have twice this axial rigidity.

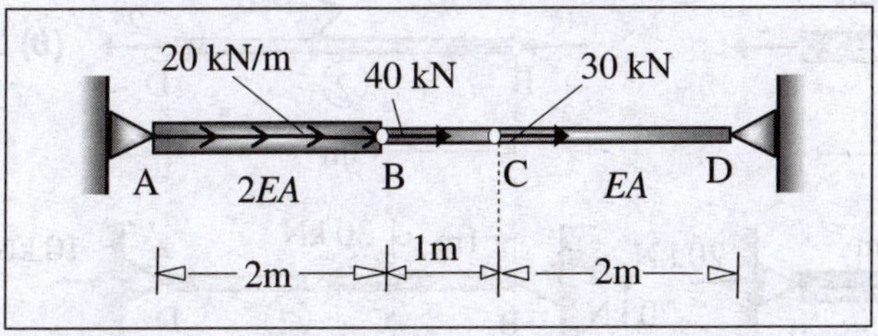

Figure 4.11 Non-prismatic axially loaded system (Example 4.1)

SOLUTION

There are two prismatic elements here ($m = 2$), with intermediate loads acting on both. There is one *active* degree of freedom ($n = 1$) and two *restrained* global coordinates ($r = 2$), as shown in Fig. 4.12a. The local coordinates for the two elements are shown in Fig. 4.12b.

Fixed End Forces and Axial Displacements

The fixed end forces are shown in Fig. 4.12c, along with the distributions of axial force $N_i(x^*)$ and axial displacement $u_i(x^*)$, for the two elements.

Element '1' (AB): (refer Eq. 4.27)

$$F^1_{1*f} = F^1_{2*f} = -\frac{P}{2} = -\frac{(20)(2)}{2} = -20.0\,\text{kN} \qquad \Rightarrow \mathbf{F}^1_{*\mathbf{f}} = \begin{bmatrix} -20 \\ -20 \end{bmatrix} \text{kN}$$

$$u_{1f,\max} = u_{1f}(x^* = L_1/2) = \frac{PL_1}{8E_1A_1} = \frac{(40\text{kN})(2\text{m})}{8(2\times5000\text{kN})} = 0.001\text{m} = 1.0\text{mm}$$

Axial force varies linearly and axial displacement varies parabolically, as shown.

Element '2' (BD): (refer Eq. 4.26)

$$F^2_{1*f} = -P\frac{b}{L_2} = -\frac{(30)(2)}{3} = -20.0\,\text{kN}; \quad F^2_{2*f} = -P\frac{a}{L_2} = -\frac{(30)(1)}{3} = -10.0\,\text{kN} \quad \Rightarrow \mathbf{F}^2_{*\mathbf{f}} = \begin{bmatrix} -20 \\ -10 \end{bmatrix} \text{kN}$$

$$u_{2f,\max} = u_{2f}(x^* = a) = \frac{Pab}{E_2A_2L_2} = \frac{(30)(1)(2)}{(5000)(3)} = 4.0\times10^{-3}\,\text{m} = 4.0\text{mm}$$

Axial force is constant on either side of the load and axial displacement varies bi-linearly, as shown.

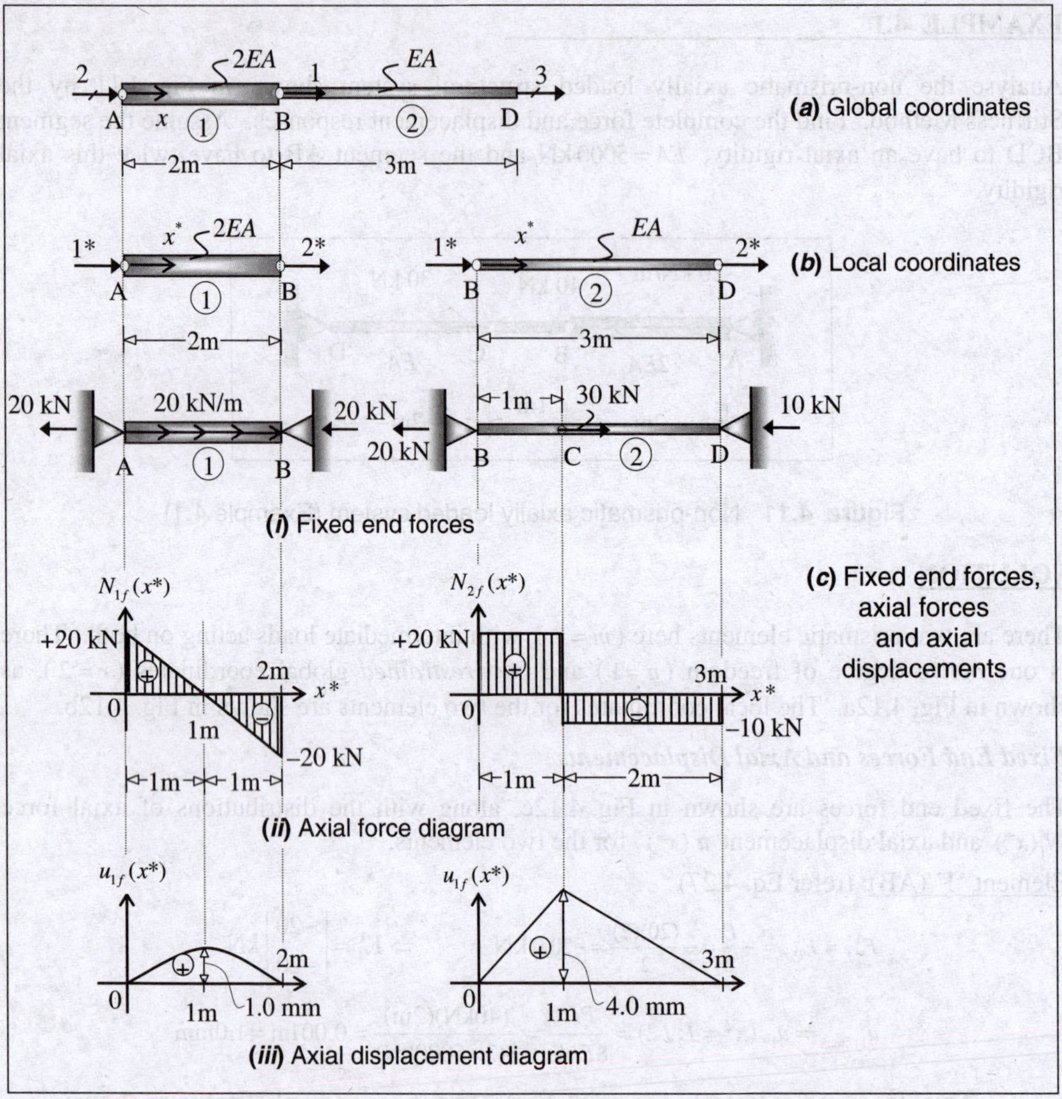

Figure 4.12 Global and local coordinates and fixed end force effects (Example 4.1)

Coordinate Transformations and Equivalent Joint Loads

The element transformation matrices $\mathbf{T}^1$ and $\mathbf{T}^2$, satisfying $\mathbf{D}_*^i = \mathbf{T}^i\mathbf{D}$ and $\mathbf{F}_*^i = \mathbf{T}^i\mathbf{F}$, take the following form (refer Figs 4.12a and b), with the linking global coordinates indicated in parentheses:

$$\mathbf{T}^1 = \overset{\quad(2)\;(1)}{\begin{bmatrix} 1 & 0 \\ 0 & 1 \end{bmatrix}}\begin{matrix}(2)\\(1)\end{matrix} \quad \text{and} \quad \mathbf{T}^2 = \overset{\quad(1)\;(3)}{\begin{bmatrix} 1 & 0 \\ 0 & 1 \end{bmatrix}}\begin{matrix}(1)\\(3)\end{matrix}$$

The fixed end forces in global axes can now be assembled, by summing up the contributions of

$$\mathbf{T}^{1^T}\mathbf{F}_{*f}^{1} = \begin{bmatrix} -20 \\ -20 \end{bmatrix}\begin{matrix}(2)\\(1)\end{matrix} \text{ kN} \quad \text{and} \quad \mathbf{T}^{2^T}\mathbf{F}_{*f}^{2} = \begin{bmatrix} -20 \\ -10 \end{bmatrix}\begin{matrix}(1)\\(3)\end{matrix} \text{ kN} \quad \Rightarrow \mathbf{F}_f = \begin{bmatrix} -40 \\ -20 \\ -10 \end{bmatrix} \text{ kN}$$

The resulting equivalent joint load at the active coordinate '1' is $F_{1e} = -F_{1f} = +40\text{kN}$, and the resultant load at this coordinate, $F_1 + F_{1e} = 40 + 40 = +80\text{kN}$, as indicated in Fig. 4.13a.

Element and Structure Stiffness Matrices

The element stiffness matrices $\mathbf{k}_*^1$ and $\mathbf{k}_*^2$, satisfying $\mathbf{F}_*^i = \mathbf{k}_*^i\mathbf{D}_*^i$, take the following form (refer Fig. 4.12b and Eq. 4.8), considering $k_1 = \frac{2EA}{2} = EA$ and $k_2 = \frac{EA}{3}$:

$$\mathbf{k}_*^1 = EA\begin{bmatrix} 1 & -1 \\ -1 & 1 \end{bmatrix} \quad \text{and} \quad \mathbf{k}_*^2 = EA\begin{bmatrix} 1/3 & -1/3 \\ -1/3 & 1/3 \end{bmatrix}$$

$$\Rightarrow \quad \mathbf{k}_*^1\mathbf{T}^1 = EA\begin{matrix}(2) \quad (1)\\ \begin{bmatrix} 1 & -1 \\ -1 & 1 \end{bmatrix}\end{matrix}\begin{matrix}(2)\\(1)\end{matrix} \quad \text{and} \quad \mathbf{k}_*^2\mathbf{T}^2 = EA\begin{matrix}(1) \quad (3)\\ \begin{bmatrix} 1/3 & -1/3 \\ -1/3 & 1/3 \end{bmatrix}\end{matrix}\begin{matrix}(1)\\(3)\end{matrix}$$

The structure stiffness matrices $\mathbf{k}$, satisfying $\mathbf{F} = \mathbf{kD}$, can now be assembled, by summing up the contributions of

$$\mathbf{T}^{1^T}\mathbf{k}_*^1\mathbf{T}^1 = EA\begin{matrix}(2) \quad (1)\\ \begin{bmatrix} 1 & -1 \\ -1 & 1 \end{bmatrix}\end{matrix}\begin{matrix}(2)\\(1)\end{matrix} \quad \text{and} \quad \mathbf{T}^{2^T}\mathbf{k}_*^2\mathbf{T}^2 = EA\begin{matrix}(1) \quad (3)\\ \begin{bmatrix} 1/3 & -1/3 \\ -1/3 & 1/3 \end{bmatrix}\end{matrix}\begin{matrix}(1)\\(3)\end{matrix}$$

$$\Rightarrow \quad \mathbf{k} = \left[\begin{array}{c|c} \mathbf{k}_{AA} & \mathbf{k}_{AR} \\ \hline \mathbf{k}_{RA} & \mathbf{k}_{RR} \end{array}\right] = EA\left[\begin{array}{c|cc} 4/3 & -1 & -1/3 \\ \hline -1 & 1 & 0 \\ -1/3 & 0 & 1/3 \end{array}\right] \quad \Rightarrow \quad [\mathbf{k}_{AA}]^{-1} = \frac{3}{4EA}$$

Displacements and Support Reactions

The structure stiffness relation to be considered here is:

$$\left[\begin{array}{c} \mathbf{F}_A \\ \hline \mathbf{F}_R \end{array}\right] - \left[\begin{array}{c} \mathbf{F}_{fA} \\ \hline \mathbf{F}_{fR} \end{array}\right] = \left[\begin{array}{c|c} \mathbf{k}_{AA} & \mathbf{k}_{AR} \\ \hline \mathbf{k}_{RA} & \mathbf{k}_{RR} \end{array}\right]\left[\begin{array}{c} \mathbf{D}_A \\ \hline \mathbf{D}_R = \mathbf{O} \end{array}\right] \Rightarrow \left[\begin{array}{c} F_1 = 40 \\ \hline F_2 = ? \\ F_3 = ? \end{array}\right] - \left[\begin{array}{c} -40 \\ \hline -20 \\ -10 \end{array}\right] = EA\left[\begin{array}{c|cc} 4/3 & -1 & -1/3 \\ \hline -1 & 1 & 0 \\ -1/3 & 0 & 1/3 \end{array}\right]\left[\begin{array}{c} D_1 = ? \\ \hline D_2 = 0 \\ D_3 = 0 \end{array}\right]$$

$$\Rightarrow \quad \mathbf{D}_A = [\mathbf{k}_{AA}]^{-1}[\mathbf{F}_A - \mathbf{F}_{fA}] \quad \Rightarrow \quad D_1 = \frac{3}{4EA}[40 + 40] = \frac{60}{EA} = \frac{60}{5000} = 0.012\text{m} = \mathbf{12.0mm}$$

The axial displacements, due to the applied loading in Fig. 4.13a, are shown in Fig. 4.13b.

Superimposing these displacements to the fixed end displacements in Fig. 4.12c(iii), the final axial displacement can be generated, as shown in Fig. 4.14b.

$$\mathbf{F_R} = \mathbf{F_{fR}} + \mathbf{k_{RA}}\mathbf{D_A} \;\Rightarrow\; \begin{Bmatrix} F_2 \\ F_3 \end{Bmatrix} = \begin{Bmatrix} -20 \\ -10 \end{Bmatrix} + EA\begin{bmatrix} -1 \\ -1/3 \end{bmatrix}\left\{ D_1 = \frac{60}{EA} \right\} = \begin{Bmatrix} -20 \\ -10 \end{Bmatrix} + \begin{Bmatrix} -60 \\ -20 \end{Bmatrix} = \begin{Bmatrix} \mathbf{-80} \\ \mathbf{-30} \end{Bmatrix}\text{kN}$$

Check $\sum F_x = 0$:

Total reaction = $F_2 + F_3 = -80 - 30 = -110\text{kN}$; Total load = $(20)(2) + 40 + 30 = +110\text{kN} \Rightarrow$ OK.

Member Forces

$$\mathbf{F_*^i} = \mathbf{F_{*f}^i} + \mathbf{k_*^i}\mathbf{T^i}\mathbf{D}$$

$$\Rightarrow \mathbf{F_*^1} = \begin{Bmatrix} F_{1*}^1 \\ F_{2*}^1 \end{Bmatrix} = \begin{Bmatrix} -20 \\ -20 \end{Bmatrix} + EA\begin{bmatrix} 1 & -1 \\ -1 & 1 \end{bmatrix}\begin{Bmatrix} D_2 = 0 \\ D_1 = 60/EA \end{Bmatrix} = \begin{Bmatrix} -20 \\ -20 \end{Bmatrix} + \begin{Bmatrix} -60 \\ +60 \end{Bmatrix} = \begin{Bmatrix} \mathbf{-80} \\ \mathbf{+40} \end{Bmatrix}\text{kN}$$

$$\mathbf{F_*^2} = \begin{Bmatrix} F_{1*}^2 \\ F_{2*}^2 \end{Bmatrix} = \begin{Bmatrix} -20 \\ -10 \end{Bmatrix} + EA\begin{bmatrix} 1/3 & -1/3 \\ -1/3 & 1/3 \end{bmatrix}\begin{Bmatrix} D_1 = 60/EA \\ D_3 = 0 \end{Bmatrix} = \begin{Bmatrix} -20 \\ -10 \end{Bmatrix} + \begin{Bmatrix} +20 \\ -20 \end{Bmatrix} = \begin{Bmatrix} \mathbf{0} \\ \mathbf{-30} \end{Bmatrix}\text{kN}$$

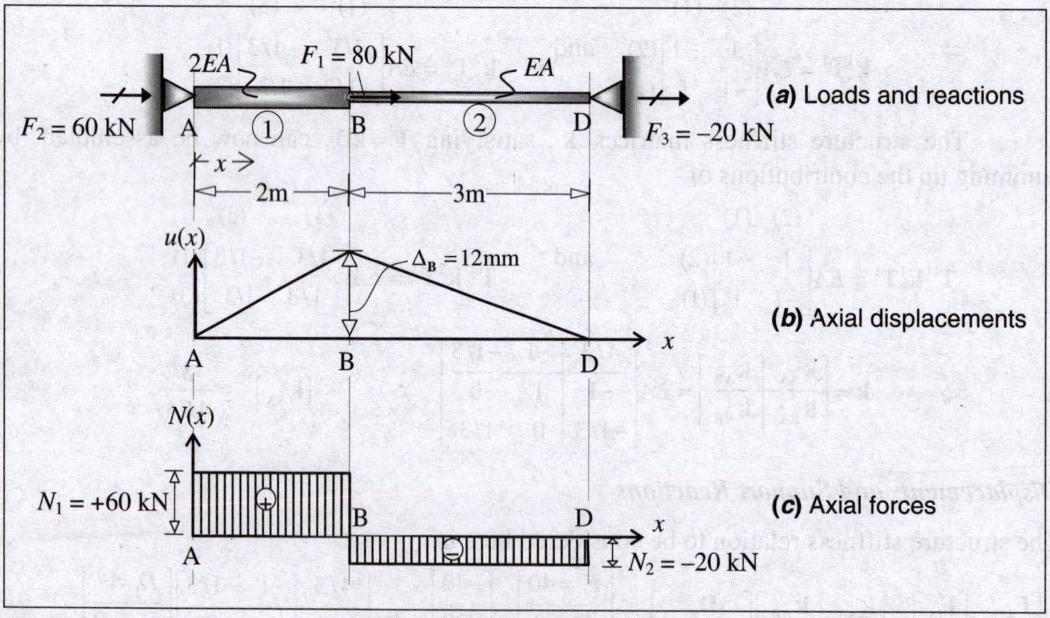

Figure 4.13 Response to nodal plus equivalent joint loads (Example 4.1)

The axial forces induced in the two elements due to the applied loading in Fig. 4.13a are shown in Fig. 4.13c. Superimposing these axial forces to the fixed end forces in Fig. 4.12c(ii), the final axial force diagram can be generated, as shown in Fig. 4.14c.

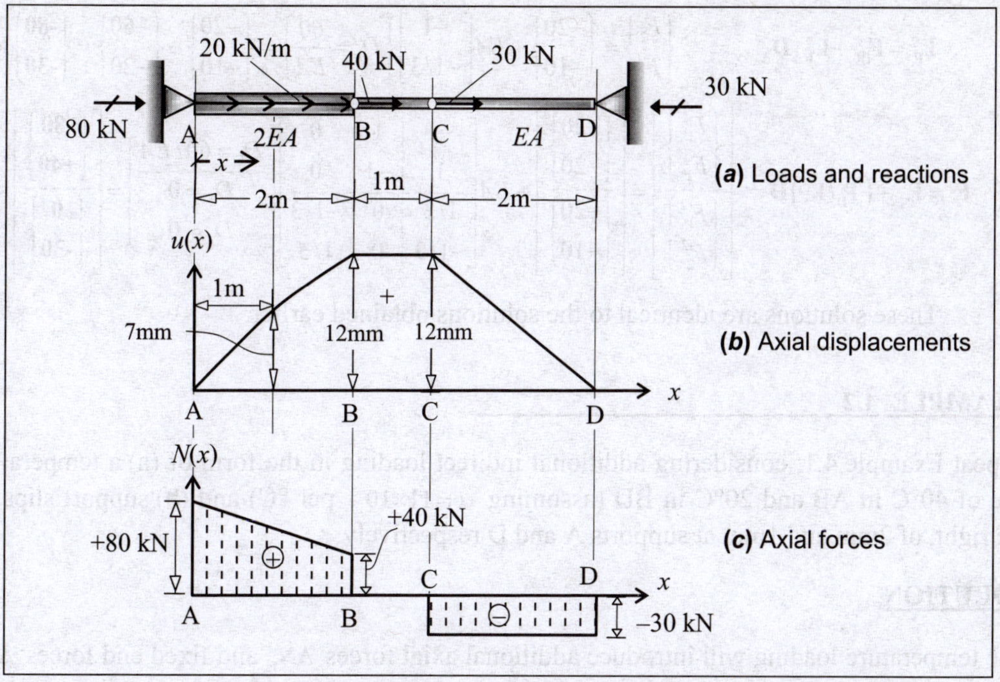

Figure 4.14 Complete force and displacement responses (Example 4.1)

Alternative Procedure Using Displacement Transformation Matrix

$$\mathbf{T}_\mathbf{D}^1 = \begin{bmatrix} 0 & 1 & 0 \\ 1 & 0 & 0 \end{bmatrix} \text{ and } \mathbf{T}_\mathbf{D}^2 = \begin{bmatrix} 1 & 0 & 0 \\ 0 & 0 & 1 \end{bmatrix}; \qquad \mathbf{D}_* = \mathbf{T}_\mathbf{D}\mathbf{D} \Rightarrow \begin{bmatrix} \begin{Bmatrix} D_{1*}^1 \\ D_{2*}^1 \end{Bmatrix} \\ \begin{Bmatrix} D_{1*}^2 \\ D_{2*}^2 \end{Bmatrix} \end{bmatrix} = \begin{bmatrix} 0 & 1 & 0 \\ 1 & 0 & 0 \\ 1 & 0 & 0 \\ 0 & 0 & 1 \end{bmatrix} \begin{bmatrix} D_1 \\ D_2 \\ D_3 \end{bmatrix}$$

$$\Rightarrow \mathbf{F}_\mathbf{f} = \mathbf{T}_\mathbf{D}{}^\mathbf{T}\mathbf{F}_{*\mathbf{f}} \Rightarrow \begin{bmatrix} F_{1f} \\ F_{2f} \\ F_{3f} \end{bmatrix} = \begin{bmatrix} 0 & 1 & 1 & 0 \\ 1 & 0 & 0 & 0 \\ 0 & 0 & 0 & 1 \end{bmatrix} \begin{bmatrix} \begin{Bmatrix} F_{1*f}^1 = -20 \\ F_{2*f}^1 = -20 \end{Bmatrix} \\ \begin{Bmatrix} F_{1*f}^2 = -20 \\ F_{2*f}^2 = -10 \end{Bmatrix} \end{bmatrix} = \begin{bmatrix} -40 \\ -20 \\ -10 \end{bmatrix} \text{kN (as obtained earlier)}$$

$$\mathbf{k} = \mathbf{T}_\mathbf{D}{}^\mathbf{T}\mathbf{k}_*\mathbf{T}_\mathbf{D} = EA\begin{bmatrix} 4/3 & -1 & -1/3 \\ -1 & 1 & 0 \\ -1/3 & 0 & 1/3 \end{bmatrix} = \begin{bmatrix} \mathbf{k}_\mathbf{AA} & \mathbf{k}_\mathbf{AR} \\ \mathbf{k}_\mathbf{RA} & \mathbf{k}_\mathbf{RR} \end{bmatrix} \text{ (as obtained earlier)}$$

$$\mathbf{D}_\mathbf{A} = [\mathbf{k}_\mathbf{AA}]^{-1}[\mathbf{F}_\mathbf{A} - \mathbf{F}_\mathbf{fA}] \Rightarrow D_1 = \frac{3}{4EA}[40+40] = \frac{60}{EA} = \frac{60}{5000} = 0.012\text{m} = \mathbf{12.0mm}$$

$$\mathbf{F_R} = \mathbf{F_{fR}} + \mathbf{k_{RA}}\mathbf{D_A} \quad \Rightarrow \quad \begin{Bmatrix} F_2 \\ F_3 \end{Bmatrix} = \begin{Bmatrix} -20 \\ -10 \end{Bmatrix} + EA\begin{bmatrix} -1 \\ -1/3 \end{bmatrix}\left\{ D_1 = \frac{60}{EA} \right\} = \begin{Bmatrix} -20 \\ -10 \end{Bmatrix} + \begin{Bmatrix} -60 \\ -20 \end{Bmatrix} = \begin{Bmatrix} \mathbf{-80} \\ \mathbf{-30} \end{Bmatrix} \mathbf{kN}$$

$$\mathbf{F_*} = \mathbf{F_{*f}} + \left(\mathbf{T_D}^{\mathbf{T}}\mathbf{k_*}\right)\mathbf{D} \Rightarrow \begin{bmatrix} \begin{Bmatrix} F_{1*}^1 \\ F_{2*}^1 \end{Bmatrix} \\ \begin{Bmatrix} F_{1*}^2 \\ F_{2*}^2 \end{Bmatrix} \end{bmatrix} = \begin{bmatrix} \begin{Bmatrix} -20 \\ -20 \end{Bmatrix} \\ \begin{Bmatrix} -20 \\ -10 \end{Bmatrix} \end{bmatrix} + EA\begin{bmatrix} -1 & 1 & 0 \\ 1 & -1 & 0 \\ 1/3 & 0 & -1/3 \\ -1/3 & 0 & 1/3 \end{bmatrix}\begin{bmatrix} D_1 = 60/EA \\ D_2 = 0 \\ D_3 = 0 \end{bmatrix} = \begin{bmatrix} \begin{Bmatrix} \mathbf{-80} \\ \mathbf{+40} \end{Bmatrix} \\ \begin{Bmatrix} \mathbf{0} \\ \mathbf{-30} \end{Bmatrix} \end{bmatrix} \mathbf{kN}$$

These solutions are identical to the solutions obtained earlier.

EXAMPLE 4.2

Repeat Example 4.1, considering additional indirect loading in the form of (a) a temperature rise of 40°C in AB and 20°C in BD (assuming $\alpha = 11\times10^{-5}$ per °C) and (b) support slips, to the right, of 2mm and 1mm at supports A and D respectively.

SOLUTION

The temperature loading will introduce additional axial forces $\mathbf{\Delta N_f}$ and fixed end forces $\mathbf{\Delta F_{*f}}$ in the elements of the primary (restrained) structure, but no axial displacements:

$$\mathbf{e_o} = \begin{Bmatrix} e_{1o} \\ e_{2o} \end{Bmatrix} = \alpha\begin{Bmatrix} L_1(\Delta T_1) \\ L_2(\Delta T_2) \end{Bmatrix} = 11\times10^{-5}\begin{Bmatrix} (2)(40) \\ (3)(20) \end{Bmatrix} = \begin{Bmatrix} 0.0088 \\ 0.0066 \end{Bmatrix} \text{m}$$

$$\Rightarrow \mathbf{\Delta N_f} = \begin{Bmatrix} \Delta N_{1f} \\ \Delta N_{2f} \end{Bmatrix} = \begin{Bmatrix} -k_1 e_{1o} \\ -k_2 e_{2o} \end{Bmatrix} = \begin{Bmatrix} -2EAe_{1o}/L_1 \\ -EAe_{2o}/L_2 \end{Bmatrix} = 5000\begin{Bmatrix} -0.0088 \\ -0.0066/3 \end{Bmatrix} = \begin{Bmatrix} -44 \\ -11 \end{Bmatrix} \text{kN}$$

$$\Rightarrow \mathbf{\Delta F_{*f}} = \begin{bmatrix} \begin{Bmatrix} F_{1*f}^1 \\ F_{2*f}^1 \end{Bmatrix} \\ \begin{Bmatrix} F_{1*f}^2 \\ F_{2*f}^2 \end{Bmatrix} \end{bmatrix} = \begin{bmatrix} \begin{Bmatrix} +44 \\ -44 \end{Bmatrix} \\ \begin{Bmatrix} +11 \\ -11 \end{Bmatrix} \end{bmatrix} \text{kN} \Rightarrow \mathbf{\Delta F_f} = \mathbf{T_D}^{\mathbf{T}}\mathbf{\Delta F_{*f}} \Rightarrow \begin{bmatrix} F_{1f} \\ F_{2f} \\ F_{3f} \end{bmatrix} = \begin{bmatrix} 0 & 1 & 1 & 0 \\ 1 & 0 & 0 & 0 \\ 0 & 0 & 0 & 1 \end{bmatrix}\begin{bmatrix} \begin{Bmatrix} +44 \\ -44 \end{Bmatrix} \\ \begin{Bmatrix} +11 \\ -11 \end{Bmatrix} \end{bmatrix} = \begin{bmatrix} -33 \\ +44 \\ -1 \end{bmatrix} \text{kN}$$

The given support movements imply $\mathbf{D_R} = \begin{Bmatrix} D_2 \\ D_3 \end{Bmatrix} = \begin{Bmatrix} 0.002 \\ 0.001 \end{Bmatrix} \text{m}$

The structure stiffness relations are given by $\begin{bmatrix} \mathbf{F_A} \\ \mathbf{F_R} \end{bmatrix} - \begin{bmatrix} \mathbf{F_{fA}} + \mathbf{\Delta F_{fA}} \\ \mathbf{F_{fR}} + \mathbf{\Delta F_{fR}} \end{bmatrix} = \begin{bmatrix} \mathbf{k_{AA}} & \mathbf{k_{AR}} \\ \mathbf{k_{RA}} & \mathbf{k_{RR}} \end{bmatrix}\begin{bmatrix} \mathbf{D_A} \\ \mathbf{D_R} \end{bmatrix}$

$$\begin{bmatrix} F_1 = 40 \\ F_2 = ? \\ F_3 = ? \end{bmatrix} - \begin{bmatrix} -40-33 \\ -20+44 \\ -10-11 \end{bmatrix} = 5000\begin{bmatrix} 4/3 & -1 & -1/3 \\ -1 & 1 & 0 \\ -1/3 & 0 & 1/3 \end{bmatrix}\begin{bmatrix} D_1 = ? \\ D_2 = 0.002 \\ D_3 = 0.001 \end{bmatrix} = 5000\begin{bmatrix} 4/3 \\ -1 \\ -1/3 \end{bmatrix} D_1 + \begin{bmatrix} -11.67 \\ 10 \\ 1.67 \end{bmatrix} \text{kN}$$

$$\mathbf{D_A} = \left[\mathbf{k_{AA}}\right]^{-1}\left[\mathbf{F_A} - \left(\mathbf{F_{fA}} + \Delta\mathbf{F_{fA}} + \mathbf{k_{AR}}\mathbf{D_R}\right)\right]$$

$$\Rightarrow D_1 = \frac{3}{4EA}\left[40-(-40-33-11.67)\right] = \frac{93.5}{EA} = \frac{93.5}{5000} = 0.0187\text{m} = \mathbf{18.7mm}$$

$$\mathbf{F_R} = \mathbf{F_{fR}} + \Delta\mathbf{F_{fR}} + \mathbf{k_{RA}}\mathbf{D_A} + \mathbf{k_{RR}}\mathbf{D_R}$$

$$\Rightarrow \begin{Bmatrix} F_2 \\ F_3 \end{Bmatrix} = \begin{Bmatrix} -20 \\ -10 \end{Bmatrix} + \begin{Bmatrix} +44 \\ -11 \end{Bmatrix} + EA\begin{bmatrix} -1 \\ -1/3 \end{bmatrix}\left\{D_1 = \frac{93.5}{EA}\right\} + \begin{Bmatrix} 10 \\ 1.67 \end{Bmatrix} = \begin{Bmatrix} \mathbf{-59.5} \\ \mathbf{-50.5} \end{Bmatrix}\textbf{kN}$$

(Check: The total support reaction is equal to the total load of 110 kN).

$$\mathbf{F_*} = \mathbf{F_{*f}} + \Delta\mathbf{F_{*f}} + \mathbf{k_*}\mathbf{T_D}\mathbf{D}$$

$$\Rightarrow \begin{bmatrix} \begin{Bmatrix} F_{1*}^1 \\ F_{2*}^1 \end{Bmatrix} \\ \begin{Bmatrix} F_{1*}^2 \\ F_{2*}^2 \end{Bmatrix} \end{bmatrix} = \begin{bmatrix} \begin{Bmatrix} -20 \\ -20 \end{Bmatrix} \\ \begin{Bmatrix} -20 \\ -10 \end{Bmatrix} \end{bmatrix} + \begin{bmatrix} \begin{Bmatrix} +44 \\ -44 \end{Bmatrix} \\ \begin{Bmatrix} +11 \\ -11 \end{Bmatrix} \end{bmatrix} + (EA)\begin{bmatrix} -1 & 1 & 0 \\ 1 & -1 & 0 \\ 1/3 & 0 & -1/3 \\ -1/3 & 0 & 1/3 \end{bmatrix}\begin{bmatrix} D_1 = 93.5/EA \\ \hline D_2 = 10/EA \\ D_3 = 5/EA \end{bmatrix} = \begin{bmatrix} \begin{Bmatrix} \mathbf{-59.5} \\ \mathbf{19.5} \end{Bmatrix} \\ \begin{Bmatrix} \mathbf{20.5} \\ \mathbf{-50.5} \end{Bmatrix} \end{bmatrix}\textbf{kN}$$

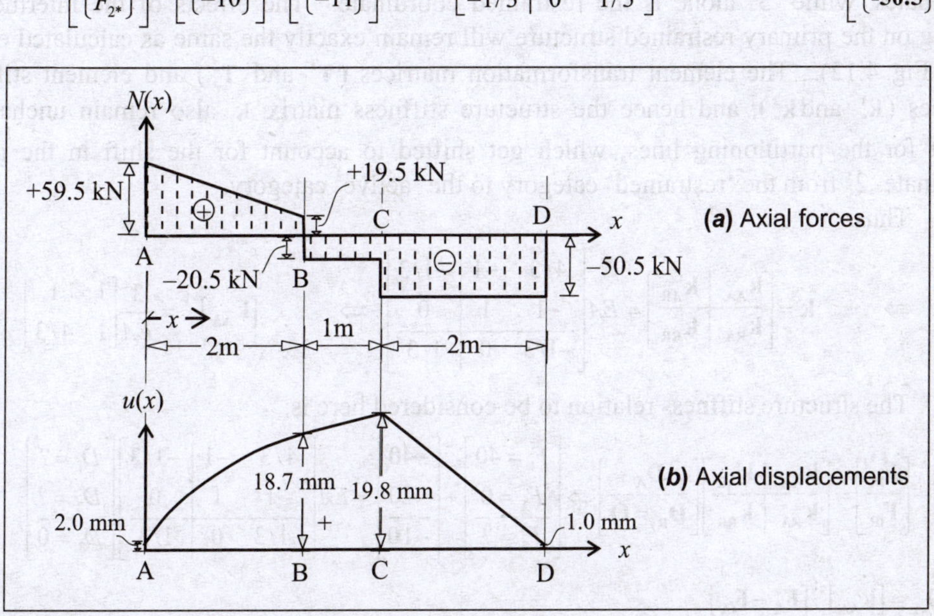

Figure 4.15 Distributions of axial force and displacement under combined loading (Example 4.2)

The distributions of axial force and axial displacement, under the combined loading are shown in Figs 4.15a and b respectively. The axial force varies linearly in the segment AB due to the action of the uniformly distributed load, causing a drop in the axial tension of 40 kN from A to B.

To get the final displacement diagram, we need to superimpose (1) the fixed end displacement diagram of Fig. 4.12c(iii) and (2) a diagram [(2mm at A) to (1mm at D)] on account of support settlements (i.e., with a value of 1.6mm at B) and (3) a linearly varying displacement diagram due to the nodal loads plus equivalent joint loads plus indirect loads, having a value of 2mm at A, 18.7 mm at B and 1mm at D. This is illustrated in Fig. 4.15b.

The temperature loading will introduce additional axial forces $\mathbf{\Delta N_f}$ and fixed end forces $\mathbf{\Delta F_{*f}}$ in the elements of the primary (restrained) structure, but no axial displacements:

EXAMPLE 4.3

Repeat Example 4.1, considering the end at A to be free (not restrained against translation).

SOLUTION

The degree of kinematic indeterminacy increases from 1 to 2. Keeping the same numbering of global coordinates as in Example 4.1, the global coordinates '1' and '2' are now the active coordinates, while '3' alone is the restrained coordinate. The effects of the intermediate loading on the primary restrained structure will remain exactly the same as calculated earlier (refer Fig. 4.12). The element transformation matrices ($\mathbf{T}^1$ and $\mathbf{T}^2$) and element stiffness matrices ($\mathbf{k}_*^1$ and $\mathbf{k}_*^2$), and hence the structure stiffness matrix $\mathbf{k}$ also remain unchanged, except for the partitioning lines, which get shifted to account for the shift in the global coordinate '2' from the 'restrained' category to the 'active' category.

Thus,

$$\Rightarrow \quad \mathbf{k} = \left[\begin{array}{c|c} \mathbf{k_{AA}} & \mathbf{k_{AR}} \\ \hline \mathbf{k_{RA}} & \mathbf{k_{RR}} \end{array}\right] = EA\left[\begin{array}{cc|c} 4/3 & -1 & -1/3 \\ -1 & 1 & 0 \\ \hline -1/3 & 0 & 1/3 \end{array}\right] \Rightarrow \quad \left[\mathbf{k_{AA}}\right]^{-1} = \frac{3}{EA}\begin{bmatrix} 1 & 1 \\ 1 & 4/3 \end{bmatrix}$$

The structure stiffness relation to be considered here is:

$$\left[\begin{array}{c} \mathbf{F_A} \\ \hline \mathbf{F_R} \end{array}\right] - \left[\begin{array}{c} \mathbf{F_{fA}} \\ \hline \mathbf{F_{fR}} \end{array}\right] = \left[\begin{array}{c|c} \mathbf{k_{AA}} & \mathbf{k_{AR}} \\ \hline \mathbf{k_{RA}} & \mathbf{k_{RR}} \end{array}\right]\left[\begin{array}{c} \mathbf{D_A} \\ \hline \mathbf{D_R} = \mathbf{O} \end{array}\right] \Rightarrow \left[\begin{array}{c} F_1 = 40 \\ F_2 = 0 \\ \hline F_3 = ? \end{array}\right] - \left[\begin{array}{c} -40 \\ -20 \\ \hline -10 \end{array}\right] = EA\left[\begin{array}{cc|c} 4/3 & -1 & -1/3 \\ -1 & 1 & 0 \\ \hline -1/3 & 0 & 1/3 \end{array}\right]\left[\begin{array}{c} D_1 = ? \\ D_2 = ? \\ \hline D_3 = 0 \end{array}\right]$$

$$\Rightarrow \quad \mathbf{D_A} = \left[\mathbf{k_{AA}}\right]^{-1}\left[\mathbf{F_A} - \mathbf{F_{fA}}\right]$$

$$\Rightarrow \quad \begin{Bmatrix} D_1 \\ D_2 \end{Bmatrix} = \frac{3}{EA}\begin{bmatrix} 1 & 1 \\ 1 & 4/3 \end{bmatrix}\begin{Bmatrix} 40+40 \\ 0+20 \end{Bmatrix} = \frac{1}{EA}\begin{Bmatrix} 300 \\ 320 \end{Bmatrix} = \frac{1}{5000}\begin{Bmatrix} 300 \\ 320 \end{Bmatrix} = \begin{Bmatrix} 0.060 \\ 0.064 \end{Bmatrix} \text{m} \Rightarrow \quad \begin{array}{l} D_1 = \mathbf{60.0mm} \\ D_2 = \mathbf{64.0mm} \end{array}$$

The axial displacements, due to the applied loading (nodal loads and equivalent joint loads) in Fig. 4.16a, are shown in Fig. 4.16b. Superimposing these displacements to the fixed end displacements in Fig. 4.12c(iii), the final axial displacement can be generated, as shown in Fig. 4.17b.

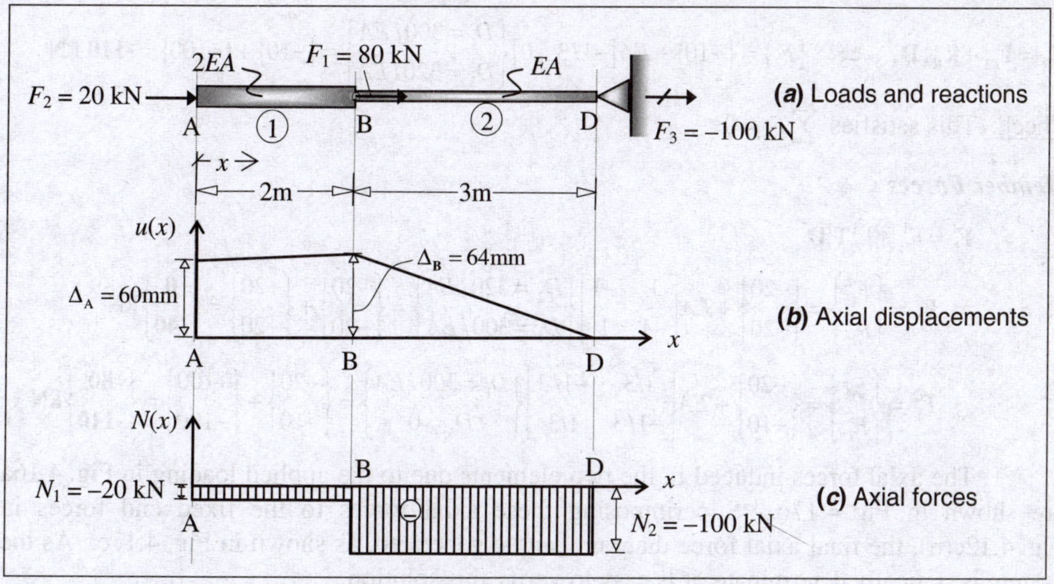

Figure 4.16 Response to nodal plus equivalent joint loads (Example 4.3)

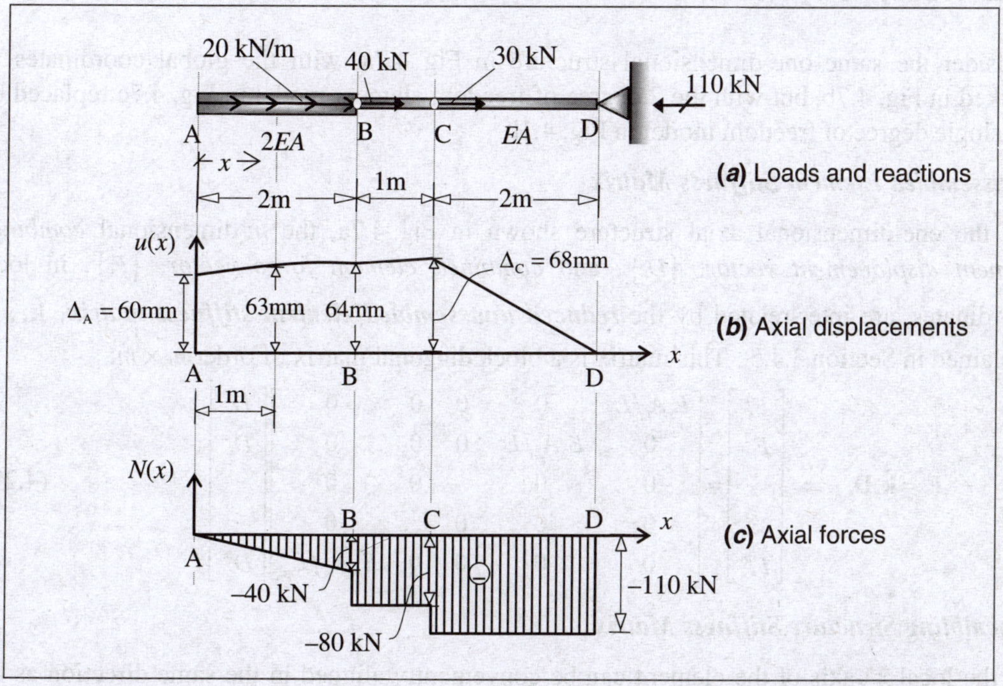

Figure 4.17 Complete force and displacement responses (Example 4.3)

$$\mathbf{F_R} = \mathbf{F_{fR}} + \mathbf{k_{RA}}\mathbf{D_A} \quad \Rightarrow \quad \{F_3\} = \{-10\} + EA[-1/3 \quad 0]\begin{Bmatrix} D_1 = 300/EA \\ D_2 = 320/EA \end{Bmatrix} = \{-10\} + \{-100\} = \mathbf{-110\ kN}$$

<u>Check:</u> This satisfies $\sum F_x = 0$.

Member Forces

$$\mathbf{F_*^i} = \mathbf{F_{*f}^i} + \mathbf{k_*^i T^i D}$$

$$\Rightarrow \mathbf{F_*^1} = \begin{Bmatrix} F_{1*}^1 \\ F_{2*}^1 \end{Bmatrix} = \begin{Bmatrix} -20 \\ -20 \end{Bmatrix} + EA\begin{bmatrix} 1 & -1 \\ -1 & 1 \end{bmatrix}\begin{Bmatrix} D_2 = 320/EA \\ D_1 = 300/EA \end{Bmatrix} = \begin{Bmatrix} -20 \\ -20 \end{Bmatrix} + \begin{Bmatrix} +20 \\ -20 \end{Bmatrix} = \begin{Bmatrix} \mathbf{0} \\ \mathbf{-40} \end{Bmatrix}\mathbf{kN}$$

$$\mathbf{F_*^2} = \begin{Bmatrix} F_{1*}^2 \\ F_{2*}^2 \end{Bmatrix} = \begin{Bmatrix} -20 \\ -10 \end{Bmatrix} + EA\begin{bmatrix} 1/3 & -1/3 \\ -1/3 & 1/3 \end{bmatrix}\begin{Bmatrix} D_1 = 300/EA \\ D_3 = 0 \end{Bmatrix} = \begin{Bmatrix} -20 \\ -10 \end{Bmatrix} + \begin{Bmatrix} +100 \\ -100 \end{Bmatrix} = \begin{Bmatrix} \mathbf{80} \\ \mathbf{-110} \end{Bmatrix}\mathbf{kN}$$

The axial forces induced in the two elements due to the applied loading in Fig. 4.16a are shown in Fig. 4.17c. Superimposing these axial forces to the fixed end forces in Fig. 4.12c(ii), the final axial force diagram can be generated, as shown in Fig. 4.17c. As the system is statically determinate, it is easy to verify this solution.

4.3.2 Analysis by Reduced Element Stiffness Method

Consider the same one-dimensional structure in Fig. 4.7a, with the global coordinates as marked in Fig. 4.7b, but with the 2 degree of freedom element model in Fig. 4.7c replaced by the single degree of freedom model in Fig. 4.1b.

Unassembled Element Stiffness Matrix

For the one-dimensional axial structure shown in Fig. 4.7a, the m-dimensional *combined element displacement vector*, $\{D_*\}$, and *combined element force vector*, $\{F_*\}$, in local coordinates, are inter-related by the ***reduced unassembled element stiffness matrix***, $\tilde{\mathbf{k}}_*$, as explained in Section 3.4.5. This matrix is a block diagonal matrix of order $m \times m$:

$$\mathbf{F_*} = \tilde{\mathbf{k}}_*\mathbf{D_*} \quad \Rightarrow \quad \begin{bmatrix} F_*^1 \\ F_*^2 \\ \vdots \\ \vdots \\ F_*^m \end{bmatrix} = \begin{bmatrix} E_1A_1/L_1 & 0 & 0 & 0 & 0 \\ 0 & E_2A_2/L_2 & 0 & 0 & 0 \\ 0 & 0 & \ddots & 0 & 0 \\ 0 & 0 & 0 & \ddots & 0 \\ 0 & 0 & 0 & 0 & E_mA_m/L_m \end{bmatrix}\begin{bmatrix} D_*^1 \\ D_*^2 \\ \vdots \\ \vdots \\ D_*^m \end{bmatrix} \qquad (4.29)$$

Assembling Structure Stiffness Matrix

As the local x^*-axis of the element can be conveniently aligned in the same direction as the global x-axis, coordinate transformations become simplified. The structure stiffness matrix $\mathbf{k}$ can be assembled by:

(i) using the element displacement transformation matrix, $\mathbf{T_D^i}$, and summing up the contributions, $\mathbf{k} = \mathbf{T_D^i}^T \tilde{\mathbf{k}}_*^i \mathbf{T_D^i}$ (direct stiffness method);

(ii) using the displacement transformation matrix $\mathbf{T_D}$, relating the displacement vector $\mathbf{D}$ to the combined element displacement vector, $\mathbf{D}_*$, and applying $\mathbf{k} = \mathbf{T_D}^T \tilde{\mathbf{k}}_* \mathbf{T_D}$;

(iii) using a 'physical approach', applying a unit displacement, $D_j = 1$, at a time on the restrained primary structure.

The final structure stiffness matrix generated will be identical to the one shown in Eq. 4.25.

Fixed End Forces

The same fixed end force effects, described earlier in the conventional stiffness method are applicable here. However, when there is a variation in the axial force diagram on account of intermediate loads, this variation **cannot** be captured through the transformation $\mathbf{F_f} = \mathbf{T_D}^F \mathbf{F_{*f}}$ in the reduced element stiffness method, as there is only one force that can be described per element. This is a shortcoming of the reduced element stiffness method, which prevents automatic generation of the equivalent joint loads. Such loads need to be manually computed. The axial forces $\mathbf{F_{*e}}$, on account of the equivalent joint loads and nodal loads, can be computed using the matrix formulation, $\mathbf{F_{*e}^i} = \left(\tilde{\mathbf{k}}_* \mathbf{T_D}\right)\mathbf{D}$. However, the superimposition with the fixed end axial forces $\mathbf{F_{*f}}$, on account of intermediate loads and temperature effects, may be done manually.

Let us now demonstrate the application of the reduced element stiffness method through Examples.

EXAMPLE 4.4

Repeat Example 4.1 (Fig. 4.11), solving by the Reduced Element Stiffness Method.

SOLUTION

There are two prismatic elements here ($m = 2$), with intermediate loads acting on both. There is one *active* degree of freedom ($n = 1$) and two *restrained* global coordinates ($r = 2$), as shown in Fig. 4.18a. The local coordinates for the two elements are shown in Fig. 4.18b.

Fixed End Forces and Axial Displacements

The fixed end forces are the same as calculated earlier (shown in Fig. 4.12c), along with the distributions of axial force $N_i(x^*)$ and axial displacement $u_i(x^*)$, for the two elements. The resulting equivalent joint load at the active coordinate '1' is $F_{1e} = -F_{1f} = +40\text{kN}$, and the resultant load at this coordinate, $F_1 + F_{1e} = 40 + 40 = +80\text{kN}$, as indicated in Fig. 4.13a.

Displacement Transformation Matrix

The element displacement transformation matrices $\mathbf{T}_\mathbf{D}^1$ and $\mathbf{T}_\mathbf{D}^2$, satisfying $\mathbf{D}_*^i = \mathbf{T}_\mathbf{D}^i\mathbf{D}$, are as follows: $\mathbf{T}_\mathbf{D}^1 = [1 \mid -1 \quad 0]$ and $\mathbf{T}_\mathbf{D}^2 = [-1 \mid 0 \quad 1]$

$$\mathbf{D}_*^i = \mathbf{T}_\mathbf{D}\mathbf{D} \Rightarrow \begin{Bmatrix} D_*^1 \\ D_*^2 \end{Bmatrix} = \left[\begin{array}{c|cc} 1 & -1 & 0 \\ -1 & 0 & 1 \end{array}\right] \begin{Bmatrix} D_1 \\ \hline D_2 \\ D_3 \end{Bmatrix}$$

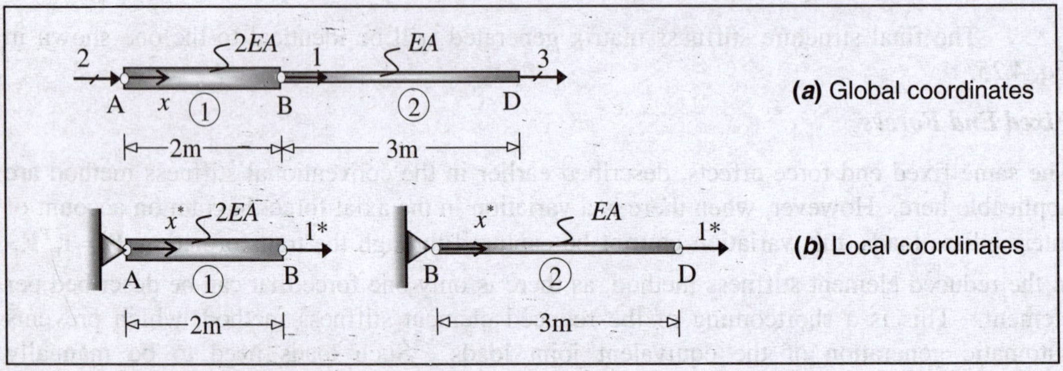

Figure 4.18 Global and local coordinates (Example 4.4)

Element and Structure Stiffness Matrices

The reduced element stiffness matrices $\tilde{\mathbf{k}}_*^1$ and $\tilde{\mathbf{k}}_*^2$, satisfying $\mathbf{F}_*^i = \mathbf{k}_*^i\mathbf{D}_*^i$, are 1×1 matrices, given by $k_1 = \frac{2EA}{2} = EA$ and $k_2 = \frac{EA}{3}$:

$$\tilde{\mathbf{k}}_*^1 = EA[1] \quad \text{and} \quad \tilde{\mathbf{k}}_*^2 = EA[1/3] \quad \Rightarrow \quad \tilde{\mathbf{k}}_* = EA\begin{bmatrix} 1 & 0 \\ 0 & 1/3 \end{bmatrix}$$

$$\Rightarrow \quad \tilde{\mathbf{k}}_*^1\mathbf{T}_\mathbf{D}^1 = EA[1 \mid -1 \quad 0] \quad \text{and} \quad \tilde{\mathbf{k}}_*^2\mathbf{T}_\mathbf{D}^2 = EA[-1/3 \mid 0 \quad 1/3]$$

$$\mathbf{F}_*^i = \left(\tilde{\mathbf{k}}_*\mathbf{T}_\mathbf{D}\right)\mathbf{D} \Rightarrow \begin{Bmatrix} F_*^1 \\ F_*^2 \end{Bmatrix} = EA\left[\begin{array}{c|cc} 1 & -1 & 0 \\ -1/3 & 0 & 1/3 \end{array}\right] \begin{Bmatrix} D_1 \\ \hline D_2 \\ D_3 \end{Bmatrix}$$

The structure stiffness matrices $\mathbf{k}$, satisfying $\mathbf{F} = \mathbf{k}\mathbf{D}$, can now be assembled by the direct stiffness method, by summing up the contributions of $\mathbf{k} = \mathbf{T}_\mathbf{D}^{i\,T}\tilde{\mathbf{k}}_*^i\mathbf{T}_\mathbf{D}^i$ (direct stiffness method):

$$\mathbf{T}_\mathbf{D}^{1T}\tilde{\mathbf{k}}_*^1\mathbf{T}_\mathbf{D}^1 = \begin{bmatrix} 1 \\ \hline -1 \\ 0 \end{bmatrix} EA[1 \mid -1 \quad 0] = EA\left[\begin{array}{c|cc} 1 & -1 & 0 \\ \hline -1 & 1 & 0 \\ 0 & 0 & 0 \end{array}\right]$$

$$\mathbf{T_D^{2T}\tilde{k}_*^2T_D^2} = \begin{bmatrix} -1 \\ \hdashline 0 \\ 1 \end{bmatrix} EA\left[-1/3 \;\vdots\; 0 \quad 1/3\right] = EA\left[\begin{array}{c|cc} 1/3 & 0 & -1/3 \\ \hline 0 & 0 & 0 \\ -1/3 & 0 & 1/3 \end{array}\right]$$

$$\Rightarrow \qquad \mathbf{k} = EA\left[\begin{array}{c|cc} (1+1/3) & (-1+0) & (0-1/3) \\ \hline (-1+0) & (1+0) & (0+0) \\ (0-1/3) & (0+0) & (0+1/3) \end{array}\right]$$

Alternatively, we may compute $\mathbf{k} = \mathbf{T_D}^{\mathbf{T}}\mathbf{\tilde{k}_*T_D}$ as follows:

$$\mathbf{k} = \mathbf{T_D}^{\mathbf{T}}\mathbf{\tilde{k}_*T_D} = \begin{bmatrix} 1 & -1 \\ \hdashline -1 & 0 \\ 0 & 1 \end{bmatrix} EA\begin{bmatrix} 1 & \vdots & -1 & 0 \\ -1/3 & \vdots & 0 & 1/3 \end{bmatrix} = EA\left[\begin{array}{c|cc} 4/3 & -1 & -1/3 \\ \hline -1 & 1 & 0 \\ -1/3 & 0 & 1/3 \end{array}\right]$$

$$\Rightarrow \qquad \mathbf{k} = \left[\begin{array}{c|c} \mathbf{k_{AA}} & \mathbf{k_{AR}} \\ \hline \mathbf{k_{RA}} & \mathbf{k_{RR}} \end{array}\right] = EA\left[\begin{array}{c|cc} 4/3 & -1 & -1/3 \\ \hline -1 & 1 & 0 \\ -1/3 & 0 & 1/3 \end{array}\right] \quad \Rightarrow \quad \left[\mathbf{k_{AA}}\right]^{-1} = \frac{3}{4EA}$$

Displacements and Support Reactions

The calculations are identical to the ones shown earlier in Example 4.1:

$$\mathbf{D_A} = \left[\mathbf{k_{AA}}\right]^{-1}\left[\mathbf{F_A} - \mathbf{F_{fA}}\right] \quad \Rightarrow \quad D_1 = \frac{3}{4EA}\left[40+40\right] = \frac{60}{EA} = \frac{60}{5000} = 0.012\text{m} = \mathbf{12.0mm}$$

$$\mathbf{F_R} = \mathbf{F_{fR}} + \mathbf{k_{RA}D_A} \quad \Rightarrow \quad \begin{Bmatrix} F_2 \\ F_3 \end{Bmatrix} = \begin{Bmatrix} -20 \\ -10 \end{Bmatrix} + EA\begin{bmatrix} -1 \\ -1/3 \end{bmatrix}\left\{D_1 = \frac{60}{EA}\right\} = \begin{Bmatrix} -20 \\ -10 \end{Bmatrix} + \begin{Bmatrix} -60 \\ -20 \end{Bmatrix} = \begin{Bmatrix} \mathbf{-80} \\ \mathbf{-30} \end{Bmatrix} \mathbf{kN}$$

Member Forces due to Nodal plus Equivalent Joint Loads

$$\mathbf{F_{*e}^i} = \left(\mathbf{\tilde{k}_*T_D}\right)\mathbf{D} \quad \Rightarrow \quad \begin{Bmatrix} F_{*e}^1 \\ F_{*e}^2 \end{Bmatrix} = EA\begin{bmatrix} 1 & \vdots & -1 & 0 \\ -1/3 & \vdots & 0 & 1/3 \end{bmatrix}\begin{Bmatrix} D_1 = 60/EA \\ \hdashline D_2 = 0 \\ D_3 = 0 \end{Bmatrix} = \begin{Bmatrix} \mathbf{60} \\ \mathbf{-20} \end{Bmatrix} \mathbf{kN}$$

These results are identical to the values obtained earlier. They are as depicted in Fig. 4.13. Superimposing these results with the fixed end effects in Fig. 4.12, the complete force and displacement responses can be generated, as shown in Fig. 4.14.

EXAMPLE 4.5

Repeat Example 4.2, solving by the Reduced Element Stiffness Method.

SOLUTION

The temperature loading will introduce additional axial forces $\mathbf{\Delta N_f} = \mathbf{\Delta F_{*f}}$ and thereby fixed end forces $\mathbf{\Delta F_{*f}}$ in the elements of the primary (restrained) structure.

But, there will be no axial displacements. As calculated in Example 4.2,

$$\mathbf{e_o} = \begin{Bmatrix} e_{1o} \\ e_{2o} \end{Bmatrix} = \begin{Bmatrix} 0.0088 \\ 0.0066 \end{Bmatrix} \text{m} \Rightarrow = \mathbf{\Delta F_{*f}} = \begin{Bmatrix} \Delta F^1_{*f} \\ \Delta F^2_{*f} \end{Bmatrix} = \begin{Bmatrix} -k_1 e_{1o} \\ -k_2 e_{2o} \end{Bmatrix} = \begin{Bmatrix} -44 \\ -11 \end{Bmatrix} \text{kN}$$

These compressive forces will induce $\mathbf{\Delta F_f} = \begin{Bmatrix} \mathbf{\Delta F_{fA}} \\ \mathbf{\Delta F_{fR}} \end{Bmatrix} = \begin{bmatrix} F_{1f} \\ F_{2f} \\ F_{3f} \end{bmatrix} = \begin{bmatrix} -44+11 \\ \hdashline +44 \\ -11 \end{bmatrix} = \begin{bmatrix} -33 \\ \hdashline +44 \\ -11 \end{bmatrix} \text{kN}$

The given support movements imply $\mathbf{D_R} = \begin{Bmatrix} D_2 \\ D_3 \end{Bmatrix} = \begin{Bmatrix} 0.002 \\ 0.001 \end{Bmatrix} \text{m}$

Displacements and Support Reactions

Solving the structure stiffness relations, as in Example 4.2,

$$\begin{bmatrix} F_1 = 40 \\ \hdashline F_2 = ? \\ F_3 = ? \end{bmatrix} - \begin{bmatrix} -40-33 \\ \hdashline -20+44 \\ -10-11 \end{bmatrix} = 5000 \left[\begin{array}{c|cc} 4/3 & -1 & -1/3 \\ \hline -1 & 1 & 0 \\ -1/3 & 0 & 1/3 \end{array}\right] \begin{bmatrix} D_1 = ? \\ \hdashline D_2 = 0.002 \\ D_3 = 0.001 \end{bmatrix} = 5000 \begin{bmatrix} 4/3 \\ \hdashline -1 \\ -1/3 \end{bmatrix} D_1 + \begin{bmatrix} -11.67 \\ \hdashline 10 \\ 1.67 \end{bmatrix} \text{kN}$$

$$\Rightarrow D_1 = \frac{3}{4EA}[40-(-40-33-11.67] = \frac{93.5}{EA} = \frac{93.5}{5000} = 0.00187\text{m} = \mathbf{1.87mm}$$

$$\begin{Bmatrix} F_2 \\ F_3 \end{Bmatrix} = \begin{Bmatrix} -20 \\ -10 \end{Bmatrix} + \begin{Bmatrix} +44 \\ -11 \end{Bmatrix} + EA \begin{bmatrix} -1 \\ -1/3 \end{bmatrix} \left\{ D_1 = \frac{93.5}{EA} \right\} + \begin{Bmatrix} 10 \\ 1.67 \end{Bmatrix} = \begin{Bmatrix} \mathbf{-59.5} \\ \mathbf{-50.5} \end{Bmatrix} \mathbf{kN}$$

Member Forces

The member forces due to the equivalent joint plus nodal loads are obtainable as follows:

$$\mathbf{F^i_{*e}} = \left(\tilde{\mathbf{k}}_* \mathbf{T_D}\right)\mathbf{D} \Rightarrow \begin{Bmatrix} F^1_{*e} \\ F^2_{*e} \end{Bmatrix} = EA \left[\begin{array}{c:cc} 1 & -1 & 0 \\ -1/3 & 0 & 1/3 \end{array}\right] \begin{Bmatrix} D_1 = 93.5/EA \\ \hdashline D_2 = 10/EA \\ D_3 = 5/EA \end{Bmatrix} = \begin{Bmatrix} \mathbf{83.5} \\ \mathbf{-29.5} \end{Bmatrix} \mathbf{kN}$$

To get the final forces, we need to superimpose the fixed end force contributions $\mathbf{F^i_{*f}}$ from the varying axial forces induced by the intermediate loads (shown in Fig. 4.12) as well as due to the temperature effects, to the values of $\mathbf{F^i_{*e}}$ generated above, as shown in Fig. 4.19. The axial forces at the two ends, A and D, match the reaction forces, F_2 and F_3, calculated earlier. The final axial force distribution in Fig. 4.19d is identical to the result obtained in Example 4.2 (Fig. 4.15b).

In a similar manner, the axial displacements can be superimposed to give the same final axial displacement diagram, under the combined loading, as shown in Fig. 4.15b.

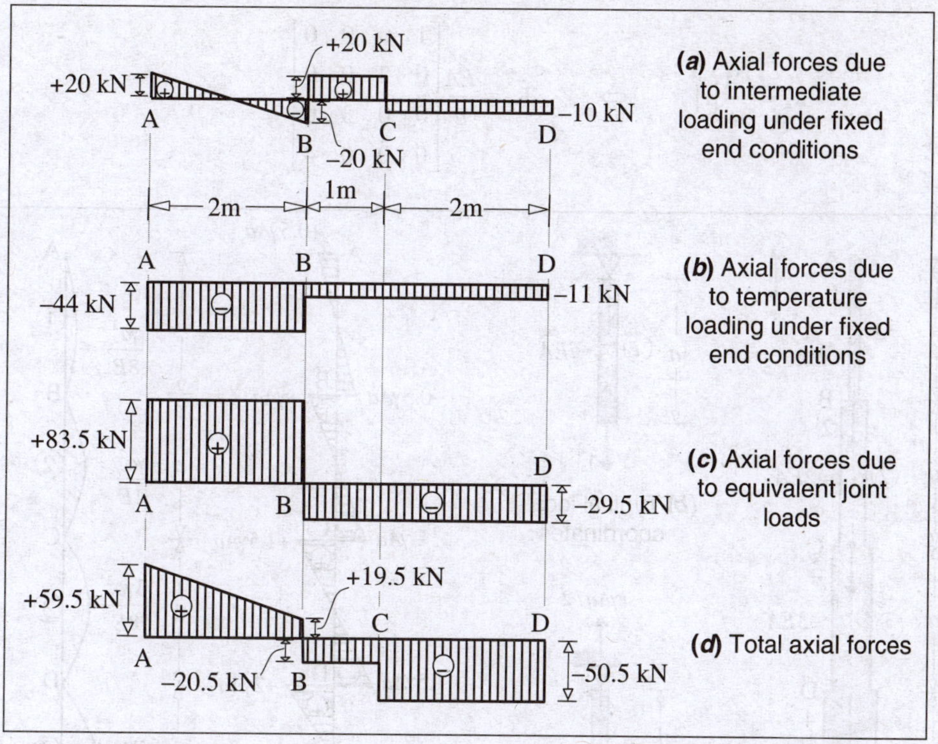

Figure 4.19 Axial forces under combined loading (Example 4.5)

EXAMPLE 4.6

Consider a stepped column ABCDE of total height $4a$, comprising four segments, each of length a, with the cross-sectional area increasing in steps of A, from A in the topmost segment AB to $4A$ in the lowermost segment DE. Analyse the complete force and displacement responses of this structure, resting on a hinged support at the base at E, subject to its own self-weight, by the Reduced Element Stiffness Method. Assume the unit weight (weight per unit volume) and elastic modulus of the column material to be γ and E respectively.

SOLUTION

The stepped column ABCDE can be modelled as a one-dimensional structure, with 4 prismatic elements, free at the top end A and on a hinged support at the base E, with the dimensions and global coordinates (4 *active* and one *restrained*), as indicated in Fig. 4.20a.

Element Stiffness Matrix

The local coordinate system, assuming one degree of freedom, for a typical element, is as shown in Fig. 4.20b, with the element stiffness matrix given by:

$$\tilde{\mathbf{k}}_*^{\mathrm{i}} = \frac{EA}{a}[i] \quad \Rightarrow \quad \tilde{\mathbf{k}}_* = \frac{EA}{a}\begin{bmatrix} 1 & 0 & 0 & 0 \\ 0 & 2 & 0 & 0 \\ 0 & 0 & 3 & 0 \\ 0 & 0 & 0 & 4 \end{bmatrix}$$

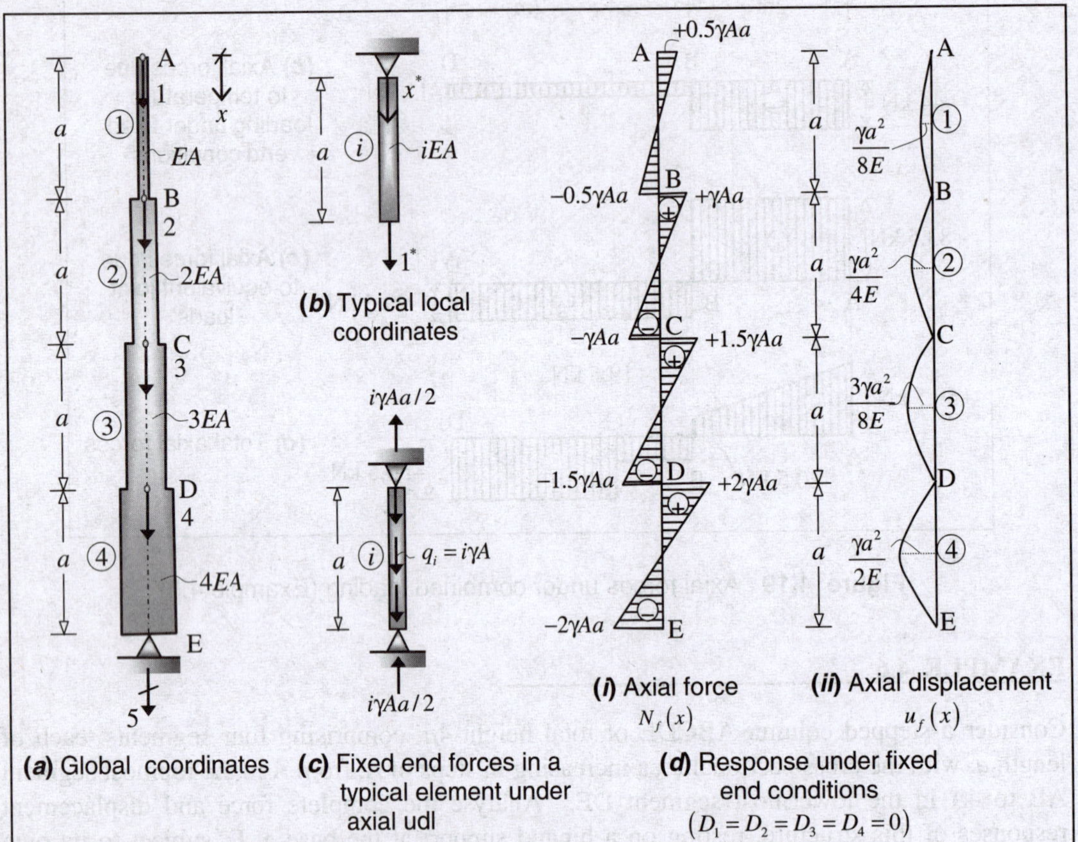

Figure 4.20 Stepped column: global coordinates and fixed end force effects (Example 4.6)

Fixed End Forces and Axial Displacements

The fixed end forces due to the action of the uniformly distributed self-weight loading of $q_i = \gamma(iA)$ are as shown in Fig. 4.20c. The total gravity load of $i\gamma Aa$ in the element i is resisted equally at the top and bottom nodes, resulting in a linearly varying axial force distribution and a parabolically varying axial displacement distribution (refer Eq. 4.27), as shown in Fig. 4.20d. The resulting fixed end forces in the structure and equivalent joint loads may be computed as follows:

$$\mathbf{F_f} = \begin{Bmatrix} \mathbf{F_{fA}} \\ \hline \mathbf{F_{fR}} \end{Bmatrix} = \begin{Bmatrix} F_{1f} \\ F_{2f} \\ F_{3f} \\ F_{4f} \\ \hline F_{5f} \end{Bmatrix} = -\gamma Aa \begin{Bmatrix} 1/2 \\ (1+2)/2 \\ (2+3)/2 \\ (3+4)/2 \\ \hline 4/2 \end{Bmatrix} = -\gamma Aa \begin{Bmatrix} 0.5 \\ 1.5 \\ 2.5 \\ 3.5 \\ \hline 2.0 \end{Bmatrix} \Rightarrow \mathbf{F_e} = -\mathbf{F_{fA}} = \begin{Bmatrix} F_{1e} \\ F_{2e} \\ F_{3e} \\ F_{4e} \end{Bmatrix} = \gamma Aa \begin{Bmatrix} 0.5 \\ 1.5 \\ 2.5 \\ 3.5 \end{Bmatrix}$$

The equivalent joint loads are shown in Fig. 4.21a.

Displacement Transformation Matrix

$$\mathbf{D_*} = \mathbf{T_D D} \quad \Rightarrow \begin{Bmatrix} D_*^1 \\ D_*^2 \\ D_*^3 \\ D_*^4 \end{Bmatrix} = \left[\begin{array}{cccc|c} -1 & 1 & 0 & 0 & 0 \\ 0 & -1 & 1 & 0 & 0 \\ 0 & 0 & -1 & 1 & 0 \\ 0 & 0 & 0 & -1 & 1 \end{array}\right] \begin{Bmatrix} D_1 \\ D_2 \\ D_3 \\ D_4 \\ \hline D_5 \end{Bmatrix}$$

Structure Stiffness Matrix

$$\tilde{\mathbf{k}}_*\mathbf{T_D} = \frac{EA}{a}\begin{bmatrix} 1 & 0 & 0 & 0 \\ 0 & 2 & 0 & 0 \\ 0 & 0 & 3 & 0 \\ 0 & 0 & 0 & 4 \end{bmatrix} \left[\begin{array}{cccc|c} -1 & 1 & 0 & 0 & 0 \\ 0 & -1 & 1 & 0 & 0 \\ 0 & 0 & -1 & 1 & 0 \\ 0 & 0 & 0 & -1 & 1 \end{array}\right] = \frac{EA}{a}\left[\begin{array}{cccc|c} -1 & 1 & 0 & 0 & 0 \\ 0 & -2 & 2 & 0 & 0 \\ 0 & 0 & -3 & 3 & 0 \\ 0 & 0 & 0 & -4 & 4 \end{array}\right]$$

$$\mathbf{k} = \mathbf{T_D}^{\mathbf{T}}\tilde{\mathbf{k}}_*\mathbf{T_D} = \left[\begin{array}{cccc} -1 & 0 & 0 & 0 \\ 1 & -1 & 0 & 0 \\ 0 & 1 & -1 & 0 \\ 0 & 0 & 1 & -1 \\ \hline 0 & 0 & 0 & 1 \end{array}\right] \frac{EA}{a}\left[\begin{array}{cccc|c} -1 & 1 & 0 & 0 & 0 \\ 0 & -2 & 2 & 0 & 0 \\ 0 & 0 & -3 & 3 & 0 \\ 0 & 0 & 0 & -4 & 4 \end{array}\right]$$

$$\Rightarrow \mathbf{k} = \left[\begin{array}{c|c} \mathbf{k_{AA}} & \mathbf{k_{AR}} \\ \hline \mathbf{k_{RA}} & \mathbf{k_{RR}} \end{array}\right] = \frac{EA}{a}\left[\begin{array}{cccc|c} 1 & -1 & 0 & 0 & 0 \\ -1 & 3 & -2 & 0 & 0 \\ 0 & -2 & 5 & -3 & 0 \\ 0 & 0 & -3 & 7 & -4 \\ \hline 0 & 0 & 0 & -4 & 4 \end{array}\right]$$

$$\Rightarrow [\mathbf{k_{AA}}]^{-1} = \frac{a}{EA}\begin{bmatrix} 2.0833 & 1.0833 & 0.5833 & 0.2500 \\ 1.0833 & 1.0833 & 0.5833 & 0.2500 \\ 0.5833 & 0.5833 & 0.5833 & 0.2500 \\ 0.2500 & 0.2500 & 0.2500 & 0.2500 \end{bmatrix}$$

[The matrix operations can be conveniently done using software like MATLAB].

Displacements and Support Reactions

There are no nodal loads prescribed here; i.e. $\mathbf{F}_A = \mathbf{O}$. The loading to be considered here consist of the equivalent joint loads shown in Fig. 4.21a. There are no support settlements prescribed, i.e. $\mathbf{D}_R = \mathbf{O}$.

$$\mathbf{D}_A = [\mathbf{k}_{AA}]^{-1}[\mathbf{F}_e] \Rightarrow \begin{Bmatrix} D_1 \\ D_2 \\ D_3 \\ D_4 \end{Bmatrix} = \frac{a}{EA}\begin{bmatrix} 2.0833 & 1.0833 & 0.5833 & 0.2500 \\ 1.0833 & 1.0833 & 0.5833 & 0.2500 \\ 0.5833 & 0.5833 & 0.5833 & 0.2500 \\ 0.2500 & 0.2500 & 0.2500 & 0.2500 \end{bmatrix} \gamma Aa \begin{Bmatrix} 0.5 \\ 1.5 \\ 2.5 \\ 3.5 \end{Bmatrix} = \frac{\gamma a^2}{E}\begin{Bmatrix} 5.0 \\ 4.5 \\ 3.5 \\ 2.0 \end{Bmatrix}$$

The axial displacements in an element has a linear variation between the end joint displacements, under the action of equivalent joint loads, as shown in Fig. 4.21b(ii). Superimposing these axial displacements with the displacements in Fig. 4.20d(ii) generated under fixed end conditions, we obtain the final displacements, as shown in Fig. 4.22c.

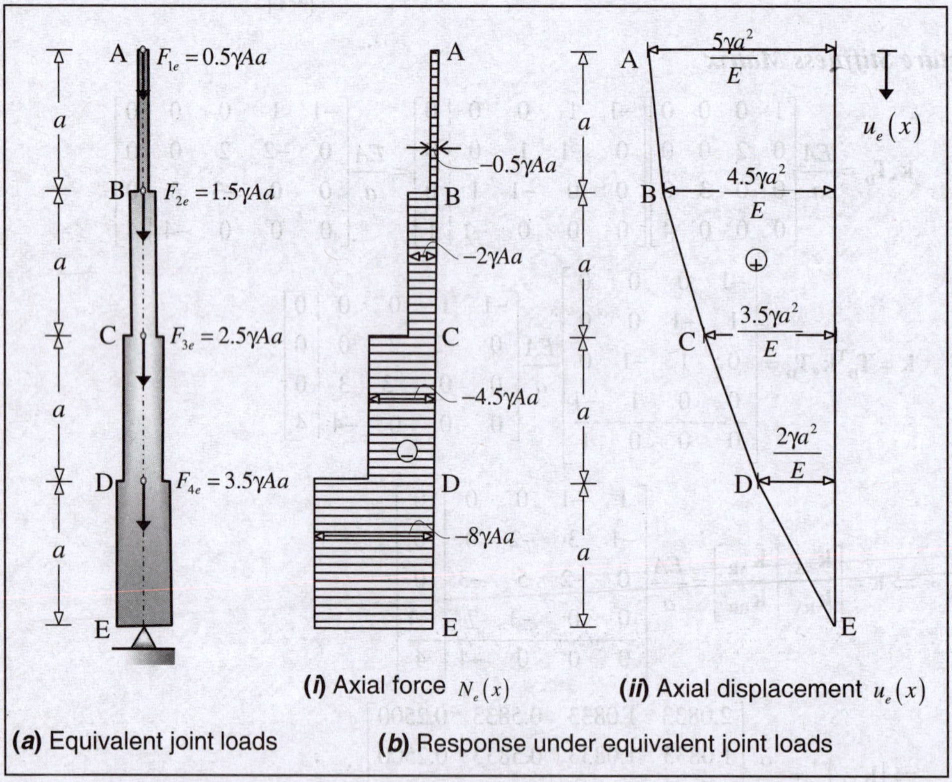

Figure 4.21 Stepped column: response to equivalent joint loads (Example 4.6)

$$\mathbf{F_R} = \mathbf{F_{fR}} + \mathbf{k_{RA}}\mathbf{D_A} \Rightarrow \{F_5\} = \{-2\gamma Aa\} + \frac{EA}{a}\begin{bmatrix}0 & 0 & 0 & -4\end{bmatrix}\frac{\gamma a^2}{E}\begin{Bmatrix}5.0\\4.5\\3.5\\2.0\end{Bmatrix} = \{-10\gamma Aa\}$$

The total upward reaction of $10\gamma Aa$ (shown in Fig. 4.22a) matches with the total downward load of $10\gamma Aa$.

Member Forces

$$\mathbf{F}^i_{*e} = \left(\tilde{\mathbf{k}}_* \mathbf{T_D}\right)\mathbf{D} \Rightarrow \begin{Bmatrix}F^1_{*e}\\F^2_{*e}\\F^3_{*e}\\F^3_{*e}\end{Bmatrix} = \frac{EA}{a}\left[\begin{array}{ccccc:c}-1 & 1 & 0 & 0 & 0\\0 & -2 & 2 & 0 & 0\\0 & 0 & -3 & 3 & 0\\0 & 0 & 0 & -4 & 4\end{array}\right]\frac{\gamma a^2}{E}\begin{Bmatrix}5.0\\4.5\\3.5\\2.0\\0\end{Bmatrix} = \gamma Aa\begin{Bmatrix}-0.5\\-2.0\\-4.5\\-8.0\end{Bmatrix}$$

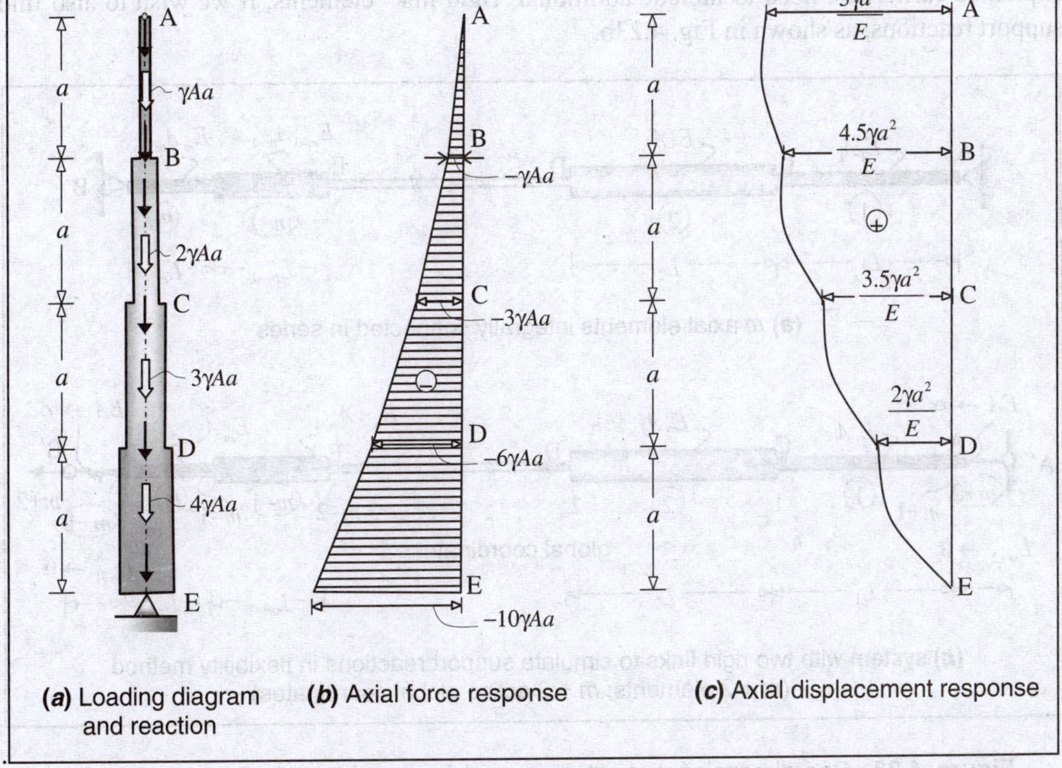

Figure 4.22 Stepped column: Total response (Example 4.6)

The axial force in an element is constant, under the action of equivalent joint loads, as shown in Fig. 4.21b(i). Superimposing these forces with the axial forces in Fig. 4.20d(i) generated under fixed end conditions, we obtain the final distribution of axial forces, as shown in Fig. 4.22b.

4.3.3 Analysis by Flexibility Method

The one-dimensional structure in Fig. 4.7a is relatively easy to analyse by the flexibility method, compared to the stiffness method, if it is statically determinate; i.e., if the restraint against translation is only at one end, as in the cases encountered in Examples 4.3 and 4.6. However, in a two-hinged situation (shown in Fig. 4.23a), the degree of static indeterminacy, equal to one, is the same as the degree of kinematic indeterminacy, and hence, both methods are comparable in terms of difficulty in analysis.

The element to be considered here is the same single degree of freedom model in Fig. 4.1b, that we had adopted for the reduced element stiffness method. However, as explained earlier, we need to include additional 'rigid link' elements, if we wish to also find support reactions, as shown in Fig. 4.23b.

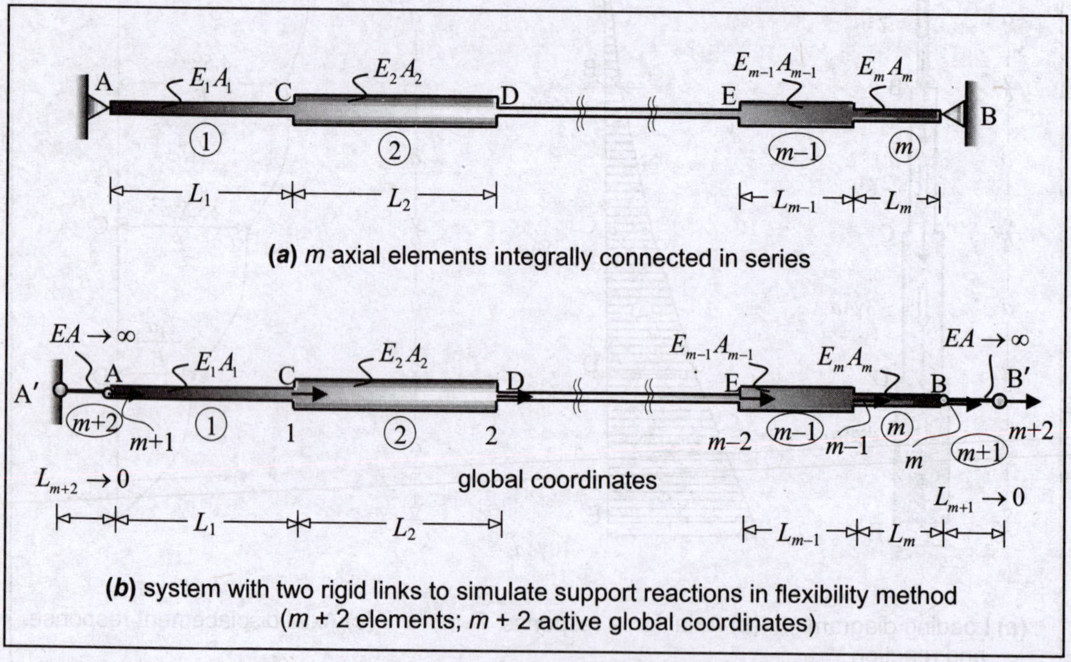

Figure 4.23 One-dimensional structure comprising *m* prismatic axial elements and additional rigid links to simulate support reactions

Unassembled Element Flexibility Matrix

For the one-dimensional axial structure shown in Fig. 4.23, the *combined element force vector*, $\{F_*\}$ and *combined element displacement vector*, $\{D_*\}$, having m (or $m+2$) dimensions, in local coordinates, are inter-related by the ***unassembled element flexibility matrix***, $\mathbf{f}_*$, as explained in Section 3.5. This matrix is a block diagonal matrix of order $m \times m$.

This flexibility matrix is a simple diagonal matrix of order $m \times m$ (which gets expanded to $(m+2)\times(m+2)$, if we include the two rigid links shown in Fig. 4.23b). The rigid links have infinite stiffness, and therefore, zero flexibility. Introducing partition lines to separate out the rigid link coordinates, the $(m+2)\times(m+2)$ matrix takes the following form:

$$\mathbf{D}_* = \mathbf{f}_*\mathbf{F}_* \Rightarrow \begin{Bmatrix} D_*^1 \\ D_*^2 \\ \vdots \\ D_*^m \\ \hline D_*^{m+1} \\ D_*^{m+2} \end{Bmatrix} = \left[\begin{array}{cccc|cc} L_1/E_1A_1 & 0 & 0 & 0 & 0 & 0 \\ 0 & L_2/E_2A_2 & 0 & 0 & 0 & 0 \\ 0 & 0 & \ddots & 0 & 0 & 0 \\ 0 & 0 & 0 & L_m/E_mA_m & 0 & 0 \\ \hline 0 & 0 & 0 & 0 & 0 & 0 \\ 0 & 0 & 0 & 0 & 0 & 0 \end{array}\right] \begin{Bmatrix} F_*^1 \\ F_*^2 \\ \vdots \\ F_*^m \\ \hline F_*^{m+1} \\ F_*^{m+2} \end{Bmatrix} \tag{4.30}$$

For convenience, we may exclude the zero rows and columns and restrict the size of the matrix to $m \times m$.

Force Transformation Matrix

If the axial structure is statically determinate (with only one end restrained against translation), we have $m + 1$ elements and $m + 1$ active global coordinates, whereby the force transformation matrix $\mathbf{T_F}$ is a square matrix, which can be easily assembled using equilibrium considerations (applying a unit load $F_j = 1$ at a time). When the structure is statically indeterminate, as in the one-dimensional axial structure shown in Fig. 4.23a, it is necessary to identify the redundant X_1, which in this case, can be one of the two support reactions ($F_{m+1} = X_1$ in Fig. 4.23b). In this case, the primary structure has $m + 2$ elements and $m + 2$ active global coordinates. It is desirable to choose the numbering of the global coordinates such that the redundant coordinate ($m + 2$) is last, while the reaction coordinate is penultimate ($m + 1$), as shown in Fig. 4.23b. For this case, the force transformation matrix $\mathbf{T_F} = [\mathbf{T_{FA}} \mid \mathbf{T_{FX}}]$ relating the element axial forces $\mathbf{F}_*$ to the nodal forces $\mathbf{F}$ is given by:

$$\mathbf{F}_* = [\mathbf{T_{FA}} \mid \mathbf{T_{FX}}]\begin{Bmatrix} \mathbf{F_A} \\ \hline \mathbf{F_X} \end{Bmatrix} \Rightarrow \begin{Bmatrix} F_*^1 \\ F_*^2 \\ \vdots \\ F_*^m \\ F_*^{m+1} \\ \hline F_*^{m+2} \end{Bmatrix} = \left[\begin{array}{cccccc|c} 1 & 1 & \cdots & 1 & 0 & & 1 \\ 0 & 1 & \cdots & 1 & 0 & & 1 \\ \vdots & \vdots & \ddots & \vdots & \vdots & & \vdots \\ 0 & 0 & \cdots & 1 & 0 & & 1 \\ 0 & 0 & \cdots & 0 & 0 & & 1 \\ \hline 1 & 1 & \cdots & 1 & 1 & & 1 \end{array}\right] \begin{Bmatrix} F_1 \\ F_2 \\ \vdots \\ F_m \\ F_{m+1} = R_A \\ \hline F_{m+2} = X_1 \end{Bmatrix} \tag{4.31}$$

When a unit load $F_j = 1$ is applied, a unit axial force (tension) gets transmitted to all the elements behind the global coordinate *j*, with elements ahead (near the free end) remaining unstressed. This is reflected in the $\mathbf{T_F}$ matrix in Eq. 4.31.

Structure Flexibility Matrix

The simplest way of assembling the structure flexibility matrix $\mathbf{f}$ is by invoking the transformation, $\mathbf{f} = \mathbf{T_F}^T \mathbf{f_*} \mathbf{T_F}$, which can be partitioned as $\mathbf{f} = \begin{bmatrix} \mathbf{f_{AA}} & \mathbf{f_{AX}} \\ \mathbf{f_{XA}} & \mathbf{f_{XX}} \end{bmatrix}$, if the structure is statically indeterminate, in which case, it is also necessary to compute $[\mathbf{f_{XX}}]^{-1}$. If there are no indirect loads involved, then it is convenient to compute $\bar{\mathbf{f}} = \mathbf{f_{AA}} - \mathbf{f_{AX}} \mathbf{f_{XX}}^{-1} \mathbf{f_{XA}}$ and $\bar{\mathbf{T}}_F^i = \mathbf{T}_{FA}^i - \mathbf{T}_{FX}^i \mathbf{f_{XX}}^{-1} \mathbf{f_{XA}}$ as the structure flexibility and force transformation matrices, suitable for handling any set of nodal loads on a statically indeterminate structure.

Fixed End Forces due to Intermediate Loads

When there are intermediate loads prescribed, these can be converted to equivalent joint loads. The same fixed end force effects (axial forces and axial displacements), described earlier in the stiffness method are applicable here. However, as in the case of the reduced element stiffness method, when there is a variation in the axial force diagram on account of intermediate loads, this variation **cannot** be captured through a matrix transformation, as there is only one force that can be described per element. For this reason, the element fixed end forces and fixed end force vector for the structure $\mathbf{F_f} = \begin{bmatrix} \mathbf{F_{fA}} \\ \mathbf{F_{fX}} \end{bmatrix}$ need to be manually computed.

Initial Displacements

Indirect loading, if any, in terms of support displacements, corresponding to the redundant coordinates ($\mathbf{D_X}$), and changes in element lengths in the 'free' condition, such as due to temperature changes or 'lack of fit' or support movements in non-redundant support locations ($\mathbf{D}_{*initial}^i$), are to be transformed to the global coordinates as $\begin{bmatrix} \mathbf{D_{A,initial}} \\ \mathbf{D_{X,initial}} \end{bmatrix} = \begin{bmatrix} \mathbf{T_{FA}^T} \\ \mathbf{T_{FX}^T} \end{bmatrix} \mathbf{D_{*initial}}$.

Force Response

In the case of statically indeterminate structures, the unknown redundant can be computed using the formula, $\mathbf{F_X} = \mathbf{F_{fX}} + [\mathbf{f_{XX}}]^{-1} \left[\mathbf{D_X} - \mathbf{D_{X,initial}} - \mathbf{f_{XA}} (\mathbf{F_A} - \mathbf{F_{fA}}) \right]$ and the element forces as $\mathbf{F_{*e}} = \mathbf{T_{FA}} \mathbf{F_A} + \mathbf{T_{FX}} \mathbf{F_X}$, which are to be combined with the fixed end axial forces computed earlier. In the absence of indirect loading, the computation of $\mathbf{F_X}$ can be avoided and $\mathbf{F_{*e}}$ can be directly computed as $\mathbf{F}_* = \overline{\mathbf{T}}_\mathbf{F}\mathbf{F}$. Similarly, in statically determinate structures ($\mathbf{F_X} = \mathbf{O}$), the element forces can be directly computed as $\mathbf{F}_{*e} = \mathbf{T}_\mathbf{F}\mathbf{F}$. The fixed end axial forces computed earlier need to be superimposed.

Displacement Response

In the case of statically indeterminate structures, the joint displacements can be computed using the formula, $\mathbf{D_A} = \mathbf{D_{A,initial}} + \mathbf{f_{AA}}(\mathbf{F_A} - \mathbf{F_{fA}}) + \mathbf{f_{AX}}(\mathbf{F_X} - \mathbf{F_{fX}})$. [Alternatively, if there are no indirect loads, to be considered, we can directly compute $\mathbf{D} = \bar{\mathbf{f}}\ (\mathbf{F} - \mathbf{F_f})$]. In statically determinate structures, this takes the form, $\mathbf{D} = \mathbf{f}(\mathbf{F} - \mathbf{F_f})$. The axial displacement in any element varies linearly between the two end joint displacements. The fixed end axial displacements computed earlier need to be superimposed.

Let us now demonstrate the application of the flexibility method through the same Examples we had used to demonstrate the Stiffness Method.

EXAMPLE 4.7

Repeat Example 4.6 (stepped column), solving by the Flexibility Method.

SOLUTION

As in Example 4.6, the stepped column ABCDE can be modelled as a one-dimensional structure, with 4 prismatic elements, free at the top end A and on a hinged support at the base E. The structure is statically determinate[†]. We may avoid modelling an additional 'rigid link' element at the bottom to model the support reaction, as this is easily determinable from the axial force in element '4'. The corresponding 4 global coordinates are as indicated earlier in Fig. 4.20a.

Element Flexibility Matrix

The local coordinate system, assuming one degree of freedom, for a typical element, is as shown in Fig. 4.20b, with the element flexibility matrix given by:

$$\mathbf{f}_* = \frac{a}{EA}\begin{bmatrix} 1 & 0 & 0 & 0 \\ 0 & 1/2 & 0 & 0 \\ 0 & 0 & 1/3 & 0 \\ 0 & 0 & 0 & 1/4 \end{bmatrix}$$

[†] Using simple statics (classical method), the force response is easy to analyse. However, using matrix methods, the computation is rendered a little difficult (and somewhat convoluted), as the distributed loading has to be first converted to equivalent joint loads. This is demonstrated here.

Fixed End Forces and Axial Displacements

The fixed end forces and axial displacements, due to the action of the uniformly distributed self-weight loading, are as calculated in Example 4.6, and as shown in Fig. 4.20d. The resulting equivalent joint loads $\mathbf{F_e}$ are shown in Fig. 4.21a.

Response to Equivalent Joint Loads

The member forces $\mathbf{F_{*e}}$ due to the action of the equivalent joint loads $\mathbf{F_e}$, can be computed by making use of the force transformation matrix $\mathbf{T_F}$:

$$\mathbf{F_{*e}} = \mathbf{T_F F_e} \Rightarrow \begin{Bmatrix} F_*^1 \\ F_*^2 \\ F_*^3 \\ F_*^4 \end{Bmatrix} = \begin{bmatrix} -1 & 0 & 0 & 0 \\ -1 & -1 & 0 & 0 \\ -1 & -1 & -1 & 0 \\ -1 & -1 & -1 & -1 \end{bmatrix} \begin{Bmatrix} F_{1e} = 0.5\gamma Aa \\ F_{2e} = 1.5\gamma Aa \\ F_{3e} = 2.5\gamma Aa \\ F_{4e} = 3.5\gamma Aa \end{Bmatrix} = \begin{Bmatrix} -0.5\gamma Aa \\ -2.0\gamma Aa \\ -4.5\gamma Aa \\ -8.0\gamma Aa \end{Bmatrix}$$

The axial forces in the various elements, due to the equivalent joint loads are identical to the ones computed earlier in Example 4.6, and as shown in Fig. 4.21b(i). Superimposing these results on the fixed end axial forces shown in Fig. 4.20d(i), we get the same final axial force distribution obtained earlier, as shown in Fig. 4.22b. The support reaction is obtained as the axial force at the base of the element '4', which is equal to $-10\gamma Aa$ (acting upwards).

Structure Flexibility Matrix and Displacements

$$\mathbf{f_* T_F} = \frac{a}{EA}\begin{bmatrix} 1 & 0 & 0 & 0 \\ 0 & 1/2 & 0 & 0 \\ 0 & 0 & 1/3 & 0 \\ 0 & 0 & 0 & 1/4 \end{bmatrix}\begin{bmatrix} -1 & 0 & 0 & 0 \\ -1 & -1 & 0 & 0 \\ -1 & -1 & -1 & 0 \\ -1 & -1 & -1 & -1 \end{bmatrix} = \frac{a}{EA}\begin{bmatrix} -1 & 0 & 0 & 0 \\ -1/2 & -1/2 & 0 & 0 \\ -1/3 & -1/3 & -1/3 & 0 \\ -1/4 & -1/4 & -1/4 & -1/4 \end{bmatrix}$$

$$\mathbf{f} = \mathbf{T_F}^{\mathbf{T}}\mathbf{f_* T_F} = \begin{bmatrix} -1 & -1 & -1 & -1 \\ 0 & -1 & -1 & -1 \\ 0 & 0 & -1 & -1 \\ 0 & 0 & 0 & -1 \end{bmatrix} \frac{a}{EA}\begin{bmatrix} -1 & 0 & 0 & 0 \\ -1/2 & -1/2 & 0 & 0 \\ -1/3 & -1/3 & -1/3 & 0 \\ -1/4 & -1/4 & -1/4 & -1/4 \end{bmatrix}$$

$$\mathbf{D} = \mathbf{f\,F_e} \Rightarrow \begin{Bmatrix} D_1 \\ D_2 \\ D_3 \\ D_4 \end{Bmatrix} = \frac{a}{EA}\begin{bmatrix} 2.0833 & 1.0833 & 0.5833 & 0.2500 \\ 1.0833 & 1.0833 & 0.5833 & 0.2500 \\ 0.5833 & 0.5833 & 0.5833 & 0.2500 \\ 0.2500 & 0.2500 & 0.2500 & 0.2500 \end{bmatrix} \gamma Aa \begin{Bmatrix} 0.5 \\ 1.5 \\ 2.5 \\ 3.5 \end{Bmatrix} = \frac{\gamma a^2}{E}\begin{Bmatrix} 5.0 \\ 4.5 \\ 3.5 \\ 2.0 \end{Bmatrix}$$

These results match exactly with the values generated in Example 4.6. Also, it may be noted that $\mathbf{f} = [\mathbf{k_{AA}}]^{-1}$. The axial displacements (linearly varying between joints) are shown in Fig. 4.21b(ii). Superimposing these axial displacements with the displacements in

Fig. 4.20d(ii) generated under fixed end conditions, we obtain the final displacements, as shown in Fig. 4.22c.

EXAMPLE 4.8

Analyse by the Flexibility Method the non-prismatic axially loaded structural system shown in Fig. 4.11 (Example 4.1), including additional indirect loading in the form of (a) a temperature rise of 40°C in AB and 20°C in BD (assuming $\alpha = 11\times10^{-5}$ per°C) and (b) support slips, to the right, of 2mm and 1mm at supports A and D respectively (Example 4.2). Assume the segment BCD to have an axial rigidity, $EA = 5000\,\text{kN}$ and the segment AB to have twice this axial rigidity. Find the complete force and displacement responses.

SOLUTION

The structure is statically indeterminate to the first degree. Let us Choose the support reaction at the right end D as the redundant X_1 and model the two supports with rigid links, as shown in Fig. 4.24a. Thus, there are 4 prismatic elements and 4 global coordinates, as shown. The local coordinates for the 4 elements (one coordinate per element) are shown in Fig. 4.24b.

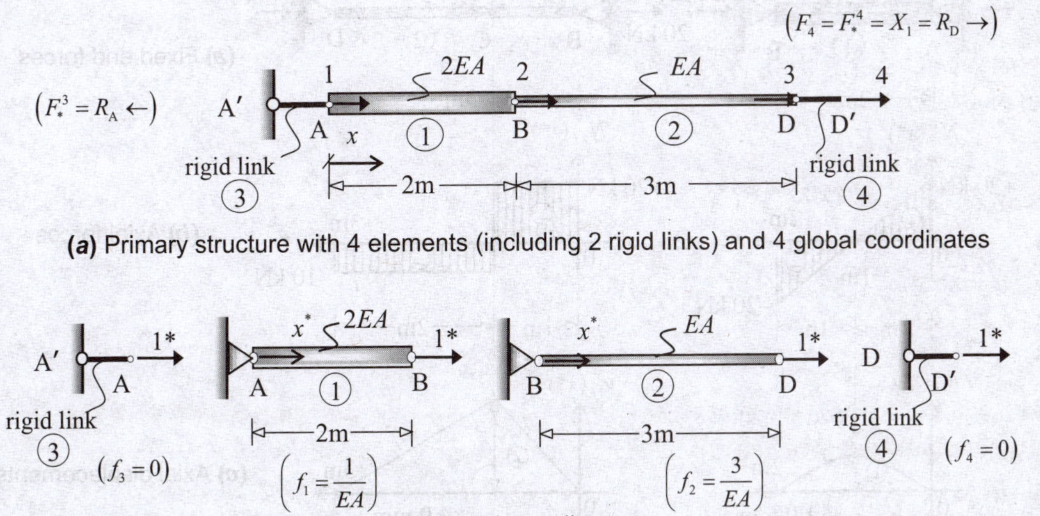

(a) Primary structure with 4 elements (including 2 rigid links) and 4 global coordinates

(b) Local coordinates

Figure 4.24 Global and local coordinates (Example 4.8)

Element Flexibility Matrix

As indicated in Fig. 4.24b, the axial flexibilities, $f_i = \frac{L_i}{E_i A_i}$, for the four elements are given by

$f_1 = \frac{2}{2EA} = \frac{1}{EA}$; $f_2 = \frac{3}{EA}$; $f_3 = f_4 = 0$ in m/kN units, where $EA = 5000$ kN. The unassembled flexibility matrix is given by:

$$\mathbf{f}_* = \frac{1}{EA}\begin{bmatrix} 1 & 0 & 0 & 0 \\ 0 & 3 & 0 & 0 \\ 0 & 0 & 0 & 0 \\ 0 & 0 & 0 & 0 \end{bmatrix}$$

Fixed End Forces and Axial Displacements

The fixed end forces and axial displacements, due to the action of the intermediate loads in elements '1' and '2', are as calculated in Example 4.1, and are shown in Fig. 4.25. There will be no axial forces and displacements induced in the rigid link elements. The fixed end force vector $\mathbf{F_f}$ can be assembled by summing up the fixed end forces at the global coordinates, and the equivalent joint loads obtained as $\mathbf{F_e} = -\mathbf{F_f}$.

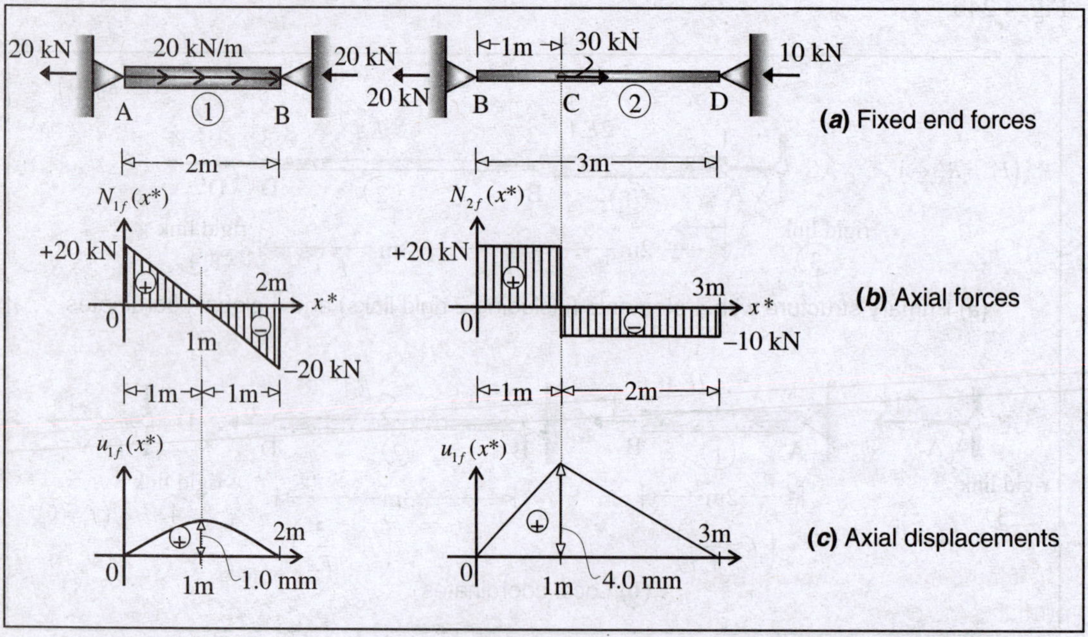

Figure 4.25 Fixed end force effects due to intermediate loading (Example 4.8)

$$\mathbf{F_f} = \begin{Bmatrix} \mathbf{F_{fA}} \\ \hline \mathbf{F_{fX}} \end{Bmatrix} = \begin{Bmatrix} F_{1f} \\ F_{2f} \\ F_{3f} \\ \hline F_{4f} \end{Bmatrix} = \begin{Bmatrix} -20 \\ -40 \\ -10 \\ \hline 0 \end{Bmatrix} \text{kN}$$

The given nodal load, $F_2 = 40\,\text{kN}$ should be added to the above equivalent joint loads to determine the load effects in terms of axial forces and axial displacements (refer Fig. 4.26).

$$\begin{Bmatrix} \mathbf{F_A - F_{fA}} \\ \hline \mathbf{F_X - F_{fX}} \end{Bmatrix} = \begin{Bmatrix} 0 \\ 40 \\ 0 \\ \hline 0 \end{Bmatrix} - \begin{Bmatrix} -20 \\ -40 \\ -10 \\ \hline 0 \end{Bmatrix} = \begin{Bmatrix} 20 \\ 80 \\ 10 \\ \hline 0 \end{Bmatrix} \text{kN}$$

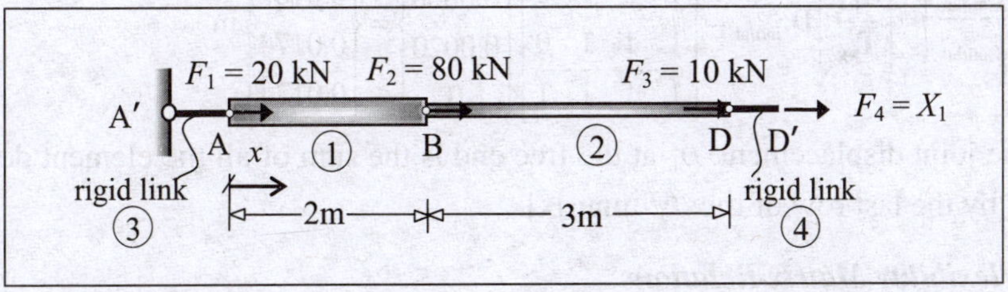

Figure 4.26 Nodal plus equivalent joint loads (Example 4.8)

Force Transformation Matrix

By applying a unit load $F_j = 1$ at a time on the primary structure, the element forces developed can be easily worked out and filled in the jth column of the $\mathbf{T_F}$ matrix. For example, when $F_4 = 1$ is applied, all the elements are subject to a unit axial force (tensile), as reflected in the fourth column of the $\mathbf{T_F}$ matrix below.

$$\mathbf{F_*} = [\mathbf{T_{FA}} \mid \mathbf{T_{FX}}] \begin{Bmatrix} \mathbf{F_A} \\ \hline \mathbf{F_X} \end{Bmatrix} \quad \Rightarrow \quad \begin{Bmatrix} F_*^1 \\ F_*^2 \\ F_*^3 \\ F_*^4 \end{Bmatrix} = \begin{bmatrix} 0 & 1 & 1 & 1 \\ 0 & 0 & 1 & 1 \\ 1 & 1 & 1 & 1 \\ 0 & 0 & 0 & 1 \end{bmatrix} \begin{Bmatrix} F_1 = 20 \\ F_2 = 40 + 40 \\ F_3 = 10 \\ \hline F_4 = X_1 \end{Bmatrix} = \begin{Bmatrix} 90 + X_1 \\ 10 + X_1 \\ 110 + X_1 \\ \hline X_1 \end{Bmatrix} \textbf{kN}$$

Indirect (Displacement) Loading

The temperature loading will introduce initial displacements in the elements '1' and '2' of the primary structure:

$$\begin{Bmatrix} e_{1o} \\ e_{2o} \end{Bmatrix} = \alpha \begin{Bmatrix} L_1(\Delta T_1) \\ L_1(\Delta T_2) \end{Bmatrix} = 11 \times 10^{-5} \begin{Bmatrix} (2)(40) \\ (3)(20) \end{Bmatrix} = \begin{Bmatrix} 0.0088 \\ 0.0066 \end{Bmatrix} \text{m (as calculated in Example 4.2)}$$

Of the two support displacements prescribed (2mm and 1mm), the one at the left support (non-redundant coordinate) can be visualised as an initial elongation in the rigid link element '3', $e_{3o} = 0.002$ mm, while the other one, which corresponds to the redundant coordinate, is to be taken as $\mathbf{D_X} = D_4 = 0.001$ mm. Thus, the initial element displacement vector is given by:

$$\mathbf{D}_{*\text{initial}} = \begin{Bmatrix} D^1_{*initial} \\ D^2_{*initial} \\ D^3_{*initial} \\ D^4_{*initial} \end{Bmatrix} = \begin{Bmatrix} 0.0088 \\ 0.0066 \\ 0.002 \\ 0 \end{Bmatrix} \text{m}$$

These changes in element lengths in the 'free' condition can be transformed to the global coordinates as follows:

$$\begin{bmatrix} \mathbf{D}_{\text{A,initial}} \\ \hline \mathbf{D}_{\text{X,initial}} \end{bmatrix} = \begin{bmatrix} \mathbf{T}^{\text{T}}_{\text{FA}} \\ \hline \mathbf{T}^{\text{T}}_{\text{FX}} \end{bmatrix} \mathbf{D}_{*\text{initial}} = \begin{bmatrix} 0 & 0 & 1 & 0 \\ 1 & 0 & 1 & 0 \\ 1 & 1 & 1 & 0 \\ \hline 1 & 1 & 1 & 1 \end{bmatrix} \begin{Bmatrix} 0.0088 \\ 0.0066 \\ 0.0020 \\ 0 \end{Bmatrix} = \begin{Bmatrix} 0.0020 \\ 0.0108 \\ 0.0174 \\ \hline 0.0174 \end{Bmatrix} \text{m}$$

[The joint displacement D_4 at the free end is the sum of all the element deformations, as reflected by the last row of the $\mathbf{T_F}^{\mathbf{T}}$ matrix].

Structure Flexibility Matrix Relations

$$\mathbf{f_* T_F} = \frac{1}{EA}\begin{bmatrix} 1 & 0 & 0 & 0 \\ 0 & 3 & 0 & 0 \\ 0 & 0 & 0 & 0 \\ 0 & 0 & 0 & 0 \end{bmatrix} \left[\begin{array}{ccc|c} 0 & 1 & 1 & 1 \\ 0 & 0 & 1 & 1 \\ 1 & 1 & 1 & 1 \\ 0 & 0 & 0 & 1 \end{array}\right] = \frac{1}{EA}\begin{bmatrix} 0 & 1 & 1 & 1 \\ 0 & 0 & 3 & 3 \\ 0 & 0 & 0 & 0 \\ 0 & 0 & 0 & 0 \end{bmatrix}$$

[The above transformation relates the axial displacements to the joint loads: $\mathbf{D}_* = (\mathbf{f}_*\mathbf{T_F})\mathbf{F}$. Thus, for example, the second row shows that element '2' will develop an axial deformation equal to $D^2_* = f_2(F_3 + F_4)$, and will remain unaffected by F_1 and F_2. The third and fourth rows are zero rows, because they correspond to the rigid links, which do not undergo any axial deformations].

$$\mathbf{f} = \mathbf{T_F}^{\mathbf{T}}(\mathbf{f}_*\mathbf{T_F}) = \begin{bmatrix} 0 & 0 & 1 & 0 \\ 1 & 0 & 1 & 0 \\ 1 & 1 & 1 & 0 \\ \hline 1 & 1 & 1 & 1 \end{bmatrix} \frac{1}{EA}\begin{bmatrix} 0 & 1 & 1 & 1 \\ 0 & 0 & 3 & 3 \\ 0 & 0 & 0 & 0 \\ 0 & 0 & 0 & 0 \end{bmatrix} = \frac{1}{EA}\left[\begin{array}{ccc|c} 0 & 0 & 0 & 0 \\ 0 & 1 & 1 & 1 \\ 0 & 1 & 4 & 4 \\ \hline 0 & 1 & 4 & 4 \end{array}\right] = \left[\begin{array}{c|c} \mathbf{f_{AA}} & \mathbf{f_{AX}} \\ \hline \mathbf{f_{XA}} & \mathbf{f_{XX}} \end{array}\right]$$

$$\begin{Bmatrix} \mathbf{D_A} \\ \hline \mathbf{D_X} \end{Bmatrix} = \begin{Bmatrix} \mathbf{D}_{\text{A,initial}} \\ \hline \mathbf{D}_{\text{X,initial}} \end{Bmatrix} + \left[\begin{array}{c|c} \mathbf{f_{AA}} & \mathbf{f_{AX}} \\ \hline \mathbf{f_{XA}} & \mathbf{f_{XX}} \end{array}\right] \begin{Bmatrix} \mathbf{F_A - F_{fA}} \\ \hline \mathbf{F_X - F_{fX}} \end{Bmatrix} \Rightarrow \begin{Bmatrix} D_1 = ? \\ D_2 = ? \\ D_3 = ? \\ \hline D_4 = 0.001 \end{Bmatrix} = \begin{Bmatrix} 0.0020 \\ 0.0108 \\ 0.0174 \\ \hline 0.0174 \end{Bmatrix} + \frac{1}{EA}\left[\begin{array}{ccc|c} 0 & 0 & 0 & 0 \\ 0 & 1 & 1 & 1 \\ 0 & 1 & 4 & 4 \\ \hline 0 & 1 & 4 & 4 \end{array}\right] \begin{Bmatrix} 20 \\ 80 \\ 10 \\ \hline X_1 \end{Bmatrix} \text{m}$$

$$[\mathbf{f_{XX}}]^{-1} = \frac{EA}{4} = \frac{5000}{4} = 1250 \text{ m/kN}$$

Redundant

$$\mathbf{F_X} = \mathbf{F_{fX}} + [\mathbf{f_{XX}}]^{-1}\left[\mathbf{D_X} - \mathbf{D_{X,initial}} - \mathbf{f_{XA}}\left(\mathbf{F_A} - \mathbf{F_{fA}}\right)\right]$$

$$\Rightarrow \; X_1 = 0 + \frac{EA}{4}\left[(0.001 - 0.0174) - \frac{1}{EA}\begin{bmatrix}0 & 1 & 4\end{bmatrix}\begin{bmatrix}20\\80\\10\end{bmatrix}\right] = \frac{EA}{4}\left[-\frac{82}{EA} - \frac{120}{EA}\right] = [-20.5 - 30] = -50.5 \text{ kN}$$

[Note that due to the direct loading, $X_1 = -30$ kN, and due to the indirect loading, $X_1 = -20.5$ kN; these results match with the solutions (support reaction at D) obtained earlier in Examples 4.1 and 4.2].

Member Forces

$$\mathbf{F}_{*e} = \left[\mathbf{T_{FA}} \mid \mathbf{T_{FX}}\right]\begin{Bmatrix}\mathbf{F_A}\\ \mathbf{F_X}\end{Bmatrix} \Rightarrow \begin{Bmatrix}F_*^1\\F_*^2\\F_*^3\\F_*^4\end{Bmatrix} = \left[\begin{array}{ccc|c}0 & 1 & 1 & 1\\0 & 0 & 1 & 1\\1 & 1 & 1 & 1\\0 & 0 & 0 & 1\end{array}\right]\begin{Bmatrix}20\\80\\10\\-50.5\end{Bmatrix} = \begin{Bmatrix}39.5\\-40.5\\59.5\\-50.5\end{Bmatrix}\text{kN}$$

The axial forces in the various elements due to the nodal plus equivalent joint loads, obtained above are shown in Fig. 4.27b (excluding the forces in the rigid links). Adding these forces to the fixed end forces in Fig. 4.27a, we get the complete axial force response, as shown in Fig. 4.27c. These results match exactly with the results in Example 4.2.

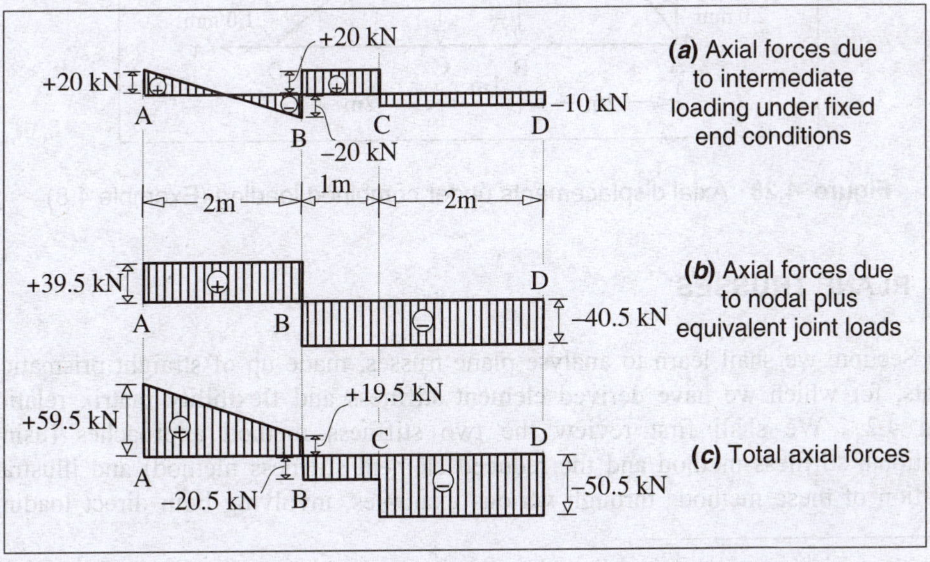

Figure 4.27 Axial forces under combined loading (Example 4.8)

Support Reactions

The final support reactions are obtained from the forces in the two rigid link elements† (refer Fig. 4.24a) as follows:

$$R_A = F_*^3 = \mathbf{59.5kN} (\leftarrow)$$

$$R_D = F_*^4 = \mathbf{-50.5kN} \; (\leftarrow)$$

The total reaction is equal and opposite to the total applied load of 110 kN.

Axial Displacements

$$\begin{Bmatrix} D_1 = ? \\ D_2 = ? \\ D_3 = ? \\ \hline D_4 = 0.001 \end{Bmatrix} = \begin{Bmatrix} 0.0020 \\ 0.0108 \\ 0.0174 \\ \hline 0.0174 \end{Bmatrix} + \frac{1}{5000}\begin{bmatrix} 0 & 0 & 0 & 0 \\ 0 & 1 & 1 & 1 \\ 0 & 1 & 4 & 4 \\ \hline 0 & 1 & 4 & 4 \end{bmatrix}\begin{Bmatrix} 20 \\ 80 \\ 10 \\ \hline -50.5 \end{Bmatrix} = \begin{Bmatrix} 0.0020 \\ 0.0187 \\ 0.0010 \\ \hline 0.0010 \end{Bmatrix} \text{m} \Rightarrow \begin{matrix} D_A = 2.00\text{mm} \\ D_B = 18.7\text{mm} \\ D_D = 1.00\text{mm} \end{matrix}$$

The axial displacements in the elements are assumed to vary linearly between the joints. Superimposing these displacements to the fixed end axial displacements shown in Fig. 4.25c, we get the complete axial force response, as shown in Fig. 4.28. These results match exactly with the results in Example 4.2.

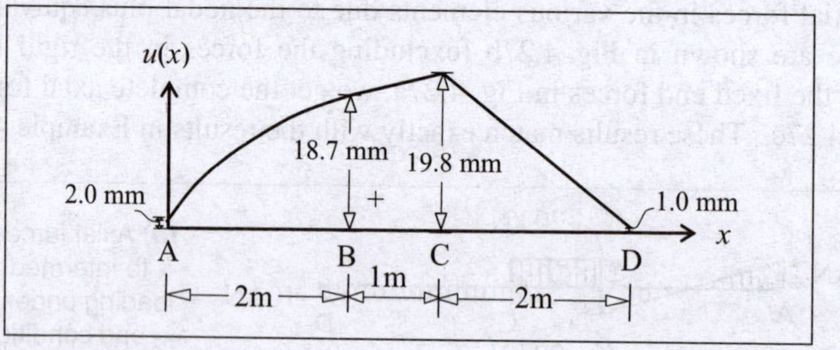

Figure 4.28 Axial displacements under combined loading (Example 4.8)

4.4 PLANE TRUSSES

In this Section, we shall learn to analyse plane trusses, made up of straight prismatic axial elements, for which we have derived element stiffness and flexibility matrix relations in Section 4.2. We shall first review the two stiffness method approaches (using the conventional stiffness method and the reduced element stiffness method) and illustrate the application of these methods through various examples, involving both direct loading and

† There are axial forces in the rigid link elements due to the intermediate loading under fixed end conditions; hence the axial forces generated in these elements due to the nodal plus equivalent joint loads give directly the final support reactions.

indirect loading. Next, we shall see how these same examples can be solved using the flexibility method. The basic steps to be followed in these matrix methods of analysis have already been outlined in Section 3.7. It may be noted that the analysis of plane trusses, we do not encounter intermediate loads, as all direct loads are assumed to act at the pinned joints. However, we do encounter fixed end forces on account of indirect loads caused by temperature effects and 'lack of fit'.

4.4.1 Analysis by Conventional Stiffness Method

Consider a plane truss, made up of *m* straight prismatic axial elements, inter-connected by *j* pin-joints, each joint having two potential degrees of freedom (in the global *x*- and *y*-directions). Thus, there are $2j$ global coordinates to be considered in the structure. Of these, let *n* be the number of *active* global coordinates, and let $r = 2j - n$ be the number of restrained global coordinates. Hence, the displacement vector, $\mathbf{D} = \begin{Bmatrix} \mathbf{D_A} \\ \mathbf{D_R} \end{Bmatrix}$ and the conjugate force vector $\mathbf{F} = \begin{Bmatrix} \mathbf{F_A} \\ \mathbf{F_R} \end{Bmatrix}$, each have a dimension of $2j$, and these two vectors are inter-related by the stiffness relation, $\begin{bmatrix} \mathbf{F_A} \\ \hline \mathbf{F_R} \end{bmatrix} = \begin{bmatrix} \mathbf{k_{AA}} & \mathbf{k_{AR}} \\ \mathbf{k_{RA}} & \mathbf{k_{RR}} \end{bmatrix} \begin{bmatrix} \mathbf{D_A} \\ \hline \mathbf{D_R} \end{bmatrix}$, in terms of the structure stiffness matrix, $\mathbf{k} = \begin{bmatrix} \mathbf{k_{AA}} & \mathbf{k_{AR}} \\ \mathbf{k_{RA}} & \mathbf{k_{RR}} \end{bmatrix}$, as mentioned earlier.

In the conventional stiffness method, we consider each plane truss element to have four degrees of freedom (with local coordinates as shown in Fig. 4.3a), with the element stiffness as defined by Eq. 4.13. The $4m$-dimensional *combined element displacement vector*, $\{D_*\}$, and *combined element force vector*, $\{F_*\}$, in local coordinates, are inter-related by the ***unassembled stiffness matrix***, $\mathbf{k}_*$, as explained in Section 3.4.4. This matrix is a block diagonal matrix of order $4m \times 4m$:

$$\mathbf{F}_* = \mathbf{k}_*\mathbf{D}_* \Rightarrow \begin{bmatrix} \begin{Bmatrix} F_{1*}^1 \\ F_{2*}^1 \\ F_{3*}^1 \\ F_{4*}^1 \end{Bmatrix} \\ \vdots \\ \begin{Bmatrix} F_{1*}^m \\ F_{2*}^m \\ F_{3*}^m \\ F_{4*}^m \end{Bmatrix} \end{bmatrix} = \begin{bmatrix} \begin{bmatrix} k_1 & 0 & -k_1 & 0 \\ 0 & 0 & 0 & 0 \\ -k_1 & 0 & k_1 & 0 \\ 0 & 0 & 0 & 0 \end{bmatrix} & \cdots & \mathbf{O} \\ \vdots & \ddots & \vdots \\ \mathbf{O} & \cdots & \begin{bmatrix} k_m & 0 & -k_m & 0 \\ 0 & 0 & 0 & 0 \\ -k_m & 0 & k_m & 0 \\ 0 & 0 & 0 & 0 \end{bmatrix} \end{bmatrix} \begin{bmatrix} \begin{Bmatrix} D_{1*}^1 \\ D_{2*}^1 \\ D_{3*}^1 \\ D_{4*}^1 \end{Bmatrix} \\ \vdots \\ \begin{Bmatrix} D_{1*}^m \\ D_{2*}^m \\ D_{3*}^m \\ D_{4*}^m \end{Bmatrix} \end{bmatrix} \quad (4.32)$$

where $k_i = \frac{E_i A_i}{L_i}$

As explained earlier (Section 4.3.1), the structure stiffness matrix $\mathbf{k}$ can be assembled in several different ways, such as by using the element transformation matrix $\mathbf{T}^i$ or the displacement transformation matrix $\mathbf{T_D}$, or even by means of a 'physical approach'. When the number of elements involved are large, it is especially convenient to adopt a 'direct stiffness' approach to assembling the $\mathbf{k}$ matrix.

Instead of applying matrix multiplication to generate first the $\mathbf{k}_*^i\mathbf{T}^i$ matrix of order 4×4 and subsequently the $\mathbf{k}^i = \mathbf{T}^{i^T}\mathbf{k}_*^i\mathbf{T}^i$ matrix for each element, we can directly generate the $\mathbf{k}_*^i\mathbf{T}^i$ and $\mathbf{k}^i = \mathbf{T}^{i^T}\mathbf{k}_*^i\mathbf{T}^i$ matrices by invoking the formulas given in Eqns 4.15 and 4.16, involving $\cos\theta_x^i$ and $\sin\theta_x^i$ terms, noting the linkages of the various rows and columns with the global coordinates. Then, summing up all the $\mathbf{k}^i$ matrix elements appropriately, the full $\mathbf{k} = \left[\begin{array}{c|c} \mathbf{k_{AA}} & \mathbf{k_{AR}} \\ \hline \mathbf{k_{RA}} & \mathbf{k_{RR}} \end{array}\right]$ matrix can be generated. The remaining procedure is as outlined in Section 3.7.

Alternatively, we may generate the $\mathbf{T_D^i}$ matrix and apply the direct stiffness approach of generating the $\mathbf{k}_*^i\mathbf{T_D^i}$ matrix of order $4 \times (n+r)$ and then the full structure stiffness matrix by the 'direct stiffness' approach. When the number of elements are relatively low, we can even generate the full $\mathbf{T_D}$ matrix, using simple compatibility relations, applying a unit displacement at a time ($D_j = 1$) on the restrained (primary) structure.

$$\mathbf{D}_* = \left[\mathbf{T_{DA}} \mid \mathbf{T_{DR}}\right]\left[\begin{array}{c}\mathbf{D_A} \\ \hline \mathbf{D_R}\end{array}\right] \Rightarrow \begin{bmatrix} \begin{Bmatrix} D_{1*}^1 \\ D_{2*}^1 \\ \hline D_{3*}^1 \\ D_{4*}^1 \end{Bmatrix} \\ \vdots \\ \begin{Bmatrix} D_{1*}^m \\ D_{2*}^m \\ \hline D_{3*}^m \\ D_{4*}^m \end{Bmatrix} \end{bmatrix}_{4m\times 1} = \left[\begin{array}{c|c} \left[\mathbf{T_{DA}^1}\right]_{4\times n} & \left[\mathbf{T_{DR}^1}\right]_{4\times r} \\ \vdots & \vdots \\ \left[\mathbf{T_{DA}^m}\right]_{4\times n} & \left[\mathbf{T_{DR}^m}\right]_{4\times r} \end{array}\right]_{4m\times(n+r)} \begin{bmatrix} D_1 \\ \vdots \\ D_n \\ \hline D_{n+1} \\ \vdots \\ D_{n+r} \end{bmatrix}_{(n+r)\times 1} \quad (4.33)$$

We can also alternatively generate the $\mathbf{T_D}^T$ matrix, by making use of the *Principle of Contra-gradient*, whereby $\left[\begin{array}{c}\mathbf{F_A} \\ \hline \mathbf{F_R}\end{array}\right] = \left[\begin{array}{c}\mathbf{T_{DA}}^T \\ \hline \mathbf{T_{DR}}^T\end{array}\right]\mathbf{F}_*$, as explained in Section 3.3.5. We next compute the product $\mathbf{k}_*\mathbf{T_D}$, which is needed later to compute the element-end forces ($\mathbf{F}_* = (\mathbf{k}_*\mathbf{T_D})\mathbf{D}$).

Finally, we generate the *structure stiffness matrix*, $\mathbf{k} = \mathbf{T_D}^\mathrm{T}\mathbf{k_*T_D}$, of order $(n+r)\times(n+r)$, making use of the $\mathbf{k_*T_D}$ matrix of order $(4m)\times(n+r)$, already generated.

When indirect loads in the form of temperature loading or 'lack of fit' are prescribed in the form of an initial (free) displacement vector $\mathbf{e_o}$ (of dimension *m*), these loads manifest as axial forces in the fixed end condition in the primary restrained structure, given by $N_{if} = -k_i e_{oi}$, which need to be expressed in the element local coordinates as follows:

$$\mathbf{F^i_{*f}} = \begin{Bmatrix} F^i_{1*f} \\ F^i_{2*f} \\ \hline F^i_{3*f} \\ F^i_{4*f} \end{Bmatrix} = \begin{Bmatrix} -N_{if} \\ 0 \\ \hline +N_{if} \\ 0 \end{Bmatrix} = \begin{Bmatrix} k_i e_{oi} \\ 0 \\ \hline -k_i e_{oi} \\ 0 \end{Bmatrix} \tag{4.34}$$

These fixed end forces can then be transformed into the global axes system by $\mathbf{F^i_f} = \mathbf{T^{i^T}} \mathbf{F^i_{*f}}$ and converted to equivalent joint loads. Alternatively, this may be done by the transformation $\mathbf{F^i_f} = \mathbf{T_D^{i\,T}} \mathbf{F^i_{*f}}$ or $\mathbf{F_f} = \mathbf{T_D}^\mathrm{T} \mathbf{F_{*f}}$. The total load vector to be considered for finding the unknown joint displacements is given by $\mathbf{F_A} - \mathbf{F_{fA}}$, as explained earlier. If there are support settlements, $\mathbf{D_R}$, these should also be accounted for in the stiffness relations. The final member end forces are obtained by $\mathbf{F^i_*} = \mathbf{F^i_{*f}} + \mathbf{k^i_* T^i D}$ or $\mathbf{F^i_*} = \mathbf{F^i_{*f}} + \mathbf{k^i_* T^i_D D}$.

The application of the conventional stiffness method, in its various forms, is demonstrated through the Examples to follow.

EXAMPLE 4.9

Analyse, by the Stiffness Method, the pin-jointed frame shown in Fig. 4.29 due to the applied direct actions as well as due to a 'lack of fit' caused by bars '1' and '4' being too short by δ. Find the complete force response and displacement response. Assume all bars to have the same axial rigidity *EA*.

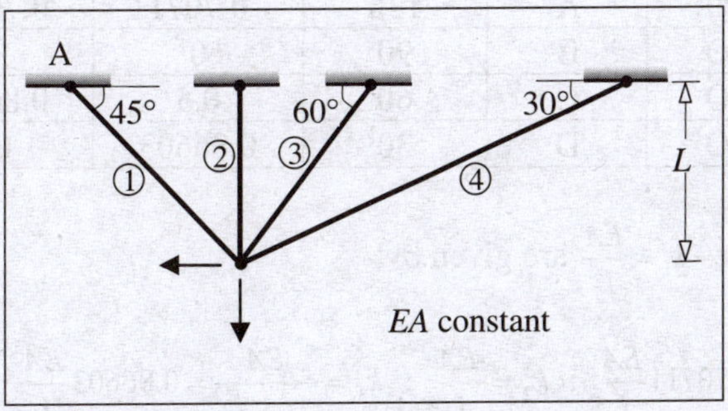

Figure 4.29 Pin-jointed frame (Example 4.9)

SOLUTION

Input Data

There are 4 prismatic elements here ($m = 4$), with 5 pin joints, i.e., $2j = 10$ global coordinates, of which 2 correspond to *active* degrees of freedom ($n = 2$) and the rest to *restrained* global coordinates ($r = 8$), as shown in Fig. 4.30a. The local coordinates for the four elements (*start node* for all at O) are shown in Fig. 4.30b.

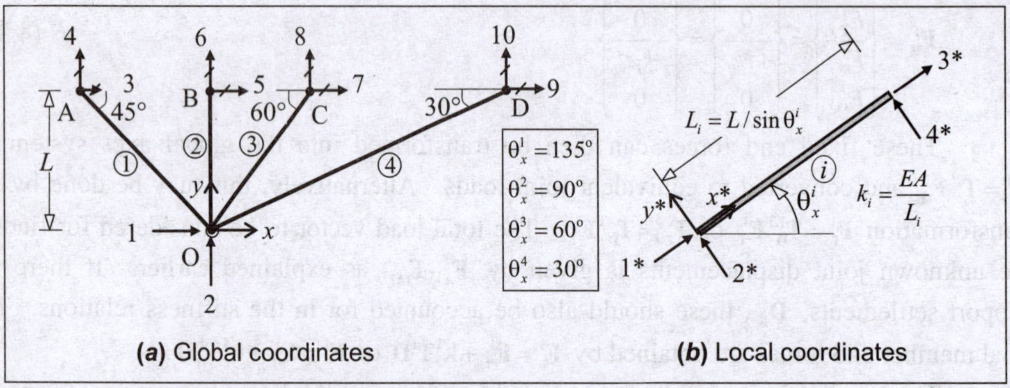

Figure 4.30 Global and local coordinates (Example 4.9)

Choosing the origin of the x-y global axes system at the node O, the start and end nodes of individual elements and their direction cosine parameters, $\cos\theta_x^i$ and $\sin\theta_x^i$, and lengths, $L_i = L/\sin\theta_x^i$, are tabulated below.

Element No.	Start Node	End Node	θ_x^i (deg)	$\cos\theta_x^i$	$\sin\theta_x^i$	$L_i = L/\sin\theta_x^i$
1	O	A	135	–0.70711	0.70711	$1.41421L$
2	O	B	90	0	1	L
3	O	C	60	0.5	0.86603	$1.15469L$
4	O	D	30	0.86603	0.5	$2L$

The element stiffnesses $k_i = \dfrac{EA}{L_i}$ are given by:

$$k_1 = \frac{EA}{\sqrt{2}L} = 0.70711\frac{EA}{L}\ ;\ k_2 = \frac{EA}{L}\ ;\ k_3 = \frac{EA}{2L/\sqrt{3}} = 0.86603\frac{EA}{L}\ ;\ k_4 = \frac{EA}{2L} = 0.5\frac{EA}{L}$$

The input load data is given in terms of the load vector, $\mathbf{F_A} = \begin{Bmatrix} F_1 \\ F_2 \end{Bmatrix} = \begin{Bmatrix} -P \\ -P \end{Bmatrix}$ (direct loading),

and the initial element displacement vector, $\mathbf{e_o} = \begin{Bmatrix} e_{o1} \\ e_{o2} \\ e_{o3} \\ e_{o4} \end{Bmatrix} = \begin{Bmatrix} -\delta \\ 0 \\ 0 \\ -\delta \end{Bmatrix}$ (indirect loading).

Element Transformation Matrices

The element transformation matrix $\mathbf{T^i}$, satisfying $\mathbf{D_*^i = T^i D}$ and $\mathbf{F_*^i = T^i F}$, take the following form (refer Figs 4.30a and b), with the linking global coordinates indicated in parentheses.

$$\begin{bmatrix} D_{1*}^i \\ D_{2*}^i \\ \hline D_{3*}^i \\ D_{4*}^i \end{bmatrix} = \begin{bmatrix} \cos\theta_x^i & \sin\theta_x^i & 0 & 0 \\ -\sin\theta_x^i & \cos\theta_x^i & 0 & 0 \\ \hline 0 & 0 & \cos\theta_x^i & \sin\theta_x^i \\ 0 & 0 & -\sin\theta_x^i & \cos\theta_x^i \end{bmatrix} \begin{bmatrix} D_1^i \\ D_2^i \\ D_3^i \\ D_4^i \end{bmatrix} \Rightarrow$$

$$\mathbf{T^1} = \begin{bmatrix} \overset{(1)}{-0.7071} & \overset{(2)}{0.7071} & \overset{(3)}{0} & \overset{(4)}{0} \\ -0.7071 & -0.7071 & 0 & 0 \\ \hline 0 & 0 & -0.7071 & 0.7071 \\ 0 & 0 & -0.7071 & -0.7071 \end{bmatrix} \begin{matrix} (1) \\ (2) \\ (3) \\ (4) \end{matrix} \qquad \mathbf{T^2} = \begin{bmatrix} \overset{(1)}{0} & \overset{(2)}{1} & \overset{(5)}{0} & \overset{(6)}{0} \\ -1 & 0 & 0 & 0 \\ \hline 0 & 0 & 0 & 1 \\ 0 & 0 & -1 & 0 \end{bmatrix} \begin{matrix} (1) \\ (2) \\ (5) \\ (6) \end{matrix}$$

$$\mathbf{T^3} = \begin{bmatrix} \overset{(1)}{0.5} & \overset{(2)}{0.8660} & \overset{(7)}{0} & \overset{(8)}{0} \\ -0.8660 & 0.5 & 0 & 0 \\ \hline 0 & 0 & 0.5 & 0.8660 \\ 0 & 0 & -0.8660 & 0.5 \end{bmatrix} \begin{matrix} (1) \\ (2) \\ (7) \\ (8) \end{matrix} \qquad \mathbf{T^4} = \begin{bmatrix} \overset{(1)}{0.8660} & \overset{(2)}{0.5} & \overset{(9)}{0} & \overset{(10)}{0} \\ -0.5 & 0.8660 & 0 & 0 \\ \hline 0 & 0 & 0.8660 & 0.5 \\ 0 & 0 & -0.5 & 0.8660 \end{bmatrix} \begin{matrix} (1) \\ (2) \\ (9) \\ (10) \end{matrix}$$

Element Stiffness Matrices

The element stiffness matrices $\mathbf{k_*^i}$ (in local coordinates), satisfying $\mathbf{F_*^i = k_*^i D_*^i}$, and the products $\mathbf{k_*^i T^i}$ satisfying $\mathbf{F_*^i = (k_*^i T^i)D}$, take the following forms (refer Figs 4.30a and b):

$$\mathbf{k_*^i} = k_i \begin{bmatrix} 1 & 0 & -1 & 0 \\ 0 & 0 & 0 & 0 \\ \hline -1 & 0 & 1 & 0 \\ 0 & 0 & 0 & 0 \end{bmatrix} \quad \text{where} \quad \begin{cases} k_1 = 0.70711\dfrac{EA}{L} \\[2mm] k_2 = \dfrac{EA}{L} \\[2mm] k_3 = 0.86603\dfrac{EA}{L} \\[2mm] k_4 = 0.5\dfrac{EA}{L} \end{cases}$$

$$\mathbf{k}_*^1\mathbf{T}^1 = \frac{EA}{L}\begin{array}{c}\begin{array}{cccc}(1) & (2) & (3) & (4)\end{array}\\ \left[\begin{array}{cc|cc} -0.5 & 0.5 & 0.5 & -0.5 \\ 0 & 0 & 0 & 0 \\ \hline 0.5 & -0.5 & -0.5 & 0.5 \\ 0 & 0 & 0 & 0 \end{array}\right]\end{array}\begin{array}{c}(1)\\(2)\\(3)\\(4)\end{array} \qquad \mathbf{k}_*^2\mathbf{T}^2 = \frac{EA}{L}\begin{array}{c}\begin{array}{cccc}(1) & (2) & (5) & (6)\end{array}\\ \left[\begin{array}{cc|cc} 0 & 1 & 0 & -1 \\ 0 & 0 & 0 & 0 \\ \hline 0 & -1 & 0 & 1 \\ 0 & 0 & 0 & 0 \end{array}\right]\end{array}\begin{array}{c}(1)\\(2)\\(5)\\(6)\end{array}$$

$$\mathbf{k}_*^3\mathbf{T}^3 = \frac{EA}{L}\begin{array}{c}\begin{array}{cccc}(1) & (2) & (7) & (8)\end{array}\\ \left[\begin{array}{cc|cc} 0.433 & 0.75 & -0.433 & -0.75 \\ 0 & 0 & 0 & 0 \\ \hline -0.433 & -0.75 & 0.433 & 0.75 \\ 0 & 0 & 0 & 0 \end{array}\right]\end{array}\begin{array}{c}(1)\\(2)\\(7)\\(8)\end{array} \qquad \mathbf{k}_*^4\mathbf{T}^4 = \frac{EA}{L}\begin{array}{c}\begin{array}{cccc}(1) & (2) & (9) & (10)\end{array}\\ \left[\begin{array}{cc|cc} 0.433 & 0.25 & -0.433 & -0.25 \\ 0 & 0 & 0 & 0 \\ \hline -0.433 & -0.25 & 0.433 & 0.25 \\ 0 & 0 & 0 & 0 \end{array}\right]\end{array}\begin{array}{c}(1)\\(2)\\(9)\\(10)\end{array}$$

Fixed End Forces and Total Load Vector

The initial element displacements due to 'lack of fit' generates fixed end forces in elements '1' and '4', when applied to the restrained (primary structure), given by Eq. 4.34:

$$\mathbf{F}_{*\mathbf{f}}^{\mathbf{i}} = \begin{Bmatrix} F_{1*f}^i \\ F_{2*f}^i \\ \hline F_{3*f}^i \\ F_{4*f}^i \end{Bmatrix} = \begin{Bmatrix} k_i e_{oi} \\ 0 \\ \hline -k_i e_{oi} \\ 0 \end{Bmatrix} \Rightarrow \begin{Bmatrix} F_{1*f}^1 \\ F_{2*f}^1 \\ \hline F_{3*f}^1 \\ F_{4*f}^1 \end{Bmatrix} = \frac{EA\delta}{L}\begin{Bmatrix} -0.7071 \\ 0 \\ \hline 0.7071 \\ 0 \end{Bmatrix};\ \begin{Bmatrix} F_{1*f}^2 \\ F_{2*f}^2 \\ \hline F_{3*f}^2 \\ F_{4*f}^2 \end{Bmatrix} = \begin{Bmatrix} F_{1*f}^3 \\ F_{2*f}^3 \\ \hline F_{3*f}^3 \\ F_{4*f}^3 \end{Bmatrix} = \begin{Bmatrix} 0 \\ 0 \\ \hline 0 \\ 0 \end{Bmatrix};\ \begin{Bmatrix} F_{1*f}^4 \\ F_{2*f}^4 \\ \hline F_{3*f}^4 \\ F_{4*f}^4 \end{Bmatrix} = \frac{EA\delta}{L}\begin{Bmatrix} -0.5 \\ 0 \\ \hline 0.5 \\ 0 \end{Bmatrix}$$

Transforming these forces to the global axes system, by applying $\mathbf{F}_\mathbf{f}^\mathbf{i} = \mathbf{T}^{\mathbf{i}^T}\mathbf{F}_{*\mathbf{f}}^\mathbf{i}$ (global coordinates marked in parenthesis), we obtain:

$$\begin{Bmatrix} F_{1f}^1 \\ F_{2f}^1 \\ \hline F_{3f}^1 \\ F_{4f}^1 \end{Bmatrix} = \frac{EA\delta}{L}\begin{Bmatrix} 0.5 \\ -0.5 \\ \hline -0.5 \\ 0.5 \end{Bmatrix}\begin{matrix}(1)\\(2)\\(3)\\(4)\end{matrix};\ \begin{Bmatrix} F_{1f}^2 \\ F_{2f}^2 \\ \hline F_{3f}^2 \\ F_{4f}^2 \end{Bmatrix} = \begin{Bmatrix} F_{1f}^3 \\ F_{2f}^3 \\ \hline F_{3f}^3 \\ F_{4f}^3 \end{Bmatrix} = \begin{Bmatrix} 0 \\ 0 \\ \hline 0 \\ 0 \end{Bmatrix};\ \begin{Bmatrix} F_{1f}^4 \\ F_{2f}^4 \\ \hline F_{3f}^4 \\ F_{4f}^4 \end{Bmatrix} = \frac{EA\delta}{L}\begin{Bmatrix} -0.433 \\ -0.25 \\ \hline 0.433 \\ 0.25 \end{Bmatrix}\begin{matrix}(1)\\(2)\\(9)\\(10)\end{matrix}$$

The resulting load vector $\mathbf{F_A} - \mathbf{F_{fA}}$ (global coordinates) is given by:

$$\mathbf{F_A} - \mathbf{F_{FA}} = \begin{Bmatrix} F_1 \\ F_2 \end{Bmatrix} - \begin{Bmatrix} F_{1f} \\ F_{2f} \end{Bmatrix} = \begin{Bmatrix} -P \\ -P \end{Bmatrix} - \frac{EA\delta}{L}\begin{Bmatrix} 0.5 - 0.433 \\ -0.5 - 0.25 \end{Bmatrix} = \begin{Bmatrix} -P - 0.067\dfrac{EA\delta}{L} \\ -P + 0.75\dfrac{EA\delta}{L} \end{Bmatrix}$$

Structure Stiffness Matrix

The structure stiffness matrix $\mathbf{k} = \left[\begin{array}{c|c} \mathbf{k_{AA}} & \mathbf{k_{AR}} \\ \hline \mathbf{k_{RA}} & \mathbf{k_{RR}} \end{array}\right]$, satisfying $\mathbf{F} = \mathbf{kD}$, can be assembled, by summing up the contributions of $\mathbf{k}^\mathbf{i} = \mathbf{T}^{\mathbf{i}^T}\mathbf{k}_*^\mathbf{i}\mathbf{T}^\mathbf{i}$ at the appropriate coordinate locations:

$$\mathbf{k}^1 = \frac{EA}{L}\begin{array}{c} \begin{array}{cccc} (1) & (2) & (3) & (4) \end{array} \\ \left[\begin{array}{cc|cc} 0.3536 & -0.3536 & -0.3536 & 0.3536 \\ -0.3536 & 0.3536 & 0.3536 & -0.3536 \\ \hline -0.3536 & 0.3536 & 0.3536 & -0.3536 \\ 0.3536 & -0.3536 & -0.3536 & 0.3536 \end{array}\right] \end{array} \begin{array}{c} \\ (1) \\ (2) \\ (3) \\ (4) \end{array} \qquad \mathbf{k}^2 = \frac{EA}{L}\begin{array}{c} \begin{array}{cccc} (1) & (2) & (5) & (6) \end{array} \\ \left[\begin{array}{cc|cc} 0 & 0 & 0 & 0 \\ 0 & 1 & 0 & -1 \\ \hline 0 & 0 & 0 & 0 \\ 0 & -1 & 0 & 1 \end{array}\right] \end{array} \begin{array}{c} \\ (1) \\ (2) \\ (5) \\ (6) \end{array}$$

$$\mathbf{k}^3 = \frac{EA}{L}\begin{array}{c} \begin{array}{cccc} (1) & (2) & (7) & (8) \end{array} \\ \left[\begin{array}{cc|cc} 0.2165 & 0.3750 & -0.2165 & -0.3750 \\ 0.3750 & 0.6495 & -0.3750 & -0.6495 \\ \hline -0.2165 & -0.3750 & 0.2165 & 0.3750 \\ -0.3750 & -0.6495 & 0.3750 & 0.6495 \end{array}\right] \end{array} \begin{array}{c} \\ (1) \\ (2) \\ (7) \\ (8) \end{array}$$

$$\mathbf{k}^4 = \frac{EA}{L}\begin{array}{c} \begin{array}{cccc} (1) & (2) & (9) & (10) \end{array} \\ \left[\begin{array}{cc|cc} 0.3750 & 0.2165 & -0.3750 & -0.2165 \\ 0.2165 & 0.1250 & -0.2165 & -0.1250 \\ \hline -0.3750 & -0.2165 & 0.3750 & 0.2165 \\ -0.2165 & -0.1250 & 0.2165 & 0.1250 \end{array}\right] \end{array} \begin{array}{c} \\ (1) \\ (2) \\ (9) \\ (10) \end{array}$$

$$k_{11} = \frac{EA}{L}(0.3536 + 0 + 0.2165 + 0.3750) = 0.9451\frac{EA}{L}$$

$$\Rightarrow \quad k_{21} = k_{12} = \frac{EA}{L}(-0.3536 + 0 + 0.3750 + 0.2165) = 0.2380\frac{EA}{L}$$

$$k_{22} = \frac{EA}{L}(0.3536 + 1 + 0.6495 + 0.1250) = 2.1281\frac{EA}{L}$$

$$\Rightarrow \quad \mathbf{k}_{AA} = \frac{L}{EA}\begin{array}{c} \begin{array}{cc} (1) & (2) \end{array} \\ \begin{bmatrix} 0.9451 & 0.2380 \\ 0.2380 & 2.1281 \end{bmatrix} \end{array} \begin{array}{c} \\ (1) \\ (2) \end{array} \quad \Rightarrow \quad [\mathbf{k}_{AA}]^{-1} = \frac{EA}{L}\begin{bmatrix} 1.0888 & -0.1217 \\ -0.1217 & 0.4835 \end{bmatrix}$$

Similarly, we can generate:

$$\mathbf{k}_{AR} = \frac{EA}{L}\begin{array}{c} \begin{array}{cccccccc} (3) & (4) & (5) & (6) & (7) & (8) & (9) & (10) \end{array} \\ \begin{bmatrix} -0.3536 & 0.3536 & 0 & 0 & -0.2165 & -0.3750 & -0.3750 & -0.2165 \\ 0.3536 & -0.3536 & 0 & -1 & -0.3750 & -0.6495 & -0.2165 & -0.1250 \end{bmatrix} \end{array} \begin{array}{c} \\ (1) \\ (2) \end{array}$$

$\mathbf{k}_{RA} = \mathbf{k}_{AR}{}^{\mathrm{T}}$. The $\mathbf{k}_{RA}$ matrix of order 8×8 can similarly be assembled; it is not shown here for convenience, as it is not required in this particular problem.

Displacements and Support Reactions

The structure stiffness relation to be considered here is:

$$\begin{bmatrix}\mathbf{F_A}\\ \hline \mathbf{F_R}\end{bmatrix}-\begin{bmatrix}\mathbf{F_{fA}}\\ \hline \mathbf{F_{fR}}\end{bmatrix}=\left[\begin{array}{c|c}\mathbf{k_{AA}} & \mathbf{k_{AR}}\\ \hline \mathbf{k_{RA}} & \mathbf{k_{RR}}\end{array}\right]\begin{bmatrix}\mathbf{D_A}\\ \hline \mathbf{D_R}=\mathbf{O}\end{bmatrix}$$

, from which the unknown displacements $\mathbf{D_A}$ and support reactions $\mathbf{F_R}$ can be computed as follows:

$$\mathbf{D_A}=[\mathbf{k_{AA}}]^{-1}[\mathbf{F_A}-\mathbf{F_{fA}}]$$

$$\Rightarrow\quad \begin{Bmatrix}D_1\\ D_2\end{Bmatrix}=\frac{L}{EA}\begin{bmatrix}1.0888 & -0.1217\\ -0.1217 & 0.4835\end{bmatrix}\begin{Bmatrix}-P-0.067\dfrac{EA\delta}{L}\\ -P+0.75\dfrac{EA\delta}{L}\end{Bmatrix}=\begin{Bmatrix}-0.9671\dfrac{PL}{EA}-0.1643\,\delta\\ -0.3618\dfrac{PL}{EA}+0.3708\,\delta\end{Bmatrix}$$

$$\mathbf{F_R}=\mathbf{F_{fR}}+\mathbf{k_{RA}}\mathbf{D_A}$$

$$\Rightarrow\quad \mathbf{F_R}=\begin{Bmatrix}F_3\\ F_4\\ F_5\\ F_6\\ F_7\\ F_8\\ F_9\\ F_{10}\end{Bmatrix}=\frac{EA\delta}{L}\begin{Bmatrix}-0.5\\ 0.5\\ 0\\ 0\\ 0\\ 0\\ 0.433\\ 0.25\end{Bmatrix}+\frac{EA}{L}\begin{bmatrix}-0.3536 & 0.3536\\ 0.3536 & -0.3536\\ 0 & 0\\ 0 & -1\\ -0.2165 & -0.3750\\ -0.3750 & -0.6495\\ -0.3750 & -0.2165\\ -0.2165 & -0.1250\end{bmatrix}\begin{Bmatrix}-0.9671\dfrac{PL}{EA}-0.1643\,\delta\\ -0.3618\dfrac{PL}{EA}+0.3708\,\delta\end{Bmatrix}$$

$$\Rightarrow\quad \mathbf{F_R}=\begin{Bmatrix}F_3\\ F_4\\ F_5\\ F_6\\ F_7\\ F_8\\ F_9\\ F_{10}\end{Bmatrix}=P\begin{Bmatrix}0.2140\\ -0.2140\\ 0\\ 0.3618\\ 0.3451\\ 0.5977\\ 0.4410\\ 0.2546\end{Bmatrix}+\frac{EA\delta}{L}\begin{Bmatrix}-0.3108\\ 0.3108\\ 0\\ -0.3708\\ -0.1035\\ -0.1792\\ 0.4143\\ 0.2392\end{Bmatrix}$$

Member Forces

$$\mathbf{F_*^i}=\mathbf{F_{*f}^i}+\mathbf{k_*^i}\mathbf{T^i}\mathbf{D}$$

For example, considering element '1',

$$\mathbf{F_*^1}=\begin{Bmatrix}F_{1*}^1\\ F_{2*}^1\\ \hline F_{3*}^1\\ F_{4*}^1\end{Bmatrix}=\frac{EA\delta}{L}\begin{Bmatrix}-0.70711\\ 0\\ \hline 0.70711\\ 0\end{Bmatrix}+\frac{EA}{L}\left[\begin{array}{cc|cc}-0.5 & 0.5 & 0.5 & -0.5\\ 0 & 0 & 0 & 0\\ \hline 0.5 & -0.5 & -0.5 & 0.5\\ 0 & 0 & 0 & 0\end{array}\right]\begin{matrix}(1)\\ (2)\\ (3)\\ (4)\end{matrix}\times\begin{Bmatrix}-0.9671\dfrac{PL}{EA}-0.1643\,\delta\\ -0.3618\dfrac{PL}{EA}+0.3708\,\delta\\ \hline 0\\ 0\end{Bmatrix}$$

$$\Rightarrow \mathbf{F}_*^1 = \begin{Bmatrix} F_{1*}^1 \\ F_{2*}^1 \\ \hdashline F_{3*}^1 \\ F_{4*}^1 \end{Bmatrix} = P\begin{Bmatrix} 0.3026 \\ 0 \\ \hdashline -0.3026 \\ 0 \end{Bmatrix} + \frac{EA\delta}{L}\begin{Bmatrix} -0.4395 \\ 0 \\ \hdashline 0.4395 \\ 0 \end{Bmatrix} \quad \Rightarrow \quad N_1 = F_{3*}^1 = -0.3026P + 0.4395\frac{EA\delta}{L}$$

We can simplify the calculations to be shown here for convenience, by observing that, for every element, there is actually only one unknown member force, namely, the axial force, given by $N_i = F_{3*}^i$, with $F_{1*}^i = -F_{3*}^i$ and $F_{2*}^i = F_{4*}^i = 0$ in all cases. Also, in the displacement vector, only D_1 and D_2 have non-zero values, as $\mathbf{D_R} = \mathbf{O}$. Thus, we need to only pick up the first two elements in the third row of the $\mathbf{k^i T^i}$ matrices derived earlier. Thus,

$$N_1 = F_{3*}^1 = \left(0.7071\frac{EA\delta}{L}\right) + \frac{EA}{L}[0.5 \quad -0.5]\begin{Bmatrix} -0.9671\dfrac{PL}{EA} - 0.1643\,\delta \\ -0.3618\dfrac{PL}{EA} + 0.3708\,\delta \end{Bmatrix} = -0.3026P + 0.4395\frac{EA\delta}{L}$$

$$N_2 = F_{3*}^2 = (0) + \frac{EA}{L}[0 \quad -1]\begin{Bmatrix} -0.9671\dfrac{PL}{EA} - 0.1643\,\delta \\ -0.3618\dfrac{PL}{EA} + 0.3708\,\delta \end{Bmatrix} = 0.3168P - 0.3708\frac{EA\delta}{L}$$

$$N_3 = F_{3*}^3 = (0) + \frac{EA}{L}[-0.433 \quad -0.75]\begin{Bmatrix} -0.9671\dfrac{PL}{EA} - 0.1643\,\delta \\ -0.3618\dfrac{PL}{EA} + 0.3708\,\delta \end{Bmatrix} = 0.6901P - 0.2070\frac{EA\delta}{L}$$

$$N_4 = F_{3*}^4 = \left(0.5\frac{EA\delta}{L}\right) + \frac{EA}{L}[-0.433 \quad -0.25]\begin{Bmatrix} -0.9671\dfrac{PL}{EA} - 0.1643\,\delta \\ -0.3618\dfrac{PL}{EA} + 0.3708\,\delta \end{Bmatrix} = 0.5092P + 0.4784\frac{EA\delta}{L}$$

The axial forces and support reactions are shown in Fig. 4.31, separately for the direct loading and indirect loading cases. It may be noted that, as expected, under the 'lack of fit' loading, elements '1' and '4' are subject to tension (as these bars, originally short in length, need to be extended in the 'force fit' condition, which pulls the joint O upward). Consequently, the other two bars ('2' and '3') are forced to reduce in length, and are hence subject to compression. We can easily verify that the forces shown in Fig. 4.31 satisfy equilibrium.

We can now easily find the various bar elongations, $e_i = \dfrac{N_i}{k_i}$, as follows:

$$e_1 = \frac{N_1}{0.7071EA/L} = -0.4279\frac{PL}{EA} + 0.6215\frac{EA\delta}{L};$$
$$e_2 = \frac{N_2}{0.5EA/L} = 0.6336\frac{PL}{EA} + 0.7416\frac{EA\delta}{L};$$
$$e_3 = \frac{N_3}{0.5EA/L} = 1.3802\frac{PL}{EA} + 0.4140\frac{EA\delta}{L};$$
$$e_4 = \frac{N_4}{1.1547EA/L} = 0.4410\frac{PL}{EA} + 0.4143\frac{EA\delta}{L}$$

Alternatively, the bar elongations can be determined from the member-end displacements given by $\mathbf{D}^i_* = \mathbf{T}^i\mathbf{D}$, with $e_i = D^i_{3*} - D^i_{1*}$. [The reader may verify this].

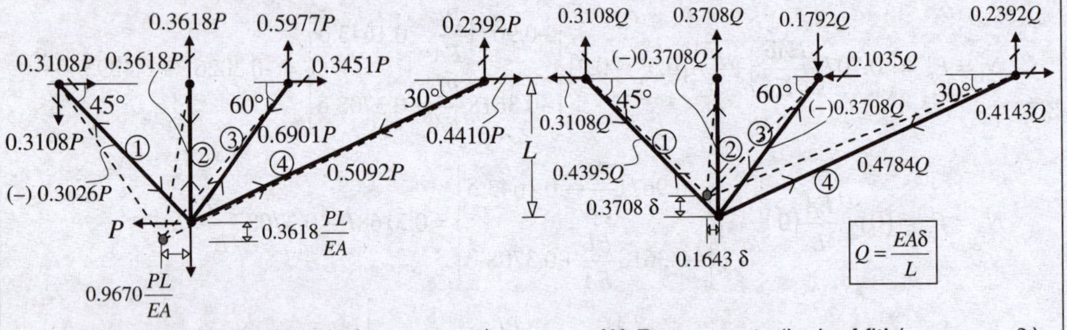

(*a*) Response to nodal loads ($F_1 = F_2 = -P$) (*b*) Response to 'lack of fit' ($e_{o1} = e_{o4} = -\delta$)

Figure 4.31 Response to direct and indirect loading (Example 4.9)

Alternative Procedure Using Displacement Transformation Matrix

The element displacement transformation matrix, $\mathbf{T}^i_{\mathbf{D}} = \left[\mathbf{T}^i_{\mathbf{DA}} \mid \mathbf{T}^i_{\mathbf{DR}}\right]$, which is used in the transformation, $\mathbf{D}^i_* = \mathbf{T}^i_{\mathbf{D}}\mathbf{D}$ and $\mathbf{F} = \mathbf{T}^{i\,T}_{\mathbf{D}}\mathbf{F}^i_*$, has an order 4×10, mostly comprising zero elements. The non-zero elements consist of $\cos\theta^i_x$ and $\sin\theta^i_x$ terms, similar to the components of the $\mathbf{T}^i$ matrix (refer Fig. 4.5), having a typical block of $\begin{bmatrix}\cos\theta^i_x & \sin\theta^i_x \\ \sin\theta^i_x & \cos\theta^i_x\end{bmatrix}$. This block is applicable at the two *start node* coordinates (corresponding to $\mathbf{T}^i_{\mathbf{DA}}$), which are common for all 4 elements and also the *end node* coordinates (corresponding to $\mathbf{T}^i_{\mathbf{DR}}$), which are different for the 4 elements. Referring to Fig. 4.30a,

$$\mathbf{T}^1_{\mathbf{D}} = \left[\mathbf{T}^1_{\mathbf{DA}} \mid \mathbf{T}^1_{\mathbf{DR}}\right] = \left[\begin{array}{cc|cccccccc} -0.7071 & 0.7071 & 0 & 0 & 0 & 0 & 0 & 0 & 0 & 0 \\ -0.7071 & -0.7071 & 0 & 0 & 0 & 0 & 0 & 0 & 0 & 0 \\ \hline 0 & 0 & -0.7071 & 0.7071 & 0 & 0 & 0 & 0 & 0 & 0 \\ 0 & 0 & -0.7071 & -0.7071 & 0 & 0 & 0 & 0 & 0 & 0 \end{array}\right]$$

$$\mathbf{T_D^2} = \left[\mathbf{T_{DA}^2} \;\vdots\; \mathbf{T_{DR}^2}\right] = \left[\begin{array}{cc|cccccccc} 0 & 1 & 0 & 0 & 0 & 0 & 0 & 0 & 0 & 0 \\ -1 & 0 & 0 & 0 & 0 & 0 & 0 & 0 & 0 & 0 \\ \hline 0 & 0 & 0 & 0 & 0 & 1 & 0 & 0 & 0 & 0 \\ 0 & 0 & 0 & 0 & -1 & 0 & 0 & 0 & 0 & 0 \end{array}\right]$$

$$\mathbf{T_D^3} = \left[\mathbf{T_{DA}^3} \;\vdots\; \mathbf{T_{DR}^3}\right] = \left[\begin{array}{cc|cccccccc} 0.5 & 0.8660 & 0 & 0 & 0 & 0 & 0 & 0 & 0 & 0 \\ -0.8660 & 0.5 & 0 & 0 & 0 & 0 & 0 & 0 & 0 & 0 \\ \hline 0 & 0 & 0 & 0 & 0 & 0 & 0.5 & 0.8660 & 0 & 0 \\ 0 & 0 & 0 & 0 & 0 & 0 & -0.8660 & 0.5 & 0 & 0 \end{array}\right]$$

$$\mathbf{T_D^4} = \left[\mathbf{T_{DA}^4} \;\vdots\; \mathbf{T_{DR}^4}\right] = \left[\begin{array}{cc|cccccccc} 0.8660 & 0.5 & 0 & 0 & 0 & 0 & 0 & 0 & 0 & 0 \\ -0.5 & 0.8660 & 0 & 0 & 0 & 0 & 0 & 0 & 0 & 0 \\ \hline 0 & 0 & 0 & 0 & 0 & 0 & 0 & 0 & 0.8660 & 0.5 \\ 0 & 0 & 0 & 0 & 0 & 0 & 0 & 0 & -0.5 & 0.8660 \end{array}\right]$$

The element stiffness matrix $\mathbf{k_*^i}$ is the same as computed earlier, whereby:

$$\mathbf{k_*^1 T_D^1} = \left[\begin{array}{cc|cccccccc} -0.5 & 0.5 & 0.5 & -0.5 & 0 & 0 & 0 & 0 & 0 & 0 \\ 0 & 0 & 0 & 0 & 0 & 0 & 0 & 0 & 0 & 0 \\ \hline 0.5 & -0.5 & -0.5 & 0.5 & 0 & 0 & 0 & 0 & 0 & 0 \\ 0 & 0 & 0 & 0 & 0 & 0 & 0 & 0 & 0 & 0 \end{array}\right]$$

$$\mathbf{k_*^2 T_D^2} = \left[\begin{array}{cc|cccccccc} 0 & 1 & 0 & 0 & 0 & -1 & 0 & 0 & 0 & 0 \\ 0 & 0 & 0 & 0 & 0 & 0 & 0 & 0 & 0 & 0 \\ \hline 0 & -1 & 0 & 0 & 0 & 1 & 0 & 0 & 0 & 0 \\ 0 & 0 & 0 & 0 & 0 & 0 & 0 & 0 & 0 & 0 \end{array}\right]$$

$$\mathbf{k_*^3 T_D^3} = \left[\begin{array}{cc|cccccccc} 0.433 & 0.75 & 0 & 0 & 0 & 0 & -0.433 & -0.75 & 0 & 0 \\ 0 & 0 & 0 & 0 & 0 & 0 & 0 & 0 & 0 & 0 \\ \hline -0.433 & -0.75 & 0 & 0 & 0 & 0 & 0.433 & 0.75 & 0 & 0 \\ 0 & 0 & 0 & 0 & 0 & 0 & 0 & 0 & 0 & 0 \end{array}\right]$$

$$\mathbf{k_*^4 T_D^4} = \left[\begin{array}{cc|cccccccc} 0.433 & 0.25 & 0 & 0 & 0 & 0 & 0 & 0 & -0.433 & -0.25 \\ 0 & 0 & 0 & 0 & 0 & 0 & 0 & 0 & 0 & 0 \\ \hline -0.433 & -0.25 & 0 & 0 & 0 & 0 & 0 & 0 & 0.433 & 0.25 \\ 0 & 0 & 0 & 0 & 0 & 0 & 0 & 0 & 0 & 0 \end{array}\right]$$

The structure stiffness matrix $\mathbf{k}$ can now be assembled by the *direct stiffness* approach, by directly summing up the contributions from all elements, $\mathbf{k} = \sum_{i=1}^{4} \mathbf{T_D^i}^{\mathbf{T}} \mathbf{k_*^i} \mathbf{T_D^i}$. Carrying out this operation, we get exactly the same $\mathbf{k}$ matrix, whose sub-matrices we had computed earlier.

The transformation of the element fixed end forces $\mathbf{F}_{*f}^{i}$ (due to 'lack of fit', calculated earlier) to global coordinates can be done by summing up the contributions from the various elements: $\mathbf{F}_f = \sum_{i=1}^{4} \mathbf{T}_D^{i\,T}\mathbf{F}_{*f}^{i} = \begin{bmatrix} \mathbf{F}_{fA} \\ \hline \mathbf{F}_{fR} \end{bmatrix}$. Only elements '1' and '4' have non-zero contributions:

$$\mathbf{T}_D^{1T}\mathbf{F}_{*f}^{1} = \left[\begin{array}{cc|cc} -0.7071 & -0.7071 & 0 & 0 \\ 0.7071 & -0.7071 & 0 & 0 \\ \hline 0 & 0 & -0.7071 & -0.7071 \\ 0 & 0 & 0.7071 & -0.7071 \\ 0 & 0 & 0 & 0 \\ 0 & 0 & 0 & 0 \\ 0 & 0 & 0 & 0 \\ 0 & 0 & 0 & 0 \\ 0 & 0 & 0 & 0 \\ 0 & 0 & 0 & 0 \end{array}\right] \frac{EA\delta}{L}\begin{Bmatrix} -0.7071 \\ 0 \\ \hline 0.7071 \\ 0 \end{Bmatrix} = \frac{EA\delta}{L}\begin{Bmatrix} 0.5 \\ -0.5 \\ \hline -0.5 \\ 0.5 \\ 0 \\ 0 \\ 0 \\ 0 \\ 0 \\ 0 \end{Bmatrix}$$

$$\mathbf{T}_D^{4T}\mathbf{F}_{*f}^{4} = \left[\begin{array}{cc|cc} 0.866 & -0.5 & 0 & 0 \\ 0.5 & 0.866 & 0 & 0 \\ \hline 0 & 0 & 0 & 0 \\ 0 & 0 & 0 & 0 \\ 0 & 0 & 0 & 0 \\ 0 & 0 & 0 & 0 \\ 0 & 0 & 0 & 0 \\ 0 & 0 & 0 & 0 \\ 0 & 0 & 0.866 & -0.5 \\ 0 & 0 & 0.5 & 0.866 \end{array}\right] \frac{EA\delta}{L}\begin{Bmatrix} -0.5 \\ 0 \\ \hline 0.5 \\ 0 \end{Bmatrix} = \frac{EA\delta}{L}\begin{Bmatrix} -0.433 \\ -0.25 \\ \hline 0 \\ 0 \\ 0 \\ 0 \\ 0 \\ 0 \\ 0.433 \\ 0.25 \end{Bmatrix}$$

$$\Rightarrow \mathbf{F}_f = \begin{bmatrix} \mathbf{F}_{fA} \\ \hline \mathbf{F}_{fR} \end{bmatrix} = \begin{Bmatrix} F_{1f} \\ F_{2f} \\ \hline F_{3f} \\ F_{4f} \\ F_{5f} \\ F_{6f} \\ F_{7f} \\ F_{8f} \\ F_{9f} \\ F_{10f} \end{Bmatrix} = \frac{EA\delta}{L}\begin{Bmatrix} 0.067 \\ -0.75 \\ \hline -0.5 \\ 0.5 \\ 0 \\ 0 \\ 0 \\ 0 \\ 0.433 \\ 0.25 \end{Bmatrix} \qquad \Rightarrow \mathbf{F}_A - \mathbf{F}_{fA} = \begin{bmatrix} -P - 0.067\dfrac{EA\delta}{L} \\ -P + 0.75\dfrac{EA\delta}{L} \end{bmatrix}$$

Solving the same stiffness equations, $\begin{bmatrix} \mathbf{F_A} \\ \hline \mathbf{F_R} \end{bmatrix} - \begin{bmatrix} \mathbf{F_{fA}} \\ \hline \mathbf{F_{fR}} \end{bmatrix} = \begin{bmatrix} \mathbf{k_{AA}} & \mathbf{k_{AR}} \\ \hline \mathbf{k_{RA}} & \mathbf{k_{RR}} \end{bmatrix} \begin{bmatrix} \mathbf{D_A} \\ \hline \mathbf{D_R} = \mathbf{O} \end{bmatrix}$, as before, we get exactly the same solutions for the unknown displacements $\mathbf{D_A}$ and support reactions $\mathbf{F_R}$. Finally, we can calculate the member forces applying $\mathbf{F_*^i} = \mathbf{F_{*f}^i} + \left(\mathbf{k_*^i T_D^i}\right)\mathbf{D}$, and get the same results. The axial forces are given by $N_i = F_{3*}^i$, and the bar elongations as $e_i = \dfrac{N_i}{k_i}$ or $e_i = D_{3*}^i - D_{1*}^i$, as indicated earlier.

EXAMPLE 4.10

Analyse, by the Stiffness Method, the simply supported truss shown in Fig. 4.32, and find the joint displacements, support reactions, bar forces and bar elongations. Assume all bars to have the same axial rigidity $EA = 6000$ kN.

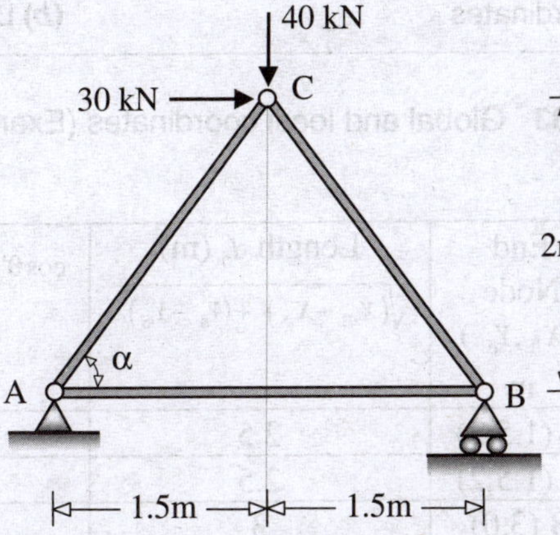

Figure 4.32 Simply supported truss (Example 4.10)

SOLUTION

Input Data

There are 3 prismatic elements here ($m = 3$), with 3 pin joints, i.e., $2j = 6$ global coordinates, of which 3 correspond to *active* degrees of freedom ($n = 3$) and the rest to *restrained* global coordinates ($r = 3$), as shown in Fig. 4.33a.

The local coordinate system for a typical element, by the reduced stiffness formulation, is as shown in Fig. 4.33b. Choosing the origin of the *x-y* global axes system at the node A, the coordinates of the start and end nodes (X_{iS}, Y_{iS} and X_{iE}, Y_{iE}) of individual elements and their lengths, $L_i = \sqrt{(X_{iE} - X_{iS})^2 + (Y_{iE} - Y_{iS})^2}$, and direction cosine parameters, $\cos\theta_x^i = \dfrac{X_{iE} - X_{iS}}{L_i}$ and $\sin\theta_x^i = \dfrac{Y_{iE} - Y_{iS}}{L_i}$, are tabulated below.

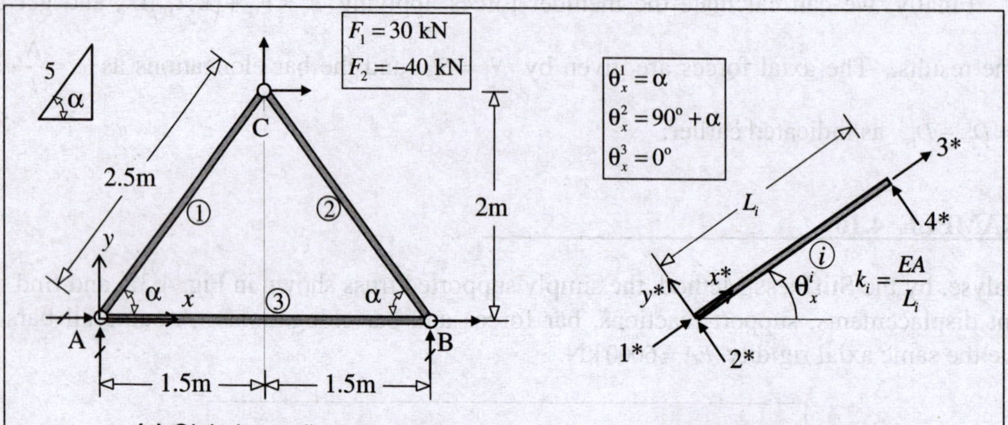

(*a*) Global coordinates (*b*) Local coordinates

Figure 4.33 Global and local coordinates (Example 4.10)

Element No.	Start Node (X_{iS}, Y_{iS}) m	End Node (X_{iE}, Y_{iE}) m	Length L_i (m) $\sqrt{(X_{iE} - X_{iS})^2 + (Y_{iE} - Y_{iS})^2}$	$\cos\theta_x^i = \dfrac{X_{iE} - X_{iS}}{L_i}$	$\sin\theta_x^i = \dfrac{Y_{iE} - Y_{iS}}{L_i}$
1	A (0,0)	C (1.5,2)	2.5	0.6	0.8
2	B (3,0)	C (1.5,2)	2.5	–0.6	0.8
3	A (0,0)	B (3,0)	3	1	0

The element stiffnesses $k_i = \dfrac{EA}{L_i}$ are given by: $k_1 = k_2 = \dfrac{6000}{2.5}$=2400 kN/m and $k_3 = \dfrac{6000}{3} = 2000$ kN/m .

The loads are specified as $\mathbf{F_A} = \begin{Bmatrix} F_1 \\ F_2 \\ F_3 \end{Bmatrix} = \begin{Bmatrix} 30 \\ -40 \\ 0 \end{Bmatrix}$ kN.

Element Transformation Matrices

The element transformation matrix $\mathbf{T^i}$, satisfying $\mathbf{D_*^i} = \mathbf{T^i D}$ and $\mathbf{F_*^i} = \mathbf{T^i F}$, takes the following form (refer Figs 4.33a and b), with the linking global coordinates indicated in parentheses.

$$\begin{bmatrix} D_{1*}^{i} \\ D_{2*}^{i} \\ \hline D_{3*}^{i} \\ D_{4*}^{i} \end{bmatrix} = \left[\begin{array}{cc|cc} \cos\theta_x^i & \sin\theta_x^i & 0 & 0 \\ -\sin\theta_x^i & \cos\theta_x^i & 0 & 0 \\ \hline 0 & 0 & \cos\theta_x^i & \sin\theta_x^i \\ 0 & 0 & -\sin\theta_x^i & \cos\theta_x^i \end{array}\right] \begin{bmatrix} D_{1}^{i} \\ D_{2}^{i} \\ \hline D_{3}^{i} \\ D_{4}^{i} \end{bmatrix} \Rightarrow \mathbf{T}^1 = \overset{\quad (5)\quad (6)\quad (1)\quad (2)}{\left[\begin{array}{cc|cc} 0.6 & 0.8 & 0 & 0 \\ -0.8 & 0.6 & 0 & 0 \\ \hline 0 & 0 & 0.6 & 0.8 \\ 0 & 0 & -0.8 & 0.6 \end{array}\right]} \begin{matrix} (5) \\ (6) \\ (1) \\ (2) \end{matrix}$$

$$\mathbf{T}^2 = \overset{\quad (3)\quad (4)\quad (1)\quad (2)}{\left[\begin{array}{cc|cc} -0.6 & 0.8 & 0 & 0 \\ -0.8 & -0.6 & 0 & 0 \\ \hline 0 & 0 & -0.6 & 0.8 \\ 0 & 0 & -0.8 & -0.6 \end{array}\right]} \begin{matrix} (3) \\ (4) \\ (1) \\ (2) \end{matrix} \qquad \mathbf{T}^3 = \overset{(5)\,(6)\,(3)\,(4)}{\left[\begin{array}{cc|cc} 1 & 0 & 0 & 0 \\ 0 & 1 & 0 & 0 \\ \hline 0 & 0 & 1 & 0 \\ 0 & 0 & 0 & 1 \end{array}\right]} \begin{matrix} (5) \\ (6) \\ (3) \\ (4) \end{matrix}$$

Element Stiffness Matrices

The element stiffness matrices $\mathbf{k}_*^i$ (in local coordinates), satisfying $\mathbf{F}_*^i = \mathbf{k}_*^i \mathbf{D}_*^i$, and the products $\mathbf{k}_*^i \mathbf{T}^i$ satisfying $\mathbf{F}_*^i = \left(\mathbf{k}_*^i \mathbf{T}^i\right)\mathbf{D}$, take the following forms (refer Figs 4.33a and b):

$$\mathbf{k}_*^i = k_i \left[\begin{array}{cc|cc} 1 & 0 & -1 & 0 \\ 0 & 0 & 0 & 0 \\ \hline -1 & 0 & 1 & 0 \\ 0 & 0 & 0 & 0 \end{array}\right] \Rightarrow \mathbf{k}_*^1 = \mathbf{k}_*^2 = \left[\begin{array}{cc|cc} 2400 & 0 & -2400 & 0 \\ 0 & 0 & 0 & 0 \\ \hline -2400 & 0 & 2400 & 0 \\ 0 & 0 & 0 & 0 \end{array}\right] \text{kN/m}$$

$$\mathbf{k}_*^3 = \left[\begin{array}{cc|cc} 2000 & 0 & -2000 & 0 \\ 0 & 0 & 0 & 0 \\ \hline -2000 & 0 & 2000 & 0 \\ 0 & 0 & 0 & 0 \end{array}\right] \text{kN/m} \Rightarrow \mathbf{k}_*^1 \mathbf{T}^1 = \overset{\quad (5)\qquad (6)\qquad (1)\qquad (2)}{\left[\begin{array}{cc|cc} 1440 & 1920 & -1440 & -1920 \\ 0 & 0 & 0 & 0 \\ \hline -1440 & -1920 & 1440 & 1920 \\ 0 & 0 & 0 & 0 \end{array}\right]} \begin{matrix} (5) \\ (6) \\ (1) \\ (2) \end{matrix} \text{kN/m}$$

$$\mathbf{k}_*^2 \mathbf{T}^2 = \overset{\quad (3)\qquad (4)\qquad (1)\qquad (2)}{\left[\begin{array}{cc|cc} -1440 & 1920 & 1440 & -1920 \\ 0 & 0 & 0 & 0 \\ \hline 1440 & -1920 & -1440 & 1920 \\ 0 & 0 & 0 & 0 \end{array}\right]} \begin{matrix} (3) \\ (4) \\ (1) \\ (2) \end{matrix} \text{kN/m} \quad \mathbf{k}_*^3 \mathbf{T}^3 = \overset{\quad (5)\quad (6)\quad (3)\quad (4)}{\left[\begin{array}{cc|cc} 2000 & 0 & -2000 & 0 \\ 0 & 0 & 0 & 0 \\ \hline -2000 & 0 & 2000 & 0 \\ 0 & 0 & 0 & 0 \end{array}\right]} \begin{matrix} (5) \\ (6) \\ (3) \\ (4) \end{matrix} \text{kN/m}$$

Structure Stiffness Matrix

The structure stiffness matrix $\mathbf{k} = \left[\begin{array}{c|c} \mathbf{k}_{AA} & \mathbf{k}_{AR} \\ \hline \mathbf{k}_{RA} & \mathbf{k}_{RR} \end{array}\right]$, satisfying $\mathbf{F} = \mathbf{kD}$, can be assembled, by summing up the contributions of $\mathbf{k}^i = \mathbf{T}^{i^T} \mathbf{k}_*^i \mathbf{T}^i$ at the appropriate coordinate locations:

$$\mathbf{k}^1 = \begin{array}{cccc} (5) & (6) & (1) & (2) \\ \left[\begin{array}{cc|cc} 864 & 1152 & -864 & -1152 \\ 1152 & 1536 & -1152 & -1536 \\ \hline -864 & -1152 & 864 & 1152 \\ -1152 & -1536 & 1152 & 1536 \end{array}\right] & \begin{array}{c} (5) \\ (6) \\ (1) \\ (2) \end{array} \end{array} \text{kN/m}$$

$$\mathbf{k}^2 = \begin{array}{cccc} (3) & (4) & (1) & (2) \\ \left[\begin{array}{cc|cc} 864 & -1152 & -864 & 1152 \\ -1152 & 1536 & 1152 & -1536 \\ \hline -864 & 1152 & 864 & -1152 \\ 1152 & -1536 & -1152 & 1536 \end{array}\right] & \begin{array}{c} (3) \\ (4) \\ (1) \\ (2) \end{array} \end{array} \text{kN/m} \qquad \mathbf{k}^3 = \begin{array}{cccc} (5) & (6) & (3) & (4) \\ \left[\begin{array}{cc|cc} 2000 & 0 & -2000 & 0 \\ 0 & 0 & 0 & 0 \\ \hline -2000 & 0 & 2000 & 0 \\ 0 & 0 & 0 & 0 \end{array}\right] & \begin{array}{c} (5) \\ (6) \\ (3) \\ (4) \end{array} \end{array} \text{kN/m}$$

Summing up the coefficients (in kN/m units) in the lower triangular portion of the symmetric **k** matrix,

$$k_{11} = (864 + 864) = 1728;$$

$$k_{21} = (1152 - 1152) = 0; \quad k_{22} = (1536 + 1536) = 3072;$$

$$k_{31} = -864; \quad k_{32} = 1152; \quad k_{33} = (864 + 2000) = 2864;$$

$$k_{41} = 1152; \quad k_{42} = -1536; \quad k_{43} = (-1152 + 0) = -1152; \quad k_{44} = (1536 + 0) = 1536;$$

$$k_{51} = -864; \quad k_{52} = -1152; \quad k_{53} = -2000; \quad k_{54} = 0; \quad k_{55} = (864 + 2000) = 2864;$$

$$k_{61} = -1152; \quad k_{62} = -1536; \quad k_{63} = 0; \quad k_{64} = 0; \quad k_{65} = (1152 + 0) = 1152; \quad k_{66} = (1536 + 0) = 1536$$

$$\Rightarrow \mathbf{k}_{AA} = \begin{array}{ccc} (1) & (2) & (3) \\ \left[\begin{array}{ccc} 1728 & 0 & -864 \\ 0 & 3072 & 1152 \\ -864 & 1152 & 2864 \end{array}\right] \end{array} \text{kN/m} \quad \Rightarrow \quad [\mathbf{k}_{AA}]^{-1} = \left[\begin{array}{ccc} 0.70370 & -0.09375 & 0.2500 \\ -0.09375 & 0.39583 & -0.18750 \\ 0.2500 & -0.18750 & 0.5000 \end{array}\right] \times 10^{-3} \text{m/kN}$$

$$[\mathbf{k}_{RA} \mid \mathbf{k}_{RR}] = \begin{array}{cccccc} (1) & (2) & (3) & (4) & (5) & (6) \\ \left[\begin{array}{ccc|ccc} 1152 & -1536 & -1152 & 1536 & 0 & 0 \\ -864 & -1152 & -2000 & 0 & 2864 & 1152 \\ -1152 & -1536 & 0 & 0 & 1152 & 1536 \end{array}\right] & \begin{array}{c} (4) \\ (5) \\ (6) \end{array} \end{array} \text{kN/m}$$

$$\mathbf{k}_{AR} = \mathbf{k}_{RA}{}^T = \begin{array}{ccc} (4) & (5) & (6) \\ \left[\begin{array}{ccc} 1152 & -864 & -1152 \\ -1536 & -1152 & -1536 \\ -1152 & -2000 & 0 \end{array}\right] & \begin{array}{c} (1) \\ (2) \\ (3) \end{array} \end{array} \text{kN/m}$$

Displacements and Support Reactions

The structure stiffness relation to be considered here is

$\begin{bmatrix} \mathbf{F_A} \\ \hline \mathbf{F_R} \end{bmatrix} - \begin{bmatrix} \mathbf{F_{fA}} = \mathbf{O} \\ \hline \mathbf{F_{fR}} = \mathbf{O} \end{bmatrix} = \begin{bmatrix} \mathbf{k_{AA}} & \mathbf{k_{AR}} \\ \hline \mathbf{k_{AR}} & \mathbf{k_{RR}} \end{bmatrix} \begin{bmatrix} \mathbf{D_A} \\ \hline \mathbf{D_R} = \mathbf{O} \end{bmatrix}$, from which the unknown joint displacements $\mathbf{D_A}$ and support reactions $\mathbf{F_R}$ can be computed as follows:

$$\mathbf{D_A} = [\mathbf{k_{AA}}]^{-1}\mathbf{F_A}$$

$$\Rightarrow \begin{Bmatrix} D_1 \\ D_2 \\ D_3 \end{Bmatrix} = \begin{bmatrix} 0.70370 & -0.09375 & 0.2500 \\ -0.09375 & 0.39583 & -0.18750 \\ 0.2500 & -0.18750 & 0.5000 \end{bmatrix} \times 10^{-3} \begin{Bmatrix} 30 \\ -40 \\ 0 \end{Bmatrix} = \begin{Bmatrix} 0.024861 \\ -0.018646 \\ 0.015000 \end{Bmatrix} \text{m} = \begin{Bmatrix} 24.86 \\ -18.65 \\ 15.00 \end{Bmatrix} \text{mm}$$

$$\mathbf{F_R} = \mathbf{k_{RA}}\mathbf{D_A}$$

$$\Rightarrow \begin{Bmatrix} F_4 \\ F_5 \\ F_6 \end{Bmatrix} = \begin{bmatrix} 1152 & -1536 & -1152 \\ -864 & -1152 & -2000 \\ -1152 & -1536 & 0 \end{bmatrix} \begin{Bmatrix} 0.024861 \\ -0.018646 \\ 0.015000 \end{Bmatrix} = \begin{Bmatrix} 40.0 \\ -30.0 \\ 0 \end{Bmatrix} \text{kN}$$

Member Forces

$$\mathbf{F_*^i} = \mathbf{F_{*f}^i} + \mathbf{k_*^i T^i D} = \mathbf{k_*^i T^i D}$$

$$\mathbf{F_*^1} = \begin{Bmatrix} F_{1*}^1 \\ F_{2*}^1 \\ F_{3*}^1 \\ F_{4*}^1 \end{Bmatrix} = \begin{bmatrix} 1440 & 1920 & -1440 & -1920 \\ 0 & 0 & 0 & 0 \\ -1440 & -1920 & 1440 & 1920 \\ 0 & 0 & 0 & 0 \end{bmatrix} \begin{matrix} (5) \\ (6) \\ (1) \\ (2) \end{matrix} \times \begin{Bmatrix} D_5 = 0 \\ D_6 = 0 \\ D_1 = 0.024861 \\ D_2 = -0.018646 \end{Bmatrix} = \begin{Bmatrix} 0 \\ 0 \\ 0 \\ 0 \end{Bmatrix} \text{kN}$$

$$\mathbf{F_*^2} = \begin{Bmatrix} F_{1*}^2 \\ F_{2*}^2 \\ F_{3*}^2 \\ F_{4*}^2 \end{Bmatrix} = \begin{bmatrix} -1440 & 1920 & 1440 & -1920 \\ 0 & 0 & 0 & 0 \\ 1440 & -1920 & -1440 & 1920 \\ 0 & 0 & 0 & 0 \end{bmatrix} \begin{matrix} (3) \\ (4) \\ (1) \\ (2) \end{matrix} \times \begin{Bmatrix} D_3 = 0.0150 \\ D_4 = 0 \\ D_1 = 0.024861 \\ D_2 = -0.018646 \end{Bmatrix} = \begin{Bmatrix} 50.0 \\ 0 \\ -50.0 \\ 0 \end{Bmatrix} \text{kN}$$

$$\mathbf{F_*^3} = \begin{Bmatrix} F_{1*}^3 \\ F_{2*}^3 \\ F_{3*}^3 \\ F_{4*}^3 \end{Bmatrix} = \begin{bmatrix} 2000 & 0 & -2000 & 0 \\ 0 & 0 & 0 & 0 \\ -2000 & 0 & 2000 & 0 \\ 0 & 0 & 0 & 0 \end{bmatrix} \begin{matrix} (5) \\ (6) \\ (3) \\ (4) \end{matrix} \times \begin{Bmatrix} D_5 = 0 \\ D_6 = 0 \\ D_3 = 0.015 \\ D_4 = 0 \end{Bmatrix} = \begin{Bmatrix} -30.0 \\ 0 \\ 30.0 \\ 0 \end{Bmatrix} \text{kN}$$

$$\Rightarrow N_1 = F_{3*}^1 = 0 \text{ kN}; \; N_2 = F_{3*}^2 = -50.0 \text{ kN}; \; N_3 = F_{3*}^3 = 30.0 \text{ kN}$$

$$\Rightarrow e_1 = \frac{N_1}{k_1} = 0 \text{ mm}; \; e_2 = \frac{N_2}{k_2} = \frac{-50.0}{2400} = 0.02083 \text{ m} = 20.83 \text{ mm};$$

$$e_3 = \frac{N_3}{k_3} = \frac{30.0}{2000} = 0.0150 \text{ m} = 15.00 \text{ mm}$$

Alternatively, the bar elongations can be determined from the member-end displacements given by $\mathbf{D_*^i} = \mathbf{T^i D}$, with $e_i = D_{3*}^i - D_{1*}^i$.

The complete response of the truss to the given loading is shown in Fig. 4.34. As the truss is statically determinate, we can easily verify (using statics) that the support reactions and bar forces are indeed the exact solutions, satisfying equilibrium.

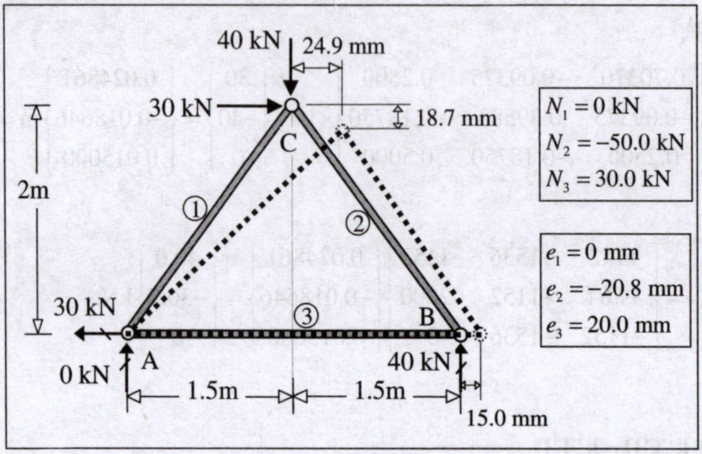

Figure 4.34 Response to loading on truss (Example 4.10)

The analysis may alternatively be done using the displacement transformation matrix, as shown in Example 4.9.

EXAMPLE 4.11

Consider the same truss shown in Fig. 4.32, but subject to arbitrary indirect loading instead of the specified direct loading. Making use of the transformation and stiffness matrices generated in Example 4.10, show that there will be no force response induced in the given 'just-rigid' truss on account of (a) support settlements and (b) temperature changes (or 'lack of fit').

SOLUTION

Indirect Loading Data

With reference to the coordinate system shown in Fig. 4.33, let us assume that the supports undergo arbitrary movements, $\mathbf{D_R} = \begin{Bmatrix} D_4 \\ D_5 \\ D_6 \end{Bmatrix}$ (in m), and that there are initial 'free' displacements in the bars, on account of either temperature rise or the bars being too long ('lack of fit'), given by $\mathbf{e_o} = \begin{Bmatrix} e_{o1} \\ e_{o2} \\ e_{o3} \end{Bmatrix}$ (in m). These initial element displacements generate fixed end forces, when applied to the restrained (primary structure), given by Eq. 4.34:

$$\mathbf{F^i_{*f}} = \begin{Bmatrix} F^i_{1*f} \\ F^i_{2*f} \\ \hline F^i_{3*f} \\ F^i_{4*f} \end{Bmatrix} = \begin{Bmatrix} k_i e_{oi} \\ 0 \\ \hline -k_i e_{oi} \\ 0 \end{Bmatrix} \Rightarrow \begin{Bmatrix} F^1_{1*f} \\ F^1_{2*f} \\ \hline F^1_{3*f} \\ F^1_{4*f} \end{Bmatrix} = \begin{Bmatrix} 2400e_{o1} \\ 0 \\ \hline -2400e_{o1} \\ 0 \end{Bmatrix}; \begin{Bmatrix} F^2_{1*f} \\ F^2_{2*f} \\ \hline F^2_{3*f} \\ F^2_{4*f} \end{Bmatrix} = \begin{Bmatrix} 2400e_{o2} \\ 0 \\ \hline -2400e_{o2} \\ 0 \end{Bmatrix}; \begin{Bmatrix} F^3_{1*f} \\ F^3_{2*f} \\ \hline F^3_{3*f} \\ F^3_{4*f} \end{Bmatrix} = \begin{Bmatrix} 2000e_{o3} \\ 0 \\ \hline -2000e_{o3} \\ 0 \end{Bmatrix}$$

Transforming these forces to the global axes system, by applying $\mathbf{F^i_f} = \mathbf{T^{i^T}} \mathbf{F^i_{*f}}$ (global coordinates marked in parenthesis), we obtain:

$$\begin{Bmatrix} F^1_{1f} \\ F^1_{2f} \\ \hline F^1_{3f} \\ F^1_{4f} \end{Bmatrix} = e_{o1} \begin{Bmatrix} 1440 \\ 1920 \\ \hline -1440 \\ -1920 \end{Bmatrix} \begin{matrix} (5) \\ (6) \\ (1) \\ (2) \end{matrix}; \quad \begin{Bmatrix} F^2_{1f} \\ F^2_{2f} \\ \hline F^2_{3f} \\ F^2_{4f} \end{Bmatrix} = e_{o2} \begin{Bmatrix} -1440 \\ 1920 \\ \hline 1440 \\ -1920 \end{Bmatrix} \begin{matrix} (3) \\ (4) \\ (1) \\ (2) \end{matrix}; \quad \begin{Bmatrix} F^3_{1f} \\ F^3_{2f} \\ \hline F^3_{3f} \\ F^3_{4f} \end{Bmatrix} = e_{o3} \begin{Bmatrix} 2000 \\ 0 \\ \hline -2000 \\ 0 \end{Bmatrix} \begin{matrix} (5) \\ (6) \\ (3) \\ (4) \end{matrix}$$

Summing up the contributions of the three elements,

$$\begin{Bmatrix} \mathbf{F_{fA}} \\ \hline \mathbf{F_{fR}} \end{Bmatrix} = \begin{Bmatrix} F_{1f} \\ F_{2f} \\ F_{3f} \\ F_{4f} \\ F_{5f} \\ F_{6f} \end{Bmatrix} = \begin{Bmatrix} -1440e_{o1} + 1440e_{o2} \\ -1920e_{o1} - 1920e_{o2} \\ -1440e_{o2} - 2000e_{o3} \\ \hline 1920e_{o2} \\ 1440e_{o1} + 2000e_{o3} \\ 1920e_{o1} \end{Bmatrix} = \begin{bmatrix} -1440 & 1440 & 0 \\ -1920 & -1920 & 0 \\ 0 & -1440 & -2000 \\ \hline 0 & 1920 & 0 \\ 1440 & 0 & 2000 \\ 1920 & 0 & 0 \end{bmatrix} \begin{Bmatrix} e_{o1} \\ e_{o2} \\ e_{o3} \end{Bmatrix}$$

Displacements

$$\begin{bmatrix} \mathbf{F_A} = \mathbf{O} \\ \hline \mathbf{F_R} \end{bmatrix} - \begin{bmatrix} \mathbf{F_{fA}} \\ \hline \mathbf{F_{fR}} \end{bmatrix} = \begin{bmatrix} \mathbf{k_{AA}} & \mathbf{k_{AR}} \\ \hline \mathbf{k_{RA}} & \mathbf{k_{RR}} \end{bmatrix} \begin{bmatrix} \mathbf{D_A} \\ \hline \mathbf{D_R} \end{bmatrix}$$

$$\mathbf{D_A} = [\mathbf{k_{AA}}]^{-1} (-\mathbf{F_{fA}} - \mathbf{k_{AR}} \mathbf{D_R})$$

$$\Rightarrow \begin{Bmatrix} D_1 \\ D_2 \\ D_3 \end{Bmatrix} = -\begin{bmatrix} 0.70370 & -0.09375 & 0.2500 \\ -0.09375 & 0.39583 & -0.18750 \\ 0.2500 & -0.18750 & 0.5000 \end{bmatrix} \times 10^{-3}$$

$$\times \left(\begin{bmatrix} -1440 & 1440 & 0 \\ -1920 & -1920 & 0 \\ 0 & -1440 & -2000 \end{bmatrix} \begin{Bmatrix} e_{o1} \\ e_{o2} \\ e_{o3} \end{Bmatrix} + \begin{bmatrix} 1152 & -864 & -1152 \\ -1536 & -1152 & -1536 \\ -1152 & -2000 & 0 \end{bmatrix} \begin{Bmatrix} D_4 \\ D_5 \\ D_6 \end{Bmatrix} \right)$$

$$\Rightarrow \begin{Bmatrix} D_1 \\ D_2 \\ D_3 \end{Bmatrix} = \begin{bmatrix} 0.8333 & -0.8333 & 0.5 \\ 0.625 & 0.625 & -0.375 \\ 0 & 0 & 1 \end{bmatrix} \begin{Bmatrix} e_{o1} \\ e_{o2} \\ e_{o3} \end{Bmatrix} + \begin{bmatrix} -0.6667 & 1 & 0.6667 \\ 0.5 & 0 & 0.5 \\ 0 & 1 & 0 \end{bmatrix} \begin{Bmatrix} D_4 \\ D_5 \\ D_6 \end{Bmatrix} \text{m}$$

Support Reactions

$$\mathbf{F_R} = \mathbf{F_{fR}} + \mathbf{k_{RA}D_A} + \mathbf{k_{RR}D_R}$$

$$\mathbf{k_{RA}D_A} = \begin{bmatrix} 1152 & -1536 & -1152 \\ -864 & -1152 & -2000 \\ -1152 & -1536 & 0 \end{bmatrix} \begin{Bmatrix} D_1 \\ D_2 \\ D_3 \end{Bmatrix}$$

$$= \begin{bmatrix} 0 & -1920 & 0 \\ -1440 & 0 & -2000 \\ -1920 & 0 & 0 \end{bmatrix} \begin{Bmatrix} e_{o1} \\ e_{o2} \\ e_{o3} \end{Bmatrix} + \begin{bmatrix} -1536 & 0 & 0 \\ 0 & -2864 & -1152 \\ 0 & -1152 & -1536 \end{bmatrix} \begin{Bmatrix} D_4 \\ D_5 \\ D_6 \end{Bmatrix}$$

$$\mathbf{k_{RR}D_R} = \begin{bmatrix} 1536 & 0 & 0 \\ 0 & 2864 & 1152 \\ 0 & 1152 & 1536 \end{bmatrix} \begin{Bmatrix} D_4 \\ D_5 \\ D_6 \end{Bmatrix} ; \quad \mathbf{F_{fR}} = \begin{bmatrix} 0 & 1920 & 0 \\ 1440 & 0 & 2000 \\ 1920 & 0 & 0 \end{bmatrix} \begin{Bmatrix} e_{o1} \\ e_{o2} \\ e_{o3} \end{Bmatrix}$$

$$\Rightarrow \quad \mathbf{F_R} = \begin{Bmatrix} F_4 \\ F_5 \\ F_6 \end{Bmatrix} = \begin{bmatrix} 0 & 0 & 0 \\ 0 & 0 & 0 \\ 0 & 0 & 0 \end{bmatrix} \begin{Bmatrix} e_{o1} \\ e_{o2} \\ e_{o3} \end{Bmatrix} + \begin{bmatrix} 0 & 0 & 0 \\ 0 & 0 & 0 \\ 0 & 0 & 0 \end{bmatrix} \begin{Bmatrix} D_4 \\ D_5 \\ D_6 \end{Bmatrix}$$

Thus, we have proved that the support reactions will always be equal to zero, regardless of the choice of the input displacement data related to the indirect loading.

Member Forces

$$\mathbf{F_*^i} = \mathbf{F_{*f}^i} + \mathbf{k_*^i T^i D}$$

$$\mathbf{F_*^1} = \begin{Bmatrix} F_{1*}^1 \\ F_{2*}^1 \\ F_{3*}^1 \\ F_{4*}^1 \end{Bmatrix} = e_{o1} \begin{Bmatrix} 2400 \\ 0 \\ -2400 \\ 0 \end{Bmatrix} + \begin{bmatrix} 1440 & 1920 & -1440 & -1920 \\ 0 & 0 & 0 & 0 \\ -1440 & -1920 & 1440 & 1920 \\ 0 & 0 & 0 & 0 \end{bmatrix} \begin{matrix} (5) \\ (6) \\ (1) \\ (2) \end{matrix} \times \begin{Bmatrix} D_5 \\ D_6 \\ D_1 \\ D_2 \end{Bmatrix}$$

Noting that $F_{2*}^1 = F_{4*}^1 = 0$ and $N_1 = F_{3*}^1 = -F_{1*}^1$,

$$N_1 = F_{3*}^1 = -2400e_{o1} - 1440D_5 - 1920D_6 + \begin{bmatrix} 1440 & 1920 \end{bmatrix} \begin{bmatrix} D_1 \\ D_2 \end{bmatrix}$$

Substituting

$$\begin{Bmatrix} D_1 \\ D_2 \end{Bmatrix} = \begin{bmatrix} 0.8333 & -0.8333 & 0.5 \\ 0.625 & 0.625 & -0.375 \end{bmatrix} \begin{Bmatrix} e_{o1} \\ e_{o2} \\ e_{o3} \end{Bmatrix} + \begin{bmatrix} -0.6667 & 1 & 0.6667 \\ 0.5 & 0 & 0.5 \end{bmatrix} \begin{Bmatrix} D_4 \\ D_5 \\ D_6 \end{Bmatrix},$$

$$N_1 = F_{3*}^1 = [-2400e_{o1} - 1440D_5 - 1920D_6] + [2400e_{o1} + 1440D_5 + 1920D_6] = 0\,\text{kN}$$

Similarly,

$$N_2 = F_{3*}^2 = 1440e_{o2} - 1920D_4 + [-1440 \quad 1920 \quad 1440]\begin{bmatrix} D_1 \\ D_2 \\ D_3 \end{bmatrix} = 0$$

$$N_3 = F_{3*}^3 = -2000e_{o3} - 2000D_5 + 2000D_3 = 0\,\text{kN}$$

Thus, it is seen that the entire force response (support reactions and member forces) will always be equal to zero, regardless of the choice of the input displacement data related to the indirect loading.

4.4.2 Analysis by Reduced Element Stiffness Method

The reduced element stiffness method is particularly advantageous in the analysis of trusses, in which there is no need to assess the fixed end forces arising from intermediate joint loads. The main advantage is that we can work with one degree of freedom per truss element, instead of four degrees of freedom. The only apparent disadvantage is that we do not get the member-end displacements in local coordinates for every element directly as output. However, this information pertaining to the displacement field is not particularly useful, as in practice, we are mainly interested in the joint displacements and bar deformations, which we get directly as output in the reduced element stiffness method. We are also able to generate the bar forces and support reactions.

In this method, we consider each plane truss element to have a single degree of freedom (with local coordinates as shown in Fig. 4.1), with the element stiffness defined as $k_*^i = k_i = \dfrac{E_i A_i}{L_i}$ (Eq. 4.3). If the truss has m elements, the ***reduced unassembled element stiffness matrix***, $\tilde{\mathbf{k}}_*$, is a simple diagonal matrix of order $m \times m$.

$$\mathbf{F}_* = \tilde{\mathbf{k}}_* \mathbf{D}_* \;\Rightarrow\; \begin{bmatrix} F_*^1 \\ F_*^2 \\ \vdots \\ \vdots \\ F_*^m \end{bmatrix} = \begin{bmatrix} k_1 & 0 & 0 & 0 & 0 \\ 0 & k_2 & 0 & 0 & 0 \\ 0 & 0 & \ddots & 0 & 0 \\ 0 & 0 & 0 & \ddots & 0 \\ 0 & 0 & 0 & 0 & k_m \end{bmatrix} \begin{bmatrix} D_*^1 \\ D_*^2 \\ \vdots \\ \vdots \\ D_*^m \end{bmatrix} \tag{4.35}$$

where $k_i = \dfrac{E_i A_i}{L_i}$.

As explained earlier (Section 4.3.2), the structure stiffness matrix $\mathbf{k}$ can be assembled by generating the displacement transformation matrix $\mathbf{T_D}$, and applying $\mathbf{k} = \mathbf{T_D}^{\mathrm{T}} \tilde{\mathbf{k}}_* \mathbf{T_D}$. For small trusses, we can also do this directly (and verify the results) by means of a 'physical approach'. When the number of elements involved is large, it is especially convenient to adopt a 'direct stiffness' approach to assembling the **k** matrix, using the element displacement transformation matrix, $\mathbf{T_D^i}$, and summing up the contributions, $\mathbf{T_D^{i\,T}\tilde{k}_*^i T_D^i}$ (direct stiffness method). The final structure stiffness matrix **k** generated will be identical to the one generated by the conventional stiffness method.

We can generate the $\mathbf{T_D^i}$ matrices for the various elements, each of order $1\times(n+r)$, or the full $\mathbf{T_D}$ matrix of order $m\times(n+r)$, using simple compatibility relations, applying a unit displacement at a time ($D_j = 1$) on the restrained (primary) structure.

$$\mathbf{D_*} = \left[\mathbf{T_{DA}} \;\vdots\; \mathbf{T_{DR}}\right]\begin{bmatrix}\mathbf{D_A}\\ \hline \mathbf{D_R}\end{bmatrix} \Rightarrow \begin{bmatrix} D_*^1 \\ D_*^2 \\ \vdots \\ \vdots \\ D_*^m \end{bmatrix}_{m\times 1} = \begin{bmatrix} \left[\mathbf{T_{DA}^1}\right]_{1\times n} & \left[\mathbf{T_{DR}^1}\right]_{1\times r} \\ \left[\mathbf{T_{DA}^2}\right]_{1\times n} & \left[\mathbf{T_{DR}^2}\right]_{1\times r} \\ \vdots & \vdots \\ \vdots & \vdots \\ \left[\mathbf{T_{DA}^m}\right]_{1\times n} & \left[\mathbf{T_{DR}^m}\right]_{1\times r} \end{bmatrix}_{m\times(n+r)} \begin{bmatrix} D_1 \\ \vdots \\ D_n \\ \hline D_{n+1} \\ \vdots \\ D_{n+r} \end{bmatrix}_{(n+r)\times 1} \qquad (4.36)$$

We can also alternatively generate the $\mathbf{T_D}^{\mathrm{T}}$ matrix, by making use of the *Principle of Contra-gradient*, whereby $\begin{bmatrix}\mathbf{F_A}\\ \hline \mathbf{F_R}\end{bmatrix} = \begin{bmatrix}\mathbf{T_{DA}}^{\mathrm{T}}\\ \hline \mathbf{T_{DR}}^{\mathrm{T}}\end{bmatrix}\mathbf{F_*}$, as explained earlier. For example, let us consider a typical truss element *i* inclined θ_x^i to the global *x*-axis, with 4 joint degrees of freedom at its two ends (global coordinates numbered '1' to '4' for convenience), as shown in Fig. 4.35a. The components of the $\mathbf{T_D^i}$ matrix can be generated from kinematic considerations (satisfying displacement compatibility), as shown in Fig. 4.35b. Alternatively, and more conveniently, the components of the $\mathbf{T_D^{i\,T}}$ matrix can be generated from static considerations (satisfying static equilibrium), as shown in Fig. 4.35c.

In practice, the global coordinates may have any set of numbers, and some of these may correspond to restrained coordinates. However, if the global *x*- and *y*- axes and θ_x^i directions are chosen as indicated in Fig. 4.35a, then invariably the block of two elements of the $\mathbf{T_D^i}$ matrix (or the $\mathbf{T_D^{i\,T}}$ matrix), corresponding to the *start node* will be negative, while the remaining two elements (corresponding to the *end node*) will be positive. The elements correspond to direction cosines, as indicated in Figs 4.35b and c.

It is convenient to also generate and store the $\mathbf{\tilde{k}_*^i T_D^i}$ matrices for the various elements (or $\mathbf{\tilde{k}_* T_D}$ for the structure) to facilitate the computation of the bar forces using the relation, $\mathbf{F_*^i} = \left(\mathbf{\tilde{k}_*^i T_D^i}\right)\mathbf{D}$ (or $\mathbf{F_*} = \left(\mathbf{\tilde{k}_* T_D}\right)\mathbf{D}$).

When indirect loads in the form of temperature loading or 'lack of fit' are prescribed in the form of an initial (free) displacement vector $\mathbf{e_o}$ (of dimension *m*), these loads manifest as axial forces in the fixed end condition in the primary restrained structure, given by: $N_{if} = -k_i e_{oi}$, which need to be expressed in the element local coordinates as follows:

$$F_{*f}^{i} = N_{if} = -k_i e_{oi} \tag{4.37}$$

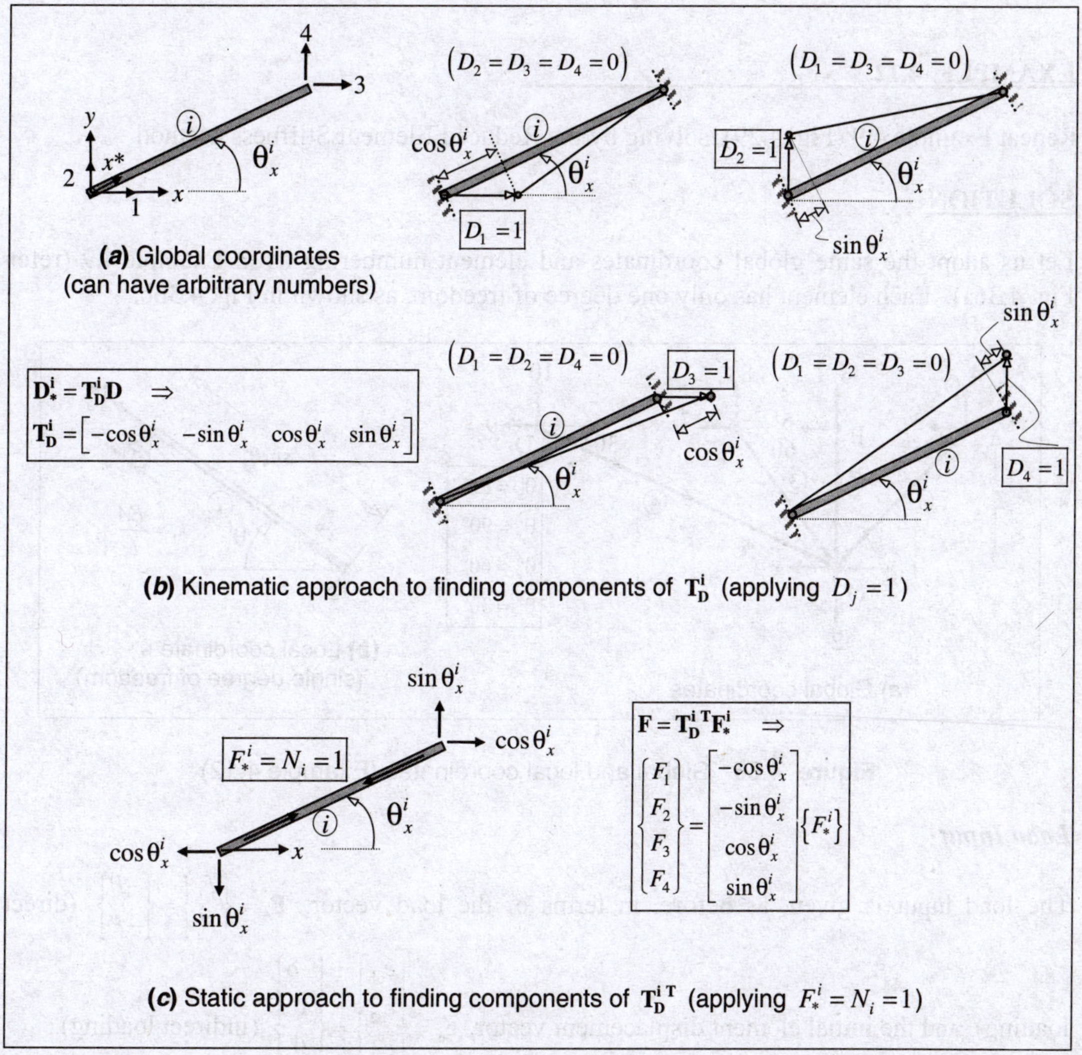

Figure 4.35 Generation of components of the $\mathbf{T}_{\mathbf{D}}^{\mathbf{i}}$ matrix for a plane truss element

These fixed end forces can then be transformed to the global coordinates by the transformation $\mathbf{F}_{\mathbf{f}}^{\mathbf{i}} = \mathbf{T}_{\mathbf{D}}^{\mathbf{i}\,\mathbf{T}}\mathbf{F}_{*\mathbf{f}}^{\mathbf{i}}$ or $\mathbf{F}_{\mathbf{f}} = \mathbf{T}_{\mathbf{D}}{}^{\mathbf{T}}\mathbf{F}_{*\mathbf{f}}$. The total load vector to be considered for finding the unknown joint displacements is given by $\mathbf{F}_{\mathbf{A}} - \mathbf{F}_{\mathbf{fA}}$, as explained earlier. If there are support settlements, $\mathbf{D}_{\mathbf{R}}$, these should also be accounted for in the stiffness relations. The final member end forces are obtained by $\mathbf{F}_*^{\mathbf{i}} = \mathbf{F}_{*\mathbf{f}}^{\mathbf{i}} + \mathbf{k}_*^{\mathbf{i}}\mathbf{T}_{\mathbf{D}}^{\mathbf{i}}\mathbf{D}$.

Let us now demonstrate the application of the reduced element stiffness method to plane trusses through Examples.

EXAMPLE 4.12

Repeat Example 4.9 (Fig. 4.29), solving by the Reduced Element Stiffness Method.

SOLUTION

Let us adopt the same global coordinates and element numbering as in Example 4.9 (refer Fig. 4.36a). Each element has only one degree of freedom, as shown in Fig. 4.36b.

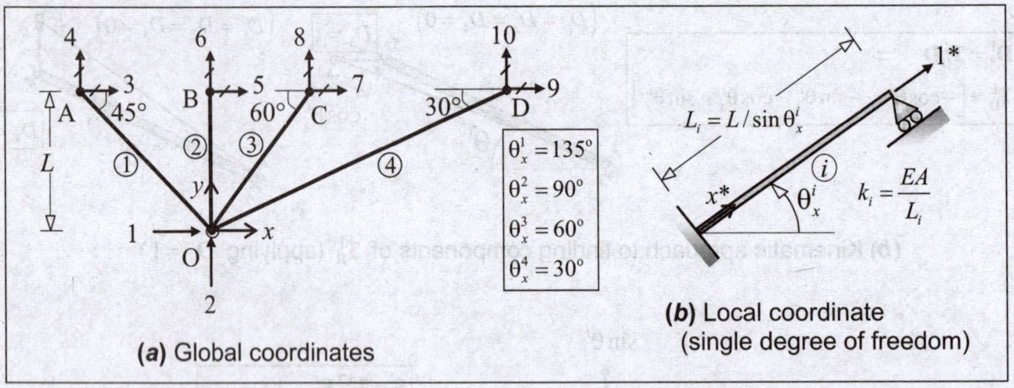

Figure 4.36 Global and local coordinates (Example 4.12)

Load Input

The load input is given, as before, in terms of the load vector, $\mathbf{F_A} = \begin{Bmatrix} F_1 \\ F_2 \end{Bmatrix} = \begin{Bmatrix} -P \\ -P \end{Bmatrix}$ (direct loading), and the initial element displacement vector, $\mathbf{e_o} = \begin{Bmatrix} e_{o1} \\ e_{o2} \\ e_{o3} \\ e_{o4} \end{Bmatrix} = \begin{Bmatrix} -\delta \\ 0 \\ 0 \\ -\delta \end{Bmatrix}$ (indirect loading).

Displacement Transformation Matrices

The element displacement transformation matrix, $\mathbf{T_D^i} = \left[\mathbf{T_{DA}^i} \mid \mathbf{T_{DR}^i} \right]$, which is used in the transformation, $\mathbf{D_*^i} = \mathbf{T_D^i D}$, has an order 1×10, mostly comprising zero elements.

Referring to Fig. 4.30a and Fig. 4.36b, and noting that $\cos\theta_x^1 = -0.7071$, $\sin\theta_x^1 = 0.7071$; $\cos\theta_x^2 = 0$, $\sin\theta_x^2 = 1$; $\cos\theta_x^3 = 0.5$, $\sin\theta_x^3 = 0.8660$; and $\cos\theta_x^4 = 0.8660$, $\sin\theta_x^4 = 0.5$.

$$\mathbf{T}_\mathbf{D}^1 = \left[\mathbf{T}_\mathbf{DA}^1 \mid \mathbf{T}_\mathbf{DR}^1\right] = [0.7071 \quad -0.7071 \mid -0.7071 \quad 0.7071 \quad 0 \quad 0 \quad 0 \quad 0 \quad 0 \quad 0]$$

$$\mathbf{T}_\mathbf{D}^2 = \left[\mathbf{T}_\mathbf{DA}^2 \mid \mathbf{T}_\mathbf{DR}^2\right] = [0 \quad -1 \mid 0 \quad 0 \quad 0 \quad 1 \quad 0 \quad 0 \quad 0 \quad 0]$$

$$\mathbf{T}_\mathbf{D}^3 = \left[\mathbf{T}_\mathbf{DA}^3 \mid \mathbf{T}_\mathbf{DR}^3\right] = [-0.5 \quad -0.8660 \mid 0 \quad 0 \quad 0 \quad 0 \quad 0.5 \quad 0.8660 \quad 0 \quad 0]$$

$$\mathbf{T}_\mathbf{D}^4 = \left[\mathbf{T}_\mathbf{DA}^4 \mid \mathbf{T}_\mathbf{DR}^4\right] = [-0.8660 \quad -0.5 \mid 0 \quad 0 \quad 0 \quad 0 \quad 0 \quad 0 \quad 0.8660 \quad 0.5]$$

We may thus generate the full displacement transformation matrix $\mathbf{T}_\mathbf{D} = \left[\mathbf{T}_\mathbf{DA} \mid \mathbf{T}_\mathbf{DR}\right]$:

$$\mathbf{T}_\mathbf{D} = \left[\mathbf{T}_\mathbf{DA} \mid \mathbf{T}_\mathbf{DR}\right] = \left[\begin{array}{cc|cccccccc} 0.7071 & -0.7071 & -0.7071 & 0.7071 & 0 & 0 & 0 & 0 & 0 & 0 \\ 0 & -1 & 0 & 0 & 0 & 1 & 0 & 0 & 0 & 0 \\ -0.5 & -0.8660 & 0 & 0 & 0 & 0 & 0.5 & 0.8660 & 0 & 0 \\ -0.8660 & -0.5 & 0 & 0 & 0 & 0 & 0 & 0 & 0.8660 & 0.5 \end{array}\right]$$

Reduced Unassembled Element Stiffness Matrix

The element stiffnesses are given by:

$$k_1 = \frac{EA}{\sqrt{2}L} = 0.70711\frac{EA}{L} \;;\; k_2 = \frac{EA}{L} \;;\; k_3 = \frac{EA}{2L/\sqrt{3}} = 0.86603\frac{EA}{L} \;;\; k_4 = \frac{EA}{2L} = 0.5\frac{EA}{L}$$

$$\mathbf{F}_* = \tilde{\mathbf{k}}_*\mathbf{D}_* \Rightarrow \begin{bmatrix} F_*^1 \\ F_*^2 \\ F_*^3 \\ F_*^4 \end{bmatrix} = \frac{EA}{L}\begin{bmatrix} 0.70711 & 0 & 0 & 0 \\ 0 & 1 & 0 & 0 \\ 0 & 0 & 0.86603 & 0 \\ 0 & 0 & 0 & 0.5 \end{bmatrix}\begin{bmatrix} D_*^1 \\ D_*^2 \\ D_*^3 \\ D_*^4 \end{bmatrix}$$

$$\tilde{\mathbf{k}}_*\mathbf{T}_\mathbf{D} = \frac{EA}{L}\left[\begin{array}{cc|cccccccc} 0.5 & -0.5 & -0.5 & 0.5 & 0 & 0 & 0 & 0 & 0 & 0 \\ 0 & -1 & 0 & 0 & 0 & 1 & 0 & 0 & 0 & 0 \\ -0.4330 & -0.75 & 0 & 0 & 0 & 0 & 0.4330 & 0.75 & 0 & 0 \\ -0.4330 & -0.25 & 0 & 0 & 0 & 0 & 0 & 0 & 0.4330 & 0.25 \end{array}\right]$$

Fixed End Forces and Total Load Vector

The initial element displacements due to 'lack of fit' generate fixed end forces in elements '1' and '4', when applied to the restrained (primary structure), given by Eq. 4.37:

$$F_{*f}^i = N_{if} = -k_i e_{oi} \Rightarrow F_{*f}^1 = 0.7071\frac{EA\delta}{L} \;;\; F_{*f}^2 = F_{*f}^3 = 0 \;;\; F_{*f}^4 = 0.5\frac{EA\delta}{L}$$

Transforming these forces to global coordinates, by applying $\mathbf{F_f} = \mathbf{T_D}^\mathbf{T}\mathbf{F}_{*\mathbf{f}}^\mathbf{i}$,

$$\begin{Bmatrix} \mathbf{F_{fA}} \\ \mathbf{F_{fR}} \end{Bmatrix} = \begin{Bmatrix} F_{1f} \\ F_{2f} \\ F_{3f} \\ F_{4f} \\ F_{5f} \\ F_{6f} \\ F_{7f} \\ F_{8f} \\ F_{9f} \\ F_{10f} \end{Bmatrix} = \begin{bmatrix} 0.5 & 0 & -0.433 & -0.433 \\ -0.5 & -1 & -0.75 & -0.25 \\ \hline -0.5 & 0 & 0 & 0 \\ 0.5 & 0 & 0 & 0 \\ 0 & 0 & 0 & 0 \\ 0 & 1 & 0 & 0 \\ 0 & 0 & 0.433 & 0 \\ 0 & 0 & 0.75 & 0 \\ 0 & 0 & 0 & 0.433 \\ 0 & 0 & 0 & 0.25 \end{bmatrix} \times \frac{EA\delta}{L} \begin{Bmatrix} 0.7071 \\ 0 \\ 0 \\ 0.5 \end{Bmatrix} = \frac{EA\delta}{L} \begin{Bmatrix} 0.067 \\ -0.75 \\ \hline -0.5 \\ 0.5 \\ 0 \\ 0 \\ 0 \\ 0 \\ 0.433 \\ 0.25 \end{Bmatrix}$$

The resulting load vector $\mathbf{F_A} - \mathbf{F_{fA}}$ (global coordinates) is given by:

$$\mathbf{F_A} - \mathbf{F_{FA}} = \begin{Bmatrix} F_1 \\ F_2 \end{Bmatrix} - \begin{Bmatrix} F_{1f} \\ F_{2f} \end{Bmatrix} = \begin{Bmatrix} -P \\ -P \end{Bmatrix} - \frac{EA\delta}{L} \begin{Bmatrix} 0.067 \\ -0.75 \end{Bmatrix} = \begin{Bmatrix} -P - 0.067\dfrac{EA\delta}{L} \\ -P + 0.75\dfrac{EA\delta}{L} \end{Bmatrix}$$

These results are identical to the ones obtained in Example 4.9.

Structure Stiffness Matrix

The structure stiffness matrix $\mathbf{k} = \begin{bmatrix} \mathbf{k_{AA}} & \mathbf{k_{AR}} \\ \hline \mathbf{k_{RA}} & \mathbf{k_{RR}} \end{bmatrix}$, satisfying $\mathbf{F} = \mathbf{kD}$, can be assembled, The structure stiffness matrix $\mathbf{k}$ can now be assembled by the *direct stiffness* approach, by directly summing up the contributions from all elements, $\mathbf{k} = \sum_{i=1}^{4} \mathbf{T_D^{i\,T}} \tilde{\mathbf{k}}_*^i \mathbf{T_D^i}$. Alternatively, we can achieve this by the transformation, $\mathbf{k} = \mathbf{T_D}^{\mathbf{T}} \left(\tilde{\mathbf{k}}_* \mathbf{T_D} \right)$. Thus, we obtain:

$$\mathbf{k_{AA}} = \frac{L}{EA} \begin{bmatrix} 0.9451 & 0.2380 \\ 0.2380 & 2.1281 \end{bmatrix} \quad \Rightarrow \quad [\mathbf{k_{AA}}]^{-1} = \frac{EA}{L} \begin{bmatrix} 1.0888 & -0.1217 \\ -0.1217 & 0.4835 \end{bmatrix}$$

$$\mathbf{k_{AR}} = \frac{EA}{L} \begin{bmatrix} -0.3536 & 0.3536 & 0 & 0 & -0.2165 & -0.3750 & -0.3750 & -0.2165 \\ 0.3536 & -0.3536 & 0 & -1 & -0.3750 & -0.6495 & -0.2165 & -0.1250 \end{bmatrix}$$

$\mathbf{k_{RA}} = \mathbf{k_{AR}}^{\mathbf{T}}$. The $\mathbf{k_{RA}}$ matrix of order 8×8 is not shown here for convenience, as it is not required in this particular problem. These results are identical to the ones obtained in Example 4.9.

Displacements and Support Reactions

$$\begin{bmatrix} \mathbf{F_A} \\ \hline \mathbf{F_R} \end{bmatrix} - \begin{bmatrix} \mathbf{F_{fA}} \\ \hline \mathbf{F_{fR}} \end{bmatrix} = \begin{bmatrix} \mathbf{k_{AA}} & \mathbf{k_{AR}} \\ \hline \mathbf{k_{RA}} & \mathbf{k_{RR}} \end{bmatrix} \begin{bmatrix} \mathbf{D_A} \\ \hline \mathbf{D_R} = \mathbf{O} \end{bmatrix}$$

$$\mathbf{D_A} = [\mathbf{k_{AA}}]^{-1}[\mathbf{F_A} - \mathbf{F_{fA}}]$$

$$\Rightarrow \begin{Bmatrix} D_1 \\ D_2 \end{Bmatrix} = \frac{L}{EA}\begin{bmatrix} 1.0888 & -0.1217 \\ -0.1217 & 0.4835 \end{bmatrix}\begin{Bmatrix} -P - 0.067\dfrac{EA\delta}{L} \\ -P + 0.75\dfrac{EA\delta}{L} \end{Bmatrix} = \begin{Bmatrix} -0.9671\dfrac{PL}{EA} - 0.1643\ \delta \\ -0.3618\dfrac{PL}{EA} + 0.3708\ \delta \end{Bmatrix}$$

$$\mathbf{F_R} = \mathbf{F_{fR}} + \mathbf{k_{RA}}\mathbf{D_A}$$

$$\Rightarrow \mathbf{F_R} = \begin{Bmatrix} F_3 \\ F_4 \\ F_5 \\ F_6 \\ F_7 \\ F_8 \\ F_9 \\ F_{10} \end{Bmatrix} = \frac{EA\delta}{L}\begin{Bmatrix} -0.5 \\ 0.5 \\ 0 \\ 0 \\ 0 \\ 0 \\ 0.433 \\ 0.25 \end{Bmatrix} + \frac{EA}{L}\begin{bmatrix} -0.3536 & 0.3536 \\ 0.3536 & -0.3536 \\ 0 & 0 \\ 0 & -1 \\ -0.2165 & -0.3750 \\ -0.3750 & -0.6495 \\ -0.3750 & -0.2165 \\ -0.2165 & -0.1250 \end{bmatrix}\begin{Bmatrix} -0.9671\dfrac{PL}{EA} - 0.1643\ \delta \\ -0.3618\dfrac{PL}{EA} + 0.3708\ \delta \end{Bmatrix}$$

$$\Rightarrow \mathbf{F_R} = \begin{Bmatrix} F_3 \\ F_4 \\ F_5 \\ F_6 \\ F_7 \\ F_8 \\ F_9 \\ F_{10} \end{Bmatrix} = P\begin{Bmatrix} 0.2140 \\ -0.2140 \\ 0 \\ 0.3618 \\ 0.3451 \\ 0.5977 \\ 0.4410 \\ 0.2546 \end{Bmatrix} + \frac{EA\delta}{L}\begin{Bmatrix} -0.3108 \\ 0.3108 \\ 0 \\ -0.3708 \\ -0.1035 \\ -0.1792 \\ 0.4143 \\ 0.2392 \end{Bmatrix}$$

Member Forces

$$\mathbf{F}_*^i = \mathbf{F}_{*f}^i + \left(\tilde{\mathbf{k}}_*^i \mathbf{T}_D^i\right)\mathbf{D} \quad \text{or} \quad \mathbf{F}_* = \mathbf{F}_{*f} + \left(\tilde{\mathbf{k}}_* \mathbf{T}_D\right)\mathbf{D}$$

We can simplify the calculations to be shown here for convenience, by noting that only D_1 and D_2 have non-zero values, as $\mathbf{D_R} = \mathbf{O}$. Thus, we need to only pick up the first two elements in the third row of the $\tilde{\mathbf{k}}_*^i \mathbf{T}_D^i$ matrices derived earlier. Thus,

$$N_1 = F_*^1 = \left(0.7071\frac{EA\delta}{L}\right) + \frac{EA}{L}[0.5 \quad -0.5]\begin{Bmatrix} -0.9671\dfrac{PL}{EA} - 0.1643\ \delta \\ -0.3618\dfrac{PL}{EA} + 0.3708\ \delta \end{Bmatrix} = -0.3026P + 0.4395\frac{EA\delta}{L}$$

$$N_2 = F_*^2 = (0) + \frac{EA}{L}[0 \quad -1]\begin{Bmatrix} -0.9671\dfrac{PL}{EA} - 0.1643\ \delta \\ -0.3618\dfrac{PL}{EA} + 0.3708\ \delta \end{Bmatrix} = 0.3168P - 0.3708\frac{EA\delta}{L}$$

$$N_3 = F_*^3 = (0) + \frac{EA}{L}[-0.433 \quad -0.75]\begin{Bmatrix} -0.9671\dfrac{PL}{EA} - 0.1643\ \delta \\ -0.3618\dfrac{PL}{EA} + 0.3708\ \delta \end{Bmatrix} = 0.6901P - 0.2070\frac{EA\delta}{L}$$

$$N_4 = F_*^4 = \left(0.5\frac{EA\delta}{L}\right) + \frac{EA}{L}[-0.433 \quad -0.25]\begin{Bmatrix} -0.9671\dfrac{PL}{EA} - 0.1643\ \delta \\ -0.3618\dfrac{PL}{EA} + 0.3708\ \delta \end{Bmatrix} = 0.5092P + 0.4784\frac{EA\delta}{L}$$

These results are identical to the ones obtained in Example 4.9. The axial forces and support reactions are as shown in Fig. 4.31, separately for the direct loading and indirect loading cases. Also, as in Example 4.9, we can now easily find the various bar elongations, $e_i = \frac{N_i}{k_i}$, if desired.

Note: If our interest is limited to finding only the bar forces, we can conveniently reduce the size of the matrices (and the computations), by ignoring the restrained degrees of freedom. Thus, we need to deal with only two active degrees of freedom, whereby:

$$\mathbf{T_D} = \mathbf{T_{DA}} = \begin{bmatrix} 0.7071 & -0.7071 \\ 0 & -1 \\ -0.5 & -0.8660 \\ -0.8660 & -0.5 \end{bmatrix} \Rightarrow \tilde{\mathbf{k}}_*\mathbf{T_D} = \frac{EA}{L}\begin{bmatrix} 0.5 & -0.5 \\ 0 & -1 \\ -0.433 & -0.75 \\ -0.433 & -0.25 \end{bmatrix}; \ \mathbf{F_f} = \mathbf{F_{fA}} = \mathbf{T_D}^{\mathbf{T}}\mathbf{F}_{*\mathbf{f}}^{\mathbf{i}}$$

$$\mathbf{k} = \mathbf{T_D}^{\mathbf{T}}\left(\tilde{\mathbf{k}}_*\mathbf{T_D}\right) = \mathbf{k_{AA}} = \frac{L}{EA}\begin{bmatrix} 0.9451 & 0.2380 \\ 0.2380 & 2.1281 \end{bmatrix}$$

Solving $\mathbf{D_A} = [\mathbf{k_{AA}}]^{-1}[\mathbf{F_A} - \mathbf{F_{fA}}]$ and applying $\mathbf{F}_* = \mathbf{F}_{*\mathbf{f}} + \left(\tilde{\mathbf{k}}_*\mathbf{T_D}\right)\mathbf{D}$, we get the bar forces directly.

EXAMPLE 4.13

Find the bar forces in the compound truss shown in Fig. 4.37 by the Stiffness Method.

SOLUTION

Input Data

In this compound truss, there are 4 prismatic elements ($m = 4$), 6 pin joints, i.e., $2j = 12$ global coordinates, of which 9 correspond to *active* degrees of freedom ($n = 9$) and the remaining 3 to *restrained* global coordinates ($r = 3$), as marked in Fig. 4.38a. The truss is to be analysed for the action of the load $F_2 = -P$, all other load components being equal to zero.

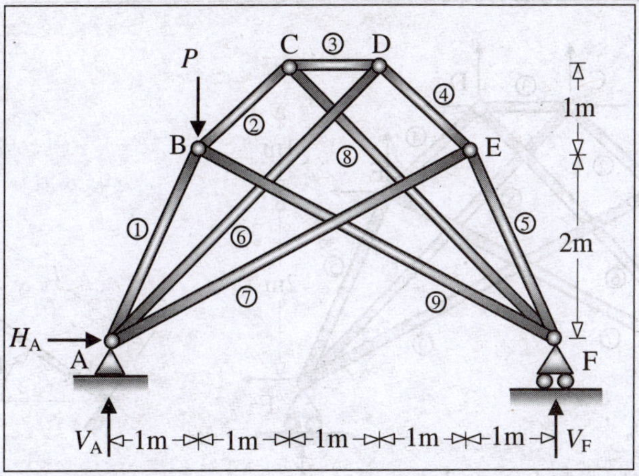

Figure 4.37 Compound truss (Example 4.13)

As the truss is statically determinate, we know that the bar forces will not be dependent on the element stiffnesses. For convenience, let us assume that all 9 bars have the same axial rigidity, *EA*. The local coordinate system for a typical element, by the reduced stiffness formulation, is as shown in Fig. 4.38b. Choosing the origin of the *x-y* global axes system at the node A, the coordinates of the start and end nodes (X_{iS}, Y_{iS} and X_{iE}, Y_{iE}) of individual elements and their lengths, $L_i = \sqrt{(X_{iE} - X_{iS})^2 + (Y_{iE} - Y_{iS})^2}$, and direction cosine parameters, $\cos\theta_x^i = \dfrac{X_{iE} - X_{iS}}{L_i}$ and $\sin\theta_x^i = \dfrac{Y_{iE} - Y_{iS}}{L_i}$, are tabulated below.

Element No.	Start Node (X_{iS}, Y_{iS}) m	End Node (X_{iE}, Y_{iE}) m	Length L_i (m) $\sqrt{(X_{iE} - X_{iS})^2 + (Y_{iE} - Y_{iS})^2}$	$\cos\theta_x^i = \dfrac{X_{iE} - X_{iS}}{L_i}$	$\sin\theta_x^i = \dfrac{Y_{iE} - Y_{iS}}{L_i}$
1	A (0,0)	B (1,2)	2.2361	0.44721	0.89443
2	B (1,2)	C (2,3)	1.4142	0.70711	0.70711
3	C (2,3)	D (3,3)	1	1	0
4	D (3,3)	E (4,2)	1.4142	–0.70711	0.70711
5	E (4,2)	F (5,0)	2.2361	–0.44721	0.89443
6	A (0,0)	D (3,3)	4.2426	0.70711	0.70711
7	A (0,0)	E (4,2)	4.4721	0.89443	0.47721
8	F (5,0)	C (2,3)	4.2426	–0.70711	0.70711
9	F (5,0)	B (1,2)	4.4721	–0.89443	0.47721

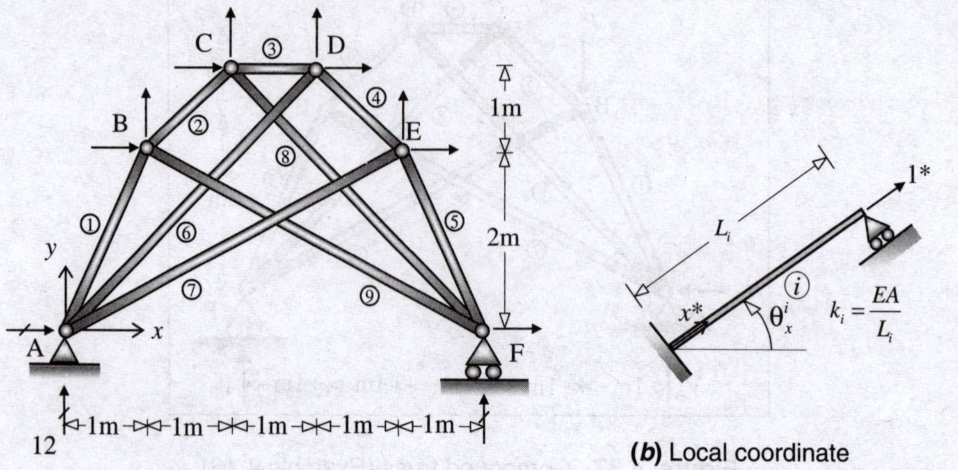

(*a*) Global coordinates

(*b*) Local coordinate (single degree of freedom)

Figure 4.38 Global and local coordinates (Example 4.13)

The element stiffnesses $k_i = \dfrac{EA}{L_i}$ are given by:

$$k_1 = k_5 = 0.44721EA\ ;\ k_2 = k_4 = 0.70711EA\ ;\ k_3 = EA\ ;\ k_6 = k_8 = 0.23570EA\ ;\ k_7 = k_9 = 0.22361EA$$

Solution Procedure

As we are required to find only the bar forces (and not the support reactions) in this problem, and as there are no support movements ($\mathbf{D_R} = \mathbf{O}$), we can ignore the restrained global coordinates, '10', '11' and '12'. We can find the bar forces by applying $\mathbf{F_*} = \left(\tilde{\mathbf{k}}_*\mathbf{T_{DA}}\right)\mathbf{D_A}$, where $\mathbf{D_A} = [\mathbf{k_{AA}}]^{-1}\,\mathbf{F_A}$. This involves inversion of a 9×9 stiffness matrix, which cannot be done manually, but can be easily generated by an appropriate software application, such as MATLAB. We can conveniently generate the $\mathbf{k_{AA}}$ matrix by the direct stiffness approach, by summing up the contributions from all elements, $\mathbf{k_{AA}} = \sum_{i=1}^{9} \mathbf{T_{DA}^{i}}^{T}\tilde{\mathbf{k}}_*^{i}\mathbf{T_{DA}^{i}}$. However, prior to this, we need to formulate the element displacement transformation matrix $\mathbf{T_{DA}^{i}}$.

Displacement Transformation Matrices

The element displacement transformation matrix, $\mathbf{T_{DA}^{i}}$, which is used in the transformation, $\mathbf{D_*^{i}} = \mathbf{T_{DA}^{i}}\mathbf{D_A}$, has an order 1×9, mostly comprising zero elements.

Referring to Fig. 4.38a and Fig. 4.36b, and making use of element connectivity data and the direction cosine parameters already computed and tabulated,

$$\mathbf{T}_{\mathbf{DA}}^{1} = [0.44721 \quad 0.89443 \quad 0 \quad 0 \quad 0 \quad 0 \quad 0 \quad 0 \quad 0]$$

$$\mathbf{T}_{\mathbf{DA}}^{2} = [-0.70711 \quad -0.70711 \quad 0.70711 \quad 0.70711 \quad 0 \quad 0 \quad 0 \quad 0 \quad 0]$$

$$\mathbf{T}_{\mathbf{DA}}^{3} = [0 \quad 0 \quad -1 \quad 0 \quad 1 \quad 0 \quad 0 \quad 0 \quad 0]$$

$$\mathbf{T}_{\mathbf{DA}}^{4} = [0 \quad 0 \quad 0 \quad 0 \quad 0.70711 \quad -0.70711 \quad -0.70711 \quad 0.70711 \quad 0]$$

$$\mathbf{T}_{\mathbf{DA}}^{5} = [0 \quad 0 \quad 0 \quad 0 \quad 0 \quad 0 \quad 0.44721 \quad -0.89443 \quad -0.44721]$$

$$\mathbf{T}_{\mathbf{DA}}^{6} = [0 \quad 0 \quad 0 \quad 0 \quad 0.70711 \quad 0.70711 \quad 0 \quad 0 \quad 0]$$

$$\mathbf{T}_{\mathbf{DA}}^{7} = [0 \quad 0 \quad 0 \quad 0 \quad 0 \quad 0 \quad 0.89443 \quad 0.47221 \quad 0]$$

$$\mathbf{T}_{\mathbf{DA}}^{8} = [0 \quad 0 \quad 0.70711 \quad -0.70711 \quad 0 \quad 0 \quad 0 \quad 0 \quad -0.70711]$$

$$\mathbf{T}_{\mathbf{DA}}^{9} = [0.89443 \quad -0.47721 \quad 0 \quad 0 \quad 0 \quad 0 \quad 0 \quad 0 \quad -0.89443]$$

Considering $\tilde{\mathbf{k}}_{*}^{\mathbf{i}} = k_i$,

$$\tilde{\mathbf{k}}_{*}^{1}\mathbf{T}_{\mathbf{DA}}^{1} = EA[0.2 \quad 0.4 \quad 0 \quad 0 \quad 0 \quad 0 \quad 0 \quad 0 \quad 0]$$

$$\tilde{\mathbf{k}}_{*}^{2}\mathbf{T}_{\mathbf{DA}}^{2} = EA[-0.5 \quad -0.5 \quad 0.5 \quad 0.5 \quad 0 \quad 0 \quad 0 \quad 0 \quad 0]$$

$$\tilde{\mathbf{k}}_{*}^{3}\mathbf{T}_{\mathbf{DA}}^{3} = EA[0 \quad 0 \quad -1 \quad 0 \quad 1 \quad 0 \quad 0 \quad 0 \quad 0]$$

$$\tilde{\mathbf{k}}_{*}^{4}\mathbf{T}_{\mathbf{DA}}^{4} = EA[0 \quad 0 \quad 0 \quad 0 \quad 0.5 \quad -0.5 \quad -0.5 \quad 0.5 \quad 0]$$

$$\tilde{\mathbf{k}}_{*}^{5}\mathbf{T}_{\mathbf{DA}}^{5} = EA[0 \quad 0 \quad 0 \quad 0 \quad 0 \quad 0 \quad 0.2 \quad -0.4 \quad -0.2]$$

$$\tilde{\mathbf{k}}_{*}^{6}\mathbf{T}_{\mathbf{DA}}^{6} = EA[0 \quad 0 \quad 0 \quad 0 \quad 0.16667 \quad 0.16667 \quad 0 \quad 0 \quad 0]$$

$$\tilde{\mathbf{k}}_{*}^{7}\mathbf{T}_{\mathbf{DA}}^{7} = EA[0 \quad 0 \quad 0 \quad 0 \quad 0 \quad 0 \quad 0.2 \quad 0.1 \quad 0]$$

$$\tilde{\mathbf{k}}_{*}^{8}\mathbf{T}_{\mathbf{DA}}^{8} = EA[0 \quad 0 \quad 0.16667 \quad -0.16667 \quad 0 \quad 0 \quad 0 \quad 0 \quad -0.16667]$$

$$\tilde{\mathbf{k}}_{*}^{9}\mathbf{T}_{\mathbf{DA}}^{9} = EA[0.2 \quad -0.1 \quad 0 \quad 0 \quad 0 \quad 0 \quad 0 \quad 0 \quad -0.2]$$

Structure Stiffness Matrix (Active Coordinates)

Using MATLAB, we can conveniently generate:

$$\mathbf{k}_{\mathbf{AA}} = \sum_{i=1}^{9} \mathbf{T}_{\mathbf{DA}}^{\mathbf{i}\,\mathrm{T}} \tilde{\mathbf{k}}_{*}^{\mathbf{i}} \mathbf{T}_{\mathbf{DA}}^{\mathbf{i}}$$

$$\mathbf{k}_{AA} = EA\begin{bmatrix} 0.6219 & 0.4430 & -0.3536 & -0.3536 & 0 & 0 & 0 & 0 & -0.1789 \\ 0.4430 & 0.7560 & -0.3536 & -0.3536 & 0 & 0 & 0 & 0 & 0.0894 \\ -0.3536 & -0.3536 & 1.4714 & 0.2357 & -1 & 0 & 0 & 0 & -0.1179 \\ -0.3536 & -0.3536 & 0.2357 & 0.4714 & 0 & 0 & 0 & 0 & 0.1179 \\ 0 & 0 & -1 & 0 & 1.4714 & -0.2357 & -0.3536 & 0.3536 & 0 \\ 0 & 0 & 0 & 0 & -0.2357 & 0.4714 & 0.3536 & -0.3536 & 0 \\ 0 & 0 & 0 & 0 & -0.3536 & 0.3536 & 0.6219 & -0.4430 & -0.0894 \\ 0 & 0 & 0 & 0 & 0.3536 & -0.3536 & -0.4430 & 0.7560 & 0.1789 \\ -0.1789 & 0.0894 & -0.1179 & 0.1179 & 0 & 0 & -0.0894 & 0.1789 & 0.3862 \end{bmatrix}$$

Displacements

$$\mathbf{F_A} = \mathbf{k_{AA} D_A} \quad \Rightarrow \quad \mathbf{D_A} = [\mathbf{k_{AA}}]^{-1}\mathbf{F_A}$$

$$\text{where } \mathbf{F_A} = \begin{Bmatrix} F_1 \\ F_2 \\ F_3 \\ F_4 \\ F_5 \\ F_6 \\ F_7 \\ F_8 \\ F_9 \end{Bmatrix} = \begin{Bmatrix} 0 \\ -P \\ 0 \\ 0 \\ 0 \\ 0 \\ 0 \\ 0 \\ 0 \end{Bmatrix} \quad \Rightarrow \quad \mathbf{D_A} = \begin{Bmatrix} D_1 \\ D_2 \\ D_3 \\ D_4 \\ D_5 \\ D_6 \\ D_7 \\ D_8 \\ D_9 \end{Bmatrix} = \frac{P}{EA}\begin{Bmatrix} 19.3511 \\ -13.5887 \\ 12.5665 \\ -8.2183 \\ 11.5665 \\ -7.3238 \\ 6.5707 \\ -10.9054 \\ 25.0274 \end{Bmatrix}$$

(Any suitable numerical scheme may be used to solve for $\mathbf{D_A}$).

Member Forces

$$\mathbf{F}_*^i = \left(\tilde{\mathbf{k}}_*^i \mathbf{T}_{DA}^i\right)\mathbf{D_A} \quad \text{and} \quad N_i = F_*^i$$

Thus,

$$N_1 = F_*^1 = -1.5652P$$

$$N_2 = F_*^2 = -0.7071P$$

$$N_3 = F_*^3 = -P$$

$$N_4 = F_*^4 = +0.7071P$$

$$N_5 = F_*^5 = +0.6708P$$

$$N_6 = F_*^6 = +0.7071P$$

$$N_7 = F_*^7 = +0.2236P$$

$N_8 = F_*^8 = -0.7071P$

$N_9 = F_*^9 = +0.2236P$

The reader may verify that these member forces satisfy equilibrium. Furthermore, by including the restrained global coordinates, the support reactions can be shown to be given by

$$\mathbf{F_R} = \mathbf{k_{RA}}\mathbf{D_A} = \begin{Bmatrix} 0.2P \\ 0 \\ 0.8P \end{Bmatrix}.$$

EXAMPLE 4.14

Consider the plane truss shown in Fig. 4.39, subject to the applied horizontal load of 100 kN at joint C combined with a 'lack of fit' on account of bar '4' being too short by 2.5mm. Find, using the Stiffness Method, all the bar forces and the force in the spring attached to the roller support at D.

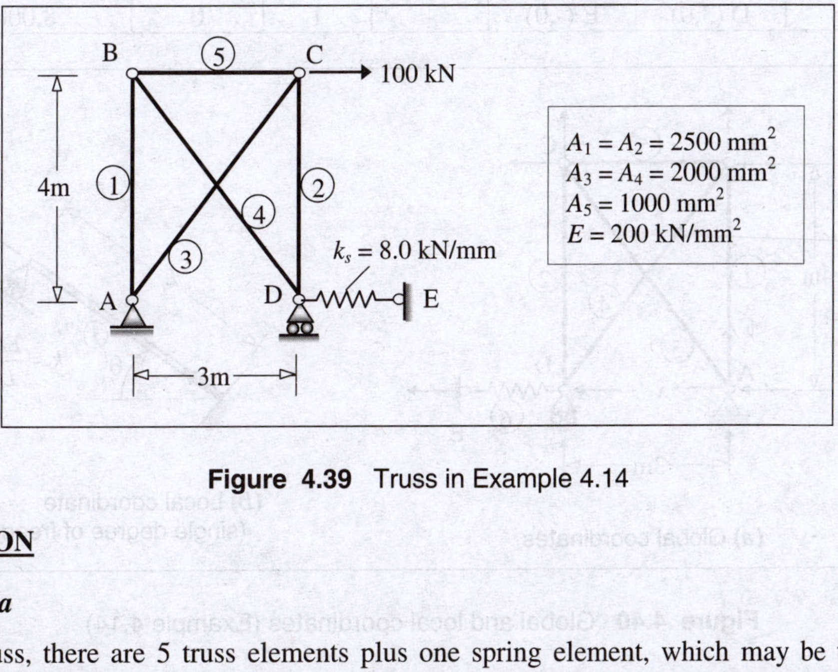

Figure 4.39 Truss in Example 4.14

SOLUTION

Input Data

In this truss, there are 5 truss elements plus one spring element, which may be treated as another truss element ($m = 6$), 9 global coordinates, of which 5 correspond to *active* degrees of freedom ($n = 5$) and 4 *restrained* global coordinates ($r = 4$), as marked in Fig. 4.40a. The truss is to be analysed for the action of (i) the load $F_3 = 100$ kN, all other load components being equal to zero and (ii) 'lack of fit' due to $e_{04} = -0.0025$ m. (As the truss is 'over rigid', this indirect loading will induce forces in the bars).

The local coordinate system for a typical element, by the reduced stiffness formulation, is as shown in Fig. 4.40b. Choosing the origin of the *x*-*y* global axes system at the node A, the coordinates of the start and end nodes (X_{iS}, Y_{iS} and X_{iE}, Y_{iE}) of individual elements and their lengths, $L_i = \sqrt{(X_{iE} - X_{iS})^2 + (Y_{iE} - Y_{iS})^2}$, and direction cosine parameters, $\cos\theta_x^i = \dfrac{X_{iE} - X_{iS}}{L_i}$ and $\sin\theta_x^i = \dfrac{Y_{iE} - Y_{iS}}{L_i}$, are tabulated below.

Element No.	Start Node (X_{iS}, Y_{iS}) m	End Node (X_{iE}, Y_{iE}) m	Length L_i (m)	$\cos\theta_x^i$	$\sin\theta_x^i$	$k_i = \dfrac{EA}{L_i}$ (kN/m)
1	A (0,0)	B (0,4)	4	0	1	125,000
2	D (3,0)	C (3,4)	4	0	1	125,000
3	A (0,0)	C (3,4)	5	0.6	0.8	80,000
4	D (3,0)	B (0,4)	5	–0.6	0.8	80,000
5	B (0,4)	C (3,4)	3	1	0	66,667
6	D (3,0)	E (-,0)	-	1	0	8,000

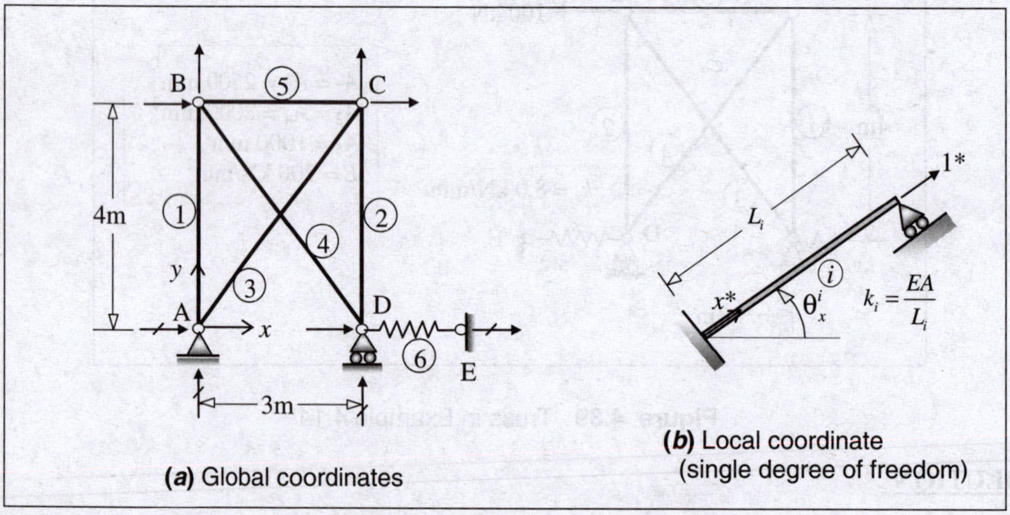

Figure 4.40 Global and local coordinates (Example 4.14)

Solution Procedure

As we are required to find only the bar forces (and not the support reactions) in this problem, and as there are no support movements ($\mathbf{D_R} = \mathbf{O}$), we can ignore the restrained global coordinates, '6', '7', '8' and '9'. The 'lack of fit' in bar '4' will generate a fixed end force in this bar, $F_{*f}^4 = N_{4f} = -k_4 e_{04}$, all other fixed end force components in $\mathbf{F_{*f}}$ being equal to zero.

We can find the bar forces by applying $\mathbf{F}_* = \mathbf{F}_{*\mathbf{f}} + \left(\tilde{\mathbf{k}}_*\mathbf{T}_{\mathbf{DA}}\right)\mathbf{D}_\mathbf{A}$, where $\mathbf{D}_\mathbf{A} = [\mathbf{k}_{\mathbf{AA}}]^{-1}[\mathbf{F}_\mathbf{A} - \mathbf{F}_{\mathbf{fA}}]$ and $\mathbf{F}_{\mathbf{fA}} = \mathbf{T}_{\mathbf{DA}}{}^{\mathbf{T}}\mathbf{F}_{*\mathbf{f}}$.

We can generate the element displacement transformation matrix, $\mathbf{T}_{\mathbf{DA}}$ and the reduced unassembled element stiffness matrix, $\tilde{\mathbf{k}}_*$, and then generate $\tilde{\mathbf{k}}_*\mathbf{T}_{\mathbf{DA}}$ and $\mathbf{k}_{\mathbf{AA}} = \mathbf{T}_{\mathbf{DA}}{}^{\mathbf{T}}\tilde{\mathbf{k}}_*\mathbf{T}_{\mathbf{DA}}$. As $n = 5$, the $\mathbf{k}_{\mathbf{AA}}$ matrix has an order of 5×5. The matrix operations can be conveniently done using MATLAB.

Displacement Transformation Matrix

The element displacement transformation matrix, $\mathbf{T}^{\mathbf{i}}_{\mathbf{DA}}$, which is used in the transformation, $\mathbf{D}^{\mathbf{i}}_* = \mathbf{T}^{\mathbf{i}}_{\mathbf{DA}}\mathbf{D}_\mathbf{A}$, has an order 1×5. Referring to Fig. 4.40a and Fig. 4.36b, and making use of element connectivity data and the direction cosine parameters already computed and tabulated,

$$\begin{aligned}
\mathbf{T}^{1}_{\mathbf{DA}} &= \begin{bmatrix} 0 & 1 & 0 & 0 & 0 \end{bmatrix} \\
\mathbf{T}^{2}_{\mathbf{DA}} &= \begin{bmatrix} 0 & 0 & 0 & 1 & 0 \end{bmatrix} \\
\mathbf{T}^{3}_{\mathbf{DA}} &= \begin{bmatrix} 0 & 0 & 0.6 & 0.8 & 0 \end{bmatrix} \\
\mathbf{T}^{4}_{\mathbf{DA}} &= \begin{bmatrix} -0.6 & 0.8 & 0 & 0 & 0.6 \end{bmatrix} \\
\mathbf{T}^{5}_{\mathbf{DA}} &= \begin{bmatrix} -1 & 0 & 1 & 0 & 0 \end{bmatrix} \\
\mathbf{T}^{6}_{\mathbf{DA}} &= \begin{bmatrix} 0 & 0 & 0 & 0 & -1 \end{bmatrix}
\end{aligned}
\quad \Rightarrow \quad
\mathbf{T}_{\mathbf{DA}} = \begin{bmatrix} 0 & 1 & 0 & 0 & 0 \\ 0 & 0 & 0 & 1 & 0 \\ 0 & 0 & 0.6 & 0.8 & 0 \\ -0.6 & 0.8 & 0 & 0 & 0.6 \\ -1 & 0 & 1 & 0 & 0 \\ 0 & 0 & 0 & 0 & -1 \end{bmatrix}$$

Reduced Unassembled Element Stiffness Matrix

Considering $\tilde{\mathbf{k}}^{\mathbf{i}}_* = k_i$,

$$\mathbf{F}_* = \tilde{\mathbf{k}}_*\mathbf{D}_* \quad \Rightarrow \quad \begin{bmatrix} F^1_* \\ F^2_* \\ F^3_* \\ F^4_* \\ F^5_* \\ F^6_* \end{bmatrix} = 10^5 \times \begin{bmatrix} 1.25 & & & & & \\ & 1.25 & & & & \\ & & 0.8 & & & \\ & & & 0.8 & & \\ & & & & 0.66667 & \\ & & & & & 0.08 \end{bmatrix} \begin{bmatrix} D^1_* \\ D^2_* \\ D^3_* \\ D^4_* \\ D^5_* \\ D^6_* \end{bmatrix}$$

$$\mathbf{F}_* = \left(\tilde{\mathbf{k}}_*\mathbf{T}_{\mathbf{DA}}\right)\mathbf{D}_\mathbf{A} \quad \Rightarrow \quad \begin{bmatrix} F^1_* \\ F^2_* \\ F^3_* \\ F^4_* \\ F^5_* \\ F^6_* \end{bmatrix} = 10^5 \times \begin{bmatrix} 0 & 1.25 & 0 & 0 & 0 \\ 0 & 0 & 0 & 1.25 & 0 \\ 0 & 0 & 0.48 & 0.64 & 0 \\ -0.48 & 0.64 & 0 & 0 & 0.48 \\ -0.6667 & 0 & 0..6667 & 0 & 0 \\ 0 & 0 & 0 & 0 & -0.08 \end{bmatrix} \begin{bmatrix} D_1 \\ D_2 \\ D_3 \\ D_4 \\ D_5 \end{bmatrix}$$

Structure Stiffness Matrix (Active Coordinates)

$$\mathbf{k}_{AA} = \mathbf{T}_{DA}{}^{T}\tilde{\mathbf{k}}_{*}\mathbf{T}_{DA} = 10^5 \times \begin{bmatrix} 0.95467 & -0.38400 & -0.6667 & 0 & -0.28800 \\ -0.38400 & 1.76200 & 0 & 0 & 0.38400 \\ -0.6667 & 0 & 0.95467 & 0.38400 & 0 \\ 0 & 0 & 0.38400 & 1.76200 & 0 \\ -0.28800 & 0.38400 & 0 & 0 & 0.36800 \end{bmatrix} \text{kN/m}$$

$$\Rightarrow [\mathbf{k}_{AA}]^{-1} = 10^{-4} \times \begin{bmatrix} 0.46756 & 0.02867 & 0.35788 & -0.07800 & 0.33600 \\ 0.02867 & 0.07522 & 0.02195 & -0.00478 & -0.05605 \\ 0.35788 & 0.02195 & 0.38874 & -0.08472 & 0.25718 \\ -0.07800 & -0.00478 & -0.08472 & 0.07522 & -0.05605 \\ 0.33600 & -0.05605 & 0.25718 & -0.05605 & 0.59318 \end{bmatrix} \text{m/kN}$$

Fixed End Forces and Total Load Vector

The initial element displacement due to 'lack of fit' generates a fixed end force in element '4', when applied to the restrained (primary structure):

$$F_{*f}^{4} = N_{4f} = -k_4 e_{o4} \Rightarrow F_{*f}^{4} = -(80000)(-0.0025) = 200\,\text{kN}$$

$$\mathbf{F}_{fA} = \mathbf{T}_{DA}{}^{T}\mathbf{F}_{*f} \Rightarrow \begin{Bmatrix} F_{f1} \\ F_{f2} \\ F_{f3} \\ F_{f4} \\ F_{f5} \end{Bmatrix} = \begin{bmatrix} 0 & 0 & 0 & -0.6 & -1 & 0 \\ 1 & 0 & 0 & 0.8 & 0 & 0 \\ 0 & 0 & 0.6 & 0 & 1 & 0 \\ 0 & 1 & 0.8 & 0 & 0 & 0 \\ 0 & 0 & 0 & 0.6 & 0 & -1 \end{bmatrix} \begin{Bmatrix} 0 \\ 0 \\ 0 \\ 200 \\ 0 \\ 0 \end{Bmatrix} = \begin{Bmatrix} -120 \\ 160 \\ 0 \\ 0 \\ 120 \end{Bmatrix} \text{kN}$$

The resulting load vector $\mathbf{F}_A - \mathbf{F}_{fA}$ (global coordinates) is given by:

$$\mathbf{F}_A - \mathbf{F}_{FA} = \begin{Bmatrix} 0 \\ 0 \\ 100 \\ 0 \\ 0 \end{Bmatrix} - \begin{Bmatrix} -120 \\ 160 \\ 0 \\ 0 \\ 120 \end{Bmatrix} = \begin{Bmatrix} 120 \\ -160 \\ 100 \\ 0 \\ -120 \end{Bmatrix} \text{kN}$$

Displacements

$$\mathbf{D}_A = [\mathbf{k}_{AA}]^{-1}[\mathbf{F}_A - \mathbf{F}_{fA}] \Rightarrow \begin{Bmatrix} D_1 \\ D_2 \\ D_3 \\ D_4 \\ D_5 \end{Bmatrix} = \begin{Bmatrix} 4.6988 \\ 0.0326 \\ 4.7477 \\ -1.0340 \\ 0.3824 \end{Bmatrix} \times 10^{-3}\,\text{m} = \begin{Bmatrix} 4.699 \\ 0.033 \\ 4.748 \\ -1.034 \\ 0.382 \end{Bmatrix} \text{mm}$$

Member Forces

$$\mathbf{F}_* = \mathbf{F}_{*f} + \left(\tilde{\mathbf{k}}_* \mathbf{T}_{DA}\right)\mathbf{D}_A \quad \text{and} \quad N_i = F_*^i$$

$$\begin{Bmatrix} N_1 \\ N_2 \\ N_3 \\ N_4 \\ N_5 \\ N_6 \end{Bmatrix} = \begin{Bmatrix} 0 \\ 0 \\ 0 \\ 200 \\ 0 \\ 0 \end{Bmatrix} + 10^5 \times \begin{bmatrix} 0 & 1.25 & 0 & 0 & 0 \\ 0 & 0 & 0 & 1.25 & 0 \\ 0 & 0 & 0.48 & 0.64 & 0 \\ -0.48 & 0.64 & 0 & 0 & 0.48 \\ -0.6667 & 0 & 0..6667 & 0 & 0 \\ 0 & 0 & 0 & 0 & -0.08 \end{bmatrix} \begin{Bmatrix} 4.6988 \\ 0.0326 \\ 4.7447 \\ -1.0340 \\ 0.3824 \end{Bmatrix} \times 10^{-3} = \begin{Bmatrix} 4.079 \\ -129.25 \\ 161.57 \\ -5.099 \\ 3.059 \\ -3.059 \end{Bmatrix} \text{kN}$$

The reader may verify that, by including the restrained global coordinates, the support reactions can be shown to be given by $\mathbf{F_R} = \mathbf{k_{RA}} \mathbf{D_A} = \begin{Bmatrix} F_6 \\ F_7 \\ F_8 \\ F_9 \end{Bmatrix} = \begin{Bmatrix} 133.33 \\ -96.94 \\ -133.33 \\ -3.06 \end{Bmatrix}$ kN. Force equilibrium can be verified.

4.4.3 Analysis by Flexibility Method

The element to be considered here is the same single degree of freedom model in Fig. 4.1b, that we had adopted for the reduced element stiffness method. However, as explained earlier, we need to include additional 'rigid link' elements, if we wish to also find support reactions. The basic structure for which we develop the force transformation matrix and the flexibility matrix is a statically determinate (just-rigid) structure. If the truss is statically indeterminate (over-rigid), then we need to select the redundants and work on the primary structure which is statically determinate. As mentioned earlier, the numbering of the global coordinates is done in such a way that the redundant coordinates are located at the end of the list. The basic force relationship, satisfying static equilibrium, is given by $\mathbf{F}_* = [\mathbf{T_{FA}} \mid \mathbf{T_{FX}}] \begin{Bmatrix} \mathbf{F_A} \\ \mathbf{F_X} \end{Bmatrix}$, in terms of the force transformation matrix $\mathbf{T_F}$. It may be noted, however, that except in the case of simple trusses, the determination of the elements of the $\mathbf{T_F}$ matrix, using simple techniques such as the *method of joints* or *method of sections*, may not be possible, and in such cases (involving compound and complex trusses), we have to take recourse to the *tension coefficient method*. In such cases, the stiffness method may be more convenient in practice (requiring less skill in analysis from the analyst), although it involves a relatively high degree of kinematic indeterminacy.

The element flexibility is defined as $f_*^i = f_i = \dfrac{L_i}{E_i A_i}$ (Eq. 4.4). If the truss has *m* elements and *r* rigid links simulating the support reactions, the *combined element force vector*, $\{F_*\}$ and *combined element displacement vector*, $\{D_*\}$, having *m* (or *m* + *r*) dimensions, in local coordinates, are inter-related by the ***unassembled element flexibility matrix***, $\mathbf{f}_*$, as explained in Section 3.5. This matrix is a simple diagonal matrix of order $m \times m$, or $(m + r)\times(m + r)$, if we wish to include support reactions using rigid links. As explained earlier, the rigid links have infinite stiffness, and therefore, zero flexibility. For convenience, we may exclude the zero rows and columns and restrict the size of the matrix to $m \times m$ (refer Eq. 4.30).

The simplest way of assembling the structure flexibility matrix $\mathbf{f}$ is by invoking the transformation, $\mathbf{f} = \mathbf{T_F}^T \mathbf{f}_* \mathbf{T_F}$, which can be partitioned as $\mathbf{f} = \left[\begin{array}{c|c} \mathbf{f_{AA}} & \mathbf{f_{AX}} \\ \hline \mathbf{f_{XA}} & \mathbf{f_{XX}} \end{array}\right]$, if the structure is statically indeterminate, in which case, it is also necessary to compute $[\mathbf{f_{XX}}]^{-1}$. As mentioned earlier, in the absence of indirect loads, it is convenient to compute $\overline{\mathbf{f}} = \mathbf{f_{AA}} - \mathbf{f_{AX}} \mathbf{f_{XX}}^{-1} \mathbf{f_{XA}}$ and $\overline{\mathbf{T}}_\mathbf{F}^\mathbf{i} = \mathbf{T_{FA}^i} - \mathbf{T_{FX}^i} \mathbf{f_{XX}}^{-1} \mathbf{f_{XA}}$ as the structure flexibility and force transformation matrices, suitable for handling any set of nodal loads on the statically indeterminate structure.

Indirect loading, if any, in terms of support displacements, corresponding to the redundant coordinates ($\mathbf{D_X}$), and changes in element lengths in the 'free' condition, such as due to temperature changes or 'lack of fit' or support movements in non-redundant support locations ($\mathbf{D^i_{*initial}}$), are to be transformed to the global coordinates as $\left[\begin{array}{c} \mathbf{D_{A,initial}} \\ \hline \mathbf{D_{X,initial}} \end{array}\right] = \left[\begin{array}{c} \mathbf{T_{FA}^T} \\ \hline \mathbf{T_{FX}^T} \end{array}\right] \mathbf{D_{*initial}}$.

Force Response

In the case of statically indeterminate trusses, the unknown redundant can be computed using the formula, $\mathbf{F_X} = [\mathbf{f_{XX}}]^{-1}\left[\mathbf{D_X} - \mathbf{D_{X,initial}} - \mathbf{f_{XA}} \mathbf{F_A}\right]$ and the element forces as $\mathbf{F_*} = \mathbf{T_{FA}} \mathbf{F_A} + \mathbf{T_{FX}} \mathbf{F_X}$. Alternatively, in the absence of indirect loading, the computation of $\mathbf{F_X}$ can be avoided and the internal forces can be directly computed as $\mathbf{F_*} = \overline{\mathbf{T}}_\mathbf{F} \mathbf{F}$. In statically determinate trusses ($\mathbf{F_X} = \mathbf{O}$), the element forces can be directly computed as $\mathbf{F_*} = \mathbf{T_F} \mathbf{F}$.

Displacement Response

In the case of statically indeterminate trusses, the joint displacements can be computed using the formula, $\mathbf{D_A} = \mathbf{D_{A,initial}} + \mathbf{f_{AA}} \mathbf{F_A} + \mathbf{f_{AX}} \mathbf{F_X}$, which simplifies to $\mathbf{D} = \mathbf{D_{initial}} + \mathbf{f}\ \mathbf{F}$ in statically determinate structures (where $\mathbf{D_{initial}} = \mathbf{T_F^T} \mathbf{D_{*initial}}$). Alternatively, if there are no indirect loads acting on the statically indeterminate truss, we can directly compute $\mathbf{D} = \overline{\mathbf{f}}\ \mathbf{F}$.

Let us now demonstrate the applications of the flexibility method to some simple plane trusses.

EXAMPLE 4.15

Consider the plane truss solved in Example 4.14 (Fig. 4.39), without the spring attached at support D. Apply the Flexibility Method to find the bar forces and joint displacements, under the application of the horizontal load of 100 kN at joint C, combined with a 'lack of fit' on account of bar '4' being too short by 2.5mm.

SOLUTION

The truss in Fig. 4.39, without the elastic support at D, is 'just-rigid' ($m=5; r=3; j=4 \Rightarrow m+r=2j$) and hence statically determinate. The global coordinates and element numbers are as indicated in Fig. 4.41a (coordinates for support reactions are not included). The local coordinate system for a typical element is as shown in Fig. 4.41b.

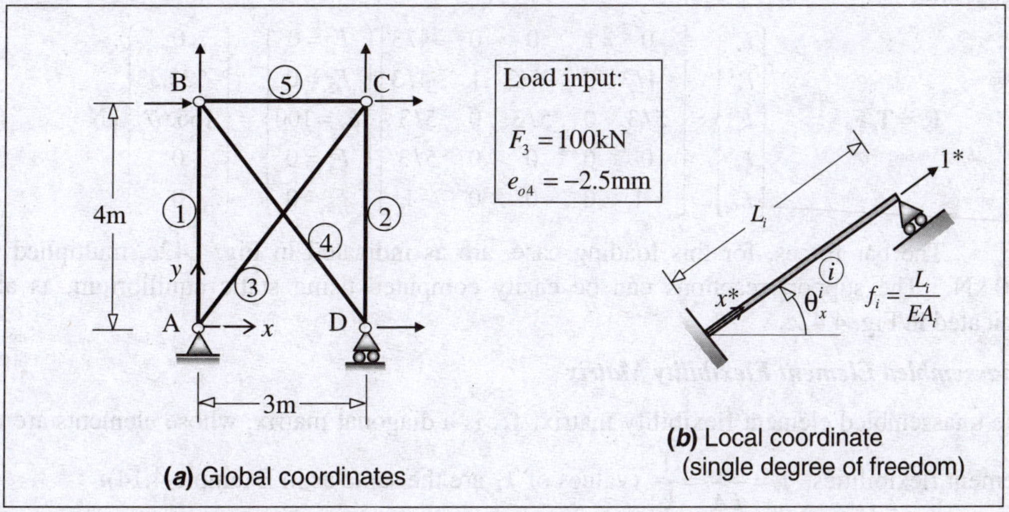

Figure 4.41 Global and local coordinates (Example 4.15)

Solution Procedure

The load input comprises a direct action, $F_3 = 100\text{kN}$, and indirect loading in the form of 'lack of fit' in bar '4', $D^4_{*\text{initial}} = e_{o4} = -2.5\text{mm}$. As the truss is 'just-rigid' (statically determinate), the indirect loading will not generate any force response. The bar forces are directly given by $\mathbf{F}_* = \mathbf{T}_\mathbf{F}\mathbf{F}$ and the joint displacements by $\mathbf{D} = \mathbf{D}_{\textbf{initial}} + \mathbf{f}\ \mathbf{F}$, where $\mathbf{D}_{\textbf{initial}} = \mathbf{T}_\mathbf{F}^\mathbf{T}\mathbf{D}_{*\textbf{initial}}$, where:

$$\mathbf{F} = \begin{Bmatrix} F_1 \\ F_2 \\ F_3 \\ F_4 \\ F_5 \end{Bmatrix} = \begin{Bmatrix} 0 \\ 0 \\ 100 \\ 0 \\ 0 \end{Bmatrix} \text{kN} \ ; \quad \mathbf{D}_{*\text{initial}} = \begin{Bmatrix} D^1_{*\text{initial}} \\ D^2_{*\text{initial}} \\ D^3_{*\text{initial}} \\ D^4_{*\text{initial}} \\ D^5_{*\text{initial}} \end{Bmatrix} = \begin{Bmatrix} 0 \\ 0 \\ 0 \\ -0.0025 \\ 0 \end{Bmatrix} \text{m}$$

We need to generate the force transformation matrix $\mathbf{T_F}$ and the unassembled flexibility matrix, $\mathbf{f}_*$, and then generate the structure flexibility matrix $\mathbf{f} = \mathbf{T_F}^{\mathrm{T}}\mathbf{f}_*\mathbf{T_F}$. As $n = 5$, these three matrices will each have an order of 5×5.

Force Transformation Matrix and Member Forces

In order to generate the force transformation matrix $\mathbf{T_F}$, we need to analyse the truss and find the bar forces, satisfying static equilibrium, due to the action of a unit load, $F_j = 1$, applied one at a time at every global coordinate location, as shown in Fig. 4.42.

Using the results shown in Fig. 4.42, and referring to the element numbers shown in Fig. 4.41a,

$$\mathbf{F}_* = \mathbf{T_F}\mathbf{F} \quad \Rightarrow \quad \begin{Bmatrix} F^1_* \\ F^2_* \\ F^3_* \\ F^4_* \\ F^5_* \end{Bmatrix} = \begin{bmatrix} 0 & 1 & 0 & 0 & -4/3 \\ -4/3 & 0 & -4/3 & 1 & -4/3 \\ 5/3 & 0 & 5/3 & 0 & 5/3 \\ 0 & 0 & 0 & 0 & 5/3 \\ -1 & 0 & 0 & 0 & -1 \end{bmatrix} \begin{Bmatrix} F_1 = 0 \\ F_2 = 0 \\ F_3 = 100 \\ F_4 = 0 \\ F_5 = 0 \end{Bmatrix} = \begin{Bmatrix} 0 \\ -133.33 \\ 166.67 \\ 0 \\ 0 \end{Bmatrix} \text{kN}$$

The bar forces, for this loading case, are as indicated in Fig. 4.42c, multiplied by 100 kN. The support reactions can be easily computed using static equilibrium, as also indicated in Fig. 4.42c.

Unassembled Element Flexibility Matrix

The unassembled element flexibility matrix, $\mathbf{f}_*$, is a diagonal matrix, whose elements are the element flexibilities, $f_i = \dfrac{L_i}{EA_i} = \dfrac{1}{k_i}$ (values of k_i are the same as in Example 4.14):

$$\mathbf{f}_* = 10^{-5} \times \begin{bmatrix} 0.8 & 0 & 0 & 0 & 0 \\ 0 & 0.8 & 0 & 0 & 0 \\ 0 & 0 & 1.25 & 0 & 0 \\ 0 & 0 & 0 & 1.25 & 0 \\ 0 & 0 & 0 & 0 & 1.5 \end{bmatrix} \text{m/kN}$$

Structure Flexibility Matrix

Carrying out the matrix operation, $\mathbf{f} = \mathbf{T_F}^{\mathrm{T}}\mathbf{f}_*\mathbf{T_F}$,

$$\mathbf{f} = \mathbf{T_F}^{\mathbf{T}} \mathbf{f_*}\, \mathbf{T_F} = 10^{-5} \times \begin{bmatrix} 6.3944 & 0 & 4.8944 & -1.0667 & 6.3944 \\ 0 & 0.8 & 0 & 0 & -1.0667 \\ 4.8944 & 0 & 4.8944 & -1.0667 & 4.8944 \\ -1.0667 & 0 & -1.0667 & 0.8 & -1.0667 \\ 6.3944 & -1.0667 & 4.8944 & -1.0667 & 11.289 \end{bmatrix} \text{m/kN}$$

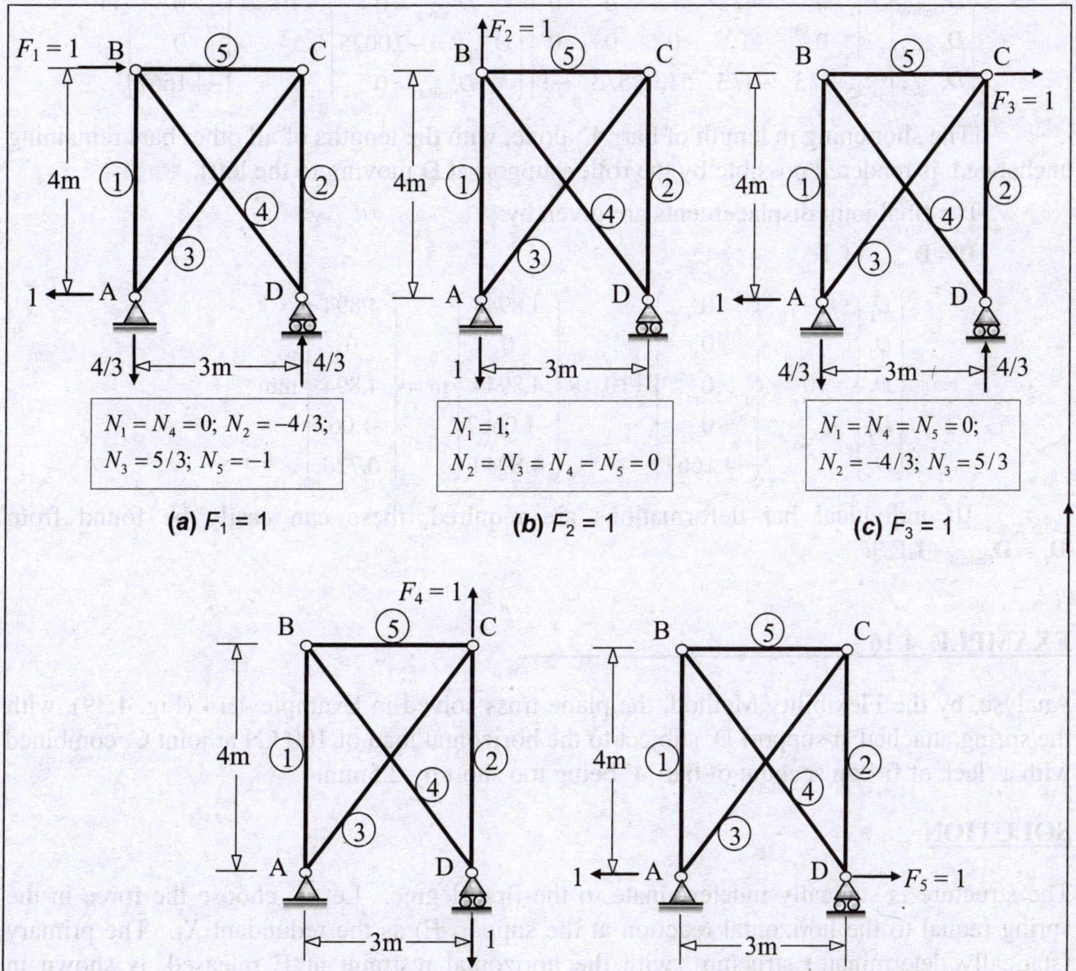

Figure 4.42 Generation of force transformation matrix, applying $F_j = 1$ (Example 4.15)

Joint Displacements

The initial joint displacements due to the indirect loading are given by:

$$\mathbf{D}_{\text{initial}} = \mathbf{T}_{\mathbf{F}}^{\mathbf{T}} \mathbf{D}_{*\text{initial}}$$

$$\Rightarrow \begin{Bmatrix} D_{1,\text{initial}} \\ D_{2,\text{initial}} \\ D_{3,\text{initial}} \\ D_{4,\text{initial}} \\ D_{5,\text{initial}} \end{Bmatrix} = \begin{bmatrix} 0 & -4/3 & 5/3 & 0 & -1 \\ 1 & 0 & 0 & 0 & 0 \\ 0 & -4/3 & 5/3 & 0 & 0 \\ 0 & 1 & 0 & 0 & 0 \\ -4/3 & -4/3 & 5/3 & 5/3 & -1 \end{bmatrix} \begin{Bmatrix} D^1_{*\text{initial}} = 0 \\ D^2_{*\text{initial}} = 0 \\ D^3_{*\text{initial}} = 0 \\ D^4_{*\text{initial}} = -0.0025 \\ D^5_{*\text{initial}} = 0 \end{Bmatrix} = 10^{-3} \times \begin{Bmatrix} 0 \\ 0 \\ 0 \\ 0 \\ -4.1667 \end{Bmatrix} \text{m}$$

[The shortening in length of bar '4' alone, with the lengths of all other bars remaining unchanged, is rendered possible by the roller support at D moving to the left].

The final joint displacements are given by:

$$\mathbf{D} = \mathbf{D}_{\text{initial}} + \mathbf{f}\ \mathbf{F}$$

$$\Rightarrow \begin{Bmatrix} D_1 \\ D_2 \\ D_3 \\ D_4 \\ D_5 \end{Bmatrix} = 10^{-3} \times \begin{Bmatrix} 0 \\ 0 \\ 0 \\ 0 \\ -4.1667 \end{Bmatrix} + 10^{-3} \times \begin{Bmatrix} 4.8944 \\ 0 \\ 4.8944 \\ -1.0667 \\ 4.8944 \end{Bmatrix} \text{m} = \begin{Bmatrix} 4.894 \\ 0 \\ 4.894 \\ -1.067 \\ 0.728 \end{Bmatrix} \text{mm}$$

[If individual bar deformations are required, these can easily be found from $\mathbf{D}_* = \mathbf{D}_{*\text{initial}} + \mathbf{f}_*\mathbf{F}_*$].

EXAMPLE 4.16

Analyse, by the Flexibility Method, the plane truss solved in Example 4.14 (Fig. 4.39), with the spring attached at support D, subject to the horizontal load of 100 kN at joint C, combined with a 'lack of fit' on account of bar '4' being too short by 2.5mm.

SOLUTION

The structure is statically indeterminate to the first degree. Let us choose the force in the spring (equal to the horizontal reaction at the support E) as the redundant X_1. The primary (statically determinate) structure, with the horizontal restraint at E released, is shown in Fig. 4.43. As in Example 4.14, we can treat the spring as another axial element (labelled '6'), having a flexibility $f_6 = \frac{1}{k_s} = \frac{1}{8} = 0.125\,\text{mm/kN}$. Ignoring the restrained coordinates, the global coordinates are as shown in Fig. 4.41, which is similar to the system in Example 4.15, except for the additional global coordinate at E, labelled '6'.

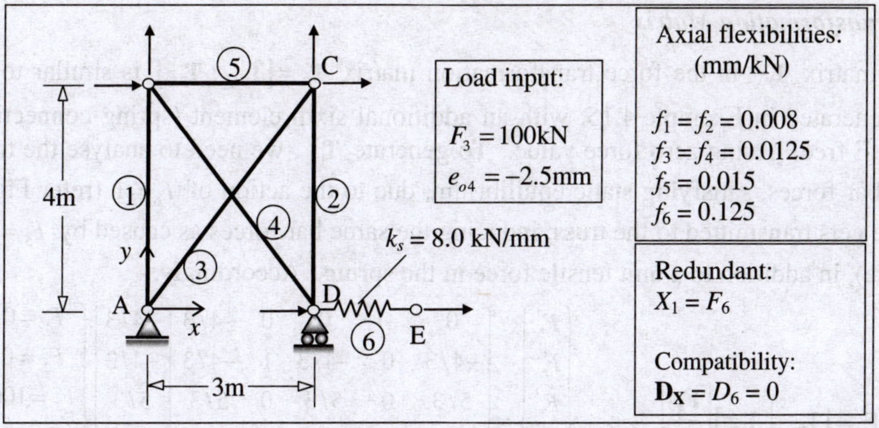

Figure 4.43 Truss in Example 4.16 (spring force as redundant)

Solution Procedure

The load input comprises a direct action, $F_3 = 100\text{kN}$, and indirect loading in the form of 'lack of fit' in bar '4', $D^4_{*\text{initial}} = e_{o4} = -2.5\text{mm}$. The bar forces are given by $\mathbf{F_* = T_{FA} F_A + T_{FX} F_X}$ and the joint displacements by $\mathbf{D_A = D_{A,initial} + f_{AA} F_A + f_{AX} F_X}$, where $\left[\begin{array}{c}\mathbf{D_{A,initial}}\\ \hline \mathbf{D_{X,initial}}\end{array}\right] = \left[\begin{array}{c}\mathbf{T^T_{FA}}\\ \hline \mathbf{T^T_{FX}}\end{array}\right]\mathbf{D_{*initial}}$, and the redundant $\mathbf{F_X} = X_1$ is obtained by solving the compatibility condition, $\mathbf{D_X = D_{X,initial} + f_{XA} F_A + f_{XX} F_X = O}$. The load input is given as follows.

$$\mathbf{F_A} = \begin{Bmatrix} F_1 \\ F_2 \\ F_3 \\ F_4 \\ F_5 \end{Bmatrix} = \begin{Bmatrix} 0 \\ 0 \\ 100 \\ 0 \\ 0 \end{Bmatrix} \text{kN} \;;\quad \mathbf{D_{*initial}} = \begin{Bmatrix} D^1_{*\text{initial}} \\ D^2_{*\text{initial}} \\ D^3_{*\text{initial}} \\ D^4_{*\text{initial}} \\ D^5_{*\text{initial}} \\ D^6_{*\text{initial}} \end{Bmatrix} = \begin{Bmatrix} 0 \\ 0 \\ 0 \\ -0.0025 \\ 0 \\ 0 \end{Bmatrix} \text{m}$$

We need to generate the force transformation matrix $\mathbf{T_F} = [\mathbf{T_{FA}} \mid \mathbf{T_{FX}}]$ and the unassembled flexibility matrix, $\mathbf{f_*}$, and then generate the structure flexibility matrix $\mathbf{f = T_F{}^T f_* T_F}$, partitioned as $\mathbf{f} = \left[\begin{array}{c|c}\mathbf{f_{AA}} & \mathbf{f_{AX}}\\ \hline \mathbf{f_{XA}} & \mathbf{f_{XX}}\end{array}\right]$. These three matrices will each have an order of 6×6. The $[\mathbf{f_{XX}}]$ sub-matrix has an order of 1×1. The unknown redundant can be computed using the formula, $\mathbf{F_X} = [\mathbf{f_{XX}}]^{-1}\left[\mathbf{D_X - D_{X,initial} - f_{XA} F_A}\right]$.

Force Transformation Matrix

The sub-matrix $\mathbf{T_{FA}}$ in the force transformation matrix $\mathbf{T_F} = [\mathbf{T_{FA}} \mid \mathbf{T_{FX}}]$ is similar to the $\mathbf{T_F}$ matrix generated in Example 4.15, with an additional sixth element (spring connecting DE, with end E free) having zero force value. To generate $\mathbf{T_{FX}}$, we need to analyse the truss and find the bar forces, satisfying static equilibrium, due to the action of $F_6 = 1$ (refer Fig. 4.43); this force gets transmitted to the truss, inducing the same bar forces as caused by $F_5 = 1$ (refer Fig. 4.42e), in addition to a unit tensile force in the spring. Accordingly,

$$\mathbf{F_*} = [\mathbf{T_{FA}} \mid \mathbf{T_{FX}}]\begin{bmatrix}\mathbf{F_A}\\ \hline \mathbf{F_X}\end{bmatrix} \Rightarrow \begin{Bmatrix}F_*^1\\F_*^2\\F_*^3\\F_*^4\\F_*^5\\F_*^6\end{Bmatrix} = \left[\begin{array}{ccccc|c} 0 & -1 & 0 & 0 & -4/3 & -4/3\\ -4/3 & 0 & -4/3 & 1 & -4/3 & -4/3\\ 5/3 & 0 & 5/3 & 0 & 5/3 & 5/3\\ 0 & 0 & 0 & 0 & 5/3 & 5/3\\ -1 & 0 & 0 & 0 & -1 & -1\\ 0 & 0 & 0 & 0 & 0 & 1 \end{array}\right]\begin{Bmatrix}F_1=0\\F_2=0\\F_3=100\\F_4=0\\F_5=0\\ \hline F_6=X_1\end{Bmatrix}\text{kN}$$

The initial joint displacements due to the indirect loading are given by:

$$\begin{bmatrix}\mathbf{D_{A,initial}}\\ \hline \mathbf{D_{X,initial}}\end{bmatrix} = \begin{bmatrix}\mathbf{T_{FA}^T}\\ \hline \mathbf{T_{FX}^T}\end{bmatrix}\mathbf{D_{*initial}}$$

$$\Rightarrow \begin{Bmatrix}D_{1,\text{initial}}\\D_{2,\text{initial}}\\D_{3,\text{initial}}\\D_{4,\text{initial}}\\D_{5,\text{initial}}\\ \hline D_{6,\text{initial}}\end{Bmatrix} = \left[\begin{array}{cccccc} 0 & -4/3 & 5/3 & 0 & -1 & 0\\ -1 & 0 & 0 & 0 & 0 & 0\\ 0 & -4/3 & 5/3 & 0 & 0 & 0\\ 0 & 1 & 0 & 0 & 0 & 0\\ -4/3 & -4/3 & 5/3 & 5/3 & -1 & 0\\ \hline -4/3 & -4/3 & 5/3 & 5/3 & -1 & 1 \end{array}\right]\begin{Bmatrix}0\\0\\0\\-0.0025\\0\\0\end{Bmatrix} = 10^{-3}\times\begin{Bmatrix}0\\0\\0\\0\\-4.1667\\ \hline -4.1667\end{Bmatrix}\text{m}$$

Unassembled Element Flexibility Matrix

The unassembled element flexibility matrix, $\mathbf{f_*}$, is identical to the one in Example 4.15, except that a sixth diagonal element with flexibility $f_6 = 0.125$ mm/kN needs to be introduced:

$$\mathbf{f_*} = 10^{-5}\times\begin{bmatrix} 0.8 & 0 & 0 & 0 & 0 & 0\\ 0 & 0.8 & 0 & 0 & 0 & 0\\ 0 & 0 & 1.25 & 0 & 0 & 0\\ 0 & 0 & 0 & 1.25 & 0 & 0\\ 0 & 0 & 0 & 0 & 1.5 & 0\\ 0 & 0 & 0 & 0 & 0 & 12.5 \end{bmatrix}\text{m/kN}$$

Structure Flexibility Matrix

Carrying out the matrix operation, $\mathbf{f} = \mathbf{T_F}^T\mathbf{f_*}\mathbf{T_F}$,

$$\mathbf{f} = \left[\begin{array}{c|c} \mathbf{f_{AA}} & \mathbf{f_{AX}} \\ \hline \mathbf{f_{XA}} & \mathbf{f_{XX}} \end{array}\right] = 10^{-5} \times \left[\begin{array}{ccccc|c} 6.3944 & 0 & 4.8944 & -1.0667 & 6.3944 & 6.3944 \\ 0 & 0.8 & 0 & 0 & -1.0667 & -1.0667 \\ 4.8944 & 0 & 4.8944 & -1.0667 & 4.8944 & 4.8944 \\ -1.0667 & 0 & -1.0667 & 0.8 & -1.0667 & -1.0667 \\ 6.3944 & -1.0667 & 4.8944 & -1.0667 & 11.289 & 11.289 \\ \hline 6.3944 & -1.0667 & 4.8944 & -1.0667 & 11.289 & 23.789 \end{array}\right] \text{m/kN}$$

$$\Rightarrow \mathbf{f_{XX}} = 23.789\times10^{-5}\ \text{m/kN}$$

(Note that the $\mathbf{f_{AA}}$ sub-matrix is identical to the $\mathbf{f}$ matrix of Example 4.15).

Redundant

$$\mathbf{F_X} = \left[\mathbf{f_{XX}}\right]^{-1}\left[\mathbf{D_X} - \mathbf{D_{X,initial}} - \mathbf{f_{XA}}\mathbf{F_A}\right]$$

$$\Rightarrow X_1 = \left[23.789\times10^{-5}\right]^{-1} \times \left[0-(-4.1667\times10^{-3})-10^{-5}\times\left[6.3944 \quad -1.0667 \quad 4.8944 \quad -1.0667 \quad 11.289\right]\begin{Bmatrix} 0 \\ 0 \\ 100 \\ 0 \\ 0 \end{Bmatrix}\right] = -3.0593\ \text{kN}$$

Member Forces

$$\mathbf{F_*} = \left[\mathbf{T_{FA}} \mid \mathbf{T_{FX}}\right]\left[\begin{array}{c} \mathbf{F_A} \\ \hline \mathbf{F_X} \end{array}\right] \Rightarrow \begin{Bmatrix} F_*^1 \\ F_*^2 \\ F_*^3 \\ F_*^4 \\ F_*^5 \\ F_*^6 \end{Bmatrix} = \left[\begin{array}{ccccc|c} 0 & -1 & 0 & 0 & -4/3 & -4/3 \\ -4/3 & 0 & -4/3 & 1 & -4/3 & -4/3 \\ 5/3 & 0 & 5/3 & 0 & 5/3 & 5/3 \\ 0 & 0 & 0 & 0 & 5/3 & 5/3 \\ -1 & 0 & 0 & 0 & -1 & -1 \\ 0 & 0 & 0 & 0 & 0 & 1 \end{array}\right]\left\{\begin{array}{c} 0 \\ 0 \\ 100 \\ 0 \\ 0 \\ \hline -3.0593 \end{array}\right\} = \begin{Bmatrix} 4.079 \\ -129.25 \\ 161.57 \\ -5.099 \\ 3.059 \\ -3.059 \end{Bmatrix} \text{kN}$$

These results match exactly with the results in Example 4.14. The support reactions can be obtained by considering joint equilibrium at supports A and D, yielding results identical to those generated in Example 4.14 by the Stiffness Method.

Joint Displacements

The final joint displacements are given by:

$$\mathbf{D_A} = \mathbf{D_{A,initial}} + \left[\mathbf{f_{AA}} \mid \mathbf{f_{AX}}\right]\left[\begin{array}{c} \mathbf{F_A} \\ \hline \mathbf{F_X} \end{array}\right]$$

$$\Rightarrow \begin{Bmatrix} D_1 \\ D_2 \\ D_3 \\ D_4 \\ D_5 \end{Bmatrix} = 10^{-3} \times \begin{Bmatrix} 0 \\ 0 \\ 0 \\ 0 \\ -4.1667 \end{Bmatrix} + 10^{-3} \times \begin{Bmatrix} 4.6988 \\ 0.0326 \\ 4.7447 \\ -1.0340 \\ 4.5491 \end{Bmatrix} \text{m} = \begin{Bmatrix} 4.699 \\ 0.033 \\ 4.745 \\ -1.034 \\ 0.382 \end{Bmatrix} \text{mm}$$

These results match exactly with the results in Example 4.14.

EXAMPLE 4.17

Analyse the pin-jointed frame in Examples 4.9 and 4.12 (Fig. 4.29) by the Flexibility Method and find the bar forces and joint displacements due to the action of the applied loads, $F_1 = F_2 = -P$, and 'lack of fit' due to bars '1' and '4' being short by δ, assuming constant EA.

SOLUTION

The structure is statically indeterminate to the second degree. Let us choose the forces in the two intermediate bars, '2' and '3', as the redundants: $X_1 = N_2 = F_*^2$ and $X_2 = N_3 = F_*^3$. The global coordinates in the primary structure (with the bars '3' and '4' cut in between[†]) are as shown in Fig. 4.44a. Each element has only one degree of freedom, as shown in Fig. 4.44b.

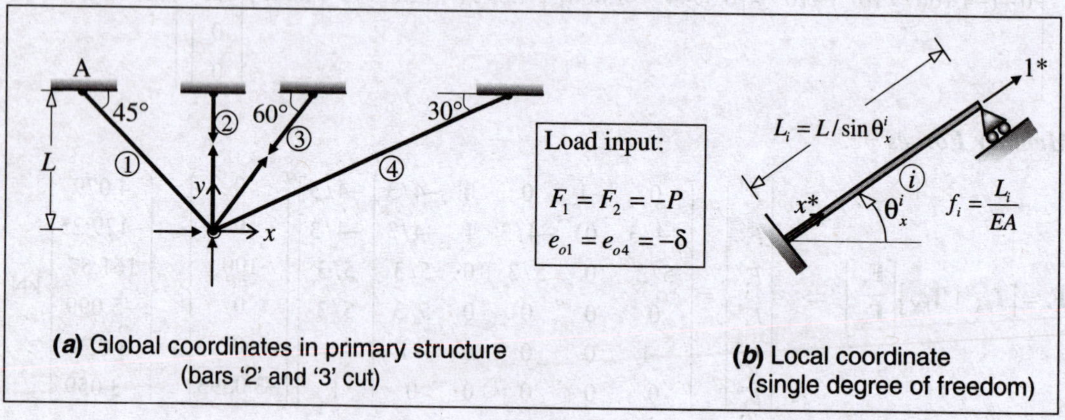

Figure 4.44 Global and local coordinates (Example 4.17)

[†] The displacements D_2 and D_3, in the primary structure, may be visualised as the relative displacements between the cut ends of the bars '2' and '3' respectively. Compatibility requires $D_2 = D_3 = 0$.

Solution Procedure

The load input is given, as in Examples 4.9 and 4.12, in terms of the load vector, $\mathbf{F_A} = \begin{Bmatrix} F_1 \\ F_2 \end{Bmatrix} = \begin{Bmatrix} -P \\ -P \end{Bmatrix}$ (direct loading), and the initial element displacement vector,

$$\mathbf{D_{*initial}} = \begin{Bmatrix} D^1_{*\text{initial}} \\ D^2_{*\text{initial}} \\ D^3_{*\text{initial}} \\ D^4_{*\text{initial}} \end{Bmatrix} = \begin{Bmatrix} e_{o1} \\ e_{o2} \\ e_{o3} \\ e_{o4} \end{Bmatrix} = \begin{Bmatrix} -\delta \\ 0 \\ 0 \\ -\delta \end{Bmatrix} \text{ (indirect loading).}$$

The bar forces are given by $\mathbf{F_* = T_{FA}F_A + T_{FX}F_X}$ and the joint displacements by $\mathbf{D_A = D_{A,initial} + f_{AA}F_A + f_{AX}F_X}$, where $\begin{bmatrix} \mathbf{D_{A,initial}} \\ \hline \mathbf{D_{X,initial}} \end{bmatrix} = \begin{bmatrix} \mathbf{T^T_{FA}} \\ \hline \mathbf{T^T_{FX}} \end{bmatrix} \mathbf{D_{*initial}}$, and the redundant $\mathbf{F_X} = \begin{Bmatrix} F_3 \\ F_4 \end{Bmatrix} = \begin{Bmatrix} X_1 \\ X_2 \end{Bmatrix}$ is obtained by solving the compatibility condition, $\mathbf{D_X = D_{X,initial} + f_{XA}F_A + f_{XX}F_X = O}$.

We need to generate the force transformation matrix $\mathbf{T_F = [T_{FA} \mid T_{FX}]}$ and the unassembled flexibility matrix, $\mathbf{f_*}$, and then generate the structure flexibility matrix $\mathbf{f = T_F{}^T f_* T_F}$, partitioned as $\mathbf{f} = \left[\begin{array}{c|c} \mathbf{f_{AA}} & \mathbf{f_{AX}} \\ \hline \mathbf{f_{XA}} & \mathbf{f_{XX}} \end{array}\right]$. These three matrices will each have an order of 4×4. The $[\mathbf{f_{XX}}]$ sub-matrix has an order of 2×2. The unknown redundants can be computed using the formula, $\mathbf{F_X = [f_{XX}]^{-1}[D_X - D_{X,initial} - f_{XA}F_A]}$.

Force Transformation Matrix

To generate the force transformation matrix $\mathbf{T_F = [T_{FA} \mid T_{FX}]}$, we need to analyse the primary structure and find the bar forces, applying a unit load $F_j = 1$ at a time, satisfying static equilibrium, as indicated in Fig. 4.45. Accordingly,

$$\mathbf{F_* = [T_{FA} \mid T_{FX}]} \begin{bmatrix} \mathbf{F_A} \\ \hline \mathbf{F_X} \end{bmatrix} \Rightarrow \begin{Bmatrix} F^1_* \\ F^2_* \\ F^3_* \\ F^4_* \end{Bmatrix} = \left[\begin{array}{cc|cc} 0.51764 & -0.89658 & -0.89658 & -0.51764 \\ 0 & 0 & 1 & 0 \\ 0 & 0 & 0 & 1 \\ -0.73205 & -0.73205 & -0.73205 & -1 \end{array}\right] \begin{Bmatrix} F_1 = -P \\ F_2 = -P \\ \hline F_3 = X_1 \\ F_4 = X_2 \end{Bmatrix}$$

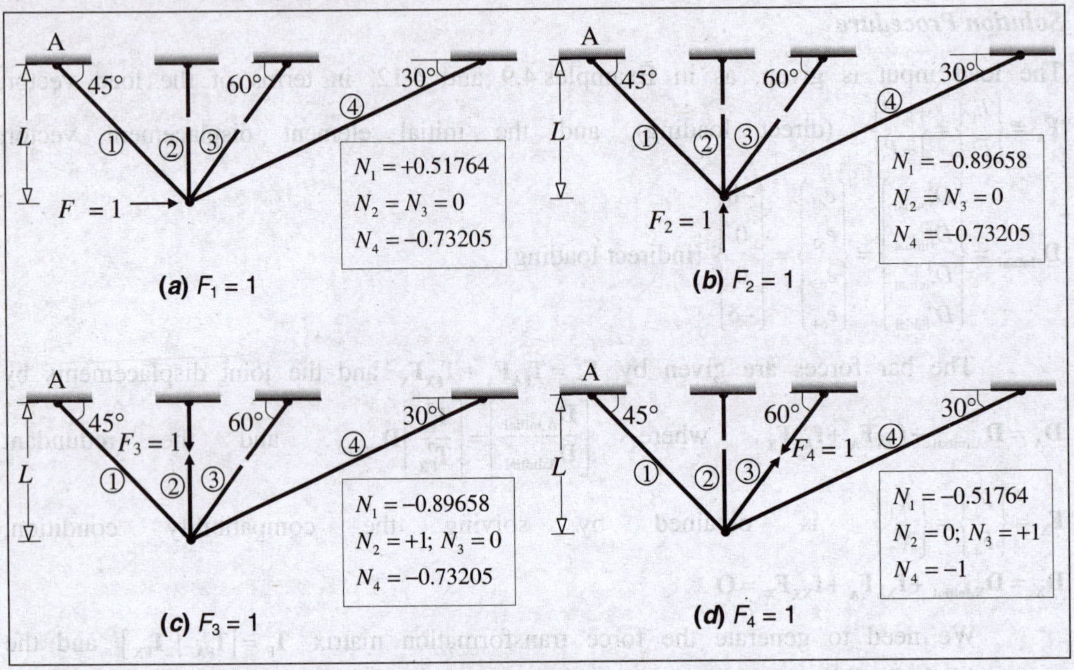

Figure 4.45 Generation of force transformation matrix, applying $F_j = 1$ (Example 4.17)

The initial joint displacements due to the indirect loading are given by:

$$\begin{bmatrix} \mathbf{D}_{\text{A,initial}} \\ \hdashline \mathbf{D}_{\text{X,initial}} \end{bmatrix} = \begin{bmatrix} \mathbf{T}_{\text{FA}}^{\text{T}} \\ \hdashline \mathbf{T}_{\text{FX}}^{\text{T}} \end{bmatrix} \mathbf{D}_{*\text{initial}} \Rightarrow \begin{Bmatrix} D_{1,\text{initial}} \\ D_{2,\text{initial}} \\ \hdashline D_{3,\text{initial}} \\ D_{4,\text{initial}} \end{Bmatrix} = \begin{Bmatrix} 0.21441 \\ 1.6286 \\ \hdashline 1.6286 \\ 1.5176 \end{Bmatrix} \delta$$

Unassembled Element Flexibility Matrix

The unassembled element flexibility matrix, $\mathbf{f}_*$, is a diagonal matrix, whose diagonal elements are given by $f_i = \dfrac{L_i}{EA}$:

$$\mathbf{f}_* = \frac{L}{EA} \begin{bmatrix} 1.4142 & 0 & 0 & 0 \\ 0 & 1 & 0 & 0 \\ 0 & 0 & 1.1547 & 0 \\ 0 & 0 & 0 & 2 \end{bmatrix}$$

Structure Flexibility Matrix

Carrying out the matrix operation, $\mathbf{f} = \mathbf{T}_\mathbf{F}{}^{\mathbf{T}} \mathbf{f}_* \mathbf{T}_\mathbf{F}$,

$$\mathbf{f} = \left[\begin{array}{c|c} \mathbf{f_{AA}} & \mathbf{f_{AX}} \\ \hline \mathbf{f_{XA}} & \mathbf{f_{XX}} \end{array}\right] = \frac{L}{EA}\left[\begin{array}{cc|cc} 1.4507 & 0.4155 & 0.4155 & 1.0852 \\ 0.4155 & 2.2086 & 2.2086 & 2.1204 \\ \hline 0.4155 & 2.2086 & 3.2086 & 2.1204 \\ 1.0852 & 2.1204 & 2.1204 & 3.5336 \end{array}\right]$$

$$\Rightarrow \quad \mathbf{f_{XX}} = \frac{L}{EA}\begin{bmatrix} 3.2086 & 2.1204 \\ 2.1204 & 3.5336 \end{bmatrix} \quad \Rightarrow \quad [\mathbf{f_{XX}}]^{-1} = \frac{EA}{L}\begin{bmatrix} 0.51648 & -0.30992 \\ -0.30992 & 0.46897 \end{bmatrix}$$

Redundants

$$\mathbf{F_X} = [\mathbf{f_{XX}}]^{-1}\left[\mathbf{D_X} - \mathbf{D_{X,initial}} - \mathbf{f_{XA}}\mathbf{F_A}\right]$$

$$\mathbf{D_X} = \mathbf{0} \;\Rightarrow\; \mathbf{F_X} = \begin{Bmatrix} X_1 \\ X_2 \end{Bmatrix} = \begin{Bmatrix} 0.3618 \\ 0.6901 \end{Bmatrix} P + \begin{Bmatrix} -0.3708 \\ -0.2070 \end{Bmatrix}\frac{EA\delta}{L}$$

Member Forces

$$\mathbf{F_*} = \left[\mathbf{T_{FA}} \mid \mathbf{T_{FX}}\right]\left[\begin{array}{c} \mathbf{F_A} \\ \hline \mathbf{F_X} \end{array}\right] \Rightarrow \begin{Bmatrix} F_*^1 \\ F_*^2 \\ F_*^3 \\ F_*^4 \end{Bmatrix} = \begin{Bmatrix} -0.3026 \\ 0.3618 \\ 0.6901 \\ 0.5092 \end{Bmatrix} P + \begin{Bmatrix} 0.4395 \\ -0.3708 \\ -0.2070 \\ 0.4784 \end{Bmatrix}\frac{EA\delta}{L}$$

These results match exactly with the results in Example 4.9 and 4.12.

Joint Displacements

The final joint displacements are given by:

$$\mathbf{D_A} = \mathbf{D_{A,initial}} + \left[\mathbf{f_{AA}} \mid \mathbf{f_{AX}}\right]\left[\begin{array}{c} \mathbf{F_A} \\ \hline \mathbf{F_X} \end{array}\right]$$

$$\Rightarrow \begin{Bmatrix} D_1 \\ D_2 \end{Bmatrix} = \begin{Bmatrix} 0.2144 \\ 1.6287 \end{Bmatrix}\delta + \begin{Bmatrix} 0.9670 \\ -0.3618 \end{Bmatrix}\frac{PL}{EA} + \begin{Bmatrix} -0.3787 \\ -1.2579 \end{Bmatrix}\delta = \begin{Bmatrix} 0.9670 \\ -0.3618 \end{Bmatrix}\frac{PL}{EA} + \begin{Bmatrix} -0.1643 \\ 0.3708 \end{Bmatrix}\delta$$

These results match exactly with the results in Example 4.9 and 4.12.

4.5 SPACE TRUSSES

In this Section, we shall learn to analyse space trusses, made up of straight prismatic axial elements, for which we have derived element stiffness and flexibility matrix relations in Section 4.2.4. The basic procedure is similar to that of plane trusses, except that the sizes of the matrices involved become larger, and the computations become relatively cumbersome to explain through a step-by-step procedure. In practice, many space trusses, such as transmission line towers are statically determinate, and the bar forces can be analysed by the *method of tension coefficients.* Nevertheless, in spite of the zero (or low degree) of static indeterminacy, it is common practice to use computer software, which is based on the stiffness method, to solve such problems, although the degree of kinematic indeterminacy is high.

In the conventional stiffness formulation, we need to deal with 6 degrees of freedom per element, and to also assign appropriate direction cosines for each element, with reference to the global axes system. The computations are greatly simplified, if we adopt the reduced element stiffness method, wherein we need to deal with only one degree of freedom per element, as in the case of plane trusses. Here, we shall demonstrate the application of this simplified stiffness method to space trusses. The detailed analysis by the conventional stiffness method and the flexibility method are not described here.

4.5.1 Analysis by Reduced Element Stiffness Method

If the space truss has *m* elements, the *reduced unassembled element stiffness matrix,* $\tilde{\mathbf{k}}_*$, is a simple diagonal matrix of order $m \times m$, as given by Eq. 4.35, with diagonal elements defined by $k_i = \dfrac{E_i A_i}{L_i}$. If there are a large number of elements involved, as is usually the case in space trusses encountered in practice, then it is convenient to work with individual elements, each with a 1×1 reduced element stiffness matrix, $\tilde{\mathbf{k}}_*^{\mathbf{i}} = k_i = \dfrac{E_i A_i}{L_i}$.

As explained earlier, the structure stiffness matrix $\mathbf{k}$ can be assembled by a 'direct stiffness' approach, using the element displacement transformation matrix, $\mathbf{T_D^i}$, and summing up the contributions, $\mathbf{T_D^i}^{\mathbf{T}} \tilde{\mathbf{k}}_*^{\mathbf{i}} \mathbf{T_D^i}$ (direct stiffness method).

We can generate the displacement transformation $\mathbf{T_D^i}$ matrices for the various elements, each of order $1 \times (n+r)$, applying a unit displacement at a time ($D_j = 1$) on the restrained (primary) structure, whereby $\mathbf{D_*^i} = \left[\mathbf{T_{DA}^i} \mid \mathbf{T_{DR}^i} \right] \begin{bmatrix} \mathbf{D_A} \\ \mathbf{D_R} \end{bmatrix}$. For example, let us consider a typical truss element *i*, with *start node* at S (X_{iS}, Y_{iS}, Z_{iS}), *end node* at E (X_{iE}, Y_{iE}, Z_{iE}) in the global axes system, with a length $L_i = \sqrt{(X_{iE} - X_{iS})^2 + (Y_{iE} - Y_{iS})^2 + (Z_{iE} - Z_{iS})^2}$, and direction cosines, $c_{ix} = \dfrac{X_{iE} - X_{iS}}{L_i}$, $c_{iy} = \dfrac{Y_{iE} - Y_{iS}}{L_i}$ and $c_{iz} = \dfrac{Z_{iE} - Z_{iS}}{L_i}$. By extending the derivation for the plane truss element in Fig. 4.36a (involving 2 global coordinates at each end), to a space truss element with 3 global coordinates at each end, we can generate the components of the $\mathbf{T_D^i}$ matrix as follows, using a global coordinate numbering system of '1', '2', '3' to define the global x-, y- and z- directions at the start node respectively, and likewise, '4', '5', and '6' at the end node of the element, for convenience.

$$\mathbf{D}_*^{\mathbf{i}} = \mathbf{T}_{\mathbf{D}}^{\mathbf{i}}\mathbf{D} \quad \Rightarrow \quad \mathbf{T}_{\mathbf{D}}^{\mathbf{i}} = \left[-c_{ix} \quad -c_{iy} \quad -c_{iz} \mid c_{ix} \quad c_{iy} \quad c_{iz}\right] \tag{4.37}$$

Alternatively, we can adopt a static approach instead of a kinematic approach, as indicated in Fig. 4.36b, to generate the $\mathbf{T}_{\mathbf{D}}^{\mathbf{i}\,\mathrm{T}}$ matrix, by making use of the *Principle of Contragradient*, whereby $\begin{bmatrix} \mathbf{F_A} \\ \hline \mathbf{F_R} \end{bmatrix} = \begin{bmatrix} \mathbf{T}_{\mathbf{DA}}^{\mathbf{i}\,\mathrm{T}} \\ \hline \mathbf{T}_{\mathbf{DR}}^{\mathbf{i}\,\mathrm{T}} \end{bmatrix} \mathbf{F}_*^{\mathbf{i}}$, as explained earlier. Thus, we can show:

$$\mathbf{F} = \mathbf{T}_{\mathbf{D}}^{\mathbf{i}\,\mathrm{T}}\mathbf{F}_*^{\mathbf{i}} \quad \Rightarrow \quad \mathbf{T}_{\mathbf{D}}^{\mathbf{i}\,\mathrm{T}} = \begin{bmatrix} -c_{ix} \\ -c_{iy} \\ -c_{iz} \\ \hline c_{ix} \\ c_{iy} \\ c_{iz} \end{bmatrix} \tag{4.38}$$

In practice, the global coordinates may have any set of numbers, some of which may correspond to restrained coordinates. However, if the global x-, y- and z-axes are chosen appropriately, then the first block of three elements of the $\mathbf{T}_{\mathbf{D}}^{\mathbf{i}}$ matrix (or the $\mathbf{T}_{\mathbf{D}}^{\mathbf{i}\,\mathrm{T}}$ matrix) with negative values should be assigned to the *start node*, while the second block of positive three elements should be assigned to the *end node*.

It is convenient to also generate and store the $\tilde{\mathbf{k}}_*^{\mathbf{i}}\mathbf{T}_{\mathbf{D}}^{\mathbf{i}}$ matrices for the various elements to facilitate the computation of the bar forces using the relation, $\mathbf{F}_*^{\mathbf{i}} = \left(\tilde{\mathbf{k}}_*^{\mathbf{i}}\mathbf{T}_{\mathbf{D}}^{\mathbf{i}}\right)\mathbf{D}$, and also to store the products, $\mathbf{T}_{\mathbf{D}}^{\mathbf{i}\,\mathrm{T}}\tilde{\mathbf{k}}_*^{\mathbf{i}}\mathbf{T}_{\mathbf{D}}^{\mathbf{i}}$, for assembling the structure stiffness matrix $\mathbf{k}$. These take the following forms, each block representing either the start node coordinates or the end node coordinates.

$$\tilde{\mathbf{k}}_*^{\mathbf{i}}\mathbf{T}_{\mathbf{D}}^{\mathbf{i}} = \frac{E_i A_i}{L_i}\left[-c_{ix} \quad -c_{iy} \quad -c_{iz} \mid c_{ix} \quad c_{iy} \quad c_{iz}\right] \tag{4.39}$$

$$\mathbf{T}_{\mathbf{D}}^{\mathbf{i}\,\mathrm{T}}\tilde{\mathbf{k}}_*^{\mathbf{i}}\mathbf{T}_{\mathbf{D}}^{\mathbf{i}} = \frac{E_i A_i}{L_i}\left[\begin{array}{ccc|ccc} c_{ix}^2 & c_{ix}c_{iy} & c_{ix}c_{iz} & -c_{ix}^2 & -c_{ix}c_{iy} & -c_{ix}c_{iz} \\ c_{ix}c_{iy} & c_{iy}^2 & c_{iy}c_{iz} & -c_{ix}c_{iy} & -c_{iy}^2 & -c_{iy}c_{iz} \\ c_{ix}c_{iz} & c_{iy}c_{iz} & c_{iz}^2 & -c_{ix}c_{iz} & -c_{iy}c_{iz} & -c_{iz}^2 \\ \hline -c_{ix}^2 & -c_{ix}c_{iy} & -c_{ix}c_{iz} & c_{ix}^2 & c_{ix}c_{iy} & c_{ix}c_{iz} \\ -c_{ix}c_{iy} & -c_{iy}^2 & -c_{iy}c_{iz} & c_{ix}c_{iy} & c_{iy}^2 & c_{iy}c_{iz} \\ -c_{ix}c_{iz} & -c_{iy}c_{iz} & -c_{iz}^2 & c_{ix}c_{iz} & c_{iy}c_{iz} & c_{iz}^2 \end{array}\right] \tag{4.40}$$

When indirect loads in the form of temperature loading or 'lack of fit' are prescribed in the form of an initial (free) displacement vector $\mathbf{e_o}$ (of dimension m), these loads manifest as axial forces in the fixed end condition in the primary restrained structure, given by: $N_{if} = -k_i e_{oi}$, which need to be expressed in the element local coordinates, as discussed in the case of the plane truss (Eq. 4.37).

EXAMPLE 4.18

Consider a simple space frame comprising 3 identical members (in a tripod arrangement), each 2m long, inter-connected to a ball-and-socket joint at O at top, 1m above ground, with their hinged bases forming an equilateral triangle on level ground. Show how the Stiffness Method can be used to find the axial forces in the 3 members, when the joint at O is subject to a gravity load of 60 kN. Also, show that no axial forces will be induced in the members on account of any 'lack of fit' or temperature effects.

SOLUTION

Preliminary Solution

In this pin-jointed space frame, there are 3 elements ($m = 3$), 4 pin joints, i.e., $3j = 12$ global coordinates, of which only 3 (at O) correspond to *active* degrees of freedom ($n = 3$) and the remaining 9 to *restrained* global coordinates ($r = 9$). As, the truss is statically determinate ($m + r = 12 = 3j$), and symmetrically loaded, all the three bars will have the same axial (compressive) force. The sum of the vertical components of the axial forces will be equal to the applied load at O. If the axial force is N and the inclination of the member to the horizontal plane is θ,

$$\sum F_y = 0 \Rightarrow 3N\sin\theta + 60 = 0 \Rightarrow N = -\frac{20}{\sin\theta}$$

Figs 4.46a and b show the plan view and the side elevation of the structure. It can be seen that $\theta = 30^\circ \Rightarrow \sin\theta = 0.5$, whereby,

$$N = -\frac{20}{0.5} = -40 \text{ kN (compression)}$$

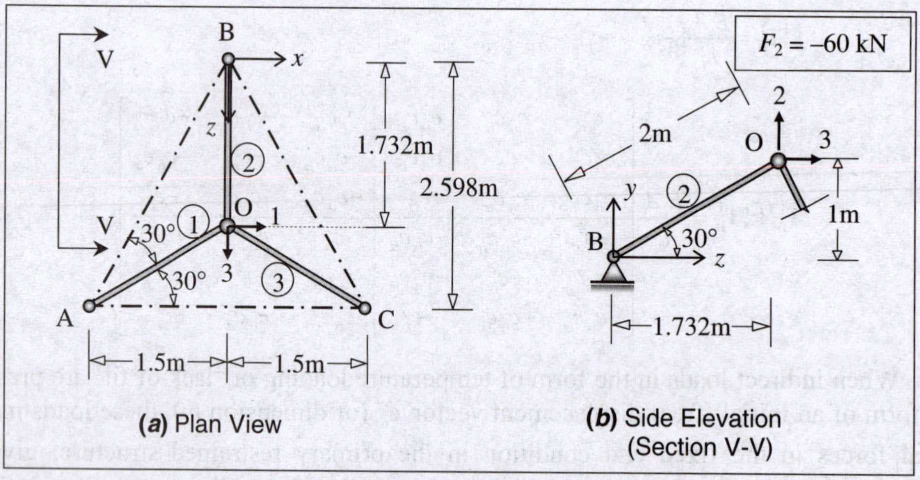

Figure 4.46 Plan view and side elevation of tripod structure (Example 4.18)

Solution by Stiffness Method (Reduced Element Stiffness)

Let us adopt the global axes system as indicated in Fig. 4.46, with the origin at B, and identifying the 3 active global coordinates at O, as shown. The coordinates of the start and end nodes, (X_{iS}, Y_{iS}, Z_{iS}) and (X_{iE}, Y_{iE}, Z_{iE}) respectively, of individual elements and their lengths, L_i, and direction cosines, $c_{ix} = \dfrac{X_{iE} - X_{iS}}{L_i}$, $c_{iy} = \dfrac{Y_{iE} - Y_{iS}}{L_i}$ and $c_{iz} = \dfrac{Z_{iE} - Z_{iS}}{L_i}$, are tabulated below.

Element No.	Start Node (X_{iS}, Y_{iS}, Z_{iS}) (m)	End Node (X_{iE}, Y_{iE}, Z_{iE}) (m)	Length L_i (m)	c_{ix}	c_{ix}	c_{iz}
1	A (–1.5,0,√6.75)	O (0,1, √3)	2	0.75	0.5	–0.43301
2	B (0,0,0)	O (0,1, √3)	2	0	0.5	0.86603
3	C (1.5,0,√6.75)	O (0,1, √3)	2	–0.75	0.5	–0.43301

Let us assume that the three identical members have the axial rigidity, EA, whereby all the bars have the same axial stiffness $k = \dfrac{EA}{2}$. The unassembled reduced element stiffness matrix is given by $\tilde{\mathbf{k}}_* = EA\begin{bmatrix} 0.5 & 0 & 0 \\ 0 & 0.5 & 0 \\ 0 & 0 & 0.5 \end{bmatrix}$ and, for the adopted member incidences (end node always at O), the element transformation matrix is given by:

$$\mathbf{T}_{\mathbf{DA}} = \begin{bmatrix} c_{1x} & c_{1y} & c_{1y} \\ c_{2x} & c_{2y} & c_{2z} \\ c_{3x} & c_{3y} & c_{3z} \end{bmatrix} = \begin{bmatrix} 0.75 & 0.5 & -0.43301 \\ 0 & 0.5 & 0.86603 \\ -0.75 & 0.5 & -0.43301 \end{bmatrix} \Rightarrow \tilde{\mathbf{k}}_*\mathbf{T}_{\mathbf{DA}} = EA\begin{bmatrix} 0.375 & 0.25 & -0.2165 \\ 0 & 0.25 & 0.4330 \\ -0.375 & 0.25 & -0.2165 \end{bmatrix}$$

$$\Rightarrow \mathbf{k}_{\mathbf{AA}} = \mathbf{T}_{\mathbf{DA}}{}^{\mathrm{T}}\tilde{\mathbf{k}}_*\mathbf{T}_{\mathbf{DA}} = EA\begin{bmatrix} 0.5625 & 0 & 0 \\ 0 & 0.375 & 0 \\ 0 & 0 & 0.5625 \end{bmatrix}$$

The truss is to be analysed for the action of the load $F_2 = -60$ kN, with $F_1 = F_3 = 0$ kN. Also, we can consider arbitrary initial 'free' deformations on account of 'lack of fit' or temperature effects in the 3 members: e_{o1}, e_{o2}, e_{o3}, which induce initial fixed end member forces in the restrained primary structure, given by $F^i_{*f} = -ke_{oi}$, and fixed end nodal forces:

$$\Rightarrow \mathbf{F}_{\mathbf{fA}} = \mathbf{T}_{\mathbf{DA}}{}^{\mathrm{T}}\mathbf{F}_{*\mathbf{f}} = \begin{bmatrix} 0.75 & 0 & -0.75 \\ 0.5 & 0.5 & 0.5 \\ -0.43301 & 0.86603 & -0.43301 \end{bmatrix} 0.5EA\begin{Bmatrix} -e_{o1} \\ -e_{o2} \\ -e_{o3} \end{Bmatrix} = EA\begin{Bmatrix} 0.375(-e_{o1} + e_{o3}) \\ 0.25(-e_{o1} - e_{o2} - e_{o3}) \\ 0.2165(e_{o1} - 2e_{o2} + e_{o3}) \end{Bmatrix}$$

$$\mathbf{D}_{\mathbf{A}} = [\mathbf{k}_{\mathbf{AA}}]^{-1}[\mathbf{F}_{\mathbf{A}} - \mathbf{F}_{\mathbf{fA}}]$$

$$\Rightarrow \begin{Bmatrix} D_1 \\ D_2 \\ D_3 \end{Bmatrix} = \frac{1}{EA}\begin{bmatrix} 1.77778 & 0 & 0 \\ 0 & 2.66667 & 0 \\ 0 & 0 & 1.77778 \end{bmatrix} \times \left[\begin{Bmatrix} 0 \\ -60 \\ 0 \end{Bmatrix} - EA \begin{Bmatrix} 0.375(-e_{o1}+e_{o3}) \\ 0.25(-e_{o1}-e_{o2}-e_{o3}) \\ 0.2165(e_{o1}-2e_{o2}+e_{o3}) \end{Bmatrix} \right]$$

$$\Rightarrow \begin{Bmatrix} D_1 \\ D_2 \\ D_3 \end{Bmatrix} = \frac{1}{EA}\begin{Bmatrix} 0 \\ -160 \\ 0 \end{Bmatrix} + e_{o1}\begin{Bmatrix} 0.6667 \\ 0.6667 \\ -0.3849 \end{Bmatrix} + e_{o2}\begin{Bmatrix} 0 \\ 0.6667 \\ 0.7698 \end{Bmatrix} + e_{o3}\begin{Bmatrix} -0.6667 \\ 0.6667 \\ -0.3849 \end{Bmatrix}$$

$$\mathbf{F}_* = \mathbf{F}_{*\mathbf{fA}} + \left(\tilde{\mathbf{k}}_* \mathbf{T}_{\mathbf{DA}}\right)\mathbf{D}_{\mathbf{A}} \Rightarrow$$

$$\begin{Bmatrix} F_*^1 \\ F_*^2 \\ F_*^3 \end{Bmatrix} = 0.5EA\begin{Bmatrix} -e_{o1} \\ -e_{o2} \\ -e_{o3} \end{Bmatrix} + EA\begin{bmatrix} 0.375 & 0.25 & -0.2165 \\ 0 & 0.25 & 0.4330 \\ -0.375 & 0.25 & -0.2165 \end{bmatrix} \times \left[\frac{1}{EA}\begin{Bmatrix} 0 \\ -160 \\ 0 \end{Bmatrix} + \begin{Bmatrix} 0.6667(e_{o1}-e_{o3}) \\ 0.6667(e_{o1}+e_{o2}+e_{o3}) \\ 0.3849(-e_{o1}+2e_{o2}-e_{o3}) \end{Bmatrix} \right]$$

$$\Rightarrow \begin{Bmatrix} N_1 \\ N_2 \\ N_3 \end{Bmatrix} = \begin{Bmatrix} F_*^1 \\ F_*^2 \\ F_*^3 \end{Bmatrix} = \begin{Bmatrix} -40 \\ -40 \\ -40 \end{Bmatrix} \text{kN}$$

The bar forces due to the direct loading match exactly with the solution obtained earlier (considering simple static equilibrium), while, as expected, the bar forces due to the indirect loading (e_{o1}, e_{o2}, e_{o3}) are all equal to zero, the frame being 'just-rigid'.

EXAMPLE 4.19

Fig. 4.47 shows the plan and elevation of a space truss, with the triangle ABC lying on the horizontal *xz* plane and the triangle DEF lying on an elevated horizontal plane at a height of 10 m above the *xz* plane. The various joint coordinates are indicated in the figure. Restraints against translation are provided along *x*, *y* and *z* directions at A, *y* and *z* directions at C and only along the *y* direction at B. Find the forces in the 12 bars, when the truss is subjected to a horizontal force *P* acting in the *z* direction at the joint F, as shown in the figure.

SOLUTION

Input Data

In this space truss, there are 12 elements ($m = 12$), 6 pin joints, i.e., $3j = 18$ global coordinates, of which 12 correspond to *active* degrees of freedom ($n = 12$) and the remaining 6 to *restrained* global coordinates ($r = 6$), as marked in Fig. 4.48. The truss is to be analysed for the action of the load $F_9 = P$, all other load components being equal to zero.

The global coordinates and member incidences (start node to end node directions) are marked in Fig. 4.48. The horizontal planes at the top and bottom are shown shaded, for convenience in visualising the space frame.

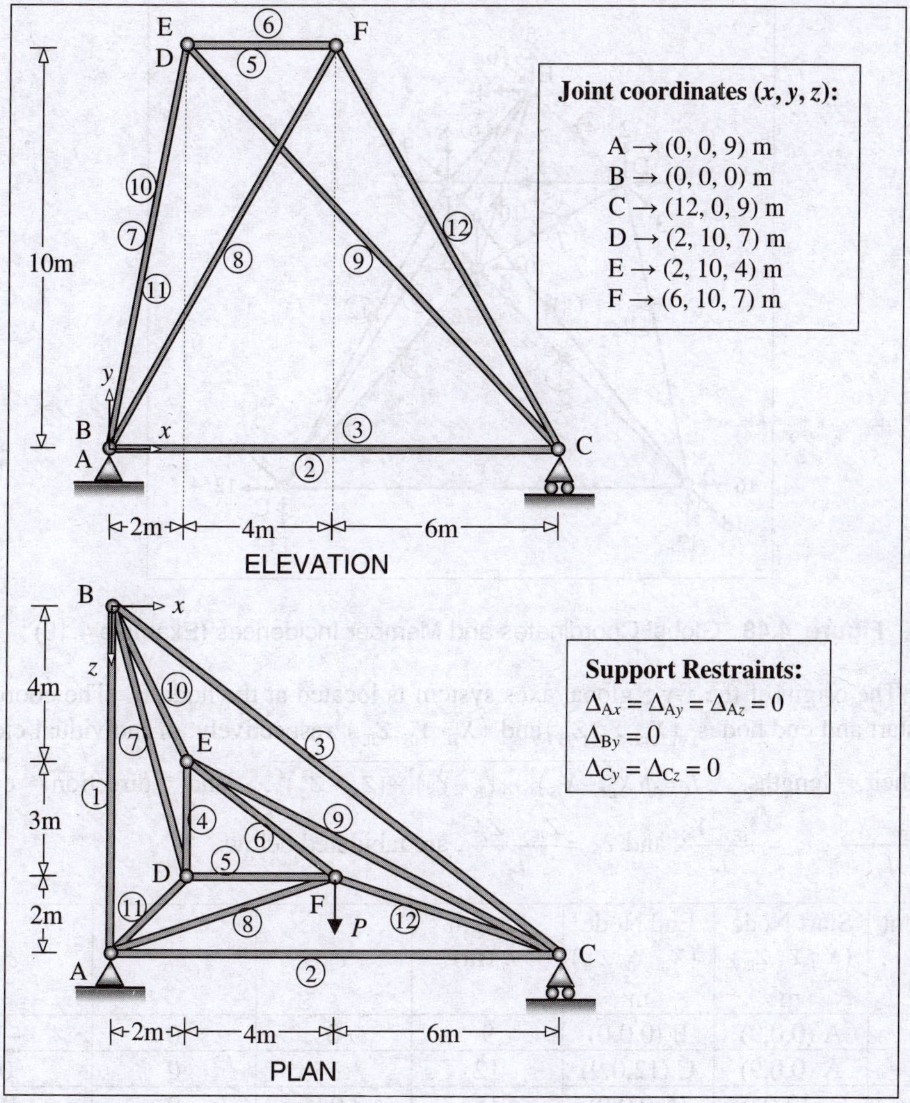

Figure 4.47 Space Truss (Example 4.19)

As the truss is statically determinate ($m + r = 18 = 3j$), we know that the bar forces will not be dependent on the element stiffnesses. For convenience, let us assume that all 12 bars have the same axial rigidity, EA.

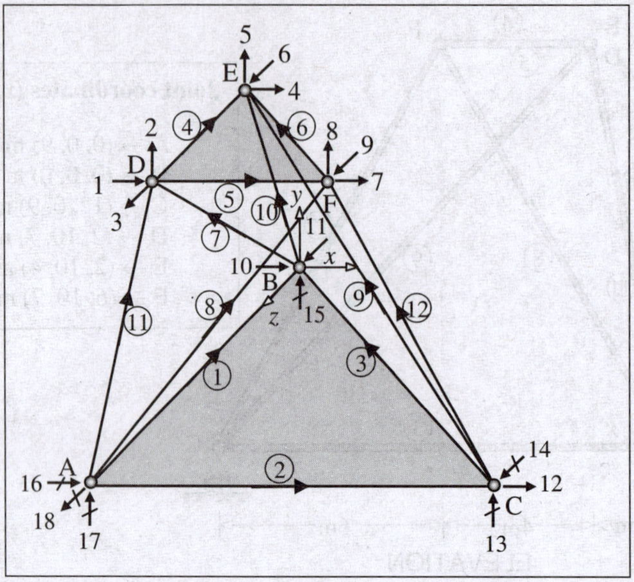

Figure 4.48 Global Coordinates and Member Incidences (Example 4.19)

The origin of the *x-y-z* global axes system is located at the node B. The coordinates of the start and end nodes, (X_{iS}, Y_{iS}, Z_{iS}) and (X_{iE}, Y_{iE}, Z_{iE}) respectively, of individual elements and their lengths, $L_i = \sqrt{(X_{iE} - X_{iS})^2 + (Y_{iE} - Y_{iS})^2 + (Z_{iE} - Z_{iS})^2}$, and direction cosines, $c_{ix} = \dfrac{X_{iE} - X_{iS}}{L_i}$, $c_{iy} = \dfrac{Y_{iE} - Y_{iS}}{L_i}$ and $c_{iz} = \dfrac{Z_{iE} - Z_{iS}}{L_i}$, are tabulated below.

Element No.	Start Node (X_{iS}, Y_{iS}, Z_{iS}) m	End Node (X_{iE}, Y_{iE}, Z_{iE}) m	Length L_i (m)	c_{ix}	c_{ix}	c_{iz}
1	A (0,0,9)	B (0,0,0)	9	0	0	–1
2	A (0,0,9)	C (12,0,9)	12	1	0	0
3	C (12,0,9)	B (0,0,0)	15	–0.8	0	–0.6
4	D (2,10,7)	E (2,10,4)	3	0	0	–1
5	D (2,10,7)	F (6,10,7)	4	1	0	0
6	F (6,10,7)	E (2,10,4)	5	–0.8	0	–0.6
7	B (0,0,0)	D (2,10,7)	12.36932	0.16169	0.80845	0.56592
8	A (0,0,9)	F (6,10,7)	11.83216	0.50709	0.84515	–0.16903
9	C (12,0,9)	E (2,10,4)	15	–0.66667	0.666667	–0.33333
10	B (0,0,0)	E (2,10,4)	10.95445	0.18257	0.91287	0.36515
11	A (0,0,9)	D (2,10,7)	10.39230	0.19245	0.96225	–0.19245
12	C (12,0,9)	F (6,10,7)	11.83216	–0.50709	0.84515	–0.16903

The element stiffnesses $k_i = \frac{EA}{L_i}$ are given by:

$k_1 = 0.11111EA;\ k_2 = 0.083333EA;\ k_3 = k_9 = 0.066667EA;\ k_4 = 0.33333EA;\ k_5 = 0.25EA;$
$k_6 = 0.2EA;\ k_7 = 0.080845EA;\ k_8 = k_{12} = 0.084515EA;\ k_{10} = 0.091287EA;\ k_{11} = 0.096225EA$

Solution Procedure

As we are required to find only the bar forces (and not the support reactions) in this problem, and as there are no support movements ($\mathbf{D_R} = \mathbf{O}$), we can ignore the restrained global coordinates, '13', to '18'. We can find the bar forces by applying $\mathbf{F_*} = \left(\tilde{\mathbf{k}}_*\mathbf{T_{DA}}\right)\mathbf{D_A}$, where $\mathbf{D_A} = [\mathbf{k_{AA}}]^{-1}\mathbf{F_A}$. This involves inversion of a 12×12 stiffness matrix, which cannot be done manually, but can be easily generated by an appropriate software application, such as MATLAB. We can conveniently generate the $\mathbf{k_{AA}}$ matrix by the direct stiffness approach, by summing up the contributions from all elements, $\mathbf{k_{AA}} = \sum_{i=1}^{12} \mathbf{T_{DA}^i}^T \tilde{\mathbf{k}}_*^i \mathbf{T_{DA}^i}$. However, prior to this, we need to formulate the element displacement transformation matrix $\mathbf{T_{DA}^i}$.

Displacement Transformation Matrices

The element displacement transformation matrix, $\mathbf{T_{DA}^i}$, which is used in the transformation, $\mathbf{D_*^i} = \mathbf{T_{DA}^i}\mathbf{D_A}$, has an order 1×12, mostly comprising zero elements.

Referring to Fig. 4.48, and making use of Eq. 4.37, along with the element connectivity data and the direction cosine parameters already computed and tabulated,

$$\mathbf{T_{DA}^1} = [0\ \ 0\ \ 0\ \ 0\ \ 0\ \ 0\ \ 0\ \ 0\ \ 0\ \ 0\ \ {-1}\ \ 0]$$

$$\mathbf{T_{DA}^2} = [0\ \ 0\ \ 0\ \ 0\ \ 0\ \ 0\ \ 0\ \ 0\ \ 0\ \ 0\ \ 0\ \ 1]$$

$$\mathbf{T_{DA}^3} = [0\ \ 0\ \ 0\ \ 0\ \ 0\ \ 0\ \ 0\ \ 0\ \ 0\ \ {-0.8}\ \ {-0.6}\ \ 0.8]$$

$$\mathbf{T_{DA}^4} = [0\ \ 0\ \ 1\ \ 0\ \ 0\ \ {-1}\ \ 0\ \ 0\ \ 0\ \ 0\ \ 0\ \ 0]$$

$$\mathbf{T_{DA}^5} = [-1\ \ 0\ \ 0\ \ 0\ \ 0\ \ 0\ \ 1\ \ 0\ \ 0\ \ 0\ \ 0\ \ 0]$$

$$\mathbf{T_{DA}^6} = [0\ \ 0\ \ 0\ \ {-0.8}\ \ 0\ \ {-0.6}\ \ 0.8\ \ 0\ \ 0.6\ \ 0\ \ 0\ \ 0]$$

$$\mathbf{T_{DA}^7} = [0.16169\ \ 0.80845\ \ 0.56592\ \ 0\ \ 0\ \ 0\ \ 0\ \ 0\ \ 0\ \ {-0.16169}\ \ {-0.56592}\ \ 0]$$

$$\mathbf{T_{DA}^8} = [0\ \ 0\ \ 0\ \ 0\ \ 0\ \ 0\ \ 0.50709\ \ 0.84515\ \ {-0.16903}\ \ 0\ \ 0\ \ 0]$$

$$\mathbf{T_{DA}^9} = [0\ \ 0\ \ 0\ \ {-0.66667}\ \ 0.66667\ \ 0.33333\ \ 0\ \ 0\ \ 0\ \ 0\ \ 0\ \ 0.66667]$$

$$\mathbf{T_{DA}^{10}} = [0\ \ 0\ \ 0\ \ 0.18257\ \ 0.91287\ \ 0.36515\ \ 0\ \ 0\ \ 0\ \ {-0.18257}\ \ {-0.36515}\ \ 0]$$

$$\mathbf{T_{DA}^{11}} = [0.19245\ \ 0.96225\ \ {-0.19245}\ \ 0\ \ 0\ \ 0\ \ 0\ \ 0\ \ 0\ \ 0\ \ 0\ \ 0]$$

$$\mathbf{T}_{\mathbf{DA}}^{12} = [-0.50709 \quad 0.84515 \quad -0.16903 \quad 0 \quad 0 \quad 0 \quad 0 \quad 0 \quad 0 \quad 0 \quad 0 \quad 0.50709]$$

Considering $\tilde{\mathbf{k}}_*^{\mathbf{i}} = k_i$, the $\tilde{\mathbf{k}}_*^{\mathbf{i}}\mathbf{T}_{\mathbf{DA}}^{\mathbf{i}} = k_i\mathbf{T}_{\mathbf{DA}}^{\mathbf{i}}$ matrices can be easily generated.

Structure Stiffness Matrix (Active Coordinates)

Using MATLAB, we can conveniently generate $\mathbf{k}_{\mathbf{AA}} = \sum_{i=1}^{12} \mathbf{T}_{\mathbf{DA}}^{\mathbf{i}\ \mathrm{T}}\tilde{\mathbf{k}}_*^{\mathbf{i}}\mathbf{T}_{\mathbf{DA}}^{\mathbf{i}}$. The order of this matrix is 12×12, whose elements are summarised below:

$k_{11} = 0.25568EA;\ k_{21}(=k_{12}) = 0.02839EA;\ k_{31}(=k_{13}) = 0.003834EA;$

$k_{41} = k_{51} = k_{61}(=k_{14} = k_{15} = k_{16}) = 0;\ k_{71}(=k_{17}) = -0.25EA;\ k_{81} = k_{91}(=k_{18} = k_{19}) = 0;$

$k_{10,1}(=k_{1,10}) = -0.002114EA;\ k_{11,1}(=k_{1,11}) = -0.007398EA;\ k_{12,1}(=k_{1,12}) = 0;$

$k_{22} = 0.14194EA;\ k_{32}(=k_{23}) = 0.01917EA;\ k_{42} = k_{52} = k_{62}(=k_{24} = k_{25} = k_{26}) = 0;$

$k_{72} = k_{82} = k_{92}(=k_{27} = k_{28} = k_{29}) = 0;$

$k_{10,2}(=k_{2,10}) = -0.01057EA;\ k_{11,2}(=k_{2,11}) = -0.03699EA;\ k_{12,2}(=k_{2,12}) = 0;$

$k_{33} = 0.36279EA;\ k_{43} = k_{53}(=k_{34} = k_{35}) = 0;\ k_{63}(=k_{36}) = -0.33333EA;$

$k_{73} = k_{83} = k_{93}(=k_{37} = k_{38} = k_{39}) = 0;$

$k_{10,3}(=k_{3,10}) = -0.007398EA;\ k_{11,3}(=k_{3,11}) = -0.02589EA;\ k_{12,3}(=k_{3,12}) = 0;$

$k_{44} = 0.16067EA;\ k_{54}(=k_{45}) = -0.01442EA;\ k_{64}(=k_{46}) = 0.11690EA;$

$k_{74}(=k_{47}) = -0.128EA;\ k_{84}(=k_{48}) = 0;\ k_{94}(=k_{49}) = -0.096EA;$

$k_{10,4}(=k_{4,10}) = -0.003043EA;\ k_{11,4}(=k_{4,11}) = -0.006086EA;\ k_{12,4}(=k_{4,12}) = -0.02963EA;$

$k_{55} = 0.10570EA;\ k_{65}(=k_{56}) = 0.01561EA;\ k_{75} = k_{85} = k_{95}(=k_{57} = k_{58} = k_{59}) = 0;$

$k_{10,5}(=k_{5,10}) = -0.01522EA;\ k_{11,5}(=k_{5,11}) = -0.03043EA;\ k_{12,5}(=k_{5,12}) = 0.02963EA;$

$k_{66} = 0.42491EA;\ k_{76}(=k_{67}) = -0.096EA;\ k_{86}(=k_{68}) = 0;\ k_{96}(=k_{69}) = -0.072EA;$

$k_{10,6}(=k_{6,10}) = -0.006086EA;\ k_{11,6}(=k_{6,11}) = -0.01217EA;\ k_{12,6}(=k_{6,12}) = -0.01482EA;$

$k_{77} = 0.42491EA;\ k_{87}(=k_{78}) = 0;\ k_{97}(=k_{79}) = 0.096EA;$

$k_{10,7} = k_{11,7}(=k_{7,10} = k_{7,11}) = 0;\ k_{12,7}(=k_{7,12}) = -0.02173EA;$

$k_{88} = 0.12074EA;\ k_{98}(=k_{89}) = -0.02415EA;$

$k_{10,8} = k_{11,8}(=k_{8,10} = k_{8,11}) = 0;\ k_{12,8}(=k_{8,12}) = 0.03622EA;$

$k_{99} = 0.07683EA;\ k_{10,9} = k_{11,9}(=k_{9,10} = k_{9,11}) = 0;\ k_{12,9}(=k_{9,12}) = -0.007244EA;$

$k_{10,10} = 0.04782EA;\ k_{11,10}(=k_{10,11}) = 0.04548;\ k_{12,10}(=k_{10,12}) = -0.04267EA;$

$k_{11,11} = 0.17317EA;\ k_{12,11}(=k_{11,12}) = -0.032EA;\ k_{12,12} = 0.17736EA$

Displacements

$$\mathbf{F_A} = \mathbf{k_{AA}}\mathbf{D_A} \quad \Rightarrow \quad \mathbf{D_A} = \left[\mathbf{k_{AA}}\right]^{-1}\mathbf{F_A}$$

$$\text{where } \mathbf{F_A} = \begin{Bmatrix} F_1 \\ F_2 \\ F_3 \\ F_4 \\ F_5 \\ F_6 \\ F_7 \\ F_8 \\ F_9 \\ F_{10} \\ F_{11} \\ F_{12} \end{Bmatrix} = \begin{Bmatrix} 0 \\ 0 \\ 0 \\ 0 \\ 0 \\ 0 \\ 0 \\ 0 \\ P \\ 0 \\ 0 \\ 0 \end{Bmatrix} \quad \Rightarrow \quad \mathbf{D_A} = \begin{Bmatrix} D_1 \\ D_2 \\ D_3 \\ D_4 \\ D_5 \\ D_6 \\ D_7 \\ D_8 \\ D_9 \\ D_{10} \\ D_{11} \\ D_{12} \end{Bmatrix} = \frac{P}{EA}\begin{Bmatrix} -26.676 \\ 6.2278 \\ 4.4630 \\ 49.463 \\ 6.1948 \\ 4.4630 \\ -26.676 \\ 21.5742 \\ 119.871 \\ 11.333 \\ 2.5 \\ 8.0 \end{Bmatrix}$$

(Any suitable numerical scheme may be used to solve for $\mathbf{D_A}$).

Member Forces

$$\mathbf{F_*^i} = \left(\tilde{\mathbf{k}}_*^i \mathbf{T_{DA}^i}\right)\mathbf{D_A} \quad \text{and} \quad N_i = F_*^i$$

Thus,

$$N_1 = F_*^1 = -0.2778P$$

$$N_2 = F_*^2 = 0.6667P$$

$$N_3 = F_*^3 = -0.2778P$$

$$N_4 = F_*^4 = 0$$

$$N_5 = F_*^5 = 0$$

$$N_6 = F_*^6 = 1.6667P$$

$$N_7 = F_*^7 = 0$$

$$N_8 = F_*^8 = -1.3147P$$

$$N_9 = F_*^9 = -1.6667P$$

$$N_{10} = F_*^{10} = 1.2172P$$

$$N_{11} = F_*^{11} = 0$$

$$N_{12} = F_*^{12} = 1.3147P$$

The reader may verify that these member forces satisfy equilibrium. Furthermore, by including the restrained global coordinates, the support reactions can be shown to be given by:

$$\mathbf{F_R} = \mathbf{k_{RA}} \mathbf{D_A} = \begin{Bmatrix} F_{13} \\ F_{14} \\ F_{15} \\ F_{16} \\ F_{17} \\ F_{18} \end{Bmatrix} = \begin{Bmatrix} 0 \\ -0.5P \\ -1.1111P \\ 0 \\ 1.1111P \\ -0.5P \end{Bmatrix}.$$

REVIEW QUESTIONS

4.1 In matrix analysis of structures with axial elements, we can consider the axial element to have one degree, two degrees, four degrees or six degrees of freedom. Explain the different circumstances, which call for such considerations.

4.2 Using an energy formulation, derive from first principles, the element stiffness matrix of a prismatic axial element, having a length L_i and axial rigidity E_iA_i, with two degrees of freedom.

4.3 Show that the 2×2 element stiffness matrix for a prismatic truss element is non-invertible, and explain, in terms of physical behaviour, why this is so.

4.4 Show how the 4×4 element stiffness matrix for a plane truss element can be transformed from local coordinates to the global axes system, in terms of the direction cosines using matrix transformation techniques. Also, show how this can be alternatively generated from first principles, using a 'physical approach'.

4.5 In a two-bar truss, the displacement transformation is given by. $\begin{Bmatrix} D_*^1 \\ D_*^2 \end{Bmatrix} = \begin{bmatrix} 10 & 5 \\ 4 & 6 \end{bmatrix} \begin{Bmatrix} D_1 \\ D_2 \end{Bmatrix}$.

where $\mathbf{D_*}$ and $\mathbf{D}$ denote the bar elongation and joint displacement vectors respectively. Find the bar forces N_1 and N_2 caused by joint loads, $F_1 = 80$ kN and $F_2 = -40$ kN.

4.6 In a simple two-bar truss, the force transformation matrix is given by $\mathbf{T_F} = \begin{bmatrix} 1 & 4/3 \\ 0 & 5/3 \end{bmatrix}$.

If the active joint displacements are given by $\begin{Bmatrix} D_1 \\ D_2 \end{Bmatrix} = \begin{Bmatrix} 2 \\ 5 \end{Bmatrix}$ mm, then find the bar elongations, D_*^1 and D_*^2.

4.7 (a) Determine the bar forces in the five-bar truss shown in Fig. 4.49, using a force transformation approach.

(b) In an unloaded condition, it is given that the five bars in Fig. 4.49 undergo thermal elongations given by $\mathbf{e_o} = [2, 2, 3, 1, 1]^T$ mm. Find the vertical deflection of joint A.

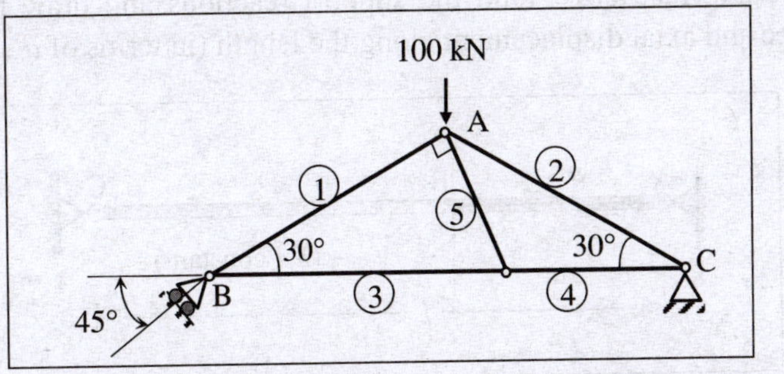

Figure 4.49 Truss in Review Question 4.7

4.8 What is meant by *rotation matrix*? Show how it is of relevance with regard to the 6 × 6 element stiffness matrix for a space truss element.

4.9 Consider a structure made up of prismatic and linearly elastic axial elements. Usually, the axial displacement will vary as a straight line from one end of the element to its other end. Under what loading conditions would this not happen?

4.10 Consider an element AB in a space truss, connecting the joint A(0,1,2) to joint B(3,4,5), where the global axes (*x*,*y*,*z*) dimensions are expressed in metres. Find the axial deformation in the bar, given that A is connected to a rigid support and B deflects by 0.5mm, –1.0mm and 1.5mm in the *x*, *y* and *z* directions respectively.

4.11 What are the advantages of adopting the reduced element stiffness method, compared to the conventional stiffness method, as applied to structures with axial elements? Are there any shortcomings?

4.12 How can you find the support reactions in structures with axial elements in (a) stiffness method and (b) flexibility method?

4.13 How do you account for the effects of 'lack of fit' in truss members in (a) stiffness method and (b) flexibility method?

4.14 How do you account for the effects of support movements in structures with axial elements in (a) stiffness method and (b) flexibility method?

4.15 Can cables be treated as truss elements? Discuss.

PROBLEMS

4.1 Consider a prismatic axially loaded system ABC of length L and axial rigidity EA, subject to a uniformly distributed axial load, q_o per unit length, over the segment AB of length a, as shown in Fig. 4.50. Analyse this system by the conventional Stiffness Method, using matrices. Find the support reactions and draw the distributions of axial force and axial displacement along the length (in terms of q_o, a, L and EA).

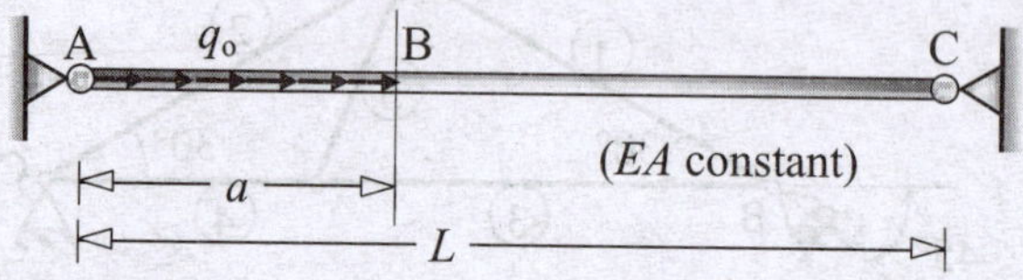

Figure 4.50 Axially loaded system in Problem 4.1

4.2 Repeat Problem 4.1, considering the end at A to be free (not restrained against translations).

4.3 Repeat Problem 4.1, considering the loading to be in the form of a temperature increase T applied on the segment AB (instead of the direct loading). Assume α to be the coefficient of thermal expansion.

4.4 Repeat Problem 4.1, using the Reduced Element Stiffness method.

4.5 Repeat Problem 4.1, using the Flexibility Method.

4.6 Analyse the non-prismatic axially loaded structural system, ABCD, shown in Fig. 4.51, with an elastic support at D, by the conventional Stiffness Method. Find the complete force and displacement responses. Assume the segment BCD to have an axial rigidity, $EA = 7500$ kN and the segment AB to have twice this axial rigidity.

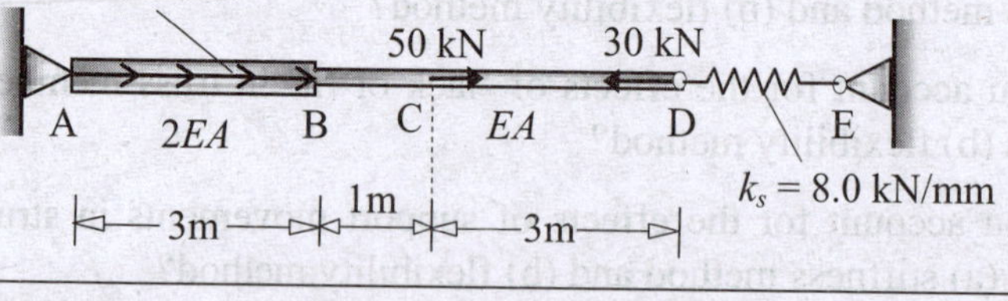

Figure 4.51 Problem 4.6

4.7 Repeat Problem 4.6, using the Reduced Element Stiffness method.

4.8 Repeat Problem 4.6, using the Flexibility Method.

4.9 Consider a stepped vertical member ABCD of total height 6m, comprising three segments, each of length 2m, with a circular cross-section having diameters of 400mm in the topmost segment AB, 300mm in the middle segment BC and 200mm in the lowermost segment CD. The member, which is suspended from a pinned ceiling support at the top end A, is subjected to a gravity load of 100 kN at the bottom end D, which is free to deflect vertically. The material has a unit weight of 60 kN/m^3 and an elastic modulus of 25 kN/m^2. Analyse the complete force and displacement responses of this structure, due to the applied loading, including its own self-weight, by the Reduced Element Stiffness Method.

4.10 Consider the symmetric pin-jointed frame AOB shown in Fig. 4.52, comprising two identical axial elements of length L and axial rigidity EA, in the vertical plane, subject to a concentrated gravity load, P at the joint O. In addition, consider a uniform temperature increase T to the two bars, which have a coefficient of thermal expansion, α. Taking advantage of symmetry, analyse this system by the conventional Stiffness Method and derive expressions for the deflection at joint O, the horizontal support reaction and the axial force in either of the two bars (in terms of P, L αT and EA).

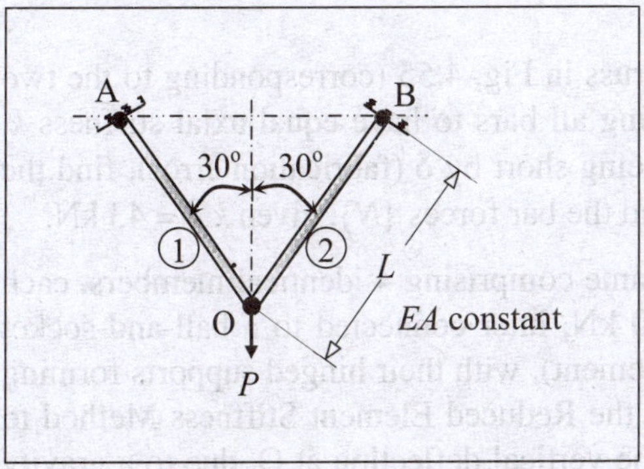

Figure 4.52 Problem 4.10

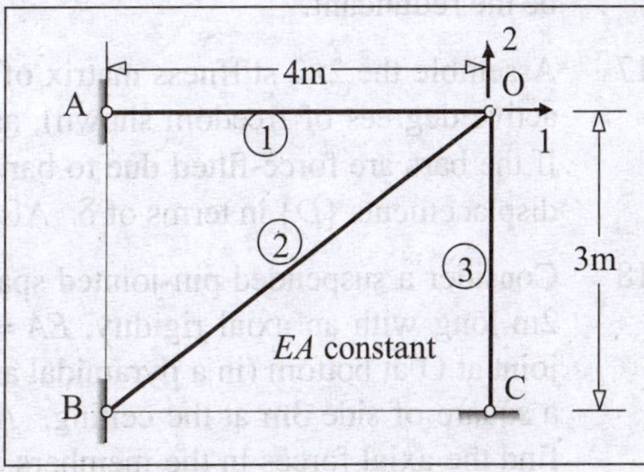

Figure 4.53 Problem 4.12

4.11 Repeat Problem 4.10, using the Reduced Element Stiffness method.

4.12 Analyse, by the Reduced Element Stiffness Method, the truss shown in Fig. 4.53 (with active global coordinates, as shown), and find the joint displacements, support reactions and bar forces, when the truss is subject to direct loads, $F_1 = 40$ kN; $F_2 = -30$ kN, plus a 'lack of fit' due to bar '2' being too long by 5mm. Assume all bars to have the same axial rigidity EA = 6000 kN.

4.13 Repeat Problem 4.12 by the Flexibility Method, considering the force in bar '3' as the redundant.

4.14 Taking advantage of symmetry, analyse, by the conventional Stiffness Method, the pin-jointed plane frame shown in Fig. 4.54, and find the vertical displacement at joint O, support reactions at A and B and axial forces in bars '1' and '2'. Assume all bars to have the same axial rigidity $EA = 6000$ kN.

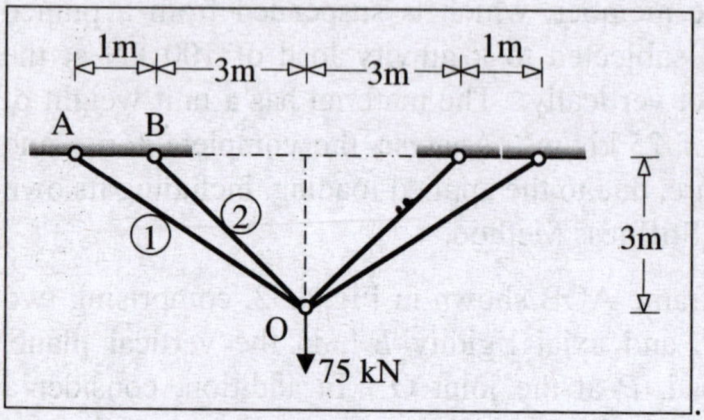

Figure 4.54 Problem 4.14

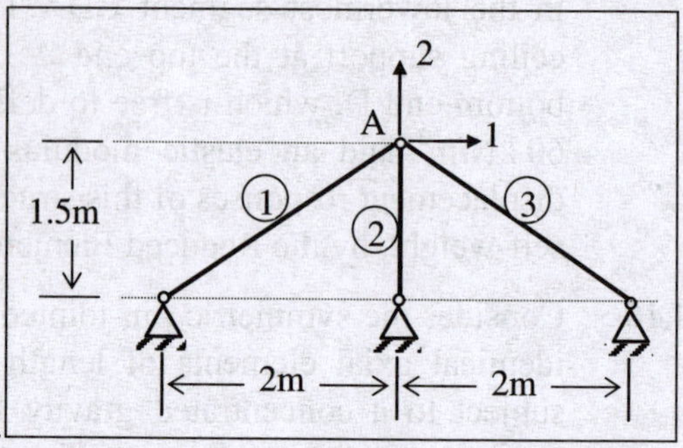

Figure 4.55 Problem 4.17

4.15 Repeat Problem 4.14, using the Reduced Element Stiffness method.

4.16 Repeat Problem 4.14, using the Flexibility Method, considering the force in bar '1' to be the redundant.

4.17 Assemble the 2×2 stiffness matrix of the truss in Fig. 4.55 (corresponding to the two active degrees of freedom shown), assuming all bars to have equal axial stiffness k. If the bars are force-fitted due to bar '2' being short by δ (fabrication error), find the displacements $\{D\}$ in terms of δ. Also, find the bar forces $\{N\}$, given $k\,\delta = 43$ kN.

4.18 Consider a suspended pin-jointed space frame comprising 4 identical members, each 2m long with an axial rigidity, $EA = 6000$ kN, inter-connected to a ball-and-socket joint at O at bottom (in a pyramidal arrangement), with their hinged supports forming a square of side 3m at the ceiling. Apply the Reduced Element Stiffness Method to find the axial forces in the members and the vertical deflection at O, due to a gravity load of 60 kN acting at O.

4.19 Repeat Problem 4.18, considering the loading to be in the form of a 'lack of fit' due to one of the bars being too long by 5mm.

5

Matrix Analysis of Beams and Grids

5.1 INTRODUCTION

The basic formulation and steps involved in the application of matrix methods (stiffness and flexibility methods) were demonstrated in the previous Chapters with reference to one-dimensional axial structures, plane trusses and space trusses. This Chapter is a continuation of the same theme, with the application directed to beams. Towards the end of the Chapter, we shall also discuss 'grids', which are spatial skeletal structures involving a planar network of beams that may resist torsion, in addition to bending and shear.

The beam is a skeletal element that resists external loads by undergoing bending (or flexure), whereby the primary internal force resultant (which can vary along the length of the element) is the *bending moment*, $M(x)$, and the corresponding deformation, over an elemental length dx, is a rotation $d\theta = \varphi\, dx$, where $\varphi(x)$ is the *curvature*. Assuming linear elastic behaviour (Euler-Bernoulli theory), the bending moment $M(x)$ varies linearly with the curvature $\varphi(x)$, and the constant of proportionality is the flexural rigidity EI of the beam element. In addition to bending moment, there is another internal force resultant, called shear force, $S(x)$, which is equal to the rate of change of bending moment; i.e., $S(x) = \dfrac{dM}{dx}$. For normal, well-proportioned beams, we assume that shear deformations are negligible. We shall see later how to account shear deformations in the analysis, wherever required.

At any point along the beam centre-line, there are two displacement quantities of interest to us, viz., the deflection $\Delta(x)$ and its derivative, the rotation $\theta(x)$. If we know the values of these two quantities at the two ends of any prismatic beam element (which may be part of a continuous beam system), then it should be possible to find $\Delta(x)$ and $\theta(x)$ at any intermediate location. This forms the basis of the stiffness formulation, whereby the number of independent displacement coordinates (i.e., *degrees of freedom*) in any prismatic beam element is limited to four. It is assumed that the loads in the form of direct actions are applied at the joints; otherwise, if there are intermediate loads, these shall be converted to *equivalent joint loads*. Thus, in the conventional stiffness formulation, the beam element has four degrees of freedom (two translational and two rotational), whereby the stiffness matrix of the element (in local coordinates) is of order 4×4. This is discussed in Section 5.2.

In the *reduced stiffness* formulation (and the *flexibility* formulation), the degrees of freedom are reduced to two, eliminating the translational degrees of freedom. In this case, the stiffness matrix of the element (in local coordinates), and its inverse, the flexibility matrix is of order 2×2. In the reduced stiffness formulation, we can also reduce the degree of kinematic indeterminacy of continuous beam systems in which the extreme ends are hinged or guided-fixed supports, by taking advantage of modified stiffnesses and also by modifying the fixed end forces appropriately. The reduced stiffness formulation is discussed in Section 5.3, while the flexibility formulation is discussed in Section 5.4.

In Section 5.5, the matrix analysis of grid structures (by the stiffness method) is discussed. It is also shown that by ignoring secondary torsion effects, the degree of kinematic indeterminacy can be significantly reduced.

5.2 CONVENTIONAL STIFFNESS METHOD APPLIED TO BEAMS

5.2.1 Four Degrees of Freedom

Consider a prismatic (initially straight) beam element *i*, having a length L_i and a cross-section with second moment of area I_i. Let the bending moment at any location x^* from the left end (start node) be $M(x^*)$, assumed sagging positive, and the corresponding displacement quantities at x^* be defined by: deflection $\Delta(x^*)$, assumed to be positive when acting upwards, and rotation $\theta(x^*) = \dfrac{d\Delta(x^*)}{dx^*}$. The curvature of the deformed centre-line of the beam at x^* is given by $\varphi(x^*) \approx \dfrac{d^2\Delta}{dx^{*2}} = \dfrac{d\theta}{dx}$ for small deformations, as explained in Section 1.5.3. Assuming linear elastic behaviour, and plane sections before bending to remain plane after bending, it can be shown (Euler-Bernoulli theory) that the curvature is linearly related to the bending moment: $\varphi(x^*) = \dfrac{M(x^*)}{EI}$, where E is the modulus of elasticity of the material.

If the beam element is subject to forces (shear forces and bending moments) only at the two ends (i.e., without any intermediate loading), then the bending moment will vary linearly along the length of the beam. This implies that the curvature $\varphi(x^*)$ will vary linearly, whereby the slope (rotation), $\theta(x^*)$, will have a quadratic variation and the deflection $\Delta(x^*)$ will have a cubic variation along the length of the beam. Thus, we can intuit that $\Delta(x^*)$ will have the following form, expressible in terms of four constants, C_o, C_1, C_2 and C_3:

$$\Delta(x^*) = C_o + C_1(x^*) + C_2(x^*)^2 + C_3(x^*)^3 \tag{5.1}$$

$$\Rightarrow \quad \theta(x^*) = \frac{d\Delta}{dx^*} = C_1 + 2C_2(x^*) + 3C_3(x^*)^2 \tag{5.2}$$

Choosing the coordinate system as shown in Fig. 5.1a, the kinematic boundary conditions at the two ends of the beam, can be expressed as follows:

$$\begin{aligned} x^* = 0 \quad &\Rightarrow \quad \Delta = D^i_{1*}; \quad \theta = D^i_{2*} \\ x^* = L_i \quad &\Rightarrow \quad \Delta = D^i_{3*}; \quad \theta = D^i_{4*} \end{aligned} \tag{5.3}$$

Applying these kinematic boundary conditions, and solving the resulting equations, we can express the four constants in Eq. 5.1 in terms of the four end displacements, D_1, D_2, D_3 and D_4:

$$\begin{aligned} C_o &= D^i_{1*} \\ C_1 &= D^i_{2*} \\ C_2 &= \frac{3\left(-D^i_{1*} + D^i_{3*}\right) - \left(2D^i_{2*} + D^i_{4*}\right)L_i}{L_i^2} \\ C_3 &= \frac{2\left(D^i_{1*} - D^i_{3*}\right) + \left(D^i_{2*} + D^i_{4*}\right)L_i}{L_i^3} \end{aligned} \tag{5.4}$$

Thus, we have established that the deflection and rotation at any location x^* in a prismatic beam element, subject to shear forces and moments at the two ends (without any intermediate loading), can be expressed in terms of the four degrees of freedom at the two ends of the beam. Applying the Principle of Superposition, we can show that the deflection $\Delta(x^*)$, rotation $\theta(x^*)$ and curvature $\varphi(x^*)$ at any distance x^* from the start node of the element can be generated as follows :

$$\begin{Bmatrix} \Delta(x^*) \\ \theta(x^*) \\ \varphi(x^*) \end{Bmatrix} = \begin{bmatrix} \Delta_1(x^*) & \Delta_2(x^*) & \Delta_3(x^*) & \Delta_4(x^*) \\ \theta_1(x^*) & \theta_2(x^*) & \theta_3(x^*) & \theta_4(x^*) \\ \varphi_1(x^*) & \varphi_2(x^*) & \varphi_3(x^*) & \varphi_4(x^*) \end{bmatrix} \begin{Bmatrix} D_{1*}^i \\ D_{2*}^i \\ D_{3*}^i \\ D_{4*}^i \end{Bmatrix} \tag{5.5}$$

where $\Delta_j(x^*)$, $\theta_j(x^*)$ and $\varphi_j(x^*)$ denote respectively, the deflection, rotation and curvature at x^*, corresponding to $D_{j*}^i = 1$ and all other end displacements restrained.

5.2.2 Stiffness Matrix of Beam Element with Four Degrees of Freedom

The stiffness matrix $\mathbf{k}_*^i$ of the beam element, i, with four degrees of freedom, as shown in Fig. 5.1a, has an order of 4×4. By definition, the first column of this stiffness matrix corresponds to the displacement condition, $D_{1*}^i = 1$ (unit vertical translation at the start node), with the other three degrees of freedom restrained. The deflected profile is as shown in Fig. 5.1b. The stiffness coefficients, k_{1*1*}^i and k_{2*1*}^i correspond respectively to the (upward) vertical force, F_{1*}^i, and the (anti-clockwise) end moment, F_{2*}^i, at the start node, while k_{3*1*}^i and k_{4*1*}^i correspond respectively to the (upward) vertical force, F_{3*}^i, and the (anti-clockwise) end moment, F_{4*}^i, at the end node. The resulting bending moment diagram is also depicted in Fig. 5.1b. Similarly, the other stiffness coefficients are defined, corresponding to $D_{2*}^i = 1$ (Fig. 5.1c), $D_{3*}^i = 1$ (Fig. 5.1d), and $D_{4*}^i = 1$ (Fig. 5.1e), with the other degrees of freedom restrained.

There are many alternative ways of generating the stiffness coefficients. We can adopt either a 'displacement approach' or a 'force approach' for this purpose.

Displacement-based Formulation

In the displacement approach, we work with the cubic deflection function defined by Eq. 5.1, and derive the following expression for curvature, $\varphi(x^*) = \Delta''(x^*)$.

$$\varphi(x^*) = \Delta''(x^*) = 2C_2 + 6C_3(x^*) \tag{5.6}$$

Now, invoking the linear relationship between bending moment and curvature,

$$M(x^*) = (EI)_i\, \varphi(x^*) = (EI)_i\,[2C_2 + 6C_3(x^*)] \tag{5.7}$$

As expected, we have a linear variation of bending moment along the length of the beam element, whereby the shear force, $S(x^*)$, will be constant, and given by:

$$S(x^*) = \frac{dM}{dx^*} = (EI)_i\,[6C_3] \tag{5.8}$$

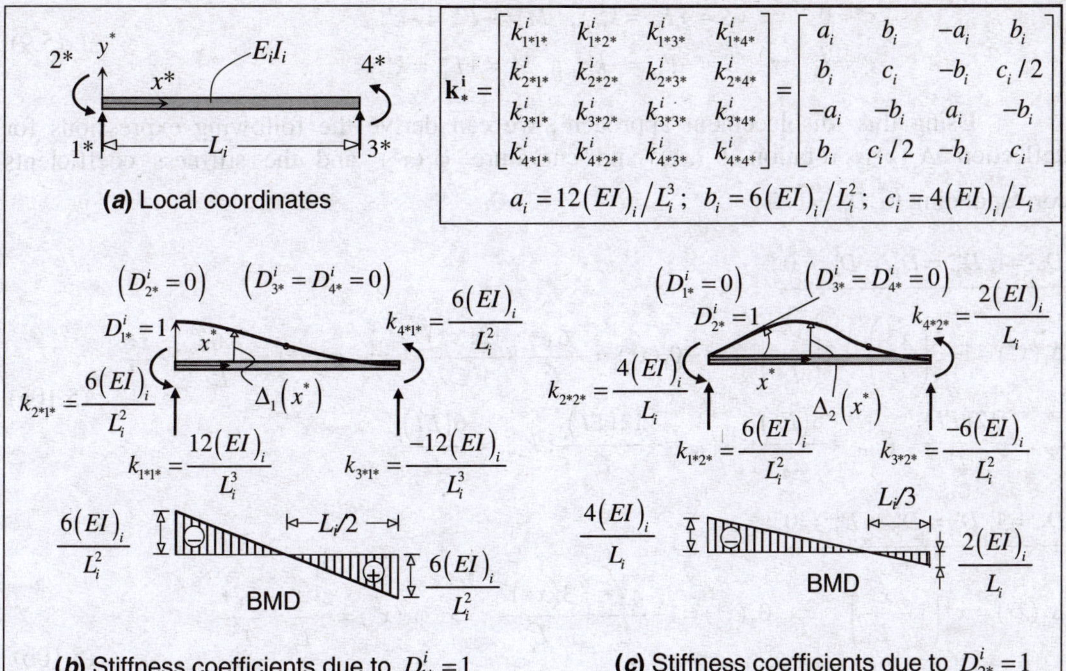

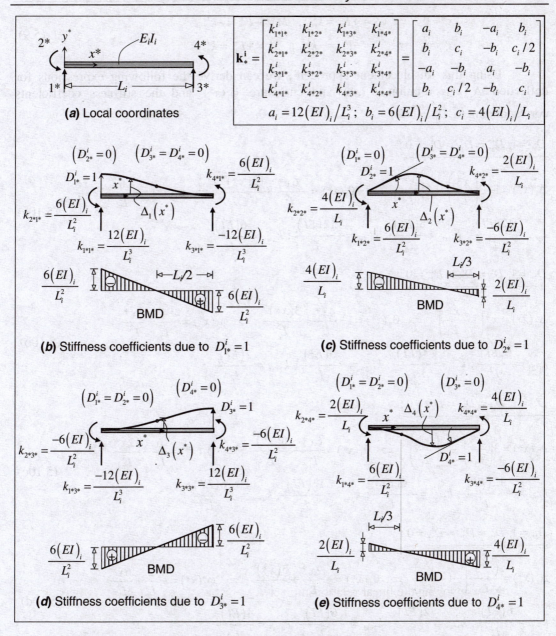

Figure 5.1 Generation of 4×4 stiffness matrix for a beam element

We can now invoke the *static* boundary conditions, relating the beam end forces to the shear force and bending moment at $x^* = 0$ and $x^* = L_i$, following the sign convention we have adopted earlier, for the condition $D^i_{j*} = 1$:

$$\begin{aligned} x^* = 0 \quad &\Rightarrow \quad S = +F_{1*} = k^i_{1*j*}; \quad M = -F_{2*} = -k^i_{2*j*} \\ x^* = L_i \quad &\Rightarrow \quad S = -F_{3*} = -k^i_{3*j*}; \quad M = +F_{4*} = k^i_{4*j*} \end{aligned} \tag{5.9}$$

Using this 'displacement approach', we can derive the following expressions for deflection $\Delta_j(x^*)$, rotation $\theta_j(x^*)$ and curvature $\varphi_j(x^*)$ and the stiffness coefficients corresponding to $D^i_{j*} = 1$:

$D^i_{1*} = 1;\ D^i_{2*} = D^i_{3*} = D^i_{4*} = 0:$

$$\begin{gathered} \Delta_1(x^*) = 1 - 3\left(\frac{x^*}{L_i}\right)^2 + 2\left(\frac{x^*}{L_i}\right)^3 \ \Rightarrow\ \theta_1(x^*) = -\frac{6x^*}{L_i^2} + \frac{12(x^*)^2}{L_i^3} \ \Rightarrow\ \varphi_1(x^*) = -\frac{6}{L_i^2} + 12\frac{x^*}{L_i^3} \\ k_{1*1*} = \frac{12(EI)_i}{L_i^3};\ k_{2*1*} = \frac{6(EI)_i}{L_i^2};\ k_{3*1*} = -\frac{12(EI)_i}{L_i^3};\ k_{4*1*} = \frac{6(EI)_i}{L_i^2} \end{gathered} \tag{5.10a}$$

$D^i_{2*} = 1;\ D^i_{1*} = D^i_{3*} = D^i_{4*} = 0:$

$$\begin{gathered} \Delta_2(x^*) = x^*\left(1 - \frac{x^*}{L_i}\right)^2 \ \Rightarrow\ \theta_2(x^*) = 1 - \frac{4x^*}{L_i} + \frac{3(x^*)^2}{L_i^2} \ \Rightarrow\ \varphi_2(x^*) = -\frac{4}{L_i} + \frac{6x^*}{L_i^2} \\ k_{1*2*} = \frac{6(EI)_i}{L_i^2};\ k_{2*2*} = \frac{4(EI)_i}{L_i};\ k_{3*2*} = -\frac{6(EI)_i}{L_i^2};\ k_{4*2*} = \frac{2(EI)_i}{L_i} \end{gathered} \tag{5.10b}$$

$D^i_{3*} = 1;\ D^i_{1*} = D^i_{2*} = D^i_{4*} = 0:$

$$\begin{gathered} \Delta_3(x^*) = 3\left(\frac{x^*}{L_i}\right)^2 + 2\left(\frac{x^*}{L_i}\right)^3 \ \Rightarrow\ \theta_3(x^*) = \frac{6x^*}{L_i^2} - \frac{6(x^*)^2}{L_i^3} \ \Rightarrow\ \varphi_3(x^*) = \frac{6}{L_i^2} - \frac{12x^*}{L_i^3} \\ k_{1*3*} = -\frac{12(EI)_i}{L_i^3};\ k_{2*3*} = -\frac{6(EI)_i}{L_i^2};\ k_{3*3*} = \frac{12(EI)_i}{L_i^3};\ k_{4*3*} = -\frac{6(EI)_i}{L_i^2} \end{gathered} \tag{5.10c}$$

$D^i_{4*} = 1;\ D^i_{1*} = D^i_{2*} = D^i_{3*} = 0:$

$$\begin{gathered} \Delta_4(x^*) = \frac{(x^*)^2}{L_i}\left(-1 + \frac{x^*}{L_i}\right) \ \Rightarrow\ \theta_4(x^*) = -\frac{2x^*}{L_i} + \frac{3(x^*)^2}{L_i^2} \ \Rightarrow\ \varphi_4(x^*) = -\frac{2}{L_i} + \frac{6x^*}{L_i^2} \\ k_{1*4*} = \frac{6(EI)_i}{L_i^2};\ k_{2*4*} = \frac{2(EI)_i}{L_i};\ k_{3*4*} = -\frac{6(EI)_i}{L_i^2};\ k_{4*4*} = \frac{4(EI)_i}{L_i} \end{gathered} \tag{5.10d}$$

These solutions are indicated in Fig. 5.1. It may be noted that the deflections at the mid-span ($x^* = 0.5L_i$), corresponding to the various displacement profiles, are given by $\Delta_1 = \Delta_3 = 0.5$ and $\Delta_2 = 0.125L_i$; $\Delta_4 = -0.125L_i$.

The resulting element stiffness matrix takes the following form:

$$\mathbf{k}_*^{\mathbf{i}} = \frac{(EI)_i}{L_i}\begin{bmatrix} 12/L_i^2 & 6/L_i & -12/L_i^2 & 6/L_i \\ 6/L_i & 4 & -6/L_i & 2 \\ -12/L_i^2 & -6/L_i & 12/L_i^2 & -6/L_i \\ 6/L_i & 2 & -6/L_i & 4 \end{bmatrix} \tag{5.11}$$

Energy Formulation

We can derive the stiffness coefficients using the energy formulation described in Section 1.7.5. This constitutes an alterative 'displacement approach'. For this purpose, we need to formulate an expression for the strain energy U_i in the element, in terms of the four end displacements, and generate $k_{l*j*} = \frac{\partial^2 U_i}{\partial D_{l*}^i \partial D_{j*}^i}$. Ignoring the contribution of shear strain energy, which is usually negligible in well-proportioned beams, the expression for strain energy is given by

$$U_i = \frac{(EI)_i}{2}\int_0^{L_i} \varphi^2(x^*)dx = \frac{(EI)_i}{2}\int_0^{L_i}\left[\varphi_1(x^*)D_{1*}^i + \varphi_2(x^*)D_{2*}^i + \varphi_3(x^*)D_{3*}^i + \varphi_4(x^*)D_{4*}^i\right]^2 dx \tag{5.12}$$

where the various expressions for $\varphi_j(x^*)$ are listed in Eq. 5.10. Expanding Eq. 5.12 and applying Eq. 1.65, it can be shown that

$$k_{l*j*} = \frac{\partial^2 U_i}{\partial D_{l*}^i \partial D_{j*}^i} = (EI)_i \int_0^{L_i} \varphi_l(x^*)\varphi_j(x^*)dx \tag{5.13}$$

For example,

$$k_{1*1*} = \frac{\partial^2 U_i}{\partial^2 D_{1*}^i} = (EI)_i \int_0^{L_i}[\varphi_1(x^*)]^2 dx = (EI)_i \int_0^{L_i}\left[\frac{-6}{L_i^2} + \frac{12x^*}{L_i^3}\right]^2 dx = \frac{(EI)_i}{L_i^3}[36-72+48] = \frac{12(EI)_i}{L_i^3}$$

$$k_{1*2*} = \frac{\partial^2 U_i}{\partial D_{1*}^i \partial D_{2*}^i} = (EI)_i \int_0^{L_i}\varphi_1(x^*)\varphi_2(x^*)dx = (EI)_i \int_0^{L_i}\left(\frac{-6}{L_i^2} + \frac{12x^*}{L_i^3}\right)\left(\frac{-4}{L_i} + \frac{6x^*}{L_i^2}\right)dx$$
$$= \frac{(EI)_i}{L_i^2}[24-42+24] = \frac{6(EI)_i}{L_i^2} = k_{2*1*}$$

Similarly, all other stiffness coefficients can be generated.

Force Approach

In this approach, we consider equilibrium of the free-body, corresponding to any of the $D_{j*}^i = 1$ situations in Fig. 5.1. By considering a section of the beam at a distance x^* from the start node, we can express the bending moment at this section as follows:

$$M(x^*) = -k_{2*j*}^i + k_{1*j*}^i(x^*) \tag{5.14}$$

We can now generate an expression for curvature $\varphi_j(x^*) = \dfrac{M(x^*)}{(EI)_i}$, and by integration, generate expressions for rotation $\theta_j(x^*)$ and deflection $\Delta_j(x^*)$:

$$\theta_j(x^*) = \frac{1}{(EI)_i}\left[-k^i_{2^*j^*}(x^*) + k^i_{1^*j^*}\frac{(x^*)^2}{2}\right] + C_{\theta j}$$
$$\Delta_j(x^*) = \frac{1}{(EI)_i}\left[-k^i_{2^*j^*}\frac{(x^*)^2}{2} + k^i_{1^*j^*}\frac{(x^*)^3}{6}\right] + C_{\theta j}(x^*) + C_{\Delta j} \tag{5.15}$$

By applying the two kinematic boundary conditions at $x^* = 0$, the two constants in Eq. 5.15 reduce to: $C_{\theta j} = \theta_j(x^*=0)$ and $C_{\Delta j} = \Delta_j(x^*=0)$. Then, the remaining two kinematic boundary conditions at $x^* = L_i$, can be solved to yield the two stiffness coefficients (at the start node):

$$\begin{Bmatrix} k^i_{2^*j^*} \\ k^i_{1^*j^*} \end{Bmatrix} = \frac{12(EI)_i}{L_i^3}\begin{bmatrix} L_i^2/6 & -L_i/2 \\ L_i/2 & -1 \end{bmatrix}\begin{Bmatrix} \theta_j(x^*=L_i) - \theta_j(x^*=0) \\ \Delta_j(x^*=L_i) - \Delta_j(x^*=0) - L_i\theta_j(x^*=0) \end{Bmatrix} \tag{5.16}$$

The remaining two stiffness coefficients (at the end node) can be obtained considering equilibrium of the overall free-body:

$$k^i_{3^*j^*} = -k^i_{1^*j^*}$$
$$k^i_{4^*j^*} = k^i_{1^*j^*}L_i - k^i_{2^*j^*} \tag{5.17}$$

For example, considering $D^i_{1^*} = 1$ (Fig. 5.1b),
$C_{\theta 1} = \theta_1(x^*=0) = 0$ and $C_{\Delta 1} = \Delta_1(x^*=0) = 1$, whereby, applying Eqns 5.16 and 5.17,

$$\begin{Bmatrix} k^i_{2^*1^*} \\ k^i_{1^*1^*} \end{Bmatrix} = \frac{12(EI)_i}{L_i^3}\begin{bmatrix} L_i^2/6 & -L_i/2 \\ L_i/2 & -1 \end{bmatrix}\begin{Bmatrix} 0-0 \\ 0-1-0 \end{Bmatrix} = \begin{Bmatrix} 6(EI)_i/L_i^2 \\ 12(EI)_i/L_i^3 \end{Bmatrix} \Rightarrow \begin{Bmatrix} k^i_{4^*1^*} \\ k^i_{3^*1^*} \end{Bmatrix} = \begin{Bmatrix} 6(EI)_i/L_i^2 \\ -12(EI)_i/L_i^3 \end{Bmatrix}$$

Substituting these values in Eq. 5.15, the expression for deflection $\Delta_1(x^*)$ can be generated. Thus, the various stiffness coefficients and expressions for deflections described in Eqns 5.10 and 5.11 can be generated using the 'force approach'.

4 × 4 Flexibility Matrix not Possible!

It is interesting to note that the determinant of the 4 × 4 stiffness matrix $\mathbf{k}^i_*$ in Eq. 5.11 is zero (rank = 2†), whereby the matrix is *not invertible*. This means that a 4 × 4 flexibility matrix (inverse of the stiffness matrix) for the element shown in Fig. 5.1a cannot be defined. This is

† For example, the third column is obtainable by multiplying the first column (or row) by –1, and thereby is linearly dependent on it. Similarly, the third row can be generated by multiplying the first row by –1.

so, because this element is *unstable* and cannot resist the action of any arbitrary combination of forces F_{1*}^i, F_{2*}^i, F_{3*}^i and F_{4*}^i. In order to have a valid flexibility matrix, it is necessary to arrest two degrees of freedom. The resulting flexibility matrix is a 2×2 matrix, to be defined later.

5.2.3 Coordinate Transformations and Structure Stiffness Matrix

Transformation Matrix

In a continuous beam system, the local coordinate *x**- and *y**-axes of any particular beam element (with 4 degrees of freedom) can be conveniently chosen to be aligned in the same direction as the global *x*- and *y*-axes of the structure. Thus, the four local coordinates, numbered 1*, 2*, 3* and 4*, as shown in Fig. 5.1a, can be directly linked in the global axes system, as 1, 2, 3 and 4, to appropriate global coordinates (say, *l*, *m*, *n* and *p*) at the same locations in the continuous beam. Compatibility of displacement requires $D_l = D_1^i = D_{1*}^i$, $D_m = D_2^i = D_{2*}^i$, $D_n = D_3^i = D_{3*}^i$ and $D_p = D_4^i = D_{4*}^i$. The 4×4 element transformation matrix $\mathbf{T^i}$, relating the displacement and force vector components, $(D_1^i, D_2^i, D_3^i, D_4^i)$ and $(F_1^i, F_2^i, F_3^i, F_4^i)$, in the global axes system to the local coordinate system, $\mathbf{D_*^i} = \mathbf{T^iD^i}$ and $\mathbf{F_*^i} = \mathbf{T^iF^i}$, as defined in Chapter 3, will be an *identity* matrix ($\mathbf{T^i} = \mathbf{I} = \mathbf{T^{i^T}}$). It is convenient to mark the global coordinates in parenthesis as follows:

$$\mathbf{T^i} = \begin{bmatrix} 1 & 0 & 0 & 0 \\ 0 & 1 & 0 & 0 \\ 0 & 0 & 1 & 0 \\ 0 & 0 & 0 & 1 \end{bmatrix} \begin{matrix} (l) \\ (m) \\ (n) \\ (p) \end{matrix} \tag{5.18}$$

The element stiffness matrix in the global axes system is given by the transformation $\mathbf{k^i} = \mathbf{T^{i^T}k_*^iT^i}$ (refer Eq. 3.32), which in this case, reduces to $\mathbf{k^i} = \mathbf{k_*^i}$, whereby:

$$\boxed{\begin{Bmatrix} F_1^i \\ F_2^i \\ F_3^i \\ F_4^i \end{Bmatrix} = \frac{(EI)_i}{L_i} \begin{bmatrix} 12/L_i^2 & 6/L_i & -12/L_i^2 & 6/L_i \\ 6/L_i & 4 & -6/L_i & 2 \\ -12/L_i^2 & -6/L_i & 12/L_i^2 & -6/L_i \\ 6/L_i & 2 & -6/L_i & 4 \end{bmatrix} \begin{Bmatrix} D_l^i \\ D_m^i \\ D_n^i \\ D_p^i \end{Bmatrix}} \tag{5.19}$$

The structure stiffness matrix $\mathbf{k} = \left[\begin{array}{c|c} \mathbf{k_{AA}} & \mathbf{k_{AR}} \\ \hline \mathbf{k_{RA}} & \mathbf{k_{RR}} \end{array}\right]$, is assembled by summing up the stiffness coefficients contributions from the various beam elements at the appropriate global coordinate locations. In a typical continuous beam system, at any joint, there will be a translational and a rotational degree of freedom. The stiffness contributions to these global coordinates will be generated only from the two beams on either side of the joint. Some of the global coordinates may be restrained, and following our earlier convention, the numbering of these coordinates must be towards the end, with the active coordinates placed at the beginning.

Displacement Transformation Matrix

As an alternative, and to avoid the additional work involved in keeping track of the various linkages with the global coordinates (*l*, *m*, *n* and *p*) connected to a particular beam element, we can generate the *displacement transformation matrix*, $\mathbf{T}_D^i$ of order $4 \times (n + r)$, where *n* is the number of active global coordinates (equal to the degree of kinematic indeterminacy) and *r* is the number of restrained coordinates. Typically, if there are *m* elements in a continuous beam system, there will be $m + 1$ joints, and hence, $n + r = 2(m + 1)$. This matrix, containing elements that are either zero or 1, relates the element local coordinates to all the global coordinates:

$$\left\{D_*^i\right\}_{4\times 1} = \left[\left[T_{DA}\right]_{4\times n} \mid \left[T_{DR}\right]_{4\times r}\right]\left[\begin{array}{c}\left\{D_A\right\}_{n\times 1} \\ \hline \left\{D_R\right\}_{r\times 1}\end{array}\right] \tag{5.20}$$

Using the element stiffness matrix $\mathbf{k}_*^i$ given in Eq. 5.11, the product $\mathbf{k}_*^i\mathbf{T}_D^i$ is first computed for each beam element and thereafter, the structure stiffness matrix $\mathbf{k}$ is assembled directly by summing up the stiffness coefficients contributions from the various beam elements using the following transformation:

$$[k]_{(n+r)\times(n+r)} = \sum_{i=1}^{m}\left[\begin{array}{c}\left[T_{DA}^i\right]^T_{n\times 4} \\ \hline \left[T_{DR}^i\right]^T_{r\times 4}\end{array}\right]\left[k_*^i\right]_{4\times 4}\left[\left[T_{DA}^i\right]_{4\times n} \mid \left[T_{DR}^i\right]_{4\times r}\right] = \left[\begin{array}{c|c}\left[k_{AA}\right]_{n\times n} & \left[k_{AR}\right]_{n\times r} \\ \hline \left[k_{RA}\right]_{r\times n} & \left[k_{RR}\right]_{r\times r}\end{array}\right] \tag{5.21}$$

5.2.4 Fixed End Forces

In case there are intermediate loads acting in any beam element, we need to find the initial force and displacement responses in the 'fixed end' element (which is part of the primary structure) and thus arrive at the *equivalent joint loads*. For this purpose, standard formulae, such as the ones listed in Table 1.3, can be conveniently used to generate the fixed end moments, and using equilibrium equations, the corresponding fixed end reactions (shear forces at the two ends) can be calculated. Care must be taken to assign the appropriate sign (positive or negative), while generating the fixed end force vector for each element,

$$\mathbf{F}_{*f}^i = \begin{Bmatrix} F_{1*f}^i \\ F_{2*f}^i \\ F_{3*f}^i \\ F_{4*f}^i \end{Bmatrix}.$$

These fixed end forces can now be converted to equivalent joint loads by invoking appropriate coordinate transformations, such as by summing up the contributions of $\mathbf{T}^{i\mathrm{T}}\mathbf{F}^{i}_{*f}$ or $\mathbf{T}^{i\,\mathrm{T}}_{\mathrm{D}}\mathbf{F}^{i}_{*f}$. The resulting fixed end force vector, in global coordinates, takes the form, $\mathbf{F}_f = \begin{bmatrix} \mathbf{F}_{fA} \\ \hline \mathbf{F}_{fR} \end{bmatrix}$. The equivalent joint loads are given by $-\mathbf{F}_{fA}$, and these are combined with the given nodal load vector $\mathbf{F}_A$.

The basic stiffness relationships take the following form:

$$\begin{bmatrix} \mathbf{F}_A \\ \hline \mathbf{F}_R \end{bmatrix} - \begin{bmatrix} \mathbf{F}_{fA} \\ \hline \mathbf{F}_{fR} \end{bmatrix} = \begin{bmatrix} \mathbf{k}_{AA} & \mathbf{k}_{AR} \\ \mathbf{k}_{RA} & \mathbf{k}_{RR} \end{bmatrix} \begin{bmatrix} \mathbf{D}_A \\ \hline \mathbf{D}_R \end{bmatrix} \qquad (5.22)$$

5.2.5 Analysis Procedure

The steps involved in the conventional stiffness method of analysis of kinematically indeterminate beams may be summarised as follows:

1. Identify the beam elements, the local and global coordinates. Arrange the numbering of the global coordinates in such a way as to position the restrained coordinates at the end of the list. Identify the input loading data at the active global coordinates, comprising direct actions, $\mathbf{F}_A$, and support displacements, $\mathbf{D}_R$, if any.
2. If there are intermediate loads acting on the beam elements, find the fixed end force vector for each element, $\mathbf{F}^{i}_{*f}$, and also the deflections at critical locations, if required.
3. Generate the transformation matrices, $\mathbf{T}^i = \mathbf{I}$, for the various elements, identifying the linking global coordinates in parenthesis. [Alternatively, generate the displacement transformation matrices, $\mathbf{T}^i_{\mathrm{D}}$]. Assemble the fixed end force vector $\mathbf{F}_f = \begin{bmatrix} \mathbf{F}_{fA} \\ \hline \mathbf{F}_{fR} \end{bmatrix}$, combining the contributions, $\mathbf{T}^{i\mathrm{T}}\mathbf{F}^{i}_{*f}$ [or $\mathbf{T}^{i\,\mathrm{T}}_{\mathrm{D}}\mathbf{F}^{i}_{*f}$] from the various elements. Generate the resultant load vector, $\mathbf{F}_A - \mathbf{F}_{fA}$.
4. Generate the element stiffness matrices $\mathbf{k}^i_*$ for all the elements, using Eq. 5.11, and hence assemble the structure stiffness matrix, $\mathbf{k} = \begin{bmatrix} \mathbf{k}_{AA} & \mathbf{k}_{AR} \\ \mathbf{k}_{RA} & \mathbf{k}_{RR} \end{bmatrix}$, combining the stiffness coefficient contributions, $\mathbf{k}^i = \mathbf{T}^{i^{\mathrm{T}}}\mathbf{k}^i_*\mathbf{T}^i = \mathbf{k}^i_*$ at the appropriate locations. [Alternatively, compute $\mathbf{k}^i_*\mathbf{T}^i_{\mathrm{D}}$ for every element and assemble $\mathbf{k} = \sum_{i=1}^{m} \mathbf{T}^{i^{\mathrm{T}}}_{\mathrm{D}}\mathbf{k}^i_*\mathbf{T}^i_{\mathrm{D}}$]. Find the inverse, $[\mathbf{k}_{AA}]^{-1}$, which is a symmetric matrix of order $n \times n$.
5. Find the unknown displacements, $\mathbf{D}_A$, by solving the following stiffness relation:

$$\mathbf{D}_{A} = \left[\mathbf{k}_{AA}\right]^{-1}\left[\left(\mathbf{F}_{A} - \mathbf{F}_{fA}\right) - \mathbf{k}_{AR}\mathbf{D}_{R}\right] \tag{5.23}$$

6. Find the support reactions by solving the following equation:

$$\mathbf{F}_{R} = \mathbf{F}_{fR} + \mathbf{k}_{RA}\mathbf{D}_{A} + \mathbf{k}_{RR}\mathbf{D}_{R} \tag{5.24}$$

   Perform a simple equilibrium check (such as $\sum F_y = 0$).

7. Find the member end forces for each element.

$$\mathbf{F}_{*}^{i} = \mathbf{F}_{*f}^{i} + \left(\mathbf{k}_{*}^{i}\mathbf{T}^{i}\right)\mathbf{D} \tag{5.25}$$

   [Alternatively, use $\mathbf{F}_{*}^{i} = \mathbf{F}_{*f}^{i} + \left(\mathbf{k}_{*}^{i}\mathbf{T}_{DA}^{i}\right)\mathbf{D}_{A}$].

8. Draw the free-body diagrams for the various beam elements, and hence, the shear force and bending moment diagrams for the beam system. Also, sketch the deflected shape (superimposing the fixed end force effects) and identify the deflections at critical locations.

EXAMPLE 5.1

Analyse the non-prismatic fixed beam in Fig. 5.2 by the conventional stiffness method (matrix approach). Find the force and displacement responses, given that the beam is subject to the loads indicated in Fig. 5.2. Assume $EI = 80000$ kNm^2.

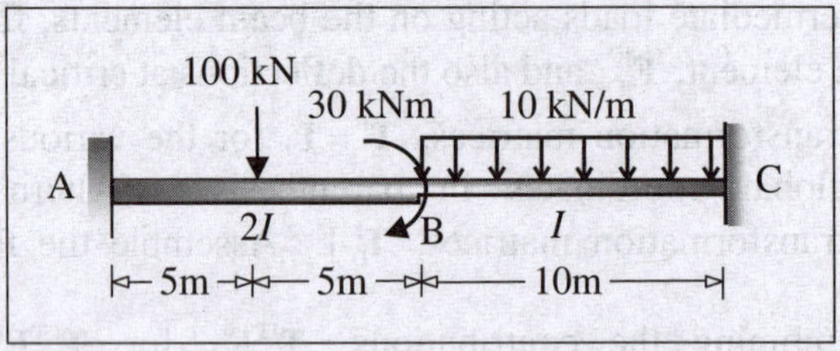

Figure 5.2 Non-Prismatic fixed beam (Example 5.1)

SOLUTION

There are two prismatic elements here ($m = 2$), with intermediate loads acting on both, and a nodal load acting at their junction. There are two *active* degrees of freedom ($n = 2$) and four *restrained* global coordinates ($r = 4$), numbered as shown in Fig. 5.3a. Accordingly, the nodal loads applied are given by $\mathbf{F}_{A} = \begin{Bmatrix} F_1 \\ F_2 \end{Bmatrix} = \begin{Bmatrix} 0 \text{ kN} \\ -30 \text{ kNm} \end{Bmatrix}$. The local coordinates for the two elements are shown in Fig. 5.3b.

Fixed End Forces

The fixed end force effects (support reactions, deflections, bending moments and shear force diagrams) on the primary structure (with $D_1 = D_2 = 0$) are shown in Fig. 5.3c.

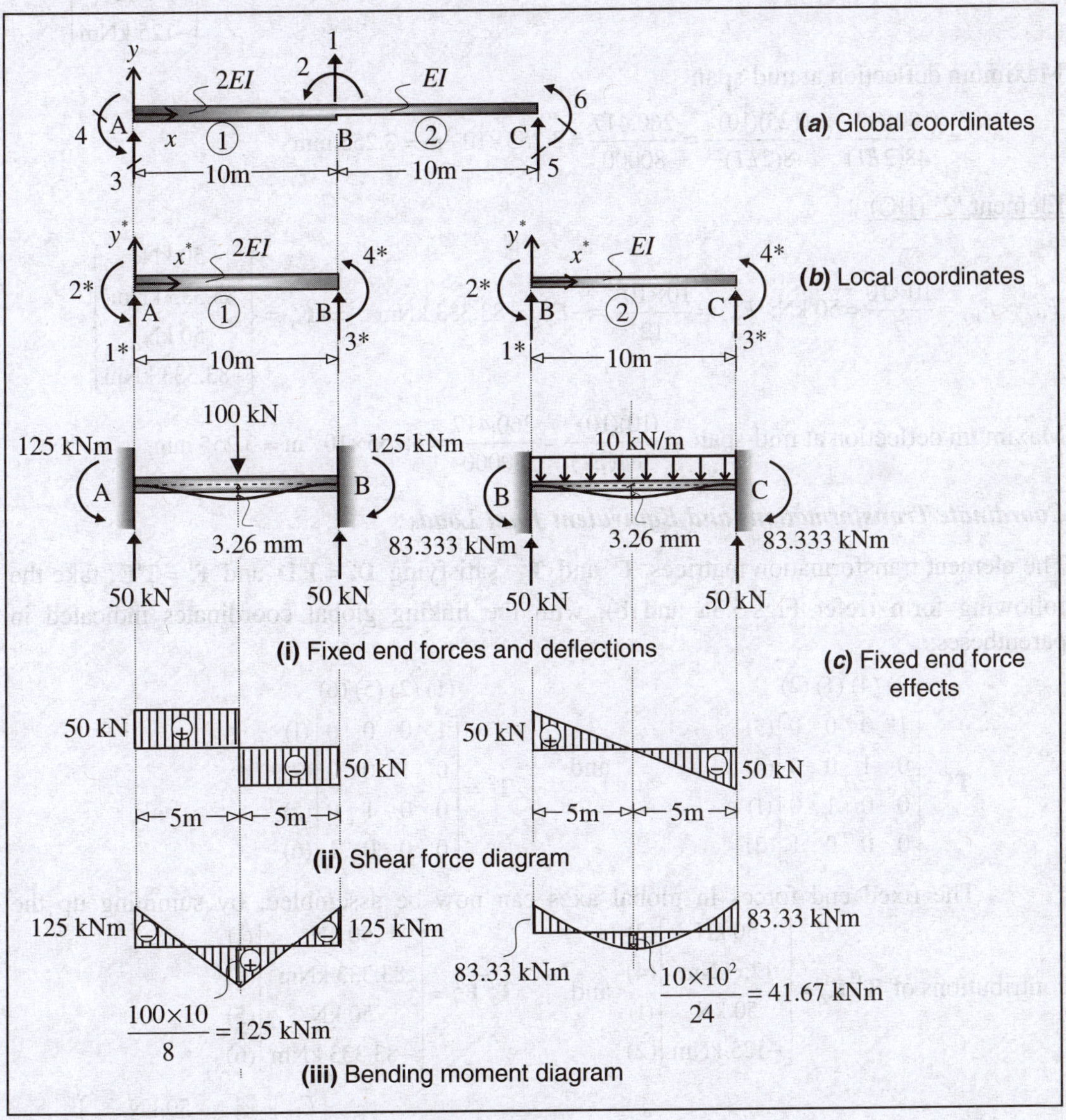

Figure 5.3 Global and local coordinates and fixed end force effects (Example 5.1)

Element '1' (AB):

$$F^1_{1*f} = F^1_{3*f} = \frac{100}{2} = 50 \text{ kN};\ F^1_{2*f} = \frac{100\times10}{8} = 125 \text{ kNm};\ F^1_{4*f} = -125 \text{ kNm} \Rightarrow \mathbf{F}^1_{*\mathbf{f}} = \begin{Bmatrix} 50 \text{ kN} \\ 125 \text{ kNm} \\ 50 \text{ kN} \\ -125 \text{ kNm} \end{Bmatrix}$$

Maximum deflection at mid-span

$$= \frac{(100)(10)^3}{48(2EI)} - \frac{(125)(10)^2}{8(2EI)} = \frac{260.417}{80000} = 3.255\times10^{-3}\text{ m} = 3.255 \text{ mm}.$$

Element '2' (BC):

$$F^2_{1*f} = F^2_{3*f} = \frac{10\times10}{2} = 50 \text{ kN};\ F^2_{2*f} = \frac{10\times10^2}{12} = -F^2_{4*f} = 83.333 \text{ kNm} \quad \Rightarrow \mathbf{F}^2_{*\mathbf{f}} = \begin{Bmatrix} 50 \text{ kN} \\ 83.333 \text{ kNm} \\ 50 \text{ kN} \\ -83.333 \text{ kNm} \end{Bmatrix}$$

Maximum deflection at mid-span $= \dfrac{(10)(10)^4}{384(EI)} = \dfrac{260.417}{80000} = 3.255\times10^{-3}\text{ m} = 3.255 \text{ mm}.$

Coordinate Transformations and Equivalent Joint Loads

The element transformation matrices $\mathbf{T}^1$ and $\mathbf{T}^2$, satisfying $\mathbf{D}^i_* = \mathbf{T}^i\mathbf{D}$ and $\mathbf{F}^i_* = \mathbf{T}^i\mathbf{F}$, take the following form (refer Figs 5.3a and b), with the linking global coordinates indicated in parentheses:

$$\mathbf{T}^1 = \begin{bmatrix} 1 & 0 & 0 & 0 \\ 0 & 1 & 0 & 0 \\ 0 & 0 & 1 & 0 \\ 0 & 0 & 0 & 1 \end{bmatrix} \begin{matrix} (3) \\ (4) \\ (1) \\ (2) \end{matrix} \text{ (columns (3) (4) (1) (2))} \quad \text{and} \quad \mathbf{T}^2 = \begin{bmatrix} 1 & 0 & 0 & 0 \\ 0 & 1 & 0 & 0 \\ 0 & 0 & 1 & 0 \\ 0 & 0 & 0 & 1 \end{bmatrix} \begin{matrix} (1) \\ (2) \\ (5) \\ (6) \end{matrix} \text{ (columns (1) (2) (5) (6))}$$

The fixed end forces in global axes can now be assembled, by summing up the contributions of

$$\mathbf{T}^{1^T}\mathbf{F}^1_{*\mathbf{f}} = \begin{bmatrix} 50 \text{ kN} \\ 125 \text{ kNm} \\ 50 \text{ kN} \\ -125 \text{ kNm} \end{bmatrix} \begin{matrix} (3) \\ (4) \\ (1) \\ (2) \end{matrix} \quad \text{and} \quad \mathbf{T}^{2^T}\mathbf{F}^2_{*\mathbf{f}} = \begin{bmatrix} 50 \text{ kN} \\ 83.333 \text{ kNm} \\ 50 \text{ kN} \\ -83.333 \text{ kNm} \end{bmatrix} \begin{matrix} (1) \\ (2) \\ (5) \\ (6) \end{matrix}$$

$$\Rightarrow \mathbf{F_f} = \begin{bmatrix} \mathbf{F_{fA}} \\ \hline \mathbf{F_{fR}} \end{bmatrix};\text{ where } \mathbf{F_{fA}} = \begin{Bmatrix} F_{f1} \\ F_{f2} \end{Bmatrix} = \begin{Bmatrix} 100 \text{ kN} \\ -41.667 \text{ kNm} \end{Bmatrix};\ \mathbf{F_{fR}} = \begin{Bmatrix} F_{3f} \\ F_{4f} \\ F_{5f} \\ F_{6f} \end{Bmatrix} = \begin{Bmatrix} 50 \text{ kN} \\ 125 \text{ kNm} \\ 50 \text{ kN} \\ -83.333 \text{ kNm} \end{Bmatrix}$$

The resultant load vector, $\mathbf{F_A} - \mathbf{F_{fA}}$, is given by:

$$\mathbf{F_A} - \mathbf{F_{fA}} = \begin{Bmatrix} 0 \text{ kN} \\ -30 \text{ kNm} \end{Bmatrix} - \begin{Bmatrix} 100 \text{ kN} \\ -41.667 \text{ kNm} \end{Bmatrix} = \begin{Bmatrix} -100 \text{ kN} \\ 11.667 \text{ kNm} \end{Bmatrix}$$

This is indicated in Fig. 5.4a. The deflections at the active coordinates, D_1 and D_2, will be identical to those in the original loading diagram (with the distributed loading) in Fig. 5.2.

Element and Structure Stiffness Matrices

The element stiffness matrices $\mathbf{k}_*^1$ and $\mathbf{k}_*^2$, satisfying $\mathbf{F}_*^i = \mathbf{k}_*^i \mathbf{D}_*^i$, take the following form (refer Fig. 5.3b), considering $\frac{E_1 I_1}{L_1} = \frac{2EI}{10}$ and $\frac{E_2 I_2}{L_2} = \frac{EI}{10}$:

$$\mathbf{k}_*^1 = \frac{2EI}{10}\begin{bmatrix} 12/10^2 & 6/10 & -12/10^2 & 6/10 \\ 6/10 & 4 & -6/10 & 2 \\ -12/10^2 & -6/10 & 12/10^2 & -6/10 \\ 6/10 & 2 & -6/10 & 4 \end{bmatrix} \Rightarrow \mathbf{k}_*^1\mathbf{T}^1 = EI \begin{matrix} (3) & (4) & (1) & (2) & \\ \end{matrix} \begin{bmatrix} 0.024 & 0.12 & -0.024 & 0.12 \\ 0.12 & 0.8 & -0.12 & 0.4 \\ -0.024 & -0.12 & 0.024 & -0.12 \\ 0.12 & 0.4 & -0.12 & 0.8 \end{bmatrix} \begin{matrix} (3) \\ (4) \\ (1) \\ (2) \end{matrix}$$

$$\mathbf{k}_*^2 = \frac{EI}{10}\begin{bmatrix} 12/10^2 & 6/10 & -12/10^2 & 6/10 \\ 6/10 & 4 & -6/10 & 2 \\ -12/10^2 & -6/10 & 12/10^2 & -6/10 \\ 6/10 & 2 & -6/10 & 4 \end{bmatrix} \Rightarrow \mathbf{k}_*^2\mathbf{T}^2 = EI \begin{matrix} (1) & (2) & (5) & (6) & \\ \end{matrix} \begin{bmatrix} 0.012 & 0.06 & -0.012 & 0.06 \\ 0.06 & 0.4 & -0.06 & 0.2 \\ -0.012 & -0.06 & 0.012 & -0.06 \\ 0.06 & 0.2 & -0.06 & 0.4 \end{bmatrix} \begin{matrix} (1) \\ (2) \\ (5) \\ (6) \end{matrix}$$

The structure stiffness matrix $\mathbf{k}$, satisfying $\mathbf{F} = \mathbf{kD}$, can now be assembled, by summing up the contributions of $\mathbf{T}^{1^T}\mathbf{k}_*^1\mathbf{T}^1$ and $\mathbf{T}^{2^T}\mathbf{k}_*^2\mathbf{T}^2$:

$$\Rightarrow \quad \mathbf{k} = \left[\begin{array}{c|c} \mathbf{k_{AA}} & \mathbf{k_{AR}} \\ \hline \mathbf{k_{RA}} & \mathbf{k_{RR}} \end{array}\right] = EI \left[\begin{array}{cc|cccc} 0.036 & -0.06 & -0.024 & -0.12 & -0.012 & 0.06 \\ -0.06 & 1.2 & 0.12 & 0.4 & -0.06 & 0.2 \\ \hline -0.024 & 0.12 & 0.024 & 0.12 & 0 & 0 \\ -0.12 & 0.4 & 0.12 & 0.8 & 0 & 0 \\ -0.012 & -0.06 & 0 & 0 & 0.012 & -0.06 \\ 0.06 & 0.2 & 0 & 0 & -0.06 & 0.4 \end{array}\right]$$

Displacements and Support Reactions

The structure stiffness relation to be considered here is:

$$\left[\begin{array}{c} \mathbf{F_A} \\ \hline \mathbf{F_R} \end{array}\right] - \left[\begin{array}{c} \mathbf{F_{fA}} \\ \hline \mathbf{F_{fR}} \end{array}\right] = \left[\begin{array}{c|c} \mathbf{k_{AA}} & \mathbf{k_{AR}} \\ \hline \mathbf{k_{RA}} & \mathbf{k_{RR}} \end{array}\right] \left[\begin{array}{c} \mathbf{D_A} \\ \hline \mathbf{D_R} = \mathbf{O} \end{array}\right]$$

$$\Rightarrow \quad \mathbf{D_A} = [\mathbf{k_{AA}}]^{-1}[\mathbf{F_A} - \mathbf{F_{fA}}] \quad \Rightarrow \quad \begin{Bmatrix} D_1 \\ D_2 \end{Bmatrix} = \frac{1}{EI}\begin{bmatrix} 30.3030 & 1.51515 \\ 1.51515 & 0.90909 \end{bmatrix}\begin{Bmatrix} -100 \\ 11.667 \end{Bmatrix} = \frac{1}{EI}\begin{Bmatrix} -3012.626 \\ -140.909 \end{Bmatrix}$$

$$\Rightarrow \quad D_1 = \frac{-3012.626}{80000} = -37.656\times10^{-3}\ \text{m}; \quad D_2 = \frac{-140.909}{80000} = -1.7614\times10^{-3}\ \text{rad}$$

The deflections, due to the applied loading in Fig. 5.4a, are shown in Fig. 5.4b. The deflection variation follows a cubic polynomial variation, as explained earlier, whereby the deflections at the mid-span locations in the two elements can be obtained:

$$\Delta_{AB}(x = L/2) = 0.5D_1 - 0.125L_1D_2 = 0.5(-37.656\times10^{-3}) - 0.125(10)(-1.7614\times10^{-3}) = -16.626\times10^{-3}\ \text{m}$$

$$\Delta_{BC}(x = L/2) = 0.5D_1 + 0.125L_1D_2 = 0.5(-37.656\times10^{-3}) + 0.125(10)(-1.7614\times10^{-3}) = -21.030\times10^{-3}\ \text{m}$$

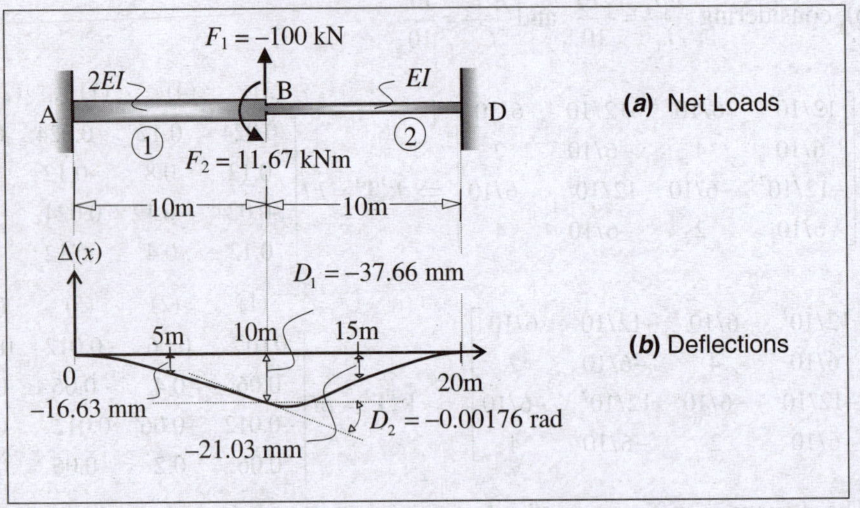

Figure 5.4 Response to nodal plus equivalent joint loads (Example 5.1)

Superimposing these displacements to the fixed end displacements in Fig. 5.3c(iii), the final deflections (at 5m intervals) can be generated, as shown in Fig. 5.5d.

$$\mathbf{F_R} = \mathbf{F_{fR}} + \mathbf{k_{RA}D_A} \Rightarrow \begin{Bmatrix} F_3 \\ F_4 \\ F_5 \\ F_6 \end{Bmatrix} = \begin{Bmatrix} 50 \\ 125 \\ 50 \\ -83.333 \end{Bmatrix} + EI\begin{bmatrix} -0.024 & 0.12 \\ -0.12 & 0.4 \\ -0.012 & -0.06 \\ 0.06 & 0.2 \end{bmatrix} \frac{1}{EI}\begin{Bmatrix} -3012.626 \\ -140.909 \end{Bmatrix} = \begin{Bmatrix} 105.394\ \text{kN} \\ 430.152\ \text{kNm} \\ 94.606\ \text{kN} \\ -292.273\ \text{kNm} \end{Bmatrix}$$

<u>Check</u> $\sum F_y = 0$:

Total reaction = $F_3 + F_5 = 105.394 + 94.606 = 200.000$ kN = Total load (downwards) ⇒ OK.

Member Forces

$$\mathbf{F_*^i} = \mathbf{F_{*f}^i} + \mathbf{k_*^i T^i D}$$

$$\Rightarrow \mathbf{F}_*^1 = \begin{Bmatrix} 50 \\ 125 \\ 50 \\ -125 \end{Bmatrix} + EI \begin{bmatrix} 0.024 & 0.12 & -0.024 & 0.12 \\ 0.12 & 0.8 & -0.12 & 0.4 \\ -0.024 & -0.12 & 0.024 & -0.12 \\ 0.12 & 0.4 & -0.12 & 0.8 \end{bmatrix} \begin{Bmatrix} D_3 = 0 \\ D_4 = 0 \\ D_1 = -3012.626/EI \\ D_2 = -140.909/EI \end{Bmatrix} = \begin{Bmatrix} 105.394 \text{ kN} \\ 430.152 \text{ kNm} \\ -5.394 \text{ kN} \\ 123.788 \text{ kNm} \end{Bmatrix}$$

$$\mathbf{F}_*^2 = \begin{Bmatrix} 50 \\ 83.333 \\ 50 \\ -83.333 \end{Bmatrix} + EI \begin{bmatrix} 0.012 & 0.06 & -0.012 & 0.06 \\ 0.06 & 0.4 & -0.06 & 0.2 \\ -0.012 & -0.06 & 0.012 & -0.06 \\ 0.06 & 0.2 & -0.06 & 0.4 \end{bmatrix} \begin{Bmatrix} D_1 = -3012.626/EI \\ D_2 = -140.909/EI \\ D_5 = 0 \\ D_6 = 0 \end{Bmatrix} = \begin{Bmatrix} 5.394 \text{ kN} \\ -153.788 \text{ kNm} \\ 94.606 \text{ kN} \\ -292.273 \text{ kNm} \end{Bmatrix}$$

The free-body, shear force and bending moment and deflection diagrams, including the fixed end force effects in Fig. 5.3, are shown in Fig. 5.5.

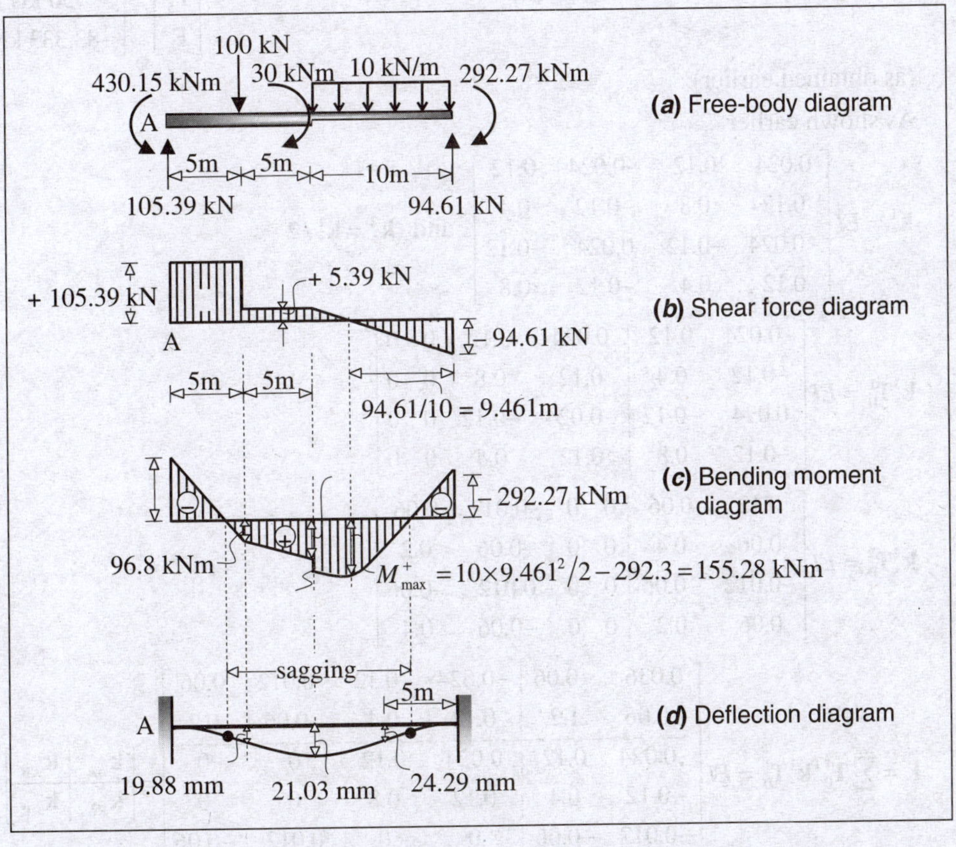

Figure 5.5 Internal force distribution and deflection profile (Example 5.1)

Alternative Procedure Using Displacement Transformation Matrix

$$\mathbf{D}_*^i = \left[\mathbf{T}_{DA}^i \mid \mathbf{T}_{DR}^i\right]\left[\frac{\mathbf{D}_A}{\mathbf{D}_R}\right] \Rightarrow \mathbf{T}_D^1 = \begin{bmatrix} 0 & 0 & 1 & 0 & 0 & 0 \\ 0 & 0 & 0 & 1 & 0 & 0 \\ 1 & 0 & 0 & 0 & 0 & 0 \\ 0 & 1 & 0 & 0 & 0 & 0 \end{bmatrix} \text{ and } \mathbf{T}_D^2 = \begin{bmatrix} 1 & 0 & 0 & 0 & 0 & 0 \\ 0 & 1 & 0 & 0 & 0 & 0 \\ 0 & 0 & 0 & 0 & 1 & 0 \\ 0 & 0 & 0 & 0 & 0 & 1 \end{bmatrix}$$

$$\mathbf{F}_{*f}^1 = \begin{Bmatrix} 50 \text{ kN} \\ 125 \text{ kNm} \\ 50 \text{ kN} \\ -125 \text{ kNm} \end{Bmatrix} \text{ and } \mathbf{F}_{*f}^2 = \begin{Bmatrix} 50 \text{ kN} \\ 83.333 \text{ kNm} \\ 50 \text{ kN} \\ -83.333 \text{ kNm} \end{Bmatrix} \Rightarrow \mathbf{F}_f = \sum_{i=1}^{2} \mathbf{T}_D^{i\,T} \mathbf{F}_{*f}^i = \left[\frac{\mathbf{F}_{fA}}{\mathbf{F}_{fR}}\right] = \begin{bmatrix} F_1 \\ F_2 \\ \hline F_3 \\ F_4 \\ F_5 \\ F_6 \end{bmatrix} = \begin{bmatrix} 100 \text{ kN} \\ -41.667 \text{ kNm} \\ \hline 50 \text{ kN} \\ 125 \text{ kNm} \\ 50 \text{ kN} \\ -83.333 \text{ kNm} \end{bmatrix}$$

(as obtained earlier)

As shown earlier,

$$\mathbf{k}_*^1 = EI \begin{bmatrix} 0.024 & 0.12 & -0.024 & 0.12 \\ 0.12 & 0.8 & -0.12 & 0.4 \\ -0.024 & -0.12 & 0.024 & -0.12 \\ 0.12 & 0.4 & -0.12 & 0.8 \end{bmatrix} \text{ and } \mathbf{k}_*^2 = \mathbf{k}_*^1/2$$

$$\Rightarrow \quad \mathbf{k}_*^1\mathbf{T}_D^1 = EI \begin{bmatrix} -0.024 & 0.12 & 0.024 & 0.12 & 0 & 0 \\ -0.12 & 0.4 & 0.12 & 0.8 & 0 & 0 \\ 0.024 & -0.12 & -0.024 & -0.12 & 0 & 0 \\ -0.12 & 0.8 & 0.12 & 0.4 & 0 & 0 \end{bmatrix}$$

$$\mathbf{k}_*^2\mathbf{T}_D^2 = EI \begin{bmatrix} 0.12 & 0.06 & 0 & 0 & -0.012 & 0.06 \\ 0.06 & 0.4 & 0 & 0 & -0.06 & 0.2 \\ -0.012 & -0.06 & 0 & 0 & 0.012 & -0.06 \\ 0.06 & 0.2 & 0 & 0 & -0.06 & 0.4 \end{bmatrix}$$

$$\Rightarrow \quad \mathbf{k} = \sum_{i=1}^{2} \mathbf{T}_D^{i\,T}\mathbf{k}_*^i\mathbf{T}_D^i = EI \begin{bmatrix} 0.036 & -0.06 & -0.024 & -0.12 & -0.012 & 0.06 \\ -0.06 & 1.2 & 0.12 & 0.4 & -0.06 & 0.2 \\ \hline -0.024 & 0.12 & 0.024 & 0.12 & 0 & 0 \\ -0.12 & 0.4 & 0.12 & 0.8 & 0 & 0 \\ -0.012 & -0.06 & 0 & 0 & 0.012 & -0.06 \\ 0.06 & 0.2 & 0 & 0 & -0.06 & 0.4 \end{bmatrix} = \begin{bmatrix} \mathbf{k}_{AA} & \mathbf{k}_{AR} \\ \hline \mathbf{k}_{RA} & \mathbf{k}_{RR} \end{bmatrix}$$

(as obtained earlier).

Thus, we get exactly the same solution for the response by applying:

$$\mathbf{D_A} = [\mathbf{k_{AA}}]^{-1}[\mathbf{F_A} - \mathbf{F_{fA}}] \quad \Rightarrow \quad \begin{Bmatrix} D_1 \\ D_2 \end{Bmatrix} = \frac{1}{EI}\begin{bmatrix} 30.3030 & 1.51515 \\ 1.51515 & 0.90909 \end{bmatrix}\begin{Bmatrix} -100 \\ 11.667 \end{Bmatrix} = \frac{1}{EI}\begin{Bmatrix} -3012.626 \\ -140.909 \end{Bmatrix},$$

$$\mathbf{F_R} = \mathbf{F_{fR}} + \mathbf{k_{RA}}\mathbf{D_A} \quad \Rightarrow \quad \begin{Bmatrix} F_3 \\ F_4 \\ F_5 \\ F_6 \end{Bmatrix} = \begin{Bmatrix} 105.394 \text{ kN} \\ 430.152 \text{ kNm} \\ 94.606 \text{ kN} \\ -292.273 \text{ kNm} \end{Bmatrix}, \text{ and}$$

$$\mathbf{F_*^i} = \mathbf{F_*^i} + \left(\mathbf{k_*^i T_D^i}\right)\mathbf{D} \Rightarrow$$

$$\begin{Bmatrix} F_{1*}^1 \\ F_{2*}^1 \\ F_{3*}^1 \\ F_{4*}^1 \end{Bmatrix} = \begin{Bmatrix} 50 \\ 125 \\ 50 \\ -125 \end{Bmatrix} + EI\left[\begin{array}{cc:cccc} -0.024 & 0.12 & 0.024 & 0.12 & 0 & 0 \\ -0.12 & 0.4 & 0.12 & 0.8 & 0 & 0 \\ 0.024 & -0.12 & -0.024 & -0.12 & 0 & 0 \\ -0.12 & 0.8 & 0.12 & 0.4 & 0 & 0 \end{array}\right]\left\{\begin{array}{c} -3012.626/EI \\ -140.909/EI \\ \hdashline 0 \\ 0 \\ 0 \\ 0 \end{array}\right\} = \begin{Bmatrix} 105.394 \text{ kN} \\ 430.152 \text{ kNm} \\ -5.394 \text{ kN} \\ 123.788 \text{ kNm} \end{Bmatrix}$$

$$\begin{Bmatrix} F_{1*}^2 \\ F_{2*}^2 \\ F_{3*}^2 \\ F_{4*}^2 \end{Bmatrix} = \begin{Bmatrix} 50 \\ 83.333 \\ 50 \\ -83.333 \end{Bmatrix} + EI\left[\begin{array}{cc:cccc} 0.12 & 0.06 & 0 & 0 & -0.012 & 0.06 \\ 0.06 & 0.4 & 0 & 0 & -0.06 & 0.2 \\ -0.012 & -0.06 & 0 & 0 & 0.012 & -0.06 \\ 0.06 & 0.2 & 0 & 0 & -0.06 & 0.4 \end{array}\right]\left\{\begin{array}{c} -3012.626/EI \\ -140.909/EI \\ \hdashline 0 \\ 0 \\ 0 \\ 0 \end{array}\right\} = \begin{Bmatrix} 5.394 \text{ kN} \\ -153.79 \text{ kNm} \\ 94.606 \text{ kN} \\ -292.27 \text{ kNm} \end{Bmatrix}$$

(exactly as obtained earlier).

EXAMPLE 5.2

Repeat Example 5.1, considering, instead of the applied loading shown in Fig. 5.2, the effect of indirect loading in the form of support movements: (i) a clockwise rotational slip of 0.002 rad at the fixed end A and (ii) a downward settlement of 10 mm at support C.

SOLUTION

Choosing the same global and local coordinates as shown in Fig. 5.3, the loading here is a *displacement loading*, given by $\mathbf{D_R} = \begin{Bmatrix} D_3 \\ D_4 \\ D_5 \\ D_6 \end{Bmatrix} = \begin{Bmatrix} 0 \text{ m} \\ -0.002 \text{ rad} \\ -0.010 \text{ m} \\ 0 \text{ rad} \end{Bmatrix}$. The direct load vector, $\mathbf{F_A} = \begin{Bmatrix} F_1 \\ F_2 \end{Bmatrix} = \begin{Bmatrix} 0 \text{ kN} \\ 0 \text{ kNm} \end{Bmatrix}$. As there are no intermediate loads, there are no fixed end forces to consider. Hence, the basic stiffness relations are given by:

$$\begin{bmatrix} \mathbf{F_A = O} \\ \hline \mathbf{F_R} \end{bmatrix} - \begin{bmatrix} \mathbf{F_{fA} = O} \\ \hline \mathbf{F_{fR} = O} \end{bmatrix} = \begin{bmatrix} \mathbf{k_{AA}} & \mathbf{k_{AR}} \\ \hline \mathbf{k_{RA}} & \mathbf{k_{RR}} \end{bmatrix} \begin{bmatrix} \mathbf{D_A} \\ \hline \mathbf{D_R} \end{bmatrix} \Rightarrow \mathbf{k_{AA} D_A + k_{AR} D_R = O}$$

$$\text{where } \mathbf{k_{AR} D_R} = (80000) \begin{bmatrix} -0.024 & -0.12 & -0.012 & 0.06 \\ 0.12 & 0.4 & -0.06 & 0.2 \end{bmatrix} \begin{Bmatrix} 0 \text{ m} \\ -0.002 \text{ rad} \\ -0.010 \text{ m} \\ 0 \text{ rad} \end{Bmatrix} = \begin{Bmatrix} 28.8 \text{ kN} \\ -16.0 \text{ kNm} \end{Bmatrix}$$

Displacements and Support Reactions

$$\mathbf{D_A} = [\mathbf{k_{AA}}]^{-1} (-\mathbf{k_{AR} D_R}) \Rightarrow \begin{Bmatrix} D_1 \\ D_2 \end{Bmatrix} = \frac{1}{EI} \begin{bmatrix} 30.3030 & 1.51515 \\ 1.51515 & 0.90909 \end{bmatrix} \begin{Bmatrix} -28.8 \\ 16.0 \end{Bmatrix} = \frac{1}{EI} \begin{Bmatrix} -848.4848 \\ -29.0909 \end{Bmatrix}$$

$$\Rightarrow D_1 = \frac{-848.4848}{80000} = -10.606 \times 10^{-3} \text{ m}; \; D_2 = \frac{-29.0909}{80000} = -0.3636 \times 10^{-3} \text{ rad}$$

$$\mathbf{F_R = k_{RA} D_A + k_{RR} D_R}$$

$$\Rightarrow \begin{Bmatrix} F_3 \\ F_4 \\ F_5 \\ F_6 \end{Bmatrix} = EI \begin{bmatrix} -0.024 & 0.12 \\ -0.12 & 0.4 \\ -0.012 & -0.06 \\ 0.06 & 0.2 \end{bmatrix} \frac{1}{EI} \begin{Bmatrix} -848.484 \\ -29.0909 \end{Bmatrix} + 80000 \begin{bmatrix} 0.024 & 0.12 & 0 & 0 \\ 0.12 & 0.8 & 0 & 0 \\ 0 & 0 & 0.012 & -0.06 \\ 0 & 0 & -0.06 & 0.4 \end{bmatrix} \begin{Bmatrix} 0 \\ -0.002 \\ -0.010 \\ 0 \end{Bmatrix}$$

$$= \begin{Bmatrix} 16.873 \\ 90.182 \\ 11.927 \\ -56.727 \end{Bmatrix} + \begin{Bmatrix} -19.20 \\ -128.0 \\ -96.00 \\ 48.00 \end{Bmatrix} = \begin{Bmatrix} -2.327 \text{ kN} \\ -37.818 \text{ kNm} \\ 2.327 \text{ kN} \\ -8.727 \text{ kNm} \end{Bmatrix}$$

<u>Check</u> $\sum F_y = 0: \; F_3 + F_5 = -2.327 + 2.327 = 0 \text{ kN} \; \Rightarrow \text{OK.}$

Member Forces

$$\mathbf{F_*^i = k_*^i T^i D}$$

$$\Rightarrow \mathbf{F_*^1} = EI \begin{bmatrix} 0.024 & 0.12 & -0.024 & 0.12 \\ 0.12 & 0.8 & -0.12 & 0.4 \\ -0.024 & -0.12 & 0.024 & -0.12 \\ 0.12 & 0.4 & -0.12 & 0.8 \end{bmatrix} \begin{Bmatrix} D_3 = 0 \\ D_4 = -0.002 \times (8 \times 10^4 / EI) \\ D_1 = -848.4848 / EI \\ D_2 = -29.0909 / EI \end{Bmatrix} = \begin{Bmatrix} -2.327 \text{ kN} \\ -37.818 \text{ kNm} \\ 2.327 \text{ kN} \\ 14.545 \text{ kNm} \end{Bmatrix}$$

$$\mathbf{F_*^2} = EI \begin{bmatrix} 0.012 & 0.06 & -0.012 & 0.06 \\ 0.06 & 0.4 & -0.06 & 0.2 \\ -0.012 & -0.06 & 0.012 & -0.06 \\ 0.06 & 0.2 & -0.06 & 0.4 \end{bmatrix} \begin{Bmatrix} D_1 = -848.4848 / EI \\ D_2 = -29.0909 / EI \\ D_5 = -0.010 \times (8 \times 10^4 / EI) \\ D_6 = 0 \end{Bmatrix} = \begin{Bmatrix} -2.327 \text{ kN} \\ -14.545 \text{ kNm} \\ 2.327 \text{ kN} \\ -8.727 \text{ kNm} \end{Bmatrix}$$

The complete response, in the form of free-body, shear force, bending moment and deflection diagrams are shown in Fig. 5.6.

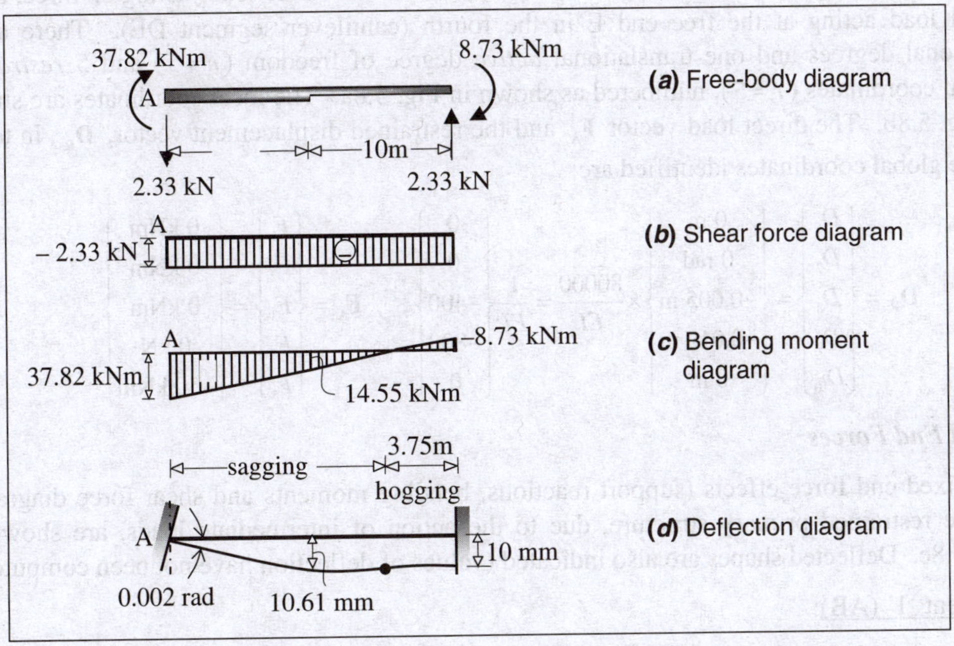

Figure 5.6 Internal force distribution and deflection profile (Example 5.2)

EXAMPLE 5.3

Analyse the continuous beam shown in Fig. 5.7, subject to the external loading and support settlements as indicated, by the Stiffness method (conventional matrix approach). Assume $EI = 80\,000$ kNm^2. Draw the shear force and bending moment diagrams.

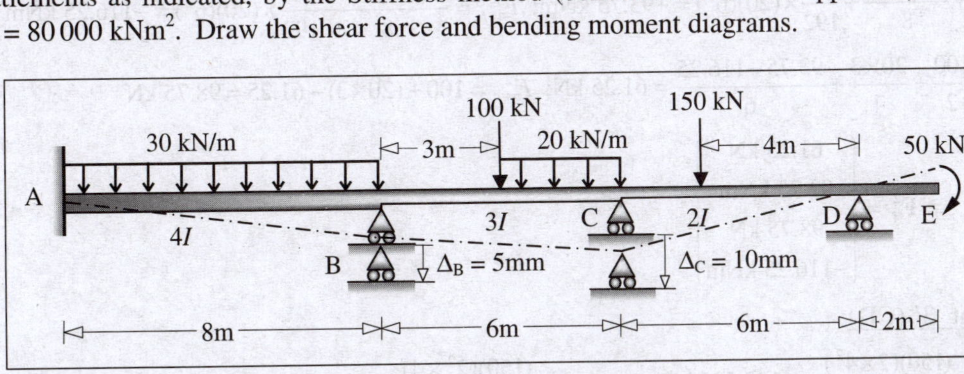

Figure 5.7 Continuous beam with gravity loading combined with support settlements (Example 5.3)

SOLUTION

There are four prismatic elements here ($m = 4$), with intermediate loads acting on three, and a nodal load acting at the free end E in the fourth (cantilever segment DE). There are 4 rotational degrees and one translational *active* degree of freedom ($n = 5$) and 5 *restrained* global coordinates ($r = 5$), numbered as shown in Fig. 5.8a. The local coordinates are shown in Fig. 5.8b. The direct load vector $\mathbf{F_A}$ and the restrained displacement vector, $\mathbf{D_R}$, in terms of the global coordinates identified are:

$$\mathbf{D_R} = \begin{Bmatrix} D_6 \\ D_7 \\ D_8 \\ D_9 \\ D_{10} \end{Bmatrix} = \begin{Bmatrix} 0 \text{ m} \\ 0 \text{ rad} \\ -0.005 \text{ m} \\ -0.010 \text{ m} \\ 0 \text{ m} \end{Bmatrix} \times \frac{80000}{EI} = \frac{1}{EI} \begin{Bmatrix} 0 \\ 0 \\ -400 \\ -800 \\ 0 \end{Bmatrix}, \quad \mathbf{F_A} = \begin{Bmatrix} F_1 \\ F_2 \\ F_3 \\ F_4 \\ F_5 \end{Bmatrix} = \begin{Bmatrix} 0 \text{ kNm} \\ 0 \text{ kNm} \\ 0 \text{ kNm} \\ 0 \text{ kN} \\ -50 \text{ kNm} \end{Bmatrix}.$$

Fixed End Forces

The fixed end force effects (support reactions, bending moments and shear force diagrams) on the restrained primary structure, due to the action of intermediate loads, are shown in Fig. 5.8c. Deflected shapes are also indicated (values of deflection have not been computed).

Element '1' (AB):

$$F^1_{1*f} = F^1_{3*f} = \frac{30 \times 8}{2} = 120 \text{ kN}; \; F^1_{2*f} = \frac{30 \times 8^2}{12} = 160 \text{ kNm}; \; F^1_{4*f} = -160 \text{ kNm} \Rightarrow \mathbf{F^1_{*f}} = \begin{Bmatrix} 120 \text{ kN} \\ 160 \text{ kNm} \\ 120 \text{ kN} \\ -160 \text{ kNm} \end{Bmatrix}$$

Element '2' (BC):

$$F^2_{2*f} = +\frac{(100)(6)}{8} + \frac{5}{192} \times (20)(6^2) = +93.75 \text{ kNm}; \; F^2_{4*f} = -\frac{(100)(6)}{8} - \frac{11}{192} \times (20)(6^2) = -116.25 \text{ kNm}$$

$$F^2_{1*f} = \frac{100}{2} + \frac{20 \times 3}{4} + \frac{93.75 - 116.25}{6} = 61.25 \text{ kN}; \; F^2_{3*f} = 100 + (20 \times 3) - 61.25 = 98.75 \text{ kN}$$

$$\Rightarrow \mathbf{F^2_{*f}} = \begin{Bmatrix} 61.25 \text{ kN} \\ 93.75 \text{ kNm} \\ 98.75 \text{ kN} \\ -116.25 \text{ kNm} \end{Bmatrix}$$

Element '3' (CD):

$$F^3_{2*f} = +\frac{(150)(2 \times 4^2)}{6^2} = +133.333 \text{ kNm}; \; F^3_{4*f} = -\frac{(150)(2^2 \times 4)}{6^2} = -66.667 \text{ kNm}$$

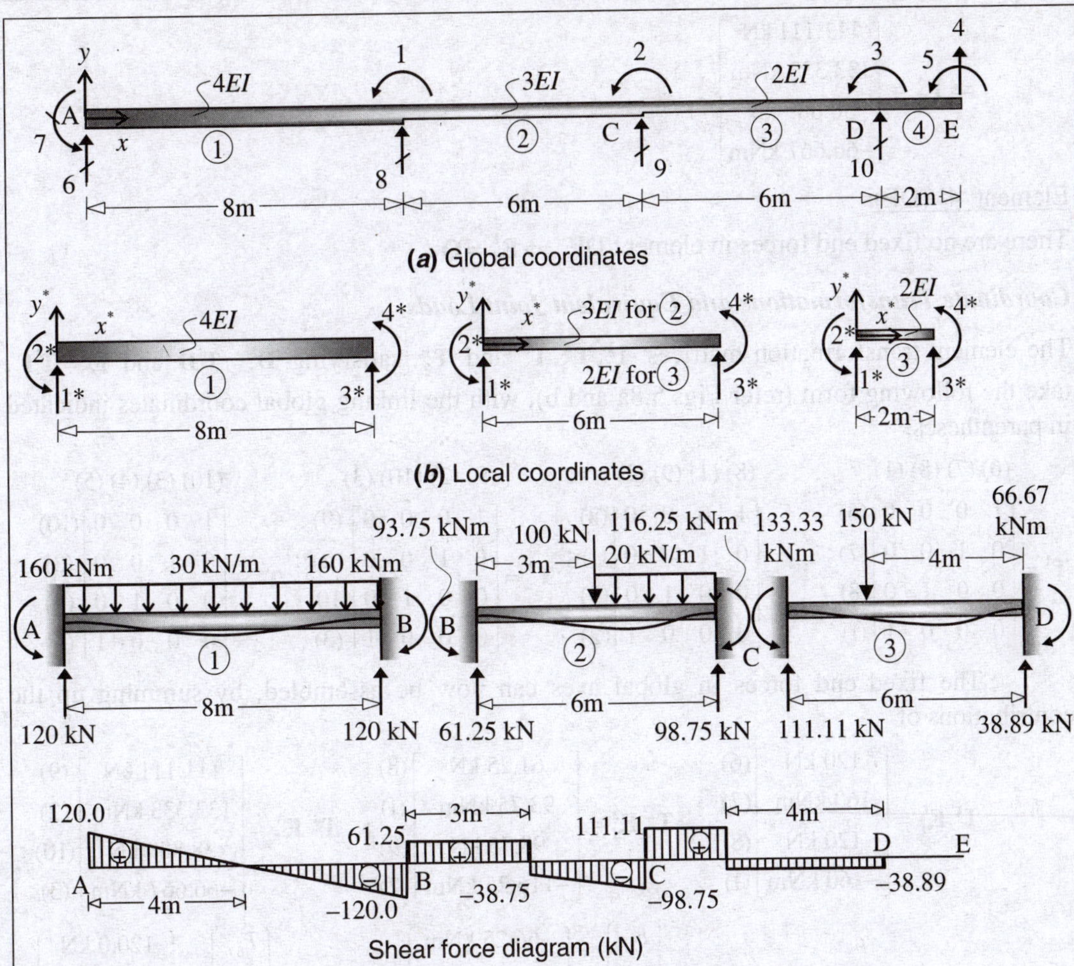

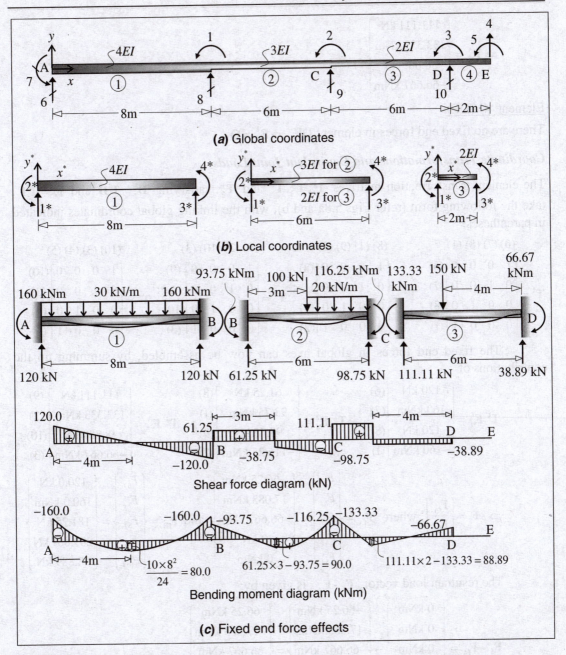

Figure 5.8 Global and local coordinates and fixed end force effects (Example 5.3)

$$F_{1^*f}^3 = \frac{150 \times 4}{6} + \frac{133.333 - 66.667}{6} = 111.111 \text{ kN}; \; F_{3^*f}^3 = 150 - 111.11 = 38.889 \text{ kN}$$

$$\Rightarrow \mathbf{F}_{*f}^{3} = \begin{Bmatrix} 111.111 \text{ kN} \\ 133.333 \text{ kNm} \\ 38.889 \text{ kN} \\ -66.667 \text{ kNm} \end{Bmatrix}$$

Element '4' (DE):

There are no fixed end forces in element DE $\Rightarrow \mathbf{F}_{*f}^{4} = \mathbf{O}$.

Coordinate Transformations and Equivalent Joint Loads

The element transformation matrices $\mathbf{T}^1, \mathbf{T}^2, \mathbf{T}^3$ and $\mathbf{T}^4$, satisfying $\mathbf{D}_*^i = \mathbf{T}^i\mathbf{D}$ and $\mathbf{F}_*^i = \mathbf{T}^i\mathbf{F}$, take the following form (refer Figs 5.8a and b), with the linking global coordinates indicated in parentheses:

$$\mathbf{T}^1 = \overset{(6)\,(7)\,(8)\,(1)}{\begin{bmatrix} 1 & 0 & 0 & 0 \\ 0 & 1 & 0 & 0 \\ 0 & 0 & 1 & 0 \\ 0 & 0 & 0 & 1 \end{bmatrix}} \begin{matrix} (6) \\ (7) \\ (8) \\ (1) \end{matrix}; \quad \mathbf{T}^2 = \overset{(8)\,(1)\,(9)\,(2)}{\begin{bmatrix} 1 & 0 & 0 & 0 \\ 0 & 1 & 0 & 0 \\ 0 & 0 & 1 & 0 \\ 0 & 0 & 0 & 1 \end{bmatrix}} \begin{matrix} (8) \\ (1) \\ (9) \\ (2) \end{matrix}; \quad \mathbf{T}^3 = \overset{(9)\,(2)\,(10)\,(3)}{\begin{bmatrix} 1 & 0 & 0 & 0 \\ 0 & 1 & 0 & 0 \\ 0 & 0 & 1 & 0 \\ 0 & 0 & 0 & 1 \end{bmatrix}} \begin{matrix} (9) \\ (2) \\ (10) \\ (3) \end{matrix}; \quad \mathbf{T}^4 = \overset{(10)\,(3)\,(4)\,(5)}{\begin{bmatrix} 1 & 0 & 0 & 0 \\ 0 & 1 & 0 & 0 \\ 0 & 0 & 1 & 0 \\ 0 & 0 & 0 & 1 \end{bmatrix}} \begin{matrix} (10) \\ (3) \\ (4) \\ (5) \end{matrix}$$

The fixed end forces in global axes can now be assembled, by summing up the contributions of

$$\mathbf{T}^{1^T}\mathbf{F}_{*f}^{1} = \begin{bmatrix} 120 \text{ kN} \\ 160 \text{ kNm} \\ 120 \text{ kN} \\ -160 \text{ kNm} \end{bmatrix} \begin{matrix} (6) \\ (7) \\ (8) \\ (1) \end{matrix}; \quad \mathbf{T}^{2^T}\mathbf{F}_{*f}^{2} = \begin{bmatrix} 61.25 \text{ kN} \\ 93.75 \text{ kNm} \\ 98.75 \text{ kN} \\ -116.25 \text{ kNm} \end{bmatrix} \begin{matrix} (8) \\ (1) \\ (9) \\ (2) \end{matrix}; \quad \mathbf{T}^{3^T}\mathbf{F}_{*f}^{3} = \begin{bmatrix} 111.111 \text{ kN} \\ 133.333 \text{ kNm} \\ 38.889 \text{ kN} \\ -66.667 \text{ kNm} \end{bmatrix} \begin{matrix} (9) \\ (2) \\ (10) \\ (3) \end{matrix}$$

$$\Rightarrow \mathbf{F}_f = \begin{bmatrix} \mathbf{F}_{fA} \\ \hline \mathbf{F}_{fR} \end{bmatrix} \text{ where } \mathbf{F}_{fA} = \begin{Bmatrix} F_{1f} \\ F_{2f} \\ F_{3f} \\ F_{4f} \\ F_{5f} \end{Bmatrix} = \begin{Bmatrix} -66.25 \text{ kNm} \\ 17.083 \text{ kNm} \\ -66.667 \text{ kNm} \\ 0 \text{ kN} \\ 0 \text{ kNm} \end{Bmatrix} \text{ and } \mathbf{F}_{fR} = \begin{Bmatrix} F_{6f} \\ F_{7f} \\ F_{8f} \\ F_{9f} \\ F_{10f} \end{Bmatrix} = \begin{Bmatrix} 120.0 \text{ kN} \\ 160.0 \text{ kNm} \\ 181.25 \text{ kN} \\ 209.861 \text{ kN} \\ 38.889 \text{ kN} \end{Bmatrix}$$

The resultant load vector, $\mathbf{F}_A - \mathbf{F}_{fA}$, is given by:

$$\mathbf{F}_A - \mathbf{F}_{fA} = \begin{Bmatrix} 0 \text{ kNm} \\ 0 \text{ kNm} \\ 0 \text{ kNm} \\ 0 \text{ kN} \\ -50 \text{ kNm} \end{Bmatrix} - \begin{Bmatrix} -66.25 \text{ kNm} \\ 17.083 \text{ kNm} \\ -66.667 \text{ kNm} \\ 0 \text{ kN} \\ 0 \text{ kNm} \end{Bmatrix} = \begin{Bmatrix} 66.25 \text{ kNm} \\ -17.083 \text{ kNm} \\ 66.667 \text{ kNm} \\ 0 \text{ kN} \\ -50 \text{ kNm} \end{Bmatrix}$$

This is indicated in Fig. 5.9. The deflections at the 5 active coordinates will be identical to those in the original loading diagram (with the distributed loading) in Fig. 5.7.

Element and Structure Stiffness Matrices

The element stiffness matrices $\mathbf{k}_*^i$ for the four elements, satisfying $\mathbf{F}_*^i = \mathbf{k}_*^i\mathbf{D}_*^i$, take the following form:

$$\mathbf{k}_*^i = \begin{bmatrix} a_i & b_i & -a_i & b_i \\ b_i & c_i & -b_i & c_i/2 \\ -a_i & -b_i & a_i & -b_i \\ b_i & c_i/2 & -b_i & c_i \end{bmatrix} \text{ where } a_i = \frac{12EI_i}{L_i^3};\ b_i = \frac{6EI_i}{L_i^2};\ c_i = \frac{4EI_i}{L_i}$$

with $L_1 = 8\text{m}$, $I_1 = 4I$; $L_2 = 6\text{m}$, $I_2 = 3I$; $L_3 = 6\text{m}$, $I_3 = 2I$; $L_4 = 2\text{m}$, $I_4 = 2I$ (refer Fig. 5.8b).

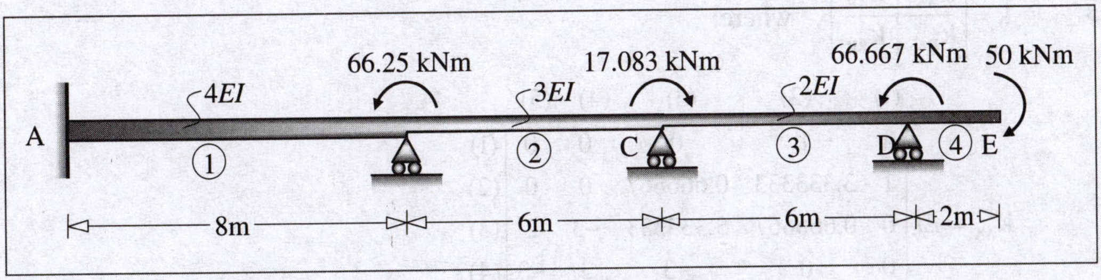

Figure 5.9 Nodal plus equivalent joint loads (Example 5.3)

$$\Rightarrow \begin{array}{l} a_1 = (0.09375)EI,\ b_1 = (0.375)EI,\ c_1 = (2)EI; \\ a_2 = (0.166667)EI,\ b_2 = (0.5)EI,\ c_2 = (2)EI; \\ a_3 = (0.111111)EI,\ b_3 = (0.333333)EI,\ c_3 = (1.333333)EI; \\ a_4 = (3)EI,\ b_4 = (3)EI,\ c_4 = (4)EI \end{array}$$

$$\Rightarrow \mathbf{k}_*^1 = EI\begin{bmatrix} 0.09375 & 0.375 & -0.09375 & 0.375 \\ 0.375 & 2 & -0.375 & 1 \\ -0.09375 & -0.375 & 0.09375 & -0.375 \\ 0.375 & 1 & -0.375 & 2 \end{bmatrix}\begin{matrix}(6)\\(7)\\(8)\\(1)\end{matrix}$$

$$\mathbf{k}_*^2 = EI\begin{bmatrix} 0.166667 & 0.5 & -0.166667 & 0.5 \\ 0.5 & 2 & -0.5 & 1 \\ -0.166667 & -0.5 & 0.166667 & -0.5 \\ 0.5 & 1 & -0.5 & 2 \end{bmatrix}\begin{matrix}(8)\\(1)\\(9)\\(2)\end{matrix}$$

$$\mathbf{k}_*^3 = EI\begin{bmatrix} 0.111111 & 0.333333 & -0.111111 & 0.333333 \\ 0.333333 & 1.333333 & -0.333333 & 0.666667 \\ -0.111111 & -0.333333 & 0.111111 & -0.333333 \\ 0.333333 & 0.666667 & -0.333333 & 1.333333 \end{bmatrix}\begin{matrix}(9)\\(2)\\(10)\\(3)\end{matrix}$$

$$\mathbf{k}_*^4 = EI\begin{bmatrix} 3 & 3 & -3 & 3 \\ 3 & 4 & -3 & 2 \\ -3 & -3 & 3 & -3 \\ 3 & 2 & -3 & 2 \end{bmatrix}\begin{matrix}(10)\\(3)\\(4)\\(5)\end{matrix}$$

As $\mathbf{T}^i$ is an identity matrix, $\mathbf{T}^{i^T}\mathbf{k}_*^i\mathbf{T}^i = \mathbf{k}_*^i$, with the relevant global coordinates (marked in parenthesis) as defined earlier. The structure stiffness matrix $\mathbf{k}$ of order 10×10, satisfying $\mathbf{F} = \mathbf{kD}$, can now be assembled, by summing up the contributions of $\mathbf{T}^{i^T}\mathbf{k}_*^i\mathbf{T}^i = \mathbf{k}_*^i$ for the 4 elements, at the appropriate coordinate locations:

$$\Rightarrow \quad \mathbf{k} = \left[\begin{array}{c|c} \mathbf{k}_{AA} & \mathbf{k}_{AR} \\ \hline \mathbf{k}_{RA} & \mathbf{k}_{RR} \end{array}\right], \text{ where:}$$

$$\Rightarrow \quad \mathbf{k}_{AA} = EI\begin{bmatrix} 4 & 1 & 0 & 0 & 0 \\ 1 & 3.333333 & 0.666667 & 0 & 0 \\ 0 & 0.666667 & 5.333333 & -3 & 2 \\ 0 & 0 & -3 & 3 & -3 \\ 0 & 0 & 2 & -3 & 4 \end{bmatrix}\begin{matrix}(1)\\(2)\\(3)\\(4)\\(5)\end{matrix} \quad \text{columns: } (1)\ (2)\ (3)\ (4)\ (5)$$

$$\mathbf{k}_{AR} = EI\begin{bmatrix} 0.375 & 1 & 0.125 & -0.5 & 0 \\ 0 & 0 & 0.5 & -0.166667 & -0.333333 \\ 0 & 0 & 0 & 0.333333 & 2.666667 \\ 0 & 0 & 0 & 0 & -3 \\ 0 & 0 & 0 & 0 & 3 \end{bmatrix}\begin{matrix}(1)\\(2)\\(3)\\(4)\\(5)\end{matrix}; \quad \mathbf{k}_{RA} = \mathbf{k}_{AR}^{T} \quad \text{columns: } (6)\ (7)\ (8)\ (9)\ (10)$$

$$\Rightarrow \quad \mathbf{k}_{RR} = EI\begin{bmatrix} 0.09375 & 0.375 & -0.09375 & 0 & 0 \\ 0.375 & 2 & -0.375 & 0 & 0 \\ -0.09375 & -0.375 & 0.260417 & -0.166667 & 0 \\ 0 & 0 & -0.166667 & 0.277778 & -0.111111 \\ 0 & 0 & 0 & -0.111111 & 3.111111 \end{bmatrix}\begin{matrix}(6)\\(7)\\(8)\\(9)\\(10)\end{matrix} \quad \text{columns: } (6)\ (7)\ (8)\ (9)\ (10)$$

The basic stiffness relations are given by:

$$\begin{bmatrix} \mathbf{F}_A \\ \hline \mathbf{F}_R \end{bmatrix} - \begin{bmatrix} \mathbf{F}_{fA} \\ \hline \mathbf{F}_{fR} \end{bmatrix} = \left[\begin{array}{c|c} \mathbf{k}_{AA} & \mathbf{k}_{AR} \\ \hline \mathbf{k}_{RA} & \mathbf{k}_{RR} \end{array}\right]\begin{bmatrix} \mathbf{D}_A \\ \hline \mathbf{D}_R \end{bmatrix}$$

$$\Rightarrow \mathbf{k}_{AA}\mathbf{D}_A = (\mathbf{F}_A - \mathbf{F}_{fA}) - \mathbf{k}_{AR}\mathbf{D}_R$$

where $\mathbf{k}_{AR}\mathbf{D}_R$, which denotes the forces at the active coordinates due to the effect of support displacements on the primary (restrained) structure.

$$\Rightarrow \mathbf{k_{AA}D_A} = \begin{Bmatrix} 66.25 \\ -17.083 \\ 66.667 \\ 0 \\ -50 \end{Bmatrix} - \begin{Bmatrix} 350 \\ -66.667 \\ -266.667 \\ 0 \\ 0 \end{Bmatrix} = \begin{Bmatrix} -283.75 \text{ kN} \\ 49.584 \text{ kNm} \\ 333.333 \text{ kN} \\ 0 \text{ kN} \\ -50 \text{ kN} \end{Bmatrix}$$

Displacements and Support Reactions

$$\mathbf{D_A} = [\mathbf{k_{AA}}]^{-1}[\mathbf{k_{AA}D_A}] \quad \Rightarrow \quad \begin{Bmatrix} D_1 \\ D_2 \\ D_3 \\ D_4 \\ D_5 \end{Bmatrix} = \frac{1}{EI}\begin{Bmatrix} -69.01517 \\ -7.68933 \\ 216.34492 \\ 382.68983 \\ 166.34492 \end{Bmatrix} = \begin{Bmatrix} -0.8627\times10^{-4} \text{ rad} \\ -0.9612\times10^{-4} \text{ rad} \\ 27.0431\times10^{-4} \text{ rad} \\ 4.7836\times10^{-3} \text{ m} \\ 20.7931\times10^{-4} \text{ rad} \end{Bmatrix}$$

$$\mathbf{F_R} = \mathbf{F_{fR}} + \mathbf{k_{RA}D_A} + \mathbf{k_{RR}D_R}$$

$$\Rightarrow \begin{Bmatrix} F_6 \\ F_7 \\ F_8 \\ F_9 \\ F_{10} \end{Bmatrix} = \begin{Bmatrix} 120.0 \text{ kN} \\ 160.0 \text{ kNm} \\ 181.25 \text{ kN} \\ 209.861 \text{ kN} \\ 38.889 \text{ kN} \end{Bmatrix} + \begin{Bmatrix} -25.881 \text{ kN} \\ -69.015 \text{ kNm} \\ -12.472 \text{ kN} \\ 107.904 \text{ kN} \\ -69.552 \text{ kN} \end{Bmatrix} + \begin{Bmatrix} 37.5 \text{ kN} \\ 150.0 \text{ kNm} \\ 29.167 \text{ kN} \\ -155.556 \text{ kN} \\ 88.889 \text{ kN} \end{Bmatrix} = \begin{Bmatrix} 131.619 \text{ kN} \\ 240.985 \text{ kNm} \\ 197.945 \text{ kN} \\ 162.210 \text{ kN} \\ 58.226 \text{ kN} \end{Bmatrix}$$

Check $\sum F_y = 0$:

Total downward load $= (30\times8)+100+(20\times3)+150 = 550$ kN

$$F_6 + F_8 + F_9 + F_{10} = 131.619+197.945+162.210+58.226 = 550.000 \text{ kN} \Rightarrow \text{OK.}$$

Member Forces

$$\mathbf{F_*^i} = \mathbf{F_{*f}^i} + \mathbf{k_*^i T^i D}$$

The fixed end forces for the four elements, $\mathbf{F_{*f}^i}$, have already been computed and the results (support reactions, shear force diagram and bending moment diagram) depicted in Fig. 5.8c. The additional forces, $\mathbf{k_*^i T^i D}$, on account of the net nodal forces (shown in Fig. 5.9) and the support settlements, for the four elements are given below and depicted in Fig. 5.10. The superposition of these forces ($\mathbf{F_*^i} = \mathbf{F_{*f}^i} + \mathbf{k_*^i T^i D}$) is also shown below, and the final support reactions, shear force diagram and bending moment diagram shown in Fig. 5.11.

$$\mathbf{k_*^1 T^1 D} = \begin{bmatrix} 0.09375 & 0.375 & -0.09375 & 0.375 \\ 0.375 & 2 & -0.375 & 1 \\ -0.09375 & -0.375 & 0.09375 & -0.375 \\ 0.375 & 1 & -0.375 & 2 \end{bmatrix} \begin{matrix} (6) \\ (7) \\ (8) \\ (1) \end{matrix} \begin{Bmatrix} 0 \\ 0 \\ -400 \\ -69.01517 \end{Bmatrix} = \begin{Bmatrix} 11.619 \text{ kN} \\ 80.985 \text{ kNm} \\ -11.619 \text{ kN} \\ 11.970 \text{ kNm} \end{Bmatrix}$$

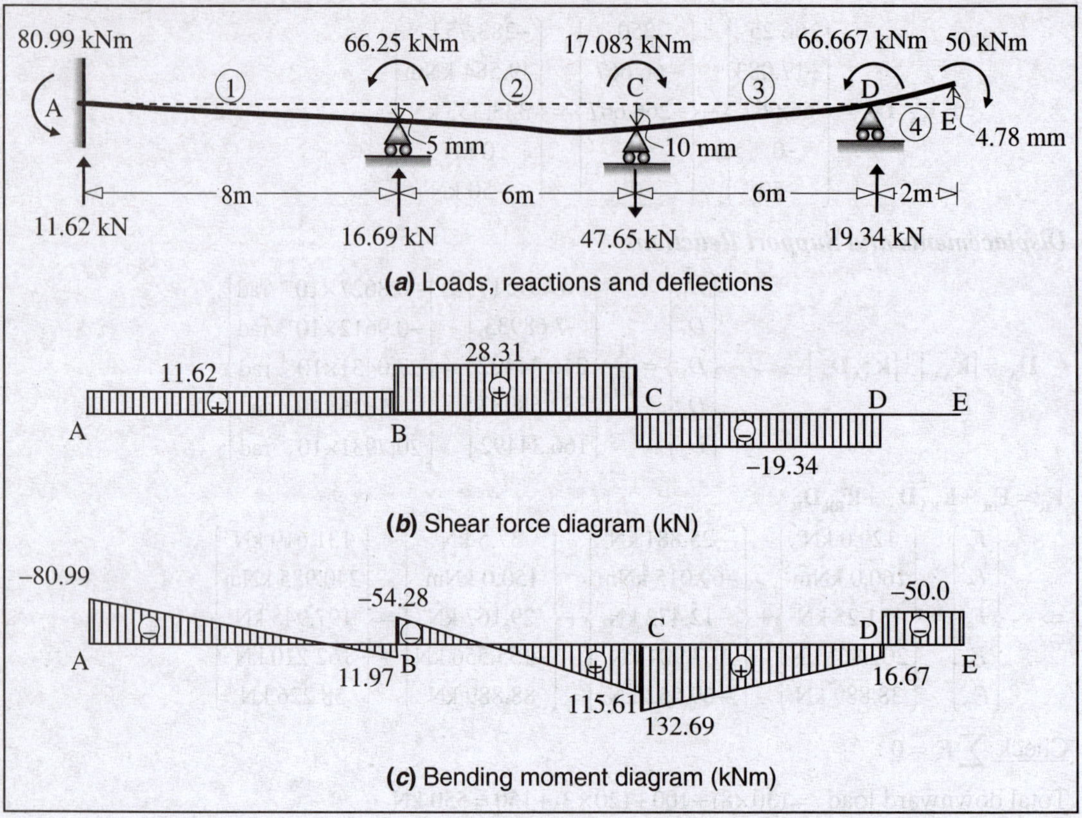

Figure 5.10 Response due to net nodal loads and support settlements (Example 5.3)

$$\Rightarrow \mathbf{F}_*^1 = \begin{Bmatrix} 120 \text{ kN} \\ 160 \text{ kNm} \\ 120 \text{ kN} \\ -160 \text{ kNm} \end{Bmatrix} + \begin{Bmatrix} 11.619 \text{ kN} \\ 80.985 \text{ kNm} \\ -11.619 \text{ kN} \\ 11.970 \text{ kNm} \end{Bmatrix} = \begin{Bmatrix} 131.619 \text{ kN} \\ 240.985 \text{ kNm} \\ 108.381 \text{ kN} \\ -148.030 \text{ kNm} \end{Bmatrix}$$

$$\mathbf{k}_*^2\mathbf{T}^2\mathbf{D} = \begin{bmatrix} 0.166667 & 0.5 & -0.166667 & 0.5 \\ 0.5 & 2 & -0.5 & 1 \\ -0.166667 & -0.5 & 0.166667 & -0.5 \\ 0.5 & 1 & -0.5 & 2 \end{bmatrix} \begin{matrix} (8) \\ (1) \\ (9) \\ (2) \end{matrix} \begin{Bmatrix} -400 \\ -69.01517 \\ -800 \\ -7.68933 \end{Bmatrix} = \begin{Bmatrix} 28.314 \text{ kN} \\ 54.280 \text{ kNm} \\ -28.314 \text{ kN} \\ 115.606 \text{ kNm} \end{Bmatrix}$$

$$\Rightarrow \mathbf{F}_*^2 = \begin{Bmatrix} 61.25 \text{ kN} \\ 93.75 \text{ kNm} \\ 98.75 \text{ kN} \\ -116.25 \text{ kNm} \end{Bmatrix} + \begin{Bmatrix} 28.314 \text{ kN} \\ 54.280 \text{ kNm} \\ -28.314 \text{ kN} \\ 115.606 \text{ kNm} \end{Bmatrix} = \begin{Bmatrix} 89.564 \text{ kN} \\ 148.030 \text{ kNm} \\ 70.436 \text{ kN} \\ -0.644 \text{ kNm} \end{Bmatrix}$$

$$\mathbf{k}_*^3\mathbf{T}^3\mathbf{D} = \begin{bmatrix} 0.111111 & 0.333333 & -0.111111 & 0.333333 \\ 0.333333 & 1.333333 & -0.333333 & 0.666667 \\ -0.111111 & -0.333333 & 0.111111 & -0.333333 \\ 0.333333 & 0.666667 & -0.333333 & 1.333333 \end{bmatrix} \begin{matrix} (9) \\ (2) \\ (10) \\ (3) \end{matrix} \begin{Bmatrix} -800 \\ -7.68933 \\ 0 \\ 216.34492 \end{Bmatrix} = \begin{Bmatrix} -19.337 \text{ kN} \\ -132.689 \text{ kNm} \\ 19.337 \text{ kN} \\ 16.667 \text{ kNm} \end{Bmatrix}$$

$$\Rightarrow \mathbf{F}_*^3 = \begin{Bmatrix} 111.111 \text{ kN} \\ 133.333 \text{ kNm} \\ 38.889 \text{ kN} \\ -66.667 \text{ kNm} \end{Bmatrix} + \begin{Bmatrix} -19.337 \text{ kN} \\ -132.689 \text{ kNm} \\ 19.337 \text{ kN} \\ 16.667 \text{ kNm} \end{Bmatrix} = \begin{Bmatrix} 91.774 \text{ kN} \\ 0.644 \text{ kNm} \\ 58.226 \text{ kN} \\ -50.000 \text{ kNm} \end{Bmatrix}$$

$$\mathbf{k}_*^4\mathbf{T}^4\mathbf{D} = \begin{bmatrix} 3 & 3 & -3 & 3 \\ 3 & 4 & -3 & 2 \\ -3 & -3 & 3 & -3 \\ 3 & 2 & -3 & 2 \end{bmatrix} \begin{matrix} (10) \\ (3) \\ (4) \\ (5) \end{matrix} \begin{Bmatrix} 0 \\ 216.34492 \\ 382.68983 \\ 166.34492 \end{Bmatrix} = \begin{Bmatrix} 0.000 \text{ kN} \\ 50.000 \text{ kNm} \\ 0.000 \text{ kN} \\ -50.000 \text{ kNm} \end{Bmatrix} \Rightarrow \mathbf{F}_*^4 = \begin{Bmatrix} 0 \\ 0 \\ 0 \\ 0 \end{Bmatrix} + \begin{Bmatrix} 0 \\ 50 \\ 0 \\ -50 \end{Bmatrix} = \begin{Bmatrix} 0 \text{ kN} \\ 50 \text{ kNm} \\ 0 \text{ kN} \\ -50 \text{ kNm} \end{Bmatrix}$$

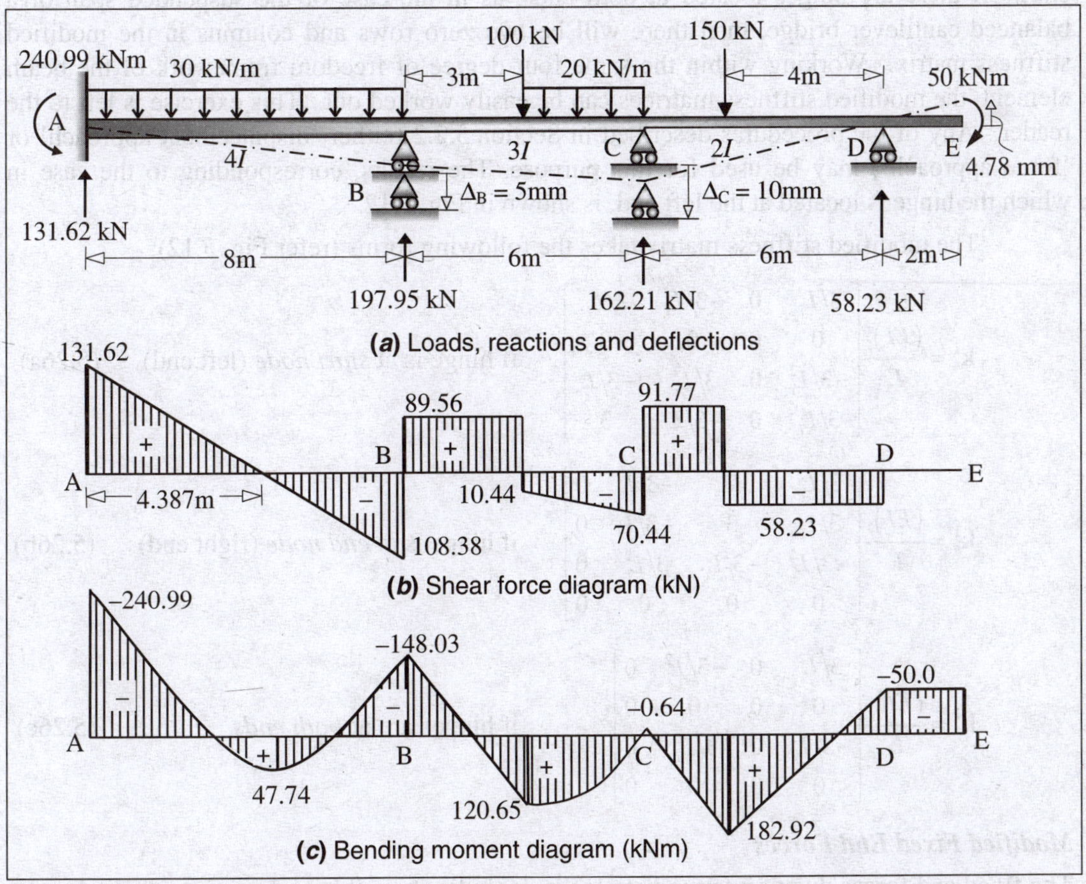

Figure 5.11 Total response due to all applied direct and indirect loading (Example 5.3)

5.2.6 Dealing with Internal Hinges

When internal hinges are present in the continuous beam system, special attention is called for. The presence of an internal hinge introduces two special features:

(i) there is no transfer of bending moment across the hinge.

(ii) there is no compatibility requirement that the rotation (slope) on the left side of the hinge should be equal to that on the right side.

We can account for the "moment release" at the hinge location by modifying the element stiffness matrix and also the computation of the fixed end force effects.

Modified Element Stiffness Matrix

The modified stiffness matrix will have a zero row and a zero column vector, corresponding the local coordinate relating to the rotational degree of freedom, which is either 2*, if the hinge is located at the start node (left end), or 4*, if the hinge is located at the end node (right end). If there are hinges located at both ends, as in the case of the suspended span of a balanced cantilever bridge, then there will be two zero rows and columns in the modified stiffness matrix. Working within the same four degree of freedom framework of the beam element, the modified stiffness matrices can be easily worked out. This exercise is left to the reader. Any of the procedures described in Section 5.2.2 (either 'displacement approach' or 'force approach') may be used for this purpose. The results, corresponding to the case in which the hinge is located at the left end, is shown in Fig. 5.12.

The modified stiffness matrix takes the following forms (refer Fig. 5.12):

$$\mathbf{k}_*^{\mathbf{i}} = \frac{(EI)_i}{L_i}\begin{bmatrix} 3/L_i^2 & 0 & -3/L_i^2 & 3/L_i \\ 0 & 0 & 0 & 0 \\ -3/L_i^2 & 0 & 3/L_i^2 & -3/L_i \\ 3/L_i & 0 & -3/L_i & 3 \end{bmatrix} \quad \text{if hinge is at } \textit{start node} \text{ (left end)} \qquad (5.26a)$$

$$\mathbf{k}_*^{\mathbf{i}} = \frac{(EI)_i}{L_i}\begin{bmatrix} 3/L_i^2 & 3/L_i & -3/L_i^2 & 0 \\ 3/L_i & 3 & -3/L_i & 0 \\ -3/L_i^2 & -3/L_i & 3/L_i^2 & 0 \\ 0 & 0 & 0 & 0 \end{bmatrix} \quad \text{if hinge is at } \textit{end node} \text{ (right end)} \qquad (5.26b)$$

$$\mathbf{k}_*^{\mathbf{i}} = \frac{(EI)_i}{L_i}\begin{bmatrix} 3/L_i^2 & 0 & -3/L_i^2 & 0 \\ 0 & 0 & 0 & 0 \\ -3/L_i^2 & 0 & 3/L_i^2 & 0 \\ 0 & 0 & 0 & 0 \end{bmatrix} \quad \text{if hinges are at } \textit{both ends} \qquad (5.26c)$$

Modified Fixed End Forces

The fixed end forces due to intermediate loads, including possible end moment applied at the hinge location, should also be suitably computed, accounting for the hinge at the member end (i.e., there should be no fixity against rotation at this end). If hinges are provided at both ends, then obviously, the beam will behave like a simply supported beam, and the fixed end effects will not induce any end moments (only vertical reactions at the ends). If the hinge is located at one end, then the beam will behave like a propped cantilever.

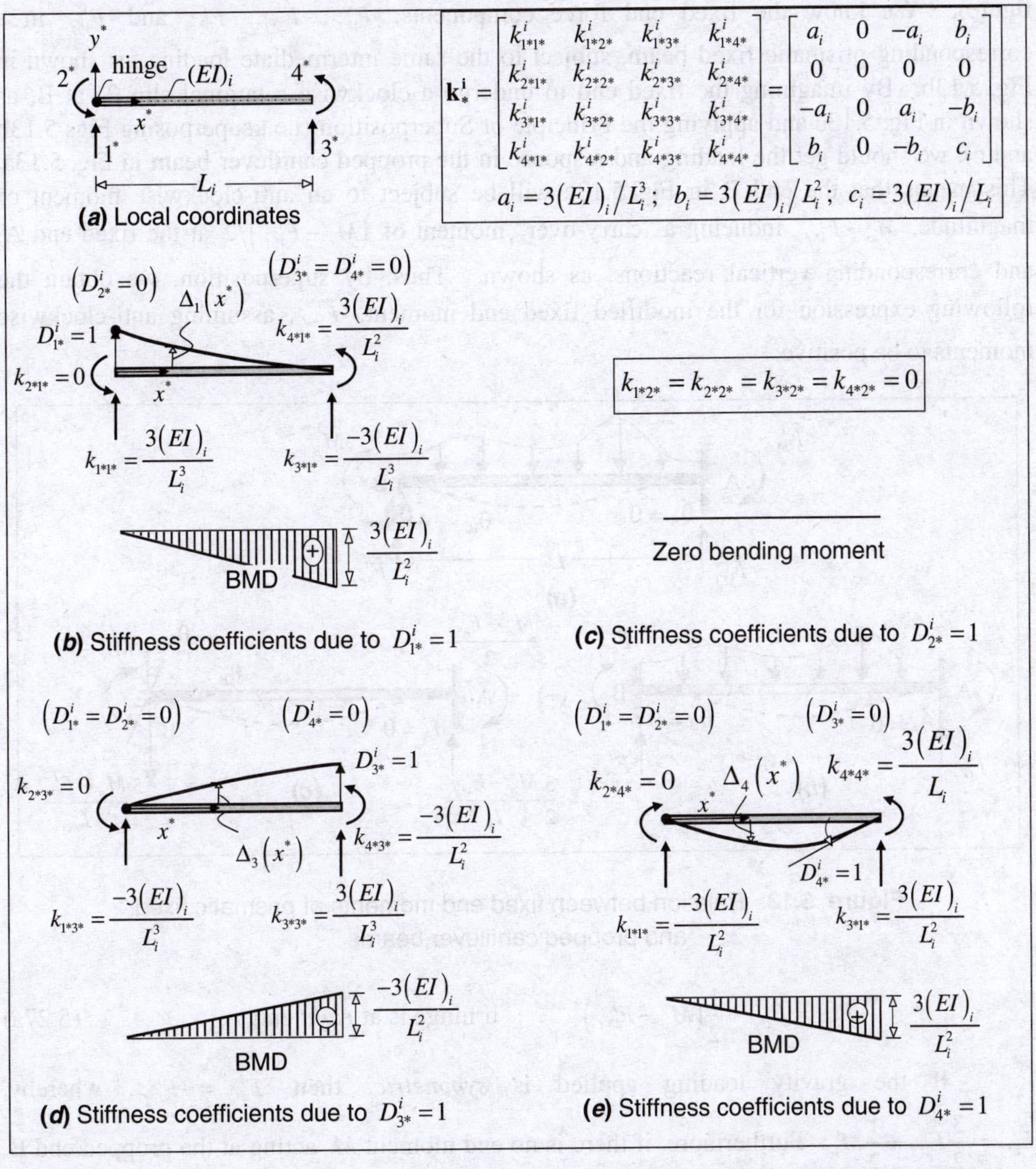

Figure 5.12 Generation of stiffness matrix for a prismatic beam element with moment release at left end

Consider, for example, the propped cantilever beam AB (with prismatic section) in Fig. 5.13a, subject to intermediate gravity loading of arbitrary magnitude and an anti-clockwise moment M_o at the propped end B, which results in an anti-clockwise rotation θ_B. We need to find the fixed end force components, $\tilde{F}^i_{1*f}$, $\tilde{F}^i_{2*f}$ and $\tilde{F}^i_{3*f}$ (as marked in the figure). We know the fixed end force components, F^i_{1*f}, F^i_{2*f}, F^i_{3*f} and F^i_{4*f} in a corresponding prismatic fixed beam, subject to the same intermediate loading, as shown in Fig. 5.13b. By imagining the fixed end to undergo a clockwise rotational slip θ_B at B, as shown in Fig. 5.13c and applying the Principle of Superposition (i.e., superposing Figs 5.13b and c), we should get the loading and response in the propped cantilever beam in Fig. 5.13a. This means that the end B in Fig. 5.13c will be subject to an anti-clockwise moment of magnitude, $M_o - F^i_{4*f}$, inducing a "carry-over" moment of $\left(M_o - F^i_{4*f}\right)/2$ at the fixed end A, and corresponding vertical reactions, as shown. Thus, by superposition, we obtain the following expression for the modified fixed end moment, $\tilde{F}^i_{2*f}$, assuming anti-clockwise moments to be positive:

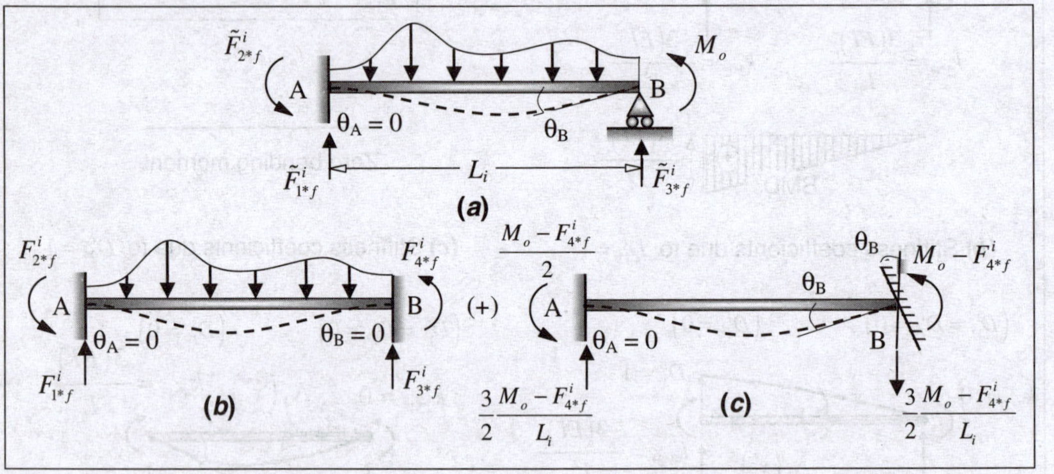

Figure 5.13 Relation between fixed end moments of prismatic fixed and propped cantilever beams

$$\tilde{F}^i_{2*f} = F^i_{2*f} + \frac{1}{2}\left(M_o - F^i_{4*f}\right) \qquad \text{if hinge is at } \textit{right} \text{ end} \qquad (5.27a)$$

If the gravity loading applied is *symmetric*, then $F^i_{4*f} = -F^i_{2*f}$, whereby, $\tilde{F}^i_{2*f} = \frac{3}{2}F^i_{2*f} + \frac{1}{2}M_o$. Furthermore, if there is no end moment M_o acting at the propped end B, it follows that $\tilde{F}^i_{2*f} = 1.5F^i_{2*f}$. If the hinge is located at the left end, Eq. 5.27a gets modified to:

$$\tilde{F}^i_{4*f} = F^i_{4*f} + \frac{1}{2}\left(M_o - F^i_{2*f}\right) \qquad \text{if hinge is at } \textit{left} \text{ end} \qquad (5.27b)$$

The vertical reaction components, $\tilde{F}^i_{1*f}$ and $\tilde{F}^i_{3*f}$, can be easily obtained by considering the equilibrium of the propped cantilever beam. Alternatively, they can be generated by superposition (Fig. 5.13):

$$\left.\begin{aligned}\tilde{F}^i_{1*f} &= F^i_{1*f} + \frac{3}{2}\frac{\left(M_o - F^i_{4*f}\right)}{L_i} \\ \tilde{F}^i_{3*f} &= F^i_{3*f} - \frac{3}{2}\frac{\left(M_o - F^i_{4*f}\right)}{L_i}\end{aligned}\right\} \quad \text{if hinge is at } \textit{right} \text{ end} \qquad (5.28a)$$

$$\left.\begin{aligned}\tilde{F}^i_{1*f} &= F^i_{1*f} + \frac{3}{2}\frac{\left(M_o - F^i_{2*f}\right)}{L_i} \\ \tilde{F}^i_{3*f} &= F^i_{3*f} - \frac{3}{2}\frac{\left(M_o - F^i_{2*f}\right)}{L_i}\end{aligned}\right\} \quad \text{if hinge is at } \textit{left} \text{ end} \qquad (5.28b)$$

Imaginary Clamp

By reducing the rotational stiffness components in the two beam elements adjoining the internal hinge location (to the left and to the right), the resultant rotational stiffness of the structure, corresponding to this rotational degree of freedom (say, global coordinate 'q'), is reduced to zero. Thus, this will appear as a zero diagonal element in the structure stiffness matrix ($k_{qq} = 0$), making the matrix $\mathbf{k}_{\mathbf{AA}}$ singular and non-invertible. We can get around this difficulty by visualising an *imaginary clamp* (see Fig. 5.14e in Example 5.4) at the internal hinge location, arresting the rotation (i.e., by setting $D_q = 0$). Although this is not physically a correct representation (rotations are possible at the internal hinge location), it serves our purpose of getting a correct solution by the stiffness method.

By arresting the rotational degree of freedom at the hinge location, we convert the coordinate q from an *active* global coordinate to a *restrained* global coordinate. In general, this will result in a zero support reaction (moment), $F_q = 0$, corresponding to this degree of freedom. However, when there is an end moment applied, at the left or right or at both sides of the internal hinge), there will be a resultant non-zero value of F_q, which may be ignored.

The application of this procedure is demonstrated in Example 5.4.

EXAMPLE 5.4

Re-analyse the non-prismatic beam in Fig. 5.2, considering an internal hinge at B. Find the force and displacement responses, given that the beam is subject to the loads indicated in Fig. 5.2, assuming the moment of 30 kNm to act at B on the segment AB (to the left of the internal hinge). Assume $EI = 80000$ kNm^2.

SOLUTION

The loading diagram is shown in Fig. 5.14a. There are two prismatic elements here ($m = 2$), with intermediate loads acting on both. As there is an internal hinge provided at the junction B, there will be no moment transfer possible across this junction, and there will be no common rotational degree of freedom at B. The stiffness matrices of the two elements can be suitably modified to account for the moment release at B, but this will result in a zero rotational stiffness at B. In order to address this problem, we can visualise an imaginary clamp at B (providing restraint against rotation at B), as indicated in Fig. 5.14e. Thus, there is only one *active* degree of freedom ($n = 1$) and five *restrained* global coordinates ($r = 5$), numbered as shown in Fig. 5.14b. There are no nodal loads applied; i.e., $\mathbf{F_A} = \{F_1\} = \{0 \text{ kN}\}$.

The local coordinates for the two elements are shown in Fig. 5.14c.

Fixed End Forces (refer Fig. 5.14d)

Element '1' (AB):

$$F^1_{4*f} = -30 \text{ kNm}; \; F^1_{2*f} = \frac{100 \times 10}{8} \times \frac{3}{2} - \frac{30}{2} = 172.5 \text{ kNm};$$

$$F^1_{1*f} = \frac{100}{2} + \frac{172.5 - 30}{10} = 64.25 \text{ kN}; \; F^1_{3*f} = 100 - 67.8125 = 35.75 \text{ kN} \;\Rightarrow \mathbf{F^1_{*f}} = \begin{Bmatrix} 64.25 \text{ kN} \\ 172.5 \text{ kNm} \\ 35.75 \text{ kN} \\ -30 \text{ kNm} \end{Bmatrix}$$

Element '2' (BC):

$$F^2_{2*f} = 0 \text{ kNm}; \; F^2_{4*f} = -\frac{10 \times 10^2}{12} \times \frac{3}{2} = 125 \text{ kNm};$$

$$F^2_{1*f} = \frac{10 \times 10}{2} - \frac{125}{10} = 37.5 \text{ kN}; \; F^2_{3*f} = 100 - 37.5 = 62.5 \text{ kN} \;\Rightarrow \mathbf{F^2_{*f}} = \begin{Bmatrix} 37.5 \text{ kN} \\ 0 \text{ kNm} \\ 62.5 \text{ kN} \\ -125 \text{ kNm} \end{Bmatrix}$$

Coordinate Transformations and Equivalent Joint Loads

The element transformation matrices $\mathbf{T^1}$ and $\mathbf{T^2}$, satisfying $\mathbf{D^i_*} = \mathbf{T^i D}$ and $\mathbf{F^i_*} = \mathbf{T^i F}$, take the following form (refer Figs 5.13b and c), with the linking global coordinates indicated in parentheses (as in Example 5.1):

$$\mathbf{T^1} = \begin{matrix} (3) & (4) & (1) & (2) & \\ \end{matrix}$$
$$\mathbf{T^1} = \begin{bmatrix} 1 & 0 & 0 & 0 \\ 0 & 1 & 0 & 0 \\ 0 & 0 & 1 & 0 \\ 0 & 0 & 0 & 1 \end{bmatrix} \begin{matrix} (3) \\ (4) \\ (1) \\ (2) \end{matrix} \quad \text{and} \quad \mathbf{T^2} = \begin{bmatrix} 1 & 0 & 0 & 0 \\ 0 & 1 & 0 & 0 \\ 0 & 0 & 1 & 0 \\ 0 & 0 & 0 & 1 \end{bmatrix} \begin{matrix} (1) \\ (2) \\ (5) \\ (6) \end{matrix}$$

(columns of $\mathbf{T^1}$: (3) (4) (1) (2); columns of $\mathbf{T^2}$: (1) (2) (5) (6))

The fixed end forces in global axes can now be assembled, by summing up the contributions of

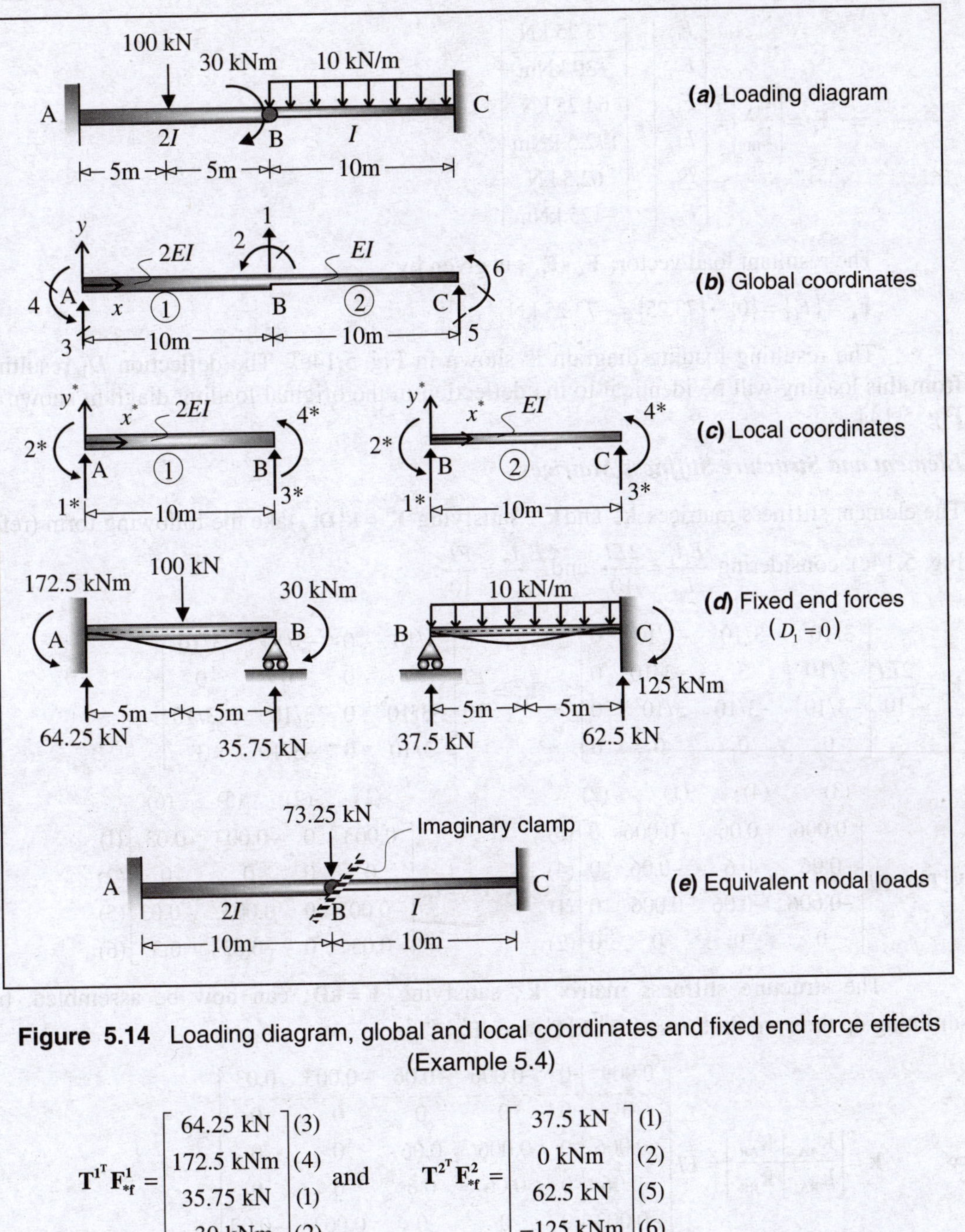

Figure 5.14 Loading diagram, global and local coordinates and fixed end force effects (Example 5.4)

$$\mathbf{T}^{1^T}\mathbf{F}^1_{*f} = \begin{bmatrix} 64.25 \text{ kN} \\ 172.5 \text{ kNm} \\ 35.75 \text{ kN} \\ -30 \text{ kNm} \end{bmatrix} \begin{matrix} (3) \\ (4) \\ (1) \\ (2) \end{matrix} \quad \text{and} \quad \mathbf{T}^{2^T}\mathbf{F}^2_{*f} = \begin{bmatrix} 37.5 \text{ kN} \\ 0 \text{ kNm} \\ 62.5 \text{ kN} \\ -125 \text{ kNm} \end{bmatrix} \begin{matrix} (1) \\ (2) \\ (5) \\ (6) \end{matrix}$$

$$\Rightarrow \mathbf{F}_f = \left[\begin{array}{c}\mathbf{F}_{fA} \\ \hline \mathbf{F}_{fR}\end{array}\right] = \begin{bmatrix} F_{1f} \\ F_{2f} \\ F_{3f} \\ F_{4f} \\ F_{5f} \\ F_{6f} \end{bmatrix} = \left[\begin{array}{c} 73.25 \text{ kN} \\ \hdashline -30 \text{ kNm} \\ 64.25 \text{ kN} \\ 172.5 \text{ kNm} \\ 62.5 \text{ kN} \\ -125 \text{ kNm} \end{array}\right]$$

The resultant load vector, $\mathbf{F}_A - \mathbf{F}_{fA}$, is given by:

$$\mathbf{F}_A = \{F_1\} = \{0\} - \{73.25\} = -73.25 \text{ kN}$$

The resulting loading diagram is shown in Fig. 5.14e. The deflection D_1 resulting from this loading will be identical to the deflection in the original loading diagram shown in Fig. 5.14a.

Element and Structure Stiffness Matrices

The element stiffness matrices $\mathbf{k}_*^1$ and $\mathbf{k}_*^2$, satisfying $\mathbf{F}_*^i = \mathbf{k}_*^i \mathbf{D}_*^i$, take the following form (refer Fig. 5.14c), considering $\frac{E_1 I_1}{L_1} = \frac{2EI}{10}$ and $\frac{E_2 I_2}{L_2} = \frac{EI}{10}$:

$$\mathbf{k}_*^1 = \frac{2EI}{10}\begin{bmatrix} 3/10^2 & 3/10 & -3/10^2 & 0 \\ 3/10 & 3 & -3/10 & 0 \\ -3/10^2 & -3/10 & 3/10^2 & 0 \\ 0 & 0 & 0 & 0 \end{bmatrix}; \quad \mathbf{k}_*^2 = \frac{EI}{10}\begin{bmatrix} 3/10^2 & 0 & -3/10^2 & 3/10 \\ 0 & 0 & 0 & 0 \\ -3/10^2 & 0 & 3/10^2 & -3/10 \\ 3/10 & 0 & -3/10 & 3 \end{bmatrix}$$

$$\mathbf{k}_*^1\mathbf{T}^1 = EI\begin{array}{c} \begin{matrix} (3) & (4) & (1) & (2) \end{matrix} \\ \begin{bmatrix} 0.006 & 0.06 & -0.006 & 0 \\ 0.06 & 0.6 & -0.06 & 0 \\ -0.006 & -0.06 & 0.006 & 0 \\ 0 & 0 & 0 & 0 \end{bmatrix} \end{array}\begin{matrix} (3) \\ (4) \\ (1) \\ (2) \end{matrix}; \quad \mathbf{k}_*^2\mathbf{T}^2 = EI\begin{array}{c} \begin{matrix} (1) & (2) & (5) & (6) \end{matrix} \\ \begin{bmatrix} 0.003 & 0 & -0.003 & 0.03 \\ 0 & 0 & 0 & 0 \\ -0.003 & 0 & 0.003 & -0.03 \\ 0.03 & 0 & -0.03 & 0.3 \end{bmatrix} \end{array}\begin{matrix} (1) \\ (2) \\ (5) \\ (6) \end{matrix}$$

The structure stiffness matrix $\mathbf{k}$, satisfying $\mathbf{F} = \mathbf{kD}$, can now be assembled, by summing up the contributions of $\mathbf{T}^{1^T}\mathbf{k}_*^1\mathbf{T}^1$ and $\mathbf{T}^{2^T}\mathbf{k}_*^2\mathbf{T}^2$:

$$\Rightarrow \quad \mathbf{k} = \left[\begin{array}{c|c} \mathbf{k}_{AA} & \mathbf{k}_{AR} \\ \hline \mathbf{k}_{RA} & \mathbf{k}_{RR} \end{array}\right] = EI\left[\begin{array}{c:ccccc} 0.009 & 0 & -0.006 & -0.06 & -0.003 & 0.03 \\ \hdashline 0 & 0 & 0 & 0 & 0 & 0 \\ -0.006 & 0 & 0.006 & 0.06 & 0 & 0 \\ -0.06 & 0 & 0.06 & 0.6 & 0 & 0 \\ -0.003 & 0 & 0 & 0 & 0.003 & -0.03 \\ 0.03 & 0 & 0 & 0 & -0.03 & 0.3 \end{array}\right]$$

$$\Rightarrow \quad [\mathbf{k}_{AA}] = EI(0.009) \quad \Rightarrow \quad [\mathbf{k}_{AA}]^{-1} = \frac{111.111}{EI}$$

Displacements and Support Reactions

The structure stiffness relation to be considered here is:

$$\begin{bmatrix} \mathbf{F_A} \\ \hline \mathbf{F_R} \end{bmatrix} - \begin{bmatrix} \mathbf{F_{fA}} \\ \hline \mathbf{F_{fR}} \end{bmatrix} = \begin{bmatrix} \mathbf{k_{AA}} & \mathbf{k_{AR}} \\ \hline \mathbf{k_{RA}} & \mathbf{k_{RR}} \end{bmatrix} \begin{bmatrix} \mathbf{D_A} \\ \hline \mathbf{D_R = O} \end{bmatrix}$$

$$\Rightarrow \quad \mathbf{D_A} = [\mathbf{k_{AA}}]^{-1}[\mathbf{F_A} - \mathbf{F_{fA}}] \quad \Rightarrow \quad D_1 = \frac{111.111}{EI}\{-73.25\} = \frac{-8138.881}{EI}$$

$$\Rightarrow \quad D_1 = \frac{-8138.881}{80000} = -101.736 \times 10^{-3} \text{ m}$$

$$\mathbf{F_R} = \mathbf{F_{fR}} + \mathbf{k_{RA}}\mathbf{D_A} \Rightarrow \quad \begin{Bmatrix} F_2 \\ F_3 \\ F_4 \\ F_5 \\ F_6 \end{Bmatrix} = \begin{Bmatrix} -30 \text{ kNm} \\ 64.25 \text{ kN} \\ 172.5 \text{ kNm} \\ 62.5 \text{ kN} \\ -125 \text{ kNm} \end{Bmatrix} + EI \begin{bmatrix} 0 \\ -0.006 \\ -0.06 \\ -0.003 \\ 0.03 \end{bmatrix} \frac{1}{EI}\{-8138.881\} = \begin{Bmatrix} -30 \text{ kNm} \\ 113.083 \text{ kN} \\ 660.833 \text{ kNm} \\ 86.917 \text{ kN} \\ -369.166 \text{ kNm} \end{Bmatrix}$$

<u>Check</u> $\sum F_y = 0$:

Total reaction = $F_3 + F_5 = 113.083 + 86.917 = 200.000$ kN = Total load (downwards) $\Rightarrow$ OK.

Member Forces

$$\mathbf{F_*^i} = \mathbf{F_{*f}^i} + \mathbf{k_*^i T^i D}$$

$$\Rightarrow \mathbf{F_*^1} = \begin{Bmatrix} 64.25 \\ 172.5 \\ 35.75 \\ -30 \end{Bmatrix} + EI \begin{bmatrix} 0.006 & 0.06 & -0.006 & 0 \\ 0.06 & 0.6 & -0.06 & 0 \\ -0.006 & -0.06 & 0.006 & 0 \\ 0 & 0 & 0 & 0 \end{bmatrix} \begin{matrix} (3) \\ (4) \\ (1) \\ (2) \end{matrix} \begin{Bmatrix} D_3 = 0 \\ D_4 = 0 \\ D_1 = -8138.881/EI \\ D_2 = 0 \end{Bmatrix} = \begin{Bmatrix} 113.083 \text{ kN} \\ 660.833 \text{ kNm} \\ -11.333 \text{ kN} \\ -30 \text{ kNm} \end{Bmatrix}$$

$$\mathbf{F_*^2} = \begin{Bmatrix} 37.5 \\ 0 \\ 62.5 \\ -125 \end{Bmatrix} + EI \begin{bmatrix} 0.003 & 0 & -0.003 & 0.03 \\ 0 & 0 & 0 & 0 \\ -0.003 & 0 & 0.003 & -0.03 \\ 0.03 & 0 & -0.03 & 0.3 \end{bmatrix} \begin{matrix} (1) \\ (2) \\ (5) \\ (6) \end{matrix} \begin{Bmatrix} D_1 = -8138.881/EI \\ D_2 = 0 \\ D_5 = 0 \\ D_6 = 0 \end{Bmatrix} = \begin{Bmatrix} 13.083 \text{ kN} \\ 0 \text{ kNm} \\ 86.917 \text{ kN} \\ -369.166 \text{ kNm} \end{Bmatrix}$$

The free-body, shear force and bending moment and deflection diagrams, including the fixed end force effects in Fig. 5.14d, are shown in Fig. 5.15.

5.2.7 Dealing with Hinged and Guided-Fixed End Supports

In the problems we have dealt with so far, with the exception of Example 5.3, the extreme end supports have been given as fixed end supports. In Example 5.3, one of the ends happened to be a free end, and we identified two active degrees of freedom at this location. We observed how the general analysis procedure described in Section 5.2.5 could be applied to this situation. In a similar way, we can deal with other boundary conditions, such as hinged or fixed-end supports. In the case of a hinged end support, we have an active rotational degree of freedom along with a restrained translational degree of freedom. In the case of a guided-fixed end support, we have an active translational degree of freedom along with a restrained rotational degree of freedom. This is demonstrated in Example 5.5.

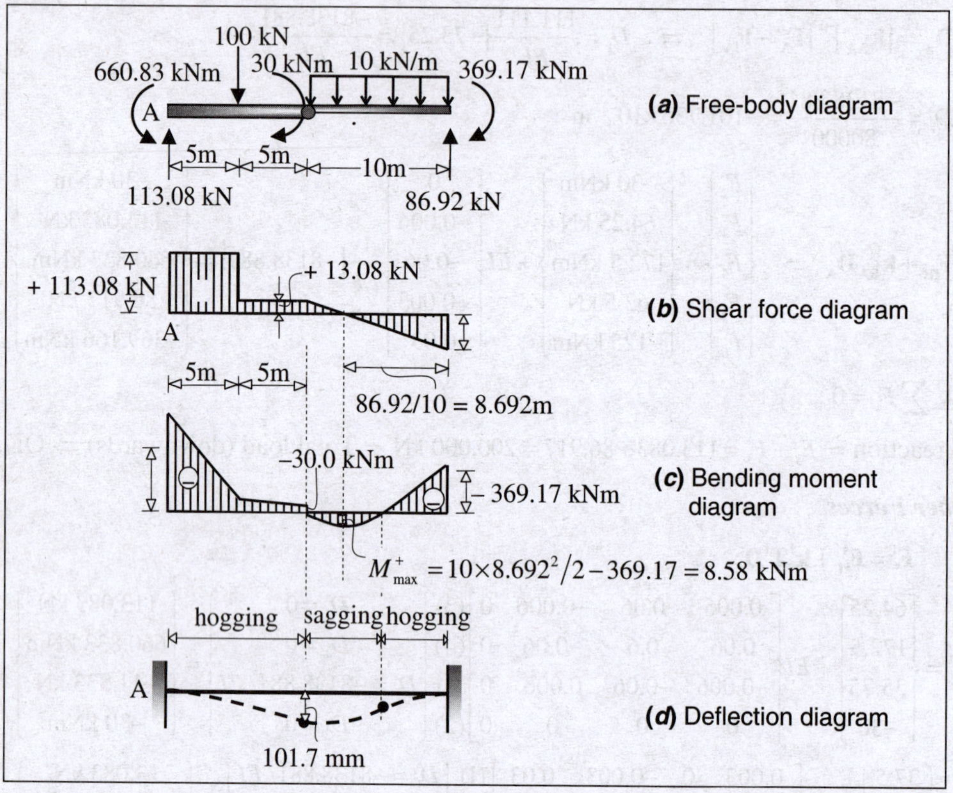

Figure 5.15 Internal force distribution and deflection profile (Example 5.4)

It may be noted, however, that when there are "member release" at an extreme end of the beam, some of the end forces turn out to be statically determinate, and we can take advantage of this. We have seen how this is possible in conventional displacement methods, such as the Slope Deflection Method and the Moment Distribution Method. We will see later, in Section 5.3, how to take advantage of a reduced degree of kinematic indeterminacy in such cases, by appropriately modifying the beam element stiffnesses and the fixed end moments.

EXAMPLE 5.5

Analyse the symmetric three-span beam shown in Fig. 5.16 by the conventional stiffness method. Draw the bending moment diagram. Assume constant flexural rigidity, *EI*.

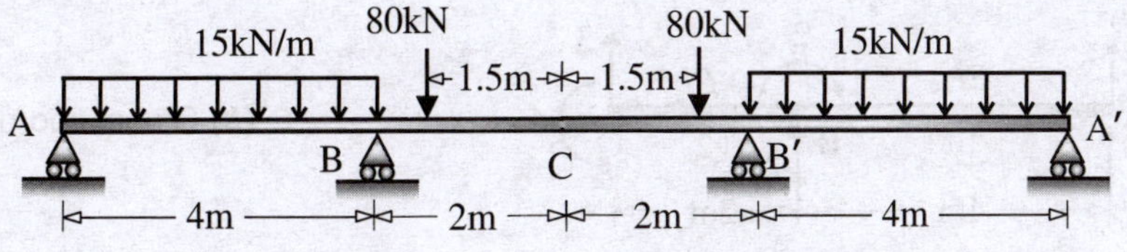

Figure 5.16 Symmetrical beam (Example 5.5)

SOLUTION

Taking advantage of symmetry, the beam can be truncated as shown in Fig. 5.17a, with a guided-fixed support at C. There are two prismatic elements in the reduced beam ($m=2$), with intermediate loads acting on both. There are three *active* degree of freedom ($n=3$) and three *restrained* global coordinates ($r=3$), numbered as shown in Fig. 5.17b. There are no nodal loads applied; i.e., $\mathbf{F_A} = \mathbf{O}$. The local coordinates for the two elements are shown in Fig. 5.17c.

Fixed End Forces (refer Fig. 5.17d)

<u>Element '1' (AB):</u>

$$F^1_{1*f} = F^1_{3*f} = \frac{(15)(4)}{2} = 30 \text{ kN};\; F^1_{2*f} = \frac{(15)(4^2)}{12} = 20.0 \text{ kNm};\; F^1_{4*f} = -20 \text{ kNm} \Rightarrow \mathbf{F^1_{*f}} = \begin{Bmatrix} 30 \text{ kN} \\ 20 \text{ kNm} \\ 30 \text{ kN} \\ -20 \text{ kNm} \end{Bmatrix}$$

<u>Element '2' (BC):</u>

$$F^2_{2*f} = \frac{(80)(0.5)(1.5)^2}{2^2} = 22.5 \text{ kNm};\; F^2_{4*f} = -\frac{(80)(0.5)^2(1.5)}{2^2} = -7.5 \text{ kNm}$$

$$F^2_{1*f} = \frac{(80)(1.5)}{2} + \frac{22.5-7.5}{2} = 67.5 \text{ kN};\; F^2_{3*f} = 80-67.5 = 12.5 \text{ kN} \quad \Rightarrow \mathbf{F^2_{*f}} = \begin{Bmatrix} 67.5 \text{ kN} \\ 22.5 \text{ kNm} \\ 12.5 \text{ kN} \\ -7.5 \text{ kNm} \end{Bmatrix}$$

The fixed end forces (support reactions) on the primary structure are shown in Fig. 5.17d.

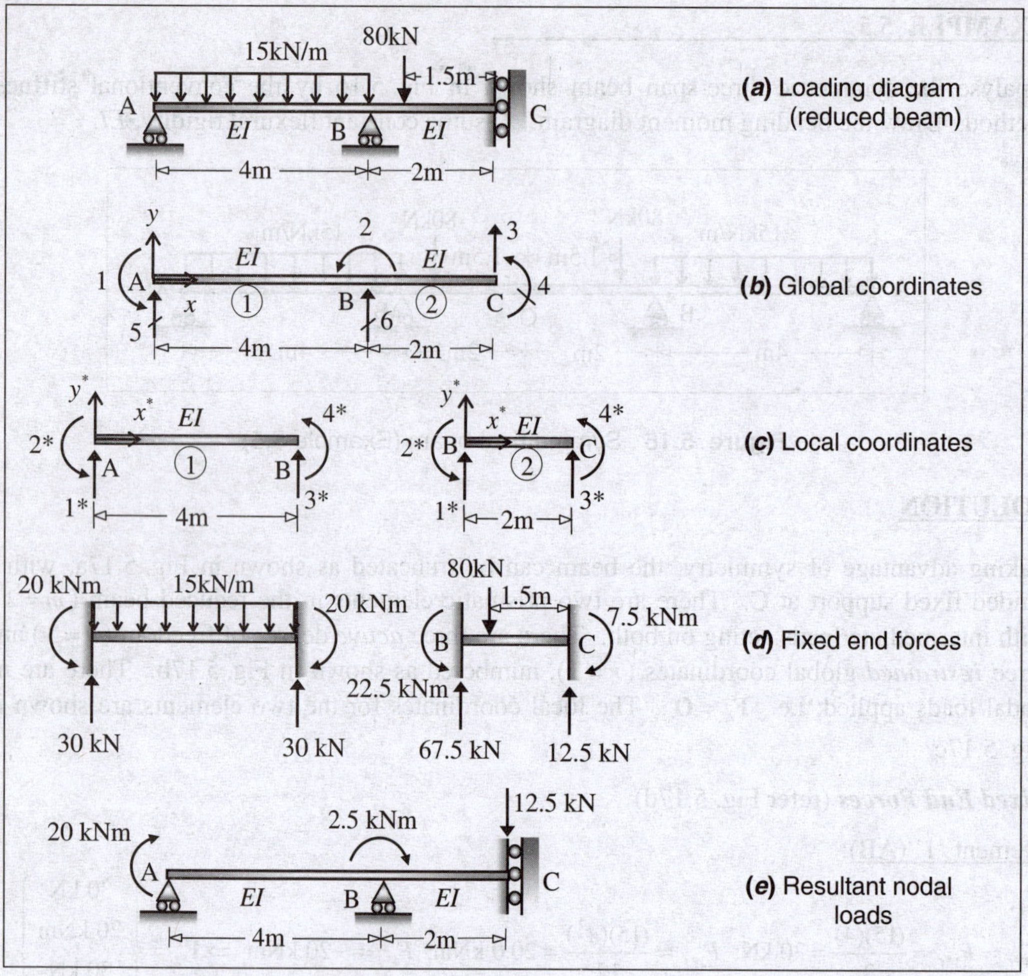

Figure 5.17 Loading diagram, coordinates and fixed end forces (Example 5.5)

Coordinate Transformations and Equivalent Joint Loads

The element transformation matrices $\mathbf{T}^1$ and $\mathbf{T}^2$, satisfying $\mathbf{D}_*^i = \mathbf{T}^i\mathbf{D}$ and $\mathbf{F}_*^i = \mathbf{T}^i\mathbf{F}$, take the following form (refer Figs 5.17b and c), with the linking global coordinates indicated in parentheses:

$$\mathbf{T}^1 = \overset{\begin{matrix}(5) & (1) & (6) & (2)\end{matrix}}{\begin{bmatrix} 1 & 0 & 0 & 0 \\ 0 & 1 & 0 & 0 \\ 0 & 0 & 1 & 0 \\ 0 & 0 & 0 & 1 \end{bmatrix}}\begin{matrix}(5)\\(1)\\(6)\\(2)\end{matrix} \quad \text{and} \quad \mathbf{T}^2 = \overset{\begin{matrix}(6) & (2) & (3) & (4)\end{matrix}}{\begin{bmatrix} 1 & 0 & 0 & 0 \\ 0 & 1 & 0 & 0 \\ 0 & 0 & 1 & 0 \\ 0 & 0 & 0 & 1 \end{bmatrix}}\begin{matrix}(6)\\(2)\\(3)\\(4)\end{matrix}$$

The fixed end forces in global axes can now be assembled, by summing up the contributions of

$$\mathbf{T}^{1^T}\mathbf{F}_{*f}^{1} = \begin{bmatrix} 30\text{ kN} \\ 20\text{ kNm} \\ 30\text{ kN} \\ -20\text{ kNm} \end{bmatrix}\begin{matrix}(5)\\(1)\\(6)\\(2)\end{matrix} \quad \text{and} \quad \mathbf{T}^{2^T}\mathbf{F}_{*f}^{2} = \begin{bmatrix} 67.5\text{ kN} \\ 22.5\text{ kNm} \\ 12.5\text{ kN} \\ -7.5\text{ kNm} \end{bmatrix}\begin{matrix}(6)\\(2)\\(3)\\(4)\end{matrix} \Rightarrow \mathbf{F}_f = \begin{bmatrix} \mathbf{F}_{fA} \\ \hline \mathbf{F}_{fR} \end{bmatrix} = \begin{bmatrix} F_{1f} \\ F_{2f} \\ F_{3f} \\ F_{4f} \\ F_{5f} \\ F_{6f} \end{bmatrix} = \begin{bmatrix} 20\text{ kNm} \\ 2.5\text{ kNm} \\ 12.5\text{ kN} \\ \hline -7.5\text{ kNm} \\ 30\text{ kN} \\ 97.5\text{ kN} \end{bmatrix}$$

$$\text{Resultant load vector, } \mathbf{F_A} - \mathbf{F_{fA}} = \begin{Bmatrix} 0\text{ kNm} \\ 0\text{ kNm} \\ 0\text{ kN} \end{Bmatrix} - \begin{Bmatrix} 20\text{ kNm} \\ 2.5\text{ kNm} \\ 12.5\text{ kN} \end{Bmatrix} = \begin{Bmatrix} -20\text{ kNm} \\ -2.5\text{ kNm} \\ -12.5\text{ kN} \end{Bmatrix}$$

This is indicated in Fig. 5.17e. The deflections at the active coordinates, D_1, D_2 and D_3, will be identical to those in the original loading diagram in Fig. 5.17a.

Element and Structure Stiffness Matrices

The element stiffness matrices $\mathbf{k}_*^1$ and $\mathbf{k}_*^2$, satisfying $\mathbf{F}_*^i = \mathbf{k}_*^i \mathbf{D}_*^i$, take the following form (refer Fig. 5.1), considering $\frac{E_1 I_1}{L_1} = \frac{EI}{4}$ and $\frac{E_2 I_2}{L_2} = \frac{EI}{2}$:

$$\mathbf{k}_*^1 = \frac{EI}{4}\begin{bmatrix} 12/4^2 & 6/4 & -12/4^2 & 6/4 \\ 6/4 & 4 & -6/4 & 2 \\ -12/4^2 & -6/4 & 12/4^2 & -6/4 \\ 6/4 & 2 & -6/4 & 4 \end{bmatrix} \quad \mathbf{k}_*^2 = \frac{EI}{2}\begin{bmatrix} 12/2^2 & 6/2 & -12/2^2 & 6/2 \\ 6/2 & 4 & -6/2 & 2 \\ -12/2^2 & -6/2 & 12/2^2 & -6/2 \\ 6/2 & 2 & -6/2 & 4 \end{bmatrix}$$

$$\mathbf{k}_*^1\mathbf{T}^1 = EI\begin{bmatrix} 0.1875 & 0.375 & -0.1875 & 0.375 \\ 0.375 & 1 & -0.375 & 0.5 \\ -0.1875 & -0.375 & 0.1875 & -0.375 \\ 0.375 & 0.5 & -0.375 & 1 \end{bmatrix}\begin{matrix}(5)\\(1)\\(6)\\(2)\end{matrix}; \quad \mathbf{k}_*^2\mathbf{T}^2 = EI\begin{bmatrix} 1.5 & 1.5 & -1.5 & 1.5 \\ 1.5 & 2 & -1.5 & 1 \\ -1.5 & -1.5 & 1.5 & -1.5 \\ 1.5 & 1 & -1.5 & 2 \end{bmatrix}\begin{matrix}(6)\\(2)\\(3)\\(4)\end{matrix}$$

(columns: (5) (1) (6) (2) for $\mathbf{k}_*^1\mathbf{T}^1$; (6) (2) (3) (4) for $\mathbf{k}_*^2\mathbf{T}^2$)

The structure stiffness matrix $\mathbf{k}$, satisfying $\mathbf{F} = \mathbf{kD}$, can now be assembled, by summing up the contributions of $\mathbf{T}^{1^T}\mathbf{k}_*^1\mathbf{T}^1$ and $\mathbf{T}^{2^T}\mathbf{k}_*^2\mathbf{T}^2$:

$$\Rightarrow \quad \mathbf{k} = \left[\begin{array}{c|c} \mathbf{k_{AA}} & \mathbf{k_{AR}} \\ \hline \mathbf{k_{RA}} & \mathbf{k_{RR}} \end{array}\right] = EI\left[\begin{array}{ccc|ccc} 1 & 0.5 & 0 & 0 & 0.375 & -0.375 \\ 0.5 & 3 & -1.5 & 1 & 0.375 & 1.125 \\ 0 & -1.5 & 1.5 & -1.5 & 0 & -1.5 \\ \hline 0 & 1 & -1.5 & 2 & 0 & 1.5 \\ 0.375 & 0.375 & 0 & 0 & 0.1875 & -0.1875 \\ -0.375 & 1.125 & -1.5 & 1.5 & -0.1875 & 0.1875 \end{array}\right]$$

Displacements and Support Reactions

The structure stiffness relation to be considered here is:

$$\begin{bmatrix} \mathbf{F_A} \\ \hline \mathbf{F_R} \end{bmatrix} - \begin{bmatrix} \mathbf{F_{fA}} \\ \hline \mathbf{F_{fR}} \end{bmatrix} = \begin{bmatrix} \mathbf{k_{AA}} & \mathbf{k_{AR}} \\ \hline \mathbf{k_{RA}} & \mathbf{k_{RR}} \end{bmatrix} \begin{bmatrix} \mathbf{D_A} \\ \hline \mathbf{D_R} = \mathbf{O} \end{bmatrix}$$

$$\mathbf{D_A} = [\mathbf{k_{AA}}]^{-1}[\mathbf{F_A} - \mathbf{F_{fA}}] \Rightarrow \begin{Bmatrix} D_1 \\ D_2 \\ D_3 \end{Bmatrix} = \frac{1}{EI}\begin{bmatrix} 1.2 & -0.4 & -0.4 \\ -0.4 & 0.8 & 0.8 \\ -0.4 & 0.8 & 1.466667 \end{bmatrix} \begin{Bmatrix} -20 \text{ kNm} \\ -2.5 \text{ kNm} \\ -12.5 \text{ kN} \end{Bmatrix} = \frac{1}{EI}\begin{Bmatrix} -18 \\ -4 \\ -12.3333 \end{Bmatrix}$$

$$\mathbf{F_R} = \mathbf{F_{fR}} + \mathbf{k_{RA}}\mathbf{D_A} \Rightarrow \begin{Bmatrix} F_4 \\ F_5 \\ F_6 \end{Bmatrix} = \begin{Bmatrix} -7.5 \text{ kNm} \\ 30 \text{ kN} \\ 97.5 \text{ kN} \end{Bmatrix} + EI\begin{bmatrix} 0 & 1 & -1.5 \\ 0.375 & 0.375 & 0 \\ -0.375 & 1.125 & -1.5 \end{bmatrix} \frac{1}{EI}\begin{Bmatrix} -18 \\ -4 \\ -12.3333 \end{Bmatrix} = \begin{Bmatrix} 7.0 \text{ kNm} \\ 21.75 \text{ kN} \\ 118.25 \text{ kN} \end{Bmatrix}$$

<u>Check</u> $\sum F_y = 0: \; F_5 + F_6 = 21.75 + 118.25 = 140.00$ kN = Total load (downwards) ⇒ OK.

Member Forces

$$\mathbf{F}_*^i = \mathbf{F}_{*f}^i + \mathbf{k}_*^i \mathbf{T}^i \mathbf{D}$$

$$\mathbf{F}_*^1 = \begin{Bmatrix} 30 \text{ kN} \\ 20 \text{ kNm} \\ 30 \text{ kN} \\ -20 \text{ kNm} \end{Bmatrix} + EI\begin{bmatrix} 0.1875 & 0.375 & -0.1875 & 0.375 \\ 0.375 & 1 & -0.375 & 0.5 \\ -0.1875 & -0.375 & 0.1875 & -0.375 \\ 0.375 & 0.5 & -0.375 & 1 \end{bmatrix} \begin{matrix} (5) \\ (1) \\ (6) \\ (2) \end{matrix} \begin{Bmatrix} D_5 = 0 \\ D_1 = -18/EI \\ D_6 = 0 \\ D_2 = -4/EI \end{Bmatrix} = \begin{Bmatrix} 21.750 \text{ kN} \\ 0.000 \text{ kNm} \\ 38.25 \text{ kN} \\ -33.000 \text{ kNm} \end{Bmatrix}$$

$$\mathbf{F}_*^2 = \begin{Bmatrix} 67.5 \text{ kN} \\ 22.5 \text{ kNm} \\ 12.5 \text{ kN} \\ -7.5 \text{ kNm} \end{Bmatrix} + EI\begin{bmatrix} 1.5 & 1.5 & -1.5 & 1.5 \\ 1.5 & 2 & -1.5 & 1 \\ -1.5 & -1.5 & 1.5 & -1.5 \\ 1.5 & 1 & -1.5 & 2 \end{bmatrix} \begin{matrix} (6) \\ (2) \\ (3) \\ (4) \end{matrix} \begin{Bmatrix} D_6 = 0 \\ D_2 = -4/EI \\ D_3 = -12.3333/EI \\ D_4 = 0 \end{Bmatrix} = \begin{Bmatrix} 80.000 \text{ kN} \\ 33.000 \text{ kNm} \\ 0.000 \text{ kN} \\ 7.000 \text{ kNm} \end{Bmatrix}$$

The free-body and bending moment diagrams, including the fixed end force effects in Fig. 5.17d, are shown in Fig. 5.18.

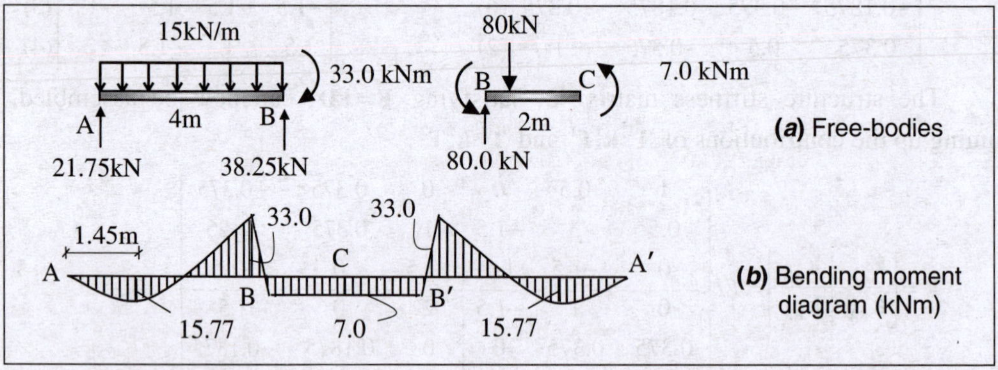

Figure 5.18 Free-bodies and bending moment diagram (Example 5.5)

5.2.8 Accounting for Shear Deformations

When the depth of the beam is relatively large (i.e., when the span/depth ratio is low), there can be significant error in employing the beam element stiffness matrix defined earlier (Eq. 5.11), which is based on the Euler-Bernoulli theory. Plane sections before bending no longer remain plane, and the associated warping of the cross-section cannot be ignored. Shear deformations contribute to additional deflections in the beam, which are not insignificant. Another way of assessing this effect of shear deformation is in terms of strain energy. The shear strain energy in a beam, which is usually negligible (less than 1 percent) in comparison to the flexural strain energy, becomes increasingly significant as the span/depth ratio of the beam reduces and approaches unity.

In such situations, the beam element stiffness matrix needs to be suitably modified. This can be done using any of the procedures adopted earlier. One of the earliest ways of incorporating the contribution of shear deformations was shown by Timoshenko, using an energy approach, on a prismatic cantilever beam (sometimes referred to as the *Timoshenko beam*). Here, we shall adopt the 'force approach' (refer Eqns 5.14 to 5.17), modifying the expression for deflection $\Delta_j(x^*)$ to include the contribution of shear deformations. Under the action of constant shear force, *S*, in a prismatic beam element [refer Fig. 1.8a], the element will undergo a constant rotation (without any curvature), whose magnitude is constant along the length of the beam, and equal to the shear strain, γ. The latter is given by the shear force *S* divided by the *shear rigidity*, $(GA')_i$, where G_i is the modulus of rigidity and A'_i is the modified area† of cross-section of the beam (modified to account for the non-uniform variation of shear stress under flexural shear). Using our earlier sign convention, corresponding to a positive shear force, this rotation will be in the clockwise direction, causing a "shear deflection", Δ_S, in the negative *y*-direction:

$$\frac{d\Delta_S}{dx} = -\gamma = -\frac{S}{GA'} \tag{5.29}$$

By considering a section of the beam at a distance x^* from the start node, in the free-bodies shown in Fig. 5.1, we observe that $S = k^i_{1^*j^*}$, whereby:

$$\Delta_S = -\frac{1}{(GA')_i}\left(k^i_{1^*j^*}\right)(x^*) + \text{constant} \tag{5.30}$$

Adding this shear deflection contribution in the expression for $\Delta_j(x^*)$ in Eq. 5.15,

$$\Delta_j(x^*) = \frac{1}{(EI)_i}\left[-\left(k^i_{2^*j^*}\right)\frac{(x^*)^2}{2} + \left(k^i_{1^*j^*}\right)\frac{(x^*)^3}{6}\right] - \frac{1}{(GA')_i}\left(k^i_{1^*j^*}\right)(x^*) + C_{\theta j}(x^*) + C_{\Delta j} \tag{5.31}$$

† For a rectangular section, with area A, $A' = A/1.2$.

Carrying out the procedure described earlier, expressions for the various stiffness influence coefficients can be derived and conveniently expressed as a function of a non-dimensional *shear deformation constant*, β_S, defined as:

$$\beta_S = \frac{12(EI)_i}{(GA')_i L_i^2} \tag{5.32}$$

Considering a rectangular section ($A' = A/1.2$) and a material with Poisson's ratio, $\nu = 0.3$ (i.e., $E/G = 2(1+0.3)$), $\beta_S = \frac{3.12}{(L/D)^2}$, taking on values in the practical range of 0.01 (for very high span/depth ratios) to 3.12 (for $L/D = 1$).

The modified element stiffness matrix, $\widehat{\mathbf{k}}_*^i$, for a prismatic beam element, including the effect of shear deformations, is given below:

$$\widehat{\mathbf{k}}_*^i = \frac{(EI)_i}{L_i(1+\beta_S)} \begin{bmatrix} 12/L_i^2 & 6/L_i & -12/L_i^2 & 6/L_i \\ 6/L_i & 4+\beta_S & -6/L_i & 2-\beta_S \\ -12/L_i^2 & -6/L_i & 12/L_i^2 & -6/L_i \\ 6/L_i & 2-\beta_S & -6/L_i & 4+\beta_S \end{bmatrix} \tag{5.33}$$

Thus, in general, the multiplication factor for most of the stiffness coefficients, accounting for the effect of shear deformations, given by $\frac{1}{1+\beta_S}$, varies from 0.99 to 0.243. The effect of shear deformations is to reduce the stiffness of the beam, and this reduction can be significant, when the span/depth ratio approaches unity. Such situations are encountered while dealing with deep beams and squat shear walls in buildings. However, for normal, well-proportioned beams, the effect of shear deformations can be considered to be negligible.

5.3 REDUCED STIFFNESS METHOD APPLIED TO BEAMS

As pointed out earlier, although the four *displacements* at the two ends of a beam element are independent, the corresponding four member end *forces* are not independent. It is for this reason, the stiffness matrix of order 4×4 does not have an inverse (i.e., an element flexibility matrix of order 4×4 does not exist). In order to generate an element flexibility matrix, we need to have a *stable* element, such as a simply supported beam or a cantilever, in which there are only two independent degrees of freedom. Thus, it is possible to model the beam element with only *two* degrees of freedom. It is usually convenient to identify the two rotational degrees of freedom at the two ends of the beam element for this purpose. We can find the shear forces (vertical reactions) at the two ends of the beam element, once we find the beam end moments, and thus we can generate the complete force response. However, unlike the four degree of freedom model, in this case, we will not be able to find directly the translations (deflections), if any, at the two member ends.

Relative translation between the two ends of the beam element induces flexure in the beam, but this can be accounted for by considering the *chord rotation* and converting this to equivalent end rotations. In the reduced element stiffness formulation, it is convenient to work with the *displacement transformation matrix* for each element, relating the local element coordinates to the global structure coordinates. However, we cannot transform the fixed end forces at the element level to the structure in this procedure, while working with two degrees of freedom; this needs to be done manually, and this is a shortcoming in the reduced stiffness method, reducing its suitability for adoption in software packages. The final structure stiffness matrix, and the final response, however, remains the same, whether we work with the conventional element stiffness matrix of order 4×4 or the reduced element stiffness matrix of order 2×2. These issues are discussed in detail in Sections 5.3.1.

In cases involving member end releases (internal hinges, hinged and guided-fixed end supports), it is possible to reduce the degree of kinematic indeterminacy of the structure. In such cases, it is possible to reduce the size of the element stiffness matrix to an order 1×1, and this can be very advantageous in terms of computational effort. This is discussed in Sections 5.3.2 and 5.3.3, and demonstrated through various examples.

5.3.1 Beam Element with Two Degrees of Freedom

Element Stiffness Matrix

The prismatic beam element with only two rotational degrees of freedom (at the two ends, with deflections restrained) is shown in Fig. 5.19a. The element stiffness matrix for this can be easily generated by eliminating the rows and columns corresponding to the translational degrees of freedom in the four degree of freedom model derived earlier (Eq. 5.11). The first column of the matrix corresponds to the condition, $\left\{D^{i}_{1*} = 1,\ D^{i}_{2*} = 0\right\}$, while the second column of the matrix corresponds to $\left\{D^{i}_{2*} = 1,\ D^{i}_{1*} = 0\right\}$. The deflected shapes and free-bodies, corresponding to these two conditions, are shown in Figs 5.19b and c respectively. Accordingly, the reduced element stiffness matrix, $\tilde{\mathbf{k}}^{\mathbf{i}}_{*}$, is obtained as follows:

$$\boxed{\tilde{\mathbf{k}}^{\mathbf{i}}_{*} = \frac{(EI)_i}{L_i}\begin{bmatrix} 4 & 2 \\ 2 & 4 \end{bmatrix}} \tag{5.34}$$

Alternatively, we can generate the element stiffness matrix as the inverse of the flexibility matrix for a simply supported prismatic beam (Fig. 5.20a). The response of such a beam, subject to two unit load conditions, $\left\{F^{i}_{1*} = 1,\ F^{i}_{2*} = 0\right\}$ and $\left\{F^{i}_{2*} = 1,\ F^{i}_{1*} = 0\right\}$, are shown in Figs 5.20b and c respectively, whereby:

$$\mathbf{f}_*^i = \frac{L_i}{6(EI)_i}\begin{bmatrix} 2 & -1 \\ -1 & 2 \end{bmatrix} \quad \Rightarrow \quad \tilde{\mathbf{k}}_*^i = \left[f_*^i \right]^{-1} = \frac{(EI)_i}{L_i}\begin{bmatrix} 4 & 2 \\ 2 & 4 \end{bmatrix} \qquad (5.35)$$

(a) Local coordinates

$$\tilde{\mathbf{k}}_*^i = \begin{bmatrix} k_{1*1*}^i & k_{1*2*}^i \\ k_{2*1*}^i & k_{2*2*}^i \end{bmatrix} = \frac{(EI)_i}{L_i}\begin{bmatrix} 4 & 2 \\ 2 & 4 \end{bmatrix}$$

(b) Stiffness coefficients due to $D_{1*}^i = 1$

(c) Stiffness coefficients due to $D_{2*}^i = 1$

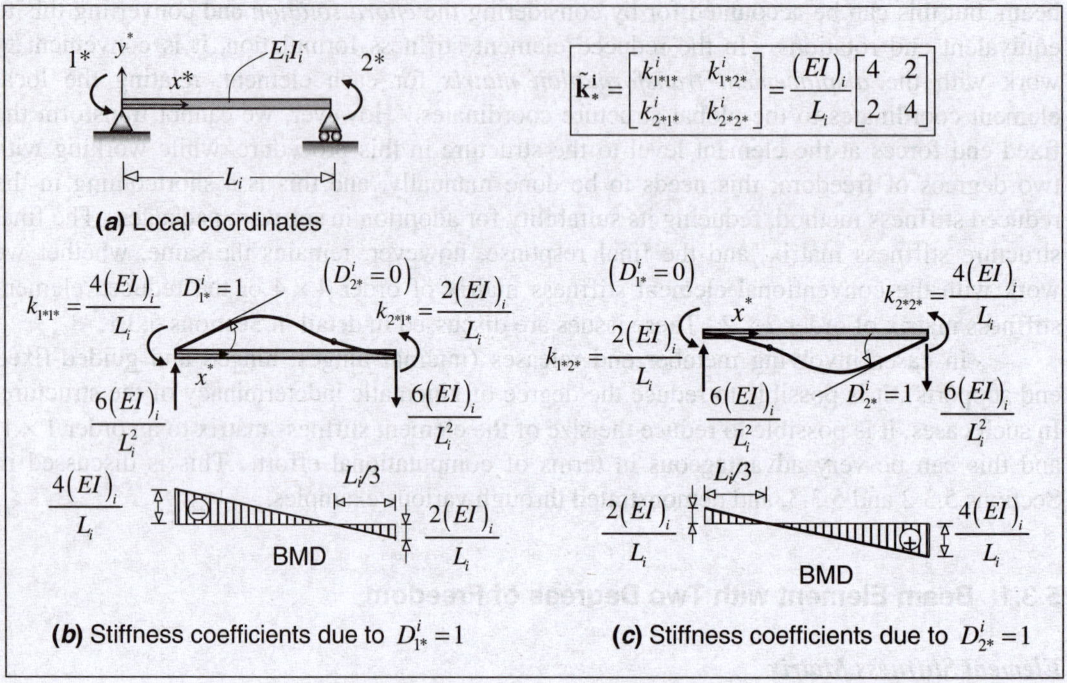

Figure 5.19 Generation of 2×2 stiffness matrix for a beam element

Displacement Transformation Matrix

In a typical continuous beam system, at every node location, two global coordinates are typically identified, corresponding to the deflection and rotation at the joint. In the four degree of freedom beam element model, it is convenient to link the element end deflection and rotation directly to the corresponding displacement in the structure, on a one-to-one basis.

In the two degree of freedom beam element model, however, this is possible only with respect to the rotational degrees of freedom. We need to devise an alternative means of linking the deflection at a node location in the structure to the end rotations in the beam element. This is possible by invoking the concept of *chord rotation*, which is defined as the ratio of the relative deflection between the two ends of the beam element AB to the span of the beam, $\phi_i = \dfrac{\Delta_B - \Delta_A}{L_i}$. Adopting our usual convention of assuming deflections acting upward to be positive, it follows that a chord rotation that acts in an anti-clockwise direction is positive. This is also consistent with our convention of assuming anti-clockwise end rotations (and moments) to be positive, for the purpose of matrix analysis.

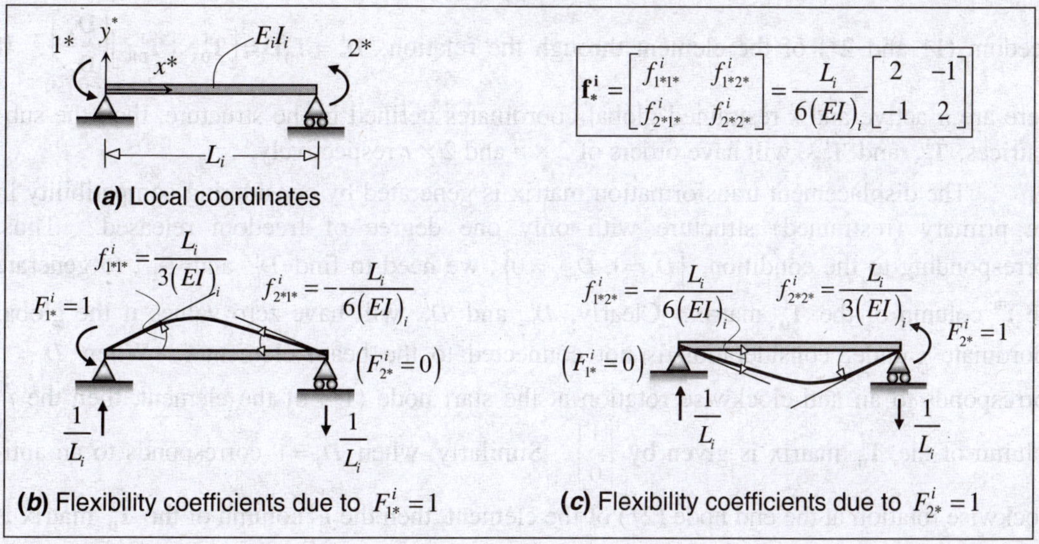

(*a*) Local coordinates

(*b*) Flexibility coefficients due to $F_{1*}^i = 1$ (*c*) Flexibility coefficients due to $F_{2*}^i = 1$

Figure 5.20 Generation of 2 × 2 flexibility matrix for a beam element

We know, from basic structural analysis, that when a prismatic fixed beam is subject to an anti-clockwise chord rotation, ϕ_i, this induces clockwise end moments of magnitude $\frac{6(EI)_i}{L_i}$ at the two ends, along with a constant negative shear force of $\frac{12(EI)_i}{L_i^2}$ [refer Fig. 5.1d]. Thus, it is possible to deal with chord rotations in the framework of the beam element with two rotational degrees of freedom, by converting the chord rotation, ϕ_i, to equivalent beam end rotations:

$$D_{1*}^i = D_{2*}^i = -\phi_i \tag{5.36}$$

Thus, for a typical prismatic beam element AB, subject to anti-clockwise end rotations, θ_A and θ_B, and upward deflections Δ_A and Δ_B, at the two ends, the relationship between the two end moments, $F_{1*}^i = M_{AB}$ and $F_{2*}^i = M_{BA}$ (anti-clockwise positive), and the various displacements can be expressed using the element stiffness matrix derived earlier (Eq. 5.34) as follows:

$$\begin{Bmatrix} F_{1*}^i \\ F_{2*}^i \end{Bmatrix} = \frac{(EI)_i}{L_i}\begin{bmatrix} 4 & 2 \\ 2 & 4 \end{bmatrix}\begin{Bmatrix} D_{1*}^i \\ D_{2*}^i \end{Bmatrix} \Rightarrow \begin{Bmatrix} M_{AB} \\ M_{BA} \end{Bmatrix} = \frac{(EI)_i}{L_i}\begin{bmatrix} 4 & 2 \\ 2 & 4 \end{bmatrix}\begin{Bmatrix} \theta_A - (\Delta_B - \Delta_A)/L_i \\ \theta_B - (\Delta_B - \Delta_A)/L_i \end{Bmatrix} \tag{5.37}$$

We are now in a position to define the *displacement transformation matrix* $\mathbf{T}_D^i$ for a beam element *i* with two degrees of freedom. This matrix defines the compatibility relations between the global degrees of freedom of the structure with the two rotational degrees of freedom (1* and 2*) of the element through the relation, $\mathbf{D}_*^i = \mathbf{T}_D^i \mathbf{D} = \left[\mathbf{T}_{DA}^i \mid \mathbf{T}_{DR}^i \right] \begin{bmatrix} \mathbf{D}_A \\ \hline \mathbf{D}_R \end{bmatrix}$. If there are n active and r restrained global coordinates defined in the structure, then the sub-matrices, $\mathbf{T}_{DA}^i$ and $\mathbf{T}_{DR}^i$, will have orders of $2 \times n$ and $2 \times r$ respectively.

The displacement transformation matrix is generated by considering compatibility in the primary (restrained) structure with only one degree of freedom released. Thus, corresponding to the condition, $\{D_j = 1, D_{l \neq j} = 0\}$, we need to find D_{1*}^i and D_{2*}^i, to generate the jth column of the $\mathbf{T}_D^i$ matrix. Clearly, D_{1*}^i and D_{2*}^i will have zero values if the global coordinate j under consideration is not connected to the beam element i. When $D_j = 1$ corresponds to an anti-clockwise rotation at the start node (1*) of the element, then the jth column of the $\mathbf{T}_D^i$ matrix is given by $\begin{bmatrix} 1 \\ 0 \end{bmatrix}$. Similarly, when $D_j = 1$ corresponds to an anti-clockwise rotation at the end node (2*) of the element, then the jth column of the $\mathbf{T}_D^i$ matrix is given by $\begin{bmatrix} 0 \\ 1 \end{bmatrix}$. When $D_j = 1$ corresponds to a vertical (upward) deflection at the start node of the element, this implies a chord rotation, $\phi_i = -1/L_i$, whereby the jth column of the $\mathbf{T}_D^i$ matrix is given by $\begin{bmatrix} 1/L_i \\ 1/L_i \end{bmatrix}$. Finally, we note that when $D_j = 1$ corresponds to a vertical (upward) deflection at the end node of the element, this implies a chord rotation, $\phi_i = +1/L_i$, whereby the jth column of the $\mathbf{T}_D^i$ matrix is given by $\begin{bmatrix} -1/L_i \\ -1/L_i \end{bmatrix}$.

We can also verify this by generating the transpose of the displacement transformation matrix, using the contra-gradient principle, whereby $\mathbf{F} = \mathbf{T}_D^{i\,T} \mathbf{F}_*^i$. We now invoke equilibrium conditions (instead of compatibility conditions), applying a unit force at a time in the simply supported beam shown in Fig. 5.20. The vertical reactions correspond to the translational degrees of freedom in the structure (global coordinates), and are correctly obtained as $\pm 1/L_i$, corresponding to the action of unit end moments, $F_{1*}^i = 1$ and $F_{2*}^i = 1$. However, it is important to note that, the use of the contra-gradient principle cannot be extended to finding the fixed end force vector through the transformation, $\mathbf{F}_f = \sum_{i=1}^{m} \mathbf{T}_D^{i\,T} \mathbf{F}_{*f}^i$, as done in the conventional stiffness method. This is so, because the two fixed end moments (F_{1*f}^i and F_{1*f}^i) in the element are not adequate by themselves to determine the vertical reactions at the beam ends on account of intermediate loading. For this reason, we cannot automate the generation of the fixed end force vector in the structure in the reduced element stiffness formulation; this must be done manually, as demonstrated in the Examples to follow.

The displacement transformation matrix is useful in generating the structure stiffness matrix by the transformation, $\mathbf{k} = \sum_{i=1}^{m} \mathbf{T}_{\mathbf{D}}^{\mathbf{i}^{\mathrm{T}}} \tilde{\mathbf{k}}_{*}^{\mathbf{i}} \mathbf{T}_{\mathbf{D}}^{\mathbf{i}}$, and for finding the internal forces, $\mathbf{F}_{*}^{\mathbf{i}} = \mathbf{F}_{*\mathbf{f}}^{\mathbf{i}} + \left(\tilde{\mathbf{k}}_{*\mathbf{f}}^{\mathbf{i}} \mathbf{T}_{\mathbf{DA}}^{\mathbf{i}}\right) \mathbf{D}_{\mathbf{A}}$.

The basic analysis procedure is as outlined earlier in Section 5.2.5.

EXAMPLE 5.6

Repeat Example 5.1 by the reduced stiffness formulation. Draw the shear force and bending moment diagrams.

SOLUTION

Here, we consider the same system of global coordinates as in Example 5.1 (shown in Fig. 5.3a) with the nodal loads given by $\mathbf{F}_{\mathbf{A}} = \begin{Bmatrix} F_1 \\ F_2 \end{Bmatrix} = \begin{Bmatrix} 0 \text{ kN} \\ -30 \text{ kNm} \end{Bmatrix}$, but with each of the two beam elements having two degrees of freedom (local coordinates as shown in Fig. 5.21b).

Fixed End Forces

The fixed end forces (support reactions) on the primary structure (with $D_1 = D_2 = 0$) are shown in Fig. 5.21c.

<u>Element '1' (AB):</u>

$$F_{1*f}^{1} = \frac{100 \times 10}{8} = 125 \text{ kNm}; \; F_{2*f}^{1} = -125 \text{ kNm} \qquad \Rightarrow \mathbf{F}_{*\mathbf{f}}^{1} = \begin{Bmatrix} 125 \text{ kNm} \\ -125 \text{ kNm} \end{Bmatrix}$$

Vertical reactions, $V_{AB} = V_{BA} = \dfrac{100}{2} = 50$ kN (upward), are induced at A and B.

<u>Element '2' (BC):</u>

$$F_{1*f}^{2} = \frac{10 \times 10^2}{12} = 83.333 \text{ kNm}; \; F_{2*f}^{2} = -83.333 \text{ kNm} \qquad \Rightarrow \mathbf{F}_{*\mathbf{f}}^{1} = \begin{Bmatrix} 83.333 \text{ kNm} \\ -83.333 \text{ kNm} \end{Bmatrix}$$

Vertical reactions, $V_{BC} = V_{CB} = \dfrac{10 \times 10}{2} = 50$ kN (upward), are induced at B and C.

Summing up the fixed end forces at the two active global coordinate locations, we note that a vertical (upward) force of 50 + 50 = 100 kN and a moment of $-125 + 83.333 = -41.667$ kNm act at B, as shown in Fig. 5.21d. Referring to the global coordinates in Fig. 5.21a, we observe that the fixed end force vector, $\mathbf{F}_{\mathbf{f}} = \left[\dfrac{\mathbf{F}_{\mathbf{fA}}}{\mathbf{F}_{\mathbf{fR}}}\right]$, is obtainable as:

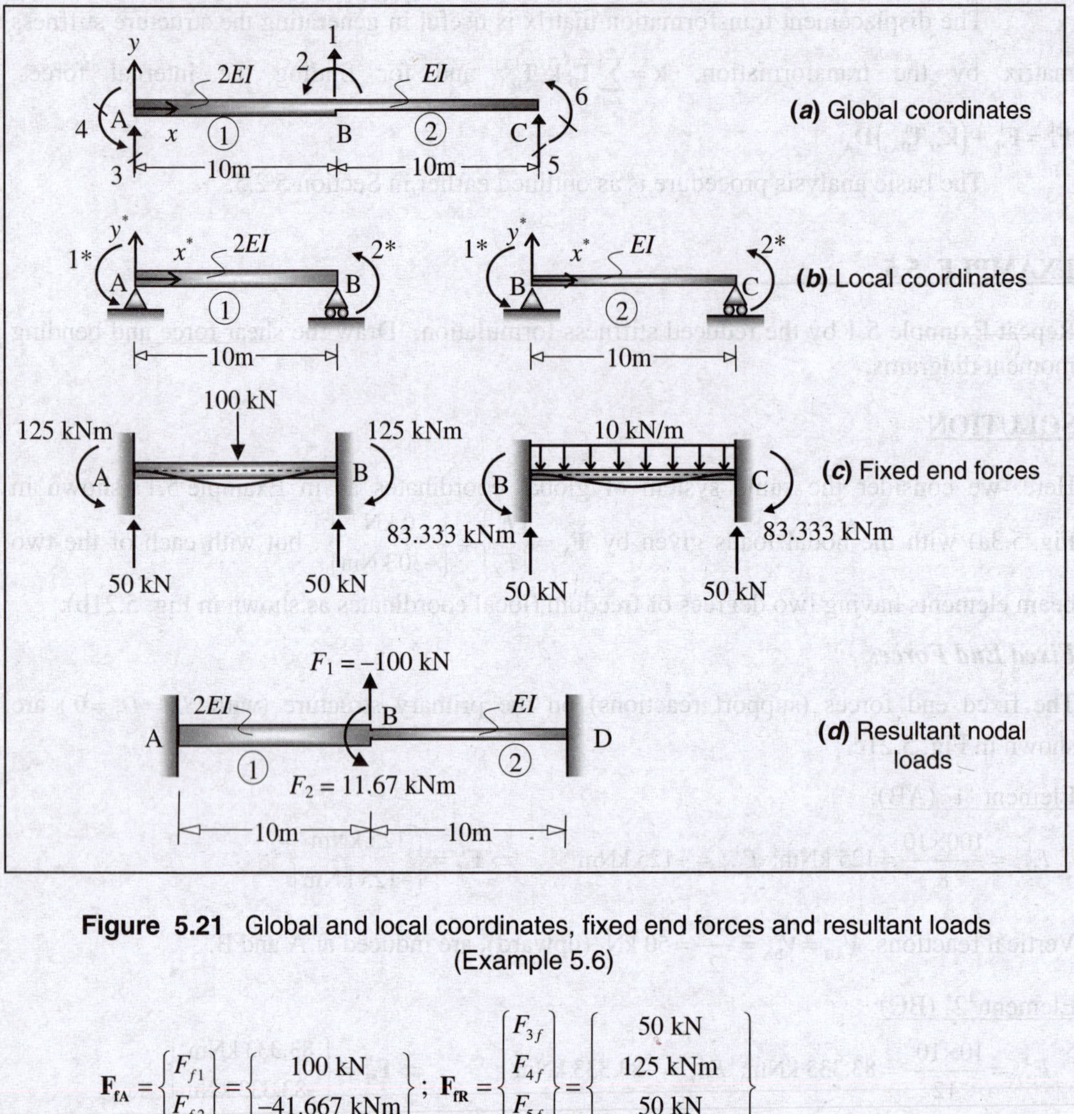

Figure 5.21 Global and local coordinates, fixed end forces and resultant loads (Example 5.6)

$$\mathbf{F_{fA}} = \begin{Bmatrix} F_{f1} \\ F_{f2} \end{Bmatrix} = \begin{Bmatrix} 100 \text{ kN} \\ -41.667 \text{ kNm} \end{Bmatrix}; \quad \mathbf{F_{fR}} = \begin{Bmatrix} F_{3f} \\ F_{4f} \\ F_{5f} \\ F_{6f} \end{Bmatrix} = \begin{Bmatrix} 50 \text{ kN} \\ 125 \text{ kNm} \\ 50 \text{ kN} \\ -83.333 \text{ kNm} \end{Bmatrix}$$

The resultant load vector, $\mathbf{F_A} - \mathbf{F_{fA}}$, is given by:

$$\mathbf{F_A} - \mathbf{F_{fA}} = \begin{Bmatrix} 0 \text{ kN} \\ -30 \text{ kNm} \end{Bmatrix} - \begin{Bmatrix} 100 \text{ kN} \\ -41.667 \text{ kNm} \end{Bmatrix} = \begin{Bmatrix} -100 \text{ kN} \\ 11.667 \text{ kNm} \end{Bmatrix}$$

The deflections at the active coordinates, D_1 and D_2, due to the above loading (shown in Fig. 5.21d) will be identical to the deflections caused by the given loads (including distributed loading) shown in Fig. 5.2.

Displacement Transformation Matrices

The element displacement transformation matrices, $\mathbf{T_D^1}$ and $\mathbf{T_D^2}$, relating the local coordinates to the global coordinates through the relation, $\mathbf{D_*^i} = \left[\mathbf{T_{DA}^i} \mid \mathbf{T_{DR}^i}\right]\begin{bmatrix}\mathbf{D_A} \\ \hline \mathbf{D_R}\end{bmatrix}$, take the following form (refer Figs 5.21a and b):

$$\begin{matrix} & (1) & (2) & (3) & (4) & (5) & (6) \end{matrix}$$

$$\mathbf{T_D^1} = \left[\begin{array}{cc|cccc} -0.1 & 0 & 0.1 & 1 & 0 & 0 \\ -0.1 & 1 & 0.1 & 0 & 0 & 0 \end{array}\right]$$

$$\mathbf{T_D^2} = \left[\begin{array}{cc|cccc} 0.1 & 1 & 0 & 0 & -0.1 & 0 \\ 0.1 & 0 & 0 & 0 & -0.1 & 1 \end{array}\right]$$

Element and Structure Stiffness Matrices

The element stiffness matrices $\mathbf{\tilde{k}_*^1}$ and $\mathbf{\tilde{k}_*^2}$, satisfying $\mathbf{F_*^i} = \mathbf{\tilde{k}_*^i D_*^i}$, take the following form (refer Fig. 5.21b), considering $\frac{E_1 I_1}{L_1} = \frac{2EI}{10}$ and $\frac{E_2 I_2}{L_2} = \frac{EI}{10}$:

$$\mathbf{\tilde{k}_*^1} = \frac{2EI}{10}\begin{bmatrix} 4 & 2 \\ 2 & 4 \end{bmatrix} = EI\begin{bmatrix} 0.8 & 0.4 \\ 0.4 & 0.8 \end{bmatrix} \quad \text{and} \quad \mathbf{\tilde{k}_*^2} = \frac{EI}{10}\begin{bmatrix} 4 & 2 \\ 2 & 4 \end{bmatrix} = EI\begin{bmatrix} 0.4 & 0.2 \\ 0.2 & 0.4 \end{bmatrix}$$

$$\Rightarrow \quad \mathbf{\tilde{k}_*^1 T_D^1} = EI\left[\begin{array}{cc|cccc} -0.12 & 0.4 & 0.12 & 0.8 & 0 & 0 \\ -0.12 & 0.8 & 0.12 & 0.4 & 0 & 0 \end{array}\right]$$

$$\mathbf{\tilde{k}_*^2 T_D^2} = EI\left[\begin{array}{cc|cccc} 0.06 & 0.4 & 0 & 0 & -0.06 & 0.2 \\ 0.06 & 0.2 & 0 & 0 & -0.06 & 0.4 \end{array}\right]$$

The structure stiffness matrix $\mathbf{k}$, satisfying $\mathbf{F} = \mathbf{kD}$, can now be assembled:

$$\mathbf{k} = \sum_{i=1}^{2} \mathbf{T_D^i}^T \mathbf{\tilde{k}_*^i T_D^i} = EI\left[\begin{array}{cc|cccc} 0.036 & -0.06 & -0.024 & -0.12 & -0.012 & 0.06 \\ -0.06 & 1.2 & 0.12 & 0.4 & -0.06 & 0.2 \\ \hline -0.024 & 0.12 & 0.024 & 0.12 & 0 & 0 \\ -0.12 & 0.4 & 0.12 & 0.8 & 0 & 0 \\ -0.012 & -0.06 & 0 & 0 & 0.012 & -0.06 \\ 0.06 & 0.2 & 0 & 0 & -0.06 & 0.4 \end{array}\right] = \left[\begin{array}{c|c} \mathbf{k_{AA}} & \mathbf{k_{AR}} \\ \hline \mathbf{k_{RA}} & \mathbf{k_{RR}} \end{array}\right]$$

This matrix is identical to the one generated in Example 5.1 (where we used a four degree of freedom beam element model).

Displacements and Support Reactions

The structure stiffness relation to be considered here is:

$$\begin{bmatrix}\mathbf{F_A} \\ \hline \mathbf{F_R}\end{bmatrix} - \begin{bmatrix}\mathbf{F_{fA}} \\ \hline \mathbf{F_{fR}}\end{bmatrix} = \left[\begin{array}{c|c} \mathbf{k_{AA}} & \mathbf{k_{AR}} \\ \hline \mathbf{k_{RA}} & \mathbf{k_{RR}} \end{array}\right]\begin{bmatrix}\mathbf{D_A} \\ \hline \mathbf{D_R} = \mathbf{O}\end{bmatrix}$$

We get exactly the same solution for the response by applying:

$$\mathbf{D_A} = [\mathbf{k_{AA}}]^{-1}[\mathbf{F_A} - \mathbf{F_{fA}}] \Rightarrow \begin{Bmatrix} D_1 \\ D_2 \end{Bmatrix} = \frac{1}{EI}\begin{bmatrix} 30.3030 & 1.51515 \\ 1.51515 & 0.90909 \end{bmatrix}\begin{Bmatrix} -100 \\ 11.667 \end{Bmatrix} = \frac{1}{EI}\begin{Bmatrix} -3012.626 \\ -140.909 \end{Bmatrix}$$

$$\mathbf{F_R} = \mathbf{F_{fR}} + \mathbf{k_{RA}}\mathbf{D_A} \Rightarrow \begin{Bmatrix} F_3 \\ F_4 \\ F_5 \\ F_6 \end{Bmatrix} = \begin{Bmatrix} 50 \\ 125 \\ 50 \\ -83.333 \end{Bmatrix} + EI\begin{bmatrix} -0.024 & 0.12 \\ -0.12 & 0.4 \\ -0.012 & -0.06 \\ 0.06 & 0.2 \end{bmatrix}\frac{1}{EI}\begin{Bmatrix} -3012.626 \\ -140.909 \end{Bmatrix} = \begin{Bmatrix} 105.394 \text{ kN} \\ 430.152 \text{ kNm} \\ 94.606 \text{ kN} \\ -292.273 \text{ kNm} \end{Bmatrix}$$

<u>Check</u> $\sum F_y = 0$:

Total reaction = $F_3 + F_5 = 105.394 + 94.606 = 200.000$ kN = Total load (downwards) $\Rightarrow$ OK.

Member Forces

$$\mathbf{F_*^i} = \mathbf{F_{*f}^i} + \tilde{\mathbf{k}}_*^i\left[\mathbf{T_{DA}^i} \mid \mathbf{T_{DR}^i}\right]\begin{Bmatrix} \mathbf{D_A} \\ \hline \mathbf{D_R} = \mathbf{O} \end{Bmatrix} = \mathbf{F_{*f}^i} + \tilde{\mathbf{k}}_*^i\mathbf{T_{DA}^i}\mathbf{D_A}$$

$$\Rightarrow \begin{Bmatrix} F_{1*}^1 \\ F_{2*}^1 \end{Bmatrix} = \begin{Bmatrix} 125 \text{ kNm} \\ -125 \text{ kNm} \end{Bmatrix} + EI\begin{bmatrix} 0.12 & 0.4 \\ 0.12 & 0.8 \end{bmatrix}\begin{Bmatrix} -3012.626/EI \\ -140.909/EI \end{Bmatrix} = \begin{Bmatrix} 430.15 \text{ kNm} \\ 123.79 \text{ kNm} \end{Bmatrix}$$

$$\begin{Bmatrix} F_{1*}^2 \\ F_{2*}^2 \end{Bmatrix} = \begin{Bmatrix} 83.333 \text{ kNm} \\ -83.333 \text{ kNm} \end{Bmatrix} + EI\begin{bmatrix} 0.06 & 0.4 \\ 0.06 & 0.2 \end{bmatrix}\begin{Bmatrix} -3012.626/EI \\ -140.909/EI \end{Bmatrix} = \begin{Bmatrix} -153.79 \text{ kNm} \\ -292.27 \text{ kNm} \end{Bmatrix}$$

The member end moments are exactly as obtained earlier. The vertical reactions (shear forces) at the member ends are not directly obtainable (as in Example 5.1), but can be easily computed from the free-bodies (shown in Fig. 5.22a), applying equilibrium equations. The shear force and bending moment diagrams are shown in Figs 5.22b and c.

5.3.2 Dealing with Moment Releases (Hinges)

We can take advantage of a reduction in the degree of kinematic indeterminacy when we encounter a "moment release" at a hinge (hinged end support or internal hinge). In such cases, we can simply ignore the degree of freedom associated with the member end release. But we need to also modify the element stiffness matrix for this element as well as the fixed end forces. Eliminating a degree of freedom from the two degree of freedom element model effectively implies that we are dealing with a beam element with a single degree of freedom.

Consider a prismatic beam element with a hinge at one end. If we ignore the rotation at this end, in addition to the deflections at the two ends, the four degree of freedom element model reduces to a single degree of freedom element model, with the rotation at the other end being the only unknown displacement. We can easily generate the element stiffness matrix of order 1×1 by eliminating the corresponding three rows and three columns from the element stiffness matrix of order 4×4 (Eq. 5.11); this yields a reduced element stiffness matrix, $\tilde{\mathbf{k}}_*^i$:

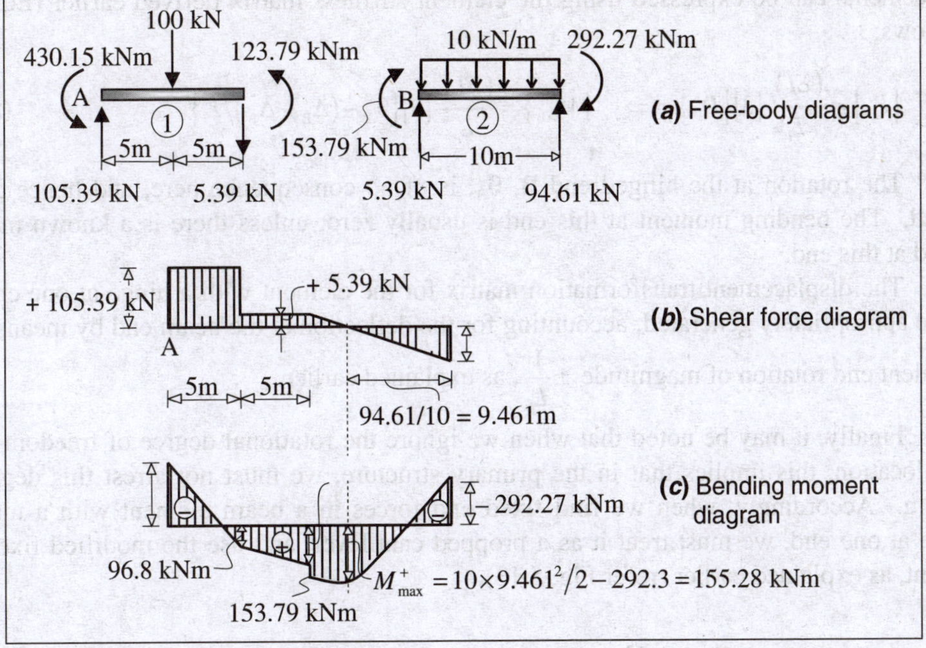

Figure 5.22 Force response (Example 5.6)

$$\boxed{\tilde{\mathbf{k}}_*^{\mathbf{i}} = \frac{(EI)_i}{L_i}[3]} \tag{5.38}$$

We can alternatively find this stiffness coefficient as the reciprocal of the corresponding flexibility coefficient, $f^i_{1*1*} = f^i_{2*2*} = \dfrac{L_i}{3(EI)_i}$, in a simply supported beam element subject to a unit end moment, as shown in Fig. 5.20.

$$\mathbf{f}_*^{\mathbf{i}} = \left[\frac{L_i}{3(EI)_i}\right] \quad \Rightarrow \quad \tilde{\mathbf{k}}_*^{\mathbf{i}} = \left[f_*^i\right]^{-1} = \left[\frac{3(EI)_i}{L_i}\right] \tag{5.39}$$

We can account for chord rotations here in the same manner as explained earlier. When a prismatic propped cantilever beam is subject to an anti-clockwise chord rotational slip, ϕ_i (on account of differential settlement), this induces a clockwise end moment of magnitude $\dfrac{3(EI)_i}{L_i}$ at the fixed end, along with a constant negative shear force of $\dfrac{3(EI)_i}{L_i^2}$.

Thus, for a typical prismatic beam element AB, with a hinge at B, subject to an anti-clockwise end rotation, θ_A at A, and upward deflections Δ_A and Δ_B, at the two ends, the relationship between the end moment, $F^i_{1*} = M_{AB}$ (anti-clockwise positive), and the various displacements can be expressed using the element stiffness matrix derived earlier (Eq. 5.38) as follows:

$$\{F_{1*}^{i}\} = \frac{(EI)_i}{L_i}[3]\{D_{1*}^{i}\} \Rightarrow \{M_{AB}\} = \frac{(EI)_i}{L_i}[3]\{\theta_A - (\Delta_B - \Delta_B)/L_i\} \tag{5.40}$$

The rotation at the hinged end B, θ_A, is of no consequence here, and hence can be ignored. The bending moment at this end is usually zero, unless there is a known moment applied at this end.

The displacement transformation matrix for the element with a hinge at one end can thus be appropriately generated, accounting for the deflection at the beam end by means of an equivalent end rotation of magnitude $\pm\frac{1}{L_i}$, as explained earlier.

Finally, it may be noted that when we ignore the rotational degree of freedom at the hinge location, this implies that in the primary structure, we must not arrest this degree of freedom. Accordingly, when we find fixed end forces in a beam element with a moment release at one end, we must treat it as a propped cantilever, and use the modified fixed end moment, as explained earlier (refer Fig. 5.13).

EXAMPLE 5.7

Repeat Example 5.3 by the reduced stiffness method and find the force response.

SOLUTION

Here, we take advantage of the fact that the overhanging portion DE of the continuous beam is statically determinate, and hence can be separated out for convenience in analysis. It is sufficient to analyse the three-span continuous beam ABCD, loaded as shown in Fig. 5.23, with an end moment at D (from overhang DE).

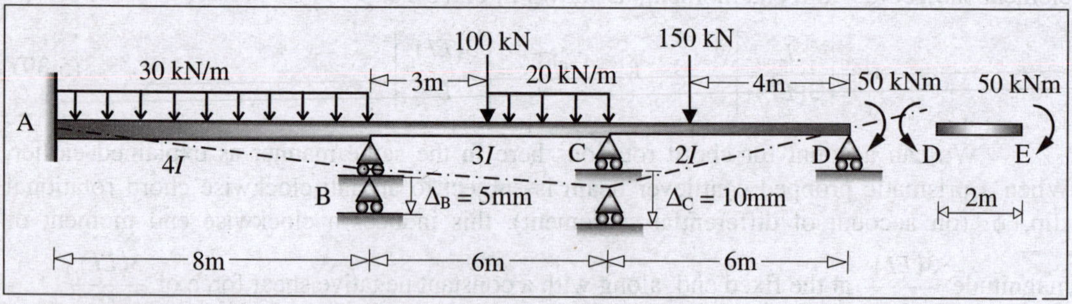

Figure 5.23 Continuous beam with gravity loading combined with support settlements (Example 5.7)

Thus, we can ignore the kinematic indeterminacy associated with the two degrees of freedom at the free end E. Furthermore, we can take advantage of the hinged end support at D and ignore the rotational degree of freedom at D. Thus, we are able to reduce the degree of kinematic indeterminacy from 5 (as solved in Example 5.3) to 2.

In the reduced (simplified) analysis, we have 3 elements, with 2 active degrees of freedom and 5 restrained degrees of freedom. The corresponding global coordinates are indicated in Fig. 5.24a. The local coordinates for the three elements are shown in Fig. 5.24b. Adopting the reduced stiffness formulation, the first two elements have two (rotational) degrees of freedom each, while the third element (CD) has only one (rotational) degree of freedom. The input loading on the structure comprises direct actions in the form of intermediate loads, with zero nodal loads. $\mathbf{F_A} = \begin{Bmatrix} F_1 \\ F_2 \end{Bmatrix} = \begin{Bmatrix} 0\text{ kNm} \\ 0\text{ kNm} \end{Bmatrix}$ and support displacements,

$$\mathbf{D_R} = \begin{Bmatrix} D_3 \\ D_4 \\ D_5 \\ D_6 \\ D_7 \end{Bmatrix} = \begin{Bmatrix} 0\text{ m} \\ 0\text{ rad} \\ -0.005\text{ m} \\ -0.010\text{ m} \\ 0\text{ m} \end{Bmatrix} \times \frac{80000}{EI} = \frac{1}{EI} \begin{Bmatrix} 0 \\ 0 \\ -400 \\ -800 \\ 0 \end{Bmatrix}.$$

Fixed End Forces

The fixed end forces (support reactions) on the primary structure (with $D_1 = D_2 = 0$) are shown in Fig. 5.24c.

<u>Element '1' (AB):</u>

$$F^1_{1*f} = \frac{30\times8^2}{12} = 160\text{ kNm};\ F^1_{2*f} = -160\text{ kNm} \qquad \Rightarrow \mathbf{F^1_{*f}} = \begin{Bmatrix} 160\text{ kNm} \\ -160\text{ kNm} \end{Bmatrix}$$

Vertical reactions, $V_{AB} = V_{BA} = \dfrac{30\times8}{2} = 120$ kN (upward), are induced at A and B.

<u>Element '2' (BC):</u>

$$F^2_{1*f} = +\frac{(100)(6)}{8} + \frac{5}{192}\times(20)(6^2) = +93.75\text{ kNm};\ F^2_{2*f} = -\frac{(100)(6)}{8} - \frac{11}{192}\times(20)(6^2) = -116.25\text{ kNm}$$

$$\Rightarrow \mathbf{F^2_{*f}} = \begin{Bmatrix} 93.75\text{ kNm} \\ -116.25\text{ kNm} \end{Bmatrix}$$

Vertical reactions:

$$V_{BC} = \frac{100}{2} + \frac{20\times3}{4} + \frac{93.75-116.25}{6} = 61.25\text{ kN};\ V_{CB} = 100 + (20\times3) - 61.25 = 98.75\text{ kN}$$

<u>Element '3' (CD):</u>

$$\tilde{F}^3_{1*f} = +\frac{(150)(2\times4)}{6^2}\left(4 + \frac{2}{2}\right) - \frac{50}{2} = +141.667\text{ kNm} \qquad \Rightarrow \mathbf{F^3_{*f}} = \{141.667\text{ kNm}\}$$

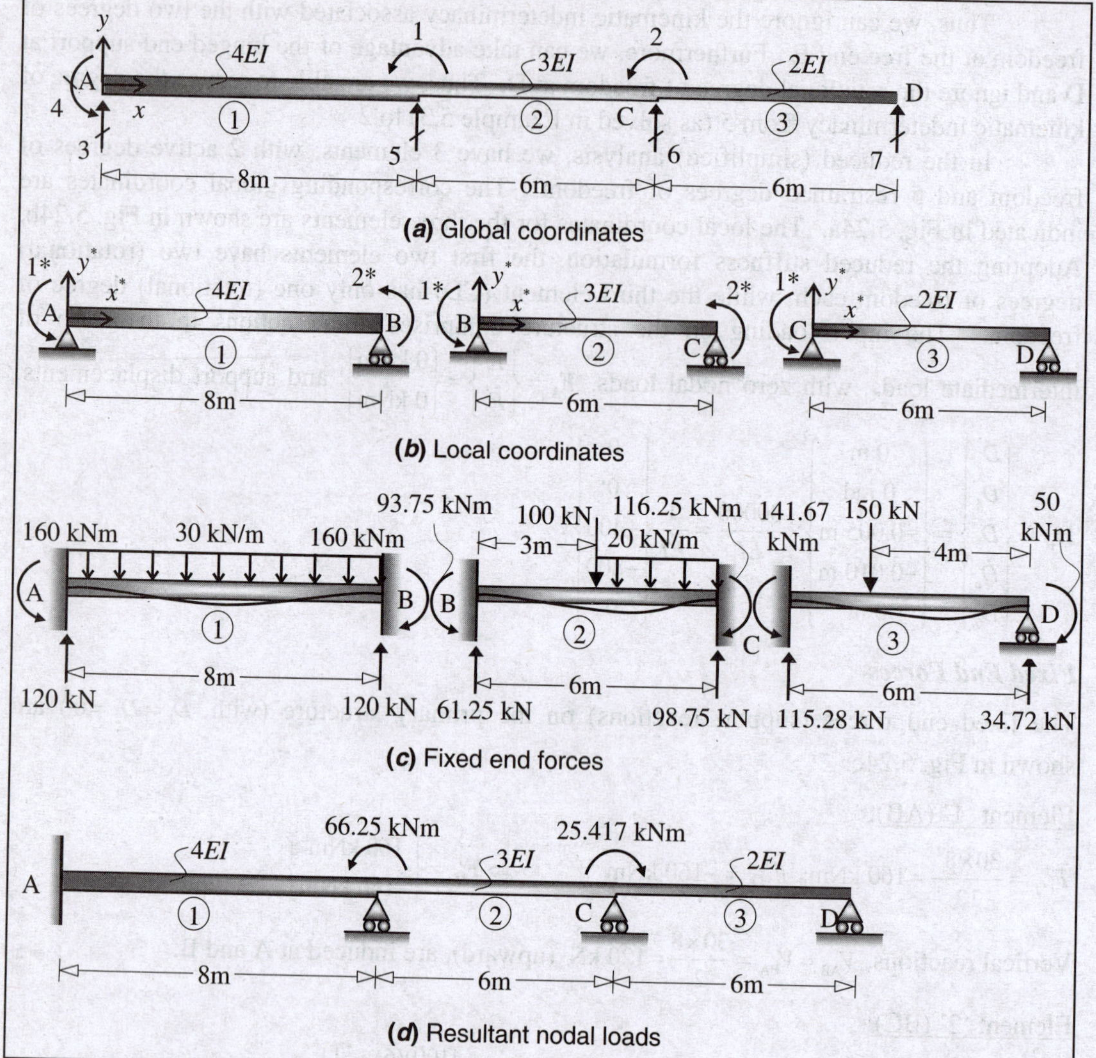

Figure 5.24 Global and local coordinates and fixed end forces (Example 5.7)

Vertical reactions:

$$V_{CD} = \frac{150\times4}{6} + \frac{141.667-50}{6} = 115.278 \text{ kN}; \; V_{DC} = 150-115.278 = 34.722 \text{ kN}$$

Summing up the fixed end moments at the two active global coordinate locations, we note that moments of $-160+93.75=-66.25$ kNm and $-116.25+141.667=25.417$ kNm act at B and C respectively, as shown in Fig. 5.24d. Referring to the global coordinates in Fig. 5.21a, we observe that the fixed end force vector is obtainable as:

$$\mathbf{F}_f = \begin{bmatrix} \mathbf{F}_{fA} \\ \hline \mathbf{F}_{fR} \end{bmatrix} \text{ where } \mathbf{F}_{fA} = \begin{Bmatrix} F_{1f} \\ F_{2f} \end{Bmatrix} = \begin{Bmatrix} -66.25 \\ 25.417 \end{Bmatrix} \text{kNm and } \mathbf{F}_{fR} = \begin{Bmatrix} F_{3f} \\ F_{4f} \\ F_{5f} \\ F_{6f} \\ F_{7f} \end{Bmatrix} = \begin{Bmatrix} 120.0 \text{ kN} \\ 160.0 \text{ kNm} \\ 181.25 \text{ kN} \\ 214.03 \text{ kN} \\ 34.72 \text{ kN} \end{Bmatrix}$$

The resultant load vector, $\mathbf{F}_A - \mathbf{F}_{fA}$, is given by:

$$\mathbf{F}_A - \mathbf{F}_{fA} = \begin{Bmatrix} 0 \\ 0 \end{Bmatrix} - \begin{Bmatrix} -66.25 \\ 25.417 \end{Bmatrix} = \begin{Bmatrix} 66.25 \\ -25.417 \end{Bmatrix} \text{kNm}$$

The deflections at the active coordinates, D_1 and D_2, due to the above loading (shown in Fig. 5.24d), combined with the given displacement loading, will be identical to the deflections caused by the given loads (including distributed loading) shown in Fig. 5.23.

Displacement Transformation Matrices

The element displacement transformation matrices, $\mathbf{T}_D^1$ and $\mathbf{T}_D^2$, relating the local coordinates to the global coordinates through the relation, $\mathbf{D}_*^i = \left[\mathbf{T}_{DA}^i \mid \mathbf{T}_{DR}^i \right] \begin{bmatrix} \mathbf{D}_A \\ \hline \mathbf{D}_R \end{bmatrix}$, take the following form (refer Figs 5.24a and b):

$$\begin{array}{c} \quad (1)\ (2)\ (3)\ \ (4)\ \ \ (5)\ \ (6)\ (7) \end{array}$$

$$\mathbf{T}_D^1 = \left[\begin{array}{cc|ccccc} 0 & 0 & 0.125 & 1 & -0.125 & 0 & 0 \\ 1 & 0 & 0.125 & 0 & -0.125 & 0 & 0 \end{array}\right]$$

$$\mathbf{T}_D^2 = \left[\begin{array}{cc|ccccc} 1 & 0 & 0 & 0 & 0.16667 & -0.16667 & 0 \\ 0 & 1 & 0 & 0 & 0.16667 & -0.16667 & 0 \end{array}\right]$$

$$\mathbf{T}_D^3 = \left[\begin{array}{cc|ccccc} 0 & 1 & 0 & 0 & 0 & 0.16667 & -0.16667 \end{array}\right]$$

Element and Structure Stiffness Matrices

The element stiffness matrices $\tilde{\mathbf{k}}_*^1$ and $\tilde{\mathbf{k}}_*^2$, satisfying $\mathbf{F}_*^i = \tilde{\mathbf{k}}_*^i \mathbf{D}_*^i$, take the following form (refer Fig. 5.24b), considering $\dfrac{E_1 I_1}{L_1} = \dfrac{4EI}{8}$, $\dfrac{E_2 I_2}{L_2} = \dfrac{3EI}{6}$ and $\dfrac{E_3 I_3}{L_3} = \dfrac{2EI}{6}$:

$$\tilde{\mathbf{k}}_*^1 = \frac{4EI}{8}\begin{bmatrix} 4 & 2 \\ 2 & 4 \end{bmatrix} = EI\begin{bmatrix} 2 & 1 \\ 1 & 2 \end{bmatrix}; \quad \tilde{\mathbf{k}}_*^2 = \frac{3EI}{6}\begin{bmatrix} 4 & 2 \\ 2 & 4 \end{bmatrix} = EI\begin{bmatrix} 2 & 1 \\ 1 & 2 \end{bmatrix}; \quad \tilde{\mathbf{k}}_*^3 = \frac{2EI}{6}[3] = EI[1]$$

$$\Rightarrow \quad \tilde{\mathbf{k}}_*^1 \mathbf{T}_D^1 = EI\left[\begin{array}{cc|ccccc} 1 & 0 & 0.375 & 2 & -0.375 & 0 & 0 \\ 2 & 0 & 0.375 & 1 & -0.375 & 0 & 0 \end{array}\right]$$

$$\tilde{\mathbf{k}}_*^2 \mathbf{T}_D^2 = EI\left[\begin{array}{cc|ccccc} 2 & 1 & 0 & 0 & 0.5 & -0.5 & 0 \\ 1 & 2 & 0 & 0 & 0.5 & -0.5 & 0 \end{array}\right]$$

$$\tilde{\mathbf{k}}_*^3\mathbf{T}_\mathbf{D}^3 = EI\begin{bmatrix}0 & 1 & \vdots & 0 & 0 & 0 & 0.16667 & -0.16667\end{bmatrix}$$

The structure stiffness matrix $\mathbf{k}$, satisfying $\mathbf{F} = \mathbf{kD}$, can now be assembled:

$$\mathbf{k} = \sum_{i=1}^{3} \mathbf{T}_\mathbf{D}^{i\,T}\tilde{\mathbf{k}}_*^i\mathbf{T}_\mathbf{D}^i \Rightarrow$$

$$\mathbf{k} = \begin{bmatrix}\mathbf{k_{AA}} & \mathbf{k_{AR}} \\ \mathbf{k_{RA}} & \mathbf{k_{RR}}\end{bmatrix} = EI\begin{bmatrix} 4 & 1 & 0.375 & 1 & 0.125 & -0.5 & 0 \\ 1 & 3 & 0 & 0 & 0.5 & -0.33333 & -0.16667 \\ 0.375 & 0 & 0.09375 & 0.375 & -0.09375 & 0 & 0 \\ 1 & 0 & 0.375 & 2 & -0.375 & 0 & 0 \\ 0.125 & 0.5 & -0.09375 & -0.375 & 0.26042 & -0.16667 & 0 \\ -0.5 & -0.33333 & 0 & 0 & -0.16667 & 0.19444 & -0.02778 \\ 0 & -0.16667 & 0 & 0 & 0 & -0.02778 & 0.02778 \end{bmatrix}$$

$$\mathbf{k_{AA}} = EI\begin{bmatrix}4 & 1 \\ 1 & 3\end{bmatrix} \Rightarrow \qquad [\mathbf{k_{AA}}]^{-1} = \frac{1}{EI}\begin{bmatrix}0.27273 & -0.09090 \\ -0.09090 & 0.36363\end{bmatrix}$$

The basic stiffness relations are given by:

$$\begin{bmatrix}\mathbf{F_A} \\ \mathbf{F_R}\end{bmatrix} - \begin{bmatrix}\mathbf{F_{fA}} \\ \mathbf{F_{fR}}\end{bmatrix} = \begin{bmatrix}\mathbf{k_{AA}} & \mathbf{k_{AR}} \\ \mathbf{k_{RA}} & \mathbf{k_{RR}}\end{bmatrix}\begin{bmatrix}\mathbf{D_A} \\ \mathbf{D_R}\end{bmatrix}$$

$$\Rightarrow \mathbf{k_{AA}D_A} = (\mathbf{F_A} - \mathbf{F_{fA}}) - \mathbf{k_{AR}D_R} \quad \text{where} \quad \mathbf{D_R^T} = \frac{1}{EI}[0 \quad 0 \quad -400 \quad -800 \quad 0]$$

where $\mathbf{k_{AR}D_R}$, which denotes the forces at the active coordinates due to the effect of support displacements on the primary (restrained) structure is given by

$$\mathbf{k_{AR}D_R} = EI\begin{bmatrix}0.375 & 1 & 0.125 & -0.5 & 0 \\ 0 & 0 & 0.5 & -0.33333 & -0.16667\end{bmatrix}\{D_R\} = \begin{Bmatrix}350.00 \\ 66.667\end{Bmatrix}\text{kNm}$$

$$\Rightarrow \mathbf{k_{AA}D_A} = \begin{Bmatrix}66.25 \\ -25.417\end{Bmatrix} - \begin{Bmatrix}350 \\ 66.667\end{Bmatrix} = \begin{Bmatrix}-283.75 \\ -92.084\end{Bmatrix}\text{kNm}$$

Displacements and Support Reactions

$$\mathbf{D_A} = [\mathbf{k_{AA}}]^{-1}[\mathbf{k_{AA}D_A}] \quad \Rightarrow \quad \begin{Bmatrix}D_1 \\ D_2\end{Bmatrix} = \frac{1}{EI}\begin{bmatrix}0.27273 & -0.09090 \\ -0.09090 & 0.36363\end{bmatrix}\begin{Bmatrix}-283.75 \\ -92.084\end{Bmatrix} = \frac{1}{EI}\begin{Bmatrix}-69.0151 \\ -7.6896\end{Bmatrix}$$

$$\mathbf{F_R} = \mathbf{F_{fR}} + \mathbf{k_{RA}D_A} + \mathbf{k_{RR}D_R}$$

$$\Rightarrow \begin{Bmatrix}F_3 \\ F_4 \\ F_5 \\ F_6 \\ F_7\end{Bmatrix} = \begin{Bmatrix}120.0\text{ kN} \\ 160.0\text{ kNm} \\ 181.25\text{ kN} \\ 214.03\text{ kN} \\ 34.72\text{ kN}\end{Bmatrix} + \begin{Bmatrix}-25.881\text{ kN} \\ -69.015\text{ kNm} \\ -12.472\text{ kN} \\ 37.071\text{ kN} \\ 1.282\text{ kN}\end{Bmatrix} + \begin{Bmatrix}37.5\text{ kN} \\ 150.0\text{ kNm} \\ 29.167\text{ kN} \\ -88.889\text{ kN} \\ 22.222\text{ kN}\end{Bmatrix} = \begin{Bmatrix}131.619\text{ kN} \\ 240.985\text{ kNm} \\ 197.945\text{ kN} \\ 162.212\text{ kN} \\ 58.224\text{ kN}\end{Bmatrix}$$

Member Forces

$$\mathbf{F}_*^i = \mathbf{F}_{*f}^i + \tilde{\mathbf{k}}_*^i \mathbf{T}_D^i \mathbf{D}, \text{ where } \mathbf{D}^T = \frac{1}{EI}[-69.0151 \quad -7.6896 \mid 0 \quad 0 \quad -400 \quad -800 \quad 0]$$

$$\Rightarrow \begin{Bmatrix} F_{1*}^1 \\ F_{2*}^1 \end{Bmatrix} = \begin{Bmatrix} 160 \text{ kNm} \\ -160 \text{ kNm} \end{Bmatrix} + EI \begin{bmatrix} 1 & 0 & 0.375 & 2 & -0.375 & 0 & 0 \\ 2 & 0 & 0.375 & 1 & -0.375 & 0 & 0 \end{bmatrix} \times \{D\} = \begin{Bmatrix} 240.98 \text{ kNm} \\ -148.03 \text{ kNm} \end{Bmatrix}$$

$$\Rightarrow \begin{Bmatrix} F_{1*}^2 \\ F_{2*}^2 \end{Bmatrix} = \begin{Bmatrix} 93.75 \text{ kNm} \\ -116.25 \text{ kNm} \end{Bmatrix} + EI \begin{bmatrix} 2 & 1 & 0 & 0 & 0.5 & -0.5 & 0 \\ 1 & 2 & 0 & 0 & 0.5 & -0.5 & 0 \end{bmatrix} \times \{D\} = \begin{Bmatrix} 148.03 \text{ kNm} \\ -0.644 \text{ kNm} \end{Bmatrix}$$

$$\{F_{1*}^3\} = \{141.667 \text{ kNm}\} + EI[0 \quad 1 \mid 0 \quad 0 \quad 0 \quad 0.16667 \quad -0.16667] \times \{D\} = \{0.644 \text{ kNm}\}$$

The member end moments are exactly as obtained earlier. The vertical reactions (shear forces) at the member ends are not directly obtainable (as in Example 5.3), but can be easily computed from the free-bodies, applying equilibrium equations. The shear force and bending moment diagrams are as shown earlier in Figs 5.11b and c.

EXAMPLE 5.8

Repeat Example 5.4 by the reduced stiffness method and find the force response.

SOLUTION

The loading diagram is shown in Fig. 5.25a. On account of the internal hinge at B, we can ignore the rotational degree of freedom at B in the reduced stiffness formulation. Thus, there is only one active degree of freedom (vertical translation at B) and four restrained global coordinates as shown in Fig. 5.25b. Each of the two prismatic elements can be considered with a single degree of freedom; the local coordinates are shown in Fig. 5.25c. [There is no need for the imaginary clamp at B, as required in the four degree of freedom model in Example 5.4].

Fixed End Forces

The fixed end forces (support reactions) on the primary structure (with $D_1 = 0$), accounting for the moment release at B, are shown in Fig. 5.26a.

Element '1' (AB):

$$F_{1*f}^1 = \frac{100 \times 10}{8} \times \frac{3}{2} - \frac{30}{2} = 172.5 \text{ kNm} \qquad \Rightarrow \mathbf{F}_{*f}^1 = \{172.5 \text{ kNm}\}$$

Vertical reactions:

$$V_{AB} = \frac{100}{2} + \frac{172.5 - 30}{10} = 64.25 \text{ kN}; \quad V_{BA} = 100 - 67.8125 = 35.75 \text{ kN}$$

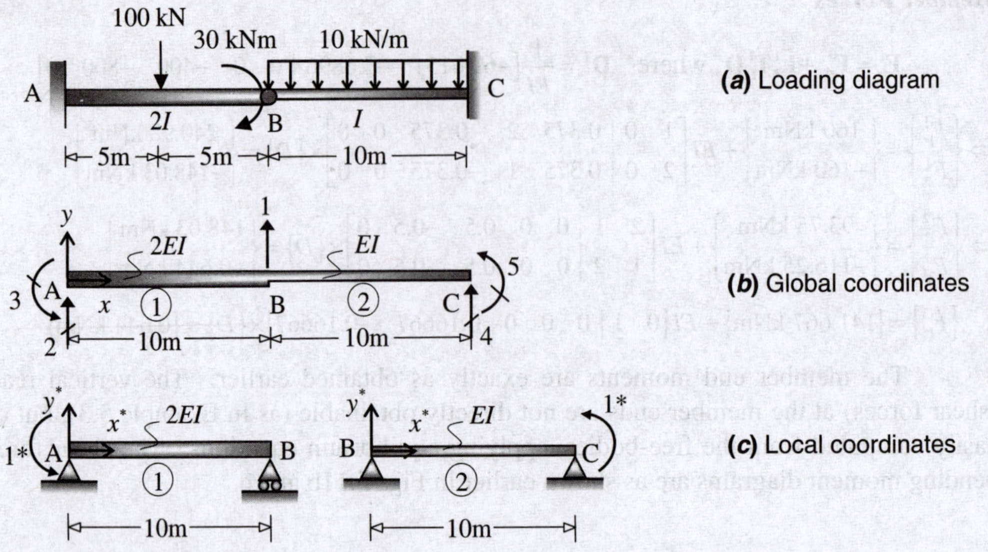

Figure 5.25 Loading diagram, global and local coordinates(Example 5.8)

<u>Element '2' (BC):</u>

$$F^2_{1^*f} = -\frac{10\times10^2}{12}\times\frac{3}{2} = -125 \text{ kNm} \Rightarrow \mathbf{F}^2_{*f} = \{-125 \text{ kNm}\}$$

Vertical reactions:

$$V_{BC} = \frac{10\times10}{2} - \frac{125}{10} = 37.5 \text{ kN}; \; V_{CB} = 100 - 37.5 = 62.5 \text{ kN}$$

The fixed end vertical force at the active global coordinate is given by

$$F_{1f} = V_{BA} + V_{BC} = 35.75 + 37.5 = 73.25 \text{ kN}$$

Referring to the global coordinates in Fig. 5.25b, the fixed end force vector is obtainable as:

$$\mathbf{F_f} = \begin{bmatrix} \mathbf{F_{fA}} \\ \hline \mathbf{F_{fR}} \end{bmatrix} = \begin{bmatrix} F_{1f} \\ \hline F_{2f} \\ F_{3f} \\ F_{4f} \\ F_{5f} \end{bmatrix} = \begin{bmatrix} 73.25 \text{ kN} \\ \hdashline 64.25 \text{ kN} \\ 172.5 \text{ kNm} \\ 62.5 \text{ kN} \\ -125 \text{ kNm} \end{bmatrix} \Rightarrow \mathbf{F_A} - \mathbf{F_{fA}} = \{0\} - \{73.25\} = \{-73.25 \text{ kN}\}$$

The resulting loading diagram is shown in Fig. 5.26b. The deflection D_1 resulting from this loading will be identical to the deflection in the original loading diagram shown in Fig. 5.25a.

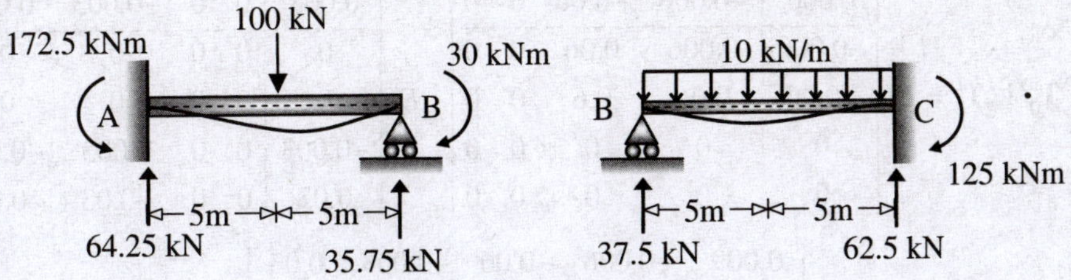

(*a*) Fixed end forces ($D_1 = 0$)

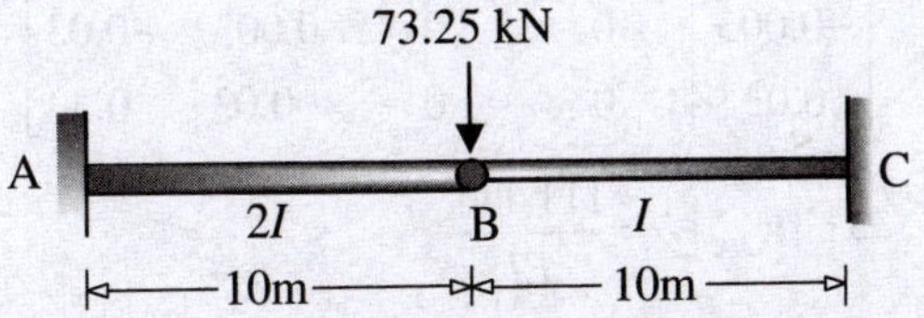

(*b*) Equivalent nodal loads

Figure 5.26 Fixed end force effects (Example 5.8)

Displacement Transformation Matrices

The element displacement transformation matrices, $\mathbf{T}_\mathbf{D}^1$ and $\mathbf{T}_\mathbf{D}^2$, relating the local coordinates to the global coordinates through the relation, $\mathbf{D}_*^i = \left[\mathbf{T}_{\mathbf{DA}}^i \mid \mathbf{T}_{\mathbf{DR}}^i\right]\begin{bmatrix}\mathbf{D}_\mathbf{A} \\ \hline \mathbf{D}_\mathbf{R}\end{bmatrix}$, take the following form (refer Figs 5.25b and c):

$$\mathbf{T}_\mathbf{D}^1 = \overset{(1)\quad(2)\quad(3)\;(4)\;(5)}{[-0.1 \mid 0.1 \quad 1 \quad 0 \quad 0]}\;;\quad \mathbf{T}_\mathbf{D}^2 = \overset{(1)\;(2)\;(3)\;(4)\;(5)}{[0.1 \mid 0 \quad 0 \quad -0.1 \quad 1]}$$

Element and Structure Stiffness Matrices

The element stiffness matrices $\tilde{\mathbf{k}}_*^1$ and $\tilde{\mathbf{k}}_*^2$, satisfying $\mathbf{F}_*^i = \tilde{\mathbf{k}}_*^i\mathbf{D}_*^i$, take the following form (refer Fig. 5.25c), considering $\frac{E_1 I_1}{L_1} = \frac{2EI}{10}$ and $\frac{E_2 I_2}{L_2} = \frac{EI}{10}$:

$$\tilde{\mathbf{k}}_*^1 = \frac{2EI}{10}[3] = EI[0.6] \quad \text{and} \quad \tilde{\mathbf{k}}_*^2 = \frac{EI}{10}[3] = EI[0.3]$$

$$\Rightarrow \quad \tilde{\mathbf{k}}_*^1\mathbf{T}_\mathbf{D}^1 = EI[-0.06 \mid 0.06 \quad 0.6 \quad 0 \quad 0];\; \tilde{\mathbf{k}}_*^2\mathbf{T}_\mathbf{D}^2 = EI[0.03 \mid 0 \quad 0 \quad -0.03 \quad 0.3]$$

The structure stiffness matrix $\mathbf{k}$, satisfying $\mathbf{F} = \mathbf{kD}$, can now be assembled:

$$\mathbf{k} = \sum_{i=1}^{2} \mathbf{T}_D^{i\,T} \tilde{\mathbf{k}}_*^i \mathbf{T}_D^i = EI \left[\begin{array}{c:cccc} 0.006 & -0.006 & -0.06 & 0 & 0 \\ \hdashline -0.006 & 0.006 & 0.06 & 0 & 0 \\ -0.06 & 0.06 & 0.6 & 0 & 0 \\ 0 & 0 & 0 & 0 & 0 \\ 0 & 0 & 0 & 0 & 0 \end{array}\right] + EI \left[\begin{array}{c:cccc} 0.003 & 0 & 0 & -0.003 & 0.03 \\ \hdashline 0 & 0 & 0 & 0 & 0 \\ 0 & 0 & 0 & 0 & 0 \\ -0.003 & 0 & 0 & 0.003 & -0.03 \\ 0.03 & 0 & 0 & -0.03 & 0.3 \end{array}\right]$$

$$\Rightarrow \quad \mathbf{k} = \left[\begin{array}{c|c} \mathbf{k}_{AA} & \mathbf{k}_{AR} \\ \hline \mathbf{k}_{RA} & \mathbf{k}_{RR} \end{array}\right] = EI \left[\begin{array}{c:cccc} 0.009 & -0.006 & -0.06 & -0.003 & 0.03 \\ \hdashline -0.006 & 0.006 & 0.06 & 0 & 0 \\ -0.06 & 0.06 & 0.6 & 0 & 0 \\ -0.003 & 0 & 0 & 0.003 & -0.03 \\ 0.03 & 0 & 0 & -0.03 & 0.3 \end{array}\right]$$

$$\Rightarrow \quad [\mathbf{k}_{AA}] = EI(0.009) \quad \Rightarrow \quad [\mathbf{k}_{AA}]^{-1} = \frac{111.111}{EI}$$

Displacements and Support Reactions

The structure stiffness relation to be considered here is:

$$\left[\begin{array}{c} \mathbf{F}_A \\ \hline \mathbf{F}_R \end{array}\right] - \left[\begin{array}{c} \mathbf{F}_{fA} \\ \hline \mathbf{F}_{fR} \end{array}\right] = \left[\begin{array}{c:c} \mathbf{k}_{AA} & \mathbf{k}_{AR} \\ \hdashline \mathbf{k}_{RA} & \mathbf{k}_{RR} \end{array}\right] \left[\begin{array}{c} \mathbf{D}_A \\ \hdashline \mathbf{D}_R = \mathbf{O} \end{array}\right]$$

$$\Rightarrow \mathbf{D}_A = [\mathbf{k}_{AA}]^{-1}[\mathbf{F}_A - \mathbf{F}_{fA}] \quad \Rightarrow D_1 = \frac{111.111}{EI}\{-73.25\} = \frac{-8138.881}{EI}$$

$$\Rightarrow \quad D_1 = \frac{-8138.881}{80000} = -101.736 \times 10^{-3} \text{ m}$$

$$\mathbf{F}_R = \mathbf{F}_{fR} + \mathbf{k}_{RA}\mathbf{D}_A \Rightarrow \begin{Bmatrix} F_2 \\ F_3 \\ F_4 \\ F_5 \end{Bmatrix} = \begin{Bmatrix} 64.25 \text{ kN} \\ 172.5 \text{ kNm} \\ 62.5 \text{ kN} \\ -125 \text{ kNm} \end{Bmatrix} + EI \begin{bmatrix} -0.006 \\ -0.06 \\ -0.003 \\ 0.03 \end{bmatrix} \frac{1}{EI}\{-8138.881\} = \begin{Bmatrix} 113.083 \text{ kN} \\ 660.833 \text{ kNm} \\ 86.917 \text{ kN} \\ -369.166 \text{ kNm} \end{Bmatrix}$$

These results match with the solutions obtained earlier in Example 5.4.

Member Forces

$$\mathbf{F}_*^i = \mathbf{F}_{*f}^i + \tilde{\mathbf{k}}_*^i \mathbf{T}_D^i \mathbf{D} \;\; ; \text{ where } \; \mathbf{D}^T = \frac{1}{EI}[-8138.881 \;\vdots\; 0 \;\; 0 \;\; 0 \;\; 0]$$

$$\Rightarrow \{F_{1*}^1\} = \{172.5 \text{ kNm}\} + EI[-0.06 \;\vdots\; 0.06 \;\; 0.6 \;\; 0 \;\; 0]\{D\} = \{660.833 \text{ kNm}\}$$

$$\Rightarrow \{F_{1*}^2\} = \{-125 \text{ kNm}\} + EI[0.03 \;\vdots\; 0 \;\; 0 \;\; -0.03 \;\; 0.3]\{D\} = \{-369.166 \text{ kNm}\}$$

The member end moments are exactly as obtained earlier. The shear force and bending moment diagrams are as shown earlier in Figs 5.15b and c.

5.3.3 Dealing with Guided-Fixed End Supports

We can take advantage of a reduction in the degree of kinematic indeterminacy when we encounter a guided-fixed end support, where there is a "shear release". We usually meet with this situation in the middle of a symmetric system of beams. At the guided-fixed end, there is usually no shear force, unless there is a concentrated external force (of known magnitude) applied here. Also, the bending moment at the guided-fixed support end can be determined by considering equilibrium of the free-body. In such cases, we can simply ignore the rotational degree of freedom in the beam element at the guided-fixed end. But we need to also modify the element stiffness matrix for this element as well as the fixed end forces. Eliminating a degree of freedom from the two degree of freedom element model effectively implies that we are dealing with a beam element with a single degree of freedom.

Consider a prismatic beam element AB, with a fixed end at A and a guided-fixed end at B. In the reduced element stiffness formulation, we consider only the rotational degree of freedom at the end A. Hence, the desired rotational stiffness is given by the moment generated at the fixed end A on account of a unit (anti-clockwise) rotational slip at A. This is depicted in Fig. 5.27a. The beam is subject to pure bending (zero shear), and it can be easily shown that the constant (hogging) moment generated has a magnitude, $(EI)_i/L_i$, as shown in Fig. 5.27b.

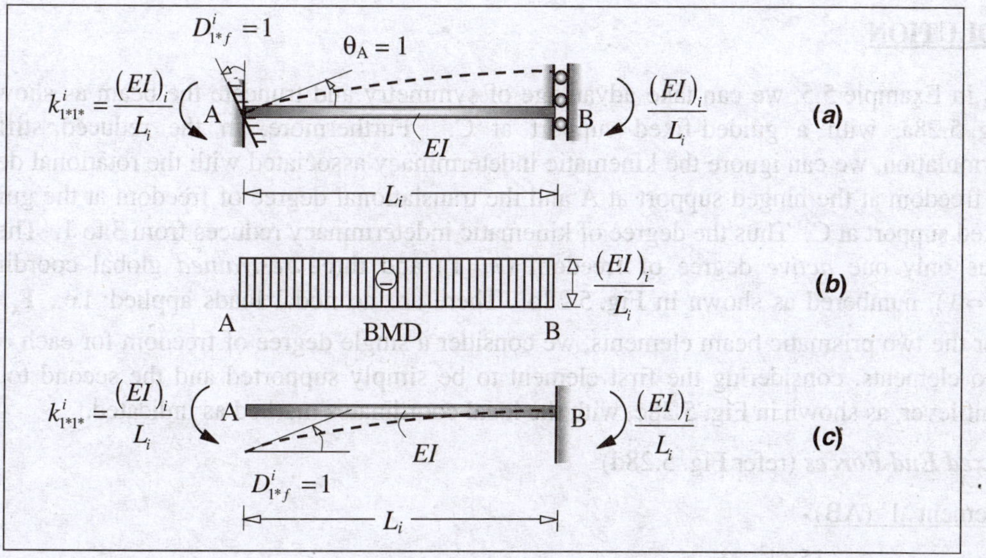

Figure 5.27 Generation of rotational stiffness for a prismatic beam element with a guided-fixed end support

Thus the reduced element stiffness matrix, of order 1×1, is given by:

$$\boxed{\tilde{\mathbf{k}}^{i}_{*} = \left[\frac{(EI)_i}{L_i}\right]} \tag{5.41}$$

The behaviour is identical to that of a prismatic cantilever beam AB (fixed at A), subject to an anti-clockwise end moment $(EI)_i/L_i$ at the free end A, causing a unit rotation at A. The relative translation between the ends A and B are of no consequence here, as there are no shear forces involved. For this reason, chord rotations in these beam elements do not have any associated flexural stiffnesses, and this fact should be accounted for, while assembling the displacement transformation matrix $\mathbf{T}^{i}_{D}$. In other words, the element in the $\mathbf{T}^{i}_{D}$ matrix, corresponding to a deflection in the structure (global coordinate), $D_j = 1$, will be zero (and not $\pm 1/L_i$, as in other cases).

Finally, care should be taken while evaluating the fixed end force effects in a beam with a guided-fixed end support. A convenient way of dealing with this situation is by considering the beam to have twice the span and subject to symmetric loading. This is demonstrated in Example 5.9.

EXAMPLE 5.9

Repeat Example 5.5 by the reduced stiffness method and find the force response.

SOLUTION

As in Example 5.5, we can take advantage of symmetry and truncate the beam as shown in Fig. 5.28a, with a guided-fixed support at C. Furthermore, in the reduced stiffness formulation, we can ignore the kinematic indeterminacy associated with the rotational degree of freedom at the hinged support at A and the translational degree of freedom at the guided-fixed support at C. Thus the degree of kinematic indeterminacy reduces from 3 to 1. There is thus only one *active* degree of freedom ($n = 1$) and three *restrained* global coordinates ($r = 3$), numbered as shown in Fig. 5.28b. There are no nodal loads applied; i.e., $\mathbf{F_A} = \mathbf{O}$. For the two prismatic beam elements, we consider a single degree of freedom for each of the two elements, considering the first element to be simply supported and the second to be a cantilever, as shown in Fig. 5.28c, with the local coordinates marked as indicated.

Fixed End Forces (refer Fig. 5.28d)

Element '1' (AB):

$$F^{1}_{1*f} = -\frac{15\times 4^2}{12}\times\frac{3}{2} = -30 \text{ kNm} \quad \Rightarrow \mathbf{F}^{1}_{*f} = \{-30 \text{ kNm}\}$$

Vertical reactions:

$$V_{AB} = \frac{(15)(4)}{2} - \frac{30}{4} = 22.5 \text{ kN}; \; V_{BA} = \frac{(15)(4)}{2} + \frac{30}{4} = 37.5 \text{ kN}$$

Element '2' (BC):

To find the fixed end moments, it is convenient to consider the full fixed beam BCB′ of span 4m, subject to two symmetrically placed loads of 80 kN:

$$F^2_{1*f} = \frac{(80)(0.5)(3.5)}{4^2}(3.5+0.5) = 35 \text{ kNm}; \; M_{CB} = (80)(0.5) - 35 = 5 \text{ kNm (sagging)}$$

$$\Rightarrow \mathbf{F}^2_{*f} = \{35 \text{ kNm}\}$$

Vertical reactions: $V_{BC} = 80$ kN; $V_{CB} = 0$ kN

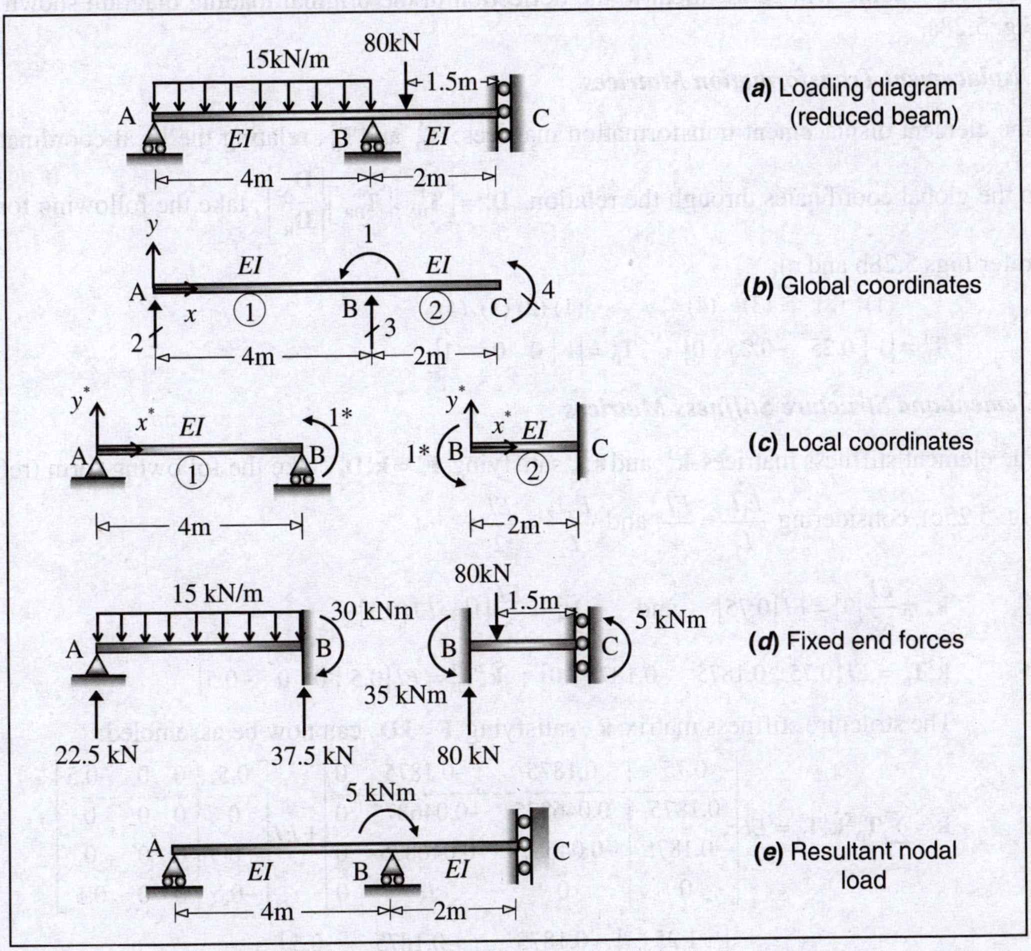

Figure 5.28 Loading diagram, coordinates and fixed end forces (Example 5.9)

The fixed end vertical force at the active global coordinate is given by

$$F_{1f} = F^1_{1*} + F^2_{1*} = -30 + 35 = 5 \text{ kNm}$$

Referring to the global coordinates in Fig. 5.28b, we observe that the fixed end force vector is obtainable as:

$$\mathbf{F_f} = \left[\begin{array}{c}\mathbf{F_{fA}} \\ \hline \mathbf{F_{fR}}\end{array}\right] = \left[\begin{array}{c}F_{1f} \\ \hline F_{2f} \\ F_{3f} \\ F_{4f}\end{array}\right] = \left[\begin{array}{c}5 \text{ kNm} \\ \hline 22.5 \text{ kN} \\ 117.5 \text{ kN} \\ 5 \text{ kNm}\end{array}\right] \Rightarrow \mathbf{F_A} - \mathbf{F_{fA}} = \{0\} - \{5\} = \{-5 \text{ kNm}\}$$

The resulting loading diagram is shown in Fig. 5.28e. The deflection D_1 resulting from this loading will be identical to the deflection in the original loading diagram shown in Fig. 5.28a.

Displacement Transformation Matrices

The element displacement transformation matrices, $\mathbf{T_D^1}$ and $\mathbf{T_D^2}$, relating the local coordinates to the global coordinates through the relation, $\mathbf{D_*^i} = \left[\mathbf{T_{DA}^i} \mid \mathbf{T_{DR}^i}\right]\left[\begin{array}{c}\mathbf{D_A} \\ \hline \mathbf{D_R}\end{array}\right]$, take the following form (refer Figs 5.28b and c):

$$\begin{matrix} & (1) & (2) & (3) & (4) \\ \mathbf{T_D^1} = [1 & \mid 0.25 & -0.25 & 0] \end{matrix} \quad ; \quad \begin{matrix} & (1) & (2) & (3) & (4) \\ \mathbf{T_D^2} = [1 & \mid 0 & 0 & -1] \end{matrix}$$

Element and Structure Stiffness Matrices

The element stiffness matrices $\tilde{\mathbf{k}}_*^1$ and $\tilde{\mathbf{k}}_*^2$, satisfying $\mathbf{F_*^i} = \tilde{\mathbf{k}}_*^i\mathbf{D_*^i}$, take the following form (refer Fig. 5.25c), considering $\frac{E_1I_1}{L_1} = \frac{EI}{4}$ and $\frac{E_2I_2}{L_2} = \frac{EI}{2}$:

$$\tilde{\mathbf{k}}_*^1 = \frac{EI}{4}[3] = EI[0.75] \quad \text{and} \quad \tilde{\mathbf{k}}_*^2 = \frac{EI}{2}[1] = EI[0.5]$$

$$\Rightarrow \quad \tilde{\mathbf{k}}_*^1\mathbf{T_D^1} = EI[0.75 \mid 0.1875 \quad -0.1875 \quad 0] \; ; \; \tilde{\mathbf{k}}_*^2\mathbf{T_D^2} = EI[0.5 \mid 0 \quad 0 \quad -0.5]$$

The structure stiffness matrix $\mathbf{k}$, satisfying $\mathbf{F} = \mathbf{kD}$, can now be assembled:

$$\mathbf{k} = \sum_{i=1}^{2} \mathbf{T_D^i}^T \tilde{\mathbf{k}}_*^i \mathbf{T_D^i} = EI\left[\begin{array}{c|ccc} 0.75 & 0.1875 & -0.1875 & 0 \\ \hline 0.1875 & 0.046875 & -0.046875 & 0 \\ -0.1875 & -0.046875 & 0.046875 & 0 \\ 0 & 0 & 0 & 0 \end{array}\right] + EI\left[\begin{array}{c|ccc} 0.5 & 0 & 0 & -0.5 \\ \hline 0 & 0 & 0 & 0 \\ 0 & 0 & 0 & 0 \\ -0.5 & 0 & 0 & 0.5 \end{array}\right]$$

$$\Rightarrow \quad \mathbf{k} = \left[\begin{array}{c|c}\mathbf{k_{AA}} & \mathbf{k_{AR}} \\ \hline \mathbf{k_{RA}} & \mathbf{k_{RR}}\end{array}\right] = EI\left[\begin{array}{c|ccc} 1.25 & 0.1875 & -0.1875 & -0.5 \\ \hline 0.1875 & 0.046875 & -0.046875 & 0 \\ -0.1875 & -0.046875 & 0.046875 & 0 \\ -0.5 & 0 & 0 & 0.5 \end{array}\right]$$

$$\Rightarrow \quad [\mathbf{k}_{AA}] = EI(1.25) \quad \Rightarrow \quad [\mathbf{k}_{AA}]^{-1} = \frac{0.8}{EI}$$

Displacements and Support Reactions

The structure stiffness relation to be considered here is:

$$\begin{bmatrix} \mathbf{F}_A \\ \mathbf{F}_R \end{bmatrix} - \begin{bmatrix} \mathbf{F}_{fA} \\ \mathbf{F}_{fR} \end{bmatrix} = \left[\begin{array}{c|c} \mathbf{k}_{AA} & \mathbf{k}_{AR} \\ \hline \mathbf{k}_{RA} & \mathbf{k}_{RR} \end{array}\right] \begin{bmatrix} \mathbf{D}_A \\ \mathbf{D}_R = \mathbf{O} \end{bmatrix}$$

$$\Rightarrow \mathbf{D}_A = [\mathbf{k}_{AA}]^{-1}[\mathbf{F}_A - \mathbf{F}_{fA}] \quad \Rightarrow \quad D_1 = \frac{0.8}{EI}\{-5\} = \frac{-4}{EI}$$

$$\mathbf{F}_R = \mathbf{F}_{fR} + \mathbf{k}_{RA}\mathbf{D}_A \quad \Rightarrow \quad \begin{Bmatrix} F_2 \\ F_3 \\ F_4 \end{Bmatrix} = \begin{Bmatrix} 22.5 \text{ kN} \\ 117.5 \text{ kN} \\ 5 \text{ kNm} \end{Bmatrix} + EI \begin{bmatrix} 0.1875 \\ -0.1875 \\ -0.5 \end{bmatrix} \frac{1}{EI}\{-4\} = \begin{Bmatrix} 21.75 \text{ kN} \\ 118.25 \text{ kN} \\ 7 \text{ kNm} \end{Bmatrix}$$

These results match with the solutions obtained earlier in Example 5.5.

Member Forces

$$\mathbf{F}_*^i = \mathbf{F}_{*f}^i + \tilde{\mathbf{k}}_*^i \mathbf{T}_D^i \mathbf{D} \text{ ; where } \mathbf{D}^T = \frac{1}{EI}[-4 \mid 0 \quad 0 \quad 0]$$

$$\Rightarrow \{F_{1*}^1\} = \{-30 \text{ kNm}\} + EI[0.75 \mid 0.1875 \quad -0.1875 \quad 0]\{D\} = \{-33 \text{ kNm}\}$$

$$\Rightarrow \{F_{1*}^2\} = \{35 \text{ kNm}\} + EI[0.5 \mid 0 \quad 0 \quad -0.5]\{D\} = \{33 \text{ kNm}\}$$

The member end moments are exactly as obtained earlier. The bending moment diagram is as shown earlier in Figs 5.18b.

5.4 FLEXIBILITY METHOD APPLIED TO BEAMS

The basic procedure underlying the flexibility method has already been introduced in Chapter 3, and demonstrated in Chapter 4 for axial elements and trusses. Here, we demonstrate the application of this method to continuous beam systems.

The loads acting on beams, unlike trusses, generally do not act at the joint locations, and invariably involve intermediate loads on the beam elements. In order to deal with the effects of such loads within the framework of matrix methods (where the global coordinates are identified at the joint locations), it is necessary to invoke the concept of *equivalent joint loads*, which is a concept that is borrowed from the stiffness method. This means that we need to know the *fixed end moments* in the various beam elements[†] in order to carry out matrix analysis by the flexibility method. In general, it is accepted that the flexibility method cannot be automated efficiently to the same extent as the stiffness method in the analysis of

[†] This may appear to be somewhat ironical in a force method framework, because the 'fixed beam' is supposed to be *statically indeterminate*, although it may be *kinematically* determinate!

beams and frames, and hence some amount of manual intervention is involved. It is not considered worthwhile to introduce additional elements to model support reactions and support movements (at non-redundant locations) in beams and frames, as this would add to the complexity of the analysis and takes away the elegance in the formulation. Support reactions can easily be computed from the final free-body diagrams and support movements at non-redundant locations can be translated into equivalent displacements at the redundant coordinates by considering their rigid body displacement effects on the primary structure.

It is important to bear in mind that all the basic formulations underlying the flexibility method (including the *force transformation matrix*) pertain to the primary structure, which is statically determinate, and on which the force and displacement loading and response are superimposed. The redundant forces are also treated as external loads (of unknown magnitude) acting on the primary structure at the redundant coordinates, along with the resultant nodal loads at the active global coordinates.

5.4.1 Force Transformation Matrix

We need to consider active global coordinates at the *node* locations on the structure, where nodal loads are applicable or where we may be interested in finding displacements. These are usually located at the joints, where there are intermediate or end supports, but may involve intermediate points, such as internal hinge locations (and occasionally, special locations where we may be interested in finding deflections or slopes). At every node location, there are potentially two global coordinates (degrees of freedom, corresponding to deflection and rotation), of which one or both may be restrained. Thus the identification and numbering of the *active* global coordinates and the *redundant* coordinates can be sequentially carried out (with the redundant coordinate numbers at the end of the list). The beam elements, between successive nodes, may also be numbered sequentially, with two rotational degrees of freedom (labelled, 1* and 2*, in local coordinates) at the ends of each element (refer Fig. 5.20).

Consider, for example, a typical continuous beam system, with *m* beam elements and a degree of static indeterminacy, n_s. Let there be *n* active global coordinates. Each beam element *i* has a force vector, $\mathbf{F}_*^i = \begin{Bmatrix} F_{1*}^i \\ F_{2*}^i \end{Bmatrix}$, comprising the two end moments (anti-clockwise positive). The *combined element force vector*, $\mathbf{F}_*$, contains all the *m* beam element vectors, and hence has a dimension, 2*m*. The components of $\mathbf{F}_*$ define the internal forces in the system, knowing which the other unknown forces, including shear forces and support reactions, as well as the shear force and bending moment diagrams, can be determined, satisfying static equilibrium.

The *external* force vector comprises forces acting at the active global coordinates, as well as the forces acting at the redundant coordinates, and hence, has a dimension, $n + n_s$. After finding the fixed end forces, $\mathbf{F}_{*f}$, in the various fixed beam elements, under the action of intermediate loads, the *resultant force vector*, $\mathbf{F}_{net} = \begin{Bmatrix} \mathbf{F}_{\mathbf{A},net} \\ \hline \mathbf{F}_{\mathbf{X},net} \end{Bmatrix} = \begin{Bmatrix} \mathbf{F}_{\mathbf{A}} - \mathbf{F}_{\mathbf{fA}} \\ \hline \mathbf{F}_{\mathbf{X}} - \mathbf{F}_{\mathbf{fX}} \end{Bmatrix}$, can be defined, in which the only unknown quantities are the redundant forces, $\mathbf{F}_{\mathbf{X}}$.

The internal force vector, $\mathbf{F}_*$, is linearly related to the resultant force vector, $\mathbf{F}_{net}$, by means of the force transformation matrix, $\mathbf{T}_{\mathbf{F}} = [\mathbf{T}_{\mathbf{FA}} \mid \mathbf{T}_{\mathbf{FX}}]$. The elements in this matrix will depend on the choice of the redundants (i.e., on the primary structure), and can be generated by applying a unit load, $F_j = 1$, at a time on the primary structure, from simple considerations of static equilibrium.

$$\mathbf{F}_* = \mathbf{F}_{*\mathbf{f}} + [\mathbf{T}_{\mathbf{FA}} \mid \mathbf{T}_{\mathbf{FX}}] \begin{Bmatrix} \mathbf{F}_{\mathbf{A},net} \\ \hline \mathbf{F}_{\mathbf{X},net} \end{Bmatrix} \tag{5.42}$$

The sub-matrix, $\mathbf{T}_{\mathbf{FA}}$, has an order $2m \times n$, while, $\mathbf{T}_{\mathbf{FX}}$ has an order $2m \times n_s$. As the primary structure is statically determinate, the full $\mathbf{T}_{\mathbf{F}} = [\mathbf{T}_{\mathbf{FA}} \mid \mathbf{T}_{\mathbf{FX}}]$ matrix will turn out to be a square matrix, of order, $2m \times 2m$; i.e., the condition, $n + n_s = 2m$, will be satisfied.

5.4.2 Flexibility Matrix

The *element flexibility matrix*, $\mathbf{f}_*^{\mathbf{i}}$, for a prismatic beam element, with two rotational degrees of freedom, satisfying the relation, $\mathbf{D}_*^{\mathbf{i}} = \mathbf{f}_*^{\mathbf{i}} \mathbf{F}_*^{\mathbf{i}}$, has already been introduced in Section 5.3.1 (refer Fig. 5.20):

$$\boxed{\mathbf{f}_*^{\mathbf{i}} = \frac{L_i}{6(EI)_i} \begin{bmatrix} 2 & -1 \\ -1 & 2 \end{bmatrix}} \tag{5.43}$$

The *combined element force vector*, $\{F_*\}$ and the conjugate *combined element displacement vector*, $\{D_*\}$, are inter-related by the ***unassembled element flexibility matrix***, $\mathbf{f}_*$, as explained in Section 3.5. This matrix is a block diagonal matrix of order $2m \times 2m$, for a continuous beam system with *m* prismatic elements.

$$\mathbf{D}_* = \mathbf{f}_* \mathbf{F}_* \Rightarrow \begin{Bmatrix} \mathbf{D}_*^1 \\ \mathbf{D}_*^2 \\ \vdots \\ \mathbf{D}_*^{\mathbf{m}} \end{Bmatrix} = \begin{bmatrix} \mathbf{f}_*^1 & \mathbf{O} & \mathbf{O} & \mathbf{O} \\ \mathbf{O} & \mathbf{f}_*^2 & \mathbf{O} & \mathbf{O} \\ \vdots & \vdots & \ddots & \vdots \\ \mathbf{O} & \mathbf{O} & \mathbf{O} & \mathbf{f}_*^{\mathbf{m}} \end{bmatrix} \begin{Bmatrix} \mathbf{F}_*^1 \\ \mathbf{F}_*^2 \\ \vdots \\ \mathbf{F}_*^{\mathbf{m}} \end{Bmatrix} \tag{5.44}$$

The full structure flexibility matrix $\mathbf{f}$, of order $2m \times 2m$, can be generated from the unassembled element flexibility matrix, $\mathbf{f}_*$, by involving the force transformation matrix, $\mathbf{T}_{\mathbf{F}}$, as described earlier.

$$\mathbf{f} = \mathbf{T_F}^T \mathbf{f_*} \mathbf{T_F} = \left[\begin{array}{c|c} \mathbf{f_{AA}} & \mathbf{f_{AX}} \\ \hline \mathbf{f_{XA}} & \mathbf{f_{XX}} \end{array}\right] \tag{5.45}$$

where, the complete force displacement relations are as described below:

$$\left[\begin{array}{c} \mathbf{D_A} \\ \hline \mathbf{D_X} \end{array}\right] = \left[\begin{array}{c} \mathbf{D_{A,initial}} \\ \hline \mathbf{D_{X,initial}} \end{array}\right] + \left[\begin{array}{c|c} \mathbf{f_{AA}} & \mathbf{f_{AX}} \\ \hline \mathbf{f_{XA}} & \mathbf{f_{XX}} \end{array}\right] \left[\begin{array}{c} \mathbf{F_A} - \mathbf{F_{fA}} \\ \hline \mathbf{F_X} - \mathbf{F_{fX}} \end{array}\right] \tag{5.46}$$

The redundants can be determined by solving the second set of (compatibility) equations, pertaining to $\mathbf{D_X}$:

$$\mathbf{F_X} = [\mathbf{f_{XX}}]^{-1}\left[(\mathbf{D_X} - \mathbf{D_{X,initial}}) - \mathbf{f_{XA}}(\mathbf{F_A} - \mathbf{F_{fA}})\right] + \mathbf{F_{fX}} \tag{5.47}$$

After finding the redundants, $\mathbf{F_X}$, the internal force vector, $\mathbf{F_*}$, can be determined by invoking Eq. 5.42, and from the free-bodies, the desired support reactions, shear force and bending moment diagrams can be generated.

Finally, the unknown displacements at the active coordinates, $\mathbf{D_A}$, can be obtained, if required, by solving the first set of equations in Eq. 5.46:

$$\mathbf{D_A} = \mathbf{D_{A,initial}} + \mathbf{f_{AA}}\mathbf{F}_{\mathbf{A},net} + \mathbf{f_{AX}}\mathbf{F}_{\mathbf{X},net} \tag{5.48}$$

5.4.3 Analysis Procedure

The analysis of continuous beam systems by the flexibility method involves the following steps:

(i) identifying the degree of static indeterminacy, selection of the redundants ($X_1, X_2, ...$); labelling the *global coordinates* (first, the *active* coordinates and then, the *redundant* coordinates) on the *primary* (statically determinate) structure;

(ii) identifying the known displacements, in terms of the displacements, $\mathbf{D_X}$, at the redundant coordinates and the initial displacement vector, $\mathbf{D_{iniital}} = \left\{\begin{array}{c} \mathbf{D_{A,iniital}} \\ \hline \mathbf{D_{X,iniital}} \end{array}\right\}$ (on account of support displacements, if any, at the non-redundant coordinates acting on the primary structure);

(iii) identifying the given nodal loads, $\mathbf{F_A}$, and finding the fixed end forces in the individual elements, $\mathbf{F_{*f}}$, and on the structure, $\mathbf{F_f} = \left\{\begin{array}{c} \mathbf{F_{fA}} \\ \hline \mathbf{F_{fX}} \end{array}\right\}$, due to intermediate loads, if any, and thus, the net force vector, $\mathbf{F}_{net} = \left\{\begin{array}{c} \mathbf{F_A} - \mathbf{F_{fA}} \\ \hline \mathbf{F_X} - \mathbf{F_{fX}} \end{array}\right\}$, on the primary structure;

(iv) setting up the force transformation matrix, $\mathbf{T_F} = [\mathbf{T_{FA}} \mid \mathbf{T_{FX}}]$, applying a unit load, $F_j = 1$, at a time on the primary structure, and satisfying equilibrium (drawing free-bodies makes this convenient);

(v) setting up the unassembled element flexibility matrix, $\mathbf{f_*}$, and generating the structure flexibility matrix, $\mathbf{f} = \mathbf{T_F}^T \mathbf{f_*} \mathbf{T_F} = \left[\begin{array}{c|c} \mathbf{f_{AA}} & \mathbf{f_{AX}} \\ \hline \mathbf{f_{XA}} & \mathbf{f_{XX}} \end{array}\right]$, and finding the inverse of the sub-matrix, $\mathbf{f_{XX}}$;

(vi) finding the unknown redundants by solving compatibility equations, $\mathbf{F_X} = [\mathbf{f_{XX}}]^{-1}\left[(\mathbf{D_X} - \mathbf{D_{X,initial}}) - \mathbf{f_{XA}}\mathbf{F_{A,\mathit{net}}}\right] + \mathbf{F_{fX}}$;

(vii) finding the beam end forces, $\mathbf{F_*} = \mathbf{F_{*f}} + [\mathbf{T_{FA}} \mid \mathbf{T_{FX}}]\left\{\begin{array}{c} \mathbf{F_{A,\mathit{net}}} \\ \hline \mathbf{F_{X,\mathit{net}}} \end{array}\right\}$, and from the free-bodies, finding the support reactions, and drawing the shear force and bending moment diagrams;

(viii) and, if required, finding the displacement response at the active coordinate locations, $\mathbf{D_A} = \mathbf{D_{A,initial}} + \mathbf{f_{AA}}\mathbf{F_{A,\mathit{net}}} + \mathbf{f_{AX}}\mathbf{F_{X,\mathit{net}}}$.

Let us now demonstrate the application of the flexibility method through a few of the Examples we had used to demonstrate the Stiffness Method.

EXAMPLE 5.10

Repeat Example 5.1/5.6 (non-prismatic fixed beam), solving by the Flexibility Method.

SOLUTION

The beam is statically indeterminate to the second degree (ignoring the horizontal reaction). Let us choose the vertical reaction, V_B, and moment M_B, at the support B as the redundants, X_1 (acting upward) and X_2 (anti-clockwise) respectively, as shown in Fig. 5.29. The primary structure, therefore, is a cantilever beam, fixed at A.

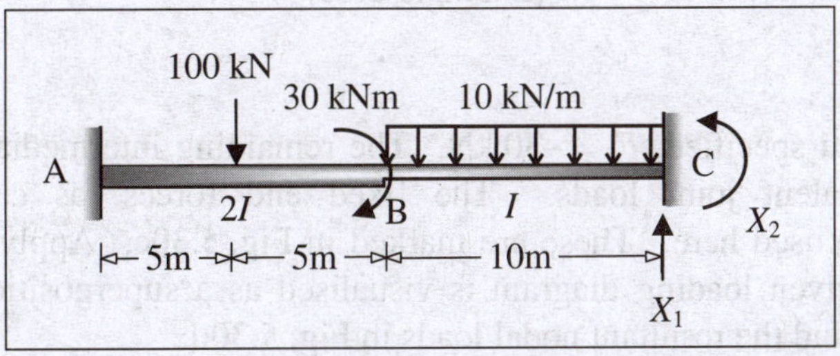

Figure 5.29 Non-prismatic fixed beam (Example 5.10)

There are two prismatic beam elements, and only two active global coordinates. The global coordinates are marked in Fig. 5.30a, and the local coordinates for the two prismatic beam elements (with two rotational degrees of freedom each) are marked in Fig. 5.30b.

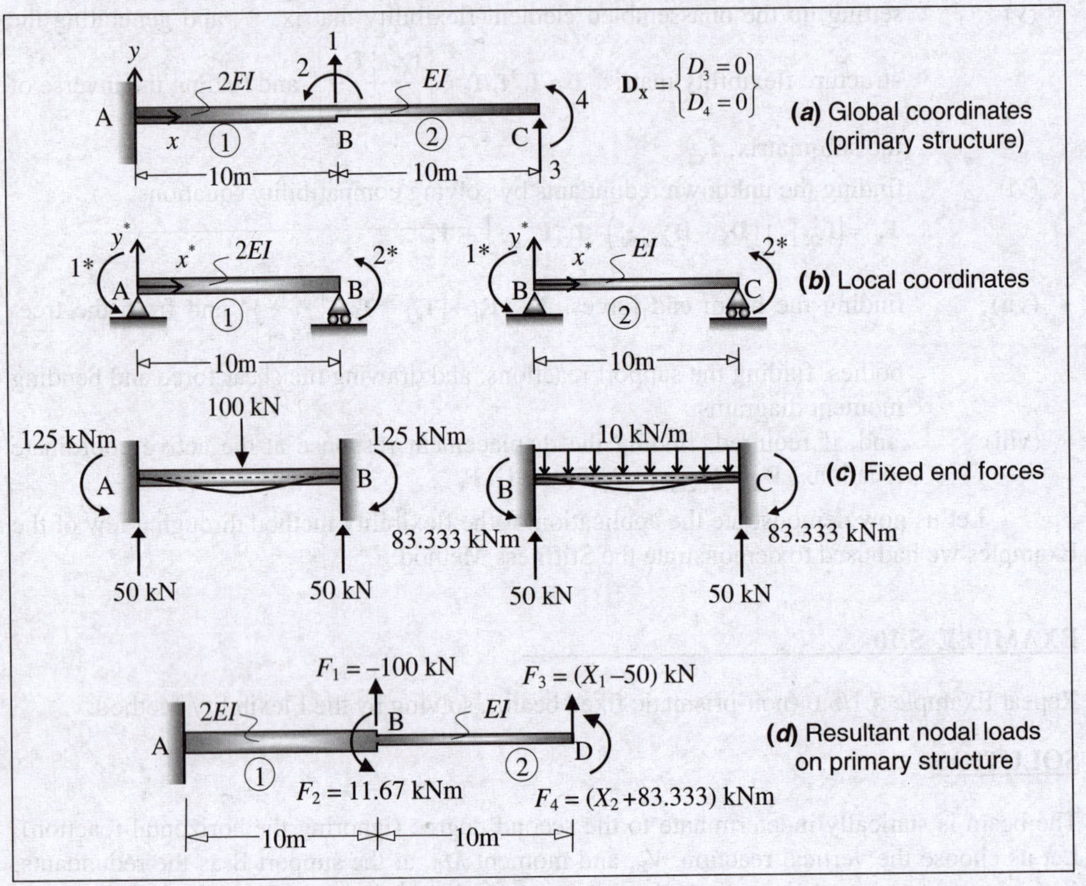

Figure 5.30 Global and local coordinates, fixed end forces and resultant loads (Example 5.10)

Fixed End Forces

There is a nodal load specified, $F_2 = -30$ kN. The remaining intermediate loads need to be converted to equivalent joint loads. The fixed end forces, as calculated earlier in Example 5.6, can be used here. These are marked in Fig. 5.30c. Applying the Principle of Superposition, the given loading diagram is visualised as a superposition of the fixed end forces in Fig. 5.30c and the resultant nodal loads in Fig. 5.30d.

$$\Rightarrow \mathbf{F}_{*f} = \begin{bmatrix} \begin{Bmatrix} F^1_{1*f} \\ F^1_{2*f} \end{Bmatrix} \\ \begin{Bmatrix} F^2_{1*f} \\ F^2_{2*f} \end{Bmatrix} \end{bmatrix} = \begin{bmatrix} \begin{Bmatrix} 125 \\ -125 \end{Bmatrix} \\ \begin{Bmatrix} 83.333 \\ -83.333 \end{Bmatrix} \end{bmatrix} \text{kNm} \Rightarrow \mathbf{F_f} = \begin{bmatrix} \mathbf{F_{fA}} \\ \mathbf{F_{fX}} \end{bmatrix} = \begin{bmatrix} \begin{Bmatrix} F_{1f} \\ F_{2f} \end{Bmatrix} \\ \begin{Bmatrix} F_{3f} \\ F_{4f} \end{Bmatrix} \end{bmatrix} = \begin{bmatrix} \begin{Bmatrix} 100 \text{ kN} \\ -41.667 \text{ kNm} \end{Bmatrix} \\ \begin{Bmatrix} 50 \text{ kN} \\ -83.333 \text{ kNm} \end{Bmatrix} \end{bmatrix}$$

The resultant load vector, $\mathbf{F}_{net}$, on the primary structure (shown in Fig. 5.30d) is given by:

$$\mathbf{F}_{net} = \begin{bmatrix} \mathbf{F_A} - \mathbf{F_{fA}} \\ \mathbf{F_X} - \mathbf{F_{fX}} \end{bmatrix} = \begin{bmatrix} \begin{Bmatrix} 0 \text{ kN} \\ -30 \text{ kNm} \end{Bmatrix} \\ \begin{Bmatrix} X_1 \\ X_2 \end{Bmatrix} \end{bmatrix} - \begin{bmatrix} \begin{Bmatrix} 100 \text{ kN} \\ -41.667 \text{ kNm} \end{Bmatrix} \\ \begin{Bmatrix} 50 \text{ kN} \\ -83.333 \text{ kNm} \end{Bmatrix} \end{bmatrix} = \begin{bmatrix} \begin{Bmatrix} -100 \text{ kN} \\ 11.667 \text{ kNm} \end{Bmatrix} \\ \begin{Bmatrix} (X_1 - 50) \text{ kN} \\ (X_2 + 83.333) \text{ kNm} \end{Bmatrix} \end{bmatrix}$$

Force Transformation Matrix

The force transformation matrix can be easily generated by analysing the primary structure under separate unit load conditions, as shown in Fig. 5.31.

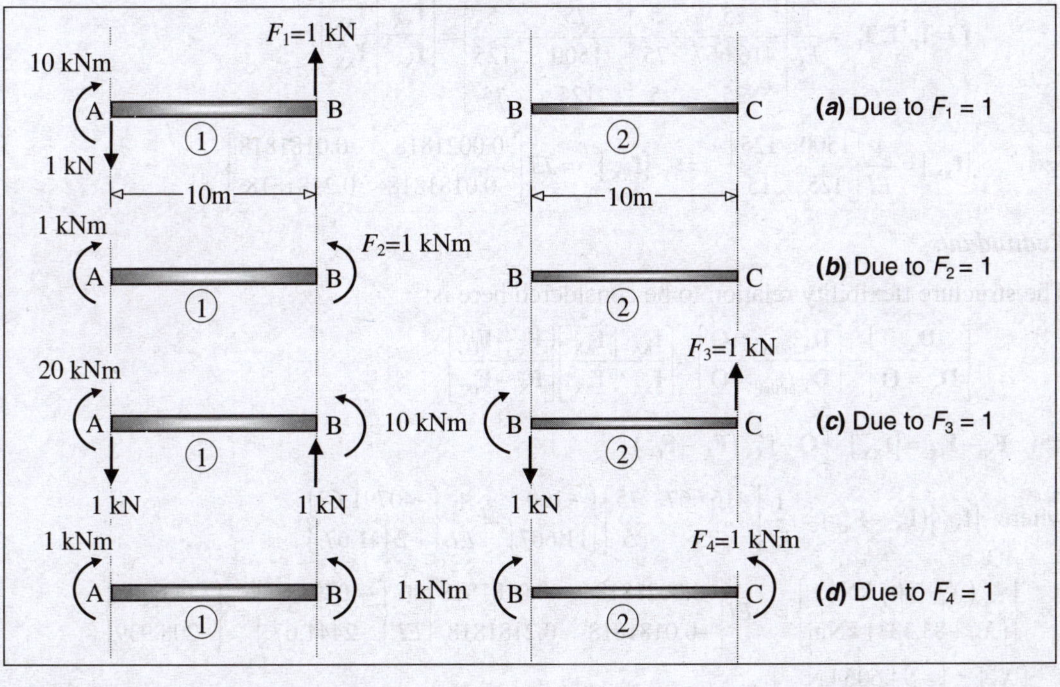

Figure 5.31 Response of primary structure to unit loads – free-bodies (Example 5.10).

Accordingly, $\mathbf{F}_* = \left[\mathbf{T}_{FA} \mid \mathbf{T}_{FX}\right]\begin{Bmatrix}\mathbf{F}_{A,net} \\ \hline \mathbf{F}_{X,net}\end{Bmatrix} \Rightarrow \begin{bmatrix}\{F_*^1\} \\ \{F_*^2\}\end{bmatrix} = \begin{bmatrix}\begin{Bmatrix}F_{1*}^1 \\ F_{2*}^1\end{Bmatrix} \\ \begin{Bmatrix}F_{1*}^2 \\ F_{2*}^2\end{Bmatrix}\end{bmatrix} = \left[\begin{array}{cc|cc}-10 & -1 & -20 & -1 \\ 0 & 1 & 10 & 1 \\ 0 & 0 & -10 & -1 \\ 0 & 0 & 0 & 1\end{array}\right]\begin{bmatrix}\begin{Bmatrix}F_1 \\ F_2\end{Bmatrix} \\ \hline \begin{Bmatrix}F_3 \\ F_4\end{Bmatrix}\end{bmatrix}$

Flexibility Matrix

The unassembled element flexibility matrix, satisfying $\mathbf{D}_* = \mathbf{f}_*\mathbf{F}_*$, may be generated as follows:

$$\begin{bmatrix}\begin{Bmatrix}D_{1*}^1 \\ D_{2*}^1\end{Bmatrix} \\ \begin{Bmatrix}D_{1*}^2 \\ D_{2*}^2\end{Bmatrix}\end{bmatrix} = \begin{bmatrix}\dfrac{10}{6(2EI)}\begin{bmatrix}2 & -1 \\ -1 & 2\end{bmatrix} & \mathbf{O} \\ \mathbf{O} & \dfrac{10}{6(EI)}\begin{bmatrix}2 & -1 \\ -1 & 2\end{bmatrix}\end{bmatrix}\begin{bmatrix}\begin{Bmatrix}F_{1*}^1 \\ F_{2*}^1\end{Bmatrix} \\ \begin{Bmatrix}F_{1*}^2 \\ F_{2*}^2\end{Bmatrix}\end{bmatrix}$$

The structure flexibility matrix $\mathbf{f}$, satisfying $\mathbf{D} = \mathbf{f}\ \mathbf{F}$, can now be assembled:

$$\mathbf{f} = \mathbf{T_F}^T\mathbf{f}_*\mathbf{T_F} = \frac{1}{EI}\left[\begin{array}{cc|cc}166.667 & 25 & 416.667 & 25 \\ 25 & 5 & 75 & 5 \\ \hline 416.667 & 75 & 1500 & 125 \\ 25 & 5 & 125 & 15\end{array}\right] = \left[\begin{array}{c|c}\mathbf{f_{AA}} & \mathbf{f_{AX}} \\ \hline \mathbf{f_{XA}} & \mathbf{f_{XX}}\end{array}\right]$$

$$\Rightarrow \quad [\mathbf{f_{XX}}] = \frac{1}{EI}\begin{bmatrix}1500 & 125 \\ 125 & 15\end{bmatrix} \quad \Rightarrow \quad [\mathbf{f_{XX}}]^{-1} = EI\begin{bmatrix}0.0021818 & -0.0181818 \\ -0.0181818 & 0.2181818\end{bmatrix}$$

Redundants

The structure flexibility relation to be considered here is:

$$\begin{bmatrix}\mathbf{D_A} \\ \hline \mathbf{D_X} = \mathbf{O}\end{bmatrix} = \begin{bmatrix}\mathbf{D_{A,initial}} = \mathbf{O} \\ \hline \mathbf{D_{X,initial}} = \mathbf{O}\end{bmatrix} + \left[\begin{array}{c|c}\mathbf{f_{AA}} & \mathbf{f_{AX}} \\ \hline \mathbf{f_{XA}} & \mathbf{f_{XX}}\end{array}\right]\begin{bmatrix}\mathbf{F_A} - \mathbf{F_{fA}} \\ \hline \mathbf{F_X} - \mathbf{F_{fX}}\end{bmatrix}$$

$$\Rightarrow \quad \mathbf{F_X} - \mathbf{F_{fX}} = [\mathbf{f_{XX}}]^{-1}\left[\mathbf{O} - \mathbf{f_{XA}}(\mathbf{F_A} - \mathbf{F_{fA}})\right]$$

where $[\mathbf{f_{XA}}](\mathbf{F_A} - \mathbf{F_{Af}}) = \dfrac{1}{EI}\begin{bmatrix}416.667 & 75 \\ 25 & 5\end{bmatrix}\begin{Bmatrix}-100 \\ 11.667\end{Bmatrix} = \dfrac{1}{EI}\begin{Bmatrix}-40791.67 \\ -2441.67\end{Bmatrix}$

$$\Rightarrow \begin{Bmatrix}(X_1 - 50)\ \text{kN} \\ (X_2 + 83.333)\ \text{kNm}\end{Bmatrix} = -EI\begin{bmatrix}0.0021818 & -0.0181818 \\ -0.0181818 & 0.2181818\end{bmatrix}\frac{1}{EI}\begin{Bmatrix}-40791.67 \\ -2441.67\end{Bmatrix} = \begin{Bmatrix}44.606 \\ -208.939\end{Bmatrix}$$

$$\Rightarrow \begin{Bmatrix}X_1 \\ X_2\end{Bmatrix} = \begin{Bmatrix}94.606\ \text{kN} \\ -292.272\ \text{kNm}\end{Bmatrix}$$

Member Forces

$$\mathbf{F}_* = \mathbf{F}_{*\mathbf{f}} + \mathbf{T_F}\mathbf{F}_{net}$$

$$\Rightarrow \mathbf{F}_* = \begin{bmatrix} \begin{Bmatrix} F_{1*}^1 \\ F_{2*}^1 \end{Bmatrix} \\ \begin{Bmatrix} F_{1*}^2 \\ F_{2*}^2 \end{Bmatrix} \end{bmatrix} = \begin{bmatrix} \begin{Bmatrix} 125 \\ -125 \end{Bmatrix} \\ \begin{Bmatrix} 83.333 \\ -83.333 \end{Bmatrix} \end{bmatrix} \text{kNm} + \left[\begin{array}{cc|cc} -10 & -1 & -20 & -1 \\ 0 & 1 & 10 & 1 \\ 0 & 0 & -10 & -1 \\ 0 & 0 & 0 & 1 \end{array}\right] \left\{\begin{array}{c} F_1 = -100 \\ F_2 = 11.667 \\ \hline F_3 = 44.606 \\ F_4 = -208.939 \end{array}\right\} = \begin{bmatrix} \begin{Bmatrix} 430.152 \\ 123.788 \end{Bmatrix} \\ \begin{Bmatrix} -153.788 \\ -292.272 \end{Bmatrix} \end{bmatrix} \text{kNm}$$

The member end moments are exactly as obtained earlier in Examples 5.1 and 5.6. The free-bodies of the two beam elements can now be drawn and the shear force and bending moment diagrams generated, exactly as shown earlier in Figs 5.22a, b and c. The support reactions are also easily obtainable from the free-bodies.

Displacements

$$\mathbf{D}_{\mathrm{A}} = \mathbf{D}_{\mathrm{A,initial}} + [\mathbf{f}_{\mathrm{AA}} \mid \mathbf{f}_{\mathrm{AX}}] \left\{\begin{array}{c} \mathbf{F}_{\mathrm{A},net} \\ \hline \mathbf{F}_{\mathrm{X},net} \end{array}\right\} \quad \text{where } \mathbf{D}_{\mathrm{A,initial}} = \mathbf{O}$$

$$\Rightarrow \begin{Bmatrix} D_1 \\ D_2 \end{Bmatrix} = \frac{1}{80000} \left[\begin{array}{cc|cc} 166.667 & 25 & 416.667 & 25 \\ 25 & 5 & 75 & 5 \end{array}\right] \left\{\begin{array}{c} F_1 = -100 \\ F_2 = 11.667 \\ \hline F_3 = 44.606 \\ F_4 = -208.939 \end{array}\right\} = \begin{Bmatrix} -37.66\times10^{-3} \text{ m} \\ -1.761\times10^{-3} \text{ rad} \end{Bmatrix}$$

These displacements at the active coordinates also match with the results obtained in Example 5.1/5.6. The deflection diagram is shown in Fig. 5.4b.

EXAMPLE 5.11

Repeat Example 5.2 (non-prismatic fixed beam subject to support movements), solving by the Flexibility Method.

SOLUTION

Choosing the same redundants and global and local coordinates as shown in Fig. 5.30, we note that the prescribed *displacement loading* comprises (i) a clockwise rotation, $\theta_A = -0.002$ rad at the fixed end A (which does not correspond to a redundant coordinate location), but which induces initial displacements in the primary structure, as shown in Fig. 5.32, $\mathbf{D}_{\mathrm{initial}} = \left\{\begin{array}{c} \mathbf{D}_{\mathrm{A,initial}} \\ \hline \mathbf{D}_{\mathrm{X,initial}} \end{array}\right\} = \left\{\begin{array}{c} D_{1,\mathrm{initial}} \\ D_{2,\mathrm{initial}} \\ \hline D_{3,\mathrm{initial}} \\ D_{4,\mathrm{initial}} \end{array}\right\} = \left\{\begin{array}{c} -0.020 \text{ m} \\ -0.002 \text{ rad} \\ \hline -0.040 \text{ m} \\ -0.002 \text{ rad} \end{array}\right\}$, and (ii) a deflection, $\Delta_C = -0.010$ m at the fixed end C (which corresponds to a redundant coordinate location), whereby, $\mathbf{D}_{\mathrm{X}} = \begin{Bmatrix} D_3 \\ D_4 \end{Bmatrix} = \begin{Bmatrix} -0.010 \text{ m} \\ 0 \text{ rad} \end{Bmatrix}$.

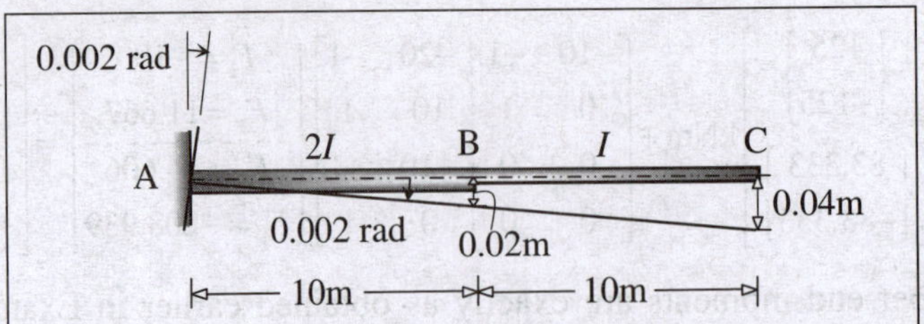

Figure 5.32 Initial displacements due to rotational slip at fixed end in primary structure (Example 5.11)

The direct load vector, $\mathbf{F_A} = \begin{Bmatrix} F_1 \\ F_2 \end{Bmatrix} = \begin{Bmatrix} 0 \text{ kN} \\ 0 \text{ kNm} \end{Bmatrix}$. As there are no intermediate loads, there are no fixed end forces to consider. Hence, the basic flexibility relations are given by:

$$\begin{bmatrix} \mathbf{D_A} \\ \mathbf{D_X} \end{bmatrix} = \begin{bmatrix} \mathbf{D_{A,initial}} \\ \mathbf{D_{X,initial}} \end{bmatrix} + \begin{bmatrix} \mathbf{f_{AA}} & \mathbf{f_{AX}} \\ \mathbf{f_{XA}} & \mathbf{f_{XX}} \end{bmatrix} \begin{bmatrix} \mathbf{F}_{\mathbf{A},net} = \mathbf{O} \\ \mathbf{F}_{\mathbf{X},net} = \mathbf{F_X} \end{bmatrix}$$

$$\Rightarrow \mathbf{F_X} = [\mathbf{f_{XX}}]^{-1} [\mathbf{D_X} - \mathbf{D_{X,initial}}]$$

where, as calculated in Example 5.10,

$$\mathbf{f} = \mathbf{T_F}^T \mathbf{f_*} \mathbf{T_F} = \frac{1}{EI} \begin{bmatrix} 166.667 & 25 & 416.667 & 25 \\ 25 & 5 & 75 & 5 \\ 416.667 & 75 & 1500 & 125 \\ 25 & 5 & 125 & 15 \end{bmatrix} = \begin{bmatrix} \mathbf{f_{AA}} & \mathbf{f_{AX}} \\ \mathbf{f_{XA}} & \mathbf{f_{XX}} \end{bmatrix}$$

$$[\mathbf{f_{XX}}] = \frac{1}{EI} \begin{bmatrix} 1500 & 125 \\ 125 & 15 \end{bmatrix} \Rightarrow [\mathbf{f_{XX}}]^{-1} = EI \begin{bmatrix} 0.0021818 & -0.0181818 \\ -0.0181818 & 0.2181818 \end{bmatrix}; \; EI = 80000 \text{ kNm}^2$$

Redundants

$$\Rightarrow \begin{Bmatrix} X_1 \\ X_2 \end{Bmatrix} = (80000) \begin{bmatrix} 0.0021818 & -0.0181818 \\ -0.0181818 & 0.2181818 \end{bmatrix} \begin{Bmatrix} -0.010 + 0.040 \text{ m} \\ 0 + 0.002 \text{ rad} \end{Bmatrix} = \begin{Bmatrix} 2.327 \text{ kN} \\ -8.727 \text{ kNm} \end{Bmatrix}$$

Member Forces

$$\mathbf{F_*} = \mathbf{F_{*f}} + \mathbf{T_F F}$$

where $\mathbf{F_{*f}} = \mathbf{O}$ and the force transformation matrix $\mathbf{T_f}$ is as calculated in Example 5.10.

$$\Rightarrow \mathbf{F_*} = \begin{bmatrix} \begin{Bmatrix} F_{1*}^1 \\ F_{2*}^1 \end{Bmatrix} \\ \begin{Bmatrix} F_{1*}^2 \\ F_{2*}^2 \end{Bmatrix} \end{bmatrix} = \begin{bmatrix} \begin{Bmatrix} 0 \\ 0 \end{Bmatrix} \\ \begin{Bmatrix} 0 \\ 0 \end{Bmatrix} \end{bmatrix} \text{kNm} + \begin{bmatrix} -10 & -1 & -20 & -1 \\ 0 & 1 & 10 & 1 \\ 0 & 0 & -10 & -1 \\ 0 & 0 & 0 & 1 \end{bmatrix} \begin{Bmatrix} F_1 = 0 \text{ kN} \\ F_2 = 0 \text{ kNm} \\ F_3 = 2.327 \text{ kN} \\ F_4 = -8.727 \text{ kNm} \end{Bmatrix} = \begin{bmatrix} \begin{Bmatrix} -37.813 \\ 14.543 \end{Bmatrix} \\ \begin{Bmatrix} -14.543 \\ -8.727 \end{Bmatrix} \end{bmatrix} \text{kNm}$$

These results match with the results obtained in Example 5.2. The free-body, shear force, bending moment and deflection diagrams are shown in Fig. 5.6.

Displacements

$$\mathbf{D}_A = \mathbf{D}_{A,initial} + \left[\mathbf{f}_{AA} \;\vdots\; \mathbf{f}_{AX}\right]\begin{bmatrix}\mathbf{F}_{A,net} = \mathbf{O} \\ \hdashline \mathbf{F}_{X,net} = \mathbf{F}_X\end{bmatrix} = \mathbf{D}_{A,initial} + \mathbf{f}_{AX}\mathbf{F}_X$$

$$\Rightarrow \begin{Bmatrix} D_1 \\ D_2 \end{Bmatrix} = \begin{Bmatrix} -0.02 \text{ m} \\ -0.002 \text{ rad} \end{Bmatrix} + \frac{1}{80000}\begin{bmatrix} 416.667 & 25 \\ 75 & 5 \end{bmatrix}\begin{Bmatrix} 2.327 \text{ kN} \\ -8.727 \text{ kNm} \end{Bmatrix} = \begin{Bmatrix} 10.607\times10^{-3} \text{ m} \\ -3.639\times10^{-4} \text{ rad} \end{Bmatrix}$$

These displacements at the active coordinates also match with the results obtained in Example 5.2. The deflection diagram is shown in Fig. 5.6d.

EXAMPLE 5.12

Repeat Example 5.3/5.7 by the flexibility method and find the force response.

SOLUTION

The beam is statically indeterminate to the third degree. Let us choose the vertical reactions, V_B, V_C and V_D at the supports B, C and D respectively as the redundants, X_1, X_2 and X_3 (all acting upward), as shown in Fig. 5.33. The primary structure, therefore, is a cantilever beam, fixed at A. We can take advantage of the fact that the overhanging segment DE is statically determinate, as in Example 5.7, thereby limiting the number of global coordinates to 6, comprising 3 active coordinates (labelled '1', '2' and '3') and 3 redundant coordinates, as shown in Fig. 5.34a.

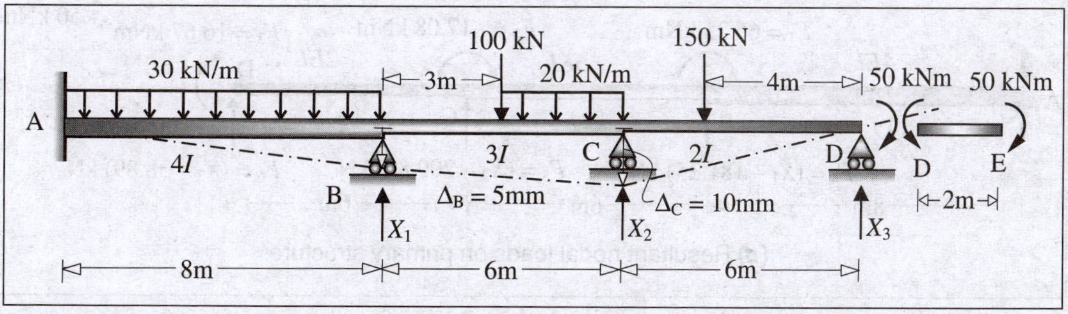

Figure 5.33 Continuous beam with gravity loading combined with support settlements (Example 5.12)

The local coordinates for the three elements (with two rotational degrees of freedom[†] each) are shown in Fig. 5.34b.

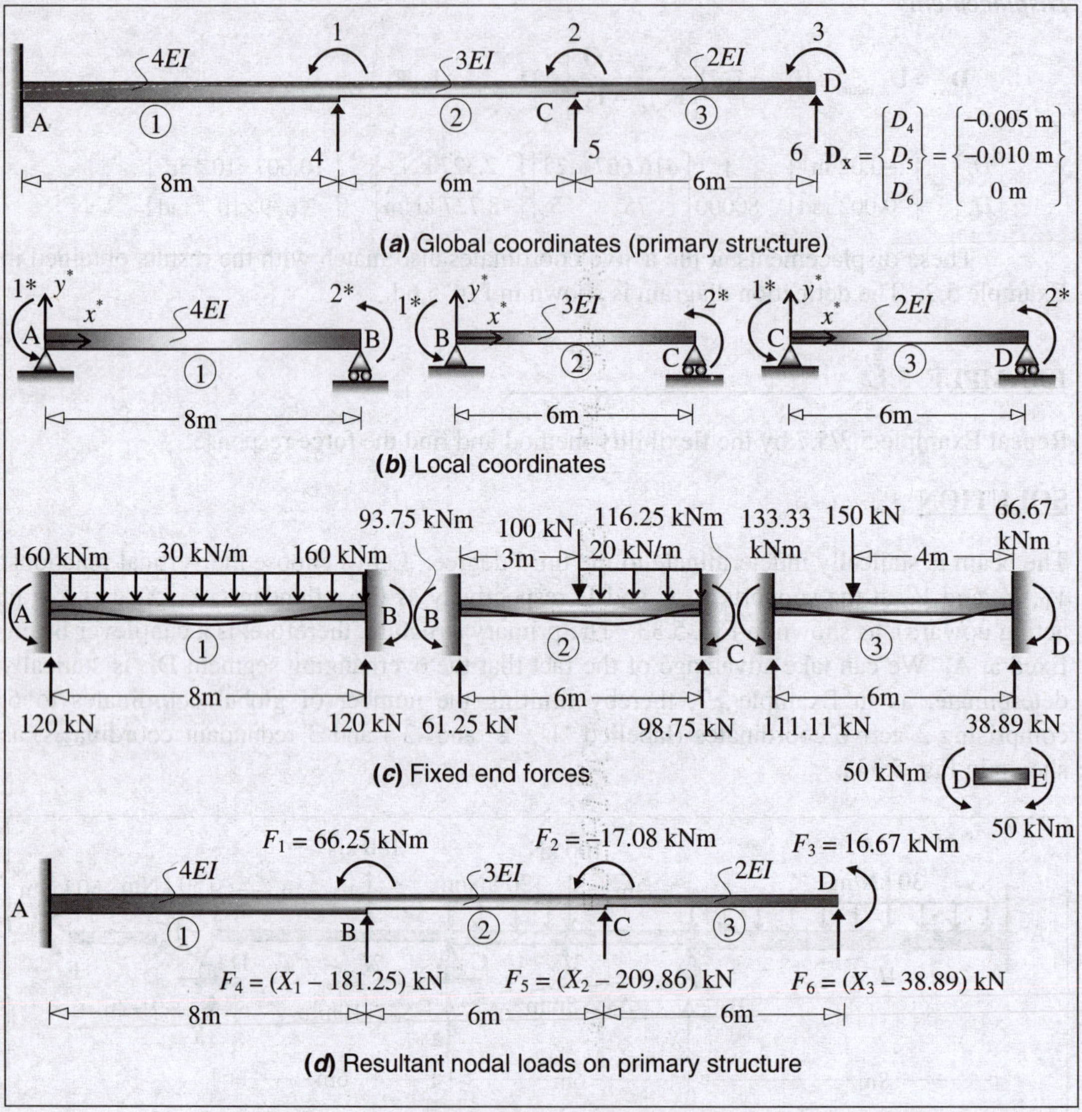

Figure 5.34 Global and local coordinates, fixed end forces and resultant nodal loads (Example 5.12)

[†] Note that this is similar to the reduced stiffness formulation in Example 5.7, except for the fact that we do not reduce the degrees of freedom (from two to one) for element '3' in the flexibility formulation. Likewise, while calculating fixed end forces, we treat this element (like all other elements) as a beam fixed at both ends, and not as a propped cantilever.

The given loading comprises direct actions as well as indirect loading due to support settlements. The direct actions due to the intermediate loads need to be converted to equivalent joint loads by considering fixed end forces, and then combined with nodal loads to generate the resultant force vector, $\mathbf{F}-\mathbf{F_f}$. The given support settlements are all at the redundant coordinate locations, with $\mathbf{D_X} = \begin{Bmatrix} D_4 \\ D_5 \\ D_6 \end{Bmatrix} \doteq \begin{Bmatrix} -0.005 \text{ m} \\ -0.010 \text{ m} \\ 0 \text{ m} \end{Bmatrix}$ and $\mathbf{D_{initial}} = \mathbf{O}$.

Fixed End Forces (Fig. 5.34c)

<u>Element '1' (AB):</u>

$$F_{1*f}^{1} = \frac{30\times 8^2}{12} = 160 \text{ kNm}; \; F_{2*f}^{1} = -160 \text{ kNm} \qquad \Rightarrow \mathbf{F_{*f}^1} = \begin{Bmatrix} 160 \text{ kNm} \\ -160 \text{ kNm} \end{Bmatrix}$$

Vertical reactions, $V_{AB} = V_{BA} = \frac{30\times 8}{2} = 120$ kN (upward), are induced at A and B.

<u>Element '2' (BC):</u>

$$F_{1*f}^{2} = +\frac{(100)(6)}{8} + \frac{5}{192}\times(20)(6^2) = +93.75 \text{ kNm}; \; F_{2*f}^{2} = -\frac{(100)(6)}{8} - \frac{11}{192}\times(20)(6^2) = -116.25 \text{ kNm}$$

$$\Rightarrow \mathbf{F_{*f}^2} = \begin{Bmatrix} 93.75 \text{ kNm} \\ -116.25 \text{ kNm} \end{Bmatrix}$$

Vertical reactions:

$$V_{BC} = \frac{100}{2} + \frac{20\times 3}{4} + \frac{93.75-116.25}{6} = 61.25 \text{ kN}; \; V_{CB} = 100 + (20\times 3) - 61.25 = 98.75 \text{ kN}$$

<u>Element '3' (CD):</u>

$$F_{1*f}^{3} = +\frac{(150)(2\times 4^2)}{6^2} = +133.333 \text{ kNm}; \; F_{2*f}^{3} = -\frac{(150)(2^2\times 4)}{6^2} = -66.667 \text{ kNm}$$

$$\Rightarrow \mathbf{F_{*f}^3} = \begin{Bmatrix} 133.333 \text{ kNm} \\ -66.667 \text{ kNm} \end{Bmatrix}$$

Vertical reactions:

$$V_{CD} = \frac{150\times 4}{6} + \frac{133.333-66.6667}{6} = 111.111 \text{ kN}; \; V_{DC} = 150 - 111.111 = 38.889 \text{ kN}$$

Summing up the fixed end moments at the three active global coordinate locations, we note that moments of $-160+93.75=-66.25$ kNm, $-116.25+133.333=17.083$ kNm and -66.667 kNm act at coordinates '1', '2' and '3' respectively. Similarly, summing up the vertical reactions at the three redundant coordinates, we note that forces of $120+61.25=181.25$ kN, $98.75+111.111=209.861$ kN and 38.889 kN act at coordinates '4', '5' and '6' respectively.

The net force vector, $\mathbf{F}_{net}$, on the primary structure (shown in Fig. 5.34d) is given by:

$$\mathbf{F}_{net} = \left[\frac{\mathbf{F}_A - \mathbf{F}_{fA}}{\mathbf{F}_X - \mathbf{F}_{fX}}\right] = \left[\frac{\begin{Bmatrix} 0 \text{ kNm} \\ 0 \text{ kNm} \\ -50 \text{ kNm} \end{Bmatrix}}{\begin{Bmatrix} X_1 \\ X_2 \\ X_3 \end{Bmatrix}}\right] - \left[\frac{\begin{Bmatrix} -66.25 \text{ kNm} \\ 17.083 \text{ kNm} \\ -66.667 \text{ kNm} \end{Bmatrix}}{\begin{Bmatrix} 181.25 \text{ kN} \\ 209.861 \text{ kN} \\ 38.889 \text{ kN} \end{Bmatrix}}\right] = \left[\frac{\begin{Bmatrix} 66.25 \text{ kNm} \\ -17.083 \text{ kNm} \\ 16.667 \text{ kNm} \end{Bmatrix}}{\begin{Bmatrix} (X_1 - 181.25) \text{ kN} \\ (X_2 - 209.861) \text{ kN} \\ (X_3 - 38.889) \text{ kN} \end{Bmatrix}}\right]$$

Force Transformation Matrix

The force transformation matrix can be easily generated by analysing the primary structure under separate unit load conditions, as shown in Fig. 5.35. Accordingly,

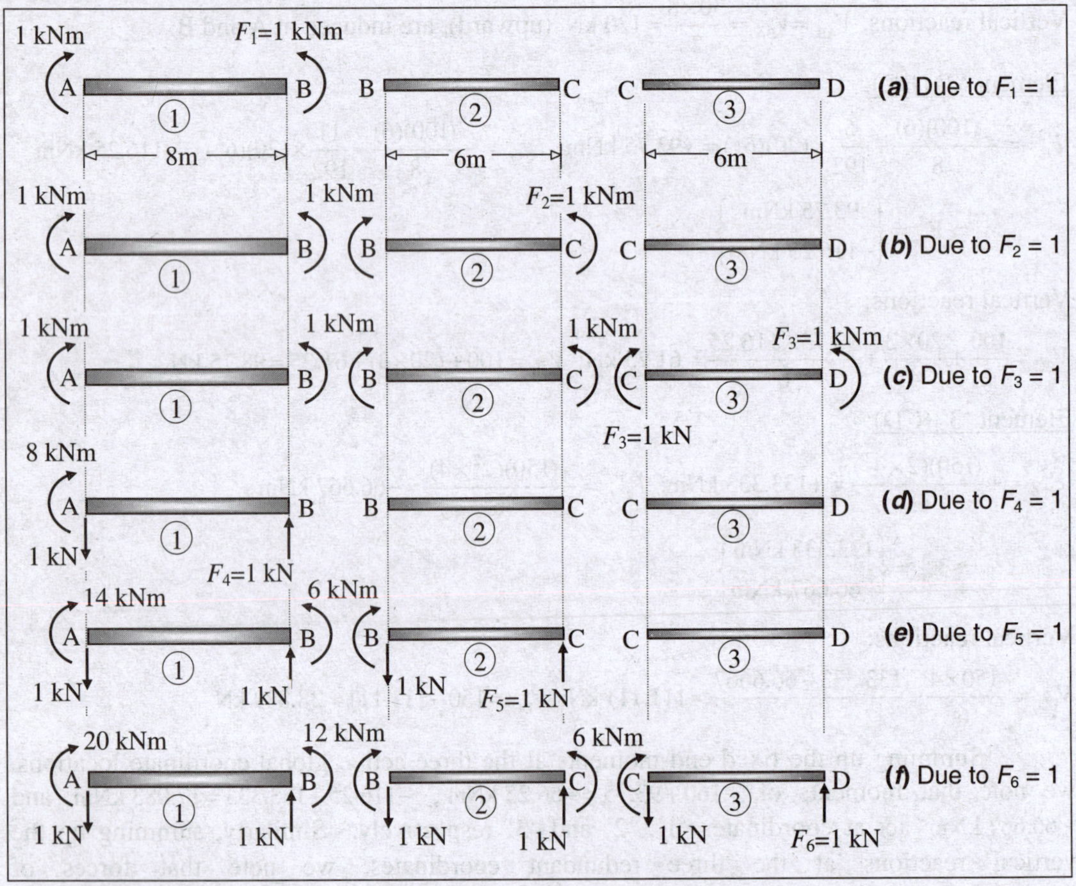

Figure 5.35 Response of primary structure to unit loads – free-bodies (Example 5.12)

$$\mathbf{F}_* = \left[\mathbf{T}_{FA} \mid \mathbf{T}_{FX}\right]\left\{\frac{\mathbf{F}_{A,net}}{\mathbf{F}_{X,net}}\right\} \Rightarrow \begin{bmatrix}\{F_*^1\}\\ \{F_*^2\}\\ \{F_*^3\}\end{bmatrix} = \begin{bmatrix}\begin{Bmatrix}F_{1*}^1\\F_{2*}^1\end{Bmatrix}\\ \begin{Bmatrix}F_{1*}^2\\F_{2*}^2\end{Bmatrix}\\ \begin{Bmatrix}F_{1*}^1\\F_{2*}^1\end{Bmatrix}\end{bmatrix} = \left[\begin{array}{ccc|ccc}-1 & -1 & -1 & -8 & -14 & -20\\ +1 & +1 & +1 & 0 & +6 & +12\\ \hline 0 & -1 & -1 & 0 & -6 & -12\\ 0 & +1 & +1 & 0 & 0 & +6\\ \hline 0 & 0 & -1 & 0 & 0 & -6\\ 0 & 0 & +1 & 0 & 0 & 0\end{array}\right]\begin{bmatrix}\begin{Bmatrix}F_1\\F_2\\F_3\end{Bmatrix}\\ \hline \begin{Bmatrix}F_4\\F_5\\F_6\end{Bmatrix}\end{bmatrix}$$

Flexibility Matrix

The unassembled element flexibility matrix may be generated as follows:

$$\mathbf{D}_* = \mathbf{f}_*\mathbf{F}_* \Rightarrow \begin{bmatrix}\begin{Bmatrix}D_{1*}^1\\D_{2*}^1\end{Bmatrix}\\ \begin{Bmatrix}D_{1*}^2\\D_{2*}^2\end{Bmatrix}\\ \begin{Bmatrix}D_{1*}^3\\D_{2*}^3\end{Bmatrix}\end{bmatrix} = \left[\begin{array}{c|c|c}\frac{8}{6(4EI)}\begin{bmatrix}2 & -1\\-1 & 2\end{bmatrix} & \mathbf{O} & \mathbf{O}\\ \hline \mathbf{O} & \frac{6}{6(3EI)}\begin{bmatrix}2 & -1\\-1 & 2\end{bmatrix} & \mathbf{O}\\ \hline \mathbf{O} & \mathbf{O} & \frac{6}{6(2EI)}\begin{bmatrix}2 & -1\\-1 & 2\end{bmatrix}\end{array}\right]\begin{bmatrix}\begin{Bmatrix}F_{1*}^1\\F_{2*}^1\end{Bmatrix}\\ \begin{Bmatrix}F_{1*}^2\\F_{2*}^2\end{Bmatrix}\\ \begin{Bmatrix}F_{1*}^3\\F_{2*}^3\end{Bmatrix}\end{bmatrix}$$

$$\Rightarrow \mathbf{f}_* = \frac{1}{EI}\begin{bmatrix}0.66667 & -0.33333 & 0 & 0 & 0 & 0\\ -0.33333 & 0.66667 & 0 & 0 & 0 & 0\\ 0 & 0 & 0.66667 & -0.33333 & 0 & 0\\ 0 & 0 & -0.33333 & 0.66667 & 0 & 0\\ 0 & 0 & 0 & 0 & 1 & -0.5\\ 0 & 0 & 0 & 0 & -0.5 & 1\end{bmatrix}$$

The structure flexibility matrix $\mathbf{f}$, satisfying $\mathbf{D} = \mathbf{f}\ \mathbf{F}$, can now be assembled:

$$\mathbf{f} = \mathbf{T_F}^T\mathbf{f}_*\mathbf{T_F} = \frac{1}{EI}\left[\begin{array}{ccc|ccc}2 & 2 & 2 & 8 & 20 & 32\\ 2 & 4 & 4 & 8 & 26 & 50\\ 2 & 4 & 7 & 8 & 26 & 59\\ \hline 8 & 8 & 8 & 42.6667 & 90.6667 & 138.6667\\ 20 & 26 & 26 & 90.6667 & 234.6667 & 390.6667\\ 32 & 50 & 59 & 138.6667 & 390.6667 & 726.6667\end{array}\right] = \left[\begin{array}{c|c}\mathbf{f}_{AA} & \mathbf{f}_{AX}\\ \hline \mathbf{f}_{XA} & \mathbf{f}_{XX}\end{array}\right]$$

Redundants

The structure flexibility relation to be considered here is:

$$\left[\frac{\mathbf{D}_A}{\mathbf{D}_X}\right] = \left[\frac{\mathbf{D}_{A,\text{initial}} = \mathbf{O}}{\mathbf{D}_{X,\text{initial}} = \mathbf{O}}\right] + \left[\begin{array}{c|c}\mathbf{f}_{AA} & \mathbf{f}_{AX}\\ \hline \mathbf{f}_{XA} & \mathbf{f}_{XX}\end{array}\right]\left[\frac{\mathbf{F}_A - \mathbf{F}_{fA}}{\mathbf{F}_X - \mathbf{F}_{fX}}\right]$$

$$\Rightarrow \quad \mathbf{F}_X - \mathbf{F}_{fX} = \left[\mathbf{f}_{XX}\right]^{-1}\left[\mathbf{D}_X - \mathbf{f}_{XA}\left(\mathbf{F}_A - \mathbf{F}_{fA}\right)\right]$$

$$\text{where}\quad \mathbf{D_X} = \begin{Bmatrix} -0.005 \text{ m} \\ -0.010 \text{ m} \\ 0 \text{ m} \end{Bmatrix} \times \frac{80000 \text{ kNm}^2}{EI} = \frac{1}{EI}\begin{Bmatrix} -400 \\ -800 \\ 0 \end{Bmatrix}$$

$$\text{and}\quad [\mathbf{f_{XA}}](\mathbf{F_A} - \mathbf{F_{fA}}) = \frac{1}{EI}\begin{bmatrix} 8 & 8 & 8 \\ 20 & 26 & 26 \\ 32 & 50 & 59 \end{bmatrix}\begin{Bmatrix} 66.25 \text{ kNm} \\ -17.083 \text{ kNm} \\ 16.667 \text{ kNm} \end{Bmatrix} = \frac{1}{EI}\begin{Bmatrix} 526.672 \\ 1314.184 \\ 2249.203 \end{Bmatrix}$$

$$\Rightarrow \begin{Bmatrix} (X_1 - 181.25) \text{ kN} \\ (X_2 - 209.861) \text{ kN} \\ (X_3 - 38.889) \text{ kN} \end{Bmatrix} = EI\begin{bmatrix} 0.176610 & -0.115530 & 0.028409 \\ -0.115530 & 0.116162 & -0.040404 \\ 0.028409 & -0.040404 & 0.017677 \end{bmatrix}\frac{1}{EI}\begin{Bmatrix} -926.672 \\ -2114.184 \\ -2249.203 \end{Bmatrix} = \begin{Bmatrix} 16.695 \\ -47.651 \\ 19.337 \end{Bmatrix}$$

$$\Rightarrow \begin{Bmatrix} X_1 \\ X_2 \\ X_3 \end{Bmatrix} = \begin{Bmatrix} 197.945 \text{ kN} \\ 162.210 \text{ kN} \\ 58.226 \text{ kN} \end{Bmatrix}$$

Member Forces

$$\mathbf{F_*} = \mathbf{F_{*f}} + \mathbf{T_F F}_{net} \quad \Rightarrow$$

$$\begin{bmatrix} \begin{Bmatrix} F^1_{1*} \\ F^1_{2*} \end{Bmatrix} \\ \begin{Bmatrix} F^2_{1*} \\ F^2_{2*} \end{Bmatrix} \\ \begin{Bmatrix} F^1_{1*} \\ F^1_{2*} \end{Bmatrix} \end{bmatrix} = \begin{bmatrix} \begin{Bmatrix} 160 \\ -160 \end{Bmatrix} \\ \begin{Bmatrix} 93.75 \\ -116.25 \end{Bmatrix} \\ \begin{Bmatrix} 133.333 \\ -66.667 \end{Bmatrix} \end{bmatrix} + \left[\begin{array}{ccc|ccc} -1 & -1 & -1 & -8 & -14 & -20 \\ +1 & +1 & +1 & 0 & +6 & +12 \\ \hline 0 & -1 & -1 & 0 & -6 & -12 \\ 0 & +1 & +1 & 0 & 0 & +6 \\ \hline 0 & 0 & -1 & 0 & 0 & -6 \\ 0 & 0 & +1 & 0 & 0 & 0 \end{array}\right]\left[\begin{array}{c} \begin{Bmatrix} F_1 = 66.25 \text{ kNm} \\ F_2 = -17.083 \text{ kNm} \\ F_3 = 16.667 \text{ kNm} \end{Bmatrix} \\ \hline \begin{Bmatrix} F_4 = 16.695 \text{ kN} \\ F_5 = -47.651 \text{ kN} \\ F_6 = 19.337 \text{ kN} \end{Bmatrix} \end{array}\right] = \begin{bmatrix} \begin{Bmatrix} 250.985 \\ -148.030 \end{Bmatrix} \\ \begin{Bmatrix} 148.030 \\ -0.644 \end{Bmatrix} \\ \begin{Bmatrix} 0.644 \\ -50.000 \end{Bmatrix} \end{bmatrix} \text{kNm}$$

The member end moments are exactly as obtained earlier in Examples 5.3 and 5.7. The free-bodies of the two beam elements can now be drawn and the shear force and bending moment diagrams generated, exactly as shown earlier in Fig 5.11. The support reactions are also easily obtainable from the free-bodies.

Displacements

$$\mathbf{D_A} = \mathbf{D_{A,initial}} + [\mathbf{f_{AA}} \mid \mathbf{f_{AX}}]\left[\begin{array}{c} \mathbf{F}_{\mathbf{A},net} \\ \hline \mathbf{F}_{\mathbf{X},net} \end{array}\right] \quad \text{where}\quad \mathbf{D_{A,initial}} = \mathbf{O}$$

$$\Rightarrow \begin{Bmatrix} D_1 \\ D_2 \\ D_3 \end{Bmatrix} = \frac{1}{80000}\left[\begin{array}{ccc|ccc} 2 & 2 & 2 & 8 & 20 & 32 \\ 2 & 4 & 4 & 8 & 26 & 50 \\ 2 & 4 & 7 & 8 & 26 & 59 \end{array}\right]\left[\begin{array}{c} \begin{Bmatrix} F_1 = 66.25 \text{ kNm} \\ F_2 = -17.083 \text{ kNm} \\ F_3 = 16.667 \text{ kNm} \end{Bmatrix} \\ \hline \begin{Bmatrix} F_4 = 16.695 \text{ kN} \\ F_5 = -47.651 \text{ kN} \\ F_6 = 19.337 \text{ kN} \end{Bmatrix} \end{array}\right] = \begin{Bmatrix} 8.627 \times 10^{-4} \text{ rad} \\ -0.961 \times 10^{-4} \text{ rad} \\ 27.043 \times 10^{-4} \text{ rad} \end{Bmatrix}$$

We can use the value of D_3 (slope at D) to find the deflection and rotation at the tip E of the cantilever segment DE, as follows. The clockwise end moment ($M_E = 50$ kNm) applied at E causes it to rotate clockwise by $M_E L_{DE}/(EI)_{DE}$ and deflect downwards by $M_E L_{DE}^2/2(EI)_{DE}$, assuming cantilever action (fixity at D), while the anti-clockwise rotation at D causes segment DE to rotate likewise, resulting in an upward deflection, $D_3 L_{DE}$. Hence, applying superposition, the net rotation and deflection at E are given by:

$$\theta_E = -\frac{(50)(2)}{(2\times 80000)} + \left(27.043\times 10^{-4}\right) = 20.793\times 10^{-4} \text{ rad (anti-clockwise)}$$

$$\Delta_E = -\frac{(50)(2)^2}{2(2\times 80000)} + \left(27.043\times 10^{-4}\right)(2) = 4.784\times 10^{-3} \text{ m} = 4.784 \text{ mm (upwards)}$$

These displacement solutions match exactly with the results obtained in Example 5.3.

5.5 ANALYSIS OF GRIDS BY STIFFNESS METHOD

A *grid* is a two-dimensional framework of interconnected prismatic elements, subjected to loads and reactions that act normal to the plane of the grid. Commonly, the plane of the grid is horizontal, and the loads applied are gravity loads. Grids are generally used for supporting long-span floors and roofs.

The joints at the intersections of elements may be rigid or provided with moment releases. Rigid joints are moment-resistant joints, whereby, moment transfer occurs in addition to shear transfer, across the joint, and the moment may be flexural (bending moment) or torsional (twisting moment) or have components of both bending and twisting. Rotational compatibility has to be maintained by all the inter-connecting members at the joint, in addition to deflection compatibility.

Consider, for example, the simple case of a grid, comprising three elements, AB, CB and DB (with fixed end supports at A, C and D), mutually intersecting at the rigid joint B, and subject to arbitrary gravity loading, as shown in Fig. 5.36a. There are three degrees of freedom in this grid system, corresponding to the vertical deflection (Δ_{By}) and rotations in two orthogonal directions (θ_{Bz} and θ_{Bx}) of the joint B, as shown in Fig. 5.36b. It may be noted that θ_{Bz} denotes the flexural rotation (slope) at B in the element AB, as well as the torsional rotation (angle of twist) at B in the elements CB and DB. Similarly, θ_{Bx} denotes the angle of twist at B in AB, and also the flexural rotation at B in CB and DB. This is the requirement of rotational compatibility at the rigid joint B. On account of this, bending moments and twisting moments are generated in the three elements at B, as depicted in the free-body diagrams shown in Fig. 5.36c. The magnitudes of these moments depend on the relative flexural and torsional stiffnesses of the elements, and must satisfy moment equilibrium at the joint B. This implies that the (hogging) bending moment at B in AB, M_{BA}, will be equal to the sum of the twisting moments, ($T_{BC} + T_{BD}$) in BC and BD, as shown in Fig. 5.36c.

Similarly, the twisting moment, T_{BA}, at B in element BA, will be equal to the difference in (sagging) bending moments ($M_{BD} - M_{BC}$) in BD and BC.

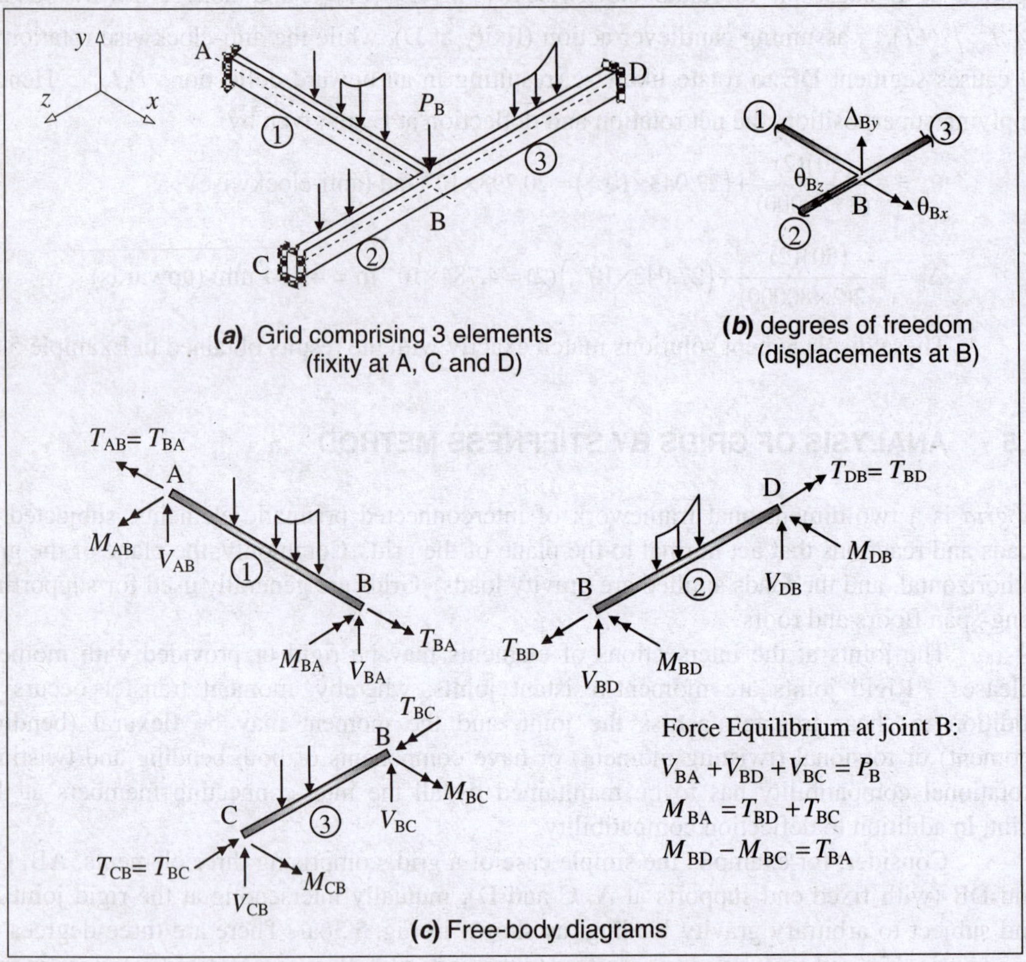

Figure 5.36 Example of a simple grid system

If the torsional stiffness values of the members are very low, as is sometimes the case in practice†, then the twisting moments generated are also low in magnitude, and can be neglected, for convenience. This can be accounted for in the modelling by assuming the torsional rigidity to be zero. Alternatively, this can be accounted for by providing 'moment releases' (internal hinges) appropriately. For example, considering the example grid in

---

† Reinforced concrete members, for example, crack under relatively low levels of torsion, and this degrades the torsional stiffness of the members considerably. Design codes generally allow the neglect of such torsion (sometimes referred to as 'secondary torsion' or 'compatibility torsion') at the analysis stage itself.

Fig. 5.36a, we can achieve moment releases by inserting a 'flexural hinge' as well as a 'torsional hinge' at the end B of element AB. The flexural hinge at B in AB is intended to release the torsion in BC and BD (and consequently M_{BA} will also be reduced to zero), while the torsional hinge at B in AB is intended to release the torsion in AB (whereby, $M_{BD} = M_{BC}$). In this situation, only a shear force can get transmitted across the joint B (satisfying force equilibrium, $V_{BA} + V_{BC} + V_{BC} = P_B$).

In design practice, the term "grid beam" is sometimes used to refer to the members in a grid. This is so, because the primary resistance to loading is attributable to bending (beam action). When torsion (twisting moment) releases are provided, then the elements involved are conventional *beam elements*, and we may use a four degree of freedom model (conventional stiffness method, refer Fig. 5.1a) or a two degree of freedom model (reduced stiffness formulation or flexibility method) for this purpose. However, when the element is subject to torsion, in addition to flexure, then we need to include additional rotational degree(s) of freedom (to account for the angle of twist and the corresponding twisting moment(s) at the end(s) of the element), as shown in Fig. 5.37. Such an element is called a *grid element*. It is assumed that the cross-section is symmetric with respect to the y^*- and z^*-axes (for example, rectangular, I-, circular or tubular sections), with the longitudinal (local x^*-axis) corresponding to the centroidal axis, so that there is no *unsymmetrical bending*.

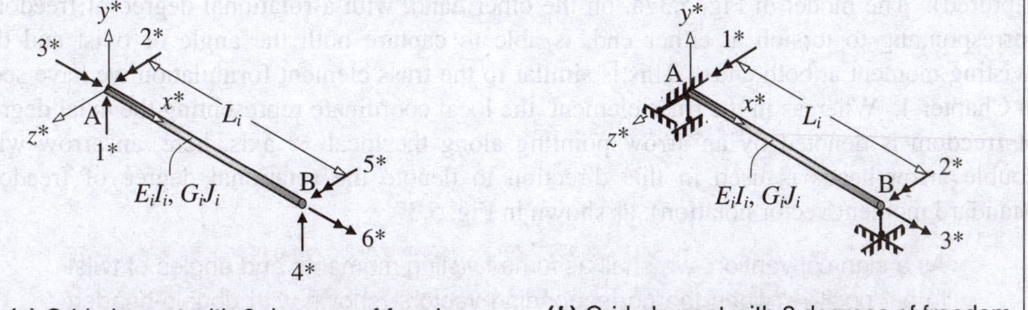

(*a*) Grid element with 6 degrees of freedom (conventional stiffness formulation)

(*b*) Grid element with 3 degrees of freedom (reduced stiffness formulation)

Figure 5.37 Prismatic grid element — local coordinates

We can consider all the elements in a typical grid system to be grid elements, but in situations where we are certain that there will be no twisting involved in some of the elements, we can adopt beam elements for these elements. Furthermore, even if there is twisting, but no associated twisting moment (on account of zero torsional stiffness specified for the element), then in the reduced stiffness formulation, it suffices to adopt the beam element. In some cases involving symmetric grids with symmetric loading, we can take advantage of symmetry and work with a reduced grid, with an appropriate mix of grid and beam elements. This is demonstrated in the examples to follow.

Here, we shall demonstrate the application of the conventional stiffness method and the reduced stiffness formulation to grid systems. Analysis by the flexibility method is not described here.

5.5.1 Torsional Stiffness

The stiffness matrix of a prismatic grid element contains the rows and columns of that of a corresponding beam element plus additional row(s) and column(s), corresponding to the torsional stiffness terms. In the reduced stiffness formulation, as indicated in Fig. 5.37b, it suffices to have only one degree of freedom corresponding to the angle of twist at one end of the element, resulting in a 3×3 reduced element stiffness matrix. Such an element stiffness matrix is invertible (the inverse being the flexibility matrix), and the corresponding element is provided with sufficient support restraints to render it stable (just-rigid), as shown. This requires the torsional degree of freedom to be restrained at one end to the simply supported beam shown we had dealt with earlier. Clearly, the twisting moment has to be constant throughout the element (with torques applied only at the two ends), and this is captured by the model shown in Fig. 5.37b, although this reduced element formulation does not capture the absolute values of the angles of twist at the two ends (only the relative angle of twist is captured). The model in Fig. 5.37a, on the other hand, with a rotational degree of freedom, corresponding to torsion at either end, is able to capture both the angle of twist and the twisting moment at both ends. This is similar to the truss element formulation we have seen in Chapter 4. Whereas in the truss element, the local coordinate representing the axial degree of freedom is denoted by an arrow pointing along the local x^* axis, here, an arrow with double arrow-heads is used in this direction to denote the rotational degree of freedom (standard moment vector notation), as shown in Fig. 5.37.

> As a sign convention, we shall assume twisting moments and angles of twist to be 'positive' when the corresponding vectors (shown with double-headed arrows) at the ends of the element are directed outward along the longitudinal axis. This is consistent with the convention of assuming axial tensile forces and elongations in truss elements to be positive (depicted using single-headed arrows at the ends of the element, directed outward).

The additional stiffness term(s) introduced on account of torsion in the grid element, is that of ***torsional stiffness***. The expression for torsional stiffness of a solid cylindrical or hollow cylindrical shaft of length L_i, can be easily shown, from basic mechanics of materials, to be given by $\frac{G_i J_i}{L_i}$, where J_i is the *polar moment of inertia* of the circular section and G_i is the *shear modulus* of the elastic material. When a torque is applied at the free end of such a shaft (fixed at the other end), it generates a constant twisting moment along the length ("pure torsion") and a linearly varying angle of twist (varying from zero at the fixed end to a maximum value at the free end). The cross-section rotates about the centroidal axis ("plane sections remain plane") without any displacement normal to the plane (no *warping*). The torsional stiffness, $k_{\phi i}$, of such an element, is, by definition, the torque required to be applied at the free end in order to generate a unit angle of twist at that end:

$$k_{\phi i} = \frac{G_i J_i}{L_i} \tag{5.49}$$

where the term, $G_i J_i$, is commonly referred to as ***torsional rigidity*** (which is similar in concept to *shear rigidity*, $G_i A_i'$, *axial rigidity*, $E_i A_i$, and *flexural rigidity*, $E_i I_i$, for a prismatic element). The expressions for polar moment of inertia for a solid circular section (with radius r_i) and a thin-walled tubular section (with a mean radius r_i and wall thickness t_i) are obtainable as follows:

$$J_i = \frac{\pi r_i^4}{2} \quad \text{for solid circular section} \tag{5.50a}$$

$$J_i = 2\pi r_i^3 t_i \quad \text{for tubular section} \tag{5.50b}$$

In the case of *non-circular* sections, however, the presence of *warping* complicates the derivation of torsional stiffness considerably. However, in cases where there is *no restraint* against warping, Eq. 5.49 can be used to approximate the torsional stiffness, provided the symbol J_i is interpreted as the *torsional constant* (St. Venant's) of the section, and not the polar moment of inertia. Expressions for J_i for various sections are available in standard texts on theory of elasticity. For a solid rectangular section (width b_i, depth d_i),

$$J_i = \frac{b_i^3 d_i}{3}\left[1 - 0.63\left(\frac{b_i}{d_i}\right)\left\{1 - \frac{1}{12}\left(\frac{b_i}{d_i}\right)^4\right\}\right] \quad \text{for } b_i \le d_i \tag{5.51}$$

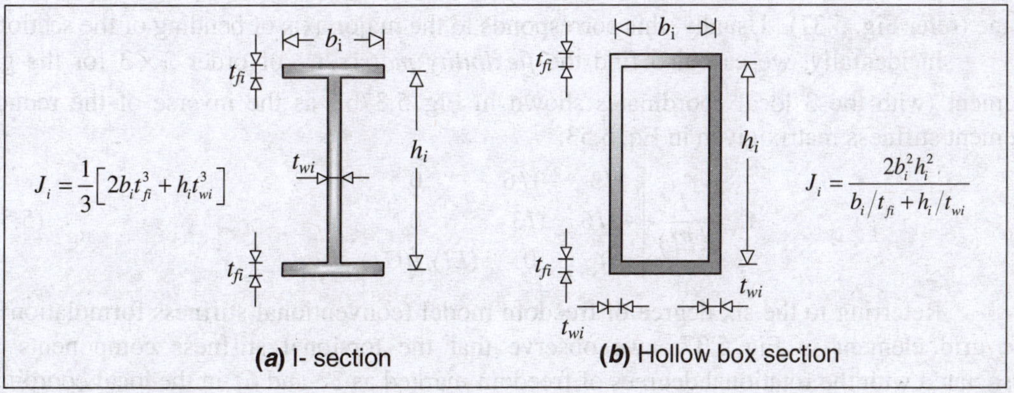

Figure 5.38 Torsional constant for I-section and hollow box section

For a very narrow rectangular section, $J_i \approx \frac{b_i^3 d_i}{3}$ can be assumed, and on this basis the value of the torsional constant for any *thin-walled open section* can be generated by summing up the contributions of the respective segments. For a typical symmetric I-section (Fig. 5.38a) and a tubular box-section (Fig. 5.38b) with height h_i, flange width b_i, flange thickness t_{fi} and web thickness t_{wi}, the expressions for J_i are obtainable as:

$$J_i = \frac{1}{3}\left[2b_i t_{fi}^3 + h_i t_{wi}^3\right] \qquad \text{for I-section} \tag{5.52a}$$

$$J_i = \frac{2b_i^2 h_i^2}{b_i/t_{fi} + h_i/t_{wi}} \qquad \text{for tubular box-section} \tag{5.52b}$$

Grid Element Stiffness Matrix

Referring to the three degree of freedom model (reduced stiffness formulation) of the grid element in Fig. 5.37b, we observe that the torsional stiffness component is associated with the rotational degree of freedom marked as 3* in the local coordinate system. Torsional effects and flexural effects can be considered to be mutually uncoupled when the cross-section is symmetric (does not undergo unsymmetrical bending), and when warping under pure torsion is absent (as in cylindrical elements) or there is freedom to warp (in non-circular sections). Under such circumstances, the reduced element stiffness matrix, $\tilde{\mathbf{k}}_*^{\mathbf{i}}$, of order 3×3 of the grid element can be easily assembled, by expanding the corresponding matrix for a beam element with two degrees of freedom (refer Eq. 5.35):

$$\tilde{\mathbf{k}}_*^{\mathbf{i}} = \frac{(EI)_i}{L_i}\begin{bmatrix} 4 & 2 & 0 \\ 2 & 4 & 0 \\ 0 & 0 & (GJ)_i/(EI)_i \end{bmatrix} \tag{5.53}$$

It may be noted that the second moment of area, I_i, to be used in the above matrix, is with respect to the centroidal axis of bending (z^*-axis), which occurs in the vertical x^*-y^* plane (refer Fig. 5.37). Usually, this corresponds to the major axis of bending of the section.

Incidentally, we can also find the *flexibility matrix* $\mathbf{f}_*^{\mathbf{i}}$, of order 3×3 for the grid element (with the 3 local coordinates shown in Fig. 5.37b), as the inverse of the reduced element stiffness matrix given in Eq. 5.53:

$$\mathbf{f}_*^{\mathbf{i}} = \frac{L_i}{(EI)_i}\begin{bmatrix} 1/3 & -1/6 & 0 \\ -1/6 & 1/3 & 0 \\ 0 & 0 & (EI)_i/(GJ)_i \end{bmatrix} \tag{5.54}$$

Referring to the six degree of freedom model (conventional stiffness formulation) of the grid element in Fig. 5.37a, we observe that the torsional stiffness components are associated with the rotational degrees of freedom marked as 3* and 6* in the local coordinate system. The grid element stiffness matrix, $\mathbf{k}_*^{\mathbf{i}}$, of order 6×6 can be easily assembled, by expanding the corresponding matrix for a beam element with two degrees of freedom (refer Eq. 5.11), and observing that when a torque is applied at one end, an equal and opposite torque must develop at the other end in order to satisfy equilibrium. Accordingly,

$$\mathbf{k}_*^{\mathbf{i}} = \frac{(EI)_i}{L_i}\begin{bmatrix} 12/L_i^2 & 6/L_i & 0 & -12/L_i^2 & 6/L_i & 0 \\ 6/L_i & 4 & 0 & -6/L_i & 2 & 0 \\ 0 & 0 & (GJ)_i/(EI)_i & 0 & 0 & -(GJ)_i/(EI)_i \\ -12/L_i^2 & -6/L_i & 0 & 12/L_i^2 & -6/L_i & 0 \\ 6/L_i & 2 & 0 & -6/L_i & 4 & 0 \\ 0 & 0 & -(GJ)_i/(EI)_i & 0 & 0 & (GJ)_i/(EI)_i \end{bmatrix} \quad (5.55)$$

5.5.2 Analysis of Grids by Conventional Stiffness Method

The basic procedure for the analysis of grids by the conventional stiffness method is similar to the one described earlier for beams. However, unlike the continuous beam system, in which all the beam elements are aligned in the same longitudinal direction, the grid elements can have any orientation, whereby the local coordinate axes in the plane of the grid (x^*-axis and z^*-axis) of the grid element will generally not be aligned in the same directions as the global axes (x-axis and z-axis) for the structure, although they are located in the same (usually horizontal) xz plane.

We need to first assemble the transformation matrix, $\mathbf{T}^{\mathbf{i}}$, for any grid element, which transforms the member end forces from local coordinates (x^*,y^*,z^*) to the global axes system (x,y,z), through the relations, $\mathbf{F}_*^{\mathbf{i}} = \mathbf{T}^{\mathbf{i}}\mathbf{F}^{\mathbf{i}}$ and $\mathbf{D}_*^{\mathbf{i}} = \mathbf{T}^{\mathbf{i}}\mathbf{D}^{\mathbf{i}}$. It is assumed that small rotations can be treated as vectors for this purpose. This transformation involves a simple rotation (anti-clockwise with reference to the positive y-axis) of the coordinate axes in the x-z plane by an angle θ_x^i, without shifting the origin (refer Figs 5.39a and b).

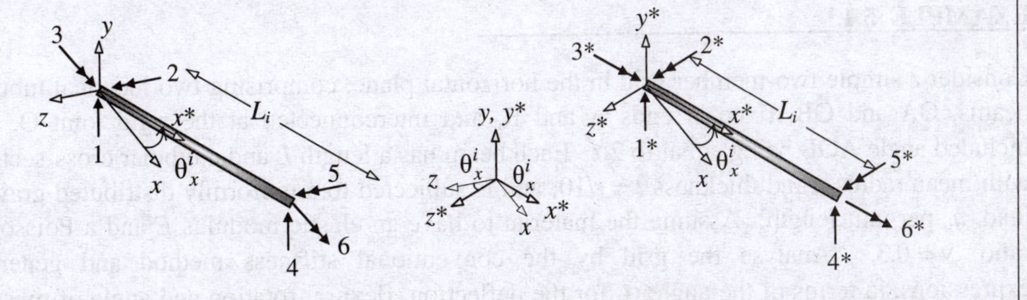

(*a*) Coordinates in global axes system (*b*) Coordinates in local coordinate system

Figure 5.39 Coordinate transformation due to rotation about the y-axis

We had derived the basis for a similar transformation earlier (refer Eqns 4.14a and b) for the plane truss element in Chapter 4. In addition, we need to include the coordinate (translation or force) in the y-direction, which remains unchanged. Accordingly, the transformation takes the following form, resulting in a square and orthogonal transformation matrix, $\mathbf{T}^{\mathbf{i}}$ (whose inverse is the same as its transpose) of order 6×6, satisfying $\mathbf{D}_{*}^{\mathbf{i}} = \mathbf{T}^{\mathbf{i}}\mathbf{D}$ and $\mathbf{F}_{*}^{\mathbf{i}} = \mathbf{T}^{\mathbf{i}}\mathbf{F}$:

$$\begin{Bmatrix} D_{1*}^i \\ D_{2*}^i \\ D_{3*}^i \\ \hline D_{4*}^i \\ D_{5*}^i \\ D_{6*}^i \end{Bmatrix} = \left[\begin{array}{ccc|ccc} 1 & 0 & 0 & 0 & 0 & 0 \\ 0 & c_i & s_i & 0 & 0 & 0 \\ 0 & -s_i & c_i & 0 & 0 & 0 \\ \hline 0 & 0 & 0 & 1 & 0 & 0 \\ 0 & 0 & 0 & 0 & c_i & s_i \\ 0 & 0 & 0 & 0 & -s_i & c_i \end{array}\right] \begin{Bmatrix} D_1^i \\ D_2^i \\ D_3^i \\ \hline D_4^i \\ D_5^i \\ D_6^i \end{Bmatrix} \; ; \; \begin{Bmatrix} F_{1*}^i \\ F_{2*}^i \\ F_{3*}^i \\ \hline F_{4*}^i \\ F_{5*}^i \\ F_{6*}^i \end{Bmatrix} = \left[\begin{array}{ccc|ccc} 1 & 0 & 0 & 0 & 0 & 0 \\ 0 & c_i & s_i & 0 & 0 & 0 \\ 0 & -s_i & c_i & 0 & 0 & 0 \\ \hline 0 & 0 & 0 & 1 & 0 & 0 \\ 0 & 0 & 0 & 0 & c_i & s_i \\ 0 & 0 & 0 & 0 & -s_i & c_i \end{array}\right] \begin{Bmatrix} F_1^i \\ F_2^i \\ F_3^i \\ \hline F_4^i \\ F_5^i \\ F_6^i \end{Bmatrix} \quad (5.56)$$

where $c_i = \cos\theta_x^i$ and $s_i = \sin\theta_x^i$, θ_x^i being the angle made by the local x^*-axis with respect to the global x-axis (measured anti-clockwise, as shown in Fig. 5.39).

It is clear that even if the element under consideration may not have any torsion, and it suffices to use a beam element (with four degrees of freedom), we must necessarily use a grid element (with six degrees of freedom) in order to enable this transformation. The moment vector will have components along the global x- and z- directions.

As explained earlier, the element stiffness matrix $\mathbf{k}^{\mathbf{i}}$ in the global axes system can be generated from the local coordinate system through the transformation, $\mathbf{T}^{\mathbf{i}\mathbf{T}}\tilde{\mathbf{k}}_{*}^{\mathbf{i}}\mathbf{T}^{\mathbf{i}}$, and by summing up all the contributions from different elements (keeping track of the appropriate linkages), the structure stiffness matrix, $\mathbf{k}$ can be generated. Alternatively, we can make use of the displacement transformation matrix, $\mathbf{T}_{\mathbf{D}}^{\mathbf{i}}$, for this purpose, as explained earlier.

The procedure is demonstrated here through a simple example.

EXAMPLE 5.13

Consider a simple two-member grid in the horizontal plane, comprising two identical tubular beams, OA and OB, fixed at ends A and B, and interconnected at the rigid joint O, the included angle AOB being equal to 2α. Each beam has a length L and a tubular cross-section with mean radius r and thickness $t = r/10$, and is subjected to a uniformly distributed gravity load, q_o per unit length. Assume the material to have an elastic modulus E and a Poisson's ratio, $\nu = 0.3$. Analyse the grid by the conventional stiffness method and generate expressions, in terms of the angle α, for the deflection, flexural rotation and angle of twist at O, and the support reactions. Hence, draw the corresponding shear force, bending moment and twisting moment diagrams of OA. Comment on the behaviour with respect to variations in the angle α.

SOLUTION

The layout plan of the grid for the three cases is shown in Fig. 5.40a. There are two prismatic grid elements here, with intermediate loads acting on both, and no nodal loads acting

$(\mathbf{F}_A = \mathbf{O})$. There are 3 *active* degrees of freedom at the rigid joint O and 6 *restrained* global coordinates (at A and B), numbered as shown in Figs 5.40a and b. [We know, in advance, that due to the symmetry in the loading and the structure, $D_2 = 0$ and $F_7 = F_4 = q_o L$; $F_8 = F_5$ and $F_9 = F_6$]. The local coordinates for the two elements are shown in Fig. 5.40c, with $\theta_x^1 = \theta_x^2 = 90° - \alpha$.

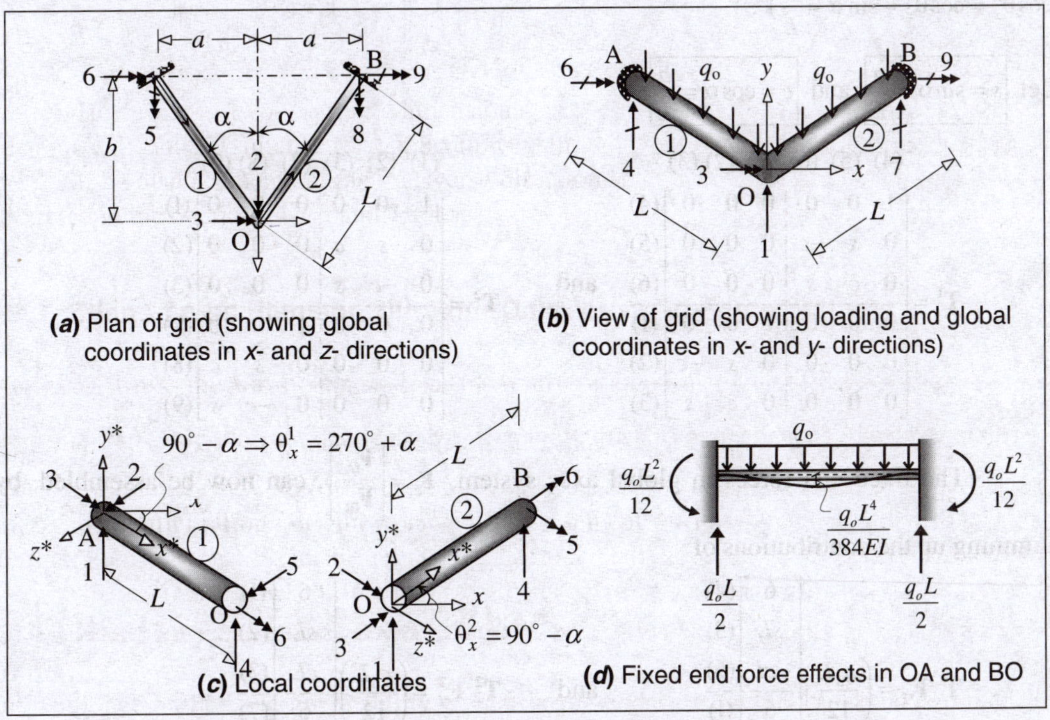
(a) Plan of grid (showing global coordinates in x- and z- directions)
(b) View of grid (showing loading and global coordinates in x- and y- directions)
90° − α ⇒ θ¹ₓ = 270° + α
θ²ₓ = 90° − α
(c) Local coordinates
q_oL²/12
q_oL⁴/384EI
q_oL/2
(d) Fixed end force effects in OA and BO

Figure 5.40 Details of grid with two elements (Example 5.13)

Fixed End Forces

The fixed end force effects (support reactions) on a typical grid element (with $D_1 = D_2 = D_3 = 0$) are shown in Fig. 5.40d.

Element '1' (AO): $F^1_{1*f} = F^1_{4*f} = \dfrac{q_o L}{2}$; $F^1_{2*f} = \dfrac{q_o L^2}{12}$; $F^1_{5*f} = -\dfrac{q_o L^2}{12}$; $F^1_{3*f} = F^1_{6*f} = 0$

Element '2' (OB): $F^2_{1*f} = F^2_{4*f} = \dfrac{q_o L}{2}$; $F^2_{2*f} = \dfrac{q_o L^2}{12}$; $F^2_{5*f} = -\dfrac{q_o L^2}{12}$; $F^2_{3*f} = F^2_{6*f} = 0$

$$\Rightarrow \mathbf{F}^1_{*f} = \mathbf{F}^2_{*f} = \left[q_o L/2 \quad q_o L^2/12 \quad 0 \;\vdots\; q_o L/2 \quad -q_o L^2/12 \quad 0 \right]^T$$

Maximum deflection at mid-span = $\dfrac{q_o L^4}{384 EI}$ in AO and OB (due to fixity at two ends)

Coordinate Transformations and Equivalent Joint Loads

The element transformation matrices $\mathbf{T}^1$ and $\mathbf{T}^2$, satisfying $\mathbf{D}_*^i = \mathbf{T}^i\mathbf{D}$ and $\mathbf{F}_*^i = \mathbf{T}^i\mathbf{F}$, take the following form (refer Fig 5.40 and Eq. 5.46), with the linking global coordinates indicated in parentheses ($\theta_x^1 = 270^\circ + \alpha;\ \theta_x^1 = 90^\circ - \alpha \quad \Rightarrow \quad \sin\theta_x^1 = -\cos\alpha = -\dfrac{b}{L};\quad \sin\theta_x^2 = \cos\alpha = \dfrac{b}{L}$ and $\cos\theta_x^1 = \cos\theta_x^2 = \sin\alpha = \dfrac{a}{L}$)[†].

Let $\boxed{s = \sin\alpha = \dfrac{a}{L}}$ and $\boxed{c = \cos\alpha = \dfrac{b}{L}}$.

$$\mathbf{T}^1 = \begin{array}{c} \begin{matrix}(4) & (5) & (6) & (1) & (2) & (3)\end{matrix} \\ \left[\begin{array}{ccc|ccc} 1 & 0 & 0 & 0 & 0 & 0 \\ 0 & s & -c & 0 & 0 & 0 \\ 0 & c & s & 0 & 0 & 0 \\ \hline 0 & 0 & 0 & 1 & 0 & 0 \\ 0 & 0 & 0 & 0 & s & -c \\ 0 & 0 & 0 & 0 & c & s \end{array}\right] \end{array} \begin{matrix}(4)\\(5)\\(6)\\(1)\\(2)\\(3)\end{matrix} \quad \text{and} \quad \mathbf{T}^2 = \begin{array}{c} \begin{matrix}(1) & (2) & (3) & (7) & (8) & (9)\end{matrix} \\ \left[\begin{array}{ccc|ccc} 1 & 0 & 0 & 0 & 0 & 0 \\ 0 & s & c & 0 & 0 & 0 \\ 0 & -c & s & 0 & 0 & 0 \\ \hline 0 & 0 & 0 & 1 & 0 & 0 \\ 0 & 0 & 0 & 0 & s & c \\ 0 & 0 & 0 & 0 & -c & s \end{array}\right] \end{array} \begin{matrix}(1)\\(2)\\(3)\\(7)\\(8)\\(9)\end{matrix}$$

The fixed end forces in global axes system, $\mathbf{F}_\mathbf{f} = \left[\begin{array}{c}\mathbf{F}_{\mathbf{fA}} \\ \hline \mathbf{F}_{\mathbf{fR}}\end{array}\right]$, can now be assembled, by summing up the contributions of:

$$\mathbf{T}^{1^T}\mathbf{F}_{*\mathbf{f}}^1 = \left(\frac{q_o L}{12}\right)\left[\begin{array}{c} 6 \\ sL \\ -cL \\ \hline 6 \\ -sL \\ cL \end{array}\right] \begin{matrix}(4)\\(5)\\(6)\\(1)\\(2)\\(3)\end{matrix} \quad \text{and} \quad \mathbf{T}^{2^T}\mathbf{F}_{*\mathbf{f}}^2 = \left(\frac{q_o L}{12}\right)\left[\begin{array}{c} 6 \\ sL \\ cL \\ \hline 6 \\ -sL \\ -cL \end{array}\right] \begin{matrix}(1)\\(2)\\(3)\\(7)\\(8)\\(9)\end{matrix}$$

$$\Rightarrow \mathbf{F}_{\mathbf{fA}} = \begin{Bmatrix} F_{1f} \\ F_{2f} \\ F_{3f} \end{Bmatrix} = \left(\frac{q_o L}{6}\right)\begin{Bmatrix} 6 \\ 0 \\ cL \end{Bmatrix} \quad ; \quad \mathbf{F}_{\mathbf{fF}} = \begin{Bmatrix} F_{4f} \\ F_{5f} \\ F_{6f} \\ F_{7f} \\ F_{8f} \\ F_{9f} \end{Bmatrix} = \left(\frac{q_o L}{12}\right)\begin{Bmatrix} 6 \\ sL \\ -cL \\ 6 \\ -sL \\ -cL \end{Bmatrix}$$

The resultant load vector, $\mathbf{F}_\mathbf{A} - \mathbf{F}_{\mathbf{fA}} = -\mathbf{F}_{\mathbf{fA}}$, is given by:

[†] Alternatively, we can make use of Eq. 5.57, considering element 1 to have the *start node* at A (-*a*,0,*b*) and *end node* at O (0,0,0) and element 2 to have the *start node* at O (0,0,0) and *end node* at B (*a*,0,*b*).

$$\mathbf{F_A} - \mathbf{F_{fA}} = \begin{Bmatrix} F_1 \\ F_2 \\ F_3 \end{Bmatrix} = \left(\frac{q_o L}{6}\right) \begin{Bmatrix} -6 \\ 0 \\ -cL \end{Bmatrix}$$

The resultant nodal loads comprise a downward force of $q_o L$ and a moment (vector in the negative x-direction) of magnitude $\dfrac{q_o c L^2}{6}$ at O. The deflections at the active coordinates, D_1, D_2 and D_3, due to these loads will be identical to those in the original loading diagram (with the distributed loading) in Fig. 5.40b.

Element and Structure Stiffness Matrices

The element stiffness matrices $\mathbf{k}_*^1$ and $\mathbf{k}_*^2$, satisfying $\mathbf{F}_*^i = \mathbf{k}_*^i \mathbf{D}_*^i$, take the following form (refer Fig. 5.40c and Eq. 5.55), considering $G = \dfrac{E}{2(1+\nu)} = \dfrac{E}{2.6}$, $I = \pi r^3 t = \pi r^4 / 10$, $\dfrac{GJ}{EI} = \dfrac{1}{2.6}\dfrac{2\pi r^3 t}{\pi r^3 t} = 0.76923$ for both the elements:

$$\mathbf{k}_*^1 = \mathbf{k}_*^2 = \frac{\pi r^4 E}{10L^3}\begin{bmatrix} 12 & 6L & 0 & -12 & 6L & 0 \\ 6L & 4L^2 & 0 & -6L & 2L^2 & 0 \\ 0 & 0 & 0.76923L^2 & 0 & 0 & -0.76923L^2 \\ -12 & -6L & 0 & 12 & -6L & 0 \\ 6L & 2L^2 & 0 & -6L & 4L^2 & 0 \\ 0 & 0 & -0.76923L^2 & 0 & 0 & 0.76923L^2 \end{bmatrix}$$

$$\Rightarrow \quad \mathbf{k}_*^1\mathbf{T}^1 = \frac{\pi r^4 E}{10L^3}\begin{bmatrix} 12 & 6Ls & -6Lc & -12 & 6Ls & -6Lc \\ 6L & 4L^2 s & -4L^2 c & -6L & 2L^2 s & -2L^2 c \\ 0 & 0.76923L^2 c & 0.76923L^2 s & 0 & -0.76923L^2 c & -0.76923L^2 s \\ -12 & -6Ls & 6Lc & 12 & -6Ls & 6Lc \\ 6L & 2L^2 s & -2L^2 c & -6L & 4L^2 s & -4L^2 c \\ 0 & -0.76923L^2 c & -0.76923L^2 s & 0 & 0.76923L^2 c & 0.76923L^2 s \end{bmatrix}\begin{matrix}(4)\\(5)\\(6)\\(1)\\(2)\\(3)\end{matrix}$$

(columns: (4) (5) (6) (1) (2) (3))

$$\mathbf{k}_*^2\mathbf{T}^2 = \frac{\pi r^4 E}{10L^3}\begin{bmatrix} 12 & 6Ls & 6Lc & -12 & 6Ls & 6Lc \\ 6L & 4L^2 s & 4L^2 c & -6L & 2L^2 s & 2L^2 c \\ 0 & -0.76923L^2 c & 0.76923L^2 s & 0 & 0.76923L^2 c & -0.76923L^2 s \\ -12 & -6Ls & -6Lc & 12 & -6Ls & -6Lc \\ 6L & 2L^2 s & 2L^2 c & -6L & 4L^2 s & 4L^2 c \\ 0 & 0.76923L^2 c & -0.76923L^2 s & 0 & -0.76923L^2 c & 0.76923L^2 s \end{bmatrix}\begin{matrix}(1)\\(2)\\(3)\\(7)\\(8)\\(9)\end{matrix}$$

(columns: (1) (2) (3) (7) (8) (9))

$$\Rightarrow \quad \mathbf{T}^{1^T}\mathbf{k}_*^1\mathbf{T}^1 = \frac{\pi r^4 E}{10L^3}\begin{bmatrix} 12 & 6sL & -6cL & -12 & 6sL & -6cL \\ 6sL & (c_1)L^2 & -3.23077scL^2 & -6sL & (c_3)L^2 & -2.76923scL \\ -6cL & -3.23077scL^2 & (c_2)L^2 & 6cL & -2.76923scL & (c_4)L^2 \\ -12 & -6sL & 6cL & 12 & -6sL & 6cL \\ 6sL & (c_3)L^2 & -2.76923scL & -6sL & (c_1)L^2 & -3.23077scL^2 \\ -6cL & -2.76923scL^2 & (c_4)L^2 & 6cL & -3.23077scL^2 & (c_2)L^2 \end{bmatrix}\begin{matrix}(4)\\(5)\\(6)\\(1)\\(2)\\(3)\end{matrix}$$

(columns: (4) (5) (6) (1) (2) (3))

$$\mathbf{T}^{2^T}\mathbf{k}_*^2\mathbf{T}^2 = \frac{\pi r^4 E}{10L^3}\begin{bmatrix} 12 & 6sL & 6cL & -12 & 6sL & 6cL \\ 6sL & (c_1)L^2 & 3.23077scL^2 & -6sL & (c_3)L^2 & 2.76923scL \\ 6cL & 3.23077scL^2 & (c_2)L^2 & -6cL & 2.76923scL & (c_4)L^2 \\ -12 & -6sL & -6cL & 12 & -6sL & -6cL \\ 6sL & (c_3)L^2 & 2.76923scL & -6sL & (c_1)L^2 & 3.23077scL^2 \\ 6cL & 2.76923scL^2 & (c_4)L^2 & -6cL & 3.23077scL^2 & (c_2)L^2 \end{bmatrix}\begin{matrix}(1)\\(2)\\(3)\\(7)\\(8)\\(9)\end{matrix}$$

(columns: (1) (2) (3) (7) (8) (9))

where $c_1 = (4s^2 + 0.76923c^2)$, $c_2 = (4c^2 + 0.76923s^2)$, $c_3 = (2s^2 - 0.76923c^2)$, $c_4 = (2c^2 - 0.76923s^2)$

The structure stiffness matrix $\mathbf{k} = \left[\begin{array}{c|c} \mathbf{k}_{AA} & \mathbf{k}_{AR} \\ \hline \mathbf{k}_{RA} & \mathbf{k}_{RR} \end{array}\right]$, satisfying $\mathbf{F} = \mathbf{kD}$, can now be assembled, by summing up the contributions of $\mathbf{T}^{1^T}\mathbf{k}_*^1\mathbf{T}^1$ and $\mathbf{T}^{2^T}\mathbf{k}_*^2\mathbf{T}^2$. It can be shown that

$$\Rightarrow \quad \mathbf{k}_{AA} = \frac{\pi r^4 E}{10L^3}\begin{bmatrix} 24 & 0 & 12cL \\ 0 & (8s^2 + 1.53846c^2)L^2 & 0 \\ 12cL & 0 & (8c^2 + 1.53846s^2)L^2 \end{bmatrix}\begin{matrix}(1)\\(2)\\(3)\end{matrix}$$

$$\mathbf{k}_{RA} = \frac{\pi r^4 E}{10L^3}\begin{bmatrix} -12 & 6sL & -6cL \\ -6sL & (2s^2 - 0.76923c^2)L^2 & -2.76923scL^2 \\ 6cL & -2.76923scL^2 & (2c^2 - 0.76923s^2)L^2 \\ -12 & -6sL & -6cL \\ 6sL & (2s^2 - 0.76923c^2)L^2 & 2.76923scL^2 \\ 6cL & 2.76923scL^2 & (2c^2 - 0.76923s^2)L^2 \end{bmatrix}\begin{matrix}(4)\\(5)\\(6)\\(7)\\(8)\\(9)\end{matrix}$$

(columns: (1) (2) (3))

An expression for the inverse of $\mathbf{k}_{AA}$, which is a symmetric matrix of order 3×3, can be easily generated, using the adjoint method (refer Eq. 2.39b).

Displacements

The structure stiffness relation to be considered here is:

$$\left[\frac{\mathbf{F_A = O}}{\mathbf{F_R}}\right] - \left[\frac{\mathbf{F_{fA}}}{\mathbf{F_{fR}}}\right] = \left[\begin{array}{c|c} \mathbf{k_{AA}} & \mathbf{k_{AR}} \\ \hline \mathbf{k_{RA}} & \mathbf{k_{RR}} \end{array}\right]\left[\frac{\mathbf{D_A}}{\mathbf{D_R = O}}\right]$$

$$\Rightarrow \quad \mathbf{D_A} = [\mathbf{k_{AA}}]^{-1}[\mathbf{F_A} - \mathbf{F_{fA}}] \quad \Rightarrow$$

$$\Rightarrow \quad \begin{Bmatrix} D_1 \\ D_2 \\ D_3 \end{Bmatrix} = \left(\frac{10L}{\pi r^4 E}\right)\frac{1}{24(c_1)(c_2 - 3c^2)}\begin{bmatrix} (c_1)(c_2)L^2 & 0 & -6(c_1)cL \\ 0 & 12\left[(c_2) - 3c^2\right] & 0 \\ -6(c_1)cL & 0 & 12(c_1) \end{bmatrix}\left(\frac{q_o L}{6}\right)\begin{Bmatrix} -6 \\ 0 \\ -cL \end{Bmatrix}$$

$$= \left(\frac{q_o}{E}\right)\left(\frac{L}{r}\right)^4 \frac{1}{2.4\pi(c_2 - 3c^2)}\begin{Bmatrix} -(c_2 - c^2) \\ 0 \\ 4c/L \end{Bmatrix}$$

$$\Rightarrow \quad \boxed{\begin{aligned} D_1 &= -\left(\frac{q_o}{E}\right)\left(\frac{L}{r}\right)^4 \frac{(3c^2 + 0.76923s^2)}{2.4\pi(c^2 + 0.76923s^2)} \\ D_2 &= 0 \\ D_3 &= \frac{1}{L}\left(\frac{q_o}{E}\right)\left(\frac{L}{r}\right)^4 \frac{4c}{2.4\pi(c^2 + 0.76923s^2)} \end{aligned}}$$

We can easily verify the accuracy of these solutions for the two extreme values of α. The deflection profile of the grid element OA (including the fixed-end force effects) is shown in Fig. 5.41a. The deflection is largest at the joint O, and at this location, Δ_O takes the largest value when $\alpha = 0^\circ$ (the two beams act independently as cantilevers, whereby $(\Delta_O)_{\alpha=0^\circ} = \frac{q_o L^4}{8EI}$). The deflection Δ_O is lowest, when $\alpha = 90^\circ$ (the two beams join together to form a fixed beam of span 2*L*, whose mid-span deflection at O is given by $\frac{q_o(2L)^4}{384EI}$). The rotation of the joint O (with reference to the global *x*-axis), likewise, reduces from the largest value, $(\theta_O)_{\alpha=0^\circ} = \frac{q_o L^3}{6EI}$, at $\alpha = 0^\circ$ to zero at $\alpha = 90^\circ$. The rotation in the perpendicular direction (about the global *z*-axis) is always zero (on account of symmetry). Hence, the components of the net rotation at O (D_3) can be easily resolved into flexural and torsional rotation components ($\theta_O = D_3 \cos\alpha$ and $\phi_O = D_3 \sin\alpha$ respectively) at the end O of the two grid elements. The variations of Δ_O, θ_O and ϕ_O, with respect to the angle α, can be conveniently expressed in terms of non-dimensional coefficients, C_Δ, C_θ and C_ϕ respectively, as shown in Fig. 5.41a, and as depicted graphically in Fig. 5.41b.

The angle of twist, ϕ_O, is zero when $\alpha = 0^\circ$ (AO behaves like a cantilever, free at O) and also when $\alpha = 90^\circ$ (AOB becomes a straight fixed beam), and takes on the largest value when $\alpha \simeq 49°$. When $\alpha = 45°$, AO and BO become mutually perpendicular at O, with the twisting moment at O in one element being equal to the bending moment at O in the other.

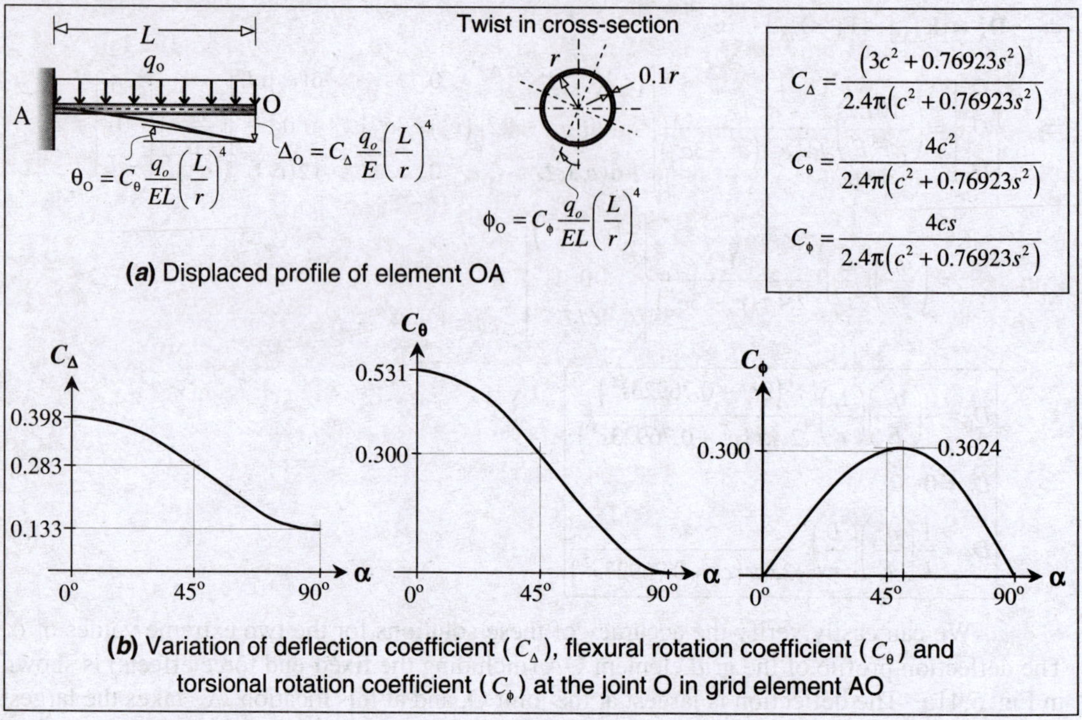

Figure 5.41 Displacement response of grid element (Example 5.13)

Support Reactions

$$\mathbf{F_R} = \mathbf{F_{fR}} + \mathbf{k_{RA}}\mathbf{D_A}$$

$$\Rightarrow \begin{Bmatrix} F_4 \\ F_5 \\ F_6 \\ F_7 \\ F_8 \\ F_9 \end{Bmatrix} = \left(\frac{q_o L}{12}\right)\begin{Bmatrix} 6 \\ sL \\ -cL \\ 6 \\ -sL \\ -cL \end{Bmatrix} + \frac{1}{2(c^2 + 0.76923s^2)}\begin{bmatrix} -12 & 6sL & -6cL \\ -6sL & (c_3)L^2 & -2.76923scL^2 \\ 6cL & -2.76923scL^2 & (c_4)L^2 \\ -12 & -6sL & -6cL \\ 6sL & (c_3)L^2 & 2.76923scL^2 \\ 6cL & 2.76923scL^2 & (c_4)L^2 \end{bmatrix}\begin{Bmatrix} -(3c^2 + 0.76923s^2) \\ 0 \\ 4c/L \end{Bmatrix}$$

$$\Rightarrow \boxed{\begin{aligned} &F_4 = F_7 = q_o L \\ &F_5 = -F_8 = \frac{q_o L^2}{12}(s)\left[1 + \frac{3.46154c^2 + 2.30769s^2}{c^2 + 0.76923s^2}\right] \\ &F_6 = F_9 = -\frac{q_o L^2}{12}(c)\left[1 + \frac{5c^2 + 3.84615s^2}{c^2 + 0.76923s^2}\right] \end{aligned}}$$

Member Forces

$$\mathbf{F}_*^i = \mathbf{F}_{*f}^i + \mathbf{k}_*^i \mathbf{T}^i \mathbf{D}$$

Noting that $\mathbf{D_R} = \mathbf{O}$, the calculations can be simplified as $\mathbf{F}_*^i = \mathbf{F}_{*f}^i + \mathbf{k}_\mathbf{A}^i \mathbf{T}_\mathbf{A}^i \mathbf{D_A}$:

$$\Rightarrow \mathbf{F}_*^1 = \begin{Bmatrix} q_oL/2 \\ q_oL^2/12 \\ 0 \\ q_oL/2 \\ -q_oL^2/12 \\ 0 \end{Bmatrix} + \frac{q_oL}{24(c_2-3c^2)} \begin{bmatrix} -12 & 6sL & -6cL \\ -6L & 2sL^2 & -2cL^2 \\ 0 & -0.76923cL^2 & -0.76923sL^2 \\ 12 & -6sL & 6cL \\ -6L & 4sL^2 & -4cL^2 \\ 0 & 0.76923cL^2 & 0.76923sL^2 \end{bmatrix} \begin{Bmatrix} -(c_2-c^2) \\ 0 \\ 4c/L \end{Bmatrix}$$

$$F_{1*}^1 = q_oL;\ F_{4*}^1 = 0$$

$$\Rightarrow F_{2*}^1 = \frac{q_oL^2}{12}\left[1 + \frac{5c^2 + 2.30769s^2}{c^2 + 0.76923s^2}\right];\ F_{5*}^1 = \frac{q_oL^2}{12}\left[-1 + \frac{c^2 + 2.30769s^2}{c^2 + 0.76923s^2}\right]$$

$$F_{3*}^1 = -\frac{q_oL^2}{12}\left[\frac{1.53846sc}{c^2 + 0.76923s^2}\right];\ F_{6*}^1 = \frac{q_oL^2}{12}\left[\frac{1.53846sc}{c^2 + 0.76923s^2}\right]$$

$$\Rightarrow \mathbf{F}_*^2 = \begin{Bmatrix} q_oL/2 \\ q_oL^2/12 \\ 0 \\ q_oL/2 \\ -q_oL^2/12 \\ 0 \end{Bmatrix} + \frac{q_oL}{24(c_2-3c^2)} \begin{bmatrix} 12 & 6sL & 6cL \\ 6L & 4sL^2 & 4cL^2 \\ 0 & -0.76923cL^2 & 0.76923sL^2 \\ -12 & -6sL & -6cL \\ 6L & 2sL^2 & 2cL^2 \\ 0 & 0.76923cL^2 & -0.76923sL^2 \end{bmatrix} \begin{Bmatrix} -(c_2-c^2) \\ 0 \\ 4c/L \end{Bmatrix}$$

$$F_{1*}^2 = 0;\ F_{4*}^2 = q_oL$$

$$\Rightarrow F_{2*}^2 = \frac{q_oL^2}{12}\left[1 - \frac{c^2 + 2.30769s^2}{c^2 + 0.76923s^2}\right];\ F_{5*}^2 = \frac{q_oL^2}{12}\left[-1 - \frac{5c^2 + 2.30769s^2}{c^2 + 0.76923s^2}\right]$$

$$F_{3*}^2 = \frac{q_oL^2}{12}\left[\frac{1.53846sc}{c^2 + 0.76923s^2}\right];\ F_{6*}^2 = -\frac{q_oL^2}{12}\left[\frac{1.53846sc}{c^2 + 0.76923s^2}\right]$$

The free-body, shear force, typical bending moment and twisting moment diagrams, including the fixed end force effects, are shown in Fig. 5.42. The grid element AO behaves essentially like a cantilever beam AO, fixed at A and connected to rotational springs (flexural and torsional) at the end O (spring stiffnesses depending on the angle α). The vertical reaction at O is always equal to zero, and hence, the shear force distribution (linearly varying along the span) does not depend on the angle α. When $\alpha = 0^\circ$ (AO behaves like a cantilever, free at O), the bending moment is fully hogging along the span (with a parabolic variation) and there is no torsion. With increasing values of α, the connecting beam at O offers rotational restraint, introducing sagging moments near O and a torque at O. This reduces the fixed end moment at O and introduces a constant twisting moment in the span of the element.

The twisting moment is directly given by the product of the torsional stiffness (GJ/L) of the element and the angle of twist, ϕ_o; the largest value occurs when $\alpha = 49°$, as explained earlier. With further increase in α, the twisting moment reduces, while the sagging moment at O increases and the hogging moment at the fixed end A decreases. When $\alpha = 90°$ (AOB becomes a straight fixed beam), the twisting moment becomes zero and the fixed end moment at A becomes equal to $\dfrac{q_o(2L)^2}{12}$.

The variations of the hogging moment, M, and the twisting moment, T, at the fixed end A, with respect to the angle α, can be conveniently expressed in terms of non-dimensional moment coefficients, C_M and C_T respectively (multiplied by $q_o L^2$), as shown in Fig. 5.42a, and as depicted graphically in Fig. 5.42e. The bending moment (sagging) at the end O is easily obtainable as $(0.5 - C_M)q_o L^2$.

$$C_M = \frac{1}{12}\left[1 + \frac{5c^2 + 2.30769s^2}{c^2 + 0.76923s^2}\right]$$

$$C_T = \frac{1}{12}\left[\frac{1.53846sc}{c^2 + 0.76923s^2}\right]$$

(a) Free-body of element OA

(b) Shear force diagram

(c) Bending moment diagram

(d) Twisting moment diagram

(e) Variation of bending moment coefficient (C_M) and twisting moment coefficient (C_T) in grid element AO

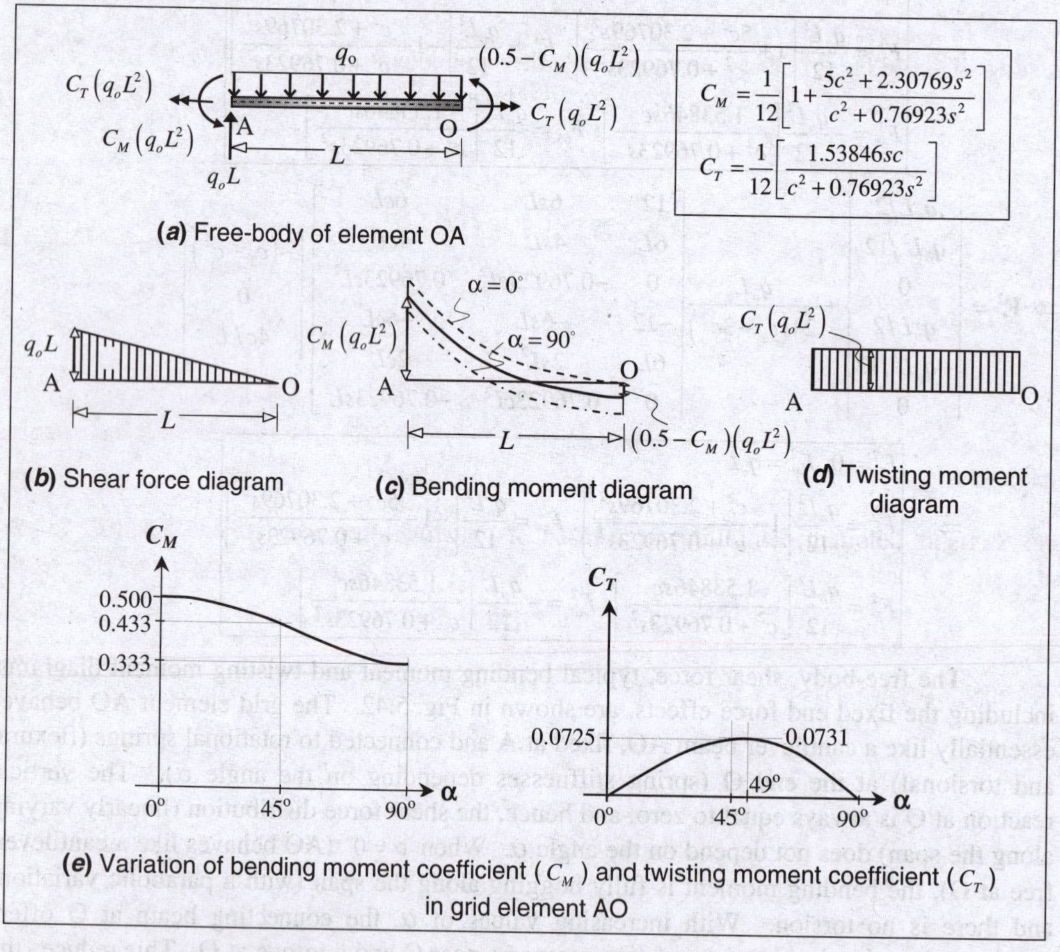

Figure 5.42 Force response of grid element (Example 5.13)

5.5.3 Analysis of Grids by Reduced Element Stiffness Method

The main advantage of the reduced element stiffness method, as applied to grids, is that we can work with three degrees of freedom per grid element, instead of six degrees of freedom, as shown in Fig. 5.37b. We can also make use of beam elements with two degrees of freedom each, when we are certain that there is no twisting moment induced in the element under the given loading. Furthermore, when we encounter hinged and guided-fixed end supports, we can further simplify and work with a single rotational degree of freedom per element, as explained in Sections 5.3.2 and 5.3.3. Thus, the overall kinematic indeterminacy of the grid structure can be considerably reduced, and this is especially possible in symmetrical grids. The only apparent disadvantage is that we do not get all the member-end displacements in local coordinates for every element directly as output. We also need to input the fixed end force effects directly, as explained earlier.

As explained earlier, the structure stiffness matrix $\mathbf{k}$ can be assembled by a 'direct stiffness' approach, using the element displacement transformation matrix, $\mathbf{T_D^i}$, and summing up the contributions, $\mathbf{T_D^i}^{\mathbf{T}}\tilde{\mathbf{k}}_*^{\mathbf{i}}\mathbf{T_D^i}$ (direct stiffness method).

We can generate the displacement transformation $\mathbf{T_D^i}$ matrices for the various elements (grid elements or beam elements) applying a unit displacement at a time ($D_j = 1$) on the restrained (primary) structure, whereby $\mathbf{D_*^i} = \left[\mathbf{T_{DA}^i} \;\vdots\; \mathbf{T_{DR}^i}\right]\begin{bmatrix}\mathbf{D_A} \\ \hline \mathbf{D_R}\end{bmatrix}$. Alternatively, we can adopt a static approach instead of a kinematic approach, to generate the $\mathbf{T_D^i}^{\mathbf{T}}$ matrix, by making use of the *Principle of Contra-gradient*, whereby $\begin{bmatrix}\mathbf{F_A} \\ \hline \mathbf{F_R}\end{bmatrix} = \begin{bmatrix}\mathbf{T_{DA}^i}^{\mathbf{T}} \\ \hline \mathbf{T_{DR}^i}^{\mathbf{T}}\end{bmatrix}\mathbf{F_*^i}$, as explained earlier. In the case of beam elements, the components of the $\mathbf{T_D^i}$ will in general be as discussed earlier in Section 5.3, except that we need to account for the fact that the local x^*-axis of the element may be inclined to the global x-axis. In such cases, and in grid elements (where twisting is involved, in addition to bending), the direction cosine components $c_i = \cos\theta_x^i$ and $s_i = \sin\theta_x^i$, need to be appropriately included, as demonstrated in Eq. 5.56 (refer Fig. 5.39). For example, if we replace the 6 degree of freedom model in Fig. 5.39b with the reduced three degree of freedom model in Fig. 5.37b, and we consider the most general case of six global degrees of freedom acting at the ends of the element, numbered as shown in Fig. 5.39a, then the transformation, $\mathbf{D_*^i} = \mathbf{T_D^i D}$, takes the following form:

$$\begin{Bmatrix} D_{1*}^i \\ D_{2*}^i \\ D_{3*}^i \end{Bmatrix} = \left[\begin{array}{ccc|ccc} 1/L_i & c_i & s_i & -1/L_i & 0 & 0 \\ 1/L_i & 0 & 0 & -1/L_i & c_i & s_i \\ 0 & -s_i & c_i & 0 & -s_i & c_i \end{array}\right]\left[\begin{array}{ccc|ccc} D_1^i & D_2^i & D_3^i & D_4^i & D_5^i & D_6^i \end{array}\right]^{\mathrm{T}} \tag{5.57}$$

where $c_i = \cos\theta_x^i$ and $s_i = \sin\theta_x^i$, θ_x^i being the angle made by the local x^*-axis with respect to the global x-axis (measured anti-clockwise, as shown in Fig. 5.39).

In practice, the global coordinates may have any set of numbers, some of which may correspond to restrained coordinates. Also, as explained earlier, it is convenient to also generate and store the $\tilde{\mathbf{k}}_*^i \mathbf{T}_D^i$ matrices for the various elements to facilitate the computation of the bar forces using the relation, $\mathbf{F}_*^i = \left(\tilde{\mathbf{k}}_*^i \mathbf{T}_D^i\right)\mathbf{D}$, and also to store the products, $\mathbf{T}_D^{i\,T}\tilde{\mathbf{k}}_*^i \mathbf{T}_D^i$, for assembling the structure stiffness matrix $\mathbf{k}$. The final structure stiffness matrix $\mathbf{k}$ generated will be identical to the one generated by the conventional stiffness method.

We will now demonstrate the application of the reduced element stiffness method to two grid example problems, in which we take advantage of symmetry in the grid. By careful examination of the structure and an understanding of the response, we can work with simple reduced models and with minimal degree of kinematic indeterminacy. The examples also demonstrate how beam elements can be effectively used instead of grid elements, when twisting moments are absent.

EXAMPLE 5.14

Fig. 5.43 shows the layout plan of an orthogonal grid of floor beams, with fixed supports at A, B, C, D, E and F. Assuming all the grid elements to have identical flexural rigidity, $EI = 5\times10^4$ kN/m^2, and torsional rigidity $GJ = 0.2EI$, analyse the grid system, given that all the grid elements are subjected to a uniformly distributed gravity load of 30 kN/m. Find the maximum deflection, and draw the shear force, bending moment and twisting moment diagrams of elements CD and AB.

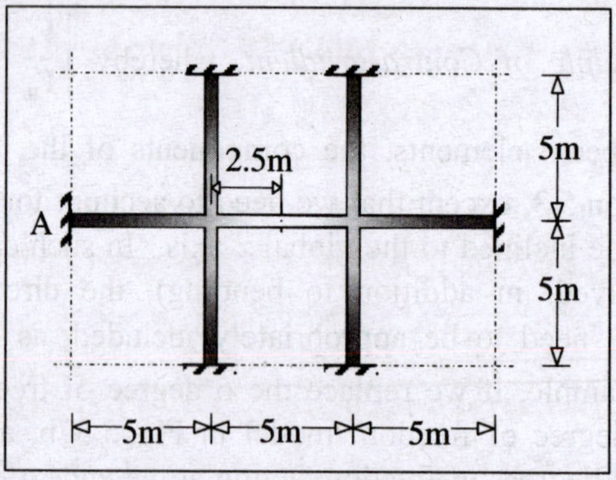

Figure 5.43 Layout plan of a grid floor (Example 5.14)

SOLUTION

As the layout plan and the loading are both symmetric (with respect to the vertical axis passing through the centre), the response will also be symmetric, and there will be no torsion in AGHB (which lies on an axis of symmetry). It is sufficient to analyse the reduced model, shown in Fig. 5.44a, truncated at J (middle of the span GH), where a guided fixed support is provided. There are two *active* and 8 *restrained* global coordinates, numbered as shown in Fig. 5.44b. Elements DG and CG ('1' and '2') are modelled as grid elements, with three degrees of freedom each. Taking advantage of the absence of torsion, elements AG ('3') and GJ ('4') are modelled as beam elements, the former considered to be simply supported with two degrees of freedom and the latter as a cantilever with one degree of freedom (refer Section 5.3.3). The local coordinates are shown in Fig. 5.44c. There are no nodal loads applied; i.e., $\mathbf{F}_A = \mathbf{O}$.

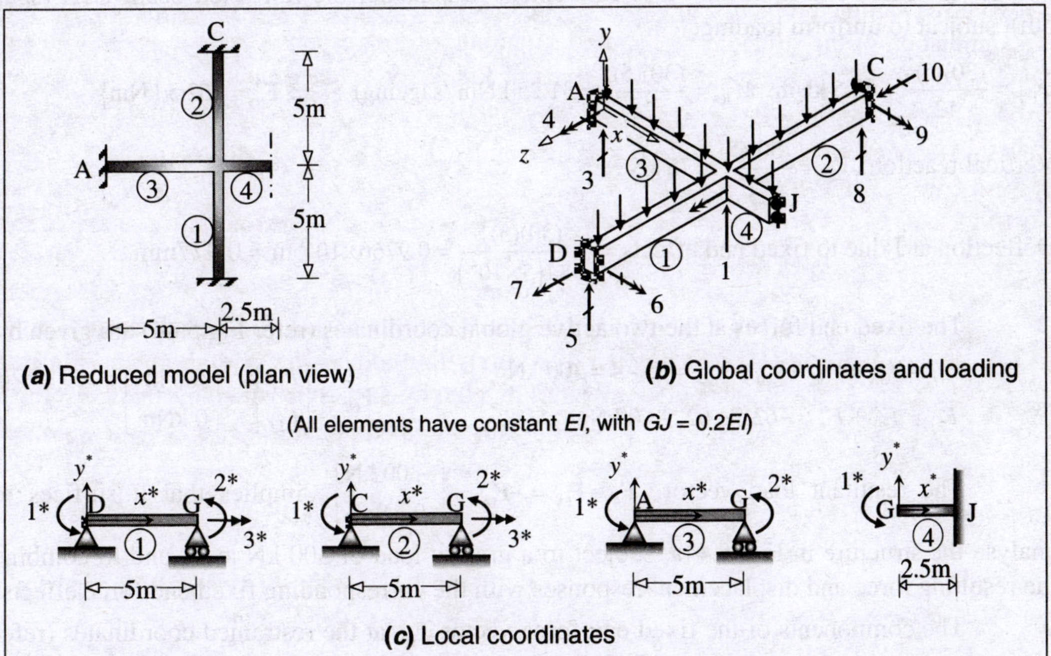

Figure 5.44 Reduced model: global and local coordinates (Example 5.14)

Fixed End Forces (due to uniformly distributed loading of 30 kN/m)

Grid Elements '1' and '2' (DG and CG):

$$F^1_{1^*f} = F^2_{1^*f} = \frac{30\times 5^2}{12} = 62.5 \text{ kNm}; \; F^1_{2^*f} = F^2_{2^*f} = -62.5 \text{ kNm}; \; F^1_{3^*f} = F^2_{3^*f} = 0 \text{ kNm}$$

$$\Rightarrow \mathbf{F}^1_{*f} = \mathbf{F}^2_{*f} = \begin{Bmatrix} 62.5 \text{ kNm} \\ -62.5 \text{ kNm} \\ 0 \text{ kNm} \end{Bmatrix}$$

Vertical reactions: $V_{CG} = V_{DG} = V_{GC} = V_{GD} = \dfrac{(30)(5)}{2} = 75 \text{ kN}$

Beam Element '3' (AG):

$$F^3_{1*f} = \frac{30\times5^2}{12} = 62.5 \text{ kNm};\ F^3_{2*f} = -62.5 \text{ kNm} \qquad \Rightarrow \mathbf{F^3_{*f}} = \begin{Bmatrix} 62.5 \text{ kNm} \\ -62.5 \text{ kNm} \end{Bmatrix}$$

Vertical reactions: $V_{AG} = V_{GA} = \frac{(30)(5)}{2} = 75 \text{ kN}$

Beam Element '4' (GJ):

To find the fixed end moments, it is convenient to consider the full fixed beam GJH (span 5m), subject to uniform loading:

$$F^4_{1*f} = \frac{(30)(5)^2}{12} = 62.5 \text{ kNm};\ M_{JG} = \frac{(30)(5)^2}{24} = 31.25 \text{ kNm (sagging)} \qquad \Rightarrow \mathbf{F^4_{*f}} = \{62.5 \text{ kNm}\}$$

Vertical reaction: $V_{GJ} = \frac{(30)(5)}{2} = 75 \text{ kN}$

Deflection at J due to fixed end effects $= \frac{(30)(5)^4}{384\left(5\times10^4\right)} = 0.9766\times10^{-3} \text{ m} = 0.977\text{mm}$

The fixed end forces at the two active global coordinates (refer Fig.5.44) are given by

$$F_{1f} = V_{GA} + V_{GC} + V_{GJ} + V_{GD} = 75\times4 = 300 \text{ kN}$$
$$F_{2f} = F^3_{2*} + F^4_{1*} = -62.5 + 62.5 = 0 \text{ kNm}$$
$$\Rightarrow \mathbf{F_{fA}} = \begin{Bmatrix} F_{1f} \\ F_{2f} \end{Bmatrix} = \begin{Bmatrix} 300 \text{ kN} \\ 0 \text{ kNm} \end{Bmatrix}$$

The resultant load vector, $\mathbf{F_A} - \mathbf{F_{fA}} = -\mathbf{F_{fA}} = \begin{Bmatrix} -300 \text{ kN} \\ 0 \text{ kNm} \end{Bmatrix}$, implies that it suffices to analyse the structure in Fig. 5.44a, subject to a gravity load of 300 kN at G, and to combine the resulting force and displacement responses with the corresponding fixed end force effects.

The components of the fixed end force vector $\mathbf{F_{fR}}$ at the restrained coordinates (refer Fig.5.44b) are given by:

$$F_{3f} = F_{5f} = F_{8f} = 75 \text{ kN};\ F_{4f} = F_{6f} = 62.5 \text{ kNm};\ F_{9f} = -62.5 \text{ kNm};\ F_{7f} = F_{10f} = 0 \text{ kNm};$$

Displacement Transformation Matrices

The element displacement transformation matrices, $\mathbf{T^1_D}$ and $\mathbf{T^2_D}$, relating the local coordinates to the global coordinates through the relation, $\mathbf{D^i_*} = \left[\mathbf{T^i_{DA}} \mid \mathbf{T^i_{DR}}\right]\left[\begin{array}{c}\mathbf{D_A} \\ \hline \mathbf{D_R}\end{array}\right]$, take the following form (refer Figs 5.44b and c):

$$\mathbf{T^1_D} = \begin{array}{c} \begin{matrix}(1) & (2) & (3) & (4) & (5) & (6) & (7) & (8) & (9) & (10)\end{matrix} \\ \left[\begin{array}{cc|cccccccc} -0.2 & 0 & 0 & 0 & 0.2 & 1 & 0 & 0 & 0 & 0 \\ -0.2 & 0 & 0 & 0 & 0.2 & 0 & 0 & 0 & 0 & 0 \\ 0 & -1 & 0 & 0 & 0 & 0 & 1 & 0 & 0 & 0 \end{array}\right] \end{array};\quad \mathbf{T^2_D} = \begin{array}{c} \begin{matrix}(1) & (2) & (3) & (4) & (5) & (6) & (7) & (8) & (9) & (10)\end{matrix} \\ \left[\begin{array}{cc|cccccccc} -0.2 & 0 & 0 & 0 & 0 & 0 & 0 & 0.2 & -1 & 0 \\ -0.2 & 0 & 0 & 0 & 0 & 0 & 0 & 0.2 & 0 & 0 \\ 0 & 1 & 0 & 0 & 0 & 0 & 0 & 0 & 0 & -1 \end{array}\right] \end{array}$$

$$\mathbf{T_D^3} = \overset{(1)\ (2)\ (3)\ (4)\ (5)\ (6)\ (7)\ (8)\ (9)\ (10)}{\begin{bmatrix} -0.2 & 0 & 0.2 & 1 & 0 & 0 & 0 & 0 & 0 & 0 \\ -0.2 & 1 & 0.2 & 0 & 0 & 0 & 0 & 0 & 0 & 0 \end{bmatrix}}$$

$$\mathbf{T_D^4} = \begin{bmatrix} 0 & 1 & 0 & 0 & 0 & 0 & 0 & 0 & 0 & 0 \end{bmatrix}$$

Element and Structure Stiffness Matrices

The element stiffness matrices $\tilde{\mathbf{k}}_*^1$ and $\tilde{\mathbf{k}}_*^2$, satisfying $\mathbf{F}_*^i = \tilde{\mathbf{k}}_*^i \mathbf{D}_*^i$, take the following form (refer Fig. 5.44c), considering $\frac{E_1I_1}{L_1} = \frac{E_2I_2}{L_2} = \frac{E_3I_3}{L_3} = \frac{EI}{5} = 0.2EI$, $\frac{E_4I_4}{L_4} = \frac{EI}{2.5} = 0.4EI$ and $\frac{G_1J_1}{L_1} = \frac{G_2J_2}{L_2} = \frac{0.2EI}{5} = 0.04EI$:

$$\tilde{\mathbf{k}}_*^1 = \tilde{\mathbf{k}}_*^2 = \frac{EI}{5}\begin{bmatrix} 4 & 2 & 0 \\ 2 & 4 & 0 \\ 0 & 0 & 0.2 \end{bmatrix} = EI\begin{bmatrix} 0.8 & 0.4 & 0 \\ 0.4 & 0.8 & 0 \\ 0 & 0 & 0.04 \end{bmatrix}$$

$$\tilde{\mathbf{k}}_*^3 = EI\begin{bmatrix} 0.8 & 0.4 \\ 0.4 & 0.8 \end{bmatrix} \quad \text{and} \quad \tilde{\mathbf{k}}_*^4 = \frac{EI}{2.5}[1] = EI[0.4]$$

$$\Rightarrow \quad \tilde{\mathbf{k}}_*^1\mathbf{T_D^1} = EI\begin{bmatrix} -0.24 & 0 & 0 & 0 & 0.24 & 0.8 & 0 & 0 & 0 & 0 \\ -0.24 & 0 & 0 & 0 & 0.24 & 0.4 & 0 & 0 & 0 & 0 \\ 0 & -0.04 & 0 & 0 & 0 & 0 & 0.04 & 0 & 0 & 0 \end{bmatrix}$$

$$\tilde{\mathbf{k}}_*^2\mathbf{T_D^2} = EI\begin{bmatrix} -0.24 & 0 & 0 & 0 & 0 & 0 & 0 & 0.24 & -0.8 & 0 \\ -0.24 & 0 & 0 & 0 & 0 & 0 & 0 & 0.24 & -0.4 & 0 \\ 0 & 0.04 & 0 & 0 & 0 & 0 & 0 & 0 & 0 & -0.04 \end{bmatrix}$$

$$\tilde{\mathbf{k}}_*^3\mathbf{T_D^3} = \begin{bmatrix} -0.24 & 0.4 & 0.24 & 0.8 & 0 & 0 & 0 & 0 & 0 & 0 \\ -0.24 & 0.8 & 0.24 & 0.4 & 0 & 0 & 0 & 0 & 0 & 0 \end{bmatrix}$$

$$\tilde{\mathbf{k}}_*^4\mathbf{T_D^4} = \begin{bmatrix} 0 & 0.4 & 0 & 0 & 0 & 0 & 0 & 0 & 0 & 0 \end{bmatrix}$$

The structure stiffness matrix $\mathbf{k} = \left[\begin{array}{c|c} \mathbf{k_{AA}} & \mathbf{k_{AR}} \\ \hline \mathbf{k_{RA}} & \mathbf{k_{RR}} \end{array}\right]$, satisfying $\mathbf{F} = \mathbf{kD}$, can now be assembled as $\mathbf{k} = \sum_{i=1}^{4} \mathbf{T_D^i}^{\mathbf{T}} \tilde{\mathbf{k}}_*^i \mathbf{T_D^i}$:

$$\Rightarrow \quad [\mathbf{k_{AA}}] = EI\begin{bmatrix} 0.288 & -0.24 \\ -0.24 & 1.28 \end{bmatrix} \quad \Rightarrow \quad [\mathbf{k_{AA}}]^{-1} = \frac{1}{EI}\begin{bmatrix} 4.11523 & 0.77160 \\ 0.77160 & 0.92593 \end{bmatrix}$$

$$\mathbf{k_{AR}} = \mathbf{k_{RA}}^{\mathbf{T}} = \begin{bmatrix} -0.096 & -0.24 & -0.096 & -0.24 & 0 & -0.096 & 0.24 & 0 \\ 0.24 & 0.40 & 0 & 0 & 0.04 & 0 & 0 & -0.04 \end{bmatrix}$$

Displacements and Support Reactions

The structure stiffness relation to be considered here is:

$$\begin{bmatrix} \mathbf{F_A} \\ \mathbf{F_R} \end{bmatrix} - \begin{bmatrix} \mathbf{F_{fA}} \\ \mathbf{F_{fR}} \end{bmatrix} = \begin{bmatrix} \mathbf{k_{AA}} & \mathbf{k_{AR}} \\ \mathbf{k_{RA}} & \mathbf{k_{RR}} \end{bmatrix} \begin{bmatrix} \mathbf{D_A} \\ \mathbf{D_R} = \mathbf{O} \end{bmatrix}$$

$$\Rightarrow \quad \mathbf{D_A} = [\mathbf{k_{AA}}]^{-1}[\mathbf{F_A} - \mathbf{F_{fA}}]$$

$$\Rightarrow \begin{Bmatrix} D_1 \\ D_2 \end{Bmatrix} = \frac{1}{EI}\begin{bmatrix} 4.11523 & 0.77160 \\ 0.77160 & 0.92593 \end{bmatrix} \begin{Bmatrix} -300 \text{ kN} \\ 0 \text{ kNm} \end{Bmatrix} = \frac{1}{EI}\begin{Bmatrix} -1234.568 \\ -231.4815 \end{Bmatrix}$$

$$\mathbf{F_R} = \mathbf{F_{fR}} + \mathbf{k_{RA}}\mathbf{D_A} \Rightarrow \begin{Bmatrix} F_3 \\ F_4 \\ F_5 \\ F_6 \\ F_7 \\ F_8 \\ F_9 \\ F_{10} \end{Bmatrix} = \begin{Bmatrix} 75 \text{ kN} \\ 62.5 \text{ kNm} \\ 75 \text{ kN} \\ 62.5 \text{ kNm} \\ 0 \text{ kNm} \\ 75 \text{ kN} \\ -62.5 \text{ kNm} \\ 0 \text{ kNm} \end{Bmatrix} + \begin{Bmatrix} 62.963 \text{ kN} \\ 203.704 \text{ kNm} \\ 118.518 \text{ kN} \\ 296.296 \text{ kNm} \\ 9.259 \text{ kNm} \\ 118.518 \text{ kN} \\ -296.296 \text{ kNm} \\ 9.259 \text{ kNm} \end{Bmatrix} = \begin{Bmatrix} 137.963 \text{ kN} \\ 266.204 \text{ kNm} \\ 193.519 \text{ kN} \\ 358.796 \text{ kNm} \\ 9.259 \text{ kNm} \\ 193.518 \text{ kN} \\ -358.796 \text{ kNm} \\ 9.259 \text{ kNm} \end{Bmatrix}$$

Member Forces

$$\mathbf{F^i_*} = \mathbf{F^i_{*f}} + \tilde{\mathbf{k}}^i_{\mathbf{DA}}\mathbf{T^i_{DA}}\mathbf{D_A}$$

$$\Rightarrow \begin{Bmatrix} F^1_{1*} \\ F^1_{2*} \\ F^1_{3*} \end{Bmatrix} = \begin{Bmatrix} 62.5 \text{ kNm} \\ -62.5 \text{ kNm} \\ 0 \text{ kNm} \end{Bmatrix} + EI\begin{bmatrix} -0.24 & 0 \\ -0.24 & 0 \\ 0 & -0.04 \end{bmatrix} \frac{1}{EI}\begin{Bmatrix} -1234.568 \\ -231.4815 \end{Bmatrix} = \begin{Bmatrix} 358.796 \text{ kNm} \\ 233.796 \text{ kNm} \\ 9.259 \text{ kNm} \end{Bmatrix}$$

$$\begin{Bmatrix} F^2_{1*} \\ F^2_{2*} \\ F^2_{3*} \end{Bmatrix} = \begin{Bmatrix} 62.5 \text{ kNm} \\ -62.5 \text{ kNm} \\ 0 \text{ kNm} \end{Bmatrix} + EI\begin{bmatrix} -0.24 & 0 \\ -0.24 & 0 \\ 0 & 0.04 \end{bmatrix} \frac{1}{EI}\begin{Bmatrix} -1234.568 \\ -231.4815 \end{Bmatrix} = \begin{Bmatrix} 358.796 \text{ kNm} \\ 233.796 \text{ kNm} \\ -9.259 \text{ kNm} \end{Bmatrix}$$

$$\begin{Bmatrix} F^3_{1*} \\ F^3_{2*} \end{Bmatrix} = \begin{Bmatrix} 62.5 \text{ kNm} \\ -62.5 \text{ kNm} \end{Bmatrix} + EI\begin{bmatrix} -0.24 & 0.4 \\ -0.24 & 0.8 \end{bmatrix} \frac{1}{EI}\begin{Bmatrix} -1234.568 \\ -231.4815 \end{Bmatrix} = \begin{Bmatrix} 266.204 \text{ kNm} \\ 48.611 \text{ kNm} \end{Bmatrix}$$

$$\{F^4_{1*}\} = \{62.5 \text{ kNm}\} + EI[0 \quad 0.4]\frac{1}{EI}\begin{Bmatrix} -1234.568 \\ -231.4815 \end{Bmatrix} = \{-30.093 \text{ kNm}\}$$

The free-body, shear force, bending moment and twisting moment diagrams are shown in Figs 5.45a, b, c and d respectively. It can be seen that the beam AGHB behaves like a three-span continuous beam, with fixed end supports at A and B and elastic supports (translational and rotational springs) at G and H. This results in a vertical force of 87.04 kN plus a twisting moment of 18.52 kNm being transmitted at the two joints G and H to the grid beams DGC and FHE.

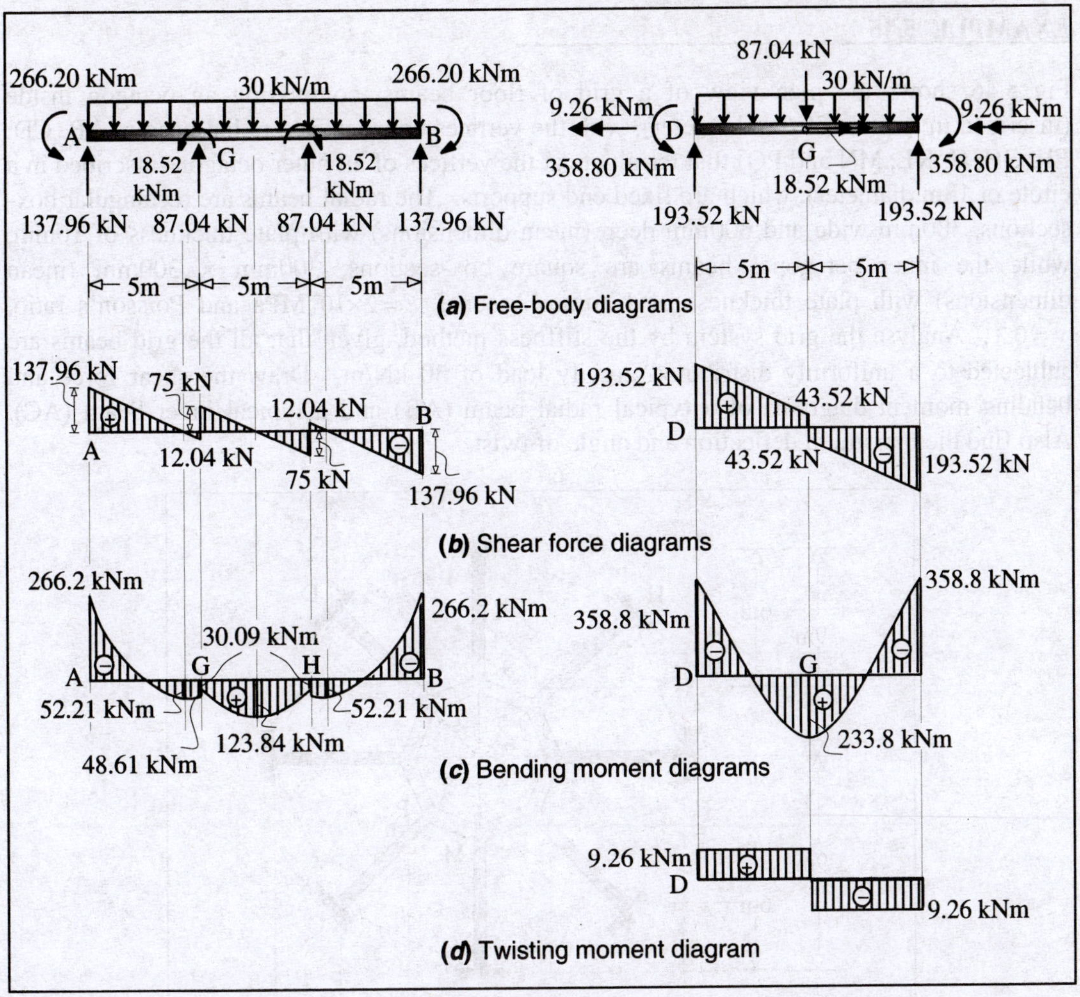

Figure 5.45 Free-bodies, shear force, bending moment and twisting moment diagrams (Example 5.14)

Maximum Deflection

The maximum deflection occurs at J (centre of the grid), and is given by the sum of (i) deflection caused by fixed end forces (calculated earlier as 0.977mm), (ii) deflections at G and H (equal to $-D_1 = \dfrac{1234.568}{5\times10^4} = 24.691\times10^{-3}$ m) and (iii) additional deflection at J due to rotation at G (given by $-D_2\dfrac{L_{GJ}}{2} = \dfrac{(231.4815)(2.5)}{2(5\times10^4)} = 5.787\times10^{-3}$ m):

$$\Delta_{max} = 0.977 + 24.691 + 5.787 = 31.455 \text{ mm}$$

EXAMPLE 5.15

Fig. 5.46 shows the plan view of a grid of floor beams, comprising an octagon inside (inscribed in a circle of 6m diameter), with the vertices connected to radial beams (AB, CD, EF, GH, IJ, KL, MN and PQ) that terminate at the vertices of an outer octagon (inscribed in a circle of 18m diameter), which are fixed end supports. The radial beams are rectangular box-sections, 300mm wide and 600mm deep (mean dimensions) with plate thickness of 16mm, while the inner octagonal beams are square box-sections, 300mm × 300mm (mean dimensions) with plate thickness of 12mm. Assume $E = 2\times10^5$ MPa and Poisson's ratio, $\nu = 0.3$. Analyse the grid system by the stiffness method, given that all the grid beams are subjected to a uniformly distributed gravity load of 30 kN/m. Draw the shear force and bending moment diagrams of a typical radial beam (AB) and a typical inner beam (AC). Also find the maximum deflection and angle of twist.

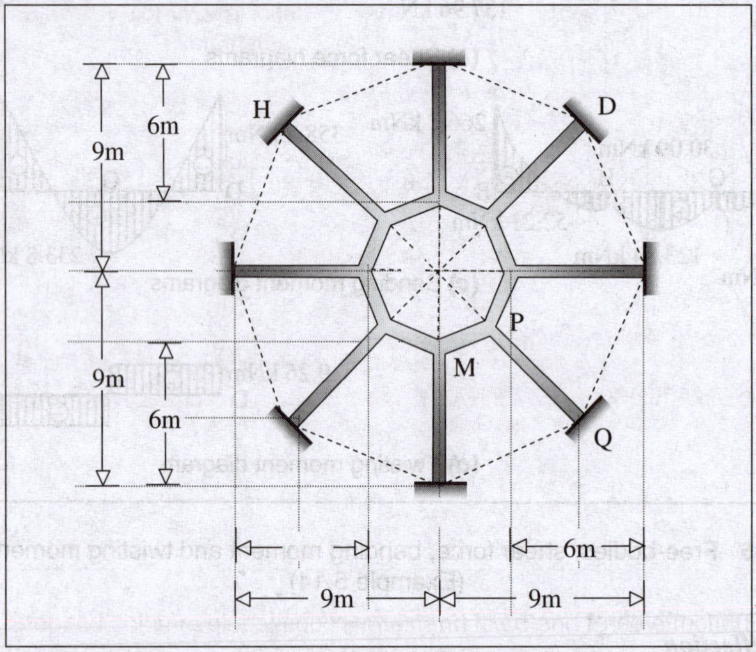

Figure 5.46 Layout plan of a symmetric grid (Example 5.15)

SOLUTION

As the layout plan and the loading are both symmetric (with respect to the vertical axis passing through the centre), the response will also be symmetric. All the radial beams (such as AB, span 6m) will behave identically, and all the inner beams (such as AC, span 2.2961m) will behave identically. The entire gravity load is symmetrically transmitted to the 8 supports at the periphery. The inner octagonal ring of elements, subject to uniformly distributed gravity loading, may be visualised as being supported on elastic supports (vertical and rotational springs) at the vertices of the octagon, transmitting a vertical force and a moment to the end of each radial beam. Thus, each radial beam (such as AB) acts like a simple cantilever beam, fixed at the support (B) at the periphery, subject to a uniformly distributed gravity loading plus the concentrated force and moment transmitted from the inner elements (such as AC and AP). Due to symmetry, there will be no translation in the horizontal plane and no rotation (twist) about the longitudinal axis of the radial beam.

From the free-bodies of the typical elements shown in Fig. 5.46, it is clear that the moment M_1 at A in the cantilever beam AB will be counter-balanced equally by the two moment vectors at A in the connecting beams, AC and AP, producing equal sagging moments (of magnitude M_2), which also occur at the ends C and P, on account of symmetry. We can also visualise twisting moment components to act in the elements AC and AP, but these would have to act in opposite directions at the two ends of each element, which is not possible. Thus, there will be no twisting moment acting in any of the elements, although there will be an angle of twist (rigid body rotation) in all the inner octagonal elements (such as AC and AP), the angle of twist being equal to $\theta_{AB} \cos 22.5^{\circ}$, where θ_{AB} is the slope at A in the cantilever beam AB. Thus, the angle of twist is independent of the torsional/flexural stiffness of the inner octagonal elements, and only depends on the flexural stiffness of the radial beam. Also, we note that, in order to satisfy equilibrium, $2M_2 \sin 22.5^{\circ} = M_1 \Rightarrow M_2 = 1.30656 M_1$, as shown in Fig. 5.47.

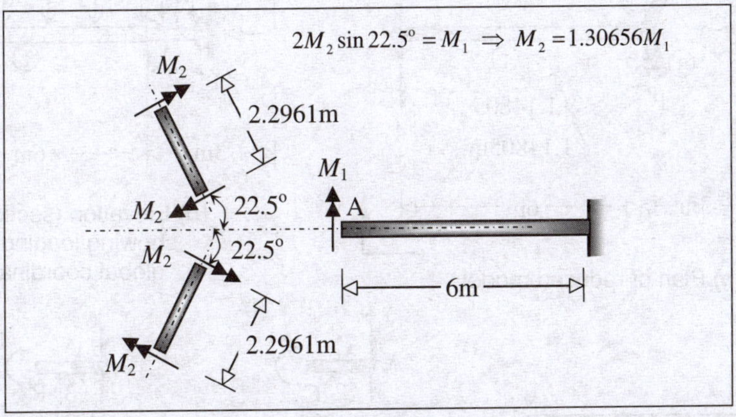

Figure 5.47 Resultant moments at end A in elements AB, AC and AP (Example 5.15)

Taking advantage of symmetry, it is sufficient to analyse the reduced model, shown in Fig. 5.48a, comprising a single radial beam AB, fixed at end B, and two inner octagonal beams, truncated at their mid-point locations (C′ and P′), where guided-fixed supports are provided (with freedom to translate vertically but not horizontally, freedom to twist, but restrained against flexural rotation). There are two *active* and two *restrained* global coordinates, numbered as shown in Fig. 5.48b. Taking advantage of the absence of torsion (due to symmetry), element AB ('1') is modelled as a beam element (simply supported with two degrees of freedom), as shown in Fig. 5.48c. Elements AC′ and AP′ ('2' and '3'), which are also free of twisting moment, are modelled as cantilever beam elements, with one degree of freedom each, as shown in Fig. 5.48d. There are no nodal loads applied; i.e., $\mathbf{F_A} = \mathbf{O}$.

We know in advance that the support reaction, F_3, is given by one-eighth the total gravity load acting on the grid; i.e., equal to the total load acting on the reduced model in Fig. 5.48a: $F_3 = (30)(6+2.2961) = 248.88$ kN. Also, as already explained, the angle of twist in the inner ring elements is given by $D_2 \cos 22.5^\circ$. The maximum deflection occurs at the mid-points of the inner beams, i.e. at C′ and P′.

Fixed End Forces (due to uniformly distributed loading of 30 kN/m)

Beam Element '1' (AB):

$$F^1_{1^*f} = \frac{30\times 6^2}{12} = 90 \text{ kNm}; \; F^1_{2^*f} = -90 \text{ kNm} \;\Rightarrow\; \mathbf{F^1_{*f}} = \begin{Bmatrix} 90 \text{ kNm} \\ -90 \text{ kNm} \end{Bmatrix}$$

$$\text{Vertical reactions: } V_{AB} = V_{BA} = \frac{(30)(6)}{2} = 90 \text{ kN}$$

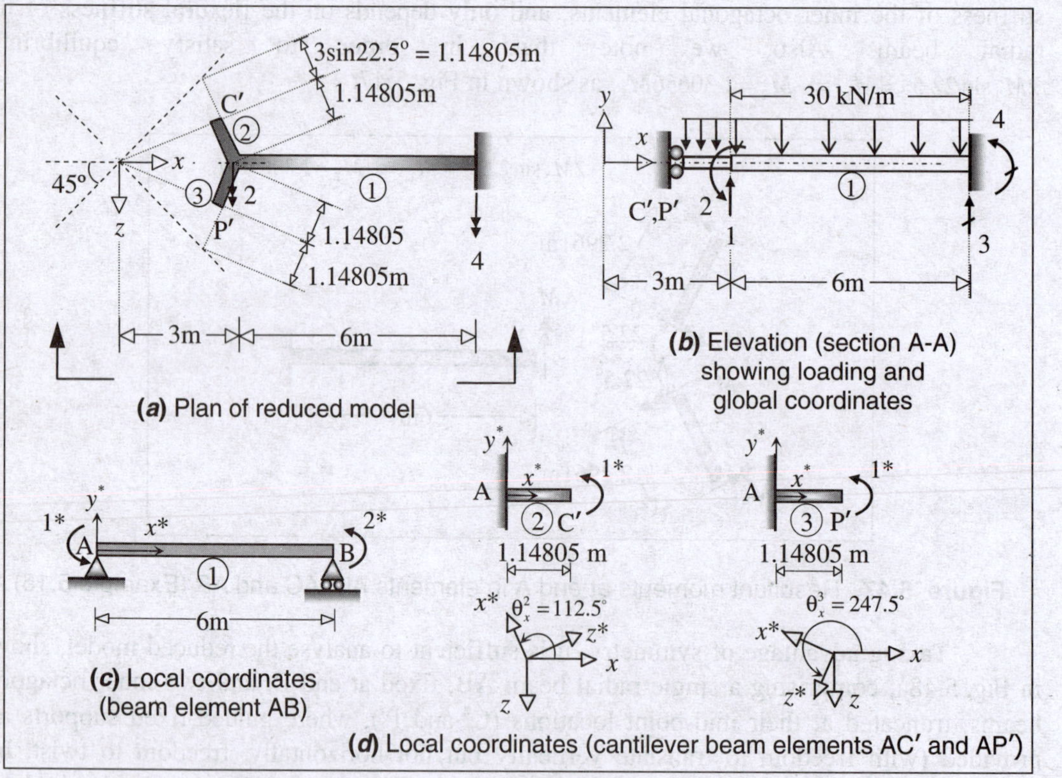

Figure 5.48 Reduced grid, global and local coordinates (Example 5.15)

Beam Elements '2' and '3' (AC′ and AP′):

To find the fixed end moments, it is convenient to consider the full fixed beam AC′C (span 2.2961m), subject to uniform loading:

$$F^2_{1*f} = F^3_{1*f} = \frac{30\times 2.2961^2}{12} = 13.180 \text{ kNm} \qquad \Rightarrow \mathbf{F}^2_{*f} = \mathbf{F}^3_{*f} = \{13.18 \text{ kNm}\}$$

Vertical reactions: $V_{AC'} = V_{AP'} = (30)(1.14805) = 34.44$ kN

Deflection at C′ and P′ due to fixed end effects = $\dfrac{(30)(2.2961)^4}{384(E_2I_2)} = \dfrac{2.17147}{(EI)_2}$

The fixed end forces at the two active global coordinates (refer Fig.5.48) are given by

$$F_{1f} = V_{AB} + V_{AC'} + V_{AP'} = 90 + (34.44\times 2) = 158.88 \text{ kN}$$

$$F_{2f} = F^1_{1*f} + F^2_{1*f}\cos\theta^2_x + F^3_{1*f}\cos\theta^3_x = 90 + 13.18(\cos 112.5^\circ + \cos 247.5^\circ) = 79.912 \text{ kNm}$$

$$\Rightarrow \mathbf{F}_{fA} = \begin{Bmatrix} F_{1f} \\ F_{2f} \end{Bmatrix} = \begin{Bmatrix} 158.88 \text{ kN} \\ 79.912 \text{ kNm} \end{Bmatrix}$$

The resultant load vector is given by $\mathbf{F}_A - \mathbf{F}_{fA} = -\mathbf{F}_{fA} = \begin{Bmatrix} -158.88 \text{ kN} \\ -79.912 \text{ kNm} \end{Bmatrix}$.

The fixed end forces at the restrained coordinates (refer Fig.5.48b) are given by

$$F_{3f} = V_{BA} = 90 \text{ kN} \qquad\qquad \Rightarrow \mathbf{F}_{fR} = \begin{Bmatrix} F_{3f} \\ F_{4f} \end{Bmatrix} = \begin{Bmatrix} 90 \text{ kN} \\ -90 \text{ kNm} \end{Bmatrix}$$

$$F_{4f} = F^1_{2*f} = -90 \text{ kNm}$$

Displacement Transformation Matrices

The element displacement transformation matrices, $\mathbf{T}^1_D$ and $\mathbf{T}^2_D$, relating the local coordinates to the global coordinates through the relation, $\mathbf{D}^i_* = \left[\mathbf{T}^i_{DA} \mid \mathbf{T}^i_{DR}\right]\begin{bmatrix} \mathbf{D}_A \\ \hline \mathbf{D}_R \end{bmatrix}$, take the following form (refer Figs 5.48b, c and d):

$$\mathbf{T}^1_D = \overset{\quad(1)\quad(2)\quad\;(3)\quad(4)}{\begin{bmatrix} 1/6 & 1 & -1/6 & 0 \\ 1/6 & 0 & -1/6 & 1 \end{bmatrix}}$$

Noting that $\theta^2_x = 112.5^\circ$ and $\theta^3_x = 247.5^\circ \quad \Rightarrow \cos\theta^2_x = \cos\theta^3_x = -0.38268$,

$$\mathbf{T}^2_D = \mathbf{T}^3_D = \overset{(1)\qquad(2)\qquad(3)\;(4)}{[0 \quad -0.38268 \mid 0 \quad 0]}$$

Element and Structure Stiffness Matrices

Section properties (box section):

Element '1': $I_1 = 2\times\left[\frac{(16)(600)^3}{12}+(300\times16)(300)^2+\frac{(300)(16)^3}{12}\right]=144.02048\times10^7\ \text{mm}^4$

$$\Rightarrow \frac{E_1 I_1}{L_1}=\frac{\left(2.0\times10^8\ \text{kN/m}^2\right)\left(144.02048\times10^{-5}\ \text{m}^4\right)}{(6\ \text{m})}=48006.83\ \text{kNm/rad}$$

Elements '2','3': $I_2 = I_3 = 2\times\left[\frac{(12)(300)^3}{12}+(300\times16)(150)^2+\frac{(150)(12)^3}{12}\right]=21.60864\times10^7\ \text{mm}^4$

$$\Rightarrow \frac{E_2 I_2}{L_2}=\frac{E_3 I_3}{L_3}=\frac{\left(2.0\times10^8\ \text{kN/m}^2\right)\left(21.60864\times10^{-5}\ \text{m}^4\right)}{(1.14805\ \text{m})}=37644.07\ \text{kNm/rad}$$

The element stiffness matrices $\tilde{\mathbf{k}}_*^1$ and $\tilde{\mathbf{k}}_*^2=\tilde{\mathbf{k}}_*^3$, satisfying $\mathbf{F}_*^i=\tilde{\mathbf{k}}_*^i\mathbf{D}_*^i$, take the following form:

$$\tilde{\mathbf{k}}_*^1=48006.83\begin{bmatrix}4 & 2\\ 2 & 4\end{bmatrix}=\begin{bmatrix}192027.3 & 96013.7\\ 96013.7 & 192027.3\end{bmatrix}\text{kNm/rad}$$

$$\tilde{\mathbf{k}}_*^2=\tilde{\mathbf{k}}_*^3=37644.07[1]=[37644.1]\ \text{kNm/rad}$$

$$\Rightarrow \quad \tilde{\mathbf{k}}_*^1\mathbf{T}_\mathbf{D}^1=\begin{bmatrix}48006.8 & 192027.3 & -48006.8 & 96013.7\\ 48006.8 & 96013.7 & -48006.8 & 192027.3\end{bmatrix}$$

$$\tilde{\mathbf{k}}_*^2\mathbf{T}_\mathbf{D}^2=\tilde{\mathbf{k}}_*^3\mathbf{T}_\mathbf{D}^3=[0\quad -14405.6\quad 0\quad 0]$$

The structure stiffness matrix $\mathbf{k}=\left[\begin{array}{c|c}\mathbf{k}_{\mathbf{AA}} & \mathbf{k}_{\mathbf{AR}}\\ \hline \mathbf{k}_{\mathbf{RA}} & \mathbf{k}_{\mathbf{RR}}\end{array}\right]$, satisfying $\mathbf{F}=\mathbf{kD}$, can now be assembled as $\mathbf{k}=\sum_{i=1}^{3}\mathbf{T}_\mathbf{D}^{i\,T}\tilde{\mathbf{k}}_*^i\mathbf{T}_\mathbf{D}^i$:

$$\Rightarrow \quad [\mathbf{k}_{\mathbf{AA}}]=\begin{bmatrix}16002.3 & 48006.8\\ 48006.8 & 203052.8\end{bmatrix}\quad \Rightarrow \quad [\mathbf{k}_{\mathbf{AA}}]^{-1}=\begin{bmatrix}214.9500 & -50.8196\\ -50.8196 & 16.9399\end{bmatrix}$$

$$\mathbf{k}_{\mathbf{RA}}=\begin{bmatrix}-16002.3 & -48006.8\\ 48006.8 & 96013.7\end{bmatrix}$$

Displacements and Support Reactions

The structure stiffness relation to be considered here is:

$$\left[\frac{\mathbf{F}_\mathbf{A}}{\mathbf{F}_\mathbf{R}}\right]-\left[\frac{\mathbf{F}_{\mathbf{fA}}}{\mathbf{F}_{\mathbf{fR}}}\right]=\left[\begin{array}{c|c}\mathbf{k}_{\mathbf{AA}} & \mathbf{k}_{\mathbf{AR}}\\ \hline \mathbf{k}_{\mathbf{RA}} & \mathbf{k}_{\mathbf{RR}}\end{array}\right]\left[\frac{\mathbf{D}_\mathbf{A}}{\mathbf{D}_\mathbf{R}=\mathbf{O}}\right]$$

$$\Rightarrow \quad \mathbf{D}_\mathbf{A}=[\mathbf{k}_{\mathbf{AA}}]^{-1}[\mathbf{F}_\mathbf{A}-\mathbf{F}_{\mathbf{fA}}]$$

$$\Rightarrow \quad \begin{Bmatrix}D_1\\ D_2\end{Bmatrix}=\begin{bmatrix}214.9500 & -50.8196\\ -50.8196 & 16.9399\end{bmatrix}\text{kNm/rad}=\begin{Bmatrix}-0.030090\ \text{m}\\ 0.0067205\ \text{rad}\end{Bmatrix}$$

$$\mathbf{F_R} = \mathbf{F_{fR}} + \mathbf{k_{RA}}\mathbf{D_A} \Rightarrow \begin{Bmatrix} F_3 \\ F_4 \end{Bmatrix} = \begin{Bmatrix} 90 \text{ kN} \\ -90 \text{ kNm} \end{Bmatrix} + \begin{Bmatrix} 158.88 \text{ kN} \\ -799.27 \text{ kNm} \end{Bmatrix} = \begin{Bmatrix} 248.88 \text{ kN} \\ -889.27 \text{ kNm} \end{Bmatrix}$$

Note: The vertical support reaction satisfies vertical force equilibrium:

$$F_3 = (30)(6+2.2961) = 248.88 \text{ kN}$$

Member Forces

$$\mathbf{F_*^i} = \mathbf{F_{*f}^i} + \mathbf{\tilde{k}_{DA}^i}\mathbf{T_{DA}^i}\mathbf{D_A} \quad \Rightarrow$$

$$\begin{Bmatrix} F_{1*}^1 \\ F_{2*}^1 \end{Bmatrix} = \begin{Bmatrix} 90 \text{ kNm} \\ -90 \text{ kNm} \end{Bmatrix} + \begin{bmatrix} 48006.8 & 192027.3 \\ 48006.8 & 96013.7 \end{bmatrix} \begin{Bmatrix} -0.030090 \\ 0.0067205 \end{Bmatrix} = \begin{Bmatrix} -64.009 \text{ kNm} \\ -889.27 \text{ kNm} \end{Bmatrix}$$

$$\{F_{1*}^2\} = \{F_{1*}^3\} = \{13.18 \text{ kNm}\} + [0 \quad -14405.6] \begin{Bmatrix} -0.030090 \\ 0.0067205 \end{Bmatrix} = \{-83.633 \text{ kNm}\}$$

Referring to Fig. 5.47, we observe that $M_1 = 64.009\,\text{kNm}$ and $M_2 = 83.633\,\text{kNm}$, satisfying the requirement, $M_2 = 1.30656 M_1$.

The loading/deflection diagrams (including support reactions) and shear force and bending moment diagrams, for a typical radial beam AB and a typical inner octagonal beam AC, are shown in Figs 5.49a, b and c respectively. It can be seen that the beam AC behaves like a beam, with elastic supports (translational and rotational springs) at A and C. The elastic support reactions are transmitted to the free end A of the cantilever beam AB.

Maximum Deflection

The maximum deflection occurs at the mid-point of the inner octagonal beam (C′), and is given by the sum of (i) deflection caused by fixed end forces (calculated earlier as $\frac{2.17147}{(EI)_2} = \frac{2.17147}{43217.3} = 0.0502 \times 10^{-3}$ m), (ii) deflection at A (equal to $-D_1 = 30.090 \times 10^{-3}$ m) and (iii) deflection at C′ due to rotation at A, given by $\left(D_2 \sin 22.5^\circ\right)\frac{L_{AC'}}{2}$

$$= \left(2.572 \times 10^{-3}\right)\left(\frac{1.14805}{2}\right) = 1.476 \times 10^{-3} \text{ m}:$$

$$\Delta_{\max} = 0.050 + 30.090 + 1.476 = 31.616 \text{ mm}$$

Angle of Twist

As mentioned earlier, all the inner octagonal elements will undergo a rigid body rotation, the angle of twist being equal to $D_2 \cos 22.5^\circ = 0.0067052 \cos 22.5^\circ = 0.006195$ rad. The radial beams will not undergo any twist.

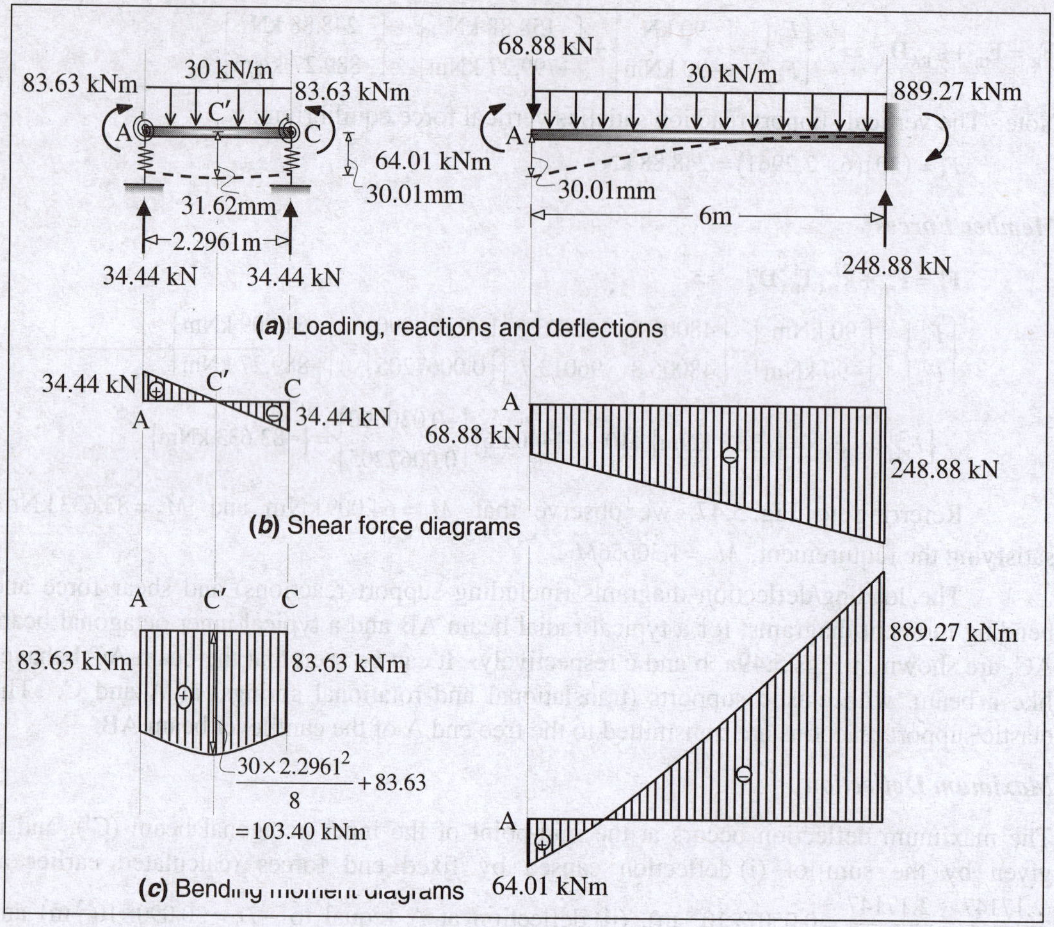

Figure 5.49 Force response in beams AC and AB (Example 5.15)

REVIEW QUESTIONS

5.1 Using a 'displacement approach', derive from first principles, the element stiffness matrix of a prismatic beam element, having a length L_i and flexural rigidity E_iI_i, with 4 degrees of freedom.

5.2 Show how an energy formulation can be used to derive the above matrix.

5.3 Show that the 4×4 element stiffness matrix for a prismatic beam element is non-invertible, and explain, in terms of physical behaviour, why this is so.

5.4 Show how the 4×4 element stiffness matrix for a continuous beam element can be easily transformed from local coordinates to the global axes system.

5.5 Explain the steps involved in the analysis of continuous beams by the conventional stiffness method, including effects of support settlements in addition to applied loading.

5.6 Consider a two-span continuous beam ABC, fixed at A and C, and with equal spans L and uniform flexural rigidity EI, subject to a uniformly distributed load q_o per unit length in the span AB alone. Define the global *active* and *restrained* coordinates appropriately. Using the conventional stiffness method (with 4 degrees of freedom per element), formulate the transformation matrices for the two elements, the resultant nodal force vector and the structure stiffness matrix.

5.7 How will you account for internal hinges in a continuous beam in the conventional stiffness method?

5.8 Derive the 4×4 element stiffness matrix for a prismatic beam element, including the effect of *shear deformations*.

5.9 Derive the reduced 2×2 element stiffness matrix for a prismatic beam element, including the effect of *shear deformations*.

5.10 What are the advantages and disadvantages in using the reduced stiffness formulation, compared to the conventional stiffness method, in the matrix analysis of continuous beams?

5.11 How can the reduced stiffness formulation be conveniently used to handle beam elements with (a) hinged end supports and (b) guided-fixed supports? Derive the element stiffness matrices.

5.12 Repeat Question 5.5, considering the reduced stiffness formulation, with a guided-fixed support at C.

5.13 Why is the flexibility method not well suited for matrix analysis of continuous beams?

5.14 Describe the steps involved in the flexibility method, as applied to continuous beams, including the effects of support settlements.

5.15 What is meant by a *grid* system? In what way is a *grid element* different from a beam element?

5.16 Can beam elements be used in matrix analysis of a grid system? Discuss.

5.17 Demonstrate how the reduced stiffness formulation can be advantageously used to solve symmetric grid problems.

PROBLEMS

5.1 Consider a two-span prismatic continuous beam, ABC, each span equal to L and having flexural rigidity EI. The beam, which is fixed at end A, is continuous over a simple support at B and is simply supported at end C. The span AB is subject to a gravity load P acting at its mid-point. Analyse the beam by the conventional Stiffness Method, and draw the shear force, bending moment and deflection diagrams.

5.2 Repeat Problem 5.1, considering the loading to be in the form of support settlements of 2δ and δ at supports B and C respectively, in lieu of the concentrated load.

5.3 Repeat (a) Problem 5.1 and (b) Problem 5.2 by the Reduced Element Stiffness Method, taking advantage of the hinged support at C.

5.4 Repeat (a) Problem 5.1 and (b) Problem 5.2 by the Flexibility Method, considering the fixed end moment at A and the vertical reaction at B to be the redundants.

5.5 Analyse the non-prismatic fixed beam with overhang shown in Fig. 5.50, by the conventional Stiffness Method. Assume $EI = 80\,000$ kNm^2. Find the support reactions and draw the shear force and bending moment diagrams.

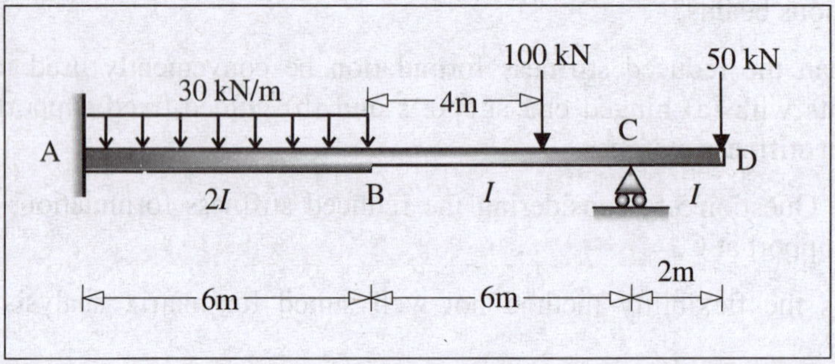

Figure 5.50 Problem 5.5

5.6 Repeat Problem 5.5 by the Reduced Element Stiffness Method, taking advantage of the hinge at C.

5.7 Consider a two-span continuous beam ABC, fixed at A and C, and with equal spans L and uniform flexural rigidity EI, subject to a uniformly distributed gravity load q_o per unit length in the span AB alone (Fig. 5.51). Define the global active and restrained coordinates appropriately. Analyse the beam using the reduced element stiffness formulation. Find the rotation at the joint B and the support reactions. Draw the bending moment diagram.

5.8 Analyse the continuous beam shown in Fig. 5.51, subject to the external loading and support settlements as indicated, by the conventional Stiffness Method. Assume

$EI = 80\,000$ kNm^2. Find the support reactions and draw the shear force and bending moment diagrams.

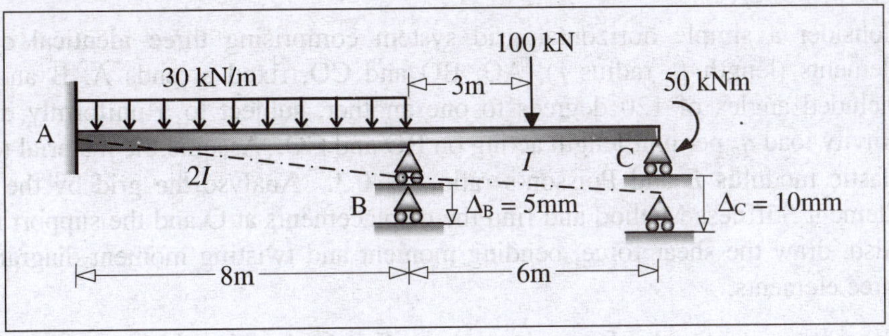

Figure 5.51 Problem 5.7

5.9 Repeat Problem 5.8 by the Reduced Element Stiffness Method, taking advantage of the hinge at C.

5.10 Repeat Problem 5.8 by the Flexibility Method.

5.11 Taking advantage of symmetry, analyse the beam shown in Fig. 5.52 (with internal hinges at B and C), by the conventional Stiffness Method. Assume $EI = 80\,000$ kNm^2. Draw the shear force, bending moment and deflection diagrams. [Check the solutions by the force method; the beam is statically determinate].

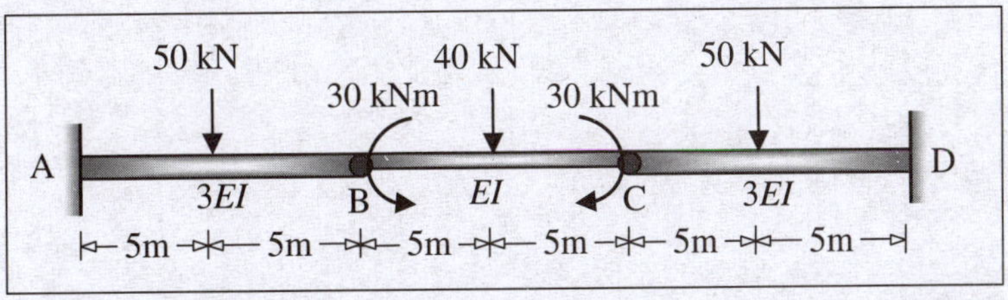

Figure 5.52 Problem 5.11

5.12 Repeat Problem 5.11 by the Reduced Element Stiffness Method.

5.13 Consider the beam in Problem 5.11 without the two internal hinges. Analyse this non-prismatic fixed beam by the Reduced Element Stiffness Method and draw the shear force and bending moment diagrams.

5.14 Repeat Problem 5.13 by the Flexibility Method.

5.15 Consider the simple two-member grid problem, solved in Example 5.13, with a concentrated gravity load P acting at the joint O, instead of the uniformly distributed loading. Analyse the grid by the conventional Stiffness Method and generate expressions, in terms of the angle α, for the deflection, flexural rotation and angle of twist at O, and the support reactions. Hence, draw the corresponding shear force, bending moment and twisting moment diagrams of OA. Comment on the behaviour with respect to variations in the angle α.

5.16 Consider a simple horizontal grid system comprising three identical cylindrical elements (length L, radius r), AO, BO and CO, fixed at ends A, B and C, with included angles of 120 degrees to one another, subject to a uniformly distributed gravity load q_o per unit length acting on BO and CO. Assume the material to have an elastic modulus E and Poisson's ratio, $\nu = 0.3$. Analyse the grid by the Reduced Element Stiffness Method and find the displacements at O and the support reactions. Also, draw the shear force, bending moment and twisting moment diagrams of the three elements.

5.17 Consider a rectangular floor plan layout, 9m×12m with grid beams forming small squares of 3m side (i.e., with 3 intermediate short beams of 9m span and two intermediate long beams of 12m span). Analyse this system by the Reduced Element Stiffness Method, considering a concentrated gravity load P to act at each intersection point, taking advantage of symmetry. Assuming all the grid elements to have identical flexural rigidity, $EI = 5\times10^4$ kN/m^2, and torsional rigidity $GJ = 0.2EI$, analyse the grid system. Find the maximum deflection, and draw the shear force, bending moment and twisting moment diagrams of a typical 12m span beam and the interior and exterior 9m spans.

6

Matrix Analysis of Plane and Space Frames

6.1 INTRODUCTION

This Chapter deals with the application of matrix methods to plane frames and space frames, comprising skeletal elements that are inter-connected with 'rigid' (moment-resistant) joints[†]. The emphasis, as in the previous Chapters, is on the *stiffness method*; the application of *flexibility method* is limited to plane frames.

The *plane frame element* (assumed to be prismatic) is a skeletal element that is a combination of the *beam element* (having flexural rigidity *EI*) and *plane truss element* (having axial rigidity *EA*) introduced earlier. The element resists external loads (acting in the plane of the frame) by undergoing bending (or flexure), shear force and axial force. The effect of axial force on flexural stiffness, accounting for slenderness effects ('P-delta effect'), is not considered in this Chapter, whose scope is limited to linear analysis; geometric nonlinearity effects are discussed in Chapter 7.

At any point along the element centre-line, there are three displacement quantities of interest to us, viz., the longitudinal translation $u(x)$, transverse deflection $\Delta(x)$ and its

[†] Some of the joints may have moment releases. If all the joints are pinned, then the frame reduces to a truss (which is discussed in Chapter 4).

derivative, the rotation (or slope) $\theta(x)$. If we know the values of these three quantities at the two ends of any prismatic plane frame element, then it should be possible to find $u(x)$, $\Delta(x)$ and $\theta(x)$ at any intermediate location. This forms the basis of the stiffness formulation, whereby the number of independent displacement coordinates (i.e., *degree of freedom*) in any prismatic frame element is limited to six. It is assumed that the loads in the form of direct actions are applied at the joints; otherwise, if there are intermediate loads, these shall be converted to *equivalent joint loads*. Thus, in the conventional stiffness formulation, the plane frame element has six degrees of freedom (four translational and two rotational), whereby the stiffness matrix of the element (in local coordinates) is of order 6×6. This is discussed in Section 6.2. For normal, well-proportioned frame elements, we assume that shear deformations are negligible. However, it is possible to account for shear deformations by suitably modifying the stiffness matrix.

In the *reduced stiffness* formulation and the *flexibility* formulation, the degrees of freedom are reduced to three. In this case, the stiffness matrix of the element (in local coordinates), and its inverse, the flexibility matrix is of order 3×3. In the reduced stiffness formulation, we can also reduce the degree of kinematic indeterminacy of plane frame systems in which the extreme ends are hinged or guided-fixed supports, by taking advantage of modified stiffnesses and also by modifying the fixed end forces appropriately, as discussed in Chapter 5. Furthermore, we can reduce the degree of kinematic indeterminacy of the plane frame by ignoring axial deformations (as is commonly done in structural analysis of small frames). However, this needs to be done cautiously, while dealing with 'sway' degrees of freedom, especially in frames with inclined members. The reduced stiffness formulation is discussed in Section 6.3, while the flexibility formulation is discussed in Section 6.4.

In three-dimensional frames, we need to consider the *space frame* element, which resists external loads by undergoing bending (or flexure) in two orthogonal planes and associated shear forces, as well as axial force and torsion. Thus, it has six degrees of freedom at each end of the element, and thus has a stiffness matrix of order 12×12. In Section 6.5, the matrix analysis of space frame structures (by the stiffness method) is discussed.

6.2 CONVENTIONAL STIFFNESS METHOD APPLIED TO PLANE FRAMES

6.2.1 Plane Frame Element with Six Degrees of Freedom

Consider a prismatic (initially straight) plane frame element i, having a length L_i and a cross-section with area A_i, second moment of area I_i, and composed of a material with an elastic modulus, E_i. If the element is subject to concentrated forces only at the two ends (i.e., without any intermediate loading), then, in order to satisfy equilibrium, the bending moment, $M(x^*)$, will, in general, vary linearly along the length of the element, while the shear force, $S(x^*)$, and axial force, $N(x^*)$, remain constant, as explained earlier. Assuming linear elastic behaviour, and ignoring the effects of axial and shear deformations on the flexural stiffness of the element, this implies that the curvature, $\varphi(x^*) = \frac{M(x^*)}{EI}$, will vary linearly, while the axial strain and shear strain remain constant along the length of the element. As explained in Chapters 4 and 5, this implies that the axial deformation $u(x^*)$ will vary linearly, while the transverse deflection $\Delta(x^*)$ will have a cubic variation along the length of the element. Thus, a complete description of the displaced geometry of the element can be generated in terms of two axial degrees of freedom (for a linear variation of $u(x^*)$) and four flexure-related degrees of freedom (for a cubic variation of $\Delta(x^*)$). This can be conveniently achieved by selecting six degrees of freedom, relating to three displacements (two translations and a rotation) at either end of the plane frame element, as indicated in Fig. 6.1a.

The axial deformation, or *longitudinal deflection*, $u(x^*)$, assumed to be positive while acting in the positive x^* direction, is easily obtainable as a linear combination of the end axial displacements, D^i_{1*} (at the start node) and D^i_{4*} (at the end node). Applying the Principle of Superposition to the displaced configurations corresponding to $D^i_{1*} = 1$ (Fig. 6.1b) and $D^i_{4*} = 1$ (Fig. 6.1e),

$$u(x^*) = \left[\eta_1(x^*) \quad \eta_4(x^*)\right] \begin{Bmatrix} D^i_{1*} \\ D^i_{4*} \end{Bmatrix} \tag{6.1}$$

where, as indicated earlier in Chapter 4 (refer Fig. 4.2), the non-dimensional 'displacement functions', called *shape functions* in finite element analysis, are given by:

$$\begin{aligned} \eta_1(x^*) &= 1 - x^*/L_i \\ \eta_4(x^*) &= x^*/L_i \end{aligned} \tag{6.2}$$

Similarly, the *transverse deflection* $\Delta(x^*)$, assumed to be positive while acting in the positive y^* direction (i.e., upwards in Fig. 6.1a), is obtainable as a linear combination of the transverse deflection D^i_{2*} and rotation D^i_{3*} at the start node, and the corresponding displacements D^i_{5*} and D^i_{6*} at the end node. Applying the Principle of Superposition to the displaced configurations corresponding to $D^i_{2*} = 1$ (Fig. 6.1c), $D^i_{3*} = 1$ (Fig. 6.1d), $D^i_{5*} = 1$ (Fig. 6.1f), and $D^i_{6*} = 1$ (Fig. 6.1g),

$$\Delta(x^*) = \left[\eta_2(x^*) \quad \eta_3(x^*) \quad \eta_5(x^*) \quad \eta_6(x^*)\right] \begin{Bmatrix} D^i_{2*} \\ D^i_{3*} \\ D^i_{5*} \\ D^i_{6*} \end{Bmatrix} \tag{6.3}$$

where, as derived earlier in Chapter 5 (refer Eq. 5.10), the non-dimensional 'displacement functions' (*shape functions*) are given by:

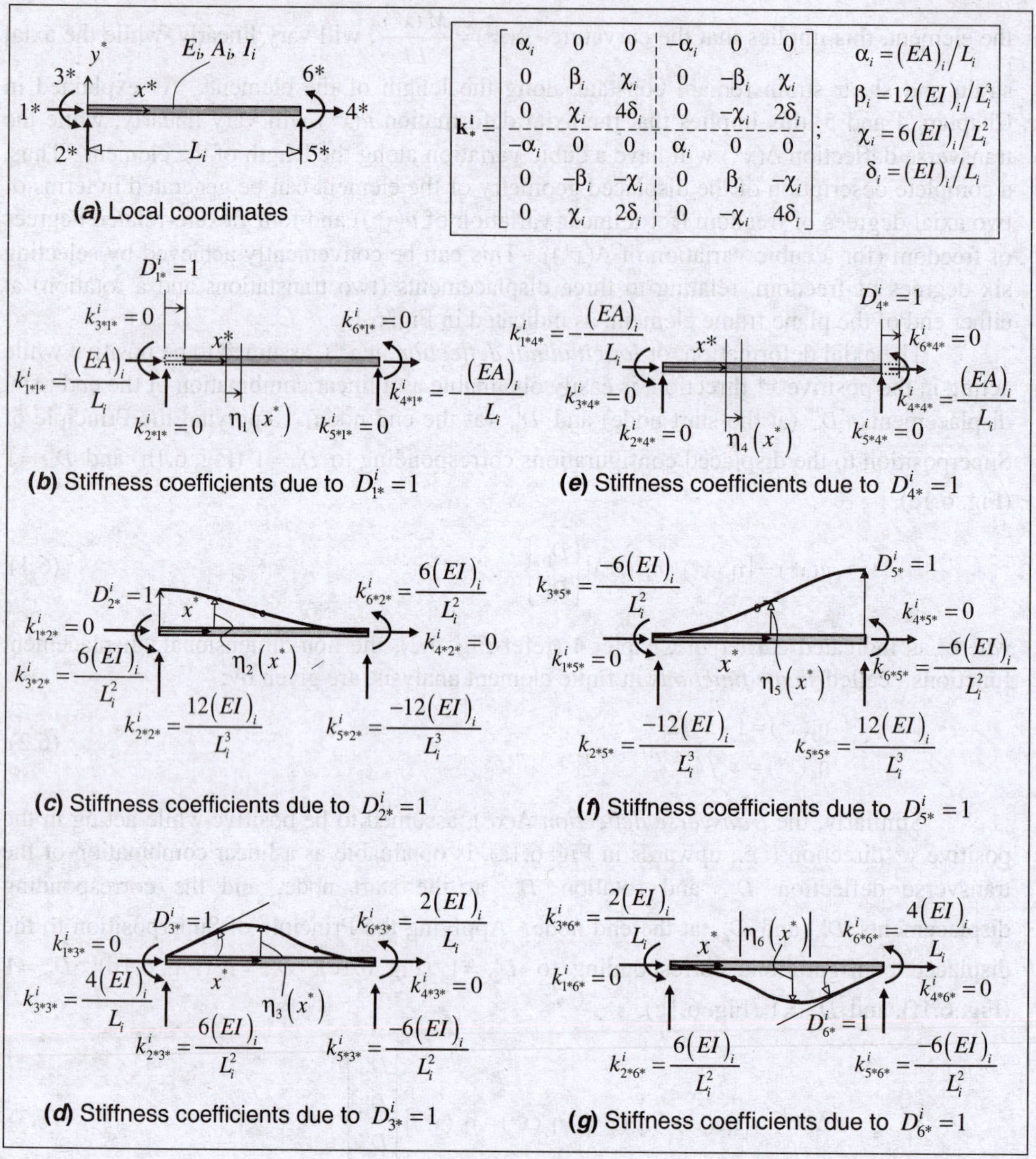

$$\mathbf{k}_*^i = \begin{bmatrix} \alpha_i & 0 & 0 & -\alpha_i & 0 & 0 \\ 0 & \beta_i & \chi_i & 0 & -\beta_i & \chi_i \\ 0 & \chi_i & 4\delta_i & 0 & -\chi_i & 2\delta_i \\ -\alpha_i & 0 & 0 & \alpha_i & 0 & 0 \\ 0 & -\beta_i & -\chi_i & 0 & \beta_i & -\chi_i \\ 0 & \chi_i & 2\delta_i & 0 & -\chi_i & 4\delta_i \end{bmatrix};$$

Figure 6.1 Generation of 6 × 6 stiffness matrix for a plane frame element

$$\eta_2(x^*) = 1 - 3(x^*/L_i)^2 + 2(x^*/L_i)^3$$
$$\eta_3(x^*) = x^*(1 - x^*/L_i)^2 \tag{6.4a}$$

$$\eta_5(x^*) = 3\left(x^*/L_i\right)^2 + 2\left(x^*/L_i\right)^3$$
$$\eta_6(x^*) = x^*\left(x^*/L_i\right)\left(-1 + x^*/L_i\right) \qquad (6.4b)$$

Expressions for the rotation $\theta(x^*)$ and curvature $\varphi(x^*)$ at any distance x^* from the start node of the element can be easily generated in terms of the derivatives of the above shape functions as follows:

$$\begin{Bmatrix} \theta(x^*) \\ \varphi(x^*) \end{Bmatrix} = \begin{bmatrix} \eta_2'(x^*) & \eta_3'(x^*) & \eta_5'(x^*) & \eta_6'(x^*) \\ \eta_2''(x^*) & \eta_3''(x^*) & \eta_5''(x^*) & \eta_6''(x^*) \end{bmatrix} \begin{Bmatrix} D_{2*}^i \\ D_{3*}^i \\ D_{5*}^i \\ D_{6*}^i \end{Bmatrix} \qquad (6.5)$$

where $\Delta_j(x^*)$, $\theta_j(x^*)$ and $\varphi_j(x^*)$ denote respectively, the deflection, rotation and curvature at x^*, corresponding to $D_{j*}^i = 1$, with all other end displacements restrained.

Stiffness Matrix

The stiffness matrix $\mathbf{k}_*^i$ of the beam element, i, with six degrees of freedom, as shown in Fig. 6.1a, has an order of 6×6. As explained earlier, elements of the jth column of this matrix can be generated by analysing the element end forces corresponding to the displacement condition, $D_{j*}^i = 1$, with the other five degrees of freedom restrained. The various deflected profiles are shown in Fig. 6.1, and the stiffness coefficients are as derived earlier, for the truss element (Fig. 4.4) and the beam element (Fig. 5.1). It is assumed that the axial and flexural stiffness components are uncoupled; i.e., axial deformations in the element do not induce bending moments and shear forces ($k_{2*1*}^i = k_{3*1*}^i = k_{5*1*}^i = k_{6*1*}^i = 0$ and $k_{2*4*}^i = k_{3*4*}^i = k_{5*4*}^i = k_{6*4*}^i = 0$), while flexural deformations do not induce axial forces ($k_{2*1*}^i = k_{3*1*}^i = k_{5*1*}^i = k_{6*1*}^i = 0$). Shear deformations are considered to be negligible[†].

There are many alternative ways of generating the stiffness coefficients. We can adopt either a 'displacement approach' or a 'force approach' for this purpose, as discussed in Chapters 4 and 5. Here, we shall take advantage of the coefficients already derived for the plane truss element and the plane frame element, as indicated in Fig. 6.1. These coefficients can be easily generated using a 'physical approach'. Accordingly, the stiffness coefficients are obtainable as follows, considering $D_{j*}^i = 1$, one at a time:

$$k_{1*1*} = \frac{(EA)_i}{L_i};\ k_{2*1*} = 0;\ k_{3*1*} = 0;\ k_{4*1*} = -\frac{(EA)_i}{L_i};\ k_{5*1*} = 0;\ k_{6*1*} = 0$$
$$k_{1*4*} = -\frac{(EA)_i}{L_i};\ k_{2*4*} = 0;\ k_{3*4*} = 0;\ k_{4*4*} = \frac{(EA)_i}{L_i};\ k_{5*4*} = 0;\ k_{6*4*} = 0 \qquad (6.6a)$$

[†] The effect of shear deformations on the flexural stiffness coefficients can be included, if desired, as shown in Section 5.2.8.

$$k_{1*2*}=0;\ k_{2*2*}=\frac{12(EI)_i}{L_i^3};\ k_{3*2*}=\frac{6(EI)_i}{L_i^2};\ k_{4*2*}=0;\ k_{5*2*}=-\frac{12(EI)_i}{L_i^3};\ k_{6*2*}=\frac{6(EI)_i}{L_i^2}$$

$$k_{1*5*}=0;\ k_{2*5*}=-\frac{12(EI)_i}{L_i^3};\ k_{3*5*}=-\frac{6(EI)_i}{L_i^2};\ k_{4*5*}=0;\ k_{5*5*}=\frac{12(EI)_i}{L_i^3};\ k_{6*5*}=-\frac{6(EI)_i}{L_i^2} \tag{6.6b}$$

$$k_{1*3*}=0;\ k_{2*3*}=\frac{6(EI)_i}{L_i^2};\ k_{3*3*}=\frac{4(EI)_i}{L_i};\ k_{4*3*}=0;\ k_{5*3*}=-\frac{6(EI)_i}{L_i^2};\ k_{6*3*}=\frac{2(EI)_i}{L_i}$$

$$k_{1*6*}=0;\ k_{2*6*}=\frac{6(EI)_i}{L_i^2};\ k_{3*6*}=\frac{2(EI)_i}{L_i};\ k_{4*6*}=0;\ k_{5*6*}=-\frac{6(EI)_i}{L_i^2};\ k_{6*6*}=\frac{4(EI)_i}{L_i} \tag{6.6c}$$

The resulting element stiffness matrix takes the following form:

$$\mathbf{k}_*^i=\left[\begin{array}{ccc|ccc}\alpha_i & 0 & 0 & -\alpha_i & 0 & 0\\ 0 & \beta_i & \chi_i & 0 & -\beta_i & \chi_i\\ 0 & \chi_i & 4\delta_i & 0 & -\chi_i & 2\delta_i\\ \hline -\alpha_i & 0 & 0 & \alpha_i & 0 & 0\\ 0 & -\beta_i & -\chi_i & 0 & \beta_i & -\chi_i\\ 0 & \chi_i & 2\delta_i & 0 & -\chi_i & 4\delta_i\end{array}\right];\quad \begin{array}{l}\alpha_i=(EA)_i/L_i\\ \beta_i=12(EI)_i/L_i^3\\ \chi_i=6(EI)_i/L_i^2\\ \delta_i=(EI)_i/L_i\end{array} \tag{6.7}$$

6 ×6 Flexibility Matrix not Possible!

It is interesting to note that the determinant of the 6×6 stiffness matrix $\mathbf{k}_*^i$ in Eq. 6.7 is zero (rank = 3), whereby the matrix is *not invertible*. This means that a 6×6 flexibility matrix (inverse of the stiffness matrix) for the element shown in Fig. 6.1a cannot be defined. This is so, because this element is *unstable* and cannot resist the action of any arbitrary combination of forces $F_{1*}^i, F_{2*}^i, \ldots, F_{6*}^i$. In order to have a valid flexibility matrix, it is necessary to arrest 3 degrees of freedom. The resulting flexibility matrix is a 3×3 matrix, to be defined later.

6.2.2 Coordinate Transformations and Structure Stiffness Matrix

Transformation Matrix

In a plane frame system, all the elements (as well as the loads) lie on the same plane (usually vertical plane). Thus, while the local coordinate x^*- and y^*-axes of an element (with six degrees of freedom) are, in general, inclined to the global x- and y-axes of the structure (defined by the angle θ_x^i, as shown in Fig. 6.2), the local z^*-axis is always aligned in the same direction as the global z-axis. The six local coordinates, numbered 1*, 2*,…, 6*, as shown in Fig. 6.1a, can be directly linked in the global axes system, as 1, 2,…, 6, to appropriate global coordinates (say, l, m, n, p, q, r) at the same locations in the plane frame. Compatibility of displacement requires $D_l=D_1^i$, $D_m=D_2^i$, $D_n=D_3^i=D_{3*}^i$, $D_p=D_4^i$, $D_q=D_5^i$ and $D_r=D_6^i=D_{6*}^i$.

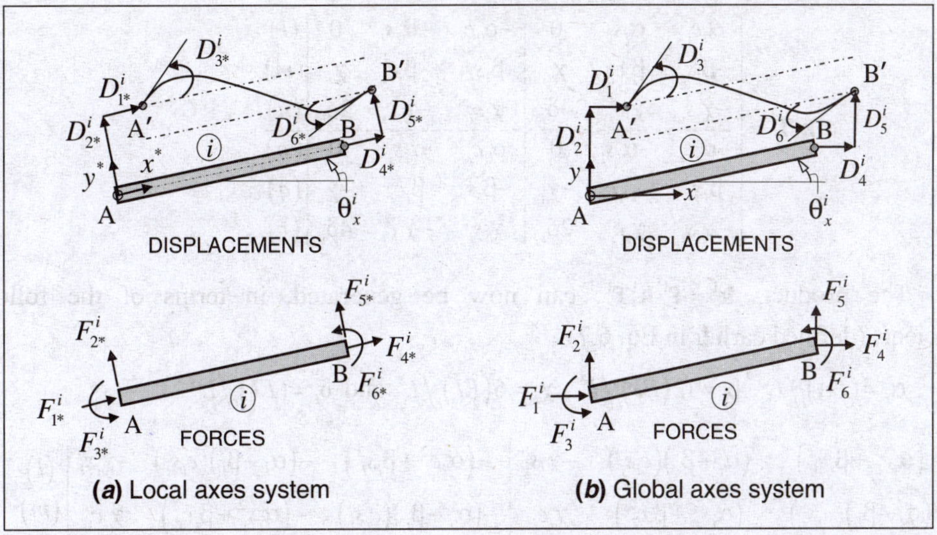

(***a***) Local axes system (***b***) Global axes system

Figure 6.2 Member-end displacements and forces in a plane frame element

The 6×6 element transformation matrix $\mathbf{T^i}$, relating the displacement and force vector components, ($D_1^i, D_2^i, D_3^i, D_4^i, D_5^i, D_6^i$) and ($F_1^i, F_2^i, F_3^i, F_4^i, F_5^i, F_6^i$), in the global axes system to the corresponding components in the local coordinate system, $\mathbf{D_*^i} = \mathbf{T^i D^i}$ and $\mathbf{F_*^i} = \mathbf{T^i F^i}$, as defined in Chapter 3, will take the following form (global coordinates in parenthesis):

$$\mathbf{T^i} = \left[\begin{array}{ccc|ccc} c_i & s_i & 0 & 0 & 0 & 0 \\ -s_i & c_i & 0 & 0 & 0 & 0 \\ 0 & 0 & 1 & 0 & 0 & 0 \\ \hline 0 & 0 & 0 & c_i & s_i & 0 \\ 0 & 0 & 0 & -s_i & c_i & 0 \\ 0 & 0 & 0 & 0 & 0 & 1 \end{array}\right] \begin{array}{c} (l) \\ (m) \\ (n) \\ (p) \\ (q) \\ (r) \end{array} \tag{6.8}$$

where $c_i = \cos\theta_x^i$ and $s_i = \sin\theta_x^i$. The elements of this transformation matrix, corresponding to the vertical and horizontal translations at the start and end nodes of the element, involving the direction cosines, are similar to those of the plane truss element (refer Eq. 4.14a); rotation (with reference to the *z*-axis) involves an identity transformation, as indicated in Fig. 6.2.

The element stiffness matrix in the global axes system is given by the transformation $\mathbf{k^i} = \mathbf{T^{i^T} k_*^i T^i}$ (refer Eq. 3.32), satisfying the relation $\mathbf{F^i} = \mathbf{k^i D^i}$. It can be computed by first finding the product, $\mathbf{k_*^i T^i}$, which takes the following form:

$$\mathbf{k}_*^{\mathbf{i}}\mathbf{T}^{\mathbf{i}} = \left[\begin{array}{ccc|ccc} \alpha_i c_i & \alpha_i s_i & 0 & -\alpha_i c_i & -\alpha_i s_i & 0 \\ -\beta_i s_i & \beta_i c_i & \chi_i & \beta_i s_i & -\beta_i c_i & \chi_i \\ -\chi_i s_i & \chi_i c_i & 4\delta_i & \chi_i s_i & -\chi_i c_i & 2\delta_i \\ \hline -\alpha_i c_i & -\alpha_i s_i & 0 & \alpha_i c_i & \alpha_i s_i & 0 \\ \beta_i s_i & -\beta_i c_i & -\chi_i & -\beta_i s_i & \beta_i c_i & -\chi_i \\ -\chi_i s_i & \chi_i c_i & 2\delta_i & \chi_i s_i & -\chi_i c_i & 4\delta_i \end{array}\right] \begin{array}{c} (l) \\ (m) \\ (n) \\ (p) \\ (q) \\ (r) \end{array} \quad (6.9)$$

The product, $\mathbf{k}^{\mathbf{i}} = \mathbf{T}^{\mathbf{i}^{\mathrm{T}}}\mathbf{k}_*^{\mathbf{i}}\mathbf{T}^{\mathbf{i}}$, can now be generated, in terms of the following coefficients (defined earlier in Eq. 6.7):

$$\alpha_i = (EA)_i/L_i,\ \beta_i = 12(EI)_i/L_i^3,\ \chi_i = 6(EI)_i/L_i^2 \text{ and } \delta_i = (EI)_i/L_i$$

$$\mathbf{k}^{\mathbf{i}} = \left[\begin{array}{ccc|ccc} (\alpha_i c_i^2 + \beta_i s_i^2) & (\alpha_i - \beta_i)(c_i s_i) & -\chi_i s_i & -(\alpha_i c_i^2 + \beta_i s_i^2) & -(\alpha_i - \beta_i)(c_i s_i) & -\chi_i s_i \\ (\alpha_i - \beta_i)(c_i s_i) & (\alpha_i s_i^2 + \beta_i c_i^2) & \chi_i c_i & -(\alpha_i - \beta_i)(c_i s_i) & -(\alpha_i s_i^2 + \beta_i c_i^2) & \chi_i c_i \\ -\chi_i s_i & \chi_i c_i & 4\delta_i & \chi_i s_i & -\chi_i c_i & 2\delta_i \\ \hline -(\alpha_i c_i^2 + \beta_i s_i^2) & -(\alpha_i - \beta_i)(c_i s_i) & \chi_i s_i & (\alpha_i c_i^2 + \beta_i s_i^2) & (\alpha_i - \beta_i)(c_i s_i) & \chi_i s_i \\ -(\alpha_i - \beta_i)(c_i s_i) & -(\alpha_i s_i^2 + \beta_i c_i^2) & -\chi_i c_i & (\alpha_i - \beta_i)(c_i s_i) & (\alpha_i s_i^2 + \beta_i c_i^2) & -\chi_i c_i \\ -\chi_i s_i & \chi_i c_i & 2\delta_i & \chi_i s_i & -\chi_i c_i & 4\delta_i \end{array}\right] \begin{array}{c} (l) \\ (m) \\ (n) \\ (p) \\ (q) \\ (r) \end{array} \quad (6.10)$$

Comparing the above transformed matrix with the original element stiffness matrix $\mathbf{k}_*^{\mathbf{i}}$ in local axes coordinates [Eq. 6.7], we observe that the transformation to the global axes has introduced coupling between the axial and flexural (translational) element stiffnesses. This coupling, however, occurs only in the case of *inclined* frame elements (i.e., when $\theta_x^i \neq 0^\circ$ and $\theta_x^i \neq 90^\circ$, whereby $c_i \neq 0$ and $s_i \neq 0$). This is so, because, in such cases, a vertical or horizontal translation at the end of the plane frame element is likely to introduce both axial deformation and bending in the element.

The structure stiffness matrix $\mathbf{k} = \left[\begin{array}{c|c} \mathbf{k}_{\mathbf{AA}} & \mathbf{k}_{\mathbf{AR}} \\ \hline \mathbf{k}_{\mathbf{RA}} & \mathbf{k}_{\mathbf{RR}} \end{array}\right]$, is assembled by summing up the stiffness coefficients contributions from the various frame elements at the appropriate global coordinate locations. The stiffness contributions to the three global coordinates at any joint in the structure will be generated only from the plane frame elements that are connected to that joint. Some of the global coordinates may be restrained, and following our earlier convention, the numbering of these coordinates must be towards the end, with the active coordinates placed at the beginning.

Displacement Transformation Matrix

As an alternative, and to avoid the additional work involved in keeping track of the various linkages with the global coordinates (*l*, *m*, *n* and *p*, *q*, *r*) connected to a particular plane frame element, we can generate the *displacement transformation matrix*, $\mathbf{T_D^i}$ of order $6 \times (n + r)$, where n is the number of active global coordinates (equal to the degree of kinematic indeterminacy) and r is the number of restrained coordinates. This matrix, whose elements comprise the same elements as the $\mathbf{T^i}$ matrix (refer Eq. 6.8) with additional zero elements, relates the frame element local coordinates to all the global coordinates:

$$\left\{D_*^i\right\}_{6\times1} = \left[\left[T_{DA}\right]_{6\times n} \;\vdots\; \left[T_{DR}\right]_{6\times r}\right]\begin{bmatrix}\left\{D_A\right\}_{n\times1} \\ \hdashline \left\{D_R\right\}_{r\times1}\end{bmatrix} \tag{6.11}$$

Using the element stiffness matrix $\mathbf{k_*^i}$ given by Eq. 6.7, the product $\mathbf{k_*^i T_D^i}$ is first computed for each frame element and thereafter, the structure stiffness matrix $\mathbf{k}$ is assembled directly by summing up the stiffness coefficients contributions from the various frame elements using the following transformation:

$$\left[k\right]_{(n+r)\times(n+r)} = \sum_{i=1}^{m}\begin{bmatrix}\left[T_{DA}^i\right]_{n\times6}^T \\ \hdashline \left[T_{DR}^i\right]_{r\times6}^T\end{bmatrix}\left[k_*^i\right]_{6\times6}\left[\left[T_{DA}^i\right]_{6\times n} \;\vdots\; \left[T_{DR}^i\right]_{6\times r}\right] = \begin{bmatrix}\left[k_{AA}\right]_{n\times n} & \left[k_{AR}\right]_{n\times r} \\ \hdashline \left[k_{RA}\right]_{r\times n} & \left[k_{RR}\right]_{r\times r}\end{bmatrix} \tag{6.12}$$

6.2.3 Fixed End Forces

In case there are intermediate loads acting in any frame element, we need to find the initial force and displacement responses in the 'fixed end' element (which is part of the primary structure) and thus arrive at the *equivalent joint loads*, as explained earlier in Section 4.3.1 (for axial loads) and Section 5.2.4 (for loads causing bending). Care must be taken to assign the appropriate sign (positive or negative), while generating the components of fixed end force vector for each element, $\mathbf{F_{*f}^i}$. These fixed end forces can now be converted to equivalent joint loads by invoking appropriate coordinate transformations, such as by summing up the contributions of $\mathbf{T^{iT}F_{*f}^i}$ or $\mathbf{T_D^{i\,T}F_{*f}^i}$. The resulting fixed end force vector, in global coordinates, takes the form, $\mathbf{F_f} = \begin{bmatrix}\mathbf{F_{fA}} \\ \hdashline \mathbf{F_{fR}}\end{bmatrix}$. The equivalent joint loads are given by $-\mathbf{F_{fA}}$, and these are combined with the given nodal load vector $\mathbf{F_A}$.

The basic stiffness relationships take the following form:

$$\begin{bmatrix}\mathbf{F_A} \\ \hdashline \mathbf{F_R}\end{bmatrix} - \begin{bmatrix}\mathbf{F_{fA}} \\ \hdashline \mathbf{F_{fR}}\end{bmatrix} = \begin{bmatrix}\mathbf{k_{AA}} & \mathbf{k_{AR}} \\ \hdashline \mathbf{k_{RA}} & \mathbf{k_{RR}}\end{bmatrix}\begin{bmatrix}\mathbf{D_A} \\ \hdashline \mathbf{D_R}\end{bmatrix} \tag{6.13}$$

6.2.4 Analysis Procedure

The steps involved in the conventional stiffness method of analysis of kinematically indeterminate frames are as discussed in Section 5.2.5 (for beams), and may be summarised as follows:

1. Identify the frame elements, the local and global coordinates, as well as the input loads at the global coordinates, comprising direct actions, $\mathbf{F_A}$, and support displacements, $\mathbf{D_R}$, if any.
2. If there are intermediate loads acting on the frame elements, find the fixed end force vector for each element, $\mathbf{F^i_{*f}}$, and if required, the deflections at critical locations.
3. Generate the transformation matrices, $\mathbf{T^i}$, for the various elements, identifying the linking global coordinates in parenthesis (refer Eq. 6.8). [Alternatively, generate the displacement transformation matrices, $\mathbf{T^i_D}$]. Assemble the fixed end force vector $\mathbf{F_f} = \begin{bmatrix} \mathbf{F_{fA}} \\ \hline \mathbf{F_{fR}} \end{bmatrix}$, combining the contributions, $\mathbf{T^{iT}F^i_{*f}}$ [or $\mathbf{T_D^{i\,T}F^i_{*f}}$] from the various elements. Generate the resultant load vector, $\mathbf{F_A} - \mathbf{F_{fA}}$.
4. Generate the element stiffness matrices $\mathbf{k^i_*}$ for all the elements, using Eq. 6.7, and compute $\mathbf{k^i_*T^i}$, and hence assemble the structure stiffness matrix, $\mathbf{k} = \left[\begin{array}{c|c} \mathbf{k_{AA}} & \mathbf{k_{AR}} \\ \hline \mathbf{k_{RA}} & \mathbf{k_{RR}} \end{array}\right]$, combining the stiffness coefficient contributions, $\mathbf{k^i} = \mathbf{T^{i^T}k^i_*T^i}$ at the appropriate locations; for convenience, use Eqns 6.9 and 6.10. [Alternatively, compute $\mathbf{k^i_*T^i_D}$ for every element and assemble $\mathbf{k} = \sum_{i=1}^{m} \mathbf{T_D^{i^T}k^i_*T^i_D}$]. Find the inverse, $[\mathbf{k_{AA}}]^{-1}$, which is a symmetric matrix of order $n \times n$.
5. Find the unknown displacements, $\mathbf{D_A}$, by solving the following stiffness relation:
$$\mathbf{D_A} = [\mathbf{k_{AA}}]^{-1}\left[(\mathbf{F_A} - \mathbf{F_{fA}}) - \mathbf{k_{AR}D_R}\right] \tag{6.14}$$
6. Find the support reactions by solving the following equation:
$$\mathbf{F_R} = \mathbf{F_{fR}} + \mathbf{k_{RA}D_A} + \mathbf{k_{RR}D_R} \tag{6.15}$$
Perform simple equilibrium checks (such as $\sum F_y = 0$).
7. Find the member end forces for each element.
$$\mathbf{F^i_*} = \mathbf{F^i_{*f}} + (\mathbf{k^i_*T^i})\mathbf{D^i} \tag{6.16}$$
[Alternatively, use $\mathbf{F^i_*} = \mathbf{F^i_{*f}} + (\mathbf{k^i_*T^i_{DA}})\mathbf{D_A}$].
8. Draw the free-body diagrams for the various frame elements, and hence, the axial force, shear force and bending moment diagrams for the frame. Also, sketch the

deflected shape (superimposing the fixed end force effects) and identify the deflections at critical locations.

EXAMPLE 6.1

Consider a simple two-member frame in the vertical plane, comprising two identical tubular elements, OA and OB, fixed at ends A and B, and interconnected at the rigid joint O, the included angle AOB being equal to 2α. Each element has a length L and a tubular cross-section with mean radius r and thickness $t = r/10$, and is subjected to a uniformly distributed gravity load, q_o per unit length. Assume the material to have an elastic modulus E. Analyse the frame by the conventional stiffness method and generate expressions, in terms of the angle α, for the deflection at O, and the support reactions. Hence, draw the corresponding axial force, shear force and bending moment diagrams of OA. Comment on the behaviour with respect to variations in the angle α. Also comment on the magnitude of error in estimating the response if axial deformations are ignored.

SOLUTION

The elevation of the frame is shown in Fig. 6.3a. Taking advantage of symmetry, we can work with the reduced frame in Fig. 6.3b, in which there is only one prismatic frame element, fixed at end A and with a guided-fixed support at O (permitting movement only in the vertical direction), as shown. Hence, there is only one *active* degree of freedom at the rigid joint O and five *restrained* global coordinates, numbered as shown in Fig 6.3b. The local coordinates for the element are shown in Fig. 6.3c, with $\theta_x^1 = 90^\circ - \alpha$. The element is subject to a uniformly distributed vertical load (having components $q_o \sin\alpha$ and $q_o \cos\alpha$ in the transverse and axial directions of the element respectively). There are no nodal loads acting on the frame $(\mathbf{F_A} = \mathbf{O})$ and there are no support displacements $(\mathbf{D_R} = \mathbf{O})$.

Fixed End Forces

The fixed end force effects (support reactions) on the frame element (with $D_1 = 0$) are shown in Fig. 6.1d. Let $\boxed{s = \sin\alpha = \frac{a}{L}}$ and $\boxed{c = \cos\alpha = \frac{b}{L}}$.

Element '1' (OA): $F^1_{1*f} = F^1_{4*f} = \frac{q_o cL}{2};\ F^1_{2*f} = F^1_{5*f} = \frac{q_o sL}{2};\ F^1_{3*f} = \frac{q_o sL^2}{12}; F^1_{6*f} = -\frac{q_o sL^2}{12}$

$$\Rightarrow \mathbf{F^1_{*f}} = \left[q_o cL/2 \quad q_o sL/2 \quad q_o sL^2/12 \;\vdots\; q_o cL/2 \quad q_o sL/2 \quad q_o sL^2/12\right]^{\mathbf{T}}$$

Coordinate Transformations and Equivalent Joint Loads

The element transformation matrix $\mathbf{T^1}$, satisfying $\mathbf{D^i_*} = \mathbf{T^i D}$ and $\mathbf{F^i_*} = \mathbf{T^i F}$, take the following form (refer Fig. 6.3 and Eq. 6.8), with the linking global coordinates indicated in parentheses.

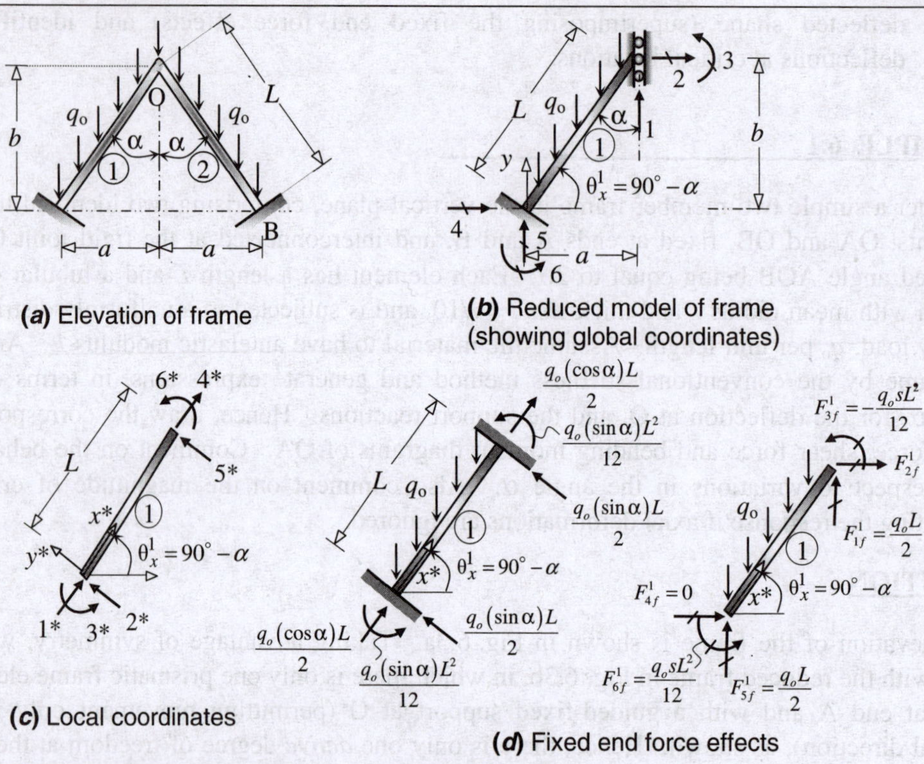

(a) Elevation of frame
(b) Reduced model of frame (showing global coordinates)
(c) Local coordinates
(d) Fixed end force effects

Figure 6.3 Details of frame with two elements (Example 6.1)

$$
\mathbf{T}^1 = \begin{array}{c} \begin{matrix} (4) & (5) & (6) & (2) & (1) & (3) \end{matrix} \\ \left[\begin{array}{ccc|ccc} s & c & 0 & 0 & 0 & 0 \\ -c & s & 0 & 0 & 0 & 0 \\ 0 & 0 & 1 & 0 & 0 & 0 \\ \hline 0 & 0 & 0 & s & c & 0 \\ 0 & 0 & 0 & -c & s & 0 \\ 0 & 0 & 0 & 0 & 0 & 1 \end{array}\right] \end{array} \begin{matrix} (4) \\ (5) \\ (6) \\ (2) \\ (1) \\ (3) \end{matrix}
$$

$$
\Rightarrow \mathbf{T}^{1^T}\mathbf{F}^1_{*f} = \left[\begin{array}{ccc|ccc} s & -c & 0 & 0 & 0 & 0 \\ c & s & 0 & 0 & 0 & 0 \\ 0 & 0 & 1 & 0 & 0 & 0 \\ \hline 0 & 0 & 0 & s & -c & 0 \\ 0 & 0 & 0 & c & s & 0 \\ 0 & 0 & 0 & 0 & 0 & 1 \end{array}\right] \begin{Bmatrix} q_o cL/2 \\ q_o sL/2 \\ q_o sL^2/12 \\ q_o cL/2 \\ q_o sL/2 \\ -q_o sL^2/12 \end{Bmatrix} = \begin{Bmatrix} 0 \\ q_o L/2 \\ q_o sL^2/12 \\ 0 \\ q_o L/2 \\ -q_o sL^2/12 \end{Bmatrix} \begin{matrix} (4) \\ (5) \\ (6) \\ (2) \\ (1) \\ (3) \end{matrix}
$$

The fixed end forces in global axes system, $\mathbf{F}_f = \begin{bmatrix} \mathbf{F}_{fA} \\ \hline \mathbf{F}_{fR} \end{bmatrix}$, can now be assembled:

$$\mathbf{F}_{fA} = \{F_{1f}\} = \left\{\frac{q_o L}{2}\right\} \quad ; \quad \mathbf{F}_{fR} = \begin{Bmatrix} F_{2f} \\ F_{3f} \\ F_{4f} \\ F_{5f} \\ F_{6f} \end{Bmatrix} = \begin{Bmatrix} 0 \\ -q_o sL^2/12 \\ 0 \\ q_o L/2 \\ q_o sL^2/12 \end{Bmatrix}$$, as shown in Fig. 6.3d.

The resultant load vector is given by:

$$\mathbf{F}_A - \mathbf{F}_{fA} = \{0\} - \{F_{1f}\} = \left\{-\frac{q_o L}{2}\right\}$$

The resultant nodal load comprises a downward force of $q_o L/2$ at O in the reduced frame (i.e., $q_o L$ in the full frame, as shown in Figs 6.4a and 6.5a). The deflections at the active coordinate, D_1, due to this load will be identical to the one in the original loading diagram (with the distributed loading) in Fig. 6.3a.

Approximate Solution: Ignoring Axial Deformations

If axial deformations are ignored, of course, $D_1 = 0$, and the problem becomes kinematically determinate. In this case, the frame in Fig. 6.4a will behave like a funicular (linear) arch, with the frame elements subject to pure axial compression (without any bending moment or shear force), as indicated in Fig. 6.4a. Superimposing this force response with the fixed end force response in Fig. 6.3d, we get an approximate solution for the total response, as shown in Fig. 6.4b.

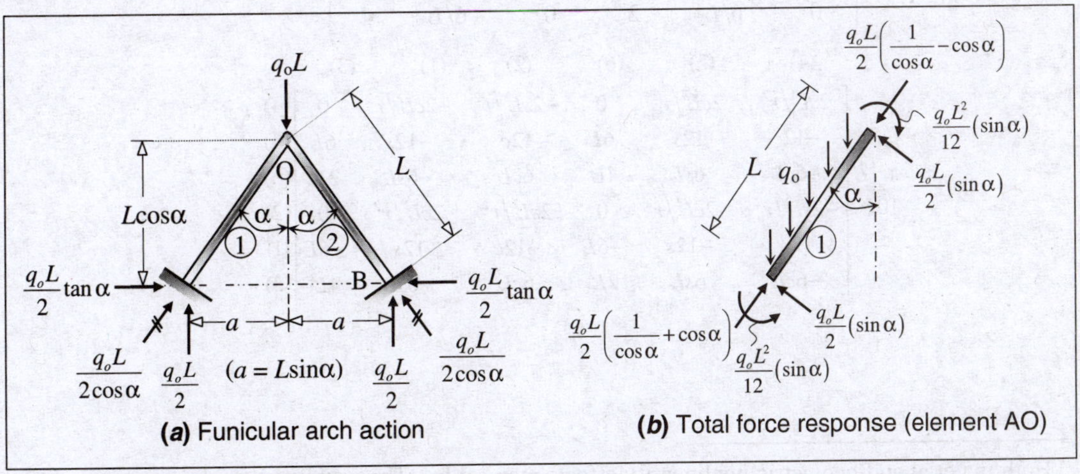

Figure 6.4 Approximate solution, ignoring axial deformations (Example 6.1)

The funicular arch assumption, however, becomes invalid at very high values of the angle α, because it would be practically impossible to sustain the applied loading through axial force alone[†]. The horizontal reaction at the support, $\frac{q_o L \tan\alpha}{2}$, approaches infinity as α approaches 90°! In the limiting case, when $\alpha = 90°$, the frame in Fig. 6.4a actually reduces to a fixed beam, where the resistance to the applied loading occurs through flexure and not by axial force; but such behaviour cannot be simulated by the simplified frame analysis (ignoring axial deformations). Furthemore, due to the beam action (at high values of α), sagging bending moments develop near the joint O, and not hogging moments (indicated in Fig. 6.4b). For these reasons, it is inappropriate to ignore axial deformations in the frame when the value of α is high, because flexural response becomes predominant in such cases. In order to activate this flexural response, the joint O has to deflect, and this is possible only if axial deformations are permitted in the structural model. The resulting stiffness matrix will involve coupling of the axial and flexural stiffnesses of the frame element.

Element and Structure Stiffness Matrices

The element stiffness matrix $\mathbf{k}_*^1$, satisfying $\mathbf{F}_*^i = \mathbf{k}_*^i \mathbf{D}_*^i$, takes the following form (refer Fig. 6.3c and Eq. 6.7), considering, $I = \pi r^3 t = \pi r^4/10$ and $\frac{EA}{EI} = \frac{2\pi r t}{\pi r^3 t} = \frac{2}{r^2}$:

$$\mathbf{k}_*^1 = \frac{\pi r^4 E}{10L}\begin{bmatrix} 2/r^2 & 0 & 0 & -2/r^2 & 0 & 0 \\ 0 & 12/L^2 & 6/L & 0 & -12/L^2 & 6/L \\ 0 & 6/L & 4 & 0 & -6/L & 2 \\ -2/r^2 & 0 & 0 & 2/r^2 & 0 & 0 \\ 0 & -12/L^2 & -6/L & 0 & 12/L^2 & -6/L \\ 0 & 6/L & 2 & 0 & -6/L & 4 \end{bmatrix}$$

$$\Rightarrow \quad \mathbf{k}_*^1\mathbf{T}^1 = \frac{\pi r^4 E}{10L^3}\begin{array}{c} \begin{array}{cccccc} (4) & (5) & (6) & (2) & (1) & (3) \end{array} \\ \left[\begin{array}{cccccc} 2sL^2/r^2 & 2cL^2/r^2 & 0 & -2sL^2/r^2 & -2cL^2/r^2 & 0 \\ -12c & 12s & 6L & 12c & -12s & 6L \\ -6cL & 6sL & 4L^2 & 6cL & -6sL & 2L^2 \\ -2sL^2/r^2 & -2cL^2/r^2 & 0 & 2sL^2/r^2 & 2cL^2/r^2 & 0 \\ 12c & -12s & -6L & -12c & 12s & -6L \\ -6cL & 6sL & 2L^2 & 6cL & -6sL & 4L^2 \end{array}\right] \end{array} \begin{array}{c} \\ (4) \\ (5) \\ (6) \\ (2) \\ (1) \\ (3) \end{array}$$

[†] Both material and geometric nonlinearity effects may set in. The computed axial compressive stress in the frame element is likely to exceed the elastic limit. Also, slenderness effects become pronounced under high axial compression. The response predicted by linear analysis can be significantly erroneous in such cases.

$$\Rightarrow \quad \mathbf{T}^{1^T}\mathbf{k}_*^1\mathbf{T}^1 = \frac{\pi r^4 E}{10L^3}\begin{bmatrix} c_1 & c_3 & -6cL & -c_1 & -c_3 & -6cL \\ c_3 & c_2 & 6sL & -c_3 & -c_2 & 6sL \\ -6cL & 6sL & 4L^2 & 6cL & -6sL & 2L^2 \\ -c_1 & -c_3 & 6cL & c_1 & c_3 & 6cL \\ -c_3 & -c_2 & -6sL & c_3 & c_2 & -6sL \\ -6cL & 6sL & 2L^2 & 6cL & -6sL & 4L^2 \end{bmatrix}\begin{matrix}(4)\\(5)\\(6)\\(2)\\(1)\\(3)\end{matrix}$$

(columns: (4) (5) (6) (2) (1) (3))

where $c_1 = \left(2s^2(L/r)^2 + 12c^2\right)$, $c_2 = \left(2c^2(L/r)^2 + 12s^2\right)$, $c_3 = sc\left(2(L/r)^2 - 12\right)$

As there is only one element, the structure stiffness matrix $\mathbf{k} = \left[\begin{array}{c|c} \mathbf{k_{AA}} & \mathbf{k_{AR}} \\ \hline \mathbf{k_{RA}} & \mathbf{k_{RR}} \end{array}\right]$, satisfying $\mathbf{F} = \mathbf{kD}$, can now be assembled by considering the contributions of $\mathbf{T}^{1^T}\mathbf{k}_*^1\mathbf{T}^1$ alone.

$$\Rightarrow \quad \mathbf{k_{AA}} = \frac{\pi r^4 E}{10L^3}[c_2] = \frac{\pi r^4 E}{10L^3}\left[2c^2(L/r)^2 + 12s^2\right]$$

$$\mathbf{k_{RA}} = \frac{\pi r^4 E}{10L^3}\begin{bmatrix} c_3 \\ -6sL \\ -c_3 \\ -c_2 \\ -6sL \end{bmatrix} = \frac{\pi r^4 E}{10L^3}\begin{bmatrix} sc\left(2(L/r)^2 - 12\right) \\ -6sL \\ -sc\left(2(L/r)^2 - 12\right) \\ -\left(2c^2(L/r)^2 + 12s^2\right) \\ -6sL \end{bmatrix}$$

The inverse of $\mathbf{k_{AA}}$ (which is a matrix of order 1×1) is given by:

$$[\mathbf{k_{AA}}]^{-1} = \left(\frac{10L^3}{\pi r^4 E}\right)\left[\frac{1}{2c^2(L/r)^2 + 12s^2}\right]$$

Displacements

The structure stiffness relation to be considered here is:

$$\left[\begin{array}{c} \mathbf{F_A = O} \\ \hline \mathbf{F_R} \end{array}\right] - \left[\begin{array}{c} \mathbf{F_{fA}} \\ \hline \mathbf{F_{fR}} \end{array}\right] = \left[\begin{array}{c|c} \mathbf{k_{AA}} & \mathbf{k_{AR}} \\ \hline \mathbf{k_{RA}} & \mathbf{k_{RR}} \end{array}\right]\left[\begin{array}{c} \mathbf{D_A} \\ \hline \mathbf{D_R = O} \end{array}\right]$$

$$\Rightarrow \quad \mathbf{D_A} = [\mathbf{k_{AA}}]^{-1}[\mathbf{F_A} - \mathbf{F_{fA}}] \quad \Rightarrow \quad \{D_1\} = \left(\frac{10L^3}{\pi r^4 E}\right)\left[\frac{1}{2(cL/r)^2 + 12s^2}\right]\left\{-\frac{q_o L}{2}\right\}$$

$$\Rightarrow \quad \boxed{D_1 = -\left(\frac{q_o}{E}\right)\left(\frac{L}{r}\right)^4 \frac{1}{0.4\pi\left((cL/r)^2 + 6s^2\right)}}$$

We can easily verify the accuracy of this solution for the two extreme values of α. The deflection $\Delta_O = -D_1$ at the joint O (refer Fig. 6.5a) takes the lowest value when $\alpha = 0^\circ$, whereby the two elements act independently as vertical columns, with

$(\Delta_O)_{\alpha=0^\circ} = \dfrac{(q_o L/2)}{EA/L} = \dfrac{0.5}{\pi}\dfrac{q_o}{E}\left(\dfrac{L}{r}\right)^2$. This deflection is very small and is practically nil for members with low *L/r* ratios.

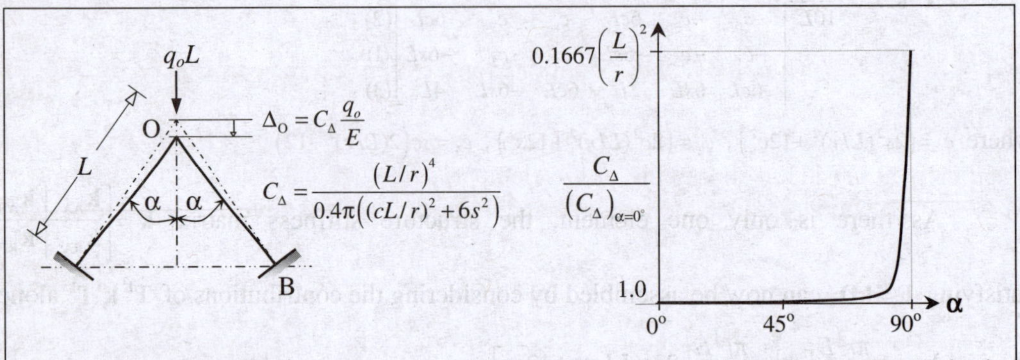

(*a*) Deflected shape (*b*) Typical variation of deflection coefficient (C_Δ)

Figure 6.5 Displacement response of frame (Example 6.1)

The variation of Δ_O, with respect to the angle α, can be conveniently expressed in terms of the non-dimensional coefficient, C_Δ, as shown in Fig. 6.5. With reference to $(\Delta_O)_{\alpha=0^\circ}$, the deflection Δ_O increases extremely slowly with increasing values of α, but starts increasing rapidly (due to the predominance of flexural action) when α exceeds about 75°, as shown in Fig. 6.5b. In the limiting case of $\alpha = 90^\circ$, the two frame elements join together to form a horizontal fixed beam of span 2*L*, whose mid-span deflection at O is given by $\dfrac{q_o(2L)^4}{384EI} = \dfrac{5}{12\pi}\dfrac{q_o}{E}\left(\dfrac{L}{r}\right)^4$, indicating $(\Delta_O)_{\alpha=90^\circ} = (\Delta_O)_{\alpha=0^\circ} \times \dfrac{(L/r)^2}{6}$. Thus, the 'beam' deflection, $(\Delta_O)_{\alpha=90^\circ}$, is 16.67 times the corresponding 'column' shortening, $(\Delta_O)_{\alpha=0^\circ}$, at O, for $L/r = 10$, but increases to as much as 1667 times $(\Delta_O)_{\alpha=0^\circ}$ for $L/r = 100$.

Support Reactions

$\mathbf{F_R} = \mathbf{F_{fR}} + \mathbf{k_{RA}}\mathbf{D_A}$

where $\mathbf{F_{fR}} = \begin{Bmatrix} F_{2f} \\ F_{3f} \\ F_{4f} \\ F_{5f} \\ F_{6f} \end{Bmatrix} = \begin{Bmatrix} 0 \\ -q_o sL^2/12 \\ 0 \\ q_o L/2 \\ q_o sL^2/12 \end{Bmatrix}$ and $\mathbf{k_{RA}}\mathbf{D_A} = \begin{bmatrix} sc\left(2(L/r)^2 - 12\right) \\ -6sL \\ -sc\left(2(L/r)^2 - 12\right) \\ -\left(2c^2(L/r)^2 + 12s^2\right) \\ -6sL \end{bmatrix} \left[\dfrac{1}{2c^2(L/r)^2 + 12s^2}\right] \left\{-\dfrac{q_o L}{2}\right\}$

$$\Rightarrow \boxed{\begin{aligned} &F_2 = -F_4 = -q_o scL\left[\frac{(L/r)^2 - 6}{2(cL/r)^2 + 12s^2}\right] \; ; \; F_5 = q_o L \\ &F_3 = \frac{q_o sL^2}{12}\left[-1 + \frac{18}{(cL/r)^2 + 6s^2}\right] \; ; \; F_6 = \frac{q_o sL^2}{12}\left[1 + \frac{18}{(cL/r)^2 + 6s^2}\right] \end{aligned}}$$

We can easily verify these solutions for the two extreme cases of $\alpha = 0^\circ$ and $\alpha = 90^\circ$. The freebody diagram for element AO is shown in Fig. 6.6a. The frame element AO behaves essentially like a fixed beam, fixed against rotations at A and O, but free to translate vertically at O (due to shortening of the element). The vertical reaction at O is always equal to zero. The horizontal reaction and the end moments can be conveniently expressed in terms of non-dimensional coefficients, as indicated in Fig. 6.6a.

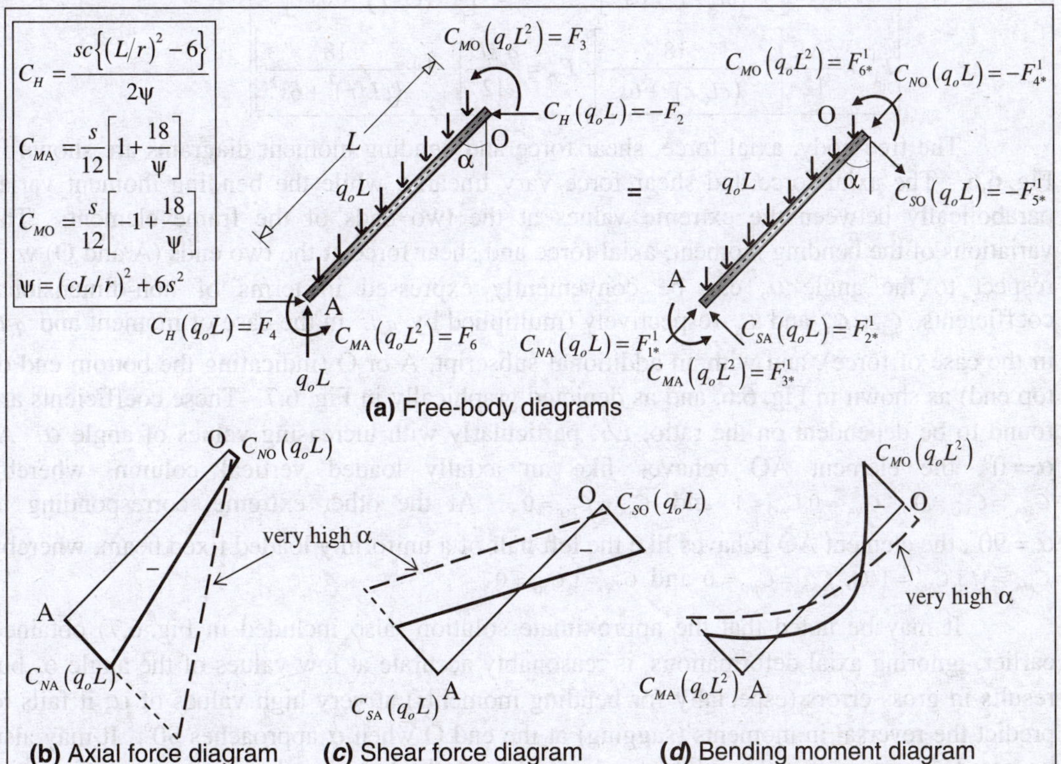

Figure 6.6 Force response of frame element (Example 6.1)

Member Forces

$$\mathbf{F}^{\mathbf{i}}_{*} = \mathbf{F}^{\mathbf{i}}_{*\mathbf{f}} + \mathbf{k}^{\mathbf{i}}\mathbf{T}^{\mathbf{i}}\mathbf{D}$$

Noting that $\mathbf{D_R} = \mathbf{O}$, the calculations can be simplified as $\mathbf{F}^{\mathbf{i}}_{*} = \mathbf{F}^{\mathbf{i}}_{*\mathbf{f}} + \mathbf{k}^{\mathbf{i}}_{\mathbf{A}}\mathbf{T}^{\mathbf{i}}_{\mathbf{A}}\mathbf{D_A}$:

$$\Rightarrow \mathbf{F}_*^1 = \begin{Bmatrix} q_o cL/2 \\ q_o sL/2 \\ q_o sL^2/12 \\ q_o cL/2 \\ q_o sL/2 \\ -q_o sL^2/12 \end{Bmatrix} + \begin{bmatrix} -2cL^2/r^2 \\ -12s \\ -6sL \\ 2cL^2/r^2 \\ 12s \\ -6sL \end{bmatrix} \left[\frac{1}{2(cL/r)^2 + 12s^2} \right] \left\{ -\frac{q_o L}{2} \right\}$$

$$\Rightarrow \boxed{\begin{array}{l} F_{1*}^1 = \dfrac{q_o cL}{2}\left[1 + \dfrac{(L/r)^2}{(cL/r)^2 + 6s^2}\right]; \; F_{4*}^1 = \dfrac{q_o cL}{2}\left[1 - \dfrac{(L/r)^2}{(cL/r)^2 + 6s^2}\right] \\ F_{2*}^1 = \dfrac{q_o sL}{2}\left[1 + \dfrac{6}{(cL/r)^2 + 6s^2}\right]; \; F_{5*}^1 = \dfrac{q_o sL}{2}\left[1 - \dfrac{6}{(cL/r)^2 + 6s^2}\right] \\ F_{3*}^1 = \dfrac{q_o sL^2}{12}\left[1 - \dfrac{18}{(cL/r)^2 + 6s^2}\right]; \; F_{6*}^1 = \dfrac{q_o sL^2}{12}\left[-1 + \dfrac{18}{(cL/r)^2 + 6s^2}\right] \end{array}}$$

The free-body, axial force, shear force and bending moment diagrams are shown in Fig. 6.6. The axial force and shear force vary linearly, while the bending moment varies parabolically between the extreme values at the two ends of the frame element. The variations of the bending moment, axial force and shear force at the two ends (A and O) with respect to the angle α, can be conveniently expressed in terms of non-dimensional coefficients, C_M, C_N and C_S respectively (multiplied by $q_o L^2$ in the case of moment and $q_o L$ in the case of force), and with an additional subscript, A or O (indicating the bottom end or top end) as shown in Fig. 6.6, and as depicted graphically in Fig. 6.7. These coefficients are found to be dependent on the ratio, *L/r*, particularly with increasing values of angle α. At $\alpha = 0^o$, the element AO behaves like an axially loaded vertical column, whereby $C_{MA} = C_{MO} = 0$, $C_{NA} = 0, C_{NO} = 1$ and $C_{SA} = C_{SO} = 0$. At the other extreme, corresponding to $\alpha = 90^o$, the element AO behaves like the left half of a uniformly loaded fixed beam, whereby $C_{MA} = 1/3, C_{MO} = 1/6$, $C_{NA} = C_{NO} = 0$ and $C_{SA} = 1, C_{SO} = 0$.

It may be noted that the approximate solution (also included in Fig. 6.7) obtained earlier, ignoring axial deformations, is reasonably accurate at low values of the angle α, but results in gross errors (especially for bending moments) at very high values of α; it fails to predict the reversal in moments (sagging) at the end O when α approaches 90°. It may also be noted, however, that the solutions obtained from the linear analysis accounting for axial deformations, at very high values of α (except at $\alpha = 90^o$), will be invalid, if the high compressive stresses exceed the elastic limit (material nonlinearity) or if the element is very slender (geometric nonlinearity, associated with high values of *L/r*).

The figure contains the following labelled boxed equations:

$$C_{MA} = \frac{\sin\alpha}{12}; \quad C_{MO} = -\frac{\sin\alpha}{12}$$

$$C_{SA} = \frac{\sin\alpha}{2}; \quad C_{SO} = -\frac{\sin\alpha}{2}$$

$$C_{NA} = \frac{q_o L}{2}\left(\frac{1}{\cos\alpha} + \cos\alpha\right)$$

$$C_{NO} = -\frac{q_o L}{2}\left(\frac{1}{\cos\alpha} - \cos\alpha\right)$$

If axial deformations are ignored,

(a) Moment coefficients

(b) Axial force coefficients

(c) Shear force coefficients

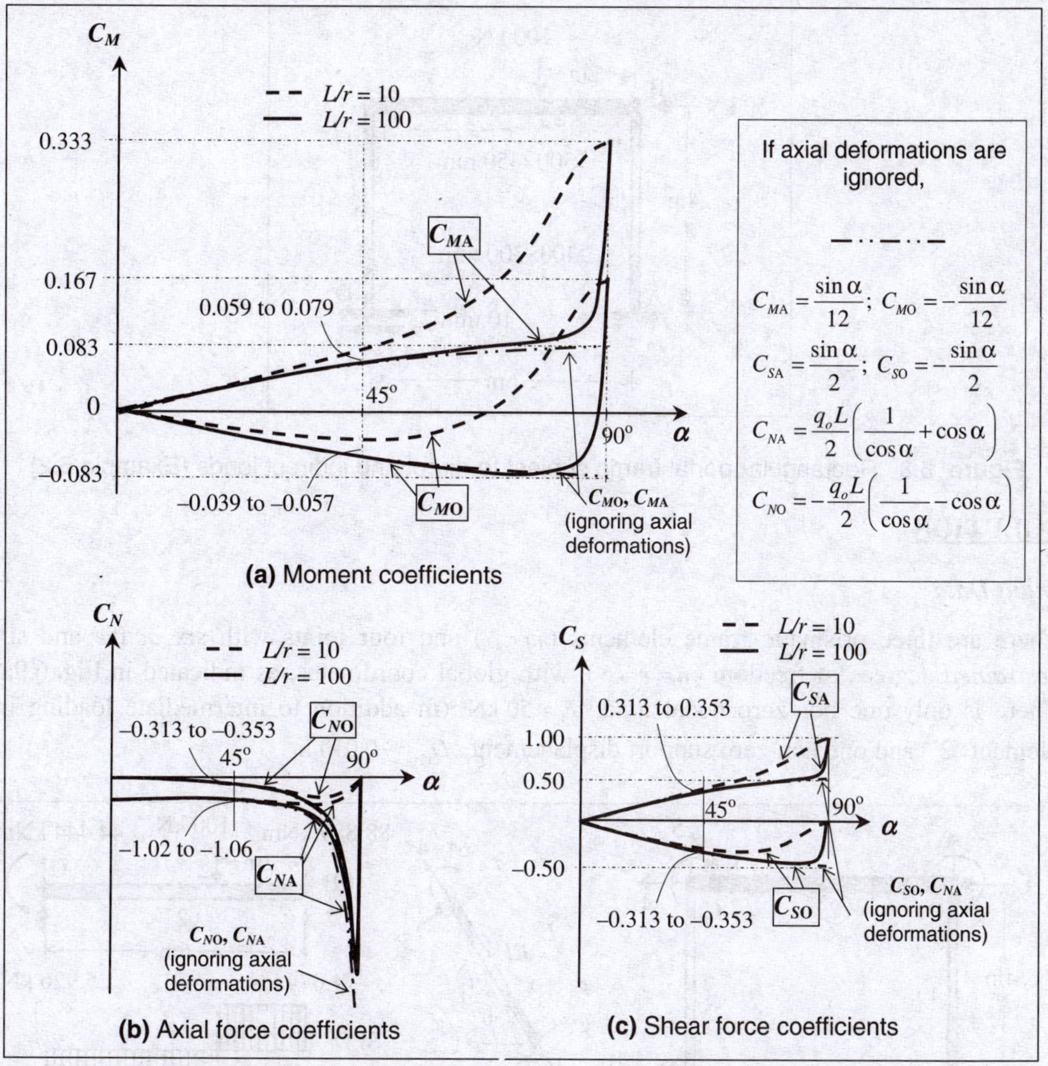

Figure 6.7 Variations of moment and force coefficients with α (Example 6.1)

EXAMPLE 6.2

Analyse the rectangular portal frame shown in Fig. 6.8 by the conventional Stiffness Method. In addition to the applied loads, consider the effect of indirect loading in the form of a settlement of 10 mm at the support D. Assume the column elements to have a square cross section, 300×300 mm and the beam to have a size, 300×450 mm. Assume all elements to have an elastic modulus, $E = 2.5 \times 10^4\,\text{N/mm}^2$. Find the complete force response and draw the deflection and bending moment diagrams.

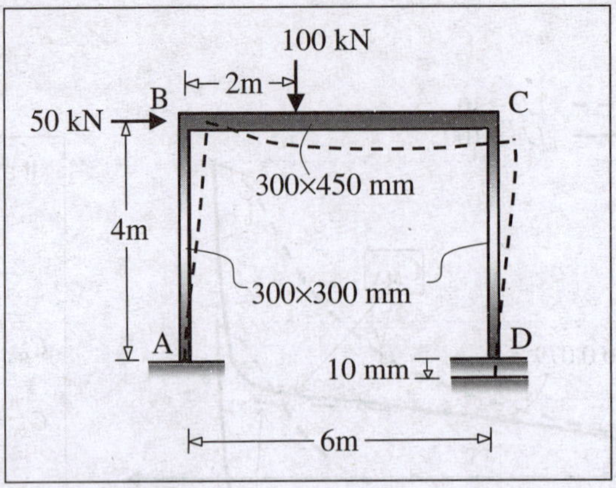

Figure 6.8 Rectangular portal frame subject to direct and indirect loads (Example 6.2)

SOLUTION

Input Data

There are three prismatic frame elements ($m = 3$) and four joints with six *active* and six *restrained* degrees of freedom ($n = r = 6$), with global coordinates, as indicated in Fig. 6.9a. There is only one non-zero nodal load, $F_1 = 50$ kN (in addition to intermediate loading in element '2') and one non-zero support displacement, $D_{11} = -0.010$ m.

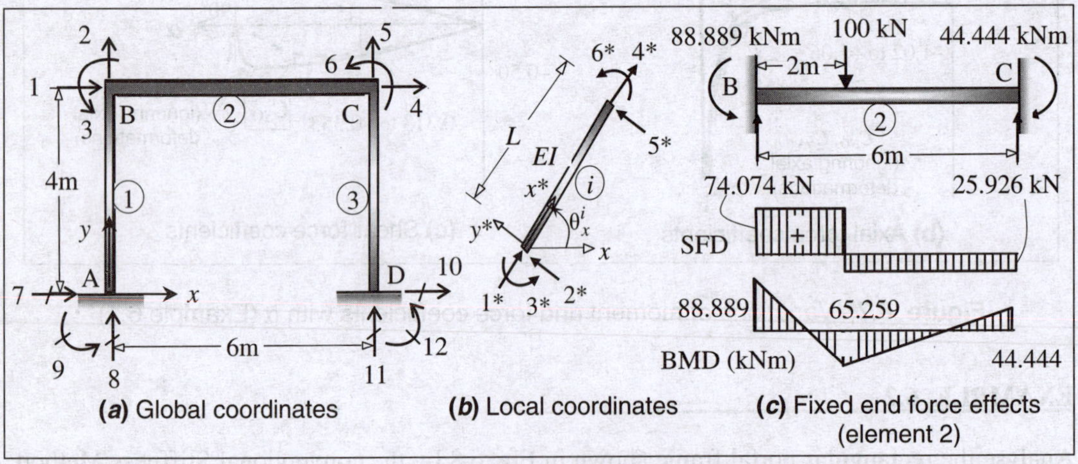

Figure 6.9 Global and local coordinates and fixed end force effects (Example 6.2)

The local coordinates for a typical frame element are as shown in Fig. 6.9b. Choosing the origin of the *x*-*y* global axes system at the joint A, the coordinates of the start and end nodes (X_{iS}, Y_{iS} and X_{iE}, Y_{iE}) of individual elements and their lengths,

$L_i = \sqrt{(X_{iE} - X_{iS})^2 + (Y_{iE} - Y_{iS})^2}$, and direction cosine parameters, $c_i = \cos\theta_x^i = \dfrac{X_{iE} - X_{iS}}{L_i}$ and $s_i = \sin\theta_x^i = \dfrac{Y_{iE} - Y_{iS}}{L_i}$, are tabulated below. The element cross-sectional properties, A_i and I_i, and the stiffness measures, $\alpha_i = (EA)_i/L_i$ and $\delta_i = (EI)_i/L_i$, are also computed, assuming $E = 2.5\times10^7$ kN/m^2.

Element No. (i)	Start Node (X_{iS}, Y_{iS}) (m)	End Node (X_{iE}, Y_{iE}) (m)	L_i (m)	c_i	s_i	A_i (m^2)	I_i (m^4) $\times10^{-4}$	$\alpha_i = (EA)_i/L_i$ (kN/m)	$\delta_i = (EI)_i/L_i$ (kNm)
1	A (0,0)	B (0,4)	4.0	0	1	0.090	6.75	562500	4218.75
2	B (0,4)	C (6,4)	6.0	1	0	0.135	2.27812	562500	9492.1875
3	D (0,6)	C (6,4)	4.0	0	1	0.090	6.75	562500	4218.75

Fixed End Forces

The fixed end force effects (support reactions, bending moment and shear force diagrams) on the element '2' in the restrained primary structure, due to the action of intermediate loading, are shown in Fig. 6.9c. There are no fixed end forces in elements '1' and '3'.

<u>Element '2' (BC):</u>

$$F^2_{1*f} = F^2_{4*f} = 0 \text{ kN};\ F^2_{3*f} = \frac{100(4)^2(2)}{(6)^2} = 88.889 \text{ kNm};\ F^2_{6*f} = -\frac{100(2)^2(4)}{(6)^2} = -44.444 \text{ kNm};$$

$$F^2_{2*f} = \frac{100\times4}{6} + \frac{44.444}{6} = 74.074 \text{ kN};\ F^2_{5*f} = \frac{100\times2}{6} - \frac{44.444}{6} = 25.926 \text{ kN}$$

$$\Rightarrow \mathbf{F}^1_{*f} = \mathbf{F}^3_{*f} = \left\{\begin{array}{c} 0 \text{ kN} \\ 0 \text{ kN} \\ 0 \text{ kNm} \\ \hdashline 0 \text{ kN} \\ 0 \text{ kN} \\ 0 \text{ kNm} \end{array}\right\};\quad \mathbf{F}^2_{*f} = \left\{\begin{array}{c} 0 \text{ kN} \\ 74.074 \text{ kN} \\ 88.889 \text{ kNm} \\ \hdashline 0 \text{ kN} \\ 25.926 \text{ kN} \\ -44.444 \text{ kNm} \end{array}\right\}$$

Coordinate Transformations and Equivalent Joint Loads

The element transformation matrices $\mathbf{T}^1$, $\mathbf{T}^2$ and $\mathbf{T}^3$, satisfying $\mathbf{D}^i_* = \mathbf{T}^i\mathbf{D}$ and $\mathbf{F}^i_* = \mathbf{T}^i\mathbf{F}$, take the following form (refer Eq. 6.8 and Figs 6.9a and b), with the linking global coordinates indicated in parentheses:

$$
\mathbf{T^1} = \begin{array}{cc} \begin{matrix} (7)\ (8)\ (9)\ (1)\ (2)\ (3) \end{matrix} \\ \left[\begin{array}{ccc|ccc} 0 & 1 & 0 & 0 & 0 & 0 \\ -1 & 0 & 0 & 0 & 0 & 0 \\ 0 & 0 & 1 & 0 & 0 & 0 \\ \hline 0 & 0 & 0 & 0 & 1 & 0 \\ 0 & 0 & 0 & -1 & 0 & 0 \\ 0 & 0 & 0 & 0 & 0 & 1 \end{array}\right] \begin{matrix} (7) \\ (8) \\ (9) \\ (1) \\ (2) \\ (3) \end{matrix} \end{array} ; \quad
\mathbf{T^2} = \begin{array}{cc} \begin{matrix} (1)\ (2)\ (3)\ (4)\ (5)\ (6) \end{matrix} \\ \left[\begin{array}{ccc|ccc} 1 & 0 & 0 & 0 & 0 & 0 \\ 0 & 1 & 0 & 0 & 0 & 0 \\ 0 & 0 & 1 & 0 & 0 & 0 \\ \hline 0 & 0 & 0 & 1 & 0 & 0 \\ 0 & 0 & 0 & 0 & 1 & 0 \\ 0 & 0 & 0 & 0 & 0 & 1 \end{array}\right] \begin{matrix} (1) \\ (2) \\ (3) \\ (4) \\ (5) \\ (6) \end{matrix} \end{array} ; \quad
\mathbf{T^3} = \begin{array}{cc} \begin{matrix} (10)\ (11)\ (12)\ (4)\ (5)\ (6) \end{matrix} \\ \left[\begin{array}{ccc|ccc} 0 & 1 & 0 & 0 & 0 & 0 \\ -1 & 0 & 0 & 0 & 0 & 0 \\ 0 & 0 & 1 & 0 & 0 & 0 \\ \hline 0 & 0 & 0 & 0 & 1 & 0 \\ 0 & 0 & 0 & -1 & 0 & 0 \\ 0 & 0 & 0 & 0 & 0 & 1 \end{array}\right] \begin{matrix} (10) \\ (11) \\ (12) \\ (4) \\ (5) \\ (6) \end{matrix} \end{array}
$$

The fixed end forces in global axes can now be assembled, by summing up the contributions of

$$
\mathbf{T^{1^T}F_{*f}^1} = \begin{Bmatrix} 0 \text{ kN} \\ 0 \text{ kN} \\ 0 \text{ kNm} \\ \hline 0 \text{ kN} \\ 0 \text{ kN} \\ 0 \text{ kNm} \end{Bmatrix} \begin{matrix} (7) \\ (8) \\ (9) \\ (1) \\ (2) \\ (3) \end{matrix} \quad ; \quad
\mathbf{T^{2^T}F_{*f}^2} = \begin{Bmatrix} 0 \text{ kN} \\ 74.074 \text{ kN} \\ 88.889 \text{ kNm} \\ \hline 0 \text{ kN} \\ 25.926 \text{ kN} \\ -44.444 \text{ kNm} \end{Bmatrix} \begin{matrix} (1) \\ (2) \\ (3) \\ (4) \\ (5) \\ (6) \end{matrix} ; \quad
\mathbf{T^{3^T}F_{*f}^3} = \begin{Bmatrix} 0 \text{ kN} \\ 0 \text{ kN} \\ 0 \text{ kNm} \\ \hline 0 \text{ kN} \\ 0 \text{ kN} \\ 0 \text{ kNm} \end{Bmatrix} \begin{matrix} (10) \\ (11) \\ (12) \\ (4) \\ (5) \\ (6) \end{matrix}
$$

$$
\Rightarrow \mathbf{F_f} = \begin{bmatrix} \mathbf{F_{fA}} \\ \hline \mathbf{F_{fR}} \end{bmatrix} \quad \text{where} \quad \mathbf{F_{fA}} = \begin{Bmatrix} F_{1f} \\ F_{2f} \\ F_{3f} \\ F_{4f} \\ F_{5f} \\ F_{6f} \end{Bmatrix} = \begin{Bmatrix} 0 \text{ kN} \\ 74.074 \text{ kN} \\ 88.889 \text{ kNm} \\ \hline 0 \text{ kN} \\ 25.926 \text{ kN} \\ -44.444 \text{ kNm} \end{Bmatrix} \quad \text{and} \quad \mathbf{F_{fR}} = \begin{Bmatrix} F_{7f} \\ F_{8f} \\ F_{9f} \\ F_{10f} \\ F_{11f} \\ F_{12f} \end{Bmatrix} = \begin{Bmatrix} 0 \text{ kN} \\ 0 \text{ kN} \\ 0 \text{ kNm} \\ 0 \text{ kN} \\ 0 \text{ kN} \\ 0 \text{ kNm} \end{Bmatrix}
$$

The resultant direct load vector, $\mathbf{F_A} - \mathbf{F_{fA}}$, and the restrained displacement vector, $\mathbf{D_R}$ are given by:

$$
\mathbf{F_A} - \mathbf{F_{fA}} = \begin{Bmatrix} 50 \text{ kN} \\ 0 \text{ kN} \\ 0 \text{ kNm} \\ \hline 0 \text{ kN} \\ 0 \text{ kN} \\ 0 \text{ kNm} \end{Bmatrix} - \begin{Bmatrix} 0 \text{ kN} \\ 74.074 \text{ kN} \\ 88.889 \text{ kNm} \\ \hline 0 \text{ kN} \\ 25.926 \text{ kN} \\ -44.444 \text{ kNm} \end{Bmatrix} = \begin{Bmatrix} 50 \text{ kN} \\ -74.074 \text{ kN} \\ -88.889 \text{ kNm} \\ \hline 0 \text{ kN} \\ -25.926 \text{ kN} \\ 44.444 \text{ kNm} \end{Bmatrix} \quad ; \quad \mathbf{D_R} = \begin{Bmatrix} D_7 \\ D_8 \\ D_9 \\ D_{10} \\ D_{11} \\ D_{12} \end{Bmatrix} = \begin{Bmatrix} 0 \text{ m} \\ 0 \text{ m} \\ 0 \text{ rad} \\ 0 \text{ m} \\ -0.01 \text{ m} \\ 0 \text{ rad} \end{Bmatrix}
$$

This is indicated in Fig. 6.10. The displacements $\mathbf{D_A}$ at the six active coordinates will be identical to those in the original loading diagram (with the intermediate loading) in Fig. 6.8.

Element and Structure Stiffness Matrices

The element stiffness matrices $\mathbf{k_*^i}$ for the three elements, satisfying $\mathbf{F_*^i} = \mathbf{k_*^i D_*^i}$, take the following form (Eq. 6.7):

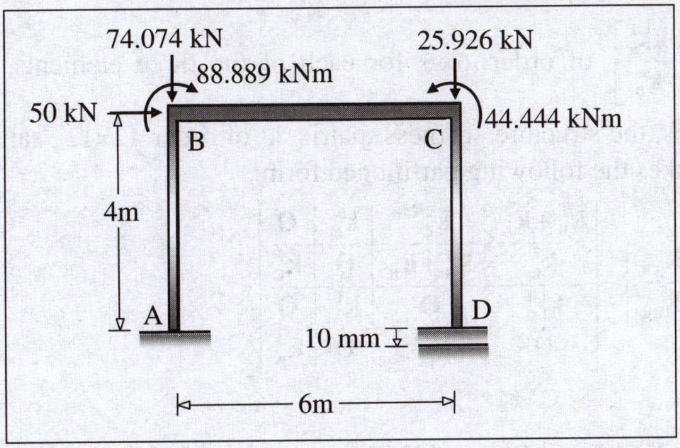

Figure 6.10 Resultant loads (Example 6.2)

$$\mathbf{k}_*^{\mathbf{i}} = \left[\begin{array}{ccc|ccc} \alpha_i & 0 & 0 & -\alpha_i & 0 & 0 \\ 0 & \beta_i & \chi_i & 0 & -\beta_i & \chi_i \\ 0 & \chi_i & 4\delta_i & 0 & -\chi_i & 2\delta_i \\ \hline -\alpha_i & 0 & 0 & \alpha_i & 0 & 0 \\ 0 & -\beta_i & -\chi_i & 0 & \beta_i & -\chi_i \\ 0 & \chi_i & 2\delta_i & 0 & -\chi_i & 4\delta_i \end{array}\right]; \quad \begin{array}{l} \alpha_i = (EA)_i / L_i \\ \beta_i = 12(\delta_i / L_i^2) \\ \chi_i = 6(\delta_i / L_i) \\ \delta_i = (EI)_i / L_i \end{array}$$

with $L_1 = L_3 = 6\text{m}$, $L_2 = 4\text{m}$; $\alpha_1 = \alpha_2 = \alpha_3 = 562500$ kN/m; $\delta_1 = \delta_3 = 4218.75$, $\delta_2 = 9492.1875$ kNm

$$\Rightarrow \mathbf{k}_*^1 = \mathbf{k}_*^3 = \left[\begin{array}{ccc|ccc} 562500 & 0 & 0 & -562500 & 0 & 0 \\ 0 & 3164.1 & 6328.1 & 0 & -3164.1 & 6328.1 \\ 0 & 6328.1 & 16875 & 0 & -6328.1 & 8437.5 \\ \hline -562500 & 0 & 0 & 562500 & 0 & 0 \\ 0 & -3164.1 & -6328.1 & 0 & 3164.1 & -6328.1 \\ 0 & 6328.1 & 8437.5 & 0 & -6328.1 & 16875 \end{array}\right]$$

$$\mathbf{k}_*^2 = \left[\begin{array}{ccc|ccc} 562500 & 0 & 0 & -562500 & 0 & 0 \\ 0 & 3164.1 & 9492.2 & 0 & -3164.1 & 9492.2 \\ 0 & 9492.2 & 37968.8 & 0 & -9492.2 & 18984.4 \\ \hline -562500 & 0 & 0 & 562500 & 0 & 0 \\ 0 & -3164.1 & -9492.2 & 0 & 3164.1 & -9492.2 \\ 0 & 9492.2 & 18984.4 & 0 & -9492.2 & 37968.8 \end{array}\right]$$

The products, $\mathbf{k}_*^{\mathbf{i}}\mathbf{T}^{\mathbf{i}}$ and $\mathbf{T}^{\mathbf{i}^{\mathrm{T}}}\mathbf{k}_*^{\mathbf{i}}\mathbf{T}^{\mathbf{i}}$, with the relevant global coordinates (marked in parenthesis), as defined earlier, may now be computed directly (using MATLAB) or generated using Eqns 6.9 and 6.10. By summing up the the contributions of

$$\mathbf{T}^{i^T}\mathbf{k}_*^i\mathbf{T}^i = \mathbf{k}^i = \begin{bmatrix} \mathbf{k}_A^i & \mathbf{k}_C^{i\,T} \\ \hline \mathbf{k}_C^i & \mathbf{k}_B^i \end{bmatrix}$$ of order 6×6 for each of the three elements, at the appropriate

coordinate locations, the structure stiffness matrix $\mathbf{k}$ of order 12×12, satisfying $\mathbf{F}=\mathbf{kD}$, can be assembled. It takes the following partitioned form:

$$\Rightarrow \qquad \mathbf{k} = \begin{bmatrix} \mathbf{k}_{AA} & \mathbf{k}_{AR} \\ \hline \mathbf{k}_{RA} & \mathbf{k}_{RR} \end{bmatrix} = \begin{bmatrix} \mathbf{k}_B^1+\mathbf{k}_A^2 & \mathbf{k}_C^{2\,T} & \mathbf{k}_C^1 & \mathbf{O} \\ \hline \mathbf{k}_C^2 & \mathbf{k}_B^2+\mathbf{k}_B^3 & \mathbf{O} & \mathbf{k}_C^3 \\ \hline \mathbf{k}_C^{1\,T} & \mathbf{O} & \mathbf{k}_A^1 & \mathbf{O} \\ \hline \mathbf{O} & \mathbf{k}_C^{3\,T} & \mathbf{O} & \mathbf{k}_A^3 \end{bmatrix},$$

where:

$$\mathbf{k}_A^1 = \mathbf{k}_A^3 = \begin{bmatrix} 3164.1 & 0 & -6328.1 \\ 0 & 562500 & 0 \\ -6328.1 & 0 & 16875 \end{bmatrix} ; \ \mathbf{k}_B^1 = \mathbf{k}_B^3 = \begin{bmatrix} 3164.1 & 0 & 6328.1 \\ 0 & 562500 & 0 \\ 6328.1 & 0 & 16875 \end{bmatrix}$$

$$\mathbf{k}_C^1 = \mathbf{k}_C^3 = \begin{bmatrix} -3164.1 & 0 & 6328.1 \\ 0 & -562500 & 0 \\ -6328.1 & 0 & 8437.5 \end{bmatrix} ; \ \mathbf{k}_C^2 = \begin{bmatrix} -562500 & 0 & 0 \\ 0 & -3164.1 & -9492.2 \\ 0 & 9492.2 & 18984.4 \end{bmatrix}$$

$$\mathbf{k}_A^2 = \begin{bmatrix} 562500 & 0 & 0 \\ 0 & 3164.1 & 9492.2 \\ 0 & 9492.2 & 37968.8 \end{bmatrix} ; \ \mathbf{k}_B^2 = \begin{bmatrix} 562500 & 0 & 0 \\ 0 & 3164.1 & -9492.2 \\ 0 & -9492.2 & 37968.8 \end{bmatrix}$$

$$\Rightarrow \quad \mathbf{k}_{AA} = \begin{array}{c} \\ \end{array} \begin{matrix} (1) & (2) & (3) & (4) & (5) & (6) \\ \begin{bmatrix} 565664.1 & 0 & 6328.1 & -562500 & 0 & 0 \\ 0 & 565664.1 & 9492.2 & 0 & -3164.1 & 9492.2 \\ 6328.1 & 9492.2 & 54843.8 & 0 & -9492.2 & 18984.4 \\ -562500 & 0 & 0 & 565664.1 & 0 & 6328.1 \\ 0 & -3164.1 & -9492.2 & 0 & 565664.1 & -9492.2 \\ 0 & 9492.2 & 18984.4 & 6328.1 & -9492.2 & 54843.8 \end{bmatrix} & \begin{matrix} (1) \\ (2) \\ (3) \\ (4) \\ (5) \\ (6) \end{matrix} \end{matrix}$$

$$\mathbf{k}_{AR} = \begin{matrix} (7) & (8) & (9) & (10) & (11) & (12) \\ \begin{bmatrix} -3164.1 & 0 & 6328.1 & 0 & 0 & 0 \\ 0 & -562500 & 0 & 0 & 0 & 0 \\ -6328.1 & 0 & 8437.5 & 0 & 0 & 0 \\ 0 & 0 & 0 & -3164.1 & 0 & 6328.1 \\ 0 & 0 & 0 & 0 & -562500 & 0 \\ 0 & 0 & 0 & -6328.1 & 0 & 8437.5 \end{bmatrix} & \begin{matrix} (1) \\ (2) \\ (3) \\ (4) \\ (5) \\ (6) \end{matrix} = \mathbf{k}_{RA}^{\ T} \end{matrix}$$

$$\mathbf{k}_{RR} = \begin{bmatrix} 3164.1 & 0 & -6328.1 & 0 & 0 & 0 \\ 0 & 562500 & 0 & 0 & 0 & 0 \\ -6328.1 & 0 & 16875 & 0 & 0 & 0 \\ \hline 0 & 0 & 0 & 3164.1 & 0 & -6328.1 \\ 0 & 0 & 0 & 0 & 562500 & 0 \\ 0 & 0 & 0 & -6328.1 & 0 & 16875 \end{bmatrix} \begin{matrix} (7) \\ (8) \\ (9) \\ (10) \\ (11) \\ (12) \end{matrix}$$

(columns (7) (8) (9) (10) (11) (12))

Displacements and Support Reactions

The basic stiffness relations are given by:

$$\left[\frac{\mathbf{F}_A}{\mathbf{F}_R}\right] - \left[\frac{\mathbf{F}_{fA}}{\mathbf{F}_{fR} = \mathbf{O}}\right] = \left[\begin{array}{c|c} \mathbf{k}_{AA} & \mathbf{k}_{AR} \\ \hline \mathbf{k}_{RA} & \mathbf{k}_{RR} \end{array}\right]\left[\frac{\mathbf{D}_A}{\mathbf{D}_R}\right]$$

$$\Rightarrow \mathbf{D}_A = [\mathbf{k}_{AA}]^{-1}\left[(\mathbf{F}_A - \mathbf{F}_{fA}) - \mathbf{k}_{AR}\mathbf{D}_R\right]$$

$$\mathbf{F}_R = \mathbf{k}_{RA}\mathbf{D}_A + \mathbf{k}_{RR}\mathbf{D}_R$$

Substituting the various matrices and solving using MATLAB,

$$\mathbf{D}_A = \begin{Bmatrix} D_1 \\ D_2 \\ D_3 \\ \hline D_4 \\ D_5 \\ D_6 \end{Bmatrix} = \begin{Bmatrix} 0.013394133 \text{ m} \\ -0.000095716 \text{ m} \\ -0.004595093 \text{ rad} \\ \hline 0.013328892 \text{ m} \\ -0.010082062 \text{ m} \\ -0.000865364 \text{ rad} \end{Bmatrix}; \quad \mathbf{F}_R = \begin{Bmatrix} F_7 \\ F_8 \\ F_9 \\ \hline F_{10} \\ F_{11} \\ F_{12} \end{Bmatrix} = \begin{Bmatrix} -13.302 \text{ kN} \\ 53.840 \text{ kN} \\ 45.988 \text{ kNm} \\ \hline -36.698 \text{ kN} \\ 46.160 \text{ kN} \\ 77.045 \text{ kNm} \end{Bmatrix}$$

Check equilibrium: $\sum F_x = 0: \; F_7 + F_{10} - 50 = -13.302 - 36.698 + 50 = 0 \text{ kN}$

$\sum F_y = 0: \; F_8 + F_{11} - 100 = 53.840 + 46.160 - 100 = 0 \text{ kN} \;\Rightarrow \text{OK.}$

Member Forces

$$\mathbf{F}_*^i = \mathbf{F}_{*f}^i + \left(\mathbf{k}_*^i\mathbf{T}^i\right)\mathbf{D}^i,$$

where

$$\mathbf{k}_*^1\mathbf{T}^1 = \mathbf{k}_*^3\mathbf{T}^3 = \begin{bmatrix} 0 & 562500 & 0 & 0 & -562500 & 0 \\ -3164.1 & 0 & 632.81 & 3164.1 & 0 & 6328.1 \\ -6328.1 & 0 & 1687.5 & 6328.1 & 0 & 8437.5 \\ \hline 0 & -562500 & 0 & 0 & 562500 & 0 \\ 3164.1 & 0 & -632.81 & -3164.1 & 0 & -6328.1 \\ -6328.1 & 0 & 843.75 & 6328.1 & 0 & 16875 \end{bmatrix}$$

$$\mathbf{k}_*^2\mathbf{T}^2 = \left[\begin{array}{ccc|ccc} 562500 & 0 & 0 & -562500 & 0 & 0 \\ 0 & 3164.1 & 6328.1 & 0 & -3164.1 & 9492.2 \\ 0 & 9492.2 & 37968.8 & 0 & -9492.2 & 18984.4 \\ \hline -562500 & 0 & 0 & 562500 & 0 & 0 \\ 0 & -3164.1 & -9492.2 & 0 & 3164.1 & -9492.2 \\ 0 & 9492.2 & 18984.4 & 0 & -9492.2 & 37968.8 \end{array}\right]$$

$$\Rightarrow \mathbf{F}_*^1 = \begin{Bmatrix} F_{1*}^1 \\ F_{2*}^1 \\ F_{3*}^1 \\ \hline F_{4*}^1 \\ F_{5*}^1 \\ F_{6*}^1 \end{Bmatrix} = \begin{Bmatrix} 0\text{ kN} \\ 0\text{ kN} \\ 0\text{ kNm} \\ \hline 0\text{ kN} \\ 0\text{ kN} \\ 0\text{ kNm} \end{Bmatrix} + \left[\mathbf{k}_*^1\mathbf{T}^1\right]\begin{Bmatrix} 0 \\ 0 \\ 0 \\ \hline D_1 \\ D_2 \\ D_3 \end{Bmatrix} = \begin{Bmatrix} 53.840\text{ kN} \\ 13.302\text{ kN} \\ 45.988\text{ kNm} \\ \hline -53.840\text{ kN} \\ -13.302\text{ kN} \\ 7.218\text{ kNm} \end{Bmatrix}$$

$$\mathbf{F}_*^2 = \begin{Bmatrix} F_{1*}^2 \\ F_{2*}^2 \\ F_{3*}^2 \\ \hline F_{4*}^2 \\ F_{5*}^2 \\ F_{6*}^2 \end{Bmatrix} = \begin{Bmatrix} 0\text{ kN} \\ 74.074\text{ kN} \\ 88.889\text{ kNm} \\ \hline 0\text{ kN} \\ 25.926\text{ kN} \\ -44.444\text{ kNm} \end{Bmatrix} + \left[\mathbf{k}_*^2\mathbf{T}^2\right]\begin{Bmatrix} D_1 \\ D_2 \\ D_3 \\ \hline D_4 \\ D_5 \\ D_6 \end{Bmatrix} = \begin{Bmatrix} 36.698\text{ kN} \\ 53.840\text{ kN} \\ -7.218\text{ kNm} \\ \hline -36.698\text{ kN} \\ 46.160\text{ kN} \\ -69.744\text{ kNm} \end{Bmatrix}$$

$$\mathbf{F}_*^3 = \begin{Bmatrix} F_{1*}^3 \\ F_{2*}^3 \\ F_{3*}^3 \\ \hline F_{4*}^3 \\ F_{5*}^3 \\ F_{6*}^3 \end{Bmatrix} = \begin{Bmatrix} 0\text{ kN} \\ 0\text{ kN} \\ 0\text{ kNm} \\ \hline 0\text{ kN} \\ 0\text{ kN} \\ 0\text{ kNm} \end{Bmatrix} + \left[\mathbf{k}_*^3\mathbf{T}^3\right]\begin{Bmatrix} 0 \\ -0.01 \\ 0 \\ \hline D_4 \\ D_5 \\ D_6 \end{Bmatrix} = \begin{Bmatrix} 46.160\text{ kN} \\ 36.697\text{ kN} \\ 77.045\text{ kNm} \\ \hline -46.160\text{ kN} \\ -36.697\text{ kN} \\ 69.744\text{ kNm} \end{Bmatrix}$$

The complete response (including the fixed end force effects), in the form of free-body, deflection and bending moment diagrams are shown in Fig. 6.11.

Alternative Procedure Using Displacement Transformation Matrix

$\mathbf{D}_*^i = \left[\mathbf{T}_{DA}^i \mid \mathbf{T}_{DR}^i\right]\begin{bmatrix} \mathbf{D}_A \\ \hline \mathbf{D}_R \end{bmatrix}$ where $\mathbf{T}_{DA}^i$ and $\mathbf{T}_{DR}^i$ are matrices of order 6×6, similar to $\mathbf{T}^i$.

$$\mathbf{T}_{DA}^1 = \left[\begin{array}{c|ccc} & 0 & 1 & 0 \\ \mathbf{O} & -1 & 0 & 0 \\ & 0 & 0 & 1 \\ \hline \mathbf{O} & & \mathbf{O} & \end{array}\right],\ \mathbf{T}_{DR}^1 = \left[\begin{array}{ccc|c} 0 & 1 & 0 & \\ -1 & 0 & 0 & \mathbf{O} \\ 0 & 0 & 1 & \\ \hline & \mathbf{O} & & \mathbf{O} \end{array}\right];\ \mathbf{T}_{DA}^3 = \mathbf{T}_{DR}^3 = \left[\begin{array}{c|ccc} \mathbf{O} & & \mathbf{O} & \\ \hline & 0 & 1 & 0 \\ \mathbf{O} & -1 & 0 & 0 \\ & 0 & 0 & 1 \end{array}\right];$$

$\mathbf{T}_{DA}^2 = \mathbf{I}$, $\mathbf{T}_{DR}^2 = \mathbf{O}$, where $\mathbf{O}$ and $\mathbf{I}$ denote the null matrix and identity matrix respectively.

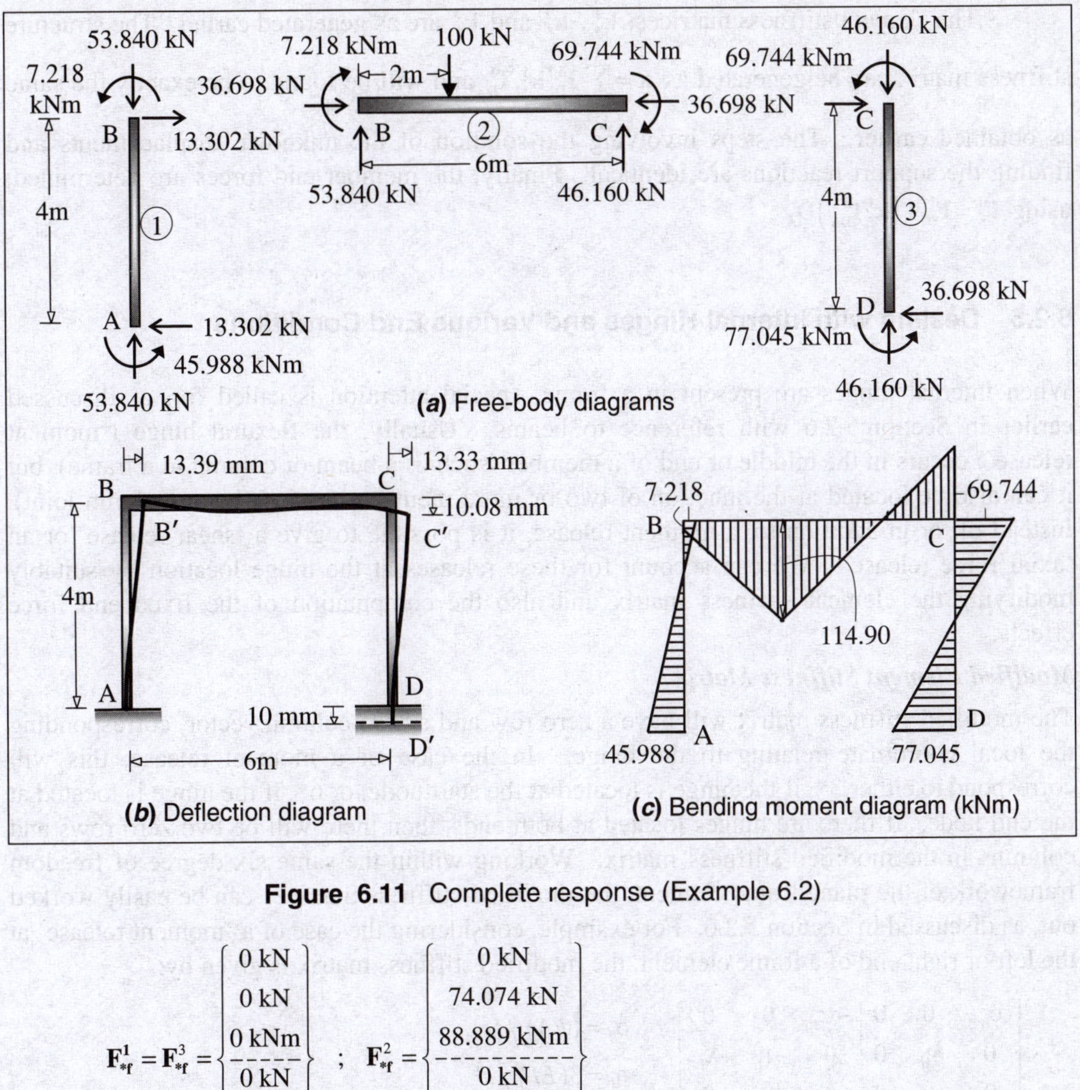

Figure 6.11 Complete response (Example 6.2)

$$\mathbf{F}^1_{*\mathbf{f}} = \mathbf{F}^3_{*\mathbf{f}} = \left\{\begin{array}{c} 0\text{ kN} \\ 0\text{ kN} \\ 0\text{ kNm} \\ \hline 0\text{ kN} \\ 0\text{ kN} \\ 0\text{ kNm} \end{array}\right\} \;;\; \mathbf{F}^2_{*\mathbf{f}} = \left\{\begin{array}{c} 0\text{ kN} \\ 74.074\text{ kN} \\ 88.889\text{ kNm} \\ \hline 0\text{ kN} \\ 25.926\text{ kN} \\ -44.444\text{ kNm} \end{array}\right\}$$

$$\Rightarrow \mathbf{F_f} = \left[\frac{\mathbf{F_{fA}}}{\mathbf{F_{fR}}}\right] = \sum_{i=1}^{3} \mathbf{T_D^{i\,T}} \mathbf{F^i_{*f}} \quad \text{where} \quad \mathbf{F_{fA}} = \left\{\begin{array}{c} F_{1f} \\ F_{2f} \\ F_{3f} \\ \hline F_{4f} \\ F_{5f} \\ F_{6f} \end{array}\right\} = \left\{\begin{array}{c} 0\text{ kN} \\ 74.074\text{ kN} \\ 88.889\text{ kNm} \\ \hline 0\text{ kN} \\ 25.926\text{ kN} \\ -44.444\text{ kNm} \end{array}\right\} \text{ and } \mathbf{F_{fR}} = \left\{\begin{array}{c} F_{7f} \\ F_{8f} \\ F_{9f} \\ \hline F_{10f} \\ F_{11f} \\ F_{12f} \end{array}\right\} = \left\{\begin{array}{c} 0\text{ kN} \\ 0\text{ kN} \\ 0\text{ kNm} \\ \hline 0\text{ kN} \\ 0\text{ kN} \\ 0\text{ kNm} \end{array}\right\}$$

(as obtained earlier).

The element stiffness matrices, $\mathbf{k}_*^1$, $\mathbf{k}_*^2$ and $\mathbf{k}_*^3$ are as generated earlier. The structure stiffness matrix can be generated as $\mathbf{k} = \sum_{i=1}^{3} \mathbf{T}_D^{i\,T} \mathbf{k}_{*f}^i \mathbf{T}_D^i$ and will be found to be exactly the same as obtained earlier. The steps involving the solution of the unknown displacements and finding the support reactions are identical. Finally, the member end forces are determined, using $\mathbf{F}_*^i = \mathbf{F}_{*f}^i + \left(\mathbf{k}_*^i \mathbf{T}_{DA}^i\right)\mathbf{D}_A$.

6.2.5 Dealing with Internal Hinges and Various End Conditions

When internal hinges are present in a frame, special attention is called for, as discussed earlier in Section 5.2.6 with reference to beams. Usually, the flexural hinge ('moment release') occurs in the middle or end of a member (such as a beam or column in a frame), but it can also be located at the junction of two or more members (as in a beam-column joint). Instead of, or in addition to, a moment release, it is possible to give a 'shear release' or an 'axial force release'. We can account for these releases at the hinge location by suitably modifying the element stiffness matrix and also the computation of the fixed end force effects.

Modified Element Stiffness Matrix

The modified stiffness matrix will have a zero row and a zero column vector, corresponding the local coordinate relating to the hinge. In the case of a moment release, this will correspond to either 3*, if the hinge is located at the start node, or 6*, if the hinge is located at the end node. If there are hinges located at both ends, then there will be two zero rows and columns in the modified stiffness matrix. Working within the same six degree of freedom framework of the plane frame element, the modified stiffness matrices can be easily worked out, as discussed in Section 5.2.6. For example, considering the case of a 'moment release' at the left or right end of a frame element, the modified stiffness matrix is given by:

$$\mathbf{k}_*^i = \left[\begin{array}{ccc|ccc} \alpha_i & 0 & 0 & -\alpha_i & 0 & 0 \\ 0 & \eta_i & 0 & 0 & -\eta_i & \lambda_i \\ 0 & 0 & 0 & 0 & 0 & 0 \\ \hline -\alpha_i & 0 & 0 & \alpha_i & 0 & 0 \\ 0 & -\eta_i & 0 & 0 & \eta_i & -\lambda_i \\ 0 & \lambda_i & 0 & 0 & -\lambda_i & 3\delta_i \end{array}\right]; \quad \begin{array}{l} \alpha_i = (EA)_i/L_i \\ \eta_i = 3(EI)_i/L_i^3 \\ \lambda_i = 3(EI)_i/L_i^2 \\ \delta_i = (EI)_i/L_i \end{array} \quad \text{if hinge is at } \textit{start node} \qquad (6.17a)$$

$$\mathbf{k}_*^{\mathbf{i}} = \left[\begin{array}{ccc|ccc} \alpha_i & 0 & 0 & -\alpha_i & 0 & 0 \\ 0 & \eta_i & \lambda_i & 0 & -\eta_i & 0 \\ 0 & \lambda_i & 3\delta_i & 0 & -\lambda_i & 0 \\ \hline -\alpha_i & 0 & 0 & \alpha_i & 0 & 0 \\ 0 & -\eta_i & -\lambda_i & 0 & \eta_i & 0 \\ 0 & 0 & 0 & 0 & 0 & 0 \end{array}\right]; \quad \begin{array}{l} \alpha_i = (EA)_i/L_i \\ \eta_i = 3(EI)_i/L_i^3 \\ \lambda_i = 3(EI)_i/L_i^2 \\ \delta_i = (EI)_i/L_i \end{array} \quad \text{if hinge is at } \textit{end node} \qquad (6.17b)$$

$$\mathbf{k}_*^{\mathbf{i}} = \left[\begin{array}{ccc|ccc} \alpha_i & 0 & 0 & -\alpha_i & 0 & 0 \\ 0 & \eta_i & 0 & 0 & -\eta_i & 0 \\ 0 & 0 & 0 & 0 & 0 & 0 \\ \hline -\alpha_i & 0 & 0 & \alpha_i & 0 & 0 \\ 0 & -\eta_i & 0 & 0 & \eta_i & 0 \\ 0 & 0 & 0 & 0 & 0 & 0 \end{array}\right]; \quad \begin{array}{l} \alpha_i = (EA)_i/L_i \\ \eta_i = 3(EI)_i/L_i^3 \\ \lambda_i = 3(EI)_i/L_i^2 \\ \delta_i = (EI)_i/L_i \end{array} \quad \text{if hinges are at } \textit{both ends} \qquad (6.17c)$$

Modified Fixed End Forces

The fixed end forces due to intermediate loads, including possible end moment applied at the hinge location, should also be suitably computed, accounting for the hinge at the member end (i.e., there should be no fixity against rotation at this end), as explained earlier. If hinges are provided at both ends, then obviously, the element will behave like a simply supported beam, and the fixed end effects will not induce any end moments (only vertical reactions at the ends). If the hinge is located at one end, then the element will behave like a propped cantilever.

Imaginary Clamp

By reducing the rotational stiffness components in the (two or more) frame elements adjoining the internal hinge location (to the left and to the right), the resultant rotational stiffness of the structure, corresponding to this rotational degree of freedom (say, global coordinate 'q'), is reduced to zero. Thus, this will appear as a zero diagonal element in the structure stiffness matrix ($k_{qq} = 0$), making the matrix $\mathbf{k_{AA}}$ singular and non-invertible. We can get around this difficulty by visualising an *imaginary clamp* at the internal hinge location, arresting the rotation (i.e., by setting $D_q = 0$), as discussed earlier in Section 5.2.6. Although this is not physically a correct representation (rotations are possible at the internal hinge location), it serves our purpose of getting a correct solution by the stiffness method.

By arresting the rotational degree of freedom at the hinge location, we convert the coordinate q from an *active* global coordinate to a *restrained* global coordinate. In general, this will result in a zero support reaction (moment), $F_q = 0$, corresponding to this degree of freedom. However, when there is an end moment applied, at the left or right or at both sides of the internal hinge), there will be a resultant non-zero value of F_q, which may be ignored.

The application of this procedure is demonstrated in Example 6.3.

Free, Hinged and Guided-Fixed Boundary Conditions

In the problems we have dealt with so far, the extreme end supports have been given as fixed end supports. The general analysis procedure described in Section 6.2.4 can be applied to any boundary condition, such as a hinged, guided-fixed support or a free end. In the case of a hinged end support, we have one active (rotational) degree of freedom (refer Example 6.4). If the support is a roller support, there is an additional active translational degree of freedom. In the case of a guided-fixed end support, we have one active translational degree of freedom. If the boundary is free, all three degrees of freedom are active.

EXAMPLE 6.3

Re-analyse the frame in Example 6.2 (Fig. 6.8), considering the frame (with the same loading) to have an internal hinge at the joint C. Draw the bending moment and deflection diagrams.

SOLUTION

Let us define the same global and local coordinates as in Example 6.2. However, as there is an internal hinge provided at the joint C, there will be no moment transfer possible across this junction. Hence, there will be no common rotational degree of freedom at C. The stiffness matrices of BC and CD can be suitably modified to account for the moment release at C, but this will result in zero rotational stiffness at C. In order to address this problem, we visualise an imaginary clamp at C (providing restraint against rotation at C), as indicated in Fig 6.12a. Thus, '6' becomes a restrained coordinate, and the number of active degrees of freedom reduces to five ($n=5$), while the *restrained* global coordinates increases to seven ($r=8$), numbered as shown in Fig. 6.12a.

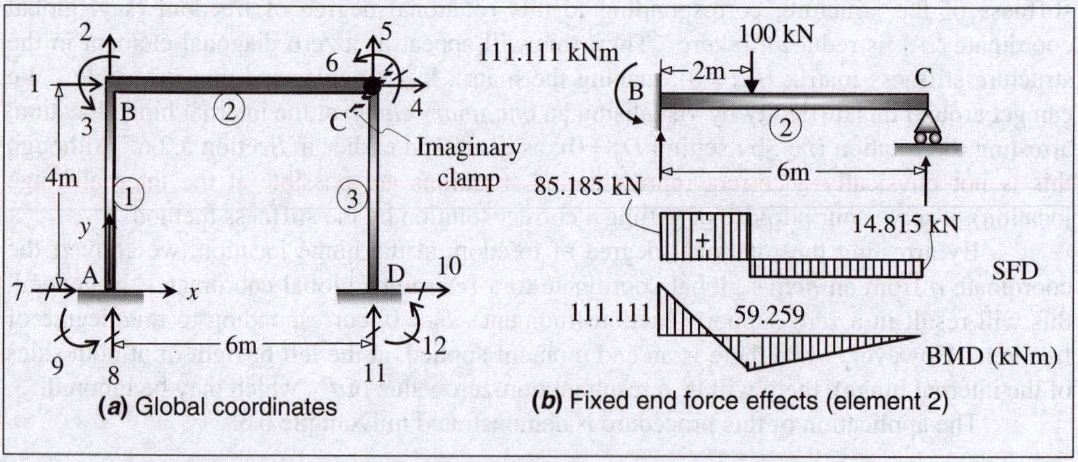

Figure 6.12 Global coordinates and fixed end force effects (Example 6.3)

Fixed End Forces and Equivalent Joint Loads

Considering the hinge at C, the element '2' is to be treated as a propped cantilever for the purpose of finding the fixed end force effects, as shown in Fig. 6.12b. There are no fixed end forces in elements '1' and '3'.

$$F_{1*f}^2 = F_{4*f}^2 = 0 \text{ kN};\ F_{3*f}^2 = \frac{100(4)^2(2)}{(6)^2} + \frac{1}{2}\frac{100(2)^2(4)}{(6)^2} = 111.111 \text{ kNm};\ F_{6*f}^2 = 0 \text{ kNm};$$

$$F_{2*f}^2 = \frac{100\times 4}{6} + \frac{111.111}{6} = 85.185 \text{ kN};\ F_{5*f}^2 = \frac{100\times 2}{6} - \frac{111.111}{6} = 14.815 \text{ kN}$$

$$\Rightarrow \mathbf{F}_{*f}^1 = \mathbf{F}_{*f}^3 = \begin{Bmatrix} 0 \text{ kN} \\ 0 \text{ kN} \\ 0 \text{ kNm} \\ \hdashline 0 \text{ kN} \\ 0 \text{ kN} \\ 0 \text{ kNm} \end{Bmatrix};\quad \mathbf{F}_{*f}^2 = \begin{Bmatrix} 0 \text{ kN} \\ 85.185 \text{ kN} \\ 111.111 \text{ kNm} \\ \hdashline 0 \text{ kN} \\ 14.815 \text{ kN} \\ 0 \text{ kNm} \end{Bmatrix}$$

Coordinate Transformations and Equivalent Joint Loads

The element transformation matrices $\mathbf{T}^1, \mathbf{T}^2$ and $\mathbf{T}^3$, satisfying $\mathbf{D}_*^i = \mathbf{T}^i\mathbf{D}$ and $\mathbf{F}_*^i = \mathbf{T}^i\mathbf{F}$, are as defined in Example 6.2. The fixed end forces in global axes can be assembled, by summing up the contributions of $\mathbf{T}^{i^T}\mathbf{F}_{*f}^i$, considering the five active and seven restrained coordinates.

$$\Rightarrow \mathbf{F}_f = \begin{bmatrix} \mathbf{F}_{fA} \\ \hdashline \mathbf{F}_{fR} \end{bmatrix} \text{ where } \mathbf{F}_{fA} = \begin{Bmatrix} F_{1f} \\ F_{2f} \\ F_{3f} \\ \hdashline F_{4f} \\ F_{5f} \end{Bmatrix} = \begin{Bmatrix} 0 \text{ kN} \\ 85.185 \text{ kN} \\ 111.111 \text{ kNm} \\ \hdashline 0 \text{ kN} \\ 14.815 \text{ kN} \end{Bmatrix} \text{ and } \mathbf{F}_{fR} = \begin{Bmatrix} F_{6f} \\ \hdashline F_{7f} \\ F_{8f} \\ F_{9f} \\ \hdashline F_{10f} \\ F_{11f} \\ F_{12f} \end{Bmatrix} = \begin{Bmatrix} 0 \text{ kNm} \\ \hdashline 0 \text{ kN} \\ 0 \text{ kN} \\ 0 \text{ kNm} \\ \hdashline 0 \text{ kN} \\ 0 \text{ kN} \\ 0 \text{ kNm} \end{Bmatrix}$$

The resultant load vector, $\mathbf{F}_A - \mathbf{F}_{fA}$, and the restrained displacement vector, $\mathbf{D}_R$ are:

$$\mathbf{F}_A - \mathbf{F}_{fA} = \begin{Bmatrix} 50 \text{ kN} \\ 0 \text{ kN} \\ 0 \text{ kNm} \\ \hdashline 0 \text{ kN} \\ 0 \text{ kN} \end{Bmatrix} - \begin{Bmatrix} 0 \text{ kN} \\ 85.185 \text{ kN} \\ 111.111 \text{ kNm} \\ \hdashline 0 \text{ kN} \\ 14.815 \text{ kN} \end{Bmatrix} = \begin{Bmatrix} 50 \text{ kN} \\ -85.185 \text{ kN} \\ -111.111 \text{ kNm} \\ \hdashline 0 \text{ kN} \\ -14.815 \text{ kN} \end{Bmatrix};\quad \mathbf{D}_R = \begin{Bmatrix} D_6 \\ \hdashline D_7 \\ D_8 \\ D_9 \\ \hdashline D_{10} \\ D_{11} \\ D_{12} \end{Bmatrix} = \begin{Bmatrix} 0 \text{ rad} \\ \hdashline 0 \text{ m} \\ 0 \text{ m} \\ 0 \text{ rad} \\ \hdashline 0 \text{ m} \\ -0.01 \text{ m} \\ 0 \text{ rad} \end{Bmatrix}$$

This loading is indicated in Fig. 6.13.

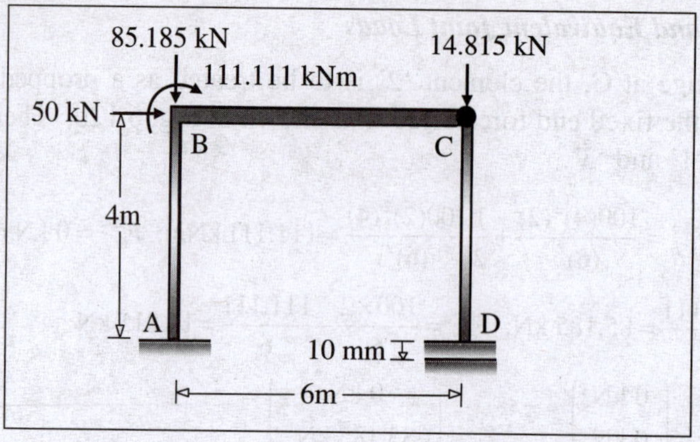

Figure 6.13 Resultant loads (Example 6.3)

The element stiffness matrices for elements '2' and '3' get modified on account of the hinge at the end node (Eq. 6.17b), while that of element '1' remains the same as in Example 6.2.

$$\Rightarrow \mathbf{k}_*^1 = \left[\begin{array}{ccc|ccc} 562500 & 0 & 0 & -562500 & 0 & 0 \\ 0 & 3164.1 & 6328.1 & 0 & -3164.1 & 6328.1 \\ 0 & 6328.1 & 16875 & 0 & -6328.1 & 8437.5 \\ \hline -562500 & 0 & 0 & 562500 & 0 & 0 \\ 0 & -3164.1 & -6328.1 & 0 & 3164.1 & -6328.1 \\ 0 & 6328.1 & 8437.5 & 0 & -6328.1 & 16875 \end{array}\right]$$

$$\mathbf{k}_*^2 = \left[\begin{array}{ccc|ccc} 562500 & 0 & 0 & -562500 & 0 & 0 \\ 0 & 791.0 & 4746.1 & 0 & -791.0 & 0 \\ 0 & 4746.1 & 28476.6 & 0 & -4746.1 & 0 \\ \hline -562500 & 0 & 0 & 562500 & 0 & 0 \\ 0 & -791.0 & -4746.1 & 0 & 791.0 & 0 \\ 0 & 0 & 0 & 0 & 0 & 0 \end{array}\right]$$

$$\mathbf{k}_*^3 = \left[\begin{array}{ccc|ccc} 562500 & 0 & 0 & -562500 & 0 & 0 \\ 0 & 791.0 & 3164.1 & 0 & -791.0 & 0 \\ 0 & 3164.1 & 12656.3 & 0 & -3164.1 & 0 \\ \hline -562500 & 0 & 0 & 562500 & 0 & 0 \\ 0 & -791.0 & -3164.1 & 0 & 791.0 & 0 \\ 0 & 0 & 0 & 0 & 0 & 0 \end{array}\right]$$

The products, $\mathbf{k_*^i T^i}$ and $\mathbf{T^{i^T} k_*^i T^i}$, with the relevant global coordinates (marked in parenthesis), as defined earlier, may now be computed directly (using MATLAB). For convenience, partitioning may be done as $\mathbf{T^{i^T} k_*^i T^i} = \mathbf{k^i} = \left[\begin{array}{c|c} \mathbf{k_A^i} & \mathbf{k_C^{i\,T}} \\ \hline \mathbf{k_C^i} & \mathbf{k_B^i} \end{array}\right]$,

where:

$$\mathbf{k_A^1} = \begin{bmatrix} 3164.1 & 0 & -6328.1 \\ 0 & 562500 & 0 \\ -6328.1 & 0 & 16875 \end{bmatrix};\quad \mathbf{k_B^1} = \begin{bmatrix} 3164.1 & 0 & 6328.1 \\ 0 & 562500 & 0 \\ 6328.1 & 0 & 16875 \end{bmatrix}$$

$$\mathbf{k_C^1} = \begin{bmatrix} -3164.1 & 0 & 6328.1 \\ 0 & -562500 & 0 \\ -6328.1 & 0 & 8437.5 \end{bmatrix};\quad \mathbf{k_A^2} = \begin{bmatrix} 562500 & 0 & 0 \\ 0 & 791.0 & 4746.1 \\ 0 & 4746.1 & 28476.6 \end{bmatrix}$$

$$\mathbf{k_B^2} = \begin{bmatrix} 562500 & 0 & 0 \\ 0 & 791.0 & 0 \\ 0 & 0 & 0 \end{bmatrix};\quad \mathbf{k_C^2} = \begin{bmatrix} -562500 & 0 & 0 \\ 0 & -791.0 & -4746.1 \\ 0 & 0 & 0 \end{bmatrix}$$

$$\mathbf{k_A^3} = \begin{bmatrix} 791.0 & 0 & -3164.1 \\ 0 & 562500 & 0 \\ -3164.1 & 0 & 12656.3 \end{bmatrix};\quad \mathbf{k_B^3} = \begin{bmatrix} 791.0 & 0 & 0 \\ 0 & 562500 & 0 \\ 0 & 0 & 0 \end{bmatrix}$$

$$\mathbf{k_C^3} = \begin{bmatrix} -791.0 & 0 & 3164.1 \\ 0 & -562500 & 0 \\ 0 & 0 & 0 \end{bmatrix}$$

By summing up the contributions of $\mathbf{T^{i^T} k_*^i T^i} = \mathbf{k^i} = \left[\begin{array}{c|c} \mathbf{k_A^i} & \mathbf{k_C^{i\,T}} \\ \hline \mathbf{k_C^i} & \mathbf{k_B^i} \end{array}\right]$ of order 6×6 for each of the three elements, at the appropriate coordinate locations, the structure stiffness matrix $\mathbf{k}$ of order 12×12, can be assembled as

$$\mathbf{k} = \left[\begin{array}{c|c|c|c} \mathbf{k_B^1 + k_A^2} & \mathbf{k_C^{2\,T}} & \mathbf{k_C^1} & \mathbf{O} \\ \hline \mathbf{k_C^2} & \mathbf{k_B^2 + k_B^3} & \mathbf{O} & \mathbf{k_C^3} \\ \hline \mathbf{k_C^{1\,T}} & \mathbf{O} & \mathbf{k_A^1} & \mathbf{O} \\ \hline \mathbf{O} & \mathbf{k_C^{3\,T}} & \mathbf{O} & \mathbf{k_A^3} \end{array}\right] = \left[\begin{array}{c|c} \mathbf{k_{AA}} & \mathbf{k_{AR}} \\ \hline \mathbf{k_{RA}} & \mathbf{k_{RR}} \end{array}\right].$$

$$\Rightarrow \quad \mathbf{k_{AA}} = \begin{array}{c} \\ \\ \\ \\ \\ \end{array}\begin{array}{ccccc} (1) & (2) & (3) & (4) & (5) \\ \end{array}$$

$$\mathbf{k_{AA}} = \begin{bmatrix} 565664.1 & 0 & 6328.1 & -562500 & 0 \\ 0 & 563291 & 4746.1 & 0 & -791.0 \\ 6328.1 & 4746.1 & 45351.6 & 0 & -4746.1 \\ -562500 & 0 & 0 & 563291 & 0 \\ 0 & -791.0 & -4746.1 & 0 & 563291 \end{bmatrix}\begin{matrix} (1) \\ (2) \\ (3) \\ (4) \\ (5) \end{matrix}$$

$$
\mathbf{k_{AR}} = \begin{bmatrix}
0 & -3164.1 & 0 & 6328.1 & 0 & 0 & 0 \\
0 & 0 & -562500 & 0 & 0 & 0 & 0 \\
0 & -6328.1 & 0 & 8437.5 & 0 & 0 & 0 \\
0 & 0 & 0 & 0 & -791.0 & 0 & 3164.1 \\
0 & 0 & 0 & 0 & 0 & -562500 & 0
\end{bmatrix}
\begin{matrix}
(1) \\ (2) \\ (3) \\ (4) \\ (5)
\end{matrix} = \mathbf{k_{RA}}^T
$$

with column headers (6) (7) (8) (9) (10) (11) (12)

$$
\mathbf{k_{RR}} = \begin{bmatrix}
0 & 0 & 0 & 0 & 0 & 0 & 0 \\
0 & 3164.1 & 0 & -6328.1 & 0 & 0 & 0 \\
0 & 0 & 562500 & 0 & 0 & 0 & 0 \\
0 & -6328.1 & 0 & 16875 & 0 & 0 & 0 \\
0 & 0 & 0 & 0 & 791.0 & 0 & -3164.1 \\
0 & 0 & 0 & 0 & 0 & 562500 & 0 \\
0 & 0 & 0 & 0 & -3164.1 & 0 & 12656.3
\end{bmatrix}
\begin{matrix}
(6) \\ (7) \\ (8) \\ (9) \\ (10) \\ (11) \\ (12)
\end{matrix}
$$

with column headers (6) (7) (8) (9) (10) (11) (12)

Displacements and Support Reactions

The basic stiffness relations are given by:

$$
\begin{bmatrix} \mathbf{F_A} \\ \hline \mathbf{F_R} \end{bmatrix} - \begin{bmatrix} \mathbf{F_{fA}} \\ \hline \mathbf{F_{fR}=O} \end{bmatrix} = \begin{bmatrix} \mathbf{k_{AA}} & \vdots & \mathbf{k_{AR}} \\ \hline \mathbf{k_{RA}} & \vdots & \mathbf{k_{RR}} \end{bmatrix} \begin{bmatrix} \mathbf{D_A} \\ \hline \mathbf{D_R} \end{bmatrix} \Rightarrow \quad
\begin{aligned}
\mathbf{D_A} &= \left[\mathbf{k_{AA}}\right]^{-1}\left[\left(\mathbf{F_A}-\mathbf{F_{fA}}\right)-\mathbf{k_{AR}D_R}\right] \\
\mathbf{F_R} &= \mathbf{k_{RA}D_A} + \mathbf{k_{RR}D_R}
\end{aligned}
$$

Substituting the various matrices and solving using MATLAB,

$$
\mathbf{D_A} = \begin{Bmatrix} D_1 \\ D_2 \\ D_3 \\ D_4 \\ D_5 \end{Bmatrix} = \begin{Bmatrix} 0.023477823 \text{ m} \\ -0.000108339 \text{ m} \\ -0.006768392 \text{ rad} \\ 0.023444854 \text{ m} \\ -0.010069439 \text{ m} \end{Bmatrix}
\qquad
\mathbf{F_R} = \begin{Bmatrix} F_6 \\ \hline F_7 \\ F_8 \\ F_9 \\ \hline F_{10} \\ F_{11} \\ F_{12} \end{Bmatrix} = \begin{Bmatrix} 0 \text{ kNm} \\ \hline -31.455 \text{ kN} \\ 60.941 \text{ kN} \\ 91.462 \text{ kNm} \\ \hline -18.545 \text{ kN} \\ 39.059 \text{ kN} \\ 74.182 \text{ kNm} \end{Bmatrix}
$$

<u>Check equilibrium:</u> $\sum F_x = 0:\ F_7 + F_{10} - 50 = -31.455 - 18.545 + 50 = 0 \text{ kN}$

$\sum F_y = 0:\ F_8 + F_{11} - 100 = 60.941 + 39.059 - 100 = 0 \text{ kN} \Rightarrow \text{OK.}$

Member Forces

$$\mathbf{F_*^i} = \mathbf{F_{*f}^i} + \left(\mathbf{k_*^i T^i}\right)\mathbf{D^i} \quad \Rightarrow$$

$$\mathbf{F}_*^1 = \begin{Bmatrix} F_{1*}^1 \\ F_{2*}^1 \\ F_{3*}^1 \\ \hline F_{4*}^1 \\ F_{5*}^1 \\ F_{6*}^1 \end{Bmatrix} = \begin{Bmatrix} 0 \\ 0 \\ 0 \\ \hline 0 \\ 0 \\ 0 \end{Bmatrix} + \left[\begin{array}{ccc|ccc} 0 & 562500 & 0 & -562500 & 0 & 0 \\ -3164.1 & 0 & 6328.1 & 3164.1 & 0 & 6328.1 \\ -6328.1 & 0 & 16875 & 6328.1 & 0 & 8437.5 \\ \hline 0 & -562500 & 0 & 0 & 562500 & 0 \\ 3164.1 & 0 & -6328.1 & -3164.1 & 0 & -6328.1 \\ -6328.1 & 0 & 8437.5 & 6328.1 & 0 & 16875 \end{array}\right] \begin{Bmatrix} 0 \\ 0 \\ 0 \\ \hline D_1 \\ D_2 \\ D_3 \end{Bmatrix} = \begin{Bmatrix} 60.941 \text{ kN} \\ 31.454 \text{ kN} \\ 91.462 \text{ kNm} \\ \hline -60.941 \text{ kN} \\ -31.454 \text{ kN} \\ 34.354 \text{ kNm} \end{Bmatrix}$$

$$\mathbf{F}_*^2 = \begin{Bmatrix} F_{1*}^2 \\ F_{2*}^2 \\ F_{3*}^2 \\ \hline F_{4*}^2 \\ F_{5*}^2 \\ F_{6*}^2 \end{Bmatrix} = \begin{Bmatrix} 0 \\ 85.185 \\ 111.111 \\ \hline 0 \\ 14.815 \\ 0 \end{Bmatrix} + \left[\begin{array}{ccc|ccc} 562500 & 0 & 0 & -562500 & 0 & 0 \\ 0 & 791.0 & 4746.1 & 0 & -791.0 & 0 \\ 0 & 4746.1 & 28476.6 & 0 & -4746.1 & 0 \\ \hline -562500 & 0 & 0 & 562500 & 0 & 0 \\ 0 & -791.0 & -4746.1 & 0 & 791.0 & 0 \\ 0 & 0 & 0 & 0 & 0 & 0 \end{array}\right] \begin{Bmatrix} D_1 \\ D_2 \\ D_3 \\ \hline D_4 \\ D_5 \\ 0 \end{Bmatrix} = \begin{Bmatrix} 18.545 \text{ kN} \\ 60.941 \text{ kN} \\ -34.353 \text{ kNm} \\ \hline -18.545 \text{ kN} \\ 39.059 \text{ kN} \\ 0 \text{ kNm} \end{Bmatrix}$$

$$\mathbf{F}_*^3 = \begin{Bmatrix} F_{1*}^3 \\ F_{2*}^3 \\ F_{3*}^3 \\ \hline F_{4*}^3 \\ F_{5*}^3 \\ F_{6*}^3 \end{Bmatrix} = \begin{Bmatrix} 0 \\ 0 \\ 0 \\ \hline 0 \\ 0 \\ 0 \end{Bmatrix} + \left[\begin{array}{ccc|ccc} 0 & 562500 & 0 & 0 & -562500 & 0 \\ -791.0 & 0 & 3164.1 & 791.0 & 0 & 0 \\ -3164.1 & 0 & 12656.3 & 3164.1 & 0 & 0 \\ \hline 0 & -562500 & 0 & 0 & 562500 & 0 \\ 791.0 & 0 & -3164.1 & -791.0 & 0 & 0 \\ 0 & 0 & 0 & 0 & 0 & 0 \end{array}\right] \begin{Bmatrix} 0 \\ -0.01 \\ 0 \\ \hline D_4 \\ D_5 \\ D_6 \end{Bmatrix} = \begin{Bmatrix} 39.059 \text{ kN} \\ 18.545 \text{ kN} \\ 74.181 \text{ kNm} \\ \hline -39.059 \text{ kN} \\ -18.545 \text{ kN} \\ 0 \text{ kNm} \end{Bmatrix}$$

The complete response (including the fixed end force effects), in the form of free-body, deflection and bending moment diagrams are shown in Fig. 6.14. Comparison with the response in Fig. 6.11 (Example 6.2) shows that the presence of the internal hinge enhances the drift in the frame significantly on account of a reduction in the lateral stiffness; the bending moments in the columns are also enhanced. The column CD behaves like a vertical cantilever, whose lateral deflection can be estimated as $\dfrac{PL_3^{\,3}}{3EI_3} = \dfrac{(18.545)(4)^3}{3(2.5\times6.75\times10^3)} = 0.02344$ m.

EXAMPLE 6.4

Analyse the two-hinged plane frame with inclined legs shown in Fig. 6.15 using the conventional stiffness method. Assume all three frame elements to have a uniform square cross-section, 300×300 mm with an elastic modulus, $E = 2.5\times10^4$ N/mm^2. Draw the axial force, shear force and bending moment diagrams. Also sketch the deflected shape.

SOLUTION

Input Data

There are three prismatic frame elements ($m = 3$) and four joints with eight *active* and four *restrained* degrees of freedom ($n = 8; r = 4$). The numbering of global coordinates (with the restrained coordinates at the end of the list) is as indicated in Fig. 6.16a.

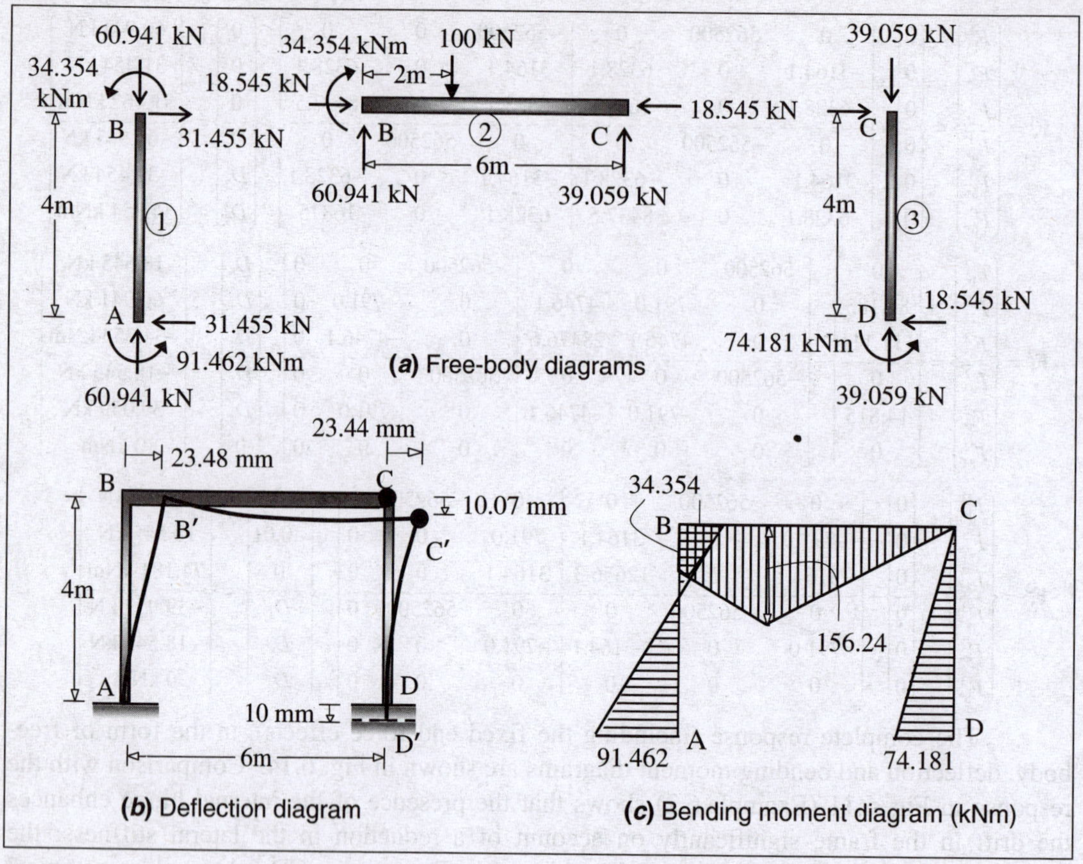

Figure 6.14 Complete response (Example 6.3)

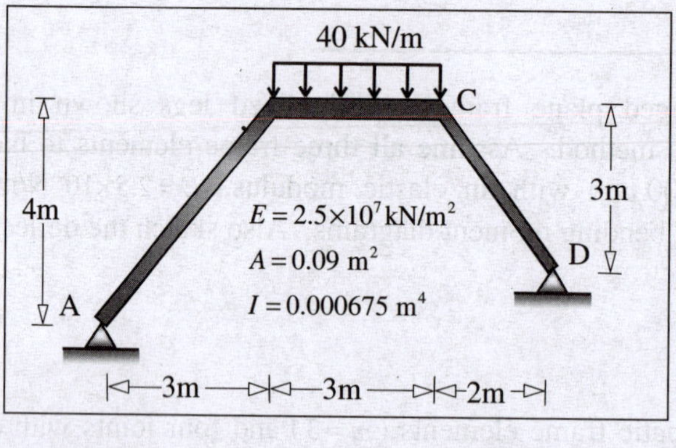

Figure 6.15 Frame with sloping legs (Example 6.4)

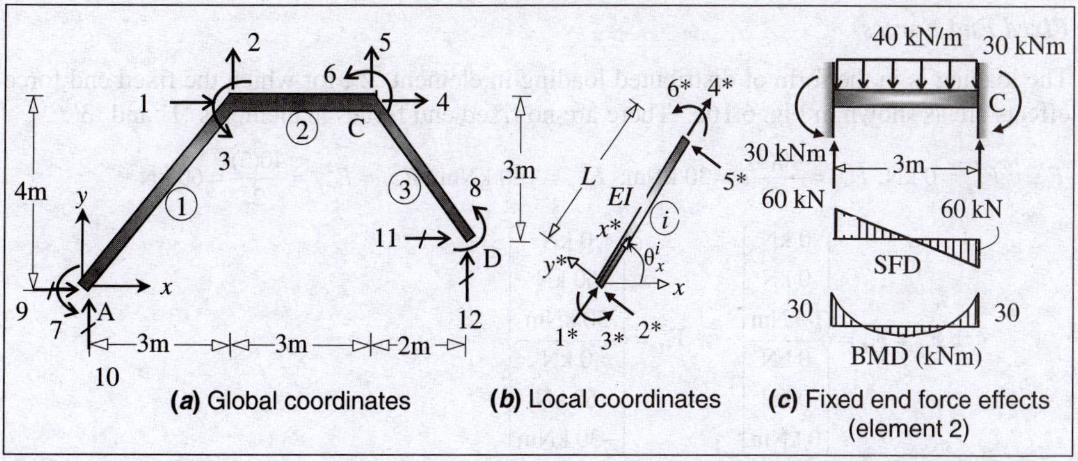

Figure 6.16 Global and local coordinates and fixed end force effects (Example 6.4)

There are no direct nodal loads and no support movements ($\mathbf{F_A} = \mathbf{D_R} = \mathbf{O}$). The typical local coordinates are shown in Fig. 6.16b. All the elements have the same sectional and material properties:

$$E = 2.5\times10^7 \text{ kN/m}^2,\ A = (0.3)(0.3) = 0.09 \text{ m}^2,\ I = (0.3)^4/12 = 0.000675 \text{ m}^4.$$

The axial and flexural rigidities are obtained as $EA = 2.25\times10^6$ kN, $EI = 16875$ kNm^2

Choosing the origin of the *x-y* global axes system at the joint A, the coordinates of the start and end nodes (X_{iS}, Y_{iS} and X_{iE}, Y_{iE}) of individual elements and their lengths, $L_i = \sqrt{(X_{iE} - X_{iS})^2 + (Y_{iE} - Y_{iS})^2}$, and direction cosine parameters, $c_i = \cos\theta_x^i = \dfrac{X_{iE} - X_{iS}}{L_i}$ and $s_i = \sin\theta_x^i = \dfrac{Y_{iE} - Y_{iS}}{L_i}$, are tabulated below. The element cross-sectional properties, A_i and I_i, and the stiffness measures, $\alpha_i = (EA)_i/L_i$ and $\delta_i = (EI)_i/L_i$, are also computed, assuming $E = 2.5\times10^7$ kN/m^2.

Element No. (i)	Start Node (X_{iS}, Y_{iS}) (m)	End Node (X_{iE}, Y_{iE}) (m)	L_i (m)	c_i	s_i	$\alpha_i = (EA)_i/L_i$ (kN/m)	$\delta_i = (EI)_i/L_i$ (kNm)
1	A (0,0)	B (3,4)	5.0	0.6	0.8	450000	3375
2	B (3,4)	C (6,4)	3.0	1	0	750000	5625
3	C (6,4)	D (8,1)	$\sqrt{13}$	0.55470	−0.83205	624037.7	4680.28

Fixed End Forces

The loading is in the form of distributed loading in element '2', for which the fixed end force effects are as shown in Fig. 6.16c. There are no fixed end forces in elements '1' and '3'.

$$F_{1*f}^2 = F_{4*f}^2 = 0 \text{ kN};\ F_{3*f}^2 = \frac{40(3)^2}{12} = 30 \text{ kNm};\ F_{6*f}^2 = -30 \text{ kNm};\ F_{2*f}^2 = F_{5*f}^2 = \frac{40(3)}{2} = 60 \text{ kN}$$

$$\Rightarrow \mathbf{F}_{*f}^1 = \mathbf{F}_{*f}^3 = \begin{Bmatrix} 0 \text{ kN} \\ 0 \text{ kN} \\ 0 \text{ kNm} \\ \hline 0 \text{ kN} \\ 0 \text{ kN} \\ 0 \text{ kNm} \end{Bmatrix};\quad \mathbf{F}_{*f}^2 = \begin{Bmatrix} 0 \text{ kN} \\ 60 \text{ kN} \\ 30 \text{ kNm} \\ \hline 0 \text{ kN} \\ 60 \text{ kN} \\ -30 \text{ kNm} \end{Bmatrix}$$

Coordinate Transformations and Equivalent Joint Loads

Let us use here the concept of displacement transformation matrix, linking all the global coordinates to the local element coordinates, defined by $\mathbf{D}_*^i = \left[\mathbf{T}_{DA}^i \mid \mathbf{T}_{DR}^i\right]\begin{bmatrix}\mathbf{D}_A \\ \hline \mathbf{D}_R\end{bmatrix}$, where $\mathbf{T}_{DA}^i$ and $\mathbf{T}_{DR}^i$ are matrices of order 6×8 and 6×4 respectively. The fixed end forces in terms of the global coordinates can now be assembled, by summing up the contributions from the three elements: $\begin{bmatrix}\mathbf{F}_{fA} \\ \hline \mathbf{F}_{fR}\end{bmatrix} = \sum_{i=1}^{3}\left[\mathbf{T}_{DA}^i \mid \mathbf{T}_{DR}^i\right]^T \mathbf{F}_{*f}^i$.

$$\mathbf{T}_{DA}^1 = \overset{\quad(1)\quad(2)\quad(3)\ (4)\ (5)\ (6)\ (7)\ (8)}{\left[\begin{array}{ccc|ccc|cc} 0 & 0 & 0 & 0 & 0 & 0 & 0 & 0 \\ 0 & 0 & 0 & 0 & 0 & 0 & 0 & 0 \\ 0 & 0 & 0 & 0 & 0 & 0 & 1 & 0 \\ \hline 0.6 & 0.8 & 0 & 0 & 0 & 0 & 0 & 0 \\ -0.8 & 0.6 & 0 & 0 & 0 & 0 & 0 & 0 \\ 0 & 0 & 1 & 0 & 0 & 0 & 0 & 0 \end{array}\right]};\quad \mathbf{T}_{DR}^1 = \overset{(9)\ (10)\ (11)\ (12)}{\left[\begin{array}{cc|cc} 0.6 & 0.8 & 0 & 0 \\ -0.8 & 0.6 & 0 & 0 \\ 0 & 0 & 0 & 0 \\ \hline 0 & 0 & 0 & 0 \\ 0 & 0 & 0 & 0 \\ 0 & 0 & 0 & 0 \end{array}\right]}$$

$$\mathbf{T}_{DA}^2 = \left[\begin{array}{ccc|ccc|cc} 1 & 0 & 0 & 0 & 0 & 0 & 0 & 0 \\ 0 & 1 & 0 & 0 & 0 & 0 & 0 & 0 \\ 0 & 0 & 1 & 0 & 0 & 0 & 0 & 0 \\ \hline 0 & 0 & 0 & 1 & 0 & 0 & 0 & 0 \\ 0 & 0 & 0 & 0 & 1 & 0 & 0 & 0 \\ 0 & 0 & 0 & 0 & 0 & 1 & 0 & 0 \end{array}\right];\quad \mathbf{T}_{DR}^2 = \left[\begin{array}{cc|cc} 0 & 0 & 0 & 0 \\ 0 & 0 & 0 & 0 \\ 0 & 0 & 0 & 0 \\ \hline 0 & 0 & 0 & 0 \\ 0 & 0 & 0 & 0 \\ 0 & 0 & 0 & 0 \end{array}\right]$$

$$\mathbf{T}_{DA}^{3}=\left[\begin{array}{ccc|ccc|cc} 0&0&0&0.55470&-0.83205&0&0&0\\ 0&0&0&0.83205&0.55470&0&0&0\\ 0&0&0&0&0&1&0&0\\ \hline 0&0&0&0&0&0&0&0\\ 0&0&0&0&0&0&0&0\\ 0&0&0&0&0&0&0&1 \end{array}\right] ; \quad \mathbf{T}_{DR}^{3}=\left[\begin{array}{cc|cc} 0&0&0&0\\ 0&0&0&0\\ 0&0&0&0\\ \hline 0&0&0.55470&-0.83205\\ 0&0&0.83205&0.55470\\ 0&0&0&0 \end{array}\right]$$

$$\mathbf{T}_{DA}^{1}\,{}^{T}\mathbf{F}_{*f}^{1}=\mathbf{T}_{DA}^{3}\,{}^{T}\mathbf{F}_{*f}^{3}=\left\{\begin{array}{c} 0\text{ kN}\\ 0\text{ kN}\\ 0\text{ kNm}\\ \hline 0\text{ kN}\\ 0\text{ kN}\\ 0\text{ kNm}\\ \hline 0\text{ kNm}\\ 0\text{ kNm} \end{array}\right\}\begin{array}{c}(1)\\(2)\\(3)\\(4)\\(5)\\(6)\\(7)\\(8)\end{array} ; \quad \mathbf{T}_{DA}^{2}\,{}^{T}\mathbf{F}_{*f}^{2}=\left\{\begin{array}{c} 0\text{ kN}\\ 60\text{ kN}\\ 30\text{ kNm}\\ \hline 0\text{ kN}\\ 60\text{ kN}\\ -30\text{ kNm}\\ \hline 0\text{ kNm}\\ 0\text{ kNm} \end{array}\right\}\begin{array}{c}(1)\\(2)\\(3)\\(4)\\(5)\\(6)\\(7)\\(8)\end{array} \Rightarrow \mathbf{F}_{fA}=\left\{\begin{array}{c} 0\text{ kN}\\ 60\text{ kN}\\ 30\text{ kNm}\\ \hline 0\text{ kN}\\ 60\text{ kN}\\ -30\text{ kNm}\\ \hline 0\text{ kNm}\\ 0\text{ kNm} \end{array}\right\}\begin{array}{c}(1)\\(2)\\(3)\\(4)\\(5)\\(6)\\(7)\\(8)\end{array}$$

$$\mathbf{T}_{DR}^{1}\,{}^{T}\mathbf{F}_{*f}^{1}=\mathbf{T}_{DR}^{2}\,{}^{T}\mathbf{F}_{*f}^{2}=\mathbf{T}_{DR}^{3}\,{}^{T}\mathbf{F}_{*f}^{3}=\left\{\begin{array}{c} 0\text{ kN}\\ 0\text{ kN}\\ \hline 0\text{ kN}\\ 0\text{ kN} \end{array}\right\}\begin{array}{c}(9)\\(10)\\(11)\\(12)\end{array} \Rightarrow \mathbf{F}_{fR}=\left\{\begin{array}{c} 0\text{ kN}\\ 0\text{ kN}\\ \hline 0\text{ kN}\\ 0\text{ kN} \end{array}\right\}\begin{array}{c}(9)\\(10)\\(11)\\(12)\end{array}$$

The resultant direct load vector is given by $\mathbf{F}_A - \mathbf{F}_{fA} = -\mathbf{F}_{fA}$, as indicated in Fig. 6.17.

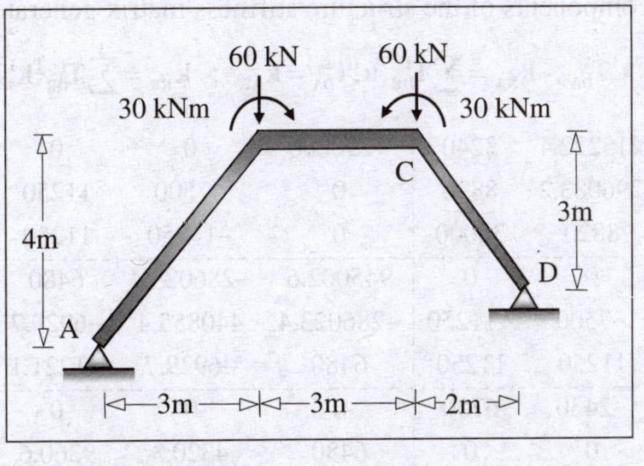

Figure 6.17 Resultant loads (Example 6.4)

Element and Structure Stiffness Matrices

The element stiffness matrices $\mathbf{k}_*^i$ for the three elements, satisfying $\mathbf{F}_*^i = \mathbf{k}_*^i \mathbf{D}_*^i$, take the form shown in Eq. 6.7. Substituting the values of L_i, α_i and δ_i listed in the element property table,

$$\mathbf{k}_*^1 = \left[\begin{array}{ccc|ccc} 450000 & 0 & 0 & -450000 & 0 & 0 \\ 0 & 1620 & 4050 & 0 & -1620 & 4050 \\ 0 & 4050 & 13500 & 0 & -4050 & 6750 \\ \hline -450000 & 0 & 0 & 450000 & 0 & 0 \\ 0 & -1620 & -4050 & 0 & 1620 & -4050 \\ 0 & 4050 & 6750 & 0 & -4050 & 13500 \end{array}\right]$$

$$\mathbf{k}_*^2 = \left[\begin{array}{ccc|ccc} 750000 & 0 & 0 & -750000 & 0 & 0 \\ 0 & 7500 & 11250 & 0 & -7500 & 11250 \\ 0 & 11250 & 22500 & 0 & -11250 & 11250 \\ \hline -750000 & 0 & 0 & 750000 & 0 & 0 \\ 0 & -7500 & -11250 & 0 & 7500 & -11250 \\ 0 & 11250 & 11250 & 0 & -11250 & 22500 \end{array}\right]$$

$$\mathbf{k}_*^3 = \left[\begin{array}{ccc|ccc} 624037.7 & 0 & 0 & -624037.7 & 0 & 0 \\ 0 & 4320.26 & 7788.46 & 0 & -4320.26 & 7788.46 \\ 0 & 7788.46 & 18721.13 & 0 & -7788.46 & 9360.57 \\ \hline -624037.7 & 0 & 0 & 624037.7 & 0 & 0 \\ 0 & -4320.26 & -7788.46 & 0 & 4320.26 & -7788.46 \\ 0 & 7788.46 & 9360.57 & 0 & -7788.46 & 18721.13 \end{array}\right]$$

The products, $\mathbf{k}_*^i\mathbf{T}_{DA}^i$, $\mathbf{k}_*^i\mathbf{T}_{DR}^i$ and $\mathbf{T}^{i^T}\mathbf{k}_*^i\mathbf{T}^i$ may now be computed directly (using MATLAB) and the components of the structure stiffness matrix generated as follows:

$$\mathbf{k}_{AA} = \sum_{i=1}^{3} \mathbf{T}_{DA}^{i\;T}\mathbf{k}_*^i\mathbf{T}_{DA}^i \; ; \quad \mathbf{k}_{RA} = \sum_{i=1}^{3} \mathbf{T}_{DR}^{i\;T}\mathbf{k}_*^i\mathbf{T}_{DA}^i = \mathbf{k}_{AR}^{\;T} \; ; \quad \mathbf{k}_{RR} = \sum_{i=1}^{3} \mathbf{T}_{DR}^{i\;T}\mathbf{k}_*^i\mathbf{T}_{DR}^i$$

$$\Rightarrow \mathbf{k}_{AA} = \left[\begin{array}{ccc|ccc|cc} 913036.8 & 215222.4 & 3240 & -750000 & 0 & 0 & 3240 & 0 \\ 215222.4 & 296083.2 & 8820 & 0 & -7500 & 11250 & -2430 & 0 \\ 3240 & 8820 & 36000 & 0 & -11250 & 11250 & 6750 & 0 \\ \hline -750000 & 0 & 0 & 945002.6 & -286023.4 & 6480 & 0 & 6480 \\ 0 & -7500 & -11250 & -286023.4 & 440855.4 & -6929.7 & 0 & 4320.3 \\ 0 & 11250 & 11250 & 6480 & -6929.7 & 41221.1 & 0 & 9360.6 \\ \hline 3240 & -2430 & 6750 & 0 & 0 & 0 & 13500 & 0 \\ 0 & 0 & 0 & 6480 & 4320.3 & 9360.6 & 0 & 18721.1 \end{array}\right]$$

$$\mathbf{k}_{RA} = \left[\begin{array}{ccc|ccc|cc} -163036.8 & -215222.4 & -3240 & 0 & 0 & 0 & -3240 & 0 \\ -215222.4 & -288583.2 & 2430 & 0 & 0 & 0 & 2430 & 0 \\ \hline 0 & 0 & 0 & -195002.6 & 286023.4 & -6480.4 & 0 & -6480.4 \\ 0 & 0 & 0 & 286023.4 & -433355.4 & -4320.3 & 0 & -4320.3 \end{array}\right]$$

$$\mathbf{k}_{RR} = \left[\begin{array}{cc|cc} 163036.8 & 215222.4 & 0 & 0 \\ 215222.4 & 288583.2 & 0 & 0 \\ \hline 0 & 0 & 195002.6 & -286023.4 \\ 0 & 0 & -286023.4 & 433355.4 \end{array}\right]$$

Displacements and Support Reactions

The basic stiffness relations are given by:

$$\left[\frac{\mathbf{F}_A = \mathbf{O}}{\mathbf{F}_R}\right] - \left[\frac{\mathbf{F}_{fA}}{\mathbf{F}_{fR} = \mathbf{O}}\right] = \left[\begin{array}{c|c} \mathbf{k}_{AA} & \mathbf{k}_{AR} \\ \hline \mathbf{k}_{RA} & \mathbf{k}_{RR} \end{array}\right]\left[\frac{\mathbf{D}_A}{\mathbf{D}_R = \mathbf{O}}\right]$$

$$\Rightarrow \mathbf{D}_A = [\mathbf{k}_{AA}]^{-1}[-\mathbf{F}_{fA}] = \begin{Bmatrix} D_1 \\ D_2 \\ D_3 \\ \hline D_4 \\ D_5 \\ D_6 \\ \hline D_7 \\ D_8 \end{Bmatrix} = \begin{Bmatrix} 0.000093379 \text{ m} \\ -0.000278113 \text{ m} \\ -0.001324194 \text{ rad} \\ \hline 0.000030697 \text{ m} \\ -0.000128508 \text{ m} \\ 0.001279619 \text{ rad} \\ \hline 0.000589626 \text{ rad} \\ -0.000620780 \text{ rad} \end{Bmatrix}$$

$$\mathbf{F}_R = \mathbf{k}_{RA}\mathbf{D}_A \Rightarrow \quad \mathbf{F}_R = \begin{Bmatrix} F_9 \\ F_{10} \\ \hline F_{11} \\ F_{12} \end{Bmatrix} = \begin{Bmatrix} 47.012 \text{ kN} \\ 58.376 \text{ kN} \\ \hline -47.012 \text{ kN} \\ 61.624 \text{ kN} \end{Bmatrix}$$

<u>Check equilibrium:</u> $\sum F_x = 0: \; F_9 + F_{11} = 47.012 - 47.012 = 0 \text{ kN}$

$\sum F_y = 0: \; F_{10} + F_{12} - (40)(3) = 58.376 + 61.624 - 120 = 0 \text{ kN} \;\; \Rightarrow \text{OK.}$

Member Forces

$$\mathbf{F}_*^i = \mathbf{F}_{*f}^i + \left(\mathbf{k}_*^i \mathbf{T}_{DA}^i\right)\mathbf{D}_A \Rightarrow$$

$$\mathbf{F}_*^1 = \begin{Bmatrix} F_{1*}^1 \\ F_{2*}^1 \\ F_{3*}^1 \\ \hline F_{4*}^1 \\ F_{5*}^1 \\ F_{6*}^1 \end{Bmatrix} = \begin{Bmatrix} 74.908 \text{ kN} \\ -2.584 \text{ kN} \\ 0 \text{ kNm} \\ \hline -74.908 \text{ kN} \\ 2.584 \text{ kN} \\ -12.918 \text{ kNm} \end{Bmatrix}; \quad \mathbf{F}_*^2 = \begin{Bmatrix} F_{1*}^2 \\ F_{2*}^2 \\ F_{3*}^2 \\ \hline F_{4*}^2 \\ F_{5*}^2 \\ F_{6*}^2 \end{Bmatrix} = \begin{Bmatrix} 47.012 \text{ kN} \\ 58.376 \text{ kN} \\ 12.918 \text{ kNm} \\ \hline -47.012 \text{ kN} \\ 61.624 \text{ kN} \\ -17.789 \text{ kNm} \end{Bmatrix}; \quad \mathbf{F}_*^3 = \begin{Bmatrix} F_{1*}^3 \\ F_{2*}^3 \\ F_{3*}^3 \\ \hline F_{4*}^3 \\ F_{5*}^3 \\ F_{6*}^3 \end{Bmatrix} = \begin{Bmatrix} 77.381 \text{ kN} \\ 4.934 \text{ kN} \\ 17.789 \text{ kNm} \\ \hline -77.381 \text{ kN} \\ -4.934 \text{ kN} \\ 0 \text{ kNm} \end{Bmatrix}$$

The complete response (including the fixed end force effects), in the form of free-body, deflection, bending moment, axial force and shear force diagrams are shown in Fig. 6.18.

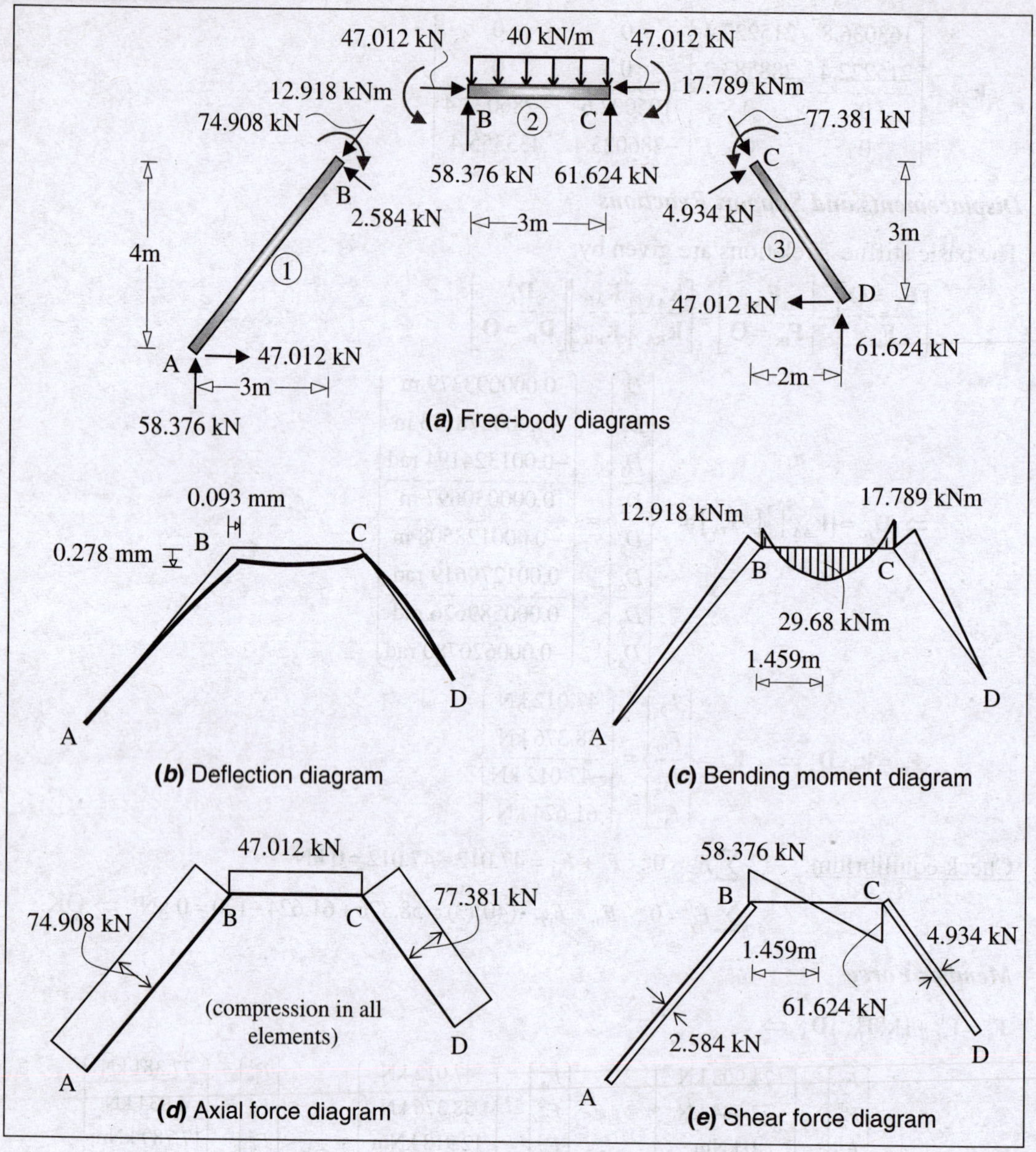

Figure 6.18 Complete response (Example 6.4)

EXAMPLE 6.5

Consider the frame in Fig. 6.15 (frame details as given in Example 6.4) to be subject to a uniform drop in temperature by 40°C. Find the response to this temperature loading by the conventional Stiffness Method, assuming the coefficient of thermal expansion, $\alpha = 11\times10^{-6}$ per °C.

SOLUTION

The global and local coordinates, transformation and stiffness matrices are as defined in Example 6.4. Due to the temperature loading, fixed end forces develop in the restrained (primary) structure, as shown in Fig. 6.19a.

Fixed End Forces

Due to restraints against axial deformation, each of the three elements will develop axial force, given by

$$N_{if} = -\left(\frac{EA}{L}\right)_i L_i\alpha(\Delta T) = -AE\alpha(\Delta T) = -(0.09)(-40)\left(2.5\times10^7\right)\left(11\times10^{-6}\right) = +990\,\text{kN}$$

$$\Rightarrow F^i_{1*f} = -N_{if} = -990\text{ kN};\; F^i_{4*f} = +N_{if} = +990\text{ kN}$$

No shear forces or bending moments develop on account of the restraints to the active degrees of freedom in the primary structure.

$$\Rightarrow \mathbf{F}^1_{*f} = \mathbf{F}^2_{*f} = \mathbf{F}^3_{*f} = \begin{Bmatrix} -990\text{ kN} \\ 0\text{ kN} \\ 0\text{ kNm} \\ \hline 990\text{ kN} \\ 0\text{ kN} \\ 0\text{ kNm} \end{Bmatrix}$$

The fixed end forces in terms of the global coordinates can now be assembled, by summing up the contributions from the three elements: $\begin{bmatrix}\mathbf{F}_{fA} \\ \hline \mathbf{F}_{fR}\end{bmatrix} = \sum_{i=1}^{3}\left[\mathbf{T}^i_{DA} \mid \mathbf{T}^i_{DR}\right]^T \mathbf{F}^i_{*f}$, using the displacement transformations derived earlier in Example 6.4.

$$\Rightarrow \mathbf{F}_{fA} = \left[\mathbf{T}^{1\ T}_{DA} + \mathbf{T}^{2\ T}_{DA} + \mathbf{T}^{3\ T}_{DA}\right]\begin{Bmatrix} -990\text{ kN} \\ 0\text{ kN} \\ 0\text{ kNm} \\ \hline 990\text{ kN} \\ 0\text{ kN} \\ 0\text{ kNm} \end{Bmatrix} = \begin{Bmatrix} -396\text{ kN} \\ 792\text{ kN} \\ 0\text{ kNm} \\ \hline 440.847\text{ kN} \\ 823.730\text{ kN} \\ 0\text{ kNm} \\ \hline 0\text{ kNm} \\ 0\text{ kNm} \end{Bmatrix} \begin{matrix} (1) \\ (2) \\ (3) \\ (4) \\ (5) \\ (6) \\ (7) \\ (8) \end{matrix}$$

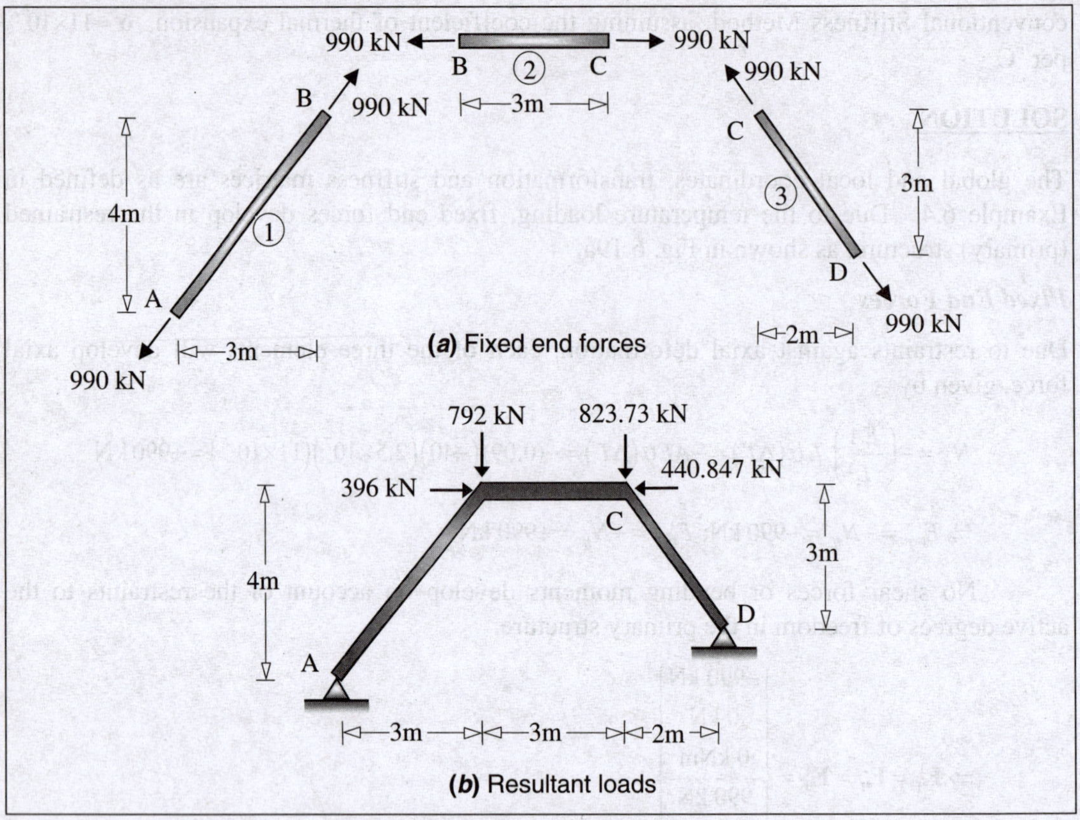

Figure 6.19 Fixed end forces and resultant loads due to temperature loading (Example 6.5)

$$\Rightarrow \mathbf{F}_{fR} = \left[\mathbf{T}_{DR}^{1\;T} + \mathbf{T}_{DR}^{2\;T} + \mathbf{T}_{DR}^{3\;T}\right]\begin{Bmatrix} -990 \text{ kN} \\ 0 \text{ kN} \\ 0 \text{ kNm} \\ \hdashline 990 \text{ kN} \\ 0 \text{ kN} \\ 0 \text{ kNm} \end{Bmatrix} = \begin{Bmatrix} -594 \text{ kN} \\ -792 \text{ kN} \\ \hdashline 549.153 \text{ kN} \\ -823.730 \text{ kN} \end{Bmatrix} \begin{matrix} (9) \\ (10) \\ (11) \\ (12) \end{matrix}$$

The resultant direct load vector is given by $\mathbf{F}_A - \mathbf{F}_{fA} = -\mathbf{F}_{fA}$, as indicated in Fig. 6.19b.

Displacements and Support Reactions

The basic stiffness relations are given by:

$$\begin{bmatrix} \mathbf{F}_A = \mathbf{O} \\ \hdashline \mathbf{F}_R \end{bmatrix} - \begin{bmatrix} \mathbf{F}_{fA} \\ \hdashline \mathbf{F}_{fR} \end{bmatrix} = \begin{bmatrix} \mathbf{k}_{AA} & \mathbf{k}_{AR} \\ \hdashline \mathbf{k}_{RA} & \mathbf{k}_{RR} \end{bmatrix} \begin{bmatrix} \mathbf{D}_A \\ \hdashline \mathbf{D}_R = \mathbf{O} \end{bmatrix}$$

$$\Rightarrow \mathbf{D_A} = [\mathbf{k_{AA}}]^{-1}[-\mathbf{F_{fA}}] = \begin{Bmatrix} D_1 \\ D_2 \\ D_3 \\ \hline D_4 \\ D_5 \\ D_6 \\ \hline D_7 \\ D_8 \end{Bmatrix} = \begin{Bmatrix} 0.00035740 \text{ m} \\ -0.00301637 \text{ m} \\ -0.00011115 \text{ rad} \\ \hline -0.00096146 \text{ m} \\ -0.0025468 \text{ m} \\ +0.00041458 \text{ rad} \\ \hline -0.00057315 \text{ rad} \\ -0.00071327 \text{ rad} \end{Bmatrix}$$

$$\mathbf{F_R} = \mathbf{F_{fR}} + \mathbf{k_{RA}}\mathbf{D_A} \Rightarrow \quad \mathbf{F_R} = \begin{Bmatrix} F_9 \\ F_{10} \\ \hline F_{11} \\ F_{12} \end{Bmatrix} = \begin{Bmatrix} -0.860 \text{ kN} \\ -0.108 \text{ kN} \\ \hline 0.860 \text{ kN} \\ 0.108 \text{ kN} \end{Bmatrix}$$

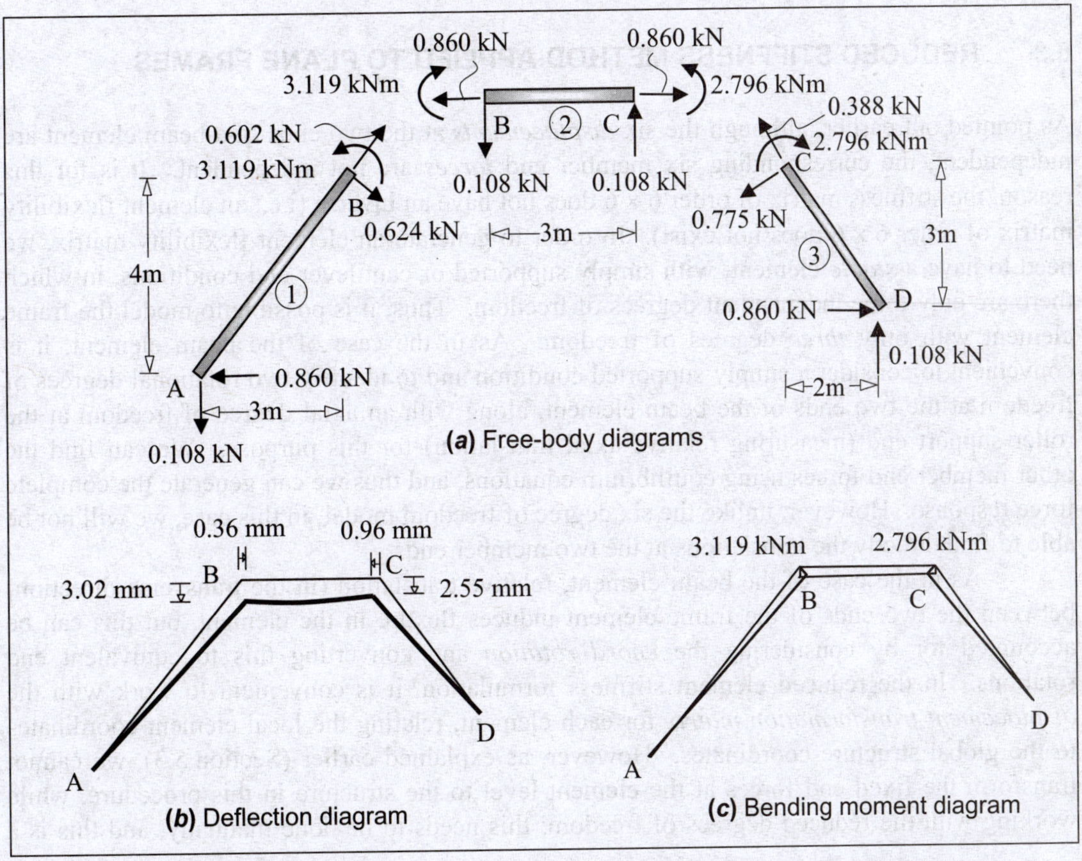

Figure 6.20 Response to temperature loading (Example 6.5)

Member Forces

$$\mathbf{F}_*^i = \mathbf{F}_{*f}^i + \left(\mathbf{k}_*^i \mathbf{T}_{DA}^i\right)\mathbf{D}_A \Rightarrow$$

$$\mathbf{F}_*^1 = \begin{Bmatrix} F_{1*}^1 \\ F_{2*}^1 \\ F_{3*}^1 \\ \hdashline F_{4*}^1 \\ F_{5*}^1 \\ F_{6*}^1 \end{Bmatrix} = \begin{Bmatrix} -0.602 \text{ kN} \\ 0.624 \text{ kN} \\ 0 \text{ kNm} \\ \hdashline 0.602 \text{ kN} \\ -0.624 \text{ kN} \\ 3.119 \text{ kNm} \end{Bmatrix}; \quad \mathbf{F}_*^2 = \begin{Bmatrix} F_{1*}^2 \\ F_{2*}^2 \\ F_{3*}^2 \\ \hdashline F_{4*}^2 \\ F_{5*}^2 \\ F_{6*}^2 \end{Bmatrix} = \begin{Bmatrix} -0.860 \text{ kN} \\ -0.108 \text{ kN} \\ -3.119 \text{ kNm} \\ \hdashline 0.860 \text{ kN} \\ 0.108 \text{ kN} \\ 2.796 \text{ kNm} \end{Bmatrix}; \quad \mathbf{F}_*^3 = \begin{Bmatrix} F_{1*}^3 \\ F_{2*}^3 \\ F_{3*}^3 \\ \hdashline F_{4*}^3 \\ F_{5*}^3 \\ F_{6*}^3 \end{Bmatrix} = \begin{Bmatrix} -0.388 \text{ kN} \\ -0.775 \text{ kN} \\ -2.796 \text{ kNm} \\ \hdashline 0.388 \text{ kN} \\ 0.775 \text{ kN} \\ 0 \text{ kNm} \end{Bmatrix}$$

The response (including the fixed end force effects), in the form of free-body, deflection and bending moment diagrams are shown in Fig. 6.20. It is interesting to note that although the fixed end forces due to the temperature loading appear to be significant, the internal forces developed are very small[†].

6.3 REDUCED STIFFNESS METHOD APPLIED TO PLANE FRAMES

As pointed out earlier, although the six *displacements* at the two ends of a beam element are independent, the corresponding six member end *forces* are not independent. It is for this reason, the stiffness matrix of order 6 × 6 does not have an inverse (i.e., an element flexibility matrix of order 6 × 6 does not exist). In order to generate an element flexibility matrix, we need to have a *stable* element, with simply supported or cantilever end conditions, in which there are only three independent degrees of freedom. Thus, it is possible to model the frame element with only *three* degrees of freedom. As in the case of the beam element, it is convenient to consider a simply supported condition and to identify two rotational degrees of freedom at the two ends of the beam element, along with an axial degree of freedom at the roller support end (measuring relative axial translation) for this purpose. We can find the other member end forces using equilibrium equations, and thus we can generate the complete force response. However, unlike the six degree of freedom model, in this case, we will not be able to find directly the translations at the two member ends.

As in the case of the beam element, relative translation (in the transverse direction) between the two ends of the frame element induces flexure in the element, but this can be accounted for by considering the *chord rotation* and converting this to equivalent end rotations. In the reduced element stiffness formulation, it is convenient to work with the *displacement transformation matrix* for each element, relating the local element coordinates to the global structure coordinates. However, as explained earlier (Section 5.3), we cannot transform the fixed end forces at the element level to the structure in this procedure, while working with the reduced degrees of freedom; this needs to be done manually, and this is a

[†] The elastic restraints offered to 'free' movements of individual elements by the connecting elements at joints B and C are negligible.

shortcoming in the reduced stiffness method, reducing its suitability for adoption in software packages. The final structure stiffness matrix, and the final response, however, remains the same, whether we work with the conventional element stiffness matrix of order 6×6 or the reduced element stiffness matrix of order 3×3.

When axial deformations are considered negligible, we can, in most cases, further simplify the analysis by eliminating the axial degree of freedom. The frame element reduces to a beam element (with two rotational degrees of freedom) in this case. This is, in most cases, a reasonably accepted simplification, which reduces the kinematic indeterminacy of the structure. The resulting error is often negligible, as demonstrated in the examples to follow.

Also, in cases involving member end releases (internal hinges, hinged and guided-fixed end supports), it is possible to further reduce the degree of kinematic indeterminacy of the structure, as explained in Section 5.3 for beams. In such cases, it is possible to reduce the size of the element stiffness matrix to an order 1×1, and this can be very advantageous in terms of computational effort. This is discussed in Sections 6.3.2 and 6.3.3, and demonstrated through various examples.

6.3.1 Plane Frame Element with Three Degrees of Freedom

Element Stiffness Matrix

The prismatic plane frame element with only three degrees of freedom is shown in Fig. 6.21a. The element stiffness matrix for this can be easily generated by eliminating the rows and columns corresponding to the relevant translational degrees of freedom in the six degree of freedom model derived earlier (Eq. 6.7). The first column and row of the matrix, which corresponds to the axial degree of freedom is uncoupled from the rotational (flexural) degrees of freedom (represented by the other two columns/rows). The deflected shapes and free-bodies, corresponding to the three unit displacement conditions are shown in Figs 6.21b, c and d. Accordingly, the reduced element stiffness matrix, $\tilde{\mathbf{k}}_*^{\mathbf{i}}$, is obtained as follows:

$$\boxed{\tilde{\mathbf{k}}_*^{\mathbf{i}} = \begin{bmatrix} \alpha_i & 0 & 0 \\ 0 & 4\delta_i & 2\delta_i \\ 0 & 2\delta_i & 4\delta_i \end{bmatrix}; \quad \begin{matrix} \alpha_i = \dfrac{E_i A_i}{L_i} \\ \delta_i = \dfrac{E_i I_i}{L_i} \end{matrix}} \tag{6.18}$$

Alternatively, we can generate the reduced element stiffness matrix $\tilde{\mathbf{k}}_*^{\mathbf{i}}$ as the inverse of the flexibility matrix $\mathbf{f}_*^{\mathbf{i}}$, as explained earlier for the truss element and the beam element (refer Fig. 5.20).

$$\mathbf{f}_*^{\mathbf{i}} = \begin{bmatrix} (L/EA)_i & 0 & 0 \\ 0 & (L/3EI)_i & -(L/6EI)_i \\ 0 & -(L/6EI)_i & (L/3EI)_i \end{bmatrix} \Rightarrow \tilde{\mathbf{k}}_*^{\mathbf{i}} = \left[f_*^i\right]^{-1} = \begin{bmatrix} (EA/L)_i & 0 & 0 \\ 0 & 4(EI/L)_i & 2(EI/L)_i \\ 0 & 2(EI/L)_i & 4(EI/L)_i \end{bmatrix} \tag{6.19}$$

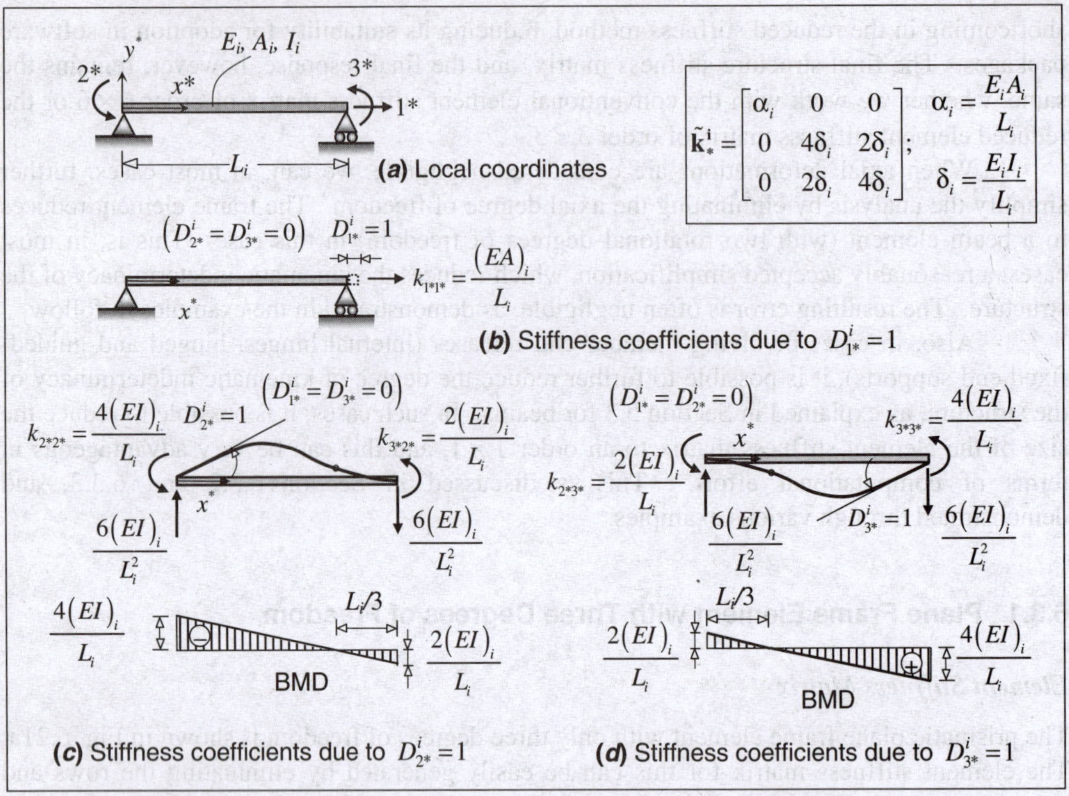

Figure 6.21 Generation of 3 × 3 stiffness matrix for a plane frame element

Displacement Transformation Matrix

In a plane frame system, at every node location, three global coordinates are typically identified, corresponding to two orthogonal translations and a rotation at the joint. In the six degree of freedom frame element model, it is convenient to link the element end deflections and rotation directly to the corresponding displacements in the structure, on a one-to-one basis.

In the three degree of freedom beam element model, however, this is possible only with respect to the rotational degrees of freedom. We need to adopt the concepts described earlier with regard to the reduced truss and beam elements in order to link the translations at a node location in the structure to the end displacements in the frame element. It is possible to account for transverse translations by converting the resulting chord rotation, ϕ_i, to equivalent beam end rotations; when the *chord rotation* acts clockwise, the beam end rotations are anti-clockwise, as explained earlier. Similarly, the displacement D_{1*}^i corresponds to the axial deformation e_i, which is equal to the difference in longitudinal translations at the two ends (positive when $e_i > 0$). These transformations are simple to carry out while dealing with vertical and horizontal members, and the displacement transformation matrix $\mathbf{T_D^i}$ can be easily generated.

However, while dealing with sloping members, special care needs to be taken. For example, let us consider a typical truss element i inclined θ_x^i to the global x-axis, with 6 joint degrees of freedom at its two ends (global coordinates numbered '1' to '6' for convenience), as shown in Fig. 6.22a. The components of the $\mathbf{T_D^i}$ matrix, linking the displacements corresponding to these degrees of freedom to the local coordinates (Fig. 6.22b) can be generated either from kinematic considerations (satisfying displacement compatibility), as shown in Fig. 6.22c. Alternatively, the transpose of this matrix can be generated using static conditions (refer Fig. 6.22c). The elements of the matrix involve direction cosines.

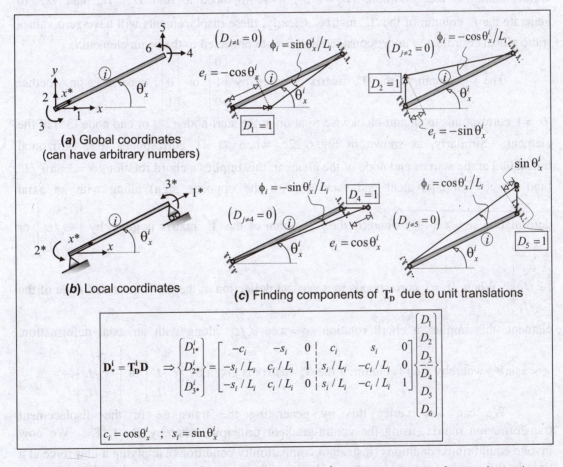

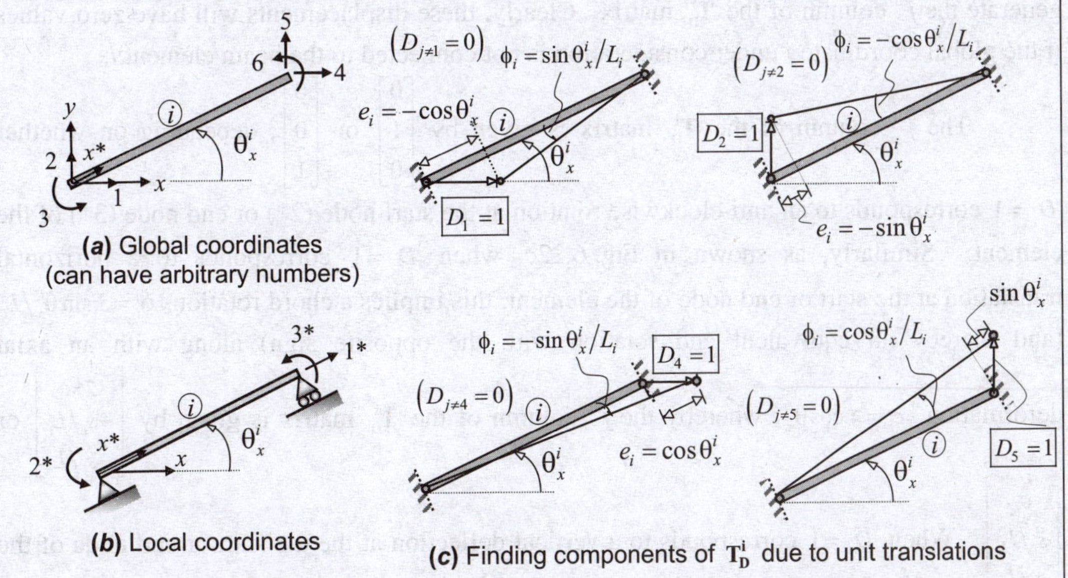

$$\mathbf{D_*^i} = \mathbf{T_D^i D} \quad \Rightarrow \begin{Bmatrix} D_{1*}^i \\ D_{2*}^i \\ D_{3*}^i \end{Bmatrix} = \left[\begin{array}{ccc|ccc} -c_i & -s_i & 0 & c_i & s_i & 0 \\ -s_i/L_i & c_i/L_i & 1 & s_i/L_i & -c_i/L_i & 0 \\ -s_i/L_i & c_i/L_i & 0 & s_i/L_i & -c_i/L_i & 1 \end{array}\right] \begin{Bmatrix} D_1 \\ D_2 \\ D_3 \\ \hline D_4 \\ D_5 \\ D_6 \end{Bmatrix}$$

$$c_i = \cos\theta_x^i \; ; \quad s_i = \sin\theta_x^i$$

Figure 6.22 Generation of components of the $\mathbf{T_D^i}$ matrix for a plane frame element (reduced element stiffness formulation)

We are now in a position to define the *displacement transformation matrix* $\mathbf{T_D^i}$ for a plane frame element *i* with three degrees of freedom. This matrix defines the compatibility relations between the global degrees of freedom of the structure with the three degrees of freedom (1*, 2* and 3*) of the element through the relation, $\mathbf{D_*^i} = \mathbf{T_D^i D} = \left[\mathbf{T_{DA}^i} \mid \mathbf{T_{DR}^i}\right]\begin{bmatrix}\mathbf{D_A} \\ \hline \mathbf{D_R}\end{bmatrix}$. If there are *n* active and *r* restrained global coordinates defined in the structure, then the sub-matrices, $\mathbf{T_{DA}^i}$ and $\mathbf{T_{DR}^i}$, will have orders of $3 \times n$ and $3 \times r$ respectively.

The displacement transformation matrix is generated by considering compatibility in the primary (restrained) structure with only one degree of freedom released. Thus, corresponding to the condition, $\left\{D_j = 1, D_{l \neq j} = 0\right\}$, we need to find D_{1*}^i, D_{2*}^i and D_{3*}^i, to generate the jth column of the $\mathbf{T_D^i}$ matrix. Clearly, these displacements will have zero values if the global coordinate *j* under consideration is not connected to the beam element *i*.

The jth column of the $\mathbf{T_D^i}$ matrix is given by $\begin{bmatrix}0\\1\\0\end{bmatrix}$ or $\begin{bmatrix}0\\0\\1\end{bmatrix}$, depending on whether $D_j = 1$ corresponds to an anti-clockwise rotation at the start node (2*) or end node (3*) of the element. Similarly, as shown in Fig. 6.22c, when $D_j = 1$ corresponds to a horizontal translation at the start or end node of the element, this implies a chord rotation, $\phi_i = \pm \sin\theta_x^i / L_i$ (and thereby an equivalent end rotation with the opposite sign) along with an axial deformation, $e_i = \mp\cos\theta_x^i$, whereby the jth column of the $\mathbf{T_D^i}$ matrix is given by $\begin{bmatrix}-c_i\\-s_i/L_i\\-s_i/L_i\end{bmatrix}$ or $\begin{bmatrix}c_i\\s_i/L_i\\s_i/L_i\end{bmatrix}$. When $D_j = 1$ corresponds to a vertical deflection at the the start or end node of the element, this implies a chord rotation, $\phi_i = \mp\cos\theta_x^i / L_i$ along with an axial deformation, $e_i = \pm\sin\theta_x^i$, whereby the jth column of the $\mathbf{T_D^i}$ matrix is given by $\begin{bmatrix}s_i\\c_i/L_i\\c_i/L_i\end{bmatrix}$ or $\begin{bmatrix}-s_i\\-c_i/L_i\\-c_i/L_i\end{bmatrix}$.

We can also verify this by generating the transpose of the displacement transformation matrix, using the contra-gradient principle, whereby $\mathbf{F} = \mathbf{T_D^{i\,T} F_*^i}$. We now invoke equilibrium conditions (instead of compatibility conditions), applying a unit force at a time in the simply supported system. However, it is important to note that, the use of the contra-gradient principle cannot be extended to finding the fixed end force vector through the transformation, $\mathbf{F_f} = \sum_{i=1}^{m} \mathbf{T_D^{i\,T} F_{*f}^i}$, as done in the conventional stiffness method. This is so, because the two fixed end moments (F^{i}_{1*f} and F^{i}_{1*f}) in the element are not adequate by themselves to determine the vertical reactions at the beam ends on account of intermediate loading. For this reason, we cannot automate the generation of the fixed end force vector in the structure in the reduced element stiffness formulation; this must be done manually, as demonstrated in the Examples to follow.

The displacement transformation matrix is useful in generating the structure stiffness matrix by the transformation, $\mathbf{k} = \sum_{i=1}^{m} \mathbf{T}^{i\mathrm{T}}_{\mathbf{D}} \tilde{\mathbf{k}}^{i}_{*} \mathbf{T}^{i}_{\mathbf{D}}$, and for finding the internal forces, $\mathbf{F}^{i}_{*} = \mathbf{F}^{i}_{*f} + \left(\tilde{\mathbf{k}}^{i}_{*f}\mathbf{T}^{i}_{\mathbf{DA}}\right)\mathbf{D}_{\mathbf{A}}$.

The basic analysis procedure is as outlined earlier in Section 6.2.4.

6.3.2 Ignoring Axial Deformations

A further simplification in analysis is possible when we ignore axial deformations in frames. This is the usual practice in basic structural analysis, and is implemented in traditional methods such as Slope Deflection method and Moment Distribution method. By means of this simplification, the axial degree of freedom in the frame element can be eliminated, whereby the frame element with three degrees of freedom reduces to a *beam element* with only two rotational degrees of freedom. The first row and column of the stiffness matrix in Eq. 6.7 get eliminated.

However, while adopting this simplification, care needs to be taken in assessing the kinematic indeterminacy of the structure, which gets reduced significantly. The translational degrees of freedom are now limited to only the 'sway' degrees of freedom (refer Fig. 6.23). In general, to identify the independent degrees of translational freedom in a plane frame, we simply have to count the total number of joint translations (typically, two per joint in a plane frame) and subtract the number of members (whose lengths are assumed to be inextensible). Thus, in the frames in Figs 6.23a, b and c, there are two joints (B and C) that can move and three members (AB, BC and CD) inter-connecting these joints, whereby the number of sway degrees of freedom is given by $2\times2-3=1$, and we may conveniently choose the horizontal deflection δ (at B or C) as the unknown translation. In order to express the various chord rotations in terms of δ, it is convenient to visualise internal hinges at all the joints and to sketch the deflection shape of the resulting mechanism. The chord rotation ϕ is treated as negative when it acts in the anti-clockwise direction, as illustrated in Fig. 6.23.

For the symmetric pitched portal frame ABCDE shown in Fig. 6.23d, we note that there are three joints and four members, whereby the number of unknown sway degrees of freedom is given by $3\times2-4=2$. We can choose the horizontal deflections at B and D as the two unknown translations (δ_B and δ_D), and the various chord rotations can be expressed as indicated in Fig. 6.23d.

In a typical multi-bay multi-storeyed frame, there will be a sway degree of freedom associated with every floor level. Thus, in a three-storeyed two-bay frame with fixed bases, there will be nine rotational degrees of freedom and three sway degrees of freedom (degree of kinematic indeterminacy is equal to 12), if we make the assumption of axial deformations being negligible. The errors involved in making this assumption are acceptably low in general, except in rare cases involving differential shortening of columns (which may occur in very tall buildings).

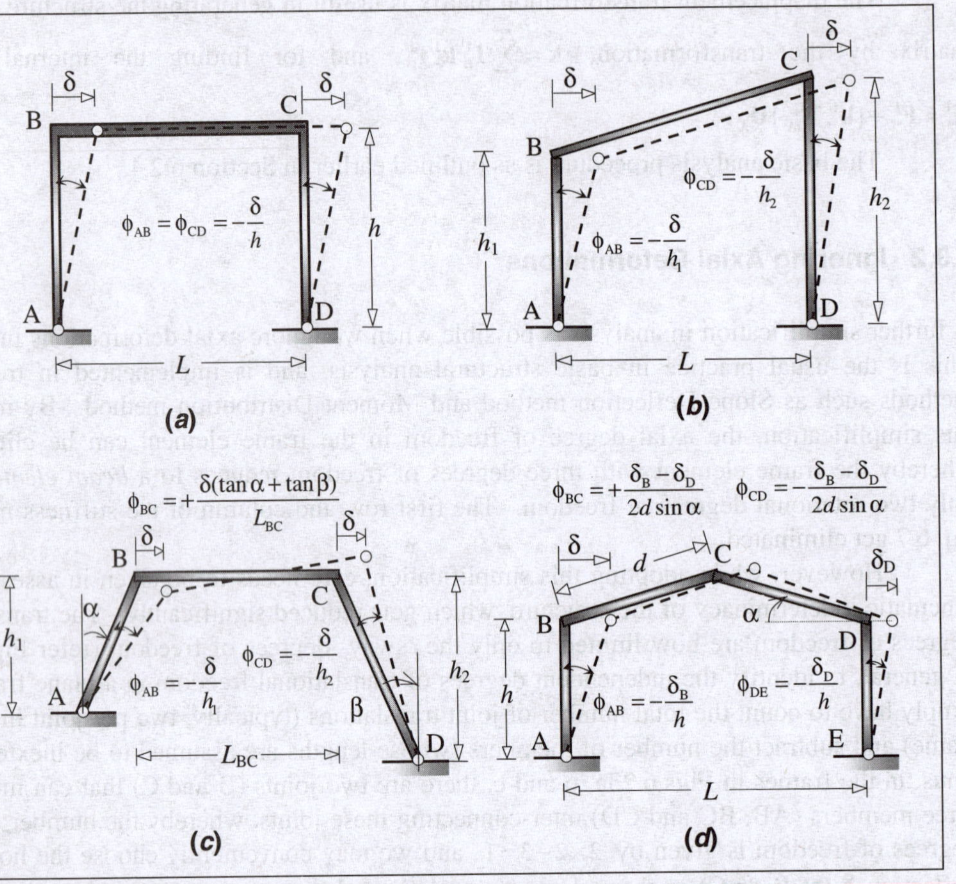

Figure 6.23 Sway degrees of freedom and chord rotations in single storey frames

Although ignoring axial deformations in the reduced stiffness formulation results in considerable savings in computational effort, it calls for supplementary manual effort and hence cannot be automated through a computer program. Furthermore, while dealing with frames with sway and sloping members, extreme caution needs to be exercised in deriving the displacement transformation matrices and the equivalent joint loads. This is demonstrated in Example 6.8.

EXAMPLE 6.6

Repeat Example 6.2 by the reduced stiffness formulation. Also, show how the analysis can be further simplified by ignoring axial deformations and find the extent of error in this approximation.

SOLUTION

Here, we consider the same system of global coordinates as in Example 6.2 (shown in Fig. 6.24a), but with each of the three frame elements having three degrees of freedom (local coordinates as shown in Fig. 6.24b). The input load data takes the following form:

$$\mathbf{F}_A = \begin{Bmatrix} 50\text{ kN} \\ 0\text{ kN} \\ 0\text{ kNm} \\ \hdashline 0\text{ kN} \\ 0\text{ kN} \\ 0\text{ kNm} \end{Bmatrix} \quad ; \quad \mathbf{D}_R = \begin{Bmatrix} D_7 \\ D_8 \\ D_9 \\ \hdashline D_{10} \\ D_{11} \\ D_{12} \end{Bmatrix} = \begin{Bmatrix} 0\text{ m} \\ 0\text{ m} \\ 0\text{ rad} \\ \hdashline 0\text{ m} \\ -0.01\text{ m} \\ 0\text{ rad} \end{Bmatrix}$$

Fixed End Forces

The fixed end forces (due to the intermediate loading in element '2') are as computed earlier (refer Fig. 6.24c).

$$\Rightarrow \mathbf{F}_{*f}^{1} = \mathbf{F}_{*f}^{3} = \begin{Bmatrix} 0\text{ kN} \\ 0\text{ kNm} \\ 0\text{ kNm} \end{Bmatrix} \quad ; \quad \mathbf{F}_{*f}^{2} = \begin{Bmatrix} 0\text{ kN} \\ 88.889\text{ kNm} \\ -44.444\text{ kNm} \end{Bmatrix}$$

Summing up the fixed end forces at the six active and six restrained global coordinate locations, we obtain the same fixed end force vector as in Example 6.2:

$$\Rightarrow \mathbf{F}_f = \begin{bmatrix} \mathbf{F}_{fA} \\ \hdashline \mathbf{F}_{fR} \end{bmatrix} \text{ where } \mathbf{F}_{fA} = \begin{Bmatrix} F_{1f} \\ F_{2f} \\ F_{3f} \\ \hdashline F_{4f} \\ F_{5f} \\ F_{6f} \end{Bmatrix} = \begin{Bmatrix} 0\text{ kN} \\ 74.074\text{ kN} \\ 88.889\text{ kNm} \\ \hdashline 0\text{ kN} \\ 25.926\text{ kN} \\ -44.444\text{ kNm} \end{Bmatrix} \text{ and } \mathbf{F}_{fR} = \begin{Bmatrix} F_{7f} \\ F_{8f} \\ F_{9f} \\ \hdashline F_{10f} \\ F_{11f} \\ F_{12f} \end{Bmatrix} = \begin{Bmatrix} 0\text{ kN} \\ 0\text{ kN} \\ 0\text{ kNm} \\ \hdashline 0\text{ kN} \\ 0\text{ kN} \\ 0\text{ kNm} \end{Bmatrix}$$

The resultant direct load vector, $\mathbf{F}_A - \mathbf{F}_{fA}$, is accordingly given by (refer Fig. 6.24d):

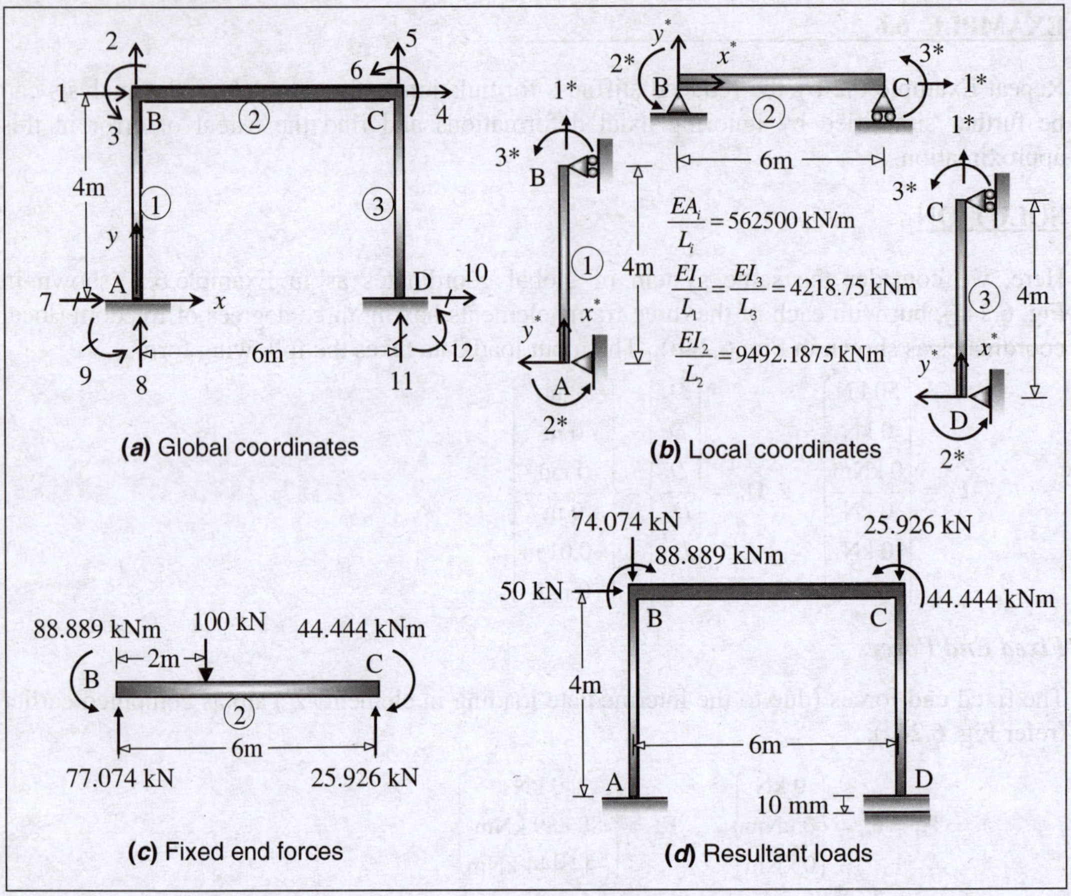

Figure 6.24 Global and local coordinates, fixed end forces and resultant loads (Example 6.6)

$$\mathbf{F_A} - \mathbf{F_{fA}} = \begin{Bmatrix} 50 \text{ kN} \\ 0 \text{ kN} \\ 0 \text{ kNm} \\ \hline 0 \text{ kN} \\ 0 \text{ kN} \\ 0 \text{ kNm} \end{Bmatrix} - \begin{Bmatrix} 0 \text{ kN} \\ 74.074 \text{ kN} \\ 88.889 \text{ kNm} \\ \hline 0 \text{ kN} \\ 25.926 \text{ kN} \\ -44.444 \text{ kNm} \end{Bmatrix} = \begin{Bmatrix} 50 \text{ kN} \\ -74.074 \text{ kN} \\ -88.889 \text{ kNm} \\ \hline 0 \text{ kN} \\ -25.926 \text{ kN} \\ 44.444 \text{ kNm} \end{Bmatrix}$$

Displacement Transformation Matrices

The element displacement transformation matrices, $\mathbf{T_D^1}$, $\mathbf{T_D^2}$ and $\mathbf{T_D^3}$, relate the local coordinates to the global coordinates through the relation, $\mathbf{D_*^i} = \left[\mathbf{T_{DA}^i} \mid \mathbf{T_{DR}^i}\right]\begin{bmatrix}\mathbf{D_A} \\ \hline \mathbf{D_R}\end{bmatrix}$.

Referring to Figs 6.24a and b,

$$\begin{array}{c} \\ \mathbf{T}_{\mathbf{D}}^{1} = \end{array} \begin{array}{c} \quad (1)\ (2)\ (3)\ (4)\ (5)\ (6)\quad (7)\ (8)\ (9)\ (10)\ (11)\ (12) \\ \left[\begin{array}{cccccc|cccccc} 0 & 1 & 0 & 0 & 0 & 0 & 0 & -1 & 0 & 0 & 0 & 0 \\ 1/4 & 0 & 0 & 0 & 0 & 0 & -1/4 & 0 & 1 & 0 & 0 & 0 \\ 1/4 & 0 & 1 & 0 & 0 & 0 & -1/4 & 0 & 0 & 0 & 0 & 0 \end{array}\right] \end{array}$$

$$\mathbf{T}_{\mathbf{D}}^{2} = \left[\begin{array}{cccccc|cccccc} -1 & 0 & 0 & 1 & 0 & 0 & 0 & 0 & 0 & 0 & 0 & 0 \\ 0 & 1/6 & 1 & 0 & -1/6 & 0 & 0 & 0 & 0 & 0 & 0 & 0 \\ 0 & 1/6 & 0 & 0 & -1/6 & 1 & 0 & 0 & 0 & 0 & 0 & 0 \end{array}\right]$$

$$\mathbf{T}_{\mathbf{D}}^{3} = \left[\begin{array}{cccccc|cccccc} 0 & 0 & 0 & 0 & 1 & 0 & 0 & 0 & 0 & 0 & -1 & 0 \\ 0 & 0 & 0 & 1/4 & 0 & 0 & 0 & 0 & 0 & -1/4 & 0 & 1 \\ 0 & 0 & 0 & 1/4 & 0 & 1 & 0 & 0 & 0 & -1/4 & 0 & 0 \end{array}\right]$$

Element and Structure Stiffness Matrices

The element stiffness matrices $\tilde{\mathbf{k}}_{*}^{1} = \tilde{\mathbf{k}}_{*}^{3}$ and $\tilde{\mathbf{k}}_{*}^{2}$, satisfying $\mathbf{F}_{*}^{i} = \tilde{\mathbf{k}}_{*}^{i}\mathbf{D}_{*}^{i}$, take the following form (refer Fig. 6.24b), considering

$$\frac{E_1A_1}{L_1} = \frac{E_2A_2}{L_2} = \frac{E_3A_3}{L_3} = 56250 \text{ kN/m}; \ \frac{E_1I_1}{L_1} = \frac{E_3I_3}{L_3} = 421.875 \text{ kNm and } \frac{E_2I_2}{L_2} = 949.21875 \text{ kNm}.$$

$$\tilde{\mathbf{k}}_{*}^{1} = \tilde{\mathbf{k}}_{*}^{3} = \begin{bmatrix} 562500 & 0 & 0 \\ 0 & 16875 & 8437.5 \\ 0 & 8437.5 & 16875 \end{bmatrix} \quad \text{and} \quad \tilde{\mathbf{k}}_{*}^{2} = \begin{bmatrix} 562500 & 0 & 0 \\ 0 & 37968.8 & 18984.4 \\ 0 & 18984.4 & 37968.8 \end{bmatrix}$$

$$\Rightarrow \quad \tilde{\mathbf{k}}_{*}^{1}\mathbf{T}_{\mathbf{D}}^{1} = \left[\begin{array}{cccccc|cccccc} 0 & 562500 & 0 & 0 & 0 & 0 & 0 & -562500 & 0 & 0 & 0 & 0 \\ 6328.1 & 0 & 8437.5 & 0 & 0 & 0 & -6328.1 & 0 & 16875 & 0 & 0 & 0 \\ 6328.1 & 0 & 16875 & 0 & 0 & 0 & -6328.1 & 0 & 8437.5 & 0 & 0 & 0 \end{array}\right]$$

$$\tilde{\mathbf{k}}_{*}^{2}\mathbf{T}_{\mathbf{D}}^{2} = \left[\begin{array}{cccccc|cccccc} -562500 & 0 & 0 & 562500 & 0 & 0 & 0 & 0 & 0 & 0 & 0 & 0 \\ 0 & 9492.2 & 37968.8 & 0 & -9492.2 & 18984.4 & 0 & 0 & 0 & 0 & 0 & 0 \\ 0 & 9492.2 & 18984.4 & 0 & -9492.2 & 37968.8 & 0 & 0 & 0 & 0 & 0 & 0 \end{array}\right]$$

$$\tilde{\mathbf{k}}_{*}^{3}\mathbf{T}_{\mathbf{D}}^{3} = \left[\begin{array}{cccccc|cccccc} 0 & 0 & 0 & 0 & 562500 & 0 & 0 & 0 & 0 & 0 & -562500 & 0 \\ 0 & 0 & 0 & 6328.1 & 0 & 8437.5 & 0 & 0 & 0 & -6328.1 & 0 & 16875 \\ 0 & 0 & 0 & 6328.1 & 0 & 16875 & 0 & 0 & 0 & -6328.1 & 0 & 8437.5 \end{array}\right]$$

The structure stiffness matrix $\mathbf{k} = \sum_{i=1}^{3} \mathbf{T}_{\mathbf{D}}^{i\,T}\tilde{\mathbf{k}}_{*}^{i}\mathbf{T}_{\mathbf{D}}^{i} = \left[\begin{array}{c|c} \mathbf{k}_{AA} & \mathbf{k}_{AR} \\ \hline \mathbf{k}_{RA} & \mathbf{k}_{RR} \end{array}\right]$ can now be assembled.

They are found to be identical to the ones generated in Example 6.2 (where we used a six degree of freedom frame element model).

Displacements and Support Reactions

The structure stiffness relation to be considered here is:

$$\begin{bmatrix} \mathbf{F_A} \\ \mathbf{F_R} \end{bmatrix} - \begin{bmatrix} \mathbf{F_{fA}} \\ \mathbf{F_{fR}} = \mathbf{O} \end{bmatrix} = \begin{bmatrix} \mathbf{k_{AA}} & \mathbf{k_{AR}} \\ \mathbf{k_{RA}} & \mathbf{k_{RR}} \end{bmatrix} \begin{bmatrix} \mathbf{D_A} \\ \mathbf{D_R} \end{bmatrix}$$

$$\Rightarrow \mathbf{D_A} = [\mathbf{k_{AA}}]^{-1}\left[(\mathbf{F_A} - \mathbf{F_{fA}}) - \mathbf{k_{AR}}\mathbf{D_R}\right]$$

$$\mathbf{F_R} = \mathbf{k_{RA}}\mathbf{D_A} + \mathbf{k_{RR}}\mathbf{D_R}$$

Substituting the various matrices and solving using MATLAB, we get exactly the same solution for the response as in Example 6.2.

Member Forces

$$\mathbf{F_*^i} = \mathbf{F_{*f}^i} + \tilde{\mathbf{k}}_*^i \mathbf{T_D^i} \mathbf{D}$$

$$\Rightarrow \mathbf{F_*^1} = \begin{Bmatrix} F_{1*}^1 \\ F_{2*}^1 \\ F_{3*}^1 \end{Bmatrix} = \begin{Bmatrix} -53.840 \text{ kN} \\ 45.988 \text{ kNm} \\ 7.218 \text{ kNm} \end{Bmatrix}; \ \mathbf{F_*^2} = \begin{Bmatrix} -36.698 \text{ kN} \\ -7.218 \text{ kNm} \\ -69.744 \text{ kNm} \end{Bmatrix}; \ \mathbf{F_*^3} = \begin{Bmatrix} -46.160 \text{ kN} \\ 77.045 \text{ kNm} \\ 69.744 \text{ kNm} \end{Bmatrix}$$

The member end moments are exactly as obtained earlier. The remaining member end forces are not directly obtainable (as in Example 6.2), but can be easily computed from the free-bodies (shown in Fig. 6.11a), applying equilibrium equations. The deflection and bending moment diagrams are as shown in Figs 6.11b and c.

<u>Simplified Analysis: Ignoring Axial Deformations</u>

If axial deformations are ignored, the kinematic indeterminacy reduces from 6 to 3, as only one translational (sway) degree of freedom of freedom is involved. The three active global coordinates are shown in Fig. 6.25a; restrained coordinates are not shown, as they cannot be included in this simplified analysis. The frame elements reduce to beam elements (with only two degrees of freedom), as shown in Fig. 6.25b.

The fixed end forces in the beam BC (element '2'), due to the intermediate loading as well as the support settlement (chord rotation, $\phi_2 = -\frac{0.010}{\cdot 6}$) are computed as follows (refer Fig. 6.25c), as done in the Slope Deflection or Moment Distribution method.

$$F_{1*f}^2 = \frac{100(4)^2(2)}{(6)^2} + 6(9492.1875)\left(\frac{0.010}{6}\right) = 88.889 + 94.922 = 183.811 \text{ kNm};$$

$$F_{2*f}^2 = -\frac{100(2)^2(4)}{(6)^2} + 94.922 = -44.444 + 94.922 = 50.478 \text{ kNm};$$

$$\Rightarrow \mathbf{F_{*f}^1} = \mathbf{F_{*f}^3} = \begin{Bmatrix} 0 \text{ kNm} \\ 0 \text{ kNm} \end{Bmatrix}; \quad \mathbf{F_{*f}^2} = \begin{Bmatrix} 183.811 \text{ kNm} \\ 50.478 \text{ kNm} \end{Bmatrix}$$

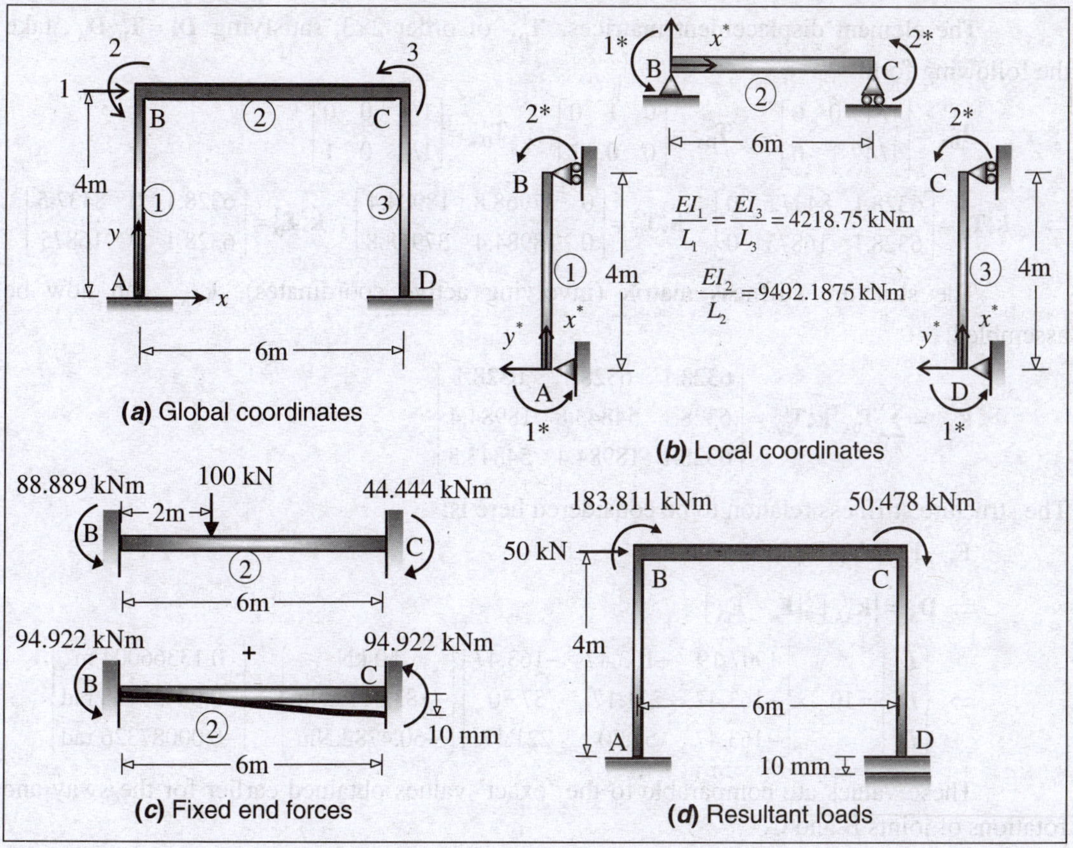

Figure 6.25 Global and local coordinates, fixed end forces and resultant loads, ignoring axial deformations (Example 6.6)

By inspection, summing up the fixed end forces at the three active global coordinate locations (refer Fig. 6.25a), we obtain the fixed end force vector, $\mathbf{F}_{\mathrm{fA}}$, and resultant direct load vector, $\mathbf{F}_{\mathrm{A}} - \mathbf{F}_{\mathrm{fA}}$, as follows (refer Fig. 6.25d):

$$\mathbf{F}_{\mathrm{A}} - \mathbf{F}_{\mathrm{fA}} = \begin{Bmatrix} 50\text{ kN} \\ 0\text{ kNm} \\ 0\text{ kNm} \end{Bmatrix} - \begin{Bmatrix} 0\text{ kN} \\ 183.811\text{ kNm} \\ 50.478\text{ kNm} \end{Bmatrix} = \begin{Bmatrix} 50\text{ kN} \\ -183.811\text{ kNm} \\ -50.478\text{ kNm} \end{Bmatrix}$$

The 2×2 element stiffness matrices, $\tilde{\mathbf{k}}_*^i$, is easily obtainable by eliminating the first row (corresponding to the axial degree of freedom) in the 3×3 $\tilde{\mathbf{k}}_*^i$ matrix derived earlier.

$$\tilde{\mathbf{k}}_*^1 = \tilde{\mathbf{k}}_*^3 = \begin{bmatrix} 16875 & 8437.5 \\ 8437.5 & 16875 \end{bmatrix} \quad ; \quad \tilde{\mathbf{k}}_*^2 = \begin{bmatrix} 37968.8 & 18984.4 \\ 18984.4 & 37968.8 \end{bmatrix}$$

The element displacement matrices, $\mathbf{T}_{DA}^{i}$ of order 2×3, satisfying $\mathbf{D}_{*}^{i} = \mathbf{T}_{DA}^{i}\mathbf{D}_{A}$, take the following form:

$$\mathbf{T}_{DA}^{1} = \begin{bmatrix} 1/4 & 0 & 0 \\ 1/4 & 1 & 0 \end{bmatrix} \; ; \; \mathbf{T}_{DA}^{2} = \begin{bmatrix} 0 & 1 & 0 \\ 0 & 0 & 1 \end{bmatrix} \; ; \; \mathbf{T}_{DA}^{3} = \begin{bmatrix} 1/4 & 0 & 0 \\ 1/4 & 0 & 1 \end{bmatrix}$$

$$\Rightarrow \tilde{\mathbf{k}}_{*}^{1}\mathbf{T}_{D}^{1} = \begin{bmatrix} 6328.1 & 8437.5 & 0 \\ 6328.1 & 16875 & 0 \end{bmatrix}; \; \tilde{\mathbf{k}}_{*}^{2}\mathbf{T}_{D}^{2} = \begin{bmatrix} 0 & 37968.8 & 18984.4 \\ 0 & 18984.4 & 37968.8 \end{bmatrix}; \; \tilde{\mathbf{k}}_{*}^{3}\mathbf{T}_{D}^{3} = \begin{bmatrix} 6328.1 & 0 & 8437.5 \\ 6328.1 & 0 & 16875 \end{bmatrix}$$

The structure stiffness matrix (involving active coordinates), $\mathbf{k}_{AA}$, can now be assembled:

$$\mathbf{k}_{AA} = \sum_{i=1}^{3} \mathbf{T}_{DA}^{i\;T}\tilde{\mathbf{k}}_{*}^{i}\mathbf{T}_{DA}^{i} = \begin{bmatrix} 6328.1 & 6328.1 & 6328.1 \\ 6328.1 & 54843.8 & 18984.4 \\ 6328.1 & 18984.4 & 54843.8 \end{bmatrix}$$

The structure stiffness relation to be considered here is:

$$\mathbf{F}_{A} - \mathbf{F}_{fA} = \mathbf{k}_{AA}\mathbf{D}_{A}$$

$$\Rightarrow \mathbf{D}_{A} = [\mathbf{k}_{AA}]^{-1}[\mathbf{F}_{A} - \mathbf{F}_{fA}]$$

$$\Rightarrow \begin{Bmatrix} D_1 \\ D_2 \\ D_3 \end{Bmatrix} = 10^{-7} \times \begin{bmatrix} 1907.19 & -163.47 & -163.47 \\ -163.47 & 221.17 & 57.70 \\ -163.47 & 57.70 & 221.17 \end{bmatrix} \begin{Bmatrix} 50 \text{ kN} \\ -183.811 \text{ kNm} \\ -50.478 \text{ kNm} \end{Bmatrix} = \begin{Bmatrix} 0.13366004 \text{ m} \\ -0.00459148 \text{ rad} \\ -0.00087326 \text{ rad} \end{Bmatrix}$$

These values are comparable to the 'exact' values obtained earlier for the sway and rotations of joints B and C.

The member end forces can now be generated.

$$\mathbf{F}_{*}^{i} = \mathbf{F}_{*f}^{i} + \tilde{\mathbf{k}}_{*}^{i}\mathbf{T}_{D}^{i}\mathbf{D}$$

$$\mathbf{F}_{*}^{1} = \begin{Bmatrix} F_{1*}^{1} \\ F_{2*}^{1} \end{Bmatrix} = \begin{Bmatrix} 0 \text{ kNm} \\ 0 \text{ kNm} \end{Bmatrix} + \begin{Bmatrix} 45.841 \text{ kNm} \\ 7.100 \text{ kNm} \end{Bmatrix} = \begin{Bmatrix} 45.841 \text{ kNm} \\ 7.100 \text{ kNm} \end{Bmatrix}$$

$$\mathbf{F}_{*}^{2} = \begin{Bmatrix} F_{1*}^{2} \\ F_{2*}^{2} \end{Bmatrix} = \begin{Bmatrix} 183.811 \text{ kNm} \\ 50.478 \text{ kNm} \end{Bmatrix} + \begin{Bmatrix} -190.911 \text{ kNm} \\ -120.323 \text{ kNm} \end{Bmatrix} = \begin{Bmatrix} -7.100 \text{ kNm} \\ -69.845 \text{ kNm} \end{Bmatrix}$$

$$\mathbf{F}_{*}^{3} = \begin{Bmatrix} F_{1*}^{3} \\ F_{2*}^{3} \end{Bmatrix} = \begin{Bmatrix} 0 \text{ kNm} \\ 0 \text{ kNm} \end{Bmatrix} + \begin{Bmatrix} 77.213 \text{ kNm} \\ 69.845 \text{ kNm} \end{Bmatrix} = \begin{Bmatrix} 77.213 \text{ kNm} \\ 69.845 \text{ kNm} \end{Bmatrix}$$

The deflection and bending moment diagrams are as shown in Figs 6.26b and c. The results obtained are very close to the 'exact' solution given in Fig. 6.11. Thus, it is seen in this Example, that the error involved in ignoring axial deformations is highly negligible (within 1 percent).

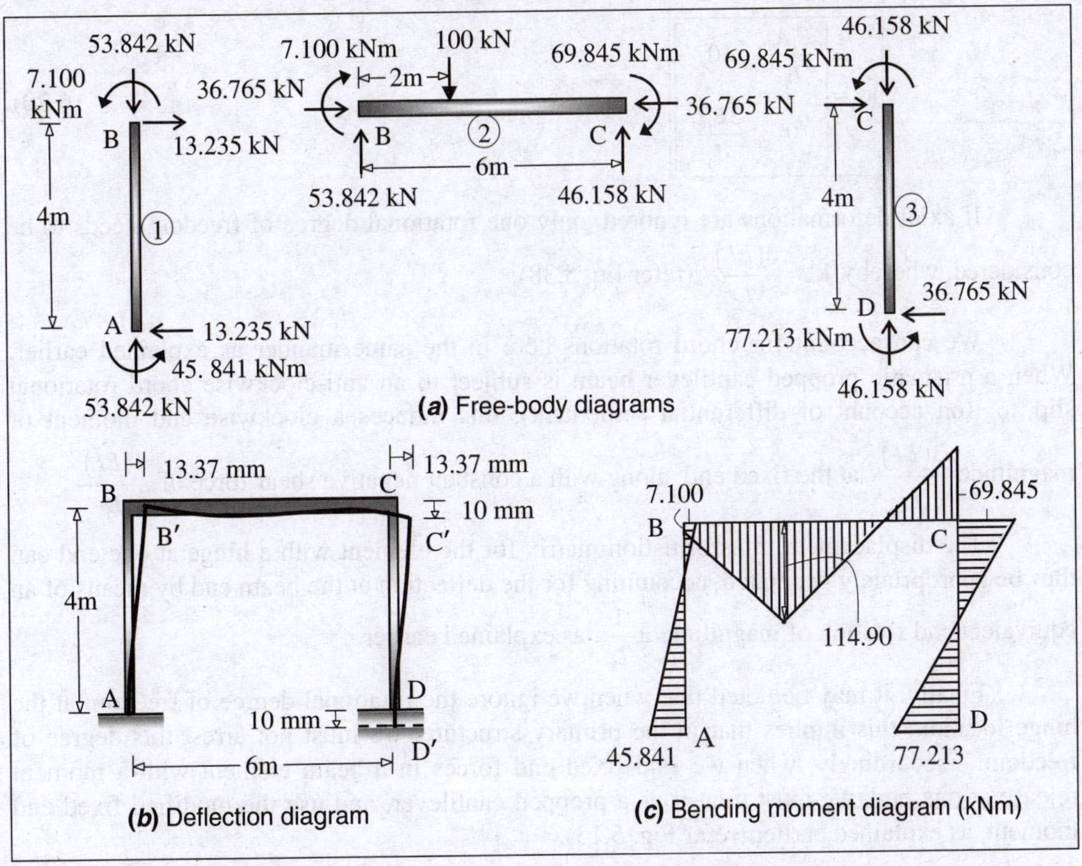

Figure 6.26 Complete response: ignoring axial deformations (Example 6.6)

6.3.3 Dealing with Moment Releases (Hinges)

We can take advantage of a reduction in the degree of kinematic indeterminacy when we encounter a "moment release" at a hinge (hinged end support or internal hinge). In such cases, we can simply ignore the degree of freedom associated with the member end release (i.e., eliminate the rotational degree of freedom). But we need to also modify the element stiffness matrix for this element as well as the fixed end forces, as explained in Section 5.3.2.

If we ignore the rotation at the hinged end, the resulting reduced element stiffness matrix, $\tilde{\mathbf{k}}_*^i$, takes the following form:

$$\boxed{\tilde{\mathbf{k}}_*^{\mathbf{i}} = \begin{bmatrix} \dfrac{E_i A_i}{L_i} & 0 \\ 0 & \dfrac{3E_i I_i}{L_i} \end{bmatrix}} \tag{6.20}$$

If axial deformations are ignored, only one rotational degree of freedom needs to be considered, whereby $\tilde{\mathbf{k}}_*^{\mathbf{i}} = \dfrac{3(EI)_i}{L_i}$ (refer Eq. 5.38).

We can account for chord rotations here in the same manner as explained earlier. When a prismatic propped cantilever beam is subject to an anti-clockwise chord rotational slip, ϕ_i (on account of differential settlement), this induces a clockwise end moment of magnitude $\dfrac{3(EI)_i}{L_i}$ at the fixed end, along with a constant negative shear force of $\dfrac{3(EI)_i}{L_i^2}$.

The displacement transformation matrix for the element with a hinge at one end can thus be appropriately generated, accounting for the deflection at the beam end by means of an equivalent end rotation of magnitude $\pm\dfrac{1}{L_i}$, as explained earlier.

Finally, it may be noted that when we ignore the rotational degree of freedom at the hinge location, this implies that in the primary structure, we must not arrest this degree of freedom. Accordingly, when we find fixed end forces in a beam element with a moment release at one end, we must treat it as a propped cantilever, and use the modified fixed end moment, as explained earlier (refer Fig. 5.13).

EXAMPLE 6.7

Repeat Example 6.3 by the reduced stiffness formulation. Ignore axial deformations.

SOLUTION

Ignoring the rotational degree of freedom at the internal hinge location at C, and ignoring axial deformations, the kinematic indeterminacy reduces to two. The two active global coordinates are shown in Fig. 6.27a; restrained coordinates are not shown, as they cannot be included in this simplified analysis. The frame elements reduce to beam elements (two degrees of freedom in element '1' and one degree of freedom in elements '2' and '3'), as shown in Fig. 6.27b.

The fixed end forces in the propped cantilever beam BC (element '2'), due to the intermediate loading as well as the support settlement (chord rotation, $\phi_2 = -\dfrac{0.010}{6}$) are computed as follows (refer Fig. 6.27c).

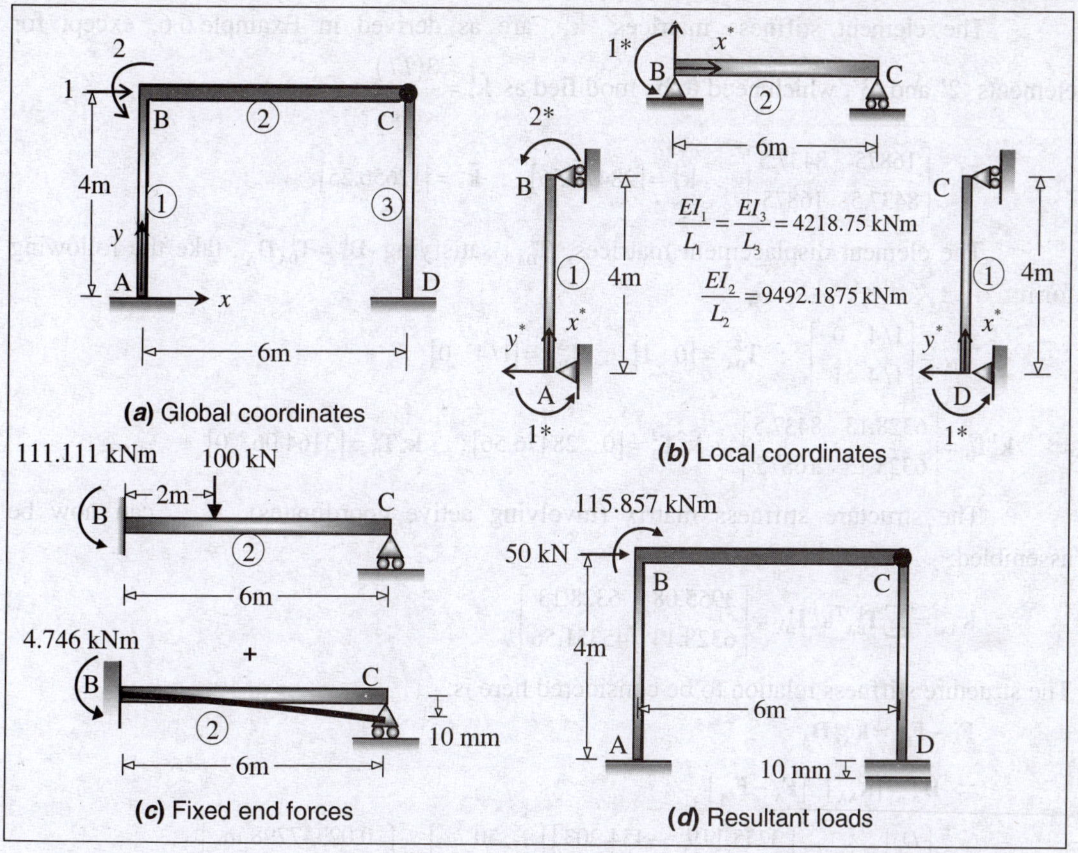

Figure 6.27 Global and local coordinates, fixed end forces and resultant loads, ignoring axial deformations (Example 6.7)

$$F_{1*f}^2 = \left[\frac{100(4)^2(2)}{(6)^2} + \frac{1}{2} \times \frac{100(2)^2(4)}{(6)^2}\right] + 3(9492.1875)\left(\frac{0.010}{6}\right) = 111.111 + 47.461 = 158.572 \text{ kNm}$$

$$\Rightarrow \mathbf{F}_{*f}^1 = \begin{Bmatrix} 0 \text{ kNm} \\ 0 \text{ kNm} \end{Bmatrix} \quad ; \quad \mathbf{F}_{*f}^2 = \{158.572 \text{ kNm}\} \quad ; \quad \mathbf{F}_{*f}^3 = \{0 \text{ kNm}\}$$

By inspection, summing up the fixed end forces at the three active global coordinate locations (refer Fig. 6.25a), we obtain the fixed end force vector, $\mathbf{F}_{fA}$, and resultant direct load vector, $\mathbf{F}_A - \mathbf{F}_{fA}$, as follows (refer Fig. 6.25d):

$$\mathbf{F}_A - \mathbf{F}_{fA} = \begin{Bmatrix} 50 \text{ kN} \\ 0 \text{ kNm} \end{Bmatrix} - \begin{Bmatrix} 0 \text{ kN} \\ 158.572 \text{ kNm} \end{Bmatrix} = \begin{Bmatrix} 50 \text{ kN} \\ -158.572 \text{ kNm} \end{Bmatrix}$$

The element stiffness matrices, $\tilde{\mathbf{k}}_*^{\mathbf{i}}$, are as derived in Example 6.6, except for elements '2' and '3', which need to be modified as $\tilde{\mathbf{k}}_*^{\mathbf{i}} = \dfrac{3(EI)_i}{L_i}$:

$$\tilde{\mathbf{k}}_*^{\mathbf{1}} = \begin{bmatrix} 16875 & 8437.5 \\ 8437.5 & 16875 \end{bmatrix} \; ; \; \tilde{\mathbf{k}}_*^{2} = [28476.56] \; ; \; \tilde{\mathbf{k}}_*^{3} = [12656.25]$$

The element displacement matrices, $\mathbf{T}_{\mathbf{DA}}^{\mathbf{i}}$, satisfying $\mathbf{D}_*^{\mathbf{i}} = \mathbf{T}_{\mathbf{DA}}^{\mathbf{i}} \mathbf{D}_{\mathbf{A}}$, take the following form:

$$\mathbf{T}_{\mathbf{DA}}^{1} = \begin{bmatrix} 1/4 & 0 \\ 1/4 & 1 \end{bmatrix} \; ; \; \mathbf{T}_{\mathbf{DA}}^{2} = [0 \quad 1] \; ; \; \mathbf{T}_{\mathbf{DA}}^{3} = [1/4 \quad 0]$$

$$\Rightarrow \quad \tilde{\mathbf{k}}_*^{1}\mathbf{T}_{\mathbf{D}}^{1} = \begin{bmatrix} 6328.13 & 8437.5 \\ 6328.13 & 16875 \end{bmatrix}; \quad \tilde{\mathbf{k}}_*^{2}\mathbf{T}_{\mathbf{D}}^{2} = [0 \quad 28476.56]; \quad \tilde{\mathbf{k}}_*^{3}\mathbf{T}_{\mathbf{D}}^{3} = [3164.06 \quad 0]$$

The structure stiffness matrix (involving active coordinates), $\mathbf{k}_{\mathbf{AA}}$, can now be assembled:

$$\mathbf{k}_{\mathbf{AA}} = \sum_{i=1}^{3} \mathbf{T}_{\mathbf{DA}}^{\mathbf{i}\;\mathrm{T}} \tilde{\mathbf{k}}_*^{\mathbf{i}} \mathbf{T}_{\mathbf{DA}}^{\mathbf{i}} = \begin{bmatrix} 3955.08 & 6328.13 \\ 6328.13 & 45351.56 \end{bmatrix}$$

The structure stiffness relation to be considered here is:

$$\mathbf{F}_{\mathbf{A}} - \mathbf{F}_{\mathbf{fA}} = \mathbf{k}_{\mathbf{AA}} \mathbf{D}_{\mathbf{A}}$$

$$\Rightarrow \mathbf{D}_{\mathbf{A}} = [\mathbf{k}_{\mathbf{AA}}]^{-1} [\mathbf{F}_{\mathbf{A}} - \mathbf{F}_{\mathbf{fA}}]$$

$$\Rightarrow \begin{Bmatrix} D_1 \\ D_2 \end{Bmatrix} = 10^{-7} \times \begin{bmatrix} 3255.119 & -454.203 \\ -454.203 & 283.877 \end{bmatrix} \begin{Bmatrix} 50 \\ -158.572 \end{Bmatrix} = \begin{Bmatrix} 0.02347798 \text{ m} \\ -0.00677251 \text{ rad} \end{Bmatrix}$$

These values are comparable to the 'exact' values obtained earlier for the sway and rotations of joints B and C.

The member end forces can now be generated.

$$\mathbf{F}_*^{\mathbf{i}} = \mathbf{F}_{*\mathbf{f}}^{\mathbf{i}} + \tilde{\mathbf{k}}_*^{\mathbf{i}} \mathbf{T}_{\mathbf{D}}^{\mathbf{i}} \mathbf{D}$$

$$\mathbf{F}_*^{1} = \begin{Bmatrix} F_{1*}^1 \\ F_{2*}^1 \end{Bmatrix} = \begin{Bmatrix} 0 \text{ kNm} \\ 0 \text{ kNm} \end{Bmatrix} + \begin{Bmatrix} 91.429 \text{ kNm} \\ 34.286 \text{ kNm} \end{Bmatrix} = \begin{Bmatrix} 91.429 \text{ kNm} \\ 34.286 \text{ kNm} \end{Bmatrix}$$

$$\mathbf{F}_*^{2} = \{F_{1*}^2\} = \{158.572 \text{ kNm}\} + \{-192.858 \text{ kNm}\} = \{-34.286 \text{ kNm}\}$$

$$\mathbf{F}_*^{3} = \{F_{1*}^3\} = \{0 \text{ kNm}\} + \{74.286 \text{ kNm}\} = \{74.286 \text{ kNm}\}$$

The complete response (including the fixed end force effects), in the form of free-body, deflection and bending moment diagrams are shown in Fig. 6.28. The results match very closely with the 'exact' solution given in Fig. 6.14 (Example 6.3).

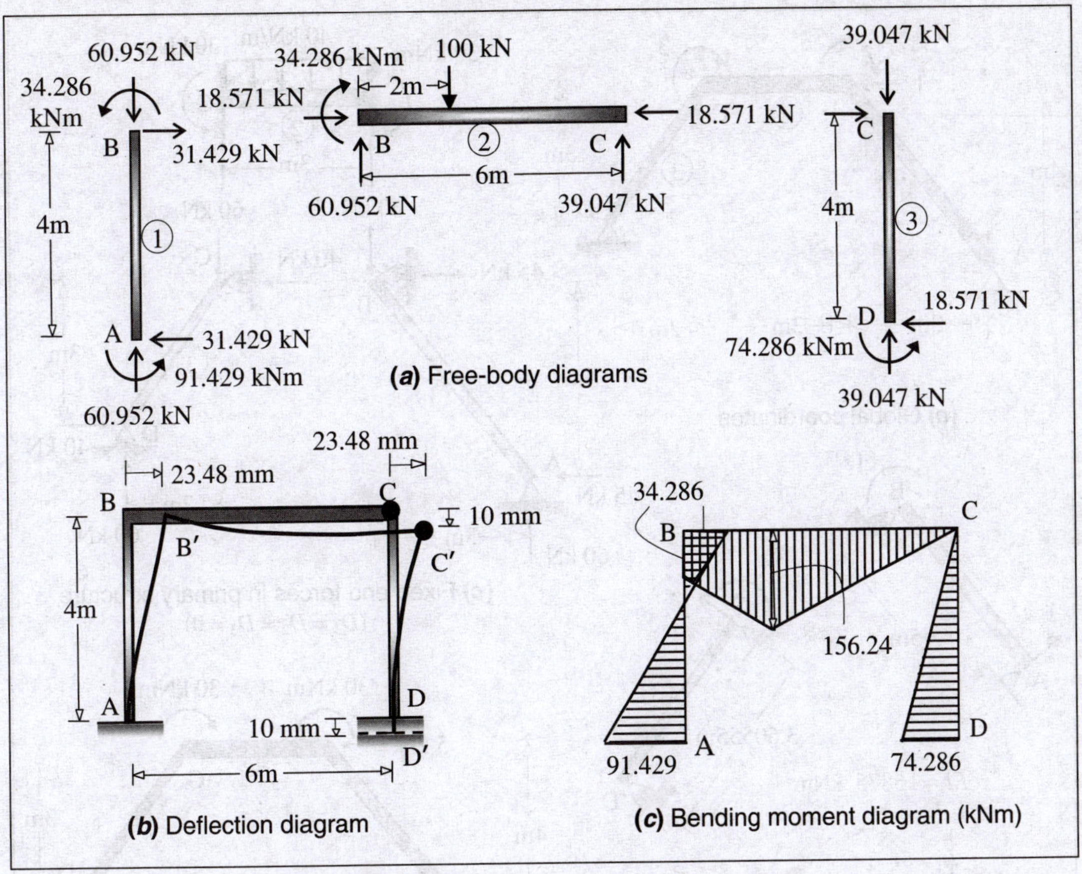

Figure 6.28 Complete response (Example 6.7)

EXAMPLE 6.8

Repeat Example 6.4 by the reduced stiffness method. Ignore axial deformations.

SOLUTION

Ignoring axial deformations and taking advantage of the hinged supports at A and D, the degree of kinematic indeterminacy can be reduced to three. The three active coordinates correspond to the displacements θ_B, θ_C, and Δ_{BC}, as shown in Fig. 6.29a. The local coordinates for the three elements (treated as beam elements) are shown in Fig. 6.29b.

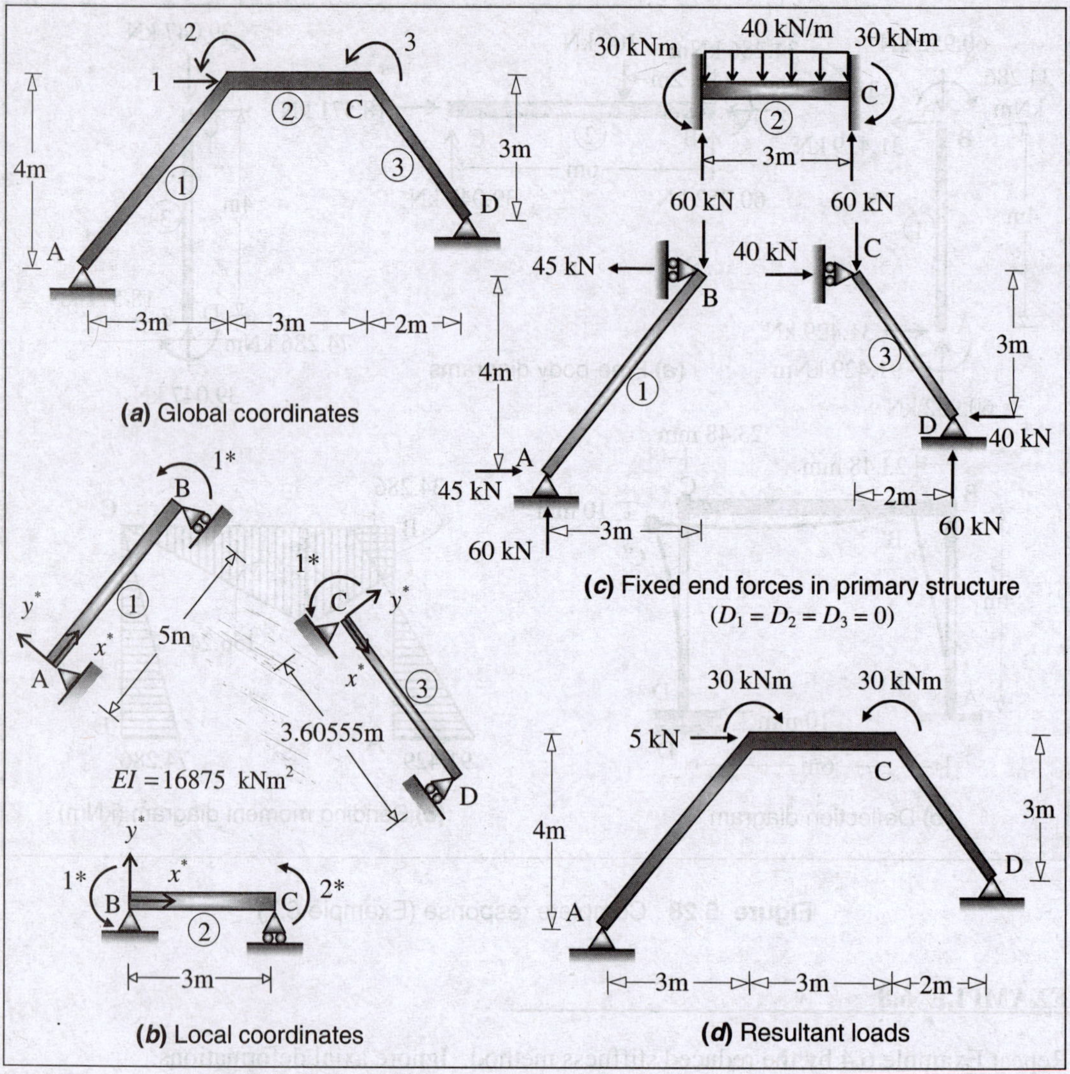

Figure 6.29 Global and local coordinates, fixed end forces and resultant loads, ignoring axial deformations (Example 6.8)

Fixed End Forces and Equivalent Joint Loads

There are no nodal loads prescribed ($\mathbf{F_A} = \mathbf{O}$). The distributed loading on element '2' (BC), however gives rise to fixed end forces on the primary structure (with $D_1 = D_2 = D_3 = 0$), as shown in Fig. 6.29c. Fixed end moments develop only in the element '2' (BC).

$$\Rightarrow \mathbf{F}^1_{*\mathbf{f}} = \{0 \text{ kNm}\} \quad ; \quad \mathbf{F}^2_{*\mathbf{f}} = \begin{Bmatrix} 30 \text{ kNm} \\ -30 \text{ kNm} \end{Bmatrix} \quad ; \quad \mathbf{F}^3_{*\mathbf{f}} = \{0 \text{ kNm}\}$$

The vertical reactions induced at B and C in the uniformly loaded beam element '2' are transmitted to the supports at A and D, inducing horizontal forces at the artificial restraint to BC, which can be computed by considering moment equilibrium of elements '1' and '3'.

<u>Element '2' (BC)</u>: Vertical reactions: $V_{BC} = V_{CB} = \frac{40(3)}{2} = 60 \text{ kN}(\uparrow)$ transmitted downward at B and C to elements '1' and '3' respectively (refer Fig. 6.29c).

<u>Element '1' (AB)</u>: Horizontal reaction: $H_B = \frac{60 \times 3}{4} = 45 \text{ kN } (\rightarrow)$

<u>Element '3' (CD)</u>: Horizontal reaction: $H_C = \frac{60 \times 2}{3} = 40 \text{ kN } (\leftarrow)$

By inspection, summing up the fixed end forces at the three active global coordinate locations, we obtain the fixed end force vector, $\mathbf{F_{fA}}$, and resultant direct load vector, $\mathbf{F_A} - \mathbf{F_{fA}}$, as follows (refer Fig. 6.29d):

$$\mathbf{F_A} - \mathbf{F_{fA}} = \begin{Bmatrix} 0 \text{ kN} \\ 0 \text{ kNm} \\ 0 \text{ kNm} \end{Bmatrix} - \begin{Bmatrix} (-45+40) \text{ kN} \\ 30 \text{ kNm} \\ -30 \text{ kNm} \end{Bmatrix} = \begin{Bmatrix} 5 \text{ kN} \\ -30 \text{ kNm} \\ 30 \text{ kNm} \end{Bmatrix}$$

The chord rotations for the three members can be expressed as $\phi_1 = -\frac{\delta}{h_1} = -\frac{D_1}{4}$, $\phi_3 = -\frac{\delta}{h_2} = -\frac{D_1}{3}$ and $\phi_3 = +\frac{\delta(\tan\alpha + \tan\beta)}{L_3} = +\frac{D_1(3/4 + 2/3)}{3} = +\frac{17}{36} D_1$ (refer Fig. 6.23c).

Accordingly, the element displacement matrices, $\mathbf{T_{DA}^i}$, satisfying $\mathbf{D_*^i} = \mathbf{T_{DA}^i D_A}$, take the following form:

$$\mathbf{T_{DA}^1} = [1/4 \quad 1 \quad 0] \; ; \; \mathbf{T_{DA}^2} = \begin{bmatrix} -0.47222 & 1 & 0 \\ -0.47222 & 0 & 1 \end{bmatrix} \; ; \; \mathbf{T_{DA}^3} = [1/3 \quad 0 \quad 1]$$

The element stiffness matrices, $\tilde{\mathbf{k}}_*^i$, can be generated, considering $\tilde{\mathbf{k}}_*^i = \frac{3(EI)_i}{L_i}$ for elements '1' and '3' (each having a single degree of freedom), and the conventional 2×2 stiffness matrix for element '2'.

$$(EI/L)_1 = 3375, \; (EI/L)_2 = 5625 \text{ and } (EI/L)_3 = 4680.2829$$

$$\Rightarrow \tilde{\mathbf{k}}_*^1 = [10125] \; ; \; \tilde{\mathbf{k}}_*^2 = \begin{bmatrix} 22500 & 11250 \\ 11250 & 22500 \end{bmatrix} ; \; \tilde{\mathbf{k}}_*^3 = [14040.85]$$

$$\Rightarrow \tilde{\mathbf{k}}_*^1 \mathbf{T_D^1} = [2531.25 \quad 10125 \quad 0]; \quad \tilde{\mathbf{k}}_*^3 \mathbf{T_D^3} = [4680.283 \quad 0 \quad 14040.849]$$

$$\tilde{\mathbf{k}}_*^2 \mathbf{T_D^2} = \begin{bmatrix} -15937.493 & 22500 & 11250 \\ -15937.493 & 11250 & 22500 \end{bmatrix}$$

The structure stiffness matrix (involving active coordinates), $\mathbf{k_{AA}}$, can now be assembled:

$$\mathbf{k_{AA}} = \sum_{i=1}^{3} \mathbf{T_{DA}^{i}}^{T}\tilde{\mathbf{k}}_{*}^{i}\mathbf{T_{DA}^{i}} = \begin{bmatrix} 17244.976 & -13406.243 & -11257.210 \\ -13406.243 & 32625 & 11250 \\ -11257.210 & 11250 & 36540.849 \end{bmatrix}$$

The structure stiffness relation to be considered here is:

$$\mathbf{F_A} - \mathbf{F_{fA}} = \mathbf{k_{AA}}\mathbf{D_A}$$

$$\Rightarrow \mathbf{D_A} = [\mathbf{k_{AA}}]^{-1}[\mathbf{F_A} - \mathbf{F_{fA}}]$$

$$\Rightarrow \begin{Bmatrix} D_1 \\ D_2 \\ D_3 \end{Bmatrix} = 10^{-7} \times \begin{bmatrix} 962.604 & 328.129 & 195.529 \\ 328.129 & 454.770 & -38.925 \\ 195.529 & -38.925 & 345.887 \end{bmatrix} \begin{Bmatrix} 5 \\ -30 \\ 30 \end{Bmatrix} = \begin{Bmatrix} 0.0000835015 \text{ m} \\ -0.001317021 \text{ rad} \\ 0.001252201 \text{ rad} \end{Bmatrix}$$

The member end forces can now be generated.

$$\mathbf{F_*^i} = \mathbf{F_{*f}^i} + \tilde{\mathbf{k}}_*^i \mathbf{T_D^i} \mathbf{D}$$

$$\mathbf{F_*^1} = \{F_{1*}^1\} = \{0 \text{ kNm}\} + \{-13.123 \text{ kNm}\} = \{-13.123 \text{ kNm}\}$$

$$\mathbf{F_*^2} = \begin{Bmatrix} F_{1*}^2 \\ F_{2*}^2 \end{Bmatrix} = \begin{Bmatrix} 30 \text{ kNm} \\ -30 \text{ kNm} \end{Bmatrix} + \begin{Bmatrix} -16.877 \text{ kNm} \\ 12.027 \text{ kNm} \end{Bmatrix} = \begin{Bmatrix} 13.123 \text{ kNm} \\ -17.973 \text{ kNm} \end{Bmatrix}$$

$$\mathbf{F_*^3} = \{F_{1*}^3\} = \{0 \text{ kNm}\} + \{17.923 \text{ kNm}\} = \{17.923 \text{ kNm}\}$$

The complete response (including the fixed end force effects), in the form of free-body, deflection and bending moment diagrams are shown in Fig. 6.30. The bending moments match closely with the 'exact' solution given in Fig. 6.18 (Example 6.4).

6.3.4 Dealing with Guided-Fixed End Supports

We can also take advantage of a reduction in the degree of kinematic indeterminacy when we encounter a guided-fixed end support, involving a "shear release", as explained in Section 5.3.3. We usually meet with this situation in the middle of a symmetric system of beams. At the guided-fixed end, there is usually no shear force, unless there is a concentrated external force (of known magnitude) applied here. Also, the bending moment at the guided-fixed support end can be determined by considering equilibrium of the free-body. In such cases, we can simply ignore the rotational degree of freedom in the beam element at the guided-fixed end. But we need to also modify the element stiffness matrix for this element as well as the fixed end forces, as explained earlier. The rotational stiffness is given by $(EI)_i/L_i$ (refer Eq. 5.41).

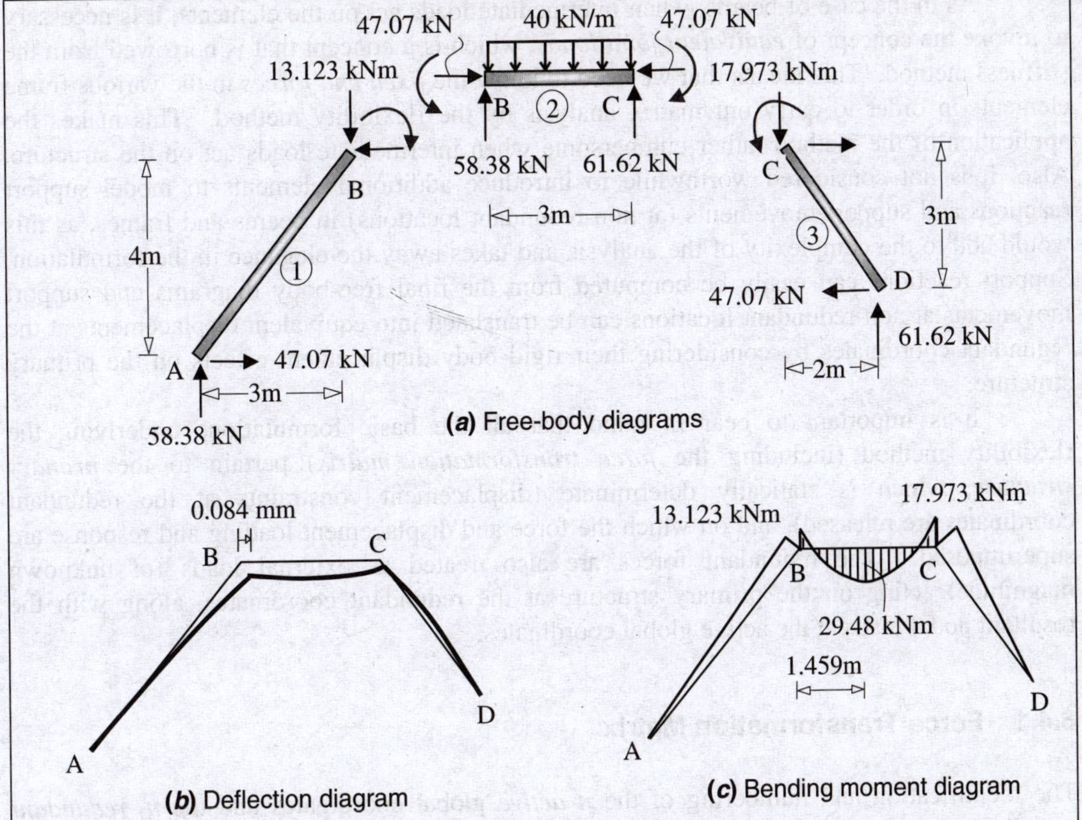

Figure 6.30 Complete response (Example 6.8)

Chord rotations in these beam elements do not have any associated flexural stiffnesses, and this fact should be accounted for, while assembling the displacement transformation matrix $\mathbf{T}_D^i$. In other words, the element in the $\mathbf{T}_D^i$ matrix, corresponding to a deflection in the structure (global coordinate), $D_j = 1$, will be zero (and not $\pm 1/L_i$, as in other cases). Finally, care should be taken while evaluating the fixed end force effects in a beam with a guided-fixed end support. A convenient way of dealing with this situation is by considering the beam to have twice the span and subject to symmetric loading (refer Example 5.9).

6.4 FLEXIBILITY METHOD APPLIED TO PLANE FRAMES

The basic procedure underlying the flexibility method has already been introduced in earlier Chapters. Here, we demonstrate the application of this method to plane frame systems.

As in the case of beams, when intermediate loads act on the elements, it is necessary to invoke the concept of *equivalent joint loads*, which is a concept that is borrowed from the stiffness method. This means that we need to know the *fixed end forces* in the various frame elements in order to carry out matrix analysis by the flexibility method. This makes the application of the method rather cumbersome when intermediate loads act on the structure. Also, it is not considered worthwhile to introduce additional elements to model support reactions and support movements (at non-redundant locations) in beams and frames, as this would add to the complexity of the analysis and takes away the elegance in the formulation. Support reactions can easily be computed from the final free-body diagrams and support movements at non-redundant locations can be translated into equivalent displacements at the redundant coordinates by considering their rigid body displacement effects on the primary structure.

It is important to bear in mind that all the basic formulations underlying the flexibility method (including the *force transformation matrix*) pertain to the *primary structure*, which is statically determinate (displacement constraints at the redundant coordinates are released), and on which the force and displacement loading and response are superimposed. The redundant forces are also treated as external loads (of unknown magnitude) acting on the primary structure at the redundant coordinates, along with the resultant nodal loads at the active global coordinates.

6.4.1 Force Transformation Matrix

The identification and numbering of the n *active* global coordinates and the n_s *redundant* coordinates in the primary structure can be sequentially carried out (with the redundant coordinate numbers at the end of the list). The local coordinates of the plane frame elements, corresponding to three degrees of freedom, are the same as in the reduced stiffness formulation (refer Fig. 6.31a). If axial deformations are ignored, the plane frame element reduces to a beam element with only two rotational degrees of freedom. Additional simplifications are possible when one end of the element is hinged.

Consider a typical plane frame system, with m elements and a degree of static indeterminacy, n_s. Let there be n active global coordinates. Each plane frame element i has a force vector, $\mathbf{F}_*^i$, comprising an axial force (tension positive) and two end moments (anti-clockwise positive). The *combined element force vector*, $\mathbf{F}_*$, contains all the m element vectors, and hence has a dimension, $3m$. The components of $\mathbf{F}_*$ define the internal forces in the system, knowing which the other unknown forces, including shear forces and support reactions, as well as the axial force, shear force and bending moment diagrams, can be determined, satisfying static equilibrium.

The *external* force vector comprises forces, $\mathbf{F}_\mathbf{A}$, acting at the active global coordinates, as well as the forces, $\mathbf{F}_\mathbf{X}$, acting at the redundant coordinates, and hence, has a dimension, $n+n_s$. After finding the fixed end forces, $\mathbf{F}_{*\mathbf{f}}$, in the various individual elements, under the action of intermediate loads, these are transferred as equivalent joint loads on the structure, and the *resultant force vector*, $\mathbf{F}_{net} = \begin{Bmatrix} \mathbf{F}_{\mathbf{A},net} \\ \hdashline \mathbf{F}_{\mathbf{X},net} \end{Bmatrix} = \begin{Bmatrix} \mathbf{F}_{\mathbf{A}} - \mathbf{F}_{\mathbf{fA}} \\ \hdashline \mathbf{F}_{\mathbf{X}} - \mathbf{F}_{\mathbf{fX}} \end{Bmatrix}$, is defined, in which the only unknown quantities are the redundant forces, $\mathbf{F}_{\mathbf{X}}$.

The internal force vector, $\mathbf{F}_*$, is linearly related to the resultant force vector, $\mathbf{F}_{net}$, by means of the force transformation matrix, $\mathbf{T}_{\mathbf{F}} = [\mathbf{T}_{\mathbf{FA}} \mid \mathbf{T}_{\mathbf{FX}}]$. The elements in this matrix will depend on the choice of the redundants (i.e., on the primary structure), and can be generated by applying a unit load, $F_j = 1$, at a time on the primary structure, from simple considerations of static equilibrium.

$$\mathbf{F}_* = \mathbf{F}_{*\mathbf{f}} + [\mathbf{T}_{\mathbf{FA}} \mid \mathbf{T}_{\mathbf{FX}}] \begin{Bmatrix} \mathbf{F}_{\mathbf{A},net} \\ \hdashline \mathbf{F}_{\mathbf{X},net} \end{Bmatrix} \tag{6.21}$$

The sub-matrix, $\mathbf{T}_{\mathbf{FA}}$, has an order, $3m \times n$, while, $\mathbf{T}_{\mathbf{FX}}$, has an order, $3m \times n_s$ when all the frame elements have three degrees of freedom each. As the primary structure is statically determinate, the full $\mathbf{T}_{\mathbf{F}} = [\mathbf{T}_{\mathbf{FA}} \mid \mathbf{T}_{\mathbf{FX}}]$ matrix will turn out to be a square matrix, of order, $3m \times 3m$; i.e., the condition, $n + n_s = 3m$, will be satisfied. In cases where we are not interested in finding the displacements at all active coordinates, we can limit the active coordinates to the load locations only.

6.4.2 Flexibility Matrix

The *element flexibility matrix*, $\mathbf{f}_*^{\mathbf{i}}$, for a prismatic plane frame element, with three degrees of freedom, satisfying the relation, $\mathbf{D}_*^{\mathbf{i}} = \mathbf{f}_*^{\mathbf{i}} \mathbf{F}_*^{\mathbf{i}}$, can be generated as the inverse of the stiffness $\tilde{\mathbf{k}}_*^{\mathbf{i}}$ of the reduced element stiffness matrix (refer Fig. 6.21). Alternatively, we can generate the matrix from first principles, applying a unit force $F_*^i = 1$ at a time, as shown in Fig. 6.31. The axial flexibility and flexural flexibility terms are assumed to be uncoupled from each other, as observed earlier.

$$\mathbf{f}_*^{\mathbf{i}} = \begin{bmatrix} \dfrac{L_i}{(EA)_i} & 0 & 0 \\ 0 & \dfrac{L_i}{3(EI)_i} & \dfrac{-L_i}{6(EI)_i} \\ 0 & \dfrac{-L_i}{6(EI)_i} & \dfrac{L_i}{3(EI)_i} \end{bmatrix} \tag{6.22}$$

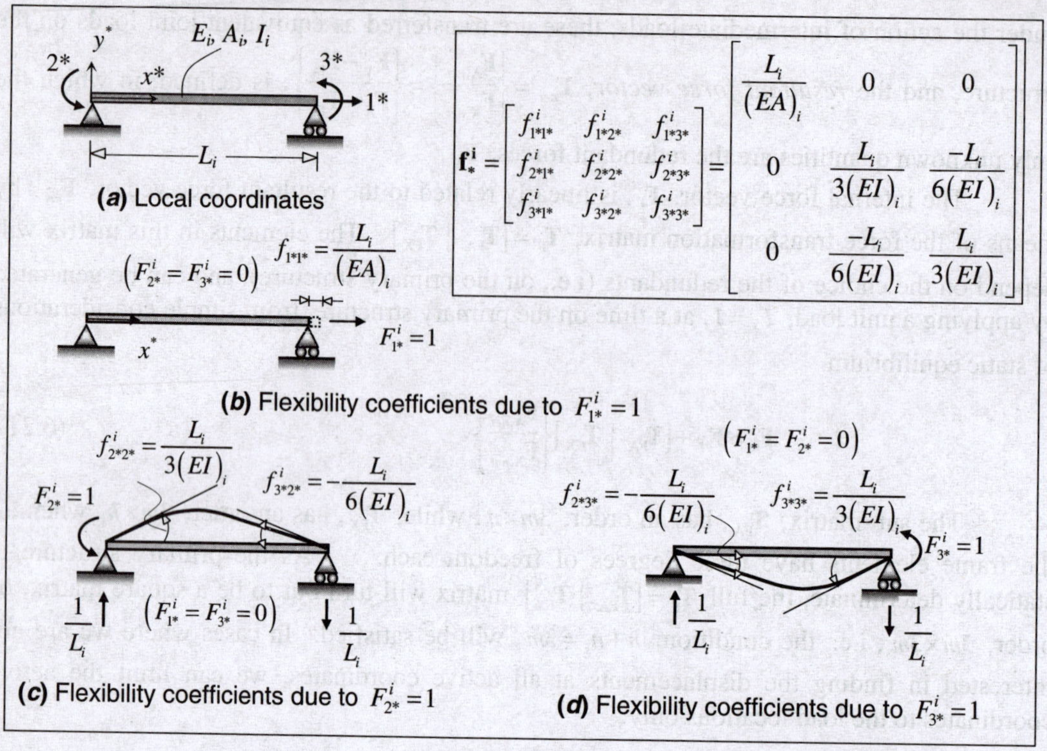

(*a*) Local coordinates

(*b*) Flexibility coefficients due to $F_{1*}^i = 1$

(*c*) Flexibility coefficients due to $F_{2*}^i = 1$

(*d*) Flexibility coefficients due to $F_{3*}^i = 1$

Figure 6.31 Generation of 3 × 3 flexibility matrix for a plane frame element

When one of the ends of the element is hinged, we can reduce the degrees of freedom from three to two, eliminating the rotational degree of freedom associated with the hinged end. The corresponding modified flexibility matrix takes the following form:

$$\tilde{\mathbf{f}}_*^{\mathbf{i}} = \begin{bmatrix} \frac{L_i}{(EA)_i} & 0 \\ 0 & \frac{L_i}{3(EI)_i} \end{bmatrix} \tag{6.22a}$$

If axial deformations are ignored, the above flexibility can further be reduced to a 1×1 order, with $\tilde{\mathbf{f}}_*^{\mathbf{i}} = \frac{L_i}{3(EI)_i}$. These simplifications are demonstrated in the Examples to follow.

The *combined element force vector*, $\{F_*\}$ and the conjugate *combined element displacement vector*, $\{D_*\}$, are inter-related by the ***unassembled element flexibility matrix***,

$\mathbf{f}_*$, as explained in Section 3.5. This matrix is a block diagonal matrix of order $3m\times3m$, for a continuous beam system with m prismatic elements.

$$\mathbf{D}_* = \mathbf{f}_*\mathbf{F}_* \Rightarrow \begin{Bmatrix} \mathbf{D}_*^1 \\ \mathbf{D}_*^2 \\ \vdots \\ \mathbf{D}_*^m \end{Bmatrix} = \begin{bmatrix} \mathbf{f}_*^1 & \mathbf{O} & \mathbf{O} & \mathbf{O} \\ \mathbf{O} & \mathbf{f}_*^2 & \mathbf{O} & \mathbf{O} \\ \vdots & \vdots & \ddots & \vdots \\ \mathbf{O} & \mathbf{O} & \mathbf{O} & \mathbf{f}_*^m \end{bmatrix} \begin{Bmatrix} \mathbf{F}_*^1 \\ \mathbf{F}_*^2 \\ \vdots \\ \mathbf{F}_*^m \end{Bmatrix} \tag{6.23}$$

The full structure flexibility matrix $\mathbf{f}$ (of order $3m\times3m$ when there are 3 degrees of freedom for each frame element), can be generated from the unassembled element flexibility matrix, $\mathbf{f}_*$, by involving the force transformation matrix, $\mathbf{T}_\mathbf{F}$, as described earlier.

$$\mathbf{f} = \mathbf{T}_\mathbf{F}{}^\mathbf{T}\mathbf{f}_*\mathbf{T}_\mathbf{F} = \left[\begin{array}{c|c} \mathbf{f}_{\mathbf{AA}} & \mathbf{f}_{\mathbf{AX}} \\ \hline \mathbf{f}_{\mathbf{XA}} & \mathbf{f}_{\mathbf{XX}} \end{array}\right] \tag{6.24}$$

where, the complete force displacement relations are as described below:

$$\left[\begin{array}{c} \mathbf{D}_\mathbf{A} \\ \hline \mathbf{D}_\mathbf{X} \end{array}\right] = \left[\begin{array}{c} \mathbf{D}_{\mathbf{A,initial}} \\ \hline \mathbf{D}_{\mathbf{X,initial}} \end{array}\right] + \left[\begin{array}{c|c} \mathbf{f}_{\mathbf{AA}} & \mathbf{f}_{\mathbf{AX}} \\ \hline \mathbf{f}_{\mathbf{XA}} & \mathbf{f}_{\mathbf{XX}} \end{array}\right] \left[\begin{array}{c} \mathbf{F}_\mathbf{A} - \mathbf{F}_{\mathbf{fA}} \\ \hline \mathbf{F}_\mathbf{X} - \mathbf{F}_{\mathbf{fX}} \end{array}\right] \tag{6.25}$$

The redundants can be determined by solving the second set of (compatibility) equations, pertaining to $\mathbf{D}_\mathbf{X}$:

$$\mathbf{F}_\mathbf{X} = [\mathbf{f}_{\mathbf{XX}}]^{-1}\left[\left(\mathbf{D}_\mathbf{X} - \mathbf{D}_{\mathbf{X,initial}}\right) - \mathbf{f}_{\mathbf{XA}}\left(\mathbf{F}_\mathbf{A} - \mathbf{F}_{\mathbf{fA}}\right)\right] + \mathbf{F}_{\mathbf{fX}} \tag{6.26}$$

After finding the redundants, $\mathbf{F}_\mathbf{X}$, the internal force vector, $\mathbf{F}_*$, can be determined by invoking Eq. 6.21, and from the free-bodies, the desired support reactions, shear force and bending moment diagrams can be generated.

Finally, the unknown displacements at the active coordinates, $\mathbf{D}_\mathbf{A}$, can be obtained, if required, by solving the first set of equations in Eq. 6.25:

$$\mathbf{D}_\mathbf{A} = \mathbf{D}_{\mathbf{A,initial}} + \mathbf{f}_{\mathbf{AA}}\mathbf{F}_{\mathbf{A},net} + \mathbf{f}_{\mathbf{AX}}\mathbf{F}_{\mathbf{X},net} \tag{6.27}$$

The steps involved in the analysis procedure are as outlined in Section 5.4.3. The procedure is demonstrated using some of the Examples solved earlier using the Stiffness Method.

EXAMPLE 6.9

Repeat Example 6.3 (portal frame with an internal hinge) by the Flexibility Method and find the complete response.

SOLUTION

The frame is statically indeterminate to the second degree. Let us choose the vertical reaction, V_D, and moment M_D, at the fixed support D as the redundants, X_1 (acting upward) and X_2 (anti-clockwise) respectively, as shown in Fig. 6.32a. The support at D is reduced to a roller support (allowing vertical deflection) in the primary structure, as shown in Fig. 6.32b, in which the global coordinates are also indicated.

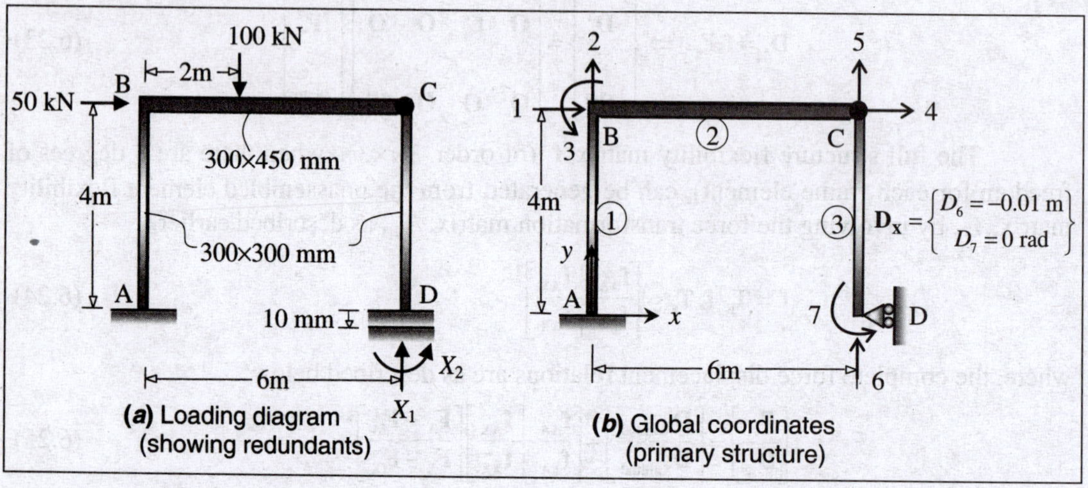

Figure 6.32 Redundants and global coordinates (Example 6.9)

There are three prismatic frame elements, whose local coordinates (with three degrees of freedom each) are marked in Fig. 6.33a.

Fixed End Forces

There is a nodal load specified, $F_1 = +50$ kN. The intermediate loading on element '2' needs to be converted to equivalent joint loads. The fixed end force effects, as calculated earlier in Example 6.3, can be used here. These are marked in Fig. 6.33b. Applying the Principle of Superposition, the given loading diagram is visualised as a superposition of the fixed end forces in Fig. 6.33b and the resultant nodal loads on the primary structure shown in Fig. 6.33c, including the displacement loading, given by the compatibility requirements associated with $\mathbf{D_X} = \begin{Bmatrix} D_6 = -0.01 \text{ m} \\ D_7 = 0 \text{ rad} \end{Bmatrix}$.

The combined element fixed end force vector, $\mathbf{F_{*f}}$, and the associated structure-level fixed end force vector, $\mathbf{F_{*f}}$, and the resultant load vector, $\mathbf{F}_{net}$, on the primary structure (shown in Fig. 6.33c) are obtained as:

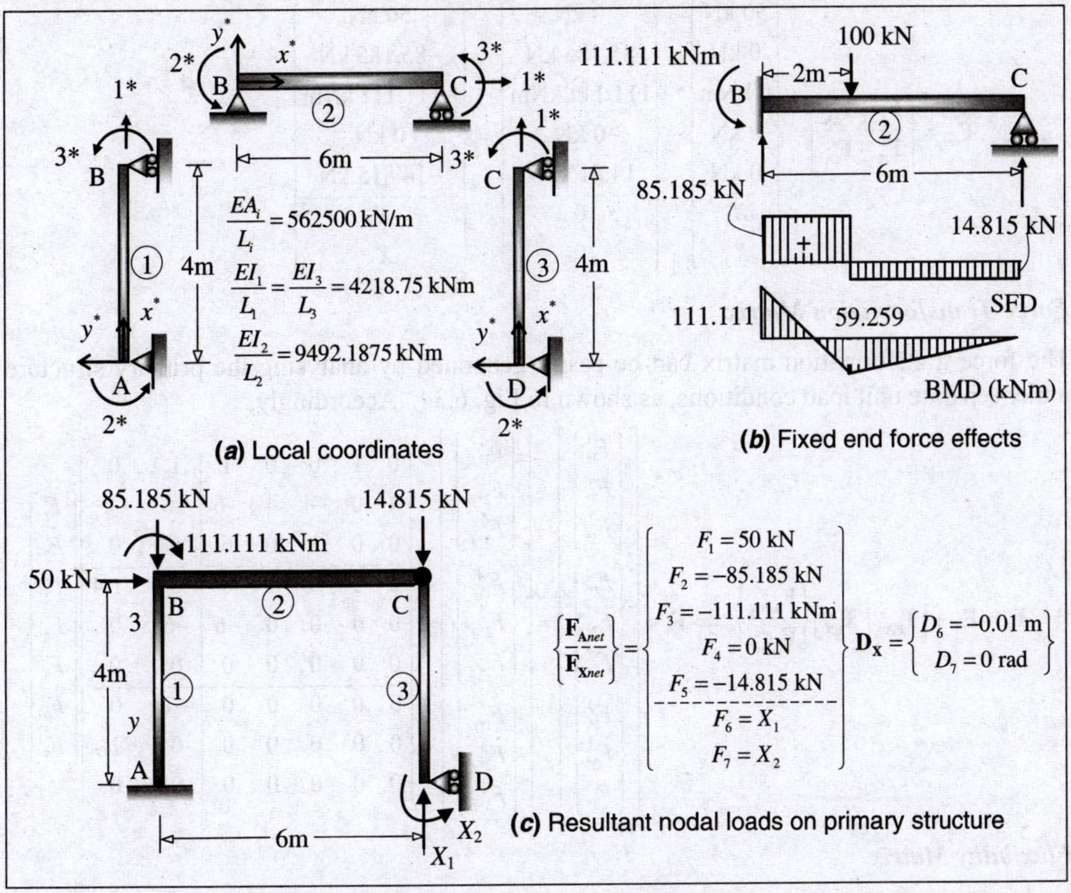

Figure 6.33 Local coordinates, fixed end forces and resultant loads (Example 5.10)

$$\mathbf{F_{*f}} = \begin{bmatrix} \begin{Bmatrix} F^1_{1*f} \\ F^1_{2*f} \\ F^1_{3*f} \end{Bmatrix} \\ \begin{Bmatrix} F^2_{1*f} \\ F^2_{2*f} \\ F^2_{3*f} \end{Bmatrix} \\ \begin{Bmatrix} F^3_{1*f} \\ F^3_{2*f} \\ F^3_{3*f} \end{Bmatrix} \end{bmatrix} = \begin{bmatrix} \begin{Bmatrix} 0 \text{ kN} \\ 0 \text{ kNm} \\ 0 \text{ kNm} \end{Bmatrix} \\ \begin{Bmatrix} 0 \text{ kN} \\ 111.111 \text{ kNm} \\ 0 \text{ kNm} \end{Bmatrix} \\ \begin{Bmatrix} 0 \text{ kN} \\ 0 \text{ kNm} \\ 0 \text{ kNm} \end{Bmatrix} \end{bmatrix} \Rightarrow \mathbf{F_f} = \begin{Bmatrix} \mathbf{F_{fA}} \\ \hline \mathbf{F_{fX}} \end{Bmatrix} = \begin{Bmatrix} F_{1f} = 0 \text{ kN} \\ F_{2f} = 85.185 \text{ kN} \\ F_{3f} = 111.111 \text{ kNm} \\ F_{4f} = 0 \text{ kN} \\ F_{5f} = 14.815 \text{ kN} \\ \hline F_{6f} = 0 \\ F_{7f} = 0 \end{Bmatrix}$$

$$\mathbf{F}_{net} = \left[\begin{array}{c}\mathbf{F_A} - \mathbf{F_{fA}} \\ \hline \mathbf{F_X} - \mathbf{F_{fX}}\end{array}\right] = \left\{\begin{array}{c} 50\text{ kN} \\ 0\text{ kN} \\ 0\text{ kNm} \\ 0\text{ kN} \\ 0\text{ kN} \\ \hline X_1 \\ X_2 \end{array}\right\} - \left\{\begin{array}{c} 0\text{ kN} \\ 85.185\text{ kN} \\ 111.111\text{ kNm} \\ 0\text{ kN} \\ 14.815\text{ kN} \\ \hline 0 \\ 0 \end{array}\right\} = \left\{\begin{array}{c} 50\text{ kN} \\ -85.185\text{ kN} \\ -111.111\text{ kNm} \\ 0\text{ kN} \\ -14.815\text{ kN} \\ \hline X_1 \\ X_2 \end{array}\right\}$$

Force Transformation Matrix

The force transformation matrix can be easily generated by analysing the primary structure under separate unit load conditions, as shown in Fig. 6.34. Accordingly,

$$\mathbf{F}_* = \mathbf{F}_{*f} + \left[\mathbf{T_{FA}} \mid \mathbf{T_{FX}}\right]\left\{\begin{array}{c}\mathbf{F}_{A,net} \\ \hline \mathbf{F}_{X,net}\end{array}\right\} \Rightarrow \mathbf{F}_* = \begin{bmatrix} \begin{Bmatrix} F_{1*}^1 \\ F_{2*}^1 \\ F_{3*}^1 \end{Bmatrix} \\ \begin{Bmatrix} F_{1*}^2 \\ F_{2*}^2 \\ F_{3*}^2 \end{Bmatrix} \\ \begin{Bmatrix} F_{1*}^3 \\ F_{2*}^3 \\ F_{3*}^3 \end{Bmatrix} \end{bmatrix} = \begin{bmatrix} \begin{Bmatrix} F_{1*f}^1 \\ F_{2*f}^1 \\ F_{3*f}^1 \end{Bmatrix} \\ \begin{Bmatrix} F_{1*f}^2 \\ F_{2*f}^2 \\ F_{3*f}^2 \end{Bmatrix} \\ \begin{Bmatrix} F_{1*f}^3 \\ F_{2*f}^3 \\ F_{3*f}^3 \end{Bmatrix} \end{bmatrix} + \left[\begin{array}{ccccc|cc} 0 & 1 & 0 & 0 & 1 & 1 & 0 \\ 4 & 0 & -1 & 4 & -6 & -6 & -1 \\ 0 & 0 & 1 & 0 & 6 & 6 & 0 \\ \hline 0 & 0 & 0 & 1 & 0 & 0 & -1/4 \\ 0 & 0 & 0 & 0 & -6 & -6 & 0 \\ 0 & 0 & 0 & 0 & 0 & 0 & 0 \\ \hline 0 & 0 & 0 & 0 & 0 & -1 & 0 \\ 0 & 0 & 0 & 0 & 0 & 0 & 1 \\ 0 & 0 & 0 & 0 & 0 & 0 & 0 \end{array}\right] \left[\begin{array}{c} F_1 \\ F_2 \\ F_3 \\ F_4 \\ F_5 \\ \hline F_6 \\ F_7 \end{array}\right]$$

Flexibility Matrix

The unassembled element flexibility matrix, satisfying $\mathbf{D}_* = \mathbf{f}_*\mathbf{F}_*$, may be generated as

$$\mathbf{f}_* = \begin{bmatrix} \mathbf{f}_*^1 & \mathbf{O} & \mathbf{O} \\ \mathbf{O} & \mathbf{f}_*^2 & \mathbf{O} \\ \mathbf{O} & \mathbf{O} & \mathbf{f}_*^3 \end{bmatrix}$$

where, considering $\frac{EA_i}{L_i} = 562500\text{ kN/m}$; $\frac{EI_1}{L_1} = \frac{EI_3}{L_3} = 4218.75$ kN/m; $\frac{EI_2}{L_2} = 9492.1875$ kN/m,

$$\mathbf{f}_*^1 = \mathbf{f}_*^3 = 10^{-6} \times \begin{bmatrix} 1.77778 & 0 & 0 \\ 0 & 79.01235 & -39.50617 \\ 0 & -39.50617 & 79.01235 \end{bmatrix};\ \mathbf{f}_*^2 = 10^{-6} \times \begin{bmatrix} 1.77778 & 0 & 0 \\ 0 & 35.11660 & -17.55830 \\ 0 & -17.55830 & 35.11660 \end{bmatrix}$$

The structure flexibility matrix $\mathbf{f} = \mathbf{T_F}^T\mathbf{f}_*\mathbf{T_F} = \left[\begin{array}{c|c}\mathbf{f_{AA}} & \mathbf{f_{AX}} \\ \hline \mathbf{f_{XA}} & \mathbf{f_{XX}}\end{array}\right]$ can now be assembled:

$$\Rightarrow \quad \mathbf{f_{XX}} = \mathbf{T_{FX}}^T\mathbf{f}_*\mathbf{T_{FX}} = 10^{-6} \times \begin{bmatrix} 9801.0864 & 711.1111 \\ 711.1111 & 158.1358 \end{bmatrix} \Rightarrow [\mathbf{f_{XX}}]^{-1} = \begin{bmatrix} 151.4386 & -680.9950 \\ -680.9950 & 9241.4361 \end{bmatrix}$$

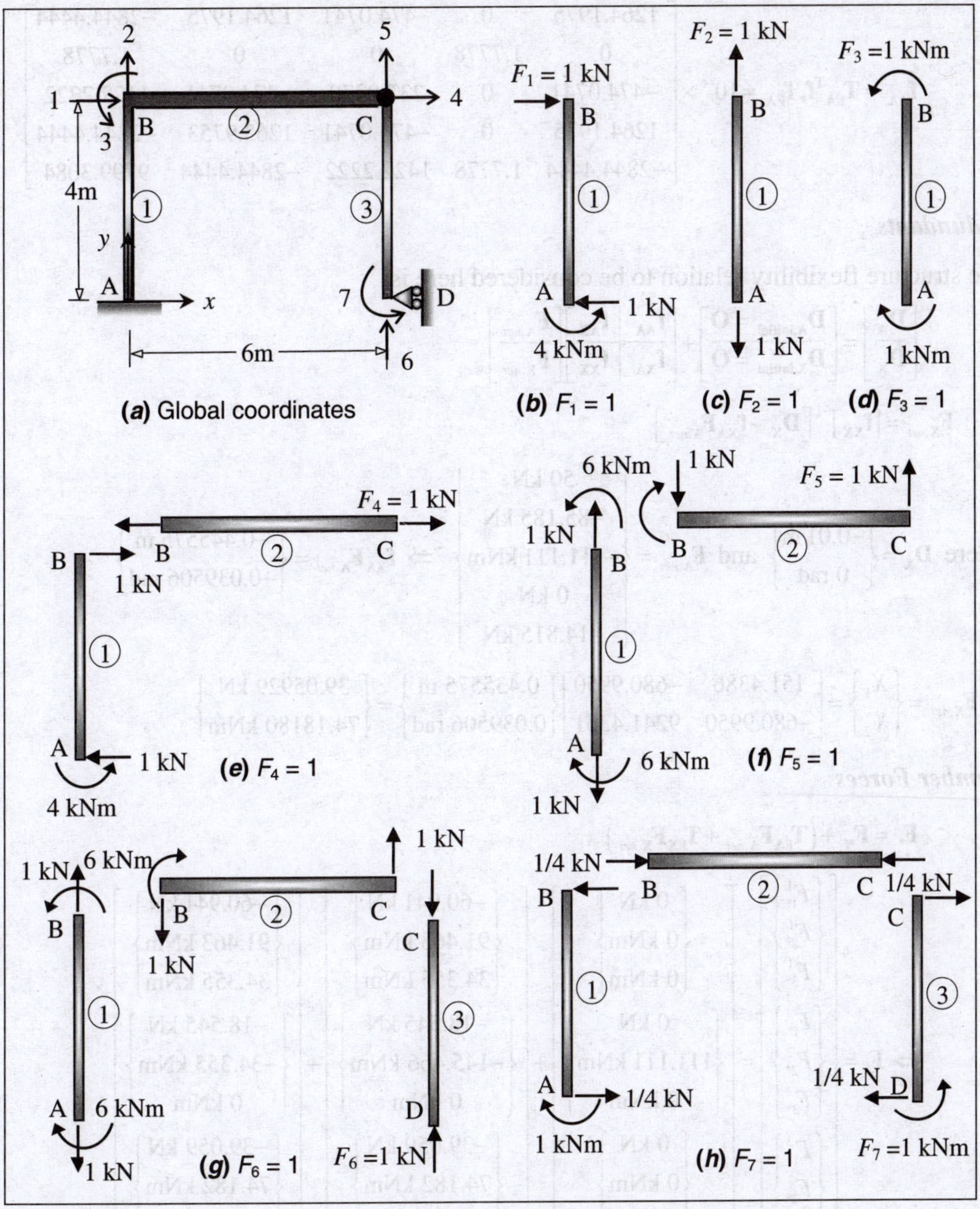

Figure 6.34 Response of primary structure to unit loads – free-bodies (Example 6.9)

$$\mathbf{f}_{\mathbf{XA}} = \mathbf{T}_{\mathbf{FX}}{}^{\mathbf{T}}\mathbf{f}_{*}\mathbf{T}_{\mathbf{FA}} = 10^{-6} \times \begin{bmatrix} -2844.4444 & 1.7778 & 1422.2222 & -2844.4444 & 9799.3086 \\ -316.0494 & 0 & 118.5185 & -316.4938 & 711.1111 \end{bmatrix} = \mathbf{f}_{\mathbf{XA}}{}^{\mathbf{T}}$$

$$\mathbf{f}_{AA} = \mathbf{T}_{FA}{}^{T}\mathbf{f}_{*}\mathbf{T}_{FA} = 10^{-6} \times \begin{bmatrix} 1264.1975 & 0 & -474.0741 & 1264.1975 & -2844.4444 \\ 0 & 1.7778 & 0 & 0 & 1.7778 \\ -474.0741 & 0 & 237.0370 & -474.0741 & 1422.2222 \\ 1264.1975 & 0 & -474.0741 & 1265.9753 & -2844.4444 \\ -2844.4444 & 1.7778 & 1422.2222 & -2844.4444 & 9799.3084 \end{bmatrix}$$

Redundants

The structure flexibility relation to be considered here is:

$$\begin{bmatrix} \mathbf{D}_{A} \\ \hline \mathbf{D}_{X} \end{bmatrix} = \begin{bmatrix} \mathbf{D}_{A,\text{initial}} = \mathbf{O} \\ \hline \mathbf{D}_{X,\text{initial}} = \mathbf{O} \end{bmatrix} + \left[\begin{array}{c|c} \mathbf{f}_{AA} & \mathbf{f}_{AX} \\ \hline \mathbf{f}_{XA} & \mathbf{f}_{XX} \end{array}\right] \begin{bmatrix} \mathbf{F}_{A,net} \\ \hline \mathbf{F}_{X,net} \end{bmatrix}$$

$$\Rightarrow \quad \mathbf{F}_{X,net} = [\mathbf{f}_{XX}]^{-1}[\mathbf{D}_{X} - \mathbf{f}_{XA}\mathbf{F}_{A,net}]$$

where $\mathbf{D}_{X} = \begin{Bmatrix} -0.01 \text{ m} \\ 0 \text{ rad} \end{Bmatrix}$ and $\mathbf{F}_{A,net} = \begin{Bmatrix} 50 \text{ kN} \\ -85.185 \text{ kN} \\ -111.111 \text{ kNm} \\ 0 \text{ kN} \\ -14.815 \text{ kN} \end{Bmatrix} \Rightarrow \mathbf{f}_{XA}\mathbf{F}_{A,net} = \begin{Bmatrix} -0.445575 \text{ m} \\ -0.039506 \text{ rad} \end{Bmatrix}$

$$\Rightarrow \mathbf{F}_{X,net} = \begin{Bmatrix} X_1 \\ X_2 \end{Bmatrix} = \begin{bmatrix} 151.4386 & -680.9950 \\ -680.9950 & 9241.4361 \end{bmatrix} \begin{Bmatrix} 0.435575 \text{ m} \\ 0.039506 \text{ rad} \end{Bmatrix} = \begin{Bmatrix} 39.05929 \text{ kN} \\ 74.18180 \text{ kNm} \end{Bmatrix}$$

Member Forces

$$\mathbf{F}_{*} = \mathbf{F}_{*f} + \left(\mathbf{T}_{FA}\mathbf{F}_{A,net} + \mathbf{T}_{FX}\mathbf{F}_{X,net}\right)$$

$$\Rightarrow \mathbf{F}_{*} = \begin{bmatrix} \begin{Bmatrix} F_{1*}^{1} \\ F_{2*}^{1} \\ F_{3*}^{1} \end{Bmatrix} \\ \begin{Bmatrix} F_{1*}^{2} \\ F_{2*}^{2} \\ F_{3*}^{2} \end{Bmatrix} \\ \begin{Bmatrix} F_{1*}^{3} \\ F_{2*}^{3} \\ F_{3*}^{3} \end{Bmatrix} \end{bmatrix} = \begin{bmatrix} \begin{Bmatrix} 0 \text{ kN} \\ 0 \text{ kNm} \\ 0 \text{ kNm} \end{Bmatrix} \\ \begin{Bmatrix} 0 \text{ kN} \\ 111.111 \text{ kNm} \\ 0 \text{ kNm} \end{Bmatrix} \\ \begin{Bmatrix} 0 \text{ kN} \\ 0 \text{ kNm} \\ 0 \text{ kNm} \end{Bmatrix} \end{bmatrix} + \begin{bmatrix} \begin{Bmatrix} -60.941 \text{ kN} \\ 91.463 \text{ kNm} \\ 34.355 \text{ kNm} \end{Bmatrix} \\ \begin{Bmatrix} -18.545 \text{ kN} \\ -145.466 \text{ kNm} \\ 0 \text{ kNm} \end{Bmatrix} \\ \begin{Bmatrix} -39.059 \text{ kN} \\ 74.182 \text{ kNm} \\ 0 \text{ kNm} \end{Bmatrix} \end{bmatrix} + \begin{bmatrix} \begin{Bmatrix} -60.941 \text{ kN} \\ 91.463 \text{ kNm} \\ 34.355 \text{ kNm} \end{Bmatrix} \\ \begin{Bmatrix} -18.545 \text{ kN} \\ -34.355 \text{ kNm} \\ 0 \text{ kNm} \end{Bmatrix} \\ \begin{Bmatrix} -39.059 \text{ kN} \\ 74.182 \text{ kNm} \\ 0 \text{ kNm} \end{Bmatrix} \end{bmatrix}$$

The member end forces are exactly as obtained earlier in Example 6.3 (Stiffness method solution). The free-bodies and the bending moment diagram can be generated, exactly as shown earlier in Fig 6.14a and c. The support reactions are also easily obtainable from the free-bodies.

Displacements

$$\mathbf{D_A} = \mathbf{D_{A,initial}} + [\mathbf{f_{AA}} \mid \mathbf{f_{AX}}]\begin{Bmatrix}\mathbf{F}_{\mathbf{A},net}\\ \mathbf{F}_{\mathbf{X},net}\end{Bmatrix} \quad \text{where} \quad \mathbf{D_{A,initial}} = \mathbf{O}$$

$$\Rightarrow \begin{Bmatrix} D_1 \\ D_2 \\ D_3 \\ D_4 \\ D_5 \end{Bmatrix} = \begin{Bmatrix} 0.023478 \text{ m} \\ -0.0001083 \text{ m} \\ -0.0067684 \text{ rad} \\ 0.023445 \text{ m} \\ -0.010069 \text{ m} \end{Bmatrix}$$

These displacements at the active coordinates also match exactly with the results obtained in Example 6.3. The deflection diagram is shown in Fig. 6.14b.

EXAMPLE 6.10

Repeat Example 6.9 by the Flexibility method, ignoring axial deformations.

SOLUTION

Let us choose the vertical reaction and moment at the support D as the two redundants, X_1 and X_2, as in Example 6.9. Ignoring the axial deformations, the plane frame element (with three degrees of freedom) reduces to a simple beam element (with two degrees of freedom in the case of element '1' and one degree of freedom in elements '2' and '3'), as shown in Fig. 6.35a (similar to Example 6.7, using the reduced stiffness formulation). We can also simplify the computations by restricting the number of active coordinates (the vertical force at B does not cause any bending, while the one at B can be included in coordinate '3'). Thus, the resultant loads that we had earlier computed (refer Fig. 6.33c) take the form shown in Fig. 6.35b. We now need to account for only two active and two restrained global coordinates, as shown in Fig. 6.35c.

The combined element fixed end force vector, $\mathbf{F_{*f}}$, and the associated structure-level fixed end force vector, $\mathbf{F_f}$, are obtained (by inspection) as:

$$\mathbf{F_{*f}} = \begin{Bmatrix} F^1_{1*f} \\ F^1_{2*f} \\ \hline F^2_{1*f} \\ \hline F^3_{1*f} \end{Bmatrix} = \begin{Bmatrix} 0 \text{ kNm} \\ 0 \text{ kNm} \\ \hline 111.111 \text{ kNm} \\ \hline 0 \text{ kNm} \end{Bmatrix} \Rightarrow \mathbf{F_f} = \begin{Bmatrix} \mathbf{F_{fA}} \\ \hline \mathbf{F_{fX}} \end{Bmatrix} = \begin{Bmatrix} F_{1f} = 0 \text{ kN} \\ F_{2f} = 111.111 \text{ kNm} \\ \hline F_{3f} = 14.815 \text{ kN} \\ F_{4f} = 0 \text{ kNm} \end{Bmatrix}$$

The resultant load vector, $\mathbf{F}_{net}$, on the primary structure (shown in Fig. 6.33c) is given by:

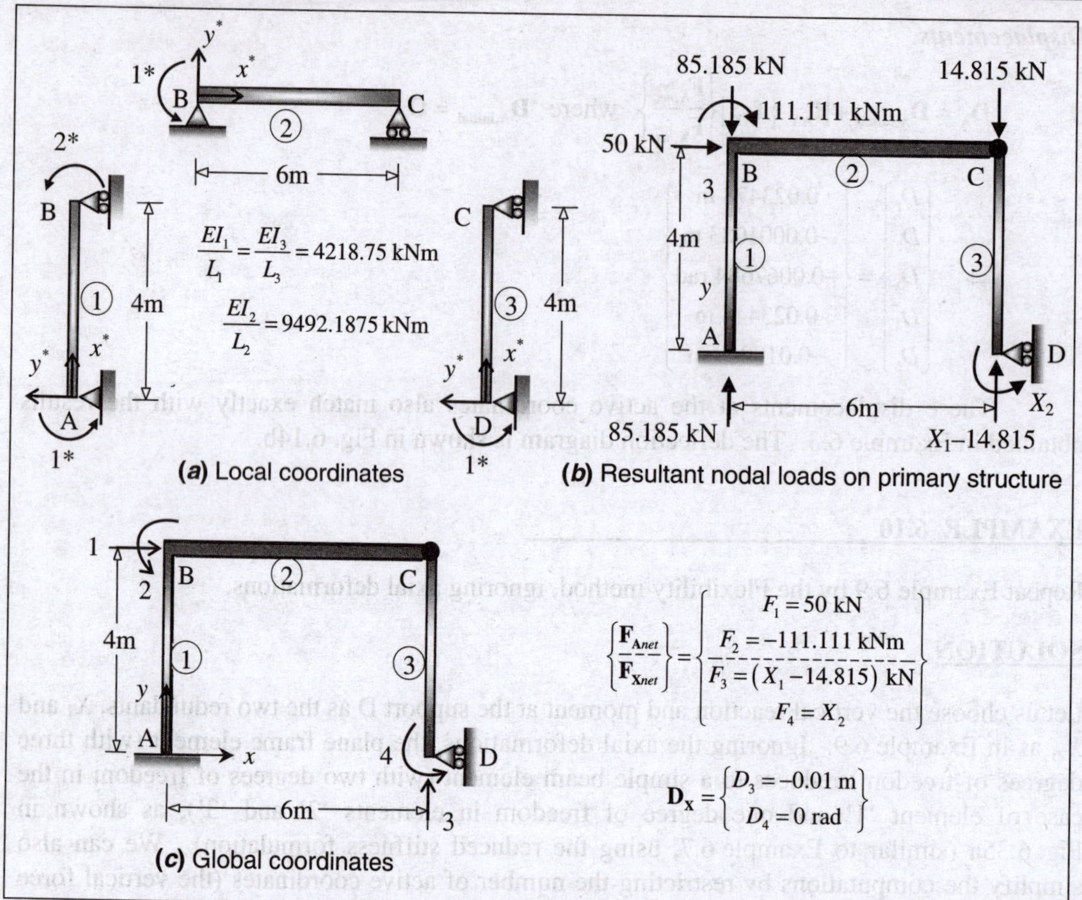

Figure 6.35 Local coordinates, resultant loads and global coordinates (Example 6.10)

$$\mathbf{F}_{net} = \left[\begin{array}{c}\mathbf{F_A} - \mathbf{F_{fA}} \\ \hline \mathbf{F_X} - \mathbf{F_{fX}}\end{array}\right] = \left\{\begin{array}{c}50\text{ kN} \\ 0\text{ kNm} \\ \hline X_1 \\ X_2\end{array}\right\} - \left\{\begin{array}{c}0\text{ kN} \\ 111.111\text{ kNm} \\ \hline 14.815\text{ kN} \\ 0\text{ kN}\end{array}\right\} = \left\{\begin{array}{c}50\text{ kN} \\ -111.111\text{ kNm} \\ \hline (X_1 - 14.815)\text{ kN} \\ X_2\end{array}\right\}$$

The force transformation matrix can be easily generated by considering the analysis of the primary structure under four separate unit load conditions (refer Fig. 6.34b, d, g and h). Accordingly,

$$\mathbf{F}_* = \left[\mathbf{T_{FA}} \mid \mathbf{T_{FX}}\right]\left\{\begin{array}{c}\mathbf{F}_{\mathbf{A},net} \\ \hline \mathbf{F}_{\mathbf{X},net}\end{array}\right\} \Rightarrow \mathbf{F}_* = \left\{\begin{array}{c}F^1_{1*} \\ F^1_{2*} \\ \hline F^2_{1*} \\ \hline F^3_{1*}\end{array}\right\} = \left\{\begin{array}{c}0\text{ kNm} \\ 0\text{ kNm} \\ \hline 111.111\text{ kNm} \\ \hline 0\text{ kNm}\end{array}\right\} + \left[\begin{array}{cc|cc}4 & -1 & -6 & -1 \\ 0 & 1 & 6 & 0 \\ \hline 0 & 0 & -6 & 0 \\ \hline 0 & 0 & 0 & 1\end{array}\right]\left[\begin{array}{c}F_1 = 50\text{ kN} \\ F_2 = -111.111\text{ kNm} \\ \hline F_3 = (X_1 - 14.815)\text{ kN} \\ F_4 = (X_2)\text{ kNm}\end{array}\right]$$

The unassembled element flexibility matrix, satisfying $\mathbf{D}_* = \mathbf{f}_*\mathbf{F}_*$, may be generated as follows, modifying the element flexibility in '2' and '3' as $\tilde{\mathbf{f}}_*^i = \left[\tilde{\mathbf{k}}_*^i\right]^{-1} = \dfrac{L}{3(EI)_i}$, and considering $\dfrac{EI_1}{L_1} = \dfrac{EI_3}{L_3} = 4218.75$ kNm; $\dfrac{EI_2}{L_2} = 9492.1875$ kNm:

$$\mathbf{f}_*^1 = 10^{-6}\times\begin{bmatrix} 79.01235 & -39.50617 \\ -39.50617 & 79.01235 \end{bmatrix}; \quad \tilde{\mathbf{f}}_*^2 = 10^{-6}\times[35.11660] \;; \quad \tilde{\mathbf{f}}_*^3 = 10^{-6}\times[79.01235]$$

$$\Rightarrow \mathbf{f}_* = \begin{bmatrix} \mathbf{f}_*^1 & \mathbf{O} & \mathbf{O} \\ \mathbf{O} & \mathbf{f}_*^2 & \mathbf{O} \\ \mathbf{O} & \mathbf{O} & \mathbf{f}_*^3 \end{bmatrix} = \mathbf{f}_*^1 = 10^{-6}\times\left[\begin{array}{cc|c|c} 79.01235 & -39.50617 & 0 & 0 \\ -39.50617 & 79.01235 & 0 & 0 \\ \hline 0 & 0 & 35.11660 & 0 \\ \hline 0 & 0 & 0 & 79.01235 \end{array}\right]$$

The structure flexibility matrix $\mathbf{f} = \mathbf{T_F}^{\mathbf{T}}\mathbf{f}_*\mathbf{T_F} = \left[\begin{array}{c|c} \mathbf{f_{AA}} & \mathbf{f_{AX}} \\ \hline \mathbf{f_{XA}} & \mathbf{f_{XX}} \end{array}\right]$ can now be assembled:

$$\Rightarrow \quad \mathbf{f_{XX}} = \mathbf{T_{FX}}^{\mathbf{T}}\mathbf{f}_*\mathbf{T_{FX}} = 10^{-6}\times\begin{bmatrix} 9797.5309 & 711.1111 \\ 711.1111 & 158.0247 \end{bmatrix} \Rightarrow [\mathbf{f_{XX}}]^{-1} = \begin{bmatrix} 151.5719 & -682.0734 \\ -682.0734 & 9397.4551 \end{bmatrix}$$

$$\mathbf{f_{XA}} = \mathbf{T_{FX}}^{\mathbf{T}}\mathbf{f}_*\mathbf{T_{FA}} = 10^{-6}\times\begin{bmatrix} -2844.4444 & 1422.2222 \\ -316.0494 & 118.5185 \end{bmatrix} = \mathbf{f_{XA}}^{\mathbf{T}}$$

$$\mathbf{f_{AA}} = \mathbf{T_{FA}}^{\mathbf{T}}\mathbf{f}_*\mathbf{T_{FA}} = 10^{-6}\times\begin{bmatrix} 1264.1975 & -474.0741 \\ -474.0741 & 237.0370 \end{bmatrix}$$

The structure flexibility relation to be considered here is:

$$\left[\begin{array}{c} \mathbf{D_A} \\ \hline \mathbf{D_X} \end{array}\right] = \left[\begin{array}{c} \mathbf{D_{A,initial}} = \mathbf{O} \\ \hline \mathbf{D_{X,initial}} = \mathbf{O} \end{array}\right] + \left[\begin{array}{c|c} \mathbf{f_{AA}} & \mathbf{f_{AX}} \\ \hline \mathbf{f_{XA}} & \mathbf{f_{XX}} \end{array}\right]\left[\begin{array}{c} \mathbf{F}_{\mathbf{A},net} \\ \hline \mathbf{F}_{\mathbf{X},net} \end{array}\right]$$

$$\Rightarrow \quad \mathbf{F}_{\mathbf{X},net} = [\mathbf{f_{XX}}]^{-1}\left[\mathbf{D_X} - \mathbf{f_{XA}}\mathbf{F}_{\mathbf{A},net}\right]$$

$$\mathbf{D_X} - \mathbf{f_{XA}}\mathbf{F}_{\mathbf{A},net} = \begin{Bmatrix} -0.01 \text{ m} \\ 0 \text{ rad} \end{Bmatrix} - 10^{-6}\times\begin{bmatrix} -2844.4444 & 1422.2222 \\ -316.0494 & 118.5185 \end{bmatrix}\begin{Bmatrix} 50 \\ -111.111 \end{Bmatrix} = \begin{Bmatrix} 0.29024676 \text{ m} \\ 0.02897118 \text{ rad} \end{Bmatrix}$$

$$\Rightarrow \mathbf{F}_{\mathbf{X},net} = \begin{Bmatrix} X_1 - 14.815 \\ X_2 \end{Bmatrix} = \begin{bmatrix} 151.5719 & -682.0734 \\ -682.0734 & 9397.4551 \end{bmatrix}\begin{Bmatrix} 0.29024676 \text{ m} \\ 0.02897118 \text{ rad} \end{Bmatrix} = \begin{Bmatrix} 24.2327695 \text{ kN} \\ 74.2857874 \text{ kNm} \end{Bmatrix}$$

$$\Rightarrow \begin{Bmatrix} X_1 \\ X_2 \end{Bmatrix} = \begin{Bmatrix} 39.048 \text{ kN} \\ 74.286 \text{ kNm} \end{Bmatrix}$$

Member Forces and Displacements

$$\mathbf{F}_* = \mathbf{F}_{*\mathbf{f}} + \left(\mathbf{T_{FA}}\mathbf{F}_{\mathbf{A},net} + \mathbf{T_{FX}}\mathbf{F}_{\mathbf{X},net}\right)$$

$$\Rightarrow \mathbf{F}_* = \begin{Bmatrix} F_{1*}^1 \\ F_{2*}^1 \\ \hline F_{1*}^2 \\ \hline F_{1*}^3 \end{Bmatrix} = \begin{Bmatrix} 0 \text{ kNm} \\ 0 \text{ kNm} \\ \hline 111.111 \text{ kNm} \\ \hline 0 \text{ kNm} \end{Bmatrix} + \left[\begin{array}{cc|c|c} 4 & -1 & -6 & -1 \\ 0 & 1 & 6 & 0 \\ \hline 0 & 0 & -6 & 0 \\ \hline 0 & 0 & 0 & 1 \end{array}\right] \begin{bmatrix} F_1 = 50 \text{ kN} \\ F_2 = -111.111 \text{ kNm} \\ \hline F_3 = 24.2327695 \text{ kN} \\ F_4 = 74.2857874 \text{ kNm} \end{bmatrix} = \begin{Bmatrix} 91.429 \text{ kNm} \\ 34.286 \text{ kNm} \\ \hline -34.286 \text{ kNm} \\ \hline 74.286 \text{ kNm} \end{Bmatrix}$$

$$\mathbf{D}_{\mathbf{A}} = \mathbf{D}_{\mathbf{A,initial}} + [\mathbf{f}_{\mathbf{AA}} \mid \mathbf{f}_{\mathbf{AX}}] \begin{Bmatrix} \mathbf{F}_{\mathbf{A},net} \\ \mathbf{F}_{\mathbf{X},net} \end{Bmatrix} \quad \text{where} \quad \mathbf{D}_{\mathbf{A,initial}} = \mathbf{O}$$

$$\Rightarrow \quad \begin{Bmatrix} D_1 \\ D_2 \end{Bmatrix} = \begin{Bmatrix} 0.023478 \text{ m} \\ -0.0067684 \text{ rad} \end{Bmatrix}$$

The member end forces and displacements (at active coordinates) are exactly as obtained earlier in Example 6.7 (reduced stiffness method solution). The free-bodies and the bending moment diagram can be generated, exactly as shown earlier in Figs 6.28a and c. The support reactions are also easily obtainable from the free-bodies.

EXAMPLE 6.11

Repeat Example 6.4 by the Flexibility method and find the complete response.

SOLUTION

The frame is statically indeterminate to the first degree. Let us choose the horizontal reaction at the hinged support D as the redundant, X_1 (acting to the right). The support at D is reduced to a roller support in the primary structure, as shown in Fig. 6.36b. There are no nodal loads prescribed ($\mathbf{F}_{\mathbf{A}} = \mathbf{O}$). The distributed loading on element '2' (BC), however, gives rise to fixed end forces on the primary structure (calculated in Example 6.4), as shown in Fig. 6.36a. The resultant nodal loads on the primary structure are as shown in Fig. 6.36b. Corresponding to these locations, four active global coordinates are chosen, as shown in Fig. 6.37a. The fifth global coordinate corresponds to the redundant location.

There are three prismatic frame elements, whose local coordinates are marked in Fig. 6.36c. Taking advantage of the hinges at A and D, we can reduce the degrees of freedom in elements '1' and '3' to two each (ignoring the rotational degree of freedom at A and D).

The combined element fixed end force vector, $\mathbf{F}_{*\mathbf{f}}$, and the associated structure-level fixed end force vector, $\mathbf{F}_{*\mathbf{f}}$, and the resultant load vector, $\mathbf{F}_{net}$, on the primary structure (shown in Fig. 6.36b) are obtained as:

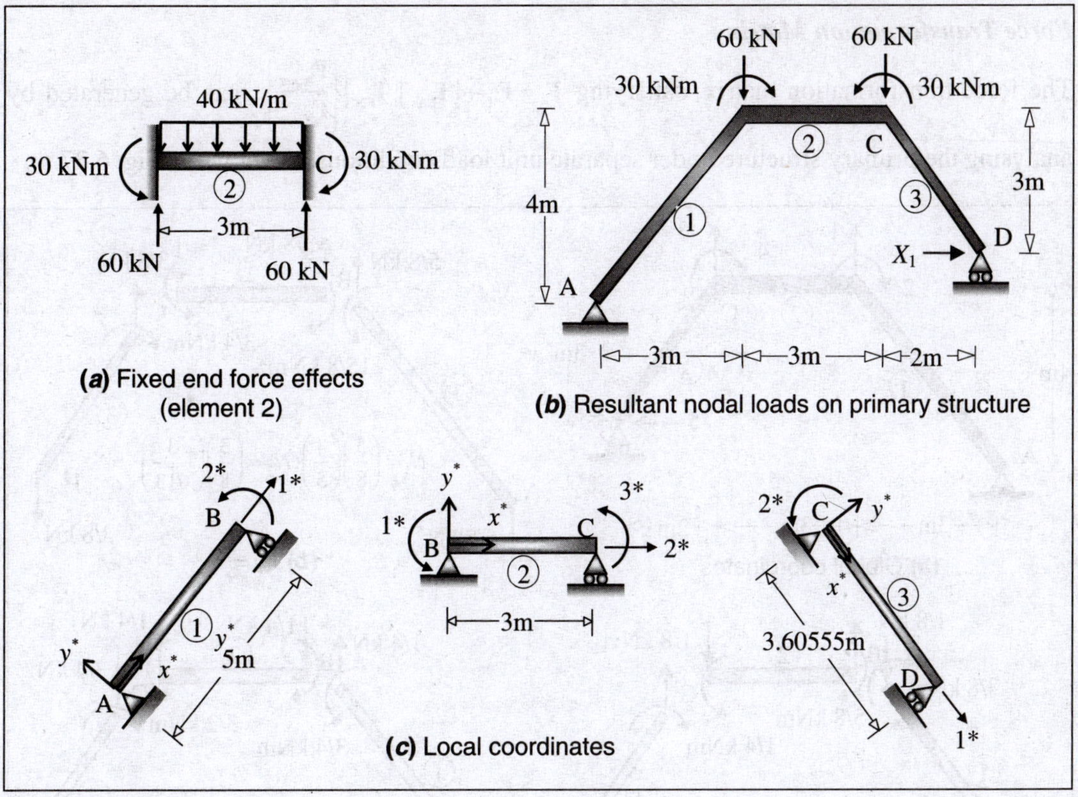

Figure 6.36 Fixed end forces, resultant loads on primary structure and local coordinates (Example 6.11)

$$\mathbf{F}_{*\mathbf{f}} = \begin{bmatrix} \begin{Bmatrix} F^1_{1*f} \\ F^1_{2*f} \end{Bmatrix} \\ \begin{Bmatrix} F^2_{1*f} \\ F^2_{2*f} \\ F^2_{3*f} \end{Bmatrix} \\ \begin{Bmatrix} F^3_{1*f} \\ F^3_{2*f} \end{Bmatrix} \end{bmatrix} = \begin{bmatrix} \begin{Bmatrix} 0 \text{ kN} \\ 0 \text{ kNm} \end{Bmatrix} \\ \begin{Bmatrix} 0 \text{ kN} \\ 30 \text{ kNm} \\ -30 \text{ kNm} \end{Bmatrix} \\ \begin{Bmatrix} 0 \text{ kN} \\ 0 \text{ kNm} \end{Bmatrix} \end{bmatrix} \Rightarrow \mathbf{F}_{\mathbf{f}} = \begin{Bmatrix} \mathbf{F}_{\mathbf{fA}} \\ \hline \mathbf{F}_{\mathbf{fX}} \end{Bmatrix} = \begin{Bmatrix} F_{1f} = 60 \text{ kN} \\ F_{2f} = 30 \text{ kNm} \\ F_{3f} = 60 \text{ kN} \\ F_{4f} = -30 \text{ kNm} \\ \hline F_{5f} = 0 \text{ kN} \end{Bmatrix}$$

$$\Rightarrow \mathbf{F}_{net} = \begin{bmatrix} \mathbf{F}_{\mathbf{A}} - \mathbf{F}_{\mathbf{fA}} \\ \hline \mathbf{F}_{\mathbf{X}} - \mathbf{F}_{\mathbf{fX}} \end{bmatrix} = \begin{Bmatrix} 0 \text{ kN} \\ 0 \text{ kNm} \\ 0 \text{ kN} \\ 0 \text{ kNm} \\ \hline X_1 \text{ kN} \end{Bmatrix} - \begin{Bmatrix} 60 \text{ kN} \\ 30 \text{ kNm} \\ 60 \text{ kN} \\ -30 \text{ kNm} \\ \hline 0 \text{ kN} \end{Bmatrix} = \begin{Bmatrix} F_1 = -60 \text{ kN} \\ F_2 = -30 \text{ kNm} \\ F_3 = -60 \text{ kN} \\ F_4 = 30 \text{ kNm} \\ \hline F_5 = X_1 \text{ kN} \end{Bmatrix}$$

Force Transformation Matrix

The force transformation matrix, satisfying $\mathbf{F}_* = \mathbf{F}_{*\mathbf{f}} + \left[\mathbf{T}_{\mathbf{FA}} \mid \mathbf{T}_{\mathbf{FX}}\right]\left\{\dfrac{\mathbf{F}_{\mathbf{A},net}}{\mathbf{F}_{\mathbf{X},net}}\right\}$, can be generated by analysing the primary structure under separate unit load conditions, as shown in Fig. 6.37.

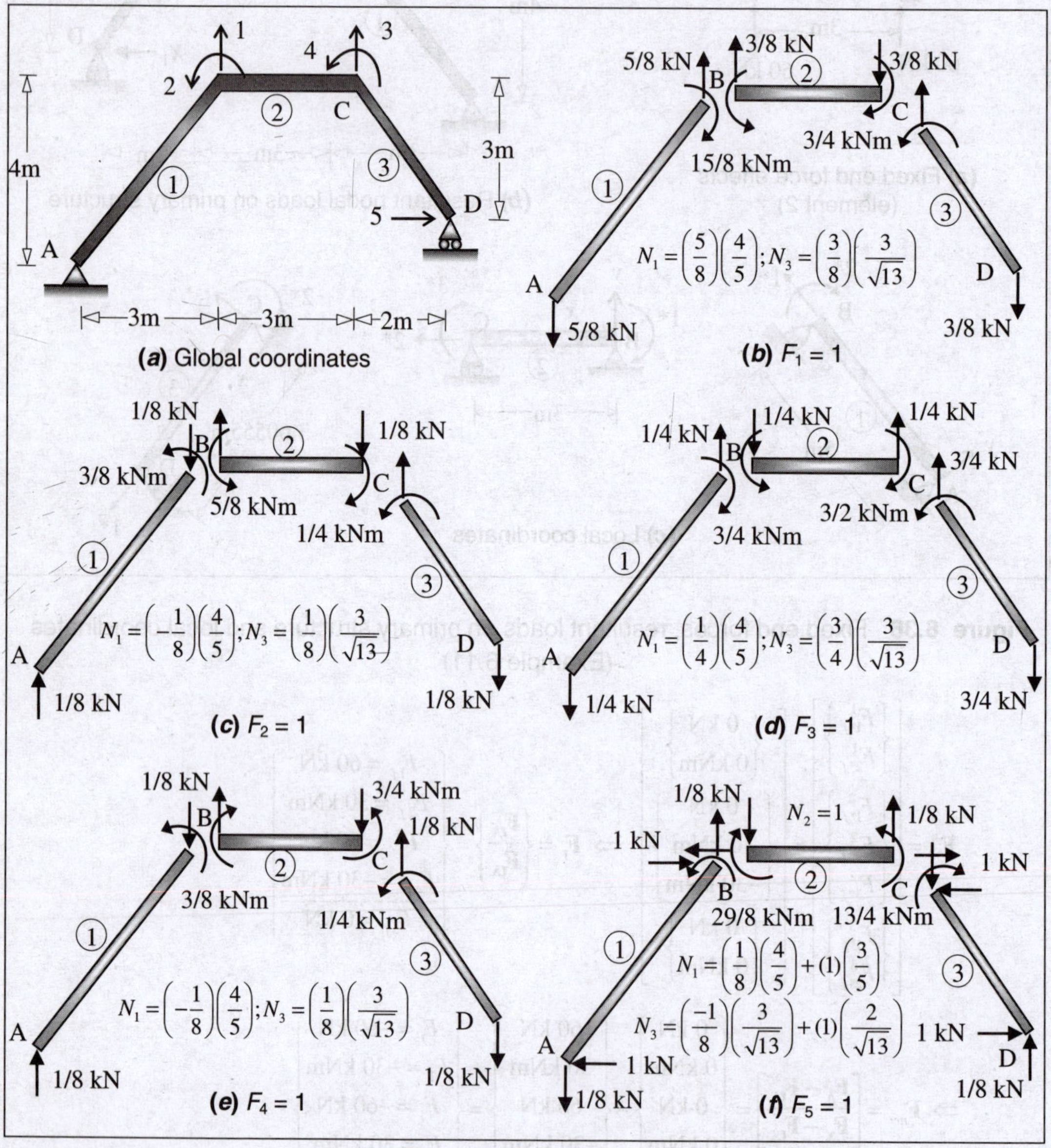

Figure 6.37 Global coordinates and free-bodies of primary structure subject to unit loads (Example 6.11)

Accordingly,

$$\mathbf{F}_* = \begin{bmatrix} \begin{Bmatrix} F_{1*}^1 \\ F_{2*}^1 \end{Bmatrix} \\ \begin{Bmatrix} F_{1*}^2 \\ F_{2*}^2 \\ F_{3*}^2 \end{Bmatrix} \\ \begin{Bmatrix} F_{1*}^3 \\ F_{2*}^3 \end{Bmatrix} \end{bmatrix} = \begin{bmatrix} \begin{Bmatrix} 0 \text{ kN} \\ 0 \text{ kNm} \end{Bmatrix} \\ \begin{Bmatrix} 0 \text{ kN} \\ 30 \text{ kNm} \\ -30 \text{ kNm} \end{Bmatrix} \\ \begin{Bmatrix} 0 \text{ kN} \\ 0 \text{ kNm} \end{Bmatrix} \end{bmatrix} + \left[\begin{array}{cccc|c} 1/2 & -1/10 & 1/5 & -1/10 & 7/10 \\ -15/8 & 3/8 & -3/4 & 3/8 & 29/8 \\ \hdashline 0 & 0 & 0 & 0 & 1 \\ 15/8 & 5/8 & 3/4 & -3/8 & -29/8 \\ -3/4 & -1/4 & -3/2 & 3/4 & 13/4 \\ \hdashline 9/8\sqrt{13} & 3/8\sqrt{13} & 9/4\sqrt{13} & 3/8\sqrt{13} & 13/8\sqrt{13} \\ 3/4 & 1/4 & 3/2 & 1/4 & -13/4 \end{array}\right] \begin{Bmatrix} F_1 \\ F_2 \\ F_3 \\ F_4 \\ F_5 \end{Bmatrix}$$

Flexibility Matrix

The unassembled element flexibility matrix, satisfying $\mathbf{D}_* = \mathbf{f}_*\mathbf{F}_*$, may be generated as $\mathbf{f}_* = \begin{bmatrix} \mathbf{f}_*^1 & \mathbf{O} & \mathbf{O} \\ \mathbf{O} & \mathbf{f}_*^2 & \mathbf{O} \\ \mathbf{O} & \mathbf{O} & \mathbf{f}_*^3 \end{bmatrix}$, where, considering values of $\frac{EA_i}{L_i}$ (in kN/m) equal to 450000, 750000 and 624037.72; $\frac{EI_i}{L_i}$ (in kNm) equal to 3375, 5625 and 4680.283 respectively for elements '1', '2' and '3',

$$\mathbf{f}_*^1 = 10^{-6} \times \begin{bmatrix} 2.22222 & 0 \\ 0 & 98.76543 \end{bmatrix} \quad ; \quad \mathbf{f}_*^3 = 10^{-6} \times \begin{bmatrix} 1.60247 & 0 \\ 0 & 71.22076 \end{bmatrix}$$

$$\mathbf{f}_*^2 = 10^{-6} \times \begin{bmatrix} 1.33333 & 0 & 0 \\ 0 & 59.25926 & -29.62963 \\ 0 & -29.62963 & 59.25926 \end{bmatrix}$$

The structure flexibility matrix $\mathbf{f} = \mathbf{T}_\mathbf{F}{}^\mathbf{T}\mathbf{f}_*\mathbf{T}_\mathbf{F} = \left[\begin{array}{c|c} \mathbf{f_{AA}} & \mathbf{f_{AX}} \\ \hline \mathbf{f_{XA}} & \mathbf{f_{XX}} \end{array}\right]$ can now be assembled:

$$\Rightarrow \quad \mathbf{f_{XX}} = \mathbf{T_{FX}}^\mathbf{T}\mathbf{f}_*\mathbf{T_{FX}} = 10^{-6} \times [4155.63435] \Rightarrow [\mathbf{f_{XX}}]^{-1} = [240.63715]$$

$$\mathbf{f_{XA}} = \mathbf{T_{FX}}^\mathbf{T}\mathbf{f}_*\mathbf{T_{FA}} = 10^{-6} \times [-1652.22712 \quad -193.13250 \quad -1298.29128 \quad 417.97861] = \mathbf{f_{XA}}^\mathbf{T}$$

$$\mathbf{f_{AA}} = \mathbf{T_{FA}}^\mathbf{T}\mathbf{f}_*\mathbf{T_{FA}} = 10^{-6} \times \begin{bmatrix} 712.99547 & 52.18367 & 46.95465 & -181.14966 \\ 52.18367 & 54.49085 & 82.32290 & -23.28692 \\ 46.95465 & 82.32290 & 449.84854 & -117.67710 \\ 181.14966 & -23.28692 & -117.67710 & 76.71308 \end{bmatrix}$$

Redundants

The structure flexibility relation to be considered here is:

$$\begin{bmatrix} \mathbf{D_A} \\ \hline \mathbf{D_X = O} \end{bmatrix} = \begin{bmatrix} \mathbf{D_{A,initial} = O} \\ \hline \mathbf{D_{X,initial} = O} \end{bmatrix} + \begin{bmatrix} \mathbf{f_{AA}} & \mathbf{f_{AX}} \\ \hline \mathbf{f_{XA}} & \mathbf{f_{XX}} \end{bmatrix} \begin{bmatrix} \mathbf{F}_{\mathbf{A},net} \\ \hline \mathbf{F}_{\mathbf{X},net} \end{bmatrix}$$

$$\Rightarrow \quad \mathbf{F}_{\mathbf{X},net} = [\mathbf{f_{XX}}]^{-1}\left[-\mathbf{f_{XA}}\mathbf{F}_{\mathbf{A},net}\right]$$

$$\text{where } \mathbf{F}_{\mathbf{A},net} = \begin{Bmatrix} -60 \text{ kN} \\ -30 \text{ kNm} \\ -60 \text{ kN} \\ 30 \text{ kNm} \end{Bmatrix} \Rightarrow \mathbf{f_{XA}}\mathbf{F}_{\mathbf{A},net} = \{0.19536444 \text{ m}\}$$

$$\Rightarrow \mathbf{F}_{\mathbf{X},net} = \{X_1\} = [240.63715]\{-0.19536444 \text{ m}\} = \{-47.0119412 \text{ kN}\}$$

Member Forces

$$\mathbf{F}_* = \mathbf{F}_{*\mathbf{f}} + \left(\mathbf{T_{FA}}\mathbf{F}_{\mathbf{A},net} + \mathbf{T_{FX}}\mathbf{F}_{\mathbf{X},net}\right)$$

$$\mathbf{F}_* = \begin{bmatrix} \begin{Bmatrix} F_{1*}^1 \\ F_{2*}^1 \end{Bmatrix} \\ \begin{Bmatrix} F_{1*}^2 \\ F_{2*}^2 \\ F_{3*}^2 \end{Bmatrix} \\ \begin{Bmatrix} F_{1*}^3 \\ F_{2*}^3 \end{Bmatrix} \end{bmatrix} = \begin{bmatrix} \begin{Bmatrix} 0 \text{ kN} \\ 0 \text{ kNm} \end{Bmatrix} \\ \begin{Bmatrix} 0 \text{ kN} \\ 30 \text{ kNm} \\ -30 \text{ kNm} \end{Bmatrix} \\ \begin{Bmatrix} 0 \text{ kN} \\ 0 \text{ kNm} \end{Bmatrix} \end{bmatrix} + \begin{bmatrix} \begin{Bmatrix} -74.908 \text{ kN} \\ -12.918 \text{ kNm} \end{Bmatrix} \\ \begin{Bmatrix} -47.012 \text{ kN} \\ -17.082 \text{ kNm} \\ 12.211 \text{ kNm} \end{Bmatrix} \\ \begin{Bmatrix} -77.351 \text{ kN} \\ 17.789 \text{ kNm} \end{Bmatrix} \end{bmatrix} = \begin{bmatrix} \begin{Bmatrix} -74.908 \text{ kN} \\ -12.918 \text{ kNm} \end{Bmatrix} \\ \begin{Bmatrix} -47.012 \text{ kN} \\ 12.918 \text{ kNm} \\ -17.789 \text{ kNm} \end{Bmatrix} \\ \begin{Bmatrix} -77.351 \text{ kN} \\ 17.789 \text{ kNm} \end{Bmatrix} \end{bmatrix}$$

The member end forces are exactly as obtained earlier in Example 6.4 (Stiffness method solution). The free-bodies and the bending moment, axial force and shear force diagrams can be generated, exactly as shown earlier in Figs 6.18a, c, d and e. The support reactions are also easily obtainable from the free-bodies.

Displacements

$$\mathbf{D_A} = \mathbf{D_{A,initial}} + [\mathbf{f_{AA}} \mid \mathbf{f_{AX}}]\begin{Bmatrix} \mathbf{F}_{\mathbf{A},net} \\ \hline \mathbf{F}_{\mathbf{X},net} \end{Bmatrix} \quad \text{where } \mathbf{D_{A,initial}} = \mathbf{O}$$

$$\Rightarrow \begin{Bmatrix} D_1 \\ D_2 \\ D_3 \\ D_4 \end{Bmatrix} = \begin{Bmatrix} -0.000278113 \text{ m} \\ -0.001324194 \text{ rad} \\ -0.000128509 \text{ m} \\ 0.001279619 \text{ rad} \end{Bmatrix}$$

These displacements at the active coordinates also match exactly with the results obtained in Example 6.4. The deflection diagram is shown in Fig. 6.18b. (The horizontal deflections at B and C are not obtainable in this analysis; if desired, these can also be generated by introducing two additional active global coordinates, as in Example 6.4).

6.5 ANALYSIS OF SPACE FRAMES BY STIFFNESS METHOD

The plane frame idealization is valid only when all the frame elements (each subject to bending moment, shear force and axial force) lie in a single plane and the loading is also on this plane. Otherwise, the frame has to be modelled as a *space frame*, in which the elements are *space frame elements*. As shown in Fig. 6.38a, at each end of the space frame element, there are six potential degrees of freedom (comprising three translations and three rotations), whereby as many as twelve degrees of freedom need to be accounted for. The element may be subject to bending in two orthogonal planes (involving two bending moments and corresponding two shear forces), axial force and twisting moment (i.e., totally six force resultants). All other elements (including plane frame element, grid element and space truss element) are special cases of the plane frame element.

The basic analysis procedure of space frames is similar to that of plane frames or grids, except that the sizes of the matrices involved become much larger, and the computations become relatively cumbersome to explain through a step-by-step procedure. In the conventional stiffness formulation, we need to deal with twelve degrees of freedom per element, and to also assign appropriate direction cosines for each element, with reference to the global axes system. The computations are greatly simplified, if we adopt the reduced element stiffness method, wherein we need to deal with only six degrees of freedom per element, as shown in Fig. 6.38b. Here, we shall demonstrate the application of the conventional stiffness method and the reduced stiffness method to space frames. The analysis by the flexibility method is not described here.

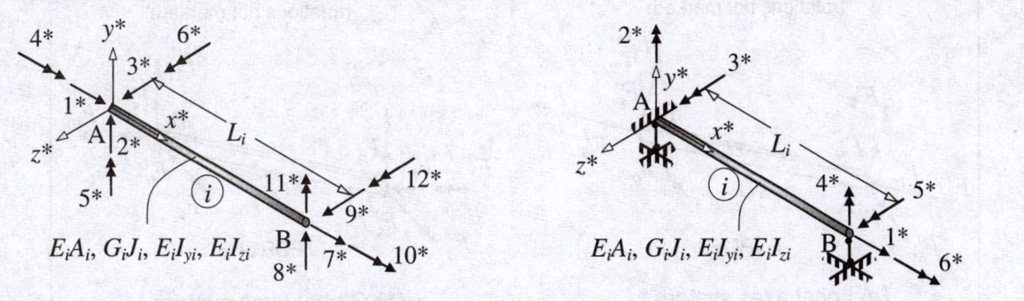

(*a*) Element with 12 degrees of freedom (conventional stiffness formulation)

(*b*) Element with 6 degrees of freedom (reduced stiffness formulation)

Figure 6.38 Prismatic space frame element — local coordinates

6.5.1 Space Frame Element with 12 Degrees of Freedom (Stiffness Method)

We can extend the formulation of the element stiffness matrix for the plane frame element to the space truss element, which can have any orientation with reference to the global axes system. As per our convention, we define the local x^*-axis in the direction of the longitudinal axis of the element, from the *start node* to the *end node* (marked as A and B in Fig. 6.39).

We are free to choose the other two local coordinate axes (y^*- and z^*-axes) as any two mutually orthogonal directions that are both perpendicular to the x^*-axis. These may be conveniently oriented along the principal axes directions of the cross-section of the element, and for this purpose, it is necessary to define the coordinates of a reference point (such as C in Fig. 6.39a, such that AC points in the positive y^*-direction. Then, the cross-product, $\hat{\mathbf{i}}^* \times \hat{\mathbf{j}}^* = \hat{\mathbf{k}}^*$ defines the direction of the base vector in the positive z^*-direction.†

In order to define the displaced configuration of the space frame element, with reference to its original configuration, it is necessary to define six degrees of freedom at each end of the element, modelled by means of translation and rotation vectors along the axial (longitudinal) x^*- direction, and transverse y^*- direction and transverse z^*- direction in the local axes system, as shown in Fig. 6.39a.

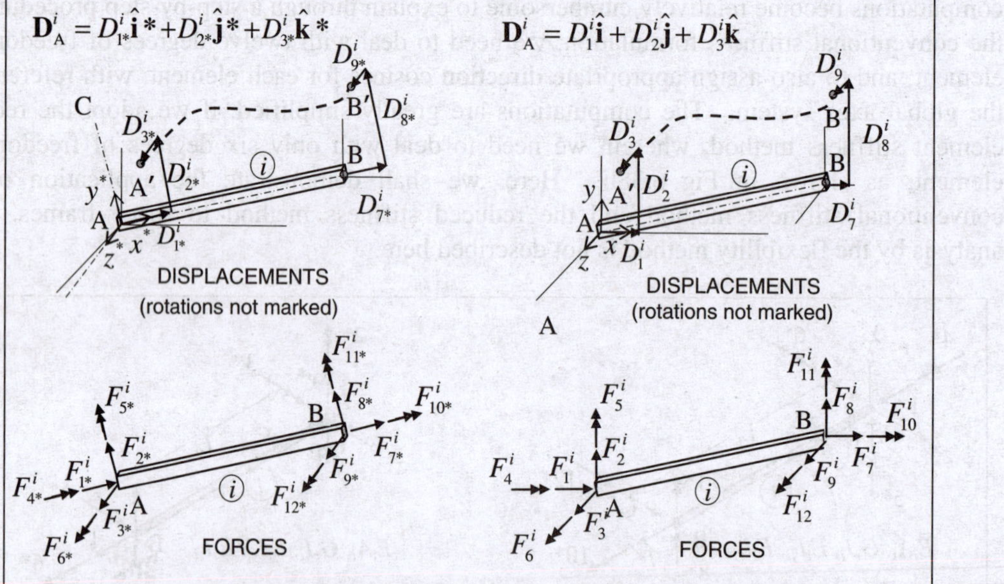

(*a*) Local axes system (*b*) Global axes system

Figure 6.39 Member-end displacements and forces in a space frame element

Element Stiffness Matrix

Extending the derivation of the element stiffness matrix for the space truss, plane truss and grid elements to the prismatic space truss element, we can easily establish the element

† Alternatively, it is convenient to choose the z^*-axis on the horizontal xz plane. Then, the y^*-axis is located in a vertical plane passing through the x^*- and y-axes.

stiffness matrix, using a 'physical approach' as follows, considering axial rigidity $(EA)_i$, flexural rigidities, $(EI_z)_i$ and $(EI_y)_i$ for bending about the local z^* and y^* local axes, and torsional rigidity $(GJ)_i$:

$$\mathbf{k}_*^\mathbf{i} = \left[\begin{array}{c|c} \mathbf{k}_{\mathbf{A*}}^\mathbf{i} & \mathbf{k}_{\mathbf{C*}}^{\mathbf{i}\ \mathbf{T}} \\ \hline \mathbf{k}_{\mathbf{C*}}^\mathbf{i} & \mathbf{k}_{\mathbf{B*}}^\mathbf{i} \end{array}\right] \tag{6.28}$$

where

$$\mathbf{k}_{\mathbf{A*}}^\mathbf{i} = \left[\begin{array}{ccc|ccc} \alpha_i & 0 & 0 & 0 & 0 & 0 \\ 0 & 12\delta_{zi}/L_i^2 & 0 & 0 & 0 & 6\delta_{zi}/L_i \\ 0 & 0 & 12\delta_{yi}/L_i^2 & 0 & -6\delta_{yi}/L_i & 0 \\ \hline 0 & 0 & 0 & \varepsilon_i & 0 & 0 \\ 0 & 0 & -6\delta_{yi}/L_i & 0 & 4\delta_{yi} & 0 \\ 0 & 6\delta_{zi}/L_i & 0 & 0 & 0 & 4\delta_{zi} \end{array}\right]; \quad \begin{array}{l} \alpha_i = (EA)_i/L_i \\ \delta_{zi} = (EI_z)_i/L_i \\ \delta_{yi} = (EI_z)_i/L_i \\ \varepsilon_i = (GJ)_i/L_i \end{array} \tag{6.28a}$$

$$\mathbf{k}_{\mathbf{B*}}^\mathbf{i} = \left[\begin{array}{ccc|ccc} \alpha_i & 0 & 0 & 0 & 0 & 0 \\ 0 & 12\delta_{zi}/L_i^2 & 0 & 0 & 0 & -6\delta_{zi}/L_i \\ 0 & 0 & 12\delta_{yi}/L_i^2 & 0 & 6\delta_{yi}/L_i & 0 \\ \hline 0 & 0 & 0 & \varepsilon_i & 0 & 0 \\ 0 & 0 & 6\delta_{yi}/L_i & 0 & 4\delta_{yi} & 0 \\ 0 & -6\delta_{zi}/L_i & 0 & 0 & 0 & 4\delta_{zi} \end{array}\right] \tag{6.28b}$$

$$\mathbf{k}_{\mathbf{C*}}^\mathbf{i} = \left[\begin{array}{ccc|ccc} -\alpha_i & 0 & 0 & 0 & 0 & 0 \\ 0 & -12\delta_{zi}/L_i^2 & 0 & 0 & 0 & -6\delta_{zi}/L_i \\ 0 & 0 & -12\delta_{yi}/L_i^2 & 0 & 6\delta_{yi}/L_i & 0 \\ \hline 0 & 0 & 0 & -\varepsilon_i & 0 & 0 \\ 0 & 0 & -6\delta_{yi}/L_i & 0 & 2\delta_{yi} & 0 \\ 0 & 6\delta_{zi}/L_i & 0 & 0 & 0 & 2\delta_{zi} \end{array}\right] \tag{6.28c}$$

Also, as assumed in the case of grids, the cross-sections are considered to be symmetric with respect to two mutually perpendicular axes, and are free to warp out of their planes under torsion. Accordingly, the bending and torsional stiffnesses get uncoupled. Furthermore, the flexural stiffnesses in the two orthogonal planes are also assumed to be uncoupled, while the axial stiffness, like the torsional stiffness, is uncoupled from all others.

Element Stiffness Matrix in Global Axes System

In order to derive the element stiffness matrix $\mathbf{k}^\mathbf{i}$ in the global axes system, we need to first generate the transformation matrix $\mathbf{T}^\mathbf{i}$ that enables a displacement or force vector, defined in the global axes system (Fig. 6.39b) to be transformed to the local coordinate system (Fig. 6.39a), without change in the length of the vector. As explained earlier with reference to the space truss element (Eq. 4.19), this can be done for the translational degrees of freedom by considering the rotation matrix, expressed in terms of the nine direction cosines. The same rotation matrix $\mathbf{R^i}$ (of order 3×3) is applicable for the rotation/moment vectors, whereby the transformation matrix $\mathbf{T^i}$ to be used in the transformations, $\mathbf{D^i_*} = \mathbf{T^i D^i}$ and $\mathbf{F^i_*} = \mathbf{T^i F^i}$, takes the following form:

$$\mathbf{T^i} = \left[\begin{array}{c|c|c|c} \mathbf{R^i} & \mathbf{O} & \mathbf{O} & \mathbf{O} \\ \hline \mathbf{O} & \mathbf{R^i} & \mathbf{O} & \mathbf{O} \\ \hline \mathbf{O} & \mathbf{O} & \mathbf{R^i} & \mathbf{O} \\ \hline \mathbf{O} & \mathbf{O} & \mathbf{O} & \mathbf{R^i} \end{array}\right] \tag{6.29}$$

where $\mathbf{R^i}$ is a 3×3 orthogonal matrix, called *rotation matrix*, comprising 9 *direction cosines*, linking each of the unit vectors, $\hat{\mathbf{i}}^*, \hat{\mathbf{j}}^*, \hat{\mathbf{k}}^*$ in the local axes system to the unit vectors $\hat{\mathbf{i}}, \hat{\mathbf{j}}, \hat{\mathbf{k}}$ in the global axes system (refer Fig. 6.39):

$$\begin{bmatrix} \hat{\mathbf{i}}^* \\ \hat{\mathbf{j}}^* \\ \hat{\mathbf{k}}^* \end{bmatrix} = \begin{bmatrix} \cos\theta^i_{x^*x} & \cos\theta^i_{x^*y} & \cos\theta^i_{x^*z} \\ \cos\theta^i_{y^*x} & \cos\theta^i_{y^*y} & \cos\theta^i_{y^*z} \\ \cos\theta^i_{z^*x} & \cos\theta^i_{z^*y} & \cos\theta^i_{z^*z} \end{bmatrix} \begin{bmatrix} \hat{\mathbf{i}} \\ \hat{\mathbf{j}} \\ \hat{\mathbf{k}} \end{bmatrix} = \begin{bmatrix} c^i_{11} \\ c^i_{21} \\ c^i_{31} \end{bmatrix} \hat{\mathbf{i}} + \begin{bmatrix} c^i_{12} \\ c^i_{22} \\ c^i_{32} \end{bmatrix} \hat{\mathbf{j}} + \begin{bmatrix} c^i_{13} \\ c^i_{23} \\ c^i_{33} \end{bmatrix} \hat{\mathbf{k}}$$

$$\Rightarrow \qquad \mathbf{R^i} = \begin{bmatrix} c^i_{11} & c^i_{12} & c^i_{13} \\ c^i_{21} & c^i_{22} & c^i_{23} \\ c^i_{31} & c^i_{32} & c^i_{33} \end{bmatrix} \tag{6.30}$$

Of the 9 direction cosines, the three in the first row, which refer to the direction cosines of the local x^*-axis, can be directly evaluated using the global axes coordinates of the start node B and end node A of any space frame element, AB:

$$c^i_{11} = \cos\theta^i_{x^*x} = \frac{x_B - x_A}{L_i} = c_{ix}$$

$$c^i_{12} = \cos\theta^i_{x^*y} = \frac{y_B - y_A}{L_i} = c_{iy} \tag{6.31a}$$

$$c^i_{13} = \cos\theta^i_{x^*z} = \frac{z_B - z_A}{L_i} = c_{iz}$$

The length L_i of the member can be easily determined:

$$L_i = \sqrt{(x_B - x_A)^2 + (y_B - y_A)^2 + (z_B - z_A)^2} \tag{6.31b}$$

The complete $\mathbf{R^i}$ matrix is required to be defined for every space frame element, and it is particularly important to generate correctly the second and third rows of the matrix, which require additional input data. The three elements of the second row of the rotation matrix $\mathbf{R^i}$ in Eq. 6.30, refer to the direction cosines of the local y^*-axis, while those of the third row refer to the direction cosines of the local z^*-axis in the global axes framework. There are several alternative ways of generating these direction cosines. A convenient way of doing this is by first describing the defining the x^*y^* plane containing the space frame element AB by defining the coordinates of a reference point Q (x_Q, y_Q, z_Q) in this plane. By normalising the position vector $\mathbf{AQ}$, directed from A to Q, vector as a unit vector $\hat{\mathbf{q}}^*$, we can generate the unit vector $\hat{\mathbf{k}}^*$ as the cross product, $\hat{\mathbf{i}}^* \times \hat{\mathbf{q}}^{\mathbf{i}*}$. Then, we can generate the unit vector $\hat{\mathbf{j}}^*$ as the cross product, $\hat{\mathbf{k}}^* \times \hat{\mathbf{i}}^*$. Carrying out these operations,

$$\hat{\mathbf{q}}^{\mathbf{i}*} = q_x^i\hat{\mathbf{i}} + q_y^i\hat{\mathbf{j}} + q_z^i\hat{\mathbf{k}} \quad ; \quad \hat{\mathbf{k}}^* = c_{31}^i\hat{\mathbf{i}} + c_{32}^i\hat{\mathbf{j}} + c_{33}^i\hat{\mathbf{k}} \quad ; \quad \hat{\mathbf{j}}^* = c_{21}^i\hat{\mathbf{i}} + c_{22}^i\hat{\mathbf{j}} + c_{23}^i\hat{\mathbf{k}} \tag{6.32}$$

where

$$\hat{\mathbf{q}}^* = \left(\frac{x_Q - x_A}{AQ}\right)\hat{\mathbf{i}} + \left(\frac{y_Q - y_A}{AQ}\right)\hat{\mathbf{j}} + \left(\frac{y_Q - y_A}{AQ}\right)\hat{\mathbf{k}} \tag{6.32a}$$

$$AQ = \sqrt{(x_Q - x_A)^2 + (y_Q - y_A)^2 + (z_Q - z_A)^2}$$

$$q_x^i = \left(\frac{x_Q - x_A}{AQ}\right) \;;\; q_y^i = \left(\frac{y_Q - y_A}{AQ}\right) \;;\; q_z^i = \left(\frac{z_Q - z_A}{AQ}\right) \tag{6.32b}$$

$$\Rightarrow c_{31}^i = \begin{vmatrix} c_{12}^i & c_{13}^i \\ q_y^i & q_z^i \end{vmatrix} \;;\; c_{32}^i = \begin{vmatrix} c_{13}^i & c_{11}^i \\ q_z^i & q_x^i \end{vmatrix} \;;\; c_{33}^i = \begin{vmatrix} c_{11}^i & c_{12}^i \\ q_x^i & q_y^i \end{vmatrix}$$

$$c_{21}^i = \begin{vmatrix} c_{32}^i & c_{33}^i \\ c_{12}^i & c_{13}^i \end{vmatrix} \;;\; c_{22}^i = \begin{vmatrix} c_{33}^i & c_{31}^i \\ c_{13}^i & c_{11}^i \end{vmatrix} \;;\; c_{23}^i = \begin{vmatrix} c_{31}^i & c_{32}^i \\ c_{11}^i & c_{12}^i \end{vmatrix} \tag{6.32c}$$

After finding the member-end displacements in the global axes system, we can find the member-end forces, using the transformation, $\mathbf{F}_*^{\mathbf{i}} = \left(\mathbf{k}_*^{\mathbf{i}}\mathbf{T}^{\mathbf{i}}\right)\mathbf{D}^{\mathbf{i}}$ and also generate the structure stiffness matrix, $\mathbf{k}^{\mathbf{i}} = \mathbf{T}^{\mathbf{i}^T}\mathbf{k}_*^{\mathbf{i}}\mathbf{T}^{\mathbf{i}}$. The other steps in the analysis are as outlined earlier for the plane frame analysis.

6.5.2 Space Frame Element with 6 Degrees of Freedom (Reduced Stiffness)

As explained earlier, in the reduced element stiffness formulation, it suffices to work with only six degrees of freedom for the space frame element, which is assumed to be simply supported in two orthogonal planes and restrained against torsion at the start node (refer Fig. 6.38b). The 6×6 reduced element stiffness matrix, $\tilde{\mathbf{k}}_*^{\mathbf{i}}$, can easily be generated from first principles (using a 'physical approach'), or by eliminating appropriately six rows from the 6×6 conventional stiffness matrix, $\mathbf{k}_*^{\mathbf{i}}$, defined by Eq. 6.28. The inverse of this matrix gives the flexibility matrix, $\mathbf{f}_*^{\mathbf{i}}$, for the space frame element.

$$\tilde{\mathbf{k}}_*^{\mathbf{i}} = \left[\begin{array}{ccc|ccc} \alpha_i & 0 & 0 & 0 & 0 & 0 \\ 0 & 4\delta_{yi} & 0 & 2\delta_{yi} & 0 & 0 \\ 0 & 0 & 4\delta_{zi} & 0 & 2\delta_{zi} & 0 \\ \hline 0 & 2\delta_{yi} & 0 & 4\delta_{yi} & 0 & 0 \\ 0 & 0 & 2\delta_{zi} & 0 & 4\delta_{zi} & 0 \\ 0 & 0 & 0 & 0 & 0 & \varepsilon_i \end{array}\right]; \quad \begin{array}{l} \alpha_i = (EA)_i / L_i \\ \delta_{zi} = (EI_z)_i / L_i \\ \delta_{yi} = (EI_z)_i / L_i \\ \varepsilon_i = (GJ)_i / L_i \end{array} \tag{6.33}$$

$$\mathbf{f}_*^{\mathbf{i}} = \left[\begin{array}{ccc|ccc} 1/\alpha_i & 0 & 0 & 0 & 0 & 0 \\ 0 & 1/(3\delta_{yi}) & 0 & -1/(6\delta_{yi}) & 0 & 0 \\ 0 & 0 & 1/(3\delta_{zi}) & 0 & -1/(6\delta_{zi}) & 0 \\ \hline 0 & -1/(6\delta_{yi}) & 0 & 1/(3\delta_{yi}) & 0 & 0 \\ 0 & 0 & -1/(6\delta_{zi}) & 0 & 1/(3\delta_{zi}) & 0 \\ 0 & 0 & 0 & 0 & 0 & 1/\varepsilon_i \end{array}\right]; \quad \begin{array}{l} \alpha_i = (EA)_i / L_i \\ \delta_{zi} = (EI_z)_i / L_i \\ \delta_{yi} = (EI_z)_i / L_i \\ \varepsilon_i = (GJ)_i / L_i \end{array} \tag{6.34}$$

As explained earlier, the structure stiffness matrix $\mathbf{k}$ can be assembled by a 'direct stiffness' approach, using the element displacement transformation matrix, $\mathbf{T_D^i}$, and summing up the contributions, $\mathbf{T_D^{i\,T}}\tilde{\mathbf{k}}_*^{\mathbf{i}}\mathbf{T_D^i}$ (direct stiffness method). We can generate the displacement transformation $\mathbf{T_D^i}$ matrices for the various elements, each of order $6\times(n+r)$, applying a unit displacement at a time ($D_j = 1$) on the restrained (primary) structure, whereby $\mathbf{D}_*^{\mathbf{i}} = \left[\mathbf{T_{DA}^i} \mid \mathbf{T_{DR}^i}\right]\left[\begin{array}{c}\mathbf{D_A} \\ \hline \mathbf{D_R}\end{array}\right]$. We need to invoke the direction cosines defined by the rotation matrix $\mathbf{R^i}$ (Eq. 6.30), and also to account for the effect of 'chord rotations', as done earlier for the plane frame element. By extending the derivation for the plane frame element (involving three global coordinates at each end), to a space frame element with six global coordinates at each end, we can generate the components of the $\mathbf{T_D^i}$ matrix as follows. Let us consider, for convenience, a global coordinate numbering system of '1', '2',..., '6' to define the global *x*-, *y*- and *z*- directions at the start node respectively, and likewise, '7', '8',..., 12' at the end node of the element. Accordingly, $\mathbf{D}_*^{\mathbf{i}} = \mathbf{T_D^i D}$ $\Rightarrow$

$$\mathbf{T_D^i} = \begin{array}{c} \begin{array}{cccccccccccc} (1) & (2) & (3) & (4) & (5) & (6) & (7) & (8) & (9) & (10) & (11) & (12) \end{array} \\ \left[\begin{array}{ccc|ccc|ccc|ccc} -c_{11}^i & -c_{12}^i & -c_{13}^i & 0 & 0 & 0 & c_{11}^i & c_{12}^i & c_{13}^i & 0 & 0 & 0 \\ -c_{31}^i/L_i & -c_{32}^i/L_i & -c_{33}^i/L_i & c_{21}^i & c_{22}^i & c_{23}^i & c_{31}^i/L_i & c_{32}^i/L_i & c_{33}^i/L_i & 0 & 0 & 0 \\ c_{21}^i/L_i & c_{22}^i/L_i & c_{23}^i/L_i & c_{31}^i & c_{32}^i & c_{33}^i & -c_{21}^i/L_i & -c_{22}^i/L_i & -c_{23}^i/L_i & 0 & 0 & 0 \\ \hline -c_{31}^i/L_i & -c_{32}^i/L_i & -c_{33}^i/L_i & 0 & 0 & 0 & c_{31}^i/L_i & c_{32}^i/L_i & c_{33}^i/L_i & c_{21}^i & c_{22}^i & c_{23}^i \\ c_{21}^i/L_i & c_{22}^i/L_i & c_{23}^i/L_i & 0 & 0 & 0 & -c_{21}^i/L_i & -c_{22}^i/L_i & -c_{23}^i/L_i & c_{31}^i & c_{32}^i & c_{33}^i \\ 0 & 0 & 0 & -c_{11}^i & -c_{12}^i & -c_{13}^i & 0 & 0 & 0 & c_{11}^i & c_{12}^i & c_{13}^i \end{array}\right] \end{array} \tag{6.35}$$

Alternatively, we can adopt a static approach instead of a kinematic approach, as indicated in Fig. 4.35b, to generate the $\mathbf{T}_{\mathbf{D}}^{\mathbf{i}\,\mathbf{T}}$ matrix, by making use of the *Principle of Contra-gradient*, whereby $\begin{bmatrix} \mathbf{F_A} \\ \hline \mathbf{F_R} \end{bmatrix} = \begin{bmatrix} \mathbf{T}_{\mathbf{DA}}^{\mathbf{i}\,\mathbf{T}} \\ \hline \mathbf{T}_{\mathbf{DR}}^{\mathbf{i}\,\mathbf{T}} \end{bmatrix} \mathbf{F}_*^{\mathbf{i}}$, as explained earlier.

In practice, the global coordinates may have any set of numbers, some of which may correspond to restrained coordinates. However, if the global *x*-, *y*- and *z*-axes are chosen appropriately, then the first 6×6 block of the $\mathbf{T}_{\mathbf{D}}^{\mathbf{i}}$ matrix (or the $\mathbf{T}_{\mathbf{D}}^{\mathbf{i}\,\mathbf{T}}$ matrix) should be assigned to the *start node*, while the second block should be assigned to the *end node*.

It is convenient to also generate and store the $\tilde{\mathbf{k}}_*^{\mathbf{i}}\mathbf{T}_{\mathbf{D}}^{\mathbf{i}}$ matrices for the various elements to facilitate the computation of the bar forces using the relation, $\mathbf{F}_*^{\mathbf{i}} = \left(\tilde{\mathbf{k}}_*^{\mathbf{i}}\mathbf{T}_{\mathbf{D}}^{\mathbf{i}}\right)\mathbf{D}$, and also to store the products, $\mathbf{T}_{\mathbf{D}}^{\mathbf{i}\,\mathbf{T}}\tilde{\mathbf{k}}_*^{\mathbf{i}}\mathbf{T}_{\mathbf{D}}^{\mathbf{i}}$, for assembling the structure stiffness matrix $\mathbf{k}$. The other steps in the analysis are as outlined earlier for the plane frame analysis.

When member end releases are encountered, the number of active degrees of freedom (kinematical indeterminacy) can be appropriately reduced and the degrees of freedom in the concerned members also reduced. The element stiffness matrix, displacement transformation matrix and fixed end forces, if any, should be appropriately computed, as discussed in the analysis of beams and plane frames.

EXAMPLE 6.12

Consider the space frame example in Chapter 4 (Example 4.18), comprising three identical members (in a tripod arrangement), each 2m long, inter-connected to a joint at O at top, 1m above ground, with their hinged bases forming an equilateral triangle on level ground. Instead of treating the joint at O as a ball-and-socket joint, treat this as a rigid joint (i.e., not as a space truss). Analyse this frame by the Stiffness Method, given that the joint at O is subject to a gravity load of 60 kN and find the complete response. Assume all bars to have a tubular cross-section, with a mean radius of 100mm and thickness of 10mm. Assume an elastic modulus $E = 2.5 \times 10^4$ MPa and a Poisson's ratio, $\nu = 0.3$.

SOLUTION

Solution by Stiffness Method (Reduced Element Stiffness)

Let us adopt the global axes system (with the origin at B) as indicated in Fig. 6.40a (similar to Fig. 4.46 in Example 4.18). Owing to symmetry in the structure and the loading, we note that there is only one active degree of freedom involved (vertical deflection at the joint O), with $\mathbf{F_A} = F_1 = -60\,\text{kN}$, as indicated in Fig. 6.40a. Also, each member will deflect and bend identically in its own vertical plane, acting like a plane frame element, with a hinged support at one end and elastic restraints against vertical translation and rotation at the other end O.

Thus, to find the support reactions, it is sufficient to define two restrained global coordinates, marked at the support B in the *y*- and *z*-axes directions, as indicated in Fig. 6.40a.

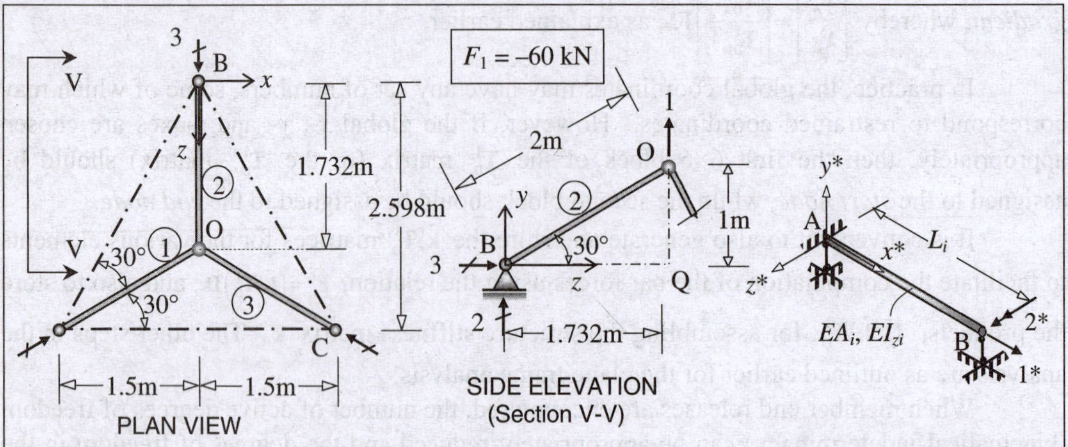

(*a*) Global coordinates (*b*) Local coordinates

Figure 6.40 Plan view and side elevation of tripod structure (Example 6.12)

It is sufficient to consider two degrees of freedom in each of the three elements in the reduced stiffness formulation, corresponding to the axial degree of freedom and the rotation at the end node as shown in Fig. 6.40b; torsional stiffness and flexural stiffness in the perpendicular direction need not be considered. All the elements have the same sectional and material properties and stiffness measures:

$$E = 2.5\times10^7 \text{ kN/m}^2;\ A_i = 2\pi(0.1)(0.01) = 2\pi\times10^{-3} \text{ m}^2;\ I_{zi} = I_{yi} = \pi(0.1)^3(0.01) = \pi\times10^{-5} \text{ m}^4$$

$$\Rightarrow \alpha_i = (EA)_i / L_i = 78539.82 \text{ kN/m};\ \delta_{zi} = (EI)_i / L_i = 392.6991 \text{ kNm}$$

The element stiffness matrix takes the following diagonal form:

$$\tilde{\mathbf{k}}_*^{\mathbf{i}} = \begin{bmatrix} \alpha_i & 0 \\ 0 & 3\delta_{zi} \end{bmatrix} \quad \Rightarrow \tilde{\mathbf{k}}_*^1 = \tilde{\mathbf{k}}_*^2 = \tilde{\mathbf{k}}_*^3 = \begin{bmatrix} 78539.82 & 0 \\ 0 & 1178.0973 \end{bmatrix}$$

The unassembled reduced element stiffness matrix of order 6×6 is given by

$$\tilde{\mathbf{k}}_* = \begin{bmatrix} \tilde{\mathbf{k}}_*^1 & \mathbf{O} & \mathbf{O} \\ \mathbf{O} & \tilde{\mathbf{k}}_*^2 & \mathbf{O} \\ \mathbf{O} & \mathbf{O} & \tilde{\mathbf{k}}_*^3 \end{bmatrix} = \left[\begin{array}{cc|cc|cc} 78539.82 & 0 & 0 & 0 & 0 & 0 \\ 0 & 1178.0973 & 0 & 0 & 0 & 0 \\ \hline 0 & 0 & 78539.82 & 0 & 0 & 0 \\ 0 & 0 & 0 & 1178.0973 & 0 & 0 \\ \hline 0 & 0 & 0 & 0 & 78539.82 & 0 \\ 0 & 0 & 0 & 0 & 0 & 1178.0973 \end{array}\right]$$

For the adopted member incidences (end node always at O), the displacement transformation matrices, satisfying $\mathbf{D}_*^i = \left[\mathbf{T}_{DA}^i \mid \mathbf{T}_{DR}^i\right]\begin{bmatrix}\mathbf{D}_A \\ \hline \mathbf{D}_R\end{bmatrix}$, can be generated, by inspection, in terms of the direction cosines (considering $\cos 30^\circ = 0.86603$ and $\sin 30^\circ = 0.5$, and converting the chord rotations to equivalent end rotations) as follows:

$$\mathbf{T}_{DA} = \begin{bmatrix}\mathbf{T}_{DA}^1 \\ \mathbf{T}_{DA}^2 \\ \mathbf{T}_{DA}^3\end{bmatrix} = \begin{bmatrix}0.5 \\ -0.433013 \\ \hdashline 0.5 \\ -0.433013 \\ \hdashline 0.5 \\ -0.433013\end{bmatrix}; \quad \mathbf{T}_{DR} = \begin{bmatrix}\mathbf{T}_{DR}^1 \\ \mathbf{T}_{DR}^2 \\ \mathbf{T}_{DR}^3\end{bmatrix} = \begin{bmatrix}-0.5 & -0.866025 \\ 0.433013 & -0.25 \\ \hdashline 0 & 0 \\ 0 & 0 \\ \hdashline 0 & 0 \\ 0 & 0\end{bmatrix}$$

$$\Rightarrow \tilde{\mathbf{k}}_*\mathbf{T}_{DA} = \begin{bmatrix}39269.91 \\ -510.1314 \\ \hdashline 39269.91 \\ -510.1314 \\ \hdashline 39269.91 \\ -510.1314\end{bmatrix} \Rightarrow \begin{matrix}\mathbf{k}_{AA} = \mathbf{T}_{DA}{}^T\tilde{\mathbf{k}}_*\mathbf{T}_{DA} = [59567.55] \\ \mathbf{k}_{RA} = \mathbf{T}_{DR}{}^T\tilde{\mathbf{k}}_*\mathbf{T}_{DA} = \begin{bmatrix}-19855.85 \\ -33881.19\end{bmatrix}\end{matrix} \text{ kN/m}$$

The solution for the vertical deflection at O, support reactions at B and element forces are given by:

$$\mathbf{D}_A = [\mathbf{k}_{AA}]^{-1}\mathbf{F}_A \quad \Rightarrow D_1 = \frac{1}{59567.55}(-60) = -1.0072599 \times 10^{-3} \text{ m}$$

$$\mathbf{F}_R = \mathbf{k}_{RA}\mathbf{D}_A = \begin{bmatrix}-19855.85 \\ -33881.19\end{bmatrix}\left\{-1.0072599 \times 10^{-3}\right\} = \begin{Bmatrix}20.000 \text{ kN} \\ 34.127 \text{ kN}\end{Bmatrix}$$

$$\mathbf{F}_*^i = \left(\tilde{\mathbf{k}}_*^i\mathbf{T}_{DA}^i\right)\mathbf{D}_A \quad \Rightarrow \mathbf{F}_*^1 = \mathbf{F}_*^2 = \mathbf{F}_*^3 = \begin{bmatrix}39269.91 \\ -510.1314\end{bmatrix}\left\{-1.0072599 \times 10^{-3}\right\} = \begin{Bmatrix}-39.555 \text{ kN} \\ 0.5138 \text{ kNm}\end{Bmatrix}$$

The above forces may be compared with the results of Example 4.18, where the axial (compressive) force in the bar is exactly 40 kN (as against 39.555 kN here). The replacement of the ball-and-socket joint at O by a rigid (moment-resistant) joint does not alter the force response significantly. The bending moments and shear forces generated in the members have negligible magnitude, as indicated in Fig. 6.41, which depicts the free-body, axial force, shear force, bending moment and deflection diagrams for a typical member (leg of the tripod).

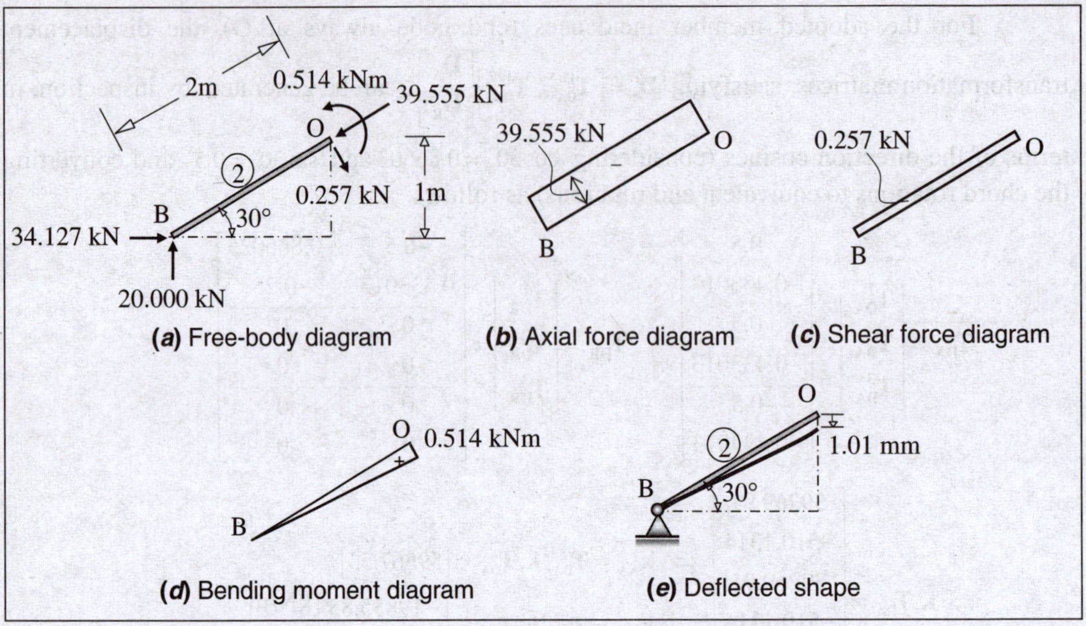

Figure 6.41 Response of a typical leg of the tripod (Example 6.12)

EXAMPLE 6.13

Analyse the space frame, comprising three members, fully fixed at supports A and D, as shown in Fig. 6.42, by the conventional stiffness method and find the complete response. Assume all members to have a tubular cross-section, with a mean radius of 150mm and thickness of 10mm. Assume elastic modulus $E = 2.5{\times}10^5$ MPa and Poisson's ratio, $\nu = 0.3$.

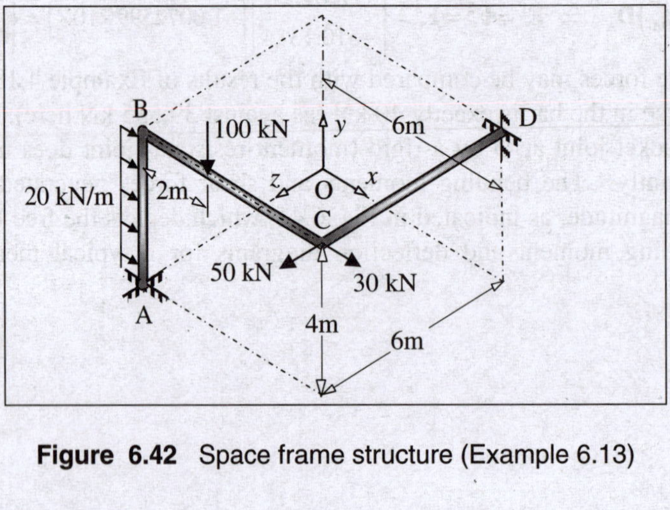

Figure 6.42 Space frame structure (Example 6.13)

SOLUTION

Input Data

There are three prismatic frame elements ($m=3$) and 4 joints with 12 *active* and 12 *restrained* degrees of freedom ($n=r=12$), with global coordinates, as indicated in Fig. 6.43a. Corresponding to these coordinates, the active load vector prescribed has only two non-zero components, $F_7 = 30$ kN and $F_9 = 50$ kN. The intermediate loads acting on elements '1' and '2' have to be converted to equivalent joint loads. There are no support displacements prescribed, whereby $\mathbf{D_R} = \mathbf{O}$.

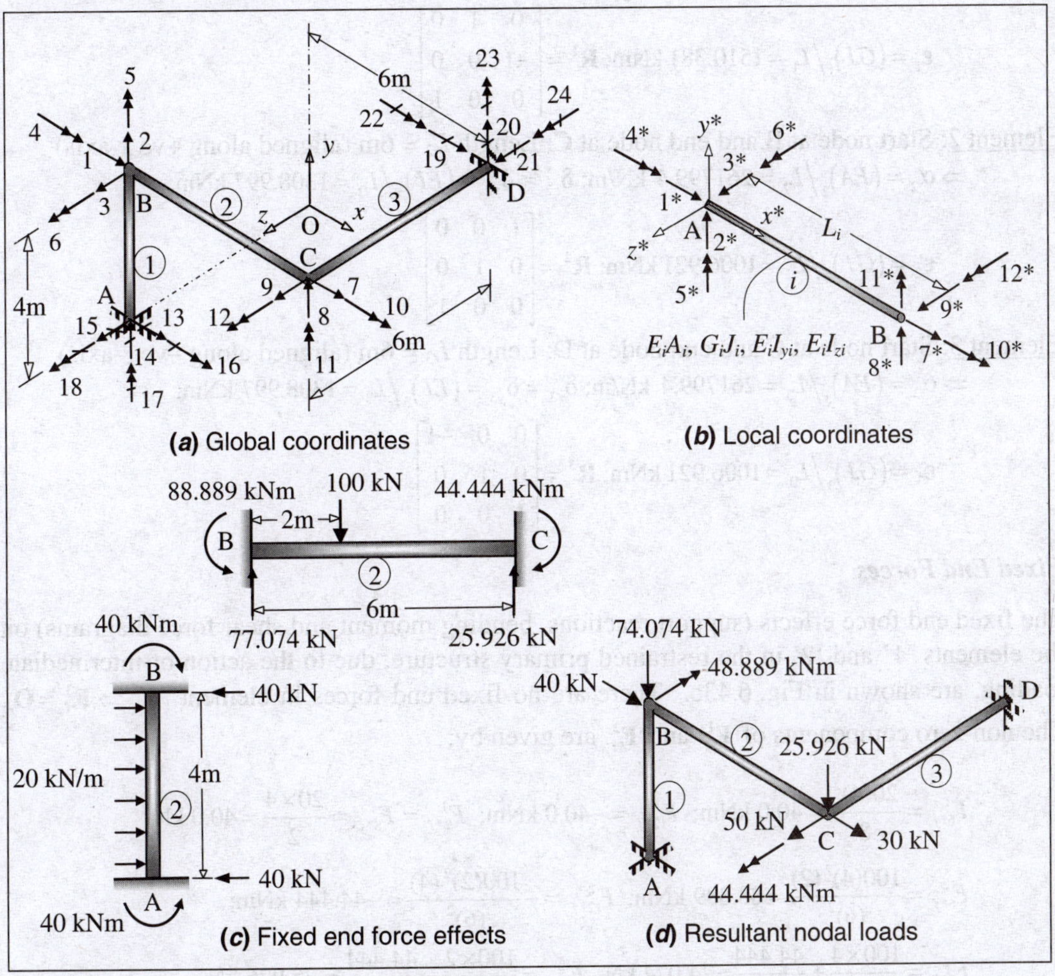

Figure 6.43 Global and local coordinates, fixed end force effects and resultant loads (Example 6.13)

The local coordinates for a typical space frame element are as shown in Fig. 6.43b. All the elements have the same sectional and material properties and stiffness measures:

$$E = 2.5\times10^8\,\text{kN/m}^2,\ \nu=0.3 \quad \Rightarrow G = \frac{E}{2(1+0.3)} = 0.9615385\times10^8\,\text{kN/m}^2$$

$$A = 2\pi(0.1)(0.01) = 2\pi\times10^{-3}\ \text{m}^2,\ I_z = I_y = \pi(0.1)^3(0.01) = \pi\times10^{-5}\ \text{m}^4,\ J = 2\pi\times10^{-5}\ \text{m}^4$$

The other element-wise details, include rotation matrix (with direction cosines) are given below:

Element 1: Start node at A and end node at B; Length L_1 = 4m (aligned along +ve y-axis)

$$\Rightarrow \alpha_1 = (EA)_1/L_1 = 392699.1\ \text{kN/m};\ \delta_{z1} = \delta_{y1} = (EI)_1/L_1 = 1963.495\ \text{kNm};$$

$$\varepsilon_1 = (GJ)_1/L_1 = 1510.381\ \text{kNm};\ \mathbf{R}^1 = \begin{bmatrix} 0 & 1 & 0 \\ -1 & 0 & 0 \\ 0 & 0 & 1 \end{bmatrix}$$

Element 2: Start node at B and end node at C; Length L_2 = 6m (aligned along +ve x-axis)

$$\Rightarrow \alpha_2 = (EA)_2/L_2 = 261799.4\ \text{kN/m};\ \delta_{z2} = \delta_{y2} = (EI)_2/L_2 = 1308.997\ \text{kNm};$$

$$\varepsilon_2 = (GJ)_2/L_2 = 1006.921\ \text{kNm};\ \mathbf{R}^2 = \begin{bmatrix} 1 & 0 & 0 \\ 0 & 1 & 0 \\ 0 & 0 & 1 \end{bmatrix}$$

Element 3: Start node at C and end node at D; Length L_3 = 6m (aligned along –ve z-axis)

$$\Rightarrow \alpha_3 = (EA)_3/L_3 = 261799.4\ \text{kN/m};\ \delta_{z3} = \delta_{y3} = (EI)_3/L_3 = 1308.997\ \text{kNm};$$

$$\varepsilon_3 = (GJ)_3/L_3 = 1006.921\ \text{kNm};\ \mathbf{R}^3 = \begin{bmatrix} 0 & 0 & -1 \\ 0 & 1 & 0 \\ 1 & 0 & 0 \end{bmatrix}$$

Fixed End Forces

The fixed end force effects (support reactions, bending moment and shear force diagrams) on the elements '1' and '2' in the restrained primary structure, due to the action of intermediate loading, are shown in Fig. 6.43c. There are no fixed end forces in element '3' $\Rightarrow \mathbf{F}^3_{*\mathbf{f}} = \mathbf{O}$. The non-zero components of $\mathbf{F}^1_{*\mathbf{f}}$ and $\mathbf{F}^2_{*\mathbf{f}}$ are given by:

$$F^1_{6*f} = \frac{20(4)^2}{8} = 40.0\ \text{kNm};\ F^1_{12*f} = -40.0\ \text{kNm};\ F^1_{2*f} = F^1_{8*f} = \frac{20\times4}{2} = 40.0\ \text{kN}$$

$$F^2_{6*f} = \frac{100(4)^2(2)}{(6)^2} = 88.889\ \text{kNm};\ F^2_{12*f} = -\frac{100(2)^2(4)}{(6)^2} = -44.444\ \text{kNm};$$

$$F^2_{2*f} = \frac{100\times4}{6} + \frac{44.444}{6} = 74.074\ \text{kN};\ F^2_{8*f} = \frac{100\times2}{6} - \frac{44.444}{6} = 25.926\ \text{kN}$$

Coordinate Transformations and Equivalent Joint Loads

The element transformation matrices $\mathbf{T}^1, \mathbf{T}^2$ and $\mathbf{T}^3$, satisfying $\mathbf{D}_*^i = \mathbf{T}^i\mathbf{D}$ and $\mathbf{F}_*^i = \mathbf{T}^i\mathbf{F}$, take the following form (refer Eq. 6.29 and Figs 6.43a and b), with the linking global coordinates indicated in parentheses.

$$
\mathbf{T}^1 = \begin{array}{c}
\scriptstyle (13)(14)(15)(16)(17)(18)\,(1)\,(2)\,(3)\,(4)\,(5)\,(6) \\
\left[\begin{array}{ccc|ccc|ccc|ccc}
0 & 1 & 0 & 0 & 0 & 0 & 0 & 0 & 0 & 0 & 0 & 0 \\
-1 & 0 & 0 & 0 & 0 & 0 & 0 & 0 & 0 & 0 & 0 & 0 \\
0 & 0 & 1 & 0 & 0 & 0 & 0 & 0 & 0 & 0 & 0 & 0 \\
\hline
0 & 0 & 0 & 0 & 1 & 0 & 0 & 0 & 0 & 0 & 0 & 0 \\
0 & 0 & 0 & -1 & 0 & 0 & 0 & 0 & 0 & 0 & 0 & 0 \\
0 & 0 & 0 & 0 & 0 & 1 & 0 & 0 & 0 & 0 & 0 & 0 \\
\hline
0 & 0 & 0 & 0 & 0 & 0 & 0 & 1 & 0 & 0 & 0 & 0 \\
0 & 0 & 0 & 0 & 0 & 0 & -1 & 0 & 0 & 0 & 0 & 0 \\
0 & 0 & 0 & 0 & 0 & 0 & 0 & 0 & 1 & 0 & 0 & 0 \\
\hline
0 & 0 & 0 & 0 & 0 & 0 & 0 & 0 & 0 & 0 & 1 & 0 \\
0 & 0 & 0 & 0 & 0 & 0 & 0 & 0 & 0 & -1 & 0 & 0 \\
0 & 0 & 0 & 0 & 0 & 0 & 0 & 0 & 0 & 0 & 0 & 1
\end{array}\right]
\end{array}
\begin{array}{c} \\ (1)\\(2)\\(3)\\(4)\\(5)\\(6)\\(7)\\(8)\\(9)\\(10)\\(11)\\(12) \end{array};
$$

$$
\mathbf{T}^2 = \begin{array}{c}
\scriptstyle (1)\,(2)\,(3)\,(4)\,(5)\,(6)\,(7)\,(8)\,(9)\,(10)\,(11)\,(12) \\
\left[\begin{array}{ccc|ccc|ccc|ccc}
1 & 0 & 0 & 0 & 0 & 0 & 0 & 0 & 0 & 0 & 0 & 0 \\
0 & 1 & 0 & 0 & 0 & 0 & 0 & 0 & 0 & 0 & 0 & 0 \\
0 & 0 & 1 & 0 & 0 & 0 & 0 & 0 & 0 & 0 & 0 & 0 \\
\hline
0 & 0 & 0 & 1 & 0 & 0 & 0 & 0 & 0 & 0 & 0 & 0 \\
0 & 0 & 0 & 0 & 1 & 0 & 0 & 0 & 0 & 0 & 0 & 0 \\
0 & 0 & 0 & 0 & 0 & 1 & 0 & 0 & 0 & 0 & 0 & 0 \\
\hline
0 & 0 & 0 & 0 & 0 & 0 & 1 & 0 & 0 & 0 & 0 & 0 \\
0 & 0 & 0 & 0 & 0 & 0 & 0 & 1 & 0 & 0 & 0 & 0 \\
0 & 0 & 0 & 0 & 0 & 0 & 0 & 0 & 1 & 0 & 0 & 0 \\
\hline
0 & 0 & 0 & 0 & 0 & 0 & 0 & 0 & 0 & 1 & 0 & 0 \\
0 & 0 & 0 & 0 & 0 & 0 & 0 & 0 & 0 & 0 & 1 & 0 \\
0 & 0 & 0 & 0 & 0 & 0 & 0 & 0 & 0 & 0 & 0 & 1
\end{array}\right]
\end{array}
$$

$$
\mathbf{T}^3 = \begin{array}{c}
\scriptstyle (7)\,(8)\,(9)(10)(11)(12)(19)(20)(21)(22)(23)(24) \\
\left[\begin{array}{ccc|ccc|ccc|ccc}
0 & 0 & -1 & 0 & 0 & 0 & 0 & 0 & 0 & 0 & 0 & 0 \\
0 & 1 & 0 & 0 & 0 & 0 & 0 & 0 & 0 & 0 & 0 & 0 \\
1 & 0 & 0 & 0 & 0 & 0 & 0 & 0 & 0 & 0 & 0 & 0 \\
\hline
0 & 0 & 0 & 0 & 0 & -1 & 0 & 0 & 0 & 0 & 0 & 0 \\
0 & 0 & 0 & 0 & 1 & 0 & 0 & 0 & 0 & 0 & 0 & 0 \\
0 & 0 & 0 & 1 & 0 & 0 & 0 & 0 & 0 & 0 & 0 & 0 \\
\hline
0 & 0 & 0 & 0 & 0 & 0 & 0 & 0 & -1 & 0 & 0 & 0 \\
0 & 0 & 0 & 0 & 0 & 0 & 0 & 1 & 0 & 0 & 0 & 0 \\
0 & 0 & 0 & 0 & 0 & 0 & 1 & 0 & 0 & 0 & 0 & 0 \\
\hline
0 & 0 & 0 & 0 & 0 & 0 & 0 & 0 & 0 & 0 & 0 & -1 \\
0 & 0 & 0 & 0 & 0 & 0 & 0 & 0 & 0 & 0 & 1 & 0 \\
0 & 0 & 0 & 0 & 0 & 0 & 0 & 0 & 0 & 1 & 0 & 0
\end{array}\right]
\end{array}
\begin{array}{c} \\ (1)\\(2)\\(3)\\(4)\\(5)\\(6)\\(7)\\(8)\\(9)\\(10)\\(11)\\(12) \end{array}
$$

The fixed end forces in global axes can now be assembled, by summing up the contributions of $\mathbf{T}^{1^T}\mathbf{F}_{*f}^1$, $\mathbf{T}^{2^T}\mathbf{F}_{*f}^2$ and $\mathbf{T}^{3^T}\mathbf{F}_{*f}^3 = \mathbf{O}$ at the appropriate locations. Thus, the components of the fixed end force vector (comprising equivalent joint loads on the structure), $\mathbf{F}_f = \left[\dfrac{\mathbf{F}_{fA}}{\mathbf{F}_{fR}}\right]$, and thereby, the resultant direct load vector, $\mathbf{F}_A - \mathbf{F}_{fA}$, can be generated.

$$\mathbf{F}^1_{*f} = \begin{bmatrix} 0\text{ kN} \\ 40\text{ kN} \\ 0\text{ kN} \\ 0\text{ kNm} \\ 0\text{ kNm} \\ 40\text{ kNm} \\ \hline 0\text{ kN} \\ 40\text{ kN} \\ 0\text{ kN} \\ 0\text{ kNm} \\ 0\text{ kNm} \\ -40\text{ kNm} \end{bmatrix} \Rightarrow \mathbf{T}^{1^T}\mathbf{F}^1_{*f} = \begin{bmatrix} -40\text{ kN} \\ 0\text{ kN} \\ 0\text{ kN} \\ 0\text{ kNm} \\ 0\text{ kNm} \\ 40\text{ kNm} \\ \hline -40\text{ kN} \\ 0\text{ kN} \\ 0\text{ kN} \\ 0\text{ kNm} \\ 0\text{ kNm} \\ -40\text{ kNm} \end{bmatrix} \begin{matrix} (13) \\ (14) \\ (15) \\ (16) \\ (17) \\ (18) \\ (1) \\ (2) \\ (3) \\ (4) \\ (5) \\ (6) \end{matrix} ; \quad \mathbf{F}^2_{*f} = \begin{bmatrix} 0\text{ kN} \\ 74.074\text{ kN} \\ 0\text{ kN} \\ 0\text{ kNm} \\ 0\text{ kNm} \\ 88.889\text{ kNm} \\ \hline 0\text{ kN} \\ 25.926\text{ kN} \\ 0\text{ kN} \\ 0\text{ kNm} \\ 0\text{ kNm} \\ -44.444\text{ kNm} \end{bmatrix} \Rightarrow \mathbf{T}^{2^T}\mathbf{F}^2_{*f} = \begin{bmatrix} 0\text{ kN} \\ 74.074\text{ kN} \\ 0\text{ kN} \\ 0\text{ kNm} \\ 0\text{ kNm} \\ 88.889\text{ kNm} \\ \hline 0\text{ kN} \\ 25.926\text{ kN} \\ 0\text{ kN} \\ 0\text{ kNm} \\ 0\text{ kNm} \\ -44.444\text{ kNm} \end{bmatrix} \begin{matrix} (1) \\ (2) \\ (3) \\ (4) \\ (5) \\ (6) \\ (7) \\ (8) \\ (9) \\ (10) \\ (11) \\ (12) \end{matrix}$$

$$\Rightarrow \mathbf{F}_{fA} = \begin{bmatrix} -40\text{ kN} \\ 74.074\text{ kN} \\ 0\text{ kN} \\ 0\text{ kNm} \\ 0\text{ kNm} \\ 48.889\text{ kNm} \\ \hline 0\text{ kN} \\ 25.926\text{ kN} \\ 0\text{ kN} \\ 0\text{ kNm} \\ 0\text{ kNm} \\ -44.444\text{ kNm} \end{bmatrix} \begin{matrix} (1) \\ (2) \\ (3) \\ (4) \\ (5) \\ (6) \\ (7) \\ (8) \\ (9) \\ (10) \\ (11) \\ (12) \end{matrix} ; \quad \mathbf{F}_{fR} = \begin{bmatrix} -40\text{ kN} \\ 0\text{ kN} \\ 0\text{ kN} \\ 0\text{ kNm} \\ 0\text{ kNm} \\ 40\text{ kNm} \\ \hline 0\text{ kN} \\ 0\text{ kN} \\ 0\text{ kN} \\ 0\text{ kNm} \\ 0\text{ kNm} \\ 0\text{ kNm} \end{bmatrix} \begin{matrix} (13) \\ (14) \\ (15) \\ (16) \\ (17) \\ (18) \\ (19) \\ (20) \\ (21) \\ (22) \\ (23) \\ (24) \end{matrix} ; \quad \mathbf{F}_A - \mathbf{F}_{fA} = \begin{bmatrix} 0 \\ 0 \\ 0 \\ 0 \\ 0 \\ 0 \\ \hline 30 \\ 0 \\ 50 \\ 0 \\ 0 \\ 0 \end{bmatrix} - \begin{bmatrix} -40 \\ 74.074 \\ 0 \\ 0 \\ 0 \\ 48.889 \\ \hline 0 \\ 25.926 \\ 0 \\ 0 \\ 0 \\ -44.444 \end{bmatrix} = \begin{bmatrix} 40\text{ kN} \\ -74.074\text{ kN} \\ 0\text{ kN} \\ 0\text{ kNm} \\ 0\text{ kNm} \\ -48.889\text{ kNm} \\ \hline 30\text{ kN} \\ -25.926\text{ kN} \\ 50\text{ kN} \\ 0\text{ kNm} \\ 0\text{ kNm} \\ 44.444\text{ kNm} \end{bmatrix} \begin{matrix} (1) \\ (2) \\ (3) \\ (4) \\ (5) \\ (6) \\ (7) \\ (8) \\ (9) \\ (10) \\ (11) \\ (12) \end{matrix}$$

The resultant loads are shown in Fig. 6.43d. The displacements $\mathbf{D}_A$ at the 12 active coordinates will be identical to those in the original loading diagram (with the intermediate loading) in Fig. 6.42.

Element and Structure Stiffness Matrices

The element stiffness matrices $\mathbf{k}^i_*$ for the three elements, satisfying $\mathbf{F}^i_* = \mathbf{k}^i_*\mathbf{D}^i_*$, take the form $\mathbf{k}^i_* = \left[\begin{array}{c|c} \mathbf{k}^i_{A*} & \mathbf{k}^{i\,T}_{C*} \\ \hline \mathbf{k}^i_{C*} & \mathbf{k}^i_{B*} \end{array}\right]$ (Eq. 6.28):

$$\mathbf{k}^1_{A*} = \left[\begin{array}{ccc|ccc} 392699.1 & 0 & 0 & 0 & 0 & 0 \\ 0 & 1472.621 & 0 & 0 & 0 & 2945.242 \\ 0 & 0 & 1472.621 & 0 & -2945.242 & 0 \\ \hline 0 & 0 & 0 & 1510.381 & 0 & 0 \\ 0 & 0 & -2945.242 & 0 & 7853.980 & 0 \\ 0 & 2945.242 & 0 & 0 & 0 & 7853.980 \end{array}\right]$$

$$\mathbf{k}^1_{B*} = \left[\begin{array}{ccc|ccc} 392699.1 & 0 & 0 & 0 & 0 & 0 \\ 0 & 1472.621 & 0 & 0 & 0 & -2945.242 \\ 0 & 0 & 1472.621 & 0 & 2945.242 & 0 \\ \hline 0 & 0 & 0 & 1510.381 & 0 & 0 \\ 0 & 0 & 2945.242 & 0 & 7853.980 & 0 \\ 0 & -2945.242 & 0 & 0 & 0 & 7853.980 \end{array}\right]$$

$$\mathbf{k}^1_{C*} = \left[\begin{array}{ccc|ccc} -392699.1 & 0 & 0 & 0 & 0 & 0 \\ 0 & -1472.621 & 0 & 0 & 0 & -2945.242 \\ 0 & 0 & -1472.621 & 0 & 2945.242 & 0 \\ \hline 0 & 0 & 0 & -151.0.381 & 0 & 0 \\ 0 & 0 & -2945.242 & 0 & 3926.990 & 0 \\ 0 & 2945.242 & 0 & 0 & 0 & 3926.990 \end{array}\right]$$

$$\mathbf{k}^2_{A*} = \mathbf{k}^3_{A*} = \left[\begin{array}{ccc|ccc} 261799.4 & 0 & 0 & 0 & 0 & 0 \\ 0 & 436.3323 & 0 & 0 & 0 & 1308.997 \\ 0 & 0 & 436.3323 & 0 & -1308.997 & 0 \\ \hline 0 & 0 & 0 & 1006.921 & 0 & 0 \\ 0 & 0 & -1308.997 & 0 & 5235.988 & 0 \\ 0 & 1308.997 & 0 & 0 & 0 & 5235.988 \end{array}\right]$$

$$\mathbf{k}^2_{B*} = \mathbf{k}^3_{B*} = \left[\begin{array}{ccc|ccc} 261799.4 & 0 & 0 & 0 & 0 & 0 \\ 0 & 436.3323 & 0 & 0 & 0 & -1308.997 \\ 0 & 0 & 436.3323 & 0 & 1308.997 & 0 \\ \hline 0 & 0 & 0 & 1006.921 & 0 & 0 \\ 0 & 0 & 1308.997 & 0 & 5235.988 & 0 \\ 0 & -1308.997 & 0 & 0 & 0 & 5235.988 \end{array}\right]$$

$$\mathbf{k}^2_{C*} = \mathbf{k}^3_{C*} = \left[\begin{array}{ccc|ccc} -261799.4 & 0 & 0 & 0 & 0 & 0 \\ 0 & -436.3323 & 0 & 0 & 0 & -1308.997 \\ 0 & 0 & -436.3323 & 0 & 1308.997 & 0 \\ \hline 0 & 0 & 0 & -1006.921 & 0 & 0 \\ 0 & 0 & -1308.997 & 0 & 2617.994 & 0 \\ 0 & 1308.997 & 0 & 0 & 0 & 2617.994 \end{array}\right]$$

The products, $\mathbf{k}^i_*\mathbf{T}^i$ and $\mathbf{T}^{i^T}\mathbf{k}^i_*\mathbf{T}^i$, with the relevant global coordinates (marked in parenthesis), as defined earlier, may now be computed directly (using MATLAB). By summing up the the contributions of $\mathbf{T}^{i^T}\mathbf{k}^i_*\mathbf{T}^i = \mathbf{k}^i = \left[\begin{array}{c|c} \mathbf{k}^i_A & \mathbf{k}^{i\,T}_C \\ \hline \mathbf{k}^i_C & \mathbf{k}^i_B \end{array}\right]$ of order 12×12 for each of the 3 elements, at the appropriate coordinate locations, the structure stiffness matrix $\mathbf{k}$ of order 24×24, satisfying $\mathbf{F} = \mathbf{kD}$, can be assembled. It takes the following partitioned form:

$$\Rightarrow \qquad \mathbf{k} = \left[\begin{array}{c|c} \mathbf{k}_{AA} & \mathbf{k}_{AR} \\ \hline \mathbf{k}_{RA} & \mathbf{k}_{RR} \end{array}\right] = \left[\begin{array}{cc|cc} \mathbf{k}_B^1 + \mathbf{k}_A^2 & \mathbf{k}_C^{2\,T} & \mathbf{k}_C^1 & \mathbf{O} \\ \mathbf{k}_C^2 & \mathbf{k}_B^2 + \mathbf{k}_A^3 & \mathbf{O} & \mathbf{k}_C^{3\,T} \\ \hline \mathbf{k}_C^{1\,T} & \mathbf{O} & \mathbf{k}_A^1 & \mathbf{O} \\ \mathbf{O} & \mathbf{k}_C^3 & \mathbf{O} & \mathbf{k}_B^3 \end{array}\right],$$

where:

$$\mathbf{k}_A^1 = \left[\begin{array}{ccc|ccc} 1472.622 & 0 & 0 & 0 & 0 & -2945.243 \\ 0 & 392699.1 & 0 & 0 & 0 & 0 \\ 0 & 0 & 1472.622 & 2945.243 & 0 & 0 \\ \hline 0 & 0 & 2945.243 & 7853.982 & 0 & 0 \\ 0 & 0 & 0 & 0 & 1510.381 & 0 \\ -2945.243 & 0 & 0 & 0 & 0 & 7853.982 \end{array}\right]$$

$$\mathbf{k}_B^1 = \left[\begin{array}{ccc|ccc} 1472.622 & 0 & 0 & 0 & 0 & 2945.243 \\ 0 & 392699.1 & 0 & 0 & 0 & 0 \\ 0 & 0 & 1472.622 & -2945.243 & 0 & 0 \\ \hline 0 & 0 & -2945.243 & 7853.982 & 0 & 0 \\ 0 & 0 & 0 & 0 & 1510.381 & 0 \\ 2945.243 & 0 & 0 & 0 & 0 & 7853.982 \end{array}\right]$$

$$\mathbf{k}_C^1 = \left[\begin{array}{ccc|ccc} -1472.621 & 0 & 0 & 0 & 0 & 2945.242 \\ 0 & -392699.1 & 0 & 0 & 0 & 0 \\ 0 & 0 & -1472.621 & -2945.242 & 0 & 0 \\ \hline 0 & 0 & 2945.242 & 3926.991 & 0 & 0 \\ 0 & 0 & 0 & 0 & -1510.381 & 0 \\ -2945.242 & 0 & 0 & 0 & 0 & 3926.991 \end{array}\right]$$

$$\mathbf{k}_A^2 = \left[\begin{array}{ccc|ccc} 261799.4 & 0 & 0 & 0 & 0 & 0 \\ 0 & 436.332 & 0 & 0 & 0 & 1308.997 \\ 0 & 0 & 436.332 & 0 & -1308.997 & 0 \\ \hline 0 & 0 & 0 & 1006.921 & 0 & 0 \\ 0 & 0 & -1308.997 & 0 & 5235988 & 0 \\ 0 & 1308.997 & 0 & 0 & 0 & 5235.988 \end{array}\right]$$

$$\mathbf{k}_B^2 = \left[\begin{array}{ccc|ccc} 261799.4 & 0 & 0 & 0 & 0 & 0 \\ 0 & 436.332 & 0 & 0 & 0 & -1308.997 \\ 0 & 0 & 436.332 & 0 & 1308.997 & 0 \\ \hline 0 & 0 & 0 & 1006.921 & 0 & 0 \\ 0 & 0 & 1308.997 & 0 & 5235988 & 0 \\ 0 & -1308.997 & 0 & 0 & 0 & 5235.988 \end{array}\right]$$

$$\mathbf{k}_C^2 = \left[\begin{array}{ccc|ccc} -261799.4 & 0 & 0 & 0 & 0 & 0 \\ 0 & -436.332 & 0 & 0 & 0 & -1308.997 \\ 0 & 0 & -436.332 & 0 & 1308.997 & 0 \\ \hline 0 & 0 & 0 & -1006.921 & 0 & 0 \\ 0 & 0 & -1308.997 & 0 & 2617.994 & 0 \\ 0 & 1308.997 & 0 & 0 & 0 & 2617.994 \end{array}\right]$$

$$\mathbf{k}_A^3 = \left[\begin{array}{ccc|ccc} 436.332 & 0 & 0 & 0 & -1308.997 & 0 \\ 0 & 436.332 & 0 & 1308.997 & 0 & 0 \\ 0 & 0 & 261799.4 & 0 & 0 & 0 \\ \hline 0 & 1308.997 & 0 & 5235.988 & 0 & 0 \\ -1308.997 & 0 & 0 & 0 & 5235.988 & 0 \\ 0 & 0 & 0 & 0 & 0 & 1006.921 \end{array}\right]$$

$$\mathbf{k}_B^3 = \left[\begin{array}{ccc|ccc} 436.332 & 0 & 0 & 0 & 1308.997 & 0 \\ 0 & 436.332 & 0 & -1308.997 & 0 & 0 \\ 0 & 0 & 261799.4 & 0 & 0 & 0 \\ \hline 0 & -1308.997 & 0 & 5235.988 & 0 & 0 \\ 1308.997 & 0 & 0 & 0 & 5235.988 & 0 \\ 0 & 0 & 0 & 0 & 0 & 1006.921 \end{array}\right]$$

$$\mathbf{k}_C^3 = \left[\begin{array}{ccc|ccc} -436.332 & 0 & 0 & 0 & 1308.997 & 0 \\ 0 & -436.332 & 0 & -1308.997 & 0 & 0 \\ 0 & 0 & -261799.4 & 0 & 0 & 0 \\ \hline 0 & 1308.997 & 0 & 2617.994 & 0 & 0 \\ -1308.997 & 0 & 0 & 0 & 2617.994 & 0 \\ 0 & 0 & 0 & 0 & 0 & -1006.921 \end{array}\right]$$

$$\Rightarrow \mathbf{k}_{AA} = \left[\begin{array}{c|c} \mathbf{k}_B^1 + \mathbf{k}_A^2 & \mathbf{k}_C^{2T} \\ \hline \mathbf{k}_C^2 & \mathbf{k}_B^2 + \mathbf{k}_A^3 \end{array}\right]; \quad \mathbf{k}_{RA} = \left[\begin{array}{c|c} \mathbf{k}_C^{1T} & \mathbf{O} \\ \hline \mathbf{O} & \mathbf{k}_C^3 \end{array}\right] \text{ (matrices too large to be shown here)}$$

Displacements and Support Reactions

The basic stiffness relations are given by:

$$\left[\begin{array}{c} \mathbf{F}_A \\ \hline \mathbf{F}_R \end{array}\right] - \left[\begin{array}{c} \mathbf{F}_{fA} \\ \hline \mathbf{F}_{fR} \end{array}\right] = \left[\begin{array}{c|c} \mathbf{k}_{AA} & \mathbf{k}_{AR} \\ \hline \mathbf{k}_{RA} & \mathbf{k}_{RR} \end{array}\right] \left[\begin{array}{c} \mathbf{D}_A \\ \hline \mathbf{D}_R = \mathbf{O} \end{array}\right] \Rightarrow \begin{array}{l} \mathbf{D}_A = [\mathbf{k}_{AA}]^{-1}(\mathbf{F}_A - \mathbf{F}_{fA}) \\ \mathbf{F}_R = \mathbf{F}_{fR} + \mathbf{k}_{RA}\mathbf{D}_A \end{array}$$

Substituting the various matrices and solving using MATLAB,

$$\mathbf{D_A} = \begin{bmatrix} 0.12342230 \text{ m} \\ -0.00019318 \text{ m} \\ 0.04434673 \text{ m} \\ \hdashline 0.01924418 \text{ rad} \\ 0.00049050 \text{ rad} \\ -0.046973467 \text{ rad} \\ \hline 0.12343532 \text{ m} \\ -0.17423226 \text{ m} \\ 0.000158029 \text{ m} \\ \hdashline 0.039636471 \text{ rad} \\ 0.020830378 \text{ rad} \\ -0.009674483 \text{ rad} \end{bmatrix} \begin{matrix} (1) \\ (2) \\ (3) \\ (4) \\ (5) \\ (6) \\ (7) \\ (8) \\ (9) \\ (10) \\ (11) \\ (12) \end{matrix} \qquad \mathbf{F_R} = \begin{bmatrix} -83.406 \text{ kN} \\ 75.861 \text{ kN} \\ -8.627 \text{ kN} \\ \hdashline -55.040 \text{ kNm} \\ -0.741 \text{ kNm} \\ 219.044 \text{ kNm} \\ \hline -26.592 \text{ kN} \\ 24.139 \text{ kN} \\ -41.372 \text{ kN} \\ \hdashline -124.301 \text{ kNm} \\ -107.043 \text{ kNm} \\ 9.741 \text{ kNm} \end{bmatrix} \begin{matrix} (13) \\ (14) \\ (15) \\ (16) \\ (17) \\ (18) \\ (19) \\ (20) \\ (21) \\ (22) \\ (23) \\ (24) \end{matrix}$$

<u>Check equilibrium:</u>

$$\sum F_x = 0: \; F_{13} + F_{19} + (20\times4) + 30 = -83.406 - 26.592 + 110 = 0 \text{ kN}$$

$$\sum F_y = 0: \; F_{14} + F_{20} - 100 = 75.861 - 24.139 - 100 = 0 \text{ kN}$$

$$\sum F_z = 0: \; F_{15} + F_{21} + 50 = -8.627 - 41.372 + 50 = 0 \text{ kN} \;\; \Rightarrow \text{OK.}$$

Member Forces

$\mathbf{F}_*^i = \mathbf{F}_{*f}^i + \left(\mathbf{k}_*^i \mathbf{T}^i\right)\mathbf{D}^i$, where considering $\mathbf{D_A}$ to be partitioned as $\mathbf{D_A} = \begin{bmatrix} \mathbf{D_{A1}} \\ \mathbf{D_{A2}} \end{bmatrix}$,

$\mathbf{D}^1 = \begin{bmatrix} \mathbf{O} \\ \mathbf{D_{A1}} \end{bmatrix}$; $\mathbf{D}^2 = \begin{bmatrix} \mathbf{D_{A1}} \\ \mathbf{D_{A2}} \end{bmatrix}$; $\mathbf{D}^3 = \begin{bmatrix} \mathbf{D_{A2}} \\ \mathbf{O} \end{bmatrix}$. Substituting the various values and solving,

$$\mathbf{F}_*^1 = \begin{bmatrix} 75.861 \text{ kN} \\ 83.406 \text{ kN} \\ -8.627 \text{ kN} \\ \hdashline -0.741 \text{ kNm} \\ 55.040 \text{ kNm} \\ 219.044 \text{ kNm} \\ \hline -75.861 \text{ kN} \\ -3.408 \text{ kN} \\ 8.628 \text{ kN} \\ \hdashline 0.741 \text{ kNm} \\ -20.533 \text{ kNm} \\ -45.425 \text{ kNm} \end{bmatrix} ; \; \mathbf{F}_*^2 = \begin{bmatrix} -3.408 \text{ kN} \\ 75.861 \text{ kN} \\ -8.628 \text{ kN} \\ \hdashline -20.533 \text{ kNm} \\ -0.741 \text{ kNm} \\ 45.425 \text{ kNm} \\ \hline 3.408 \text{ kN} \\ 24.139 \text{ kN} \\ 8.628 \text{ kN} \\ \hdashline 20.533 \text{ kNm} \\ 52.509 \text{ kNm} \\ 9.741 \text{ kNm} \end{bmatrix} ; \; \mathbf{F}_*^3 = \begin{bmatrix} -41.372 \text{ kN} \\ -24.139 \text{ kN} \\ 26.592 \text{ kN} \\ \hdashline 9.741 \text{ kNm} \\ -52.509 \text{ kNm} \\ -20.533 \text{ kNm} \\ \hline 41.372 \text{ kN} \\ 24.139 \text{ kN} \\ -26.592 \text{ kN} \\ \hdashline -9.741 \text{ kNm} \\ -107.043 \text{ kNm} \\ -124.301 \text{ kNm} \end{bmatrix}$$

The complete force response (including the fixed end force effects), in the form of free-body diagrams are shown in Fig. 6.44a. The bending moment diagrams in the local x^*y^* and x^*z^* planes are also shown in Figs 6.44b and c respectively. Similarly, shear force diagrams in these two planes are shown in Figs 6.44d and e. Twisting moments of 0.74 kNm, 20.53 kNm and 9.74 kNm act on elements '1', '2' and '3' respectively. Element '1' is subject to an axial compressive force of 75.86 kN, while elements '2' and '3' are subject to tensile forces of 3.41 kN and 41.37 kN respectively.

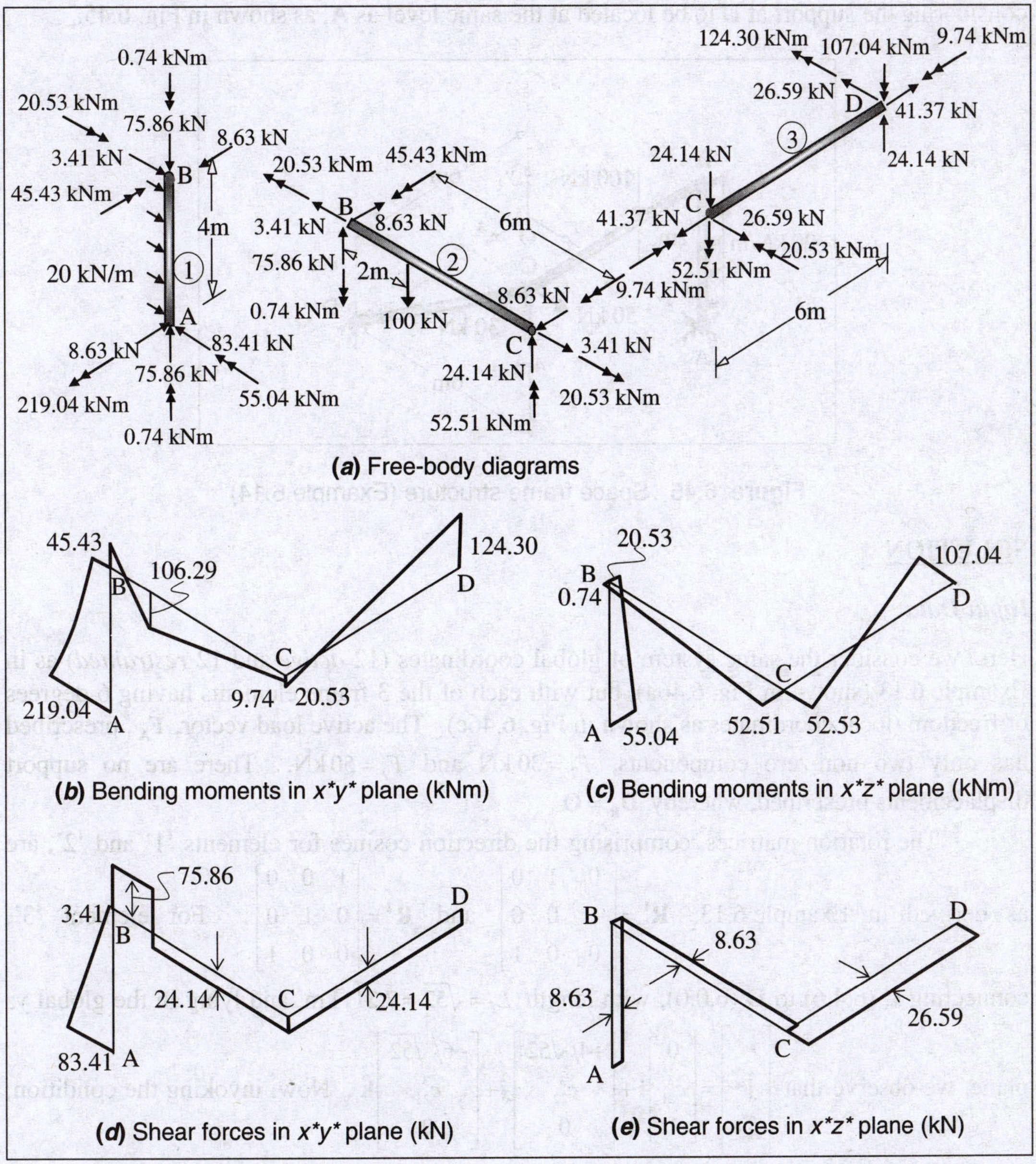

Figure 6.44 Free-body and bending moment diagrams (Example 6.13)

EXAMPLE 6.14

Re-analyse the space frame of Example 6.14 by the reduced stiffness formulation, considering the support at D to be located at the same level as A, as shown in Fig. 6.45.

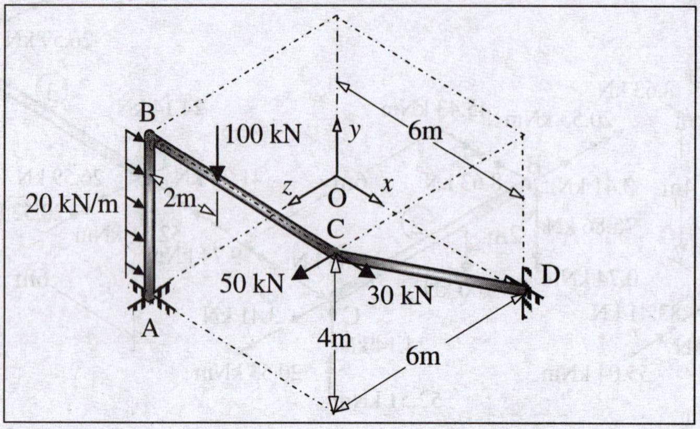

Figure 6.45 Space frame structure (Example 6.14)

SOLUTION

Input Data

Here, we consider the same system of global coordinates (12 *active* and 12 *restrained*) as in Example 6.13 (shown in Fig. 6.46a), but with each of the 3 frame elements having 6 degrees of freedom (local coordinates as shown in Fig. 6.46c). The active load vector, $\mathbf{F_A}$, prescribed has only two non-zero components, $F_7 = 30\,\text{kN}$ and $F_9 = 50\,\text{kN}$. There are no support displacements prescribed, whereby $\mathbf{D_R} = \mathbf{O}$.

The rotation matrices, comprising the direction cosines for elements '1' and '2', are as derived in Example 6.13: $\mathbf{R}^1 = \begin{bmatrix} 0 & 1 & 0 \\ -1 & 0 & 0 \\ 0 & 0 & 1 \end{bmatrix}$ and $\mathbf{R}^2 = \begin{bmatrix} 1 & 0 & 0 \\ 0 & 1 & 0 \\ 0 & 0 & 1 \end{bmatrix}$. For element '3', connecting C (6,4,6) to D (6,0,0), with length $L_2 = \sqrt{52} = 7.2111\,\text{m}$, and lying in the global *yz* plane, we observe that $\begin{bmatrix} \hat{\mathbf{i}}* \\ \hat{\mathbf{j}}* \\ \hat{\mathbf{k}}* \end{bmatrix} = \begin{bmatrix} 0 \\ c_{21}^3 \\ 1 \end{bmatrix} \hat{\mathbf{i}} + \begin{bmatrix} -4/\sqrt{52} \\ c_{22}^3 \\ 0 \end{bmatrix} \hat{\mathbf{j}} + \begin{bmatrix} -6/\sqrt{52} \\ c_{23}^3 \\ 0 \end{bmatrix} \hat{\mathbf{k}}$. Now, invoking the condition,

$\hat{\mathbf{j}}^* = \hat{\mathbf{k}}^* \times \hat{\mathbf{i}}^*$ (refer Eq. 6.32c), $c^3_{21} = \begin{vmatrix} 0 & 0 \\ -4/\sqrt{52} & -6/\sqrt{52} \end{vmatrix} = 0$, $c^3_{22} = \begin{vmatrix} 0 & 1 \\ -6/\sqrt{52} & 0 \end{vmatrix} = 6/\sqrt{52}$ and

$c^3_{23} = \begin{vmatrix} 1 & 0 \\ 0 & -4/\sqrt{52} \end{vmatrix} = -4/\sqrt{52}$, whereby $\mathbf{R}^3 = \begin{bmatrix} 0 & -4/\sqrt{52} & -6/\sqrt{52} \\ 0 & 6/\sqrt{52} & -4/\sqrt{52} \\ 1 & 0 & 0 \end{bmatrix}$.

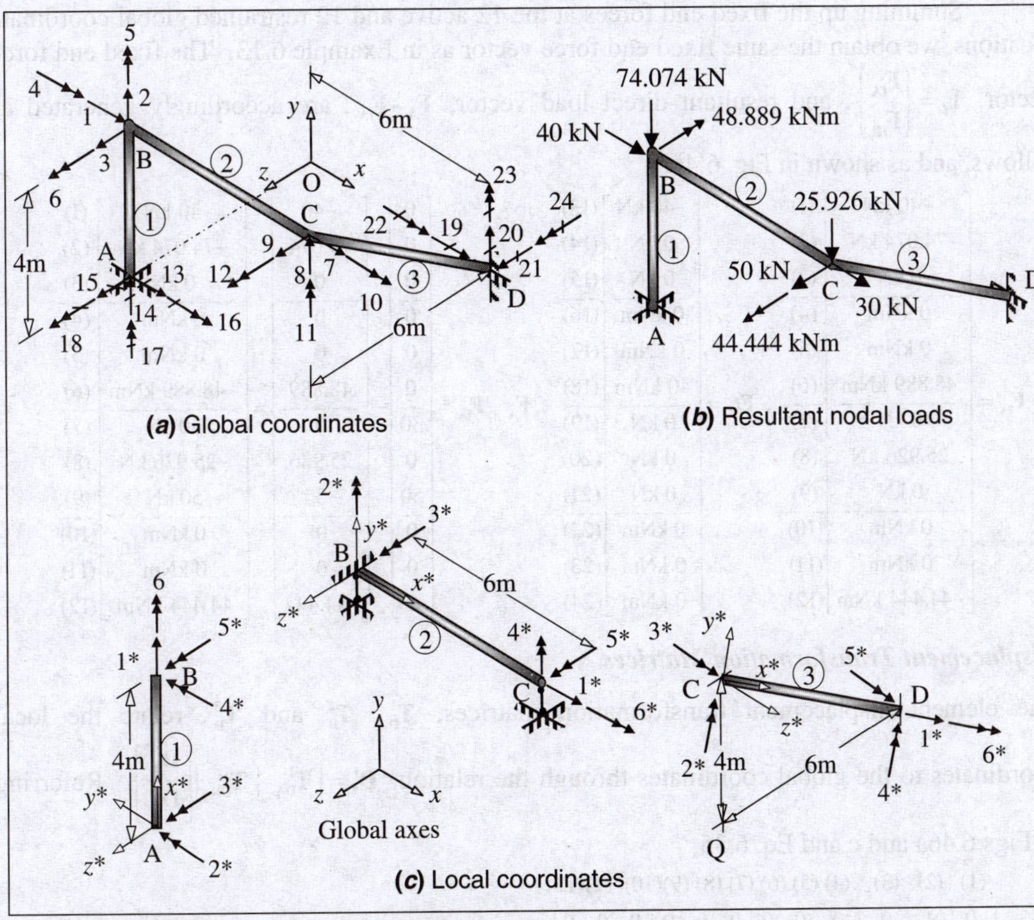

(*a*) Global coordinates (*b*) Resultant nodal loads (*c*) Local coordinates

Figure 6.46 Global and local coordinates and resultant loads (Example 6.14)

Fixed End Forces

The fixed end forces (due to the intermediate loading in elements '1' and '2') are as computed earlier (refer Fig. 6.44c). Referring to the local coordinates in Fig. 6.46b.

$$\Rightarrow \mathbf{F}_{*f}^{1} = \begin{Bmatrix} 0\text{ kN} \\ \hdashline 0\text{ kNm} \\ 40\text{ kNm} \\ \hdashline 0\text{ kNm} \\ -40\text{ kNm} \\ \hdashline 0\text{ kNm} \end{Bmatrix}\ ;\ \mathbf{F}_{*f}^{2} = \begin{Bmatrix} 0\text{ kN} \\ \hdashline 0\text{ kNm} \\ 88.889\text{ kNm} \\ \hdashline 0\text{ kNm} \\ -44.444\text{ kNm} \\ \hdashline 0\text{ kNm} \end{Bmatrix}\ ;\ \mathbf{F}_{*f}^{3} = \begin{Bmatrix} 0\text{ kN} \\ \hdashline 0\text{ kNm} \\ 0\text{ kNm} \\ \hdashline 0\text{ kNm} \\ 0\text{ kNm} \\ \hdashline 0\text{ kNm} \end{Bmatrix}$$

Summing up the fixed end forces at the 12 active and 12 restrained global coordinate locations, we obtain the same fixed end force vector as in Example 6.13. The fixed end force vector, $\mathbf{F}_f = \begin{Bmatrix} \mathbf{F}_{fA} \\ \mathbf{F}_{fR} \end{Bmatrix}$, and resultant direct load vector, $\mathbf{F}_A - \mathbf{F}_{fA}$, are accordingly generated as follows, and as shown in Fig. 6.46b.

$$\Rightarrow \mathbf{F}_{fA} = \begin{bmatrix} -40\text{ kN} \\ 74.074\text{ kN} \\ 0\text{ kN} \\ \hdashline 0\text{ kNm} \\ 0\text{ kNm} \\ 48.889\text{ kNm} \\ \hline 0\text{ kN} \\ 25.926\text{ kN} \\ 0\text{ kN} \\ \hdashline 0\text{ kNm} \\ 0\text{ kNm} \\ -44.444\text{ kNm} \end{bmatrix} \begin{matrix} (1) \\ (2) \\ (3) \\ (4) \\ (5) \\ (6) \\ (7) \\ (8) \\ (9) \\ (10) \\ (11) \\ (12) \end{matrix}\ ;\ \mathbf{F}_{fR} = \begin{bmatrix} -40\text{ kN} \\ 0\text{ kN} \\ 0\text{ kN} \\ \hdashline 0\text{ kNm} \\ 0\text{ kNm} \\ 40\text{ kNm} \\ \hline 0\text{ kN} \\ 0\text{ kN} \\ 0\text{ kN} \\ \hdashline 0\text{ kNm} \\ 0\text{ kNm} \\ 0\text{ kNm} \end{bmatrix} \begin{matrix} (13) \\ (14) \\ (15) \\ (16) \\ (17) \\ (18) \\ (19) \\ (20) \\ (21) \\ (22) \\ (23) \\ (24) \end{matrix}\ ;\ \mathbf{F}_A - \mathbf{F}_{fA} = \begin{bmatrix} 0 \\ 0 \\ 0 \\ \hdashline 0 \\ 0 \\ 0 \\ \hline 30 \\ 0 \\ 50 \\ \hdashline 0 \\ 0 \\ 0 \end{bmatrix} - \begin{bmatrix} -40 \\ 74.074 \\ 0 \\ \hdashline 0 \\ 0 \\ 48.889 \\ \hline 0 \\ 25.926 \\ 0 \\ \hdashline 0 \\ 0 \\ -44.444 \end{bmatrix} = \begin{bmatrix} 40\text{ kN} \\ -74.074\text{ kN} \\ 0\text{ kN} \\ \hdashline 0\text{ kNm} \\ 0\text{ kNm} \\ -48.889\text{ kNm} \\ \hline 30\text{ kN} \\ -25.926\text{ kN} \\ 50\text{ kN} \\ \hdashline 0\text{ kNm} \\ 0\text{ kNm} \\ 44.444\text{ kNm} \end{bmatrix} \begin{matrix} (1) \\ (2) \\ (3) \\ (4) \\ (5) \\ (6) \\ (7) \\ (8) \\ (9) \\ (10) \\ (11) \\ (12) \end{matrix}$$

Displacement Transformation Matrices

The element displacement transformation matrices, $\mathbf{T}_D^1$, $\mathbf{T}_D^2$ and $\mathbf{T}_D^3$, relate the local coordinates to the global coordinates through the relation, $\mathbf{D}_*^i = \left[\mathbf{T}_{DA}^i \mid \mathbf{T}_{DR}^i\right]\begin{bmatrix} \mathbf{D}_A \\ \hline \mathbf{D}_R \end{bmatrix}$. Referring to Figs 6.46a and c and Eq. 6.35,

$$\mathbf{T}_{DA}^1 = \begin{matrix} (1) & (2) & (3) & (4) & (5) & (6) & (7) & (8) & (9) & (10) & (11) & (12) \\ \end{matrix}$$
$$\mathbf{T}_{DA}^1 = \left[\begin{array}{ccc:ccc:ccc:ccc} 0 & 1 & 0 & 0 & 0 & 0 & 0 & 0 & 0 & 0 & 0 & 0 \\ 0 & 0 & 1/4 & 0 & 0 & 0 & 0 & 0 & 0 & 0 & 0 & 0 \\ 1/4 & 0 & 0 & 0 & 0 & 0 & 0 & 0 & 0 & 0 & 0 & 0 \\ \hdashline 0 & 0 & 1/4 & -1 & 0 & 0 & 0 & 0 & 0 & 0 & 0 & 0 \\ 1/4 & 0 & 0 & 0 & 0 & 1 & 0 & 0 & 0 & 0 & 0 & 0 \\ 0 & 0 & 0 & 0 & 1 & 0 & 0 & 0 & 0 & 0 & 0 & 0 \end{array}\right]$$

(13) (14) (15) (16)(17)(18)(19)(20)(21)(22)(23)(24)

$$\mathbf{T}_{\mathbf{DR}}^{\mathbf{1}} = \left[\begin{array}{ccc:ccc:ccc:ccc} 0 & -1 & 0 & 0 & 0 & 0 & 0 & 0 & 0 & 0 & 0 & 0 \\ 0 & 0 & -1/4 & -1 & 0 & 0 & 0 & 0 & 0 & 0 & 0 & 0 \\ -1/4 & 0 & 0 & 0 & 0 & 1 & 0 & 0 & 0 & 0 & 0 & 0 \\ \hdashline 0 & 0 & -1/4 & 0 & 0 & 0 & 0 & 0 & 0 & 0 & 0 & 0 \\ -1/4 & 0 & 0 & 0 & 0 & 0 & 0 & 0 & 0 & 0 & 0 & 0 \\ 0 & 0 & 0 & 0 & -1 & 0 & 0 & 0 & 0 & 0 & 0 & 0 \end{array}\right]$$

(1) (2) (3) (4) (5) (6) (7) (8) (9) (10)(11)(12)

$$\mathbf{T}_{\mathbf{DA}}^{\mathbf{2}} = \left[\begin{array}{ccc:ccc:ccc:ccc} -1 & 0 & 0 & 0 & 0 & 0 & 1 & 0 & 0 & 0 & 0 & 0 \\ 0 & 0 & -1/6 & 0 & 1 & 0 & 0 & 0 & 1/6 & 0 & 0 & 0 \\ 0 & 1/6 & 0 & 0 & 0 & 1 & 0 & -1/6 & 0 & 0 & 0 & 0 \\ \hdashline 0 & 0 & -1/6 & 0 & 0 & 0 & 0 & 0 & 1/6 & 0 & 1 & 0 \\ 0 & 1/6 & 0 & 0 & 0 & 0 & 0 & -1/6 & 0 & 0 & 0 & 1 \\ 0 & 0 & 0 & -1 & 0 & 0 & 0 & 0 & 0 & 1 & 0 & 0 \end{array}\right]; \quad \mathbf{T}_{\mathbf{DR}}^{\mathbf{2}} = \mathbf{O}$$

(1) (2) (3) (4) (5) (6) (7) (8) (9) (10) s(11) (12)

$$\mathbf{T}_{\mathbf{DA}}^{\mathbf{3}} = \left[\begin{array}{ccc:ccc:ccc:ccc} 0 & 0 & 0 & 0 & 0 & 0 & 0 & 4/a & 6/a & 0 & 0 & 0 \\ 0 & 0 & 0 & 0 & 0 & 0 & -1/a & 0 & 0 & 0 & 6/a & -4/a \\ 0 & 0 & 0 & 0 & 0 & 0 & 0 & 3/26 & -1/13 & 1 & 0 & 0 \\ \hdashline 0 & 0 & 0 & 0 & 0 & 0 & -1/a & 0 & 0 & 0 & 0 & 0 \\ 0 & 0 & 0 & 0 & 0 & 0 & 0 & 3/26 & -1/13 & 0 & 0 & 0 \\ 0 & 0 & 0 & 0 & 0 & 0 & 0 & 0 & 0 & 0 & 4/a & 6/a \end{array}\right], \text{ where } a = \sqrt{52}, \text{ and}$$

(13)(14)(15)(16)(17)(18)(19) (20) (21) (22) (23) (24)

$$\mathbf{T}_{\mathbf{DR}}^{\mathbf{3}} = \left[\begin{array}{ccc:ccc:ccc:ccc} 0 & 0 & 0 & 0 & 0 & 0 & 0 & -4/a & -6/a & 0 & 0 & 0 \\ 0 & 0 & 0 & 0 & 0 & 0 & 1/a & 0 & 0 & 0 & 0 & 0 \\ 0 & 0 & 0 & 0 & 0 & 0 & 0 & -3/26 & 1/13 & 0 & 0 & 0 \\ \hdashline 0 & 0 & 0 & 0 & 0 & 0 & 1/a & 0 & 0 & 0 & 6/a & -4/a \\ 0 & 0 & 0 & 0 & 0 & 0 & 0 & -3/26 & 1/13 & 1 & 0 & 0 \\ 0 & 0 & 0 & 0 & 0 & 0 & 0 & 0 & 0 & 0 & -4/a & -6/a \end{array}\right]$$

Element and Structure Stiffness Matrices

The element stiffness matrices $\tilde{\mathbf{k}}_*^1 = \tilde{\mathbf{k}}_*^3$ and $\tilde{\mathbf{k}}_*^2$, satisfying $\mathbf{F}_*^{\mathbf{i}} = \tilde{\mathbf{k}}_*^{\mathbf{i}}\mathbf{D}_*^{\mathbf{i}}$, take the following form (refer Fig. 6.46b and Eq. 6.33), considering $L_1 = 4.0\,\text{m}$, $L_2 = 6.0\,\text{m}$ and $L_3 = \sqrt{52}$ m and values of $\alpha_i = (EI)_i/L_i$; $\delta_{zi} = \delta_{yi} = (EI)_i/L_i$; $\varepsilon_i = (GJ)_i/L_i$ as follows:

$$E = 2.5\times10^8\,\text{kN/m}^2,\ \nu = 0.3 \ \Rightarrow G = \frac{E}{2(1+0.3)} = 0.9615385\times10^8\,\text{kN/m}^2$$

$$A = 2\pi(0.1)(0.01) = 2\pi\times10^{-3}\ \text{m}^2,\ I_z = I_y = \pi(0.1)^3(0.01) = \pi\times10^{-5}\ \text{m}^4,\ J = 2\pi\times10^{-5}\ \text{m}^4$$

$\Rightarrow \alpha_1 = 392699.1$ kN/m; $\alpha_2 = 261799.4$ kN/m; $\alpha_3 = 217830.3$ kN/m;

$\delta_{z1} = \delta_{y1} = 1963.495$ kNm; $\delta_{z2} = \delta_{y2} = 1308.997$ kNm; $\delta_{z3} = \delta_{y3} = 1089.151$ kNm;

$\varepsilon_1 = 1510.381$ kNm; $\varepsilon_2 = 1006.921$ kNm; $\varepsilon_3 = 837.8086$ kNm

$$\Rightarrow \tilde{\mathbf{k}}_*^1 = \left[\begin{array}{ccc|ccc} 392699.1 & 0 & 0 & 0 & 0 & 0 \\ 0 & 7853.982 & 0 & 3926.991 & 0 & 0 \\ 0 & 0 & 7853.982 & 0 & 3926.991 & 0 \\ \hline 0 & 3926.991 & 0 & 7853.982 & 0 & 0 \\ 0 & 0 & 3926.991 & 0 & 7853.982 & 0 \\ 0 & 0 & 0 & 0 & 0 & 1510.381 \end{array}\right]$$

$$\tilde{\mathbf{k}}_*^2 = \left[\begin{array}{ccc|ccc} 261799.4 & 0 & 0 & 0 & 0 & 0 \\ 0 & 5235.988 & 0 & 2617.994 & 0 & 0 \\ 0 & 0 & 5235.988 & 0 & 2617.994 & 0 \\ \hline 0 & 2617.994 & 0 & 5235.988 & 0 & 0 \\ 0 & 0 & 2617.994 & 0 & 5235.988 & 0 \\ 0 & 0 & 0 & 0 & 0 & 1006.921 \end{array}\right]$$

$$\tilde{\mathbf{k}}_*^3 = \left[\begin{array}{ccc|ccc} 217830.3 & 0 & 0 & 0 & 0 & 0 \\ 0 & 4356.605 & 0 & 2178.303 & 0 & 0 \\ 0 & 0 & 4356.605 & 0 & 2178.303 & 0 \\ \hline 0 & 2178.303 & 0 & 4356.605 & 0 & 0 \\ 0 & 0 & 2178.303 & 0 & 4356.605 & 0 \\ 0 & 0 & 0 & 0 & 0 & 837.8086 \end{array}\right]$$

We can now compute and generate the $\tilde{\mathbf{k}}_*^i \mathbf{T}_D^i$ matrices as well as the components of the structure stiffness matrix $\mathbf{k} = \sum_{i=1}^{3} \mathbf{T}_D^{i\,T} \tilde{\mathbf{k}}_*^i \mathbf{T}_D^i = \left[\begin{array}{c|c} \mathbf{k}_{AA} & \mathbf{k}_{AR} \\ \hline \mathbf{k}_{RA} & \mathbf{k}_{RR} \end{array}\right]$. We need to find $\mathbf{k}_{AA} = \sum_{i=1}^{3} \mathbf{T}_{DA}^{i\,T} \tilde{\mathbf{k}}_*^i \mathbf{T}_{DA}^i$ and $\mathbf{k}_{RA} = \sum_{i=1}^{3} \mathbf{T}_{RA}^{i\,T} \tilde{\mathbf{k}}_*^i \mathbf{T}_{DA}^i$, which are as follows: $\mathbf{k}_{AA} = \left[\begin{array}{c|c} \mathbf{k}_{AA}^a & \mathbf{k}_{AA}^{c\,T} \\ \hline \mathbf{k}_{AA}^c & \mathbf{k}_{AA}^b \end{array}\right]$ and

$$\mathbf{k}_{RA} = \left[\begin{array}{c|c} \mathbf{k}_{RA}^a & \mathbf{O} \\ \hline \mathbf{O} & \mathbf{k}_{RA}^b \end{array}\right], \text{ where } \mathbf{k}_{AA}^a = \left[\begin{array}{ccc|ccc} 263272.0 & 0 & 0 & 0 & 0 & 2945.243 \\ 0 & 393135.4 & 0 & 0 & 0 & 1308.997 \\ 0 & 0 & 1908.954 & -2945.243 & -1308.997 & 0 \\ \hline 0 & 0 & -2945.243 & 8860.902 & 0 & 0 \\ 0 & 0 & -1308.997 & 0 & 6746.369 & 0 \\ 2945.243 & 1308.997 & 0 & 0 & 0 & 1308.997 \end{array}\right];$$

$$\mathbf{k}_{AA}^{b}=\left[\begin{array}{ccc|ccc} 263272.0 & 0 & 0 & 0 & -754.029 & 502.685 \\ 0 & 67635.033 & 100421.038 & 754.029 & 0 & -1308.997 \\ 0 & 100421.038 & 151319.2 & -502.685 & 1308.997 & 0 \\ \hline 0 & 754.029 & -502.685 & 5363.526 & 0 & 0 \\ -754.029 & 0 & 1308.997 & 0 & 8509.886 & -1624.060 \\ 502.685 & -1308.997 & 0 & 0 & -1624.060 & 7156.503 \end{array}\right];$$

$$\mathbf{k}_{AA}^{c}=\left[\begin{array}{ccc|ccc} -261799.4 & 0 & 0 & 0 & 0 & 0 \\ 0 & -436.332 & 0 & 0 & 0 & -1308.997 \\ 0 & 0 & -436.332 & 0 & 1308.997 & 0 \\ \hline 0 & 0 & 0 & -1006.921 & 0 & 0 \\ 0 & 0 & -1308.997 & 0 & 2617.994 & 0 \\ 0 & 1308.997 & 0 & 0 & 0 & 2617.994 \end{array}\right];$$

$$\mathbf{k}_{RA}^{a}=\left[\begin{array}{ccc|ccc} -1472.622 & 0 & 0 & 0 & 0 & -2945.243 \\ 0 & -392699.1 & 0 & 0 & 0 & 0 \\ 0 & 0 & -1472.622 & 2945.243 & 0 & 0 \\ \hline 0 & 0 & -2945.243 & 3926.991 & 0 & 0 \\ 0 & 0 & 0 & 0 & -1510.381 & 0 \\ 2945.243 & 0 & 0 & 0 & 0 & 3926.699 \end{array}\right]$$

$$\mathbf{k}_{RA}^{b}=\left[\begin{array}{ccc|ccc} -2513.426 & 0 & 0 & 0 & 754.028 & -502.685 \\ 0 & -67198.701 & -100421.0 & -754.028 & 0 & 0 \\ 0 & -100421.0 & -150882.9 & 502.685 & 0 & 0 \\ \hline 0 & 754.028 & -502.685 & 2178.303 & 0 & 0 \\ -754.028 & 0 & 0 & 0 & 1250.268 & -1392.051 \\ 502.685 & 0 & 0 & 0 & -1392.051 & 90.226 \end{array}\right]$$

Displacements and Support Reactions

The basic stiffness relations are given by:

$$\left[\begin{array}{c}\mathbf{F_A}\\ \hline \mathbf{F_R}\end{array}\right]-\left[\begin{array}{c}\mathbf{F_{fA}}\\ \hline \mathbf{F_{fR}}\end{array}\right]=\left[\begin{array}{c|c}\mathbf{k_{AA}} & \mathbf{k_{AR}}\\ \hline \mathbf{k_{RA}} & \mathbf{k_{RR}}\end{array}\right]\left[\begin{array}{c}\mathbf{D_A}\\ \hline \mathbf{D_R}=\mathbf{O}\end{array}\right] \Rightarrow \begin{array}{l}\mathbf{D_A}=\left[\mathbf{k_{AA}}\right]^{-1}\left(\mathbf{F_A}-\mathbf{F_{fA}}\right)\\ \mathbf{F_R}=\mathbf{F_{fR}}+\mathbf{k_{RA}}\mathbf{D_A}\end{array}$$

Substituting the various matrices and solving using MATLAB,

$$\mathbf{D}_A = \begin{bmatrix} 0.163780620\text{ m} \\ -0.000200638\text{ m} \\ 0.09173483\text{ m} \\ \hdashline 0.03947646\text{ rad} \\ -0.02429291\text{ rad} \\ -0.06632591\text{ rad} \\ \hline 0.16380293\text{ m} \\ -0.35284041\text{ m} \\ 0.23530515\text{ m} \\ \hdashline 0.07906843\text{ rad} \\ -0.009184204\text{ rad} \\ -0.047617748\text{ rad} \end{bmatrix} \begin{matrix} (1) \\ (2) \\ (3) \\ (4) \\ (5) \\ (6) \\ (7) \\ (8) \\ (9) \\ (10) \\ (11) \\ (12) \end{matrix} \quad ; \quad \mathbf{F}_R = \begin{bmatrix} -85.841\text{ kN} \\ 78.790\text{ kN} \\ -18.823\text{ kN} \\ \hdashline -115.158\text{ kNm} \\ 36.692\text{ kNm} \\ 261.912\text{ kNm} \\ \hline -24.160\text{ kN} \\ 21.210\text{ kN} \\ -31.177\text{ kN} \\ \hdashline -212.101\text{ kNm} \\ -68.708\text{ kNm} \\ 90.830\text{ kNm} \end{bmatrix} \begin{matrix} (13) \\ (14) \\ (15) \\ (16) \\ (17) \\ (18) \\ (19) \\ (20) \\ (21) \\ (22) \\ (23) \\ (24) \end{matrix}$$

Check equilibrium:

$$\sum F_x = 0: \ F_{13} + F_{19} + (20\times4) + 30 = -85.841 - 24.160 + 110 = 0\text{ kN}$$

$$\sum F_y = 0: \ F_{14} + F_{20} - 100 = 78.790 - 21.210 - 100 = 0\text{ kN}$$

$$\sum F_z = 0: \ F_{15} + F_{21} + 50 = -18.823 - 31.177 + 50 = 0\text{ kN} \ \Rightarrow \text{OK.}$$

Member Forces

$$\mathbf{F}_*^i = \mathbf{F}_{*f}^i + \tilde{\mathbf{k}}_*^i \mathbf{T}_{DA}^i \mathbf{D}_A$$

$$\Rightarrow \mathbf{F}_*^1 = \begin{Bmatrix} F_{1*}^1 \\ F_{2*}^1 \\ F_{3*}^1 \\ F_{4*}^1 \\ F_{5*}^1 \\ F_{6*}^1 \end{Bmatrix} = \begin{Bmatrix} -78.790\text{ kN} \\ 115.158\text{ kNm} \\ 261.912\text{ kNm} \\ -39.866\text{ kNm} \\ -78.549\text{ kNm} \\ -36.692\text{ kNm} \end{Bmatrix}; \ \mathbf{F}_*^2 = \begin{Bmatrix} F_{1*}^2 \\ F_{2*}^2 \\ F_{3*}^2 \\ F_{4*}^2 \\ F_{5*}^2 \\ F_{6*}^2 \end{Bmatrix} = \begin{Bmatrix} 5.841\text{ kN} \\ 36.692\text{ kNm} \\ 78.549\text{ kNm} \\ 76.246\text{ kNm} \\ -5.806\text{ kNm} \\ 39.866\text{ kNm} \end{Bmatrix}; \ \mathbf{F}_*^3 = \begin{Bmatrix} F_{1*}^3 \\ F_{2*}^3 \\ F_{3*}^3 \\ F_{4*}^3 \\ F_{5*}^3 \\ F_{6*}^3 \end{Bmatrix} = \begin{Bmatrix} 14.176\text{ kN} \\ -66.661\text{ kNm} \\ -39.866\text{ kNm} \\ -107.552\text{ kNm} \\ -212.101\text{ kNm} \\ -37.463\text{ kNm} \end{Bmatrix}$$

The complete force response (including the fixed end force effects), in the form of free-body diagrams (with the forces missing in the reduced $\mathbf{F}_*^i$ vector calculated, satisfying equilibrium) are shown in Fig. 6.47a. The bending moment diagrams in the local x^*y^* and x^*z^* planes are also shown in Figs 6.47b and c respectively. Similarly, shear force diagrams in these two planes are shown in Figs 6.47d and e. Twisting moments of 36.69 kNm, 39.87 kNm and 37.46 kNm act on elements '1', '2' and '3' respectively. Element '1' is subject to an axial compressive force of 78.79 kN, while elements '2' and '3' are subject to tensile forces of 5.84 kN and 14.18 kN respectively. It may be observed that some of these forces are significantly different from the values generated in Example 6.13, due to the change in the configuration of element '3'.

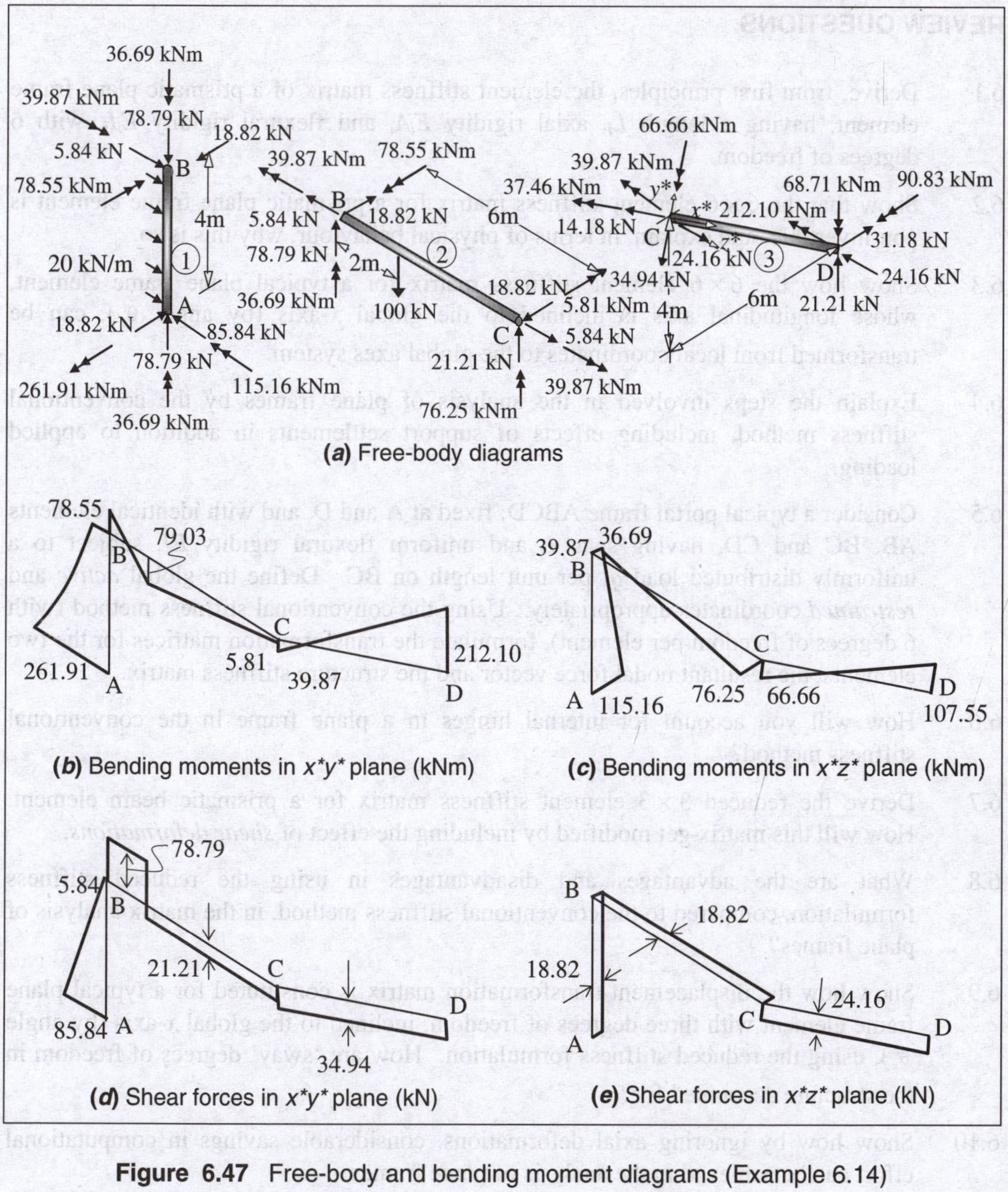

Figure 6.47 Free-body and bending moment diagrams (Example 6.14)

REVIEW QUESTIONS

6.1 Derive, from first principles, the element stiffness matrix of a prismatic plane frame element, having a length L_i, axial rigidity E_iA_i and flexural rigidity E_iI_i, with 6 degrees of freedom.

6.2 Show that the 6×6 element stiffness matrix for a prismatic plane frame element is non-invertible, and explain, in terms of physical behaviour, why this is so.

6.3 Show how the 6×6 element stiffness matrix for a typical plane frame element, whose longitudinal axis is inclined to the global *x*-axis (by angle θ_x^i), can be transformed from local coordinates to the global axes system.

6.4 Explain the steps involved in the analysis of plane frames by the conventional stiffness method, including effects of support settlements in addition to applied loading.

6.5 Consider a typical portal frame ABCD, fixed at A and D, and with identical elements AB, BC and CD, having span *L* and uniform flexural rigidity *EI*, subject to a uniformly distributed load q_o per unit length on BC. Define the global *active* and *restrained* coordinates appropriately. Using the conventional stiffness method (with 6 degrees of freedom per element), formulate the transformation matrices for the two elements, the resultant nodal force vector and the structure stiffness matrix.

6.6 How will you account for internal hinges in a plane frame in the conventional stiffness method?

6.7 Derive the reduced 3×3 element stiffness matrix for a prismatic beam element. How will this matrix get modified by including the effect of *shear deformations.*

6.8 What are the advantages and disadvantages in using the reduced stiffness formulation, compared to the conventional stiffness method, in the matrix analysis of plane frames?

6.9 Show how the displacement transformation matrix is constituted for a typical plane frame element with three degrees of freedom, inclined to the global *x*-axis (by angle θ_x^i), using the reduced stiffness formulation. How are 'sway' degrees of freedom in the structure accounted for?

6.10 Show how by ignoring axial deformations, considerable savings in computational effort can be achieved in the analysis of plane frames.

6.11 How can the reduced stiffness formulation be conveniently used to handle beam elements with (a) hinged end supports and (b) guided-fixed supports? Derive the element stiffness matrices.

6.12 Repeat Question 6.5, considering the reduced stiffness formulation, and taking advantage of symmetry (i.e., with a guided-fixed support at the middle of BC).

6.13 Why is the flexibility method not well suited for matrix analysis of plane frames?

6.14 Describe the steps involved in the flexibility method, as applied to plane frames, including the effects of support settlements.

6.15 In what way is a *space frame* different from (i) a *plane frame*, (ii) a *grid* system and (iii) a *space truss*? Show how the *plane frame element*, the *grid element* and the *space truss element* are special cases of the *space frame element*, considering (a) the conventional stiffness formulation and (b) the reduced stiffness formulation.

6.16 Under what situations can plane frame elements be used in the matrix analysis of a space frame system? Discuss.

6.17 Generate the stiffness matrix of a typical prismatic space frame element with (i) 12 degrees of freedom and (ii) 6 degrees of freedom (reduced stiffness formulation).

6.18 Consider an element AB in a space frame, connecting the joint A(0,1,2) to joint B(3,4,5), where the global axes (x,y,z) dimensions are expressed in metres. Find all the member end forces, given that A is connected to a rigid support and B deflects by 0.5mm, –1.0mm and 1.5mm, and rotates by 0.001, –0.001 and 0.002 rad in the x, y and z directions respectively. Assume that the member is tubular, having a mean radius of 150mm, shell thickness of 15mm, with modulus of elasticity, $E = 2\times10^8\,\text{kN/m}^2$ and Poisson's ratio, $\nu = 0.3$.

PROBLEMS

6.1 Re-analyse the plane frame in Example 6.1 (Fig. 6.3a), considering the uniformly distributed loading to be replaced by a concentrated load Q acting downwards at the mid-point of member AO. Analyse by the conventional Stiffness Method, and draw the axial force, shear force, bending moment and deflection diagrams.

6.2 Repeat Problem 6.1, using the reduced stiffness formulation, and verify the solution.

6.3 Analyse the rectangular portal frame shown in Fig. 6.48 by the conventional Stiffness Method, taking advantage of symmetry. Assume the column elements to have a square cross section, 300×300 mm and the beam to have a size, 300×450 mm. Assume all elements to have an elastic modulus, $E = 2.5\times10^4\,\text{N/mm}^2$. Find the complete force response and draw the deflection and bending moment diagrams.

6.4 Applying the reduced stiffness formulation (a) repeat Problem 6.1, and verify the solution. (b) Re-analyse, considering axial deformations to be negligible, and comment on the magnitude of error due to this.

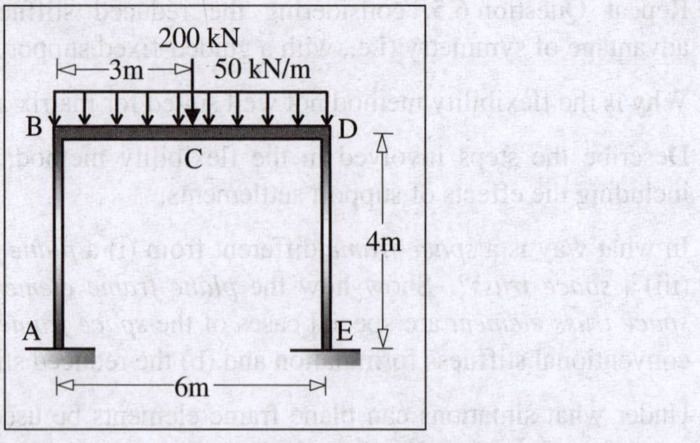

Figure 6.48 Symmetric portal frame (Problem 6.3)

6.5 Repeat Problem 6.4(a) and (b) by the Flexibility Method.

6.6 Analyse the unsymmetric portal frame shown in Fig. 6.49 by the conventional Stiffness Method, subject to the given applied loads as well as a settlement of 10 mm at the support D. Assume all elements to have a size 300×300 mm and an elastic modulus, $E = 2.5\times10^4\ \text{N/mm}^2$. Find the complete force response and draw the deflection, bending moment, shear force and axial force diagrams.

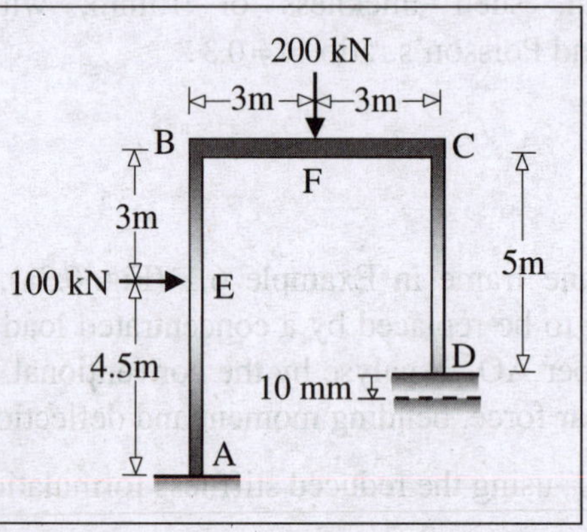

Figure 6.49 Unsymmetric rigid-jointed frame (Problem 6.6)

6.7 Applying the reduced stiffness formulation (a) repeat Problem 6.6, and verify the solution. (b) Re-analyse, considering axial deformations to be negligible, and comment on the magnitude of error due to this.

6.8 Repeat Problem 6.7(b) by the Flexibility Method.

6.9 Repeat Problem 6.6 by the conventional stiffness method, considering an internal hinge to be located at joint C.

6.10 Repeat Problem 6.9 by the reduced stiffness formulation, also considering axial deformations to be negligible.

6.11 Analyse the anti-symmetric frame shown in Fig. 6.50 by the conventional Stiffness Method. Assume all elements to have a size 300×300 mm and an elastic modulus, $E = 2.5 \times 10^4$ N/mm^2. Find the complete force response and draw the deflection, bending moment, shear force and axial force diagrams.

6.12 Repeat Problem 6.11 by the conventional stiffness method, considering internal hinges to be located at joints B and C.

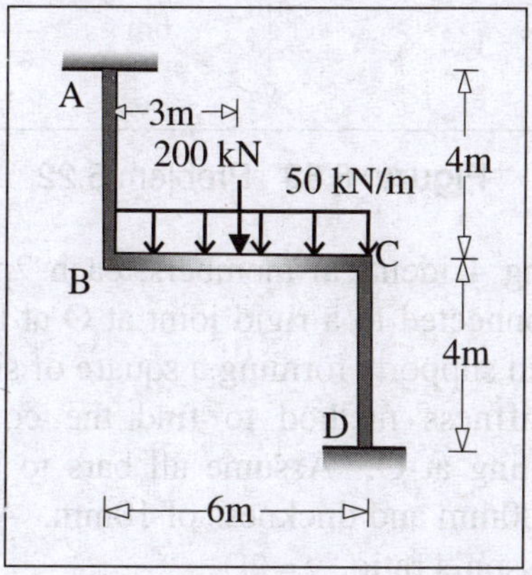

Figure 6.50 Problem 6.11

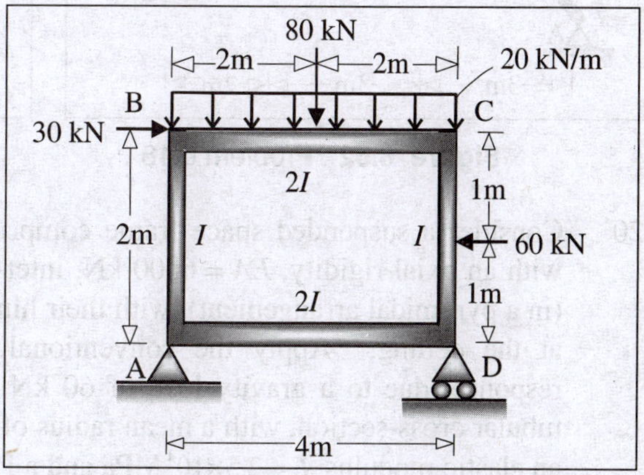

Figure 6.51 Problem 6.15

6.13 Repeat Problem 6.11 by the reduced stiffness formulation, also considering axial deformations to be negligible.

6.14 Repeat Problem 6.11 by the Flexibility Method, ignoring axial deformations.

6.15 Analyse the box frame shown in Fig. 6.51 by the reduced stiffness formulation, considering axial deformations to be negligible, and draw the axial force, shear force and bending moment diagrams. Also, sketch the deflected shape.

6.16 Repeat Problem 6.15, without including the lateral loads of 30 kN and 60 kN in Fig. 6.51. Take advantage of symmetry.

6.17 Repeat Problem 6.16 by the conventional stiffness method.

6.18 Analyse the frame with inclined members shown in Fig. 6.52 by the conventional stiffness method. Assume all elements to have a size 300×450 mm and an elastic modulus, $E = 2.5 \times 10^4$ N/mm^2. Find the complete force response and draw the deflection, bending moment, shear force and axial force diagrams.

6.19 Repeat Problem 6.18 by the reduced stiffness formulation.

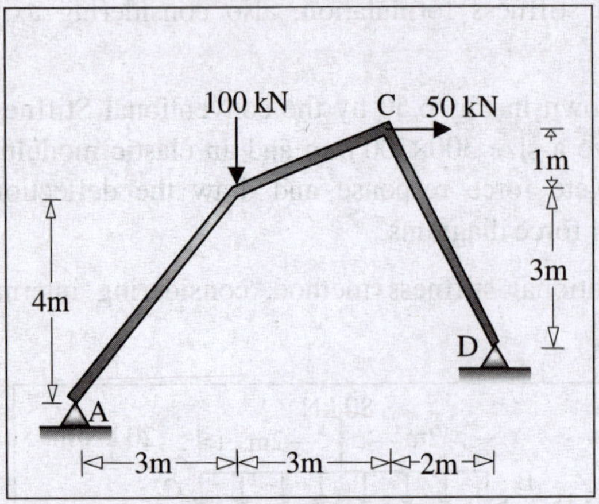

Figure 6.52 Problem 6.18

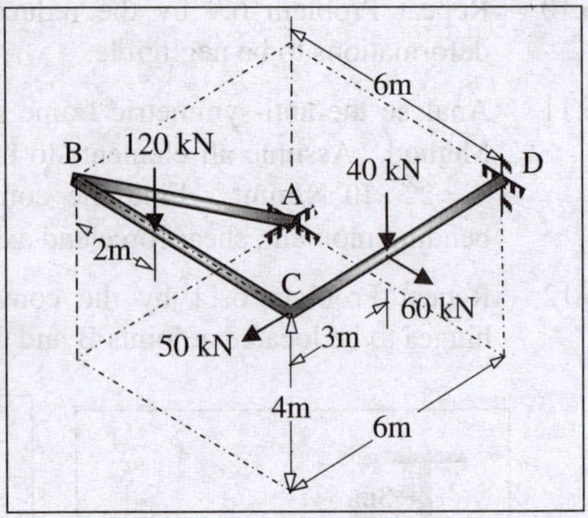

Figure 6.53 Problem 6.22

6.20 Consider a suspended space frame comprising 4 identical members, each 2m long with an axial rigidity, $EA = 6000$ kN, inter-connected to a rigid joint at O at bottom (in a pyramidal arrangement), with their hinged supports forming a square of side 3m at the ceiling. Apply the conventional stiffness method to find the complete response, due to a gravity load of 60 kN acting at O. Assume all bars to have a tubular cross-section, with a mean radius of 100mm and thickness of 10mm. Assume an elastic modulus $E = 2.5\times10^4$ MPa and a Poisson's ratio, $\nu = 0.3$.

6.21 Repeat Problem 6.20, using the reduced stiffness formulation.

6.22 Analyse the space frame, comprising three members, fully fixed at supports A and D, as shown in Fig. 6.53, by the conventional stiffness method and find the complete response. Assume all members to have a tubular cross-section, with a mean radius of 150mm and thickness of 10mm. Assume elastic modulus $E = 2.5\times10^5$ MPa and Poisson's ratio, $\nu = 0.3$.

6.23 Repeat Problem 6.22, using the reduced stiffness formulation.

7

Analysis of Elastic Instability and Second-order Response

7.1 INTRODUCTION

In the methods of structural analysis of skeletal structures we have studied so far (in the previous Chapters), we have tacitly assumed that the deformations are small, and for this reason, we have conveniently considered the equilibrium of the structure in the *undeformed* configuration. In reality, the equilibrium should be considered in the *deformed* configuration, which is something that we do not know in advance. This introduces *geometric nonlinearity* in the analysis, and we need to carry out a step-by-step (iterative) analysis in order to converge on the response in the deformed configuration. At every stage in the analysis, we may observe that the stiffness of the structure (or components of the structure), and sometimes even the external loading, depend on the deformations and changed geometry of the structure. In addition, *material nonlinearity* effects may have to be considered. This is usually a tedious exercise, which is normally carried out, only when really warranted, i.e., when the nonlinear effects are significant and cannot be reasonably captured by an equivalent linear analysis. Furthermore, we note that the Principle of Superposition is not applicable when the behaviour of the structure is nonlinear.

In this Chapter, we limit considerations to some aspects of geometric nonlinearity in frame elements, related to *slenderness effects*. In particular, when we consider the effect of an axial force on a beam or plane frame element in its deflected configuration, the contribution of the so-called *P-delta moment* (arising out of the eccentricity in the axial loading) needs to be accounted for. The presence of axial compression reduces the flexural stiffness of the element, and conversely, the presence of axial tension enhances the flexural stiffness. A critical condition of *buckling instability* is likely to occur when the flexural stiffness of the 'beam-column' element reduces to zero under axial compression. The axial compression load that leads to such instability is referred to as the *critical buckling load,* P_{cr}. This kind of instability is referred to as *elastic instability*, because, if the material behaviour remains elastic, the structure is expected to regain its original configuration upon unloading†. In the analysis here, we make a further assumption of *linear elastic* behaviour.

Whether or not *secondary* (*P-delta*) effects in a frame are significant depends on the *slenderness* of the element, the boundary conditions and the level of axial loading. When the element slenderness is high, the deformations are likely to be large (enhancing the *P-delta moment*) and the *critical buckling load* will be low. If the axial force, P, acting on the element is compressive and comparable to the buckling load, P_{cr}, then the secondary effects are likely to be significant. As a rule of thumb, we can expect the response of this element (calculated on the basis of a 'first-order' analysis) to be amplified by a factor, approximately given by $\frac{1}{1-P/P_{cr}}$. Thus, there is a clear need to explicitly account for the amplification in the response, when the ratio of P/P_{cr} in a "beam-column" exceeds about 0.15. Moreover, there is a possibility of instability of the structure as a whole, and it is necessary to determine the loading at which such an event can occur. This cannot be done by conventional 'first-order' structural analysis, assuming linear behaviour.

A rigorous treatment of 'second-order effects' induced by geometric nonlinearity calls for nonlinear analysis, and this is often rendered complex by coupling with material nonlinearity. Such analysis lies outside the scope of this book. It is possible to carry out analysis of geometric nonlineariy in a matrix framework, using the concept of *geometric stiffness*. However, by making certain simplifying assumptions, it is possible to work within the framework of conventional linear analysis. Such simplified ('quasi-static') second-order analysis, including estimation of buckling loads in beam-columns and plane frames (considering elastic instability), is discussed in this Chapter. It is assumed that the elements are adequately braced against out-of-plane and torsional modes of buckling.

To begin with, the fundamentals of buckling behaviour of columns and beam-columns, and the influence of axial force on rotational and translational stiffnesses of a prismatic plane frame element are described in Section 7.2. This is followed, in Section 7.3, by a detailed description and demonstration of the Slope-deflection method to find buckling loads and second-order response of continuous beams and plane frames; the analysis usually calls for an iterative solution procedure. In Section 7.4, it is shown how the geometric nonlinearity can be approximated in the form of a *geometric stiffness matrix*, for application in matrix methods using the stiffness method.

† For example, in the design of 'cross-bracing' in a pin-jointed steel portal frame, the diagonal bracing elements are usually slender, and one of them is expected to buckle under compression under incidental lateral (wind) loading, while the other diagonal is designed to resist the tension developed, assuming the compression diagonal to be ineffective.

7.2 EFFECTS OF AXIAL FORCE ON FLEXURAL STIFFNESS

7.2.1 Buckling of Ideal Columns

It is instructive to begin with a review of the basic concepts related to the buckling of *ideal columns*, i.e., *initially straight* prismatic frame elements, subject to axial compressive forces. Consider, for example, the behaviour of such an element subject to a gradually increasing axial compression P. At low load levels, the element will remain straight. However, at a certain load, referred to as the *critical buckling load*, P_{cr}, the column deflects all of a sudden in the lateral direction. This phenomenon is known as *buckling*, which results in an *instability* failure of the column.

As observed first by Euler, the condition of buckling can be established theoretically by considering the equilibrium† of the column in a deformed configuration, as shown in Figs 7.1a and b, for two different boundary conditions of the column (cantilever and simply supported conditions). We observe that in the assumed deflected configuration, the initially straight column is now bent, and the bending curvature $\varphi(x)$ at any section x from a column end location (which is assumed to be restrained against deflection) can be expressed in terms of the deflection at x as $\varphi(x) \approx \Delta''(x)$, considering the deflections to be very small. Assuming linear elastic behaviour and assuming plane sections before bending to remain plane (i.e., ignoring shear deformations), the resulting bending moment at this section is given by $M(x) = EI\varphi(x) = EI\Delta''(x)$, where EI is the flexural rigidity of the prismatic element. The distribution of bending moments due to the so-called *P-delta* effect depends on the shape of the lateral deflection profile, as shown in Fig. 7.1.

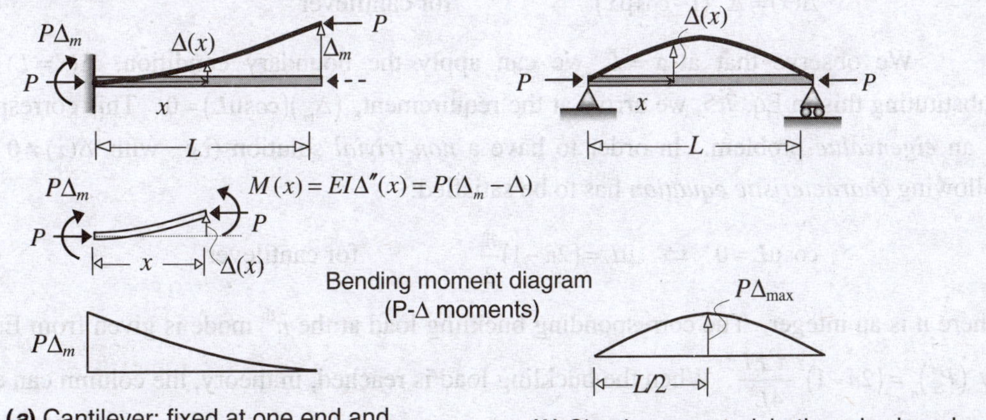

(*a*) Cantilever: fixed at one end and free at the other

(*b*) Simply supported: both ends pinned

Figure 7.1 Bending moments in columns due to P-delta effect

† As an alternative to the 'direct equilibrium' approach, we can adopt an 'energy' approach.

Moment equilibrium of the free-body at x gives us a relationship between the deflected shape and the applied compressive load P, as indicated in Fig. 7.1a:

$$EI\Delta''(x) + P\Delta(x) = P\Delta_m \tag{7.1}$$

where Δ_m is the relative deflection, if any, between the two ends of the element (equal to zero in columns that are braced against sidesway, as in Fig. 7.1b).

Defining the parameter μ as:

$$\mu = \sqrt{\frac{P}{EI}} \tag{7.2}$$

and substituting in Eq. 7.1,

$$\Delta''(x) + \mu^2\Delta(x) = \mu^2\Delta_m \tag{7.3}$$

The solution of the above second-order differential equation is given by:

$$\Delta(x) = A\sin\mu x + B\cos\mu x + \Delta_m \tag{7.4}$$

where A and B are constants that can be determined by applying appropriate boundary condtions.

Cantilever condition: fixed at one end and free at the other

For the cantilever condition shown in Fig. 7.1a, the boundary conditions are given by $\Delta(x=0) = \Delta'(x=0) = 0$, whereby $A = 0$ and $B = \Delta_m$, and the deflected shape, known as *mode shape*, is given by:

$$\Delta(x) = \Delta_m(1 - \cos\mu x) \quad \text{for cantilever} \tag{7.5}$$

We observe that at $x = L$, we can apply the boundary condition, $\Delta(x = L) = \Delta_m$. Substituting this in Eq. 7.5, we arrive at the requirement, $(\Delta_m)(\cos\mu L) = 0$. This corresponds to an *eigenvalue* problem. In order to have a *non-trivial* solution (i.e., with $\Delta(x) \neq 0$), the following *characteristic equation* has to be satisfied:

$$\cos\mu L = 0 \;\Rightarrow\; \mu L = (2n-1)\frac{\pi}{2} \quad \text{for cantilever} \tag{7.6}$$

where n is an integer. The corresponding buckling load at the nth mode is given from Eq. 7.2 by $(P_{cr})_n = (2n-1)^2\dfrac{\pi^2 EI}{4L^2}$. When the buckling load is reached, in theory, the column can either continue to remain straight (trivial solution) or buckle, in order to satisfy equilibrium. Hence, this critical point of loading, in which two alternative equilibrium states are possible, is referred to as a *bifurcation point*. In practice, the buckling will occur invariably at the lowest (first) mode (i.e., $n = 1$), corresponding to which the buckling load (eigenvalue) and mode shape (eigenfunction) are given by:

$$P_{cr} = \frac{\pi^2 EI}{4L^2} \quad ; \quad \Delta(x) = (\Delta_m)\left(1 - \cos\frac{\pi x}{2L}\right) \qquad \text{for cantilever} \qquad (7.7)$$

Note that the mode shape defines only the relative ordinates of different points along the longitudinal axis, and the absolute value of the deflection Δ_m remains undefined. This is a basic feature of the eigenvalue solution. According to this theory, once the buckling has taken place the deflections may go on increasing indefinitely, without any increase in axial load, in a condition of apparent neutral equilibrium, as indicated in Fig. 7.2. In reality, of course, this does not happen. The post-buckling formulation based on small deflection theory is not valid, and it can be shown that the post-buckling equilibrium condition is a relatively stable condition, with Δ_m having a unique value† corresponding to $P > P_{cr}$, provided the elastic behaviour can be sustained (i.e., the material does not become inelastic). If the material undergoes plastification, the post-buckling strength can drop down. These possibilities are indicated in Fig. 7.2.

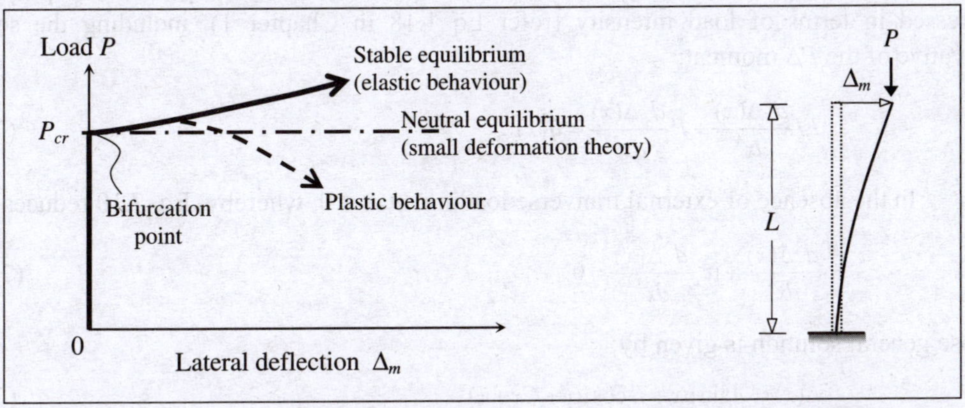

Figure 7.2 Bifurcation and post-buckling behaviour in an ideal column

Simply supported condition: pinned at both ends

For the simply supported condition shown in Fig. 7.1b, by substituting $\Delta_m = 0$, and applying the boundary condition, $\Delta(x=0)=0$ in Eq. 7.4, we obtain $B = 0$. Applying the second boundary condition, $\Delta(x=L)=0$, we arrive at the requirement, $(A)(\sin\mu L)=0$. Corresponding to the non-trivial solution, the characteristic equation is obtained as:

$$\sin\mu L = 0 \quad \Rightarrow \quad \mu L = n\pi \qquad \text{for both ends pinned} \qquad (7.8)$$

† One can observe this in the legendary portraits of Charlie Chaplin leaning against his buckled cane; the more he presses down the cane, the more the cane bends. Of course, only a marginal increase in load above the critical buckling load is possible; otherwise, the cane will get damaged or will break. If the behaviour is still within the elastic buckling load, the cane bounces back to its initial straight configuration.

$$\sin \mu L = 0 \quad \Rightarrow \quad \mu L = n\pi \qquad \text{for both ends pinned} \tag{7.8}$$

Corresponding to the first mode (i.e., $n = 1$), the buckling load (eigenvalue) and mode shape (eigenfunction) are given by:

$$P_{cr} = \frac{\pi^2 EI}{L^2} \quad ; \quad \Delta(x) = (\Delta_{\max})\left(\sin\frac{\pi x}{L}\right) \qquad \text{for both ends pinned} \tag{7.9}$$

The above expression for the critical buckling load of an ideal column, with the two ends pinned, is also popularly referred to as the *Euler buckling load*, and may be denoted by the notation, P_{Eo}, for the standard pinned-pinned case.

General approach for any set of four boundary conditions

It is possible to deal with any set of boundary conditions at the ends of a prismatic beam-column subject to an axial compressive force P and arbitrary transverse loading of intensity $q(x)$ per unit length, by invoking the basic fourth-order differential equation of equilibrium expressed in terms of load intensity (refer Eq. 1.18 in Chapter 1), including the second derivative of the $P\Delta$ moment:

$$EI\frac{d^4\Delta(x)}{dx^4} + P\frac{d^2\Delta(x)}{dx^2} = q(x) \tag{7.10}$$

In the absence of external tranverse loading, $q(x) = 0$, whereby, Eq. 7.10 reduces to:

$$\frac{d^4\Delta(x)}{dx^4} + \mu^2\frac{d^2\Delta(x)}{dx^2} = 0 \tag{7.10a}$$

whose general solution is given by:

$$\Delta(x) = A\sin\mu x + B\cos\mu x + Cx + D \tag{7.11}$$

By substituting four appropriate boundary conditions (kinematic and static), we can solve for the four constants, A, B, C and D in Eq. 7.11. For example, considering the *propped cantilever* condition, with one end hinged and the other fixed, the boundary conditions are given by (i) $\Delta(x=0) = 0$, whereby $B + D = 0$; (ii) $\Delta'(x=0) = 0$, whereby $\mu A + C = 0$; (iii) $\Delta(x=L) = 0$, whereby $CL + D = 0$; and (iv) $\Delta''(x=L) = 0$, whereby $A\sin\mu L + B\cos\mu L = 0$. This can be conveniently expressed in matrix form as follows:

$$\begin{bmatrix} 0 & 1 & 0 & 1 \\ \mu & 0 & 1 & 0 \\ 0 & 0 & L & 1 \\ \sin\mu L & \cos\mu L & 0 & 0 \end{bmatrix}\begin{Bmatrix} A \\ B \\ C \\ D \end{Bmatrix} = \begin{Bmatrix} 0 \\ 0 \\ 0 \\ 0 \end{Bmatrix} \qquad \text{for propped cantilever} \tag{7.12}$$

$$\tan \mu L = \mu L \quad \Rightarrow \quad \mu L = 4.493 \quad \text{for propped cantilever} \qquad (7.13)$$

Corresponding to the first mode, the buckling load (lowest eigenvalue) is given by:

$$P_{cr} = \frac{20.19EI}{L^2} = \frac{\pi^2 EI}{(0.699L)^2} \quad \text{for propped cantilever} \qquad (7.14)$$

Following a similar approach, it can be shown that for a prismatic column with fixed-fixed boundary conditions, the buckling load is given by:

$$P_{cr} = \frac{4\pi^2 EI}{L^2} = \frac{\pi^2 EI}{(0.5L)^2} \quad \text{for fixed-fixed end conditions} \qquad (7.15)$$

Effective Length

A generalised expression for the *Euler buckling load* P_E of a prismatic column with any set of boundary conditions can be conveniently expressed as follows:

$$P_E = \frac{\pi^2 EI}{L_e^2} = \left(\frac{1}{k_e}\right)^2 P_{Eo} \qquad (7.16)$$

where $L_e = k_e L$ is called the *effective length* of the column and k_e is called the *effective length ratio* or *effective length factor*. As established in Eq. 7.9, k_e has a value of unity when the two ends of the column are pinned, and this is the standard case referred to in Euler buckling ($P_E = P_{Eo}$). Also, comparing Eq. 7.16 with Eq. 7.7, we observe that k_e has a value of 2.0 when one end of the column is free while the other end is fixed (cantilever condition), whereby $P_E = 0.25P_{Eo}$. Similarly, comparing Eq. 7.16 with Eq. 7.14, we observe that k_e has a value of 0.699 when one end of the column is pinned while the other end is fixed (propped cantilever condition), whereby $P_E = 2.047P_{Eo}$. Finally, comparing Eq. 7.16 with Eq. 7.15, we observe that k_e has a value of 0.5 when both ends of the column are restrained against rotation and also relative translation (fixed-fixed condition), whereby $P_E = 4P_{Eo}$.

It can be shown that the value of the effective length ratio lies between 0.5 and 1.0 (i.e., $1 \le P_E / P_{Eo} \le 4$) when the column is braced against sidesway, i.e., no relative translation is permitted between the two ends of the column (refer Fig. 7.3). The lowest value of $k_e = 0.5$ (whereby, $P_E = 4P_{Eo}$) is encountered when both ends are fully fixed against rotation, as shown in Fig. 7.3a. When one end is fixed and the other is pinned (i.e. free to rotate), $k_e = 0.699$ (whereby, $P_E = 2.047P_{Eo}$), as shown in Fig. 7.3b. When both ends are pinned, $k_e = 1$ (whereby, $P_E = P_{Eo}$).

The *effective length* L_e can be interpreted as the distance between the points of contraflexure in the buckled configuration. This is indicated in the four cases shown in Fig. 7.3. For a typical column in a multi-storey frame that is braced against sidesway (by stiffening elements such as shear walls), the effective length will never exceed the actual (so-called 'unsupported') length, L, as indicated in Fig. 7.3d.

It can be shown that the value of the effective length ratio exceeds (or at best, equals to) 1.0 (i.e., $P_E \leq P_{Eo}$) when the column is *unbraced* against sidesway, i.e., when relative lateral deflection is permitted between the two ends of the column (refer Fig. 7.4). When both ends are fixed against rotation, but at one end translation is allowed, the buckled shape will be as shown in Fig. 7.4a. In order to get the distance between the points of contraflexure, we need to extrapolate the curve beyond the length of the column as shown. In this case, we notice that the curve between the two points of contraflexure in Fig. 7.4a resembles the pinned-pinned braced case in Fig. 7.3c, with $L_e = L$, whereby $P_E = P_{Eo}$. As the rotational fixity at the two ends is released, the effective length increases. For the cantilever condition shown in Fig. 7.4b, we observe, by extrapolation, that $L_e = 2L$, whereby $P_E = 0.25P_{Eo}$.

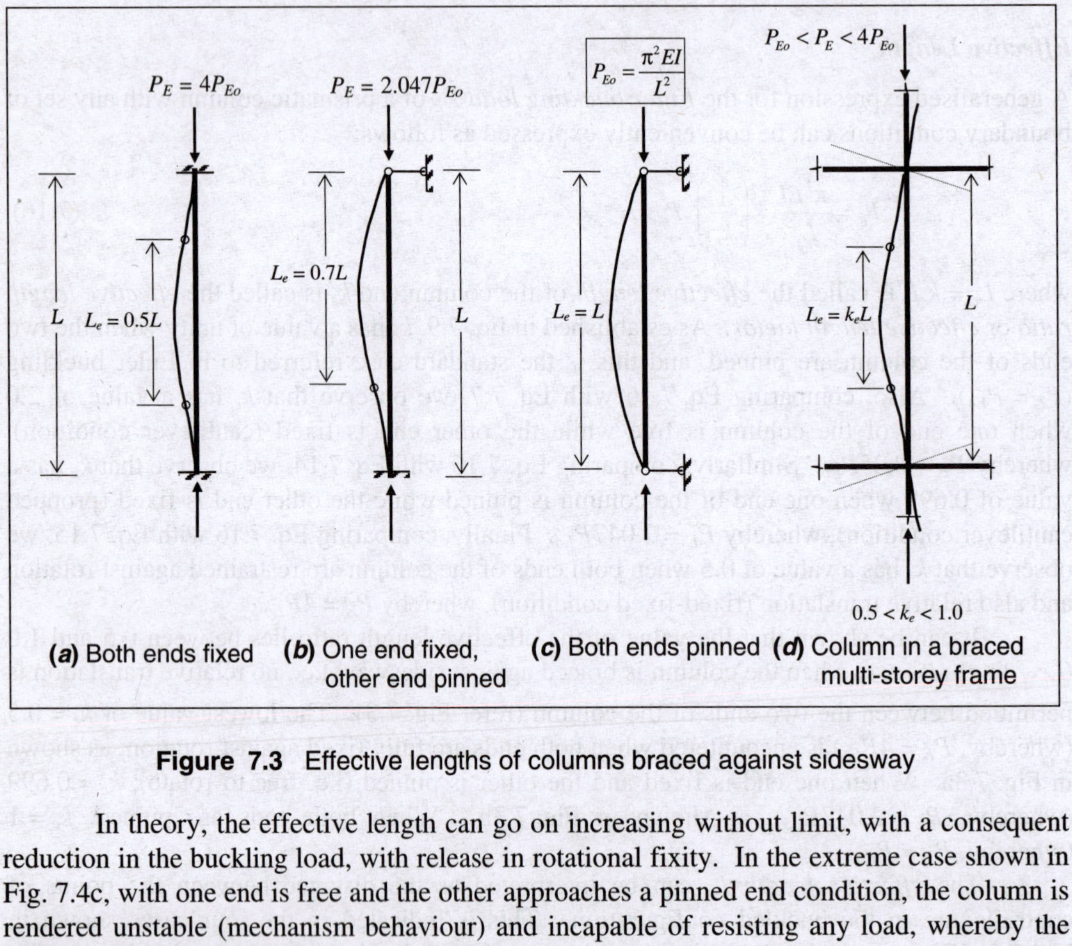

Figure 7.3 Effective lengths of columns braced against sidesway

In theory, the effective length can go on increasing without limit, with a consequent reduction in the buckling load, with release in rotational fixity. In the extreme case shown in Fig. 7.4c, with one end is free and the other approaches a pinned end condition, the column is rendered unstable (mechanism behaviour) and incapable of resisting any load, whereby the critical buckling load approaches zero and the effective length becomes infinite. For a typical column in a multi-storey frame that is unbraced against sidesway, the effective length will exceed the actual (so-called 'unsupported') length, L, by some amount, as indicated in Fig. 7.4d, and the Euler buckling load will be less than P_{Eo}. In practice, it is important to assess the critical buckling load as accurately as possible and to design the structure so as to prevent a buckling failure.

While applying the Euler buckling formula to a column in a plane frame, the second moment of area, I, to be considered, is with respect to bending in the plane of the frame. In the more general case of a space frame element, where buckling can occur in any direction, it is necessary to explore the two *principal* moments of inertia (I_{xx} and I_{yy}, or I_{uu} and I_{vv} in the case of unsymmetrical bending), along with the corresponding effective lengths (such as L_{ex} and L_{ey}) in the formula (Eq. 7.16), in order to find the lowest possible value of the critical buckling load.

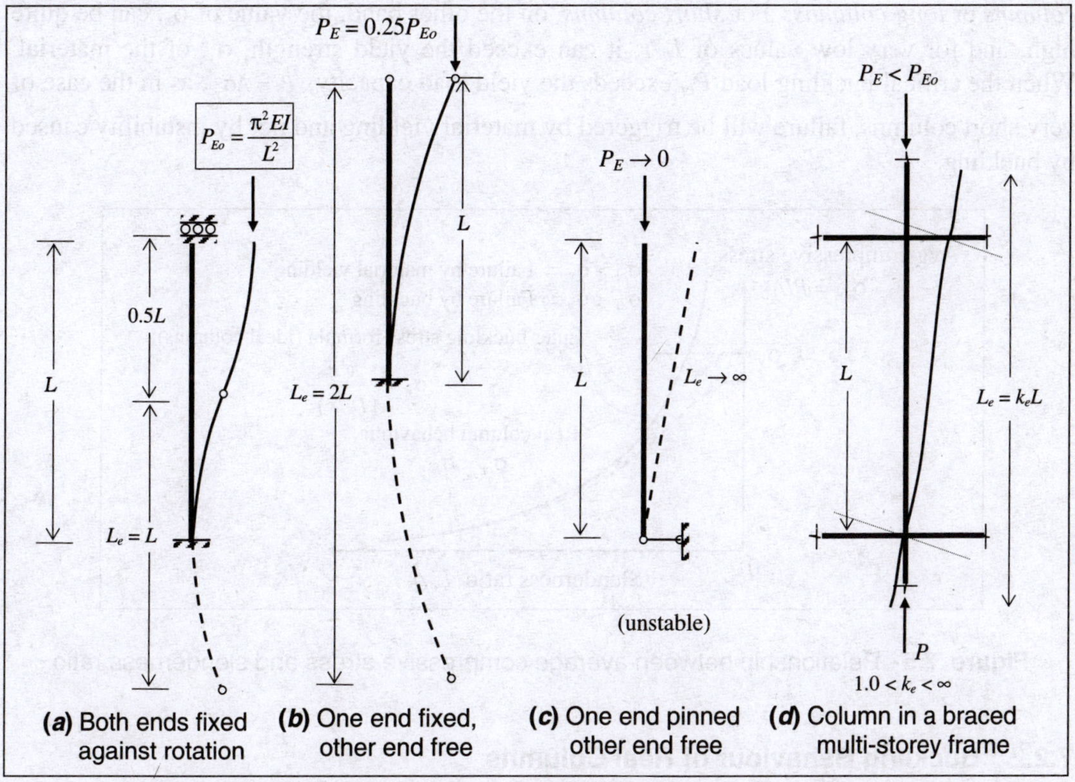

Figure 7.4 Effective lengths of columns unbraced against sidesway

Slenderness Ratio

When the column is made of a homogeneous material like steel, it is meaningful to refer to the critical buckling stress, σ_{cr}, which is obtained by dividing the critical buckling load, P_{cr}, by the cross-sectional area, A, of the column. While using the Euler buckling load formula (Eq. 7.16), it is appropriate to refer to this stress as the *Euler buckling stress*, σ_E. Thus, dividing the expression in Eq. 7.16 by A, and observing that $I = Ar^2$, where r is the radius of gyration of the column section with respect to the axis of bending,

$$\sigma_E = \frac{\pi^2 E}{(L_e/r)^2} \tag{7.17}$$

Thus, for a given material, the Euler buckling stress is found to depend only on the ratio of the effective length to the radius of gyration. This ratio, L_e/r, is often referred to as slenderness ratio. The relationship between σ_{cr} and L_e/r is depicted graphically in Fig. 7.5. The value of σ_{cr} drops drastically at high values of slenderness ratios, i.e., for *slender columns* or *long columns*. For *short columns*, on the other hand, the value of σ_{cr} can be quite high, and for very low values of L_e/r, it can exceed the yield strength, σ_y, of the material. When the critical buckling load P_{cr} exceeds the yield load capacity, $P_y = A\sigma_y$, as in the case of very short columns, failure will be triggered by material yielding and not by instability caused by buckling.

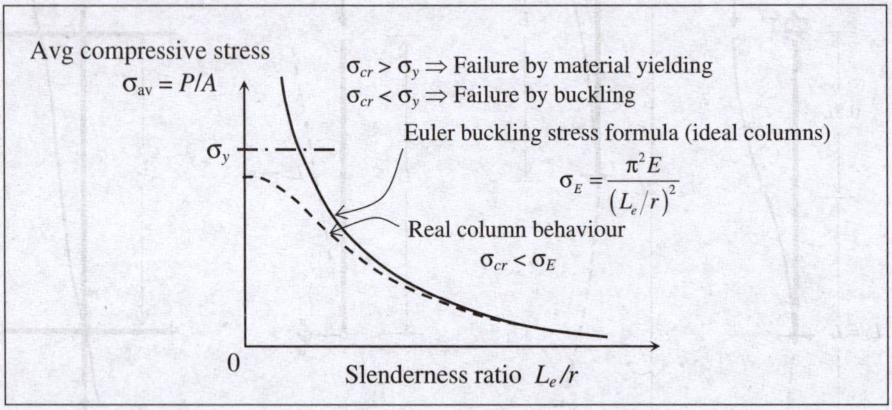

Figure 7.5 Relationship between average compressive stress and slenderness ratio

7.2.2 Buckling Behaviour of Real Columns

Experimental studies have shown that *bifurcation instability* rarely occurs in real columns. When a slender column is subject to an axial compressive load, it will start showing lateral deflections well before the critical buckling load is reached, and the value of P_{cr} is found to be invariably less than the Euler buckling load P_E computed using Eq. 7.16. This is attributable to the fact that, in reality, it is difficult to simulate conditions of perfectly concentric loading, particularly when the column is long. Eccentricity between the line of action of the applied load and the centroidal axis of the column occurs due to *initial imperfections* in the column as well as due to possible eccentricity in the loading itself. This "accidental eccentricity" introduces additional curvatures, and hence bending moments in the column, thereby reducing the buckling strength. A realistic relationship between $\sigma_{cr} = P_{cr}/A$ and L_e/r is shown by means of a dashed line in Fig. 7.5.

Eccentric Loading on Slender Columns

Let us consider the case of an eccentrically loaded slender column, with both ends pinned. Let the initial eccentricity in loading be e, and the lateral deflection be $\Delta(x)$, as shown in Fig. 7.6a. By considering moment equilibrium of a free-body at a section at x, and considering $M(x) = EI\Delta''(x)$, and substituting $\mu = \sqrt{\frac{P}{EI}}$,

$$\Delta''(x) + \mu^2\Delta(x) + \mu^2 e = 0 \tag{7.18}$$

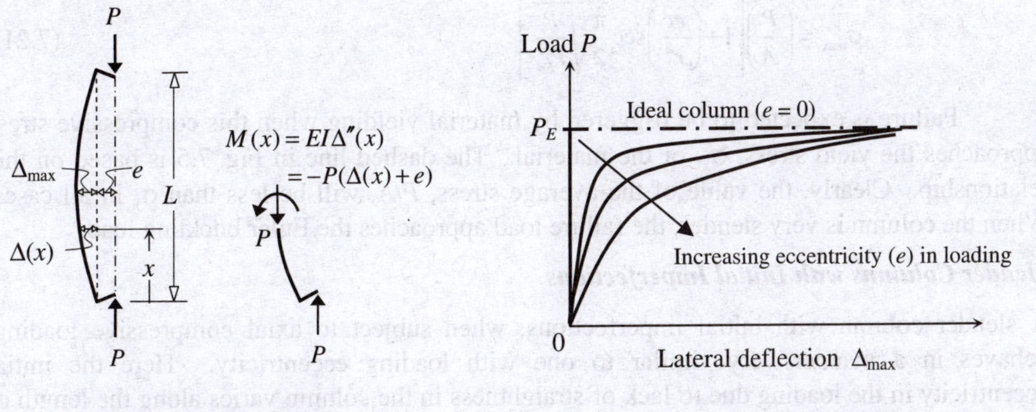

(a) Freebody diagrams (both ends pinned) **(b)** Load-deflection curves

Figure 7.6 Behaviour of a slender column with initial eccentricity in loading

The solution of the differential equation (Eq. 7.18) is as given earlier by Eq. 7.4. Applying the boundary conditions, $\Delta(x=0) = \Delta(x=L) = 0$, we obtain the expressions for the two constants, A and B as follows: $B = e$ and $A = \frac{e(1-\cos\mu L)}{\sin\mu L}$. Accordingly, the expression for the deflected shape is given by:

$$\Delta(x) = e\left[(1-\cos\mu L)\frac{\sin\mu x}{\sin\mu L} + \cos\mu x - 1\right] \tag{7.19}$$

Clearly, this is not an eigenvalue problem, and finite values of deflection are obtainable, which depend on the magnitude of the applied loading, eccentricity and the slenderness of the column. The expression for the maximum deflection, which occurs at $x = L/2$, can be simplified into the following form, known as the *secant formula*:

$$\Delta_{max} = e\left[\sec\frac{\pi}{2}\sqrt{\frac{P}{P_{Eo}}} - 1\right] \tag{7.20}$$

The relationship between the applied loading P and the maximum deflection for various eccentricities is depicted schematically in Fig. 7.6b. Clearly, with increasing eccentricity in loading, the maximum deflection due to the slenderness of the column, also increases. Also, when the eccentricity approaches zero, the maximum deflection becomes unbounded and the condition of Euler buckling is approached. The maximum compressive stress, due to the combined effect of axial compression and bending moment is given by $\sigma_{max} = \frac{P}{A} + \frac{P(e+\Delta_{max})}{I/c}$, where c is the distance from the centroidal axis to the extreme fibre in the cross-section. Simplifying this expression, it can be shown that

$$\sigma_{max} = \left(\frac{P}{A}\right)\left[1 + \left(\frac{ec}{r^2}\right)\sec\frac{\pi}{2}\sqrt{\frac{P}{P_{Eo}}}\right] \tag{7.21}$$

Failure is expected to be triggered by material yielding when this compressive stress approaches the yield stress, σ_y, of the material. The dashed line in Fig. 7.5 is based on this relationship. Clearly, the value of the average stress, P/A, will be less than σ_y in all cases. When the column is very slender, the failure load approaches the Euler buckling load.

Slender Columns with Initial Imperfections

A slender column with initial imperfections, when subject to axial compressive loading, behaves in a manner very similar to one with loading eccentricity. Here the initial eccentricity in the loading due to lack of straightness in the column varies along the length of the column. For the most adverse effects, we can assume the initial profile of the column to be similar to the first buckling mode shape. For example, for the simply supported case shown in Fig. 7.6a, it is appropriate to consider a sinusoidal initial shape with a maximum deviation of e_o at the mid-point of the column: $e(x) = e_o \sin\frac{\pi x}{L}$. Substituting this expression for e in Eq. 7.18, and solving the differential equation for the given boundary conditions, the expression for the additional deflection $\Delta(x)$ due to the applied loading takes the following sinusoidal form:

$$\Delta(x) = e_o\left(\frac{P/P_E}{1-P/P_E}\right)\sin\frac{\pi x}{L} \tag{7.22}$$

The total deflection with respect to the straight line joining the two pinned ends is therefore obtained as an amplification of the initial imperfection by an amplification factor, ω, given by:

$$\omega = \frac{1}{1-P/P_E} \tag{7.23}$$

This amplification factor is applicable also to the bending moment $M(x)$ at any section. In other words, the intial deflection (due to imperfection) and consequent bending moment (due to eccentricity) both get magnified, with the application of the load P by the factor, ω. As the applied load increases in magnitude, the deflections and bending moments also increase nonlinearly and get unbounded as the load approaches the Euler buckling load. The expression for maximum compressive stress in the column (similar to Eq. 7.21) takes the following form:

$$\sigma_{max} = \left(\frac{P}{A}\right)\left[1+\left(\frac{e_o c}{r^2}\right)\omega\right] \tag{7.24}$$

Based on the above discussions on the behaviour of real columns subject to axial compressive loading, the following conclusions can be drawn.

Bifurcation instability of columns, as predicted by the Euler buckling analysis (eigenvalue analysis), does not usually occur in columns in practice, due to the presence of eccentricity in loading and initial imperfections. Lateral deflections and bending begin to manifest at early stages of loading, and these get amplified with increase in loading, eccentricity and slenderness. With gradual increase in loading, the Euler buckling load is asymptotically reached in the case of slender columns, with the lateral deflections becoming unbounded. Buckling may be interpreted as a condition of instability that occurs when the flexural stiffness of the "beam-column" reduces to zero.

7.2.3 Flexural Behaviour of Beam-Columns

Beam-columns are frame elements that are subject to bending combined with axial compression and whose flexural response (deflections, bending moments) gets amplified on account of the reduction in flexural stiffness caused by the axial compression. From the previous discussions, we know that this effect becomes very pronounced when the axial compression approaches the Euler buckling load of the element (high value of P/P_E ratio). This occurs when the slenderness ratio ($k_e L/r$) is high and the applied compression P is also relatively high. In elements that have very low slenderness ratio and subject to relatively low axial compression, the amplification in the response is likely to be negligible. Hence, in such situations, the conventional 'first-order' frame analysis (as discussed in Chapter 6) is applicable without resulting in any significant error. Let us demonstrate this with a simple example.

Example: Simply supported beam-column

Consider the example of a simply supported beam-column (span L, flexural rigidity EI) subject to a central concentrated load W and an axial compression P, as shown in Fig. 7.7a.

The beam-column is assumed to be restrained against out-of-plane deflections. If we ignore the influence of the axial compression and treat the element as a simple beam (assuming $P = 0$), we know that the maximum values of bending moment and deflection at the mid-span are given by:

$$M_o = \frac{WL}{4}; \quad \Delta_o = \frac{WL^3}{48EI} \tag{7.25}$$

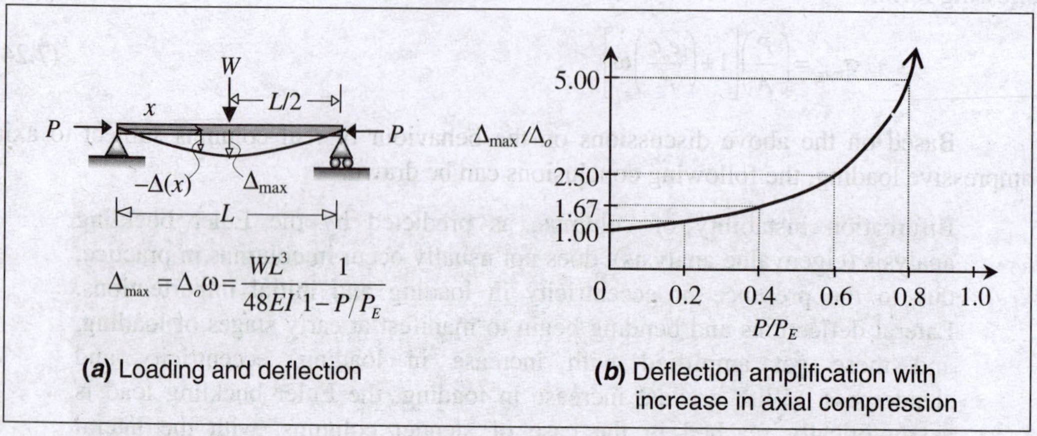

Figure 7.7 Amplification of response in a beam-column due to axial compression

Also, considering ideal column behaviour (assuming $W = 0$), we know that the element will become unstable when the axial compression P approaches the Euler buckling load, given by: $P_{Eo} = \frac{\pi^2 EI}{L^2}$.

$$P_{Eo} = \frac{\pi^2 EI}{L^2} \Rightarrow \frac{L^2}{EI} = \frac{\pi^2}{P_{Eo}} \tag{7.26}$$

By considering moment equilibrium of a free-body at a section at x (Fig. 7.7a), and considering $M(x) = EI\Delta''(x)$ (considering upward $\Delta(x)$ and sagging $M(x)$ to be positive), and substituting $\mu = \sqrt{\frac{P}{EI}}$,

$$\Delta''(x) + \mu^2 \Delta(x) - \frac{Wx}{2EI} = 0 \tag{7.27}$$

The solution to the differential equation (Eq. 7.27) is given by:

$$\Delta(x) = A \sin \mu x + B \cos \mu x + \frac{Wx}{2P} \tag{7.28}$$

Applying the boundary conditions, $\Delta(x=0)=\Delta'(x=L/2)=0$, we obtain the expressions for the two constants, *A* and *B* as follows: $B=0$ and $A=-\frac{W}{2\mu P}\frac{1}{\cos(\mu L/2)}$.

Accordingly, the expression for the deflected shape is given by:

$$\Delta(x)=\frac{-W}{2\mu P}\left[\frac{\sin\mu x}{\cos(\mu L/2)}-\mu x\right] \tag{7.29}$$

The expression for the maximum deflection, $\Delta_{max}=|\Delta(x=L/2)|$ can be derived from the above. Dividing this expression with $\Delta_o=\frac{WL^3}{48EI}$ (refer Eq. 7.25), and simplifying, we can generate the following expression for the amplification factor for maximum deflection as follows:

$$\frac{\Delta_{max}}{\Delta_o}=3\left[\frac{\tan(\mu L/2)-(\mu L/2)}{(\mu L/2)^3}\right] \tag{7.30}$$

Considering the series expansion, $\tan\theta=\theta+\frac{\theta^3}{3}+\frac{2\theta^5}{15}+\frac{17\theta^7}{315}+\ldots$, the above expression simplifies to $\frac{\Delta_{max}}{\Delta_o}=1+\frac{2}{5}(\mu L/2)^2+\frac{17}{105}(\mu L/2)^4+\cdots$. Invoking $\frac{L^2}{EI}=\frac{\pi^2}{P_{Eo}}$ (refer Eq. 7.26), whereby $\left(\frac{\mu L}{2}\right)^2=\frac{PL^2}{4EI}=\frac{\pi^2}{4}\frac{P}{P_{Eo}}$, we obtain: $\frac{\Delta_{max}}{\Delta_o}=1+0.987\left(\frac{P}{P_{Eo}}\right)+0.986\left(\frac{P}{P_{Eo}}\right)^2+\cdots$, which may be approximated as the series expansion of $\frac{1}{1-P/P_{Eo}}$. Thus, we observe that the deflection amplification factor is the same as the one derived earlier (ω) in Eq. 7.23 (for initial imperfections):

$$\Delta_{max}=\Delta_o\omega=\frac{\Delta_o}{1-P/P_{Eo}} \tag{7.31}$$

The variation of the deflection amplification factor with the load ratio, P/P_{Eo}, is shown in Fig. 7.7b. The midspan deflection gets amplified by 1.25 times when $P=0.2P_{Eo}$ and by 5 times when $P=0.8P_{Eo}$, and becomes unbounded as the Euler buckling load is approached.

The amplification factor for maximum bending moment at midspan (refer Eq. 7.26) is given by:

$$\frac{M_{max}}{M_o}=\frac{\frac{WL}{4}+P\omega\frac{PL^3}{48EI}}{WL/4}=\frac{1+0.1775P/P_{Eo}}{1-P/P_{Eo}} \tag{7.32}$$

This works out to be slightly less than the deflection amplification factor. The midspan moment gets amplified by 1.206 times when $P = 0.2P_{Eo}$ and by 4.29 times when $P = 0.8P_{Eo}$, and becomes unbounded as the Euler buckling load is approached.

In general, it can be shown that for a simply supported beam subject to any symmetric loading, the amplification in the values of maximum deflection and maximum bending moment (at midspan) can be expressed as follows:

$$\Delta_{max} = \omega_\Delta \Delta_o = \left(\frac{1}{1-P/P_E}\right)\Delta_o$$
$$M_{max} = \omega_M M_o = \left(\frac{1+\psi P/P_E}{1-P/P_E}\right)M_o \tag{7.33}$$

where

$$\Delta_o = \frac{WL^3}{48EI} \; ; \; M_o = \frac{WL}{4} \; ; \; \psi = +0.178 \quad \text{for point load } W \text{ at midspan}$$
$$\Delta_o = \frac{5q_o L^4}{384EI} \; ; \; M_o = \frac{qL^2}{8} \; ; \; \psi = +0.028 \quad \text{for udl } q_o \text{ per unit length} \tag{7.34}$$
$$\Delta_o = \frac{M_o L^2}{8EI} \; ; \; \psi = +0.233 \quad \text{for uniform moment } M_o$$

Clearly, the amplification in the flexural response due to slenderness effects in a beam-column needs to be explicitly accounted for when the P/P_E ratio is not small (say, more than 10 percent).

Linear or nonlinear analysis?

The analysis carried out so far has tacitly assumed *linear* behaviour. However, we must be cautious in applying the Principle of Superposition, while dealing with beam-columns. This Principle, based on assumed linear behaviour, is valid only when we assume that the axial compression P remains unchanged during the process of loading. If we plot the variation of the applied transverse load W with the midspan deflection for different values of the axial compression P, we will get straight-line plots, as shown in Fig. 7.8a (which can be generated using the results in Fig. 7.7b). The slope of the W-Δ_{max} line provides a measure of the flexural stiffness of the beam, which is seen to drop significantly with increase in the P/P_E ratio. The flexural stiffness reduces to zero when this ratio approaches unity, i.e. when the axial compression P is equal to the Euler buckling load, P_E. Of course, it should be noted that the stiffness measure, as calculated above, is subject to error under large deflections, because it is based on small deformation theory and also does not account for possible material nonlinearity.

Situations, where P remains more-or-less constant while W changes, are commonly encountered in columns in framed structures where the axial compression is induced due to permanent gravity loads while the bending is caused by incidental lateral loads (wind or earthquake). In such situations, we can indeed assume the load-response behaviour of the beam-column to be linear during the application of the flexural loading and thereby invoke the Principle of Superposition while considering various load combinations, provided the axial compressive forces remain practically unchanged in all the load cases. Thus, it is possible to extend the conventional matrix methods of frame analysis to include the slenderness effects (on flexural stiffness) in beam-columns. This is discussed in the Sections to follow.

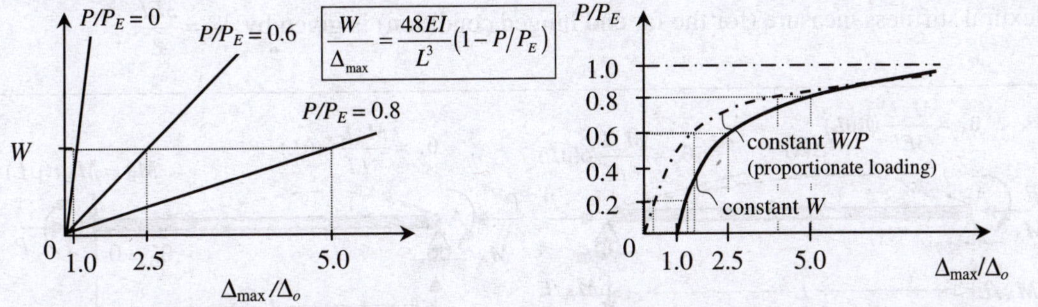

(*a*) Linear relationship between W and Δ_{max} for constant P

(*b*) Nonlinear relationship between P and Δ_{max} for constant W or proportionate loading

Figure 7.8 Relationship between loading and deflection in a beam-column (simply supported with central point load)

However, in other situations where the axial compression P changes in magnitude during the loading process, while W remains constant or also changes along with P (usually proportionately), the load-response behaviour is nonlinear, as indicated in Fig. 7.8b. Although at any stage of loading (any point on the load-deflection curve), the analysis of the beam-column can be done by assuming linear elastic behaviour (Eq. 7.33 is valid), the analysis has to be carried out separately at every stage of loading and the overall load-response behaviour turns out to be nonlinear, whereby the Principle of Superposition is not valid.

7.2.4 Flexural Stiffness Measures for Braced Prismatic Beam-Columns

In order to formulate the stiffness matrix for a prismatic beam-column element, including slenderness effects, we need to establish the basic moment-rotation relations for the element. Let us consider a prismatic element AB with span L and flexural rigidity EI, subject to a constant axial compression P, as shown in Fig. 7.9. Such a beam-column is said to be braced, in the absence of any relative deflection between the two ends (i.e., no chord rotation). Let us consider the application of a moment M_A at end A of the element, which is connected to a hinged support, and let the resulting rotation at this end be θ_A. The other end, B, of the element may be connected to a pinned (roller) support (Fig. 7.9a) or a fixed support (Fig. 7.9b).

We seek the flexural stiffness measure, defined by the ratio, M_A/θ_A, or alternatively, the flexibility measure, θ_A/M_A. We know the standard solutions, from basic structural analysis of beams, ignoring the "beam-column" interaction effects. For example, for the simply supported condition shown in Fig. 7.9a, $\theta_A = \frac{M_A L}{3EI}$ and $\theta_B = \frac{M_A L}{6EI}$, whereby the flexural stiffness measure (for the far end hinged condition) is given by $S^o = \frac{3EI}{L}$.

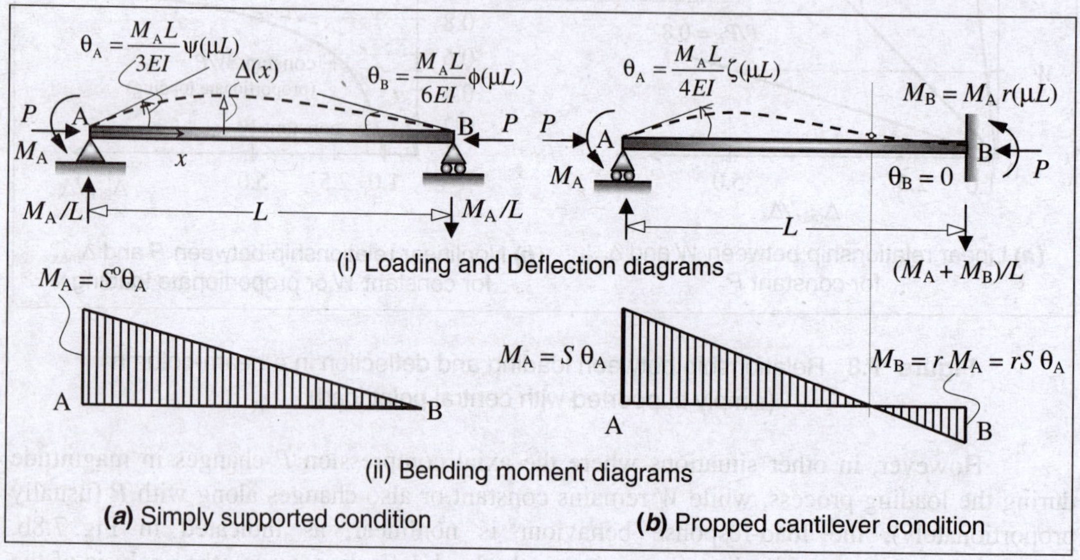

Figure 7.9 Response of a braced beam-column subject to end moment

Similarly, for the propped cantilever condition shown in Fig. 7.9b, $\theta_A = \frac{M_A L}{4EI}$, whereby the flexural stiffness measure (for the far end fixed condition) is given by $S = \frac{4EI}{L}$, and there is a "carry-over moment" induced at the fixed end B, given by $M_B = r\,M_A$, the carry-over factor r being equal to 0.5. We know that owing to the reduction in stiffness due to the influence of the axial compression P, the rotations will increase. Appropriate modifications have to be introduced, which can be shown to be functions of the parameter μL, where $\mu = \sqrt{\frac{P}{EI}}$, as defined in Eq. 7.2, and as indicated by the factors $\psi(\mu L)$ and $\phi(\mu L)$ in Fig. 7.9a and $\zeta(\mu L)$ and $r(\mu L)$ in Fig. 7.9b.

For the simply supported condition shown in Fig. 7.9a, the differential equation of equilibrium is given by:

$$EI\Delta''(x)+P\Delta(x)+M_A(1-x/L)=0 \qquad (7.35)$$

$$\Rightarrow \quad \Delta''(x)+\mu^2\Delta(x)+\frac{M_A}{EI}(1-x/L)=0$$

The solution takes the following form:

$$\Delta(x)=A\sin\mu x+B\cos\mu x-\frac{M_A}{P}\left(1-\frac{x}{L}\right)$$

Applying the boundary conditions, $\Delta(x=0)=\Delta(x=L)=0$, we obtain $B=M_A/P$ and $A=-M_A/(P\tan\mu L)$, whereby the solution is given by:

$$\Delta(x)=\frac{M_A}{P}\left[\frac{\sin\mu(L-x)}{\sin\mu L}-\left(1-\frac{x}{L}\right)\right] \qquad (7.36)$$

whereby

$$\theta_A=\Delta'(0)=\frac{M_A}{P}\left[\frac{-\mu}{\tan\mu L}+\frac{1}{L}\right]=\frac{M_A L}{3EI}\psi(\mu L) \qquad (7.37)$$

where

$$\psi(\mu L)=\frac{3}{\mu L}\left(\frac{1}{\mu L}-\frac{1}{\tan\mu L}\right) \qquad (7.37a)$$

Similarly,

$$\theta_B=-\Delta'(L)=\frac{M_A}{P}\left[\frac{\mu}{\sin\mu L}-\frac{1}{L}\right]=\frac{M_A L}{6EI}\phi(\mu L) \qquad (7.38)$$

where

$$\phi(\mu L)=\frac{6}{\mu L}\left(\frac{1}{\sin\mu L}-\frac{1}{\mu L}\right) \qquad (7.38a)$$

Thus, the flexural stiffness measure, defined by the ratio, $S^o = M_A/\theta_A$, for the simply supported condition shown in Fig. 7.9a, is given by:

$$\boxed{S^o=\frac{3EI}{L}\frac{1}{\psi(\mu L)}} \qquad (7.39)$$

The propped cantilever condition shown in Fig. 7.9b can be viewed as a superposition of (i) an end moment M_A applied at end A and (ii) an end moment M_B applied at end B on a simply supported beam, with $\theta_B = 0$. The former is depicted in Fig. 7.9a, while the latter is depicted in Fig. 7.10. The equation for the deflected shape in Fig. 7.10 can be easily shown to take the following form:

$$\Delta(x)=-\frac{M_B}{P}\left[\frac{\sin\mu x}{\sin\mu L}-\frac{x}{L}\right] \qquad (7.40)$$

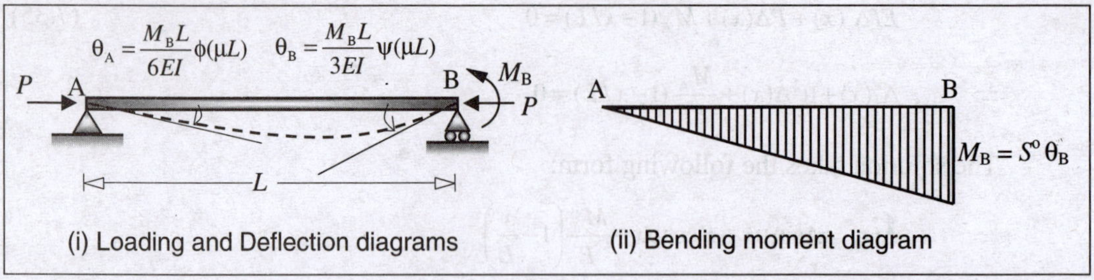

Figure 7.10 Response of simply supported beam-column subject to moment at far end

Thus, the expression for deflection $\Delta(x)$ in the propped cantilever, subject to anti-clockwise end moments M_A and M_B, shown in Fig. 7.9b is given by superposing Eqns 7.36 and 7.40. Setting the compatibility requirement, $\theta_B = \Delta'(x = L) = 0$, we obtain:

$$\theta_B = -\Delta'(L) = \frac{M_A L}{6EI}\phi(\mu L) - \frac{M_B L}{3EI}\psi(\mu L) = 0$$

whereby, the *carry-over factor*, *r*, is given by:

$$\boxed{r = \frac{M_B}{M_A} = \frac{\phi(\mu L)}{2\psi(\mu L)}} \qquad (7.41)$$

In the absence of any axial force ($\mu L = 0$), the functions, ψ and ϕ, are both equal to one, whereby the carry-over factor takes the standard value of 0.5 (applicable for prismatic beams). Furthermore, on simplification,

$$\theta_A = \frac{M_A L}{4EI}\left[\frac{3\psi}{4\psi^2 - \phi^2}\right] = \frac{M_A L}{4EI}\zeta(\mu L) \qquad (7.42)$$

where

$$\zeta(\mu L) = \frac{3\psi}{4\psi^2 - \phi^2} \qquad (7.42a)$$

Thus, the flexural stiffness measure, defined by the ratio, $S = M_A/\theta_A$, for the propped cantilever condition shown in Fig. 7.9b, is given by:

$$\boxed{S = \frac{4EI}{L}\frac{1}{\zeta(\mu L)}} \qquad (7.43)$$

Expanding and simplifying Eqns 7.43, 7.41 and 7.39, we can establish:

$$\boxed{S = \frac{EI}{L}\frac{(\mu L)(\sin\mu L - \mu L\cos\mu L)}{(2 - 2\cos\mu L - \mu L\sin\mu L)}} \qquad (7.44)$$

$$\boxed{r = \frac{\mu L - \sin\mu L}{\sin\mu L - \mu L\cos\mu L}} \qquad (7.45)$$

$$\boxed{S^{o} = \frac{EI}{L}\left[\frac{(\mu L)^2 \tan\mu L}{\tan\mu L - \mu L}\right] = S(1-r^2)} \tag{7.46}$$

The variations of the flexural stiffness measures (end moment required to cause a unit end rotation), S^{o} and S, normalised with respect to EI/L, for the far end pinned and fixed conditions respectively, with respect to the parameter μL, under axial compression, are shown in Fig. 7.11. The variation of the carry-over stiffness, rS, normalised with respect to EI/L, with respect to μL, is also depicted in Fig. 7.11. These normalised functions are sometimes referred to as *stability functions*.

From the trends in Fig. 7.11, we observe that the stiffness measures, S^{o} and S, reduce monotonically with increase in axial compression from their initial values of $3EI/L$ and $4EI/L$ respectively (at $P = 0$). The rate of fall in stiffness is gradual at low levels of axial loading, but becomes steep at higher load levels. The carry-over stiffness, rS, however, increases from the initial value of $2EI/L$ with increase in μL, very slowly initially and more steeply at higher values of μL.

The stiffness of the beam-column drops to zero as the value of P approaches the Euler buckling load, P_E. At this load level, the end rotation occurs freely as the beam-column becomes fully flexible. The 'critical' condition $S^{o} = 0$ corresponds to $\mu L = \pi$ ($P_{Eos} = \frac{\pi^2 EI}{L^2}$) for the both-ends-pinned beam-column (Fig. 7.9a), while the condition $S = 0$ corresponds to $\mu L = 4.493$ ($P_E = \frac{20.19EI}{L^2} = 2.045P_{Eo}$) for the pinned-fixed beam-column (Fig. 7.9b).

The axial compression level can exceed the Euler buckling load, P_E, corresponding to $S^{o} = 0$ or $S = 0$, when there is some additional rotational restraint at the end A (represented by a rotational spring with stiffness k_θ, as shown in Figs 7.11a and b), inducing a restraining end moment at A. This situation is commonly encountered when there is one or more connecting beam or frame elements at the end A, thereby enhancing the rotational stiffness of the joint A. The total rotational stiffness at A is now given by $S^{o} + k_\theta$ in the case shown in Fig. 7.11a and by $S + k_\theta$ in the case shown in Fig. 7.11b. Buckling instability is encountered under a critical load, P_{cr}, when this total stiffness is equal to zero, i.e., corresponding to the condition, $S^{o} = -k_\theta$ in Fig. 7.11a, and $S = -k_\theta$ in Fig. 7.11b. This *negative* stiffness is reflected in the plots shown in Fig. 7.11, and each value of negative stiffness corresponds effectively to a magnitude of the spring stiffness, k_θ. The lowest value of k_θ is zero, corresponding to the hinged condition at A, while the upper bound of k_θ can be infinite, corresponding to the condition of full rotational fixity at A.

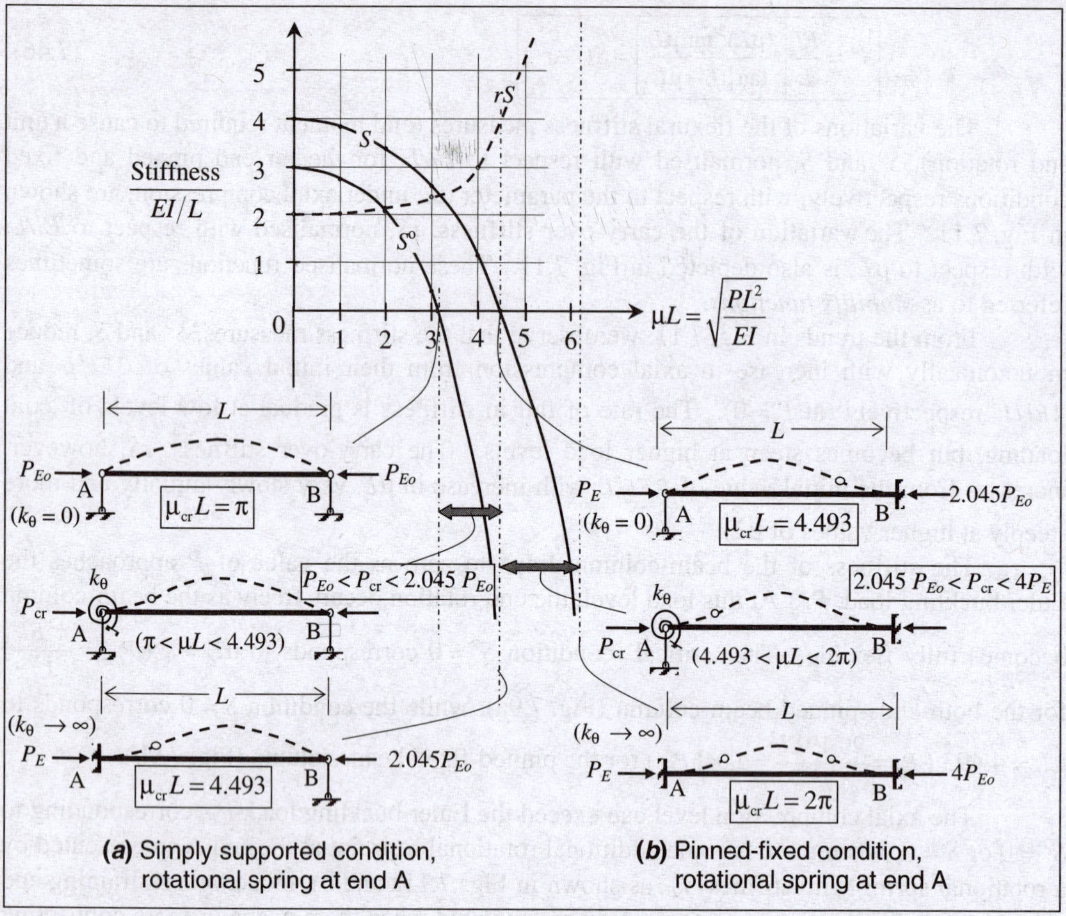

Figure 7.11 Variation of various beam-column stiffness measures with μL (axial compression)

The buckled configurations corresponding to these four limiting cases (pinned-pinned, fixed-pinned, pinned-fixed and fixed-fixed) are also shown in Figs 7.11a and b. Thus, we observe that buckling instability can be encountered in braced prismatic beam-columns, when $\pi < \mu L < 4.493$ ($P_{Eo} < P_{cr} < 2.045P_{Eo}$) for the 'end B hinged' condition in Fig. 7.11a, and $4.493 < \mu L < 2\pi$ ($2.045P_{Eo} < P_{cr} < 4P_{Eo}$) for the 'end B fixed' condition in Fig. 7.11b. With increase in stiffness of the rotational spring at end A, the stiffness measure, S or S, increases sharply and asymptotically approaches infinity when the end A becomes fully fixed against rotation, as indicated in the plots in Fig. 7.11.

The calculation of P_{cr} for simple frame problems (with one degree of freedom and known k_θ), using the concept, $S^o = -k_\theta$ and $S = -k_\theta$, is demonstrated in Example 7.1. It may be noted that the solutions correspond to values of $\mu L = \sqrt{\dfrac{PL^2}{EI}}$, which can be obtained by solving Eq. 7.39 or 7.43, by using refined tables or charts (similar to Fig. 7.11). These are transcendental equations, and for accurate solutions, it is desirable to apply a suitable iterative numerical procedure to find the root. One such procedure to solve transcendental equations of the form, $f(x) = 0$, is the *bisection method*. According to this method, initial bounds, x_1 and x_2, of the root are estimated, such that $g(x_1) < 0$ and $g(x_2) > 0$, and the root is tentatively taken as the average, x_{av}, of these two values. The value, $g(x_{av})$, is computed and if it is positive, the trial value x_2 is replaced by x_{av}; otherwise, if it is negative, x_1 is replaced by x_{av} for the next iteration. The next bisection of the interval between x_1 and x_2 will yield an improved solution, and after repeated trials, convergence is attained.

In situations involving rotational spring stiffnesses that are dependent on the buckling load and multiple degrees of freedom, the Slope-deflection method can be conveniently used, as described in Section 7.3. Alternatively, the matrix method described in Section 7.4 can be used for this purpose. Furthermore, it may be noted that for a given k_θ, if the applied axial compression P is less than the corresponding buckling load P_{cr}, i.e., $S^o + k_\theta > 0$ or $S + k_\theta > 0$, buckling will not take place and the system is capable of resisting an applied end moment M_A, undergoing a finite end rotation, θ_A. The response can be easily computed, using the stiffness measure[†], S or S, as demonstrated in Example 7.2 for simple frame problems. In general, there will be multiple degrees of freedom, and the Slope-deflection method or matrix method, can be used to carry out this 'second-order' analysis.

The values of the non-dimensional stiffness measures, S^o and S, and the carry-over factor, r, are tabulated for various values of μL, under axial compression, in Table 7.1. Tables of this kind (preferably tabulated at very close intervals of μL) can be conveniently used for solving problems by the Slope-deflection method. It may be noted that the carry-over factor r increases with increase in μL, very slowly initially, starting with 0.5 at $\mu L = 0$, and steeply subsequently, asymptotically becoming infinite when the critical value, $\mu L = 4.493$, is approached. For $\mu L > 4.493$, the carry-over factor r turns negative, reducing from $-\infty$ to -1.0 at the maximum (critical) value, $\mu L = 2\pi$. However, the product, rS, denoting the carry-over moment, always remains positive, and follows a smooth trend, as shown in Fig. 7.11.

[†] Observe the point of contra-flexure near the end A in the deflected shapes in Figs 7.11a and b, in the presence of the rotational spring at A. This implies that the bending moment close to the end A (shown sagging) indicates a direction of the end moment at A (clockwise moment M_A) that is opposite to the direction of the end rotation (anti-clockwise θ_A); this explains the negative nature of the stiffness measure.

Table 7.1 Values of beam-column stiffness measures with μL (under axial compression)

$\mu L = \sqrt{\frac{PL^2}{EI}}$	0	0.5	1.0	1.5	2.0	2.5	3.0	3.5	4.0	4.493
$\frac{S^o}{EI/L}$	3	2.950	2.794	2.518	2.088	1.438	0.4082	−1.468	−6.518	−∞
$\frac{S}{EI/L}$	4	3.967	3.865	3.691	3.436	3.088	2.624	2.008	1.173	0
r	0.5	0.5063	0.5264	0.5637	0.6263	0.7310	0.9189	1.3157	2.5605	+∞

$\mu L = \sqrt{\frac{PL^2}{EI}}$	4.5	5	5.5	6	2π
$\frac{S}{EI/L}$	−0.0191	−1.9807	−5.6726	−20.634	−∞
r	13.357	−2.5067	−1.3481	−1.0396	0

EXAMPLE 7.1

Consider a T-shaped frame, comprising a horizontal beam ABC of total length 6m, fixed at A and with a roller support at C, integrally connected at the mid-point B to a vertical column BD below, of length 4.5m, and with a point load P acting downward at B. Assume BD and ABC to have flexural rigidities, 1.5EI and 2EI respectively, where EI =1200 kNm^2. Estimate the critical value of the load (P_{cr}) that would cause buckling instability of the frame, considering the base at D to be (a) a fixed support and (b) a hinged support.

SOLUTION

The elevation of the frames, with fixed base at D and hinged support at D, are shown in Figs 7.12a and b respectively. The buckled configurations of these braced frames are also shown (with the rotation of the joint B assumed to be clockwise). The element BD, which is subject to axial compression, can be isolated as a braced beam-column, and the connection with the elements AB and BC can be modelled, as shown, by means of a rotational spring at B, whose stiffness is given by:

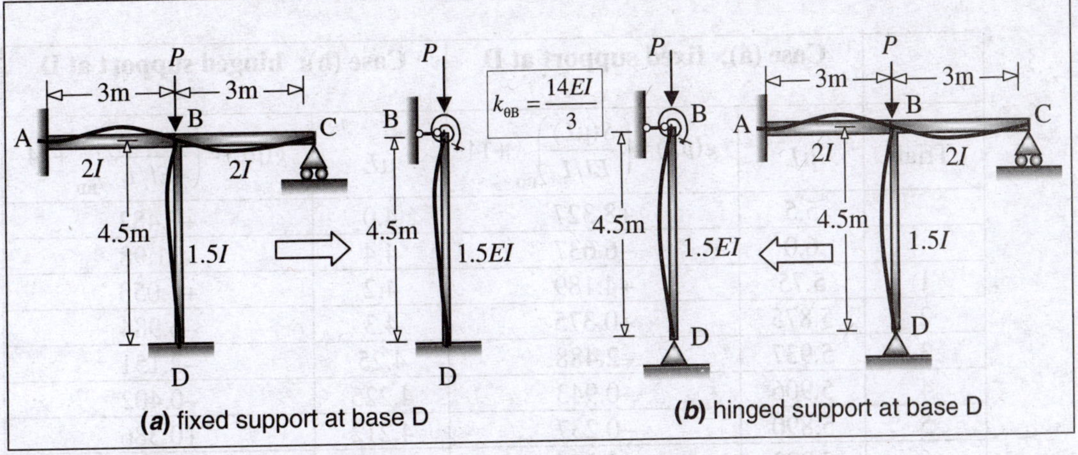

Figure 7.12 Estimation of buckling load in a braced frame (Example 7.1)

$$k_{\theta B} = \frac{4E(2I)}{3} + \frac{3E(2I)}{3} = \frac{14EI}{3}$$

The buckling instability condition corresponds to zero rotational stiffness at the joint B, i.e., $(S_{BD} + k_{\theta B})\theta_B = 0$ for the fixed base case (Fig. 7.12a) and $(S^o_{BD} + k_{\theta B})\theta_B = 0$ for the hinged base case (Fig. 7.12b). These represent the characteristic equation of an eigenvalue problem and for a non-trivial solution ($\theta_B \neq 0$), it follows that the critical buckling load corresponds to the conditions:

(a) base D fixed: $S_{BD} = -k_{\theta B} = -\frac{14EI}{3} \Rightarrow \left(\frac{S}{EI/L}\right)_{BD} = \frac{-14EI}{3}\frac{4.5}{1.5EI} = -14$

(b) base D hinged: $S^o_{BD} = -k_{\theta B} = -\frac{14EI}{3} \Rightarrow \left(\frac{S^o}{EI/L}\right)_{BD} = -14$

Referring to Table 7.1, we observe that $\mu L = \sqrt{\frac{PL^2}{EI}}$ lies between 5.5 and 6.0 for the base D fixed case, and between 4.0 and 4.4 for the base D hinged case. Using these as initial estimates, and applying the bisection method to solve $g(\mu L = 0)$, invoking Eqns 7.43 and 7.39, the solutions are obtained, as summarised below.

	Case (a): fixed support at D		Case (b): hinged support at D	
Trial	μL	$g(\mu L)=\left(\frac{S(\mu L)}{EI/L}\right)_{BD}+14$	μL	$g(\mu L)=\left(\frac{S^o(\mu L)}{EI/L}\right)_{BD}+14$
	5.5	+8.327	4.0	+7.482
	6.0	–6.637	4.4	–31.98
1	5.75	+4.189	4.2	+1.053
2	5.875	+0.375	4.3	–6.984
3	5.937	–2.488	4.25	–2.151
4	5.906	–0.943	4.225	–0.402
5	5.890	–0.237	4.212	+0.386
6	5.882	+0.095	4.218	+0.031
7	5.886	–0.070	4.221	–0.152
8	**5.884**	+0.013	**4.219**	–0.029

Case (a): fixed support at D:

$$\mu_{cr}L=\sqrt{\frac{P_{cr}}{1.5EI}}L=5.884\Rightarrow\ P_{cr}=\left(\frac{5.884}{4.5}\right)^2(1.5\times1200)=\mathbf{3077.5\ kN}$$

Case (b): hinged support at D:

$$\mu_{cr}L=\sqrt{\frac{P_{cr}}{1.5EI}}L=4.219\Rightarrow\ P_{cr}=\left(\frac{4.219}{4.5}\right)^2(1.5\times1200)=\mathbf{1582.2\ kN}$$

EXAMPLE 7.2

Consider the two frames in Example 7.1 to be subjected to a load, P = 1000 kN and a moment of 120 kNm acting clockwise at the joint B. Analyse the frames and draw the bending moment diagrams, including the second-order effects.

SOLUTION

As the applied compressive load of 1000 kN is less than the critical buckling load (computed in Example 7.1), the beam-column BD has a finite end-rotational stiffness, whereby the frame is rendered capable of resisting the applied loading. The applied moment of 120 kNm will be apportioned to the elements BA, BC and BD in proportion to their relative stiffnesses. The stiffness measures of element BD can be determined using Eqns 7.44-7.46.

Case (a): fixed support at D:

$$\mu L=\sqrt{\frac{1000}{1.5\times1200}}(4.5)=3.354\quad\Rightarrow\left(\frac{S}{EI/L}\right)_{BD}=2.20703;\ r_{BD}=1.1620$$

$$\Rightarrow M_{BA} : M_{BC} : M_{BD} = \frac{4E(2I)}{3} : \frac{3E(2I)}{3} : \frac{2.20703E(1.5I)}{4.5} = 0.4936 : 0.3702 : 0.1362$$

$$\Rightarrow M_{BA} = 0.4936 \times 120 = 59.23 \text{ kNm}; M_{BC} = 0.3702 \times 120 = 44.42 \text{ kNm}; M_{AB} = 0.5 \times 59.23 = 29.62 \text{ kNm};$$

$$M_{BD} = 0.1362 \times 120 = 16.34 \text{ kNm}; M_{DB} = 1.1620 \times 16.34 = 18.99 \text{ kNm}$$

(all clockwise)

<u>*Case (b): hinged support at D*</u>:

$$\mu L = \sqrt{\frac{1000}{1.5 \times 1200}}(4.5) = 3.354 \quad \Rightarrow \left(\frac{S^o}{EI/L}\right)_{BD} = -0.77303$$

(The negative sign implies that M_{BD} acts in an anti-clockwise direction).

$$\Rightarrow M_{BA} : M_{BC} : M_{BD} = \frac{4E(2I)}{3} : \frac{3E(2I)}{3} : \frac{-0.77303E(1.5I)}{4.5} = 0.6048 : 0.4536 : -0.05844$$

$$\Rightarrow M_{BA} = 0.6048 \times 120 = 72.58 \text{ kNm}; M_{BC} = 0.4536 \times 120 = 54.43 \text{ kNm}; M_{AB} = 0.5 \times 72.58 = 36.29 \text{ kNm};$$

$$M_{BD} = -0.05844 \times 120 = -7.01 \text{ kNm}; M_{DB} = 0 \text{ kNm}$$

(all clockwise, except M_{BD})

The bending moment diagrams (drawn on the tension side) for the two cases are shown in Fig. 7.13. It may be noted that these results include the second-order effects associated with the action of the load *P* in the deformed configuration of the frame.

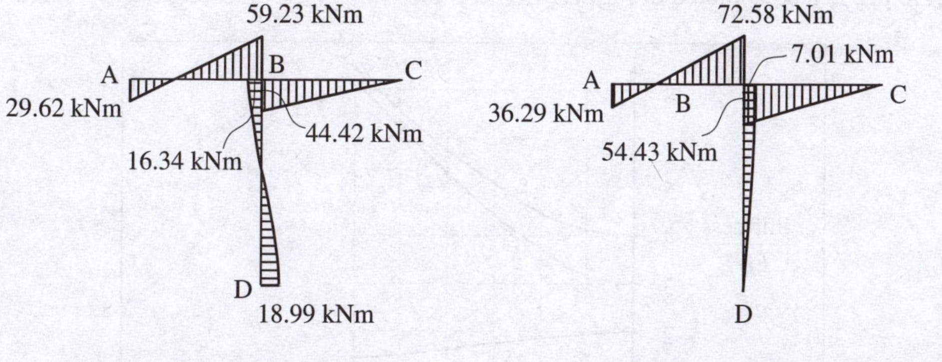

(*a*) fixed support at base D (*b*) hinged support at base D

Figure 7.13 Bending moment diagrams, including second-order effects (Example 7.2)

7.2.5 Effect of Axial Tension

If the axial force is tensile, instead of compressive, it has the beneficial effect of enhancing the flexural stiffness of the beam-column. The flexural stiffness measures can be easily derived by replacing P with $-P$ in the basic equations already derived for the case of axial compression. This implies that the basic parameter, $\mu L = \sqrt{\frac{PL^2}{EI}}$, is now to be taken as $\sqrt{\frac{-PL^2}{EI}} = i\mu L$, where $i = \sqrt{-1}$. Substituting this in Eqns 7.44-7.46, and also noting that $\sinh \mu L = -i \sin i\mu L$, $\cosh \mu L = i \cos i\mu L$ and $\tanh \mu L = -i \tan i\mu L$, we obtain the following modified expressions for S, r and S^o:

$$S = \frac{EI}{L} \frac{(\mu L)(\mu L \cosh \mu L - \sinh \mu L)}{(2 - 2\cosh \mu L + \mu L \sinh \mu L)} \quad (7.44a)$$

$$r = \frac{\sinh \mu L - \mu L}{\mu L \cosh \mu L - \sinh \mu L} \quad (7.45a)$$

$$S^o = \frac{EI}{L}\left[\frac{(\mu L)^2 \tanh \mu L}{\mu L - \tanh \mu L}\right] = S(1 - r^2) \quad (7.46a)$$

The variations of the flexural stiffness measures (end moment required to cause a unit end rotation), S^o and S, normalised with respect to *EI/L*, for the far end pinned and fixed conditions respectively, with respect to the parameter μL, under axial tension, are shown in Fig. 7.14. The variation of the carry-over stiffness, *rS*, normalised with respect to *EI/L*, with respect to μL, is also depicted in Fig. 7.14.

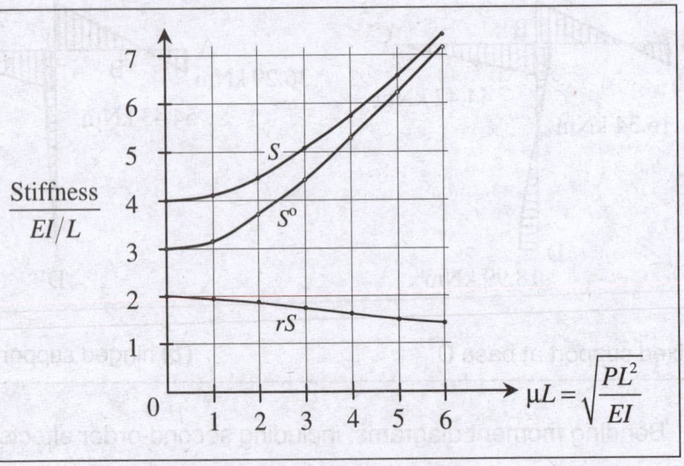

Figure 7.14 Variation of various beam-column stiffness measures with μL (axial tension)

These normalised functions may be compared with the corresponding *stability functions* for the case of axial compression shown in Fig. 7.11. From the trends in Fig. 7.14, we observe that the stiffness measures, S^o and S, increase monotonically with increase in axial tension from their initial values of $3EI/L$ and $4EI/L$ respectively (at $P = 0$). The rate of rise in stiffness is very gradual at low levels of axial loading. The carry-over stiffness, rS, however, decreases from the initial value of $2EI/L$ with increase in μL. These trends stand out in contrast with the trends in Fig. 7.11.

The values of the non-dimensional stiffness measures and the carry-over factor, r, are tabulated for various values of μL, under axial tension, in Table 7.2. All the values are positive (unlike the case of axial compression, shown in Table 7.1).

Table 7.2 Values of beam-column stiffness measures with μL (under axial tension)

$\mu L = \sqrt{\frac{PL^2}{EI}}$	0	0.5	1.0	1.5	2.0	2.5	3.0	4.0	5.0	6.0
$\frac{S^o}{EI/L}$	3	3.050	3.195	3.424	3.722	4.075	4.467	5.329	5.329	7.200
$\frac{S}{EI/L}$	4	4.033	4.132	4.292	4.508	4.773	5.081	5.797	5.797	7.482
r	0.5	0.4938	0.4762	0.4497	0.4174	0.3825	0.3477	0.2842	0.2842	0.1940

EXAMPLE 7.3

Repeat Example 7.2, considering the load $P = 1000$ kN to act in an upward direction (causing tension in BD), along with the moment of 120 kNm acting clockwise at the joint B. Analyse the frames and draw the bending moment diagrams, including the second-order effects. Also sketch the deflected shapes. Compare the results obtained with those in Example 7.2.

SOLUTION

The loading and probable deflected shapes are shown in Figs 7.15a and b, for the fixed base and hinged base conditions. The applied moment of 120 kNm will be apportioned to the elements BA, BC and BD in proportion to their relative stiffnesses. The stiffness measures of the element BD, under axial tension, with an equivalent rotational spring at end B (stiffness as calculated in Example 7.1) can be determined using Eqns 7.44a-7.46a.

Case (a): fixed support at D:

$$\mu L = \sqrt{\frac{1000}{1.5\times 1200}}(4.5) = 3.354 \Rightarrow \left(\frac{S}{EI/L}\right)_{\mathbf{BD}} = 5.3207;\ r_{\mathrm{BD}} = 0.3240$$

$$\Rightarrow M_{\mathrm{BA}} : M_{\mathrm{BC}} : M_{\mathrm{BD}} = \frac{4E(2I)}{3} : \frac{3E(2I)}{3} : \frac{5.3207E(1.5I)}{4.5} = 0.4141 : 0.3105 : 0.2754$$

$\Rightarrow M_{BA} = 0.4141 \times 120 = 49.69$ kNm; $M_{BC} = 0.3105 \times 120 = 37.26$ kNm; $M_{AB} = 0.5 \times 49.69 = 24.85$ kNm;
$M_{BD} = 0.2754 \times 120 = 33.05$ kNm; $M_{DB} = 0.3240 \times 33.05 = 10.71$ kNm
(all clockwise)

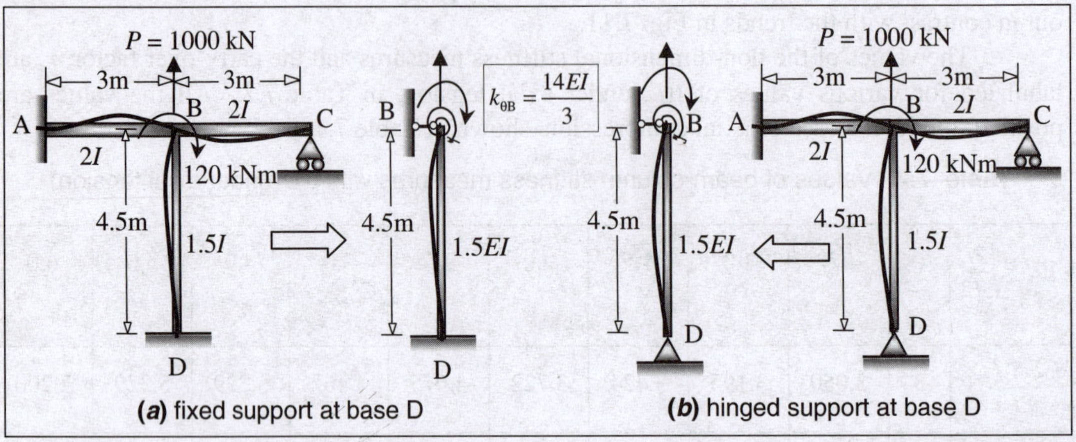

Figure 7.15 Loading and deflection diagrams in a braced frame (Example 7.3)

Case (b): hinged support at D:

$$\mu L = \sqrt{\frac{1000}{1.5 \times 1200}}(4.5) = 3.354 \quad \Rightarrow \left(\frac{S^{\circ}}{EI/L}\right)_{BD} = 4.7622$$

$$\Rightarrow M_{BA} : M_{BC} : M_{BD} = \frac{4E(2I)}{3} : \frac{3E(2I)}{3} : \frac{4.7622E(1.5I)}{4.5} = 0.4264 : 0.3198 : 0.2538$$

$\Rightarrow M_{BA} = 0.4264 \times 120 = 51.17$ kNm; $M_{BC} = 0.3198 \times 120 = 38.38$ kNm; $M_{AB} = 0.5 \times 51.17 = 25.58$ kNm;
$M_{BD} = 0.2538 \times 120 = 30.46$ kNm; $M_{DB} = 0$ kNm
(all clockwise)

The bending moment diagrams (drawn on the tension side) for the two cases are shown in Fig. 7.16. It may be noted that these results include the second-order effects associated with the action of the tensile load P in the deformed configuration of the frame. Comparison with the compression load case in Example 7.2 (refer Fig. 7.13) shows that the bending moment values are considerably different, especially for the beam-column BD. The presence of axial tension enhances the stiffness of BD considerably, thereby attracting a higher moment at the end B, compared to the case of axial compression. Furthermore, the stiffness values of the beam-column are invariably positive under tension, implying that the end moment always acts in the same direction as the joint (notice the difference in the bending moments in element BD in Figs 7.13b and 7.16b).

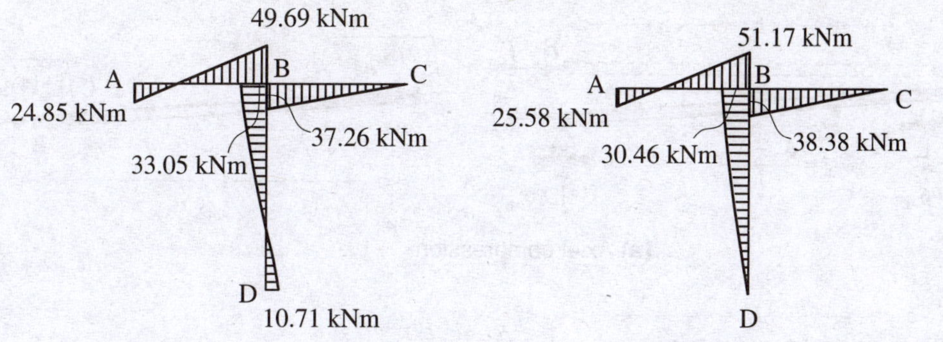

(*a*) fixed support at base D (*b*) hinged support at base D

Figure 7.16 Bending moment diagrams associated with axial tension in beam-column BD (Example 7.3)

7.2.6 Flexural Stiffness Measures for Unbraced Prismatic Beam-Columns

In unbraced beam-columns, relative deflection (sidesway) between the two ends of the element is possible. This differential deflection can be visualised as a *chord rotation*, $\phi_{AB} = \frac{\Delta_B - \Delta_A}{L}$, and the response depends on the degree of rotational (and translational) restraints provided at the two ends. Typical cases are discussed below. It is assumed that the pair of equal and opposite forces applied at the two ends of the element remain unchanged in magnitude (P), as well as direction (aligned along the longitudinal axis in the undeflected configuration). Furthermore, it is assumed that the element is axially rigid.

Case 1: Pinned-pinned condition

If both ends are pinned, then the chord rotation will not result in any bending (curvature) of the element, in the presence of axial force. However, the *P-delta* effect will generate a resisting couple in the form of a pair of equal and opposite reactions, $P\phi_{AB}$, acting perpendicular to the pair of P forces at the two ends, as shown in Fig. 7.17, in order to satisfy equilibrium. This causes an enhancement in the axial force. The element will essentially behave like a truss element, with a resultant axial force (compressive or tensile) equal to $P\sqrt{1+(\phi_{AB})^2}$. The cases of axial compression and axial tension are illustrated in Figs 7.17a and b respectively, under a clockwise chord rotation. When the chord rotation acts anti-clockwise, the directions of the reactions, $P\phi_{AB}$, at the two ends, get reversed.

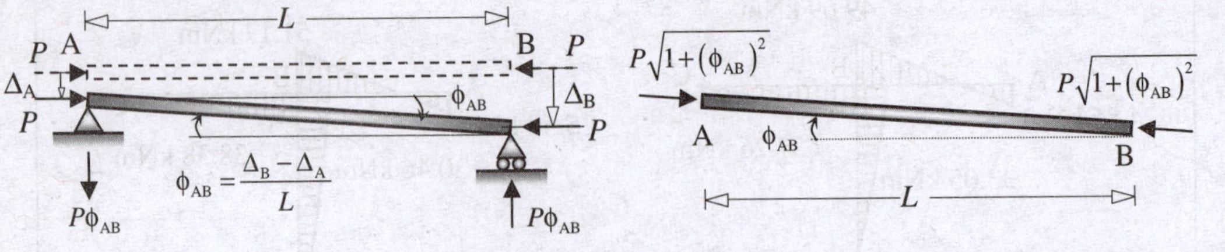

(*a*) Axial compression

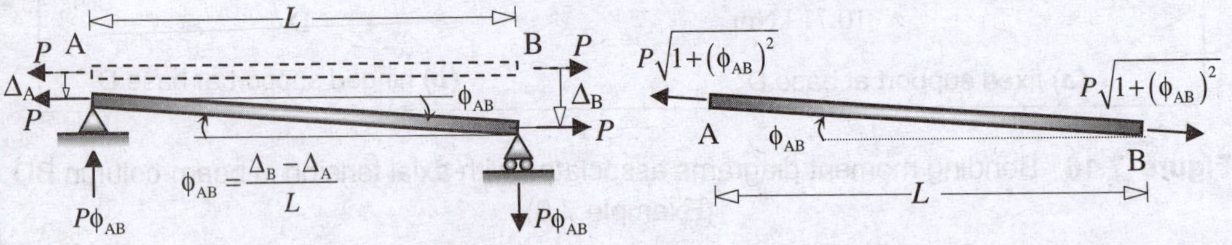

(*b*) Axial tension

Figure 7.17 Effect of chord rotation in an unbraced beam-column with both ends pinned (axial force gets enhanced, but no bending)

This concept of lateral stiffness in a braced pinned-pinned element can be usefully applied to determine the minimum stiffness requirement of lateral bracing for a long column with both ends pinned. In the absence of the brace, the critical buckling load of the ideal column AB (of length *L*), is given by $P_{cr} = P_{Eo} = \frac{\pi^2 EI}{L^2}$, and the buckled mode shape (sinusoidal) is as shown in Fig. 7.18a. If the column is braced with an elastic spring at the mid-height location C (as shown in Fig. 7.18b), the flexural stiffness of the column gets enhanced, thereby increasing its buckling load capacity ($P_{cr} > P_{Eo}$), which depends on the stiffness of the brace, k_b. Clearly, if the brace acts like a rigid link, with its stiffness, $k_b \to \infty$, the point C is prevented from deflecting laterally. The buckled mode shape, comprising two sine waves, with a point of contraflexure at C, is shown in Fig. 7.18b(ii). Comparison with Fig. 7.18a shows that the effective length of the column is halved (second mode in Eq. 7.8), thereby enhancing the critical load capacity to $P_{cr} = \frac{4\pi^2 EI}{L^2} = 4P_{Eo}$. However, it is not necessary for the brace to be infinitely stiff for this to happen, and the minimum value, $(k_b)_{min}$ required for this can be calculated by equating the total lateral force requirement at C to the spring force in the buckled configuration.

We can visualise an internal hinge at the point of contraflexure at C in Fig. 7.18b(ii). Thus, the column (assumed to be axially rigid) can be visualised as being made up of two identical rigid links (with pinned-pinned ends), connected in series, with the same critical buckling load of $4P_{Eo}$. Bifurcation buckling is expected to occur when the applied load *P* reaches this value, and the buckled configuration of this system is expected to be as shown in Fig. 7.18c. If the lateral translation at C is Δ, the chord rotation in each element is given by $\phi = 2\Delta / L$, thereby generating a horizontal force of $P\phi = 8P_{Eo}\Delta / L$ at the ends of each element. Equating the total horizontal force requirement of $2P\phi = 16P_{Eo}\Delta / L$ to the spring force, $k_b\Delta$, we observe that the desired minimum value of k_b is given by $(k_b)_{\min} = 16P_{Eo} / L$. If the brace has a lateral stiffness exceeding $(k_b)_{\min}$, the column will have a critical buckling load equal to $4P_{Eo}$, as shown in Fig. 7.18b(ii). Otherwise, if $k_b < (k_b)_{\min}$, the critical buckling load will lie in between P_{Eo} and $4P_{Eo}$, as shown in Fig. 7.18b(i); the relationship between k_b and P_{cr} is derived in Example 7.4.

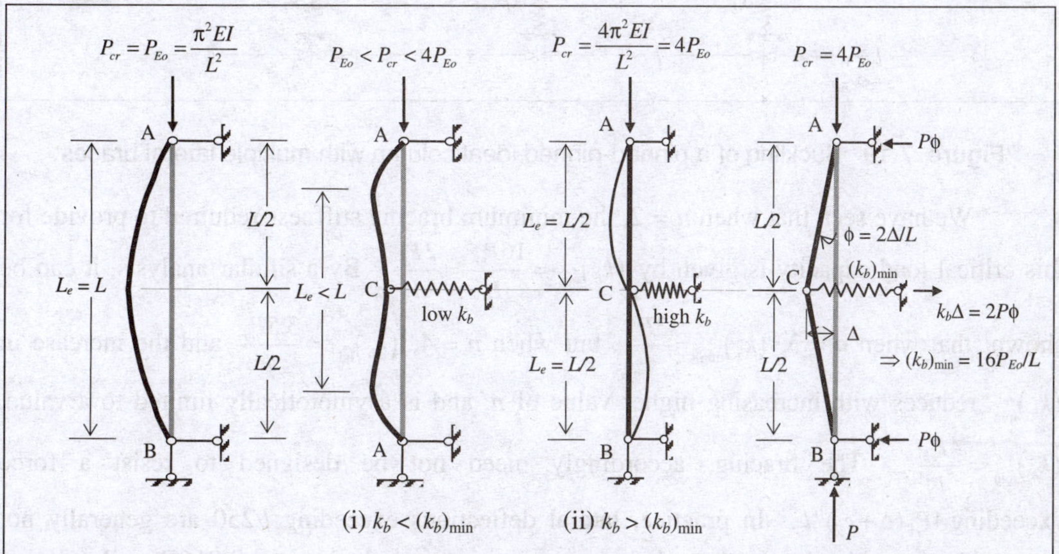

(a) Column without ***(b)*** Column with brace at mid-height ***(c)*** Rigid links with pinned joint and spring at C

Figure 7.18 Buckling of a pinned-pinned ideal column with a lateral brace (Example 7.4)

In reality, there are likely to be initial imperfections (lack of straightness) in the column, and bifurcation (sudden) buckling does not take place. The calculated critical buckling load is reached only asymptotically, with increasing lateral deflections, as discussed in Section 7.2.2. The brace, therefore, is required to resist a force that is given by $(k_b)(\Delta + e_o)$, where e_o is the assumed initial eccentricity at C due to imperfections. Furthermore, it is intuitively obvious that the critical buckling load can be enhanced further by providing additional lateral bracing along the height of the long column. If the column is divided into n equal segments, each of length, $l = L / n$, and adequately braced at the $n-1$ intermediate locations between the two extreme pinned ends, it will behave like a continuous beam-column with n equal spans. The buckled mode shape will be in the form of n sine waves, as shown in Fig. 7.19, and the corresponding critical buckling load is given by:

$$P_{cr} = \frac{\pi^2 EI}{l^2} = \frac{(n\pi)^2 EI}{L^2} = n^2 P_{Eo} \tag{7.47}$$

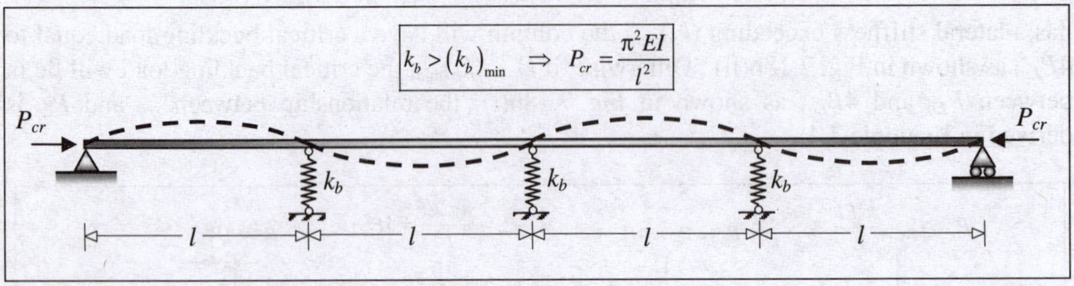

Figure 7.19 Buckling of a pinned-pinned ideal column with multiple lateral braces

We have seen that when $n = 2$, the minimum bracing stiffness required to provide for this critical load capacity is given by $(k_b)_{\min} = \frac{16P_{Eo}}{L} = \frac{2P_{cr}}{l}$. By a similar analysis, it can be shown[†] that when $n = 3$, $(k_b)_{\min} = \frac{3P_{cr}}{l}$, but when $n = 4$, $(k_b)_{\min} = \frac{3.5P_{cr}}{l}$ and the increase in $(k_b)_{\min}$ reduces with increasing higher value of n, and is asymptotically limited to a value, $(k_b)_{\min} = \frac{4P_{cr}}{l}$. The bracing, accordingly, need not be designed to resist a force exceeding $4P_{cr}(\Delta + e_o)/l$. In practice, lateral deflections exceeding $l/250$ are generally not acceptable; accordingly the lateral force requirement works out to $0.016P_{cr}$. In design practice (such as design of lacing in steel columns), it is usually estimated as about 2 percent of the applied axial load.

Case 2: Pinned-fixed condition

If one end is pinned and the other fixed, then the chord rotation will induce bending (curvature) of the element, whether or not an axial force is present. In the absence of axial force, we know from basic structural analysis, that the deflected shape of the beam corresponds to that of a cantilever, with a moment equal to $\frac{3EI}{L}\phi_{AB}$ generated at the fixed end (in a direction that opposes the chord rotation) and a pair of equal and opposite reactions of magnitude, $\frac{3EI}{L^2}\phi_{AB}$, acting at the two ends. This curvature diagram is identical to that of a

[†] Reference may be made to the paper titled "Lateral bracing of columns and frames" by George Winter, published in *Transactions of ASCE*, Vol. 125, Part 1, p. 807, 1960.

pinned-fixed beam undergoing a rotational slip ϕ_{AB} at the fixed end, in a direction that is opposed to the chord rotation. For this reason, we can treat the effect of a clockwise chord rotation ϕ_{AB} as that of an end rotation, $-\phi_{AB}$, at the fixed end, and make use of this principle to determine the force response in the presence of the axial force P (refer Fig. 7.20).

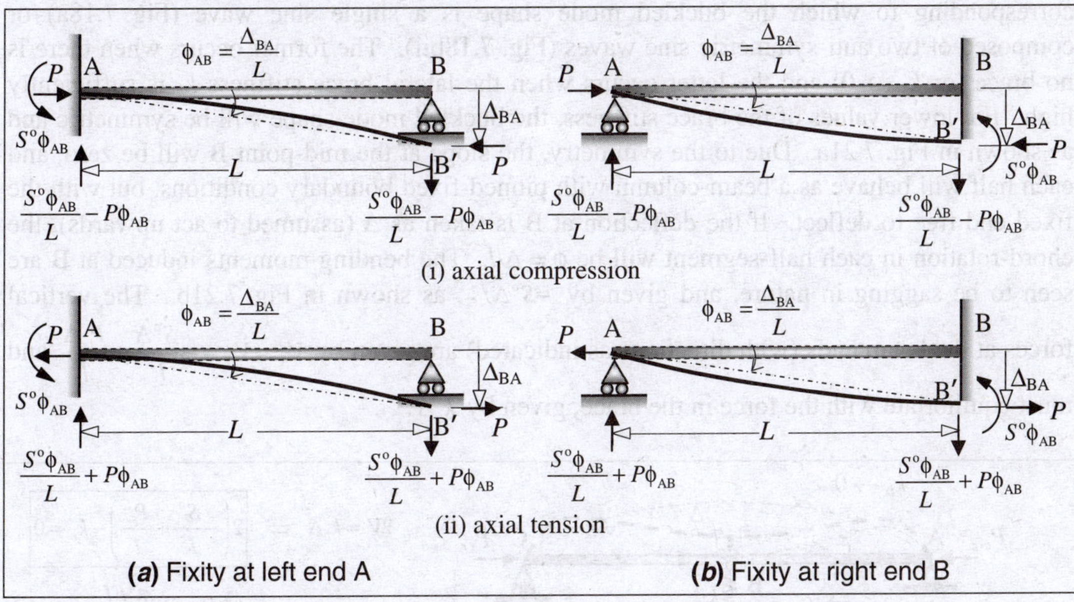

Figure 7.20 Effect of chord rotation in an unbraced beam-column with fixed-pinned ends

The moment at the fixed end will, accordingly, be given by $S^o\phi_{AB}$, and will act in a direction that is opposed to the chord rotation. To satisfy equilibrium in the deformed configuration, the reactions at the two ends, will be equal to $S^o\phi_{AB}/L \pm P\phi_{AB}$, with directions as shown in Fig. 7.20 (the positive sign is applicable when the axial force is tensile and the negative sign when the axial force is compressive).

These stiffness measures can be used (in elements with a pinned support at one end) to find the critical buckling load and second-order effects in continuous beams and frames.

EXAMPLE 7.4

Find the critical buckling load, P_{cr}, in an ideal braced column, with length $L = 2l$ and flexural rigidity EI, and with both ends pinned, and with a lateral brace with stiffness k_b at the mid-height location (as shown in Fig. 7.18b). subject to an axial compressive load P. Show how P_{cr} is related to k_b.

SOLUTION

The lower and upper bounds for the critical buckling load are given by

$$(P_{cr})_{min} = \frac{\pi^2 EI}{L^2} = 0.25 \frac{\pi^2 EI}{l^2} \quad \text{and} \quad (P_{cr})_{max} = \frac{\pi^2 EI}{l^2}$$

respectively (as discussed earlier), corresponding to which the buckled mode shape is a single sine wave (Fig. 7.18a) or composed of two anti-symmetric sine waves (Fig. 7.18bii). The former occurs when there is no brace (or $k_b \to 0$) and the latter occurs when the lateral brace stiffness k_b is sufficiently high. For lower values of the brace stiffness, the buckled mode shape will be symmetric and as shown in Fig. 7.21a. Due to the symmetry, the slope at the mid-point B will be zero, and each half will behave as a beam-column with pinned-fixed boundary conditions, but with the fixed end free to deflect. If the deflection at B is taken as Δ (assumed to act upwards), the chord-rotation in each half-segment will be $\phi = \Delta/l$. The bending moments induced at B are seen to be sagging in nature, and given by $-S^\circ \Delta/l$, as shown in Fig. 7.21b. The vertical forces at the beam ends (with directions as indicated) are given by $V_A = V_B = -\dfrac{S^\circ \Delta}{l^2} + \dfrac{P_{cr}\Delta}{l}$ and must equilibrate with the force in the brace, given by $k_b \Delta$.

$$2V = k_b\Delta \quad \Rightarrow \quad \boxed{2\left(-\frac{S^\circ}{l^2} + \frac{P_{cr}}{l}\right) - k_b = 0}$$

$$(P_{cr})_{max} = \frac{\pi^2 EI}{l^2}$$

(a) Loading diagram under symmetrical buckled shape

$$V = -\frac{S^\circ \Delta}{l^2} + \frac{P_{cr}\Delta}{l}$$

(b) Free-bodies in deformed configuration

(c) Variation of critical load with brace stiffness (normalised)

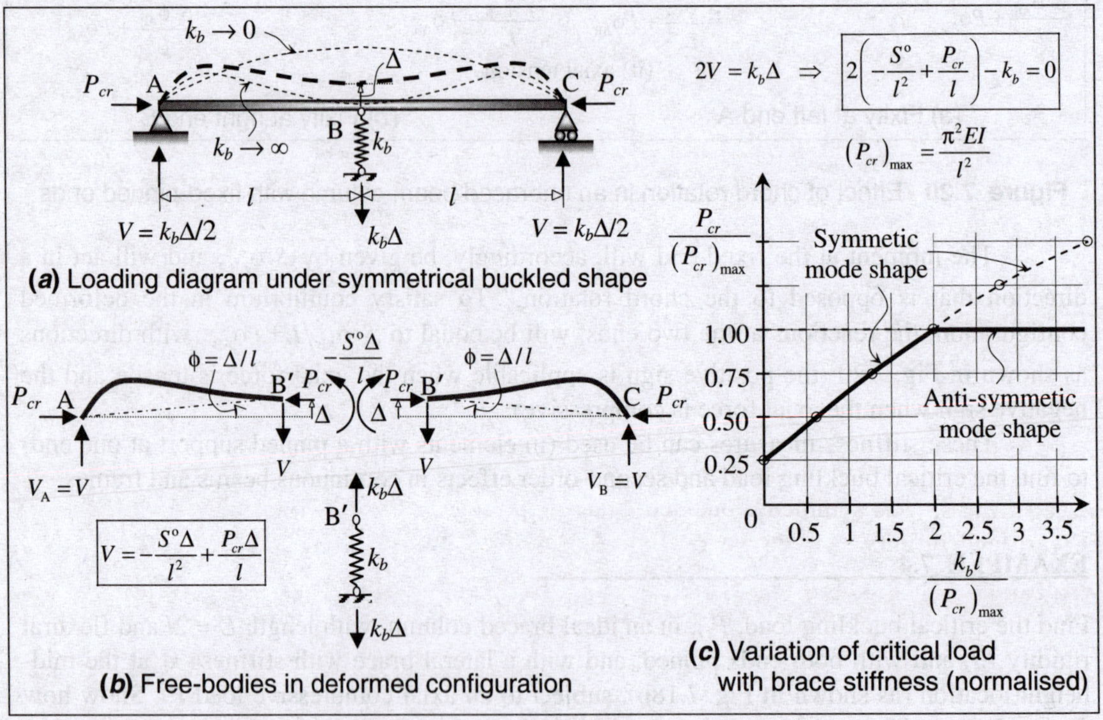

Figure 7.21 Buckling of a pinned-pinned ideal column with a middle brace (Example 7.4)

Accordingly,

$$\left[2\left(-\frac{S^o}{l^2}+\frac{P_{cr}}{l}\right)-k_b\right](\Delta)=0$$

For a non-trivial solution (i.e, $\Delta \neq 0$),

$$2\left(-\frac{S^o}{l^2}+\frac{P_{cr}}{l}\right)-k_b=0 \qquad \text{(a)}$$

Dividing the above characteristic equation throughout by $(P_{cr})_{\max}/l=\pi^2 EI/l^3$, and transposing the non-dimensional terms,

$$\frac{k_b l}{(P_{cr})_{\max}}=2\left[\frac{P_{cr}}{(P_{cr})_{\max}}-\frac{1}{\pi^2}\left(\frac{S^o}{EI/l}\right)\right]=2\left[A-B/\pi^2\right] \qquad \text{(b)}$$

where $A=\dfrac{P_{cr}}{(P_{cr})_{\max}}$ and $B=\dfrac{S^o}{EI/l}$ is obtainable from Eq. 7.46 in terms of $(\mu_{cr}l)=\sqrt{\dfrac{P_{cr}l^2}{EI}}=\pi\sqrt{A}$.

The results are tabulated below for typical values of $A=\dfrac{P_{cr}}{(P_{cr})_{\max}}$:

$A=\dfrac{P_{cr}}{(P_{cr})_{\max}}$	$(\mu_{cr}l)=\pi\sqrt{A}$	$B=\dfrac{S^o}{EI/l}$	$\dfrac{k_b l}{(P_{cr})_{\max}}$
0.25	1.57080	2.46740	0
0.50	2.22144	1.83381	0.6284
0.75	2.72070	1.04583	1.2881
1.00	π	0	**2.0000**
1.25*	3.51241	–1.53561	2.8112
1.50*	3.84765	–4.21497	3.85413

* not actually realizable in practice, , as the anti-symmetric mode with lower P_{cr} will govern

Thus, Eq. (b) predicts that the critical buckling load increases with increase in the brace stiffness. The symmetric buckled mode shapes corresponding to the extreme values of k_b (zero and infinity) are shown by light dashed lines in Fig. 7.21a, and the two extreme critical buckling loads are given by $\dfrac{\pi^2 EI}{(2l)^2}$ and $\dfrac{\pi^2 EI}{(0.699l)^2}$ (refer Eq. 7.14).

The variation of $\dfrac{P_{cr}}{(P_{cr})_{\max}}$ with $\dfrac{k_b l}{(P_{cr})_{\max}}$ is illustrated in Fig. 7.21c, in which the predictions from the assumed symmetric buckled mode shape (Fig. 7.21a) as well as the anti-symmetric buckled mode shape (Fig. 7.18bii) are shown. The two estimates of critical buckling load intersect at $\frac{k_b l}{(P_{cr})_{max}} = 2.0$, which corresponds to the $(k_b)_{min}$ value referred to in Fig. 7.18c. Buckling instability will occur at the lowest predicted P_{cr}. Accordingly,

$$\begin{aligned} 0.0 \le \frac{k_b l}{(P_{cr})_{max}} \le 2.0 \quad &\Rightarrow \quad 0.25 \le \frac{P_{cr}}{(P_{cr})_{max}} \le 1.0 \\ \frac{k_b l}{(P_{cr})_{max}} > 2.0 \quad &\Rightarrow \quad P_{cr} = (P_{cr})_{max} = \frac{\pi^2 EI}{l^2} \end{aligned} \tag{7.48}$$

Case 3: Fixed-fixed condition

If both ends are fixed, then the chord rotation will induce anti-symmetric bending (curvature) of the element (with a point of contra-flexure at the mid-point), whether or not an axial force is present. In the absence of axial force, the deflected shape of each half of the beam corresponds to that of a cantilever, with a moment equal to $\frac{3EI}{L/2}\phi_{AB}$ generated at the fixed end (in a direction that opposes the chord rotation) and a constant shear force of magnitude, $\frac{12EI}{L^2}\phi_{AB}$. This curvature diagram is identical to that of a fixed-fixed beam undergoing rotational slips ϕ_{AB} at both fixed ends, in a direction that is opposed to the chord rotation. For this reason, we can treat the effect of a clockwise chord rotation ϕ_{AB} as that of end rotations, $-\phi_{AB}$, at the two fixed ends, and make use of this principle to determine the force response in the presence of the axial force P (refer Fig. 7.22). The contribution of the *carry-over moment* also needs to be included.

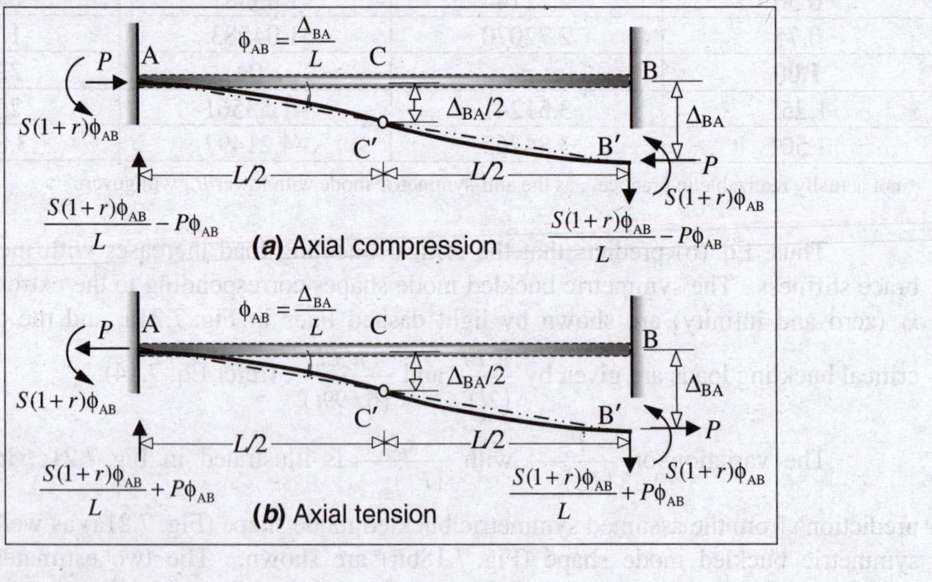

Figure 7.22 Effect of chord rotation in an unbraced beam-column with fixed-fixed ends

The moment at the fixed end (at A and B) will, accordingly, be given by $S(1+r)\phi_{AB}$, and will act in a direction that is opposed to the chord rotation. To satisfy equilibrium in the deformed configuration, the reactions at the two ends, will be equal to $S(1+r)\phi_{AB}/L \pm P\phi_{AB}$, with directions as shown in Fig. 7.22 (the positive sign is applicable when the axial force is tensile and the negative sign when the axial force is compressive).

These stiffness measures can be used to find the critical buckling load and second-order effects in continuous beams and frames.

EXAMPLE 7.5

Repeat Example 7.4, considering the two extreme ends (A and C) of the ideal braced column ABC (length $L = 2l$ and flexural rigidity EI) to be fixed against rotation.

SOLUTION

The lower bound for the critical buckling load, $(P_{cr})_{\min}$, occurs when the bracing stiffness is zero ($k_b = 0$), corresponding to which the effective length of the fixed-fixed column is $0.5L = l$, whereby $(P_{cr})_{\min} = \dfrac{\pi^2 EI}{(0.5L)^2} = \dfrac{\pi^2 EI}{l^2}$. The upper bound value corresponds to the anti-symmetric buckled mode shape, with a point of contraflexure at the midpoint B, whereby $L_e = 0.699l$ and $(P_{cr})_{\max} = \dfrac{\pi^2 EI}{(0.699l)^2} = 2.047(P_{cr})_{\min}$. The latter occurs when the lateral brace stiffness k_b is sufficiently high (see Fig. 7.24). For lower values of the brace stiffness, the buckled mode shape will be symmetric and as shown in Fig. 7.23a. Due to the symmetry, the slope at the mid-point B will be zero, and each half will behave as a beam-column with fixed-fixed boundary conditions, but with the fixed end at the middle free to deflect. If the deflection at B is taken as Δ (assumed to act upwards), the chord-rotation in each half-segment will be $\phi = \Delta/l$. The bending moments induced at B are seen to be sagging in nature, and given by $-S(1+r)\Delta/l$, with identical sagging moments at A and C, as shown in Fig. 7.23b. The vertical forces at the beam ends (with directions as indicated) are given by $V_A = V_B = -\dfrac{2S(1+r)\Delta}{l^2} + \dfrac{P_{cr}\Delta}{l}$ and must equilibriate with the force in the brace, given by $k_b\Delta$.

Accordingly, $\left[2\left(-\dfrac{2S(1+r)}{l^2} + \dfrac{P_{cr}}{l}\right) - k_b\right](\Delta) = 0$

For a non-trivial solution (i.e, $\Delta \neq 0$),

$$2\left(-\frac{2S(1+r)}{l^2} + \frac{P_{cr}}{l}\right) - k_b = 0 \qquad \text{(a)}$$

Dividing the above characteristic equation throughout by $(P_{cr})_{\min}/l = \pi^2 EI/l^3$, and transposing the non-dimensional terms,

$$\frac{k_b l}{(P_{cr})_{\min}} = 2\left[\frac{P_{cr}}{(P_{cr})_{\min}} - \frac{2}{\pi^2}\left(\frac{S(1+r)}{EI/l}\right)\right] = 2\left[A - 2B/\pi^2\right] \qquad \text{(b)}$$

where $A = \frac{P_{cr}}{(P_{cr})_{\min}}$ and $B = \frac{S(1+r)}{EI/l}$ is obtainable from Eqns 7.44 and 7.45 in terms of

$$(\mu_{cr} l) = \sqrt{\frac{P_{cr} l^2}{EI}} = \pi\sqrt{A}.$$

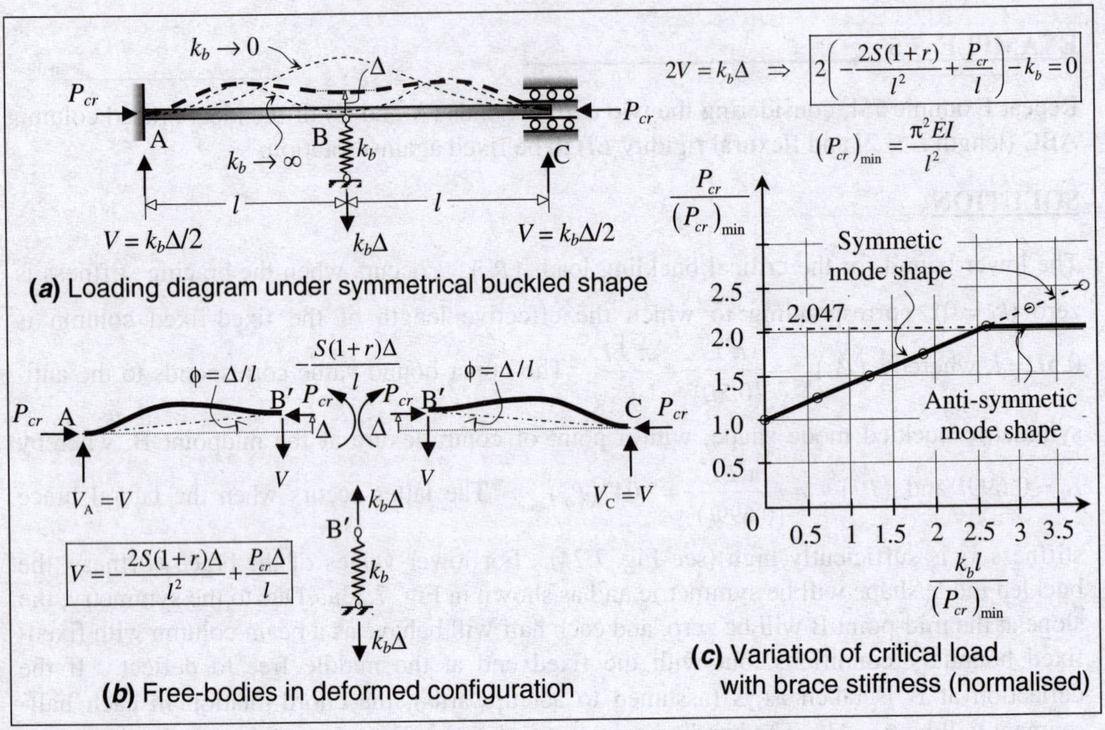

Figure 7.23 Buckling of a fixed-fixed ideal column with a middle brace (Example 7.5)

The results are tabulated below for typical values of $A = \frac{P_{cr}}{(P_{cr})_{\min}}$:

$A = \dfrac{P_{cr}}{(P_{cr})_{min}}$	$(\mu_{cr} l) = \pi\sqrt{A}$	$B = \dfrac{S(1+r)}{EI/l}$	$\dfrac{k_b l}{(P_{cr})_{min}}$
1.00	π	$\pi^2/2$	0
1.25	3.51241	4.64008	0.6194
1.50	3.84765	4.33166	1.2444
1.75	4.15594	4.00808	1.8756
2.047	4.49478	3.60157	**2.6343**
2.25*	4.71239	3.30830	3.1592
2.50*	4.96729	2.92780	3.8134

* not actually realizable in practice, as the anti-symmetric mode with lower P_{cr} will govern

Eq. (b) predicts that the critical buckling load increases with increase in the brace stiffness. The symmetric buckled mode shapes corresponding to the extreme values of k_b (zero and infinity) are shown by light dashed lines in Fig. 7.23a, and the two extreme critical buckling loads are given by $\dfrac{\pi^2 EI}{l^2}$ and $\dfrac{4\pi^2 EI}{l^2}$ (with effective lengths of l and $0.5l$ respectively). The variation of $\dfrac{P_{cr}}{(P_{cr})_{min}}$ with $\dfrac{k_b l}{(P_{cr})_{min}}$ is illustrated in Fig. 7.23c, in which the predictions from the assumed symmetric buckled mode shape (Fig. 7.23a) as well as the anti-symmetric buckled mode shape (Fig. 7.24) are shown. The two estimates of critical buckling load intersect at $\dfrac{P_{cr}}{(P_{cr})_{min}} = 2.047$ and $\dfrac{k_b l}{(P_{cr})_{min}} = 2.6343$. Buckling instability will occur at the lowest predicted P_{cr}. Accordingly,

$$0.0 \le \frac{k_b l}{(P_{cr})_{min}} \le 2.6343 \quad \Rightarrow \quad 1.0 \le \frac{P_{cr}}{(P_{cr})_{min}} \le 2.047$$
$$\frac{k_b l}{(P_{cr})_{min}} > 2.6343 \quad \Rightarrow \quad P_{cr} = (P_{cr})_{max} = 2.047(P_{cr})_{min} = \frac{2.047\pi^2 EI}{l^2} \tag{7.49}$$

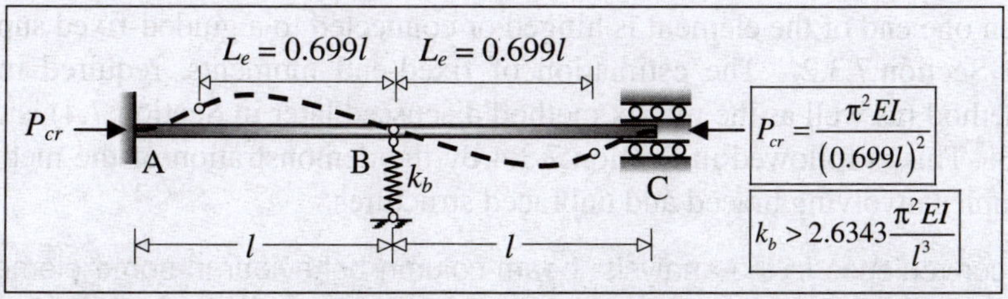

Figure 7.24 Anti-symmetric buckling mode (Example 7.5)

7.3 SLOPE-DEFLECTION METHOD OF ANALYSIS

The slope-deflection method, developed by George Maney in 1915, is perhaps the earliest displacement method that served to provide a solution to the response of continuous beams and plane frames in traditional 'first-order' structural analysis. This method, in its original form, cannot be used to find critical buckling loads and second-order effects, because it does not account for possible *P-delta* moments in the presence of axial forces.

We can easily extend this method to systems involving beam-columns, by modifying the expressions for the various stiffness measures to include the effect of the axial force in the deflected configuration. The fixed-end moments also need to be suitably modified to include the effect of the axial force. The effects of axial and shear deformations are, however, ignored. This method provides a convenient alternative to the classical procedure of solving the differential equation of equilibrium in the deformed configuration.

It may be noted that, unlike trusses, continuous beams and rigid-jointed frames involve rotational compatibility at the inter-connecting joints. Thus, whereas a truss member can undergo elastic buckling as a local phenonmenon, this is not possible in flexural members that are forced to bend on account of rotational (or translational) compatibility requirements.

> Individual beam-column instability cannot generally be studied in isolation in continuous beams and frames and the buckling of the structure as a whole has to be investigated under different possible modes in order to find the lowest possible critical buckling load. This condition of elastic instability is associated by a complete loss of flexural stiffness in the structure, which is reflected by the structure stiffness matrix becoming singular at the critical load (assuming the condition of bifurcation buckling).

Such analysis can be conveniently achieved in the flexural stiffness formulation underlying displacement methods such as the slope-deflection method. In order to do this, as well as to perform a second-order analysis in situations when the stiffness matrix remains non-singular, we need to write down the slope-deflection equations for a typical prismatic beam-column element that is subject to axial force. This is described in Section 7.3.1 for the standard case. Simplifications, in terms of reduced degree of kinematic indeterminacy, are possible when one end of the element is hinged or connected to a guided-fixed support; this is discussed in Section 7.3.2. The estimation of fixed-end moments, required in the slope-deflection method (as well as the matrix method discussed later in Section 7.4) is described in Section 7.3.3. This is followed in Section 7.3.4 by the demonstration of the method through various examples involving braced and unbraced structures.

> For convenience in the analysis, beam-column behaviour in some elements of the structure (such as beams in portal frames with low axial forces) may be ignored; similarly, the effect of minor variations in the axial force P (of known magnitude) on the beam-column stiffness may be ignored.

7.3.1 Slope-deflection Equations for Prismatic Beam-Columns

Consider a typical prismatic beam-column element AB (which may be part of a continuous beam or rigid jointed frame), subject to any arbitrary loading, as well as an axial force, P, as shown in Fig. 7.25a. Let the rotation (slope) at A be denoted as θ_A and the rotation (slope) at B be denoted as θ_B, as shown. Let us assume all end rotations and end moments to be positive when they act in a clockwise manner (as traditionally assumed in the slope-deflection method). Let there also be support settlements Δ_A and Δ_B at A and B respectively, resulting in a clockwise chord rotation $\phi_{AB} = \dfrac{\Delta_B - \Delta_A}{L}$.

The following slope-deflection equations (assuming all end moments and rotations to be positive when acting clockwise) can be easily derived for this beam-column element, by superposing the various effects, as shown in Fig. 7.25, assuming the axial force to act constantly in all cases. These equations are similar to Eq. 1.85 for the beam element, except that the stiffness measures are now expressed as *stability functions* (in terms of $\mu L = \sqrt{\dfrac{PL^2}{EI}}$):

$$\begin{Bmatrix} M_{AB} \\ M_{BA} \end{Bmatrix} = \begin{Bmatrix} M^F_{AB} \\ M^F_{BA} \end{Bmatrix} + S \begin{bmatrix} 1 & r \\ r & 1 \end{bmatrix} \begin{Bmatrix} \theta_A - \phi_{AB} \\ \theta_B - \phi_{AB} \end{Bmatrix} \tag{7.50}$$

where $M^F{}_{AB}$ and $M^F{}_{BA}$ are appropriate *fixed-end moments* (which include the effect of the axial force) and the stiffness measures S and r are as defined by Eqns 7.44-7.45 (under axial compression) and Eqns 7.44a-7.45a (under axial tension). In the absence of the axial force P, of course, Eq. 7.44 reduces to the conventional pair of slope-deflection equations, given by Eq. 1.85 (with $S = 4EI/L$ and $r = 0.5$).

7.3.2 Modified Slope-deflection Equations

The basic slope-deflection equations (Eq. 7.50) are comprehensive in the sense that they can accommodate all possible displacements at the beam ends, and hence can be applied to solve practically any type of beam or framed structure involving prismatic beam-column elements. However, in some instances, it is advantageous to ignore an unknown displacement (end rotation or deflection), thereby reducing the degree of kinematic indeterminacy in the structure. Such situations are encountered when the extreme end of a beam or frame element in the structure has freedom to rotate or translate. A commonly encountered situation is when the extreme end has a hinged or roller support, as shown in Fig. 7.26. The bending moment in the beam at this end is known (usually zero, unless there is a concentrated moment applied at this end).

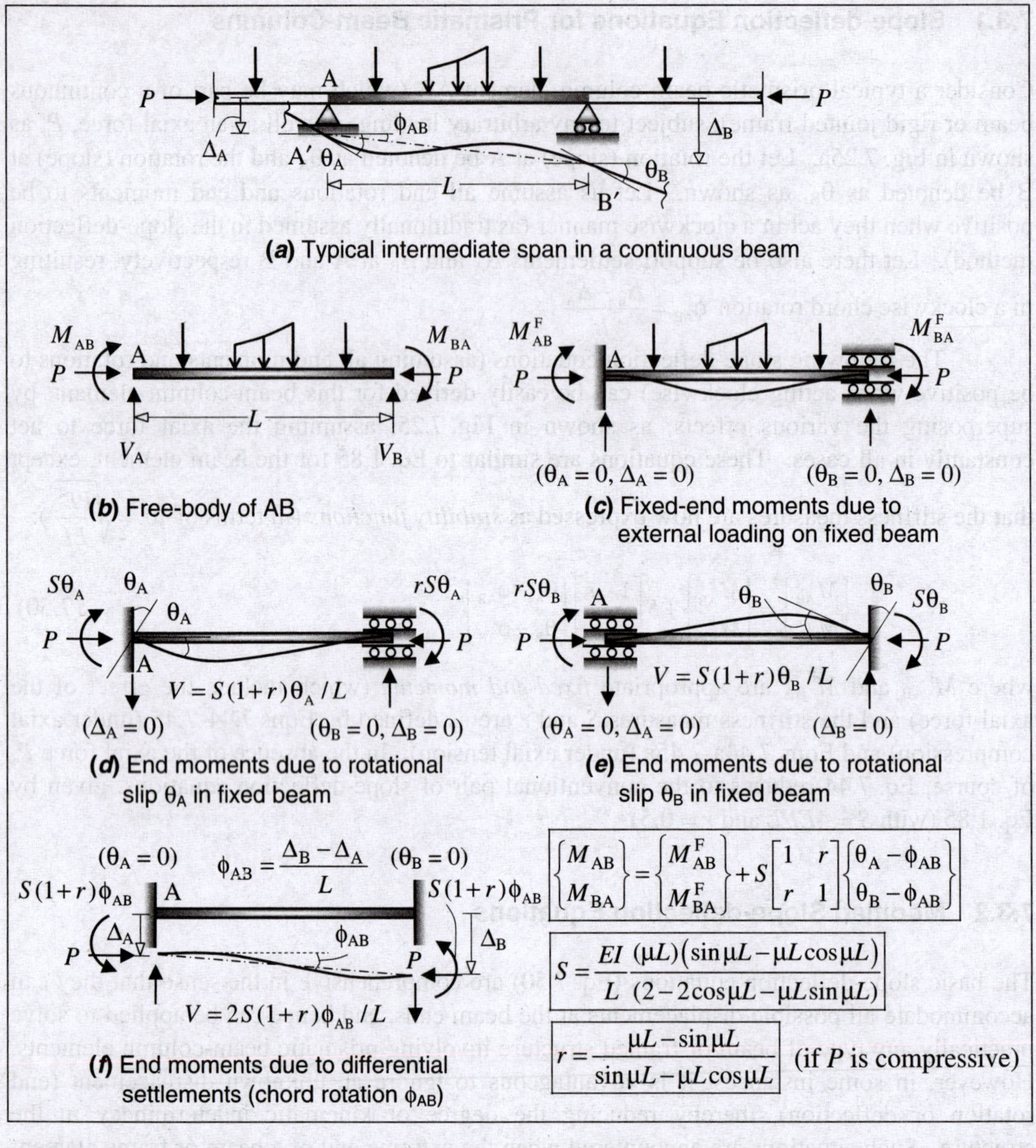

Figure 7.25 Basic slope-deflection equations for a prismatic beam-column element

Beam-column with pinned end at B

This case is depicted in Fig. 7.26a. The unknown displacements are the end rotation θ_A and chord rotation ϕ_{AB}. The rotation at the pinned end, θ_B, is also unknown, but we can afford to ignore this rotation, if our objective is limited to finding the force response. By applying the Principle of superposition, as shown earlier in Fig. 7.25, the slope-deflection equation takes the following modified form, considering the modified stiffness of the beam-column and modified fixed end moment $M^{F_o}_{AB}$ for the propped cantilever beam (including the effect of P):

$$\boxed{M_{AB} = M^{F_o}_{AB} + S^o(\theta_A - \phi_{AB})} \tag{7.51a}$$

where the stiffness measure S^o is defined by Eq. 7.46 (under axial compression) and Eq. 7.46a (under axial tension); it is equal to $3EI/L$ in the absence of axial force.

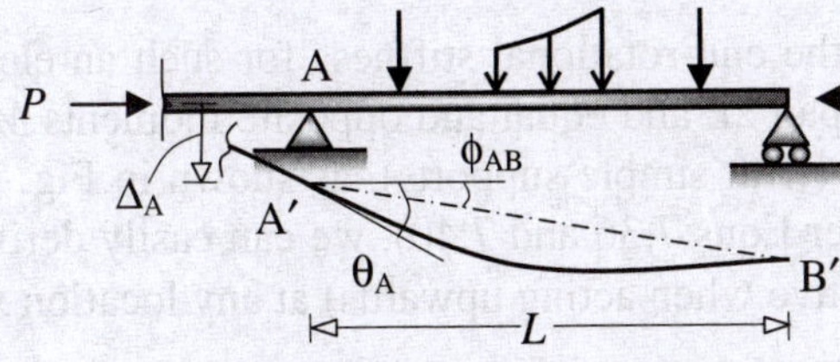

(*a*) Pinned support at end B

$$\boxed{M_{AB} = M^{F_o}_{AB} + S^o(\theta_A - \phi_{AB})}$$

(Ignore θ_B)

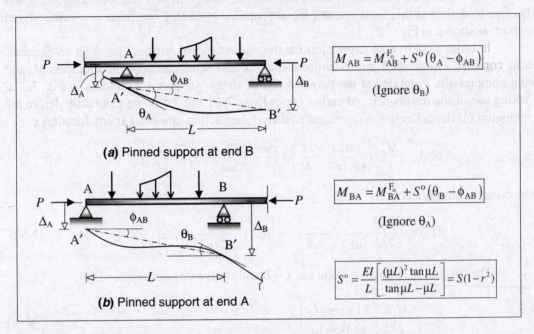

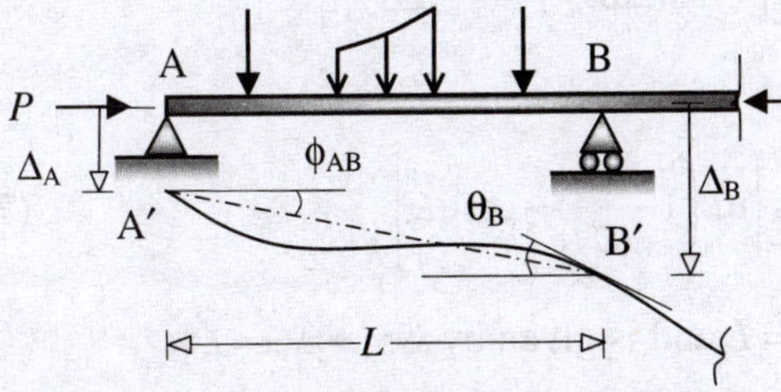

(*b*) Pinned support at end A

$$\boxed{M_{BA} = M^{F_o}_{BA} + S^o(\theta_B - \phi_{AB})}$$

(Ignore θ_A)

$$\boxed{S^o = \frac{EI}{L}\left[\frac{(\mu L)^2 \tan\mu L}{\tan\mu L - \mu L}\right] = S(1-r^2)}$$

Figure 7.26 Modified slope-deflection equations for a prismatic beam-column element with pinned support at one end

Beam-column with pinned end at A

This case is depicted in Fig. 7.26b. The unknown displacements are the end rotation θ_B and chord rotation ϕ_{AB}. The rotation at the pinned end, θ_A, is also unknown, but we can afford to ignore this rotation, if our objective is limited to finding the force response. By applying the Principle of superposition, the slope-deflection equation takes the following modified form, considering the modified stiffness of the beam and modified fixed end moment $M^{F_o}_{BA}$ for the propped cantilever beam (including the effect of the axial force):

$$\boxed{M_{BA} = M^{F_o}_{BA} + S^o(\theta_B - \phi_{AB})} \tag{7.51b}$$

Beam-column with symmetry

In some cases, we encounter symmetric frames and continuous beams, where we can take advantage of the symmetry in the loading and response, and thereby reduce the model of the structure to one-half (either left-half or right-half). The slope of any element intersected by the axis of symmetry is invariably equal to zero and hence, while truncating the structure along the axis of symmetry, we should impose the condition of rotational fixity at this location. If the axis of symmetry passes through the mid-point of a beam-column, then it is also free to deflect at this location, and the appropriate boundary condition is a guided-fixed support, as shown in Fig. 7.27.

In order to derive an expression for the end-rotational stiffness for such an element, let us consider a prismatic beam-column of span $2L$ and equal and opposite moments M_o and axial compression P applied at the two ends (with simple supports), as shown in Fig. 7.27a. Making use of the results derived earlier (refer Eqns 7.36 and 7.40), we can easily derive the expression for the deflection Δ (assumed positive when acting upwards) at any location x:

$$\Delta(x) = -\frac{M_o}{P}\left[\left\{\frac{\sin \mu x}{\sin 2\mu L} - \frac{x}{2L}\right\} + \left\{\frac{\sin \mu(2L-x)}{\sin 2\mu L} - \left(1 - \frac{x}{2L}\right)\right\}\right]$$

which can be simplified to

$$\Delta(x) = -\frac{M_o L^2}{2EI}\frac{2}{(\mu L)^2 \cos \mu L}\left[\cos\left\{\mu L\left(1 - \frac{x}{L}\right)\right\} - \cos \mu L\right] \tag{7.52}$$

The maximum deflection occurs at $x = L$ and is given by $\Delta_{max} = |\Delta(x = L)|$:

$$\Delta_{max} = \frac{M_o L^2}{2EI}\left[\frac{2(1-\cos \mu L)}{(\mu L)^2 \cos \mu L}\right] \tag{7.53a}$$

The slope, θ_o, at each of the two ends, is given by the first derivative of $\Delta(x)$:

$$\theta_o = |\Delta'(0)| = \frac{M_o L}{EI}\left[\frac{\tan \mu L}{\mu L}\right] \tag{7.53b}$$

whereby the end rotational stiffness measure, M_o/θ_o, which we may denote as $\tilde{S}$, is given by:

$$\boxed{\tilde{S} = \frac{EI}{L}\left(\frac{\mu L}{\tan \mu L}\right)} \tag{7.54}$$

Furthermore, by taking the second derivative of $\Delta(x)$ at $x = L$, we obtain an expression for the moment at the symmetry location (guided-fixed support in the truncated beam) as $M_o \sec \mu L$, acting in a direction opposite to that of the applied end moment. Thus, we can define the carry-over factor, $\tilde{r}$, for a beam-column with a guided-fixed support, as:

$$\boxed{\tilde{r} = -\sec \mu L} \tag{7.55}$$

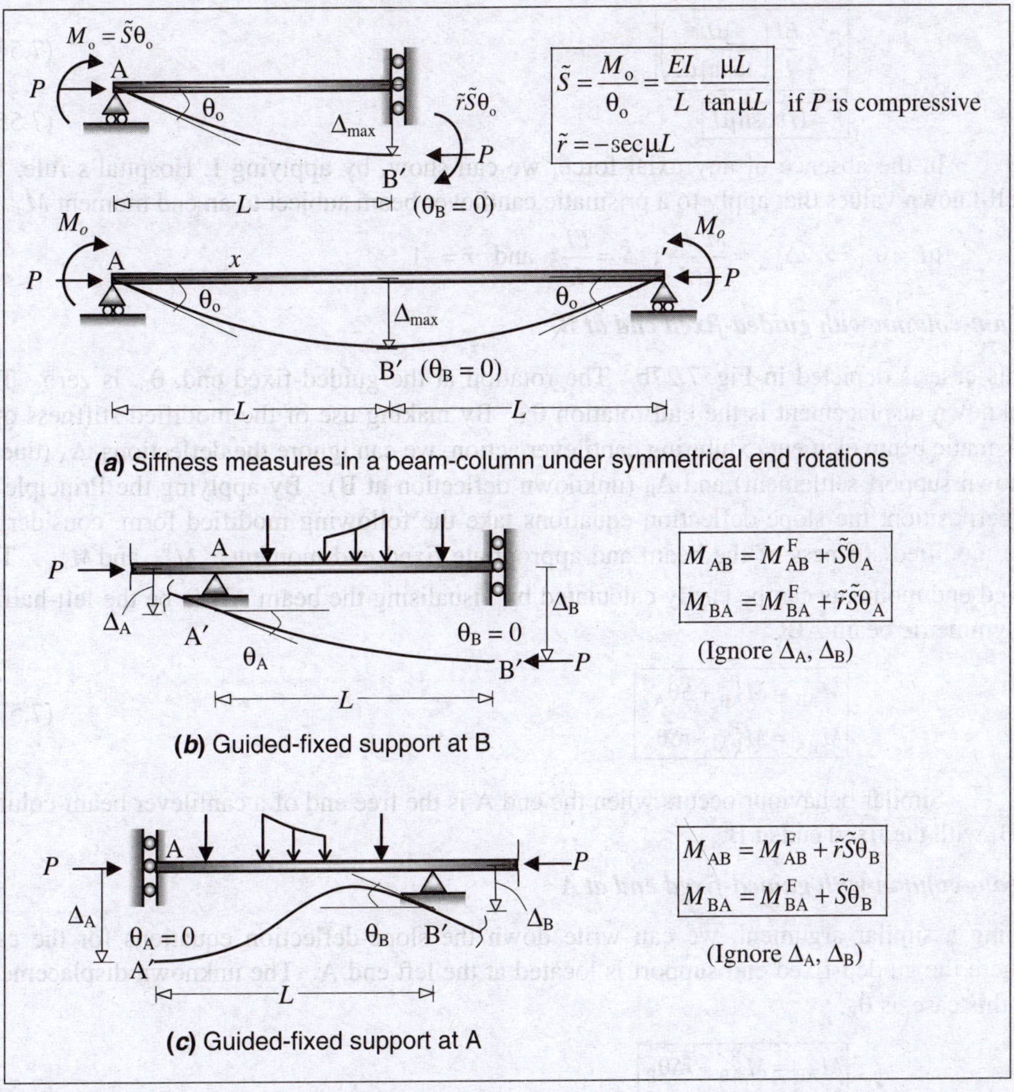

Figure 7.27 Modified slope-deflection equations for a prismatic beam-column element with guided-fixed support at one end

Alternatively, we can establish inter-relationships with the stability functions, S and r, by applying the principle of superposition to a propped cantilever beam of span $2L$, such that the rotations at the two ends are equal and opposite. Thus, it can be easily shown:

$$\tilde{S}(\mu L) = \frac{1}{2} S(2\mu L)\left[1 - r(2\mu L)\right] \tag{7.56}$$

If the beam-column is subject to axial tension instead of axial compression, we can show that:

$$\boxed{\tilde{S} = \frac{EI}{L}\left(\frac{\mu L}{\tanh \mu L}\right)} \tag{7.54a}$$

$$\boxed{\tilde{r} = 1/\cosh \mu L} \tag{7.55a}$$

In the absence of any axial force, we can show, by applying L'Hospital's rule, the well-known values that apply to a prismatic cantilever beam subject to an end moment M_o:

$$\mu L = 0 \quad \Rightarrow \quad \Delta_{max} = \frac{M_o L^2}{2EI}; \quad \tilde{S} = \frac{EI}{L}; \text{ and } \tilde{r} = -1.$$

Beam-column with guided-fixed end at B

This case is depicted in Fig. 7.27b. The rotation at the guided-fixed end, θ_B, is zero. The unknown displacement is the end rotation θ_A. By making use of the modified stiffness of a prismatic beam element exhibiting cantilever action, we can ignore the deflections, Δ_A (due to known support settlement) and Δ_B (unknown deflection at B). By applying the Principle of superposition, the slope-deflection equations take the following modified form, considering the modified stiffness of the beam and appropriate fixed end moments, M_{AB}^F and M_{BA}^F. The fixed end moments can be easily calculated by visualising the beam AB to be the left-half of a symmetric beam ABC.

$$\boxed{\begin{aligned} M_{AB} &= M_{AB}^F + \tilde{S}\theta_A \\ M_{BA} &= M_{BA}^F + \tilde{r}\tilde{S}\theta_A \end{aligned}} \tag{7.57}$$

Similar behaviour occurs when the end A is the free end of a cantilever beam-column AB, with the fixed end at B.

Beam-column with guided-fixed end at A

Using a similar argument, we can write down the slope-deflection equations for the case where the guided-fixed end support is located at the left end A. The unknown displacement, in this case, is θ_B.

$$\boxed{\begin{aligned} M_{AB} &= M_{AB}^F + \tilde{r}\tilde{S}\theta_B \\ M_{BA} &= M_{BA}^F + \tilde{S}\theta_B \end{aligned}} \tag{7.57a}$$

Similar behaviour occurs when the end B is the free end of a cantilever beam-column AB, with the fixed end at A.

7.3.3 Fixed-end Moments in Beam-Columns

In laterally loaded beam-columns, we need to find fixed-end moments in the displacement method of analysis, in order to arrive at equivalent joint loads. The moments generated at the fixed ends of such a prismatic element AB are affected by the magnitude of the axial force P.

The beam with fixed end conditions, subject to any arbitrary loading, can be analysed from first principles, by solving the basic fourth-order differential equation of equilibrium (Eq. 7.10) and applying the four kinematic boundary conditions (deflection and slope at $x = 0$ and $x = L$ are zero). Using the resulting expression for deflection $\Delta(x)$, the bending moment at any location, $M(x) = EI\Delta''(x)$, can be generated and the fixed end moments easily obtained as the values of $M(x)$ at $x = 0$ and $x = L$.

Alternatively, these end moments can be determined by applying the Principle of consistent deformations, considering the simply supported beam-column as the primary structure subject to the applied axial force P. We need to first find out the end rotations, θ_A and θ_B, due to the given lateral loading, and then find out the desired fixed-end moments as the end moments that need to be applied to nullify these end rotations. Using the flexibility measures shown in Fig. 7.9, the compatibility equations can be expressed in the following form, assuming the the end moments and rotations to be positive when acting in a clockwise direction, in accordance with the convention of the slope-deflection method:

$$\frac{L}{6EI}\begin{bmatrix} 2\psi(\mu L) & -\phi(\mu L) \\ -\phi(\mu L) & 2\psi(\mu L) \end{bmatrix}\begin{Bmatrix} M_{AB}^{F} \\ M_{BA}^{F} \end{Bmatrix} + \begin{Bmatrix} \theta_A \\ \theta_B \end{Bmatrix} = \begin{Bmatrix} 0 \\ 0 \end{Bmatrix} \tag{7.58}$$

where the functions, $\psi(\mu L)$ and $\phi(\mu L)$, are as defined in Eqns 7.37a and 7.38a respectively. Solving Eq. 7.57, the fixed end moments can be obtained.

When the applied transverse loading is symmetrical, we can take advantage of the symmetry in the response ($\theta_B = -\theta_A$ and $M_{AB}^{F} = -M_{BA}^{F}$). Making use of the stiffness measures in Eqns 7.54/7.54a (refer Fig. 7.27a), and considering the half-span as $L/2$, the desired fixed end moment is directly obtainable as:

$$M_{AB}^{F} = -\tilde{S}\left(\frac{\mu L}{2}\right)\theta_A \tag{7.59}$$

For example, let us consider a prismatic beam-column with fixed ends, subject axial compression P and a concentrated load, W, at mid-point, as shown in Fig. 7.28a. We have analysed the simply supported condition earlier (refer Fig. 7.7) and have derived an expression for the deflection $\Delta(x)$, as given by Eq. 7.29. The rotation, θ_A, at the left end can be easily obtained as:

$$\theta_A = |\Delta'(x=0)| = \frac{W}{2P}\left[\frac{1}{\cos\mu L/2} - 1\right]$$

Applying Eqns 7.58 and 7.54, and simplifying, we obtain:

$$M_{AB}^{F} = -\frac{WL}{8}\left[\frac{2\left(1-\cos\frac{\mu L}{2}\right)}{\frac{\mu L}{2}\sin\frac{\mu L}{2}}\right] = -M_{BA}^{F} \quad \text{if } P \text{ is compressive} \tag{7.59a}$$

The term in parenthesis is a measure of the amplification in the 'primary' fixed end moment (i.e., with $P = 0$) on account of the *P-delta* effect. The support reactions, equal to $W/2$, remain unchanged by this effect, as shown in Fig. 7.28a.

(a) Concentrated load at mid-span

$$M_{AB}^{F} = -\frac{WL}{8}\left[\frac{2\left(1-\cos\frac{\mu L}{2}\right)}{\frac{\mu L}{2}\sin\frac{\mu L}{2}}\right] = -M_{BA}^{F}$$

if P is compressive

(b) Concentrated load at any location

$$M_{AB}^{F} = -WL\left[\frac{\mu b\cos\mu L - \sin\mu L + \sin\mu a + \sin\mu b - \mu L\cos\mu b + \mu a}{\mu L(2 - 2\cos\mu L - \mu L\sin\mu L)}\right]$$

if P is compressive

(c) Uniformly distributed load

$$M_{AB}^{F} = -\frac{q_o L^2}{12}\left[\frac{3}{(\mu L/2)^2}\left(1-\frac{\mu L}{2}\cot\frac{\mu L}{2}\right)\right] = -M_{BA}^{F}$$

if P is compressive

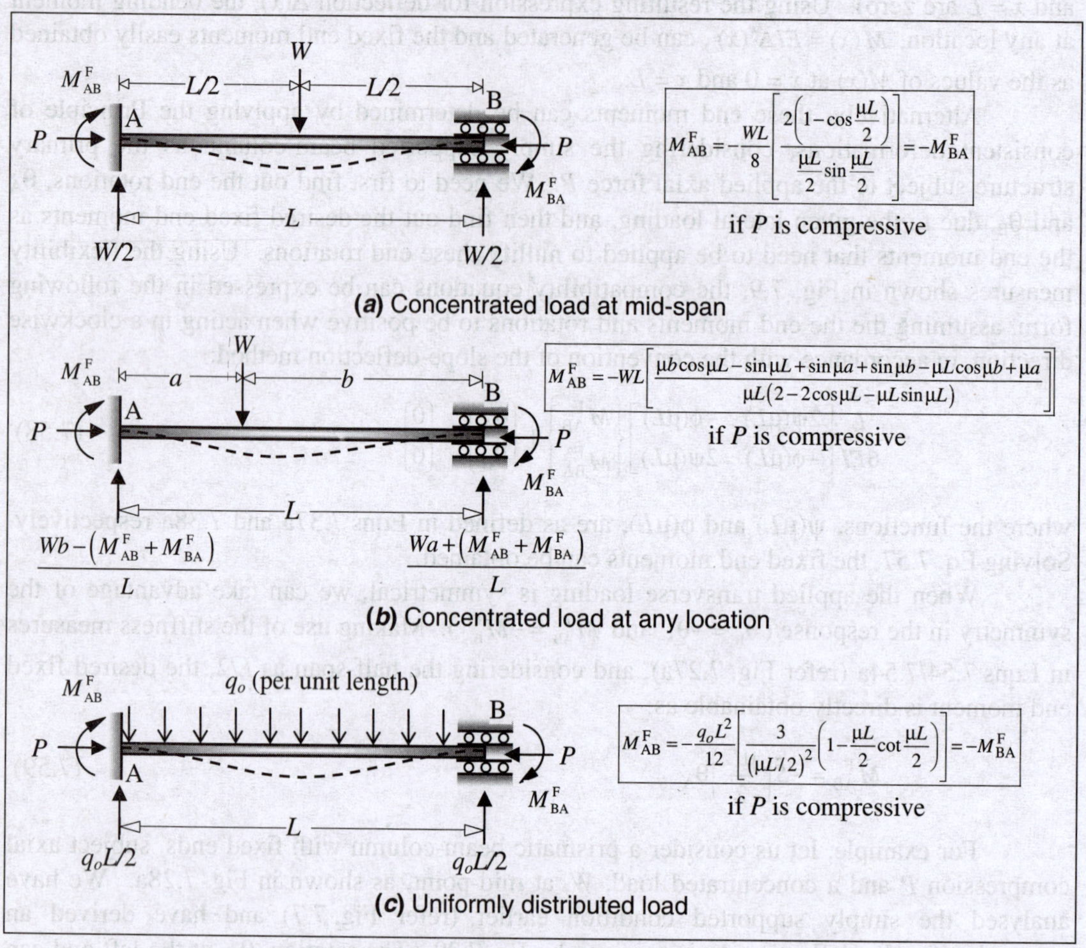

Figure 7.28 Fixed-end forces in prismatic beam-columns

In a similar way, it can be shown that when P is tensile, under a concentrated load W at midspan,

$$M_{AB}^{F} = -\frac{WL}{8}\left[\frac{-2\left(1-\cosh\frac{\mu L}{2}\right)}{\frac{\mu L}{2}\sinh\frac{\mu L}{2}}\right] = M_{BA}^{F} \qquad \text{if } P \text{ is tensile} \qquad (7.59b)$$

Axial compression causes an increase in the fixed end moment, while axial tension causes a reduction. In the case of axial compression, the rate of increase is gradual at low values of μL but steep at higher values (approaching infinity as $\mu \rightarrow \mu_{cr} = 2\pi$).

Two other commonly encountered loading cases, viz., eccentrically applied point load W and uniformly distributed load q_o (per unit length) are shown in Figs 7.28b and c respectively. The fixed-end moments (assumed clockwise positive) for these two cases are as follows:

Concentrated load W applied at a distance a from A (b from B):

$$M_{AB}^{F} = -WL\left[\frac{\mu b\cos\mu L - \sin\mu L + \sin\mu a + \sin\mu b - \mu L\cos\mu b + \mu a}{\mu L(2 - 2\cos\mu L - \mu L\sin\mu L)}\right] \quad \text{if } P \text{ is compressive} \quad (7.60)$$

$$M_{AB}^{F} = -WL\left[\frac{\mu b\cosh\mu L - \sinh\mu L + \sinh\mu a + \sinh\mu b - \mu L\cosh\mu b + \mu a}{\mu L(2 - 2\cosh\mu L + \mu L\sinh\mu L)}\right] \quad \text{if } P \text{ is tensile} \quad (7.60a)$$

The above expressions can also be used to generate values of M_{BA}^{F} by interchanging the terms a and b, and applying an overall positive sign.

Uniformly distributed load q_o per unit length:

$$M_{AB}^{F} = -\frac{q_o L^2}{12}\left[\frac{3}{(\mu L/2)^2}\left(1 - \frac{\mu L}{2}\cot\frac{\mu L}{2}\right)\right] = -M_{BA}^{F} \quad \text{if } P \text{ is compressive} \quad (7.61)$$

$$M_{AB}^{F} = -\frac{q_o L^2}{12}\left[\frac{3}{(\mu L/2)^2}\left(-1 + \frac{\mu L}{2}\coth\frac{\mu L}{2}\right)\right] = -M_{BA}^{F} \quad \text{if } P \text{ is tensile} \quad (7.61a)$$

Modified Fixed-end Moment in a Propped Cantilever

When we wish to take advantage of reduced kinematic indeterminacy in beam-columns that are pinned at one end, we need to apply the fixed-end moment corresponding to the 'propped cantilever' condition. The expression for this can be easily derived from the appropriate fixed end moment expression (Eqns 7.59-7.61), by applying the principle of superposition (eliminating the fixed end moment at the propped end), using the carry-over stiffness measure, r. Thus, if the propped end is located at B, the modified fixed end moment at A, denoted as M_{AB}^{Fo}, is given by:

$$M_{AB}^{Fo} = M_{AB}^{F} - rM_{BA}^{F} \quad \text{if prop is at end B} \quad (7.62a)$$

Similarly, if the prop is at end A,

$$M_{BA}^{Fo} = M_{BA}^{F} - rM_{AB}^{F} \quad \text{if prop is at end A} \quad (7.62b)$$

If the loading on the fixed-fixed beam is symmetrical, then for the propped cantilever case, Eq. 7.62a reduces to $M_{AB}^{Fo} = M_{AB}^{F}(1+r)$, as shown in Fig. 7.29 for two typical load cases.

$$M_{BA}^{Fo} = \frac{WL}{8}\left[\frac{2\left(1-\cos\frac{\mu L}{2}\right)}{\frac{\mu L}{2}\sin\frac{\mu L}{2}}\right](1+r)$$

if P is compressive

$$r = \frac{\mu L - \sin\mu L}{\sin\mu L - \mu L\cos\mu L}$$

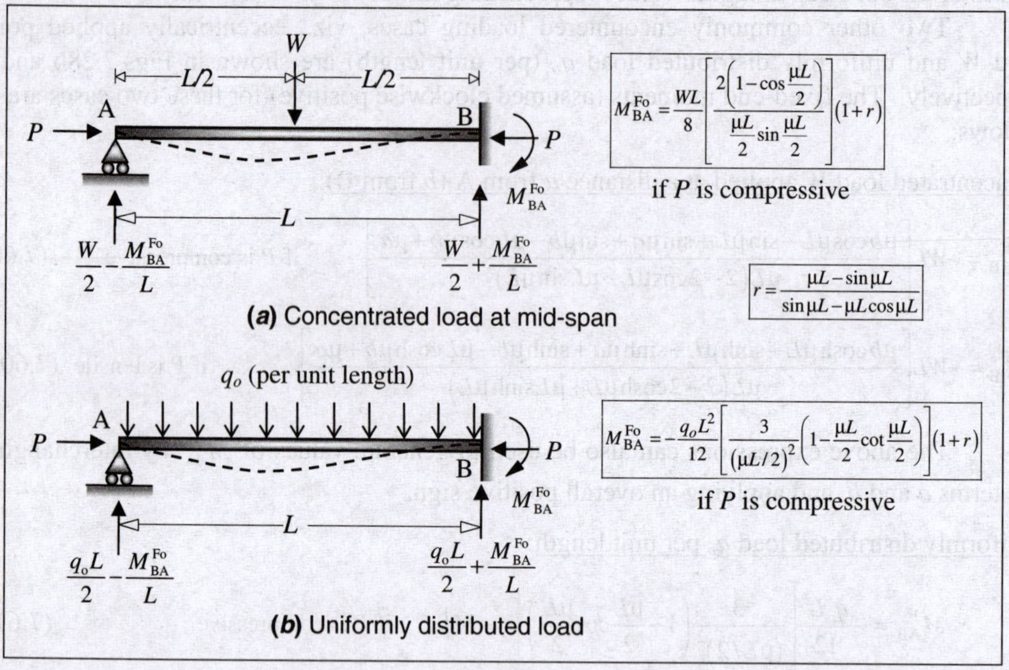

(a) Concentrated load at mid-span

$$M_{BA}^{Fo} = -\frac{q_o L^2}{12}\left[\frac{3}{(\mu L/2)^2}\left(1-\frac{\mu L}{2}\cot\frac{\mu L}{2}\right)\right](1+r)$$

if P is compressive

(b) Uniformly distributed load

Figure 7.29 Modified fixed-end forces in propped cantilever (prismatic) beam-columns

7.3.4 Examples

We now have the tools, in terms of relevant expressions of stiffness measures and fixed-end moments, to analyse continuous beams and frames with beam-column elements (subject to axial compression or tension). We intend to do this in the framework of linear analysis, and hence we need to make appropriate assumptions in order to avoid the need to carry out a nonlinear analysis. This is demonstrated in the Examples to follow.

Broadly, there are two classes of problems: (i) finding the critical buckling load and (ii) finding the response (especially bending moment diagram), inclusive of second-order effects. In the former category, the objective is to find the lowest buckling load (usually triggered by the buckling of the beam-column element with the lowest flexural stiffness) that would induce elastic instability in the structure, when it is subject to a given pattern of loading. Usually, the loading pattern chosen is simple, involving only concentrated forces (P) applied at one or more joints, so as to induce axial compressive forces in the beam-column elements, without any initial curvatures, so that a condition of bifurcation† buckling is rendered possible. We seek a non-trivial solution to the response, which is associated with a sudden loss in stiffness of the entire structure. This is obtained by solving the *characteristic equation* generated by setting the determinant of the stiffness matrix equal to zero. The stiffness matrix gets generated while applying the relevant equilibrium equations, expressed in terms of the slope-deflection equations, as in the conventional Slope-deflection method. The solution procedure is generally iterative in nature, and we shall make use of the bisection method described earlier to solve transcendental equations of the form, $g(P_{cr}) = 0$, which can be conveniently done using a simple spreadsheet (like EXCEL). One begins this numerical procedure by making a reasonable guess of the upper and lower bound values of P_{cr}.

In the second category, involving given axial forces that may be compressive or tensile, the stiffness matrix is non-singular, and by inverting it, the unknown displacements (slopes and deflections) can be determined and the end moments obtained in the various elements. However, bending moments at intermediate locations cannot be estimated conveniently, as we need to know the deflections at these locations in order to account for the moment contribution of the eccentrically acting axial load. This can be done by including additional nodes at one or more intermediate locations; this, however, increases the degree of kinematic indeterminacy. The effort required in the analysis increases significantly with increase in the degree of kinematic indeterminacy, and it would be more appropriate to resort to computer-based matrix methods (discussed in Section 7.4) in such cases.

Wherever possible, we should take advantage of conditions of symmetry (or anti-symmetry), and other situations involving end supports that are pinned or guided-fixed, in order to reduce the degree of kinematic indeterminacy.

Problems involving braced beam-columns ("non-sway") are relatively simpler, as they involve only rotational degrees of freedom (moment equilibrium equations). "Sway" type problems are encountered while dealing with unbraced beam-columns. Such problems involve unknown translations, in addition to unknown rotations, and call for additional consideration of equilibrium of member end shear forces (which are expressed in terms of the member end moments).

EXAMPLE 7.6

Find the critical buckling load, P_{cr}, for the two-span continuous beam-column system ABC, with a fixed support at A and a roller support at C, as shown in Fig. 7.30. Assume constant flexural rigidity *EI*.

† This is not possible if there are intermediate loads acting in the beam-column elements, as such loads would induce initial curvatures (similar to initial imperfections), which get enhanced gradually with increase in the applied loading. As discussed in Section 7.2.2, the same buckling load capacity in a beam-column element is reached asymptotically, even in the presence of initial curvatures.

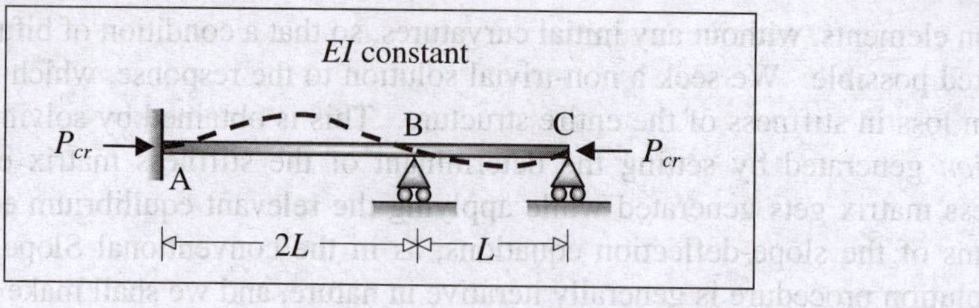

Figure 7.30 Buckling of a continuous beam-column (Example 7.6)

SOLUTION

Taking advantage of the pinned support at C, the degree of kinematic indeterminacy may be taken as one (unknown rotation θ_B at B) in this braced beam-column system. The condition of moment equilibrium at the joint B in the buckled configuration (shown in Fig. 7.30) is given by:

$$\sum M_B = 0 \Rightarrow M_{BA} + M_{BC} = 0 \Rightarrow \left(S_{BA} + S^o_{BC}\right)\theta_B = 0$$

For a non-trivial solution ($\theta_B \neq 0$), the *characteristic equation*, which renders the total rotational stiffness at the joint B equal to zero, is given by:

$$S_{BA} + S^o_{BC} = 0$$

$$\Rightarrow \frac{S_{BA}}{EI/2L}\left(\frac{1}{2}\right) + \frac{S^o_{BC}}{EI/L} = 0$$

$$\Rightarrow g\left(P_{cr}\right) = \frac{1}{2}\left(\frac{S}{EI/L}\right)_{BA} + \left(\frac{S^o}{EI/L}\right)_{BC} = 0 \qquad \text{(a)}$$

Clearly, the beam-column AB is less stiff than BC, and hence the critical buckling load is likely to be governed by it. Assuming the braced beam-column AB to be stiffened by a rotational spring with stiffness $k_{\theta B}$ (representing BC) at B, the lower and upper bounds for the critical buckling load are accordingly given by $(P_{cr})_{min} = \frac{\pi^2 EI}{(0.699 \times 2L)^2}$ and $(P_{cr})_{max} = \frac{\pi^2 EI}{(0.5L)^2}$, corresponding to $k_{\theta B} = 0$ yand $k_{\theta B} \to \infty$ respectively. The corresponding initial bounds of the parameter $\mu_{cr} L = \sqrt{\frac{P_{cr}L^2}{EI}}$ are accordingly given by $\frac{\pi}{0.699} < (\mu_{cr}L)_{BA} < \frac{\pi}{0.5}$, with $(\mu_{cr}L)_{BC} = \frac{1}{2}(\mu_{cr}L)_{BA}$. Making use of the stiffness measures given in Eqns 7.44 and 7.46, and solving Eq. (a) iteratively by the *bisection method*, the results are tabulated below.

Trial	$(\mu_{cr}L)_{BA}$	$\left(\frac{S}{EI/L}\right)_{BA}$	$\left(\frac{S^o}{EI/L}\right)_{BC}$	$g(P_{cr}) = \frac{1}{2}\left(\frac{S}{EI/L}\right)_{BA} + \left(\frac{S^o}{EI/L}\right)_{BC}$
	4.49	0.010	1.804	+1.809
	6.28	–1971.6	0.005	–985.8
1	5.385	–4.486	1.101	+0.680
2	4.938	–1.613	1.487	–6.984
3	5.162	–2.808	1.303	–0.101
4	5.050	–2.165	1.397	+0.315
5	5.106	–2.474	1.351	+0.114
6	5.134	–2.637	1.327	+0.009
7	5.137	–2.655	1.325	–0.003
8	**5.136**	–2.649	1.326	+0.001

$$\Rightarrow (\mu_{cr}L)_{BA} = 5.136 = \frac{\pi}{0.6117} \qquad \Rightarrow P_{cr} = \frac{\pi^2 EI}{(0.6117\times 2L)^2} = \boxed{\frac{6.595EI}{L^2}}$$

$$\text{Also, } (\mu_{cr}L)_{BC} = \frac{5.136}{2} = \frac{\pi}{1.223} \qquad \Rightarrow P_{cr} = \frac{\pi^2 EI}{(1.223L)^2} = \frac{6.595EI}{L^2}$$

[Note that the effective length ratio for element BC is 1.223, which is higher than the maximum value of 1.0 for any braced beam-column. This shows that the integral connection at B ends up weakening the stiffer element BC. However, we can also view this positively, as a means of stiffening the weaker element AB.]

EXAMPLE 7.7

Consider a uniformly distributed gravity load of 30 kN/m acting on the continuous beam-column system in Fig. 7.30, along with an axial load of P = 1000 kN. Analyse the structure, assuming L = 3m and EI = 5000 kNm^2, and draw the bending moment diagram (including second-order effects) for the following two cases: (a) P is compressive and (b) P induces axial tension.

SOLUTION

The degree of kinematic indeterminacy, n_k = 1. The unknown displacement is θ_B.
From Example 7.6, we note that the critical buckling load is given by

$$P_{cr} = \frac{6.595EI}{L^2} = \frac{6.595(5000)}{(3.0)^2} = 3663.9 \text{ kN}$$

The applied axial load P = 1000 kN is clearly less than P_{cr}, and hence the system is capable of resisting the applied loading.

Stiffness Measures for Beam-Column Elements:

$$\mu = \sqrt{\frac{P}{EI}} = \sqrt{\frac{1200}{5000}} = 0.4472 \text{ m}^{-1} \quad \Rightarrow (\mu L)_{AB} = (0.4472)(6) = 2.6833; \quad (\mu L)_{BC} = (0.4472)(3) = 1.3416$$

(a) *P* is compressive:

$$S_{AB} = \frac{EI}{L}\left[\frac{\mu L(\sin\mu L - \mu L\cos\mu L)}{2 - 2\cos\mu L - \mu L\sin\mu L}\right] = \frac{EI}{6}[2.93279] = 0.4888EI$$

$$r_{AB} = \left[\frac{\mu L - \sin\mu L}{\sin\mu L - \mu L\cos\mu L}\right] = 0.78660 \quad \Rightarrow (rS)_{AB} = 0.3845EI$$

$$S^{o}_{BC} = \frac{EI}{L}\left[\frac{(\mu L)^2\tan\mu L}{\tan\mu L - \mu L}\right] = \frac{EI}{3}[2.61989] = 0.8733EI$$

(b) *P* is tensile:

$$S_{AB} = \frac{EI}{L}\left[\frac{\mu L(-\sinh\mu L + \mu L\cosh\mu L)}{2 - 2\cosh\mu L + \mu L\sinh\mu L}\right] = \frac{EI}{6}[4.88132] = 0.8136EI$$

$$r_{AB} = \left[\frac{-\mu L + \sinh\mu L}{-\sinh\mu L + \mu L\cosh\mu L}\right] = 0.36965 \quad \Rightarrow (rS)_{AB} = 0.3007EI$$

$$S^{o}_{BC} = \frac{EI}{L}\left[\frac{(\mu L)^2\tanh\mu L}{-\tanh\mu L + \mu L}\right] = \frac{EI}{3}[3.34283] = 1.1143EI$$

Fixed End Moments: (with $\theta_B = 0$): Using Eqns 7.61 and 7.61a,

Case (a): *P* is compressive

$$M^{F}_{AB} = -\frac{q_o L^2}{12}\left[\frac{3}{(\mu L/2)^2}\left(1 - \frac{\mu L}{2}\cot\frac{\mu L}{2}\right)\right] = -\frac{(30)(6^2)}{12}[1.14510] = -103.06 \text{ kNm}$$

$$M^{F}_{BA} = +103.06 \text{ kNm}$$

$$M^{Fo}_{BC} = -\frac{q_o L^2}{12}\left[\frac{3}{(\mu L/2)^2}\left(1 - \frac{\mu L}{2}\cot\frac{\mu L}{2}\right)\right]\left(1 + \frac{\mu L - \sin\mu L}{\sin\mu L - \mu L\cos\mu L}\right) = -\frac{(30)(3^2)}{12}[1.03134](1.54967)$$

$$= -35.96 \text{ kNm}$$

Case (b): *P* is tensile

$$M^{F}_{AB} = -\frac{q_o L^2}{12}\left[\frac{3}{(\mu L/2)^2}\left(-1 + \frac{\mu L}{2}\coth\frac{\mu L}{2}\right)\right] = -\frac{(30)(6^2)}{12}[0.89744] = -80.77 \text{ kNm}$$

$$M^{F}_{BA} = +80.77 \text{ kNm}$$

$$M_{BC}^{Fo} = -\frac{q_o L^2}{12}\left[\frac{3}{(\mu L/2)^2}\left(-1+\frac{\mu L}{2}\coth\frac{\mu L}{2}\right)\right]\left(1+\frac{\sinh\mu L - \mu L}{\mu L\cosh\mu L - \sinh\mu L}\right) = -\frac{(30)(3^2)}{12}[0.97123](1.45889)$$

$$= -31.88 \text{ kNm}$$

Slope-deflection Equations:

Beam AB ($L = 6\text{m}$; $\theta_A = 0$; $\theta_B = ?$; $\phi_{AB} = 0$)

$$M_{AB} = M_{AB}^F + (rS)_{AB}\,\theta_B = \begin{cases}(-103.06 + 0.3845 EI\theta_B)\text{ kNm (Case 'a')}\\(-80.77 + 0.3007 EI\theta_B)\text{ kNm (Case 'b')}\end{cases}\text{kNm}$$

$$M_{BA} = M_{BA}^F + S_{AB}\theta_B = \begin{cases}(+103.06 + 0.4888 EI\theta_B)\text{ kNm (Case 'a')}\\(+80.77 + 0.8136 EI\theta_B)\text{ kNm (Case 'b')}\end{cases}\text{kNm}$$

Beam BC ($L = 3\text{m}$; $\theta_B = ?$; θ_C not reqd ; $\phi_{BC} = 0$)

$$M_{BC} = M_{BC}^{Fo} + S_{BC}^{o}\theta_B = \begin{cases}(-35.96 + 0.8733 EI\theta_B)\text{ kNm (Case 'a')}\\(-31.88 + 1.1143 EI\theta_B)\text{ kNm (Case 'b')}\end{cases}\text{kNm}$$

Equilibrium Equation:

$$M_{BA} + M_{BC} = M_B = 0$$

Case (a): P is compressive

$$67.10 + 1.3621 EI\theta_B = 0 \quad \Rightarrow EI\theta_B = -49.2622 \text{ kNm}$$

Case (b): P is tensile

$$48.89 + 1.9279 EI\theta_B = 0 \quad \Rightarrow EI\theta_B = -25.3592 \text{ kNm}$$

Final End Moments:

Case (a): P is compressive

$$\begin{Bmatrix} M_{AB}\\ M_{BA}\\ M_{BC}\end{Bmatrix} = \begin{Bmatrix}-103.06\\+103.06\\-35.96\end{Bmatrix} + \begin{bmatrix}0.3845\\0.4888\\0.8733\end{bmatrix}\{EI\theta_B = -49.2622\} = \begin{Bmatrix}-122.00\\+78.98\\-78.98\end{Bmatrix}\text{kNm}$$

Case (b): P is tensile

$$\begin{Bmatrix} M_{AB}\\ M_{BA}\\ M_{BC}\end{Bmatrix} = \begin{Bmatrix}-80.77\\+80.77\\-31.88\end{Bmatrix} + \begin{bmatrix}0.3007\\0.8136\\1.1143\end{bmatrix}\{EI\theta_B = -25.3592\} = \begin{Bmatrix}-88.33\\+60.14\\-60.14\end{Bmatrix}\text{kNm}$$

The loading, free-body and bending moment diagrams (including second-order effects) for the two cases are shown in Fig. 7.31. The bending moments in the axial compression case are significantly more than the axial tension case, as expected. It may be noted that the bending moment at any intermediate location can be computed accurately, using the free-bodies, only if the *P-delta* moment is included, and this requires an estimation of the value of the deflection, which calls for additional computation. This can be done using the Principle of superposition, by using appropriate formulae, combining the deflection responses due to the applied loading and the end moments on a simply supported beam-column. Thus, for example, the bending moment at the midspan of AB in the axial compression case may be computed as follows:

$$\Delta_{mid} = \frac{5q_oL^4}{384EI}\left[\frac{12\left(2\sec(\mu L/2)-2-(\mu L/2)^2\right)}{5(\mu L/2)^4}\right] - \frac{\left(|M_{AB}|+|M_{BA}|\right)L^2}{16EI}\left[\frac{8}{(\mu L)^2}\left(\frac{1}{\cos(\mu L/2)}-1\right)\right]$$

$$= \frac{5(30)(6)^4}{384(5000)}[3.7072] - \frac{(97.17+82.83)(6)^2}{16(5000)}[3.7805] = 0.06914 \text{ m}$$

$$\Rightarrow M_{AB,\text{ midspan}} = (97.17\times 3)-\left(30\times 3^2/2\right)-122.00+(1000)(0.06914) = 103.53 \text{ kNm (Case 'a')}$$

Thus, it is seen that the axial compression enhances both the negative and positive moments considerably. (The absolute maximum sagging moment location is likely to be slightly away from the mid-span location). In the case of axial tension, the bending moments get reduced in second-order analysis, compared to first-order analysis (with $P = 0$).

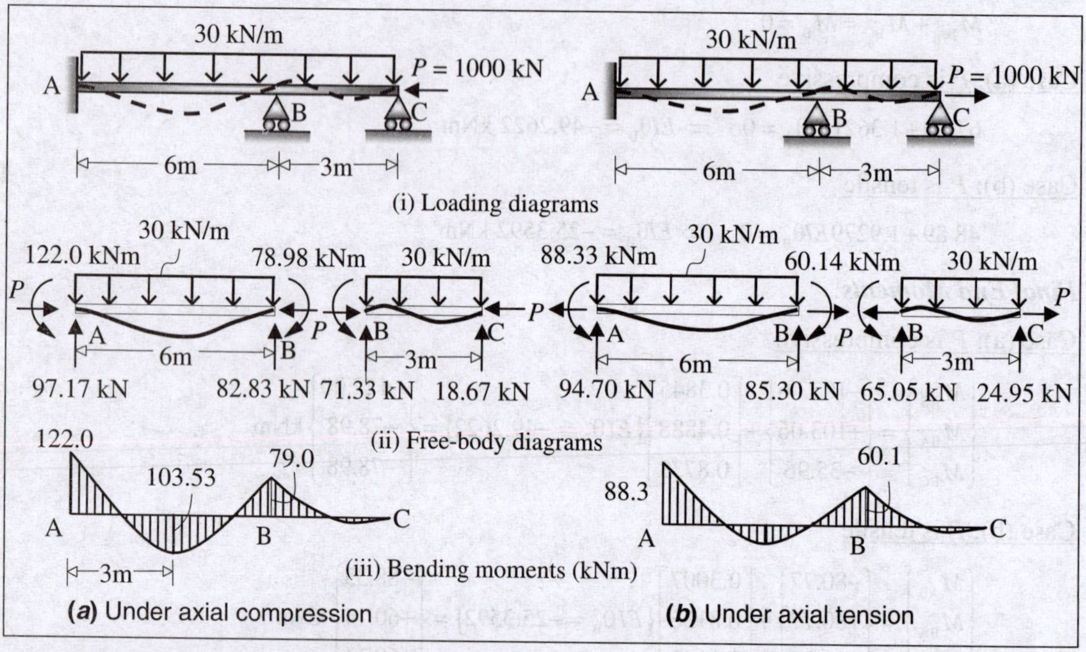

Figure 7.31 Loading, free-body and bending moment diagrams including second-order effects (Example 7.7)

EXAMPLE 7.8

Find the critical buckling load, W_{cr}, for the two-member frame ABC, loaded as shown in Fig. 7.32, with hinged supports at A and C, considering joint B to be (a) pinned and (b) rigid. Neglect axial deformations and assume both members to have the same flexural rigidity *EI*. Express W_{cr} in terms of *EI* and *L*.

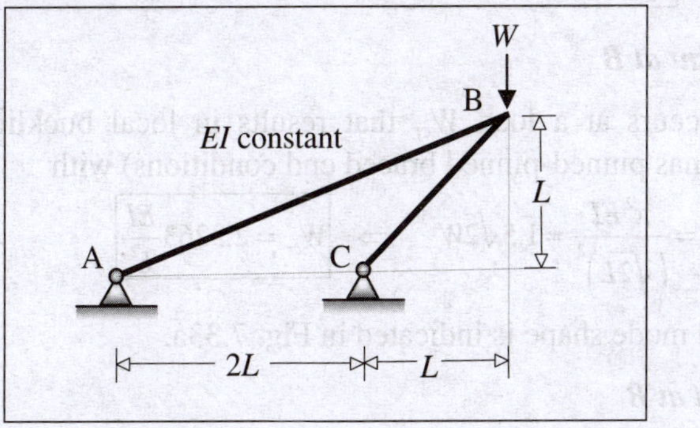

Figure 7.32 Two-member frame (Example 7.8)

SOLUTION

As the structure is *funicular* for the given loading, it will behave like a pin-jointed truss, with only axial forces in the two members and no bending (even if joint B is rigid, assuming axial deformations to be negligible). The bar forces (tension in BA and compression in BC) can be easily determined from equilibrium conditions (refer Fig. 7.33a).

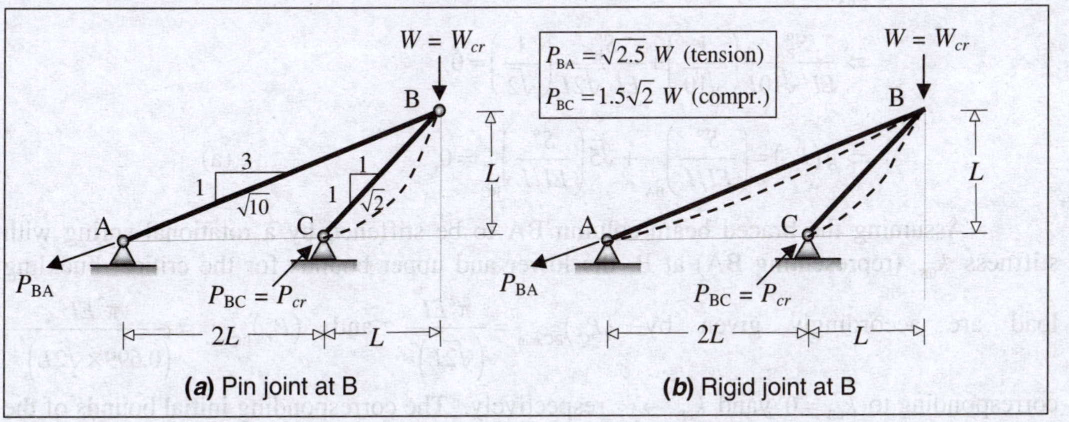

Figure 7.33 Buckling of a two-member frame (Example 7.8)

$$\sum F_x = 0 \quad \Rightarrow \quad \frac{3}{\sqrt{10}} P_{BA} = \frac{1}{\sqrt{2}} P_{BC} \quad \Rightarrow \quad P_{BC} = \frac{3}{\sqrt{5}} P_{BA}$$

$$\sum F_y = 0 \quad \Rightarrow \quad \frac{1}{\sqrt{10}} P_{BA} + W = \frac{1}{\sqrt{2}} P_{BC} = \frac{3}{\sqrt{10}} P_{BA} \quad \Rightarrow \quad P_{BA} = \frac{\sqrt{10}}{2} W$$

$$\Rightarrow \quad \boxed{P_{BA} = \sqrt{2.5}\ W \ \text{(tension)}} \quad \text{and} \quad \boxed{P_{BC} = 1.5\sqrt{2}\ W \ \text{(compression)}}$$

Case (a): Pinned joint at B

Elastic instability occurs at a load W_{cr} that results in local buckling in the compression member BC (which has pinned-pinned braced end conditions) with

$$\left(P_{BC}\right)_{cr} = P_{Eo} = \frac{\pi^2 EI}{\left(\sqrt{2}L\right)^2} = 1.5\sqrt{2}W_{cr} \quad \Rightarrow \quad \boxed{W_{cr} = 2.3263\frac{EI}{L^2}}$$

The buckled mode shape is indicated in Fig. 7.33a.

Case (b): Rigid joint at B

Here, the rigid joint at B provides some rotational restraint at one end of the compression member BC, i.e., the element AB serves to enhance the local buckling strength of the BC, compared to Case (a). The condition of moment equilibrium at the joint B in the buckled configuration (shown in Fig. 7.33b) is given by:

$$\sum M_B = 0 \quad \Rightarrow \quad M_{BA} + M_{BC} = 0 \quad \Rightarrow \quad \left(S^o_{BA} + S^o_{BC}\right)\theta_B = 0$$

For a non-trivial solution ($\theta_B \neq 0$), the *characteristic equation*, which renders the total rotational stiffness at the joint B equal to zero, is given by:

$$S^o_{BA} + S^o_{BC} = 0$$

$$\Rightarrow \frac{S^o_{BA}}{EI/\sqrt{10}L}\left(\frac{1}{\sqrt{10}}\right) + \frac{S^o_{BC}}{EI/\sqrt{2}L}\left(\frac{1}{\sqrt{2}}\right) = 0$$

$$\Rightarrow g\left(P_{cr}\right) = \left(\frac{S^o}{EI/L}\right)_{BA} + \sqrt{5}\left(\frac{S^o}{EI/L}\right)_{BC} = 0 \qquad \text{(a)}$$

Assuming the braced beam-column BA to be stiffened by a rotational spring with stiffness $k_{\theta B}$ (representing BA) at B, the lower and upper bounds for the critical buckling load are accordingly given by $\left(P_{cr}\right)_{BC,min} = \dfrac{\pi^2 EI}{\left(\sqrt{2}L\right)^2}$ and $\left(P_{cr}\right)_{BC,max} = \dfrac{\pi^2 EI}{\left(0.699\times\sqrt{2}L\right)^2}$, corresponding to $k_{\theta B} = 0$ yand $k_{\theta B} \to \infty$ respectively. The corresponding initial bounds of the parameter $\mu_{cr}L = \sqrt{\dfrac{P_{cr}L^2}{EI}} = \dfrac{\pi}{k_e}$ are accordingly given by $\dfrac{\pi}{1} < \left(\mu_{cr}L\right)_{BC} < \dfrac{\pi}{0.699}$, with

$\frac{(\mu_{cr}L)_{BA}}{(\mu_{cr}L)_{BC}} = \left(\sqrt{\frac{P_{BA}}{P_{BC}}}\right)\frac{L_{BA}}{L_{BC}} = \left(\sqrt{\frac{\sqrt{5}}{3}}\right)\frac{\sqrt{10}}{\sqrt{2}} = 1.9305$. Making use of the stiffness measures given in Eqns 7.46 (for BC in compression) and 7.46a (for BA in tension), and solving Eq. (a) iteratively by the *bisection method*, the results are tabulated below.

Trial	$(\mu_{cr}L)_{BC}$	$\left(\frac{S^o}{EI/L}\right)_{BC}$	$\left(\frac{S^o}{EI/L}\right)_{BA}$	$g(P_{cr}) = \left(\frac{S^o}{EI/L}\right)_{BA} + \sqrt{5}\left(\frac{S^o}{EI/L}\right)_{BC}$
	3.14	0.0050	4.5821	+4.593
	4.49	–1315.9	5.7747	–2937
1	3.815	–3.8485	5.1364	–3.442
2	3.478	–1.3522	4.8685	+1.845
3	3.646	–2.3719	5.0145	–0.289
4	3.562	–1.8210	4.9412	+0.869
5	3.604	–2.0848	4.9778	+0.316
6	3.625	–2.2252	4.9961	+0.020
7	3.627	–2.2389	4.9979	–0.008
8	**3.626**	–2.2320	4.9970	+0.006

$$\Rightarrow (\mu_{cr}L)_{BC} = 3.626 = \frac{\pi}{0.8664} \quad \text{and} \quad (\mu_{cr}L)_{BA} = 3.626 \times 1.9305 = \frac{\pi}{0.4488}$$

$$\Rightarrow (P_{BC})_{cr} = \frac{\pi^2 EI}{\left(0.8664 \times \sqrt{2}L\right)^2} = 1.5\sqrt{2}W_{cr} \Rightarrow \boxed{W_{cr} = 3.0990\frac{EI}{L^2}}$$

$$\text{Alternatively, } (P_{BA})_{cr} = \frac{\pi^2 EI}{\left(0.4488 \times \sqrt{10}L\right)^2} = \sqrt{2.5}W_{cr} \Rightarrow \boxed{W_{cr} = 3.0990\frac{EI}{L^2}}$$

Comparing this result with that of Case (a), we observe that by making the joint B rigid, the critical buckling load is enhanced by 33.2 percent.

EXAMPLE 7.9

Find the critical buckling load, P_{cr}, for the braced frame shown in Fig. 7.34. Express P_{cr} in terms of *EI* and *L*. Ignore axial deformations.

SOLUTION

The buckling of the frame will be governed by the buckling of the compression member AB, which will behave like a braced column with fixity at end A and a rotational spring at the end B.

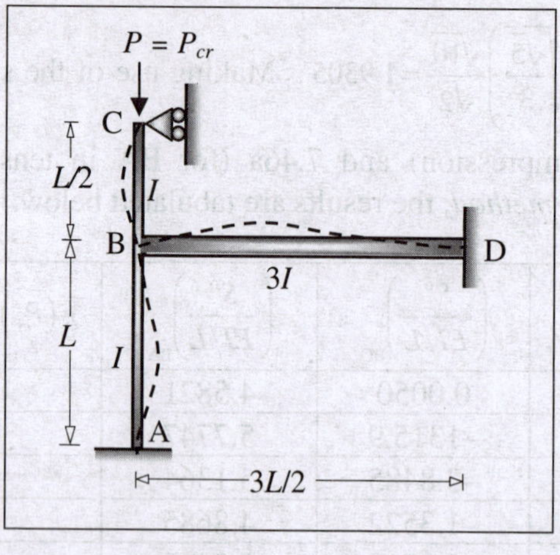

Figure 7.34 Braced frame (Example 7.9)

The condition of moment equilibrium at the joint B in the buckled configuration (shown in Fig. 7.34) is given by:

$$\sum M_{\text{B}}=0 \quad \Rightarrow \quad M_{\text{BA}}+M_{\text{BC}}+M_{\text{BD}}=0 \quad \Rightarrow \quad \left(S_{\text{BA}}+S_{\text{BC}}^{\text{o}}+\frac{4E(3I)}{(3/2)L}\right)\theta_{\mathbf{B}}=0$$

(neglecting the marginal effect of axial force in the beam BD due to the transmission of horizontal shear from AB and BC to BD).

For a non-trivial solution ($\theta_{\text{B}} \neq 0$), the *characteristic equation*, which renders the total rotational stiffness at the joint B equal to zero, is given by:

$$S_{\text{BA}}+S_{\text{BC}}^{\text{o}}+\frac{8EI}{L}=0$$

$$\Rightarrow \frac{S_{\text{BA}}}{EI/L}+\frac{S_{\text{BC}}^{\text{o}}}{EI/(L/2)}\left(\frac{1}{2}\right)+8=0$$

$$\Rightarrow g(P_{\text{cr}})=\left(\frac{S}{EI/L}\right)_{\text{BA}}+\frac{1}{2}\left(\frac{S^{\text{o}}}{EI/L}\right)_{\text{BC}}+8=0 \qquad \text{(a)}$$

Assuming the braced beam-column BA to be stiffened by a rotational spring with stiffness $k_{\theta\text{B}}$ at B, the lower and upper bounds for the critical buckling load are accordingly given by $(P_{cr})_{\text{BA,min}}=\dfrac{\pi^2 EI}{(0.699\times L)^2}$ and $(P_{cr})_{\text{BA,max}}=\dfrac{\pi^2 EI}{(0.5\times L)^2}$, corresponding to $k_{\theta\text{B}}=0$ and $k_{\theta\text{B}}\to\infty$ respectively. The corresponding initial bounds of the parameter $\mu_{\text{cr}}L=\sqrt{\dfrac{P_{cr}L^2}{EI}}=\dfrac{\pi}{k_e}$ are accordingly given by $\dfrac{\pi}{0.699}<(\mu_{\text{cr}}L)_{\text{BA}}<\dfrac{\pi}{0.5}$, with $\dfrac{(\mu_{\text{cr}}L)_{\text{BC}}}{(\mu_{\text{cr}}L)_{\text{BA}}}=\dfrac{L_{\text{BC}}}{L_{\text{BA}}}=\dfrac{L/2}{L}=0.5$. Making use of the stiffness measures given in Eqns 7.44 (for BA) and 7.46 (for BC), and solving Eq. (a) iteratively by the *bisection method*, the results are tabulated below.

Trial	$(\mu_{cr}L)_{BA}$	$\left(\frac{S}{EI/L}\right)_{BA}$	$\left(\frac{S^o}{EI/L}\right)_{BC}$	$g(P_{cr}) = \left(\frac{S}{EI/L}\right)_{BA} + \frac{1}{2}\left(\frac{S^o}{EI/L}\right)_{BC} + 8$
	4.49	0.0098	1.8039	+8.912
	6.28	−1971.6	0.0050	−1964
1	5.385	−4.4862	1.1006	+4.064
2	5.833	−12.118	0.6192	−3.809
3	5.609	−7.1284	0.8737	+1.308
4	5.721	−9.1567	0.7502	−0.782
5	5.665	−8.0574	0.8129	+0.349
6	5.693	−8.5827	0.7818	−0.192
7	5.686	−8.4471	0.7896	−0.052
8	**5.683**	−8.3899	0.7930	+0.007

$$\Rightarrow (\mu_{cr}L)_{BA} = 5.683 = \frac{\pi}{0.5528} \quad \text{and} \quad (\mu_{cr}L)_{BC} = 5.683 \times 0.5 = \frac{\pi}{1.1056}$$

$$\Rightarrow (P_{BA})_{cr} = \frac{\pi^2 EI}{(0.5528 \times L)^2} = P_{cr} \quad \Rightarrow \quad \boxed{P_{cr} = 32.297\frac{EI}{L^2}}$$

$$\text{Alternatively, } (P_{BC})_{cr} = \frac{\pi^2 EI}{(1.1056 \times 0.5L)^2} = P_{cr} \Rightarrow \boxed{P_{cr} = 32.297\frac{EI}{L^2}}$$

EXAMPLE 7.10

Find the critical buckling load, P_{cr}, for the unbraced frame shown in Fig. 7.35 (where the support at D is fixed against rotation and vertical deflection but free to translate horizontally). Assume all members to have uniform flexural rigidity *EI*. Express P_{cr} in terms of *EI* and *L*. Ignore axial deformations.

SOLUTION

The buckled configuration of the unbraced frame, as indicated in Fig. 7.35, has two degrees of freedom, θ_B and Δ. As the elements BC and CD are free to translate horizontally, under the given loading, there will be no axial forces in these elements, which will behave as ordinary beam elements. The buckling of the frame will be governed by the buckling of the compression member AB, which will behave like an unbraced column with fixity at end A and a rotational spring at the end B.

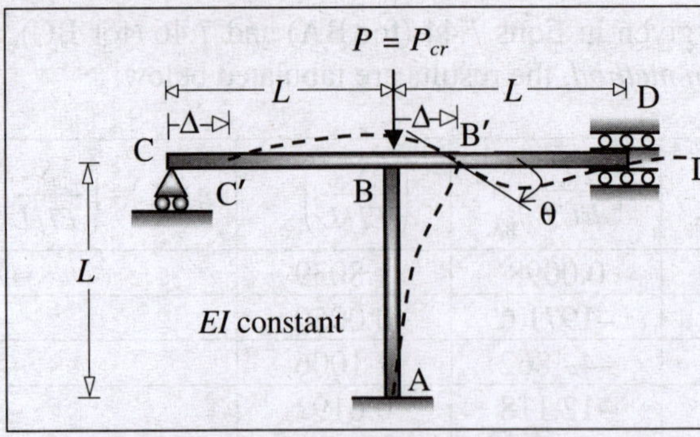

Figure 7.35 Unbraced frame (Example 7.10)

The conditions of moment equilibrium and horizontal force equilibrium (shear in AB) at the joint B in the buckled configuration (shown in Fig. 7.35) are given by:

$$\sum M_B = 0 \quad \Rightarrow \quad M_{BA} + M_{BC} + M_{BD} = 0$$

$$\sum S_{AB} = 0 \quad \Rightarrow \quad \frac{M_{BA} + M_{AB} + P\Delta}{L} = 0 \quad \Rightarrow M_{BA} + M_{AB} + \frac{EI}{L}(\mu L)^2_{AB}\frac{\Delta}{L} = 0 \qquad \text{(a)}$$

The end moments shown in Eq. (a) can be expressed in terms of the unknown displacements, θ_B and Δ, using slope-deflection equations for the beam-column AB, with zero fixed-end moments, $\theta_A = 0$ and $\phi_{AB} = \Delta/L$ (refer Eq. 7.50), and for beam elements BC and BD are as follows:

$$\begin{Bmatrix} M_{AB} \\ M_{BA} \\ M_{BC} \\ M_{BD} \end{Bmatrix} = \begin{bmatrix} (rS)_{AB} & -S_{AB}(1+r_{AB}) \\ S_{AB} & -S_{AB}(1+r_{AB}) \\ 3EI/L & 0 \\ 4EI/L & 0 \end{bmatrix} \begin{Bmatrix} \theta_B \\ \Delta/L \end{Bmatrix} \qquad \text{(b)}$$

Substituting Eq. (b) in (a), and simplifying to generate a symmetric coefficient matrix (which, in fact, is the stiffness matrix),

$$\begin{bmatrix} S_{AB} + \dfrac{7EI}{L} & -S_{AB}(1+r_{AB}) \\ -S_{AB}(1+r_{AB}) & 2S_{AB}(1+r_{AB}) - \dfrac{EI}{L}(\mu L)^2_{AB} \end{bmatrix} \begin{Bmatrix} \theta_B \\ \Delta/L \end{Bmatrix} = \begin{Bmatrix} 0 \\ 0 \end{Bmatrix}$$

$$\Rightarrow \left(\frac{EI}{L}\right) \begin{bmatrix} \left[\left(\dfrac{S}{EI/L}\right)_{AB} + 7\right] & \left[-\left(\dfrac{S}{EI/L}\right)_{AB}(1+r_{AB})\right] \\ \left[-\left(\dfrac{S}{EI/L}\right)_{AB}(1+r_{AB})\right] & \left[2\left(\dfrac{S}{EI/L}\right)_{AB}(1+r_{AB}) - (\mu L)^2_{AB}\right] \end{bmatrix} \begin{Bmatrix} \theta_B \\ \Delta/L \end{Bmatrix} = \begin{Bmatrix} 0 \\ 0 \end{Bmatrix} \qquad \text{(c)}$$

The above formulation can be shown to conform to an eigenvalue problem formulation (refer Eq. 2.50), in which the coefficients in the stiffness matrix are functions of the parameter, P (expressed in terms of μL). For a non-trivial solution, the determinant of the stiffness matrix must vanish (implying instability in the system). This condition yields the *characteristic equation*, whose roots provide the eigenvalues (critical buckling load), of which the lowest value (P_{cr}) is of practical interest. Thus, setting the determinant of the coefficient marix in Eq. (c) to zero, and simplifying, the characteristic equation reduces to:

$$g(P_{cr}) = \left\{\left(\frac{S}{EI/L}\right)_{AB} + 7\right\}\left[2\left(\frac{S}{EI/L}\right)_{AB}(1+r_{AB}) - (\mu_{cr}L)^2_{AB}\right] - \left[\left\{\left(\frac{S}{EI/L}\right)_{AB}(1+r_{AB})\right\}^2\right] = 0 \quad \text{(d)}$$

Assuming the cantilever beam-column AB to be stiffened by a rotational spring with stiffness $k_{\theta B}$ at B, the lower and upper bounds for the critical buckling load are accordingly given by $(P_{cr})_{min} = \frac{\pi^2 EI}{(2\times L)^2}$ and $(P_{cr})_{max} = \frac{\pi^2 EI}{(1\times L)^2}$, corresponding to $k_{\theta B} = 0$ and $k_{\theta B} \to \infty$ respectively. The corresponding initial bounds of the parameter $\mu_{cr}L = \sqrt{\frac{P_{cr}L^2}{EI}} = \frac{\pi}{k_e}$ are accordingly given by $0.5\pi < \mu_{cr}L < \pi$. Making use of the stiffness measures given in Eqns 7.44 and 7.45, and solving Eq. (d) iteratively by the *bisection method*, the results are tabulated below.

Trial	$(\mu_{cr}L)_{AB}$	$\left(\frac{S}{EI/L}\right)_{AB}$	$\left(\frac{r}{EI/L}\right)_{AB}$	$g(P_{cr})$
	1.57	3.6601	0.5707	+63.24
	3.14	2.4692	0.9990	–24.25
1	2.355	3.1995	0.6946	+24.64
2	2.747	2.8750	0.8090	+1.152
3	2.551	3.0463	0.7453	+13.18
4	2.845	2.7821	0.8472	–5.044
5	2.698	2.9196	0.7916	+4.298
6	2.771	2.8527	0.8179	–0.355
7	2.766	2.8574	0.8160	–0.040
8	**2.765**	2.8584	0.8157	+0.023

$$\Rightarrow (\mu_{cr}L)_{AB} = 2.765 = \frac{\pi}{1.1362}$$

$$\Rightarrow (P_{AB})_{cr} = \frac{\pi^2 EI}{(1.1362\times L)^2} = P_{cr} \quad \Rightarrow \quad \boxed{P_{cr} = 7.645\frac{EI}{L^2}}$$

Alternative Solution

Observing the cantilever behaviour of the element AB, it is possible to take advantage of the modified stiffness measure given by Eq. 7.54, and thereby reduce the degree of kinematic indeterminacy from two to one (i.e., treat only θ_B as the unknown). Accordingly,

$$M_{\text{BA}} = \frac{EI}{L}\left(\frac{\mu L}{\tan \mu L}\right)$$

$$\sum M_{\text{B}} = 0 \quad \Rightarrow \quad M_{\text{BA}} + M_{\text{BC}} + M_{\text{BD}} = 0$$

$$\Rightarrow \quad \frac{EI}{L}\left(\frac{(\mu L)_{\text{AB}}}{\tan(\mu L)_{\text{AB}}} + 3 + 4\right)(\theta_{\text{B}}) = 0$$

For a non-trivial solution ($\theta_{\text{B}} \neq 0$), the *characteristic equation* reduces to:

$$\boxed{(\mu L)_{\text{AB}} + 7\tan(\mu L)_{\text{AB}} = 0} \qquad \text{(e)}$$

Clearly, this is a much simpler equation to solve than Eq.(d). Substituting the value, $(\mu_{\text{cr}} L)_{\text{AB}} = 2.765$, obtained earlier in Eq.(e), we observe that the solution is correct. A more accurate solution can be generated as $(\mu_{\text{cr}} L)_{\text{AB}} = 2.76536$, whereby $\boxed{P_{cr} = 7.6472\frac{EI}{L^2}}$.

EXAMPLE 7.11

Consider the same frame as in Example 7.10, but with the single point load at B, replaced by two point loads, each of magnitude, $P = \frac{4EI}{L^2}$, applied at the mid-points of beams BC and BD, as shown in Fig. 7.36. Analyse the structure (including second-order effects) and compare the bending moments obtained from second-order analysis with the first-order analysis.

SOLUTION

Taking advantage of the cantilever behaviour of the beam-column AB, we can limit the degree of kinematic indeterminacy, $n_k = 1$. The unknown displacement is θ_{B} †.

From Example 7.10, we note that elastic instability of the frame occurs through buckling of the beam-column AB at a critical buckling load, $P_{cr} = 7.6472\frac{EI}{L^2}$. The axial compression in AB due to the loading shown in Fig. 7.36 is clearly less than this (approximately $P = \frac{4EI}{L^2}$), and hence the frame is capable of resisting the applied loading. The actual value of P_{cr} can be estimated from a detailed analysis of the frame using the Slope deflection method. Referring to the free-bodies in Fig. 7.37, and considering vertical force equilibrium at the joint B, $P_{\text{AB}} = V_{\text{BC}} + V_{\text{BD}}$. For an exact solution, we need to consider the slope deflection equations and solve for the unknown rotation θ_{B}.

† The direction of θ_{B} is assumed to be clockwise positive, as per the convention of the Slope deflection method. For the given loading, it is likely to be anti-clockwise; i.e., the frame is likely to sway to the left, and not to the right (as assumed in the deflected shape shown in Fig. 7.36).

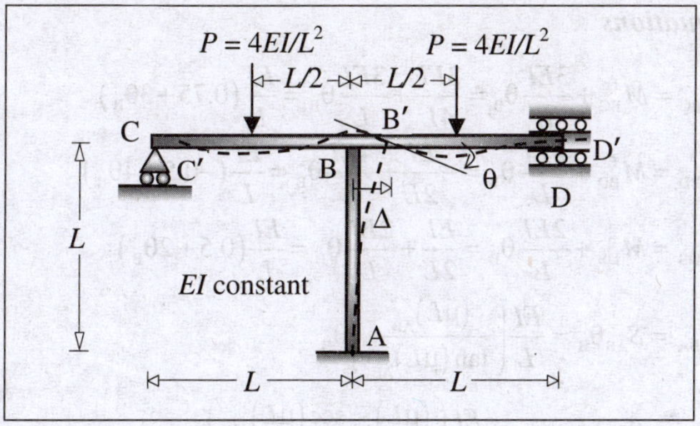

Figure 7.36 Second-order analysis of unbraced frame (Example 7.11)

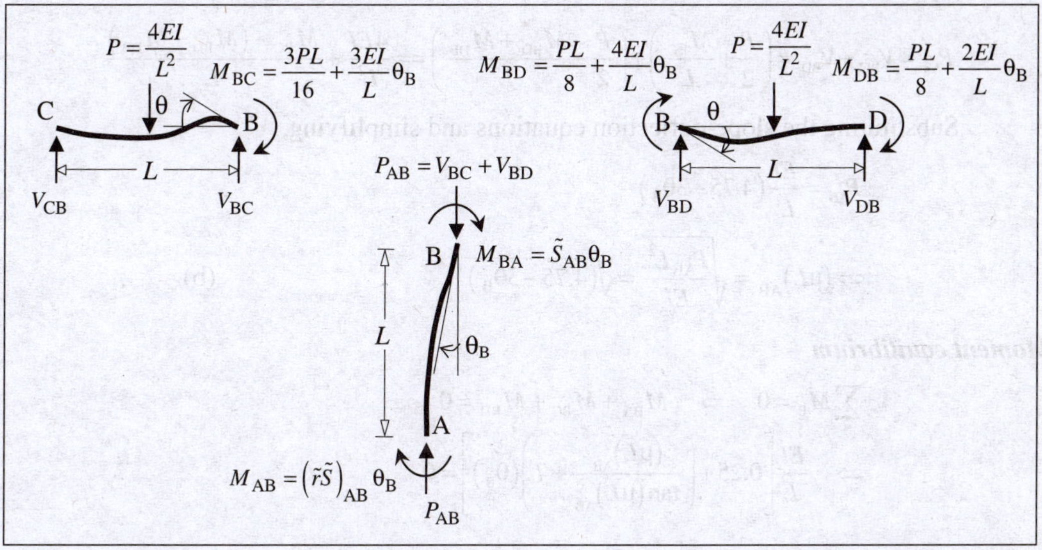

Figure 7.37 Free-bodies of elements in unbraced frame (Example 7.11)

Fixed end moments

$$M^{Fo}_{BC} = \frac{PL}{8} \times \frac{3}{2} = \frac{4EI}{L^2} \times \frac{3L}{16} = \frac{3}{4}\left(\frac{EI}{L}\right)$$

$$M^{F}_{BD} = -\frac{PL}{8} = -\frac{4EI}{L^2} \times \frac{L}{8} = -\frac{1}{2}\left(\frac{EI}{L}\right)$$

$$\Rightarrow M^{F}_{DB} = +\frac{PL}{8} = +\frac{1}{2}\left(\frac{EI}{L}\right)$$

Slope deflection equations

$$M_{BC} = M_{BC}^{Fo} + \frac{3EI}{L}\theta_B = \frac{3EI}{4L} + \frac{3EI}{L}\theta_B = \frac{EI}{L}(0.75 + 3\theta_B)$$

$$M_{BD} = M_{BD}^{F} + \frac{4EI}{L}\theta_B = -\frac{EI}{2L} + \frac{4EI}{L}\theta_B = \frac{EI}{L}(-0.5 + 4\theta_B)$$

$$M_{DB} = M_{DB}^{F} + \frac{2EI}{L}\theta_B = \frac{EI}{2L} + \frac{2EI}{L}\theta_B = \frac{EI}{L}(0.5 + 2\theta_B)$$

$$M_{BA} = \tilde{S}_{AB}\theta_B = \frac{EI}{L}\left(\frac{(\mu L)_{AB}}{\tan(\mu L)_{AB}}\right)\theta_B \tag{a}$$

$$M_{AB} = \left(\tilde{r}\tilde{S}\right)_{AB}\theta_B = -\frac{EI}{L}\left(\frac{(\mu L)_{AB}\sec(\mu L)_{AB}}{\tan(\mu L)_{AB}}\right)\theta_B$$

Axial compression in beam-column

$$P_{AB} = V_{BC} + V_{BD} = \left(\frac{P}{2} + \frac{M_{BC}}{L}\right) + \left(\frac{P}{2} - \frac{M_{BD} + M_{DB}}{L}\right) = \frac{4EI}{L^2} + \frac{M_{BC} - (M_{BD} + M_{DB})}{L}$$

Substituting the slope-deflection equations and simplifying,

$$P_{AB} = \frac{EI}{L^2}(4.75 - 3\theta_B)$$

$$\Rightarrow (\mu L)_{AB} = \sqrt{\frac{P_{AB}L^2}{EI}} = \sqrt{(4.75 - 3\theta_B)} \tag{b}$$

Moment equilibrium

$$\sum M_B = 0 \quad \Rightarrow \quad M_{BA} + M_{BC} + M_{BD} = 0$$

$$\Rightarrow \quad \frac{EI}{L}\left[0.25 + \left(\frac{(\mu L)_{AB}}{\tan(\mu L)_{AB}} + 7\right)(\theta_B)\right] = 0$$

$$\Rightarrow \quad \theta_B = -0.25\left[\frac{(\mu L)_{AB}}{\tan(\mu L)_{AB}} + 7\right]^{-1} \tag{c}$$

First-order Analysis

In first-order analysis, we ignore the change in stiffness due to the axial compression, whereby $\tilde{S}_{AB} = \frac{EI}{L}$ and $\left(\tilde{r}\tilde{S}\right)_{AB} = -\frac{EI}{L}$, and hence Eq.(a) reduces to:

$$\Rightarrow M_{BA} = \frac{EI}{L}\theta_B \tag{d}$$

$$M_{AB} = -\frac{EI}{L}\theta_B$$

Thus, Eq. (c) reduces to:

$$\theta_B = -0.25[1+7]^{-1} = -\frac{1}{32} = -0.03125 \text{ rad.}$$

Substituting this value in the slope-deflection equations, and considering $\frac{EI}{L} = \frac{PL}{4}$,

$$M_{BA} = \frac{EI}{L}\theta_B = -0.03125\frac{EI}{L} = -0.00781(PL)$$

$$M_{AB} = -\frac{EI}{L}\theta_B = +0.03125\frac{EI}{L} = +0.00781(PL)$$

$$M_{BC} = \frac{EI}{L}(0.75+3\theta_B) = +0.65625\frac{EI}{L} = +0.16406(PL)$$

$$M_{BD} = \frac{EI}{L}(-0.5+4\theta_B) = -0.62500\frac{EI}{L} = -0.15625(PL)$$

$$M_{DB} = \frac{EI}{L}(0.5+2\theta_B) = 0.43750\frac{EI}{L} = 0.109375(PL)$$

The resulting bending moment diagram is shown in Fig. 7.38a.

Second-order Analysis

For an 'exact' solution, we need to solve Eqns (b) and (c) iteratively till the results converge. We can get a reasonably accurate solution even if we limit the iteration to a single step, using the value of axial compression obtainable from the results of first-order analysis ($\theta_B = -\frac{1}{32}$).

$$\Rightarrow (\mu L)_{AB} = \sqrt{(4.75 - 3\theta_B)} = 2.200852$$

$$\Rightarrow \quad \theta_B = -0.25\left[\frac{(\mu L)_{AB}}{\tan(\mu L)_{AB}} + 7\right]^{-1} = -0.046338$$

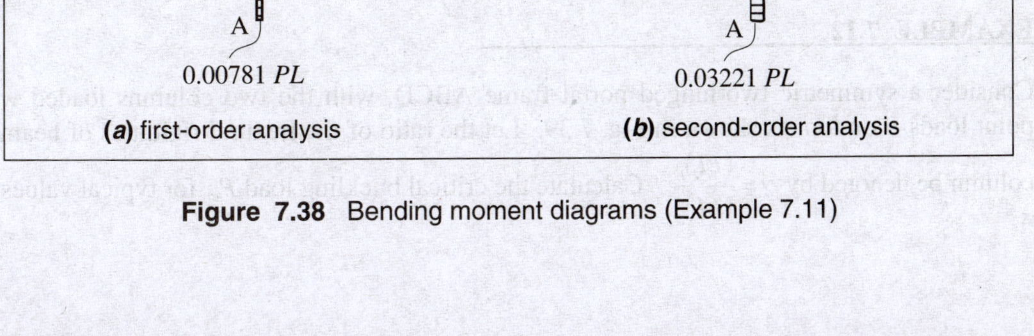

(*a*) first-order analysis (*b*) second-order analysis

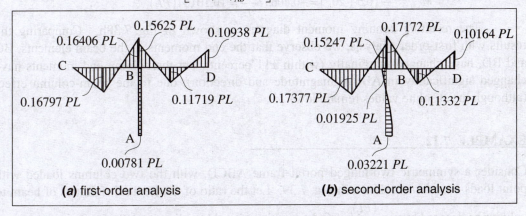

Figure 7.38 Bending moment diagrams (Example 7.11)

Second trial:

$$(\mu L)_{AB} = \sqrt{(4.75 - 3\theta_B)} = 2.211111$$

$$\Rightarrow \ \theta_B = -0.25\left[\frac{(\mu L)_{AB}}{\tan(\mu L)_{AB}} + 7\right]^{-1} = -0.046706$$

Third trial:

$$(\mu L)_{AB} = \sqrt{(4.75 - 3\theta_B)} = 2.211361$$

$$\Rightarrow \ \theta_B = -0.25\left[\frac{(\mu L)_{AB}}{\tan(\mu L)_{AB}} + 7\right]^{-1} = -0.046715$$

(converged to sufficient accuracy).

$$\Rightarrow \tilde{S}_{AB} = \frac{EI}{L}\left(\frac{(\mu L)_{AB}}{\tan(\mu L)_{AB}}\right) = -1.64842\frac{EI}{L} \quad \text{and} \quad \tilde{r}_{AB} = -\sec(\mu L)_{AB} = 1.67321$$

Substituting in the slope-deflection equations, and considering $\frac{EI}{L} = \frac{PL}{4}$,

$$M_{BA} = \tilde{S}_{AB}\theta_B = +0.07701\frac{EI}{L} = +0.01925(PL)$$

$$M_{AB} = \left(\tilde{r}\tilde{S}\right)_{AB}\theta_B = +0.12885\frac{EI}{L} = +0.03221(PL)$$

$$M_{BC} = \frac{EI}{L}(0.75 + 3\theta_B) = +0.60985\frac{EI}{L} = +0.15246(PL)$$

$$M_{BD} = \frac{EI}{L}(-0.5 + 4\theta_B) = -0.68686\frac{EI}{L} = -0.17172(PL)$$

$$M_{DB} = \frac{EI}{L}(0.5 + 2\theta_B) = +0.40657\frac{EI}{L} = +0.10164(PL)$$

The resulting bending moment diagram is shown in Fig. 7.38b. Comparing the results with first-order analysis, we observe that the end moments in the beam elements, BC and BD, have changed marginally (within ±11 percent), but the column end moments have changed significantly in AB (in magnitude and direction), due to the beam-column effect (although the absolute values remain low).

EXAMPLE 7.12

Consider a symmetric two-hinged portal frame ABCD, with the two columns loaded with point loads P on top, as shown in Fig. 7.39. Let the ratio of the flexural rigidities of beam to column be denoted by $\gamma = \frac{(EI)_b}{(EI)_c}$. Calculate the critical buckling load P_{cr} for typical values of

γ (equal to 0.5, 1, 5 and 10), considering the frame to be (a) braced and (b) unbraced against sidesway. Hence, plot the variations of P_{cr} (normalised with respect to $P_{Eo} = \frac{\pi^2 EI}{L^2}$) with γ.

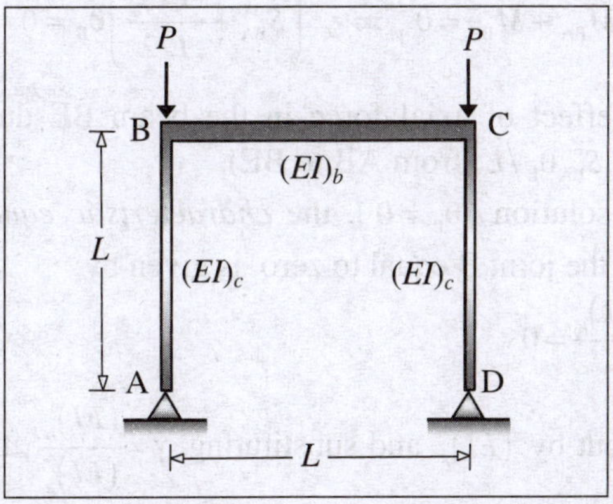

Figure 7.39 Buckling of symmetric portal frame (Example 7.12)

SOLUTION

Case (a): Frame braced against sidesway

In this case, the buckled mode shape of the frame is expected to be symmetric, as shown in Fig. 7.40a. Hence, it is sufficient to consider one-half of the frame, and providing a guided-fixed support at the truncated location (mid-point E of BC), as shown in Fig. 7.40b. The buckling of the frame ABE will be governed by the buckling of the compression member AB, which will behave like a braced column with a hinged support at end A and a rotational spring at the end B, as shown in Fig. 7.40c.

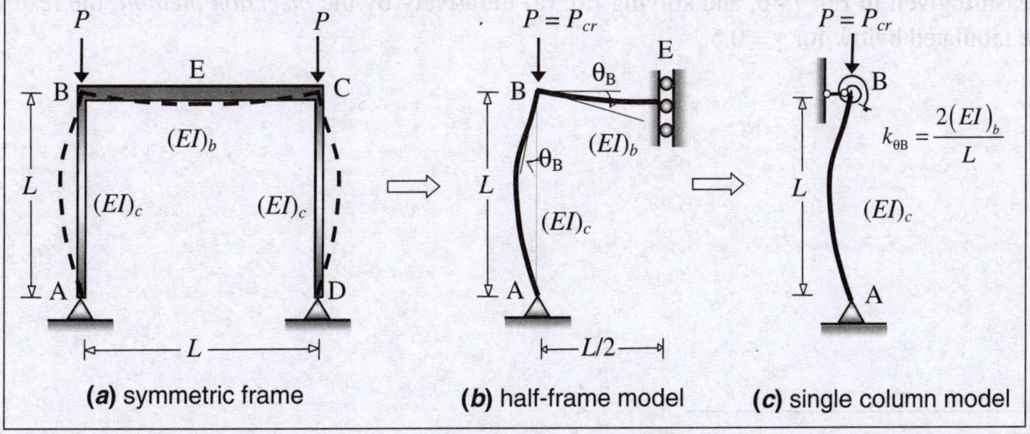

Figure 7.40 Modelling of symmetric portal frame braced against sidesway (Example 7.12)

There is only one degree of freedom involved here, viz., the rotation θ_B. The condition of moment equilibrium at the joint B in the buckled configuration (shown in Fig. 7.40b) is given by:

$$\sum M_B = 0 \quad \Rightarrow \quad M_{BA} + M_{BE} = 0 \quad \Rightarrow \quad \left(S^o_{BA} + \frac{(EI)_b}{L/2} \right) \theta_B = 0$$

(neglecting the marginal effect of axial force in the beam BE due to the transmission of horizontal shear[†], equal to $S^o_{BA}\theta_B / L$, from AB to BE).

For a non-trivial solution ($\theta_B \neq 0$), the *characteristic equation*, which renders the total rotational stiffness at the joint B equal to zero, is given by:

$$S^o_{BA} + \frac{2(EI)_b}{L} = 0$$

Dividing throughout by $(EI)_c$ and substituting $\gamma = \frac{(EI)_b}{(EI)_c}$, the required condition for elastic instability is given by:

$$g(P_{cr}) = \left(\frac{S^o}{EI/L} \right)_{BA} + 2\gamma = 0 \qquad \text{(a)}$$

The lower and upper bounds for the critical buckling load are given by $P_{cr} = \frac{\pi^2 EI}{L^2} = P_{Eo}$ and $P_{cr} = \frac{\pi^2 EI}{(0.699 \times L)^2} = 2.047 P_{Eo}$, corresponding to $k_{\theta B} = 0 \ (\Rightarrow \gamma = 0)$ and $k_{\theta B} \to \infty \ (\Rightarrow \gamma \to \infty)$ respectively. The corresponding initial bounds of the parameter $\mu_{cr} L = \sqrt{\frac{P_{cr} L^2}{EI}} = \frac{\pi}{k_e}$ are accordingly given by $\frac{\pi}{1} < (\mu_{cr} L)_{BA} < \frac{\pi}{0.699}$. Making use of the stiffness measure given in Eq. 7.46, and solving Eq. (a) iteratively by the *bisection method*, the results are tabulated below for $\gamma = 0.5$.

[†] The corresponding axial force in BE may be compressive or tensile, depending on whether θ_B is clockwise or anti-clockwise (of which the one causing a lower critical load is more likely). The magnitude of this force is very small in any case, for low values of θ_B, and hence can be neglected.

$\gamma = 0.5$			
Trial	$(\mu_{cr}L)_{\text{BA}}$	$\left(\dfrac{S^{o}}{EI/L}\right)_{\text{BA}}$	$g(P_{\text{cr}}) = \left(\dfrac{S^{o}}{EI/L}\right)_{\text{BA}} + 1$
	3.14	0.0050	+1.005
	4.49	−1316	−1315
1	3.815	−3.8485	−2.848
2	3.477	−1.3471	−0.347
3	3.308	−0.5853	+0.415
4	3.392	−0.9383	+0.062
5	3.434	−1.133	−0.133
6	3.413	−1.034	−0.034
7	3.402	−0.9835	+0.016
8	**3.406**	−1.002	−0.002

$$\gamma = 0.5 \;\Rightarrow (\mu_{\text{cr}}L)_{\text{BA}} = 3.406 = \frac{\pi}{0.9224}$$

$$\Rightarrow (P_{\text{BA}})_{cr} = P_{cr} = \frac{\pi^2 EI}{(0.9224L)^2} = \frac{P_{Eo}}{(0.9224)^2} = 1.175 P_{Eo}$$

Similarly, this procedure can be repeated for other values of γ (iteration tables not shown here). The results for $\gamma = 1$, $\gamma = 5$ and $\gamma = 10$ are as follows:

$$\gamma = 1 \;\Rightarrow (\mu_{\text{cr}}L)_{\text{BA}} = 3.591 = \frac{\pi}{0.8749} \quad\Rightarrow\quad P_{cr} = \frac{P_{Eo}}{0.8749^2} = 1.307 P_{Eo}$$

$$\gamma = 5 \;\Rightarrow (\mu_{\text{cr}}L)_{\text{BA}} = 4.132 = \frac{\pi}{0.7603} \quad\Rightarrow\quad P_{cr} = \frac{P_{Eo}}{0.7603^2} = 1.730 P_{Eo}$$

$$\gamma = 10 \;\Rightarrow (\mu_{\text{cr}}L)_{\text{BA}} = 4.2915 = \frac{\pi}{0.7321} \quad\Rightarrow\quad P_{cr} = \frac{P_{Eo}}{0.7321^2} = 1.866 P_{Eo}$$

Case (b): Frame unbraced against sidesway

In this case, the buckled mode shape of the frame is expected to be anti-symmetric, as shown in Fig. 7.41a (with no deflection and a point of contraflexure at E). Hence, it is sufficient to consider one-half of the frame, by providing a roller support at the truncated location (mid-point E of BC), as shown in Fig. 7.41b. The buckling of the frame ABE will be governed by the buckling of the compression member AB, which will behave like an unbraced column with a hinged support at end A and a rotational spring at the end B (which is free to deflect laterally), as shown in Fig. 7.41c. The axial force in AB may be taken as P, although in the buckled configuration, it will change[†] marginally by $P\Delta/L$.

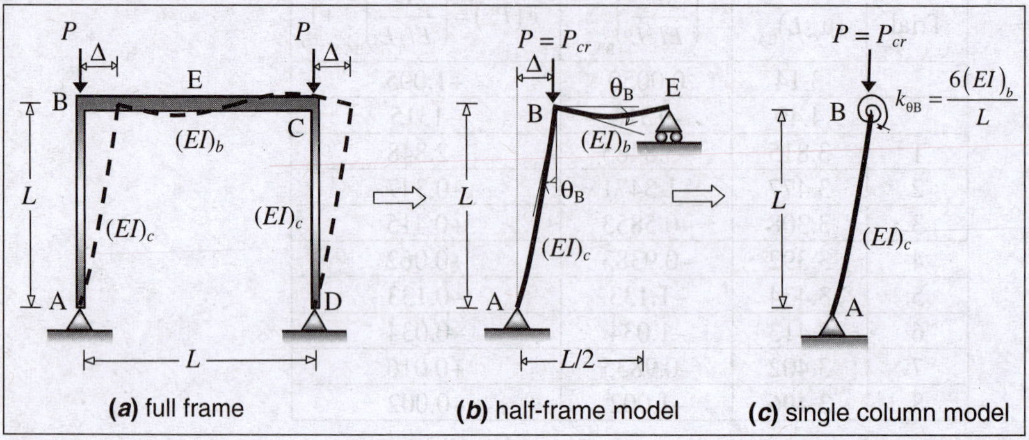

Figure 7.41 Modelling of unbraced portal frame, buckling in sway mode (Example 7.12)

The degree of kinematic indeterminacy is equal to two, considering the rotation θ_B and the lateral drift Δ as the unknown displacements. The slope-deflection equations in the buckled configuration (shown in Fig. 7.41b) are given by:

$$M_{BA} = \left(S^{o}_{BA}\right)\left(\theta_B - \frac{\Delta}{L}\right) = S^{o}_{BA}\theta_B - S^{o}_{BA}\frac{\Delta}{L}$$

$$M_{BE} = \left(\frac{3(EI)_b}{L/2}\right)\theta_B = \frac{6(EI)_b}{L}\theta_B$$

The equilibrium conditions correspond to (i) moment equilibrium at the joint B and (ii) zero horizontal force at the support A:

$$\sum M_B = 0 \;\Rightarrow\; M_{BA} + M_{BE} = 0 \;\Rightarrow\; \left[S^{o}_{BA} + \frac{6(EI)_b}{L}\right]\theta_B + \left(S^{o}_{BA}\right)\frac{\Delta}{L} = 0$$

$$H_A = 0 \;\Rightarrow\; \frac{M_{BA} + P\Delta}{L} = 0 \;\Rightarrow\; \left(S^{o}_{BA}\right)\theta_B - \left(S^{o}_{BA} - PL\right)\frac{\Delta}{L} = 0$$

Dividing both equations throughout by $\frac{(EI)_c}{L}$ and substituting $\gamma = \frac{(EI)_b}{(EI)_c}$ and $\frac{PL}{(EI)_c/L} = (\mu L)^2_{BA}$, we obtain the following matrix form:

[†] In the deformed configuration shown in Fig. 7.41a, the axial forces P in AB and CD will reduce and increase respectively by $P\Delta/L$, to satisfy moment equilibrium. This quantity is considered very small in comparison with P_{cr} and hence is ignored, for convenience in calculations.

$$\begin{bmatrix} \left[\left(\dfrac{S^{o}}{(EI)_c/L}\right)_{BA} + 6\gamma\right] & -\left(\dfrac{S^{o}}{(EI)_c/L}\right)_{BA} \\ -\left(\dfrac{S^{o}}{(EI)_c/L}\right)_{BA} & \left[\left(\dfrac{S^{o}}{(EI)_c/L}\right)_{BA} - (\mu L)^2_{BA}\right] \end{bmatrix} \begin{Bmatrix} \theta_B \\ \dfrac{\Delta}{L} \end{Bmatrix} = 0$$

where the coefficient matrix represents the stiffness matrix of the frame.

For a non-trivial solution, the *characteristic equation*, which renders the stiffness matrix equal to zero at $P = P_{cr}$ is obtained by setting the determinant of the matrix equal to zero. On simplification, the required condition for elastic instability is given by:

$$g(P_{cr}) = \left(\frac{S^{o}}{EI/L}\right)_{BA}\left[6\gamma - (\mu_{cr}L)^2_{BA}\right] - 6\gamma(\mu_{cr}L)^2_{BA} = 0 \qquad \text{(a)}$$

The lower and upper bounds for the critical buckling load are given by $P_{cr} = 0$ and $P_{cr} = \dfrac{\pi^2 EI}{(2L)^2} = 0.25P_{Eo}$, corresponding to $k_{\theta B} = 0$ $(\Rightarrow \gamma = 0)$ and $k_{\theta B} \to \infty$ $(\Rightarrow \gamma \to \infty)$ respectively (refer Fig. 7.4). The corresponding initial bounds of the parameter $\mu_{cr}L = \sqrt{\dfrac{P_{cr}L^2}{EI}} = \dfrac{\pi}{k_e}$ are accordingly given by $0 < (\mu_{cr}L)_{BA} < \dfrac{\pi}{2}$. Making use of the stiffness measure given in Eq. 7.46, and solving Eq. (a) iteratively by the *bisection method*, the results are tabulated below for $\gamma = 0.5$.

$\gamma = 0.5$			
Trial	$(\mu_{cr}L)_{BA}$	$\left(\dfrac{S^{o}}{EI/L}\right)_{BA}$	$g(P_{cr}) = \left(\dfrac{S^{o}}{EI/L}\right)_{BA} + 1$
	0	3	+9
	1.57	+2.4680	–6.074
1	0.785	+2.8745	+5.003
2	1.178	+2.7107	+0.208
3	1.374	+2.6002	–2.772
4	1.276	+2.6850	–1.238
5	1.202	+2.6983	–0.138
6	1.190	+2.7045	+0.035
7	1.196	+2.7014	–0.051
8	**1.192**	+2.7035	+0.007

$$\gamma = 0.5 \;\Rightarrow (\mu_{cr}L)_{BA} = 1.192 = \frac{\pi}{2.6356}$$

$$\Rightarrow (P_{\text{BA}})_{cr} = P_{cr} = \frac{\pi^2 EI}{(2.6356L)^2} = \frac{P_{Eo}}{(2.6356)^2} = 0.144P_{Eo}$$

Similarly, this procedure can be repeated for other values of γ (iteration tables not shown here). The results for γ = 1, γ = 5 and γ = 10 are as follows:

$$\gamma = 1 \;\Rightarrow (\mu_{cr}L)_{\text{BA}} = 1.350 = \frac{\pi}{2.3271} \;\Rightarrow\; P_{cr} = \frac{P_{Eo}}{2.3271^2} = 0.185P_{Eo}$$

$$\gamma = 5 \;\Rightarrow (\mu_{cr}L)_{\text{BA}} = 1.520 = \frac{\pi}{2.0668} \;\Rightarrow\; P_{cr} = \frac{P_{Eo}}{2.0668^2} = 0.234P_{Eo}$$

$$\gamma = 10 \;\Rightarrow (\mu_{cr}L)_{\text{BA}} = 1.545 = \frac{\pi}{2.0333} \;\Rightarrow\; P_{cr} = \frac{P_{Eo}}{2.0333^2} = 0.242P_{Eo}$$

The value of the critical load will asymptotically reach $0.25P_{Eo}$ as γ approaches infinity, in the sway buckling mode (unbraced behaviour), as indicated in the plot shown in Fig. 7.42. Similarly, as illustrated, in the non-sway buckling mode (braced behaviour), the maximum value of $2.047P_{Eo}$ is asymptotically reached as γ approaches infinity. In the absence of external bracing, it is obvious that the elastic instability will occur in the sway buckling mode, which predicts a much lower buckling load.

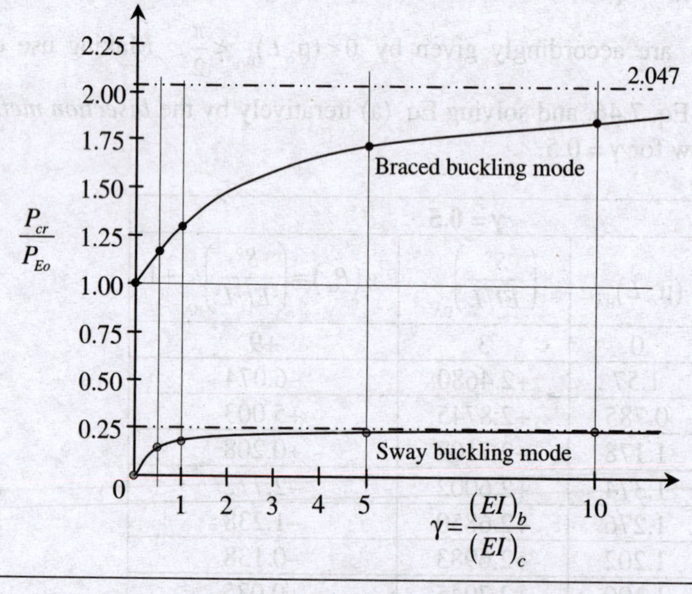

Figure 7.42 Variation of critical buckling load with beam-column stiffness (Example 7.12)

EXAMPLE 7.13

Consider a rectangular frame ABCD, with fixed bases at A and D, and pinned joints at B and C, and with columns AB and CD having flexural rigidities equal to *EI* and 4*EI* respectively, as shown in Fig. 7.43. Two point loads, P and λP, applied to the top ends of columns AB and CD respectively, are increased proportionately and simultaneously until elastic instability occurs (corresponding to which $P = P_{cr}$). Estimate the total load capacity of the frame, W_{cr}, under this loading arrangement, considering (a) $\lambda = 4$ and (b) $\lambda = 1$. Assume all members to be axially rigid.

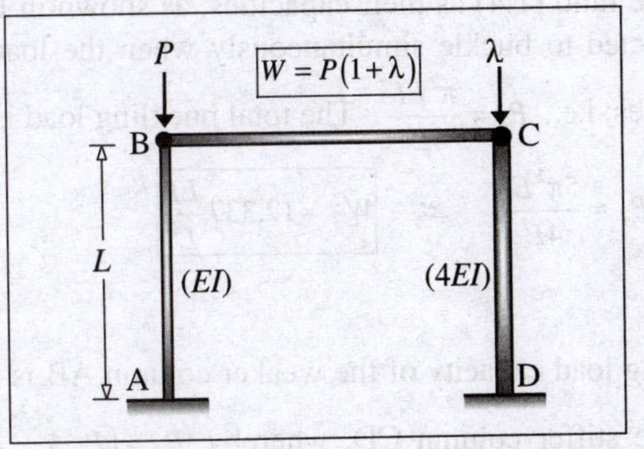

Figure 7.43 Unsymmetric portal frame with internal hinges (Example 7.13)

SOLUTION

In this case, the element BC with pinned joints at the two ends, will not undergo any curvature, and will merely act as a tie between the two columns, AB and DC, ensuring that both columns will deflect by the same amount, Δ, at the top, in the sway buckling mode (refer Fig. 7.44). The sway mode is, without doubt, more critical than the non-sway (braced) buckling mode, and each column will essentially behave like a cantilever, with single curvature in the fundamental buckling mode.

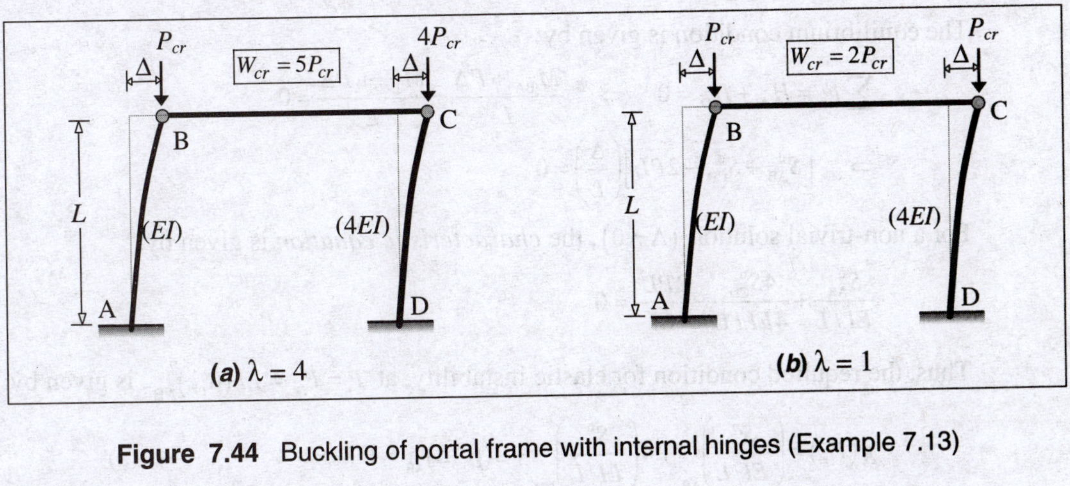

Figure 7.44 Buckling of portal frame with internal hinges (Example 7.13)

Case (a): **$\lambda = 4$**

The individual buckling load capacities of the two cantilever columns are $(P_{AB})_{max} = \dfrac{\pi^2 EI}{(2L)^2} = \dfrac{\pi^2 EI}{4L^2}$ and $(P_{CD})_{max} = \dfrac{\pi^2 E(4I)}{4L^2} = 4(P_{AB})_{max}$. The loads applied to the columns are in the same ratio (1:4) as their capacities, as shown in Fig. 7.44a. Thus, clearly both columns are expected to buckle simultaneously when the loads acting on them attain their respective capacities; i.e., $P_{cr} = \dfrac{\pi^2 EI}{4L^2}$. The total buckling load is accordingly given by:

$$W_{cr} = 5P_{cr} = \frac{5\pi^2 EI}{4L^2} \quad \Rightarrow \quad \boxed{W_{cr} = 12.337\frac{EI}{L^2}}$$

Case (b): **$\lambda = 1$**

In this case, the buckling load capacity of the weaker column AB is likely to be enhanced by the connection with the stiffer column CD, whereby $P_{cr} > (P_{AB})_{max} = \dfrac{\pi^2 EI}{4L^2}$. In any case, it cannot exceed the capacity of the column CD, whereby $P_{cr} < (P_{CD})_{max} = \dfrac{\pi^2 EI}{L^2}$. We can find the true value of P_{cr} by applying the slope-deflection method.

The degree of kinematic indeterminacy is one, considering the lateral drift Δ as the unknown displacement. The slope-deflection equations in the buckled configuration (shown in Fig. 7.44b) are given by (refer Fig. 7.20a):

$$M_{AB} = -S_{AB}^{o}\frac{\Delta}{L}$$

$$M_{DC} = -S_{CD}^{o}\frac{\Delta}{L}$$

The equilibrium condition is given by:

$$\sum F_x = H_A + H_D = 0 \quad \Rightarrow \quad \frac{M_{BA} + P\Delta}{L} + \frac{M_{DC} + P\Delta}{L} = 0$$

$$\Rightarrow \quad \left[S_{AB}^{o} + S_{CD}^{o} - 2PL\right]\left(\frac{\Delta}{L}\right) = 0$$

For a non-trivial solution $(\Delta \neq 0)$, the *characteristic equation* is given by:

$$\frac{S_{AB}^{o}}{EI/L} + \frac{4S_{CD}^{o}}{4EI/L} - \frac{2PL^2}{EI} = 0$$

Thus, the required condition for elastic instability, at $P = P_{cr} = EI(\mu_{cr})_{AB}^2$, is given by:

$$g(P_{cr}) = \left(\frac{S^{o}}{EI/L}\right)_{AB} + 4\left(\frac{S^{o}}{EI/L}\right)_{CD} - 2(\mu_{cr}L)_{AB}^2 = 0 \qquad \text{(a)}$$

As $\dfrac{\pi^2 EI}{4L^2} < P_{cr} < \dfrac{\pi^2 EI}{L^2}$, the lower and upper bounds of the parameter $\mu_{cr} L = \sqrt{\dfrac{P_{cr} L^2}{EI}} = \dfrac{\pi}{k_e}$ are accordingly given by $\dfrac{\pi}{2} < (\mu_{cr} L)_{AB} < \dfrac{\pi}{1}$, with $(\mu_{cr} L)_{CD} = 0.5(\mu_{cr} L)_{AB}$.

Making use of the stiffness measure given in Eq. 7.46, and solving Eq. (a) iteratively by the *bisection method*, the results obtained are as tabulated below.

Trial	$(\mu_{cr} L)_{AB}$	$\left(\dfrac{S^o}{EI/L}\right)_{AB}$	$\left(\dfrac{S^o}{EI/L}\right)_{CD}$	$g(P_{cr}) = \left(\dfrac{S^o}{EI/L}\right)_{AB} + 4\left(\dfrac{S^o}{EI/L}\right)_{CD} - 2(\mu_{cr} L)^2_{AB}$
	1.57	+2.4680	+2.8745	+9.036
	3.14	+0.0050	+2.4680	−9.842
1	2.355	+1.6558	+2.7110	+1.408
2	2.747	+0.9934	+26005	−3.698
3	2.551	+1.3543	+2.6583	−1.028
4	2.453	+1.5116	+2.6853	+0.218
5	2.502	+1.4347	+2.6720	−0.398
6	2.477	+1.4744	+2.6788	−0.082
7	2.470	+1.4853	+2.6807	+0.006
8	**2.4705**	+1.4845	+1.2353	0.000

$$\Rightarrow (\mu_{cr} L)_{AB} = 2.4705 = \frac{\pi}{1.2716}$$

$$\Rightarrow (P_{AB})_{cr} = P_{cr} = \frac{\pi^2 EI}{(1.2716L)^2} = \frac{6.104EI}{L^2}$$

$$W_{cr} = 2P_{cr} \quad \Rightarrow \quad \boxed{W_{cr} = \frac{12.208EI}{L^2}}$$

Thus, we observe that the buckling capacity of the weaker column is significantly enhanced, and that the final load capacity of the frame is only about 10 percent less than the sum of the individual buckling capacities of the two columns (refer to Case (a)).

7.4 MATRIX METHOD OF ANALYSIS

The slope-deflection method, although accurate, is not a convenient method to apply when we deal with structures that have a high degree of kinematic indeterminacy. In such cases, it is appropriate that we resort to a computer-based solution. The matrix method is ideally suited for this. We can conveniently extend the stiffness formulation of 'first-order' analysis (discussed in earlier chapters) to include the second-order effects induced by *P-delta* effects. However, as in the case of the slope-deflection method, it is desirable to work within the framework of linear analysis. We can simply extend the same stiffness formulation underlying the slope-deflection method in a more direct manner in the matrix method. As in the slope-deflection method, we can adopt an iterative approach in order to find the lowest critical buckling load that renders the stiffness matrix singular (inducing elastic instability).

However, as we shall see shortly, the matrix method offers a simpler solution that completely avoids iteration in the solution procedure. For this, we need to make an approximation of the stability functions discussed earlier, by considering their first-order series expansions. The approximation does not result in significant error if the value of the parameter, $\mu L = \sqrt{\frac{PL^2}{EI}}$, is kept low, and this can always be achieved by introducing intermediate nodes in the elements, if necessary, thereby reducing the element length *L*.

In Section 7.4.1, the expressions for the coefficients of the stiffness matrix of a beam-column element with four degrees of freedom (as in a beam element) are generated, and it is shown how this matrix can be decomposed into a *primary stiffness* matrix (free of *P-delta* effects) and a *geometric stiffness* matrix that is a linear function of the axial load *P* on the element. The effects of axial deformations on the flexural stiffness measures of the beam-column element are ignored, as in the slope-deflection method (otherwise, a nonlinear analysis is called for). However, we need to include the axial degrees of freedom while dealing with plane frame elements. It is shown how the 6×6 element stiffness matrix can be generated for such an element. The effect of axial deformations is included only in the axial stiffness formulation, and the coupling with the flexural stiffness is ignored.

In Section 7.4.2, the application of the matrix method is demonstrated, through various examples, for finding the critical load causing elastic instability in continuous beams and plane frames. This is followed in Section 7.4.3 with a demonstration of the matrix method to carrying out a second-order analysis of continuous beams and plane frames.

We shall limit considerations to the conventional stiffness method, and not to any reduced stiffness formulation (involving less degrees of freedom in the basic element stiffness formulation, as shown in earlier Chapters).

<u>Note on sign convention for rotations and moments</u>: In the slope-deflection method, the element end rotations and moments were assumed to be positive, if acting clockwise (in conformity with tradition). In the matrix method formulations (in this book), however, we maintain a sign convention that is consistent with vector algebra $(\hat{\mathbf{i}} \times \hat{\mathbf{j}} = \hat{\mathbf{k}})$. Assuming the local x^*-axis to act in the longitudinal direction of the element (pointing from the start node to the end node), the end rotation vector will point along the local z^*-axis (normal to the x^*-y^* plane of the element) in such a way that the rotation (or moment) will be positive, if acting in an anti-clockwise direction.

7.4.1 Stiffness Matrix for Prismatic Beam-Column Elements

Beam Element with Four Degrees of Freedom

The element stiffness matrix for a beam-column element with four degrees of freedom (for application in continuous beam systems) can be easily generated, by making use of the stiffness measures (involving *stability functions*), previously derived, that incorporate the *P-delta* effect. This is illustrated in Fig. 7.45, which is similar to Fig. 5.1, except for the addition of the axial force P_i (indicated as compressive) on the element.

The element stiffness matrix takes the following form:

$$\mathbf{k}_*^\mathbf{i} = \begin{bmatrix} k_{1*1*}^i & k_{1*2*}^i & k_{1*3*}^i & k_{1*4*}^i \\ k_{2*1*}^i & k_{2*2*}^i & k_{2*3*}^i & k_{2*4*}^i \\ k_{3*1*}^i & k_{3*2*}^i & k_{3*3*}^i & k_{3*4*}^i \\ k_{4*1*}^i & k_{4*2*}^i & k_{4*3*}^i & k_{4*4*}^i \end{bmatrix} = \begin{bmatrix} \tilde{\beta}_i & \tilde{\chi}_i & -\tilde{\beta}_i & \tilde{\chi}_i \\ \tilde{\chi}_i & S_i & -\tilde{\chi}_i & r_i S_i \\ -\tilde{\beta}_i & -\tilde{\chi}_i & \tilde{\beta}_i & -\tilde{\chi}_i \\ \tilde{\chi}_i & r_i S_i & -\tilde{\chi}_i & S_i \end{bmatrix} \tag{7.63}$$

where

$$\begin{aligned} \tilde{\beta}_i &= 2S_i(1+r_i)/L_i^2 \\ \tilde{\chi}_i &= S_i(1+r_i)/L_i \end{aligned} \tag{7.64}$$

$$S_i = \frac{(EI)_i}{L_i}\frac{(\mu L)_i\left[\sin(\mu L)_i - (\mu L)_i\cos(\mu L)_i\right]}{2-2\cos(\mu L)_i - (\mu L)_i\sin(\mu L)_i} \quad \text{if } P_i \text{ is compressive} \tag{7.65}$$

$$r_i = \frac{(\mu L)_i - \sin(\mu L)_i}{\sin(\mu L)_i - (\mu L)_i\cos(\mu L)_i} \quad \text{if } P_i \text{ is compressive} \tag{7.66}$$

$$S_i = \frac{(EI)_i}{L_i}\frac{(\mu L)_i\left[-\sinh(\mu L)_i + (\mu L)_i\cosh(\mu L)_i\right]}{2-2\cosh(\mu L)_i + (\mu L)_i\sinh(\mu L)_i} \quad \text{if } P_i \text{ is tensile} \tag{7.65a}$$

$$r_i = \frac{-(\mu L)_i + \sinh(\mu L)_i}{-\sinh(\mu L)_i + (\mu L)_i\cosh(\mu L)_i} \quad \text{if } P_i \text{ is tensile} \tag{7.66a}$$

$$(\mu L)_i = \sqrt{\frac{P_i L_i^2}{(EI)_i}} \tag{7.67}$$

The element stiffness matrix (Eq. 7.63) reduces to the 'first-order' form given in Eq. 5.11 when P_i is equal to zero (whereby $(\mu L)_i = 0$). If the value of the parameter $(\mu L)_i$ is small, we can simplify the formulation by considering the first two terms of the Taylor series expansion of the trigonometric functions, as an approximation:

$$\begin{aligned} \sin(\mu L)_i &\approx (\mu L)_i - (\mu L)_i^3/6 \\ \cos(\mu L)_i &\approx 1-(\mu L)_i/2 \end{aligned} \tag{7.68}$$

$$\begin{aligned} \sinh(\mu L)_i &\approx (\mu L)_i + (\mu L)_i^3/6 \\ \cosh(\mu L)_i &\approx 1+(\mu L)_i/2 \end{aligned} \tag{7.68a}$$

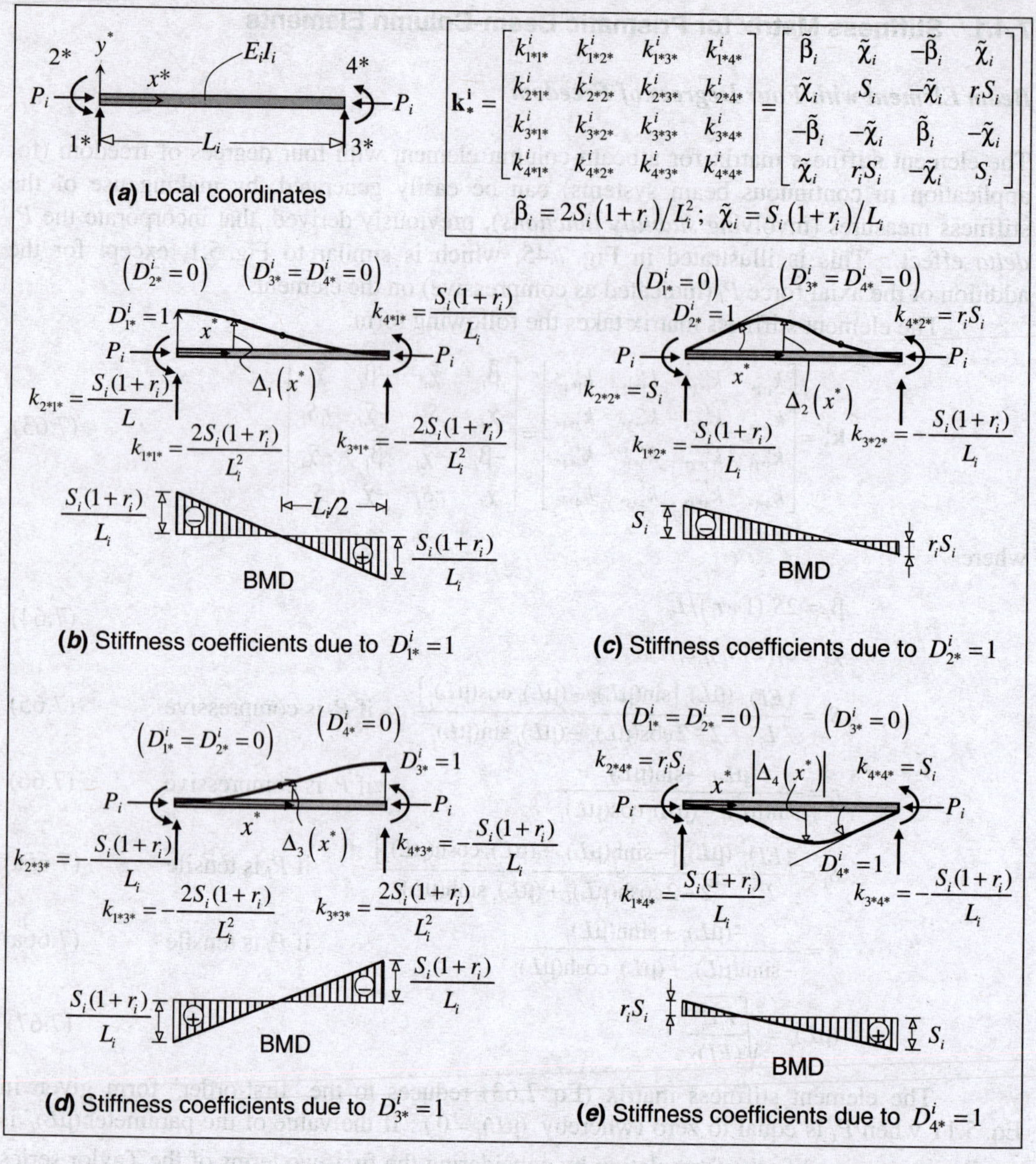

Figure 7.45 Generation of 4 × 4 stiffness matrix for a beam-column element

Substituting the above expressions in the stability functions and simplifying, it can be shown that the stiffness measures in Eqns 7.65-7.66a reduce to the following compact form:

$$S_i \approx \frac{4(EI)_i}{L_i} - \frac{2P_iL_i}{15}, \quad r_iS_i \approx \frac{2(EI)_i}{L_i} + \frac{P_iL_i}{30} \qquad \text{if } P_i \text{ is compressive} \tag{7.69}$$

$$S_i \approx \frac{4(EI)_i}{L_i} + \frac{2P_iL_i}{15}, \quad r_iS_i \approx \frac{2(EI)_i}{L_i} - \frac{P_iL_i}{30} \qquad \text{if } P_i \text{ is tensile} \tag{7.69a}$$

The above stiffness measures are the stiffness coefficients applicable to *braced* beam-columns (corresponding to Figs 7.45c and e), as indicated in Fig. 7.46a. The simplified forms suggest that they are linear functions of the axial load P_i. The end rotational stiffness, S_i, is equal to the 'first-order' value, $4(EI)_i/L_i$, minus a 'second-order' contribution, $2P_iL_i/15$ (due to an axial compression P_i). Similarly, the carry-over stiffness, r_iS_i, is equal to the 'first-order' value, $4(EI)_i/L_i$, plus a 'second-order' contribution, $P_iL_i/30$. If the axial force is tensile, instead of compressive, the signs of the second-order contributions get reversed, as reflected in Eq. 7.69a.

Similarly, on substitution and simplification, the expressions in Eq. 7.64 reduce to:

$$\tilde{\beta}_i \approx \frac{12(EI)_i}{L_i^3} - \frac{6P_iL_i}{5}, \quad \tilde{\chi}_i \approx \frac{6(EI)_i}{L_i^2} - \frac{P_i}{10} \qquad \text{if } P_i \text{ is compressive} \tag{7.70}$$

$$\tilde{\beta}_i \approx \frac{12(EI)_i}{L_i^3} + \frac{6P_iL_i}{5}, \quad \tilde{\chi}_i \approx \frac{6(EI)_i}{L_i^2} + \frac{P_i}{10} \qquad \text{if } P_i \text{ is tensile} \tag{7.70a}$$

The above stiffness measures are the stiffness coefficients applicable to *unbraced* beam-columns (corresponding to Figs 7.45b and d), as indicated in Fig. 7.46b. The simplified forms suggest that they are also linear functions of the axial load P_i. The stiffness measure, $\tilde{\chi}_i$, is not only a measure of the end moment caused by unit relative deflection in an unbraced element (refer Fig. 7.46b), but is also a measure of the shear force caused by unit end rotation in a braced element, satisfying equilibrium (refer Fig. 7.46a). The value of $\tilde{\chi}_i$ is equal to the 'first-order' value, $6(EI)_i/L_i^2$, minus a 'second-order' contribution, $P_i/10$ (due to an axial compression P_i). Similarly, the translational stiffness, $\tilde{\beta}_i$, is equal to the 'first-order' value, $12(EI)_i/L_i^3$, plus a 'second-order' contribution, $6P_iL_i/5$. If the axial force is tensile, instead of compressive, the signs of the second-order contributions get reversed, as reflected in Eq. 7.70a.

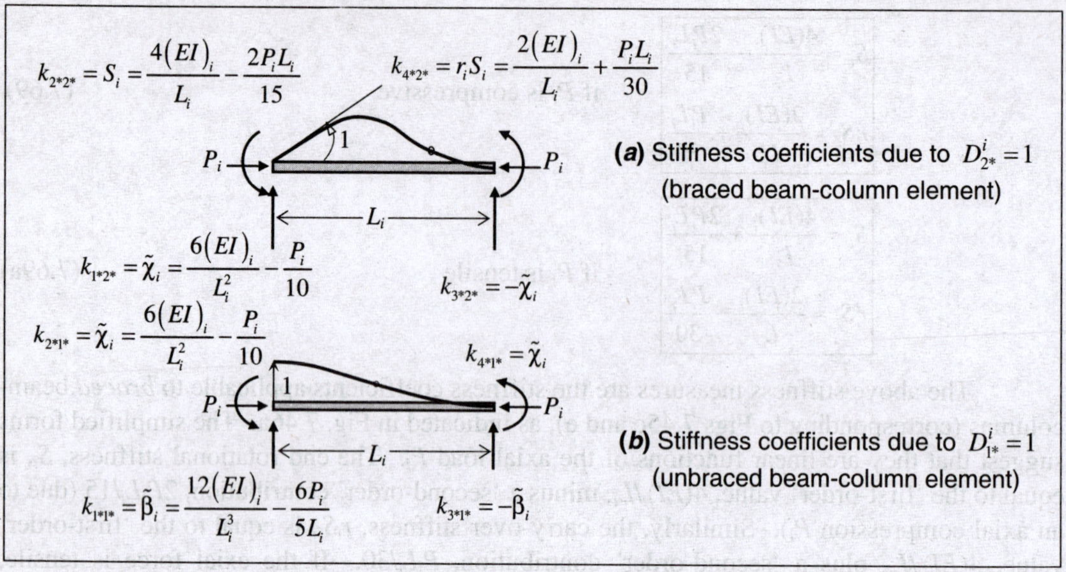

Figure 7.46 Simplified stiffness coefficients (for small $(\mu L)_i$; axial compression)

The expressions given in Eqns 7.69 and 7.70 can also be derived from first principles, by assuming the 'first-order' (cubic) *shape functions* (refer Eqns 6.4a and 6.4b) to approximate the deflected shapes and including the additional curvatures induced by the *P-delta* effects.

Thus, applying these simplifications, the stiffness matrix (Eq. 7.63) can be decomposed into two parts: (i) the 'first-order' (primary) stiffness matrix (denoted as $\mathbf{k}^{\mathbf{i}}_{\mathbf{O}*}$ and given by Eq. 5.11) and (ii) the 'geometric stiffness' matrix, which accounts for the *P-delta* effects (denoted as $\mathbf{k}^{\mathbf{i}}_{\mathbf{G}*}$):

$$\mathbf{k}^{\mathbf{i}}_* = \mathbf{k}^{\mathbf{i}}_{\mathbf{O}*} + \mathbf{k}^{\mathbf{i}}_{\mathbf{G}*} \quad \Rightarrow$$

$$\mathbf{k}^{\mathbf{i}}_* = \frac{(EI)_i}{L_i}\begin{bmatrix} 12/L_i^2 & 6/L_i & -12/L_i^2 & 6/L_i \\ 6/L_i & 4 & -6/L_i & 2 \\ -12/L_i^2 & -6/L_i & 12/L_i^2 & -6/L_i \\ 6/L_i & 2 & -6/L_i & 4 \end{bmatrix} \mp P_i \begin{bmatrix} 1.2/L_i & 0.1 & -1.2/L_i & 0.1 \\ 0.1 & 2L_i/15 & -0.1 & -L_i/30 \\ -1.2/L_i & -0.1 & 1.2/L_i & -0.1 \\ 0.1 & -L_i/30 & -0.1 & 2L_i/15 \end{bmatrix} \qquad (7.71)$$

where the negative sign with regard to P_i is applicable when the axial force is compressive and the positive sign is applicable when the axial force is tensile.

Clearly, under axial compression, the stiffness of the element gets reduced. When the stiffnesses of various elements are combined to form the structure stiffness matrix, the latter too can be expressed in a form, $\mathbf{k} = \mathbf{k_O} + \mathbf{k_G}$. When elements in the structure are subject to axial compression, the structure is vulnerable to elastic instability when the structure stiffness matrix becomes singular; i.e., when the determinant, $|\mathbf{k_O} + \mathbf{k_G}|$, approaches zero.

However, the estimate of the critical buckling load is likely to be approximate, owing to the approximations involved in Eq. 7.68. The accuracy can be significantly enhanced by reducing the value of the parameter $(\mu L)_i$, which is achieved by introducing intermediate nodes in the beam-column elements (thereby reducing L_i).

Plane Frame Element with Six Degrees of Freedom

The element stiffness matrix for a beam-column element with six degrees of freedom (for application in plane frames) can be easily generated, by including an axial degree of freedom at each end to the beam element model considered earlier. This is indicated in Fig. 7.47. The axial stiffness of the element is given by $(EA)_i/L_i$, where $(EA)_i$ is the axial rigidity of the element, and it is assumed that there is no coupling between the axial stiffness and the flexural stiffness terms. It may be noted that the flexural stiffness terms derived earlier (based on *stability functions*) are based on the assumption of axial rigidity (no axial deformations). However, here it is meaningful to include the axial degrees of freedom (labelled '1*' and '3*' in Fig. 7.47 (identical to Fig. 6.1a), although in most practical cases, assuming the condition of axial rigidity in the member will usually not result in significant loss in accuracy.

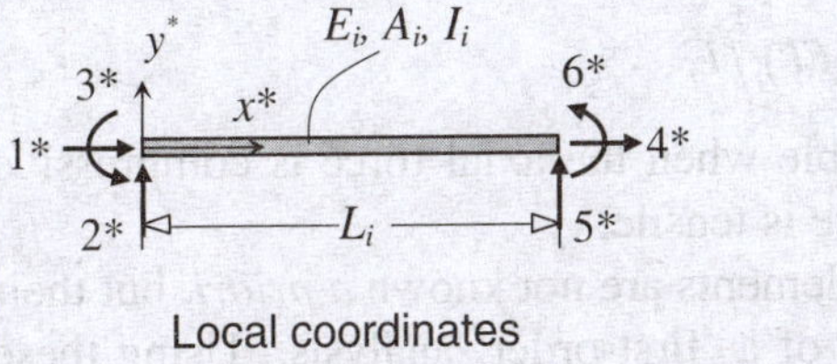

$$\mathbf{k}_*^i = \left[\begin{array}{ccc|ccc} \alpha_i & 0 & 0 & -\alpha_i & 0 & 0 \\ 0 & \tilde{\beta}_i & \tilde{\chi}_i & 0 & -\tilde{\beta}_i & \tilde{\chi}_i \\ 0 & \tilde{\chi}_i & S_i & 0 & -\tilde{\chi}_i & r_i S_i \\ \hline -\alpha_i & 0 & 0 & \alpha_i & 0 & 0 \\ 0 & -\tilde{\beta}_i & -\tilde{\chi}_i & 0 & \tilde{\beta}_i & -\tilde{\chi}_i \\ 0 & \tilde{\chi}_i & r_i S_i & 0 & -\tilde{\chi}_i & S_i \end{array}\right]; \quad \begin{array}{l} \alpha_i = (EA)_i/L_i \\ \tilde{\beta}_i = 2S_i(1+r_i)/L_i^2 \\ \tilde{\chi}_i = S_i(1+r_i)/L_i \end{array}$$

$$\mathbf{k}_*^i = \mathbf{k}_{O*}^i + \mathbf{k}_{G*}^i$$

$$\mathbf{k}_*^i \approx \left[\begin{array}{ccc|ccc} \alpha_i & 0 & 0 & -\alpha_i & 0 & 0 \\ 0 & \beta_i & \chi_i & 0 & -\beta_i & \chi_i \\ 0 & \chi_i & 4\delta_i & 0 & -\chi_i & 2\delta_i \\ \hline -\alpha_i & 0 & 0 & \alpha_i & 0 & 0 \\ 0 & -\beta_i & -\chi_i & 0 & \beta_i & -\chi_i \\ 0 & \chi_i & 2\delta_i & 0 & -\chi_i & 4\delta_i \end{array}\right] \mp P_i \left[\begin{array}{ccc|ccc} 0 & 0 & 0 & 0 & 0 & 0 \\ 0 & 1.2/L_i & 0.1 & 0 & -1.2/L_i & 0.1 \\ 0 & 0.1 & 2L_i/15 & 0 & -0.1 & -L_i/30 \\ \hline 0 & 0 & 0 & 0 & 0 & 0 \\ 0 & -1.2/L_i & -0.1 & 0 & 1.2/L_i & -0.1 \\ 0 & 0.1 & -L_i/30 & 0 & -0.1 & 2L_i/15 \end{array}\right]$$

$$\alpha_i = (EA)_i/L_i;\ \beta_i = 12(EI)_i/L_i^3;\ \chi_i = 6(EI)_i/L_i^2;\ \delta_i = (EI)_i/L_i$$

Figure 7.47 6×6 stiffness matrix for a plane frame element, including *P-delta* effects

The element stiffness matrix takes the following form, in terms of stability functions, for the coordinates shown in Fig. 7.47:

$$\mathbf{k}_*^{\mathbf{i}} = \left[\begin{array}{ccc|ccc} \alpha_i & 0 & 0 & -\alpha_i & 0 & 0 \\ 0 & \tilde{\beta}_i & \tilde{\chi}_i & 0 & -\tilde{\beta}_i & \tilde{\chi}_i \\ 0 & \tilde{\chi}_i & S_i & 0 & -\tilde{\chi}_i & r_i S_i \\ \hline -\alpha_i & 0 & 0 & \alpha_i & 0 & 0 \\ 0 & -\tilde{\beta}_i & -\tilde{\chi}_i & 0 & \tilde{\beta}_i & -\tilde{\chi}_i \\ 0 & \tilde{\chi}_i & r_i S_i & 0 & -\tilde{\chi}_i & S_i \end{array}\right]; \quad \begin{array}{l} \alpha_i = (EA)_i / L_i \\ \tilde{\beta}_i = 2S_i(1+r_i)/L_i^2 \\ \tilde{\chi}_i = S_i(1+r_i)/L_i \end{array} \qquad (7.72)$$

As indicated earlier (refer Eq. 7.71), we can decompose this stiffness matrix into a 'first-order' (primary) stiffness matrix, $\mathbf{k}_{\mathbf{O}*}^{\mathbf{i}}$, (given by Eq. 6.7) and a 'geometric stiffness' matrix, $\mathbf{k}_{\mathbf{G}*}^{\mathbf{i}}$, which accounts for the *P-delta* effects:

$$\mathbf{k}_*^{\mathbf{i}} = \mathbf{k}_{\mathbf{O}*}^{\mathbf{i}} + \mathbf{k}_{\mathbf{G}*}^{\mathbf{i}}$$

$$\mathbf{k}_*^{\mathbf{i}} \approx \left[\begin{array}{ccc|ccc} \alpha_i & 0 & 0 & -\alpha_i & 0 & 0 \\ 0 & \beta_i & \chi_i & 0 & -\beta_i & \chi_i \\ 0 & \chi_i & 4\delta_i & 0 & -\chi_i & 2\delta_i \\ \hline -\alpha_i & 0 & 0 & \alpha_i & 0 & 0 \\ 0 & -\beta_i & -\chi_i & 0 & \beta_i & -\chi_i \\ 0 & \chi_i & 2\delta_i & 0 & -\chi_i & 4\delta_i \end{array}\right] \mp P_i \left[\begin{array}{ccc|ccc} 0 & 0 & 0 & 0 & 0 & 0 \\ 0 & 1.2/L_i & 0.1 & 0 & -1.2/L_i & 0.1 \\ 0 & 0.1 & 2L_i/15 & 0 & -0.1 & -L_i/30 \\ \hline 0 & 0 & 0 & 0 & 0 & 0 \\ 0 & -1.2/L_i & -0.1 & 0 & 1.2/L_i & -0.1 \\ 0 & 0.1 & -L_i/30 & 0 & -0.1 & 2L_i/15 \end{array}\right] \qquad (7.73)$$

$$\alpha_i = (EA)_i/L_i;\ \beta_i = 12(EI)_i/L_i^3;\ \chi_i = 6(EI)_i/L_i^2;\ \delta_i = (EI)_i/L_i$$

where the negative sign with regard to P_i is applicable when the axial force is compressive and the positive sign is applicable when the axial force is tensile.

In general, the axial forces P_i in the various elements are not known *a priori*, but their approximate values can be deduced from the results of a 'first-order' analysis. Using these values (compressive or tensile), the stiffness matrices of the individual elements (including effect of P_i on flexural stiffness terms) can be generated and the stiffness method applied to get a solution. Further iterations can be carried out, if deemed necessary, till satisfactory convergence in the results is obtained.

Dealing with Internal Hinges

When internal hinges are present in a frame, special attention is called for, as discussed earlier in Sections 5.2.6 and 6.2.5 with reference to beams and plane frames respectively. Usually, the flexural hinge ('moment release') occurs in the middle or end of a member (such as a beam or column in a frame), but it can also be located at the junction of two or more members (as in a beam-column joint). We can account for this moment release at the hinge location by suitably modifying the element stiffness matrix and also the computation of the fixed end force effects, if any (refer Eqns 7.62a and b).

The modified stiffness matrix will have a zero row and a zero column vector, corresponding the local coordinate relating to the hinge, as explained earlier. If there are hinges located at both ends, then there will be two zero rows and columns in the modified stiffness matrix.

The modified stiffness matrix (including *P-delta* effects) takes the following forms for a four degree of freedom element:

$$\mathbf{k}_*^{\mathbf{i}} = \frac{3(EI)_i}{L_i}\begin{bmatrix} 1/L_i^2 & 0 & -1/L_i^2 & 1/L_i \\ 0 & 0 & 0 & 0 \\ -1/L_i^2 & 0 & 1/L_i^2 & -1/L_i \\ 1/L_i & 0 & -1/L_i & 1 \end{bmatrix} \mp \frac{P_i L_i}{8}\begin{bmatrix} 1/L_i^2 & 0 & -1/L_i^2 & 1/L_i \\ 0 & 0 & 0 & 0 \\ -1/L_i^2 & 0 & 1/L_i^2 & -1/L_i \\ 1/L_i & 0 & -1/L_i & 1 \end{bmatrix}$$

if hinge is at *start node* (7.74a)

$$\mathbf{k}_*^{\mathbf{i}} = \frac{3(EI)_i}{L_i}\begin{bmatrix} 1/L_i^2 & 1/L_i & -1/L_i^2 & 0 \\ 1/L_i & 1 & -1/L_i & 0 \\ -1/L_i^2 & -1/L_i & 1/L_i^2 & 0 \\ 0 & 0 & 0 & 0 \end{bmatrix} \mp \frac{P_i L_i}{8}\begin{bmatrix} 1/L_i^2 & 1/L_i & -1/L_i^2 & 0 \\ 1/L_i & 1 & -1/L_i & 0 \\ -1/L_i^2 & -1/L_i & 1/L_i^2 & 0 \\ 0 & 0 & 0 & 0 \end{bmatrix}$$

if hinge is at *end node* (7.74b)

$$\mathbf{k}_*^{\mathbf{i}} = \frac{3(EI)_i}{L_i^3}\begin{bmatrix} 1 & 0 & -1 & 0 \\ 0 & 0 & 0 & 0 \\ -1 & 0 & 1 & 0 \\ 0 & 0 & 0 & 0 \end{bmatrix} \mp \frac{P_i}{8L_i^2}\begin{bmatrix} 1 & 0 & -1 & 0 \\ 0 & 0 & 0 & 0 \\ -1 & 0 & 1 & 0 \\ 0 & 0 & 0 & 0 \end{bmatrix}$$

if hinges are at *both ends* (7.74c)

where the negative sign associated with the axial force P_i should be used when the force is compressive.

Similarly, for a six degree of freedom element:

$$\mathbf{k}_*^{\mathbf{i}} = \frac{3(EI)_i}{L_i}\left[\begin{array}{ccc|ccc} A_i/(3I_i) & 0 & 0 & -A_i/(3I_i) & 0 & 0 \\ 0 & 1/L_i^2 & 0 & 0 & -1/L_i^2 & 1/L_i \\ 0 & 0 & 0 & 0 & 0 & 0 \\ \hline -A_i/(3I_i) & 0 & 0 & A_i/(3I_i) & 0 & 0 \\ 0 & -1/L_i^2 & 0 & 0 & 1/L_i^2 & -1/L_i \\ 0 & 1/L_i & 0 & 0 & -1/L_i & 1 \end{array}\right] \mp \frac{P_i L_i}{8}\left[\begin{array}{ccc|ccc} 0 & 0 & 0 & 0 & 0 & 0 \\ 0 & 1/L_i^2 & 0 & 0 & -1/L_i^2 & 1/L_i \\ 0 & 0 & 0 & 0 & 0 & 0 \\ \hline 0 & 0 & 0 & 0 & 0 & 0 \\ 0 & -1/L_i^2 & 0 & 0 & 1/L_i^2 & -1/L_i \\ 0 & 1/L_i & 0 & 0 & -1/L_i & 1 \end{array}\right]$$

if hinge is at *start node* (7.75a)

$$\mathbf{k}_*^{\mathbf{i}} = \frac{3(EI)_i}{L_i}\left[\begin{array}{ccc|ccc} A_i/(3I_i) & 0 & 0 & -A_i/(3I_i) & 0 & 0 \\ 0 & 1/L_i^2 & 1/L_i & 0 & -1/L_i^2 & 0 \\ 0 & 1/L_i & 1 & 0 & -1/L_i & 0 \\ \hline -A_i/(3I_i) & 0 & 0 & A_i/(3I_i) & 0 & 0 \\ 0 & -1/L_i^2 & -1/L_i & 0 & 1/L_i^2 & 0 \\ 0 & 0 & 0 & 0 & 0 & 0 \end{array}\right] \mp \frac{P_i L_i}{8}\left[\begin{array}{ccc|ccc} 0 & 0 & 0 & 0 & 0 & 0 \\ 0 & 1/L_i^2 & 1/L_i & 0 & -1/L_i^2 & 0 \\ 0 & 1/L_i & 1 & 0 & -1/L_i & 0 \\ \hline 0 & 0 & 0 & 0 & 0 & 0 \\ 0 & -1/L_i^2 & -1/L_i & 0 & 1/L_i^2 & 0 \\ 0 & 0 & 0 & 0 & 0 & 0 \end{array}\right]$$

if hinge is at *end node* (7.75b)

$$\mathbf{k}_*^{\mathrm{i}} = \frac{3(EI)_i}{L_i^3}\left[\begin{array}{ccc|ccc} A_iL_i^2/(3I_i) & 0 & 0 & -A_iL_i^2/(3I_i) & 0 & 0 \\ 0 & 1 & 0 & 0 & -1 & 0 \\ 0 & 0 & 0 & 0 & 0 & 0 \\ \hline -A_iL_i^2/(3I_i) & 0 & 0 & A_iL_i^2/(3I_i) & 0 & 0 \\ 0 & -1 & 0 & 0 & 1 & 0 \\ 0 & 0 & 0 & 0 & 0 & 0 \end{array}\right] \mp \frac{P_i}{8L_i^2}\left[\begin{array}{ccc|ccc} 0 & 0 & 0 & 0 & 0 & 0 \\ 0 & 1 & 0 & 0 & -1 & 0 \\ 0 & 0 & 0 & 0 & 0 & 0 \\ \hline 0 & 0 & 0 & 0 & 0 & 0 \\ 0 & -1 & 0 & 0 & 1 & 0 \\ 0 & 0 & 0 & 0 & 0 & 0 \end{array}\right]$$

if hinges are at *both ends* (7.75c)

The use of the imaginary clamp (using the procedure discussed earlier in Chapters 5 and 6) is demonstrated in Example 7.18.

7.4.2 Estimation of Critical Elastic Buckling Loads

In this case, the loading prescribed on the structure is such that it is applied directly at the nodes in such a manner as to induce axial compressive forces in the critical elements (such as columns in a frame). The objective is to arrive at a condition of bifurcation instability. As mentioned earlier, other loads causing bending in the elements are usually not applied, as they induce pre-buckling moments, which are ignored in the eigenvalue analysis for bifurcation instability (refer Example 7.19).

When the applied loading on the structure is such that it induces bending in the various members, a condition of bifurcation instability cannot occur. To find the critical buckling load in such cases, a 'first-order' analysis is required to be carried out to determine the axial forces in the various members (for use in the assessment of geometric stiffness), and then a conventional instability analysis is carried out, ignoring the effect of pre-buckling moments. This critical load is expected to be reached asymptotically, with increasing loading on the structure[†].

Analysis Procedure

The steps involved in the analysis procedure may be summarised as follows, and is based on the methods described earlier in Chapter 5 (for beams) and Chapter 6 (for frames):

1. Identify the various elements and express the axial compressive forces P_i in them in terms of a force P, whose lowest critical value we seek. For improved accuracy, the critical elements may be segmented by introducing intermediate nodes. Also,

[†] This is accompanied by large deformations, and in reality, failure is likely to occur prior to reaching the estimated critical buckling load. Moreover, other nonlinear effects, due to large deformations and material nonlinearity, come into play during the process of loading to failure. These are not accounted for in the simple elastic instability analysis discussed here.

identify the various local and global coordinates (*n* active and *r* restrained coordinates).

2. Generate the transformation matrices, $\mathbf{T^i}$, for the various elements, identifying the linking global coordinates in parenthesis (use Eq. 6.8, if inclined members are involved). [Alternatively, generate the displacement transformation matrices, $\mathbf{T_D^i}$].

3. Generate the element stiffness matrices $\mathbf{k_*^i}$ for all the elements, using Eq. 7.71 or Eq. 7.73, and assemble the structure stiffness matrix, $\mathbf{k} = \left[\begin{array}{c|c} \mathbf{k_{AA}} & \mathbf{k_{AR}} \\ \hline \mathbf{k_{RA}} & \mathbf{k_{RR}} \end{array}\right]$, combining the stiffness coefficient contributions, $\mathbf{k^i = T^{i^T} k_*^i T^i}$ at the appropriate locations. [Alternatively, compute $\mathbf{k_*^i T_D^i}$ for every element and assemble $\mathbf{k} = \sum_{i=1}^{m} \mathbf{T_D^{i^T} k_*^i T_D^i}$]. It is sufficient to assemble $\mathbf{k_{AA}}$ in the form, $\mathbf{k_{O,AA} + k_{G,AA}}$, where the elements of the geometric stiffness matrix $\mathbf{k_{G,AA}}$ are linear functions of *P*.

4. Set up the equilibrium equation in the deformed configuration at the active coordinates:

$$\mathbf{k_{AA} D_A = F_A = O} \qquad \Rightarrow \left[\mathbf{k_{O,AA} + k_{G,AA}}\right]\mathbf{D_A = O} \tag{7.76}$$

For a non-trivial solution, the *characteristic equation*, given by setting the determinant of the symmetric matrix $\mathbf{k_{AA}}$ equal to zero:

$$\left|\mathbf{k_{O,AA} + k_{G,AA}}\right| = 0 \tag{7.77}$$

Use a suitable numerical technique to find the lowest root (eigenvalue) of the equation, which is a polynomial function of *P*. The lowest root gives the critical value, P_{cr}. In an iterative procedure (like the bisection method), it is important to begin with a lower bound estimate of P_{cr}.

5. If required, the corresponding mode shape can be generated from the eigenvalue analysis.

EXAMPLE 7.14

Apply the matrix method to derive the critical buckling load of an ideal braced column (length *L* and flexural rigidity *EI*), considering (a) both ends to be restrained against rotation and (b) both ends to be pinned. Sketch the corresponding mode shapes for the two cases.

SOLUTION

Let us consider an intermediate node at the mid-point of the axially loaded braced member, as shown in Fig. 7.48. Thus, we have two prismatic elements, each of length $l = L/2$, subject to a constant axial compression *P*. In the fixed-fixed case, there are two active degrees of freedom and four restrained global coordinates (numbered as shown in Fig. 7.48a), while in the pinned-pinned case, there are four active degrees of freedom and two restrained global coordinates (numbered as shown in Fig. 7.48b). The local coordinates for the two beam-column elements, with four degrees of freedom each (assuming axial rigidity), are shown in Fig. 7.48c.

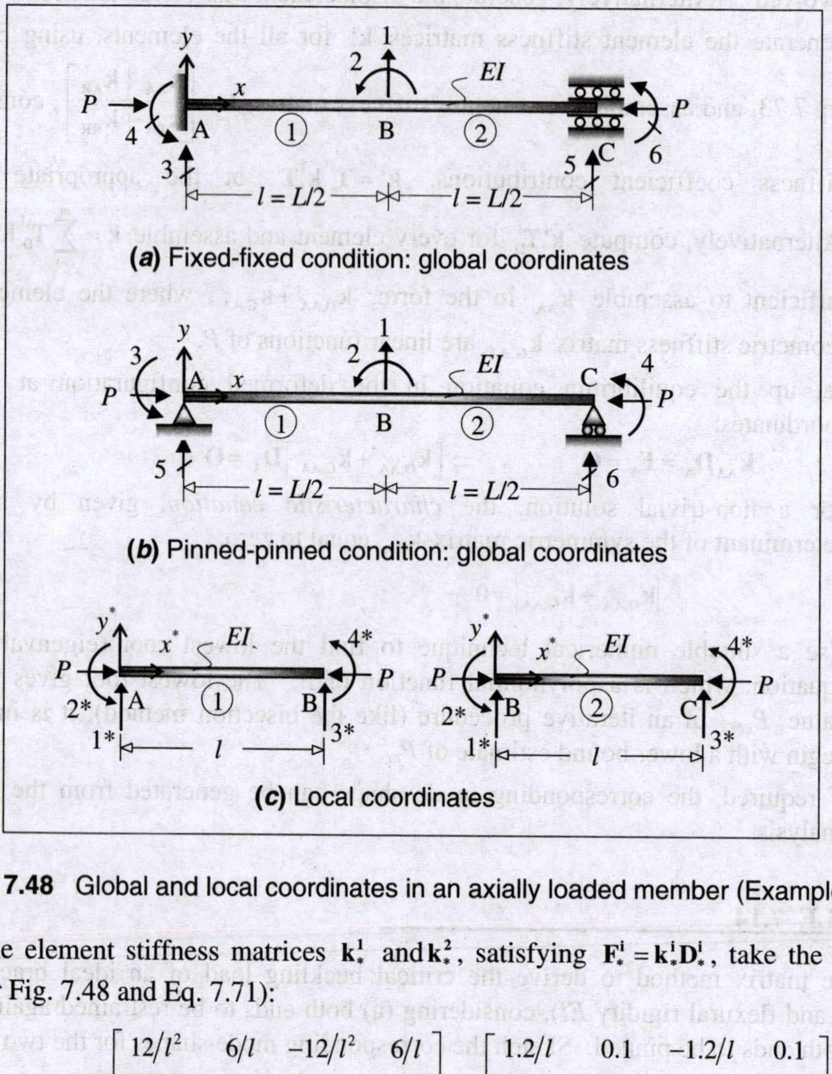

Figure 7.48 Global and local coordinates in an axially loaded member (Example 7.14)

The element stiffness matrices $\mathbf{k}_*^1$ and $\mathbf{k}_*^2$, satisfying $\mathbf{F}_*^i = \mathbf{k}_*^i \mathbf{D}_*^i$, take the following form (refer Fig. 7.48 and Eq. 7.71):

$$\mathbf{k}_*^1 = \mathbf{k}_*^2 = \frac{EI}{l}\begin{bmatrix} 12/l^2 & 6/l & -12/l^2 & 6/l \\ 6/l & 4 & -6/l & 2 \\ -12/l^2 & -6/l & 12/l^2 & -6/l \\ 6/l & 2 & -6/l & 4 \end{bmatrix} - P\begin{bmatrix} 1.2/l & 0.1 & -1.2/l & 0.1 \\ 0.1 & 2l/15 & -0.1 & -l/30 \\ -1.2/l & -0.1 & 1.2/l & -0.1 \\ 0.1 & -l/30 & -0.1 & 2l/15 \end{bmatrix}$$

Case (a): both ends fixed (refer Fig 7.48a):

The element transformation matrices $\mathbf{T}^1$ and $\mathbf{T}^2$, satisfying $\mathbf{D}_*^i = \mathbf{T}^i\mathbf{D}$ and $\mathbf{F}_*^i = \mathbf{T}^i\mathbf{F}$, take the following form, with the linking global coordinates indicated in parentheses:

$$\mathbf{T}^1 = \overset{(3)\ (4)\ (1)\ (2)}{\begin{bmatrix} 1 & 0 & 0 & 0 \\ 0 & 1 & 0 & 0 \\ 0 & 0 & 1 & 0 \\ 0 & 0 & 0 & 1 \end{bmatrix}} \begin{matrix} (3) \\ (4) \\ (1) \\ (2) \end{matrix} \quad \text{and} \quad \mathbf{T}^2 = \overset{(1)\ (2)\ (5)\ (6)}{\begin{bmatrix} 1 & 0 & 0 & 0 \\ 0 & 1 & 0 & 0 \\ 0 & 0 & 1 & 0 \\ 0 & 0 & 0 & 1 \end{bmatrix}} \begin{matrix} (1) \\ (2) \\ (5) \\ (6) \end{matrix}$$

As the transformation matrices, $\mathbf{T}^1$ and $\mathbf{T}^2$, are identity matrices, $\mathbf{T}^{1^T}\mathbf{k}_*^1\mathbf{T}^1 = \mathbf{k}_*^1$ and $\mathbf{T}^{2^T}\mathbf{k}_*^2\mathbf{T}^2 = \mathbf{k}_*^2$. Summing up these contributions, the structure stiffness matrix, $\mathbf{k} = \mathbf{k}_O + \mathbf{k}_G = \left[\begin{array}{c|c} \mathbf{k}_{AA} & \mathbf{k}_{AR} \\ \hline \mathbf{k}_{RA} & \mathbf{k}_{RR} \end{array}\right]$, satisfying $\mathbf{F} = \mathbf{kD}$, can be assembled as follows:

$$\mathbf{k} = \frac{EI}{l}\left[\begin{array}{cc|cccc} 24/l^2 & 0 & -12/l^2 & -6/l & -12/l^2 & 6/l \\ 0 & 8 & 6/l & 2 & -6/l & 2 \\ \hline -12/l^2 & 6/l & 12/l^2 & 6/l & 0 & 0 \\ -6/l & 2 & 6/l & 4 & 0 & 0 \\ -12/l^2 & -6/l & 0 & 0 & 12/l^2 & -6/l \\ 6/l & 2 & 0 & 0 & -6/l & 4 \end{array}\right] - P\left[\begin{array}{cc|cccc} 2.4/l & 0 & -1.2/l & -0.1 & -1.2/l & 0.1 \\ 0 & 4l/15 & 0.1 & -l/30 & -0.1 & -l/30 \\ \hline -1.2/l & 0.1 & 1.2/l & 0.1 & 0 & 0 \\ -0.1 & -l/30 & 0.1 & 2l/15 & 0 & 0 \\ -1.2/l & -0.1 & 0 & 0 & 1.2/l & -0.1 \\ 0.1 & -l/30 & 0 & 0 & -0.1 & 2l/15 \end{array}\right] \begin{matrix} (1) \\ (2) \\ (3) \\ (4) \\ (5) \\ (6) \end{matrix}$$

$$\mathbf{k}_{AA}\mathbf{D}_A = \mathbf{O} \Rightarrow \begin{bmatrix} \left(24\frac{EI}{l^3} - \frac{2.4P}{l}\right) & 0 \\ 0 & \left(8\frac{EI}{l} - \frac{4Pl}{15}\right) \end{bmatrix} \begin{Bmatrix} D_1 \\ D_2 \end{Bmatrix} = \begin{Bmatrix} 0 \\ 0 \end{Bmatrix} \qquad \text{(a)}$$

For a non-trivial solution to the equilibrium equation, the *characteristic equation* is obtained by setting the determinant of $\mathbf{k}_{AA}$ equal to zero.

$$|\mathbf{k}_{AA}| = 0 \Rightarrow \begin{vmatrix} \left(24\frac{EI}{l^3} - \frac{2.4P}{l}\right) & 0 \\ 0 & \left(8\frac{EI}{l} - \frac{4Pl}{15}\right) \end{vmatrix} = 0 \Rightarrow \left(24\frac{EI}{l^3} - \frac{2.4P}{l}\right)\left(8\frac{EI}{l} - \frac{4Pl}{15}\right) = 0$$

Of the two eigenvalues, the critical (lower) value is given by:

$$P_{cr} = 10\frac{EI}{l^2} \Rightarrow \boxed{P_{cr} = 40\frac{EI}{L^2}}.$$

This compares very well with the exact solution, $P_{cr} = \dfrac{4\pi^2 EI}{L^2} = 39.478\dfrac{EI}{L^2}$.

The corresponding eigenvector is given by $\begin{Bmatrix} D_1 = \Delta \\ D_2 = 0 \end{Bmatrix}$, indicating a symmetric mode shape, as shown in Fig. 7.49a. (The second eigenvector will be anti-symmetric, given by $\begin{Bmatrix} D_1 = 0 \\ D_2 = \theta \end{Bmatrix}$).

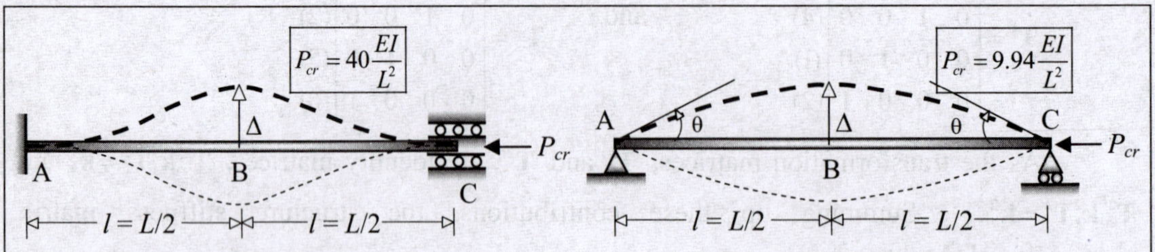

(*a*) Fixed-fixed condition (*b*) pinned-pinned condition

Figure 7.49 Critical buckling mode shapes in an axially loaded member (Example 7.14)

Case (b): both ends pinned (refer Fig 7.48b):

$$\mathbf{T}^1 = \overset{\quad(5)\ (3)\ (1)\ (2)}{\begin{bmatrix} 1 & 0 & 0 & 0 \\ 0 & 1 & 0 & 0 \\ 0 & 0 & 1 & 0 \\ 0 & 0 & 0 & 1 \end{bmatrix}} \begin{matrix} (3) \\ (5) \\ (1) \\ (2) \end{matrix} \quad \text{and} \quad \mathbf{T}^2 = \overset{\quad(1)\ (2)\ (6)\ (4)}{\begin{bmatrix} 1 & 0 & 0 & 0 \\ 0 & 1 & 0 & 0 \\ 0 & 0 & 1 & 0 \\ 0 & 0 & 0 & 1 \end{bmatrix}} \begin{matrix} (1) \\ (2) \\ (6) \\ (4) \end{matrix}$$

Summing up the contributions of $\mathbf{T}^{1^T}\mathbf{k}_*^1\mathbf{T}^1 = \mathbf{k}_*^1$ and $\mathbf{T}^{2^T}\mathbf{k}_*^2\mathbf{T}^2 = \mathbf{k}_*^2$, the structure stiffness matrix, $\mathbf{k} = \mathbf{k_O} + \mathbf{k_G} = \left[\begin{array}{c|c} \mathbf{k_{AA}} & \mathbf{k_{AR}} \\ \hline \mathbf{k_{RA}} & \mathbf{k_{RR}} \end{array}\right]$, satisfying $\mathbf{F} = \mathbf{kD}$, can be assembled as follows:

$$\mathbf{k} = \frac{EI}{l}\left[\begin{array}{cccc|cc} 24/l^2 & 0 & -6/l & 6/l & -12/l^2 & -12/l^2 \\ 0 & 8 & 2 & 2 & 6/l & -6/l \\ -6/l & 2 & 4 & 0 & 6/l & 0 \\ 6/l & 2 & 0 & 4 & 0 & -6/l \\ \hline -12/l^2 & 6/l & 6/l & 0 & 12/l^2 & 0 \\ -12/l^2 & -6/l & 0 & -6/l & 0 & 12/l^2 \end{array}\right] - P\left[\begin{array}{cccc|cc} 2.4/l & 0 & -0.1 & 0.1 & -1.2/l & -1.2/l \\ 0 & 4l/15 & -l/30 & -l/30 & 0.1 & -0.1 \\ -0.1 & -l/30 & 2l/15 & 0 & 0.1 & 0 \\ 0.1 & -l/30 & 0 & 2l/15 & 0 & -0.1 \\ \hline -1.2/l & 0.1 & 0.1 & 0 & 1.2/l & 0 \\ -1.2/l & -0.1 & 0 & -0.1 & 0 & 1.2/l \end{array}\right] \begin{matrix} (1) \\ (2) \\ (3) \\ (4) \\ (5) \\ (6) \end{matrix}$$

$$\mathbf{k}_{AA}\mathbf{D}_A = \mathbf{O} \Rightarrow \begin{bmatrix} \left(24\frac{EI}{l^3}-\frac{2.4P}{l}\right) & 0 & -\left(6\frac{EI}{l^2}-0.1P\right) & \left(6\frac{EI}{l^2}-0.1P\right) \\ 0 & \left(8\frac{EI}{l}-\frac{4Pl}{15}\right) & \left(2\frac{EI}{l}+\frac{Pl}{30}\right) & \left(2\frac{EI}{l}+\frac{Pl}{30}\right) \\ -\left(6\frac{EI}{l^2}-0.1P\right) & \left(2\frac{EI}{l}+\frac{Pl}{30}\right) & \left(4\frac{EI}{l}-\frac{2Pl}{15}\right) & 0 \\ \left(6\frac{EI}{l^2}-0.1P\right) & \left(2\frac{EI}{l}+\frac{Pl}{30}\right) & 0 & \left(4\frac{EI}{l}-\frac{2Pl}{15}\right) \end{bmatrix} \begin{Bmatrix} D_1 \\ D_2 \\ D_3 \\ D_4 \end{Bmatrix} = \begin{Bmatrix} 0 \\ 0 \\ 0 \\ 0 \end{Bmatrix} \quad \text{(b)}$$

We can reduce the kinematic indeterminacy by recognising that the mode shape corresponding to the critical load is symmetric, whereby $D_2 = 0$ and $D_4 = -D_3$ (refer Fig. 7.49b). Thus, eliminating the second row and column in the stiffness matrix in Eq. (b), and condensing out the fourth row (by subtracting the elements in this column to the elements in the third column), Eq. (b) reduces to:

$$\mathbf{k}_{AA}\mathbf{D}_A = \mathbf{O} \Rightarrow \begin{bmatrix} \left(24\frac{EI}{l^3}-\frac{2.4P}{l}\right) & -\left(12\frac{EI}{l^2}-0.2P\right) \\ -\left(6\frac{EI}{l^2}-0.1P\right) & \left(4\frac{EI}{l}-\frac{2Pl}{15}\right) \end{bmatrix} \begin{Bmatrix} D_1 \\ D_3 \end{Bmatrix} = \begin{Bmatrix} 0 \\ 0 \end{Bmatrix} \quad \text{(c)}$$

For a non-trivial solution, considering $\lambda = \frac{P_{cr}l^2}{EI}$

$$\begin{vmatrix} (24-2.4\lambda) & -(12-0.2\lambda) \\ -(6-0.1P\lambda) & \left(4-\frac{2\lambda}{15}\right) \end{vmatrix} = 0 \quad \Rightarrow \quad 0.3\lambda^2 - 10.4\lambda + 24 = 0$$

Of the two eigenvalues, the critical (lower) value is given by:

$$\lambda = \frac{P_{cr}(L/2)^2}{EI} = 2.486 \quad \Rightarrow \quad \boxed{P_{cr} = 9.944\frac{EI}{L^2}}.$$

This is comparable to the exact solution, $P_{cr} = \frac{\pi^2 EI}{L^2} = 9.869\frac{EI}{L^2}$. (For improved accuracy, the element may be divided into three or more equal segments). The critical mode shape is described by the elements of the eigenvector, $\begin{Bmatrix} D_1 \\ D_2 \\ D_3 \\ D_4 \end{Bmatrix} = \begin{Bmatrix} \Delta \\ 0 \\ \theta \\ -\theta \end{Bmatrix}$, as shown in Fig. 7.49b†.

† A relationship between θ and Δ can be established by solving Eq. (c); however, this is likely to be approximate, and for improved accuracy, we must divide the member into many more elements. (However, this academic exercise has hardly any practical significance).

EXAMPLE 7.15

Repeat Example 7.14 for the fixed-fixed case, considering the stiffening effect of a lateral brace (modelled as a spring with stiffness k_b) at the middle of the member. Derive an expression for the critical buckling load P_{cr}, considering two elements only, each of length $l = L/2$, and show how the mode of buckling is dependent on the magnitude of k_b.

SOLUTION

Let us consider the same system of global coordinates as in Example 7.14, but with an additional coordinate ('7') at the base of the spring BD, as shown in Fig. 7.50a. The spring can be modelled as an axial element with two degrees of freedom, one of which is restrained (at D). Its contribution to the stiffness matrix $\mathbf{k}_{AA}$ in Example 7.14 is by incrementing k_{11} by k_b.

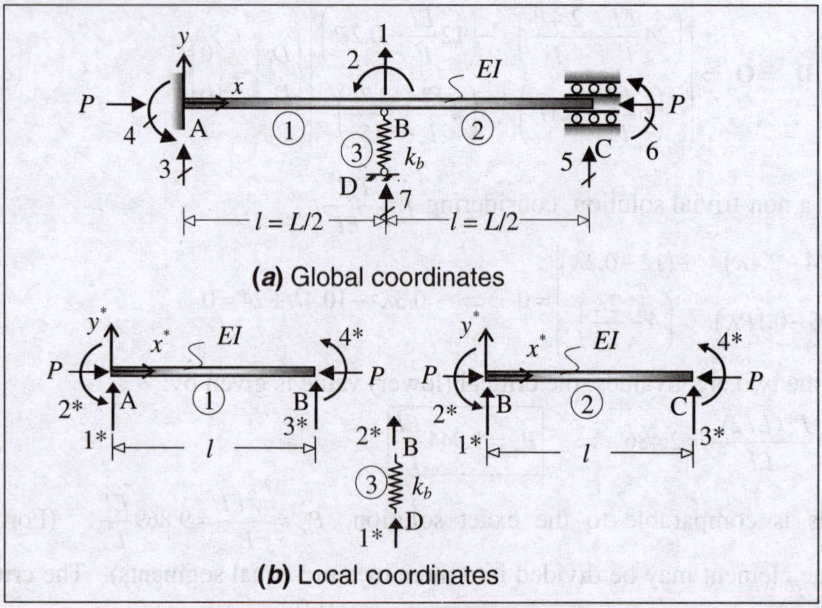

Figure 7.50 Global and local coordinates in a braced compression member (Example 7.15)

Thus, using the results of Example 7.14 for the pinned-pinned case, and including the contribution of the spring stiffness k_b,

$$\mathbf{k}_{AA}\mathbf{D}_A = \mathbf{O} \Rightarrow \begin{bmatrix} \left(24\dfrac{EI}{l^3} - \dfrac{2.4P}{l} + k_b\right) & 0 \\ 0 & \left(8\dfrac{EI}{l} - \dfrac{4Pl}{15}\right) \end{bmatrix} \begin{Bmatrix} D_1 \\ D_2 \end{Bmatrix} = \begin{Bmatrix} 0 \\ 0 \end{Bmatrix}$$

For a non-trivial solution to the equilibrium equation, the *characteristic equation* is given by setting the determinant of $\mathbf{k}_{AA}$ equal to zero.

$$|\mathbf{k}_{AA}| = 0 \Rightarrow \left(24\frac{EI}{l^3} - \frac{2.4P}{l} + k_b\right)\left(8\frac{EI}{l} - \frac{4Pl}{15}\right) = 0 \quad \text{(a)}$$

The two roots (eigenvalues) and eigenvectors of the above equation are given by:

(1) $P_{cr} = 10\frac{EI}{l^2} + \frac{k_b l}{2.4}$; $\begin{Bmatrix} D_1 = \Delta \\ D_2 = 0 \end{Bmatrix}$ symmetric mode (Fig. 7.51a)

(2) $P_{cr} = 30\frac{EI}{l^2}$; $\begin{Bmatrix} D_1 = 0 \\ D_2 = \theta \end{Bmatrix}$ anti-symmetric mode (Fig. 7.51b)

Clearly, for low values of k_b, the first critical load will govern (with a symmetric mode shape, as shown in Fig. 7.51a), and the lowest critical load of $P_{cr} = 10\frac{EI}{l^2}$ is predicted when $k_b = 0$ is very close to the exact solution, $P_{cr} = \frac{4\pi^2 EI}{L^2} = 9.87\frac{EI}{l^2}$ (corresponding to which the mode shape is as shown earlier in Fig. 7.49a). The buckling load is seen to increase linearly with increase in k_b (which is nearly true, as indicated in Fig. 7.23c), but obviously cannot exceed the value, $P_{cr} = 30\frac{EI}{l^2}$, associated with the anti-symmetric mode shape shown in Fig. 7.51b. This occurs when $\frac{k_b l}{2.4} > 20\frac{EI}{l^2} \Rightarrow k_b > 96\frac{EI}{l^3}$ (The exact solution, however, for this condition, as proved in Example 7.5, is given by $k_b > 2.6343\pi^2\frac{EI}{l^3} = 26\frac{EI}{l^3}$, and the corresponding $P_{cr} = 2.047\frac{\pi^2 EI}{l^2} = 20.203\frac{EI}{l^2}$. Thus, we observe that there is considerable error with regard to the second (anti-symmetric) mode; the error can be reduced by increasing the number of elements in the analysis. The error is attributable to the approximation made in the shape function in the element stiffness formulation.

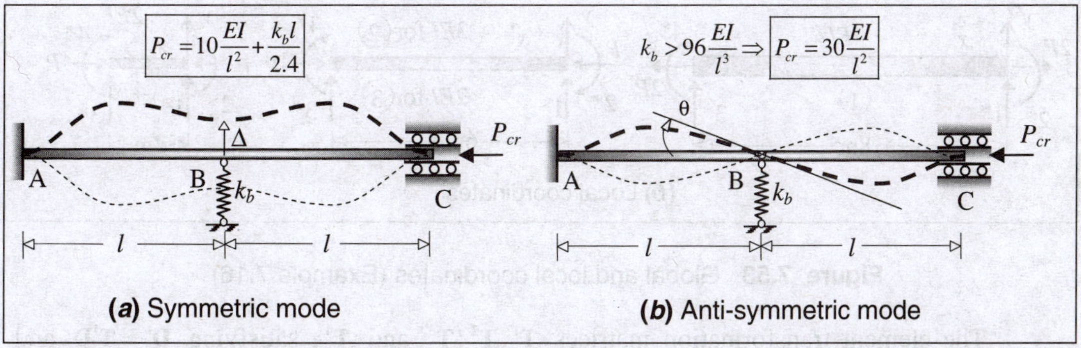

Figure 7.51 Critical buckling mode shapes in a member with middle brace (Example 7.15)

EXAMPLE 7.16

Find the critical buckling load, $P = P_{cr}$, in the continuous beam ABCDE shown in Fig. 7.52, subject to axial forces, $2P$ and P at B and E, as indicated. Assume $EI = 8000$ kNm^2 and that the members are axially rigid.

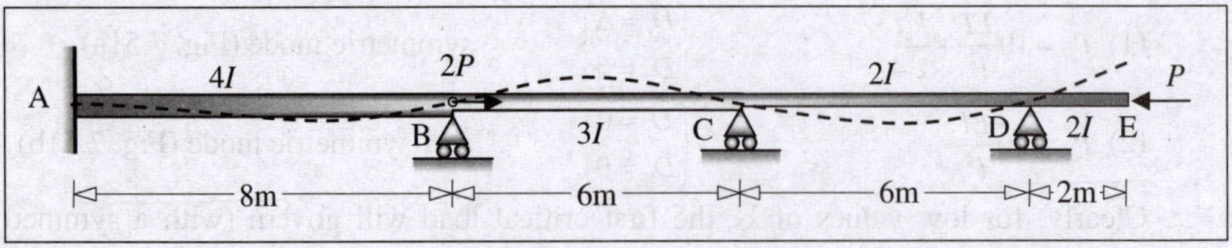

Figure 7.52 Instability in continuous beam system subject to axial forces (Example 7.16)

SOLUTION

There are four prismatic elements, of which AB (element '1') is subject to an axial tension $2P$, while the other three elements are subject to axial compression, P. There are four rotational degrees and one translational *active* degree of freedom ($n = 5$) and five *restrained* global coordinates ($r = 5$), numbered as shown in Fig. 7.53a (similar to Fig. 5.8a in Example 5.3). The local coordinates are shown in Fig. 7.53b.

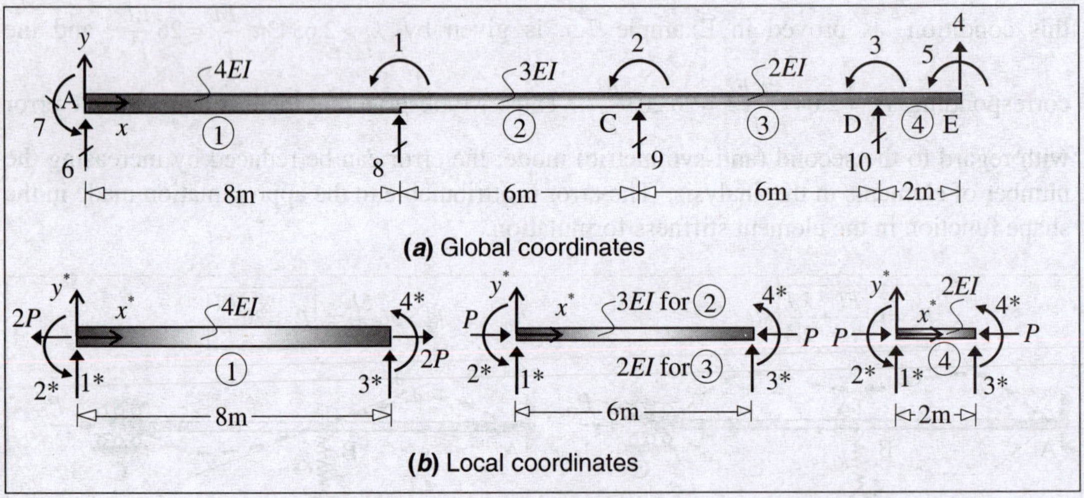

Figure 7.53 Global and local coordinates (Example 7.16)

The element transformation matrices $\mathbf{T}^1, \mathbf{T}^2, \mathbf{T}^3$ and $\mathbf{T}^4$, satisfying $\mathbf{D}_*^i = \mathbf{T}^i\mathbf{D}$ and $\mathbf{F}_*^i = \mathbf{T}^i\mathbf{F}$, take the following form (refer Figs 7.53a and b), with the linking global coordinates indicated in parentheses:

$$\mathbf{T}^1 = \begin{matrix} (6) & (7) & (8) & (1) \\ \end{matrix}\begin{bmatrix} 1 & 0 & 0 & 0 \\ 0 & 1 & 0 & 0 \\ 0 & 0 & 1 & 0 \\ 0 & 0 & 0 & 1 \end{bmatrix}\begin{matrix} (6) \\ (7) \\ (8) \\ (1) \end{matrix}; \quad \mathbf{T}^2 = \begin{matrix} (8) & (1) & (9) & (2) \\ \end{matrix}\begin{bmatrix} 1 & 0 & 0 & 0 \\ 0 & 1 & 0 & 0 \\ 0 & 0 & 1 & 0 \\ 0 & 0 & 0 & 1 \end{bmatrix}\begin{matrix} (8) \\ (1) \\ (9) \\ (2) \end{matrix}; \quad \mathbf{T}^3 = \begin{matrix} (9) & (2) & (10) & (3) \\ \end{matrix}\begin{bmatrix} 1 & 0 & 0 & 0 \\ 0 & 1 & 0 & 0 \\ 0 & 0 & 1 & 0 \\ 0 & 0 & 0 & 1 \end{bmatrix}\begin{matrix} (9) \\ (2) \\ (10) \\ (3) \end{matrix}; \quad \mathbf{T}^4 = \begin{matrix} (10) & (3) & (4) & (5) \\ \end{matrix}\begin{bmatrix} 1 & 0 & 0 & 0 \\ 0 & 1 & 0 & 0 \\ 0 & 0 & 1 & 0 \\ 0 & 0 & 0 & 1 \end{bmatrix}\begin{matrix} (10) \\ (3) \\ (4) \\ (5) \end{matrix}$$

The element stiffness matrices, $\mathbf{k}_*^i = \mathbf{k}_{O*}^i + \mathbf{k}_{G*}^i$, for the four elements, satisfying $\mathbf{F}_*^i = \mathbf{k}_*^i \mathbf{D}_*^i$, take the following form (refer Fig. 7.53b and Eq. 7.71):

$$\mathbf{k}_*^i = \frac{EI_i}{L_i}\begin{bmatrix} 12/L_i^2 & 6/L_i & -12/L_i^2 & 6/L_i \\ 6/L_i & 4 & -6/L_i & 2 \\ -12/L_i^2 & -6/L_i & 12/L_i^2 & -6/L_i \\ 6/L_i & 2 & -6/L_i & 4 \end{bmatrix} - P_i \begin{bmatrix} 1.2/L_i & 0.1 & -1.2/L_i & 0.1 \\ 0.1 & 2L_i/15 & -0.1 & -L_i/30 \\ -1.2/L_i & -0.1 & 1.2/L_i & -0.1 \\ 0.1 & -L_i/30 & -0.1 & 2L_i/15 \end{bmatrix}$$

with $L_1 = 8\text{m},\ I_1 = 4I;\ L_2 = 6\text{m},\ I_2 = 3I;\ L_3 = 6\text{m},\ I_3 = 2I;\ L_4 = 2\text{m},\ I_4 = 2I$

and $P_1 = -2P$, $P_2 = P_3 = P_4 = P$.

$$\mathbf{k}_*^1 = EI\begin{bmatrix} 0.09375 & 0.375 & -0.09375 & 0.375 \\ 0.375 & 2 & -0.375 & 1 \\ -0.09375 & -0.375 & 0.09375 & -0.375 \\ 0.375 & 1 & -0.375 & 2 \end{bmatrix} + 2P\begin{bmatrix} 0.15 & 0.1 & -0.15 & 0.1 \\ 0.1 & 1.06667 & -0.1 & -0.26667 \\ -0.15 & -0.1 & 0.15 & -0.1 \\ 0.1 & -0.26667 & -0.1 & 1.06667 \end{bmatrix}\begin{matrix} (6) \\ (7) \\ (8) \\ (1) \end{matrix}$$

$$\mathbf{k}_*^2 = EI\begin{bmatrix} 0.166667 & 0.5 & -0.166667 & 0.5 \\ 0.5 & 2 & -0.5 & 1 \\ -0.166667 & -0.5 & 0.166667 & -0.5 \\ 0.5 & 1 & -0.5 & 2 \end{bmatrix} - P\begin{bmatrix} 0.2 & 0.1 & -0.2 & 0.1 \\ 0.1 & 0.8 & -0.1 & -0.2 \\ -0.2 & -0.1 & 0.2 & -0.1 \\ 0.1 & -0.2 & -0.1 & 0.8 \end{bmatrix}\begin{matrix} (8) \\ (1) \\ (9) \\ (2) \end{matrix}$$

$$\mathbf{k}_*^3 = EI\begin{bmatrix} 0.11111 & 0.33333 & -0.11111 & 0.33333 \\ 0.33333 & 1.33333 & -0.33333 & 0.66667 \\ -0.11111 & -0.33333 & 0.11111 & -0.33333 \\ 0.33333 & 0.66667 & -0.33333 & 1.33333 \end{bmatrix} - P\begin{bmatrix} 0.2 & 0.1 & -0.2 & 0.1 \\ 0.1 & 0.8 & -0.1 & -0.2 \\ -0.2 & -0.1 & 0.2 & -0.1 \\ 0.1 & -0.2 & -0.1 & 0.8 \end{bmatrix}\begin{matrix} (9) \\ (2) \\ (10) \\ (3) \end{matrix}$$

$$\mathbf{k}_*^4 = EI\begin{bmatrix} 3 & 3 & -3 & 3 \\ 3 & 4 & -3 & 2 \\ -3 & -3 & 3 & -3 \\ 3 & 2 & -3 & 2 \end{bmatrix} - P\begin{bmatrix} 0.6 & 0.1 & -0.6 & 0.1 \\ 0.1 & 0.26667 & -0.1 & -0.06667 \\ -0.6 & -0.1 & 0.6 & -0.1 \\ 0.1 & -0.06667 & -0.1 & 0.26667 \end{bmatrix}\begin{matrix} (10) \\ (3) \\ (4) \\ (5) \end{matrix}$$

As $\mathbf{T}^i$ is an identity matrix, $\mathbf{T}^{i^T}\mathbf{k}_*^i\mathbf{T}^i = \mathbf{k}_*^i$, with the relevant global coordinates (marked in parenthesis) as defined earlier. The structure stiffness matrix $\mathbf{k}$ of order 10×10, satisfying $\mathbf{F} = \mathbf{kD}$, can now be assembled, by summing up the contributions of $\mathbf{T}^{i^T}\mathbf{k}_*^i\mathbf{T}^i = \mathbf{k}_*^i$ for the four elements, at the appropriate coordinate locations:

$$\Rightarrow \quad \mathbf{k} = \left[\begin{array}{c|c} \mathbf{k}_{AA} & \mathbf{k}_{AR} \\ \hline \mathbf{k}_{RA} & \mathbf{k}_{RR} \end{array}\right], \text{ where it is sufficient to generate } \mathbf{k}_{AA} \text{ to solve } [\mathbf{k}_{AA}]\begin{Bmatrix} D_1 \\ D_2 \\ D_3 \\ D_4 \\ D_5 \end{Bmatrix} = \begin{Bmatrix} 0 \\ 0 \\ 0 \\ 0 \\ 0 \end{Bmatrix}.$$

$$\mathbf{k}_{AA} = \begin{bmatrix} (4EI+1.33333P) & (EI+0.2P) & 0 & 0 & 0 \\ (EI+0.2P) & (3.33333EI-1.6P) & (0.66667EI+0.2P) & 0 & 0 \\ 0 & (0.66667EI+0.2P) & (5.33333EI-1.06667P) & (-3EI+0.1P) & (2EI+0.06667P) \\ 0 & 0 & (-3EI+0.1P) & (3EI-0.6P) & (-3EI+0.1P) \\ 0 & 0 & (2EI+0.06667P) & (-3EI+0.1P) & (4EI+0.26667P) \end{bmatrix}$$

(columns (1), (2), (3), (4), (5))

For a non-trivial solution to the equilibrium equation, the *characteristic equation* is obtained by setting the determinant of $\mathbf{k}_{AA}$ (which is a symmetric banded matrix) equal to zero. Setting $\lambda = \frac{P}{EI}$, $|\mathbf{k}_{AA}| = 0 \Rightarrow g(\lambda) = 0$, where

$$g(\lambda) = \begin{vmatrix} (4+1.33333\lambda) & (1+0.2\lambda) & 0 & 0 & 0 \\ (1+0.2\lambda) & (3.33333-1.6\lambda) & (0.66667+0.2\lambda) & 0 & 0 \\ 0 & (0.66667+0.2\lambda) & (5.33333-1.06667\lambda) & (-3+0.1\lambda) & (2+0.06667\lambda) \\ 0 & 0 & (-3+0.1\lambda) & (3-0.6\lambda) & (-3+0.1\lambda) \\ 0 & 0 & (2+0.06667\lambda) & (-3+0.1\lambda) & (4+0.26667\lambda) \end{vmatrix}$$

Of the five eigenvalues possible that will render $g(\lambda) = 0$, we are interested in the lowest value of λ, and we can get this by adopting a numerical procedure like the bisection method, using MATLAB to conveniently compute the determinant, and starting with the lowest possible value of λ. We note that element DE will behave like an unbraced element (cantilever with a rotational spring at D), with $(P_{cr})_{DE} < \frac{\pi^2 E(2I)}{(2\times 2.0)^2} \Rightarrow 0 < \lambda < 4.935$. Thus, $\lambda = 0$ provides an absolute lowest bound. Using MATLAB (or any other tool),

$\lambda = 0 \Rightarrow g(\lambda) = +44.000$

Now, we should gradually increase λ in small increments (say, steps of 0.1) and observe when $g(\lambda)$ turns from a positive value to a negative value. This occurs at $\lambda = 0.5$:

$\lambda = 0.5 \Rightarrow g(\lambda) = -6.180$

Now, applying the bisection method, we can easily converge on the 'exact' value, which turns out to be $\lambda_{cr} = 0.4283$, corresponding to which, $g(\lambda) = 0.0020$.

Substituting EI = 8000 kNm^2, the desired critical buckling load is obtained as:

P_{cr} = 0.4283×8000 = **3426 kN**

EXAMPLE 7.17

Find the critical buckling load, $P = P_{cr}$, in the portal frame ABCD shown in Fig. 7.54, with the columns AB and CD subject to axial compressive forces, $2P$ and P, as indicated. Assume the column elements to have a square cross section, 300×300 mm and the beam to have a size, 300×450 mm. Assume all elements to have an elastic modulus, $E = 2.5\times10^4$ N/mm^2.

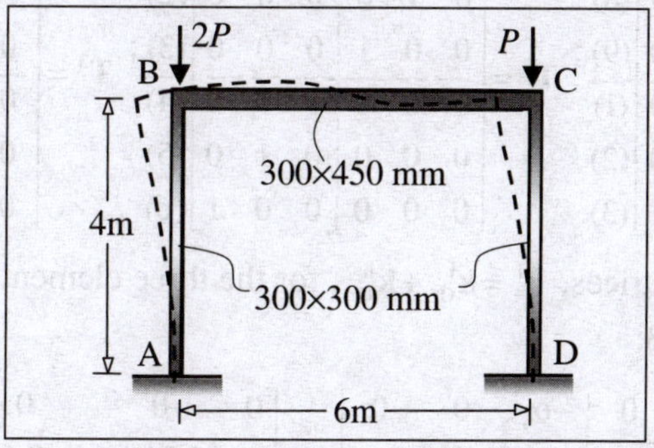

Figure 7.54 Elastic buckling of a portal frame (Example 7.17)

SOLUTION

There are three prismatic frame elements and four joints with six *active* and six *restrained* degrees of freedom ($n = r = 6$), with global coordinates, as indicated in Fig. 7.55a; local coordinates are shown in Fig. 7.55b (similar to Fig. 6.9 in Example 6.2)

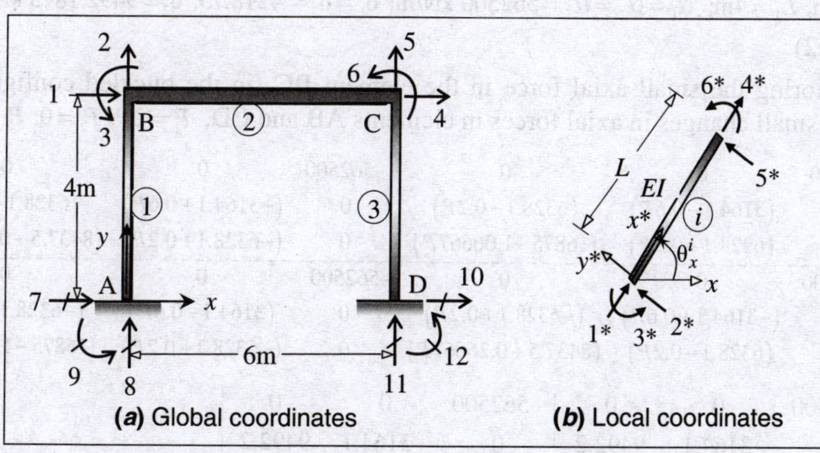

Figure 7.55 Global and local coordinates (Example 7.17)

The element transformation matrices $\mathbf{T}^1, \mathbf{T}^2$ and $\mathbf{T}^3$, satisfying $\mathbf{D}_*^i = \mathbf{T}^i\mathbf{D}$ and $\mathbf{F}_*^i = \mathbf{T}^i\mathbf{F}$, take the following form (refer Eq. 6.8 and Figs 7.55a and b), with the linking global coordinates indicated in parentheses:

$$\mathbf{T}^1 = \begin{array}{c} (7)\,(8)\,(9)\,(1)\,(2)\,(3) \\ \left[\begin{array}{ccc|ccc} 0 & 1 & 0 & 0 & 0 & 0 \\ -1 & 0 & 0 & 0 & 0 & 0 \\ 0 & 0 & 1 & 0 & 0 & 0 \\ \hline 0 & 0 & 0 & 0 & 1 & 0 \\ 0 & 0 & 0 & -1 & 0 & 0 \\ 0 & 0 & 0 & 0 & 0 & 1 \end{array}\right] \end{array} \begin{array}{c} \\ (7) \\ (8) \\ (9) \\ (1) \\ (2) \\ (3) \end{array}; \quad \mathbf{T}^2 = \begin{array}{c} (1)\,(2)\,(3)\,(4)\,(5)\,(6) \\ \left[\begin{array}{ccc|ccc} 1 & 0 & 0 & 0 & 0 & 0 \\ 0 & 1 & 0 & 0 & 0 & 0 \\ 0 & 0 & 1 & 0 & 0 & 0 \\ \hline 0 & 0 & 0 & 1 & 0 & 0 \\ 0 & 0 & 0 & 0 & 1 & 0 \\ 0 & 0 & 0 & 0 & 0 & 1 \end{array}\right] \end{array} \begin{array}{c} \\ (1) \\ (2) \\ (3) \\ (4) \\ (5) \\ (6) \end{array}; \quad \mathbf{T}^3 = \begin{array}{c} (10)\,(11)\,(12)\,(4)\,(5)\,(6) \\ \left[\begin{array}{ccc|ccc} 0 & 1 & 0 & 0 & 0 & 0 \\ -1 & 0 & 0 & 0 & 0 & 0 \\ 0 & 0 & 1 & 0 & 0 & 0 \\ \hline 0 & 0 & 0 & 0 & 1 & 0 \\ 0 & 0 & 0 & -1 & 0 & 0 \\ 0 & 0 & 0 & 0 & 0 & 1 \end{array}\right] \end{array} \begin{array}{c} \\ (10) \\ (11) \\ (12) \\ (4) \\ (5) \\ (6) \end{array}$$

The element stiffness matrices, $\mathbf{k}_*^i = \mathbf{k}_{O*}^i + \mathbf{k}_{G*}^i$, for the three elements, satisfying $\mathbf{F}_*^i = \mathbf{k}_*^i\mathbf{D}_*^i$ can be derived using Eq. 7.73:

$$\mathbf{k}_*^i = \left[\begin{array}{ccc|ccc} \alpha_i & 0 & 0 & -\alpha_i & 0 & 0 \\ 0 & \beta_i & \chi_i & 0 & -\beta_i & \chi_i \\ 0 & \chi_i & 4\delta_i & 0 & -\chi_i & 2\delta_i \\ \hline -\alpha_i & 0 & 0 & \alpha_i & 0 & 0 \\ 0 & -\beta_i & -\chi_i & 0 & \beta_i & -\chi_i \\ 0 & \chi_i & 2\delta_i & 0 & -\chi_i & 4\delta_i \end{array}\right] - P_i \left[\begin{array}{ccc|ccc} 0 & 0 & 0 & 0 & 0 & 0 \\ 0 & 1.2/L_i & 0.1 & 0 & -1.2/L_i & 0.1 \\ 0 & 0.1 & 2L_i/15 & 0 & -0.1 & -L_i/30 \\ \hline 0 & 0 & 0 & 0 & 0 & 0 \\ 0 & -1.2/L_i & -0.1 & 0 & 1.2/L_i & -0.1 \\ 0 & 0.1 & -L_i/30 & 0 & -0.1 & 2L_i/15 \end{array}\right]$$

$$\alpha_i = (EA)_i/L_i;\ \beta_i = 12(EI)_i/L_i^3;\ \chi_i = 6(EI)_i/L_i^2;\ (EI)_i/L_i$$

$L_1 = L_3 = 6\text{m},\ L_2 = 4\text{m};\ \alpha_1 = \alpha_2 = \alpha_3 = 562500$ kN/m; $\delta_1 = \delta_3 = 4218.75,\ \delta_2 = 9492.1875$ kNm (refer Example 6.2).

Ignoring the small axial force in the element BC (in the buckled configuration) as well as the small changes in axial forces in elements AB and CD, $P_1 = 2P;\ P_2 = 0;\ P_3 = P$,

$$\mathbf{k}_*^1 = \left[\begin{array}{ccc|ccc} 562500 & 0 & 0 & -562500 & 0 & 0 \\ 0 & (3164.1-0.6P) & (6328.1-0.2P) & 0 & (-3164.1+0.6P) & (6328.1-0.2P) \\ 0 & (6328.1-0.2P) & (16875-1.06667P) & 0 & (-6328.1+0.2P) & (8437.5+0.26667P) \\ \hline -562500 & 0 & 0 & 562500 & 0 & 0 \\ 0 & (-3164.1+0.6P) & (-6328.1+0.2P) & 0 & (3164.1-0.6P) & (-6328.1+0.2P) \\ 0 & (6328.1-0.2P) & (8437.5+0.26667P) & 0 & (-6328.1+0.2P) & (16875-1.06667P) \end{array}\right]$$

$$\mathbf{k}_*^2 = \left[\begin{array}{ccc|ccc} 562500 & 0 & 0 & -562500 & 0 & 0 \\ 0 & 3164.1 & 9492.2 & 0 & -3164.1 & 9492.2 \\ 0 & 9492.2 & 37968.8 & 0 & -9492.2 & 18984.4 \\ \hline -562500 & 0 & 0 & 562500 & 0 & 0 \\ 0 & -3164.1 & -9492.2 & 0 & 3164.1 & -9492.2 \\ 0 & 9492.2 & 18984.4 & 0 & -9492.2 & 37968.8 \end{array}\right]$$

$$\mathbf{k}_*^3 = \left[\begin{array}{ccc|ccc} 562500 & 0 & 0 & -562500 & 0 & 0 \\ 0 & (3164.1-0.3P) & (6328.1-0.1P) & 0 & (-3164.1+0.3P) & (6328.1-0.1P) \\ 0 & (6328.1-0.1P) & (16875-0.53333P) & 0 & (-6328.1+0.1P) & (8437.5+0.13333P) \\ \hline -562500 & 0 & 0 & 562500 & 0 & 0 \\ 0 & (-3164.1+0.3P) & (-6328.1+0.1P) & 0 & (3164.1-0.3P) & (-6328.1+0.1P) \\ 0 & (6328.1-0.1P) & (8437.5+0.13333P) & 0 & (-6328.1+0.1P) & (16875-0.53333P) \end{array}\right]$$

By summing up the the contributions of $\mathbf{T}^{i^T}\mathbf{k}_*^i\mathbf{T}^i = \mathbf{k}^i = \left[\begin{array}{c|c} \mathbf{k}_A^i & \mathbf{k}_C^{i\,T} \\ \hline \mathbf{k}_C^i & \mathbf{k}_B^i \end{array}\right]$ of order 6×6 for each of the three elements (using Eq. 6.10), at the appropriate coordinate locations, the structure stiffness matrix $\mathbf{k}$ of order 12×12, satisfying $\mathbf{F}=\mathbf{kD}$, can be assembled, as done in Example 6.2. It takes the following partitioned form:

$$\mathbf{k} = \left[\begin{array}{c|c} \mathbf{k}_{AA} & \mathbf{k}_{AR} \\ \hline \mathbf{k}_{RA} & \mathbf{k}_{RR} \end{array}\right] = \left[\begin{array}{cc|cc} \mathbf{k}_B^1+\mathbf{k}_A^2 & \mathbf{k}_C^{2\,T} & \mathbf{k}_C^1 & \mathbf{O} \\ \mathbf{k}_C^2 & \mathbf{k}_B^2+\mathbf{k}_B^3 & \mathbf{O} & \mathbf{k}_C^3 \\ \hline \mathbf{k}_C^{1\,T} & \mathbf{O} & \mathbf{k}_A^1 & \mathbf{O} \\ \mathbf{O} & \mathbf{k}_C^{3\,T} & \mathbf{O} & \mathbf{k}_A^3 \end{array}\right]$$, of which only $\mathbf{k}_{AA}$ is of interest here.

$$\mathbf{k}_{AA} = \left[\begin{array}{c|c} \mathbf{k}_B^1+\mathbf{k}_A^2 & \mathbf{k}_C^{2\,T} \\ \hline \mathbf{k}_C^2 & \mathbf{k}_B^2+\mathbf{k}_B^3 \end{array}\right]$$

where:

$$\mathbf{k}_B^1 = \begin{bmatrix} (3164.1-0.6P) & 0 & (6328.1-0.2P) \\ 0 & 562500 & 0 \\ (6328.1-0.2P) & 0 & (16875-1.06667P) \end{bmatrix}$$

$$\mathbf{k}_A^2 = \begin{bmatrix} 562500 & 0 & 0 \\ 0 & 3164.1 & 9492.2 \\ 0 & 9492.2 & 37968.8 \end{bmatrix};\ \mathbf{k}_B^2 = \begin{bmatrix} 562500 & 0 & 0 \\ 0 & 3164.1 & -9492.2 \\ 0 & -9492.2 & 37968.8 \end{bmatrix}$$

$$\mathbf{k}_C^2 = \begin{bmatrix} -562500 & 0 & 0 \\ 0 & -3164.1 & -9492.2 \\ 0 & 9492.2 & 18984.4 \end{bmatrix}$$

$$\mathbf{k}_B^3 = \begin{bmatrix} (3164.1-0.3P) & 0 & (6328.1-0.1P) \\ 0 & 562500 & 0 \\ (6328.1-0.1P) & 0 & (16875-0.53333P) \end{bmatrix}$$

Substituting,

$$\mathbf{k}_{AA}=\left[\begin{array}{ccc|ccc}(565664.1-0.6P) & 0 & (6328.1-0.2P) & -562500 & 0 & 0\\ 0 & 565664.1 & 9492.2 & 0 & -3164.1 & 9492.2\\ (6328.1-0.2P) & 9492.2 & (54843.8-1.06667P) & 0 & -9492.2 & 18984.4\\ \hline -562500 & 0 & 0 & (565664.1-0.3P) & 0 & (6328.1-0.1P)\\ 0 & -3164.1 & -9492.2 & 0 & 565664.1 & -9492.2\\ 0 & 9492.2 & 18984.4 & (6328.1-0.1P) & -9492.2 & (54843.8-0.53333P)\end{array}\right]$$

For a non-trivial solution to the equilibrium equation $\mathbf{k}_{AA}\mathbf{D}_A=\mathbf{O}$, the *characteristic equation* is obtained by setting the determinant of $\mathbf{k}_{AA}$ (which is a symmetric matrix) equal to zero. For convenience, in order to avoid dealing with large numbers, let us set $\lambda=\dfrac{P}{EI_3}$, where $EI_3=EI_1=(2.5\times10^7)(6.75\times10^{-4})=16875\text{ kNm}^2$.

$|\mathbf{k}_{AA}|=0\Rightarrow g(\lambda)=0$, where

$$g(\lambda)=\left|\begin{array}{ccc|ccc}(33.5208-0.6\lambda) & 0 & (0.37500-0.2\lambda) & -33.3333 & 0 & 0\\ 0 & 33.5208 & 0.56250 & 0 & -0.18750 & 0.56250\\ (0.37500-0.2\lambda) & 0.56250 & (3.25000-1.06667\lambda) & 0 & -0.56250 & 1.12500\\ \hline -33.3333 & 0 & 0 & (33.5208-0.3\lambda) & 0 & (0.37500-0.1\lambda)\\ 0 & -0.18750 & -0.56250 & 0 & 33.5208 & -0.56250\\ 0 & 0.56250 & 1.12500 & (0.37500-0.1\lambda) & -0.56250 & (3.25000-0.53333\lambda)\end{array}\right|\begin{array}{c}(1)\\(2)\\(3)\\(4)\\(5)\\(6)\end{array}$$

Of the six eigenvalues possible that will render $g(\lambda)=0$, we are interested in the lowest value of λ. We note that the buckling will be governed by the more heavily loaded column AB, and the lower bound solution is given by considering its cantilever buckling behaviour, whereby $\dfrac{(P_{cr})_{AB}}{EI_3}=2\lambda_{cr}>\dfrac{\pi^2}{(2\times4.0)^2}\Rightarrow\lambda_{cr}>0.077$. Beginning with this solution, the value of $g(\lambda)$ is computed (using MATLAB to find the determinant of the 6×6 stiffness matrix):

$$\lambda=0.077\ \Rightarrow\ g(\lambda)=+80681$$

Now, we should gradually increase λ in small increments (say, steps of 0.1, starting with 0.2) and observe when $g(\lambda)$ turns from a positive value to a negative value. This occurs at $\lambda=0.4$:

$$\lambda=0.4\ \Rightarrow\ g(\lambda)=-9403.6$$

Now, applying the bisection method, we can easily converge on the 'exact' value, which turns out to be $\lambda_{cr}=0.3598$, corresponding to which, $g(\lambda)=+7.3^{\dagger}$.

Substituting EI = 16875 kNm^2, the desired critical buckling load is obtained as:

† This is as close to $g(\lambda)=0$ as possible, for the level of accuracy we need. Note that the 'exact' value lies between $\lambda=0.3598$ and $\lambda=0.3599$, corresponding to which, $g(\lambda)=-16.7$).

$P_{cr} = 0.3598 \times 16875 =$ **6072 kN**

Note:

A more accurate solution is possible by introducing nodes in the middle of the columns AB and CD. (However, such an exercise will reveal that the change (reduction) in P_{cr} is very small, in this case). Also, it is possible to reduce the kinematic indeterminacy in the structure, from six to three, by assuming all the members to be axially rigid. With this assumption, $D_2 = D_5 = 0$ and $D_1 = D_4$ (refer Fig. 7.55a), whereby the stiffness matrix can be condensed into an order 3×3.

EXAMPLE 7.18

Re-analyse the frame in Example 7.17 (Fig. 7.54), considering the frame (with the same loading) to have an internal hinge at the joint C.

SOLUTION

Let us define the same global and local coordinates as in Example 7.17. However, as there is an internal hinge provided at the joint C (refer Fig. 7.56a), there will be no moment transfer possible across this junction. Hence, there will be no common rotational degree of freedom at C. The stiffness matrices of BC and CD can be suitably modified to account for the moment release at C, but this will result in zero rotational stiffness at C. In order to address this problem, we visualise an imaginary clamp at C (providing restraint against rotation at C), as indicated in Fig 7.56b (as done earlier in Example 6.3). Thus, '6' becomes a restrained coordinate, and the number of active degrees of freedom reduces to five ($n = 5$), while the *restrained* global coordinates increases to seven ($r = 8$), numbered as shown in Fig. 7.56b.

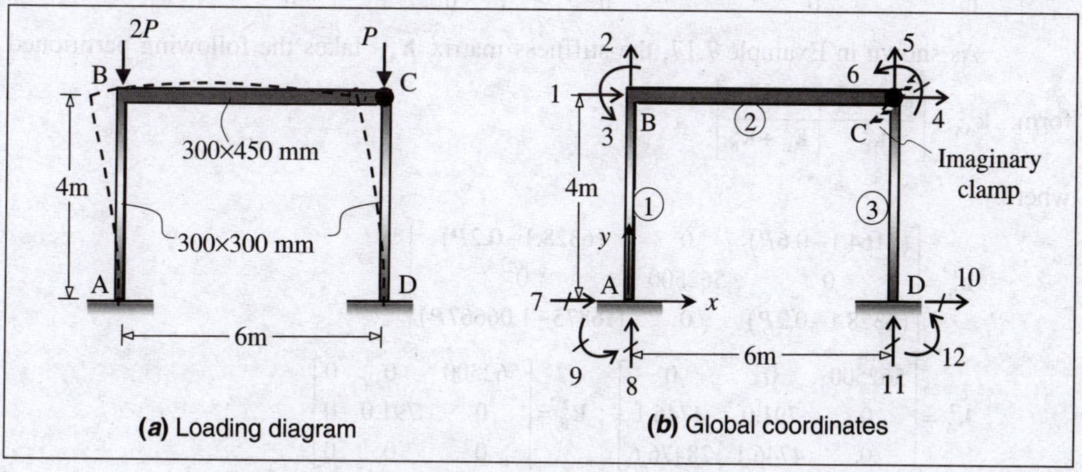

Figure 7.56 Loading diagram and global coordinates (Example 7.18)

The element transformation matrices $\mathbf{T}^1, \mathbf{T}^2$ and $\mathbf{T}^3$, satisfying $\mathbf{D}_*^i = \mathbf{T}^i\mathbf{D}$ and $\mathbf{F}_*^i = \mathbf{T}^i\mathbf{F}$, are as defined in Example 7.17.

The element stiffness matrices for elements '2' and '3' get modified on account of the hinge at the end node (refer Eqns 6.17b and 7.75b), while that of element '1' remains the same as in Example 7.17.

$$\mathbf{k}_*^1 = \left[\begin{array}{ccc|ccc} 562500 & 0 & 0 & -562500 & 0 & 0 \\ 0 & (3164.1-0.6P) & (6328.1-0.2P) & 0 & (-3164.1+0.6P) & (6328.1-0.2P) \\ 0 & (6328.1-0.2P) & (16875-1.06667P) & 0 & (-6328.1+0.2P) & (8437.5+0.26667P) \\ \hline -562500 & 0 & 0 & 562500 & 0 & 0 \\ 0 & (-3164.1+0.6P) & (-6328.1+0.2P) & 0 & (3164.1-0.6P) & (-6328.1+0.2P) \\ 0 & (6328.1-0.2P) & (8437.5+0.26667P) & 0 & (-6328.1+0.2P) & (16875-1.06667P) \end{array}\right]$$

$$\mathbf{k}_*^2 = \left[\begin{array}{ccc|ccc} 562500 & 0 & 0 & -562500 & 0 & 0 \\ 0 & 791.0 & 4746.1 & 0 & -791.0 & 0 \\ 0 & 4746.1 & 28476.6 & 0 & -4746.1 & 0 \\ \hline -562500 & 0 & 0 & 562500 & 0 & 0 \\ 0 & -791.0 & -4746.1 & 0 & 791.0 & 0 \\ 0 & 0 & 0 & 0 & 0 & 0 \end{array}\right] \quad \text{(as in Example 6.3)}$$

$$\mathbf{k}_*^3 = \left[\begin{array}{ccc|ccc} 562500 & 0 & 0 & -562500 & 0 & 0 \\ 0 & (791.0-0.03125P) & (3164.1-0.125P) & 0 & (-791.0+0.03125P) & 0 \\ 0 & (3164.1-0.125P) & (12656.3-0.5P) & 0 & (-3164.1+0.125P) & 0 \\ \hline -562500 & 0 & 0 & 562500 & 0 & 0 \\ 0 & (-791.0+0.03125P) & (-3164.1+0.125P) & 0 & (791.0-0.03125P) & 0 \\ 0 & 0 & 0 & 0 & 0 & 0 \end{array}\right]$$

As shown in Example 7.17, the stiffness matrix $\mathbf{k}_{AA}$ takes the following partitioned form: $\mathbf{k}_{AA} = \left[\begin{array}{c|c} \mathbf{k}_B^1 + \mathbf{k}_A^2 & \mathbf{k}_C^{2\,T} \\ \hline \mathbf{k}_C^2 & \mathbf{k}_B^2 + \mathbf{k}_B^3 \end{array}\right]$

where:

$$\mathbf{k}_B^1 = \begin{bmatrix} (3164.1-0.6P) & 0 & (6328.1-0.2P) \\ 0 & 562500 & 0 \\ (6328.1-0.2P) & 0 & (16875-1.06667P) \end{bmatrix}$$

$$\mathbf{k}_A^2 = \begin{bmatrix} 562500 & 0 & 0 \\ 0 & 791.0 & 4746.1 \\ 0 & 4746.1 & 28476.6 \end{bmatrix}; \quad \mathbf{k}_B^2 = \begin{bmatrix} 562500 & 0 & 0 \\ 0 & 791.0 & 0 \\ 0 & 0 & 0 \end{bmatrix}$$

$$\mathbf{k}_{\mathbf{C}}^{2} = \begin{bmatrix} -562500 & 0 & 0 \\ 0 & -791.0 & -4746.1 \\ 0 & 0 & 0 \end{bmatrix}$$

$$\mathbf{k}_{\mathbf{B}}^{3} = \begin{bmatrix} (791.0-0.03125P) & 0 & 0 \\ 0 & 562500 & 0 \\ 0 & 0 & 0 \end{bmatrix}$$

$$\Rightarrow \mathbf{k}_{\mathbf{AA}} = \begin{bmatrix} (565664.1-0.6P) & 0 & (6328.1-0.2P) & -562500 & 0 \\ 0 & 563291 & 4746.1 & 0 & -791.0 \\ (6328.1-0.2P) & 4746.1 & (45351.6-1.06667P) & 0 & -4746.1 \\ -562500 & 0 & 0 & (563291-0.03125P) & 0 \\ 0 & -791.0 & -4746.1 & 0 & 563291 \end{bmatrix} \begin{matrix} (1) \\ (2) \\ (3) \\ (4) \\ (5) \end{matrix}$$

For a non-trivial solution to the equilibrium equation $\mathbf{k}_{\mathbf{AA}}\mathbf{D}_{\mathbf{A}} = \mathbf{O}$, the *characteristic equation* is obtained by setting the determinant of $\mathbf{k}_{\mathbf{AA}}$ equal to zero. As in Example 7.17, let us set $\lambda = \dfrac{P}{EI_3}$, where $EI_3 = EI_1 = (2.5\times10^7)(6.75\times10^{-4}) = 16875 \text{ kNm}^2$.

$|\mathbf{k}_{\mathbf{AA}}| = 0 \Rightarrow g(\lambda) = 0$, where

$$g(\lambda) = \begin{vmatrix} (33.5208-0.6\lambda) & 0 & (0.375-0.2\lambda) & -33.3333 & 0 \\ 0 & 33.3802 & 0.28125 & 0 & -0.04687 \\ (0.375-0.2\lambda) & 0.28125 & (2.6875-1.06667\lambda) & 0 & -0.28125 \\ -33.3333 & 0 & 0 & (33.3802-0.03125\lambda) & 0 \\ 0 & -0.04687 & -0.28125 & 0 & 33.3802 \end{vmatrix} \begin{matrix} (1) \\ (2) \\ (3) \\ (4) \\ (5) \end{matrix}$$

Of the five eigenvalues possible that will render $g(\lambda) = 0$, we are interested in the lowest value of λ. As in Example 7.18, let us begin with the lower bound solution (cantilever behaviour of column AB): $\lambda = 0.077$. (The frame is less stiff than the one in Example 7.17, and hence we know that $\lambda < 0.36$). Using MATLAB to find the determinant of the 5×5 matrix,

$\lambda = 0.077 \Rightarrow g(\lambda) = +13155$

Now, we should gradually increase λ in small increments (say, steps of 0.1, starting with 0.2) and observe when $g(\lambda)$ turns from a positive value to a negative value.

$\lambda = 0.3 \Rightarrow g(\lambda) = +262.4$

$\lambda = 0.31 \Rightarrow g(\lambda) = -260.7$

Thus, we quickly converge on the solution: $\lambda_{cr} = 0.3050$, corresponding to which, $g(\lambda) = +0.3$. Substituting EI = 16875 kNm^2, the desired critical buckling load is obtained as:

$P_{cr} = 0.3050\times16875 =$ **5147 kN**

(which, as expected, is less than the value of 6072 kN obtained in Example 7.17).

EXAMPLE 7.19

Find the critical buckling load, $P = P_{cr}$, in the two-hinged plane frame with inclined legs ABCD shown in Fig. 7.57, with vertical loads equal to P applied at B and C, as indicated. Assume all three frame elements to have a uniform square cross-section, 300×300 mm, with an elastic modulus, $E = 2.5\times10^4$ N/mm^2. Also sketch the deflected shape as $P \rightarrow P_{cr}$.

SOLUTION

There are three prismatic frame elements and four joints with eight *active* and four *restrained* degrees of freedom ($n = 8; r = 4$). The numbering of global coordinates (with the restrained coordinates at the end of the list) is as indicated in Fig. 7.58a; local coordinates are shown in Fig. 7.58b (similar to Fig. 6.16a in Example 6.4).

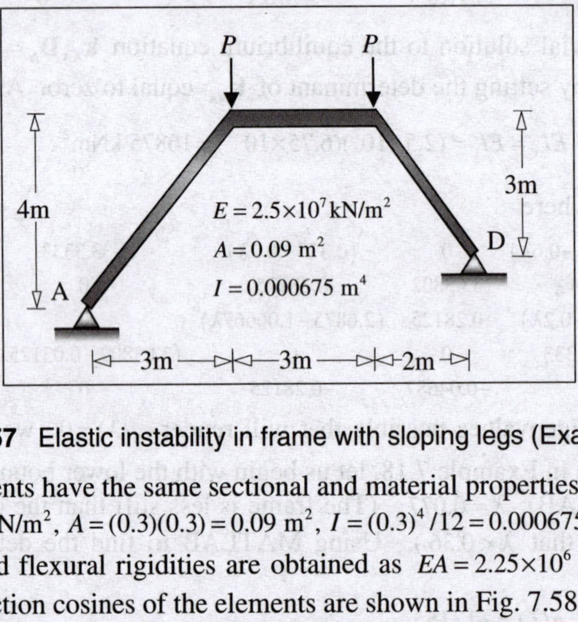

Figure 7.57 Elastic instability in frame with sloping legs (Example 7.19)

All the elements have the same sectional and material properties:
$E = 2.5\times10^7\,\text{kN/m}^2$, $A = (0.3)(0.3) = 0.09\ \text{m}^2$, $I = (0.3)^4/12 = 0.000675\ \text{m}^4$.

The axial and flexural rigidities are obtained as $EA = 2.25\times10^6$ kN, $EI = 16875$ kNm^2. The lengths and direction cosines of the elements are shown in Fig. 7.58.

We observe that the nodal loads applied on the frame in Fig. 7.57 will induce bending in the frame, prior to the onset of elastic instability, and that the axial forces in the three members cannot be known without carrying out a first-order analysis of the frame, similar to the one carried out in Example 6.4. There are no fixed-end moments involved, as the loads applied are nodal loads, defined by $F_2 = F_5 = -P$ (with the other six elements in the load vector $\mathbf{F}_A$ equal to zero).

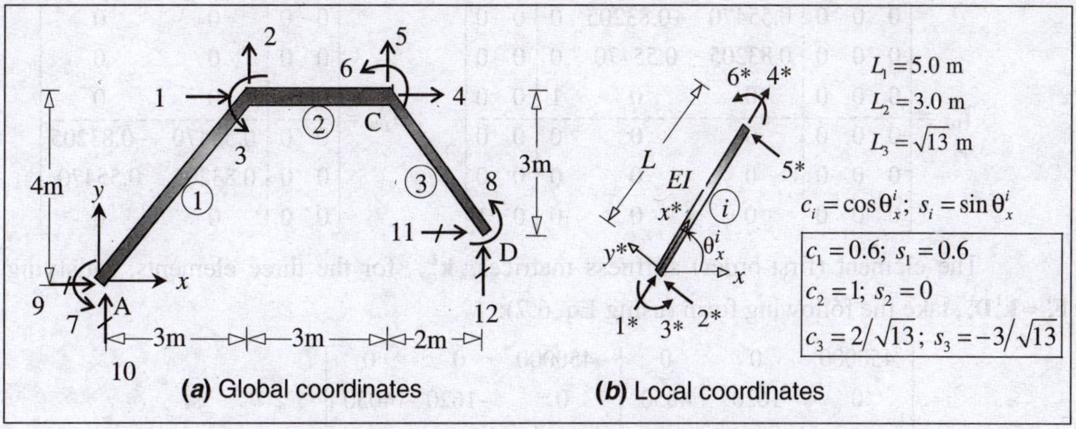

Figure 7.58 Global and local coordinates (Example 7.19)

The following transformation and 'first-order' stiffness matrices derived in Example 6.4 are summarised below for the three elements, with the linking global coordinates indicated in parentheses. The concept of displacement transformation matrix, defined by $\mathbf{D}_*^i = \left[\mathbf{T}_{DA}^i \mid \mathbf{T}_{DR}^i\right]\left[\dfrac{\mathbf{D}_A}{\mathbf{D}_R}\right]$, has been used here.

$$\mathbf{T}_{DA}^1 = \begin{array}{c} \begin{matrix}(1) & (2) & (3) & (4) & (5) & (6) & (7) & (8)\end{matrix} \\ \left[\begin{array}{ccc|ccc|cc} 0 & 0 & 0 & 0 & 0 & 0 & 0 & 0 \\ 0 & 0 & 0 & 0 & 0 & 0 & 0 & 0 \\ 0 & 0 & 0 & 0 & 0 & 0 & 1 & 0 \\ \hline 0.6 & 0.8 & 0 & 0 & 0 & 0 & 0 & 0 \\ -0.8 & 0.6 & 0 & 0 & 0 & 0 & 0 & 0 \\ 0 & 0 & 1 & 0 & 0 & 0 & 0 & 0 \end{array}\right]\end{array} \qquad \mathbf{T}_{DR}^1 = \begin{array}{c} \begin{matrix}(9) & (10) & (11) & (12)\end{matrix} \\ \left[\begin{array}{cc|cc} 0.6 & 0.8 & 0 & 0 \\ -0.8 & 0.6 & 0 & 0 \\ 0 & 0 & 0 & 0 \\ \hline 0 & 0 & 0 & 0 \\ 0 & 0 & 0 & 0 \\ 0 & 0 & 0 & 0 \end{array}\right]\end{array}$$

$$\mathbf{T}_{DA}^2 = \left[\begin{array}{ccc|ccc|cc} 1 & 0 & 0 & 0 & 0 & 0 & 0 & 0 \\ 0 & 1 & 0 & 0 & 0 & 0 & 0 & 0 \\ 0 & 0 & 1 & 0 & 0 & 0 & 0 & 0 \\ \hline 0 & 0 & 0 & 1 & 0 & 0 & 0 & 0 \\ 0 & 0 & 0 & 0 & 1 & 0 & 0 & 0 \\ 0 & 0 & 0 & 0 & 0 & 1 & 0 & 0 \end{array}\right]; \qquad \mathbf{T}_{DR}^2 = \left[\begin{array}{cc|cc} 0 & 0 & 0 & 0 \\ 0 & 0 & 0 & 0 \\ 0 & 0 & 0 & 0 \\ \hline 0 & 0 & 0 & 0 \\ 0 & 0 & 0 & 0 \\ 0 & 0 & 0 & 0 \end{array}\right]$$

$$\mathbf{T}_{DA}^{3}=\left[\begin{array}{ccc|ccc|cc} 0 & 0 & 0 & 0.55470 & -0.83205 & 0 & 0 & 0 \\ 0 & 0 & 0 & 0.83205 & 0.55470 & 0 & 0 & 0 \\ 0 & 0 & 0 & 0 & 0 & 1 & 0 & 0 \\ \hline 0 & 0 & 0 & 0 & 0 & 0 & 0 & 0 \\ 0 & 0 & 0 & 0 & 0 & 0 & 0 & 0 \\ 0 & 0 & 0 & 0 & 0 & 0 & 0 & 1 \end{array}\right] ; \quad \mathbf{T}_{DR}^{3}=\left[\begin{array}{cc|cc} 0 & 0 & 0 & 0 \\ 0 & 0 & 0 & 0 \\ 0 & 0 & 0 & 0 \\ \hline 0 & 0 & 0.55470 & -0.83205 \\ 0 & 0 & 0.83205 & 0.55470 \\ 0 & 0 & 0 & 0 \end{array}\right]$$

The element (first-order) stiffness matrices, $\mathbf{k}_{O*}^{i}$, for the three elements, satisfying $\mathbf{F}_{*}^{i}=\mathbf{k}_{*}^{i}\mathbf{D}_{*}^{i}$, take the following form (using Eq. 6.7):

$$\mathbf{k}_{O*}^{1}=\left[\begin{array}{ccc|ccc} 450000 & 0 & 0 & -450000 & 0 & 0 \\ 0 & 1620 & 4050 & 0 & -1620 & 4050 \\ 0 & 4050 & 13500 & 0 & -4050 & 6750 \\ \hline -450000 & 0 & 0 & 450000 & 0 & 0 \\ 0 & -1620 & -4050 & 0 & 1620 & -4050 \\ 0 & 4050 & 6750 & 0 & -4050 & 13500 \end{array}\right]$$

$$\mathbf{k}_{O*}^{2}=\left[\begin{array}{ccc|ccc} 750000 & 0 & 0 & -750000 & 0 & 0 \\ 0 & 7500 & 11250 & 0 & -7500 & 11250 \\ 0 & 11250 & 22500 & 0 & -11250 & 11250 \\ \hline -750000 & 0 & 0 & 750000 & 0 & 0 \\ 0 & -7500 & -11250 & 0 & 7500 & -11250 \\ 0 & 11250 & 11250 & 0 & -11250 & 22500 \end{array}\right]$$

$$\mathbf{k}_{O*}^{3}=\left[\begin{array}{ccc|ccc} 624037.7 & 0 & 0 & -624037.7 & 0 & 0 \\ 0 & 4320.26 & 7788.46 & 0 & -4320.26 & 7788.46 \\ 0 & 7788.46 & 18721.13 & 0 & -7788.46 & 9360.57 \\ \hline -624037.7 & 0 & 0 & 624037.7 & 0 & 0 \\ 0 & -4320.26 & -7788.46 & 0 & 4320.26 & -7788.46 \\ 0 & 7788.46 & 9360.57 & 0 & -7788.46 & 18721.13 \end{array}\right]$$

The products, $\mathbf{k}_{O*}^{i}\mathbf{T}_{DA}^{i}$, $\mathbf{k}_{O*}^{i}\mathbf{T}_{DR}^{i}$ and $\mathbf{T}^{i^{T}}\mathbf{k}_{O*}^{i}\mathbf{T}^{i}$, may now be computed directly (using MATLAB) and the components of the structure stiffness matrix generated as follows:

$$\mathbf{k}_{O,AA}=\sum_{i=1}^{3}\mathbf{T}_{DA}^{i\ T}\mathbf{k}_{O*}^{i}\mathbf{T}_{DA}^{i}\,;\quad \mathbf{k}_{O,RA}=\sum_{i=1}^{3}\mathbf{T}_{DR}^{i\ T}\mathbf{k}_{O*}^{i}\mathbf{T}_{DA}^{i}=\mathbf{k}_{O,AR}^{\ T}\,;\quad \mathbf{k}_{O,RR}=\sum_{i=1}^{3}\mathbf{T}_{DR}^{i\ T}\mathbf{k}_{O*}^{i}\mathbf{T}_{DR}^{i}$$

$$\Rightarrow \mathbf{k}_{\mathbf{O,AA}} = \left[\begin{array}{ccc|ccc|cc}
913036.8 & 215222.4 & 3240 & -750000 & 0 & 0 & 3240 & 0 \\
215222.4 & 296083.2 & 8820 & 0 & -7500 & 11250 & -2430 & 0 \\
3240 & 8820 & 36000 & 0 & -11250 & 11250 & 6750 & 0 \\ \hline
-750000 & 0 & 0 & 945002.6 & -286023.4 & 6480 & 0 & 6480 \\
0 & -7500 & -11250 & -286023.4 & 440855.4 & -6929.7 & 0 & 4320.3 \\
0 & 11250 & 11250 & 6480 & -6929.7 & 41221.1 & 0 & 9360.6 \\ \hline
3240 & -2430 & 6750 & 0 & 0 & 0 & 13500 & 0 \\
0 & 0 & 0 & 6480 & 4320.3 & 9360.6 & 0 & 18721.1
\end{array}\right]$$

$$\mathbf{k}_{\mathbf{O,RA}} = \left[\begin{array}{ccc|ccc|cc}
-163036.8 & -215222.4 & -3240 & 0 & 0 & 0 & -3240 & 0 \\
-215222.4 & -288583.2 & 2430 & 0 & 0 & 0 & 2430 & 0 \\ \hline
0 & 0 & 0 & -195002.6 & 286023.4 & -6480.4 & 0 & -6480.4 \\
0 & 0 & 0 & 286023.4 & -433355.4 & -4320.3 & 0 & -4320.3
\end{array}\right]$$

$$\mathbf{k}_{\mathbf{O,RR}} = \left[\begin{array}{cc|cc}
163036.8 & 215222.4 & 0 & 0 \\
215222.4 & 288583.2 & 0 & 0 \\ \hline
0 & 0 & 195002.6 & -286023.4 \\
0 & 0 & -286023.4 & 433355.4
\end{array}\right]$$

The basic stiffness relations are given by:

$$\left[\begin{array}{c}\mathbf{F_A}\\ \hline \mathbf{F_R}\end{array}\right] - \left[\begin{array}{c}\mathbf{F_{fA}=O}\\ \hline \mathbf{F_{fR}=O}\end{array}\right] = \left[\begin{array}{c|c}\mathbf{k_{AA}} & \mathbf{k_{AR}}\\ \hline \mathbf{k_{RA}} & \mathbf{k_{RR}}\end{array}\right]\left[\begin{array}{c}\mathbf{D_A}\\ \hline \mathbf{D_R=O}\end{array}\right]$$

The *'first-order' response* is, accordingly, given (with P in kN units) by:

$$\Rightarrow \mathbf{D_{A,O}} = \left[\mathbf{k_{AA,O}}\right]^{-1}\left[-\mathbf{F_A}\right] = \left[\mathbf{k_{AA}}\right]^{-1}\left\{\begin{array}{c}0\\P\\0\\ \hline 0\\P\\0\\ \hline 0\\0\end{array}\right\} \quad\Rightarrow\quad \left\{\begin{array}{c}D_1\\D_2\\D_3\\ \hline D_4\\D_5\\D_6\\ \hline D_7\\D_8\end{array}\right\} = (P\times10^{-5})\left\{\begin{array}{c}-0.817138\text{ m}\\0.945354\text{ m}\\-0.261833\text{ rad}\\ \hline -0.722470\text{ m}\\-0.239741\text{ m}\\-0.206013\text{ rad}\\ \hline 0.497193\text{ rad}\\0.408417\text{ rad}\end{array}\right\}$$

$$\mathbf{F_{R,O}} = \mathbf{k_{RA,O}}\mathbf{D_{A,O}} \Rightarrow \quad \mathbf{F_{R,O}} = \left\{\begin{array}{c}F_9\\F_{10}\\ \hline F_{11}\\F_{12}\end{array}\right\} = \left\{\begin{array}{c}-0.710004\,P\\-0.963751\,P\\ \hline +0.710004\,P\\-1.036249\,P\end{array}\right\}$$

<u>Check equilibrium:</u> $\sum F_x = 0:\ F_9 + F_{11} = 0$

$$\sum F_y = 0:\ F_{10} + F_{12} - 2P = 0 \ \Rightarrow \text{OK.}$$

The 'first-order' member end forces are given by

$$\mathbf{F}^{\mathbf{i}}_{\mathbf{O*}} = \mathbf{F}^{\mathbf{i}}_{\mathbf{O*f}} + \left(\mathbf{k}^{\mathbf{i}}_{\mathbf{O*}}\mathbf{T}^{\mathbf{i}}_{\mathbf{DA}}\right)\mathbf{D}_{\mathbf{A,O}} = \left(\mathbf{k}^{\mathbf{i}}_{\mathbf{O*}}\mathbf{T}^{\mathbf{i}}_{\mathbf{DA}}\right)\mathbf{D}_{\mathbf{A,O}}:$$

$$\mathbf{F}^{1}_{\mathbf{O*}} = \begin{Bmatrix} F^1_{1*} \\ F^1_{2*} \\ F^1_{3*} \\ \hline F^1_{4*} \\ F^1_{5*} \\ F^1_{6*} \end{Bmatrix} = P\begin{Bmatrix} -1.19700 \text{ kN} \\ -0.01025 \text{ kN} \\ 0 \text{ kNm} \\ \hline 1.19700 \text{ kN} \\ 0.01025 \text{ kN} \\ -0.05123 \text{ kNm} \end{Bmatrix}; \quad \mathbf{F}^{2}_{\mathbf{O*}} = \begin{Bmatrix} F^2_{1*} \\ F^2_{2*} \\ F^2_{3*} \\ \hline F^2_{4*} \\ F^2_{5*} \\ F^2_{6*} \end{Bmatrix} = P\begin{Bmatrix} -0.71000 \text{ kN} \\ 0.03625 \text{ kN} \\ 0.05123 \text{ kNm} \\ \hline 0.71000 \text{ kN} \\ -0.03625 \text{ kN} \\ 0.05751 \text{ kNm} \end{Bmatrix}; \quad \mathbf{F}^{3}_{\mathbf{O*}} = \begin{Bmatrix} F^3_{1*} \\ F^3_{2*} \\ F^3_{3*} \\ \hline F^3_{4*} \\ F^3_{5*} \\ F^3_{6*} \end{Bmatrix} = P\begin{Bmatrix} -1.25605 \text{ kN} \\ -0.01595 \text{ kN} \\ -0.05751 \text{ kNm} \\ \hline 1.25605 \text{ kN} \\ 0.01595 \text{ kN} \\ 0 \text{ kNm} \end{Bmatrix}$$

Thus, from a 'first-order' analysis, we obtain the axial forces in the three members as:

$$\boxed{P_1 = 1.19700P;\quad P_2 = 0.71000P;\quad P_3 = 1.25605P} \text{ (all compressive)}$$

We will now use these values of axial forces to find the element geometric stiffness matrices, and then proceed to find the critical buckling loads (ignoring the effect of the curvatures introduced due to pre-buckling moments). As shown in Eq. 7.73,

$$\mathbf{k}^{\mathbf{i}}_{\mathbf{G*}} = -P_i\left[\begin{array}{ccc|ccc} 0 & 0 & 0 & 0 & 0 & 0 \\ 0 & 1.2/L_i & 0.1 & 0 & -1.2/L_i & 0.1 \\ 0 & 0.1 & 2L_i/15 & 0 & -0.1 & -L_i/30 \\ \hline 0 & 0 & 0 & 0 & 0 & 0 \\ 0 & -1.2/L_i & -0.1 & 0 & 1.2/L_i & -0.1 \\ 0 & 0.1 & -L_i/30 & 0 & -0.1 & 2L_i/15 \end{array}\right], \text{ with } \begin{array}{l} P_1 = 1.197P;\ P_2 = 0.710P;\ P_3 = 1.25605P \\ L_1 = 5\text{m};\ L_2 = 3\text{m};\ L_2 = \sqrt{13}\text{m} \end{array}$$

$$\mathbf{k}_{\mathbf{G,AA}} = \sum_{i=1}^{3} \mathbf{T}^{\mathbf{i}\ \mathrm{T}}_{\mathbf{DA}}\mathbf{k}^{\mathbf{i}}_{\mathbf{G*}}\mathbf{T}^{\mathbf{i}}_{\mathbf{DA}} \quad \text{(using MATLAB)}$$

$$\mathbf{k}_{\mathbf{G,AA}} = (-P)\left[\begin{array}{ccc|ccc|cc} 0.183859 & -0.137894 & 0.095760 & 0 & 0 & 0 & 0.095760 & 0 \\ -0.137894 & 0.387421 & -0.000820 & 0 & -0.284000 & 0.071000 & -0.071820 & 0 \\ 0.095760 & -0.000820 & 1.082000 & 0 & -0.071000 & -0.071000 & -0.199500 & 0 \\ \hline 0 & 0 & 0 & 0.289411 & 0.192941 & 0.104510 & 0 & 0.104510 \\ 0 & -0.284000 & -0.071000 & 0.192941 & 0.412627 & -0.001327 & 0 & 0.069673 \\ 0 & 0.071000 & -0.071000 & 0.104510 & -0.001327 & 0.887834 & 0 & -0.150958 \\ \hline 0.095760 & -0.071820 & -0.199500 & 0 & 0 & 0 & 0.798000 & 0 \\ 0 & 0 & 0 & 0.104510 & 0.069673 & -0.150958 & 0 & 0.603834 \end{array}\right]$$

For a non-trivial solution to the equilibrium equation $\mathbf{k}_{\mathbf{AA}}\mathbf{D}_{\mathbf{A}} = \mathbf{O}$, the *characteristic equation* is obtained by setting the determinant of $\mathbf{k}_{\mathbf{AA}} = \mathbf{k}_{\mathbf{O,AA}} + \mathbf{k}_{\mathbf{G,AA}}$ (which is a symmetric matrix) equal to zero. For convenience, in order to avoid dealing with large numbers, let us set $\lambda = \dfrac{P}{EI}$, where $EI = (2.5\times10^7)(6.75\times10^{-4}) = 16875 \text{ kNm}^2$, and write an expression for $g(\lambda)$:

$g(\lambda)=\left|\left(\frac{1}{16875}\right)\mathbf{k}_{\mathbf{O,AA}}+\left(\frac{\lambda}{P}\right)\mathbf{k}_{\mathbf{G,AA}}\right|$ (which is too large to be shown here, but can be easily set up in MATLAB and computed for any input value of λ).

Of the eight eigenvalues possible that will render $g(\lambda)=0$, we are interested in the lowest value of λ. It is safe to begin by assigning a zero value to λ.

$\lambda=0.0 \;\Rightarrow\; g(\lambda)=+39729$

Now, we should gradually increase λ in small increments (steps of 0.1) and observe when $g(\lambda)$ turns from a positive value to a negative value. This occurs at $\lambda=0.3$:

$\lambda=0.3 \;\Rightarrow\; g(\lambda)=-1396.5$

Now, applying the bisection method, we can easily converge on the 'exact' value, which turns out to be $\lambda_{cr}=0.2742$, corresponding to which, $g(\lambda)=+2.96$[†].

Substituting EI = 16875 kNm^2, the desired critical buckling load is obtained as:
P_{cr} = 0.2742×16875 = **4627 kN**

This value predicted by eigenvalue analysis, assuming bifurcation instability, is likely to over-estimate slightly the actual critical load in the presence of initial curvatures. Although, it is possible to find the eigenvector corresponding to the first mode, it is more meaningful to use the result obtained earlier (displacement vector $\mathbf{D}_{\mathbf{A,O}}$) from the first-order analysis of the frame. This vector has been generated as a linear function of P, and although it does not include the contribution of geometric stiffness, it represents approximately the deflected shape as P asymptotically approaches P_{cr} = 4627 kN. This is depicted in Fig. 7.59.

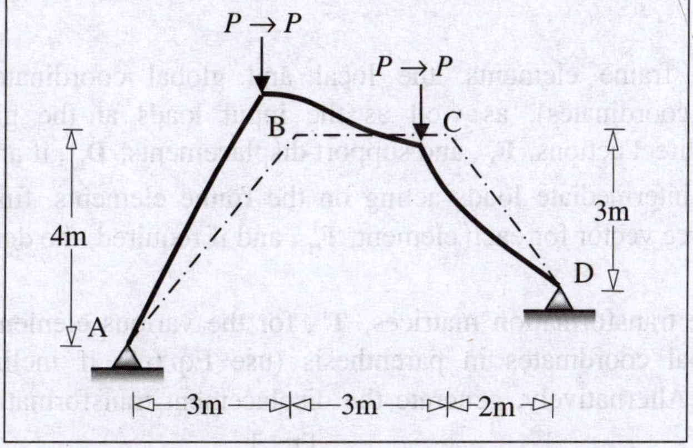

Figure 7.59 Deflected shape as $P \rightarrow P_{cr}$ (Example 7.19)

[†] This is as close to $g(\lambda)=0$ as possible, for the level of accuracy we need. Note that the 'exact' value lies between $\lambda=0.2742$ and $\lambda=0.2743$, corresponding to which, $g(\lambda)=-3.04$.

7.4.3 Second-Order Analysis

In this case, the objective is to determine the response of the structure, including the *P-delta* effects. As mentioned earlier, this strictly requires a nonlinear approach. However, working within the framework of linear analysis, we can get a reasonable solution, by considering the geometric stiffness matrix, provided the deformations under the loading are small, axial deformations are negligible and the material behaviour remains linear elastic. The second-order analysis can be carried out in a single stage if the axial forces in the various members are known *a priori*, as in continuous beam systems, in which the axial loads are independent of the lateral loads acting on the various elements.

However, in plane frames, in general, the axial forces in the various elements are dependent on the loading on the structure, and, in order to determine these forces, we need to first carry out a conventional 'first-order' analysis, as demonstrated in Example 7.19. Then, applying the modifications to the stiffness matrix (geometric stiffness matrix), we carry out a second-stage analysis (so-called 'second-order' analysis) of the structure, and thereby determine the response which now includes the *P-delta* effects. If the axial forces resulting from this analysis are significantly different from the previous step, we can repeat the analysis one more time (or as many times as desired, until satisfactory convergence is attained in the response). Usually, one or two cycles are adequate.

Analysis Procedure

The steps involved in the analysis procedure may be summarised as follows, for the more general case of loading in which the axial forces in the various members are not known *a priori*. It is based on the methods described earlier in Chapter 5 (for beams) and Chapter 6 (for frames):

1. Identify the frame elements, the local and global coordinates (n active and r restrained coordinates), as well as the input loads at the global coordinates, comprising direct actions, $\mathbf{F_A}$, and support displacements, $\mathbf{D_R}$, if any.
2. If there are intermediate loads acting on the frame elements, find the 'first-order' fixed end force vector for each element, $\mathbf{F^i_{*f}}$, and if required, the deflections at critical locations.
3. Generate the transformation matrices, $\mathbf{T^i}$, for the various elements, identifying the linking global coordinates in parenthesis (use Eq. 6.8, if inclined members are involved). [Alternatively, generate the displacement transformation matrices, $\mathbf{T^i_D}$].

   Assemble the fixed end force vector $\mathbf{F_f} = \begin{bmatrix} \mathbf{F_{fA}} \\ \hline \mathbf{F_{fR}} \end{bmatrix}$, combining the contributions, $\mathbf{T^{iT}F^i_{*f}}$ [or $\mathbf{T_D^{i\,T}F^i_{*f}}$] from the various elements. Generate the resultant load vector, $\mathbf{F_A} - \mathbf{F_{fA}}$.
4. Generate the element stiffness matrices $\mathbf{k^i_*}$ for all the elements, using Eq. 7.69 or Eq. 7.71, assuming $P_i = 0$ for all elements in the first-order analysis (i.e., ignoring geometric stiffness). Compute $\mathbf{k^i_*T^i}$, and hence assemble the structure stiffness

matrix, $\mathbf{k} = \left[\begin{array}{c|c} \mathbf{k_{AA}} & \mathbf{k_{AR}} \\ \hline \mathbf{k_{RA}} & \mathbf{k_{RR}} \end{array}\right]$, combining the stiffness coefficient contributions, $\mathbf{k^i} = \mathbf{T^{i^T}} \mathbf{k_*^i} \mathbf{T^i}$ at the appropriate locations; for convenience, use Eqns 6.9 and 6.10. [Alternatively, compute $\mathbf{k_*^i T_D^i}$ for every element and assemble $\mathbf{k} = \sum_{i=1}^{m} \mathbf{T_D^{i^T}} \mathbf{k_*^i T_D^i}$]. Find the inverse, $[\mathbf{k_{AA}}]^{-1}$, which is a symmetric matrix of order $n \times n$.

5. Find the unknown 'first-order' displacements, $\mathbf{D_A} = \mathbf{D_{A,O}}$, by solving the following stiffness relation:

$$\mathbf{D_A} = [\mathbf{k_{AA}}]^{-1}\left[(\mathbf{F_A} - \mathbf{F_{fA}}) - \mathbf{k_{AR}} \mathbf{D_R}\right] \tag{7.78}$$

6. Find the 'first-order' support reactions, $\mathbf{F_R} = \mathbf{F_{R,O}}$, by solving the following equation:

$$\mathbf{F_R} = \mathbf{F_{fR}} + \mathbf{k_{RA}} \mathbf{D_A} + \mathbf{k_{RR}} \mathbf{D_R} \tag{7.79}$$

Perform simple equilibrium checks (such as $\sum F_y = 0$).

7. Find the 'first-order' member end forces, $\mathbf{F_*^i} = \mathbf{F_{O*}^i}$, for each element.

$$\mathbf{F_*^i} = \mathbf{F_{*f}^i} + \left(\mathbf{k_*^i T^i}\right)\mathbf{D^i} \tag{7.80}$$

[Alternatively, use $\mathbf{F_*^i} = \mathbf{F_{*f}^i} + \left(\mathbf{k_*^i T_{DA}^i}\right)\mathbf{D_A}$].

8. From the above output, determine the axial forces, P_i (compressive or tensile), in the various elements, and update the element stiffness matrices, including the geometric stiffness components, as well as the fixed-end forces (wherever relevant).
9. Repeat steps 4-7, and find the 'second-order' response.
10. Inspect the results obtained (compare the 'second-order' P_i values with the 'first-order' results), and if required (i.e., if it is felt that there is significant disparity in the results), repeat the steps involved in the second-order analysis once more.
11. Draw the free-body diagrams for the various elements, and hence, the axial force, shear force and bending moment diagrams. Also, sketch the deflected shape (superimposing the fixed end force effects) and identify the deflections at critical locations.

EXAMPLE 7.20

Analyse the braced beam-column system shown in Fig. 7.60, including *P-delta* effects. Assume the brace in the middle to have a stiffness, k_b = 480 kN/m. Assume the beam-column to be axially rigid and to have a uniform flexural rigidity, EI = 8000 kNm^2. Find the maximum deflection and draw the shear force bending moment diagrams. Compare the results with those of first-order analysis.

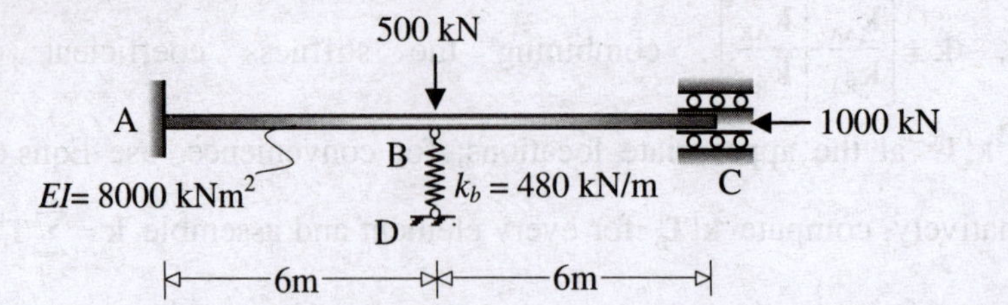

Figure 7.60 Loading diagram in a braced beam-column system (Example 7.20)

SOLUTION

Let us consider the same system of global coordinates as in Example 7.15, as shown in Fig. 7.61a. There are two active degrees of freedom and four restrained global coordinates (numbered as shown). The local coordinates for the two beam-column elements, with four degrees of freedom each (assuming axial rigidity), are shown in Fig. 7.61b. The load vector is given by $\mathbf{F_A} = \begin{Bmatrix} -500 \text{ kN} \\ 0 \text{ kNm} \end{Bmatrix}$; there are no fixed end forces to calculate ($\mathbf{F_f} = \mathbf{O}$) and no support movements have been prescribed ($\mathbf{D_R} = \mathbf{O}$).

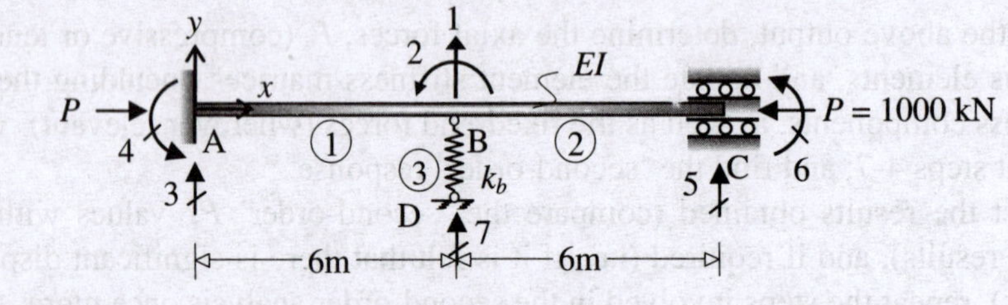

(*a*) Global coordinates

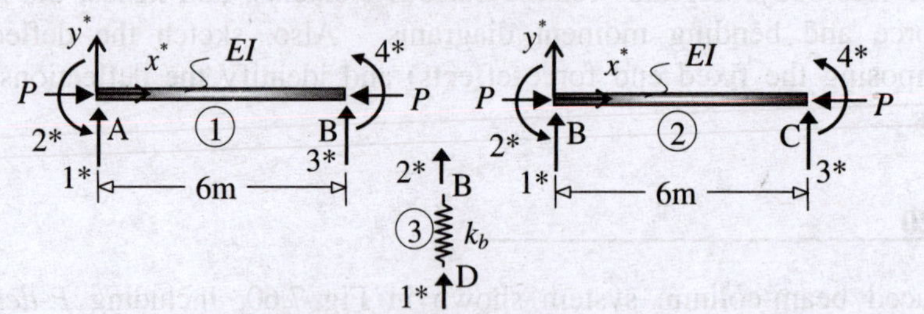

(*b*) Local coordinates

Figure 7.61 Global and local coordinates (Example 7.20)

The element stiffness matrices $\mathbf{k}_*^1$ and $\mathbf{k}_*^2$, satisfying $\mathbf{F}_*^i = \mathbf{k}_*^i\mathbf{D}_*^i$, take the following form (refer Fig. 7.61b and Eq. 7.71):

$$\mathbf{k}_*^1 = \mathbf{k}_*^2 = \frac{EI}{l}\begin{bmatrix} 12/l^2 & 6/l & -12/l^2 & 6/l \\ 6/l & 4 & -6/l & 2 \\ -12/l^2 & -6/l & 12/l^2 & -6/l \\ 6/l & 2 & -6/l & 4 \end{bmatrix} - P\begin{bmatrix} 1.2/l & 0.1 & -1.2/l & 0.1 \\ 0.1 & 2l/15 & -0.1 & -l/30 \\ -1.2/l & -0.1 & 1.2/l & -0.1 \\ 0.1 & -l/30 & -0.1 & 2l/15 \end{bmatrix}$$

Substituting $l = 6$m, $EI = 8000$ kNm^2 and $P = 1000$ kN,

$$\Rightarrow \mathbf{k}_*^1 = \mathbf{k}_*^2 = \begin{bmatrix} 244.444 & 1233.333 & -244.444 & 1233.333 \\ 1233.333 & 4533.333 & -1233.333 & 2866.667 \\ -244.444 & -1233.333 & 244.444 & -1233.333 \\ 1233.333 & 2866.667 & -1233.333 & 4533.333 \end{bmatrix}$$

The element stiffness matrix of the spring element (with local coordinates as shown in Fig. 7.61b), considering $k_b = 480$ kN/m, is given by:

$$\mathbf{k}_*^3 = \begin{bmatrix} 480 & -480 \\ -480 & 480 \end{bmatrix}$$

The element transformation matrices $\mathbf{T}^1$, $\mathbf{T}^2$ and $\mathbf{T}^3$, satisfying $\mathbf{D}_*^i = \mathbf{T}^i\mathbf{D}$ and $\mathbf{F}_*^i = \mathbf{T}^i\mathbf{F}$, take the following form, with the linking global coordinates indicated in parentheses:

$$\mathbf{T}^1 = \begin{array}{c} \begin{matrix}(3) & (4) & (1) & (2)\end{matrix} \\ \begin{bmatrix} 1 & 0 & 0 & 0 \\ 0 & 1 & 0 & 0 \\ 0 & 0 & 1 & 0 \\ 0 & 0 & 0 & 1 \end{bmatrix} \end{array} \begin{matrix} \\ (3) \\ (4) \\ (1) \\ (2) \end{matrix}; \quad \mathbf{T}^2 = \begin{array}{c} \begin{matrix}(1) & (2) & (5) & (6)\end{matrix} \\ \begin{bmatrix} 1 & 0 & 0 & 0 \\ 0 & 1 & 0 & 0 \\ 0 & 0 & 1 & 0 \\ 0 & 0 & 0 & 1 \end{bmatrix} \end{array} \begin{matrix} \\ (1) \\ (2) \\ (5) \\ (6) \end{matrix}; \quad \mathbf{T}^3 = \begin{array}{c} \begin{matrix}(7) & (1)\end{matrix} \\ \begin{bmatrix} 1 & 0 \\ 0 & 1 \end{bmatrix} \end{array} \begin{matrix} \\ (7) \\ (1) \end{matrix}$$

As all the transformation matrices are identity matrices, $\mathbf{T}^{i^T}\mathbf{k}_*^i\mathbf{T}^i = \mathbf{k}_*^i$. Summing up these contributions, the structure stiffness matrix, $\mathbf{k} = \mathbf{k_O} + \mathbf{k_G} = \left[\begin{array}{c|c} \mathbf{k_{AA}} & \mathbf{k_{AR}} \\ \hline \mathbf{k_{RA}} & \mathbf{k_{RR}} \end{array}\right]$, satisfying $\mathbf{F} = \mathbf{kD}$, can be assembled and partitioned as follows:

$$\mathbf{k} = \frac{EI}{l}\left[\begin{array}{cc|cccc} 24/l^2 & 0 & -12/l^2 & -6/l & -12/l^2 & 6/l \\ 0 & 8 & 6/l & 2 & -6/l & 2 \\ \hline -12/l^2 & 6/l & 12/l^2 & 6/l & 0 & 0 \\ -6/l & 2 & 6/l & 4 & 0 & 0 \\ -12/l^2 & -6/l & 0 & 0 & 12/l^2 & -6/l \\ 6/l & 2 & 0 & 0 & -6/l & 4 \end{array}\right] - P\left[\begin{array}{cc|cccc} 2.4/l & 0 & -1.2/l & -0.1 & -1.2/l & 0.1 \\ 0 & 4l/15 & 0.1 & -l/30 & -0.1 & -l/30 \\ \hline -1.2/l & 0.1 & 1.2/l & 0.1 & 0 & 0 \\ -0.1 & -l/30 & 0.1 & 2l/15 & 0 & 0 \\ -1.2/l & -0.1 & 0 & 0 & 1.2/l & -0.1 \\ 0.1 & -l/30 & 0 & 0 & -0.1 & 2l/15 \end{array}\right] \begin{matrix} (1) \\ (2) \\ (3) \\ (4) \\ (5) \\ (6) \end{matrix}$$

$$\mathbf{k}_{AA} = \begin{bmatrix} 968.889 & 0 \\ 0 & 9066.667 \end{bmatrix} \begin{matrix} (1) \\ (2) \end{matrix}$$

$$\mathbf{k}_{AR} = \begin{bmatrix} 244.444 & -1233.333 & -244.444 & 1233.333 & -480.000 \\ 1233.333 & 2866.667 & -1233.333 & 2866.667 & 0 \end{bmatrix} \begin{matrix} (1) \\ (2) \end{matrix}; \ \mathbf{k}_{RA} = \mathbf{k}_{AR}^{T}$$

($\mathbf{k}_{RR}$ is not required here, as there are no support settlements prescribed).

Displacements and Support Reactions

The basic stiffness relations are given by:

$$\begin{bmatrix} \mathbf{F}_A \\ \mathbf{F}_R \end{bmatrix} - \begin{bmatrix} \mathbf{F}_{fA} = \mathbf{O} \\ \hline \mathbf{F}_{fR} = \mathbf{O} \end{bmatrix} = \begin{bmatrix} \mathbf{k}_{AA} & \mathbf{k}_{AR} \\ \hline \mathbf{k}_{RA} & \mathbf{k}_{RR} \end{bmatrix} \begin{bmatrix} \mathbf{D}_A \\ \hline \mathbf{D}_R = \mathbf{O} \end{bmatrix}$$

$$\mathbf{D}_A = [\mathbf{k}_{AA}]^{-1} \mathbf{F}_A$$

$$\Rightarrow \begin{Bmatrix} D_1 \\ D_2 \end{Bmatrix} = 10^{-4} \times \begin{bmatrix} 10.3211 & \\ & 1.10294 \end{bmatrix} \begin{Bmatrix} -500 \\ 0 \end{Bmatrix} = \begin{Bmatrix} -0.516055 \text{ m} \\ 0 \text{ rad} \end{Bmatrix}$$

Thus, the maximum deflection predicted (under the load) is given by $\mathbf{D_1 = -0.516}$ **m.**

$$\mathbf{F}_R = \mathbf{k}_{RA} \mathbf{D}_A \quad \Rightarrow \quad \mathbf{F}_R = \begin{Bmatrix} F_3 \\ F_4 \\ F_5 \\ F_6 \\ F_7 \end{Bmatrix} = \begin{Bmatrix} 126.147 \text{ kN} \\ 636.468 \text{ kNm} \\ 126.147 \text{ kN} \\ -636.468 \text{ kNm} \\ 247.706 \text{ kN} \end{Bmatrix}$$

Check equilibrium:

$$\sum F_y = 0: \ F_3 + F_5 + F_7 - 500 = (126.147) \times 2 + 247.706 - 500 = 0 \text{ kN} \ \Rightarrow \text{OK}$$

The force response is symmetric, as expected.

Member Forces

$$\mathbf{F}_*^i = \mathbf{F}_{*f}^i + \left(\mathbf{k}_*^i \mathbf{T}^i\right) \mathbf{D}^i = \mathbf{k}_*^i \mathbf{D}^i$$

$$\Rightarrow \mathbf{F}_*^1 = \begin{Bmatrix} F_{1*}^1 \\ F_{2*}^1 \\ F_{3*}^1 \\ F_{4*}^1 \end{Bmatrix} = \begin{bmatrix} 244.444 & 1233.333 & -244.444 & 1233.333 \\ 1233.333 & 4533.333 & -1233.333 & 2866.667 \\ -244.444 & -1233.333 & 244.444 & -1233.333 \\ 1233.333 & 2866.667 & -1233.333 & 4533.333 \end{bmatrix} \begin{Bmatrix} D_3 = 0 \\ D_4 = 0 \\ \hline D_1 = -0.516055 \\ D_2 = 0 \end{Bmatrix} = \begin{Bmatrix} 126.147 \text{ kN} \\ 636.468 \text{ kNm} \\ \hline -126.147 \text{ kN} \\ 636.468 \text{ kNm} \end{Bmatrix}$$

$$\Rightarrow \mathbf{F}_*^2 = \begin{Bmatrix} F_{1*}^2 \\ F_{2*}^2 \\ F_{3*}^2 \\ F_{4*}^2 \end{Bmatrix} = \begin{bmatrix} 244.444 & 1233.333 & -244.444 & 1233.333 \\ 1233.333 & 4533.333 & -1233.333 & 2866.667 \\ -244.444 & -1233.333 & 244.444 & -1233.333 \\ 1233.333 & 2866.667 & -1233.333 & 4533.333 \end{bmatrix} \begin{Bmatrix} D_1 = -0.516055 \\ D_2 = 0 \\ \hline D_5 = 0 \\ D_6 = 0 \end{Bmatrix} = \begin{Bmatrix} -126.147 \text{ kN} \\ -636.468 \text{ kNm} \\ \hline 126.147 \text{ kN} \\ -636.468 \text{ kNm} \end{Bmatrix}$$

Comparison with first-order response:

The first-order response is easy to compute in this case. The applied load is shared by the fixed beam ABC (with span L = 12m) and the spring BD in proportion to their relative stiffnesses. The beam stiffness works out to $192EI/L^3 = 888.888$ kN/m, compared to $k_b = 480$ kN/m. The total stiffness is therefore equal to 1368.889 kN/m, whereby the deflection at B is given by: $D_1 = \dfrac{500 \text{ kN}}{1368.889 \text{ kN/m}} = 0.36526 \text{ m}$, whereby the spring force is given by $k_b D_1 = 175.325\,\text{kN}$, and the shear force in the beam is given by $(500-175.325)/2 = 162.338\,\text{kN}$. The fixed-end and middle moments are given by $(500-175.325)(12)/8 = 487.01\,\text{kNm}$. Thus, we observe, from the second-order analysis, an enhancement in the midspan deflection by 41.3 percent and an enhancement in the bending moments by 30.7 percent, compared to first-order analysis. However, there is a reduction in shear force by 28.7 percent.

The deflection, shear force and bending moment diagrams, from first-order and second-order analyses are shown in Fig. 7.62.

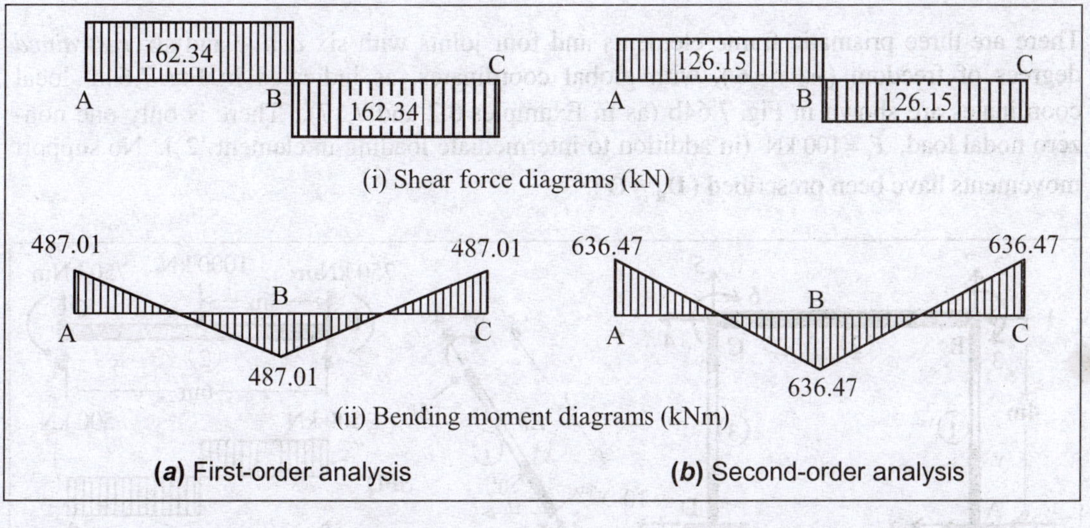

Figure 7.62 Shear force and bending moment diagrams (Example 7.20)

EXAMPLE 7.21

Analyse the rectangular portal frame shown in Fig. 7.63, including *P-delta* effects. Assume the column elements to have a square cross section, 300×300 mm and the beam to have a size, 300×450 mm. Assume all elements to have an elastic modulus, $E = 2.5\times10^4\,\text{N/mm}^2$. Find the complete force response and draw the deflection and bending moment diagrams. Compare the results with those of first-order analysis.

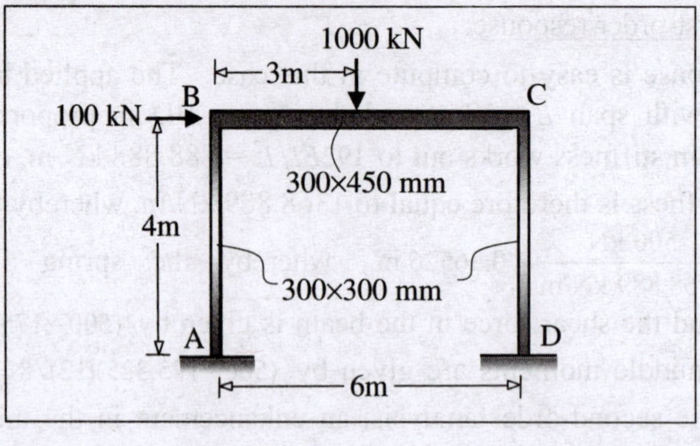

Figure 7.63 Second-order analysis of a portal frame (Example 7.21)

SOLUTION

Input Data

There are three prismatic frame elements and four joints with six *active* and six *restrained* degrees of freedom ($n = r = 6$), with global coordinates, as indicated in Fig. 7.64a; local coordinates are shown in Fig. 7.64b (as in Examples 6.2 and 7.17). There is only one non-zero nodal load, $F_1 = 100$ kN (in addition to intermediate loading in element '2'). No support movements have been prescribed ($\mathbf{D_R} = \mathbf{O}$).

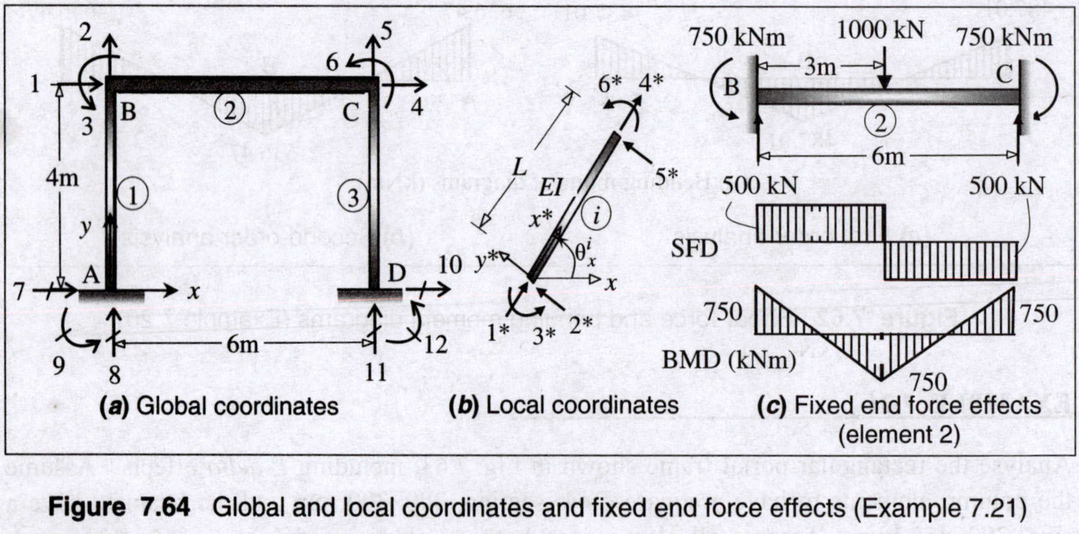

Figure 7.64 Global and local coordinates and fixed end force effects (Example 7.21)

Fixed End Forces

The fixed end force effects (support reactions, bending moment and shear force diagrams) on the element '2' in the restrained primary structure, due to the action of intermediate loading, are shown in Fig. 7.64c (ignoring the *P-delta* effect of the small axial force in this element). There are no fixed end forces in elements '1' and '3'.

Element '2' (BC):

$$F_{1*f}^2 = F_{4*f}^2 = 0 \text{ kN};\ F_{3*f}^2 = \frac{1000\times 6}{8} = 750.0 \text{ kNm};\ F_{6*f}^2 = -750.0 \text{ kNm};\ F_{2*f}^2 = F_{5*f}^2 = \frac{1000}{2} = 500.0 \text{ kN}$$

$$\Rightarrow \mathbf{F}_{*f}^1 = \mathbf{F}_{*f}^3 = \begin{Bmatrix} 0 \text{ kN} \\ 0 \text{ kN} \\ 0 \text{ kNm} \\ \hline 0 \text{ kN} \\ 0 \text{ kN} \\ 0 \text{ kNm} \end{Bmatrix};\quad \mathbf{F}_{*f}^2 = \begin{Bmatrix} 0 \text{ kN} \\ 500 \text{ kN} \\ 750 \text{ kNm} \\ \hline 0 \text{ kN} \\ 500 \text{ kN} \\ -750 \text{ kNm} \end{Bmatrix}$$

(The deflection under the load is given by $\dfrac{WL_2^3}{192EI_2} = \dfrac{(1000)(6.0)^3}{192(56953.1)} = 0.01975 \text{ m}$)

Coordinate Transformations and Equivalent Joint Loads

The element transformation matrices $\mathbf{T}^1$, $\mathbf{T}^2$ and $\mathbf{T}^3$, satisfying $\mathbf{D}_*^i = \mathbf{T}^i\mathbf{D}$ and $\mathbf{F}_*^i = \mathbf{T}^i\mathbf{F}$, take the following form (as in Examples 6.2 and 7.17), with the linking global coordinates indicated in parentheses:

$$\mathbf{T}^1 = \overset{(7)\ (8)\ (9)\ (1)\ (2)\ (3)}{\begin{bmatrix} 0 & 1 & 0 & 0 & 0 & 0 \\ -1 & 0 & 0 & 0 & 0 & 0 \\ 0 & 0 & 1 & 0 & 0 & 0 \\ \hline 0 & 0 & 0 & 0 & 1 & 0 \\ 0 & 0 & 0 & -1 & 0 & 0 \\ 0 & 0 & 0 & 0 & 0 & 1 \end{bmatrix}} \begin{matrix} (7) \\ (8) \\ (9) \\ (1) \\ (2) \\ (3) \end{matrix};\quad \mathbf{T}^2 = \overset{(1)\ (2)\ (3)\ (4)\ (5)\ (6)}{\begin{bmatrix} 1 & 0 & 0 & 0 & 0 & 0 \\ 0 & 1 & 0 & 0 & 0 & 0 \\ 0 & 0 & 1 & 0 & 0 & 0 \\ \hline 0 & 0 & 0 & 1 & 0 & 0 \\ 0 & 0 & 0 & 0 & 1 & 0 \\ 0 & 0 & 0 & 0 & 0 & 1 \end{bmatrix}} \begin{matrix} (1) \\ (2) \\ (3) \\ (4) \\ (5) \\ (6) \end{matrix};\quad \mathbf{T}^3 = \overset{(10)\ (11)\ (12)\ (4)\ (5)\ (6)}{\begin{bmatrix} 0 & 1 & 0 & 0 & 0 & 0 \\ -1 & 0 & 0 & 0 & 0 & 0 \\ 0 & 0 & 1 & 0 & 0 & 0 \\ \hline 0 & 0 & 0 & 0 & 1 & 0 \\ 0 & 0 & 0 & -1 & 0 & 0 \\ 0 & 0 & 0 & 0 & 0 & 1 \end{bmatrix}} \begin{matrix} (10) \\ (11) \\ (12) \\ (4) \\ (5) \\ (6) \end{matrix}$$

The fixed end forces in global axes can now be assembled, by summing up the contributions of

$$\mathbf{T}^{1^T}\mathbf{F}_{*f}^1 = \begin{Bmatrix} 0 \text{ kN} \\ 0 \text{ kN} \\ 0 \text{ kNm} \\ \hline 0 \text{ kN} \\ 0 \text{ kN} \\ 0 \text{ kNm} \end{Bmatrix} \begin{matrix} (7) \\ (8) \\ (9) \\ (1) \\ (2) \\ (3) \end{matrix};\quad \mathbf{T}^{2^T}\mathbf{F}_{*f}^2 = \begin{Bmatrix} 0 \text{ kN} \\ 500 \text{ kN} \\ 750 \text{ kNm} \\ \hline 0 \text{ kN} \\ 500 \text{ kN} \\ -750 \text{ kNm} \end{Bmatrix} \begin{matrix} (1) \\ (2) \\ (3) \\ (4) \\ (5) \\ (6) \end{matrix};\quad \mathbf{T}^{3^T}\mathbf{F}_{*f}^3 = \begin{Bmatrix} 0 \text{ kN} \\ 0 \text{ kN} \\ 0 \text{ kNm} \\ \hline 0 \text{ kN} \\ 0 \text{ kN} \\ 0 \text{ kNm} \end{Bmatrix} \begin{matrix} (10) \\ (11) \\ (12) \\ (4) \\ (5) \\ (6) \end{matrix}$$

$$\Rightarrow \mathbf{F_f} = \left[\begin{array}{c}\mathbf{F_{fA}}\\ \hline \mathbf{F_{fR}}\end{array}\right] \quad \text{where } \mathbf{F_{fA}} = \begin{Bmatrix} F_{1f} \\ F_{2f} \\ F_{3f} \\ F_{4f} \\ F_{5f} \\ F_{6f} \end{Bmatrix} = \begin{Bmatrix} 0\text{ kN} \\ 500\text{ kN} \\ 750\text{ kNm} \\ \hline 0\text{ kN} \\ 500\text{ kN} \\ -750\text{ kNm} \end{Bmatrix} \text{ and } \mathbf{F_{fR}} = \begin{Bmatrix} F_{7f} \\ F_{8f} \\ F_{9f} \\ F_{10f} \\ F_{11f} \\ F_{12f} \end{Bmatrix} = \begin{Bmatrix} 0\text{ kN} \\ 0\text{ kN} \\ 0\text{ kNm} \\ \hline 0\text{ kN} \\ 0\text{ kN} \\ 0\text{ kNm} \end{Bmatrix}$$

The resultant direct load vector, $\mathbf{F_A} - \mathbf{F_{fA}}$, is given by:

$$\mathbf{F_A} - \mathbf{F_{fA}} = \begin{Bmatrix} 100\text{ kN} \\ 0\text{ kN} \\ 0\text{ kNm} \\ \hline 0\text{ kN} \\ 0\text{ kN} \\ 0\text{ kNm} \end{Bmatrix} - \begin{Bmatrix} 0\text{ kN} \\ 500\text{ kN} \\ 750\text{ kNm} \\ \hline 0\text{ kN} \\ 500\text{ kN} \\ -750\text{ kNm} \end{Bmatrix} = \begin{Bmatrix} 100\text{ kN} \\ -500\text{ kN} \\ -750\text{ kNm} \\ \hline 0\text{ kN} \\ -500\text{ kN} \\ 750\text{ kNm} \end{Bmatrix}$$

This is indicated in Fig. 7.65a. The displacements $\mathbf{D_A}$ at the six active coordinates will be identical to those in the original loading diagram (with the intermediate loading) in Fig. 7.63. This loading will produce axial compressive forces of 500 kN in elements '1' and '3', which will get modified by an amount Q on account of the action of the 100 kN lateral load, as shown.

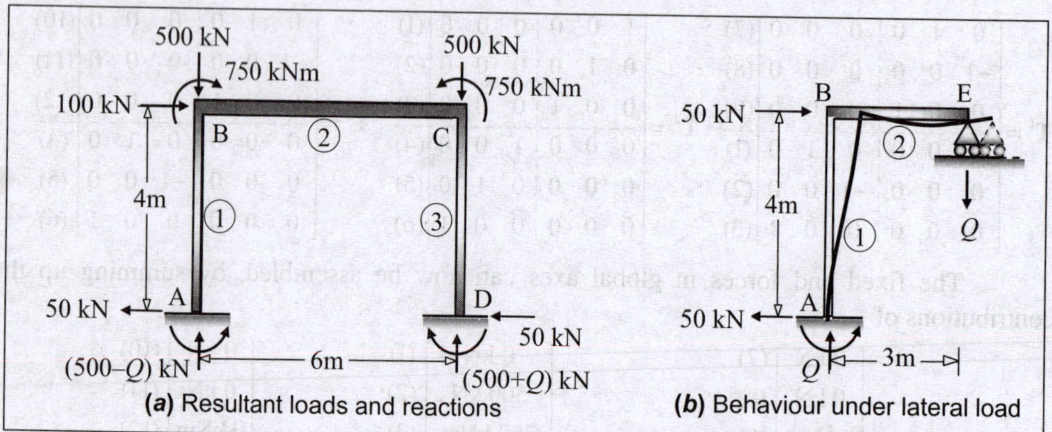

Figure 7.65 Resultant loads and 'first-order' column axial forces (Example 7.21)

We can easily determine this quantity Q, by analysing the half-frame shown in Fig. 7.65b (where the roller support at the mid-point E of BC is used to take advantage of the anti-symmetric response under lateral loading). A simple analysis of this frame by moment distribution shows that the unbalanced moment at joint B, $50 \times 4.0 / 2 = 100$ kNm (fixed-end moment in BA at B) gets apportioned to BA and BE in the ratio of their relative stiffnesses:

$k_{BA} : k_{BE} = \frac{EI_1}{L_1} : \frac{3EI_2}{L_2/2} = 4218.75 : (6 \times 9492.1875) = 1 : 13.5$. The bending moment at B in BE is accordingly given by $100 \times \frac{13.5}{14.5} = 93.1$ kNm, whereby $Q = \frac{93.1}{3.0} = 31.03$ kN.

Thus the axial compressive forces in elements '1' and '3' are obtained from first-order analysis as:

$$P_1 = 500.0 - 31.03 = 468.97 \text{ kN}$$
$$P_3 = 500.0 + 31.03 = 531.03 \text{ kN}$$

The element '2' will also be subject to a nominal axial compressive force:

$$P_2 = 50.0 \text{ kN}$$

Element and Structure Stiffness Matrices

The element stiffness matrices, $\mathbf{k}_*^\mathbf{i} = \mathbf{k}_{O*}^\mathbf{i} + \mathbf{k}_{G*}^\mathbf{i}$, for the three elements (including second-order effects), satisfying $\mathbf{F}_*^\mathbf{i} = \mathbf{k}_*^\mathbf{i} \mathbf{D}_*^\mathbf{i}$ can be derived using Eq. 7.73, as follows:

$$\mathbf{k}_*^\mathbf{i} = \begin{bmatrix} \alpha_i & 0 & 0 & -\alpha_i & 0 & 0 \\ 0 & \beta_i & \chi_i & 0 & -\beta_i & \chi_i \\ 0 & \chi_i & 4\delta_i & 0 & -\chi_i & 2\delta_i \\ -\alpha_i & 0 & 0 & \alpha_i & 0 & 0 \\ 0 & -\beta_i & -\chi_i & 0 & \beta_i & -\chi_i \\ 0 & \chi_i & 2\delta_i & 0 & -\chi_i & 4\delta_i \end{bmatrix} - P_i \begin{bmatrix} 0 & 0 & 0 & 0 & 0 & 0 \\ 0 & 1.2/L_i & 0.1 & 0 & -1.2/L_i & 0.1 \\ 0 & 0.1 & 2L_i/15 & 0 & -0.1 & -L_i/30 \\ 0 & 0 & 0 & 0 & 0 & 0 \\ 0 & -1.2/L_i & -0.1 & 0 & 1.2/L_i & -0.1 \\ 0 & 0.1 & -L_i/30 & 0 & -0.1 & 2L_i/15 \end{bmatrix}$$

$$\alpha_i = (EA)_i/L_i; \ \beta_i = 12(EI)_i/L_i^3; \ \chi_i = 6(EI)_i/L_i^2; \ \delta_i = (EI)_i/L_i$$

$L_1 = L_3 = 4$m, $L_2 = 6$m; $\alpha_1 = \alpha_2 = \alpha_3 = 562500$ kN/m; $\delta_1 = \delta_3 = 4218.75$, $\delta_2 = 9492.1875$ kNm (refer Example 6.2); $P_1 = 468.97$ kN; $P_2 = 50.0$ kN; $P_3 = 531.03$ kN. Accordingly,

$$\mathbf{k}_*^\mathbf{1} = \begin{bmatrix} 562500 & 0 & 0 & -562500 & 0 & 0 \\ 0 & 3023.4 & 6281.2 & 0 & -3023.4 & 6281.2 \\ 0 & 6281.2 & 16625 & 0 & -6281.2 & 8187.4 \\ -562500 & 0 & 0 & 562500 & 0 & 0 \\ 0 & -3023.4 & -6281.2 & 0 & 3023.4 & -6281.2 \\ 0 & 6281.2 & 8187.4 & 0 & -6281.2 & 16625 \end{bmatrix}$$

$$\mathbf{k}_*^\mathbf{2} = \begin{bmatrix} 562500 & 0 & 0 & -562500 & 0 & 0 \\ 0 & 3154.06 & 9487.19 & 0 & -3154.06 & 9487.19 \\ 0 & 9487.19 & 37928.75 & 0 & -9487.19 & 18994.37 \\ -562500 & 0 & 0 & 562500 & 0 & 0 \\ 0 & -3154.06 & -9487.19 & 0 & 3154.06 & -9487.19 \\ 0 & 9487.19 & 18994.37 & 0 & -9487.19 & 37928.75 \end{bmatrix}$$

$$\mathbf{k}_*^3 = \left[\begin{array}{ccc|ccc} 562500 & 0 & 0 & -562500 & 0 & 0 \\ 0 & 3004.8 & 6275 & 0 & -3004.8 & 6275 \\ 0 & 6275 & 16592 & 0 & -6275 & 8154.3 \\ \hline -562500 & 0 & 0 & 562500 & 0 & 0 \\ 0 & -3004.8 & -6275 & 0 & 3004.8 & -6275 \\ 0 & 6275 & 8154.3 & 0 & -6275 & 16592 \end{array}\right]$$

The products, $\mathbf{k}_*^i\mathbf{T}^i$ and $\mathbf{T}^{i^T}\mathbf{k}_*^i\mathbf{T}^i$, with the relevant global coordinates (marked in parenthesis), as defined earlier, may now be computed directly (using MATLAB). By summing up the the contributions of $\mathbf{T}^{i^T}\mathbf{k}_*^i\mathbf{T}^i = \mathbf{k}^i = \left[\begin{array}{c|c} \mathbf{k}_A^i & \mathbf{k}_C^{i\,T} \\ \hline \mathbf{k}_C^i & \mathbf{k}_B^i \end{array}\right]$ of order 6×6 for each of the three elements, at the appropriate coordinate locations, the structure stiffness matrix $\mathbf{k}$ of order 12×12, satisfying $\mathbf{F} = \mathbf{kD}$, can be assembled. It takes the following partitioned form:

$$\Rightarrow \qquad \mathbf{k} = \left[\begin{array}{c|c} \mathbf{k}_{AA} & \mathbf{k}_{AR} \\ \hline \mathbf{k}_{RA} & \mathbf{k}_{RR} \end{array}\right] = \left[\begin{array}{cc|cc} \mathbf{k}_B^1 + \mathbf{k}_A^2 & \mathbf{k}_C^{2\,T} & \mathbf{k}_C^1 & \mathbf{O} \\ \mathbf{k}_C^2 & \mathbf{k}_B^2 + \mathbf{k}_B^3 & \mathbf{O} & \mathbf{k}_C^3 \\ \hline \mathbf{k}_C^{1\,T} & \mathbf{O} & \mathbf{k}_A^1 & \mathbf{O} \\ \mathbf{O} & \mathbf{k}_C^{3\,T} & \mathbf{O} & \mathbf{k}_A^3 \end{array}\right],$$

where:

$$\mathbf{k}_A^1 = \begin{bmatrix} 3023.4 & 0 & -6281.2 \\ 0 & 562500 & 0 \\ -6281.2 & 0 & 16625 \end{bmatrix}; \quad \mathbf{k}_B^1 = \begin{bmatrix} 3023.4 & 0 & 6281.2 \\ 0 & 562500 & 0 \\ 6281.2 & 0 & 16625 \end{bmatrix};$$

$$\mathbf{k}_C^1 = \begin{bmatrix} -3023.4 & 0 & 6281.2 \\ 0 & -562500 & 0 \\ -6281.2 & 0 & 8187.4 \end{bmatrix}$$

$$\mathbf{k}_A^2 = \begin{bmatrix} 562500 & 0 & 0 \\ 0 & 3154.06 & 9487.19 \\ 0 & 9487.19 & 37928.75 \end{bmatrix}; \quad \mathbf{k}_B^2 = \begin{bmatrix} 562500 & 0 & 0 \\ 0 & 3154.06 & -9487.19 \\ 0 & -9487.19 & 37928.75 \end{bmatrix};$$

$$\mathbf{k}_C^2 = \begin{bmatrix} -562500 & 0 & 0 \\ 0 & -3154.06 & -9487.19 \\ 0 & 9487.19 & 18994.38 \end{bmatrix}$$

$$\mathbf{k}_A^3 = \begin{bmatrix} 3004.8 & 0 & -6275 \\ 0 & 562500 & 0 \\ -6275 & 0 & 16592 \end{bmatrix}; \quad \mathbf{k}_B^3 = \begin{bmatrix} 3004.8 & 0 & 6275 \\ 0 & 562500 & 0 \\ 6275 & 0 & 16592 \end{bmatrix};$$

$$\mathbf{k}_C^3 = \begin{bmatrix} -3004.8 & 0 & 6275 \\ 0 & -562500 & 0 \\ -6275 & 0 & 8154.3 \end{bmatrix}$$

$$\Rightarrow \quad \mathbf{k}_{AA} = \begin{array}{cccccc} (1) & (2) & (3) & (4) & (5) & (6) \end{array}$$

$$\mathbf{k}_{AA} = \left[\begin{array}{ccc:ccc} 565523.4 & 0 & 6281.2 & -562500 & 0 & 0 \\ 0 & 565654.06 & 9487.19 & 0 & -3154.06 & 9487.19 \\ 6281.2 & 9487.19 & 54553.75 & 0 & -9487.19 & 18994.38 \\ \hdashline -562500 & 0 & 0 & 565504.8 & 0 & 6275 \\ 0 & -3154.06 & -9487.19 & 0 & 565654.06 & -9487.19 \\ 0 & 9487.19 & 18994.38 & 6275 & -9487.19 & 54520.75 \end{array}\right] \begin{array}{l} (1) \\ (2) \\ (3) \\ (4) \\ (5) \\ (6) \end{array}$$

$$\begin{array}{cccccc} (7) & (8) & (9) & (10) & (11) & (12) \end{array}$$

$$\mathbf{k}_{AR} = \left[\begin{array}{ccc:ccc} -3023.4 & 0 & 6281.2 & 0 & 0 & 0 \\ 0 & -562500 & 0 & 0 & 0 & 0 \\ -6281.2 & 0 & 8187.4 & 0 & 0 & 0 \\ \hdashline 0 & 0 & 0 & -3004.8 & 0 & 6275 \\ 0 & 0 & 0 & 0 & -562500 & 0 \\ 0 & 0 & 0 & -6275 & 0 & 8154.3 \end{array}\right] \begin{array}{l} (1) \\ (2) \\ (3) \\ (4) \\ (5) \\ (6) \end{array} = \mathbf{k}_{RA}^{\ \mathbf{T}}$$

$$\begin{array}{cccccc} (7) & (8) & (9) & (10) & (11) & (12) \end{array}$$

$$\mathbf{k}_{RR} = \left[\begin{array}{ccc:ccc} 3023.4 & 0 & -6281.2 & 0 & 0 & 0 \\ 0 & 562500 & 0 & 0 & 0 & 0 \\ -6281.2 & 0 & 16625 & 0 & 0 & 0 \\ \hdashline 0 & 0 & 0 & 3004.8 & 0 & -6275 \\ 0 & 0 & 0 & 0 & 562500 & 0 \\ 0 & 0 & 0 & -6275 & 0 & 16592 \end{array}\right] \begin{array}{l} (7) \\ (8) \\ (9) \\ (10) \\ (11) \\ (12) \end{array}$$

Displacements and Support Reactions

The basic stiffness relations are given by:

$$\left[\frac{\mathbf{F}_A}{\mathbf{F}_R}\right] - \left[\frac{\mathbf{F}_{fA}}{\mathbf{F}_{fR} = \mathbf{O}}\right] = \left[\begin{array}{c:c} \mathbf{k}_{AA} & \mathbf{k}_{AR} \\ \hdashline \mathbf{k}_{RA} & \mathbf{k}_{RR} \end{array}\right] \left[\frac{\mathbf{D}_A}{\mathbf{D}_R = \mathbf{O}}\right]$$

$$\Rightarrow \mathbf{D}_A = \left[\mathbf{k}_{AA}\right]^{-1}\left[\mathbf{F}_A - \mathbf{F}_{fA}\right]$$

$$\mathbf{F}_R = \mathbf{k}_{RA}\mathbf{D}_A$$

Substituting the various matrices and solving using MATLAB,

$$\mathbf{D_A} = \begin{Bmatrix} D_1 \\ D_2 \\ D_3 \\ \hline D_4 \\ D_5 \\ D_6 \end{Bmatrix} = \begin{Bmatrix} 0.02039041 \text{ m} \\ -0.00083093 \text{ m} \\ -0.02286814 \text{ rad} \\ \hline 0.02006687 \text{ m} \\ -0.00094684 \text{ m} \\ 0.01939347 \text{ rad} \end{Bmatrix} \quad ; \quad \mathbf{F_R} = \begin{Bmatrix} F_7 \\ F_8 \\ F_9 \\ \hline F_{10} \\ F_{11} \\ F_{12} \end{Bmatrix} = \begin{Bmatrix} 81.991 \text{ kN} \\ 467.400 \text{ kN} \\ -59.154 \text{ kNm} \\ \hline -181.991 \text{ kN} \\ 532.599 \text{ kN} \\ 284.059 \text{ kNm} \end{Bmatrix}$$

Check equilibrium:

$$\sum F_x = 0: \; F_7 + F_{10} - 100 = 81.991 - 181.991 + 100 = 0 \text{ kN}$$

$$\sum F_y = 0: \; F_8 + F_{11} - 1000 = 467.400 + 532.599 - 1000 = 0 \text{ kN} \;\; \Rightarrow \text{OK.}$$

Member Forces

$$\mathbf{F_*^i} = \mathbf{F_{*f}^i} + \left(\mathbf{k_*^i T^i}\right)\mathbf{D^i},$$

$$\Rightarrow \mathbf{F_*^1} = \begin{Bmatrix} F_{1*}^1 \\ F_{2*}^1 \\ F_{3*}^1 \\ \hline F_{4*}^1 \\ F_{5*}^1 \\ F_{6*}^1 \end{Bmatrix} = \begin{Bmatrix} 0 \text{ kN} \\ 0 \text{ kN} \\ 0 \text{ kNm} \\ \hline 0 \text{ kN} \\ 0 \text{ kN} \\ 0 \text{ kNm} \end{Bmatrix} + \left[\mathbf{k_*^1 T^1}\right] \begin{Bmatrix} 0 \\ 0 \\ 0 \\ \hline D_1 \\ D_2 \\ D_3 \end{Bmatrix} = \begin{Bmatrix} 467.400 \text{ kN} \\ -81.991 \text{ kN} \\ -59.154 \text{ kNm} \\ \hline -467.400 \text{ kN} \\ 81.991 \text{ kN} \\ -252.106 \text{ kNm} \end{Bmatrix}$$

$$\mathbf{F_*^2} = \begin{Bmatrix} F_{1*}^2 \\ F_{2*}^2 \\ F_{3*}^2 \\ \hline F_{4*}^2 \\ F_{5*}^2 \\ F_{6*}^2 \end{Bmatrix} = \begin{Bmatrix} 0 \text{ kN} \\ 500 \text{ kN} \\ 750 \text{ kNm} \\ \hline 0 \text{ kN} \\ 500 \text{ kN} \\ -750 \text{ kNm} \end{Bmatrix} + \left[\mathbf{k_*^2 T^2}\right] \begin{Bmatrix} D_1 \\ D_2 \\ D_3 \\ \hline D_4 \\ D_5 \\ D_6 \end{Bmatrix} = \begin{Bmatrix} 181.991 \text{ kN} \\ 467.400 \text{ kN} \\ 252.106 \text{ kNm} \\ \hline -181.991 \text{ kN} \\ 532.599 \text{ kN} \\ -447.695 \text{ kNm} \end{Bmatrix}$$

$$\mathbf{F_*^3} = \begin{Bmatrix} F_{1*}^3 \\ F_{2*}^3 \\ F_{3*}^3 \\ \hline F_{4*}^3 \\ F_{5*}^3 \\ F_{6*}^3 \end{Bmatrix} = \begin{Bmatrix} 0 \text{ kN} \\ 0 \text{ kN} \\ 0 \text{ kNm} \\ \hline 0 \text{ kN} \\ 0 \text{ kN} \\ 0 \text{ kNm} \end{Bmatrix} + \left[\mathbf{k_*^3 T^3}\right] \begin{Bmatrix} 0 \\ 0 \\ 0 \\ \hline D_4 \\ D_5 \\ D_6 \end{Bmatrix} = \begin{Bmatrix} 532.599 \text{ kN} \\ 181.991 \text{ kN} \\ 291.059 \text{ kNm} \\ \hline -532.599 \text{ kN} \\ -181.991 \text{ kN} \\ 447.696 \text{ kNm} \end{Bmatrix}$$

The complete 'second-order' response (including the fixed end force effects), in the form of free-body, deflection and bending moment diagrams are shown in Fig. 7.66. The maximum sagging moment under the load, considering a mid-span deflection of approximately 0.04m, is given by $(467.4 \times 3.0) - 252.11 + (182 \times 0.04) = 1157.37\,\text{kNm}$.

Note that the axial forces obtained from the above 'second-order' analysis, given by $P_1 = 467.4$ kN; $P_2 = 182$ kN; $P_3 = 532.60$ kN, are comparable to the initially assumed 'first-order' values, $P_1 = 468.97$ kN; $P_2 = 50.0$ kN; $P_3 = 531.03$ kN (except for the axial force in element '2', which may not be critical). However, if greater accuracy in the results is required, another iteration may be carried out.

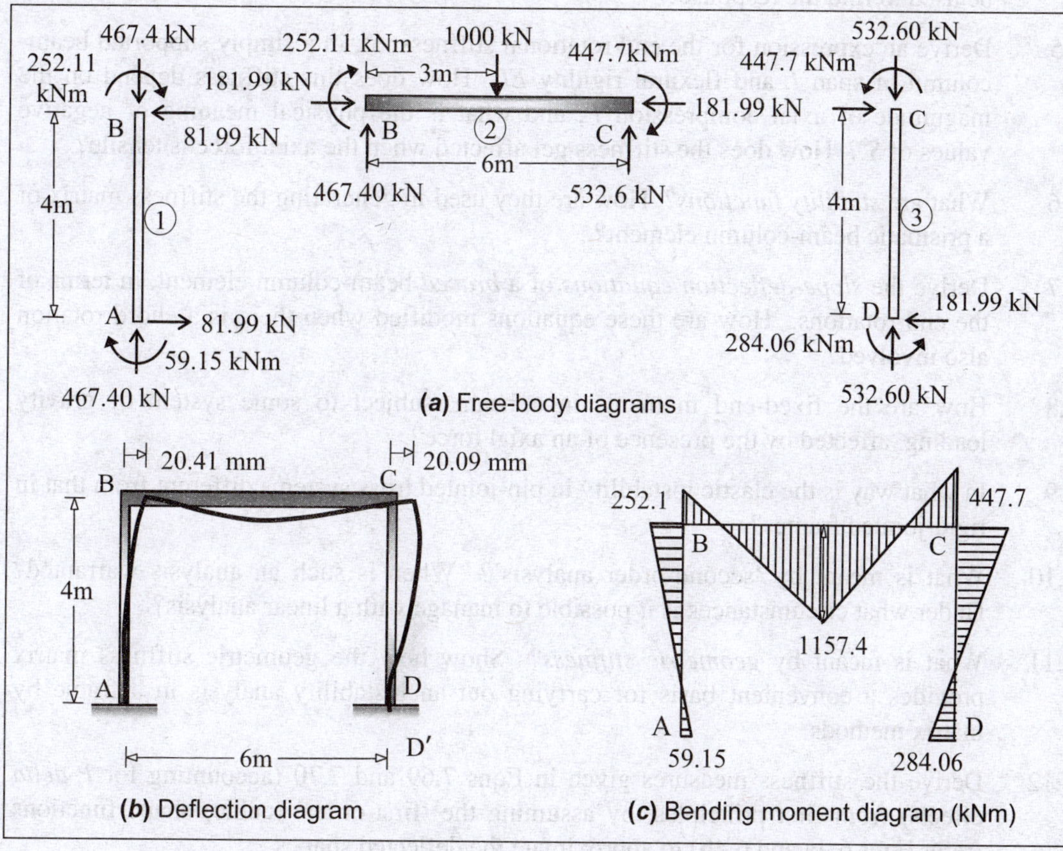

Figure 7.66 Complete ('second-order' response (Example 7.21)

REVIEW QUESTIONS

7.1 What is meant by *bifurcation instability*? How does an eigenvalue formulation provide a mathematical basis for it?

7.2 How is the buckling behaviour of *real columns* different from *ideal columns*, subject to axial compression?

7.3 What is meant by *beam-column*? How is the element stiffness matrix of a beam-column element different from a conventional beam element?

7.4 Consider a simply supported beam, made of a linear elastic material, subject to a uniformly distributed load of total magnituded W and an axial compressive force P.

Under what circumstances is it adequate to carry out a *linear* analysis to find the response? If additional gravity loads are applied, can the principle of superposition be used to find the response?

7.5 Derive an expression for the end rotational stiffness, S^o, in a simply supported beam-column of span L and flexural rigidity EI. How does this stiffness depend on the magnitude of axial compression P, and what is the physical meaning of negative values of S^o? How does the stiffness get affected when the axial force is tensile?

7.6 What are *stability functions*? How are they used in generating the stiffness matrix of a prismatic beam-column element?

7.7 Derive the *slope-deflection equations* of a *braced* beam-column element, in terms of the end rotations. How are these equations modified when there is a chord rotation also involved?

7.8 How are the fixed-end moments in a beam, subject to some system of gravity loading, affected by the presence of an axial force?

7.9 In what way is the elastic instability in pin-jointed truss systems different from that in rigid-jointed frames?

7.10 What is meant by 'second-order analysis'? When is such an analysis warranted? Under what circumstances is it possible to manage with a linear analysis?

7.11 What is meant by *geometric stiffness*? Show how the geometric stiffness matrix provides a convenient basis for carrying out an instability analysis in a frame by matrix methods.

7.12 Derive the stiffness measures given in Eqns 7.69 and 7.70 (accounting for *P-delta* effects) from first principles, by assuming the 'first-order' (cubic) shape functions (refer Eqns 6.4a and 6.4b) to approximate the deflected shapes.

7.13 Why are intermediate loads generally not included in investigating the critical load in a frame?

7.14 How will you carry out an instability analysis when the axial compressive forces in the various members in the frame are not known *a priori*?

7.15 How will you account for internal hinges in a plane frame in carrying out an instability or second-order analysis?

PROBLEMS

7.1 Derive, from first principles (by solving the differential equation of equilibrium), an expression for the mid-span deflection Δ_{max} in a prismatic beam-column, with a simply supported span L and flexural rigidity EI and axial compression P, when it is subject to a uniformly distributed load q_o per unit length. Show how this deflection (normalised against its 'first-order' value) varies with the axial load P (normalised against the Euler buckling load, P_{Eo}) by means of a graph.

7.2 Repeat Problem 7.1, considering the ends to be fixed against rotation. Also, derive an expression for the fixed-end moment.

7.3 Repeat Example 7.4, considering the end A to be fixed against rotation.

7.4 Estimate the critical buckling load in the continuous beam ABC shown in Fig. 7.67 using the slope deflection method, and draw the mode shape. Assume constant $EI = 8000$ kNm^2.

7.5 (a) Repeat Problem 7.4 by the matrix method of analysis.
(b) Considering the applied axial load to be 50 percent of the computed critical load, carry out a second-order analysis and draw the bending moment diagram.

7.6 Repeat Problem 7.4, considering a hinged support at end A (instead of the fixed support shown in Fig. 7.67.

7.7 Repeat Problem 7.4, considering an additional horizontal force, $P/2$, applied at B, pointing towards A.

7.8 Repeat Problem 7.7 by the matrix method.

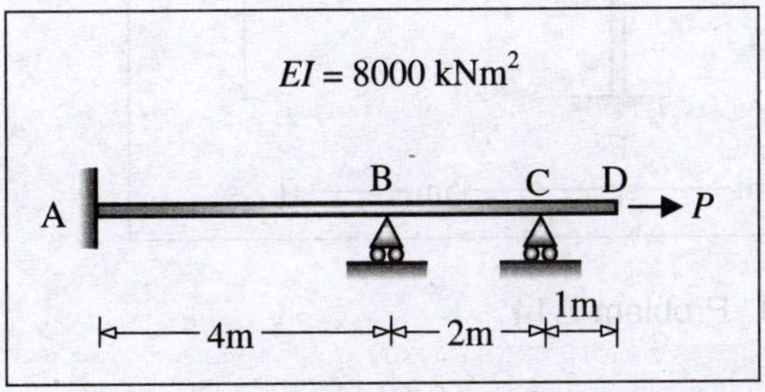

Figure 7.67 Problem 7.4

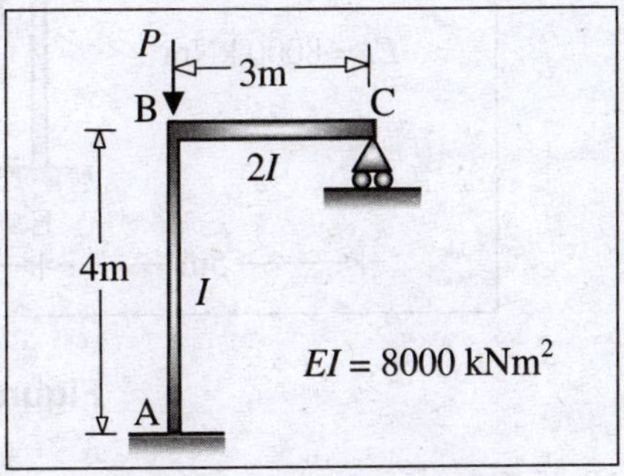

Figure 7.68 Problem 7.9

7.9 Estimate the critical buckling load in the frame ABC shown in Fig. 7.68, using the slope deflection method. Assume $EI = 8000$ kNm^2.

7.10 Repeat Problem 7.9 by the matrix method of analysis.

7.11 Repeat Problem 7.10, considering a hinged support at end A.

7.12 Carry out a second-order analysis, by the slope-deflection method, on the frame in Fig. 7.68, considering an additional uniformly distributed lateral load of 30 kN/m acting on AB from left to right. Sketch the deflected shape and draw the bending moment diagram.

7.13 Repeat Problem 7.12 by the matrix method of analysis.

7.14 Estimate the critical load P causing instability in the frame shown in Fig. 7.69 by the slope deflection method. Assume $EI = 8000$ kNm^2. Sketch the mode shape.

7.15 Consider the fixed end A in Fig. 7.69 to be replaced with a roller support. Carry out a second-order analysis, considering $P = 750$ kN, and draw the bending moment diagram.

7.16 Consider the end D in Fig. 7.69 to be restrained against rotation. Taking advantage of symmetry, sketch a reduced model of the frame. Apply the matrix method of analysis to find the critical load P causing instability in the frame.

7.17 For the symmetric frame in Example 7.16, carry out a second-order analysis, $P = 1000$ kN. Find the complete force response and draw the deflection and bending moment diagrams.

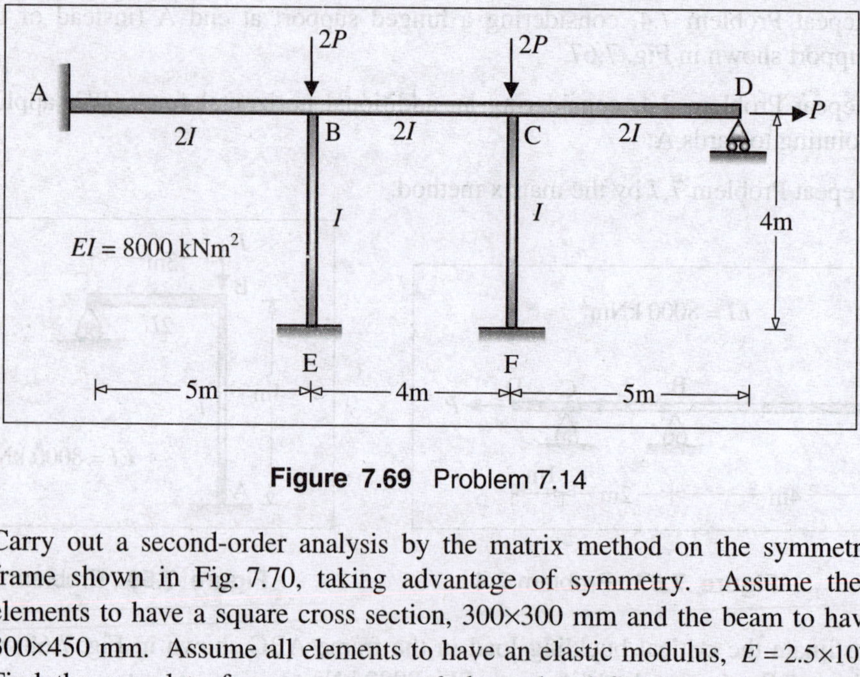

Figure 7.69 Problem 7.14

7.18 Carry out a second-order analysis by the matrix method on the symmetric portal frame shown in Fig. 7.70, taking advantage of symmetry. Assume the column elements to have a square cross section, 300×300 mm and the beam to have a size, 300×450 mm. Assume all elements to have an elastic modulus, $E = 2.5 \times 10^4$ N/mm^2. Find the complete force response and draw the deflection and bending moment diagrams.

7.19 Repeat Problem 7.18 (Fig. 7.70), considering pinned joints at B and D.

7.20 Apply the slope deflection method to find the critical buckling load, W_{cr}, for the three-member frame ABC, loaded as shown in Fig. 7.71, considering joint B to be pinned. Neglect axial deformations and assume all members to have the same flexural rigidity EI. Express W_{cr} in terms of EI and L.

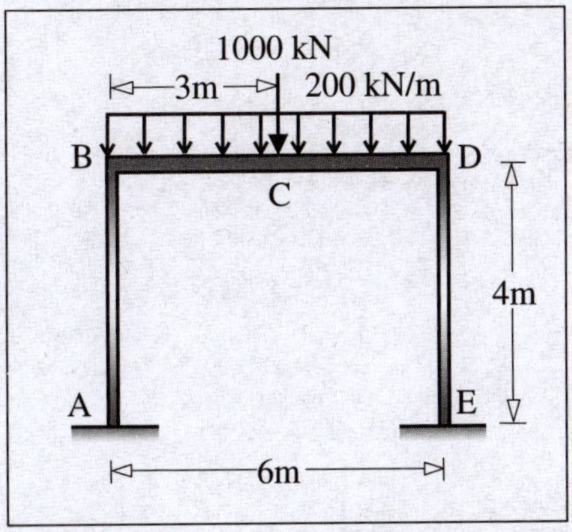

Figure 7.70 Symmetric portal frame (Problem 7.18)

7.21 Repeat Problem 7.20 by the matrix method of analysis, considering the joint at B to be rigid.

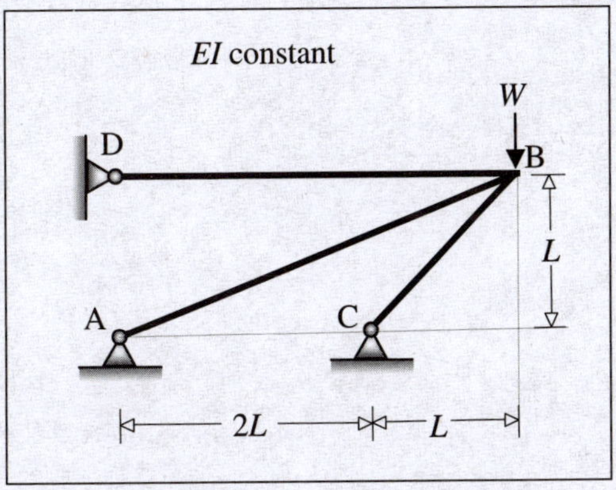

Figure 7.71 Problem 7.20

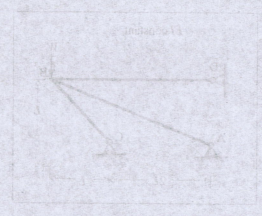

Index